Physical Constants

Quantity	Symbol	Value	SI unit
Speed of light in vacuum	c	3.00×10^8	m/s
Permittivity of vacuum	ϵ_0	8.85×10^{-12}	F/m
Coulomb constant, $1/4\pi\epsilon_0$	k	9.0×10^9	$N \cdot m^2/C^2$
Permeability of vacuum	μ_0	1.26×10^{-6}	N/A^2
		($4\pi \times 10^{-7}$ exactly)	
Elementary charge	e	1.60×10^{-19}	C
Planck's constant	h	6.63×10^{-34}	$J \cdot s$
	$\hbar = h/2\pi$	1.05×10^{-34}	$J \cdot s$
Electron rest mass	m_e	9.11×10^{-31}	kg
		5.49×10^{-4}	u
Proton rest mass	m_p	1.67265×10^{-27}	kg
		1.007276	u
Neutron rest mass	m_n	1.67495×10^{-27}	kg
		1.008665	u
Avogadro constant	N_A	6.02×10^{-23}	$(gmol)^{-1}$
Molar gas constant	R	8.31	$J/mol \cdot K$
Boltzmann's constant	k	1.38×10^{-23}	J/K
Stefan-Boltzmann constant	σ	5.67×10^{-8}	$W/m^2 \cdot K^4$
Molar volume of ideal gas at STP	V	22.4	liters/mol
		2.24×10^{-2}	m^3/mol
Rydberg constant	R	1.10×10^{-7}	m^{-1}
Bohr radius	a_0	5.29×10^{-11}	m
Electron Compton wavelength	$h/m_e c$	2.43×10^{-12}	m
Gravitational constant	G	6.67×10^{-11}	$m^3/kg \cdot s^2$
Standard gravity	g	9.80	m/s^2
Radius of earth (at equator)	R_e	6.38×10^6	m
Mass of earth	M_e	5.98×10^{24}	kg
Radius of moon	R_m	1.74×10^6	m
Mass of moon	M_m	7.36×10^{22}	kg

The values presented in this table are those used in computations in the text. Generally, the physical constants are known to much better precision.

COLLEGE PHYSICS

Third Edition

COLLEGE PHYSICS

Third Edition

Raymond A. Serway
James Madison University

Jerry S. Faughn
Eastern Kentucky University

Saunders Golden Sunburst Series

Saunders College Publishing
A Harcourt Brace Jovanovich Publisher

Fort Worth	San Antonio
Philadelphia	Toronto
San Diego	Montreal
New York	London
Orlando	Sydney
Austin	Tokyo

Requests for permission to make copies of any part of the work should be mailed to: Permissions Department, Harcourt Brace Jovanovich, Publishers, 8th floor, Orlando, Florida 32887.

Text Typeface: Caledonia
Compositor: Progressive Typographers
Acquisitions Editor: John Vondeling
Developmental Editor: Ellen Newman
Senior Project Manager: Sally Kusch
Copy Editor: Charlotte Nelson
Managing Editor: Carol Field
Manager of Art and Design: Carol Bleistine
Text Designer: Tracy Baldwin
Cover Designer: Lawrence R. Didona
Text Artwork: Rolin Graphics
Layout Artists: Tracy Baldwin; Dorothy Chattin
Production Manager: Tim Frelick
Cover: Front: Fireworks, © Guy Motil/West Light
        Back: Fireworks, © Ron Watts/West Light
Frontispiece: Bullet traveling faster than the speed of sound.
© Harold E. Edgerton, courtesy of Palm Press, Inc.

Printed in the United States of America

COLLEGE PHYSICS, 3rd edition

ISBN 0-03-073331-6

Library of Congress Catalog Card Number: 91-53139

234   032   987654321

Preface

This textbook, now in its third edition, is appropriate for a one-year course in introductory physics commonly taken by students majoring in biology and pre-medicine, as well as other disciplines such as health, environmental, earth, and social sciences. The mathematical techniques used in the book include algebra and trigonometry, but not calculus.

© Tom Branch 1984/Photo Researchers, Inc.

Changes in the Third Edition

A number of changes and improvements have been made in preparing the third edition of this text. Many of these changes are in response to comments and suggestions offered by users of the second edition and reviewers of the manuscript. The following represent the major changes in the third edition:

1. Substantial revision of the problems to include more at the intermediate level (about 30% of the problems are new or revised). In addition, the end-of-chapter questions have been revised and increased by 50%. *All problems in the text have been reviewed by several professors to ensure accuracy in the text as well as in the Instructor's Manual.*

2. Addition of Problem-Solving Strategies in appropriate sections throughout the text provides additional information and guidelines to help students solve problems.

3. Biographical sketches of many important physicists have been added to give the third edition a more historical emphasis.

4. New full-color design, which is used to highlight important concepts throughout the text and in the figures. In addition, approximately 300 color photographs have been carefully selected, and their accompanying captions have been written to serve as an added instructional tool. A chart explaining the pedagogical use of color is included on the inside front cover for easy reference.

5. Physics in Action boxes feature photographs of interesting demonstrations and phenomena in physics, accompanied by detailed explanations. Students learn from these special features that physics is a science based on experimental observations.

6. Conversations with prominent physicists, appearing in each of the text's six parts, introduce students to individuals who have achieved international recognition in their fields. The conversations cover specialized scientific topics and are intended as optional reading.

7. The guest essays have been revised and updated; two new essays—on the topics of chaos and the third law of thermodynamics—have been included.

8. Some sections have been rewritten to improve clarity, and several optional topics (e.g., Gauss' law, equipotential surfaces, and the twin paradox) have been added according to reviewers' suggestions.

9. The mechanics section of the text has been reorganized so that Chapter 2 now deals only with one-dimensional motion, and Chapter 3 is an introduction to vectors and a treatment of two-dimensional motion, including projectiles and relative motion (new). In addition, Chapter 4 has been combined with Chapter 8 of the second edition: the new Chapter 8 is now entitled "Objects at Rest and Rotational Dynamics."

Objectives

The main objectives of this introductory physics textbook are twofold: to provide the student with a clear and logical presentation of the basic concepts and principles of physics, and to strengthen an understanding of the concepts and principles through a

v

broad range of interesting applications to the real world. In order to meet these objectives, emphasis is placed on sound physical arguments and discussions of everyday experiences. At the same time, we have attempted to motivate the student through practical examples that demonstrate the role of physics in other disciplines.

Coverage

The material covered in this book is concerned with standard topics in classical physics and modern physics. The book is divided into six parts: Part 1 (Chapters 1–10) deals with the fundamentals of Newtonian mechanics and the physics of fluids; Part 2 (Chapters 11–13) is concerned with heat and thermodynamics; Part 3 (Chapters 14–15) covers wave motion and sound; Part 4 (Chapters 16–23) is concerned with electricity and magnetism; Part 5 (Chapters 24–27) treats the properties of light and the field of geometric and wave optics; and Part 6 (Chapters 28–34) represents an introduction to the theory of relativity and modern physics.

Features

Most instructors would agree that the textbook assigned in a course should be the student's major "guide" for understanding and learning the subject matter. With this in mind, we have included many pedagogic features in the textbook that enhance its usefulness to both the student and instructor. These are as follows:

Style We have attempted to write the book in a style that is clear, relaxed, and pleasing to the reader. New terms are carefully defined, and we have tried to avoid jargon.

Organization The book is divided into six parts: mechanics, thermodynamics, vibrations and wave motion, electricity and magnetism, light and optics, and modern physics. Each part includes an overview of the subject matter to be covered in that part and some historical perspectives.

Introductory Chapter An introductory chapter is included to "set the stage" for the text and to discuss the units of physical quantities, order-of-magnitude calculations, dimensional analysis, significant figures, and mathematical notation.

Units The international system of units (SI) is used throughout the book. The British engineering system of units (conventional system) is used only to a limited extent in the problem sets of the early chapters on mechanics.

Previews Most chapters begin with a chapter preview, which includes a brief discussion of the chapter objectives and content.

Marginal Notes and Equations Important equations are highlighted with a light tan screen, and marginal notes set in blue type are often used to describe their meaning. Marginal notes are also used to locate specific definitions and important statements.

Biographical Sketches Throughout the text we have added short biographies of important scientists to give the third edition a more historical and human emphasis.

Problem-Solving Strategies As an added feature of the third edition, we have included general strategies for solving the types of problems featured in both the examples and end-of-chapter problems. It is our hope that this feature will help the students identify the steps in solving a problem and eliminate any uncertainty they might have. This feature is highlighted by a light tan screen for emphasis, which helps the student locate the strategies quickly.

Physics in Action   A new feature of the third edition, these boxes contain interesting demonstrations and phenomena in physics, accompanied by detailed explanations. The boxes can also serve as a source of information for initiating classroom discussions.

Conversations with Prominent Scientists   As an additional feature of the third edition, we have included interviews with physicists who have achieved international recognition in their fields. Appearing in each of the text's six parts, the conversations cover specialized scientific topics as well as current concerns such as science education and the role of science in society.

Worked Examples   A large number of worked examples (300) are presented as an aid in understanding concepts. In many cases, these examples serve as models for solving end-of-chapter problems. The examples are set off from the text by blue rules for ease of location, and most examples are given titles to describe their content. The solution answer is highlighted with a light blue screen.

Worked Example Exercises   Many of the worked examples (over 100) are followed immediately by exercises with answers. These examples are intended to make the textbook more interactive with the student, and to test the student's understanding of problem-solving techniques.

Special Topics   Many chapters include special topic sections which are intended to expose the student to various practical and interesting applications of physical principles.

Guest Essays   In this third edition, we have included 12 essays written by guest authors. Two of the essays are new, while the remaining ten are revisions of essays that appeared in the previous edition. The topics covered in the essays include the Big Bang, arch structures, the circulatory system, the nervous system, applications of lasers, exponential growth, general relativity, superconductivity, chaos, the third law of thermodynamics, vacuum tunneling spectroscopy, and alternate energy sources. Although the essays are intended as supplemental reading for the student, many include questions and problems for added flexibility in covering these topics. The essays have been placed at the ends of the chapters to avoid interrupting the main textual material.

Jack Finch/Science Photo Library

Important Statements   Many important statements and definitions are set in blue type for added emphasis and ease of review.

Illustrations and Photographs   The readability and effectiveness of the text material and worked examples are enhanced by the large number of figures, diagrams, photographs, and tables. Full color is used to add clarity to the artwork and to make it as realistic as possible. For example, vectors are color-coded, and curves in xy-plots are drawn in color. Three-dimensional effects are produced with the use of color airbrushed areas, where appropriate. Approximately 300 color photographs have been carefully selected, and their accompanying captions have been written to serve as an added instructional tool. A complete description of the pedagogical use of color appears on the inside front cover.

Summaries   Each chapter contains a summary that reviews the important concepts and equations in that chapter.

Readings   Each chapter contains a set of suggested additional readings that have been selected on the basis of their level and the degree to which they supplement the text.

Thought Questions   The list of questions requiring verbal answers given at the end of each chapter has been expanded by more than 50% for a total of 846. Some questions provide the student with a means of self-testing the concepts presented in the chapter. Others could serve as a basis for initiating classroom discussions. Answers to selected questions are included in the Student Study Guide accompanying the text.

Problems   An extensive set of problems is included at the end of each chapter. In this third edition, many new problems have been added to the text, and most of the old problems have been revised or rewritten. The text contains 2300 problems, which is an increase of over 30% from the previous edition. Answers to odd-numbered problems are given at the end of the book; these pages have colored edges to help locate them easily. For the convenience of both the student and the instructor, most problems are keyed to specific sections of the chapter; the remaining problems, labeled "Additional Problems," are not keyed to specific sections.

In general, the problems within a given section are presented so that the straightforward problems are first, followed by problems of increasing difficulty. For ease in identifying the problems of intermediate difficulty, the problem number is printed in blue. We have also included a small number of more challenging problems, which are indicated by a problem number printed in magenta. In our opinion, assignments should mainly consist of problems from the first two categories to help build self-confidence. As mentioned earlier, most of the guest essays now include several questions and problems for those instructors who choose to cover these topics.

Appendices   Several appendices are provided at the end of the text. Most of the appendix material represents a review of mathematical techniques used in the text, including scientific notation, algebra, geometry, and trigonometry. Reference to these appendices is made throughout the text. Most of the mathematical review sections include worked examples and exercises with answers. The last few appendices contain useful tables that supplement textual information. For easy reference, the front endpapers contain a chart explaining the pedagogical use of color throughout the text and a list of physical constants, and the back endpapers provide a periodic table of the elements in full color.

Ancillaries

The following ancillaries are available with this text:

Instructor's Manual with Full Solutions   This manual consists of complete, worked-out solutions to *all* the problems in the text, as well as answers to the even-numbered problems.

Student Study Guide with Selected Solutions   The Study Guide contains chapter objectives, a skills section that reviews mathematical techniques, and suggested approaches to problem-solving methodology. Answers to selected end-of-chapter questions, as well as full, worked-out solutions to selected end-of-chapter problems from the text (approximately 10%) are included. The Study Guide also provides several review problems and notes from important sections of each chapter, which include a glossary of important terms, theorems, and concepts.

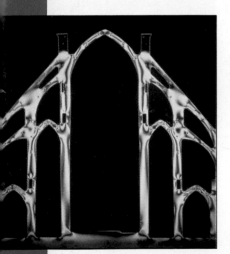

© *Sepp Seitz 1981*

Printed Test Bank The printed test bank contains approximately 2300 problems, both open-ended and multiple choice, generated from the Computerized Test Bank. The format allows instructors to duplicate pages for distribution to students.

ExaMaster™ Computerized Test Bank Available for the IBM and Macintosh computers, the test bank contains approximately 2300 open-ended and multiple-choice problems, representing every chapter of the text. The test bank allows instructors to customize tests by rearranging, editing, or adding questions. The problems are graded in level of difficulty for the instructor's convenience; the software program solves all problems and prints the answers on a separate grading key.

Overhead Transparency Acetates This collection of transparencies consists of more than 300 full-color figures and photographs from the text to enhance lectures; they feature large print for easy viewing in the classroom.

Computer Exercises with Courseware Disk A number of interactive computer programs (presented in special computer modules) accompany a set of printed exercises. The student's input has direct and immediate effect on the output, which enables the student to work through many challenging numerical problems while experiencing the power of the computer in scientific work. The printed exercises direct the student's use of the programs contained on the Courseware Disk, which is available for Apple, IBM, and Macintosh computers.

Guy Marche/FPG International

Physics Laboratory Manual To supplement the learning of basic physical principles while introducing laboratory procedures and equipment, each chapter of the laboratory manual includes a pre-laboratory assignment, objectives, an equipment list, the theory behind the experiment, experimental procedure, calculations, graphs, and questions. A laboratory report is provided for each experiment so the student can record data, calculations, and experimental results. The manual contains 45 standard experiments that can be used for either algebra-based or calculus-based courses.

Instructor's Manual for Physics Laboratory Manual Each chapter contains a discussion of the experiment, teaching hints, answers to selected questions, and a post-laboratory quiz with short answers and essay questions. We have also included a list of the suppliers of scientific equipment and a summary of the equipment needed for all the laboratory experiments in the manual.

Physics Demonstration Videotapes A unique collection of 60 physics demonstrations is provided to supplement classroom presentations and to help motivate the students.

Teaching Options

This book contains more than enough material for a one-year course in introductory physics. This serves two purposes. First, it gives the instructor more flexibility in choosing topics for a specific course. Second, the book becomes more useful as a resource for students. On the average, it should be possible to cover about one chapter each week. Many sections containing interesting applications are considered optional and are therefore marked with an asterisk(*). Some of these could be covered to suit the needs of the students and instructors. For shorter courses, instructors may choose to omit several chapters without loss of continuity. The chapters that we suggest could be omitted in such circumstances are Chapter 8 (Objects at Rest and Rotational Dynamics), Chapter 22 (Alternating Current Circuits), Chapter 23 (Electromagnetic Waves), Chapter 27 (Optical Instruments), Chapter 31 (Molecules and Solids), and Chapter 33 (Nuclear Physics Applications).

Acknowledgements

In preparing the third edition of this textbook, we have been guided by the expertise of many people who reviewed part or all of the manuscript. We wish to acknowledge the following scholars and express our sincere appreciation for their helpful suggestions, criticisms, and encouragement: Ralph Barnett, Florissant Valley Community College; Paul Feldker, Florissant Valley Community College; Robert Hudson, Roanoke College; Frank Kolp, Trenton State College; Karl F. Kuhn, Eastern Kentucky University; Joel Levine, Orange Coast College; David Lamp, Texas Tech University; David Markowitz, University of Connecticut; Ralph V. McGrew, Broome Community College; Brooke M. Pridmore, Clayton State College; Virginia Roundy, California State University, Fullerton; Reinhard A. Schumacher, Carnegie Mellon University; Jerry H. Wilson, Metropolitan State College; and Clyde A. Zaidins, University of Colorado at Denver.

We wish to thank the following scholars for their comments and suggestions during the revision of the second edition: John Anderson, University of Pittsburgh; Neil W. Ashcroft, Cornell University; Gordon Aubrecht, Ohio State University; Charles Bacon, Ferris State University; Louis Barrett, Western Washington University; Jeffrey Braun, University of Evansville; John Brennan, University of Central Florida; Joseph Catanzarite, Cypress College; Roger W. Clapp, University of South Florida; Giuseppe Colaccico, University of South Florida; Jorge Cossio, Miami-Dade Community College; Terry T. Crow, Mississippi State College; Albert Thomas Fromhold, Jr., Auburn University; Teymoor Gedayloo, California Polytechnic State University; Wlodzimierz Guryn, Brookhaven National Laboratory; James Harmon, Oklahoma State University; Fred Inman, Mankato State University; Drasko Jovanovic, Fermilab; Ronald E. Jodoin, Rochester Institute of Technology; Joan P. S. Kowalski, George Mason University; Sol Krasner, University of Chicago; James Linblad, Saddleback Community College; Michael Lieber, University of Arkansas; Jeffrey V. Mallow, Loyola University of Chicago; Lewis J. Oakland, University of Minnesota; Louis O'Kelly, Memphis State University; T. A. K. Pillai, University of Wisconsin—La Crosse; William D. Ploughe, Ohio State University; Joseph Priest, Miami University; Kurt Reibel, Ohio State University; Donald D. Snyder, Indiana University at South Bend; Carey E. Stronach, Virginia State University; Thomas W. Taylor, Cleveland State University; Larry Weaver, Kansas State University; and L. L. Van Zandt, Purdue University.

For their suggestions during the various stages of development of the first edition, we thank the following scholars: Albert Altman, University of Lowell; Paul Bender, Washington State University; Roger W. Clapp, Jr., University of South Florida; John R. Gordon, James Madison University; Karl F. Kuhn, Eastern Kentucky University; Harvey S. Leff, California State Polytechnic University; Bill Lochslet, Pennsylvania State University; David Markowitz, The University of Connecticut; Joe McCauley, Jr., University of Houston; Bill F. Melton, The University of North Carolina at Charlotte; H. Kent Moore, James Madison University; Carl R. Nave, Georgia State University; Blaine Norum, University of Virginia; M. E. Oakes, The University of Texas at Austin; James Purcell, Georgia State University; William R. Savage, The University of Iowa; Howard G. Voss, Arizona State University; Donald H. White, Western Oregon State College; George A. Williams, The University of Utah; Robert M. Wood, University of Georgia; and Peter D. Zimmerman, Louisiana State University.

We wish to express our gratitude to Linda Miller, who typed parts of the manuscript and the Student Study Guide. We thank Christine Livecchi, Janet Nuciforo, Henry Leap, and Jim Lehman for locating and providing many excellent photographs.

Ward's Natural Science

We are especially indebted to John R. Gordon for his fine work on the Student Study Guide, to Charles Teague for checking solutions to all the problems and for his assistance in preparing the Instructor's Manual, and to Steve Van Wyk for accuracy review of all the problems in the text and for his help in revising the Test Bank.

We thank the following individuals for writing the interesting essays that appear throughout the text: Isaac D. Abella, University of Chicago; Albert A. Bartlett, University of Colorado at Boulder; Gordon Batson, Clarkson University; Paul Davidovits, Boston College; Roger A. Freedman, University of California; Paul K. Hansma, University of California; Laurent Hodges, Iowa State University; David Markowitz, University of Connecticut; J. Clint Sprott, University of Michigan; Hans C. von Baeyer, College of William and Mary; Clifford Will, Washington University; and Sidney Wolff, Kitt Peak National Observatory.

Special thanks to the following scholars who were kind enough to be interviewed for our text: Hans A. Bethe, Cornell University; Steven Chu, Stanford University; Freeman J. Dyson, The Institute for Advanced Study; Leon M. Lederman, University of Chicago; M. Brian Maple, University of California at San Diego; and Heinrich Rohrer, IBM Research Laboratory in Zurich. We appreciate the expertise of Janice K. Mandel in conducting the interviews that appear in the text.

We wish to thank the professional staff at Saunders College Publishing for their expertise and perseverance in the project. In particular, we thank Ellen Newman, who worked so diligently on all aspects of the manuscript during its development; Sally Kusch, who was most instrumental in turning the manuscript into a most attractive book; and Charlotte Nelson for her excellent copyediting. We thank John Vondeling for his expert guidance in this revision and Carol Bleistine for her superb work as Manager of Art and Design. Special thanks also go to Rolin Graphics for the new text art.

© *Eric Meola/The IMAGE Bank*

Finally, we dedicate this book to our wives and children for their love, support and long-term sacrifices.

Raymond A. Serway
James Madison University
Harrisonburg, Virginia

Jerry S. Faughn
Eastern Kentucky University
Richmond, Kentucky

July 1991

To the Student

Peter Aprahamian/Science Photo Library

We feel it is appropriate to offer some words of advice that should be of benefit to you, the student. Before doing so, we shall assume that you have read the preface, which describes the various features of the text that will help you through the course.

How to Study

Very often we are asked "How should I study physics and prepare for examinations?" There is no simple answer to this question, but we would like to offer some suggestions based on our own experiences in learning and teaching over the years.

First and foremost, maintain a positive attitude towards the subject matter, keeping in mind that physics is the most fundamental of all natural sciences. Other science courses that follow will use the same physical principles, so it is important that you understand and be able to apply the various concepts and theories discussed in the text.

Concepts and Principles

It is essential that you understand the basic concepts and principles *before* attempting to solve assigned problems. This is best accomplished through a careful reading of the textbook before attending your lecture on that material. In the process, it is useful to jot down certain points that are not clear to you. Take careful notes in class, and then ask questions pertaining to those ideas that require clarification. Keep in mind that few people are able to absorb the full meaning of scientific material after one reading. Several readings of the text and notes may be necessary. Your lectures and laboratory work should supplement the text and clarify some of the more difficult material. You should reduce memorization of material to a minimum. Memorizing passages from a text, equations, and derivations does not necessarily mean you understand the material. Your understanding of the material will be enhanced through a combination of efficient study habits, discussions with other students and instructors, and your ability to solve the problems in the text. Ask questions whenever you feel it is necessary. If you are reluctant to ask questions in class, seek private consultation or initiate discussions with your classmates. Many individuals are able to speed up the learning process when the subject is discussed on a one-to-one basis.

Study Schedule

It is important to set up a regular study schedule, preferably on a daily basis. Make sure to read the syllabus for the course and adhere to the schedule set by your instructor. The lectures will be much more meaningful if you read the corresponding textual material *before* attending the lecture. As a general rule, you should devote about two hours of study time for every hour in class. If you are having trouble with the course, seek the advice of the instructor or students who have already taken the course. You may find it necessary to seek further instruction from experienced students. Very often, instructors offer review sessions in addition to regular class periods. It is important that you avoid the practice of delaying study until a day or two before an exam. More often than not, this will lead to disastrous results. Rather than an all-night study session before an exam, it is better to briefly review the basic concepts and equations, followed by a good night's rest. If you feel you need additional help in understanding the concepts, preparing for exams, or in problem-solving, we suggest that you acquire a copy of the student study guide that accompanies the text, which should be available at your college bookstore.

Use the Features

You should make *full* use of the various features of the text discussed in the preface. For example, marginal notes are useful for locating and describing important equations, while important statements and definitions are highlighted in blue type. Many useful tables are contained in the appendices, but most are incorporated into the text where they are used most often. Appendix A is a convenient review of mathematical techniques. Answers to odd-numbered problems are given at the end of the text, and answers to most end-of-chapter questions are provided in the study guide. Exercises (with answers), which follow some worked examples, represent extensions of those examples, and in most cases you are expected to perform a simple calculation. Their purpose is to test your problem-solving skills as you read through the text. Problem-solving strategies are included in selected chapters throughout the text to give you additional information to help you solve problems. An overview of the entire text is given in the table of contents, while the index will enable you to locate specific material quickly. Footnotes are sometimes used to supplement the discussion or to cite other references on the subject. A list of suggested additional readings is given at the end of each chapter.

After reading a chapter, you should be able to define any new quantities introduced in that chapter and to discuss the principles and assumptions that were used to arrive at certain key relations. The chapter summaries and the review sections of the study guide should help you in this regard. In some cases, it will be necessary to refer to the index of the text to locate certain topics. You should be able to correctly associate with each physical quantity a symbol used to represent that quantity and the unit in which the quantity is specified. Furthermore, you should be able to express each important relation in a concise and accurate prose statement.

Problem Solving

R. P. Feynman, Nobel laureate in physics, once said, "You do not know anything until you have practiced." In keeping with this statement, we strongly advise that you develop the skills necessary to solve a wide range of problems. Your ability to solve problems will be one of the main tests of your knowledge of physics, and therefore you should try to solve as many problems as possible. It is essential that you understand basic concepts and principles before attempting to solve problems. It is good practice to try to find alternate solutions to the same problem. For example, problems in mechanics can be solved using Newton's laws, but very often an alternative method using energy considerations is more direct. You should not deceive yourself into thinking you understand the problem after seeing its solution in class. You must be able to solve the problem and similar problems on your own.

The method of solving problems should be carefully planned. A systematic plan is especially important when a problem involves several concepts. First, read the problem several times until you are confident you understand what is being asked. Look for any key words that will help you interpret the problem, and perhaps allow you to make certain assumptions. Your ability to interpret the question properly is an integral part of problem solving. You should acquire the habit of writing down the information given in a problem, and decide what quantities need to be found. You might want to construct a table listing quantities given, and quantities to be found. This procedure is sometimes used in the worked examples of the text. After you have decided on the method you feel is appropriate for the situation, proceed with your solution. General problem-solving strategies of this type are included in the text and are highlighted by a light tan screen.

We often find that students fail to recognize the limitations of certain formulas or physical laws in a particular situation. It is very important that you understand and

Courtesy of CENCO

remember the assumptions underlying a particular theory or formalism. For example, we shall find that certain equations in kinematics apply only to a particle moving with constant acceleration. They are not valid for situations in which the acceleration is not constant, as in the cases of the motion of an object connected to a spring and the motion of an object through a fluid.

Experiments

Physics is a science based upon experimental observations. In view of this fact, we recommend that you make every effort to supplement the text through various types of "hands-on" experiments, either in the laboratory or at home. Many simple experiments can be used to test ideas and models discussed in class or in the text. For example, an object swinging on the end of a long string together with a wristwatch can be used to investigate pendulum motion; an object attached to the end of a vertical spring or a rubber band can be used to determine the nature of restoring forces and to investigate periodic motion; collisions between equal masses can be observed while playing billiards; an approximate value for the acceleration due to gravity can be obtained by dropping an object from a known height and simply measuring the time of its fall with a stopwatch; traveling waves can be investigated with the aid of a stretched rope or the common "Slinky" toy (a stretched spring); a pair of Polaroid sunglasses and some discarded lenses or a magnifying glass can be used to perform various experiments in optics. The list is endless. When physical models are not available, be imaginative and try to develop models of your own.

Joseph Drivas/The IMAGE Bank

Closing Comments

Someone once said that there are only two professions in which people truly enjoy what they are doing: professional sports and physics. Although this statement is most likely an exaggeration, both professions are truly exciting and stretch your skills to the limit. It is our sincere hope that you too will find physics exciting and that you will benefit from this experience, regardless of your chosen profession.

Welcome to the exciting world of physics.

To see a World in a Grain of Sand
And a Heaven in a Wild Flower,
Hold infinity in the palm of your hand
And Eternity in an hour.

W. Blake, "Auguries of Innocence"

Introducing the Book

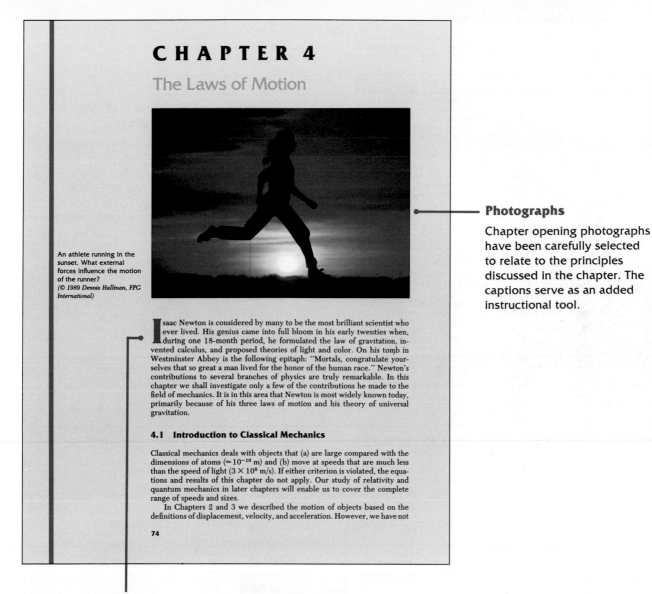

CHAPTER 4

The Laws of Motion

An athlete running in the sunset. What external forces influence the motion of the runner?
(© 1989 Dennis Hallinan, FPG International)

Isaac Newton is considered by many to be the most brilliant scientist who ever lived. His genius came into full bloom in his early twenties when, during one 18-month period, he formulated the law of gravitation, invented calculus, and proposed theories of light and color. On his tomb in Westminster Abbey is the following epitaph: "Mortals, congratulate yourselves that so great a man lived for the honor of the human race." Newton's contributions to several branches of physics are truly remarkable. In this chapter we shall investigate only a few of the contributions he made to the field of mechanics. It is in this area that Newton is most widely known today, primarily because of his three laws of motion and his theory of universal gravitation.

4.1 Introduction to Classical Mechanics

Classical mechanics deals with objects that (a) are large compared with the dimensions of atoms ($\approx 10^{-10}$ m) and (b) move at speeds that are much less than the speed of light (3×10^8 m/s). If either criterion is violated, the equations and results of this chapter do not apply. Our study of relativity and quantum mechanics in later chapters will enable us to cover the complete range of speeds and sizes.

In Chapters 2 and 3 we described the motion of objects based on the definitions of displacement, velocity, and acceleration. However, we have not

74

Photographs

Chapter opening photographs have been carefully selected to relate to the principles discussed in the chapter. The captions serve as an added instructional tool.

Chapter Previews

Include a brief discussion of chapter objectives and content.

Objects in Equilibrium and Newton's First Law

Objects that are at rest or that are moving with constant velocity are said to be in equilibrium, and Newton's first law is a statement of one condition that must be true for equilibrium conditions to prevail. In equation form, this can be expressed as

$$\sum \mathbf{F} = 0 \qquad (4.7)$$

This statement signifies that the *vector* sum of all the forces (the net force) acting on an object is zero. Usually, the problems we will encounter in our study of equilibrium will be more easily solved if we work with Equation 4.7 in terms of the components of the external forces acting on an object. By this, we mean that, in a two-dimensional problem, the sum of all the external forces in the x and y directions must separately equal zero, that is

$$\sum F_x = 0$$
$$\sum F_y = 0 \qquad (4.8)$$

This set of equations is often referred to as the **first condition for equilibrium**. We shall not consider three-dimensional problems in this text, but the extension of Equation 4.8 to a three-dimensional situation can be made by adding a third equation, $\sum F_z = 0$.

Problem-Solving Strategy:
Objects in Equilibrium

The following procedure is recommended for problems involving objects in equilibrium:

1. Make a sketch of the object under consideration.
2. Draw a free-body diagram and label all external forces acting on the object. Assume a direction for each force. If you select a direction that leads to a negative sign in your solution for a force, do not be alarmed; this merely means that the direction of the force is the opposite of what you assumed.
3. Resolve all forces into x and y components, choosing a convenient coordinate system.
4. Use the equations $\sum F_x = 0$ and $\sum F_y = 0$. Remember to keep track of the signs of the various force components.
5. Application of Step 4 above leads to a set of equations with several unknowns. Solve the simultaneous equations for the unknowns in terms of the known quantities.

Example 4.1 Tug-a-Boat

Two fishermen are pulling a boat through the water as in Figure 4.7. Each exerts a force of 600 N directed at a 30° angle relative to the forward motion of the boat. If the boat moves with constant velocity, find the resistive force **F** of the water on the boat.

Problem-Solving Strategies

Outlines general strategies for solving the types of problems featured in both the examples and the end-of-chapter problems. These steps are highlighted with a light tan screen so they can be easily located.

ng that the period of a

e period of a simple

$$\qquad (14.14)$$

that the period of a
od of a simple pendu-
e to gravity. Further-
ng as we restrict the
e motion of a simple
Figure 14.13.

The period of a simple pendulum depends only on L and g

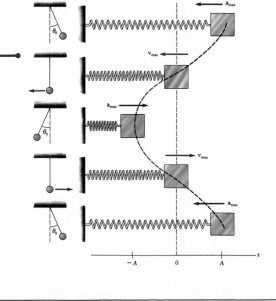

Figure 14.13
Simple harmonic motion for a mass-spring system and its analogy to the motion of a simple pendulum.

Color-Coded Line Art

Color is used to add clarity and three-dimensional effects. A complete description of the color system used appears on the inside front cover.

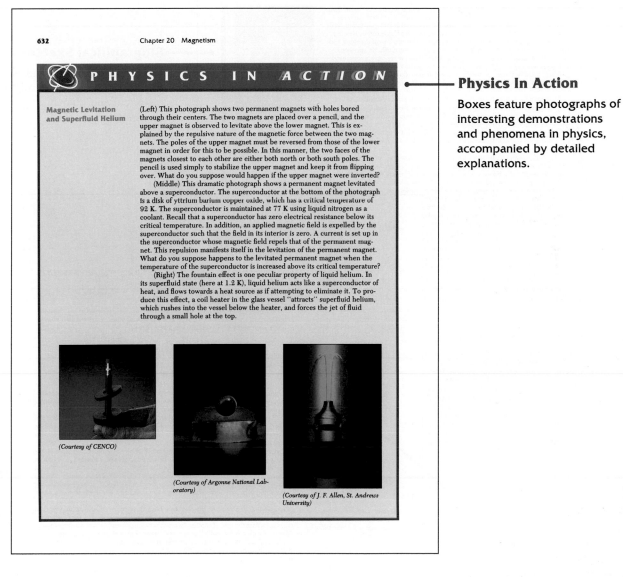

632   Chapter 20   Magnetism

PHYSICS IN ACTION

Magnetic Levitation and Superfluid Helium

(Left) This photograph shows two permanent magnets with holes bored through their centers. The two magnets are placed over a pencil, and the upper magnet is observed to levitate above the lower magnet. This is explained by the repulsive nature of the magnetic force between the two magnets. The poles of the upper magnet must be reversed from those of the lower magnet in order for this to be possible. In this manner, the two faces of the magnets closest to each other are either both north or both south poles. The pencil is used simply to stabilize the upper magnet and keep it from flipping over. What do you suppose would happen if the upper magnet were inverted?

(Middle) This dramatic photograph shows a permanent magnet levitated above a superconductor. The superconductor at the bottom of the photograph is a disk of yttrium barium copper oxide, which has a critical temperature of 92 K. The superconductor is maintained at 77 K using liquid nitrogen as a coolant. Recall that a superconductor has zero electrical resistance below its critical temperature. In addition, an applied magnetic field is expelled by the superconductor such that the field in its interior is zero. A current is set up in the superconductor whose magnetic field repels that of the permanent magnet. This repulsion manifests itself in the levitation of the permanent magnet. What do you suppose happens to the levitated permanent magnet when the temperature of the superconductor is increased above its critical temperature?

(Right) The fountain effect is one peculiar property of liquid helium. In its superfluid state (here at 1.2 K), liquid helium acts like a superconductor of heat, and flows towards a heat source as if attempting to eliminate it. To produce this effect, a coil heater in the glass vessel "attracts" superfluid helium, which rushes into the vessel below the heater, and forces the jet of fluid through a small hole at the top.

(Courtesy of CENCO)

(Courtesy of Argonne National Laboratory)

(Courtesy of J. F. Allen, St. Andrews University)

Physics In Action

Boxes feature photographs of interesting demonstrations and phenomena in physics, accompanied by detailed explanations.

4.4   Newton's Second Law   **79**

Isaac Newton, a British physicist and mathematician, is regarded as one of the greatest scientists in history. Before the age of 30, he formulated the basic concepts and laws of motion, discovered the universal law of gravitation, and formulated the methods of calculus. As a consequence of his theories, Newton was able to explain the motion of the planets, the ebb and flow of the tides, and many special features of the motion of the Moon and the Earth. He also made many fundamental discoveries in optics, showing that white light is composed of a spectrum of colors. His contributions to physical theories dominated scientific thought for two centuries and remain important today.

Newton was born prematurely on Christmas Day in 1642 shortly after his father's death. When he was three, his mother remarried and he was left in his grandmother's care. Because he was small in stature as a child, he was bullied by other children and would take refuge in mechanical things such as building waterclocks, kites carrying fiery lanterns, sundials, and model windmills powered by mice. His mother withdrew him from school at the age of 12 with the intention of making him into a farmer. Fortunately, his uncle recognized his scientific and mathematical abilities and helped send him to Trinity College in Cambridge.

In 1665, Newton completed his Bachelor of Arts degree and the university was closed because of the bubonic plague that raged through England. During this period of turmoil, Newton returned to the family farm at Woolsthorpe to study. This was an especially creative period of his life during which time he laid the foundations of his work in mathematics, optics, the laws of motion, celestial mechanics, and his famous law of gravit

Newton was a very private person who stud night in his laboratory conducting experiments, immersing himself in theological studies. There and he never married. His greatest single work *Natural Philosophy,* was published in 1687. In of his time quarreling with other eminent minds, Gottfried Leibnitz, who worked independently calculus; Christian Huygens, who worked on th Robert Hooke, who supported Huygens' theory his studies and his work on alchemy, which invol him to suffer severe depression in 1692. He was Society in 1703, and he retained that office unti

Isaac Newton
(1642–1727)

● ── **Biographical Sketches**

Feature prominent scientists, giving the third edition added historical and human emphasis.

4.4   Newton's Second Law

Newton's first law explains what happens to an ob acting on it is zero. In such instances we find that or moves with constant velocity. Newton's secon what happens to an object that has a nonzero r

In order to gain some insight into what happe to an object, consider a situation in which you ar a polished surface, smooth enough to make frict you exert a certain force on the block of ice, y acceleration of, say, 2 m/s². If you push twice as h ation also doubles. Pushing three times as hard t on. From observations such as this, we can concl

Special Topics

Special topics sections expose the student to interesting applications of physical principles.

---

332   Chapter 11   Thermal Physics

(b) Determine the rms speed of the atoms.

Solution
Because the molecular mass of helium is 4 g/mol = 4×10^{-3} kg/mol, the rms speed is

$$v_{rms} = \sqrt{\frac{3RT}{M}} = \sqrt{\frac{3(8.317\ \text{J/mol·K})(293\ \text{K})}{4 \times 10^{-3}\ \text{kg/mol}}} = \boxed{1.35 \times 10^3\ \text{m/s}}$$

The same result is obtained using $v_{rms} = \sqrt{3kT/m_0}$. Remember that m_0, the mass of one molecule, is obtained by dividing the molecular mass by the number of particles in 1 mole of the gas (Avogadro's number). Try it!

▶ **SPECIAL TOPIC**
Evaporation, Humidity, and Boiling

If a container of water is left open, the water slowly evaporates. That is, it spontaneously goes from the liquid state to the gaseous state. The kinetic theory of gases enables us to explain why this phenomenon occurs.

The molecules of a liquid are in continuous random motion. Occasionally, the upward component of velocity of a molecule near the surface of the liquid will be sufficiently large to enable the molecule to escape from the liquid and become a gaseous molecule.

Equation 11.18 shows that the temperature of an ideal gas is a measure of the average kinetic energy of its molecules. This concept can be extended to liquids as well. The molecules that evaporate from the liquid are those that have sufficient energy (greater than the average kinetic energy) to break free of their neighbors on the surface. Since the evaporating molecules are very energetic, the molecules remaining in the liquid must have lower-than-average kinetic energy. For this reason, the liquid becomes cooler as a result of evaporation. You may have noticed this cooling due to evaporation after a swim or shower. Your body feels cooler when you are wet than when you are dry because the water is evaporating. As the rate of evaporation increases, so does the rate of cooling. In another familiar example, when a doctor swabs your arm with alcohol (which evaporates faster than water) before giving you an injection, the alcohol serves two purposes. It sterilizes the area to be injected, and it slightly numbs the area as a result of its rapid evaporation and subsequent cooling of the skin.

Humidity

The rate of evaporation is enhanced if air is moved rapidly over the surface of the liquid. This occurs because the liquid molecules that vaporize are removed by the moving air before they have time to undergo collisions with air molecules near the surface that could deflect them back into the liquid (Fig. 11.13). For the same reason, fanning yourself on a hot day helps you cool down.

Let us now look at the exact opposite of fanning. Let us assume that water is placed in a sealed container such that the gaseous molecules cannot escape, as in Figure 11.14. In this case, the air above the liquid will absorb water molecules until it can hold no more. At this point, the air is said to be **saturated** with water vapor. Saturation occurs when the number of water molecules returning to the liquid in some time interval is equal to the number leaving the liquid in the same time interval. The amount of water vapor that air can hold increases as the temperature of the air increases.

Figure 11.13
The evaporation rate increases as air moves over the surface of the liquid.

Figure 11.14
A liquid placed in a sealed container. The air above the liquid becomes saturated when the number of molecules leaving the liquid equals the number returning to the liquid in some time interval.

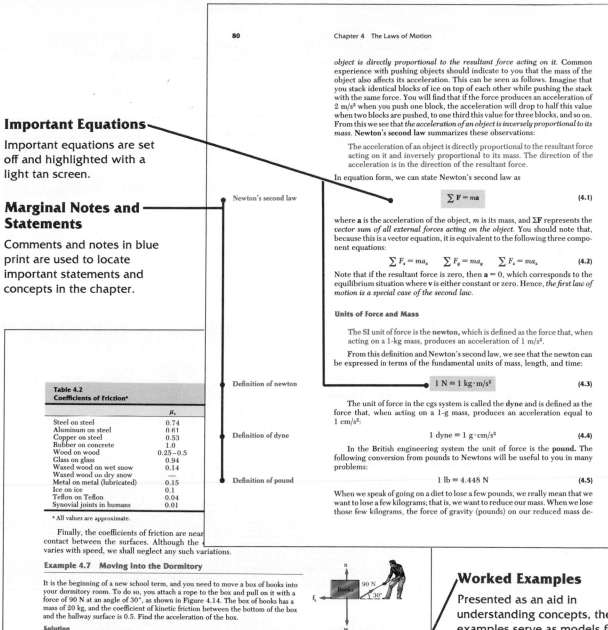

Important Equations

Important equations are set off and highlighted with a light tan screen.

Marginal Notes and Statements

Comments and notes in blue print are used to locate important statements and concepts in the chapter.

Table 4.2
Coefficients of Friction*

	μ_s
Steel on steel	0.74
Aluminum on steel	0.61
Copper on steel	0.53
Rubber on concrete	1.0
Wood on wood	0.25–0.5
Glass on glass	0.94
Waxed wood on wet snow	0.14
Waxed wood on dry snow	—
Metal on metal (lubricated)	0.15
Ice on ice	0.1
Teflon on Teflon	0.04
Synovial joints in humans	0.01

* All values are approximate.

Finally, the coefficients of friction are near contact between the surfaces. Although the varies with speed, we shall neglect any such variations.

Example 4.7   Moving Into the Dormitory

It is the beginning of a new school term, and you need to move a box of books into your dormitory room. To do so, you attach a rope to the box and pull on it with a force of 90 N at an angle of 30°, as shown in Figure 4.14. The box of books has a mass of 20 kg, and the coefficient of kinetic friction between the bottom of the box and the hallway surface is 0.5. Find the acceleration of the box.

Solution
The box is not accelerating in the vertical direction, and so we find the normal force from $\Sigma F_y = 0$. The forces in the y direction are the weight, w, the normal force, n, and the vertical component of your 90-N pull. This vertical component has a magnitude equal to (90 N)(sin 30°). We find that

$$\Sigma F_y = n + (90\ \text{N})(\sin 30°) - (20\ \text{kg})(9.80\ \text{m/s}^2) = 0$$

$$n = 151\ \text{N}$$

Note that in this case the normal force is *not* equal to the weight of the box of books because the vertical component of your pull is helping to support some of the weight of the box.

Since the normal force is known, we can find the force of kinetic friction:

$$f_k = \mu_k n = (0.5)(151\ \text{N}) = 75.5\ \text{N} \quad \text{(to the left)}$$

Finally, we determine the horizontal acceleration using Newton's second law:

$$\Sigma F_x = (90\ \text{N})(\cos 30°) - 75.5\ \text{N} = (20\ \text{kg})(a_x)$$

$$a_x = \ +0.122\ \text{m/s}^2$$

Exercise If the initial speed of the box is zero, what is its speed after it has traveled 2 m? How long does it take you to pull it a distance of 2 m?

Answer 0.699 m/s, 5.73 s.

---

object is directly proportional to the resultant force acting on it. Common experience with pushing objects should indicate to you that the mass of the object also affects its acceleration. This can be seen as follows. Imagine that you stack identical blocks of ice on top of each other while pushing the stack with the same force. You will find that if the force produces an acceleration of 2 m/s² when you push one block, the acceleration will drop to half this value when two blocks are pushed, to one third this value for three blocks, and so on. From this we see that *the acceleration of an object is inversely proportional to its mass.* **Newton's second law** summarizes these observations:

> The acceleration of an object is directly proportional to the resultant force acting on it and inversely proportional to its mass. The direction of the acceleration is in the direction of the resultant force.

In equation form, we can state Newton's second law as

Newton's second law

$$\Sigma \mathbf{F} = m\mathbf{a} \tag{4.1}$$

where $\mathbf{a}$ is the acceleration of the object, m is its mass, and $\Sigma\mathbf{F}$ represents the *vector sum of all external forces acting on the object*. You should note that, because this is a vector equation, it is equivalent to the following three component equations:

$$\Sigma F_x = ma_x \qquad \Sigma F_y = ma_y \qquad \Sigma F_z = ma_z \tag{4.2}$$

Note that if the resultant force is zero, then $\mathbf{a} = 0$, which corresponds to the equilibrium situation where $\mathbf{v}$ is either constant or zero. Hence, *the first law of motion is a special case of the second law.*

Units of Force and Mass

The SI unit of force is the **newton**, which is defined as the force that, when acting on a 1-kg mass, produces an acceleration of 1 m/s².

From this definition and Newton's second law, we see that the newton can be expressed in terms of the fundamental units of mass, length, and time:

Definition of newton

$$1\ \text{N} \equiv 1\ \text{kg} \cdot \text{m/s}^2 \tag{4.3}$$

The unit of force in the cgs system is called the **dyne** and is defined as the force that, when acting on a 1-g mass, produces an acceleration equal to 1 cm/s²:

Definition of dyne

$$1\ \text{dyne} \equiv 1\ \text{g} \cdot \text{cm/s}^2 \tag{4.4}$$

In the British engineering system the unit of force is the **pound**. The following conversion from pounds to Newtons will be useful to you in many problems:

Definition of pound

$$1\ \text{lb} = 4.448\ \text{N} \tag{4.5}$$

When we speak of going on a diet to lose a few pounds, we really mean that we want to lose a few kilograms; that is, we want to reduce our mass. When we lose those few kilograms, the force of gravity (pounds) on our reduced mass de-

Figure 4.14 (Example 4.7) A box of books being pulled to the right at an angle of 30° with a rough horizontal surface.

Worked Examples

Presented as an aid in understanding concepts, these examples serve as models for solving the end-of-chapter problems. The examples are set between blue rules, and the solution answer is highlighted with a light blue screen.

Worked Example Exercises

Many of the worked examples are followed immediately by exercises with answers to make the textbook more interactive and to test the student's understanding of problem-solving techniques.

Summary

The **linear momentum** of an object of mass m moving with a velocity $\mathbf{v}$ is defined to be

$$\mathbf{p} = m\mathbf{v} \qquad (6.1)$$

The **impulse** of a force $\mathbf{F}$ acting on an object is equal to the product of the force and the time interval during which the force acts:

$$\text{Impulse} = \mathbf{F}\,\Delta t$$

The **impulse-momentum theorem** states that the impulse of a force on an object is equal to the change in momentum of the object:

$$\mathbf{F}\,\Delta t = \Delta \mathbf{p} = m\mathbf{v}_f - m\mathbf{v}_i \qquad (6.4)$$

The **law of conservation of momentum** of two interacting objects states that, if the two objects form an isolated system, their total momentum is conserved regardless of the nature of the forces between them:

$$m_1\mathbf{v}_{1i} + m_2\mathbf{v}_{2i} = m_1\mathbf{v}_{1f} + m_2\mathbf{v}_{2f} \qquad (6.5)$$

An **inelastic collision** is one in which mome[...]
energy is not. **A perfectly inelastic collision** [...]
objects stick together after the collision. An e[...]
both momentum and kinetic energy are conse[...]
In glancing collisions, conservation of mo[...]
two perpendicular directions, that is, along an [...]
The x and y coordinates of the center of grav[...]
object are

$$x_c = \frac{\sum m_i x_i}{\sum m_i} \qquad y_c = \frac{\sum ...}{...}$$

Newton's second law applied to a system o[...]

$$\mathbf{F}_{ext} = M\mathbf{a}_c$$

where $\mathbf{a}_c$ is the acceleration of the center of mas[...]
external force on the system. This equation says[...]
like an imaginary particle of mass M under t[...]
external force on the system.

Summary

Reviews important concepts and equations discussed in the chapter.

37. A 1500-kg car traveling east with a speed of 20 m/s collides at an intersection with a 2500-kg van traveling south at a speed of 15 m/s. The vehicles undergo a perfectly inelastic collision, and the wreckage slides 6 m before coming to rest. Find the magnitude and direction of the constant force that has decelerated them.

38. A neutron in a reactor makes an elastic head-on collision with a carbon atom initially at rest. (The mass of the carbon nucleus is about 12 times the mass of the neutron.) (a) What fraction of the neutron's kinetic energy is transferred to the carbon nucleus? (b) If the initial kinetic energy of the neutron is 1.6×10^{-13} J, find its final kinetic energy and the kinetic energy of the carbon nucleus after the collision.

39. Consider the ballistic pendulum described in Example 6.7 and shown in Figure 6.12. If the mass of the bullet is 8 g and the mass of the pendulum is 2 kg, find the ratio of the kinetic energy after the collision to the kinetic energy before collision. What accounts for this missing energy?

40. A cue ball traveling at a speed of 4 m/s makes a glancing, elastic collision with a target ball of equal mass initially at rest. It is found that the cue ball is deflected such that it makes an angle of 30° with respect to its original direction of travel. Find (a) the angle the second ball will move and (b) the speed of each ball after the collision.

Section 6.5 The Center of Gravity and The Center of Mass

41. Four objects are located along the y axis as follows. A 2-kg object is at $+3$ m, a 3-kg object is at $+2.5$ m, a 2.5-kg object is at the origin, and a 4-kg object is at -0.5 m. Where is the center of gravity of these objects located?

42. Where would a fifth object of mass 5 kg have to be placed along the y axis so that the center of gravity of the objects in Problem 41 is at the origin?

43. Three masses are located in a rectangular coordinate system as follows: a 2-kg mass is at $(3, -2)$, a 3-kg mass is at $(-2, 4)$, and a 1-kg mass is at $(2, 2)$, where all distances are in meters. Find the coordinates of the center of gravity for the three masses.

44. A uniform T square used by a draftsman has dimensions as shown in Figure 6.24. Locate the center of gravity with respect to point O. (*Hint:* Note that the mass of each rectangular part is proportional to its area.)

45. A uniform carpenter's square has the shape of an L, as in Figure 6.25. Locate the center of gravity relative to an origin at the lower left corner. (Note the hint for Problem 44.)

Figure 6.24 Problem 44

Figure 6.25 Problem 45

46. The mass of the Earth is 5.98×10^{24} kg and the mass of the Moon is 7.36×10^{22} kg. The distance of separation measured from their centers is 3.84×10^8 m. Determine the location of the center of mass of the Earth-Moon system as measured from the center of the Earth.

47. The nucleus of a hydrogen atom is a single proton of mass 1.67×10^{-27} kg. The mass of the electron which circles the proton at a radius of about 0.51×10^{-10} m is 9.1×10^{-31} kg. Find the center of mass of the proton-electron system as measured from the proton.

48. Three masses located in the xy plane have the following coordinates: a 2-kg mass has coordinates $(3, -2)$ m, a 3-kg mass has coordinates $(-2, 4)$ m, and a 1-kg mass has coordinates $(2, 2)$ m. Find the coordinates of the center of mass of the system.

49. An 80.0-kg astronaut is on a space "walk" when the engine on his back fails. He is at rest with respect to

Problems Graded By Level of Difficulty

Intermediate-level problems are indicated by printing the problem number in blue print; challenging problems have a magenta problem number.

Problems Keyed To Sections

Allows students to refer back to the sections to look at a worked example for help in solving problems.

Newton's universal law of gravitation—today it is taught in every introductory physics course and is so familiar that it is easy to forget how profoundly this simple equation changed the nature of science. Newton not only offered a precise mathematical description of the motions of material objects, he also showed that this description applied far beyond the boundaries of the Earth. The same force that causes apples to fall to the ground also keeps the Moon revolving about the Earth and holds the planets in their orbits around the Sun. The heavens were transformed; no longer mysterious and unknowable, the Moon, the planets, and potentially even the stars, all subject to the same physical laws that are valid on Earth, became legitimate objects of scientific inquiry.

But is Newton's law of gravitation truly universal? Does it apply to the stars? Newton assumed that it did, but measurements available during his lifetime were not precise enough to offer proof. The influence of gravity is detected by analyzing the motions of objects, and apart from their rising and setting as a consequence of the Earth's rotation, stars did not appear to the astronomers of Newton's time to have motions of their own. The fact that there are clusters or groupings of stars, like the Pleiades and Hyades, suggested that attractive forces do influence stars, but it was meticulous measurements by Sir William Herschel, best known for his discovery of the planet Uranus, that offered the first concrete evidence.

In January 1782, Herschel published a catalog of 269 pairs of stars that appear to be very close together in the sky. At that time, he did not know whether these double stars form physical systems or simply happen to appear in very nearly the same direction but at very different distances from the Earth. About 20 years later, Herschel repeated the observations of the double stars that he had found, and discovered that some pairs had changed position relative to each other in a way that could only be explained if they were revolving around each other. Several more decades of measurements by other astronomers proved that these stars do indeed follow orbits that can be described by the inverse square force of Newton's law of gravitation.

The conclusion that the motions of stars are c̶ problem that was recognized by Newton. Up until century it was assumed that the Universe is static, t panding nor contracting. If the theory of gravitati in a static Universe, and if the Universe is finite in why the Universe has not already collapsed to a si are attracted to one another. Newton thought that perhaps be stable, since gravitational forces of all throughout space might possibly cancel their mutu

The next major advance in gravitational theor general theory of relativity. In 1917, Einstein appl verse as a whole and found that there were no solu a homogeneous static Universe, even if the Univers Einstein consequently introduced a new term, call which represents a repulsion that can balance gravi distances and permit a static Universe. There is, ho repulsion in nature; the cosmological constant was general relativity conform to the preconceived noti

Subsequent theoretical work by the Russian m mann showed that general relativity is compatible expanding Universe for all values of the cosmologi At about the same time, observations proved that t expanding. American physicist George Gamow rep ing that his relativity theory is compatible with an

The Big Bang

Essay

Sidney C. Wolff
*National Optical
Astronomy Observatory*

Guest Essays

Intended as supplemental reading for the student, the essays have been placed at the ends of chapters to avoid interrupting the main textual material. The color bar at the edges of the pages helps students locate them easily.

With time, the picture presented here of the origin, evolution, and ultimate fate of the material Universe may undergo a comparably radical transformation. Indeed, astronomers are attracted to this area of research in substantial numbers precisely because the subject is not fully understood and seems likely to yield its secrets to a systematic scientific attack. To learn something new, something never known before, is the ultimate challenge and satisfaction of scientific research.

Suggested Reading

T. Ferris, *The Red Limit*, 2nd ed., New York, Morrow, Quill, 1983.
E. R. Harrison, *Cosmology*, Cambridge, England, Cambridge University Press, 1981.
R. Wagoner and D. Goldsmith, *Cosmic Horizons: Understanding the Universe*, New York, W. H. Freeman, 1983.

Essay Questions

1. When Hubble and Humason first reported their results for the relation between distance and velocity for galaxies, they thought that the constant of proportionality H was equal to about 200 km/s per million lightyears. How did their estimate for the age of the Universe differ from the modern value based on $H = 20$ km/s per million lightyears? The oldest stars in the Universe are about 15 billion years old. Is this consistent with Hubble and Humason's value for the Hubble constant H?
2. In the text, it was explained that it was possible to estimate the masses of two galaxies orbiting about each other by measuring their separations and velocities and applying Kepler's third law. Could this be done by measuring the orbital period directly? Why or why not? (Remember that galaxies contain billions of stars and have diameters of about 100,000 lightyears.)
3. Explain why the observation that all galaxies are moving away from us does not mean that we are at the center of the Universe.

Essay Problems

1. Plot the "velocity-distance relation" for the raisins-in-the-bread analogy from the data given in Figure 3. Repeat but use some other raisin than A for a reference. Is your new plot the same as the last one?
2. Current estimates for the Hubble constant H range from 15 to 30 km/s per million lightyears, where one lightyear is the distance that light travels in one year. What is the corresponding range in the estimated age of the universe?
3. Derive an expression for escape velocity in terms of mean density. Now apply this result to two galaxies, separated by a distance R and moving away from each other at just the escape velocity and derive an expression for the critical density at which the escape and expansion velocities are equal. Assume that $H = 20$ km/s per million lightyears. How does your value for the critical density compare with the observed density, which is currently estimated to be less than about 10^{-30} g/cm^3. What does your calculation indicate about the ultimate fate of the Universe?

Suggested Readings, Essay Questions, and Problems

Most of the essays contain suggested readings, questions, and problems for added flexibility in covering these topics.

210

Conversations With Prominent Scientists

Appearing in each of the text's six parts, the interviews introduce students to individuals who have achieved international recognition in their fields. The conversations cover specialized scientific topics as well as current concerns such as science education and the role of science in society.

Interview

Conversation with
Hans Bethe

Born in Strasbourg in 1906, Hans Albrecht Bethe studied physics, chemistry, and mathematics at the University of Frankfurt and passed his doctoral exam in physics at the University of Munich in 1928. There he studied with Arnold Sommerfeld, an influential theoretical physicist whose students included Wolfgang Pauli, Werner Heisenberg, Peter Debye, and Max von Laue, all of whom later won the Nobel Prize. He collaborated with Sommerfeld on a long article containing the basis for the modern solid-state theory of metals in 1932, and he was subsequently the author of two other important articles. One was on atomic physics, and the other was the first comprehensive treatment of nuclear physics.

As the Nazi movement grew, Bethe lost his job as an assistant professor at Tübingen because his mother was Jewish; he was given a fellowship by Sommerfeld, who aided many physicists of Jewish ancestry. Bethe spent the years 1930 and 1931 at Cambridge University on a Rockefeller Foundation fellowship and at the University of Rome, where he worked with Enrico Fermi. In 1935 he joined the faculty of Cornell University in Ithaca, New York, where he remains today as professor emeritus.

Interview conducted by Janice K. Mandel

872

Bethe's many honors include receiving the President's Medal of Merit in 1946 for his work as the head of the Theoretical Division of the Los Alamos atomic bomb project (known as the Manhattan Project), and receiving the 1967 Nobel Prize in Physics for his work on the energy production in stars.

Why did you decide to become a physicist?

I was always interested in numbers as a child. My mother found a little booklet in which I had done all sorts of exercises with numbers like powers of 2, 3, and 5 when I was about 7 years old. Then [numbers] seemed a little too narrow a subject, and so, as I learned later in high school about physics, I saw that the real place I could use my interest in numbers was physics, and that was it.

How did you become involved in astrophysics?

I got into this by good luck. The Carnegie Foundation sponsored a conference in Washington in which astrophysicists were discussing solar energy and reported on their investigations of the interior of stars. They were convinced that the energy in stars comes from nuclear reactions, and so I said to myself, "I know nuclear reactions; I have written a long article about it in *The Review of Modern Physics*, so I probably could answer that question on just what nuclear reactions are responsible."

I didn't want to go to that conference. I was invited by my friend Edward Teller, but it was not a subject along the main line of my interest, so I was very hesitant in accepting and only after some persuasion did I go. When I came home after that conference I set to work. Karl Weizsäcker had previously made the suggestion that the source of stellar energy was the collision between two protons (hydrogen nuclei) making the nucleus of heavy hydrogen. Following this suggestion, a young graduate student in Washington, Critchfield, put some actual numbers into the model. Later, Critchfield and I wrote a paper together on this reaction.

However, I now wanted to know whether this was the only nuclear reaction that can take place in stars, and so I searched through the periodic table of the elements. Nothing worked until I came to carbon, and then it worked beautifully. Carbon adds four protons one after the other; of course, that takes millions of years in the star. Then a beautiful thing happens. The protons combine to make a helium nucleus, which releases a great amount of energy, and the car-

Contents Overview

Conversation with
Leon M. Lederman   *22*

Courtesy of Fermilab

Conversation with
Steven Chu   *308*

Courtesy of Frans Alkemade

Conversation with
Freeman J. Dyson   *416*

Courtesy of Mia Dyson

Conversation with
Heinrich Rohrer   *496*

Courtesy of Chuck Slade and IBM Research

Conversation with
M. Brian Maple   *744*

Conversation with
Hans Bethe   *872*

Contents

Francois Dardelet/The IMAGE Bank

* These sections are optional.

* These sections are optional.

* These sections are optional.

Hank Delespinasse/The IMAGE Bank

Courtesy of IBM Research

* These sections are optional.

© *Michael Keller/FPG
International*

* These sections are optional.

© Orion 1988/FPG International

* These sections are optional.

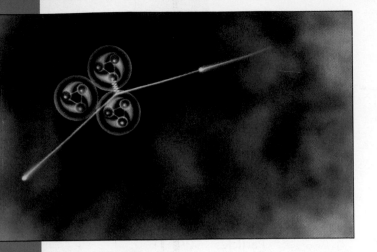

Courtesy of CEBAF

* These sections are optional.

PART 1
Mechanics

Physics, the most fundamental science, is concerned with the basic principles of the Universe. It is one of the foundations upon which the other physical sciences—astronomy, chemistry, and geology—are based. The beauty of physics lies in the simplicity of the fundamental physical theories and in the way just a small number of fundamental concepts, equations, and assumptions can alter and expand our view of the world around us.

The myriad physical phenomena in our world are a part of one or more of the following five areas of physics:

1. Mechanics, which is concerned with the motion of material objects
2. Thermodynamics, which deals with heat, temperature, and the behavior of a large number of particles
3. Electromagnetism, which deals with charges, currents, and electromagnetic fields
4. Relativity, which is a theory describing particles moving at any speed
5. Quantum mechanics, a theory dealing with the behavior of particles at the submicroscopic level as well as the macroscopic world

The first part of this textbook deals with mechanics, sometimes referred to as classical mechanics or Newtonian mechanics. This is an appropriate place to begin an introductory text since many of the basic principles used to understand mechanical systems can later be used to describe such natural phenomena as waves and heat transfer. Furthermore, the laws of conservation of energy and momentum introduced in mechanics retain their importance in the fundamental theories that follow, including the theories of modern physics.

The first serious attempts to develop a theory of motion were provided by the Greek astronomers and philosophers. Although they devised a complex model to describe the motions of heavenly bodies, their model lacked correlation between such motions and the motions of objects on Earth. The study of mechanics was enhanced by a number of careful astronomical investigations by Copernicus, Brahe, and Kepler in the 16th century. In the 16th and 17th centuries, Galileo attempted to relate the motion of falling bodies and projectiles to the motion of planetary bodies, and Sevin and Hooke were studying forces and their relation to motion. A major development in the theory of mechanics was provided by Newton in 1687 when he published his *Principia*. Newton's elegant theory, which remained unchallenged for more than 200 years, was based on contributions made by Galileo and others, together with Newton's hypothesis of universal gravitation.

Today, mechanics is of vital importance to students from all disciplines. It is highly successful in describing the motions of material bodies, such as planets, rockets, and baseballs. In the first ten chapters of the text, we shall describe the laws of mechanics and examine a wide range of phenomena which can be understood with these fundamental ideas.

Facts which at first seem improbable will, even in scant explanation, drop the cloak which has hidden them and stand forth in naked and simple beauty.
GALILEO GALILEI

Nature and Nature's laws lay hid in night;
God said, Let Newton be.
and all was light.
ALEXANDER POPE

1

CHAPTER 1

Introduction

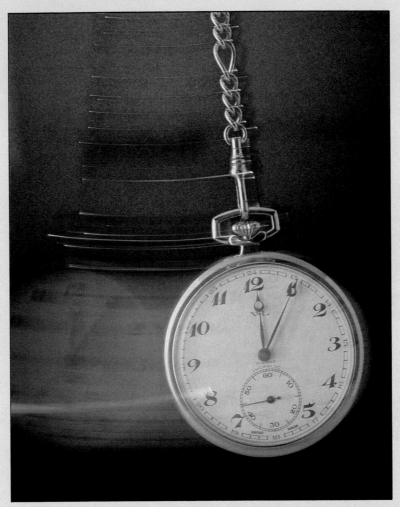

A timepiece is commonly used to measure time intervals in sporting events, laboratory environments, and in industry.
(Don Allen Sparks, The Image Bank)

The goal of physics is to provide a quantitative understanding of certain basic phenomena that occur in our Universe. Physics is a science based on experimental observations and mathematical analyses. The main objective behind such experiments and analyses is to develop theories that explain the phenomenon being studied and to relate these theories to other established theories. Fortunately, it is possible to explain the behavior of various physical systems using relatively few fundamental laws. Analytical procedures require expressing these laws in the language of mathematics, the tool that provides a bridge between theory and experiment. In this chapter we shall discuss a few mathematical concepts and techniques that will be used throughout the text.

Since future chapters will be concerned with the laws of physics, it is necessary to provide a clear definition of the basic quantities involved in these laws. For example, such physical quantities as force, velocity, volume, and acceleration can be described in terms of more fundamental quantities. In the next several chapters we shall encounter three basic quantities: **length** (L), **time** (T), and **mass** (M). In later chapters we shall have to add two other standard units to our list, one for temperature (the kelvin), the other for electric current (the ampere). In our study of mechanics, however, we shall be concerned only with the units of mass, length, and time.

1.1 Standards of Length, Mass, and Time

If we are to report the results of a measurement of a certain quantity to someone who wishes to reproduce this measurement, a unit for this quantity must be defined. For example, it would be meaningless for a visitor from another planet to talk to us about a length of 8 "gliches" if we do not know the meaning of the unit "glich." On the other hand, if someone familiar with our system of measurement and weights reports that a wall is 2.0 meters high and our unit of length is defined to be 1.0 meter, we then know that the height of the wall is twice our fundamental unit of length. Likewise, if we are told that a person has a mass of 75 kilograms and our unit of mass is defined as 1.0 kilogram, then that person has a mass 75 times greater than our fundamental unit of mass. In 1960, an international committee agreed on a set of definitions and standards to describe these fundamental physical quantities. The system that was established is called the **SI system** (Système International) of units. In this system, the units of mass, length, and time are the kilogram, meter, and second, respectively.

Length

In A.D. 1120 the king of England decreed that the standard of length in his country would be the yard and that the yard would be precisely equal to the distance from the tip of his nose to the end of his outstretched arm. Similarly, the original standard for the foot adopted by the French was the length of the royal foot of King Louis XIV. This standard prevailed until 1799, when the legal standard of length in France became the meter, defined as one ten-millionth of the distance from the equator to the north pole.

There have been many other systems developed in addition to those discussed above, but the advantages of the French system have caused it to become the prevailing system in most countries and in scientific circles everywhere. As recently as 1960, the length of the meter was preserved as the distance between two lines on a specific bar of platinum-iridium alloy stored under controlled conditions. This standard was abandoned for several reasons, a principal one being that the limited accuracy with which the separation between the lines can be determined does not meet the present requirements of science and technology. Until recently, the meter was defined as 1,650,763.73 wavelengths of orange-red light emitted from a krypton-86 lamp. However, in October 1983, the **meter** was redefined to be *the distance traveled by light in a vacuum during a time of 1/299,792,458 second*. In effect, this latest definition establishes that the speed of light in a vacuum is 299,792,458 meters per second.

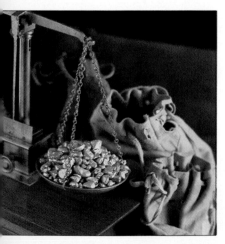

Quantitative measurement is one of the tools of physicists. Here the mass of gold nuggets is shown being measured on a scale. (© R. Pleasant, FPG International)

(Left) The National Standard Kilogram No. 20, an accurate copy of the International Standard Kilogram kept at Sèvres, France, is housed under a double bell jar in a vault at the National Institute of Standards and Technology. (Right) The primary frequency standard (an atomic clock) at the National Institute of Standards and Technology. This device keeps time with an accuracy of about 3 millionths of a second per year. (Photos courtesy of National Institute of Standards and Technology, U.S. Dept. of Commerce)

Mass

The SI unit of mass, the **kilogram,** is defined as *the mass of a specific platinum-iridium alloy cylinder kept at the International Bureau of Weights and Measures at Sèvres, France.* At this point, we should add a word of caution. Most beginning students of physics tend to confuse the quantities called weight and mass. For the present, we shall not discuss the distinction between these two quantities, which will be clearly defined in later chapters. For now, however, you should note that these are two distinctly different physical quantities.

Time

Before 1960, the standard of time was defined in terms of the average length of a solar day for the year 1900. (A solar day is the time interval between successive appearances of the sun at the highest point it reaches in the sky each day.) The basic unit of time, the second, was defined to be $(1/60)(1/60)(1/24) = 1/86,400$ of the average solar day. In 1967, the second was redefined to take advantage of the high precision obtainable with a device known as an atomic clock, which uses the characteristic frequency of the cesium-133 atom as the "reference clock." The **second** is now defined as *9,192,631,770 times the period of one oscillation of the cesium atom.*

Approximate Values for Length, Mass, and Time

Approximate values of various lengths, masses, and time intervals are presented in Tables 1.1, 1.2, and 1.3, respectively. Note the wide range of values for these quantities. You should study these tables and get a feel for

Table 1.1
Approximate Values of Some Measured Lengths

	Length (m)
Distance from Earth to most remote known quasar	1×10^{26}
Distance from Earth to most remote known normal galaxies	4×10^{25}
Distance from Earth to nearest large galaxy (M31 in Andromeda)	2×10^{22}
Distance from Earth to nearest star (Proxima Centauri)	4×10^{16}
One lightyear	9×10^{15}
Mean orbit radius of the Earth about the Sun	2×10^{11}
Mean distance from Earth to Moon	4×10^{8}
Mean radius of the Earth	6×10^{6}
Typical altitude of orbiting Earth satellite	2×10^{5}
Length of a football field	9×10^{1}
Length of a housefly	5×10^{-3}
Size of smallest dust particles	1×10^{-4}
Size of cells of most living organisms	1×10^{-5}
Diameter of a hydrogen atom	1×10^{-10}
Diameter of an atomic nucleus	1×10^{-14}
Diameter of a proton	1×10^{-15}

Table 1.2
Approximate Values of Various Masses

	Mass (kg)
Universe	10^{52}
Milky Way galaxy	7×10^{41}
Sun	2×10^{30}
Earth	6×10^{24}
Moon	7×10^{22}
Shark	1×10^{2}
Human	7×10^{1}
Frog	1×10^{-1}
Mosquito	1×10^{-5}
Bacterium	1×10^{-15}
Hydrogen atom	2×10^{-27}
Electron	9×10^{-31}

what is meant by a kilogram of mass (this book has a mass of about 2 kilograms) or a time interval of 10^{10} seconds (one year is about 3×10^{7} seconds). If you need to learn or review the powers of 10 notation, such as the designation of the number 50,000 in the form 5×10^{4}, you should study Appendix A.

Systems of units commonly used are the *SI system*, in which the units of mass, length, and time are the kilogram (kg), meter (m), and second (s), respectively; the *cgs* or *gaussian system*, in which the units of mass, length, and time are the gram (g), centimeter (cm), and second, respectively; and the *British engineering system* (sometimes called the conventional system), in which the units of mass, length, and time are the slug, foot (ft), and second, respectively. Throughout most of this text we shall use SI units since they are almost universally accepted in science and industry. We shall make some limited use of British engineering units in the study of mechanics.

Table 1.3
Approximate Values of Some Time Intervals

	Time Interval (s)
Age of the Universe	5×10^{17}
Age of the Earth	1×10^{17}
Average age of a college student	6×10^{8}
One year	3×10^{7}
One day (time for one revolution of Earth about its axis)	9×10^{4}
Time between normal heartbeats	8×10^{-1}
Period[a] of audible sound waves	1×10^{-3}
Period of typical radio waves	1×10^{-6}
Period of vibration of an atom in a solid	1×10^{-13}
Period of visible light waves	2×10^{-15}
Duration of a nuclear collision	1×10^{-22}
Time for light to cross a proton	3×10^{-24}

[a] Period is defined as the time interval of one complete vibration.

Table 1.4
Some Prefixes for Powers of Ten Used in the SI System

Power	Prefix	Abbreviation
10^{-18}	atto	a
10^{-15}	femto	f
10^{-12}	pico	p
10^{-9}	nano	n
10^{-6}	micro	μ
10^{-3}	milli	m
10^{-2}	centi	c
10	deca	d
10^3	kilo	k
10^6	mega	M
10^9	giga	G
10^{12}	tera	T
10^{15}	peta	P
10^{18}	exa	E

Some of the most frequently used prefixes defined within the SI system for the various powers of 10 and their abbreviations are listed in Table 1.4. For example, 10^{-3} m is equivalent to 1 millimeter (mm), and 10^3 m is 1 kilometer (km). Likewise, 1 kg is equal to 10^3 g, and 1 megavolt (MV) is equivalent to 10^6 volts (V).

1.2   Dimensional Analysis

The word *dimension* has a special meaning in physics. It denotes the qualitative nature of a physical quantity. Whether the separation between two points is measured in units of feet or meters or furlongs, the measured quantity is a distance. We say that the dimension of distance is *length.*

The symbols that will be used to specify length, mass, and time are L, M, and T, respectively. We shall often use brackets [ ] to denote the dimensions of a physical quantity. For example, in this notation the dimensions of velocity, v, are written $v = [L]/[T]$, and the dimensions of area, A, are $A = [L^2]$. The dimensions of area, volume, velocity, and acceleration are listed in Table 1.5, along with their units in the three common systems. The dimensions of other quantities, such as force and energy, will be described as they are introduced in the text.

Table 1.5
Dimensions of Area, Volume, Velocity, and Acceleration

System	Area (L^2)	Volume (L^3)	Velocity (L/T)	Acceleration (L/T^2)
SI	m^2	m^3	m/s	m/s^2
cgs	cm^2	cm^3	cm/s	cm/s^2
British engineering (conventional)	ft^2	ft^3	ft/s	ft/s^2

In many situations, you may be faced with having to derive or check a specific formula. Although you may have forgotten the details of the derivation, there is a useful and powerful procedure called **dimensional analysis** that can be used to assist in the derivation or to check your final expression. Dimensional analysis makes use of the fact that *dimensions can be treated as algebraic quantities*. That is, quantities can be added or subtracted only if they have the same dimensions. Furthermore, the quantities on each side of an equation must have the same dimensions. By following these simple rules, you can use dimensional analysis to help determine whether or not you have the correct form of an expression, because the relationship can be correct only if the dimensions on each side of the equation are the same.

To illustrate this procedure, suppose that during an examination you wish to derive the formula for the distance x traveled by a car in a time t if the car starts from rest and moves with constant acceleration a. In Chapter 2 we shall find that the correct expression for this situation is $x = \frac{1}{2}at^2$. Suppose you suspect that this equation is in error and you wish to check its validity from a dimensional analysis approach.

We know that the quantity x on the left side has a dimension of length. In order for the equation to be dimensionally correct, the quantity on the right side must also have the dimension of length. To make a dimensional check, we substitute in the basic dimensions for acceleration, $[L]/[T^2]$, and time, $[T]$. That is, the equation $x = \frac{1}{2}at^2$ can be written dimensionally as

$$[L] = \frac{[L]}{[T^2]} \cdot [T^2] = [L]$$

The dimensions of time cancel as shown, leaving us with the dimensions of length. Note that the factor $\frac{1}{2}$ was ignored because it has no dimensions.

Example 1.1   Check the Dimensions

Show that the expression $v = v_0 + at$ is dimensionally correct, where v and v_0 represent velocities, a is acceleration, and t is a time interval.

Solution

Since the dimensions of velocity are $[L]/[T]$ and those of acceleration are $[L]/[T^2]$, the equation $v = v_0 + at$ can be written in dimensional terms as

$$\frac{[L]}{[T]} = \frac{[L]}{[T]} + \frac{[L]}{[T^2]} \cdot [T]$$

$$= \frac{[L]}{[T]} + \frac{[L]}{[T]}$$

$$= \frac{[L]}{[T]}$$

In the second step, the two quantities to the right of the equal sign have the same dimensions; they can be added (or subtracted) to give a single quantity with those dimensions. Thus we see in the last step that $[L]/[T] = [L]/[T]$, and the expression is dimensionally correct. On the other hand, if the expression were $v = v_0 + at^2$, it would be dimensionally *incorrect*. Try it and see!

Example 1.2   The Clever Student

In a desperate attempt to come up with an equation to use during an examination, a student tries the equation $v^2 = ax$. Use dimensional analysis to determine if the equation might be valid.

Solution
The dimensions on the left side are $[L^2]/[T^2]$, so the dimensions on the right side must be the same. Let us check to see if this is the case:

$$\frac{[L^2]}{[T^2]} = \frac{[L]}{[T^2]} \cdot [L] = \frac{[L^2]}{[T^2]}$$

The dimensions do check, but the equation is incorrect nonetheless. The correct form is $v^2 = 2ax$. Dimensional analysis can tell us nothing about the numerical constants in an equation. You should bear in mind that even though dimensional analysis is a powerful tool, it does have its limitations.

1.3   Significant Figures

This photograph shows a flange being measured with a device called a micrometer. A typical micrometer can measure lengths to a precision of about 0.01 mm. *(Jeffrey Coolidge, Stockphotos, Inc.)*

When one performs measurements on certain quantities, the measured values are only known to within the limits of the experimental uncertainty. The value of the uncertainty can depend on various factors such as the quality of the apparatus, the skill of the experimenter, and the number of measurements performed. Suppose that in a laboratory experiment we are asked to measure the area of a rectangular plate using a meter stick as a measuring instrument. Let us assume that the accuracy to which we can measure a particular dimension of the plate is ± 0.1 cm. If the length of the plate is measured to be 16.3 cm, we can only claim that its length lies somewhere between 16.2 cm and 16.4 cm. In this case, we say that the measured value has three significant figures. Likewise, if its width is measured to be 4.5 cm, the actual value lies between 4.4 cm and 4.6 cm. (This measured value has only two significant figures.) Note that the significant figures include the first estimated digit. Thus, we could write the measured values as 16.3 ± 0.1 cm and 4.5 ± 0.1 cm.

Suppose now that we would like to find the area of the plate by multiplying the two measured values together. If we were to claim that the area is $(16.3 \text{ cm})(4.5 \text{ cm}) = 73.35 \text{ cm}^2$, our answer would be unjustifiable since it contains four significant figures, which is greater than the number of significant figures in either of the measured lengths. A good "rule of thumb" to use as a guide in determining the number of significant figures that can be claimed is as follows:

When multiplying several quantities, the number of significant figures in the final answer is the same as the number of significant figures in the *least* accurate of the quantities being multiplied, where "least accurate" means "having the lowest number of significant figures." The same rule applies to division.

Applying this rule to the multiplication example above, we see that the answer for the area can have only two significant figures since the dimension of 4.5 cm has only two significant figures. Thus, we can only claim the area to be

Could this be the result of poor data analysis? Perhaps the improper handling of significant figures caused this disaster. What do you think? *(Roger-Viollet. © ND-Viollet)*

73 cm^2, realizing that the value can range between $(16.2 \text{ cm})(4.4 \text{ cm}) = 71 \text{ cm}^2$ and $(16.4 \text{ cm})(4.6 \text{ cm}) = 75 \text{ cm}^2$.

The presence of zeros in an answer may be misinterpreted. For example, suppose the mass of an object is measured to be 1500 g. This value is ambiguous because it is not known whether the last two zeros are being used to locate the decimal point or whether they represent significant figures in the measurement. In order to remove this ambiguity, it is common to use scientific notation to indicate the number of significant figures. In this case, we would express the mass as 1.5×10^3 g if there are two significant figures in the measured value and 1.50×10^3 g if there are three significant figures. Likewise, a number such as 0.00015 should be expressed in scientific notation as 1.5×10^{-4} if it has two significant figures or as 1.50×10^{-4} if it has three significant figures. The three zeros between the decimal point and the digit 1 in the number 0.00015 are not counted as significant figures because they are present only to locate the decimal point. In general, a **significant figure** is a reliably known digit (other than a zero used to locate the decimal point).

For addition and subtraction, the number of decimal places must be considered.

When numbers are added (or subtracted), the number of decimal places in the result should equal the smallest number of decimal places of any term in the sum.

For example, if we wish to compute $123 + 5.35$, the answer would be 128 and not 128.35.

Throughout this book, *we shall generally assume that the given data are precise enough to yield an answer having three significant figures.* Thus, if we state that a jogger runs a distance of 5 m, it is to be understood that the distance covered is 5.00 m. Likewise, if the speed of a car is given as 23 m/s, its value is understood to be 23.0 m/s.

Example 1.3   Installing a Carpet

A carpet is to be installed in a room whose length is measured to be 12.71 m (four significant figures), and whose width is measured to be 3.46 m (three significant figures). Find the area of the room.

Solution
If you multiply 12.71 m by 3.46 m on your calculator, you will get an answer of 43.9766 m². How many of these numbers should you claim? Our rule of thumb for multiplication tells us that you can only claim the number of significant figures in the least accurate of the quantities being measured. In this example, we have only three significant figures in our least accurate measurement, so we should express our final answer as 44.0 m². Note that in the answer given, we used a general rule for rounding off numbers which states that the last digit retained is to be increased by 1 if the first digit dropped was equal to 5 or greater.

Example 1.4   Time to Go Fishing

A lucky fisherman catches two sturgeon. The smaller fish has a measured length of 93.46 cm (two decimal places, four significant figures), while the larger fish has a measured length of 135.3 cm (one decimal place, four significant figures). What is the total length of fish caught for the day?

Solution
Adding the two lengths together we find an answer of 228.76 cm. However, the length known to the smallest number of decimal places is the 135.3 cm length, which has one decimal place. Thus, according to our rule of thumb for adding numbers, you can only claim your answer for the total length to be 228.8 cm.

1.4   Conversion of Units

Sometimes it is necessary to convert units from one system to another. Conversion factors between the SI and conventional systems for units of length are as follows:

$$1 \text{ mile} = 1609 \text{ m} = 1.609 \text{ km} \qquad 1 \text{ ft} = 0.3048 \text{ m} = 30.48 \text{ cm}$$

$$1 \text{ m} = 39.37 \text{ in.} = 3.281 \text{ ft} \qquad 1 \text{ in.} = 0.0254 \text{ m} = 2.54 \text{ cm}$$

A more complete list of conversion factors can be found in Appendix C of this book. Units can be treated as algebraic quantities that can cancel each other. For example, suppose we wish to convert 15.0 in. to centimeters. From the list of conversion factors above, we find that 1 in. = 2.54 cm.[1] We can now convert 15.0 in. to centimeters by multiplying 15.0 in. by 2.54 cm/1 in. The units cancel as follows:

$$15.0 \text{ in.} = 15.0 \text{ in.} \times 2.54 \, \frac{\text{cm}}{\text{in.}} = 38.1 \text{ cm}$$

---

[1] The conversion 1 in. = 2.54 cm is *exact* and is the legal conversion in the United States.

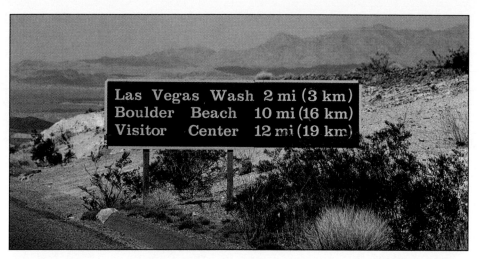

Conversion of miles to kilometers.

Example 1.5 The Volume of a Cube

A cube is 5 in. on a side. What is its volume in SI units?

Solution

The conversion factor to use is 1 in. = 0.0254 m. Thus, the length of a side in meters is

$$L = (5 \text{ in.}) \left(\frac{0.0254 \text{ m}}{1 \text{ in.}} \right) = 0.127 \text{ m}$$

and the volume is

$$V = L^3 = (0.127 \text{ m})^3 = 2.05 \times 10^{-3} \text{ m}^3$$

Conversions of units from the labels of some grocery store items. You should check the accuracy of some of these conversions. *(Courtesy of Henry Leap and Jim Lehman)*

1.5 Order-of-Magnitude Calculations

It is often useful to estimate an approximate answer to a problem in which information is available. Such results can then be used to determine whether or not a more precise calculation is necessary. These approximations are usually based on certain assumptions, which must be modified if more precision is needed. Thus, we shall sometimes refer to the **order of magnitude** of a certain quantity as the power of ten that is closest to the quantity. For example, the mass of a person might be 75 kg $\approx 10^2$ kg. Thus, we would say that the person's mass is of the order of 10^2 kg. Usually, when an order-of-magnitude calculation is made, the results are reliable to within a factor of 10.

Example 1.6 How Much Gasoline Do We Use?

Estimate the number of gallons of gasoline used annually by all cars in the United States.

Solution

Since there are about 200 million people in the United States, an estimate of the number of cars in the country is 40 million (assuming one car and five people per family). We shall also estimate that the average distance traveled per year is 10,000 miles. If we assume a gasoline consumption of 20 miles/gal, each car uses about 500 gal/year. Multiplying this by the total number of cars in the United States gives an estimated total consumption of 2×10^{10} gal. This corresponds to a yearly consumer expenditure of over 20 billion dollars! This is probably a low estimate since we haven't accounted for commercial consumption and for such factors as two-car families.

1.6 Mathematical Notation

Many mathematical symbols will be used throughout this book, some of which you are aware of, such as the symbol = to denote the equality of two quantities.

The symbol $\propto$ denotes a proportionality. For example, $y \propto x^2$ means that y is proportional to the square of x.

The symbol $<$ means *less than,* and $>$ means *greater than.* For example, $x > y$ means x is greater than y.

The symbol $\ll$ means *much less than,* and $\gg$ means *much greater than.*

The symbol $\approx$ indicates that two quantities are *approximately equal* to each other.

The symbol $\equiv$ means *is defined as.* This is a stronger statement than a simple =.

It is convenient to use a symbol to indicate the *change in a quantity.* For example, Δx (read as "delta x") means "the change in the quantity x." (It does not mean "the product of Δ and x.") For example, suppose that a person out for a morning stroll starts measuring his distance away from home when he is 10 m from his doorway. He then walks along a straight-line path and stops strolling when he is 50 m from the door. His change in position during his walk is $\Delta x = 50$ m $-$ 10 m $= 40$ m, or in symbolic form

$$\Delta x = x_{\mathrm{f}} - x_{\mathrm{i}}$$

In this equation x_{f} is the *final position* and x_{i} is the *initial position.*

We shall often have occasion to add several quantities. A useful abbreviation for representing such a sum is the Greek Letter Σ (capital sigma). Suppose we wish to add a set of five numbers represented by $x_1, x_2, x_3, x_4,$ and x_5. In the abbreviated notation, we would write the sum as

$$x_1 + x_2 + x_3 + x_4 + x_5 = \sum_{i=1}^{5} x_i$$

where the subscript i on x represents any one of the numbers in the set. For example, if there are five masses in a system, $m_1, m_2, m_3, m_4,$ and m_5, the total mass of the system $M = m_1 + m_2 + m_3 + m_4 + m_5$ could be expressed

$$M = \sum_{i=1}^{5} m_i$$

Finally, the magnitude of a quantity x, written $|x|$, is simply the absolute value of that quantity. The sign of $|x|$ is always positive, regardless of the sign of x. For example, if $x = -5$, $|x| = 5$; if $x = 8$, $|x| = 8$.

1.7 Coordinate Systems

In many of the interesting applications and problems in physics, it is necessary to specify the locations of objects. For example, the description of the motion of an airplane requires a method for describing the position of the plane. Locating an object in space is accomplished by means of a system of coordinates.

A **coordinate system** used to specify locations in space consists of:

1. A fixed reference point, called the origin
2. A set of specified axes or directions
3. Instructions that tell us how to label a point in space relative to the origin and axes.

One convenient coordinate system that we shall use frequently is the *cartesian coordinate system,* sometimes called the rectangular coordinate system. Such a system is illustrated in Figure 1.1. The point *P* shown in the figure has coordinates (5, 3). This notation means that if we start at the origin (*O*), we can reach *P* by moving 5 units along the positive *x* axis and then 3 units along the positive *y* axis. (Positive *x* is usually selected to the right of the origin and positive *y* upward from the origin. Negative *x* is usually to the left of the origin and negative *y* downward from the origin. This convention is not an absolute necessity, however. There may be instances in which you would like to take positive *x* to the left of the origin and negative *x* to the right. Feel free to do so.) The point *Q* has coordinates (−3, 4), corresponding to going 3 units in the negative *x* direction and 4 units in the positive *y* direction.

The cartesian coordinate system will be the one used far more often than any other in this text. However, sometimes it is more convenient to locate a point in space by its plane polar coordinates, (*r*, *θ*), as shown in Figure 1.2. When using this system, an origin and a reference line are selected as shown. A point is then specified by the distance *r* from the origin to the point and by the angle *θ* between *r* and the reference line. (Frequently the reference line is selected to be the positive *x* axis of a cartesian coordinate system, although this choice is not necessary.) The angle *θ* is considered to be positive when measured counterclockwise from the reference line and negative when measured clockwise. For example, if a point is specified by the polar coordinates 3 m and 60°, one locates this point by moving out 3 m from the origin at an angle of 60° above (or counterclockwise from) the reference line. A point specified by polar coordinates 3 m and −60° is located 3 m out from the origin and 60° below (or clockwise from) the reference line.

1.8 Trigonometry

That portion of mathematics based on the special properties of a right triangle is called trigonometry. Many of the ideas and concepts of this branch of mathematics are of utmost importance in the study of physics. We will now review some of the more inherent fundamental definitions and ideas that will be necessary for you to know.

Consider the right triangle shown in Figure 1.3, where side *a* is opposite the angle *θ*, side *b* is adjacent to the angle *θ*, and side *c* is the hypotenuse of the triangle. The basic trigonometry functions defined by such a triangle are the

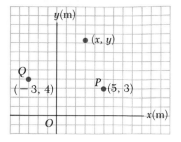

Figure 1.1
Designation of points in a cartesian coordinate system. Every point is labeled with coordinates (*x*, *y*).

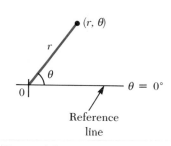

Figure 1.2
A polar coordinate system.

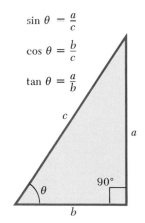

$$\sin \theta = \frac{a}{c}$$

$$\cos \theta = \frac{b}{c}$$

$$\tan \theta = \frac{a}{b}$$

Figure 1.3
Certain trigonometric functions for a right triangle.

ratios of the lengths of certain sides of the triangle. Specifically, the most important relationships are called the sine (sin), cosine (cos), and tangent (tan) functions. In terms of the angle θ, these functions are[2]

$$\sin \theta = \frac{\text{side opposite } \theta}{\text{hypotenuse}} = \frac{a}{c}$$

$$\cos \theta = \frac{\text{side adjacent to } \theta}{\text{hypotenuse}} = \frac{b}{c} \tag{1.1}$$

$$\tan \theta = \frac{\text{side opposite to } \theta}{\text{side adjacent to } \theta} = \frac{a}{b}$$

For example, if the angle θ is equal to $30°$, it is found that the ratio of a to c is always 0.50, that is, $\sin 30° = 0.50$. Note that the sin, cos, or tan are quantities without units, since they represent the ratio of two lengths.

Another important relationship, called the **Pythagorean theorem**, exists between the lengths of the sides of a right triangle; it is

$$c^2 = a^2 + b^2 \tag{1.2}$$

Finally, it will often be necessary to find the value of **inverse** relationships. For example, suppose you know the value of the sine of an angle, but you need to know the value of the angle itself. The inverse sine function may be expressed as $\sin^{-1}(0.866)$, which is a shorthand way of asking the question, "What angle has a sine of 0.866?" Punching a couple of buttons on your calculator reveals that this angle is $60°$. Try it for yourself and show that $\tan^{-1}(0.400) = 21.8°$.

The following example problems will give you a little practice in working with these ideas from trigonometry and will also show you how they can be used in practical situations.

Example 1.7   How High Is the Building?

A man attempts to measure the height of a building by walking out a distance of 46.0 m from the base and shining a flashlight beam toward its top. He finds that when the beam is elevated at an angle of $39°$ with respect to the horizontal, as shown in Figure 1.4, the beam just strikes the top of the building. Find the height of the building and the distance that the flashlight beam has to travel before it strikes the top of the building.

Solution
We know the length of the adjacent side of the red triangle shown in the figure, and we know the value of the angle. Thus, we can use the definition of the tangent function to find the height of the building, which is the opposite side of the triangle.

$$\tan 39° = \frac{\text{height of building}}{46.0 \text{ m}}$$

or

$$\text{height of building} = (\tan 39°)(46.0 \text{ m}) = (0.810)(46.0 \text{ m}) = \boxed{37.3 \text{ m}}$$

---

[2] In order to recall these trigonometric relationships, you should consider using the following "memory device." Use the Indian sounding name "SOHCAHTOA" whose letters stand for Sine = Opposite/Hypotenuse, Cosine = Adjacent/Hypotenuse, and Tangent = Opposite/Adjacent. (Thanks go to Professor Don Chodrow for pointing this out to one of the authors, RAS.)

Height

39°

46.0 m

Figure 1.4
Example 1.7

Now that we know the length of both the adjacent side and the opposite side of the triangle, we can find the length of the hypotenuse, which is the distance c that the beam travels before it strikes the top of the building.

$$c = \sqrt{a^2 + b^2} = \sqrt{(37.3)^2 + (46.0)^2} = \boxed{59.2 \text{ m}}$$

Exercise Try finding the distance the light traveled by a method other than that used above. Would another trigonometric function such as the sin or cos give you the answer? Try it.

Example 1.8 Things To Do When the CB Goes Bad

A truck driver moves up a straight mountain highway, as shown in Figure 1.5. Elevation markers at the beginning and ending points of the trip show that he has risen vertically 0.53 km, and the mileage indicator on the truck shows that he has traveled a total distance of 3.00 km during the ascent. Find the angle of incline of the hill.

Solution
This is an example in which we must use an inverse trigonometric relationship. From the sine of the angle, defined as the opposite side over the hypotenuse, we find

$$\sin \theta = \frac{0.53}{3.00} = 0.177$$

To find θ, we use the inverse sine relationship.

$$\theta = \sin^{-1}(0.177) = \boxed{10.2°}$$

3.00 km

0.53 km

θ

Figure 1.5
Example 1.8

1.9 Problem-Solving Strategy

Most courses in general physics require the student to learn the skills of problem solving, and examinations are largely composed of problems that test such skills. This brief section describes some useful ideas that will enable you to increase your accuracy in solving problems, enhance your understanding of physical concepts, eliminate initial panic or lack of direction in approaching a problem, and organize your work. One way to help accomplish these goals is to adopt a problem-solving strategy. Many chapters in this text will include a section labeled "Problem-Solving Strategies and Hints" that should help you through the "rough spots."

In developing problem-solving strategies, six basic steps are commonly used.

| Read Problem |

| Draw Diagram |

| Identify Data |

| Choose Equation(s) |

| Solve Equation(s) |

| Evaluate and Check Answer |

A menu for problem solving.

1. Read the problem carefully at least twice. Be sure you understand the nature of the problem before proceeding further.
2. Draw a suitable diagram with appropriate labels and coordinate axes if needed.
3. As you examine what is being asked in the problem, identify the basic physical principle (or principles) that are involved, listing the knowns and unknowns.
4. Select a basic relationship or derive an equation that can be used to find the unknown, and solve the equation for the unknown symbolically.
5. Substitute the given values along with the appropriate units into the equation.
6. Obtain a numerical value for the unknown. The problem is verified and receives a check mark if the following questions can be properly answered: Do the units match? Is the answer reasonable? Is the plus or minus sign proper or meaningful?

One of the purposes of this strategy is to promote accuracy. Properly drawn diagrams can eliminate many sign errors. Diagrams also help to isolate the physical principles of the problem. Symbolic solutions and carefully labeled knowns and unknowns will help eliminate other careless errors. The use of symbolic solutions should help you think in terms of the physics of the problem. A check of units at the end of the problem can indicate a possible algebraic error. The physical layout and organization of your problem will make the final product more understandable and easier to follow. Once you have developed an organized system for examining problems and extracting relevant information, you will become a more confident problem solver.

The following example illustrates this procedure.

Example 1.9   A Round Trip by Air

An airplane travels 450 km due east and then travels due north through an unknown distance. Finally, the airplane returns to its starting point by traveling a distance of 525 km. How far did the airplane travel in the northerly direction?

Solution
After reading the problem carefully, you should draw a diagram of the situation as in Figure 1.6. Since the path of the airplane forms a right triangle, you can use the

Pythagorean theorem, which relates the three sides of the triangle:

$$c^2 = a^2 + b^2$$

In this problem, the hypotenuse c and the side a are known quantities. Hence, we must solve for the unknown quantity b. Subtracting a^2 from both sides of the equation above gives

$$b^2 = c^2 - a^2$$

We now take the square root of both sides to find the solution for b:

$$b = \sqrt{c^2 - a^2}$$

Note that the negative solution has been disregarded since it is not physically meaningful.

Finally, we substitute the given values for the quantities c and a to find a numerical value for b:

$$b = \sqrt{(525\text{ km})^2 - (450\text{ km})^2} = \boxed{270\text{ km}}$$

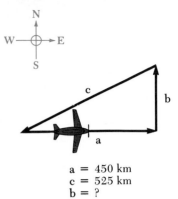

$a = 450$ km
$c = 525$ km
$b = ?$

Figure 1.6
Example 1.9

Summary

The physical quantities we shall encounter in our study of mechanics can be expressed in terms of three fundamental quantities, mass, length, and time, which have the units kilograms (kg), meters (m), and seconds (s), respectively, in the SI system. It is often helpful to use dimensional analysis to check equations and to assist in deriving equations.

When inserting numerical factors into equations, you must be sure that the dimensions of these factors are consistent throughout the equation. In many cases it will be necessary to use the table of conversion factors given in Appendix C of this text in order to convert from one system of units to another.

It is often useful to estimate an approximate answer to a problem in which little or no information is given. Such estimates are called **order-of-magnitude calculations.**

The three most basic trigonometric functions for a right triangle are the sine, cosine, and tangent defined as

$$\sin \theta = \frac{\text{side opposite }\theta}{\text{hypotenuse}}$$

$$\cos \theta = \frac{\text{side adjacent to }\theta}{\text{hypotenuse}}$$

$$\tan \theta = \frac{\text{side opposite to }\theta}{\text{side adjacent to }\theta}$$

Another important relationship, called the **Pythagorean theorem,** exists between the lengths of the sides of a right triangle; it is

$$c^2 = a^2 + b^2$$

where c is the hypotenuse of the triangle and a and b are its other two sides.

Additional Reading

A. V. Astin, "Standards of Measurements," *Sci. American,* June 1968, p. 50.

H. Butterfield, "The Scientific Revolution," *Sci. American,* September 1960, p. 173.

J. Friberg, "Numbers and Measures in the Earliest Written Records," *Sci. American,* February 1984, p. 110.

G. Goth, "Dimensional Analysis by Computer," *The Physics Teacher,* February 1986, p. 75.

L. M. Lederman, "The Value of Fundamental Science," *Sci. American,* November 1984, p. 40.

Philip and Phylis Morrison and the office of Charles and Ray Eames, *The Powers of Ten,* The Sci. American Library, New York, W. H. Freeman, 1982.

Lord Ritchie-Calder, "Conversion to the Metric System," *Sci. American,* July 1970, p. 17.

Questions

1. The height of a horse is sometimes given in units of "hands." Why is this a poor standard of length?
2. Does a dimensional analysis give any information on constants of proportionality that may appear in an algebraic expression? Explain.
3. Suppose that two quantities A and B have different dimensions. Determine which of the following arithmetic operations *could* be physically meaningful: (a) A + B, (b) A/B, (c) B − A, (d) AB.
4. Would a person with a height of 1.85 m be considered tall? Discuss.
5. Estimate the order of magnitude of the length in meters of each of the following: (a) a ladybug; (b) a femur bone; (c) your campus administration building; (d) a giraffe; (e) a city block; (f) a football field.

6. How many of the lengths or time intervals given in Tables 1.2 and 1.3 could you verify using only equipment that might be found in a typical dormitory room?
7. An ancient unit of length called the cubit was equal to six palms, where a palm was the width of the four fingers of an open hand. Noah's ark was 300 cubits long, 50 cubits wide, and 30 cubits high. Estimate the volume of the ark in cubic feet. Also, estimate the volume of a typical home and compare it to the volume of the ark.
8. If an equation is dimensionally correct, does this mean that the equation must be true?

Problems

Section 1.2   Dimensional Analysis

1. Show that the equation $v^2 = v_0^2 + 2ax$ is dimensionally correct, where v and v_0 represent velocities, a is acceleration, and x is a distance.
2. (a) Suppose that the displacement of an object is related to the time according to the expression $x = ct^2$. What are the dimensions of c? (b) A displacement is related to the time as $x = A \sin (2\pi ft)$, where A and f are constants. Find the dimensions of A. (*Hint:* A trigo-

nometric function appearing in an equation must be dimensionless.)
3. (a) One of the fundamental laws of motion states that the acceleration of an object is directly proportional to the resultant force on it and inversely proportional to its mass. From this statement, determine the dimensions of force. (b) The newton is the SI unit of force. According to the results for (a), how can you express a force having the units of newtons using the fundamental units of mass, length, and time?

4. Which of the equations below are dimensionally correct? (a) $v = v_0 + ax$ (b) $y = (2\text{ m}) \cos(kx)$, where $k = 2\text{ m}^{-1}$.

5. A spherical droplet of blood is drawn into a pipette, as in Figure 1.7. Write a formula for the height h of the column of blood in the pipette in terms of the radius r of the pipette and the radius R of the original droplet. Check the dimensions of your formula.

Figure 1.7
Problem 5

6. The period of a simple pendulum, which is defined as the time for one complete oscillation, is measured in time units and is given by

$$T = 2\pi \sqrt{\frac{\ell}{g}}$$

where ℓ is the length of the pendulum and g is the acceleration due to gravity in units of length divided by the square of time. Show that this equation is dimensionally consistent.

Section 1.3   Significant Figures

7. The value of the speed of light is now known to be 2.99792458×10^8 m/s. Express the speed of light to (a) three significant figures, (b) five significant figures, and (c) seven significant figures.

8. How many significant figures are there in (a) 78.9 ± 0.2, (b) 3.788×10^9, (c) 2.46×10^{-6}, (d) 0.0032?

9. Carry out the following arithmetic operations: (a) the sum of the numbers 756, 37.2, 0.83, and 2.5; (b) the product 3.2×3.563; (c) the product $5.67 \times \pi$.

10. Carry out the following arithmetic operations: (a) 756 minus 37.2, (b) 3.2 divided by 1.4577.

11. Calculate (a) the circumference of a circle of radius 3.5 cm and (b) the area of a circle of radius 4.65 cm.

12. A farmer measures the distance around a rectangular field. The length of the long sides of the rectangle is found to be 38.44 m, and the length of the short sides is found to be 19.5 m. What is the total distance around the field?

Section 1.4   Conversion of Units

13. How many meters are there in the hundred-yard dash?

14. A quart container of ice cream is to be made in the form of a cube. What should be the length of a side in centimeters? (Use the conversion 1 gallon = 3.786 liter.)

15. Estimate the age of the Earth in years using the data in Table 1.3 and the appropriate conversion factors.

16. Estimate the distance to the nearest star in feet using the data in Table 1.1 and the appropriate conversion factors.

17. A house is 50 ft long, 26 ft wide, and has 8-ft high ceilings. What is the volume of the house in cubic meters and in cubic centimeters?

18. (a) How many seconds are there in a year? (b) If one micrometeorite (a sphere with a diameter of 10^{-6} m) strikes each square meter of the moon each second, how many years would it take to cover the moon to a depth of 1 m? (*Hint*: Consider a cubic box on the moon 1 m on a side, and find how long it will take to fill the box.)

19. A painter is to cover the walls in a room 8 ft high and 12 ft along a side. What surface area in square meters must he cover?

20. Assume an oil slick consists of a single layer of molecules and that each molecule occupies a cube 1.0 μm on a side. Determine the area of an oil slick formed by 1.0 m^3 of oil.

21. One cubic meter (1.0 m^3) of aluminum has a mass of 2.70×10^3 kg, and 1.0 m^3 of iron has a mass of 7.86×10^3 kg. Find the radius of an aluminum sphere whose mass is the same as that of an iron sphere of radius 2.0 cm. (*Note*: Density is defined as the mass of an object divided by its volume.)

22. One cubic centimeter (1.0 cm^3) of water has a mass of 1.0×10^{-3} kg. (a) Determine the mass of 1 m^3 of water. (b) Assuming biological substances are 98% water, estimate the mass of a cell with a diameter of 1.0 μm, a human kidney, and a fly. Assume a kidney is roughly a sphere with a radius of 4.0 cm, and a fly is roughly a cylinder 4.0 mm long and 2.0 mm in diameter. (See hint in Problem 21.)

23. (a) Find a conversion factor to convert from mi/h to km/h. (b) Until recently, federal law mandated that maximum highway speeds would be 55 mi/h. Use the conversion factor of part (a) to find the speed in km/h. (c) The maximum highway speed has been raised to

65 mi/h in some places. In km/h, how much increase is this over the 55 mi/h limit?

24. The speed of light is 3.00×10^8 m/s. Convert this to miles per hour.

25. Use the fact that the speed of light in free space is about 3.00×10^8 m/s to determine how many miles a pulse from a laser beam will travel in 1 hour.

26. The base of a pyramid covers an area of 13 acres (1 acre = 43,560 ft^2) and has a height of 481 ft. If the volume of a pyramid is given by the expression $V = (1/3)bh$, where b is the area of the base and h is the height, find the volume of this pyramid in cubic meters.

27. The pyramid described in Problem 26 contains approximately two million stone blocks that average 2.5 tons each. Find the weight of this pyramid in pounds and in newtons.

Section 1.5   Order-of-Magnitude Calculations

Note: In arriving at answers to the problems in this section, you should state your important assumptions, including the numerical values assigned to parameters used in the solution. Since only order-of-magnitude results are expected, do not be surprised if your results differ from those given in the answer section of the text.

28. Imagine yourself to be the equipment manager of a professional baseball team. One of your jobs is to keep baseballs on hand for games. One of the ways by which balls are lost is by players hitting them into the stands either as home runs or foul balls. Estimate how many baseballs you have to buy per season in order to account for such losses. Assume your team plays an 81-game home schedule in a season.

29. A particular hamburger chain advertises that it has sold more than 50 billion hamburgers. Estimate how many pounds of hamburger meat must have been used by the restaurant chain and how many head of cattle were required to furnish the meat.

30. Estimate the number of piano tuners living in New York City. This question was raised by the physicist Enrico Fermi, who was well known for making order-of-magnitude calculations.

31. Assume that you watch every pitch of every game of a 162-game major league baseball season. Approximately how many pitches would you see thrown?

Section 1.7   Coordinate Systems

32. A point is located in a polar coordinate system by the coordinates $r = 2.5$ m and $\theta = 35°$. Find the x and y coordinates of this point, assuming the two coordinate systems have the same origin.

33. A certain corner of a room is selected as the origin of a rectangular coordinate system. If a fly is crawling on an adjacent wall at a point having coordinates (2.0, 1.0), where the units are in meters, what is the distance of the fly from the corner of the room?

34. Express the location of the fly in Problem 33 in polar coordinates.

35. Two points in a rectangular coordinate system have the coordinates (5, 3) and (−3, 4), where the units are in centimeters. Determine the distance between these points.

Section 1.8   Trigonometry

36. In Figure 1.8, find (a) the side opposite θ, (b) the side adjacent to ϕ, (c) cos θ, (d) sin ϕ, and (e) tan ϕ.

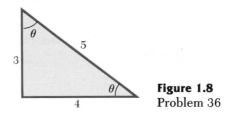

Figure 1.8
Problem 36

37. In a certain right triangle, the two sides that are perpendicular to each other are 5 m and 7 m long. What is the length of the third side of the triangle?

38. In Problem 37, what is the tangent of the angle for which 5 is the opposite side?

39. A right triangle has a hypotenuse of length 3 m, and one of its angles is 30°. What is the length of (a) the side opposite the 30° angle and (b) the side adjacent to the 30° angle?

40. For the triangle shown in Figure 1.9, what is (a) the length of the unknown side, (b) the tangent of θ, and (c) the sin of ϕ?

Figure 1.9
Problem 40

Additional Problems

41. Use the standard prefixes in Table 1.4 and the well-established prefixes for the numbers 1 to 10 to determine appropriate words for the following conversions: (a) 10^{12} microphones, (b) 10^{21} picolos, (c) 10 rations, (d) 10^6 bicycles, (e) 10^{12} pins, (f) $3\frac{1}{3}$ tridents, (g) 2000 mockingbirds, (h) 10^{-12} boo, (i) 10^{-9} goat. (Some of these amusing conversions are from a table published by Solomon W. Golomb of UCLA.)

42. The radius of the planet Saturn is 5.85×10^7 m and its mass is 5.68×10^{26} kg. (a) Find the density of Saturn (its mass divided by its volume) in g/cm^3. (The volume of a sphere is given by $(4/3)\pi r^3$.) (b) Find the surface area of Saturn in square feet. (The surface area of a sphere is given by $4\pi r^2$.)

43. The consumption of natural gas by a company satisfies the empirical equation $V = 1.5t + 0.008t^2$, where V is the volume in millions of cubic feet and t the time in months. Express this equation in cubic feet and seconds. Put the proper units on the coefficients. Assume a month is 30 days.

44. Soft drinks are commonly sold in aluminum containers. Estimate the number of such containers thrown away each year by U.S. consumers. Approximately how many tons of aluminum does this represent?

45. You close out your $1,000,000 checking account at a bank, and take your money all in one-dollar bills. Estimate the height of your stack of money.

46. The displacement of an object moving under uniform acceleration is some function of time and the acceleration. Suppose we write this displacement as $s = k\,a^m t^n$, where k is a dimensionless constant. Show by dimensional analysis that this expression is satisfied if $m = 1$ and $n = 2$. Can this analysis give the value of k?

47. Two points in a rectangular coordinate system have coordinates $(2.0, -4.0)$ and $(-3.0, 3.0)$, where the units are in meters. Determine the distance between these points.

METRIC CLOCK

Conversation with
Leon M. Lederman

(Courtesy of Fermilab)

Leon M. Lederman began his career in physics in 1951 as a research associate at Columbia University's Nevis Laboratories, the physics department's center for experimental research in high energy physics. Earlier that year at Columbia, Dr. Lederman had received his Ph.D. in physics, preceded by a master's degree he earned there in 1946.

Dr. Lederman served as director of the Nevis Labs from 1960 to 1979, where the work he conducted earlier in collaboration with Columbia colleagues Melvin Schwartz and Jack Steinberger led to his sharing the 1988 Nobel Prize in physics for his role in detecting a subatomic particle, the muon neutrino. That discovery and other key clues to the subatomic mystery are components of the Standard Model of the structure of matter. While at Columbia, Dr. Lederman also taught physics. In addition to receiving several other scientific honors, he was awarded the National Medal of Science in 1965.

From 1979 to 1989 Dr. Lederman served as director of Fermilab, a U.S. Department of Energy facility with a staff of 2000 and an annual budget of about $175 million. Since retiring as Fermilab Director in 1989, Dr.

Interview conducted by
Janice K. Mandel

Lederman continues to work at the laboratory and serves as professor of physics at the University of Chicago. He is a driving force behind bringing progressive science education to Chicago's public schools.

What inspired your interest in physics?

Einstein wrote a big-print book about science for kids *(Einstein for Children)* that my father gave me when I was 10 that started off by saying: "Think of a detective story, one that has a lot of clues as to how nature works, how the world works, and by looking at these clues, the ques-

tion is: Can you find out who did it or how it works?" I found that idea of the mystery and the clues intriguing.

You received a B.S. degree in Chemistry from City College in 1943. Why did you switch to physics?

In order to understand chemistry, I had to take some physics classes, and I think I enjoyed physics more, and for some reason—maybe an accident—found the kids that were in the physics classes and who were physics majors were a lot of fun and I wanted to hang around with them.

Were you an A student?

No. I was a B student, and I did lots of other things besides that, such as reading many books of various kinds. I wasn't really a hotshot student, and some of it was very hard. When I spent three years in the army and went back to graduate school, I found it very difficult.

What kept you going?

After my first year of graduate school, I was getting discouraged, and my friends encouraged me. They said, "Stick it out, stick it out, it won't be that bad." And then, occasionally, through the drudgery of trying to understand things and get good grades on exams and pass, there were some sparks in some of the subjects that recalled to me the kind of fun of reading about how things work and how different things surprised you and how you would see different connections.

In 1988, you were approached by the Department of Energy to help improve science education in Chicago's troubled public school system. What are your thoughts on this project?

To get students involved in science, their teachers have to be much more skillful in general than they are trained to be in how to teach kids math and science. There have been a lot of developments in how to teach math and science, especially to children, and if we can get a whole city activated and improved in their teaching of math and science, there will be tens of thousands of kids who might be turned on by this. And once students find pleasure in science,

they tend to become better students in other areas, such as language and history. It's really a process of educating that the country needs badly. Especially the kids in big cities who have never had the kind of opportunities they should have.

If students really want to know what physics is all about, what do you recommend they do outside of reading this book?

The textbook is only one step. You've got to read *Scientific American* and *Omni* and *Discover* and the magazines that talk about science. You've got to insist that your teacher bring people into the classroom or take you to some place at a university where other things are going on. There are also some very good TV programs on science to be aware of watching, such as *Nova*. Go to museums and look at what they're doing.

Do you recall reading anything that influenced your professional life?

I remember reading an article when I was 12 about a professor who took some apparatus on top of a high mountain in Colorado, trapped some cosmic rays, and discovered the positive electron. I thought that was the most adventurous thing you could possibly imagine doing.

Congratulations on sharing the Nobel Prize in physics in 1988. Can you explain in the simplest terms what you discovered?

My field is particle physics and, in fact, what we discovered was a particle no one knew existed.

It's called a neutrino. But, it wasn't just another particle, it was a very spooky particle because, unlike standard particles like the electron or the proton that have many properties you can study, this particle didn't have an electric charge and, in fact as far as we know, the mass of this particle is practically zero, if not zero. Not only didn't it have any electric charge or mass, but it seemed to ignore all the other particles. You can see an electron and a proton attract each other when you make collisions. The neutrino is uninterested in all the other particles.

So, if you wanted to put up a wall of lead that would stop a neutrino, the wall would have to be billions and billions and billions of miles thick, and we couldn't get enough money to make such a wall. But we found a trick way of actually seeing these neutrinos and measuring what properties they have left. And that was very impressive because it was like the button on which eventually an overcoat would be sewn.

What impact did discovering the muon neutrino and measuring its properties have on science?

Our discovery of this particular kind of neutrino led to a way of organizing the basic particles of nature into a simple pattern which we now call the Standard Model—it's a nice arrangement of basic particles called quarks and leptons. It turns out there is more than one kind of neutrino. This is the key ingredient to the construction of the Universe.

CHAPTER 2

Motion in One Dimension

An apple and a feather, released from rest in a four-foot vacuum chamber, fall at the same rate, regardless of their mass. Neglecting air resistance, all objects fall to the earth with the same acceleration. *(© James Sugar, Black Star)*

In some fashion, people have always been concerned with motion. For example, in today's world you consider motion when you describe to someone how fast your new car will go or how much pickup it has. Likewise, Neanderthal people must have, in their own way, pondered about motion as they devised methods for capturing a rapidly moving antelope. *The branch of physics concerned with the study of the motion of an object and the relationship of this motion to such physical concepts as force and mass is called* **mechanics.** *That part of mechanics which describes motion without regard to the causes of the motion is called* **kinematics.** In this chapter we shall focus on kinematics and on one-dimensional motion, that is, motion along a straight line. We shall start by discussing the concepts of displacement, velocity, and acceleration. These concepts allow us to study the motion of objects undergoing constant acceleration. In Chapter 3 we shall discuss motion of objects in two dimensions.

Before we begin our study, it would be well to consider the long history of mechanics. The first recorded evidence of the study of mechanics can be traced to ancient Babylonian and Egyptian civilizations. The primary concern in those days was to understand the motion of heavenly bodies. The most systematic and detailed early studies of the heavens, however, were conducted by the Greeks during the period from about 300 B.C. to A.D. 300. For hundreds of years, scientists and lay people regarded the Earth as the center of the Universe. This geocentric model was accepted by such notables as Aristotle (384 to 322 B.C.) and Claudius Ptolemy (about A.D. 140). Largely because of the authority of Aristotle, the geocentric model became the officially accepted theory of the Universe and maintained this position until the seventeenth century. The Polish astronomer Nicolaus Copernicus (1473–1543) suggested that the Earth and the other planets revolved in circular orbits around the Sun (the heliocentric model). This early knowledge formed the foundation for the work of Galileo Galilei (1564–1642). Galileo stands out as perhaps the dominant figure in leading the world of physics into the modern era. In 1609 he became one of the first to make astronomical observations with a telescope. He observed mountains on the Moon, the larger satellites of Jupiter, the rings of Saturn, and spots on the Sun. His observations convinced him of the correctness of the Copernican theory. Galileo's work with motion is particularly well known, and because of his leadership, experimentation has become an important part of our search for knowledge.

2.1 Displacement

In order to describe the motion of an object, it is necessary to be able to specify its position at all times using some convenient coordinate system and a specified origin. For example, consider a sprinter moving along the x axis from an initial position x_i to some final position x_f as in Figure 2.1. The **displacement** of an object, defined as *the change in position of the object,* is given by the difference between its final and initial coordinates, or $x_f - x_i$. As mentioned in Chapter 1, we use the Greek letter delta (Δ) to denote the change in a quantity. Therefore, we write the displacement, or change in position of the object, as

$$\Delta x \equiv x_f - x_i \qquad (2.1)$$

Figure 2.1
A sprinter moving along the x axis from x_i to x_f undergoes a displacement $\Delta x = x_f - x_i$.

Displacement

From this definition, we see that Δx is positive if x_f is greater than x_i and negative if x_f is less than x_i. If the sprinter moves from an initial position of $x_i = -3$ m to a final position of $x_f = 5$ m, his displacement is 8 m. [That is, $\Delta x = 5$ m $- (-3$ m$) = 5$ m $+ 3$ m $= 8$ m.]

It is important to note that displacement is an example of a vector quantity since it has both a magnitude and a direction associated with it. Many physical quantities that we shall encounter in this text, including displacement, velocity, and acceleration, are vectors. In general, a **vector** is a physical quantity that requires the specification of both direction and magnitude. In contrast, a **scalar** is a quantity that has magnitude and no direction. Scalar quantities, such as mass and temperature, can be specified by a number with appropriate units.

In the future we shall designate vector quantities with a boldface type. For example, **v** will denote a velocity vector, while **a** will denote an acceleration

A vector has both direction and magnitude

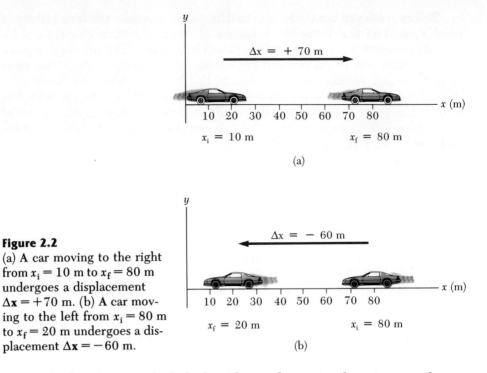

Figure 2.2
(a) A car moving to the right from $x_i = 10$ m to $x_f = 80$ m undergoes a displacement $\Delta\mathbf{x} = +70$ m. (b) A car moving to the left from $x_i = 80$ m to $x_f = 20$ m undergoes a displacement $\Delta\mathbf{x} = -60$ m.

vector. In this chapter which deals with one-dimensional motion, we choose not to use this notation. Rather, we shall specify the direction of vector quantities with positive and negative values. For example, if a car moves from an initial position of 10 m to a final position of 80 m as in Figure 2.2a, its displacement is

$$\Delta x = x_f - x_i = 80 \text{ m} - 10 \text{ m} = +70 \text{ m}$$

In this case, the magnitude of the displacement is 70 m and is directed in the positive x direction as indicated by the plus sign of the result. (Sometimes the plus sign is omitted, but a result of say 70 m for the displacement is understood to be the same as $+70$ m.) The displacement vector is usually represented by an arrow as in Figure 2.2a. The length of the arrow represents the magnitude of the displacement, while the head of the arrow indicates its direction.

Now suppose the car moves to the left from an initial position of 80 m to a final position of 20 m as in Figure 2.2b. In this situation, its displacement is

$$\Delta x = x_f - x_i = 20 \text{ m} - 80 \text{ m} = -60 \text{ m}$$

The minus sign in this result indicates that the displacement is in the negative x direction. Likewise, the arrow representing the displacement vector is to the left as in Figure 2.2b.

2.2   Average Velocity

Consider a car moving along a highway (the x axis), as in Figure 2.2. Let the car's position be x_i at some time t_i, and let its position be x_f at time t_f. (The indices i and f refer to the initial and final locations, respectively.) In the time interval $\Delta t = t_f - t_i$, the displacement of the car is $\Delta x = x_f - x_i$.

The **average velocity**, $\bar{v}$, is defined as the displacement, Δx, divided by the time interval during which the displacement occurred:

$$\bar{v} = \frac{\Delta x}{\Delta t} = \frac{x_f - x_i}{t_f - t_i}$$

(2.2)   Average velocity

It is important to recognize that velocity is a vector quantity. The average velocity of an object can be either positive or negative, depending on the sign of the displacement. (The time interval, Δt, is always positive.) For example, if we select a coordinate system such that a car moves from a position 100 m from the origin to a point 50 m from that origin in a time interval of 2 s, the average velocity is -25 m/s. In this case, the minus sign indicates motion to the left. If an object moves in the negative x direction during some time interval, its velocity in the x direction during that interval is necessarily negative. Similarly, an object moving toward negative y values has a negative y component of velocity.

In order to understand the vector nature of velocity, consider the following situation. Suppose a friend tells you that she will be taking a trip in her car and will travel at a constant rate of 55 mi/h in a straight line for 1 h. If she starts from her home, where will she be at the end of the trip? Obviously, you cannot answer this question because the direction in which she will travel is not specified. All you can say is that she will be located 55 mi from her starting point. However, if she tells you that she will be driving at the rate of 55 mi/h directly northward, then her final location will be known exactly. Therefore, the velocity of an object is known only if its direction and its magnitude (speed) are specified. In general, *any physical quantity that is a vector must be characterized by both a magnitude and a direction.*

As an example of the use of Equation 2.2, suppose the car in Figure 2.2 moves 100 m to the right in 5 s. The substitution of these values into Equation 2.2 gives the average velocity in this time interval as

$$\bar{v} = \frac{\Delta x}{\Delta t} = \frac{100 \text{ m}}{5 \text{ s}} = +20 \text{ m/s}$$

Note that the units of average velocity are units of length divided by units of time, which is meters per second (m/s) in SI units. Other units for velocity might be feet per second (ft/s) in the conventional system or centimeters per second (cm/s) if we decide to measure distances in centimeters. Even furlongs per fortnight would be acceptable (but not very common); the only requirement is that velocity must have units of length divided by units of time.

Let us assume we are watching a drag race from the Goodyear blimp. In one race we see a car follow the straight-line path from P to Q shown in Figure 2.3 during a time interval Δt, while in a second run a car follows the curved path shown in Figure 2.3 for the same time interval. If you examine our definition of average velocity carefully, you will see that both cars have the same average velocity. This is because both cars have the same displacement $(x_f - x_i)$ during the same time interval (Δt), as indicated in Figure 2.3.

Figure 2.4 shows the unusual path of a confused football player. He receives a kickoff at his own goal, runs downfield to within inches of a touchdown, and then reverses his direction to race backward until he is tackled at the exact location where he first caught the ball. What is his average velocity

Figure 2.3
The view of a drag race from a blimp. One car follows the red straight-line path from P to Q, and a second car follows the blue curved path.

Figure 2.4
The path followed by a confused football player.

Figure 2.5
Position-time graph for an object moving along the x axis. The average velocity $\bar{v}$ in the time interval $\Delta t = t_f - t_i$ is the slope of the blue straight line connecting the points P and Q.

during this run? From the definition of average velocity (Eq. 2.2), we see that his average velocity is zero because his displacement is zero. In other words, since x_i and x_f have the same value, $\bar{v} = \Delta x / \Delta t = 0$. From this we see that displacement should not be confused with distance traveled. The confused football player clearly traveled a long distance, almost 200 yards, yet his displacement is zero.

Graphical Interpretation of Velocity

Figure 2.5 is a graph drawn for the motion of an object moving along a straight-line path from the position x_i at time t_i to the position x_f at time t_f. (Note that the motion is along a straight line, yet the position-time graph is *not* a straight line. Why is that?) The straight line connecting points P and Q provides us with a geometric interpretation of average velocity. As indicated on the graph, the slope of the line is $\Delta x \, (= x_f - x_i)$ divided by the time interval for the motion, $\Delta t \, (= t_f - t_i)$. Therefore,

> the average velocity of an object during the time interval t_i to t_f is equal to the slope of the straight line joining the initial and final points on a graph of the position of the object plotted versus time.

2.3   Instantaneous Velocity

Let's imagine that you take a trip in your car along a perfectly straight highway. At the end of your journey, it is a relatively simple task to calculate your average velocity. The car's odometer would give you the distance traveled, a compass would give you direction, and a watch could supply the time interval. However, such a calculation would omit a great deal of information about what actually occurred on your trip. For example, if your calculation indicates that your average velocity was 55 mi/h, this does not necessarily mean that your velocity at every instant was 55 mi/h. Instead, you might have traveled at 55 mi/h for a short distance, stopped for lunch, and then made up some time by traveling at 70 mi/h, paused again as a police officer wrote out a ticket, and then raced forward again when the coast was clear.

More precisely, the instantaneous velocity is defined as the limit of the average velocity as the time interval Δt becomes infinitesimally short. In mathematical language this is written as

Definition of instantaneous velocity

$$v \equiv \lim_{\Delta t \to 0} \frac{\Delta x}{\Delta t} \qquad \text{(2.3)}$$

The notation $\lim \Delta t \to 0$ means that the ratio $\Delta x / \Delta t$ is to be evaluated as the time interval Δt approaches zero. Note that we denote instantaneous velocities by the symbol v without a bar across the top.

To better understand the meaning of instantaneous velocity as expressed by Equation 2.3, consider the data shown in Table 2.1. Assume you have been observing a runner racing along a track. At a time 1 s after starting into motion, the runner has moved to a position 1 m from the starting point, at $t = 1.5$ s, the runner is 2.25 m from the starting point, and so on. After collecting these data,

Table 2.1 Position of a Runner at Specific Instants of Time	
t(s)	x(m)
1.00	1.00
1.01	1.02
1.10	1.21
1.20	1.44
1.50	2.25
2.00	4.00
3.00	9.00

Table 2.2 Calculated Values of the Time Intervals, Displacements, and Average Velocities for the Runner Using the Data in Table 2.1			
Time Interval(s)	Δt(s)	Δx(m)	v(m/s)
1 to 1.01	0.01	0.02	2.0
1 to 1.10	0.10	0.21	2.1
1 to 1.20	0.20	0.44	2.2
1 to 1.50	0.50	1.25	2.5
1 to 2.00	1.00	3.00	3.0
1 to 3.00	2.00	8.00	4.0

suppose you wish to determine the velocity of the runner at the time $t = 1$ s. Table 2.2 represents some of the calculations you might perform to determine the velocity in question. Let us use $t = 1$ s as our initial time, and first consider the bottom row of the table. For the observed portion of the run (from 1 s to 3 s), the time interval $\Delta t = 2$ s, and the displacement $\Delta x = +8$ m. Thus, the average velocity in this interval is $\Delta x/\Delta t = +4$ m/s. This gives only a rough approximation to the instantaneous velocity at $t = 1$ s. According to Equation 2.3, we can find a more reliable answer by letting the time interval become smaller and smaller. Therefore, consider the next to the last entry in Table 2.2, from 1 s to 2 s, corresponding to a time interval of $\Delta t = 1$ s. In this interval, $\Delta x = +3$ m, so the average velocity in this interval is $+3$ m/s. This is closer to the correct answer because the time interval is smaller than the first chosen interval. Now, let us consider a very short time interval indicated by the first entry in Table 2.2. In this case, $\Delta t = 0.01$ s and the displacement $\Delta x = +0.02$ m. Thus, the average velocity in this interval is $+2$ m/s. We could improve the reliability of our calculation by allowing the time interval to become even smaller, but we can state with some degree of confidence that the instantaneous velocity of the runner was $+2$ m/s at the time $t = 1$ s.

In day-to-day usage, the terms *speed* and *velocity* are used interchangeably. In physics, however, there is a clear distinction between these two quantities.

The **instantaneous speed** of an object, which is a scalar quantity, is defined as the magnitude of the instantaneous velocity. Hence, by definition, speed can never be negative.

*Graphical Interpretation of Instantaneous Velocity

Figure 2.6 is a repeat of the graph drawn earlier (Fig. 2.5) for the position of an object versus time. To find the instantaneous velocity of the object at point P, we must find the average velocity during an infinitesimally short time interval. This means that point Q on the curve must be brought closer and closer to point P until the two points are nearly overlapping each other. From this construction we see that the line joining P and Q is a line that is approaching the line tangent to the curve at point P.

The slope of the line tangent to the position-time curve at P is defined to be the instantaneous velocity at that time.

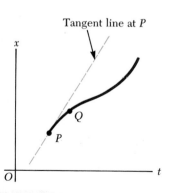

Figure 2.6
Geometric construction for obtaining the instantaneous velocity from the x versus t curve. The instantaneous velocity at P is defined as the slope of the line tangent to the curve at P.

Figure 2.7   Example 2.1

Example 2.1   A Toy Train

A toy train moves slowly along a straight portion of track according to the graph of position versus time shown in Figure 2.7. Find (a) the average velocity for the total trip, (b) the average velocity during the first 4 s of motion, (c) the average velocity during the next 4 s of motion, (d) the instantaneous velocity at $t = 2$ s, and (e) the instantaneous velocity at $t = 5$ s.

Solution
(a) The slope of the line joining the starting point and end point on the graph (the dashed line) gives us the average velocity for the total trip. A measurement of its slope gives

$$\bar{v} = \frac{\Delta x}{\Delta t} = \frac{10 \text{ cm}}{12 \text{ s}} = \boxed{+0.83 \text{ cm/s}}$$

(b) The slope of the line joining the starting point to the point on the curve at $t = 4$ s gives us the average velocity during the first 4 s:

$$v = \frac{\Delta x}{\Delta t} = \frac{4.0 \text{ cm}}{4.0 \text{ s}} = \boxed{+1.0 \text{ cm/s}}$$

(c) Following the same procedure for the next 4-s interval, we see that the slope of the line between points A and B is zero. During this time interval, the train has remained at the same location, 4 cm from the starting point.

(d) A line drawn tangent to the curve at the point corresponding to $t = 2$ s has the same slope as that found for the line in part (b). Thus, the instantaneous velocity at this time is $+1.0$ cm/s. This has to be true because the graph indicates that during the first 4 s of motion, the train is covering equal distances in equal intervals of time. In other words, the train is moving at a constant velocity during the first 4 s. Under these conditions the average velocity and the instantaneous velocity at all times are identical.

(e) At $t = 5$ s, the slope of the position-time curve is zero. Therefore, the instantaneous velocity is zero at this instant. In fact, the train is at rest in the entire time interval between 4 s and 8 s.

2.4   Acceleration

As you travel from place to place in your car, you normally do not travel long distances at a constant velocity. The velocity of the car will increase when you step on the gas, and it will decrease when you apply the brakes. Furthermore, the velocity of the car changes when you round a curve, corresponding to a change in direction of motion. The rate of change of velocity is referred to as acceleration. However, we need a more precise definition of acceleration than that.

Suppose a car moves along a straight highway as in Figure 2.8. At time t_i it has a velocity v_i, and at time t_f its velocity is v_f.

The **average acceleration** during this time interval is defined as the change in velocity divided by the time interval during which this change occurs:

Figure 2.8
A car moving to the right accelerates from a velocity v_i to a velocity v_f in the time interval $\Delta t = t_f - t_i$.

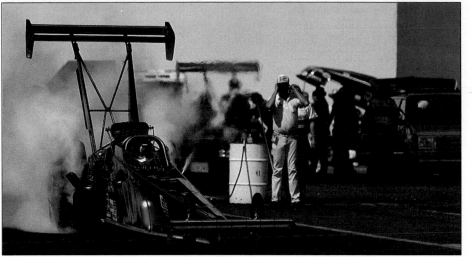

A car in a drag race undergoes large accelerations and decelerations during a race. (© Jeffrey Sylvester, FPG International)

$$\bar{a} \equiv \frac{\Delta v}{\Delta t} = \frac{v_f - v_i}{t_f - t_i}$$

(2.4) Definition of average acceleration

As an example of the computation of acceleration, suppose the car shown in Figure 2.8 accelerates from an initial velocity of $v_i = +10$ m/s to a final velocity of $v_f = +30$ m/s in a time interval of 2 s. (Note that both velocities are toward the right, the direction selected as the positive direction.) These values can be inserted into Equation 2.4 to give the average acceleration:

$$\bar{a} = \frac{30 \text{ m/s} - 10 \text{ m/s}}{2.0 \text{ s}} = +10 \text{ m/s}^2$$

Note that the dimensions of acceleration are those of length divided by (time)². Some units of acceleration are meters per second per second (m/s²), feet per second per second (ft/s²), and miles per hour per second (mi/h · s). The acceleration we calculated in the above example was $+10$ m/s². This notation means that, on the average, the car was moving such that its velocity increased at a rate of 10 m/s every second.

As a second example of the computation of acceleration, consider the car pictured in Figure 2.9. In this case, the velocity of the car has changed from an initial value of $+30$ m/s to a final value of $+10$ m/s in a time interval of 2 s. The average acceleration during this time interval is

$$\bar{a} = \frac{10 \text{ m/s} - 30 \text{ m/s}}{2.0 \text{ s}} = -10 \text{ m/s}^2$$

Because acceleration is a vector quantity, the minus sign in the above example indicates that the acceleration vector is in the negative x direction (to the left). For the case of motion in a straight line, the direction of the velocity of an object and the direction of its acceleration are related as follows. *When the object's velocity and acceleration are in the same direction, the object is speeding up in that direction.* (The first example demonstrates this situation.)

$t_i = 0$ $t_f = 2$ s
$+30$ m/s $+10$ m/s

Figure 2.9
The velocity of the car decreases from $+30$ m/s to $+10$ m/s in a time interval of 2 s.

When the object's velocity and acceleration are in opposite directions, the object is slowing down. (Decreases in speed are sometimes called decelerations.)

To clarify this point, consider the following situation. Suppose the velocity of a car changes from −10 m/s to −30 m/s in a time interval of 2 s. The minus signs here indicate that the motion of the car is in the negative *x* direction. The average acceleration of the car in this time interval is

$$\bar{a} = \frac{-30 \text{ m/s} - (-10 \text{ m/s})}{2.0 \text{ s}} = -10 \text{ m/s}^2$$

The minus sign indicates that the acceleration is also in the negative *x* direction. Since the velocity and acceleration are in the same direction, the car must be speeding up as it moves to the left. Note that a negative value for the acceleration does not always indicate a deceleration.

Instantaneous Acceleration

In some situations, the value of the average acceleration may be different over different time intervals. It is therefore useful to define **instantaneous acceleration** as the *acceleration of an object at some particular instant of time.* This concept is analogous to the definition of instantaneous velocity discussed in Section 2.3. In mathematical terms, the instantaneous acceleration, *a*, is defined as the limit of the average acceleration as the time interval Δ*t* becomes infinitesimally short. That is,

Definition of instantaneous acceleration

$$a \equiv \lim_{\Delta t \to 0} \frac{\Delta v}{\Delta t} \qquad (2.5)$$

Here again, the notation lim Δ*t* → 0 means that the ratio Δ*v*/Δ*t* is to be evaluated as the time interval Δ*t* approaches zero.

Figure 2.10 is a useful diagram for understanding the concept of instantaneous acceleration. Plotted here is the velocity of an object versus time. This could represent, for example, the motion of a car along a busy street. The average acceleration of the car between times t_i and t_f can be found by finding the slope of the line joining points *P* and *Q*. If we imagine that point *Q* is brought closer and closer to point *P*, the value that we find for the average acceleration between these points approaches the value of the acceleration of the car at point *P*. You should be able to see that the instantaneous acceleration at point *P*, for example, is the slope of the graph at the point. That is,

> the instantaneous acceleration of an object at a certain time equals the slope of the velocity-time graph at that instant of time.

From now on we shall use the term *acceleration* to mean "instantaneous acceleration."

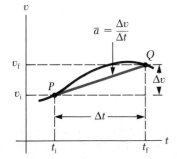

Figure 2.10
Velocity-time graph for an object moving in a straight line. The slope of the blue line connecting points *P* and *Q* is defined as the average acceleration in the time interval $\Delta t = t_f - t_i$.

Example 2.2   A Fly Ball Is Caught

A baseball player moves in a straight-line path in order to catch a fly ball hit to the outfield. His velocity as a function of time is shown in Figure 2.11. Find his instantaneous acceleration at points *A*, *B*, and *C* on the curve.

Solution

The instantaneous acceleration at any time is the slope of the velocity-time curve at that instant. The slope of the curve at point A is

$$a = \frac{\Delta v}{\Delta t} = \frac{4 \text{ m/s}}{2 \text{ s}} = \boxed{+2 \text{ m/s}^2}$$

At point B, the slope of the line is zero and hence the instantaneous acceleration at this time is also zero. You should note that even though the instantaneous acceleration has dropped to zero at this instant, the velocity of the player is not zero. (In general, if $a = 0$, the velocity is constant in time but *not necessarily* zero.) Instead, he is running at a constant velocity of $+4$ m/s. A value of zero for an instantaneous acceleration means that at the particular instant the velocity of the object is not changing.

Finally, at point C we calculate the slope of the velocity-time curve and find that the instantaneous acceleration is

$$a = \frac{\Delta v}{\Delta t} = \frac{-2 \text{ m/s}}{1 \text{ s}} = \boxed{-2 \text{ m/s}^2}$$

The minus sign here indicates that the player is decelerating as he approaches the location where he will attempt to catch the baseball.

Figure 2.11 Example 2.2

2.5 Motion Diagrams

The concepts of velocity and acceleration are often confused with each other, but in fact they are quite different quantities. It is instructive to make use of motion diagrams to describe the velocity and acceleration vectors as time progresses during the motion of an object. In order not to confuse these two vector quantities in Figure 2.12, we use red for velocity vectors and violet for acceleration vectors. The vectors are sketched at several instants during the motion of the object and the time intervals between adjacent positions are assumed to be equal. Figure 2.12a represents a car moving to the right with a constant positive acceleration. In this case, note that the velocity vector increases in time. Because the car moves faster as it travels to the right, its displacement between adjacent positions increases as time progresses. If the car moves initially to the right with a constant negative acceleration (or deceleration) as in Figure 2.12b, the velocity vector decreases in time and eventually reaches zero. (This type of motion is exhibited by a car that skids to a stop

(a)

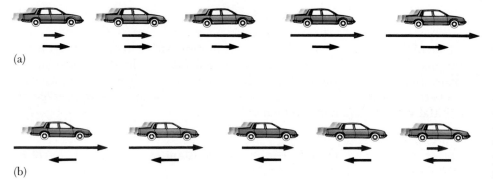

(b)

Figure 2.12
(a) Motion diagram for an object whose constant acceleration is in the direction of its velocity. The velocity vectors at each instant are in red, while the constant acceleration vector is violet. (b) Motion diagram for an object whose constant acceleration points *opposite* the velocity at each instant.

after applying its brakes.) In this case, the car moves slower as it moves to the right, so its displacement between adjacent positions decreases as time progresses. From this diagram, we see that the acceleration and velocity vectors are *not* in the same direction.

You should be able to construct a motion diagram corresponding to a particle moving initially to the left with a constant positive or negative acceleration. You should also construct appropriate motion diagrams after completing the mathematical solution to kinematic problems to see if your answers are consistent with the diagrams.

2.6   One-Dimensional Motion with Constant Acceleration

Most of the applications in this text will be concerned with objects moving with *constant acceleration*. This type of motion is important because many objects in nature move with constant acceleration. For example, an object in free fall near the Earth's surface moves in the vertical direction with constant acceleration, assuming that air resistance can be neglected. Furthermore, in many situations where the acceleration is not constant, an approximate solution can be obtained by assuming the acceleration to be constant and equal to its average value over some time interval. In almost all situations, the actual motion by a series of motions all having a constant acceleration can be approximated. When an object moves with constant acceleration, *the average acceleration equals the instantaneous acceleration.* Consequently, the velocity increases or decreases at the same rate throughout the motion.

Since the average acceleration equals the instantaneous acceleration when $a = $ constant, we can eliminate the bar used to denote average values from our defining equation for acceleration. That is, since $\bar{a} = a$, we can write Equation 2.4 as

$$a = \frac{v_f - v_i}{t_f - t_i}$$

For convenience, let $t_i = 0$ and t_f be any arbitrary time t. Also, we shall let $v_i = v_0$ (the initial velocity at $t = 0$) and $v_f = v$ (the velocity at any arbitrary time t). With this notation, we can express the acceleration as

$$a = \frac{v - v_0}{t}$$

or

$$v = v_0 + at \tag{2.6}$$

One of the features of one-dimensional motion with constant acceleration is the manner in which the initial, final, and average velocities are related. Because the velocity is increasing or decreasing *uniformly* with time, we can express the **average velocity** in any time interval as the arithmetic average of the initial velocity, v_0, and the final velocity, v:

$$\bar{v} = \frac{v_0 + v}{2} \qquad \text{(for constant } a\text{)} \tag{2.7}$$

Note that *this expression is valid only when the acceleration is constant, that is, when the velocity changes uniformly with time.* The graphical interpretation of $\bar{v}$ is shown in Figure 2.13. As you can see, the velocity varies linearly with time according to Equation 2.6.

We can now use this result along with the defining equation for average velocity, Equation 2.2, to obtain an expression for the displacement of an object as a function of time. Again we choose $t_i = 0$, and for convenience we shall also assume that we have selected our coordinate system such that the initial position of the object under consideration is at the origin, that is, $x_i = 0$. This gives

$$x = \bar{v}t = \left(\frac{v_0 + v}{2}\right) t$$

or

$$x = \frac{1}{2}\,(v + v_0)t \qquad \textbf{(2.8)}$$

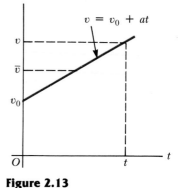

Figure 2.13
The velocity varies linearly with time for a particle moving with constant acceleration. The average velocity is just the mean value of the initial and final velocities.

We can obtain another useful expression for displacement by substituting the equation for v (Eq. 2.6) into Equation 2.8:

$$x = \frac{1}{2}\,(v_0 + at + v_0)t$$

or

$$x = v_0 t + \frac{1}{2}\,at^2 \qquad \textbf{(2.9)}$$

Finally, we can obtain an expression that does not contain time by substituting the value of t from Equation 2.6 into Equation 2.8. This gives

$$x = \frac{1}{2}\,(v + v_0)\left(\frac{v - v_0}{a}\right) = \frac{v^2 - v_0{}^2}{2a}$$

or

$$v^2 = v_0{}^2 + 2ax \qquad \textbf{(2.10)}$$

Equations 2.6 through 2.10 may be used to solve any problem in one-dimensional motion with constant acceleration. Keep in mind that *these relationships were derived from the definitions of velocity and acceleration,* together with some algebraic manipulations and the requirement that the acceleration be constant. The four equations that are used most often are listed in Table 2.3 for convenience.

The best way to gain confidence in the use of these equations is through working a number of problems. Many times you will discover that there is more than one method for obtaining a solution to a given problem.

Table 2.3
Equations for Motion in a Straight Line Under Constant Acceleration

Equation	Information Given by Equation
$v = v_0 + at$	Velocity as a function of time
$x = \frac{1}{2}(v + v_0)t$	Displacement as a function of velocity and time
$x = v_0 t + \frac{1}{2}at^2$	Displacement as a function of time
$v^2 = v_0^2 + 2ax$	Velocity as a function of displacement

Note: Motion is along the x axis. At $t = 0$, the particle is at the origin ($x_0 = 0$) and its velocity is v_0.

Problem-Solving Strategy:
Accelerated Motion

The following procedure is recommended when solving problems involving acceleration motion:

1. Make sure all the units in the problem are consistent. That is, if distances are measured in meters, be sure that velocities have units of m/s and accelerations have units of m/s².
2. Choose a coordinate system.
3. Make a list of all the quantities given in the problem and a separate list of those to be determined.
4. Select from the list of kinematic equations the one or ones that will enable you to determine the unknowns.
5. Construct an appropriate motion diagram and check to see if your answers are consistent with the diagram.

Example 2.3   The Indianapolis 500

A racing car starting from rest accelerates at a rate of 5 m/s². What is the velocity of the car after it has traveled 100 ft?

Solution

Refer to the Problem-Solving Strategy section to see how the steps indicated there are applied to this example. First, you must be sure that the units in the problem are consistent. As stated, the units in this problem are not consistent. If we choose to leave the distance as 100 ft, we must change the length dimension of the units of acceleration from m/s² to ft/s². As an alternative, we can leave the units of acceleration above as m/s² and convert the distance traveled to meters. Let's do the latter. The table of conversion factors in Appendix C of this text gives 1 ft = 0.305 m; thus 100 ft = 30.5 m.

Second, you must choose a coordinate system. In Figure 2.14, a convenient one is shown. The origin of the coordinate system is at the initial location of the car, and the positive direction is to the right. Using this convention, we require that velocities, accelerations, and displacements to the right are positive and vice versa.

$v_0 = 0$ $v = ?$

$x = 0$ $x = 30.5$ m

Figure 2.14   Example 2.3

Next, you will find it convenient to make a list of the quantities given in the problem and a separate list of those to be determined:

Given	To be Determined
$v_0 = 0$	v
$a = +5 \text{ m/s}^2$	
$x = +30.5 \text{ m}$	

The final step is to select from kinematic equations (Table 2.3) those that will allow you to determine the unknowns. In our present case, the equation

$$v^2 = v_0{}^2 + 2ax$$

is our best choice since it will give us a value for v directly:

$$v^2 = (0)^2 + 2(5 \text{ m/s}^2)(30.5 \text{ m}) = 305 \text{ m}^2/\text{s}^2$$

from which,

$$v = \sqrt{305 \text{ m}^2/\text{s}^2} = \pm 17.5 \text{ m/s}$$

Since the car is moving to the right, we choose $+17.5$ m/s as the correct solution for v. Alternatively, the problem can be solved by using $x = v_0 t + \frac{1}{2}at^2$ to find t and then using the expression $v = v_0 + at$ to find v. Try it!

Example 2.4 The Supercharged Sportscar

A certain automobile manufacturer claims that its super-deluxe sportscar will accelerate uniformly from rest to a speed of 87.0 mi/h in 8 s. (a) Determine the acceleration of the car.

Solution
First note that $v_0 = 0$ and the velocity after 8 s is 87.0 mi/h = 38.9 m/s, where the conversion 1 mi/h = 0.447 m/s has been used. Using $v = v_0 + at$, we can find the acceleration:

$$a = \frac{v - v_0}{t} = \frac{38.9 \text{ m/s}}{8 \text{ s}} = \boxed{+4.86 \text{ m/s}^2}$$

(b) Find the distance the car travels in the first 8 s.

Solution
The distance traveled by the car can be found from Equation 2.8:

$$x = \frac{1}{2}(v_0 + v)t = \frac{1}{2}(38.9 \text{ m/s})(8 \text{ s}) = \boxed{156 \text{ m}}$$

Many examples in this text will be followed by an exercise. The purpose of these exercises is to test your understanding of the example by asking you a question related to it. Answers to these exercises will be provided at the end of the exercise, when appropriate. Here is your first exercise, which relates to Example 2.4.

Exercise What is the velocity of the car 10 s after it begins its motion, assuming it continues to accelerate at the rate of $+4.86$ m/s^2?

Answer 48.6 m/s, or 109 mi/h.

Two teams of skydivers forming two rings. Skydivers reach a terminal speed of about 160 km/h in the "spread-eagle" position. They can control their speed by adjusting the area they present to the atmosphere. (© Vandystadt/Photo Researchers, Inc.)

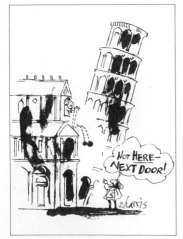

NOT HERE—
NEXT DOOR!

2.7 Freely Falling Bodies

It is well known that, in the absence of air resistance, all objects when dropped fall toward the Earth with the same constant acceleration. However, it was not until about 1600 that this conclusion came to be accepted. Prior to that time, the teachings of the great philosopher Aristotle (384 — 322 B.C.) had held that heavier objects fell faster than lighter ones. It was Galileo, often called the father of modern science, who was responsible for our present ideas concerning falling bodies. There is a legendary story that he discovered the law of falling bodies by observing that two different weights dropped simultaneously from the Leaning Tower of Pisa hit the ground at approximately the same time. Although there is some doubt that this particular experiment was carried out, it is well established that Galileo did perform many systematic experiments on objects moving on inclined planes. In his experiments he rolled balls down a slight incline and measured the distance they moved in successive time intervals. The purpose of the incline was to produce a smaller acceleration and enable him to make accurate measurements of the time intervals. (Some people refer to this experiment as "diluting gravity.") By gradually increasing the slope of the incline, he was finally able to draw conclusions about freely falling objects because a falling ball is equivalent to a ball falling down a vertical incline. Galileo's achievements in the science of mechanics paved the way for Newton in his development of the laws of motion, which we shall study in Chapter 4.

You might want to try the following experiment. Drop a coin and a crumpled-up piece of paper simultaneously from the same height. If the effects of air friction are negligible, both will have the same motion and hit the floor at the same time. In the idealized case, where air resistance is absent, such motion is referred to as free fall. If this same experiment could be conducted in a vacuum, where air friction is truly negligible, the paper and coin would fall with the same acceleration, regardless of the shape of the paper. (See photographs in "Physics in Action.") On August 2, 1971, such a demonstration was conducted on the Moon by astronaut David Scott. He simultaneously released a hammer and a feather, and they fell with the same acceleration to the lunar surface. This demonstration would have surely pleased Galileo!

Freely Falling Bodies

On the left is a multiflash photograph of a falling apple. The time interval between successive images is (1/60) s. As the apple falls, the spacing between successive images increases, indicating that the apple accelerates downward. Can you estimate **g** from this photograph?

The middle photograph shows a multiflash image of two freely falling balls released simultaneously. The ball on the left side is solid, while the ball on the right is a hollow Ping-Pong ball. The time interval between flashes is (1/30) s and the scale is in cm. Can you determine **g** from these data? Note that the effect of air resistance is greater for the smaller hollow ball on the right, as indicated by the smaller separation between successive images.

In the photograph on the right a feather and an apple are released from rest in a 4-ft vacuum chamber. The trap door that released the two objects was opened with an electronic switch at the same instant the camera shutter was opened. The two objects fall at the same rate as indicated by the horizontal alignment of the multiple images. This shows that when air resistance is negligible, all objects fall with the same acceleration regardless of their mass.

(© Harold E. Edgerton. Courtesy of Palm Press Inc.)

(Education Development Center, Newton, MA)

(© James Sugar, Black Star)

Galileo Galilei
(1564 – 1642)

Galileo Galilei, an Italian physicist and astronomer, was born into the family of a Florentine cloth merchant in Pisa. As a young boy, Galileo was sent to school at the local monastery until his father, who was interested in many things including mathematics and music, decided to teach him. The young Galileo entered medical school, but soon withdrew because he could not accept many beliefs of the profession. Because of his mathematical skills, and through the influence of his father's friends, he received the chair in mathematics at the University of Pisa at age 25.

Although Galileo was an exceedingly gifted person, he was very determined and often spiteful. According to one story, Galileo published a pamphlet explaining his invention of a particular type of military compass. Shortly thereafter, a young student published another pamphlet describing a similar compass. Galileo took the young student to court, claiming plagiarism, and won easily. Following the decision, he wrote and distributed an account of the incident, which ruined the student's reputation.

Another colorful story asserts that Galileo climbed the Leaning Tower of Pisa and simultaneously dropped a bullet and a cannonball to prove that they fell at the same rate. Although there is no real evidence to support this story, Galileo formulated the laws that govern the motion of objects in free fall (including projectiles). He also investigated the motion of an object on an inclined plane, established the concept of relative motion, invented the thermometer, and noted that the motion of a swinging pendulum could be used to measure time intervals. Some claim that he discovered the properties of a pendulum by using his pulse to time the motion of a swinging lamp in the Pisa Cathedral.

In 1609, Galileo became aware of the invention of the telescope in Holland, and immediately designed and constructed his own telescope, which he then used to make several major discoveries in astronomy: He discovered four of Jupiter's moons, found that the Moon's surface was rough (contrary to Aristotle's view), discovered sunspots and the phases of Venus, and showed that the Milky Way consists of an enormous number of stars.

As a result of his observations in astronomy, Galileo publicly defended the view of Nicholas Copernicus, which asserted that the Sun was at the center of the universe (the heliocentric system). At the time, the Church upheld Aristotle's viewpoint that the Universe was fixed and that the Earth was at its center. Following many skirmishes with the Church, which declared the Copernican view to be heretical, Galileo published *Dialogue Concerning The Two World System* to support this view. He was taken to Rome in 1633 on a charge of heresy and was sentenced to life imprisonment. The sentence was commuted to house arrest and he was confined to his villa at Arcetri near Florence for the rest of his life. In his last years, he summed up his life's work in a manuscript entitled *Discourses and Mathematical Discoveries Concerning Two New Sciences*, which was smuggled out of Italy and published in Holland in 1638. After completing this work, he became blind and died in Arcetri in 1642.

(Courtesy of AIP Niels Bohr Library)

We shall denote the acceleration due to gravity by the symbol **g**. The magnitude of **g** decreases with increasing altitude. Furthermore, there are slight variations in the magnitude of **g** with latitude. However, at the surface of the Earth the magnitude of **g** is approximately 9.8 m/s^2, or 980 cm/s^2, or 32 ft/s^2. Unless stated otherwise, we shall use the value 9.80 m/s^2 when doing calculations. Furthermore, we shall assume that the vector **g** is directed downward toward the center of the Earth.

When we use the expression *freely falling body,* we do not necessarily refer to an object dropped from rest.

A freely falling body is an object moving freely under the influence of gravity only, regardless of its initial motion. Objects thrown upward or downward and those released from rest are all falling freely once they are released!

It is important to emphasize that any freely falling object experiences an acceleration directed downward. This is true regardless of the initial motion of the object. An object thrown upward (or downward) will experience the same acceleration as an object released from rest.

Once they are in free fall, all objects have an acceleration downward, equal to the acceleration due to gravity.

If we neglect air resistance and assume that the gravitational acceleration does not vary with altitude, then the motion of a freely falling body is equivalent to motion in one dimension under constant acceleration. Therefore, the equations developed in Section 2.6 for objects moving with constant acceleration can be applied. The only modification that we need to make in these equations for freely falling bodies is to note that the motion is in the vertical direction (the y direction) rather than along the horizontal (x direction), and that the acceleration is downward and has a magnitude of 9.80 m/s². Thus, we always take $a = -g = -9.80$ m/s², where the minus sign means that the acceleration of a freely falling body is downward. We shall study how to deal with variations in g with altitude in Chapter 7, and the effects of air resistance on motion will be discussed in Chapter 10.

Example 2.5 Look Out Below!

A golf ball is released from rest from the top of a very tall building. Neglecting air resistance, calculate the position and velocity of the ball after 1, 2, and 3 s.

Solution

We choose our coordinates such that the starting point of the ball is at the origin ($y_0 = 0$ at $t = 0$) and remember that we have defined y to be positive upward. Since $v_0 = 0$, and $a = -g = -9.80$ m/s², Equations 2.6 and 2.9 become

$$v = at = (-9.80 \text{ m/s}^2)t$$

$$y = \frac{1}{2} at^2 = \frac{1}{2}(-9.80 \text{ m/s}^2)t^2$$

where t is in s, v is in m/s, and y is in m. These expressions give the velocity and displacement at any time t after the ball is released. Therefore, at $t = 1$ s,

$$v = (-9.80 \text{ m/s}^2)(1 \text{ s}) = \boxed{-9.80 \text{ m/s}}$$

$$y = \frac{1}{2}(-9.80 \text{ m/s}^2)(1 \text{ s})^2 = \boxed{-4.90 \text{ m}}$$

Likewise, at $t = 2$ s, we find that $v = -19.6$ m/s and $y = -19.6$ m. Finally, at $t = 3$ s, $v = -29.4$ m/s and $y = -44.1$ m. The minus signs for v indicate that the velocity vector is directed downward, and the minus signs for y indicate displacement in the negative y direction.

Exercise   Calculate the position and velocity of the ball after 4 s.

Answer -78.4 m, -38.4 m/s.

Example 2.6   Not a Bad Throw for a Rookie!

A stone is thrown from the top of a building with an initial velocity of 20 m/s straight upward. The building is 50 m high, and the stone just misses the edge of the roof on its way down, as in Figure 2.15. Determine (a) the time needed for the stone to reach its maximum height, (b) the maximum height, (c) the time needed for the stone to return to the level of the thrower, (d) the velocity of the stone at this instant, and (e) the velocity and position of the stone at $t = 5$ s.

Solution

(a) To find the time necessary to reach the maximum height, use Equation 2.6, $v = v_0 + at$, noting that $v = 0$ at maximum height:

$$20 \text{ m/s} + (-9.80 \text{ m/s}^2)t_1 = 0$$

$$t_1 = \frac{20 \text{ m/s}}{9.80 \text{ m/s}^2} = \boxed{2.04 \text{ s}}$$

(b) This value of time can be substituted into Equation 2.9, $y = v_0 t + \frac{1}{2}at^2$, to give the maximum height as measured from the position of the thrower:

$$y_{\text{max}} = (20 \text{ m/s})(2.04 \text{ s}) + \tfrac{1}{2}(-9.80 \text{ m/s}^2)(2.04 \text{ s})^2 = \boxed{20.4 \text{ m}}$$

(c) When the stone is back at the height of the thrower, the y coordinate is zero. From the expression $y = v_0 t + \frac{1}{2}at^2$ (Eq. 2.9), with $y = 0$, we obtain the expression

$$20t - 4.9t^2 = 0$$

This is a quadratic equation and has two solutions for t. The equation can be factored to give

$$t(20 - 4.9t) = 0$$

One solution is $t = 0$, corresponding to the time the stone starts its motion. The other solution is $t = 4.08$ s, which is the solution we are after.

(d) The value for t found in (c) can be inserted into $v = v_0 + at$ (Eq. 2.6) to give

$$v = 20 \text{ m/s} + (-9.80 \text{ m/s}^2)(4.08 \text{ s}) = \boxed{-20.0 \text{ m/s}}$$

Note that the velocity of the stone when it arrives back at its original height is equal in magnitude to its initial velocity but opposite in direction. This indicates that the motion is symmetric.

(e) From $v = v_0 + at$ (Eq. 2.6), the velocity after 5 s is

$$v = 20 \text{ m/s} + (-9.80 \text{ m/s}^2)(5 \text{ s}) = \boxed{-29.0 \text{ m/s}}$$

We can use $y = v_0 t + \frac{1}{2}at^2$ (Eq. 2.9) to find the position of the particle at $t = 5$ s:

$$y = (20 \text{ m/s})(5 \text{ s}) + \tfrac{1}{2}(-9.80 \text{ m/s}^2)(5 \text{ s})^2 = \boxed{-22.5 \text{ m}}$$

Exercise   Find the velocity of the stone just before it hits the ground.

Answer -37 m/s.

$t_1 = 2.04$ s
$y_{\text{max}} = 20.4$ m
$v_y = 0$

$t = 0, y_0 = 0$
$v_0 = 20$ m/s

$t = 4.08$ s
$y = 0$
$v = -20$ m/s

$t = 5$ s
$y = -22.5$ m
$v = -29$ m/s

$t_2 = 5.8$ s
$y = -50$ m
$v = -37$ m/s

50 m

Figure 2.15   Example 2.6 Position and velocity versus time for a freely falling object thrown initially upward with a velocity $v_0 = +20$ m/s.

Summary

The **displacement** of an object moving along the x axis is defined as the change in position of the object, and is given by

$$\Delta x \equiv x_f - x_i \qquad\qquad (2.1)$$

where x_i is the initial position of the object and x_f is its final position.

The **average velocity** of an object moving along the x axis during some time interval is equal to the displacement of the object, Δx, divided by the time interval, Δt, during which the displacement occurred:

$$\bar{v} \equiv \frac{\Delta x}{\Delta t} = \frac{x_f - x_i}{t_f - t_i} \qquad\qquad (2.2)$$

The average velocity is equal to the slope of the straight line joining the initial and final points on a graph of the displacement of the object versus time.

The velocity of an object at any instant of time is called the **instantaneous velocity.** The slope of the line tangent to the position-time curve at some point is defined to be the instantaneous velocity at that time. The **instantaneous speed** of an object is defined as the magnitude of the instantaneous velocity.

The **average acceleration** of an object during some time interval is defined as the change in velocity, Δv, divided by the time interval, Δt, during which the change occurred:

$$\bar{a} \equiv \frac{\Delta v}{\Delta t} = \frac{v_f - v_i}{t_f - t_i} \qquad\qquad (2.4)$$

The **instantaneous acceleration** of an object is its acceleration at some particular instant of time. The instantaneous acceleration of an object at a certain time equals the slope of a velocity time graph at that instant of time.

A **vector** quantity is any quantity that is characterized by both a magnitude and a direction. In contrast, a **scalar** quantity such as mass has a magnitude only. Displacement, velocity, and acceleration are all vector quantities. In the case of motion along a straight line, we use algebraic signs to describe the direction of the vectors. For example, an object moving to the right with a speed of 8 m/s would have a velocity of $+8$ m/s. If the object were moving to the left, its velocity would be -8 m/s.

The equations that describe the motion of an object moving with constant acceleration along the x axis are

$$v = v_0 + at \qquad\qquad (2.6)$$

$$x = \frac{1}{2}(v + v_0)t \qquad\qquad (2.8)$$

$$x = v_0 t + \frac{1}{2}at^2 \qquad\qquad (2.9)$$

$$v^2 = v_0^2 + 2ax \qquad\qquad (2.10)$$

An object falling in the presence of the Earth's gravity experiences a gravitational acceleration directed toward the center of the Earth. If air friction is neglected and if the altitude of the falling object is small compared with the Earth's radius, then one can assume that the *acceleration of gravity*, **g**, is constant over the range of motion, where g is equal to 9.80 m/s^2, or 32 ft/s^2.

Assuming the positive direction for y is chosen to be upward, the acceleration is $-g$ (downward) and the equations describing the motion of the falling object are the same as the equations above, with the substitutions $x \to y$ and $a \to -g$.

Additional Reading

I. B. Cohen, "Galileo," *Sci. American,* August 1949, p. 40.

S. Drake, "Galileo's Discovery of the Law of Free Fall," *Sci. American,* May 1973, p. 84.

Galileo Galilei, "Dialogues Concerning Two New Sciences," translated by H. Crew and A. de Salvio, Evanston, IL, Northwestern University Press, 1939.

O. Gingerich, "The Galileo Affair," *Sci. American,* August 1982, p. 132.

Questions

1. The speed of sound in air is 331 m/s. During the next thunderstorm try to estimate your distance from a lightning bolt by measuring the time lag between the flash and thunderclap. You can ignore the time for the light flash to reach you. Why?

2. If the average velocity of an object is zero in some time interval, what can you say about the displacement of the object for that interval?

3. Average velocity and instantaneous velocity are generally different quantities. Can they ever be equal for a specific type of motion?

4. Can the instantaneous velocity of an object ever be greater in magnitude than the average velocity? Can it ever be less?

5. An object would travel the same distance if it moved with a constant velocity equal to the average velocity. Explain.

6. If the average velocity is nonzero for some time interval, does this mean that the instantaneous velocity is never zero during this interval? Explain.

7. If an object is stationary at some instant, is its acceleration necessarily zero at this time?

8. At the end of its arc, the velocity of a pendulum is zero. Is its acceleration also zero at this point?

9. If the velocity of a particle is not zero, can its acceleration ever be zero?

10. A ball is thrown vertically upward. What are its velocity and acceleration when it reaches its maximum altitude? What is its acceleration just before it hits the ground?

11. If a car is traveling eastward, can its acceleration be westward? Explain.

12. Can the sign of the acceleration ever be positive for an object that is slowing down? Explain.

13. An object is thrown upward. Discuss the sign of the acceleration of gravity relative to the velocity while the object is in the air.

14. Car A traveling from New York to Miami has a speed of 25 m/s. Car B traveling from New York to Chicago also has a speed of 25 m/s. Are their velocities equal? Explain.

15. Two cars are moving in the same direction in parallel lanes along a highway. At some instant, the velocity of car A exceeds the velocity of car B. Does this mean that the acceleration of A is greater than that of B? Explain.

16. Suppose you are guilty of a speeding violation and are presented with a traffic ticket. Will your fine be based on your average speed or on your instantaneous speed?

17. A rule of thumb for driving is that a separation of one car length for each 10 miles per hour of speed should be maintained between moving vehicles. Assuming a constant reaction time, discuss the relevance of this rule for (a) motion with constant velocity and (b) motion with constant acceleration.

18. A stone is thrown upward from the top of a building. Does the stone's displacement depend on the location of the origin of the coordinate system? Does the stone's velocity depend on the origin?

19. A child throws a marble into the air with an initial velocity v_0. Another child drops a ball at the same instant. Compare the accelerations of the two objects while they are in flight.

20. A baseball is thrown vertically upwards into the air and returns to the starting point. Sketch graphs of (a) its velocity versus time and (b) its acceleration versus time.

21. Two children stand at the edge of a tall building. One drops a rock over the edge, while simultaneously the second throws a rock downward such that it has an initial speed of 10 m/s. Compare the accelerations of the two objects while in flight.

22. A ball is thrown vertically upward with an initial speed of 8 m/s. What will its speed be when it returns to its starting point?

23. A student at the top of a building of height h throws one ball upward with a speed v_0 and then throws a second ball downward with the same initial speed v_0. How do the final velocities of the balls compare when they reach the ground?

24. A book is moved once around the perimeter of a table of dimensions 1 m $\times$ 2 m. If the book ends up at its initial position, what is its displacement?

25. A space shuttle takes off from Florida and circles the globe several times, finally landing in California. While the shuttle is in flight, the wife of an astronaut flies from Florida to California so she can greet her husband when he steps off the shuttle. Who undergoes the greater displacement, the wife or the astronaut?

26. Does the displacement between two objects locate the objects? Discuss in terms of the information carried by a vector.

Problems

Section 2.2   Average Velocity and
Section 2.3   Instantaneous Velocity

1. A football player makes a touchdown run of 100 yards in a time of 15 seconds. What was his average velocity in m/s during his run?

2. If the average speed of an orbiting space shuttle is 19,800 miles per hour, find the time required for it to circle the Earth. Use the distance around the Earth in your calculation and neglect the fact that the shuttle is about 200 miles above the Earth's surface.

3. A certain bacterium swims with a speed of 3.5 μm/s. How long would it take this bacterium to swim across a petri dish having a diameter of 8.4 cm?

4. A person makes a trip by car between two cities. A record of the trip follows: 30.0 min at 80.0 km/h, 12.0 min at 100 km/h, 45.0 min at 40.0 km/h, and 15.0 min for lunch break and gas-up. (a) Determine the average speed for the trip. (b) Determine the distance between the cities.

5. An athlete swims the length of a 50-m pool in 20 s and makes the return trip to the starting position in 22 s. Determine her average velocity in (a) the first half of the swim, and (b) the second half of the swim, and (c) the round trip.

6. A person takes a trip, driving with a constant speed of 89.5 km/h except for a 22.0-min rest stop. If the person's average speed is 77.8 km/h, how far is the trip?

7. A jogger runs eastward in a straight line with an average speed of 2 m/s for 5 min and then continues with an average speed of 1.5 m/s for 2 min. (a) What is her total displacement? (b) What is her average velocity during this time?

8. Runner A is initially 4.0 mi west of a flagpole and is running with a constant velocity of 6.0 mi/h due east. Runner B is initially 3.0 mi east of the flagpole and is running with a constant velocity of 5.0 mi/h due west.

How far are the runners from the flagpole when their paths cross?

9. The position-time graph for a bug crawling along the x axis is shown in Figure 2.16. Determine whether the velocity is positive, negative, or zero for the times (a) t_1, (b) t_2, (c) t_3, (d) t_4.

Figure 2.16
Problem 9

10. A runner moves such that his position at certain time intervals is given by the data in the table below. Use these data to construct a table like that given in the text as Table 2.2. From this table, find (a) the average velocity during the complete interval and (b) the instantaneous velocity of $t = 2$ s.

t(s)	x(m)	t(s)	x(m)
2	5.66	2.5	6.32
2.01	5.674	3	6.92
2.2	5.93	4	8.00

11. As a race car pulls away from the starting point, its position fits the relation

$$x = (5.0 \text{ m/s})t + (0.75 \text{ m/s}^3)t^3$$

if x is measured in meters and t in seconds. (a) Plot a graph of position versus time. (b) Determine the instantaneous velocity at $t = 4.0$ s using time intervals of 0.4 s, 0.2 s, and 0.1 s. (c) Compare the average velocity during the first 4.0 s with the results of (b).

12. It is found that the position of a model airplane as a function of time is given for a portion of time by $x = 2t^2$. Plot a graph of this equation between $t = 0$ and $t = 3$ s. From your graph find (a) the average velocity during this 3-s interval and (b) the instantaneous velocity at 2 s.

13. A tennis player moves in a straight-line path as shown in Figure 2.17. Find her average velocity in the time interval (a) 0 to 1 s, (b) 0 to 4 s, (c) 1 s to 5 s, (d) 0 to 5 s.

Figure 2.17
Problems 13 and 14

14. Find the instantaneous velocity of the tennis player described in Figure 2.17 at (a) 0.5 s, (b) 2 s, (c) 3 s, (d) 4.5 s.

15. A shopper in a supermarket is in a great hurry. Plot a position-time graph for him as he moves along an aisle in a straight-line path. Use the following data and assume the origin of coordinates is at the initial position of the shopper. He moves from position 0 to -3 m at constant velocity in 1 s. He then moves from this position to $+3$ m at a constant velocity in 2 s. Finally, he pauses to catch his breath for 1 s. Use the graph you have plotted to find the average velocity during the total time interval, and the instantaneous velocity at 0.5 s, 2 s, and 3.5 s.

Section 2.4   Acceleration

16. A car traveling in a straight line has a velocity of $+5$ m/s at some instant. After 4 s, its velocity is equal to $+8$ m/s. What is its average acceleration in this time interval?

17. A car traveling in a straight-line path has a velocity of $+10$ m/s at some instant. After 3 s, its speed is $+6$ m/s. What is the average acceleration in this time interval?

18. A tennis ball with a speed of 10.0 m/s is thrown perpendicularly at a wall. After striking the wall, the ball rebounds in the opposite direction with a speed of 8.0 m/s. If the ball is in contact with the wall for 0.012 s, what is the average acceleration of the ball while in contact with the wall?

19. A car traveling initially at $+7$ m/s accelerates at the rate of $+0.8$ m/s^2 for a time of 2 s. What is its velocity at the end of the acceleration?

20. The velocity versus time graph for an object moving along a straight path is shown in Figure 2.18 (a) Find the average acceleration of this object during the time intervals 0 to 5 s, 5 s to 15 s, and 0 to 20 s. (b) Find the instantaneous acceleration at 2 s, 10 s, and 18 s.

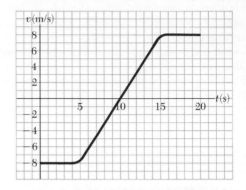

Figure 2.18
Problem 20

21. A record of travel along a straight path is as follows: (1) Start from rest with constant acceleration of 2.77 m/s^2 for 15.0 s. (2) Constant velocity for the next 2.05 min. (3) Constant negative acceleration -9.47 m/s^2 for 4.39 s. (a) What was the total displacement for the complete trip? (b) What was the average speed during legs (1), (2), and (3) of the trip as well as for the complete trip?

22. A certain car is capable of accelerating at a rate of $+0.6$ m/s^2. How long a time is taken for this car to go from a speed of 55 mi/h to a speed of 60 mi/h?

23. The engine of a model rocket accelerates the rocket vertically upward for 2 s such that its speed is given by the following data. At $t = 0$, its speed is zero; at $t = 1$ s, its speed is 5 m/s; at $t = 2$ s, its speed is 16 m/s. Plot a velocity-time graph for this motion and from it determine (a) the average acceleration during the 2-s interval and (b) the instantaneous acceleration at $t = 1.5$ s.

Section 2.6   One-Dimensional Motion with Constant Acceleration

24. A speedboat increases its speed from 20 m/s to 30 m/s in a distance of 200 m. Find (a) the magnitude of its acceleration and (b) the time it takes the boat to travel this distance.

25. A car starts from rest and accelerates at 0.3 m/s^2. What is the speed of the car after it has traveled 25 m?

26. A car accelerates uniformly from rest to a speed of 40 mi/h in 12 s. (a) Find the distance the car travels during this time and (b) the constant acceleration of the car.

27. A car, initially traveling at 20 m/s, accelerates at a uniform rate of 4 m/s^2 for a distance of 50 m. How much time is required to cover this distance?

28. A racing car reaches a speed of 40 m/s. At this instant, it begins a uniform deceleration using a parachute and a braking system and comes to rest 5 s later. (a) Determine the deceleration of the car. (b) How far does the car travel after deceleration starts?

29. A bicyclist starts down a hill with an initial speed of 2.0 m/s. She moves down the hill with constant acceleration, arriving at the bottom with a speed of 8.0 m/s. If the hill is 12.0 m long, what is the acceleration of the bicyclist on the hill?

30. An electron moving in a straight line has an initial speed of 3.0×10^5 m/s. If it undergoes an acceleration of 8.0×10^{14} m/s^2, (a) how long will it take to reach a speed of 5.4×10^5 m/s and (b) how far will it have traveled in this time?

31. A jet plane lands with a velocity of $+100$ m/s and can accelerate at a maximum rate of -5.0 m/s^2 as it comes to rest. (a) From the instant it touches the runway, what is the minimum time needed before it comes to rest? (b) Can this plane land on a small tropic island airport where the runway is 0.80 km long?

32. A train 400 m long is moving on a straight track with a speed of 82.4 km/h. The engineer applies the brakes at a crossing, and later the caboose passes the crossing with a speed of 16.4 km/h. Assuming constant acceleration, determine how long the train blocks the crossing. Disregard the width of the crossing.

33. A car starts from rest and travels for 5.0 s with a uniform acceleration of $+1.5$ m/s^2. The driver then applies the brakes, causing a uniform negative acceleration of -2.0 m/s^2. If the brakes are applied for 3.0 s, how fast is the car going at the end of the braking period and how far has it gone?

34. In order to pass a physical education class at a university, a student must run 1 mile in 12 minutes. After running for 10 min, she still has 500 yd to go. If her maximum acceleration is 0.15 m/s^2 can she make it? If the answer is no, determine what acceleration she would have to have to be successful.

35. A hockey player is standing on his skates on a frozen pond when an opposing player moving with a uniform speed of 12 m/s skates by with the puck. After 3 s, the first player makes up his mind to chase his opponent. If he accelerates uniformly at 4 m/s^2, (a) how long does it take him to catch his opponent and (b) how far has he traveled in this time? (Assume the player with the puck remains in motion at constant speed.)

Section 2.7   Freely Falling Bodies

36. A ball is thrown vertically upward with a speed of 25 m/s. (a) How high does it rise? (b) How long does it take to reach its highest point? (c) How long does it take to hit the ground from its highest point? (d) What is its speed when it returns to the level from which it started?

37. A pebble is dropped into a deep water well, and the sound of the splash as the pebble reaches the bottom of the well is heard 3 s later. From this information, estimate the depth of the well. (Neglect the time required for the sound to travel the length of the well.)

38. When a nurse squeezes a syringe, the liquid squirts 3.43 cm into the air. With what speed does the liquid emerge from the syringe?

39. A foul ball is hit into the stands at a baseball game. The ball rises to a height of 40 m and is caught by a fan at a height of 30 m as it drops back toward the field. What is its velocity in the vertical direction just before it is caught?

40. A juggler performs in a room with a ceiling 2 m above hand level. (a) What is the maximum upward speed she can give a ball without letting the ball hit the ceiling? (b) How long is the ball in the air before it is caught?

41. A peregrine falcon dives at a pigeon. The falcon starts downward from rest and falls with the acceleration of gravity. If the pigeon is 76.0 m below the initial position of the falcon, how long does it take the falcon to reach the pigeon? Assume the pigeon remains at rest.

42. A parachutist descending at a speed of 10 m/s drops a camera from an altitude of 50 m. (a) How long does it take the camera to reach the ground? (b) What is the velocity of the camera just before it hits the ground?

43. A rocket moves upward starting from rest with an acceleration of $+29.4$ m/s^2 for 4 s. It runs out of fuel at the end of this 4 s and continues to move upward, acted upon only by gravity. How high does it rise?

44. The tallest volcano in the solar system is the 24-km tall Martian volcano, Olympus Mons. Assume an astronaut drops a ball off the rim of the crater and that the acceleration of gravity remains constant throughout its 24-km fall at a value of 0.38 the acceleration due to gravity on Earth. (We assume the crater is as deep as the volcano is tall, which is not the case for real volcanoes.) Find (a) the time for the ball to reach the crater floor and (b) the velocity with which it hits. (In light of your answer for the velocity, does it seem reasonable that air resistance, even in Mars' thin atmosphere, can really be neglected in this problem?)

45. A "superball" is dropped from a height of 2 m above the ground. On the first bounce the ball reaches a height of 1.85 m, where it is caught. Find the velocity of the ball (a) just as it makes contact with the ground and (b) just as it leaves the ground on the bounce. (c) Neglecting the time the ball is in contact with the ground, find the total time required for the ball to move from the dropping point to the point where it is caught.

Additional Problems

46. A bullet is fired through a board 10 cm thick in such a way that the bullet's line of motion is perpendicular to the face of the board. If the initial speed of the bullet is 400 m/s and it emerges from the other side of the board with a speed of 300 m/s, find (a) the acceleration of the bullet as it passes through the board and (b) the total time the bullet is in contact with the board.

47. A ball is thrown downward from the top of a tall cliff with an initial speed of 10 m/s. Determine the velocity and acceleration of the ball at $t = 2$ s.

48. The velocity of an object is shown as a function of time in Figure 2.19. Find the average acceleration of the object in the intervals (a) 0 to 1 s, (b) 1 s to 3 s, and (c) 3 s to 4 s. (d) Find the instantaneous acceleration at 0.5 s, 2 s, and 3.5 s.

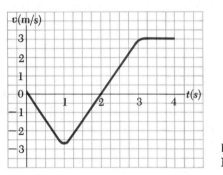

Figure 2.19
Problem 48

49. One swimmer in a relay race has a 0.5-s lead and is swimming at a constant speed of 4 m/s. He has 50 m to swim before reaching the end of the pool. A second swimmer moves in the same direction as the leader. What constant speed must the second swimmer have in order to catch up to the leader at the end of the pool?

50. A woman is reported to have fallen 44 m from the 17th floor of a building and to have landed on a metal ventilator box, which she crushed to a depth of 46 cm. She suffered only minor injuries. Neglecting air resistance, calculate (a) the speed of the woman just before she collided with the box, (b) her acceleration while in contact with the box, and (c) the time it took to crush the box.

51. A ball is thrown upward from the ground with an initial speed of 25 m/s at the same instant that a ball is dropped from a building 15 m high. At what time will the balls be at the same height?

52. A ranger in a National Park is driving at 35 mph when a deer jumps into the road 200 ft ahead of her posi-tion. After a reaction time of t s, she applies the brakes and decelerates at $a = -9.0$ ft/s^2. What is the maxi-mum reaction time allowed if she is to avoid hitting the deer?

53. A mountain climber stands at the top of a 50-m cliff that overhangs a calm pool of water. He throws two stones vertically downward 1 s apart and observes that they cause a single splash. The first stone has an initial velocity of $+2$ m/s. (a) At what time after re-lease of the first stone will the two stones hit the water? (b) What initial velocity must the second stone have if they are to hit simultaneously? (c) What will the velocity of each stone be at the instant they hit the water?

54. A person sees a lightning bolt passing close to an air-plane flying off in the distance. The person hears thunder 5 s after seeing the bolt and sees the airplane overhead 10 s after hearing the thunder. If the speed of sound in air is 1100 ft/s, (a) find the distance the airplane is from the person at the instant of the bolt. (Neglect the time it takes the light to travel from the bolt to the eye.) (b) Assuming the plane travels with a constant speed toward the person, find the velocity of the airplane. (c) Look up the speed of light in air and defend the approximation used in (a).

55. A sportscar enthusiast buys a super deluxe machine that can accelerate at the rate of $+16$ ft/s^2. She de-cides to test her car in a drag race with another speed-ster in a souped-up stock car. Both start from rest, but the experienced stock car driver leaves 1 s before the sportscar. If the stock car moves with a constant accel-eration of $+12$ ft/s^2, find (a) the time it takes the sportscar to overtake it, (b) the distance the two travel before they are side by side, and (c) the velocities of both cars at the instant they are side by side.

56. A certain cable car in San Francisco can stop in 10 s when traveling at maximum speed. On one occasion, the driver sees a dog a distance d m in front of the car and slams on the brakes instantly. The car reaches the dog 8 s later, and the dog jumps off the track just in time. If the car travels 4 m beyond the position of the dog before coming to a stop, how far was the car from the dog? (*Hint:* You will need three equations.)

57. An ice sled powered by a rocket engine starts from rest on a large frozen lake and accelerates at $+40$ ft/s^2. After some time t_1 the rocket engine is shut down and the sled moves with constant velocity **v** for a time t_2. If the total distance traveled by the sled is 17,500 ft and the total time is 90 s, find (a) the times t_1 and t_2 and (b) the velocity **v**. If, at the 17,500-ft mark, the sled begins to decelerate at -20 ft/s^2, (c) what is the final position of the sled when it comes to rest and (d) how long does it take to come to rest?

CHAPTER 3

Vectors and Two-Dimensional Motion

A dramatic fireworks display photographed in the night sky. Each fragment of the explosion follows a unique parabolic trajectory.
(© John Gillmoure, Photo Researchers, Inc.)

In our discussion of one-dimensional motion in the preceding chapter, we used the concept of vectors only to a very limited extent; however, as we move further into our study of motion, the ability to manipulate vector quantities will become increasingly important to us. As a result, much of this present chapter will be devoted to techniques for adding vectors, subtracting them, and so forth. We will then apply our newfound skills to a special case of two-dimensional motion, that of the motion of a projectile. We shall also see that an understanding of vector manipulation is necessary in order to work with and to understand relative motion.

3.1 Vectors and Scalars Revisited

The physical quantities that we shall encounter in this text can be placed in one of two categories: they are either scalar or vector quantities. A scalar is a quantity that can be completely specified by its magnitude only with appropriate units. That is,

a **scalar** has only magnitude and no direction. On the other hand, a **vector** is a physical quantity that requires the specification of both direction and magnitude.

51

Jennifer pointing in the right direction. *(Photo by Raymond A. Serway)*

Temperature is an example of a scalar quantity. If someone tells you that the temperature of an object is $-5°C$, that information completely specifies the temperature of the object; no direction is required. Other examples of scalars are mass, time intervals, and the number of pages in this textbook. The rules of ordinary arithmetic are used to manipulate scalar quantities. For example, if you have 2 liters of water in a container (where 1 liter is defined to be 1000 cm³) and you add 3 more liters, by ordinary arithmetic the amount of water you have in the container is 5 liters.

An example of a vector quantity is force. If you are told that someone is going to exert a force of 10 lb on an object, that is not enough information to let you know what will happen to the object. The effect of a force of 10 lb exerted horizontally is different from the effect of a force of 10 lb exerted vertically. In other words, you need to know the direction of the force as well as its magnitude.

Velocity is another example of a vector quantity. If we wish to describe the velocity of a moving vehicle, we must specify both its speed (say 30 m/s) and the direction in which the vehicle is moving (say northeast). Other examples of vector quantities include displacement and acceleration defined in Chapter 2.

3.2 Some Properties of Vectors

Adding Vectors

When two or more vectors are added together, they must all have the same units. For example, it would be meaningless to add a velocity vector to a displacement vector because they are different physical quantities. Scalars also obey this same rule. For example, it would be meaningless to add temperatures and areas. (Remember dimensional analysis in Chapter 1?)

In this printed text, a vector quantity will be represented by boldface roman type **A**. The magnitude of a vector **A** will be represented by standard italic type *A*. Likewise, italic type will be used to represent scalars. Blackboard presentations or other handwritten work often use a letter with an arrow over it ($\vec{A}$) to represent a vector quantity.

The rules for adding vectors are conveniently described by geometric methods. You should be aware, however, that adding vectors geometrically is presented here to introduce you to the concepts involved. Later we shall develop a more precise technique for adding vectors that is much more convenient and will be the method used throughout the remainder of this text.

To add vector **B** to vector **A**, first draw **A** on a piece of graph paper to some scale, such as 1 cm equals 1 m. Also, the vector **A** must be drawn such that its direction is specified relative to a coordinate system. Then draw vector **B** using this same scale and with the tail of **B** starting at the tip of **A**, as in Figure 3.1.

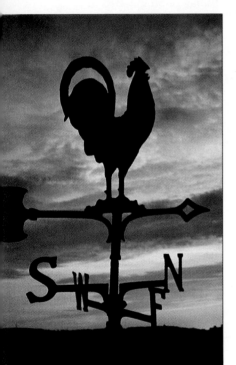

Weathervanes can be used to determine the direction of the wind velocity vector at any instant. *(© Pat Le Croix, The IMAGE Bank)*

Figure 3.1
When vector **B** is added to vector **A**, the vector sum **R** is the vector that runs from the tail of **A** to the tip of **B**.

Figure 3.2
This construction shows that
A + **B** = **B** + **A**. Note that the
resultant vector **R** is the
diagonal of a parallelogram
with sides *A* and *B*.

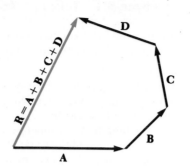

Figure 3.3
Geometric construction for
summing four vectors. The
resultant vector **R** is the vector
that completes the polygon.

Vector **B** must be drawn along the direction that makes the proper angle relative to vector **A**. The resultant vector **R**, given by **R** = **A** + **B**, is the vector drawn from the tail of **A** to the tip of **B**. This is known as the *triangle method of addition*.

An alternative graphical procedure for adding two vectors, known as the *parallelogram rule of addition*, is shown in Figure 3.2. In this construction, the tails of the two vectors **A** and **B** are joined together and the resultant vector **R** is the diagonal of the parallelogram formed with **A** and **B** as its sides.

This same general approach can also be used to add more than two vectors. This is shown in Figure 3.3 for four vectors. The resultant vector sum **R** = **A** + **B** + **C** + **D** is the vector drawn from the tail of the first vector to the tip of the last vector. The order in which you add vectors is unimportant.

Negative of a Vector The negative of the vector **A** is defined as the vector that when added to **A** gives a zero vector. This means that **A** and −**A** must have the same magnitude but point in opposite directions.

Subtraction of Vectors The operation of vector subtraction makes use of the definition of the negative of a vector. We define the operation **A** − **B** as vector −**B** added to vector **A**:

$$\mathbf{A} - \mathbf{B} = \mathbf{A} + (-\mathbf{B}) \tag{3.1}$$

The geometric construction for subtracting two vectors is shown in Figure 3.4.

Multiplication and Division of Vectors by Scalars

The multiplication or division of a vector by a scalar gives a vector. For example, if a vector **A** is multiplied by the scalar number 3, written as 3**A**, the result is a vector whose magnitude is three times that of the original vector **A** and pointing in the same direction as **A**. On the other hand, if we multiply a vector **A** by the scalar −3, the result is a vector whose magnitude is three times that of **A** but pointing in the direction opposite **A** (because of the negative sign).

Figure 3.4
This construction shows how
to subtract vector **B** from
vector **A**. The vector −**B** has
the same magnitude as the
vector **B** but points in the
opposite direction.

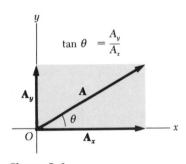

Figure 3.5 (Example 3.1) Graphical method for finding the resultant displacement vector $\mathbf{R} = \mathbf{A} + \mathbf{B}$.

Figure 3.6
Any vector $\mathbf{A}$ lying in the xy plane can be represented by its rectangular components, $\mathbf{A}_x$ and $\mathbf{A}_y$.

Example 3.1   Taking a Trip

A car travels 20.0 km due north and then 35.0 km in a direction 60° west of north, as in Figure 3.5. Find the magnitude and direction of the car's resultant displacement.

Solution
The problem can be solved geometrically using graph paper and a protractor, as shown in Figure 3.5. The resultant displacement $\mathbf{R}$ is the sum of the two individual displacements $\mathbf{A}$ and $\mathbf{B}$.

The length of $\mathbf{R}$, using the same scale factor used to draw $\mathbf{A}$ and $\mathbf{B}$, indicates that the displacement of the car is 48.2 km, and a measurement of the angle β shows that the displacement is approximately 39° west of north.

3.3   Components of a Vector

As noted earlier, the geometric method of adding vectors is not the recommended procedure in situations where high precision is required. In this section, we describe a method of adding vectors that makes use of the projections of a vector along the axes of a rectangular coordinate system. These projections are called the **components** of the vector. Any vector can be completely described by its components.

Consider a vector $\mathbf{A}$ in a rectangular coordinate system, as shown in Figure 3.6. Note that the vector $\mathbf{A}$ can be expressed as the sum of two vectors, $\mathbf{A}_x$ parallel to the x axis, and $\mathbf{A}_y$ parallel to the y axis. That is,

$$\mathbf{A} = \mathbf{A}_x + \mathbf{A}_y$$

where $\mathbf{A}_x$ and $\mathbf{A}_y$ are called the component vectors of $\mathbf{A}$. The projection of $\mathbf{A}$ along the x axis, A_x, is called the x component of $\mathbf{A}$, and the projection of $\mathbf{A}$ along the y axis, A_y, is called the y component of $\mathbf{A}$. *These scalar components can be either positive or negative numbers with units.* From the definitions of sine and cosine of an angle, we see that $\cos\theta = A_x/A$ and $\sin\theta = A_y/A$. Hence, the **scalar components** of $\mathbf{A}$ are given by

$$A_x = A\cos\theta$$
$$A_y = A\sin\theta \tag{3.2}$$

These components form two sides of a right triangle, the hypotenuse of which has a magnitude A. Thus, it follows that the magnitude of $\mathbf{A}$ and its direction are related to its scalar components through the Pythagorean theorem and the definition of the tangent

$$A = \sqrt{A_x^2 + A_y^2} \tag{3.3}$$

and

$$\tan\theta = \frac{A_y}{A_x} \tag{3.4}$$

To solve for the angle θ, we can invert Equation 3.4 and write the expression as $\theta = \tan^{-1}(A_y/A_x)$.

If you choose a coordinate system other than the one shown in Figure 3.6, the components of the vector must be modified accordingly. In many applications it is more convenient to express the components of a vector in a coordinate system having axes that are not horizontal and vertical but still perpendicular to each other. Suppose a vector **B** makes an angle θ with the x' axis defined in Figure 3.7. The rectangular components of **B** along the axes of Figure 3.7 are given by $B_{x'} = B \cos \theta$ and $B_{y'} = B \sin \theta$, as in Equation 3.2. The magnitude and direction of **B** are obtained from expressions equivalent to Equations 3.3 and 3.4. Thus, we can express the components of a vector in any coordinate system that is convenient for a particular situation.

If you wish to add or subtract vectors analytically rather than graphically, first find the components of the vectors. Note that the sum of the x components of the vectors equals the x component of the resultant vector. Likewise, the sum of the y components of the vectors equals the y component of the resultant vector.

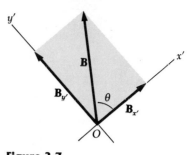

Figure 3.7
The components of vector **B** in a tilted coordinate system.

Problem-Solving Strategy:
Adding Vectors

When two or more vectors are to be added, the following step-by-step procedure is recommended:

1. Select a coordinate system.
2. Draw a sketch of the vectors to be added (or subtracted), with a label on each vector.
3. Find the x and y components of all vectors.
4. Find the algebraic sum of the components in both the x and y directions.
5. Use the Pythagorean theorem to find the magnitude of the resultant vector.
6. Use a suitable trigonometric function to find the angle the resultant vector makes with the x axis.

Example 3.2   Help Is on the Way!

Find the horizontal and vertical components of the 100-m displacement of a super-hero who flies from the top of a tall building following the path shown in Figure 3.8a.

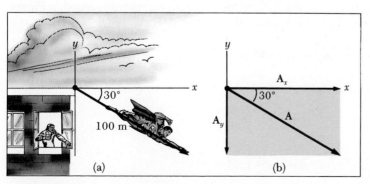

Figure 3.8
Example 3.2

Solution
The triangle formed by the displacement and its components is shown in Figure 3.8b. Since $A = 100$ m and $\theta = -30°$ (θ is negative because it is measured clockwise from the x axis), we have

$$A_y = A \sin \theta = (100 \text{ m}) \sin(-30°) = \boxed{-50.0 \text{ m}}$$

Note that $\sin(-\theta) = -\sin \theta$. The negative sign for A_y corresponds to the fact that displacement in the y direction is *downward* from the origin.

The x component of displacement is

$$A_x = A \cos \theta = (100 \text{ m}) \cos(-30°) = \boxed{+86.6 \text{ m}}$$

Note that $\cos(-\theta) = \cos \theta$. Also, from an inspection of the figure, you should be able to see that A_x is positive in this case.

Example 3.3   Go and Take a Hike

On an extended hike, a camper follows the path shown in Figure 3.9a. The total trip consists of four straight-line paths. At the end of the hike, what is the hiker's resultant displacement from the tent site?

Solution
A summary of the steps for approaching problems of this nature is given in the problem-solving section that precedes these examples. The coordinate system we shall use is shown in Figure 3.9a. The 100-m displacement has only an x component, and the 300-m displacement has only a y component (which is negative). You should use the procedures for finding components of vectors to show that the table below has the proper components for all the displacements.

Displacement	x Component	y Component
100 m	100 m	0
300 m	0	−300 m
150 m	−130 m	−75 m
200 m	−100 m	173 m
Total	−130 m	−202 m

The table also shows that the resultant x component is −130 m and the resultant y component is −202 m, as shown in Figure 3.9b.

Applying the Pythagorean theorem to the resultant components found in Figure 3.9b gives the magnitude of the resultant displacement.

$$A = \sqrt{A_x^2 + A_y^2} = \sqrt{(-130 \text{ m})^2 + (-202 \text{ m})^2} = \boxed{240 \text{ m}}$$

The direction is found by noting that

$$\tan \theta = \frac{-202 \text{ m}}{-130 \text{ m}} = 1.55$$

and that

$$\theta = \tan^{-1} 1.55 = \boxed{57°}$$

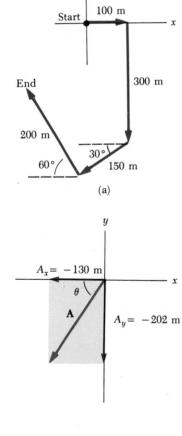

(a)

(b)

Figure 3.9
Example 3.3

The direction of the resultant is frequently specified as an angle measured relative to the positive x direction. If we follow this procedure for the above, the angle relative to the positive x direction is $180° + 57° = 237°$. You should be able to show that the total distance traveled is 750 m.

3.4   Velocity and Acceleration in Two Dimensions

In the case of one-dimensional motion discussed in Chapter 2, the vector nature of displacement, velocity, and acceleration was taken into account by using positive and negative signs. In order to completely describe the motion of an object in two or three dimensions, we must make use of vectors.

Consider an object moving along some curve in space as shown in Figure 3.10. When the object is at some point P at time t_i, its position is described by the position vector $\mathbf{r}_i$ drawn from the origin to P. Likewise, when the particle has moved to some other point Q at time t_f, its position vector is $\mathbf{r}_f$. As you can see from the vector diagram in Figure 3.10, the final position vector is the sum of the initial position vector and $\Delta\mathbf{r}$. Since $\mathbf{r}_f = \mathbf{r}_i + \Delta\mathbf{r}$, the displacement of the object is defined as the change in the position vector

$$\Delta\mathbf{r} \equiv \mathbf{r}_f - \mathbf{r}_i \qquad (3.5)$$

Note that the magnitude of the displacement vector is *not* equal to the distance traveled measured along the curved path. In fact, the displacement is less than this distance when the path is curved.

In analogy with our one-dimensional case, the average velocity of the object in the time interval $\Delta t = t_f - t_i$ is given by

$$\bar{\mathbf{v}} \equiv \frac{\Delta\mathbf{r}}{\Delta t} \qquad (3.6)$$

We now define the instantaneous velocity vector $\mathbf{v}$ as the limiting value of the average velocity as the time interval approaches zero. With reference to Figure 3.11, we see that as the time interval gets smaller and smaller, and point Q approaches P, the direction of $\Delta\mathbf{r}$ approaches that of the line tangent to the curve at P. Thus, the direction of the instantaneous velocity vector at any point is along the line tangent to the path of the motion at that point.

The average acceleration of an object whose velocity changes by $\Delta\mathbf{v}$ in the time interval Δt is a vector defined as the ratio $\Delta\mathbf{v}/\Delta t$.

$$\bar{\mathbf{a}} \equiv \frac{\Delta\mathbf{v}}{\Delta t} \qquad (3.7)$$

The instantaneous acceleration vector $\mathbf{a}$ is defined as the limit of the average acceleration vector as the time interval approaches zero.

You should note that an object is accelerating whenever the velocity vector changes in magnitude, direction, or both. The most familiar example of

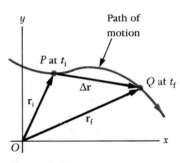

Figure 3.10
An object moving along some curved path between points P and Q. The displacement vector $\Delta\mathbf{r}$ is the difference in the position vectors. That is, $\Delta\mathbf{r} = \mathbf{r}_f - \mathbf{r}_i$.

Average velocity vector

Figure 3.11
As the time interval between the final and initial positions approaches zero, the direction of the displacement vector approaches that of the line tangent to the path at P.

acceleration is motion in one dimension when the magnitude of the object's velocity (speed) changes. On the other hand,

> an object moving with constant speed is accelerating if the direction of the velocity vector is changing.

One example of this is an object moving in a circular path with constant speed. In this case, the acceleration is directed towards the center of the circle. We shall return to this type of motion in Chapter 7.

3.5   Projectile Motion

The situations we considered in Chapter 2 were those in which an object moves along a straight-line path, such as the x axis. We now look at some cases in which the object moves in a plane. By this we mean that the object may move in both the x and the y direction simultaneously, corresponding to motion in two dimensions. The particular type of motion we shall concentrate on is called **projectile motion.** Anyone who has observed a baseball in motion (or for that matter, any object thrown into the air) has observed projectile motion. This very common form of two-dimensional motion is surprisingly simple to analyze if the following three assumptions are made:

1. The acceleration due to gravity, **g**, has a magnitude of 9.80 m/s², is constant over the range of motion and is directed downward.
2. The effect of air resistance is negligible.
3. The rotation of the Earth does not affect the motion.

With these assumptions, we shall find that the path of a projectile is curved as shown in Figure 3.12. (Such a curve is called a parabola.)

Let us choose our coordinate system such that the y direction is vertical and positive upward. In this case

> the acceleration in the y direction is $-g$, just as in free fall, and the acceleration in the x direction is 0 (because air friction is neglected).

Figure 3.12
The parabolic trajectory of a particle that leaves the origin with a velocity $\mathbf{v}_0$. Note that the velocity, $\mathbf{v}$, changes with time. However, the x component of the velocity, v_x, remains constant in time. Also, $v_y = 0$ at the peak, but the acceleration is always equal to the acceleration of gravity, and acts vertically downwards.

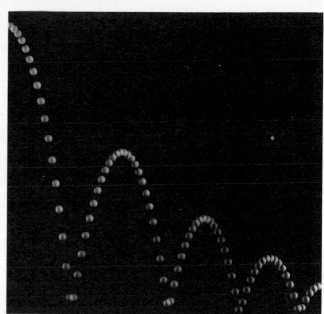

In a volcanic eruption in Mt. Etna, Sicily, the lava particles from the eruption follow parabolic paths as one would expect, since they are projectiles falling in the presence of gravity. *(© Otto Hahn, Peter Arnold, Inc.)*

Multiflash photograph of a ball undergoing several bounces off a hard surface. Note the parabolic path of the ball following each bounce. *(Photo courtesy of Educational Development Center, Newton, MA)*

Furthermore, let us assume that at $t = 0$, the projectile leaves the origin with a velocity $\mathbf{v}_0$, as shown in Figure 3.12. If the velocity vector makes an angle θ_0 with the horizontal, as shown, then the initial x and y components of velocity are given by

$$v_{x0} = v_0 \cos \theta_0 \qquad \text{and} \qquad v_{y0} = v_0 \sin \theta_0$$

In order to analyze projectile motion, we shall separate the motion into two parts, the x (or horizontal) motion and the y (or vertical) motion, and solve each part separately. We shall look first at the x motion. As noted before, motion along the x direction occurs with $a_x = 0$. This means that *the velocity component along the x direction never changes.* Thus, if the initial value of the velocity component in the x direction is $v_{x0} = v_0 \cos \theta_0$, this is also the value of the velocity at any later time. That is,

$$v_x = v_{x0} = v_0 \cos \theta_0 = \text{constant} \qquad (3.8)$$

Equation 3.8 can be substituted into the defining equation for velocity (Eq. 2.2) to give us an expression for the horizontal position of the projectile as a function of time:

$$x = v_{x0} t = (v_0 \cos \theta_0)t \qquad (3.9)$$

These equations tell us all we need to know about the motion in the x direction. Let us now consider the y motion. Since the motion in the y direction is motion with constant acceleration, the equations developed in Section

2.6 can be used. In these equations, we shall denote v_{y0} as the initial velocity in the y direction and $-g$ as the acceleration due to gravity. The negative sign for g assumes that the positive direction for the vertical motion is upward. With this choice of signs we have

$$v_y = v_{y0} - gt \tag{3.10}$$

$$y = v_{y0}t - \frac{1}{2}gt^2 \tag{3.11}$$

$$v_y{}^2 = v_{y0}{}^2 - 2gy \tag{3.12}$$

where $v_{y0} = v_0 \sin\theta$. Note that it is not necessary to assume that the upward direction is positive. If you choose the positive direction as downward, simply replace $-g$ with g in the equations above. The speed v of the projectile at any instant can be calculated from the components of velocity at that instant using the Pythagorean theorem:

$$v = \sqrt{v_x{}^2 + v_y{}^2}$$

Before we look at some numerical examples dealing with projectile motion, let us pause to summarize what we have learned so far about this kind of motion:

1. Provided air resistance is negligible, the horizontal component of velocity, v_x, remains constant since there is no horizontal component of acceleration.
2. The vertical component of acceleration is equal to the acceleration due to gravity, **g**.
3. The vertical component of velocity, v_y, and the displacement in the y direction are identical to those of a freely falling body.
4. Projectile motion can be described as a superposition of the two motions in the x and y directions.

Problem-Solving Strategy:
Projectile Motion

We suggest that you use the following approach to solving projectile motion problems:

1. Select a coordinate system.
2. Resolve the initial velocity vector into x and y components.
3. Treat the horizontal motion and the vertical motion independently.
4. Follow the techniques for solving problems with constant velocity to analyze the horizontal motion of the projectile.
5. Follow the techniques for solving problems with constant acceleration to analyze the vertical motion of the projectile.

Example 3.4   The Stranded Explorers

An Alaskan rescue plane drops a package of emergency rations to a stranded party of explorers, as shown in Figure 3.13. If the plane is traveling horizontally at 40 m/s at a height of 100 m above the ground, (a) where does the package strike the ground relative to the point at which it was released?

Solution

The coordinate system for this problem is selected as shown in Figure 3.13, with the positive x direction to the right and the positive y direction upward.

Consider first the horizontal motion of the package. The only equation available to us is

$$x = v_{x0}t$$

The initial x component of the package velocity is the same as the velocity of the plane when the package was released, 40 m/s. Thus, we have

$$x = (40 \text{ m/s})t$$

If we know t, the length of time the package is in the air, we can determine x, the distance traveled by the package along the horizontal. To find t, we move to the equations for the vertical motion of the package. We know that at the instant the package hits the ground its y coordinate is -100 m. We also know that the initial velocity of the package in the vertical direction, v_{y0}, is zero because the package was released with only a horizontal component of velocity. From Equation 3.11, we have

$$y = -\frac{1}{2} gt^2$$

$$-100 \text{ m} = -\frac{1}{2} (9.80 \text{ m/s}^2)t^2$$

$$t^2 = 20.4 \text{ s}^2$$

$$t = 4.51 \text{ s}$$

This value for the time of flight substituted into the equation for the x coordinate gives

$$x = (40 \text{ m/s})(4.51 \text{ s}) = \boxed{180 \text{ m}}$$

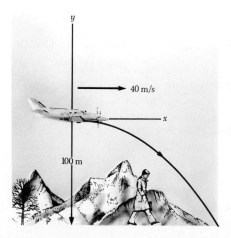

Figure 3.13   Example 3.4
To an observer on the ground, a package released from the rescue plane travels along the path shown. (This figure is not drawn to scale.)

(b) What are the horizontal and vertical components of the velocity of the package just before it hits the ground?

Solution
We already know the horizontal component of the velocity of the package just before it hits, because the velocity in the horizontal direction remains constant at 40 m/s throughout the flight.

The vertical component of the velocity just before the package hits the ground may be found by using $v_y = v_0 \sin \theta_0 - gt$ (Eq. 3.10):

$$v_y = -(9.80 \text{ m/s}^2)(4.51 \text{ s}) = \boxed{-44.1 \text{ m/s}}$$

Example 3.5   The Long Jump

A long jumper leaves the ground at an angle of 20° to the horizontal and at a speed of 11 m/s, as in the photograph at the left. (a) How far does he jump? (Assume that the motion of the long jumper is equivalent to that of a particle.)

Solution
His horizontal motion is described by using $x = (v_0 \cos \theta_0)t$ (Eq. 3.9), or

$$x = (11 \text{ m/s})(\cos 20°)t$$

The value of x can be found if t, the total time of the jump, is known. We are able to find t using $v_y = v_0 \sin \theta_0 - gt$ (Eq. 3.10) by noting that at the top of the jump the vertical component of velocity goes to zero:

$$v_y = v_0 \sin \theta_0 - gt$$
$$0 = (11 \text{ m/s}) \sin 20° - (9.80 \text{ m/s}^2)t_1$$
$$t_1 = 0.384 \text{ s}$$

Note that t_1 is the time interval to reach the *top* of the jump. Because of the symmetry of the vertical motion, an identical time interval passes before the jumper returns to the ground. Therefore, the *total time* in the air is $t = 2t_1 = 0.768$ s and the distance jumped is

$$x = (11 \text{ m/s})(\cos 20°)(0.768 \text{ s}) = \boxed{7.94 \text{ m}}$$

(b) What is the maximum height reached?

Solution
The maximum height reached is found using $y = (v_0 \sin \theta_0)t - \frac{1}{2}gt^2$ (Eq. 3.11) with $t = t_1 = 0.384$ s.

$$y_{\max} = (11 \text{ m/s})(\sin 20°)(0.384 \text{ s}) - \frac{1}{2}(9.80 \text{ m/s}^2)(0.384 \text{ s})^2 = \boxed{0.722 \text{ m}}$$

The assumption that the motion of the long jumper is that of a projectile is an oversimplification of the situation. Nevertheless, the values obtained are reasonable.

Example 3.6   That's Quite an Arm

A stone is thrown upward from the top of a building at an angle of 30° to the horizontal and with an initial speed of 20 m/s, as in Figure 3.14. If the height of the building is 45 m, (a) how long is the stone "in flight"?

A track star in a long-jump event can leap through horizontal distances of at least 8 meters. (© R. Mackson/FPG International)

On the left is a multiflash photograph of a popular laboratory demonstration in which a projectile is fired at a falling target. The conditions of the experiment are that the gun is aimed at the target and that the projectile leaves the gun at the same instant that the target is released from rest. Under these conditions the projectile will hit the target, as shown in the photo, independent of the initial velocity of the projectile. The reason for this is that they fall through the same vertical distance in the same time interval. The middle photograph shows an evening view of an illuminated water fountain in La Place de la Concorde, Paris, France. The individual water streams follow a parabolic trajectory. The photograph shows that the horizontal range and maximum height of a given stream depend on the elevation angle of its initial velocity. On the right, a welder is able to cut holes through a heavy metal construction beam with a hot torch. Note that the sparks generated in the process follow parabolic paths.

Parabolic Paths

(Courtesy of CENCO)

(© The Telegraph Colour Library/ FPG International)

(© The Telegraph Colour Library/ FPG International)

Solution

The initial x and y components of the velocity are

$$v_{x0} = v_0 \cos \theta_0 = (20 \text{ m/s})(\cos 30°) = +17.3 \text{ m/s}$$

$$v_{y0} = v_0 \sin \theta_0 = (20 \text{ m/s})(\sin 30°) = +10 \text{ m/s}$$

To find t, we can use $y = v_{y0}t - \frac{1}{2}gt^2$ (Eq. 3.11) with $y = -45$ m and $v_{y0} = 10$ m/s (we have chosen the top of the building as the origin, as shown in Fig. 3.14):

$$-45 \text{ m} = (10 \text{ m/s})t - \frac{1}{2}(9.80 \text{ m/s}^2)t^2$$

Solving the quadratic equation (see Appendix A) for t gives, for the positive root, the value $t = 4.22$ s. Does the negative root have any physical meaning? (Why not think of another way of finding t from the information given?)

(b) What is the speed of the stone just before it strikes the ground?

Solution

The y component of the velocity just before the stone strikes the ground can be obtained using the equation $v_y = v_{y0} - gt$ (Eq. 3.10) with $t = 4.22$ s:

$$v_y = 10 \text{ m/s} - (9.80 \text{ m/s}^2)(4.22 \text{ s}) = -31.4 \text{ m/s}$$

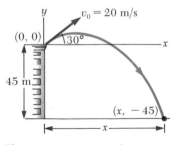

Figure 3.14 Example 3.5

Since $v_x = v_{x0} = 17.3$ m/s, the required speed is given by

$$v = \sqrt{v_x{}^2 + v_y{}^2} = \sqrt{(17.3)^2 + (-31.4)^2} \text{ m/s} = \boxed{35.9 \text{ m/s}}$$

Exercise   Where does the stone strike the ground?

Answer   73.0 m from the base of the building.

---

*3.6   Relative Velocity

In this section, we shall describe how observations made by different observers in different frames of reference are related to each other. We shall find that observers in different frames of reference may measure different displacements or velocities for an object in motion. That is, two observers moving with respect to each other will generally not agree on the outcome of a measurement.

For example, if two cars are moving in the same direction with speeds of 50 mi/h and 60 mi/h, a passenger in the slower car will claim that the speed of the faster car relative to the slower car is 10 mi/h. Of course, a stationary observer will measure the speed of the faster car to be 60 mi/h. This simple example demonstrates that velocity measurements differ in different frames of reference.

Another simple example is to imagine a package being dropped from an airplane flying parallel to the Earth with a constant velocity. An observer from the airplane would describe the motion of the package as a straight line toward the Earth. On the other hand, an observer on the ground would view the trajectory of the package as that of a projectile. Relative to the ground, the package has a vertical component of velocity (resulting from the acceleration due to gravity and equal to the velocity measured by the observer in the airplane) *and* a horizontal component of velocity (given to it by the airplane's motion). If the airplane continues to move horizontally with the same velocity, the package will hit the ground directly beneath the airplane (assuming that air resistance is negligible).

In order to develop a general approach to solving relative-motion problems, let us return to the example given earlier of the fast car and the slower car. The situation described was that of a fast car moving at 60 mi/h, a slower car moving at 50 mi/h, and we wanted to know the speed of the fast car with respect to the slower car. We solved such a problem with a minimum of thought and effort, but there are many situations that you will meet in the future in which a more systematic method for attacking such problems will be beneficial. To develop this method, let us write down all that we are given and that which we want to know by writing down our velocities with subscripts appended. We have

$\mathbf{v}_{se} = +50$ mi/h   (Here the subscript se means the velocity of the *slower* car with respect to the *Earth*.)

$\mathbf{v}_{fe} = +60$ mi/h   (The subscript fe means the velocity of the *fast* car with respect to the *Earth*.)

We want to know $\mathbf{v}_{fs}$, which is the velocity of the *fast* car with respect to the *slower* car. To find this, we write an equation for $\mathbf{v}_{fs}$ in terms of the other velocities, such that the subscripts on the right side of the equation start with f and eventually end with s. Also, each velocity on the right has a subscript that starts with the same letter that ended the velocity subscript which preceded it in the equation. For our case, we have

$$\mathbf{v}_{fs} = \mathbf{v}_{fe} + \mathbf{v}_{es}$$

The boldface notation is, of course, indicative of the fact that velocity is a vector quantity, and, as we shall see in the example problems below, this vector nature of the velocity will be of paramount importance in certain instances. (Note that in the terms on the right, we must start our velocity subscripts with f and end with s, because the subscripts on $\mathbf{v}_{fs}$ start and end with these letters. Also note that the subscripts on the second term on the right, $\mathbf{v}_{es}$, must start with e since the preceding velocity subscript, $\mathbf{v}_{fe}$, ended with e.)

We know that $\mathbf{v}_{es} = -\mathbf{v}_{se}$, so

$$\mathbf{v}_{fs} = +60 \text{ mi/h} - 50 \text{ mi/h} = +10 \text{ mi/h}$$

There is no general equation for you to memorize in order to work relative velocity problems; instead, you should develop the necessary equations on your own by following the technique for writing subscripts demonstrated above. We suggest that you practice this technique for the example problems which follow.

Example 3.7   Where Does the Ball Land?

A passenger at the rear of a train, traveling at 15 m/s relative to the Earth, throws a baseball with a speed of 15 m/s in the opposite direction to the motion of the train. What is the velocity of the baseball relative to the Earth?

Solution
We first write down our knowns and unknowns with appropriate subscripts, as

$\mathbf{v}_{te} = +15 \text{ m/s}$ (This is the velocity of the *train*, t, relative to the *Earth*, e.)
$\mathbf{v}_{bt} = -15 \text{ m/s}$ (This is the velocity of the *baseball*, b, relative to the *train*, t.)

We need $\mathbf{v}_{be}$, the velocity of the *baseball* relative to the *Earth*. Following the strategy for watching the subscripts while we write down the equation for our unknown, we have

$$\mathbf{v}_{be} = \mathbf{v}_{bt} + \mathbf{v}_{te}$$

or

$$\mathbf{v}_{be} = +15 \text{ m/s} - 15 \text{ m/s} = 0$$

Example 3.8   Crossing a River

A boat heading due north crosses a wide river with a velocity of 10 km/h relative to the water. The river has a uniform velocity of 5 km/h due east. Determine the velocity of the boat with respect to an observer on the riverbank.

Solution
We have

$\mathbf{v}_{br} = 10 \text{ km/h due north}$ (The velocity of the *boat*, b, with respect to the *river*, r.)
$\mathbf{v}_{re} = 5 \text{ km/h due east}$ (The velocity of the *river*, r, with respect to the *Earth*, e.)

Figure 3.15 Example 3.7

and we want, $\mathbf{v_{be}}$, the velocity of the *boat* with respect to the *Earth*. Our equation becomes

$$\mathbf{v_{be}} = \mathbf{v_{br}} + \mathbf{v_{re}}$$

Since the various velocities are not along the same direction as in our previous examples, the terms in the equation must be manipulated as vector quantities, where the vectors are shown in Figure 3.15. The quantity $\mathbf{v_{br}}$ is due north, $\mathbf{v_{re}}$ is due east, and the vector sum of the two, $\mathbf{v_{be}}$, is at an angle θ, as defined in Figure 3.15a. Thus, we see that the velocity of the boat with respect to the Earth can be found from the Pythagorean theorem as

$$v_{be} = \sqrt{(v_{br})^2 + (v_{re})^2} = \sqrt{(10 \text{ km/h})^2 + (5 \text{ km/h})^2} = 11.2 \text{ km/h}$$

and the direction of $\mathbf{v_{be}}$ is

$$\theta = \tan^{-1}\left(\frac{v_{re}}{v_{br}}\right) = \tan^{-1}\frac{5}{10} = 26.6°$$

Therefore, the boat will be traveling at a speed of 11.2 km/h in the direction 63.4° north of east with respect to the Earth.

Example 3.9 Which Way Should the Boat Head?

If the boat of the preceding example travels with the same speed of 10 km/h relative to the water and is to travel due north, as in Figure 3.16, what should be its heading?

Solution
We know

$\mathbf{v_{br}}$ = the velocity of the *boat*, b, with respect to the *river*, r
$\mathbf{v_{re}}$ = the velocity of the *river*, r, with respect to the *Earth*, e

and we need, $\mathbf{v_{be}}$, the velocity of the *boat*, b, with respect to the *Earth*, e. We have

$$\mathbf{v_{be}} = \mathbf{v_{br}} + \mathbf{v_{re}}$$

where the relationship between these three quantities is shown in Figure 3.16, which agrees with our intuitive guess that the boat must head upstream in order to be pushed directly northward across the water. The speed $\mathbf{v_{be}}$ can be found from the Pythagorean theorem as

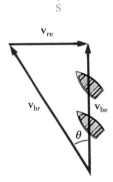

Figure 3.16 Example 3.8

$$v_{be} = \sqrt{(v_{br})^2 - (v_{re})^2} = \sqrt{(10 \text{ km/h})^2 - (5 \text{ km/h})^2} = 8.66 \text{ km/h}$$

and the direction of v_{be} is

$$\theta = \tan^{-1}\frac{v_{re}}{v_{be}} = \tan^{-1}\left(\frac{5}{8.66}\right) = 30°$$

where θ is west of north.

▶ S P E C I A L T O P I C
Relative Motion at High Speeds

As we mentioned in the previous section, we can transform the velocity of a particle as measured in the Earth's reference frame to those measured in a frame moving with uniform motion with respect to the Earth. However, it is important to note that these transformations are valid *only* at particle speeds (relative to both observers) that are small compared with the speed of light, c (where $c \approx 3 \times 10^8$ m/s). When the particle speed according to either observer approaches the speed of light, these transformation equations must be replaced by the equations used by Einstein in his special theory of relativity. Although we shall discuss the theory of relativity in Chapter 28, a forecast of some of its predictions is in order. As it turns out, the relativistic transformation equations for the velocity of a particle reduce to those of the last section when the particle speed is small compared with the speed of light. This is in keeping with the **correspondence principle** first proposed by Niels Bohr, which in effect states that if an old theory accurately describes a number of physical phenomena, then any new theory *must* account for the same phenomena over the range of validity of the old theory.

You might wonder how one can test the validity of the transformation equations. The velocity transformation equations given in the preceding section are used in Newtonian mechanics, while Einstein's theory uses relativistic transformations. From experiments on high-speed particles such as electrons and protons in particle accelerators, one finds that Newtonian mechanics *fails* at particle speeds approaching the speed of light. On the other hand, Einstein's theory of special relativity is in agreement with experiment at *all* speeds. Finally, Newtonian mechanics places no upper limit on speed of a particle. In contrast, the relativistic velocity transformation equations predict particle speeds that *can never exceed the speed of light*. Electrons and protons accelerated through very high voltages can acquire speeds close to the speed of light, but never reach this value. Hence, experimental results are in complete agreement with the theory of relativity. ◄

Summary

Two vectors **A** and **B** can be added geometrically by either the triangle method or the parallelogram rule. In the triangle method, the two vectors are drawn to scale, on graph paper, such that the tail of the second vector starts at the tip of the first. The resultant vector is the vector drawn from the tail of the first to the tip of the second. In the parallelogram method, the vectors are drawn to scale on graph paper with the tails of the two vectors joined together. The resultant vector is the diagonal of the parallelogram formed with the two vectors as its sides.

The negative of a vector **A** is a vector having the same magnitude as **A** but pointing in the direction opposite the direction of **A**.

The x component of a vector is equivalent to its projection along the x axis of a coordinate system. Likewise, the y component is the projection of the vector along the y axis of this coordinate system. The resultant of two or more vectors can be found mathematically by resolving all vectors into their x and y components, finding the resultant x and y components, and then using the Pythagorean theorem to find the resultant vector. The angle of the resultant vector with respect to the x axis can be found by use of a suitable trigonometric function.

An object moving above the surface of the Earth such that it moves in both the x and the y direction simultaneously is said to be undergoing **projectile motion**. Projectile motion is equivalent to two different motions occurring at the same time: (1) the object moves along the horizontal (x) direction such that its velocity in this direction, v_x, is a constant and (2) the object moves in the vertical (y) direction with a constant downward acceleration of magnitude $g = 9.80$ m/s^2. The equations describing the motion of a projectile are

$$x = v_{x0}t \tag{3.9}$$

$$v_y = v_{y0} - gt \tag{3.10}$$

$$y = v_{y0}t - \frac{1}{2}gt^2 \tag{3.11}$$

$$v_y{}^2 = v_{y0}{}^2 - 2gy \tag{3.12}$$

Projectile motion

where $v_{y0} = v_0 \sin \theta_0$ is the initial vertical component of the velocity and $v_{x0} = v_0 \cos \theta_0$ is the initial horizontal component of the velocity.

Observations made by observers in different frames of reference can be related to one another through the techniques of the transformation of relative velocities.

Additional Reading

P. Brancazio, "The Trajectory of a Fly Ball," *The Physics Teacher*, January 1985, p. 20.

S. Drake, and J. MacLachlan, "Galileo's Discovery of the Parabolic Trajectory," *Sci. American*, March 1975, p. 102.

W. F. Magie, *Source Book in Physics*, Cambridge, MA, Harvard University Press, 1963. Contains many excerpts from Galileo on projectile motion.

Questions

1. If **B** is added to **A**, under what conditions does the resultant vector have a magnitude equal to $A + B$? Under what conditions is the resultant vector equal to zero?

2. A student accurately uses the method for combining vectors. The two vectors that he combines have magnitudes of 50 and 25. The answer he gets is either 80, 20, or 55. Pick the correct answer and tell why it is the only one of the three that can be correct.

3. Two vectors have unequal magnitudes. Can their sum be zero? Explain.

4. Vector **A** lies in the xy plane. For what orientations will both of its rectangular components be negative? For what orientations will its components have opposite signs?

5. Can a vector have a component greater than its magnitude?

6. Under what circumstances would a vector have components which are equal in magnitude?

7. If the component of vector **A** along the direction of vector **B** is zero, what can you conclude about these two vectors?

8. Can a vector have a component equal to zero and still have a nonzero magnitude? Explain.

9. If $\mathbf{A} + \mathbf{B} = 0$, what can you say about the components of the two vectors?

10. Can a force directed vertically on an object ever cancel a force directed horizontally?

11. Make a list detailing a procedure for reconstructing a vector from a set of components. Sketch any appropriate diagrams to accompany the list.

12. If a set of vectors laid tail to head forms a closed polygon, the resultant is zero. Is this statement true? Discuss.

13. The magnitude of a vector is a scalar. Explain this statement.

14. Can the magnitude of a vector have a negative value?

15. Which of the following are vectors and which are not: force, temperature, the amount of water in a can, the weight of a book, the height of a building, the velocity of a sports car, the age of the Universe?

16. Is it possible to add a vector quantity to a scalar quantity? Explain.

17. Does a ball dropped out of the window of a moving car take longer to reach the ground than one dropped from a car at rest?

18. If a rock is dropped from the top of a sailboat's mast, will it hit the deck at the same point whether or not the boat is at rest or in motion at constant velocity?

19. A rock is dropped at the same instant a ball, at the same initial elevation, is thrown horizontally. Which will have the greater velocity when they reach ground level?

Problems

Section 3.2   Some Properties of Vectors

1. While traveling along a straight interstate highway you notice that the mile marker reads 260. You travel until you reach the 150-mile marker and note that you are low on gas. You then retrace your path to the 175-mile marker. What is the magnitude of your resultant displacement from the 260-mile marker?
2. (a) What is the resultant displacement of a walk of 80 m followed by a walk of 125 m when both displacements are in the eastward direction? (b) What is the resultant displacement in a situation in which the 125-m walk is in the direction opposite the 80-m walk?
3. Vector **A** is 3 units in length and points along the positive x axis. Vector **B** is 4 units in length and points along the negative y axis. Use graphical methods to find the magnitude and direction of the vectors (a) **A** + **B** and (b) **A** − **B**.
4. Each of the displacement vectors **A** and **B** shown in Figure 3.17 has a magnitude of 3 m. Find graphically (a) **A** + **B**, (b) **A** − **B**, (c) **B** − **A**, (d) **A** − 2**B**.

Figure 3.17
Problem 4

5. A roller coaster moves 200 ft horizontally, then rises 135 ft at an angle of 30° above the horizontal. It then travels 135 ft at an angle of 40° downward. What is its displacement from its starting point at the end of this movement? Use graphical techniques.
6. A dog searching for a bone walks 3.5 m south, then 8.2 m at an angle 30° north of east, and finally 15 m west. Find the dog's resultant displacement vector using graphical techniques.
7. A jogger runs 100 m due west, then changes direction for the second leg of the run. At the end of the run, she is 175 m away from the starting point at an angle of 15° north of west. What was the direction and length of her second displacement? Use graphical techniques.
8. A man lost in a maze makes three consecutive displacements such that at the end of the walks he is right back where he started. The first displacement is 8 m westward, and the second is 13 m northward. Find the magnitude and direction of the third displacement using the graphical method.

Section 3.3   Components of a Vector

9. A roller coaster travels 135 ft at an angle of 40° above the horizontal. How far does it move horizontally and vertically?
10. A submarine dives at an angle of 30° with the horizontal and follows a straight-line path for a total distance of 50 m. How far is the submarine below the surface of the water?
11. A quarterback takes the ball from the line of scrimmage, runs backward for 10 yards, then sideways parallel to the line of scrimmage for 15 yards. At this point, he throws a 50-yard forward pass straight downfield perpendicular to the line of scrimmage. What is the magnitude of the football's resultant displacement?
12. A person going for a walk follows the path shown in Figure 3.18. The total trip consists of four straight-line paths. At the end of the walk, what is the person's resultant displacement measured from the starting point?

Figure 3.18
Problem 12

13. A shopper pushing a cart through a store moves a distance 40 m down one aisle then makes a 90° turn and moves 15 m. He then makes another 90° turn and moves 20 m. How far is the shopper away from his original position in magnitude and direction? Note that you are not given the direction moved in any of the 90° turns. As a result, there could be more than one answer.
14. An airplane flies from city A to city B in a direction due east for 800 miles. In the next part of the trip the

airplane flies from city B to city C in a direction 40° north of east for 600 miles. What is the resultant displacement of the airplane between city A and city C?

15. Two people pull on a stubborn mule as seen from the helicopter view shown in Figure 3.19. Find (a) the single force which is equivalent to the two forces shown, and (b) the force that a third person would have to exert on the mule to make the net force equal to zero.

Figure 3.19
Problem 15

16. A man pushing a mop across a floor causes it to undergo two displacements. The first has a magnitude of 150 cm and makes an angle of 120° with the positive x axis. The resultant displacement has a magnitude of 140 cm and is directed at an angle of 35° to the positive x axis. Find the magnitude and direction of the second displacement.

Section 3.4   Velocity and Acceleration in Two Dimensions and
Section 3.5   Projectile Motion

17. A punter kicks a football at an angle of 30° with the horizontal at an initial speed of 20 m/s. Where should a punt returner position himself to catch the ball just before it strikes the ground?

18. Tom the cat is chasing Jerry the mouse across a table surface 1.5 m high. Jerry steps out of the way at the last second, and Tom slides off the edge of the table at a speed of 5 m/s. Where will Tom strike the floor and what velocity components will he have just before he hits?

19. In an ideal punt, the football has a "hangtime" (total time in the air) of 5 s. If a punter kicks the ball at an angle of 45° with the horizontal, what must be the initial velocity of the ball to achieve this?

20. A brick is thrown upward from the top of a building at an angle of 25° to the horizontal and with an initial speed of 15 m/s. If the brick is in flight for 3 s, how tall is the building?

21. A ball is thrown straight upward and returns to the thrower's hand after 3 s in the air. A second ball is thrown at an angle of 30° with the horizontal. At what speed must the second ball be thrown so that it reaches the same height as the one thrown vertically?

22. Two objects are initially the same height above the ground. Simultaneously, one is released from rest and the other is shot off horizontally with an initial speed of 2.5 m/s. The two objects collide after falling 20.0 m. How far apart were the objects initially?

23. A mountain climber is stranded on a ledge 300 m above the ground. Rescuers on the ground want to shoot a projectile to him with a rope tied to it. If the projectile is directed upward at an initial angle of 35° and fired at the ledge from a horizontal distance of 500 m, determine the initial speed the projectile must have to hit the ledge.

24. A car is parked on a cliff overlooking the ocean on an incline that makes an angle of 37° below the horizontal. The negligent driver leaves the car in neutral, and the emergency brakes are defective. The car rolls from rest down the incline with a constant acceleration of 4 m/s² and travels 50 m to the edge of the cliff. The cliff is 30 m above the ocean. Find (a) the position of the car relative to the base of the cliff when it lands in the ocean, and (b) the time the car is in the air.

Section 3.6   Relative Velocity

25. Two rowboats are heading toward one another. The speed of rowboat 1 is 0.511 km/h. The speed of rowboat 2 is 0.433 km/h. Assuming the direction of motion of rowboat 1 is positive, find the velocity of rowboat 2 as seen from rowboat 1.

26. A motorist traveling west on Interstate Route 80 at 80 km/h is being chased by a police car traveling at 95 km/h. (a) What is the velocity of the motorist relative to the police car? (b) What is the velocity of the police car relative to the motorist?

27. An escalator is 20.0 m long. If a person stands on the "up" side, it takes 50.0 s to ride to the top. (a) If a person walks up the moving escalator with a speed of 0.5 m/s relative to the escalator, how long does it take to get to the top? (b) If a person walks down the "up" escalator with the same relative speed as in (a), how long does it take to reach the bottom?

28. A boat moves at 10 m/s relative to the water regardless of which direction it travels through the water. If

the water in a river is flowing at 1.5 m/s, how long does it take the boat to make a round trip consisting of a 300-m displacement downstream followed by a 300-m displacement upstream?

29. A skater (water spider) maintains an average position on the surface of a stream by darting upstream (against the current), then drifting downstream (with the current) to its original position. The current in the stream is 0.50 m/s relative to the shore, and the skater darts upstream 0.56 m (relative to shore) in 0.8 s during the first part of its motion. Take upstream as the positive direction and determine: (a) the velocity of the skater relative to the water (i) during its dash upstream and (ii) during its drift downstream. (b) How far upstream relative to the water does the skater move during one cycle of this motion? (c) What is the average velocity of the skater relative to the water?

30. Two canoeists in identical canoes exert the same effort paddling and hence maintain the same speed relative to the water. One paddles directly upstream (and moves upstream), whereas the other paddles directly downstream. If downstream is the positive direction, an observer on shore determines the velocities of the two canoes to be −1.2 m/s and +2.9 m/s respectively. (a) What is the speed of the water relative to shore? (b) What is the velocity of each canoe relative to the water?

31. A river flows due east at 1.5 m/s. A boat crosses the river from the south shore to the north shore by maintaining a constant velocity of 10.0 m/s due north relative to the water. (a) What is the velocity of the boat relative to shore? (b) If the river is 300 m wide, how far downstream has the boat moved by the time it reaches the north shore?

32. A rowboat crosses a river with a velocity of 3.30 mi/h at an angle 62.5° north of west relative to the water. The river is 0.505 mi wide and carries an eastward current of 1.25 mi/h. How far downstream is the boat when it reaches the opposite shore?

33. The pilot of an aircraft wishes to fly due west in a wind blowing at 50 km/h toward the south. If the speed of the aircraft in the absence of a wind is 200 km/h, (a) in what direction should the aircraft head and (b) what should its speed be relative to the ground?

34. A car travels due east with a speed of 50 km/h. Rain is falling vertically with respect to the Earth. The traces of the rain on the side windows of the car make an angle of 60° with the vertical. Find the velocity of the rain with respect to (a) the car and (b) the Earth.

35. A science student is riding on a flatcar of a train traveling along a straight horizontal track at a constant speed of 10 m/s. The student throws a ball along a path that he judges to make an initial angle of 60° with the horizontal and to be in line with the track. The

student's professor, who is standing on the ground nearby, observes the ball to rise vertically. How high does the ball rise?

Additional Problems

36. A baseball is thrown by an outfielder at a speed of 20 m/s at an angle of 35° above the horizontal. Find (a) the component of velocity along the ground at the instant the ball is released and (b) the upward component of velocity at this instant.

37. A river has a steady speed of 0.5 m/s. A student swims upstream a distance of 1 km and returns to the starting point. If the student can swim at a speed of 1.2 m/s in still water, how long does the trip take? Compare this with the time the trip would take if the water were still.

38. A shopper in a department store can walk up a stationary (stalled) escalator in 30 s. If the normally functioning escalator can carry the standing shopper to the next floor in 20 s, how long would it take the shopper to walk up the moving escalator? Assume the same walking effort for the shopper whether the escalator is stalled or moving.

39. A taxi driver travels due south for 10 km and then moves 6 km in a direction 30° north of east. Find the magnitude and direction of the car's resultant displacement.

40. The multiflash photograph of Figure 3.20 shows two golf balls released simultaneously. The time interval between flashes is (1/30) s, and the white parallel lines were placed 15 cm apart. (a) Find the speed at which the right ball was projected and (b) show that both balls should be expected to reach the floor simultaneously.

Figure 3.20   Problem 40

41. If a person can jump a horizontal distance of 3 m on the Earth, how far could the person jump on the Moon, where the acceleration due to gravity is g/6, where g = 9.80 m/s²? Repeat for Mars, where the acceleration due to gravity is 0.38g.

42. A baseball is hit in such a way that the ball just clears a wall 3 m tall, 130 m from home plate. The ball leaves the bat at an angle of 35° to the horizontal and 1 m above the ground. Find its initial velocity.

43. Towns A and B in Figure 3.21 are 80.0 km apart. A couple driving from town A arranges to meet a couple driving from town B at the lake L. They leave simultaneously and drive for 2.5 h in the directions shown. Car 1 has a speed of 90.0 km/h. If the cars arrive simultaneously at the lake, what is the speed of car 2?

Figure 3.21
Problem 43

44. A daredevil is shot out of a cannon at 45° to the horizontal with an initial speed of 25 m/s. A net is located at a horizontal distance of 50 m from the cannon. At what height above the cannon should the net be placed in order to catch the daredevil?

45. A daredevil decides to jump a canyon of width 10 m. To do so, he drives a motorcycle up an incline sloped at an angle of 15°. What minimum speed must he have in order to clear the canyon?

46. A rocket is launched at an angle of 53° above the horizontal with an initial speed of 100 m/s. It moves along its initial line of motion with an acceleration of 30 m/s² for 3 s. At this time its engines fail and the rocket proceeds to move as a free body. Find (a) the maximum altitude reached by the rocket, (b) its total time of flight, and (c) its horizontal range.

47. A dart gun is fired while being held horizontally at a height of 1 m above ground level. With the gun at rest relative to the ground, the dart from the gun travels a horizontal distance of 5 m. A child holds the same gun in a horizontal position while sliding down a 45° incline at a constant speed of 2 m/s. How far will the dart travel if the gun is fired when it is 1 m above the ground?

48. The determined coyote is out once more to try to capture the elusive roadrunner. The coyote wears a new pair of Acme power roller skates, which provide a constant horizontal acceleration of 15 m/s², as shown in Figure 3.22. The coyote starts off at rest 70 m from the edge of a cliff at the instant the roadrunner zips by in the direction of the cliff. (a) If the roadrunner moves with constant speed, determine the minimum speed he must have in order to reach the cliff before the coyote. (b) If the cliff is 100 m above the base of a canyon, determine where the coyote lands in the canyon. (Assume that his skates are still in operation when he is in "flight" and that his horizontal component of acceleration remains constant at 15 m/s².)

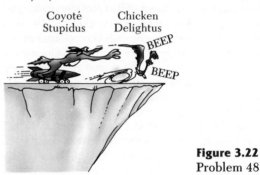

Figure 3.22
Problem 48

49. A projectile is fired with an initial speed of v_0 at an angle, θ_0, to the horizontal, as in Figure 3.12. When it reaches its peak, it has (x, y) coordinates given by $(R/2, h)$, and when it strikes the ground, its coordinates are $(R, 0)$, where R is called the horizontal range. (a) Show that it reaches a maximum height, h, given by

$$h = \frac{v_0^2 \sin^2 \theta_0}{2g}$$

(b) Show that its horizontal range is given by

$$R = \frac{v_0^2 \sin 2\theta_0}{g}$$

50. In a very popular lecture demonstration, a projectile is fired at a falling target as in Figure 3.23. The projectile leaves the gun at the same instant that the target is dropped from rest. Assuming that the gun is initially aimed at the target, show that the projectile will hit the target. (One restriction of this experiment is that the projectile must reach the target before the target strikes the floor.)

Figure 3.23
Problem 50

Figure 3.24
Problem 51

51. An enemy ship is on the east side of a mountainous island as shown in Figure 3.24. The enemy ship can maneuver to within 2500 m of the 1800-m high mountain peak and can shoot projectiles with an initial speed of 250 m/s. If the western shoreline is horizontally 300 m from the peak, what are the distances from the western shore at which a ship can be safe from the bombardment of the enemy ship?

52. A vector **R** has a magnitude of 100 units. Find its components in magnitude, including sign, when it makes the following angles with the $+x$ axis. (a) 60°, (b) 130°, (c) 200°, (d) 290°.

53. While exploring a cave, a spelunker starts at the entrance and moves the following distances. She goes 75 m north, 250 m east, 125 m at an angle 30° north of east, and 150 m south. Find the resultant displacement from the cave entrance.

54. Figure 3.25 illustrates the difference in proportions of the male and female anatomy. The displacements $\mathbf{d}_{1m}$ and $\mathbf{d}_{1f}$ from the bottom of the feet to the navel have magnitudes of 104 cm and 84.0 cm, respectively. The displacements $\mathbf{d}_{2m}$ and $\mathbf{d}_{2f}$ have magnitudes of 50.0 cm and 43.0 cm, respectively. (a) Find

the vector sum of the displacements $\mathbf{d}_1$ and $\mathbf{d}_2$ in each case. (b) The male figure is 180 cm tall, the female 168 cm. Normalize the displacements of each figure to a common height of 200 cm and reform the vector sums as in part a. Then find the vector difference between the two sums.

55. Determine the height h in Figure 3.26.

Figure 3.26
Problem 55

56. Find the height h, the length of the side L, and the angle α of the triangle in Figure 3.27.

Figure 3.25
Problem 54

Figure 3.27
Problem 56

CHAPTER 4

The Laws of Motion

An athlete running in the sunset. What external forces influence the motion of the runner?
(© 1989 Dennis Hallinan, FPG International)

Isaac Newton is considered by many to be the most brilliant scientist who ever lived. His genius came into full bloom in his early twenties when, during one 18-month period, he formulated the law of gravitation, invented calculus, and proposed theories of light and color. On his tomb in Westminster Abbey is the following epitaph: "Mortals, congratulate yourselves that so great a man lived for the honor of the human race." Newton's contributions to several branches of physics are truly remarkable. In this chapter we shall investigate only a few of the contributions he made to the field of mechanics. It is in this area that Newton is most widely known today, primarily because of his three laws of motion and his theory of universal gravitation.

4.1 Introduction to Classical Mechanics

Classical mechanics deals with objects that (a) are large compared with the dimensions of atoms ($\approx 10^{-10}$ m) and (b) move at speeds that are much less than the speed of light (3×10^8 m/s). If either criterion is violated, the equations and results of this chapter do not apply. Our study of relativity and quantum mechanics in later chapters will enable us to cover the complete range of speeds and sizes.

In Chapters 2 and 3 we described the motion of objects based on the definitions of displacement, velocity, and acceleration. However, we have not

74

yet been able to answer specific questions related to the nature of motion, questions such as "What causes motion?" and "Why do some objects accelerate at a greater rate than others?" In this chapter, we shall use the concepts of force and mass to describe the change in motion of objects. We shall see that it is possible to describe the acceleration of an object in terms of its mass and the external force acting on it. This force represents the interaction of the object with its environment. The mass of an object is a measure of the object's inertia, that is, the tendency of the object to resist an acceleration when a force acts on it.

Force is a vector quantity which has both magnitude and direction. We shall represent a force on a diagram by an arrow whose length represents the magnitude of the force and whose tip points in the direction of the force. The rules of vector addition and the use of components of a vector discussed in Chapter 3 will be used extensively to analyze a number of mechanical systems.

4.2   The Concept of Force

Everyone has an idea of the concept of force from everyday experiences. When you push or pull an object, you exert a force on it. For example, you exert a force on a ball when you throw it or kick it. In each of these instances the word *force* is associated with the result of muscular activity, leading to some form of motion. Although forces can cause motion, it does not necessarily follow that forces acting on an object will always cause it to move. For example, as you sit reading this book, the force of gravity acts on your body and yet you remain stationary. Also, you can push on a wall and not move it.

What force (if any) causes a distant star to move freely through space? Newton answered such questions by stating that the change in velocity of an object is caused by forces. Therefore, if an object moves with uniform motion (constant velocity), no force is required to maintain the motion. Since only a force can cause a change in velocity, a force can be viewed as that which causes an object to accelerate.

Now consider a situation in which several forces act simultaneously on an object. In this situation, the object will accelerate only if the *net* force acting on it is not equal to zero. We shall often refer to the net force as the *resultant* force or the unbalanced force. *If the net force on an object is zero, the acceleration is zero and the velocity of the object remains constant.* That is, if the net force acting on the object is zero, either the object will be at rest or it will move with constant velocity. *When the velocity of an object is constant, or if the object is at rest, it is said to be in equilibrium.*

Kinds of Forces

Whenever a force is exerted on an object, its shape can change. For example, when you squeeze a rubber ball or strike a punching bag with your fist, the object will deform to some extent. Even more rigid objects, such as automobiles, are deformed under the action of external forces. Often the deformations are permanent as in the case of a collision between vehicles. If you pull on a coiled spring, as in Figure 4.1a, the spring stretches. If the spring is calibrated, the distance that it stretches can be used to measure the strength of the

Many forces influence the motion of this beautiful sailing ship. Can you identify some of them? *(© Bruce Beyers, FPG International)*

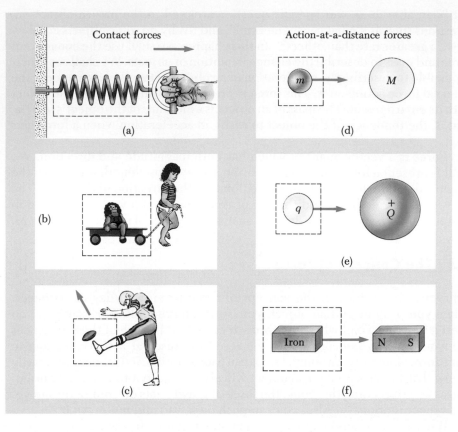

Figure 4.1
Some examples of forces applied to various objects. In each case a force is exerted on the particle or object within the dashed boxed area. Some agent or object external to the boxed area provides the force on the object.

Compression of a football as kicker's foot sets it in motion. (© *Harold E. Edgerton, Courtesy of Palm Press, Inc.*)

force. If you pull hard enough on a cart, as in Figure 4.1b, the cart will move. Finally, when a football is kicked, as in Figure 4.1c, it is deformed and set into motion. These are all examples of a class of forces called **contact forces** because they arise as the result of physical contact between two objects. Other examples of contact forces are the force that a gas exerts on the walls of a container (the result of gas molecules colliding with the walls) and the force of our feet on the floor.

Another class of forces is known as **field forces.** These forces do not involve physical contact between an object and its surroundings, but act through space. The force of gravitational attraction between two objects is an example of this class of force. Early scientists, including Newton himself, were uneasy with the concept of a force acting between two disconnected objects. To overcome this conceptual problem, Michael Faraday (1791–1867) introduced the concept of a *field*. According to this approach, when a mass m_1 is placed at some point P near a mass m_2, one can say that m_1 interacts with the gravitational field that exists at P produced by m_2. In Chapter 16, we shall see that the field concept is also useful in describing electric interactions between charged particles.

We should mention that the distinction between contact forces and field forces is not as sharp as you may have been led to believe by the above discussion. At the atomic level, the so-called contact forces are actually due to repulsive electric forces between charges, which themselves are field forces.

Now let us consider the force of attraction between any two objects. This force of attraction is called the gravitational force and is pictured in Figure 4.1d. This force keeps objects bound to the Earth and gives rise to what we commonly call the weight of an object. The planets of our Solar System move in their elliptical orbits under the action of gravitational forces exerted on them by the Sun. Another common example of a field force is the electric force that one electric charge exerts on another electric charge, as in Figure 4.1e. These charges might be an electron and proton pair forming the hydrogen atom. A third example of a field force is the force that a bar magnet exerts on a piece of iron, as shown in Figure 4.1f.

4.3 Newton's First Law

Before we begin our discussion, try the following simple experiment: slide a book across a table and observe what happens to it. You know what happens before you do the experiment: the book stops sliding after only a very short distance. Before about 1600, scientists chose to concentrate on this aspect of matter. They felt that the natural state of matter was the state of rest, a reasonable assumption in light of our experiment. Now, however, imagine sliding this same book across a smooth, highly polished floor. The book will again come to rest, but not as quickly as before. Finally, let your mind leap to the possibility of a floor so highly polished that friction is completely absent. Under these conditions, could it be that the book would slide forever? It was Galileo who developed such thought experiments as these and concluded that scientists were considering motion from an unproductive viewpoint. His conclusion was that it is not the nature of an object to come to rest once set in motion; instead, the object tends to resist a change in velocity.

This new approach to motion was later formalized by Newton in a form that has come to be known as **Newton's first law of motion:**

> An object at rest will remain at rest and an object in motion will continue in motion with a constant velocity (that is, constant speed in a straight line) unless it experiences a net external force (called the resultant force).

A statement of Newton's first law

In simpler terms, you can say that when the resultant force on a body is zero, its acceleration is zero. That is, when $\Sigma \mathbf{F} = 0$, then $\mathbf{a} = 0$. From the first law, we conclude that an isolated body (a body that does not interact with its environment) is either at rest or moving with constant velocity. Actually, Newton was not the first to state this law. Several decades earlier, in his publication *The New Sciences,* Galileo wrote, "Any velocity once imparted to a moving body will be rigidly maintained as long as the external causes of retardation are removed."

A good example of uniform motion on a nearly frictionless surface is the motion of a light disk on a column of air, as in Figure 4.2. If the disk is given an initial velocity, it will coast a great distance before coming to rest. This idea is used in the game of air hockey, in which the disk makes many collisions with the sides of the table before coming to rest.

Finally, consider a moving spaceship far from any planets or other matter. The spaceship requires some propulsion system to change its velocity. However, if the propulsion system is turned off when the spaceship reaches a

Figure 4.2
A disk moving on a column of air is an example of uniform motion, that is, motion in which the acceleration is zero and the velocity remains constant.

velocity **v**, the spaceship will "coast" in space with the same velocity and the astronauts get a "free ride," that is, no propulsion system is required to keep them moving at the velocity **v**.

Mass and Inertia

If you attempt to change the state of motion of any object, the object will resist the change.

> The tendency of an object to either remain at rest or to maintain a state of uniform motion in a straight line is called **inertia**.

For this reason, Newton's first law is sometimes called the *law of inertia.* For example, imagine striking two objects with a golf club: a golf ball and a bowling ball. Both are initially at rest and, according to the first law, they will remain at rest unless a net force is applied. The golf club supplies this net force, but obviously the net force will have a much greater effect on the golf ball than on the bowling ball. The golf ball, if hit solidly, will fly many meters, whereas the bowling ball will barely move if the same force is applied. We conclude that the two balls have different amounts of inertia.

Mass is a term used to measure inertia, and the SI unit of mass is the kilogram. The greater the mass of a body, the less it will accelerate (change its state of motion) under the action of an applied force. For example, if a given force acting on a 3-kg mass produces an acceleration of 4 m/s², the same force applied to a 6-kg mass will produce an acceleration of only 2 m/s². This idea will be used shortly to obtain a quantitative description of the concept of mass.

It is important to point out that mass should not be confused with *weight*. Mass and weight are two different quantities and have different units.

> The **weight** of a body, a vector quantity, is equal to the force of gravity acting on the body and varies with location.

For example, a person who weighs 180 lb on Earth weighs only about 30 lb on the Moon. On the other hand, the mass of a body is the same everywhere, regardless of location. An object having a mass of 2 kg on Earth also has a mass of 2 kg on the Moon.

Mass is an inherent property of a body and is independent of the body's surroundings and of the method used to measure the mass. It is an experimental fact that mass is a scalar quantity, since its value is independent of the direction of motion and independent of the coordinate system used in describing the motion. Finally, mass is a quantity that obeys the rules of ordinary arithmetic. That is, several masses can be combined in a simple numerical fashion. For example, if you combine a 3-kg mass with a 5-kg mass, the total mass is 8 kg. This can be verified experimentally by comparing the acceleration of each object produced by a known force with the acceleration of the combined system using the same force.

In the introduction to this chapter, we pointed out that much of the present discussion must be modified when we treat objects moving at speeds approaching the speed of light. One of these modifications is concerned with the nature of mass, which we assume to be constant in Newtonian mechanics. In our study of special relativity (Chapter 28), we shall find that the mass of an object increases with increasing speed. This effect, however, is perceptible only for speeds approaching 3×10^8 m/s, the speed of light.

Inertia

Astronaut Edwin E. Aldrin, Jr., walking on the Moon after the Apollo 11 lunar landing. The weight of this astronaut on the Moon is less than it is on Earth, but his mass remains the same. *(Courtesy of NASA)*

Isaac Newton, a British physicist and mathematician, is regarded as one of the greatest scientists in history. Before the age of 30, he formulated the basic concepts and laws of motion, discovered the universal law of gravitation, and formulated the methods of calculus. As a consequence of his theories, Newton was able to explain the motion of the planets, the ebb and flow of the tides, and many special features of the motion of the Moon and the Earth. He also made many fundamental discoveries in optics, showing that white light is composed of a spectrum of colors. His contributions to physical theories dominated scientific thought for two centuries and remain important today.

Newton was born prematurely on Christmas Day in 1642 shortly after his father's death. When he was three, his mother remarried and he was left in his grandmother's care. Because he was small in stature as a child, he was bullied by other children and would take refuge in mechanical things such as building waterclocks, kites carrying fiery lanterns, sundials, and model windmills powered by mice. His mother withdrew him from school at the age of 12 with the intention of making him into a farmer. Fortunately, his uncle recognized his scientific and mathematical abilities and helped send him to Trinity College in Cambridge.

In 1665, Newton completed his Bachelor of Arts degree and the university was closed because of the bubonic plague that raged through England. During this period of turmoil, Newton returned to the family farm at Woolsthorpe to study. This was an especially creative period of his life during which time he laid the foundations of his work in mathematics, optics, the laws of motion, celestial mechanics, and his famous law of gravity.

Newton was a very private person who studied alone and labored day and night in his laboratory conducting experiments, performing calculations, or immersing himself in theological studies. There was little happiness in his life and he never married. His greatest single work, *Mathematical Principles of Natural Philosophy*, was published in 1687. In his later years, he spent much of his time quarreling with other eminent minds, including the mathematician Gottfried Leibnitz, who worked independently on the development of calculus; Christian Huygens, who worked on the wave theory of light; and Robert Hooke, who supported Huygens' theory. These disputes, the strain of his studies and his work on alchemy, which involved mercury (a poison) caused him to suffer severe depression in 1692. He was elected President of the Royal Society in 1703, and he retained that office until his death in 1727.

Isaac Newton
(1642 – 1727)

(Courtesy AIP Niels Bohr Library)

4.4 Newton's Second Law

Newton's first law explains what happens to an object when the resultant force acting on it is zero. In such instances we find that the object either stays at rest or moves with constant velocity. Newton's second law answers the question of what happens to an object that has a nonzero resultant force acting on it.

In order to gain some insight into what happens when a net force is applied to an object, consider a situation in which you are pushing a block of ice across a polished surface, smooth enough to make frictional forces negligible. When you exert a certain force on the block of ice, you find that it moves with an acceleration of, say, 2 m/s². If you push twice as hard, you find that the acceleration also doubles. Pushing three times as hard triples the acceleration, and so on. From observations such as this, we can conclude that *the acceleration of an object is directly proportional to the resultant force acting on it.* Common experience with pushing objects should indicate to you that the mass of the object also affects its acceleration. This can be seen as follows. Imagine that you stack identical blocks of ice on top of each other while pushing the stack with the same force. You will find that if the force produces an acceleration of 2 m/s² when you push one block, the acceleration will drop to half this value when two blocks are pushed, to one third this value for three blocks, and so on. From this we see that *the acceleration of an object is inversely proportional to its mass.* **Newton's second law** summarizes these observations:

> The acceleration of an object is directly proportional to the resultant force acting on it and inversely proportional to its mass. The direction of the acceleration is in the direction of the resultant force.

In equation form, we can state Newton's second law as

Newton's second law

$$\sum \mathbf{F} = m\mathbf{a} \tag{4.1}$$

where **a** is the acceleration of the object, m is its mass, and $\Sigma\mathbf{F}$ represents the *vector sum of all external forces acting on the object.* You should note that, because this is a vector equation, it is equivalent to the following three component equations:

$$\sum F_x = ma_x \qquad \sum F_y = ma_y \qquad \sum F_z = ma_z \tag{4.2}$$

Note that if the resultant force is zero, then $\mathbf{a} = 0$, which corresponds to the equilibrium situation where **v** is either constant or zero. Hence, *the first law of motion is a special case of the second law.*

Units of Force and Mass

The SI unit of force is the **newton,** which is defined as the force that, when acting on a 1-kg mass, produces an acceleration of 1 m/s².

From this definition and Newton's second law, we see that the newton can be expressed in terms of the fundamental units of mass, length, and time:

Definition of newton

$$1 \text{ N} \equiv 1 \text{ kg} \cdot \text{m/s}^2 \tag{4.3}$$

The unit of force in the cgs system is called the **dyne** and is defined as the force that, when acting on a 1-g mass, produces an acceleration equal to 1 cm/s²:

Definition of dyne

$$1 \text{ dyne} \equiv 1 \text{ g} \cdot \text{cm/s}^2 \tag{4.4}$$

In the British engineering system the unit of force is the **pound.** The following conversion from pounds to Newtons will be useful to you in many problems:

Definition of pound

$$1 \text{ lb} \equiv 4.448 \text{ N} \tag{4.5}$$

When we speak of going on a diet to lose a few pounds, we really mean that we want to lose a few kilograms; that is, we want to reduce our mass. When we lose those few kilograms, the force of gravity (pounds) on our reduced mass de-

Table 4.1
Units of Mass, Acceleration, and Force

System	Mass	Acceleration	Force
SI	kg	m/s²	$N = kg \cdot m/s^2$
cgs	g	cm/s²	$dyne = g \cdot cm/s^2$

Note: $1 \text{ N} = 10^5 \text{ dyne} = 0.225 \text{ lb.}$

creases and that is how we "lose a few pounds." Most of the calculations we shall carry out in our study of classical mechanics will be in SI units, and therefore the kilogram will be the unit of mass used most often in this text.

Since $1 \text{ kg} = 10^3 \text{ g}$ and $1 \text{ m} = 10^2 \text{ cm}$, it follows that $1 \text{ N} = 10^5 \text{ dynes}$. It is left as an exercise to show that $1 \text{ N} = 0.225 \text{ lb}$. The units of mass, acceleration, and force in the SI and cgs system are summarized in Table 4.1.

Weight

We are well aware of the fact that bodies are attracted to the Earth. The force exerted by the Earth on a body is called the **weight** of the body, **w**. This force is directed toward the center of the Earth. Thus, weight is a vector while mass is a scalar.

We have seen that a freely falling body experiences an acceleration **g** acting toward the center of the Earth. Applying Newton's second law to the freely falling body, shown in Figure 4.3, with $\mathbf{a} = \mathbf{g}$ and $\mathbf{F} = \mathbf{w}$, gives

$$\mathbf{w} = m\mathbf{g} \tag{4.6}$$

Weight

Since it depends on **g**, weight varies with geographic location. Bodies weigh less at higher altitudes than at sea level. This is because g decreases with

Figure 4.3
The only force acting on an object in free fall is the weight, **w**. Newton's second law applied in the vertical direction gives $w = mg$.

An astronaut "floating" in a gravity-free environment of a space shuttle cannot tell whether he is floating in free space or is in free-fall. *(Courtesy of NASA)*

increasing distance from the center of the Earth. Hence, weight, unlike mass, is not an inherent property of a body. For example, if a body has a mass of 70 kg, then the magnitude of its weight in a location where $g = 9.80$ m/s² is $mg = 686$ N (about 154 lb). At the top of a mountain, where g might be 9.76 m/s², say, this weight would be 683 N. This corresponds to a decrease in weight of about 0.4 lb. Therefore, if you want to lose weight without going on a diet, move to the top of a mountain or weigh yourself at an altitude of 30,000 ft during a flight on a jet airplane.

4.5   Newton's Third Law

In Section 4.2 we found that a force is exerted on an object when it comes into contact with some other object. For example, consider the task of driving a nail into a block of wood as illustrated in Figure 4.4. In order to accelerate the nail and drive it into the block, a net force must be supplied to the nail by the hammer. Newton, however, recognized that a single isolated force (such as the force on the nail by the hammer) cannot exist. Instead, *forces in nature always exist in pairs.* According to Newton, the hammer does exert a force on the nail, but the nail also exerts a force on the hammer. Furthermore, there is clearly a net force on the hammer because it rapidly slows down (decelerates) after coming into contact with the nail. Newton stated this type of situation in terms of his third law of motion, which may be stated as follows:

Newton's third law

> If two bodies interact, the force exerted on body 1 by body 2 is equal in magnitude and opposite in direction to the force exerted on body 2 by body 1.

An alternative statement of this law is that *for every action there is an equal and opposite reaction.* It is important to note that action-reaction pairs are always exerted on *different* objects. In our hammer-nail example above, we could call the force the hammer exerts on the nail the action and the force that the nail exerts on the hammer the reaction. However, there is nothing special about this assignment of action-reaction pairs. If you choose to call the force the hammer exerts on the nail the reaction and the force that the nail exerts on the hammer the action, it is perfectly satisfactory to do so. Let us discuss some other examples of action-reaction pairs.

The force acting on a freely falling object is its weight **w**. Let us call this the action. What is the reaction? Since the weight is the force exerted on the falling object by the Earth, the reaction is the force exerted on the Earth by the falling object. Thus, as the falling object accelerates toward the Earth, the Earth accelerates toward the object. However, since the Earth has such a large mass compared to the object's mass, the acceleration of the Earth due to this reaction force is negligibly small.

Figure 4.5a shows a block at rest on a table. There are two forces acting on the block indicated by **n** and **w** in the figure. The force **w** is the weight of the block and the force **n** is the upward force exerted on the block by the table. The force **n** is called the **normal** force. The word normal is used because the direction of **n** is perpendicular to the table surface and normal is the word for perpendicular. Because the block is not accelerating, we know from the first law that **w** and **n** must be equal in magnitude and opposite in direction. How-

Figure 4.4
The force of the hammer on the nail is equal in magnitude and opposite in direction to the force of the nail on the hammer.

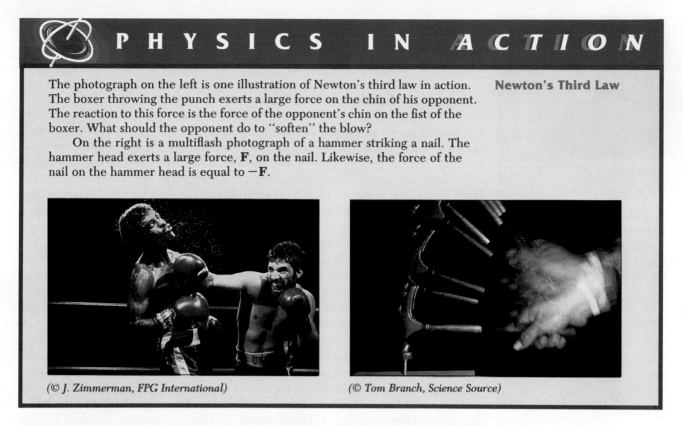

PHYSICS IN ACTION

Newton's Third Law

The photograph on the left is one illustration of Newton's third law in action. The boxer throwing the punch exerts a large force on the chin of his opponent. The reaction to this force is the force of the opponent's chin on the fist of the boxer. What should the opponent do to "soften" the blow?

On the right is a multiflash photograph of a hammer striking a nail. The hammer head exerts a large force, **F**, on the nail. Likewise, the force of the nail on the hammer head is equal to $-\mathbf{F}$.

(© J. Zimmerman, FPG International)   (© Tom Branch, Science Source)

ever, you should note that these two forces do not constitute an action-reaction pair because such forces always act on different objects. In our case, both **w** and **n** are exerted on the same object, the block. If we choose to call **w** the action, the reaction to this force is the force denoted as **w′** in Figure 4.5a, which is the force the block exerts on the Earth. The reaction to **n**, denoted by **n′** in the figure, is the force exerted on the table by the block. Note that the forces acting on the block are **w** and **n** as indicated in Figure 4.5b. When examining the motion of an object, we shall be interested only in such external forces.

(a)

(b)

Figure 4.5
When a block is lying on a table, the forces acting on the block are the normal force **n**, and the force of gravity, **w**, as illustrated in (b). The reaction to **n** is the force of the block on the table, **n′**. The reaction to **w** is the force of the block on the Earth, **w′**.

Multiflash exposure of a tennis player executing a backhand swing. The force of the tennis racket on the ball is equal and opposite to the force of the ball on the tennis racket.
(© J. Zimmerman, FPG International)

You experience Newton's third law directly when you slam your fist against a wall or kick a football. You should try to identify the action and reaction forces in these cases.

4.6   Some Applications of Newton's Laws

In this section we present some simple applications of Newton's laws to bodies moving under the action of constant external forces. As our model, we shall assume that the bodies behave as particles so that we need not worry about rotational motion. We shall also neglect the effects of friction for those problems involving motion. This is equivalent to stating that the surfaces are smooth. Finally, we shall neglect the mass of any ropes involved in a particular problem. In this approximation, the magnitude of the force exerted at any point along a rope is the same at all points along the rope.

When we apply Newton's laws to a body, we shall be interested only in those external forces that act on the body. For example, in Figure 4.5b the only external forces acting on the block are **n** and **w**. The reactions to these forces, **n'** and **w'**, act on the table and on the Earth, respectively, and do not appear in Newton's second law as applied to the block.

Consider a block being pulled to the right on the smooth horizontal surface of a table, as in Figure 4.6a. Suppose you are asked to find the acceleration of the block and the force of the table on it. First, note that the horizontal force being applied to the block acts through the string. The force that the string exerts on the block is denoted by the symbol **T**. The magnitude of **T** is called the **tension** in the string. A dashed circle is drawn around the block in Figure 4.6a to remind you to isolate the block from its surroundings. Since we are interested only in the motion of the block, we must be able to identify all external forces acting on it. These are illustrated in Figure 4.6b. In addition to the force **T**, the force diagram for the block includes the weight, **w**, and the normal force, **n**. As before, **w** corresponds to the force of gravity pulling down on the block and **n** to the upward force of the table on the block. Such a force diagram is referred to as a **free-body diagram**. The construction of such a diagram is an important step in applying Newton's laws. The reactions to the forces we have listed, namely, the force of the string on the hand doing the pulling, the force of the block on the Earth, and the force of the block on the table, are not included in the free-body diagram because they act on other bodies and not on the block.

We are now in a position to apply Newton's second law to the block. First, we must choose an appropriate coordinate system. In this case it is convenient to use the coordinate system shown in Figure 4.6b, with the x axis horizontal and the y axis vertical. We are now ready to apply Newton's second law in the x direction, y direction, or both, depending on what we are asked to find in the problem. In addition, you may be able to make use of the equations of motion for constant acceleration found in Chapter 2. However, you should note that these equations can be used only when the acceleration is a constant. For example, if the force **T** in Figure 4.6 is constant, then it follows that the acceleration in the x direction is also a constant because $a_x = T/m$. Hence, if you need to find the displacement or the velocity of the object at some instant of time, the equations of motion with constant acceleration can be used.

Figure 4.6
(a) A block being pulled to the right on a smooth surface.
(b) The free-body diagram that represents the external forces on the block.

Objects in Equilibrium and Newton's First Law

Objects that are at rest or that are moving with constant velocity are said to be in equilibrium, and Newton's first law is a statement of one condition that must be true for equilibrium conditions to prevail. In equation form, this can be expressed as

$$\sum \mathbf{F} = 0 \tag{4.7}$$

This statement signifies that the *vector* sum of all the forces (the net force) acting on an object is zero. Usually, the problems we will encounter in our study of equilibrium will be more easily solved if we work with Equation 4.7 in terms of the components of the external forces acting on an object. By this, we mean that, in a two-dimensional problem, the sum of all the external forces in the x and y directions must separately equal zero, that is

$$\sum F_x = 0$$
$$\sum F_y = 0 \tag{4.8}$$

This set of equations is often referred to as the **first condition for equilibrium.** We shall not consider three-dimensional problems in this text, but the extension of Equation 4.8 to a three-dimensional situation can be made by adding a third equation, $\sum F_z = 0$.

Problem-Solving Strategy:
Objects in Equilibrium

The following procedure is recommended for problems involving objects in equilibrium:

1. Make a sketch of the object under consideration.
2. Draw a free-body diagram and label all external forces acting on the object. Assume a direction for each force. If you select a direction that leads to a negative sign in your solution for a force, do not be alarmed; this merely means that the direction of the force is the opposite of what you assumed.
3. Resolve all forces into x and y components, choosing a convenient coordinate system.
4. Use the equations $\sum F_x = 0$ and $\sum F_y = 0$. Remember to keep track of the signs of the various force components.
5. Application of Step 4 above leads to a set of equations with several unknowns. Solve the simultaneous equations for the unknowns in terms of the known quantities.

Example 4.1 Tug-a-Boat

Two fishermen are pulling a boat through the water as in Figure 4.7. Each exerts a force of 600 N directed at a 30° angle relative to the forward motion of the boat. If the boat moves with constant velocity, find the resistive force **F** of the water on the boat.

Figure 4.7 (Example 4.2)
Two fishermen pulling a boat.

Solution

If the boat moves with constant velocity, the first condition for equilibrium tells us that the resultant force on the boat in the x direction must be zero. The x component of each man's force is (600 N)(cos 30°), and the force of the water on the boat is **F**. Hence, the net force on the boat in the x direction is

$$\sum F_x = (600 \text{ N})(\cos 30°) + (600 \text{ N})(\cos 30°) - F = 0$$

$$F = 2(600 \text{ N})(\cos 30°) = \boxed{1.04 \times 10^3 \text{ N}}$$

Example 4.2   A Traffic Light at Rest

A traffic light weighing 100 N hangs from a cable tied to two other cables fastened to a support, as in Figure 4.8a. The upper cables make angles of 37° and 53° with the horizontal. Find the tension in each of the three cables.

Solution

Take note of the techniques indicated in the preceding Problem-Solving Strategy section as they are applied to this problem. First we construct a free-body diagram for the traffic light, as in Figure 4.8b. The tension in the vertical cable, T_3, supports the light, and so we see that $T_3 = w = 100$ N. Now we construct a free-body diagram for the knot that holds the three cables together, as in Figure 4.8c. This is a convenient point to choose because all forces in question act at this point. We choose the

Figure 4.8 (Example 4.1)
(a) A traffic light suspended by cables. (b) Free-body diagram for the traffic light. (c) Free-body diagram for the knot in the cable.

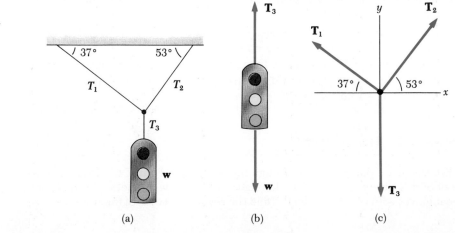

(a)   (b)   (c)

coordinate axes as shown in Figure 4.8c and resolve the forces into their x and y components:

Force	x Component	y Component
$\mathbf{T}_1$	$-T_1 \cos 37°$	$T_1 \sin 37°$
$\mathbf{T}_2$	$T_2 \cos 53°$	$T_2 \sin 53°$
$\mathbf{T}_3$	0	-100 N

The first condition for equilibrium gives us the equations

(1) $\qquad \sum F_x = T_2 \cos 53° - T_1 \cos 37° = 0$

(2) $\qquad \sum F_y = T_1 \sin 37° + T_2 \sin 53° - 100 \text{ N} = 0$

From (1) we see that the horizontal components of $\mathbf{T}_1$ and $\mathbf{T}_2$ must be equal in magnitude, and from (2) we see that the sum of the vertical components of $\mathbf{T}_1$ and $\mathbf{T}_2$ must balance the weight of the light. We can solve (1) for T_2 in terms of T_1 to give

$$T_2 = T_1 \left(\frac{\cos 37°}{\cos 53°} \right) = T_1 \left(\frac{0.800}{0.602} \right) = 1.33 T_1$$

This value for T_2 can be substituted into (2) to give

$$T_1 \sin 37° + (1.33 T_1)(\sin 53°) - 100 \text{ N} = 0$$

$$T_1[(0.602) + 1.33(0.800)] = 100 \text{ N}$$

$$T_1 = \boxed{60.0 \text{ N}}$$

$$T_2 = 1.33 T_1 = 1.33(60 \text{ N}) = \boxed{79.8 \text{ N}}$$

Example 4.3   Sled on a Slick Hill

A child holds a sled at rest on a frictionless snow-covered hill, as shown in Figure 4.9a. If the sled weighs 100 N, find the force the child must exert on the rope and the force the hill exerts on the sled.

(a)   (b)

Figure 4.9 (Example 4.3) (a) A child holding a sled on a frictionless hill. (b) Free-body diagram for the sled.

Solution

Figure 4.9b shows the forces acting on the sled and a convenient coordinate system to use for this type of problem. Note that **n**, the force the ground exerts on the sled, is perpendicular to the hill. The hill can exert a component of force along the incline only if there is friction between the sled and the hill.

Applying the first condition for equilibrium to the sled, we find

$$\sum F_x = T - (100 \text{ N})(\sin 40°) = 0$$

$$T = \boxed{64.3 \text{ N}}$$

$$\sum F_y = n - (100 \text{ N})(\cos 40°) = 0$$

$$n = \boxed{76.6 \text{ N}}$$

Note that n is *less* than the weight of the sled in this case because the sled is on an incline and **n** is equal to and opposite the component of weight perpendicular to the incline.

Exercise   What happens to the normal force as the angle of incline increases?

Answer   It decreases.

Exercise   Under what conditions would the normal force equal the weight of the sled?

Answer   If the sled were on a horizontal surface and the applied force were either zero or along the horizontal.

Accelerating Objects and Newton's Second Law

In those situations in which there is a net force acting on an object, the object will be accelerated, and we will have to use Newton's second law in order to determine the various features of its motion. The representative problems that follow, along with the suggestions below, should be helpful to you in indicating the approach to solving problems of this kind.

Problem-Solving Strategy: Newton's Second Law

The following procedure is recommended when dealing with problems involving the application of Newton's law:

1. Draw a simple, neat diagram of the system.
2. Isolate the object of interest whose motion is being analyzed. Draw a free-body diagram for this object, that is, a diagram showing *all external forces acting on the object.* For systems containing more than one object, draw *separate* diagrams for each object. Do not include forces that the object exerts on its surroundings.
3. Establish convenient coordinate axes for each body and find the components of the forces along these axes. Now, apply Newton's second law, $\Sigma \mathbf{F} = m\mathbf{a}$ in the x and y directions for each object under consideration.

4. Solve the component equations for the unknowns. Remember that you must have as many independent equations as you have unknowns in order to obtain a complete solution.
5. Often in solving such problems, one must also use the equations of kinematics (motion with constant acceleration) to find all the unknowns.

Example 4.4 A Boy and His Wagon

A boy pulls with a force of 20 N on the 300-N wagon of Figure 4.10. What acceleration does the wagon have, and how far will it move in 2 s? Assume that the wagon starts from rest and assume there are no frictional forces opposing the motion of the wagon.

Solution
In order to apply Newton's second law to the wagon, we must first know its mass:

$$m = \frac{w}{g} = \frac{300\ \text{N}}{9.80\ \text{m/s}^2} = 30.6\ \text{kg}$$

The acceleration can now be found from the second law:

$$a_x = \frac{F_x}{m} = \frac{20\ \text{N}}{30.6\ \text{kg}} = 0.654\ \text{m/s}^2$$

Since the acceleration of the wagon is constant, we can find the distance it moves in 2 s using the relation $x = v_0t + \frac{1}{2}at^2$ with $v_0 = 0$:

$$x = \frac{1}{2}\,at^2 = \frac{1}{2}\,(0.654\ \text{m/s}^2)(2\ \text{s})^2 = \boxed{1.31\ \text{m}}$$

It is important to note that the constant applied force of 20 N is assumed to act on the wagon all during its motion. If the force is removed at some instant, the wagon would continue to move with constant velocity and hence zero acceleration.

Figure 4.10 (Example 4.4)

Example 4.5 Getting a Free Ride

An airplane of mass 1.5×10^4 kg tows a glider of mass 0.5×10^4 kg, as in Figure 4.11a. If the propellers produce a forward force of 7.5×10^4 N, what is the tension in the connecting cable (neglecting air resistance)?

Solution
Problems involving connected bodies are usually most easily handled by considering the objects separately. In the present example, if we first focus our attention on the airplane, we find that two horizontal forces act on it, as shown in Figure 4.11b: **F**, the forward force of the propellers, and **T**, the tension in the connecting cable. Let us apply Newton's second law to the airplane, using m_p to denote its mass:

$$F - T = m_p a_x$$

$$(1) \qquad (7.5 \times 10^4\ \text{N}) - T = (1.5 \times 10^4\ \text{kg})a_x$$

With two unknowns and only a single equation, we must seek additional information in the form of another independent equation. This can be obtained by applying

(a)

(b)

(c)

Figure 4.11 (Example 4.5) (a) A glider of mass m_g being towed by an airplane of mass m_p. (b) Horizontal forces on the airplane. (c) Horizontal forces on the glider. The vertical forces (not shown) on both airplane and glider are the weight (downward) and the net lift (upward). These vertical forces balance each other.

Newton's second law to the glider, as shown in Figure 4.11c. Since the tension, **T**, in the cable is the only horizontal force acting on the glider, we get

$$(2) \qquad T = m_g a_x = (0.5 \times 10^4 \text{ kg}) a_x$$

The equations for the plane and glider form a set of simultaneous equations that can be solved for T and a_x. Let us substitute the expression given in (2) for T into (1) to get

$$(7.5 \times 10^4 \text{ N}) - (0.5 \times 10^4 \text{ kg}) a_x = (1.5 \times 10^4 \text{ kg}) a_x$$

When the terms in a_x are combined, we get

$$(1.5 \times 10^4 \text{ kg} + 0.5 \times 10^4 \text{ kg}) a_x = 7.5 \times 10^4 \text{ N}$$

$$a_x = \frac{7.5 \times 10^4 \text{ N}}{2.0 \times 10^4 \text{ kg}} = 3.75 \text{ m/s}^2$$

Substituting this value into (2) gives

$$T = (0.5 \times 10^4 \text{ kg})(3.75 \text{ m/s}^2) = \boxed{1.88 \times 10^4 \text{ N}}$$

Example 4.6   An Object on an Incline

Two objects of masses 10 kg and 5 kg are connected by a light string that passes over a light frictionless pulley, as in Figure 4.12a. The 5-kg object lies on a smooth incline of angle $\theta = 40°$. Find the acceleration of the two objects and the tension in the string.

Solution

Since the two objects are connected by a string (which we assume doesn't stretch), they must have the same acceleration. The free-body diagrams for the two are shown in Figures 4.12b and 4.12c. Newton's second law applied to the 10-kg object in component form gives

$$(1) \qquad \sum F_x = 0$$
$$(2) \qquad \sum F_y = T - (10 \text{ kg}) g = (10 \text{ kg}) a$$

Note that a is positive (upward) only if $T > 10g$.

In describing the motion of the 5-kg object, we shall choose the positive x axis along the incline, as in Figure 4.12c. (We can always choose different coordinate systems for different objects.) Newton's second law applied to this object in component form gives

$$(3) \qquad \sum F_x = (5 \text{ kg}) g \sin 40° - T = (5 \text{ kg}) a$$
$$(4) \qquad \sum F_y = n - (5 \text{ kg}) g \cos 40° = 0$$

Figure 4.12   (Example 4.6) (a) Two objects connected by a light string passing over a frictionless pulley. (b) Free-body diagram for the 10-kg object. (c) Free-body diagram for the 5-kg object.

Equations (1) and (4) give no information on the acceleration. However, we can solve (2) and (3) simultaneously for a and T. By adding (2) and (3), we get

$$5g \sin 40° - 10g = 15a$$

$$a = \frac{(5 \text{ kg})(9.80 \text{ m/s}^2)(0.643) - (10 \text{ kg})(9.80 \text{ m/s}^2)}{15 \text{ kg}}$$

$$= -4.43 \text{ m/s}^2$$

The negative sign for a simply means that the acceleration is in the negative x direction. That is, the 10-kg object accelerates *downward*, whereas the 5-kg one accelerates *up* the incline.

Now that we know the acceleration, we can use Equation (2) to determine that $T = 53.7$ N.

4.7 Forces of Friction

When a body is in motion on a rough surface or moves through a viscous medium such as air or water, there is resistance to the motion because of the interaction between the body and its surroundings. We call such resistance a **force of friction**. Forces of friction are very important in our everyday lives. They allow us to walk or run and are necessary for the motion of wheeled vehicles.

Consider a block on a table, as in Figure 4.13a. If we apply an external horizontal force **F** to the block, acting to the right, the block will remain

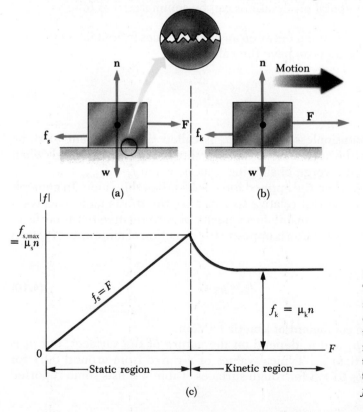

(a)   (b)

(c)

Figure 4.13
The force of friction, **f**, between a block and a rough surface is opposite the applied force, **F**. (a) The force of static friction equals the applied force. (b) When the magnitude of the applied force exceeds the maximum static friction force, $f_{s,max}$, the block accelerates to the right. (c) A graph of the applied force versus the magnitude of the frictional force. Note that $f_{s,max} > f_k$.

stationary if **F** is not too large. The force that keeps the block from moving acts to the left and is called the **frictional force, f**. As long as the block is not moving, $f = F$. Since the block is stationary, we call this frictional force the **force of static friction, f_s**. Thus, if F is increased, the magnitude of the frictional force f also increases. Likewise, if F decreases, f also decreases. Experiments show that this force arises from the nature of the two surfaces; because of the roughness of the surfaces, contact is made only at a few points, as shown in the "magnified" view of the surfaces in Figure 4.13a. Actually, the frictional force is much more complicated than presented here since it ultimately involves forces between atoms or molecules where the surfaces are in contact.

If we increase the magnitude of **F** enough, as in Figure 4.13b, the block will eventually slip. When the block is on the verge of slipping, f_s is a maximum. When F exceeds $f_{s,max}$, the block moves and accelerates to the right. When the block is in motion, the retarding frictional force becomes less than $f_{s,max}$ (Fig. 4.13c). We call the retarding force for an object in motion the **force of kinetic friction, f_k**. The unbalanced force in the x direction, $\mathbf{F} - \mathbf{f_k}$, produces an acceleration to the right. If $F = f_k$, the block moves to the right with constant speed. If the applied force is removed, then the frictional force acting to the left decelerates the block and eventually brings it to rest.

The force of kinetic friction is less than $f_{s,max}$ for the following reason. When the object is stationary, the contact points between the object and the surface are said to be cold-welded. While the object is in motion, these small welds can no longer form and the frictional force decreases.

Experimentally, one finds that both f_s and f_k are proportional to the normal force acting on the block and depend on the nature of the two surfaces in contact. The experimental observations can be summarized as follows:

1. The force of static friction between any two surfaces in contact is opposite the applied force and can have the values

$$f_s \leq \mu_s n \qquad (4.9)$$

where the dimensionless constant μ_s is called the **coefficient of static friction** and n is the normal force. The equality in Equation 4.9 holds when the block is on the verge of slipping, that is, when $f_s = f_{s,max} \equiv \mu_s n$. The inequality holds when the applied force is less than this value. In general, when an object is at rest relative to a surface, the frictional force always acts in such a way as to maintain a velocity of zero relative to the surface.
2. The force of kinetic friction is opposite the direction of motion and is given by

$$f_k = \mu_k n \qquad (4.10)$$

where μ_k is the **coefficient of kinetic friction**.
3. The values of μ_k and μ_s depend on the nature of the surfaces, but μ_k is generally less than μ_s. Typical values of μ range from around 0.01 for smooth surfaces to 1.5 for rough surfaces. Table 4.2 lists some reported values.

Table 4.2
Coefficients of Friction[a]

	μ_s	μ_k
Steel on steel	0.74	0.57
Aluminum on steel	0.61	0.47
Copper on steel	0.53	0.36
Rubber on concrete	1.0	0.8
Wood on wood	0.25–0.5	0.2
Glass on glass	0.94	0.4
Waxed wood on wet snow	0.14	0.1
Waxed wood on dry snow	—	0.04
Metal on metal (lubricated)	0.15	0.06
Ice on ice	0.1	0.03
Teflon on Teflon	0.04	0.04
Synovial joints in humans	0.01	0.003

[a] All values are approximate.

Finally, the coefficients of friction are nearly independent of the area of contact between the surfaces. Although the coefficient of kinetic friction varies with speed, we shall neglect any such variations.

Example 4.7   Moving Into the Dormitory

It is the beginning of a new school term, and you need to move a box of books into your dormitory room. To do so, you attach a rope to the box and pull on it with a force of 90 N at an angle of 30°, as shown in Figure 4.14. The box of books has a mass of 20 kg, and the coefficient of kinetic friction between the bottom of the box and the hallway surface is 0.5. Find the acceleration of the box.

Solution
The box is not accelerating in the vertical direction, and so we find the normal force from $\Sigma F_y = 0$. The forces in the y direction are the weight, w, the normal force, n, and the vertical component of your 90-N pull. This vertical component has a magnitude equal to $(90 \text{ N})(\sin 30°)$. We find that

$$\Sigma F_y = n + (90 \text{ N})(\sin 30°) - (20 \text{ kg})(9.80 \text{ m/s}^2) = 0$$

$$n = 151 \text{ N}$$

Note that in this case the normal force is *not* equal to the weight of the box of books because the vertical component of your pull is helping to support some of the weight of the box.

Since the normal force is known, we can find the force of kinetic friction:

$$f_k = \mu_k n = (0.5)(151 \text{ N}) = 75.5 \text{ N} \qquad \text{(to the left)}$$

Finally, we determine the horizontal acceleration using Newton's second law:

$$\Sigma F_x = (90 \text{ N})(\cos 30°) - 75.5 \text{ N} = (20 \text{ kg})(a_x)$$

$$a_x = \;+0.122 \text{ m/s}^2$$

Exercise   If the initial speed of the box is zero, what is its speed after it has traveled 2 m? How long does it take you to pull it a distance of 2 m?

Answer   0.699 m/s, 5.73 s.

Figure 4.14   (Example 4.7) A box of books being pulled to the right at an angle of 30° with a rough horizontal surface.

Motion

n

f_k

mg

Figure 4.15 (Example 4.8) *After* the puck is given an initial velocity to the right, the external forces acting on it are its weight, *mg*, the normal force, **n**, and the force of kinetic friction, f_k.

Example 4.8   The Sliding Hockey Puck

A hockey puck is given an initial speed of 20 m/s on a frozen pond as in Figure 4.15. The puck remains on the ice and slides 120 m before coming to rest. Determine the coefficient of friction between the puck and the ice.

Solution
You can find the acceleration of the puck from $v^2 = v_0^2 + 2ax$, with the final speed, v equal to zero, the initial speed $v_0 = 20$ m/s, and the distance traveled $x = 120$ m.

$$v^2 = v_0^2 + 2ax$$

$$0 = (20 \text{ m/s})^2 + 2a(120 \text{ m})$$

$$a = -1.67 \text{ m/s}^2$$

The negative sign means that the acceleration is to the left, *opposite* the direction of the velocity.

The magnitude of the force of kinetic friction is found from $f_k = \mu_k n$, where n is found from $\Sigma F_y = 0$ as follows:

$$\sum F_y = n - w = 0$$

$$n = w = mg$$

Thus,

$$f_k = \mu_k n = \mu_k mg$$

Now we apply Newton's second law along the horizontal direction. We shall take the positive direction toward the right.

$$\sum F_x = -f_k = ma$$

$$-\mu_k mg = m(-1.67 \text{ m/s}^2)$$

$$\mu_k = \frac{1.67 \text{ m/s}^2}{9.80 \text{ m/s}^2} = \boxed{0.170}$$

Example 4.9   Connected Objects

Two objects are connected by a light string that passes over a frictionless pulley, as in Figure 4.16a. The coefficient of sliding friction between the 4-kg object and the surface is 0.30. Find the acceleration of the two objects and the tension in the string.

Solution
First, let us isolate each object in Figure 4.16a and determine the external forces on each. Newton's second law applied to the 4-kg object in component form with the positive x direction to the right gives

$$\sum F_x = T - f = (4 \text{ kg})(a)$$

$$\sum F_y = n - (4 \text{ kg})(g) = 0$$

Since $f = \mu_k n$ and $n = 4g = (4 \text{ kg})(9.80 \text{ m/s}^2) = 39.2$ N, we have $f = (0.30)(39.2 \text{ N}) = 11.8$ N. Therefore,

$$(1) \qquad T = f + (4 \text{ kg})(a) = 11.8 \text{ N} + (4 \text{ kg})(a)$$

Now we apply Newton's second law to the 7-kg object moving in the vertical direction, where the downward direction is selected as the positive direction:

(a)

(b)

Figure 4.16 (Example 4.9) (a) Two objects connected by a light string that passes over a frictionless pulley. The surface is rough. (b) Free-body diagrams for the objects.

$$\sum F_y = (7 \text{ kg})(g) - T = (7 \text{ kg})(a)$$

or

(2) $T = 68.6 \text{ N} - (7 \text{ kg})(a)$

Subtracting (1) from (2) eliminates T:

$$56.8 \text{ N} - (11 \text{ kg})(a) = 0$$

$$a = \boxed{5.16 \text{ m/s}^2}$$

When this value for the acceleration is substituted into (1), we get

$$T = \boxed{32.5 \text{ N}}$$

Example 4.10   Experimental Determination of μ_s and μ_k

In this example we describe a simple method of measuring the coefficients of friction between an object and a rough surface. Suppose the object is a small block placed on a surface inclined with respect to the horizontal, as in Figure 4.17. The angle of the inclined plane is increased until the block slips. By measuring the critical angle, θ_c, at which this slipping occurs, we obtain μ_s directly.

Figure 4.17 (Example 4.10) The external forces acting on a block on a rough incline are the weight of the block, $m\mathbf{g}$, the normal force, $\mathbf{n}$, and the force of friction, $\mathbf{f}$. Note that the weight vector is resolved into a component along the incline, $mg \sin \theta$, and a component perpendicular to the incline, $mg \cos \theta$.

Block in motion

We note that the only forces acting on the block are its weight, $m\mathbf{g}$, the normal force, $\mathbf{n}$, and the force of static friction, $\mathbf{f}_s$. With x parallel to the plane and y perpendicular to the plane, Newton's second law applied to the block gives

$$(1) \qquad \sum F_x = mg \sin \theta - f_s = 0$$

$$(2) \qquad \sum F_y = n - mg \cos \theta = 0$$

We can eliminate mg by solving (2) for mg to get $mg = n/\cos \theta$ and substituting this value into (1) to get

$$(3) \qquad f_s = mg \sin \theta = \left(\frac{n}{\cos \theta} \right) (\sin \theta) = n \tan \theta$$

When the inclined plane is at the critical angle, $f_s = f_{s,\text{max}} = \mu_s n$, and so at this angle, (3) becomes

$$\mu_s n = n \tan \theta_c$$

$$\mu_s = \tan \theta_c$$

For example, if we find that the block just slips at $\theta_c = 20°$, then the value of μ_s is given by $\mu_s = \tan 20° = 0.364$. Once the block starts to move at $\theta \geq \theta_c$, it will accelerate down the incline and the force of friction is $f_k = \mu_k n$. However, once the block is in motion, and θ is reduced to a value less than θ_c, an angle θ_c' can be found such that the block moves down the incline with constant speed ($a_x = 0$). In this case, (1) and (2) with f_s replaced by f_k give

$$\mu_k = \tan \theta_c'$$

where $\theta_c' < \theta_c$.

You should try this simple experiment using a coin as the block and a book as the inclined plane. Also, you can try taping two coins together to prove that you still get the same critical angles as with one coin.

Summary

Newton's first law states that an object at rest will remain at rest and an object in motion will continue in motion with a constant velocity unless it experiences a net force.

An **object in equilibrium** has no net force acting on it, and the first law, in component form, is $\Sigma F_x = 0$ and $\Sigma F_y = 0$.

The resistance of an object to a change in its state of motion is called **inertia.** Mass is a physical quantity which is a measure of inertia.

Newton's second law states that the resultant force acting on an object is equal to the product of the mass of the object and its acceleration.

$$\sum \mathbf{F} = m\mathbf{a} \qquad \qquad (4.1)$$

The **weight** of an object is equal to the product of its mass and the acceleration due to gravity:

$$\mathbf{w} = m\mathbf{g} \qquad \qquad (4.6)$$

Newton's **third law** states that, if two bodies exert force on each other, the force of body 1 on body 2 is equal in magnitude and opposite in direction to the force of body 2 on body 1. Thus, an isolated force can never occur in nature.

The maximum force of static friction, $\mathbf{f}_{s,max}$, between an object and a rough surface is proportional to the normal force acting on the object. This maximum force occurs when the object is on the verge of slipping. In general,

$$f_s \le \mu_s n \tag{4.9}$$

where μ_s is the **coefficient of static friction.** When a body slides over a rough surface, the force of kinetic friction, $\mathbf{f}_k$, is opposite the motion and is also proportional to the normal force. The magnitude of this force is

$$f_k = \mu_k n \tag{4.10}$$

Here, μ_k is the **coefficient of kinetic friction.** In general, $\mu_k < \mu_s$.

Additional Reading

P. Brancazio, "The Physics of Kicking a Football," *The Physics Teacher*, October 1985, p. 403.

B. I. Cohen, "Isaac Newton," *Sci. American*, December 1955, p. 73.

S. Drake, "Galileo's Discovery of the Law of Free Fall," *Sci American*, May 1973, p. 17.

G. Gamow, "Gravity," *Sci. American*, March 1961, p. 94.

G. Gamow, *Gravity*, Science Study Series, Garden City, N.Y., Doubleday, 1962.

W. F. Magie, *Source Book in Physics*, Cambridge, MA, Harvard University Press, 1963. Excerpts from Newton on the laws of motion.

M. McCloskey, "Intuitive Physics," *Sci. American*, April 1983, p. 122.

M. McCloskey, A. Caramaza, and B. Gross, "Curvilinear Motion in the Absence of External Forces: Naive Beliefs About the Motion of Objects," *Science*, December 1980, p. 1139.

I. Newton, *Mathematical Principles of Natural Philosophy (Principia)*, translated by A. Motte, revised by F. Cajori, Berkeley, University of California Press, 1947.

F. Palmer, "Friction," *Sci. American*, February 1951, p. 54.

R. Zimmerer, "The Measurement of Mass," *The Physics Teacher*, September 1983, p. 354.

Questions

1. If an object is at rest, can we conclude that there are no external forces acting on it?
2. If gold were sold by weight, would you rather buy it in Denver or in Death Valley? If sold by mass, in which of the two locations would you prefer to buy it? Why?
3. A passenger sitting in the rear of a bus claims that he was injured when the driver slammed on the brakes, causing a suitcase to come flying toward him from the front of the bus. If you were the judge in this case, what disposition would you make? Why?
4. A space explorer is moving through space far from any planet or star. He notices a large rock, taken as a specimen from an alien planet, floating around the cabin of his ship. Should he push it gently toward a storage compartment or kick it toward the compartment? Why?
5. How much does an astronaut weigh out in space, far from any planets?
6. Although the frictional force between two surfaces may decrease as the surfaces are smoothed, the force

will again increase if the surfaces are made extremely smooth and flat. How do you explain this?

7. Why is it that the frictional force involved in the rolling of a body over another is less than for sliding motion?

8. In discussing friction, the forces exerted by a surface have been treated as though there are two separate forces, a normal force and a friction force. Is this really necessary? Explain.

9. A massive metal object on a rough metal surface may actually undergo contact welding to that surface. Discuss how this affects the frictional forces that arise between the object and the surface.

10. Analyze the motion of a rock dropped in water in terms of its speed and acceleration as it falls. Assume that there is a resistive force acting on the rock that increases as the velocity increases.

11. Identify the action-reaction pairs in the following situations: a man takes a step; a snowball hits a girl in the back; a baseball player catches a ball; a gust of wind strikes a window.

12. While a football is in flight, what forces act on it? What are the action-reaction pairs while the football is being kicked, and while it is in flight?

13. A ball is held in a person's hand. (a) Identify all the external forces acting on the ball and the reaction to each of these forces. (b) If the ball is dropped, what force is exerted on it while it is falling? Identify the reaction force in this case. (Neglect air resistance.)

14. Identify all the action-reaction pairs that exist for a horse pulling on a cart. Include the Earth in your examination.

15. If a car is traveling with a constant speed of 20 m/s in the westward direction, what is the resultant force acting on it?

16. A large crate is placed on the bed of a truck without being tied to the truck. (a) As the truck accelerates forward, the crate remains at rest relative to the truck. What force causes the crate to accelerate forward? (b) If the truck driver slams on his brakes, what could happen to the crate?

17. A child pulls a wagon with some force, causing it to accelerate. Newton's third law says that the wagon exerts an equal and opposite reaction force on the child. How can the wagon accelerate?

18. Explain why a rope climber must pull downward on the rope in order to move upward. Discuss the force exerted by the arms in relation to the weight of the person during the various stages of one "step" up the rope.

19. In a tug-of-war between two athletes, each pulls on the rope with a force of 200 N. What is the tension in the rope? What horizontal force does each athlete exert against the ground?

20. A rubber ball is dropped onto a floor. What force causes the ball to bounce back into the air?

21. What is wrong with the following statement. "Since the car is at rest, there are no forces acting on it." How would you restate this sentence to make it correct?

22. Suppose you are driving a car along a highway at a high speed. Why should you avoid "slamming on" your brakes if you want to stop in the shortest distance?

23. The driver of a speeding empty truck slams on the brakes and skids to a stop through a distance d. (a) If the truck carried a heavy load such that its mass were doubled, what would be its "skidding distance"? (b) If the initial speed of the truck is halved, what would be its "skidding distance"?

24. A child in a car holds a helium-filled balloon attached to a string. What happens to the balloon when the car accelerates forward?

25. Is it possible to have motion in the absence of a force? Explain.

26. What causes a rotary lawn sprinkler to turn?

27. A janitor finds it quite easy to sweep a floor using a long-handled broom when the angle between the handle and floor is small. However, if the angle is large, it becomes very difficult to push the broom. Explain why this occurs.

28. The force of gravity is twice as great on a 20-N rock as it is on a 10-N rock. Why doesn't the 20-N rock have a greater free-fall acceleration?

29. The Earth is attracted to each and every object on it with a force equal and opposite to the force the Earth exerts on the object. Explain why the Earth's acceleration is not equal and opposite to that of the object.

30. The ropes and strings used in the text examples are all tacitly assumed to be massless. Actually, however, they must have mass. Discuss the effect of the mass of the rope on a typical example of an acceleration problem involving a rope. Draw diagrams of the forces.

31. Can an object be in equilibrium if only one force acts on it?

32. It is often said that Newton's first law is really a special case of his second law. Explain that statement.

33. An object thrown into the air comes to rest at the highest point in its path. Is it in equilibrium at this point? Explain this carefully.

34. Is the Moon in equilibrium as it moves in an elliptical orbit about the Earth?

35. Under what conditions would a body falling through a viscous fluid be in equilibrium?

36. Can a body be in equilibrium if it is in motion? If so, give an example.

37. The mayor of a city decides to fire some city employees because they will not remove the sag from the cables that support the city traffic lights. If you were a lawyer, what defense would you give on behalf of the employees? Who do you think will win the case in court?

Problems

Section 4.2 through Section 4.5

1. What is the weight in newtons of a 2-kg cannonball? What is its weight in pounds?

2. (a) A car with a mass of 850 kg is moving to the right with a constant speed of 1.44 m/s. What is the force on the car? (b) What is the force on the car if it is moving to the left?

3. If a man weighs 900 N on Earth, what would he weigh on Jupiter, where the acceleration of gravity is 25.9 m/s²?

4. A bag of sugar weighs 5 lb on Earth. What should it weigh in newtons on the Moon where the acceleration due to gravity is 1/6 that on Earth? Repeat for Jupiter where g is 2.64 times Earth gravity. Find the mass in kilograms at all three locations.

5. A person weighs 120 lb. Determine (a) her weight in newtons and (b) her mass in kilograms.

6. (a) Draw a free-body diagram indicating all the forces acting on a freely falling baseball. For each force drawn, indicate the reaction force to it. (b) Repeat part (a) for the baseball while moving as a projectile toward an outfielder.

7. A 6-kg object undergoes an acceleration of 2 m/s². (a) What is the magnitude of the resultant force acting on it? (b) If this same force is applied to a 4-kg object, what acceleration will be produced?

8. A box weighing 20 N is to be pushed across a smooth floor with a horizontal force of 5 N. What acceleration will be produced?

9. A football punter accelerates a football from rest to a speed of 10 m/s during the time in which his toe is in contact with the ball (about 0.2 s). If the football has a mass of 0.5 kg, what average force does the punter exert on the ball?

10. The engine on a 0.2-kg model airplane exerts a forward force on the plane of 10 N. If the plane accelerates at 2 m/s², what is the magnitude of the resistive force due to the wind acting on the airplane?

11. A 5.0-g bullet leaves the muzzle of a rifle with a speed of 320 m/s. What force (assumed constant) is exerted on the bullet while it is traveling down the 0.82-m long barrel of the rifle?

12. The force of the wind on the sails of a sailboat is 390 N north. The water exerts a force of 180 N east. If the boat including crew has a mass of 270 kg, what is the magnitude and direction of the acceleration of the sailboat?

13. Two forces are applied to a car in an effort to move it, as shown in Figure 4.18. (a) What is the resultant of these two forces? (b) If the car has a mass of 3000 kg, what acceleration does it have?

Figure 4.18 Problem 13

14. A boat moves through the water with two forces acting on it. One is a 2000-N forward push by the motor, while the second is an 1800-N resistive force due to the water. (a) What is the acceleration of the 1000-kg boat? (b) If it starts from rest, how far will it move in 10 s? (c) What will be its velocity at the end of this time?

Section 4.6   Some Applications of Newton's Laws

Newton's First Law and the First Condition of Equilibrium

15. It is amateur night at the tightrope walkers' convention, and a 600-N performer finds himself in the awkward position shown in Figure 4.19. If the angle between the rope and the horizontal is 8°, find the tension in the rope on either side of the performer.

Figure 4.19 Problem 15

16. Two horses are pulling a sled along an icy road; assume that friction between the sled and the road is negligible. A farmer keeps the sled along the proper

path by pulling on it with a rope attached at the point where the cables from the horses are attached. A noise startles the horses and they move toward opposite sides of the road as shown in Figure 4.20. What are the magnitude and direction of the force that the farmer will have to exert on the sled to keep it moving along the road at constant velocity?

Figure 4.20   Problem 16

17. Find the tension in the two wires that support the 100-N light fixture shown in Figure 4.21.

Figure 4.21   Problem 17

18. A 150-N bird feeder is supported by three cables as shown in Figure 4.22. Find the tension in each cable.

Figure 4.22   Problem 18

19. Assuming the hand grips the rope tightly in Figure 4.23, determine the force of the hand on the cord.

Figure 4.23   Problem 19

20. The leg in the cast in Figure 4.24 weighs 220 N (w_1). Determine the weight w_2 and the angle α in order that there be no force exerted on the hip joint J by the leg in the cast.

Figure 4.24   Problem 20

Newton's Second Law

21. A shopper in a supermarket pushes a loaded cart with a horizontal force of 10 N. If the cart has a mass of 30 kg, how far will it move in 3 s, starting from rest? (Ignore friction.)
22. How far will the cart in Problem 21 move in the 3-s interval if the shopper places her 30-N child in the cart before she begins to push it?
23. A 0.15-kg baseball moving at 20 m/s strikes the glove of a catcher. The glove recoils a distance of 8 cm. What is the average force applied on the ball by the glove? What is the average force applied on the glove by the ball?
24. A train has a mass of 5.22×10^6 kg and is moving with a speed of 90.0 km/h. The engineer applies the brakes, which results in a net backward force of 1.87×10^6 N on the train. The brakes are held on for 30.0 s. (a) What is the new speed of the train? (b) How far does the train travel during this period?
25. A roller coaster starts its descent with an initial speed of 4 m/s. It moves through a distance of 135 ft along

an incline that makes an angle of 40° with the horizontal. Neglect friction and find its speed at the bottom of the incline.

26. A 2-kg mass starts from rest and slides down an inclined plane 80 cm long in 0.5 s. What is the *net force* along the incline acting on the mass?

27. A 40.0-kg wagon is towed up a hill which is inclined at 18.5° with respect to the horizontal. The tow rope is parallel to the incline and has a tension of 140 N in it. Assume that the wagon starts from rest at the bottom of the hill and neglect friction. How fast is the wagon going after moving 80.0 m up the hill?

28. A shopper pushes a 40-kg cart up a 20° smooth ramp. What force directed along the incline must the shopper exert to give the cart an acceleration of 0.5 m/s²?

29. The parachute on a race car of weight 8820 N opens at the end of a quarter-mile run when the car is traveling at 35 m/s. What total retarding force must be supplied by the parachute to stop the car in a distance of 1000 m?

30. A 5-kg bucket of water is raised from a well by a rope attached to the bucket. If the upward acceleration of the bucket is 3 m/s², find the force exerted on the bucket by the rope.

31. Two people pull as hard as they can on ropes attached to a boat that has a mass of 200 kg. If they pull in the same direction, the boat has an acceleration of 1.52 m/s² to the right. If they pull in opposite directions, the boat has an acceleration of 0.518 m/s² to the left. What is the force exerted by each person on the boat? (Disregard any other forces on the boat.)

32. On takeoff, the combined action of the engines and the wings of an airplane exerts a force of 8000 N on the plane upward at an angle of 65.0° above the horizontal. The plane rises with constant velocity in the vertical direction while continuing to accelerate in the horizontal direction. (a) What is the weight of the plane? (b) What is the horizontal acceleration of the plane?

33. A 8000-kg tow truck pulls a 3000-kg car horizontally into a garage. If the motor of the tow truck exerts a forward thrust on the truck of 800 N, what is the tension in the cable connecting the truck to the car? (Neglect any forces of friction.)

34. A car with a mass of 1000 kg is pulling a trailer with a mass of 300 kg. Together, the car and trailer have an acceleration of 2.15 m/s² in the forward direction. Neglecting frictional forces on the trailer, determine (a) the net force on the car; (b) the net force on the trailer; (c) the force of the trailer on the car; (d) the force of the car on the road.

35. Two blocks are fastened to the ceiling of an elevator as in Figure 4.25. The elevator accelerates upward at the rate of 2 m/s². Find the tension in each rope.

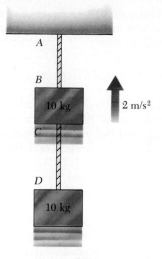

Figure 4.25   Problems 35 and 91

36. A 5-kg mass resting on a smooth horizontal table is connected to a cable that passes over a pulley and then is fastened to a hanging 10-kg mass as in Figure 4.26. Find the acceleration of the two objects and the tension in the string.

Figure 4.26   Problems 36 and 55

37. Two masses of 3 kg and 5 kg are connected by a light string that passes over a frictionless pulley, as in Figure 4.27. Determine (a) the tension in the string,

Figure 4.27   Problem 37

(b) the acceleration of each mass, and (c) the distance each mass will move in the first second of motion if they start from rest.

Section 4.7   Forces of Friction

38. (a) A 10-kg box of books rests on a horizontal surface. If the coefficient of static friction between box and floor is 0.4, find the maximum horizontal force that can be applied to the box before it slips. (b) The coefficient of kinetic friction is 20% less than that of static friction. If the force found in part (a) is exerted on the box once it is in motion, find the acceleration of the box.

39. A 1000-N bobsled leaves a horizontal track with a speed of 30 m/s. It then moves onto a horizontal track lightly sprinkled with sawdust and comes to a stop in a distance of 150 m. Find the coefficient of kinetic friction between the runners of the sled and the sawdust-covered surface.

40. A dockworker loading crates on a ship finds that a 20-kg crate, initially at rest on a rough horizontal surface, requires a 75-N horizontal force to set it into motion. However, after it is in motion, a horizontal force of 60 N is required to keep the crate moving with a constant speed. Find the coefficients of static and kinetic friction between the crate and the floor.

41. A box of books weighing 300 N is shoved across the floor of an apartment by exerting a force of 400 N on it downward at an angle of 35.2° below the horizontal. If the coefficient of kinetic friction between the box and the floor is 0.57, how long does it take to move the box 4.0 m starting from rest?

42. A man doing his spring cleaning pulls a 150-N vacuum cleaner across the floor at constant velocity by exerting a force on it at an angle of 30° with the horizontal. If he must pull with a force of 35 N to move the vacuum cleaner, what is the coefficient of friction between the vacuum cleaner and the floor?

43. A boy drags a 100-N sled up a 20° slope at constant velocity. If the coefficient of friction between sled and hill is 0.2, what force must he exert at an angle of 35° with respect to the hill?

44. A crate is held at rest on a frictionless 53° ramp by a rope parallel to the incline. If the tension in the rope is 7000 N, what is the mass of the crate?

45. The center board sandwiched between the other two boards in Figure 4.28 weighs 95.5 N. If the coeffi-

Figure 4.28   Problem 45

cient of friction between the boards is 0.663, what must be the magnitude of the compression forces (assume horizontal) acting on either side of the board in the center to keep it from slipping?

46. A person standing on crutches in Figure 4.29 weighs 170 lb. The crutches each make an angle of 22° with the vertical (as seen from the front). Half of the person's weight is supported by the crutches. The other half is supported by the vertical forces of the ground on the feet. Assuming the person is at rest and the force of the ground on the crutches acts along the crutches, determine (a) the coefficient of friction of the crutches on the ground and (b) the magnitude of the compression force supported by each crutch.

Figure 4.29   Problem 46

47. (a) A student decides to move a box of books into her dormitory room by pulling on a rope attached to the box with a force of 80 N at an angle of 25° with the horizontal. The box of books has a mass of 25 kg, and the coefficient of kinetic friction between box and floor is 0.3. Find the acceleration of the box. (b) Along the way, the student must move the box up a ramp inclined at 10° with the horizontal. If the box starts from rest at the bottom of the incline and is pulled at an angle of 25° with respect to the incline and with the same force, can it be moved up the incline? If so, what is the acceleration up the ramp?

48. A box slides down a 30° ramp with an acceleration of 1.2 m/s². Determine the coefficient of kinetic friction between the box and the ramp.

49. A car is traveling at 50 km/h on a flat highway. (a) If the coefficient of friction between the road and tires on a rainy day is 0.1, what is the minimum distance in which the car will stop? (b) What is the stopping distance when the surface is dry and the coefficient of friction is 0.6?

50. A racing car accelerates uniformly from 0 to 80 mph in 8 s. The external force that accelerates the car is the frictional force between the tires and the road. If the tires do not spin, determine the *minimum* coefficient of friction between the tires and the road.

51. A car with a speed of 40.0 km/h approaches the bottom of an icy hill. The hill has an angle of inclination of 10.5°. The driver applies the brakes, which makes the car skid up the hill. If the coefficient of kinetic friction between the ice and the tires is 0.153, how high, measured along the incline, is the car on the hill when it comes to rest?

52. A box weighing 5400 N is being pulled up an inclined plane that rises 1.3 m for every 7.5 m of length measured along the incline. If the coefficient of friction is 0.6, determine the force applied parallel to the incline necessary to move it up the incline at constant speed.

53. A furniture crate of mass 60 kg is at rest on a loading ramp that makes an angle of 25° with the horizontal. The coefficient of kinetic friction between the crate and the ramp is 0.2. What force, applied parallel to the incline, is required to push the crate up the incline at constant speed?

54. In order to determine the coefficients of friction between rubber and various surfaces, a student uses a rubber eraser and an inclined board. In one experiment, the eraser slips down the incline when the angle of inclination is 36°, and then moves down the incline with constant speed when the angle is reduced to 30°. From these data, determine the coefficients of static and kinetic friction for this experiment.

55. Find the acceleration experienced by the two masses shown in Figure 4.26, if the coefficient of friction between the 5-kg mass and the table is 0.25.

56. Find the acceleration experienced by the two masses shown in Figure 4.30 if the coefficient of friction between the 5-kg mass and the plane is 0.25.

Figure 4.30   Problem 56

57. A hockey puck is hit on a frozen lake and starts moving with a velocity of 12.0 m/s. Five seconds later, its velocity is 6.0 m/s. (a) What is the average acceleration of the puck? (b) What is the average value of the coefficient of friction between the puck and the ice? (c) How far does the puck travel during this 5-s interval?

58. A tractor is pulling two logs in tandem. The logs are fastened together by a chain. Each log weighs 1750 N, and the logs are moving with constant velocity. The tractor exerts a force of 1600 N on the front log, and this same log is acted on by a force of friction of 900 N. Find the tension in the chain between the logs and the force of friction on the back log.

59. A sled weighing 60 N is pulled horizontally across snow such that the coefficient of kinetic friction between sled and snow is 0.1. A penguin weighing 70 N rides on the sled. (See Fig. 4.31.) If the coefficient of static friction between penguin and sled is 0.7, find the maximum horizontal force that can be exerted on the sled before the penguin begins to slide off.

Figure 4.31   Problem 59

Additional Problems

60. A man attempting to train his dog is shown in Figure 4.32 pulling on him with a force of 70 N at an angle of 30° to the horizontal. Find the *x* and *y* components of this force.

Figure 4.32   Problem 60

61. Four forces act on a boat shown from a top view in Figure 4.33. Find the magnitude and direction of the resultant force on the boat.

Figure 4.33   Problem 61

62. (a) What is the resultant force exerted by the two cables supporting the traffic light shown in Figure 4.34? (b) What is the weight of the light?

45°   45°

60 N   60 N

Figure 4.34   Problem 62

63. The graph in Figure 4.35 shows the speed of a person's body while doing a chin-up. Assuming the motion is vertical and the mass of the person (excluding the arms) is 64.0 kg, determine the magnitude of the force exerted on the body by the arms at various stages of the motion.

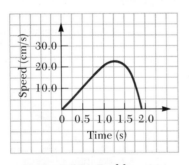

Figure 4.35   Problem 63

64. Some baseball pitchers are capable of throwing a fastball at 100 mi/h. The pitcher achieves this speed by moving his arm through a distance of about 1.5 m. What average force must he exert on the 0.15-kg ball during this time?

65. A baseball pitcher throws a ball straight up into the air at a speed of 20 m/s as a protest against the umpire's calls. In the process, he moves his hand through a distance of 1.5 m. If the ball has a mass of 0.15 kg, find the force he exerts on the ball to give it this upward speed. Note that the force of gravity is acting against the motion of the ball.

66. A car is at rest at the top of a driveway that has a slope of 20°. If the brake of the car is released, find (a) the acceleration of the car down the drive and (b) the time it takes for the car to reach the street 10 m away.

67. A 2000-kg car is slowed down uniformly from a speed of 20 m/s to 5 m/s in a time of 4 s. What average force acted on the car during this time and how far did the car travel during the deceleration?

68. A girl coasts down a hill on a sled, reaching the bottom of the hill with a speed of 7 m/s. If the coefficient of friction between the steel runners and the snow is 0.05 and the girl and sled together weigh 600 N, how far does the sled travel on the level surface before coming to rest?

69. An elevator accelerates upward at 1.5 m/s². If the elevator has a mass of 200 kg, find the tension in the supporting cable.

70. Two barrels, of masses 4 kg and 7 kg, are pulled across the frictionless surface of a frozen pond by an ice fisherman. If he exerts a force of 30 N on the first barrel as shown in Figure 4.36, determine the acceleration of the system and the tension in the cord connecting the barrels.

4 kg   7 kg

30 N

Figure 4.36   Problem 70

71. A student slides a 2000-N crate across a floor by pulling with a force of 750 N at an angle of 30° with the horizontal. If the crate moves at a constant speed, find the coefficient of sliding friction between the crate and the floor.

72. A 1200-N go-cart is being pulled up a 25° incline by a rope that makes an angle of 35° with the horizontal. Neglecting all frictional effects, determine the tension in the rope necessary to pull the cart up the incline at constant speed.

73. A tightrope walker weighing 600 N stands at the center of a rope. (a) If the ends of the rope make an angle of 10° with the horizontal, what is the tension in the rope? (b) If the tightrope can withstand a tension of 2000 N without breaking, what is the minimum angle the rope can make with the horizontal without breaking?

74. A solid sphere of radius R and mass M is placed in a wedge as shown in Figure 4.37. The inner surfaces of the wedge are frictionless. Determine the forces exerted by the wedge on the sphere at the two contact points.

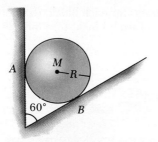

Figure 4.37   Problem 74

75. Find the tension in each cable supporting the 600-N cat burglar in Figure 4.38.

Figure 4.38   Problem 75

76. (a) What is the minimum force of friction required to hold the system of Figure 4.39 in equilibrium? (b) What coefficient of static friction between the 100-N block and the table will ensure equilibrium? (c) If the coefficient of kinetic friction between the 100-N block and the table is 0.25, what hanging weight should replace the 50-N weight to allow the system to move at a constant speed when set in motion?

Figure 4.39   Problem 76

77. A box rests on the back of a truck. The coefficient of friction between the box and the surface is 0.3. (a) When the truck accelerates forward, what force

accelerates the box? (b) Find the maximum acceleration the truck can have before the box slides.

78. A 10,000-N force is exerted on a car by a cable pulling the car up a loading ramp which is inclined at 30° with the horizontal. If the car has a mass of 1500 kg, determine the acceleration of the system. (Ignore any effects of friction.)

79. The car of Problem 78 is pulled up the same ramp, but this time the pull is supplied by a 10,000-N counterweight passing over a pulley as shown in Figure 4.40. Find (a) the tension in the cable, and (b) the acceleration of the system. (c) What mass should the counterweight have in order for the car to move down the incline at an acceleration of 2 m/s²? (Ignore any effects of friction.)

Figure 4.40   Problem 79

80. A 3-kg block starts from rest at the top of a 30° incline and slides a distance of 2 m down the incline in 1.5 s. Find (a) the acceleration of the block, (b) the coefficient of kinetic friction between the block and the plane, (c) the frictional force acting on the block, and (d) the speed of the block after it has slid 2 m.

81. A 72-kg man stands on a spring scale in an elevator. Starting from rest, the elevator ascends, attaining its maximum velocity of 1.2 m/s in 0.8 s. It travels with this constant velocity for the next 5.0 s. The elevator then undergoes a uniform *negative* acceleration for 1.5 s and comes to rest. What does the spring scale register (a) before the elevator starts to move? (b) during the first 0.8 s? (c) while the elevator is traveling at constant velocity? (d) during the negative acceleration period?

82. A 3-kg mass hangs at one end of a rope that is attached to a support on a railroad car. When the car accelerates to the right, the cord makes an angle of 4° with the vertical as shown in Figure 4.41. Find the acceleration of the car.

Figure 4.41   Problem 82

83. A bowling ball attached to a spring scale is suspended from the ceiling of an elevator. The scale reads 70 N when the elevator is at rest. (a) What will the scale read when the elevator accelerates upward at the rate of 3 m/s²? (b) What will the scale read when the elevator accelerates downward at the rate of 3 m/s²?

84. Two blocks on a frictionless horizontal surface are connected by a light string, as in Figure 4.42, where $m_1 = 10$ kg and $m_2 = 20$ kg. A force of 50 N is applied to the 20-kg block. (a) Determine the acceleration of each block and the tension in the connecting cable, assuming the surface is frictionless. (b) Repeat the problem, assuming the coefficient of kinetic friction between each block and the surface is 0.1.

Figure 4.42   Problem 84

85. A 5-kg penguin sits on top of a 10-kg sled, as in Figure 4.43. A horizontal force of 45 N is applied to the sled, but the penguin attempts to impede the motion by holding onto a cord attached to a wall. The coefficient of kinetic friction between the moving surfaces is 0.2. (a) Draw force diagrams for the penguin and the sled, and identify the reaction force for each force you include. (b) Determine the tension in the string held by the penguin and the acceleration of the sled.

Figure 4.43   Problem 85

86. A book of mass 0.9 kg is projected up a 30° ramp with an initial speed of 3 m/s. The coefficient of a kinetic friction between the ramp and the book is 0.2. Determine (a) how far up the incline the book will go before coming to rest, (b) its acceleration as it slides back down the incline, and (c) its speed when it reaches the bottom of the incline.

87. In Figure 4.44, the coefficient of kinetic friction between the 2-kg and 3-kg blocks is 0.3. The horizontal surface and the pulleys are frictionless. (a) Draw free-body diagrams for each block. (b) Determine the acceleration of each block. (c) Find the tension in the strings.

Figure 4.44   Problem 87

88. A horizontal force **F** is applied to a frictionless pulley of mass m_2 as in Figure 4.45. The horizontal surface is smooth. (a) Show that the acceleration of the block of mass m_1 is *twice* the acceleration of the pulley. Find (b) the acceleration of the pulley and the block and (c) the tension in the string. A constant supporting force is applied to the axle of the pulley equal to its weight.

Figure 4.45   Problem 88

89. Three blocks are in contact with each other on a frictionless, horizontal surface as in Figure 4.46. A horizontal force **F** is applied to m_1. If $m_1 = 2$ kg, $m_2 = 3$ kg, $m_3 = 4$ kg, and **F** = 18 N, find (a) the acceleration of the blocks, (b) the resultant force on each block, and (c) the magnitude of the contact forces between the blocks.

Figure 4.46   Problem 89

90. Repeat Problem 89 given that the coefficient of kinetic friction between the blocks and the surface is 0.1. Use the data given in Problem 89.

91. Two blocks are fastened to the top of an elevator as in Figure 4.25. The mass of each rope is 1 kg. The elevator accelerates upward at 4 m/s². Find the tensions in the ropes at (a) the ceiling of the elevator, (b) the point of contact of the top rope with the top block, (c) the point at which the lower rope attaches to the top block, and (d) the point of contact of the lower rope to the lower block.

92. The three blocks in Figure 4.47 are connected by light strings that pass over frictionless pulleys. The acceleration of the system is 2 m/s² to the left and the surfaces are rough. Find (a) the tension in each string and (b) the coefficient of kinetic friction between blocks and surface. (Assume the same μ for both blocks.)

93. Two blocks on a rough incline are connected by a light string that passes over a frictionless pulley as in Figure 4.48. Assuming $m_1 > m_2$ and taking the coefficients of kinetic friction for each block to be μ, determine expressions for (a) the acceleration of the blocks and (b) the tension in the string. (Assume the system is in motion.)

Figure 4.47   Problem 92

Figure 4.48   Problem 93

CHAPTER 5

Work and Energy

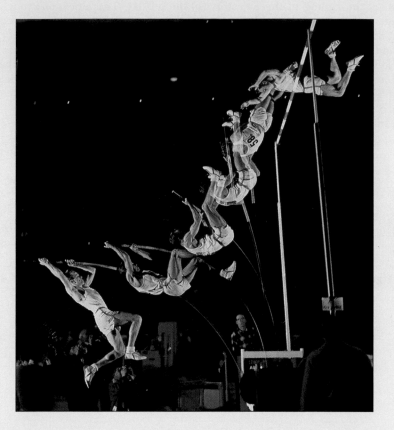

This multiflash photo of a pole-vaulter illustrates various forms of energy, including energy of motion (kinetic energy) and energy associated with position in space (potential energy). *(© Harold E. Edgerton. Courtesy of Palm Press, Inc.)*

The concept of energy is one of the most important in the world of science. In everyday usage, we think of energy in terms of the cost of fuel for transportation and heating, electricity for lights and appliances, and the foods we consume. However, these ideas do not really define energy. They tell us only that fuels are needed to do a job and that those fuels provide us with something we call energy.

Energy is present in various forms, including mechanical energy, chemical energy, electromagnetic energy, heat energy, and nuclear energy. The various forms are related to each other such that, when energy is transformed from one form to another, the total amount of energy remains the same. This is the point that makes the energy concept so useful. If an isolated system loses energy in some form, then, by the law of conservation of energy, the system must gain an equal amount of energy in other forms. For example, when an electric motor is connected to a battery, chemical energy is converted to electrical energy, which in turn is converted to mechanical energy. The transformation of energy from one form to another is an essential part of the study of physics, chemistry, biology, geology, and astronomy.

5.1   Introduction

In this chapter, we shall be concerned only with the mechanical forms of energy. We shall introduce the concept of *kinetic energy*, the energy associated with the motion of an object, and the concept of *potential energy*, the energy associated with the position of an object. We shall see that the ideas of work and energy can be used in place of Newton's laws to solve certain problems. In a complex situation, the "energy approach" can often provide a much simpler analysis than the direct application of Newton's second law.

We begin by defining *work*, a concept that provides a link between force and energy. With this as a foundation, we shall then be able to discuss the law of conservation of energy and apply it to various problems.

5.2   Work

Almost all of the terms we have used thus far in our study have conveyed the same meaning in physics as they do in everyday life. Terms such as *velocity*, *acceleration*, and *force* have been used in virtually the same context here as you would use them if talking to a friend. We now encounter a term whose meaning in physics is distinctly different from its meaning in our day-to-day affairs. This new term is **work**, which can be defined with the help of Figure 5.1. Here we see an object that undergoes a displacement **s** along a straight line while acted on by a constant force **F**, which makes an angle θ with **s**.

Figure 5.1
If an object undergoes a displacement **s**, the work done by the force **F** is $(F \cos \theta)s$.

The work done, W, by the constant force is defined as the product of the component of the force along the direction of displacement and the magnitude of the displacement:

$$W = (F \cos \theta)s \qquad (5.1)$$

As an example of the distinction between this definition of *work* and what we mean in our everyday usage of the word, consider holding a heavy chair at arm's length for 10 min. At the end of this time interval, your tired arms might lead you to express the opinion that you have done a considerable amount of work. However, according to our definition, no work has been done on the chair.[1] You have exerted a force in order to support the chair, but you have not moved it. A force does no work on an object if the object does not move. This can be seen by noting that if $s = 0$, Equation 5.1 gives $W = 0$. Also note from Equation 5.1 that the work done by a force is zero when the force is perpendicular to the displacement. That is, if $\theta = 90°$, then $W = 0$ because $\cos 90° = 0$. For example, no work is done when a bucket of water is carried horizontally at constant velocity because the upward force exerted to support the bucket is perpendicular to the displacement of the bucket, as shown in Figure 5.2. Likewise, the work done by the force of gravity is also zero for the same reason.

Figure 5.2
No work is done when a bucket of water is moved horizontally because the applied force **F** is perpendicular to the displacement.

---

[1] Actually, you do burn calories while holding a chair at arm's length. This is explained by the fact that the muscles are continually contracting and relaxing while the chair is being supported. Thus, work is being done on our body, but not on the chair.

(a)   (b)

Figure 5.3
(a) Positive work is done by the person when the box is lifted because the applied force **F** is in the same direction as the displacement. (b) Negative work is done by the frictional force, **f**, which is opposite the displacement.

Does the weight lifter do any work as he holds the weight over his head? (© *M. Brittan/Index Stock International*)

The sign of the work depends on the direction of **F** relative to **s**. The work done by an applied force is positive when the component $F \cos \theta$ is in the same direction as the displacement. For example, when you lift a box as in Figure 5.3a, the work done by the force you exert on the box is positive because the lifting force is upward, that is, in the same direction as the displacement. Work is negative when the component of the applied force is in the direction opposite the displacement. A common example of a situation in which work is always negative is the work done by a frictional force when a body slides over a rough surface. In Figure 5.3b, the work done by the frictional force on the sliding base runner is negative because the direction of the force is opposite the displacement. The negative sign comes from the fact that $\theta = 180°$ and $\cos 180° = -1$, which, from Equation 5.1, gives a negative value for W.

Finally, if an applied force acts along the direction of the displacement, then $\theta = 0°$. Since $\cos 0° = 1$, in this case Equation 5.1 becomes

$$W = Fs \qquad (5.2)$$

Work is a scalar quantity, and its units are force multiplied by length. Therefore, the SI unit of work is the **newton-meter** (N·m). Another name for the newton-meter is the **joule** (J). The unit of work in the cgs system is the **dyne-centimeter** (dyne·cm), which is also called the **erg,** and the unit in the conventional (British engineering) system is the **foot-pound** (ft·lb). These are summarized in Table 5.1. Note that $1 \text{ J} = 10^7$ ergs.

Table 5.1
Units of Work in the Three Common Systems of Measurement

System	Unit of Work	Name of Combined Unit
SI	Newton-meter (N·m)	Joule (J)
cgs	Dyne-centimeter (dyne·cm)	Erg
British engineering (conventional)	Foot-pound (ft·lb)	Foot-pound

Example 5.1   Mr. Clean

A man cleaning his apartment pulls a vacuum cleaner with a force of 50 N at an angle of 30°, as shown in Figure 5.4. A frictional force of magnitude 40 N retards the motion, and the vacuum is pulled a distance of 3 m. Calculate (a) the work done by the 50-N pull, (b) the work done by the frictional force, and (c) the net work done on the vacuum by all forces acting on it.

Solution
(a) We can use $W = (F \cos \theta)s$ with $F = 50$ N, $\theta = 30°$, and $s = 3$ m to get

$$W_F = (50 \text{ N})(\cos 30°)(3 \text{ m}) = \boxed{130 \text{ J}}$$

(b) In this case, we use $W_f = (f \cos \theta)s$ and take $f = 40$ N and $\theta = 180°$ to get

$$W_f = (40 \text{ N})(\cos 180°)(3 \text{ m}) = \boxed{-120 \text{ J}}$$

(c) The normal force, **n**, the weight $m\mathbf{g}$, and the upward component of the applied force, 50 sin 30°, do *no* work because they are perpendicular to the displacement. Thus, the net work done is

$$W_{net} = W_F + W_f = \boxed{10 \text{ J}}$$

Exercise   Find the net work done on the vacuum cleaner if the man pulls it horizontally with a force of 50 N, assuming the frictional force is 40 N.

Answer   30 J.

Figure 5.4   (Example 5.1) A vacuum cleaner being pulled at an angle of 30° with the horizontal.

5.3   Work and Kinetic Energy

Figure 5.5 shows an object of mass m moving to the right under the action of a constant net force **F**. Because the force is constant, we know from Newton's second law that the object will move with a constant acceleration **a**. If the object is displaced a distance s, the work done by the net force **F** is

$$W_{net} = Fs = mas \qquad \textbf{(5.3)}$$

However, in Chapter 2 we found that the following relationship holds when an object undergoes constant acceleration:

$$v^2 = v_0^2 + 2as \qquad \text{or} \qquad as = \frac{v^2 - v_0^2}{2}$$

This expression is now substituted into Equation 5.3 to give

$$W_{net} = m\left(\frac{v^2 - v_0^2}{2}\right)$$

$$\boxed{W_{net} = \frac{1}{2}mv^2 - \frac{1}{2}mv_0^2} \qquad \textbf{(5.4)}$$

Figure 5.5
An object undergoing a displacement and change in velocity under the action of a constant net force **F**.

The **kinetic energy,** *KE,* of an object of mass *m* and speed *v* is defined as

Kinetic energy

$$KE \equiv \frac{1}{2}\, mv^2 \qquad \text{(5.5)}$$

Kinetic energy is a scalar quantity and has the same units as work. For example, a 1-kg mass moving with a speed of 4.0 m/s has a kinetic energy of 8.0 J. We can think of kinetic energy as the energy associated with the motion of an object. It is often convenient to write Equation 5.4 as

Work-energy theorem

$$W_{net} = KE_f - KE_i \qquad \text{(5.6)}$$

This equation says that, when work is done on an object, the effect of the work done by the *net* force is to change the kinetic energy from some initial value KE_i to some final value KE_f. Equation 5.6 is an important result known as the **work-energy theorem.** Thus, we conclude that

> the net work done on an object by the force or forces acting on it is equal to the change in the kinetic energy of the object.

The work-energy theorem also says that the speed of the object will increase if the net work done on it is positive because the final kinetic energy will be greater than the initial kinetic energy. Its speed will decrease if the net work is negative because the final kinetic energy will be less than the initial kinetic energy. Notice that the speed and kinetic energy of an object will change only if work is done on the object by some external force or forces.

Consider the relationship between the work done on an object and the change in its kinetic energy as expressed by Equation 5.4. Because of this connection, we can also think of kinetic energy as the work the object can do in coming to rest. For example, suppose a hammer is on the verge of striking a nail as in Figure 5.6. The moving hammer has kinetic energy and is able to do work on the nail once it has struck it. The work done on the nail appears as the product *Fs,* where *F* is the average force exerted on the nail by the hammer and *s* is the distance it is driven into the wall.

For convenience, Equation 5.6 was derived under the assumption that the net force acting on the object was constant. A more general derivation would show that this equation is valid under all circumstances, including that of a variable force.

Example 5.2   Towing a Car

A 1400-kg car has a net forward force of 4500 N applied to it. The car starts from rest and travels down a horizontal highway. What are its kinetic energy and speed after it has traveled 100 m? (Ignore losses in energy due to friction, such as that caused by air resistance.)

Solution
With the initial velocity given as zero, Equation 5.4 reduces to

$$(1) \qquad W_{net} = \frac{1}{2}\, mv^2$$

Figure 5.6
The moving hammer has kinetic energy and thus is able to do work on the nail, driving it into the wall.

Handwritten notes in margin:

$W = \Delta KE$

- after striking nail hammer moves through distance $\mathfrak{1}s$

$\Delta KE = W = Fs$ (in coming to rest)

- F is force imparted to nail

The work done by the net force on the car is

$$W_{net} = Fs = (4500 \text{ N})(100 \text{ m}) = 4.50 \times 10^5 \text{ J}$$

This work has all gone into changing the kinetic energy of the car; thus the final value of the kinetic energy, from (1), is also 4.50×10^5 J.

The speed of the car can be found from (1) as follows:

$$\frac{1}{2} mv^2 = 4.50 \times 10^5 \text{ J}$$

$$v^2 = \frac{2(4.50 \times 10^5 \text{ J})}{1400 \text{ kg}} = 643 \text{ m}^2/\text{s}^2$$

$$v = \boxed{25.4 \text{ m/s}}$$

5.4 Potential Energy

In the previous section, we saw that an object with kinetic energy is able to do work on another object as illustrated by the moving hammer driving a nail into the wall. In this section, we shall see that an object can also do work on another object because of the energy it has due to its *position* in space.

Gravitational Potential Energy

In order to understand how an object can do work by virtue of its position in space, consider Figure 5.7 which shows a brick of mass m at a height h above a nail in a horizontal board lying on the ground. When the brick is released, it falls towards the ground, does work on the nail, and drives it into the plank.

> The energy that an object has due to its position in space is called **gravitational potential energy.**

Let us now derive an expression for the gravitational potential energy of an object at a given location in space. To do this consider a block of mass m at an initial height y_i above the ground as in Figure 5.8. Neglecting air resistance, as the block falls the only force that does work on it is the gravitational force, mg. The work done by the gravitational force as the block undergoes a downward displacement $s = y_i - y_f$ is given by the product of the downward force mg times the displacement, or

$$W_g = mgs = mgy_i - mgy_f$$

We now define the gravitational potential energy, PE, as

$$PE \equiv mgy \qquad (5.7)$$

If we substitute this expression for PE into the expression for W_g, we have

$$W_g = PE_i - PE_f \qquad (5.8)$$

Brick

Figure 5.7
The brick at the top can do work on the nail because of its position in space.

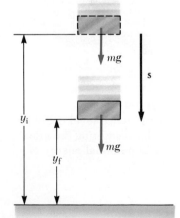

Figure 5.8
The work done by the gravitational force as the block falls from y_i to y_f is equal to $mgy_i - mgy_f$.

From this result we see that

> The work done by the force of gravity is equal to the initial value of the potential energy minus the final value of the potential energy.

Note that the units of gravitational potential energy must be the same as those of work. That is, potential energy has units of joules, ergs, or foot-pounds. Potential energy, like work and kinetic energy, is a scalar quantity.

Also, note that the gravitational potential energy of an object depends only on the vertical height of an object above the surface of the Earth. From this result, we see that an object will do the same amount of work if it falls vertically to the Earth as it will if it starts at the same point and slides down a frictionless incline to the Earth. Finally, you should note that Equation 5.7 is valid only for objects near the surface of the Earth where **g** is approximately constant. We shall derive a more general expression for potential energy in Chapter 7.

Reference Levels for Gravitational Potential Energy

In working problems involving the concept of gravitational potential energy, it is always necessary to choose a location at which we set the gravitational potential energy equal to zero. We choose to call this the zero level for the gravitational potential energy. The choice of location of this zero level is completely arbitrary. This is true because the quantity of importance is the *difference* in potential energy, and the difference in potential energy between two points is independent of the choice of zero level.

It is often convenient to choose the surface of the Earth as the reference position for zero potential energy, but again this is not essential. Often, the statement of a problem will suggest a convenient level to use. As an example, consider a book at several possible locations as in Figure 5.9. When the book is at A, above the surface of a desk, a logical zero level for potential energy could be taken at the desk surface. When the book is at B, however, the floor surface might be a more appropriate zero reference level. Finally, a location such as C, where the book is flying out a window, would suggest choosing the ground level as the zero level of potential energy. The choice, however, makes no difference. Any of the reference levels mentioned could be used as the zero level, regardless of whether the book is at A, B, or C.

Figure 5.9
Any reference level could be used for measuring the gravitational potential energy of the book.

Example 5.3   Let's Go Skiing

A novice skier of mass 60 kg is at the top of a slope as shown in Figure 5.10a. At the initial point A, the skier is 10 m vertically above point B. (a) Choose the zero level for gravitational potential energy to be at B and find the gravitational potential energy of the skier at A and B, and the difference in potential energy between these two points.

Solution
The gravitational potential energy at the final point, B, is zero by choice. Hence, the potential energy at the initial point is

$$PE_i = mgy_i = (60 \text{ kg})(9.80 \text{ m/s}^2)(10 \text{ m}) = \boxed{5880 \text{ J}}$$

(a)

(b)

Figure 5.10 (Example 5.3).
(Photo © François Dardelet/THE IMAGE BANK)

Since $PE_f = 0$, the difference in potential energy is

$$\Delta PE = PE_i - PE_f = 5880 \text{ J} - 0 \text{ J} = \boxed{5880 \text{ J}}$$

(b) Repeat this problem with the zero level at point A.

Solution
In this case, the initial potential energy is zero because of the choice of the reference level. The final potential energy is

$$PE_f = mgy_f = (60 \text{ kg})(9.80 \text{ m/s}^2)(-10 \text{ m}) = \boxed{-5880 \text{ J}}$$

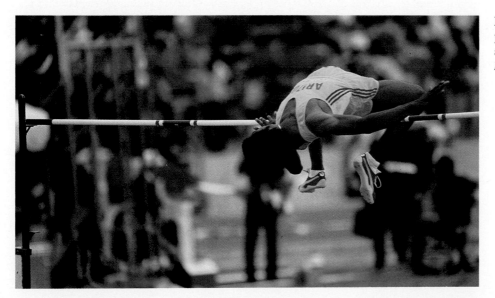

Estimate the gravitational potential energy of this athlete at the top of his jump. (© R. Mackson/FPG International)

Note that the distance y_f is -10 m because the final point is 10 m *below* the zero reference level. The difference in potential energy is

$$\Delta PE = PE_i - PE_f = 0 \text{ J} - (-5880 \text{ J}) = \boxed{5880 \text{ J}}$$

These calculations show that the potential energy of the skier at the top of the slope is greater than the potential energy at the bottom by 5880 J, *regardless of the zero level selected*.

Exercise   If the zero level for gravitational potential energy is selected to be midway down the slope, at a height of 5 m, find the initial potential energy, the final potential energy, and the difference in potential energy between points A and B.

Answer   2940 J, -2940 J, 5880 J

5.5   Conservative and Nonconservative Forces

Conservative Forces

In Section 5.4 we found that the work done by the gravitational force acting on an object is given by

$$W_g = mgy_i - mgy_f$$

In other words, *the work done by gravity depends only on the initial and final coordinates. The amount of work done does not depend on the path selected to move between these points.* When a force exhibits this property, it is called a **conservative force**. In addition to the gravitational force, other examples of conservative forces, which we shall encounter later, are the electrostatic force and the forces exerted by springs. In general,

Definition of a conservative force

a force is conservative if the work it does on an object moving between two points is independent of the path the object takes between the points. The work done on an object by a conservative force depends only on the initial and final positions of the object.

Conservative forces are important because we can always associate a potential energy function with them. For the gravitational force, we found the potential energy to be

$$PE = mgy$$

We are able to define potential energy functions only for conservative forces. In general, one always finds that the work done on an object by a conservative force, W_c, is given by the initial potential energy of the object minus its final potential energy. That is,

$$W_c = PE_i - PE_f \qquad\qquad \textbf{(5.9)}$$

Nonconservative Forces

A force is nonconservative if the work it does on an object moving between two points depends on the path taken.

Definition of a
nonconservative force

The force of sliding friction is a good example of a nonconservative force. If an object is moved over a rough horizontal surface between two points along various paths, the work done by the frictional force certainly depends on the path. The negative work done by the frictional force along any particular path between two points equals the force of friction multiplied by the length of the path. Different paths involve different amounts of work. The least amount of work will be done by the frictional force along a straight-line path between the two points. For example, suppose you were to displace a book between two points on a rough horizontal surface, such as a table. If the book is displaced in a straight line along the blue path between points A and B in Figure 5.11, the work done by friction is simply $-fd$, where d is the distance between the two points. However, if the book is moved along any other path between the two points, the work done by friction is greater (in absolute magnitude) than $-fd$. For example, the work done by friction along the red semicircular path in Figure 5.11 is equal to $-f(\pi d/2)$, where d is the diameter of the circle. Potential energy functions cannot be defined for nonconservative forces.

Figure 5.11
The work done by the force of friction depends on the path taken as the book is moved from A to B.

5.6 Conservation of Mechanical Energy

Conservation laws play a very important role in physics, and we shall encounter a number of these laws as we proceed through the remainder of this course. The first of these that we meet, the law of conservation of energy, is one of the most important. Before we describe the mathematical details of the law of conservation of energy, we must pause briefly to examine what is meant when we say something is conserved. When we say that a physical quantity is *conserved*, we simply mean that the value of this quantity does not change with time. Although the form of the quantity may change in some manner, its final value will be the same as its initial value. We shall find, for example, that the energy in a system may change from gravitational potential energy to kinetic energy or to one of a variety of other forms that we shall encounter soon, but energy is never lost from the system.

In order to develop the principle of conservation of energy, let us return to the work-energy relation, Equation 5.6, which says that the net work done on a system equals the change in the system's kinetic energy.

$$W_{net} = KE_f - KE_i$$

Now, let us assume that the only force doing work on the system is a conservative force. In this case the net work on the system is equal to W_c, and from Equation 5.9 we have

$$W_{net} = W_c = PE_i - PE_f$$

We now substitute this expression into the work energy equation for W_{net} to find

$$PE_i - PE_f = KE_f - KE_i$$

or

$$KE_i + PE_i = KE_f + PE_f \qquad \textbf{(5.10)}$$

Conservation of mechanical energy

The sum of the kinetic energy plus the potential energy is called the **total mechanical energy**. Equation 5.10 is a statement of **the law of conservation of mechanical energy**. We can describe this law as follows:

The law of conservation of mechanical energy states that the mechanical energy of a system remains constant if the only forces that do work on the system are conservative forces.

Therefore if a conservative system (one subject only to conservative forces) has some initial amount of kinetic and potential energy, the energy may later transform from one kind into the other, but the final total mechanical energy will remain the same. This is equivalent to saying that, if the kinetic energy of a conservative system increases (or decreases) by some amount, the potential energy of the system must decrease (or increase) by the same amount.

So far, we have encountered only one conservative force, and as a result we have only one potential energy function, that of gravitational potential energy. Thus, *if the force of gravity is the only force doing work on an object, then the total mechanical energy of the object is conserved*, and the law of conservation of mechanical energy takes the form

$$\frac{1}{2} m v_i{}^2 + mgy_i = \frac{1}{2} m v_f{}^2 + mgy_f \qquad \textbf{(5.11)}$$

As we go further into this course, we shall discover that there are other kinds of energy besides kinetic energy and gravitational potential energy. As a result, you should be aware of the fact that we shall have to modify Equation 5.11 to include these new forms of energy as they occur.

Example 5.4   The Daring Diver

A diver weighing 750 N (mass = w/g = 77 kg) drops from a board 10 m above the surface of a pool of water as in Figure 5.12. (a) Use the conservation of mechanical energy to find his speed at a point 5 m above the water surface.

Solution
As the diver falls toward the water, only one force acts on him: the force of gravity. (This assumes that air resistance can be neglected.) Therefore, we can be assured that no forces other than the force of gravity do any work on the diver and that mechanical energy is conserved. In order to find the diver's speed at the 5-m mark, let us choose the zero level for potential energy to be at the surface of the water. Also,

10 m

$KE_i = 0$
$PE_i = 7500$ J

5 m

$KE = 3500$ J
$PE = 3750$ J

$KE_f = 7500$ J
$PE_f = 0$

0

Figure 5.12   (Example 5.4) The kinetic energy and potential energy of a diver at various heights. The zero of potential energy is taken to be at the surface of the pool.

note that the diver drops from the board, that is, he leaves with zero velocity and zero kinetic energy. Equation 5.11 gives

$$\frac{1}{2} m v_i{}^2 + mgy_i = \frac{1}{2} m v_f{}^2 + mgy_f$$

$$0 + (750 \text{ N})(10 \text{ m}) = \frac{1}{2} (77 \text{ kg}) v_f{}^2 + (750 \text{ N})(5 \text{ m})$$

$$v_f = \boxed{9.86 \text{ m/s}}$$

(b) Find the speed of the diver just before he strikes the water.

Solution
With the final position of the diver at the surface of the pool, Equation 5.11 gives

$$0 + (750 \text{ N})(10 \text{ m}) = \frac{1}{2} (77 \text{ kg}) v_f{}^2 + 0$$

$$v_f = \boxed{13.9 \text{ m/s}}$$

Exercise   If the diver pushes off such that he leaves the board with an initial speed of 2 m/s, find his speed when he strikes the water. Use conservation of energy.

Answer   14.1 m/s

Example 5.5   Sliding Down a Slick Hill

A sled and its rider together weigh 800 N. They move down a frictionless hill through a vertical distance of 10 m, as shown in Figure 5.15a. Use conservation of mechanical energy to find the speed of the sled-rider system at the bottom of the hill, assuming that the rider pushes off with an initial speed of 5 m/s.

Solution
The forces acting on the sled and rider as they move down the hill are shown in Figure 5.13b. In the absence of a frictional force, the only forces acting are the normal force, **n**, and the gravitational force. At all points along the path, **n** is perpendicular to the direction of travel and hence does no work. Likewise, the component of weight perpendicular to the incline ($w \cos \theta$) does no work. The only force that does any work is the component of the gravitational force **F** along the slope of the hill. As a result, the only force doing any work is the gravitational force and we are justified in using Equation 5.11. Note that in this case the initial energy includes kinetic energy because of the initial speed:

$$\frac{1}{2} m v_i{}^2 + mgy_i = \frac{1}{2} m v_f{}^2 + mgy_f$$

or, after canceling m throughout the equation,

$$\frac{1}{2} v_i{}^2 + gy_i = \frac{1}{2} v_f{}^2 + gy_f$$

Figure 5.13   (Example 5.5)
(a) A sled and rider start from the top of a frictionless hill with a speed of 5 m/s.
(b) Free-body diagram for the system (sled plus rider).

If we select the origin at the bottom of the incline, we see that the initial and final y coordinates are $y_i = 10$ m and $y_f = 0$. Thus we get

$$\frac{1}{2}(5 \text{ m/s})^2 + (9.80 \text{ m/s}^2)(10 \text{ m}) = \frac{1}{2} v_f^2 + (9.80 \text{ m/s}^2)(0)$$

$$v_f = \boxed{14.9 \text{ m/s}}$$

Exercise   If the sled and rider start at the bottom of the incline and are given an initial velocity of 5 m/s up the incline, how high will they rise vertically and what will their speed be when they return to the bottom of the hill?

Answer   1.28 m; 5 m/s.

Potential Energy Stored in a Spring

We have found the concept of gravitational potential energy to be of tremendous value in the description of certain types of mechanical motion. As we continue with our study we will find that there are other types of systems that can be described using the concept of potential, or stored, energy. One of these, the energy stored in a compressed or stretched spring, will be discussed more completely in Chapter 14 but it is instructive to examine this type of energy briefly here.

Consider Figure 5.14a, which shows a spring, which has been neither compressed nor stretched, at the end of a frictionless horizontal surface. In Figure 5.14b, a block has been pushed against the spring, compressing it a distance x. In order to compress the spring, we would find that we must exert a

Figure 5.14
A block of mass m on a horizontal, frictionless surface is pushed against a spring and released from rest. If x is the compression in the spring as in (b), the potential energy stored in the spring is $\frac{1}{2}kx^2$. This energy is transferred to the block in the form of kinetic energy as in (c).

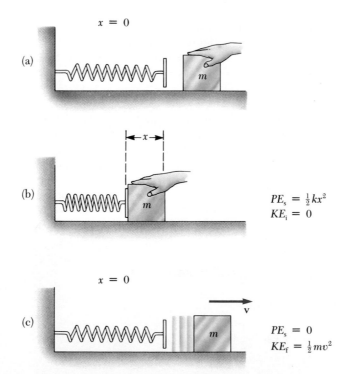

force on the block whose magnitude is given by $F = kx$, where k is a constant for a particular spring called the **spring constant.** For a flexible spring, k is a small number, whereas for a stiff spring, k is large. The equation $F = kx$ says that the more we compress or stretch the spring, the larger is the force that we must exert. Since the block is in equilibrium at the position shown in Figure 5.14b, we see that the spring must be exerting a force on it given by $F = -kx$. This equation is often called **Hooke's law,** after Sir Robert Hooke, who discovered the relationship between F and x.

To find an expression for elastic potential energy, let us find the work required to compress a spring from its equilibrium position to some final arbitrary position x. The force that must be applied to compress the spring varies from $F = 0$ to $F = kx$ at maximum compression. Because the force increases linearly with position (that is, $F \propto x$), the average force that must be applied is

$$\overline{F} = \frac{F_0 + F_x}{2} = \frac{0 + kx}{2} = \frac{1}{2}\,kx$$

Therefore, the work *done by the applied force* is

$$W = \overline{F}x = \frac{1}{2}\,kx^2$$

This work is stored in the spring as elastic potential energy. Thus, we define the elastic potential energy, PE_s, as

$$PE_s \equiv \frac{1}{2}\,kx^2 \qquad \textbf{(5.12)}$$

Figure 5.14c shows how this stored elastic potential energy can be recovered. When the block is released, the spring snaps back to its original length and the stored elastic potential energy is converted to kinetic energy of the block. Note that the elastic potential energy stored in the spring is zero when the spring is unstretched or uncompressed ($x = 0$). Energy is stored in the spring only when it is either stretched or compressed. Furthermore, the elastic potential energy is a maximum when the spring has reached its maximum compression or extension. Finally, the potential energy in the spring is always positive because it is proportional to x^2. This new form of energy is included as another term in our equation for conservation of mechanical energy:

$$(KE + PE_g + PE_s)_i = (KE + PE_g + PE_s)_f$$

where PE_g is the gravitational potential energy.

The following examples show some applications of the conservation of energy for a few systems containing springs. Additional problems will be found in Chapter 14.

Example 5.6 Block Projected on a Smooth Surface

A block of mass 0.5 kg rests on a horizontal, frictionless surface, as shown in Figure 5.14. The block is pressed against a light spring having a spring constant $k = 80$ N/m. The spring is compressed a distance of 2 cm and released. Consider the system to be the spring and block, and find the speed of the block at the instant it loses contact with the spring at the $x = 0$ position.

Solution

We use the expression for the conservation of mechanical energy as

$$(KE + PE_g + PE_s)_i = (KE + PE_g + PE_s)_f$$

The initial kinetic energy of the block is zero, and since the block remains at the same level throughout its motion, the gravitational potential energy terms on each side of the equation for the conservation of mechanical energy are equal. Thus,

$$PE_{si} = KE_f$$

or

$$\frac{1}{2} kx_i{}^2 = \frac{1}{2} mv_f{}^2$$

$$\frac{1}{2} (80 \text{ N/m})(2 \times 10^{-2} \text{ m})^2 = \frac{1}{2} (0.5 \text{ kg})v_f{}^2$$

$$v_f = \boxed{0.253 \text{ m/s}}$$

Example 5.7   The Spring-Loaded Popgun

Figure 5.15
(Example 5.7)

The launching mechanism of a toy gun consists of a spring of unknown spring constant as shown in Figure 5.15. By compressing the spring a distance of 0.12 m, the gun is able to launch a 20-g projectile to a maximum height of 20 m when fired vertically from rest. Neglecting all resistive forces, (a) determine the value of the spring constant.

Solution

Since the projectile starts at rest, the initial kinetic energy in the system is zero. If the reference of gravitational potential energy is taken at the lowest position of the projectile, then its initial gravitational potential energy is also zero. Hence, the total initial energy of the system is the elastic potential energy stored in the spring, which is $kx^2/2$, where $x = 0.12$ m. Since the projectile rises to a maximum height $h = 20$ m, its final gravitational potential energy is equal to mgh, its final kinetic energy is zero, and the final elastic potential energy is zero. Conservation of mechanical energy gives

$$\tfrac{1}{2}kx^2 = mgh$$

$$\tfrac{1}{2}k(0.12 \text{ m})^2 = (0.02 \text{ kg})(9.80 \text{ m/s}^2)(20 \text{ m})$$

or

$$k = \boxed{544 \text{ N/m}}$$

(b) Find the speed of the projectile as it moves through the equilibrium position of the spring (where $x = 0$) as shown in Figure 5.15b.

Solution

Using the same reference level for the gravitational potential energy as in part (a), we see that the initial energy of the system is still the elastic potential energy $kx^2/2$. The final energy of the system when the projectile moves through the unstretched position of the spring consists of the kinetic energy of the projectile, $mv^2/2$, and the gravitational potential energy of the projectile, mgx. Hence, conservation of energy in this case gives

$$\tfrac{1}{2}kx^2 = \tfrac{1}{2}mv^2 + mgx$$

Solving for v gives

$$v = \sqrt{\frac{kx^2}{m} - 2gx}$$

$$v = \sqrt{\frac{(544 \text{ N/m})(0.12 \text{ m})^2}{(0.02 \text{ kg})} - 2(9.80 \text{ m/s}^2)(0.12 \text{ m})}$$

$$= \boxed{19.7 \text{ m/s}}$$

Exercise What is the speed of the projectile when it is at a height of 10 m?

Answer 14.0 m/s

5.7 Nonconservative Forces and the Work-Energy Theorem

In realistic situations, nonconservative forces, such as friction, are usually present. In such situations, the total mechanical energy of the system is not constant, and one cannot apply Equation 5.10. This is due to the fact that the law of conservation of mechanical energy holds only when the forces that do work on the system are conservative forces. In order to be able to account for nonconservative forces, let us return to the work-energy theorem.

$$W_{\text{net}} = \frac{1}{2}mv_{\text{f}}^2 - \frac{1}{2}mv_{\text{i}}^2$$

Let us separate the net work, W_{net}, into two parts, that due to the nonconservative forces, W_{nc}, and the work done by conservative forces, W_{c}. Thus, the work-energy relation becomes

$$W_{\text{nc}} + W_{\text{c}} = \frac{1}{2}mv_{\text{f}}^2 - \frac{1}{2}mv_{\text{i}}^2$$

Let us consider the special case in which the conservative force is the gravitational force. In this case,

$$W_{\text{c}} = mgy_{\text{i}} - mgy_{\text{f}} = PE_{\text{i}} - PE_{\text{f}}$$

and we have

$$W_{\text{nc}} = \left(\frac{1}{2}mv_{\text{f}}^2 - \frac{1}{2}mv_{\text{i}}^2\right) + (mgy_{\text{f}} - mgy_{\text{i}}) \qquad (5.13)$$

or

$$W_{\text{nc}} = (KE_{\text{f}} - KE_{\text{i}}) + (PE_{\text{f}} - PE_{\text{i}}) \qquad (5.14)$$

That is, the work done by all nonconservative forces equals the change in kinetic energy plus the change in potential energy.

According to Equation 5.13, the work done by nonconservative forces can be positive, negative, or zero. An example of a situation in which W_{nc} is positive is that of an object being pushed along the floor by an applied force which is greater than the force of friction. In this case the applied force and the displacement are in the same direction, and the work done by the applied force is positive. In such situations, the nonconservative force is adding mechanical energy to the system. The work done by a nonconservative force is negative when the direction of the force is opposite that of the displacement. Forces of sliding friction are nonconservative in nature. When nonconservative forces act on a system, they remove mechanical energy from the system. In reality, energy is not lost in the process; it is simply being transformed into some form of energy other than kinetic or potential energy. As we shall see in a later chapter, this other form of energy is usually heat energy. Because frictional forces produce a decrease in the total mechanical energy of a system, they are called *dissipative* forces.

Note that when there are no nonconservative forces present, $W_{nc} = 0$, and Equation 5.13 reduces to Equation 5.11. This is to be expected on the basis of the discussion in Section 5.6.

Finally, you should note that when Equation 5.13 is used, the term W_{nc} must involve only nonconservative forces. *This does not mean that conservative forces such as gravity are being ignored; they are already taken into account by the presence of the potential energy terms in Equation 5.13.* For example, the work done by the force of gravity appears in the terms mgy_i and mgy_f.

Conservation of Energy as an "Accounting" Procedure

As we have seen, the work done by nonconservative forces can either add energy to an object or remove energy from the object. For example, a jet engine adds energy to an airplane whereas the drag force due to air resistance removes energy from the airplane. Thus, we can consider the work done by nonconservative forces as

$$W_{nc} = \text{energy added} - \text{energy removed}$$

This expression enables us to write Equation 5.14 as follows:

$$KE_i + PE_i + \text{energy added} = KE_f + PE_f + \text{energy removed} \quad \textbf{(5.15)}$$

When the equation is written in this form, we see that conservation of energy is simply a balance sheet for energy. If we add the energy of the system before some event takes place to the energy added by nonconservative forces, the sum must equal the energy of the system after the event occurs plus any energy that may have been removed by nonconservative forces. Thus, solving problems using conservation of energy is essentially an exercise in accounting procedures.

Finally, it should be noted that, when Equation 5.15 is used, only nonconservative forces need to be considered in determining the energy added or removed. This does not mean that conservative forces, such as gravity, are being ignored, however; they are already taken into account by the presence of the potential energy terms in Equation 5.15. For example, the work added or removed by the force of gravity appears in the terms mgy_i and mgy_f.

Problem-Solving Strategy:
Conservation of Energy

The following steps should be followed in applying the principle of the conservation of energy:

1. Define your system, which may consist of more than one object.
2. Select a reference position for the zero point of gravitational potential energy.
3. Determine whether or not nonconservative forces are present. If so, the work done by these forces must be included in the conservation of energy expression; if not, mechanical energy is conserved.
4. If mechanical energy is conserved, then you can proceed to write the total initial energy at some point as the sum of the kinetic and potential energy at that point. Then, write an expression for the total final energy $KE_f + PE_f$ at the final point that is of interest. Since mechanical energy is conserved, you can equate the two total energies and solve for the quantity that is unknown.
5. If nonconservative forces are present, you should write expressions for the total initial and total final energies. In this case, the difference between the two is equal to the work done by the nonconservative force(s). That is, $W_{nc} = KE_f + PE_f - KE_i - PE_i$.

Example 5.8 Crate on a Rough Ramp

A 3-kg crate slides down a rough ramp at a loading dock. The ramp is 1 m in length and is inclined at an angle of 30°, as shown in Figure 5.16. The crate starts from rest at the top and experiences a constant frictional force of magnitude 5 N. Use energy methods to determine the speed of the crate when it reaches the bottom of the ramp.

Solution
Because $v_i = 0$, the initial kinetic energy is zero. If the y coordinate is measured from the bottom of the incline, then $y_i = 0.50$ m. Therefore, the total mechanical energy of the crate at the top is all potential energy, given by

$$PE_i = mgy_i = (3\text{ kg})(9.80\text{ m/s}^2)(0.50\text{ m}) = 14.7\text{ J}$$

When the crate reaches the bottom, its potential energy is zero because its elevation is $y_f = 0$. Therefore, the total mechanical energy at the bottom is all kinetic energy:

$$KE_f = \frac{1}{2}mv_f^2$$

However, we cannot say that $PE_i = KE_f$ in this case because there is a nonconservative force that does work on the crate: the force of friction. In this case, $W_{nc} = -fs$, where s is the displacement along the ramp. (Recall that the forces normal to the ramp do no work on the crate because they are perpendicular to the displacement.) With $f = 5$ N and $s = 1$ m, we have

$$W_{nc} = -fs = (-5\text{ N})(1\text{ m}) = -5\text{ J}$$

Figure 5.16 (Example 5.8) A crate slides down a rough incline under the influence of gravity. Its potential energy decreases while its kinetic energy increases.

This says that some mechanical energy is lost because of the presence of the retarding frictional force. Applying the work-energy theorem in the form of Equation 5.13 gives

$$-fs = \frac{1}{2} mv_f{}^2 - mgy_i$$

$$\frac{1}{2} mv_f{}^2 = 14.7 \text{ J} - 5 \text{ J} = 9.7 \text{ J}$$

$$v_f{}^2 = \frac{19.4 \text{ J}}{3 \text{ kg}} = 6.47 \text{ m}^2/\text{s}^2$$

$$v_f = \boxed{2.54 \text{ m/s}}$$

Exercise   Find the final speed at the bottom of the ramp if the plane is assumed to be frictionless.

Answer $v_f = 3.13$ m/s.

Example 5.9   Fun on the Slide

A child of mass 20 kg takes a ride on an irregularly curved slide of height 6 m, as in Figure 5.17. The child starts from rest at the top. Determine the speed of the child at the bottom, assuming there is no friction present.

Solution
Note that the normal force, **n**, does no work on the child because this force is always perpendicular to the displacement. Furthermore, because there is no friction, $W_{nc} = 0$ and we can apply the law of conservation of mechanical energy. If we measure the y coordinate from the bottom of the slide, then $y_i = 6$ m, $y_f = 0$, and we get

$$KE_i + PE_i = KE_f + PE_f$$

$$0 + (20 \text{ kg})(9.80 \text{ m/s}^2)(6 \text{ m}) = \frac{1}{2} (20 \text{ kg})(v_f{}^2) + 0$$

$$v_f = \boxed{10.8 \text{ m/s}}$$

Note that this speed is the same as if the child had fallen vertically through a distance of 6 m! From the free-fall equation, $v_f{}^2 = v_i{}^2 - 2gy$, we get the same result:

$$v_f = \sqrt{2(9.80 \text{ m/s}^2)(6 \text{ m})} = 10.8 \text{ m/s}$$

The effect of the slide is to direct the motion at the end into a horizontal motion instead of the vertical motion of free fall.

Exercise   If the speed of the child at the bottom of the slide is 8 m/s rather than 10.8 m/s found above, how much work would have been done by the force of friction?

Answer -536 J.

Example 5.10   Fasten Your Seat Belts

An airplane of mass 1.5×10^4 kg is moving at a speed of 60 m/s. The pilot then revs up the engine such that the forward thrust of the propeller becomes 7.5×10^4 N. If the force of air resistance on the airplane is 4×10^4 N, find the speed of the air-

Figure 5.17   (Example 5.9) If the slide is very smooth, the speed of the child at the bottom of the slide depends essentially on the height of the slide, and is independent of its shape.

plane after it has traveled 500 m. Assume the airplane is in level flight throughout this motion.

Solution

There are two nonconservative forces doing work on the airplane. One of these is the forward thrust of the propeller on the airplane. The work done by this force, W_{nc1}, is

$$W_{nc1} = Fs = (7.5 \times 10^4 \text{ N})(500 \text{ m}) = 3.75 \times 10^7 \text{ J}$$

The second nonconservative force acting on the airplane is the force of air resistance. The work done by this force, W_{nc2}, is negative because this force and the displacement are in opposite directions. Hence, we have

$$W_{nc2} = -fs = -(4 \times 10^4 \text{ N})(500 \text{ m}) = -2 \times 10^7 \text{ J}$$

We choose the zero level of potential energy to be at the level of the airplane. Because the airplane remains in level flight during the motion, we see that $mgy_i = mgy_f = 0$. Therefore, Equation 5.13, together with the above values, gives

$$W_{nc1} + W_{nc2} = \frac{1}{2} mv_f^2 - \frac{1}{2} mv_i^2$$

$$3.75 \times 10^7 \text{ J} - 2 \times 10^7 \text{ J} = \frac{1}{2}(1.5 \times 10^4 \text{ kg})v_f^2 - \frac{1}{2}(1.5 \times 10^4 \text{ kg})(60 \text{ m/s})^2$$

$$v_f = \boxed{77 \text{ m/s}}$$

5.8   Power

Suppose you wish to climb several flights of stairs to the top of a tall building. You may be able to perform this task with little difficulty if you make the climb slowly. However, if you decide to sprint up the stairs, you will probably find yourself taking a rest after climbing a few flights. Although the total work done in both situations is the same, the *rate* at which work is being done is quite different. In fact, the rate at which work is being done is the key factor in such situations. **Power** is defined as the rate at which work is done, or as the rate at which energy is changed from one form to another:

$$\overline{P} = \frac{W}{\Delta t} \qquad \textbf{(5.16)} \qquad \text{Average power}$$

$\overline{P}$ is the average power, W is the work done, and Δt is the time interval during which work is done. It is sometimes useful to rewrite Equation 5.16 by substituting $W = F \Delta s$ and noting that $\Delta s/\Delta t$ is the average velocity of the object during the time Δt:

$$\overline{P} = \frac{W}{\Delta t} = \frac{F \Delta s}{\Delta t} = F\overline{v} \qquad \textbf{(5.17)}$$

This result says that the average power either delivered to an object or delivered by the object is equal to the product of the force acting on the object during some time interval and its average velocity during this time interval. In Equation 5.17, F is the component of force in the direction of the average velocity.

A result of the same form as Equation 5.17 is obtained for instantaneous values. That is, the instantaneous power delivered to an object is equal to the product of the force on the object at that instant and the instantaneous velocity, or $P = Fv$.

The unit of power in the SI system is joules per second, which is also called the **watt** (W), after James Watt:

$$1 \text{ W} = 1 \text{ J/s} = 1 \text{ kg} \cdot \text{m}^2/\text{s}^3 \tag{5.18}$$

The unit of power in the British engineering system is the horsepower (hp), where

$$1 \text{ hp} = 550 \, \frac{\text{ft} \cdot \text{lb}}{\text{s}} = 746 \text{ W} \tag{5.19}$$

The watt is commonly used in electrical applications, but it can be used in other scientific areas. For example, an automobile engine can be rated in watts as well as in horsepower. Likewise, the power consumption of an electric appliance can be expressed in horsepower. Use your skills to calculate the horsepower of a 100-W lightbulb and the wattage rating of a car engine.

When one refers to a car engine of 100 hp, one means that the engine is capable of changing 55,000 ft·lb of chemical or electrical energy to mechanical energy each second. Likewise, in 1 s, a 100-W lightbulb transforms 100 J of electrical energy into other forms, primarily heat and light. Returning to our stair-climbing example at the opening of this section, the reason that one may be unable to race up a flight of stairs is because the body may be unable to convert chemical energy to mechanical energy rapidly enough.

Example 5.11   Drag Racing

A 2000-kg car starts from rest and accelerates to a final velocity of $+20$ m/s in a time of 15 s. Assume that the force of air resistance remains constant at a value of -500 N during this time. (a) Find the average power developed by the engine of the car.

Solution
The equations of linear motion with constant acceleration allow us to find the acceleration of the car as follows:

$$v = v_0 + at$$

Since $v_0 = 0$, we find

$$20 \text{ m/s} = a(15 \text{ s})$$

$$a = +1.33 \text{ m/s}^2$$

The forces acting on the car along the direction of motion are the forward thrust of the engine F, and the retarding 500-N force of air resistance. Let us now use Newton's second law as follows:

$$\sum F = ma$$

$$F - 500 \text{ N} = (2000 \text{ kg})(1.33 \text{ m/s}^2)$$

$$F = +3160 \text{ N}$$

The average velocity of the car during this time interval is

$$\bar{v} = \frac{v_f + v_i}{2} = \frac{20 \text{ m/s} + 0}{2} = +10 \text{ m/s}$$

Therefore, the average power can be found using Equation 5.17:

$$\bar{P} = F\bar{v} = (3160 \text{ N})(10 \text{ m/s}) = 3.16 \times 10^4 \text{ W} = \boxed{42.4 \text{ hp}}$$

(b) Find the instantaneous power being developed by the engine when the car reaches its final velocity.

Solution
The forward thrust of the car engine remains constant at the value calculated in Part (a), $F = +3160$ N, throughout the motion. Since the final velocity is $+20$ m/s, and the instantaneous power is given by $P = Fv$, we find

$$P = Fv = (3160 \text{ N})(20 \text{ m/s}) = 6.32 \times 10^4 \text{ W} = \boxed{84.8 \text{ hp}}$$

Exercise This example is somewhat unrealistic because all of the power developed by an engine does not actually appear as a forward thrust on the car. In reality, one must account for frictional and mechanical losses within the engine. Assuming that 75% of the power developed by the engine reaches the wheels, find the average power the engine would have to develop to achieve the acceleration calculated in Part (a).

Answer 56.5 hp.

Example 5.12   Power Delivered by an Elevator Motor

An elevator has a mass of 1000 kg and carries a maximum load of 800 kg. A constant frictional force of 4000 N retards its motion upward, as in Figure 5.18. (a) What minimum power must the motor deliver to lift the fully loaded elevator at a constant speed of 3 m/s?

Solution
The motor must supply the force **T** that pulls the elevator upward. From Newton's second law and from the fact that $\mathbf{a} = 0$ because **v** is constant, we get

$$T - f - Mg = 0$$

where M is the total mass (elevator plus load), equal to 1800 kg. Therefore,

$$T = f + Mg$$
$$= 4 \times 10^3 \text{ N} + (1.8 \times 10^3 \text{ kg})(9.80 \text{ m/s}^2)$$
$$= 2.16 \times 10^4 \text{ N}$$

From $P = Fv$ and the fact that **T** is in the same direction as **v**, we have

$$P = Tv$$
$$= (2.16 \times 10^4 \text{ N})(3 \text{ m/s}) = 6.48 \times 10^4 \text{ W}$$
$$= 64.8 \text{ kW} = \boxed{86.9 \text{ hp}}$$

Figure 5.18 (Example 5.12) The motor provides a force **T** upward on the elevator. A frictional force **f** and the total weight Mg act downward.

5.9   Conservation of Energy in General

We have seen that the total mechanical energy of a system is conserved when only conservative forces act on the system. Furthermore, we have been able to associate a potential energy with each conservative force. In other words, mechanical energy is lost when nonconservative forces, such as friction, are present.

We can generalize the energy conservation principle to include all forces acting on the system, both conservative and nonconservative. In our study of heat, we shall find that mechanical energy can be transformed to thermal energy. For example, when a block slides over a rough surface, the mechanical energy lost is transformed to thermal energy stored in the block, as evidenced by a measurable increase in its temperature. On a submicroscopic scale, we shall see that this internal thermal energy is associated with the vibration of atoms about their equilibrium positions. Since this internal atomic motion has kinetic and potential energy, one can say that frictional forces arise fundamentally from conservative atomic forces. Therefore, if we include this increase in the internal energy of the system in our work-energy theorem, the total energy is conserved.

This is just one example of how you can analyze an isolated system and always find that the total energy does not change as long as you account for all forms of energy. That is, *energy can never be created or destroyed*. Energy may be transformed from one form to another, but *the total energy of an isolated system is always constant*. From a universal point of view, we believe that the total energy of the universe is constant. Therefore, if one part of the universe gains energy in some form, another part must lose an equal amount of energy. No violation of this principle has been found.

Other examples of energy transformations are the energy carried by sound waves resulting from the collision of two objects, the energy radiated by an accelerating charge in the form of electromagnetic waves (a radio antenna), and the elaborate sequence of energy conversions in a thermonuclear reaction.

In subsequent chapters, we shall see that the various parts of the energy concept, especially transformations of energy between various forms, join together the various branches of physics. In other words, one cannot completely separate the subjects of mechanics, thermodynamics, and electromagnetism. Finally, from a practical viewpoint, all mechanical and electronic devices rely on some form of energy transformation.

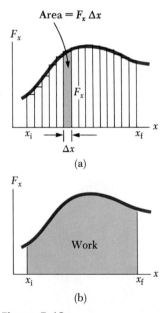

Figure 5.19
(a) The work done by a force F_x for the small displacement Δx is $F_x \, \Delta x$, which equals the area of the blue-shaded rectangle. The total work done for the displacement x_i to x_f is approximately equal to the sum of the areas of all the rectangles. (b) The work done by a variable force F_x as the particle moves from x_i to x_f is *exactly* equal to the area under this curve shaded in blue.

*5.10   Work Done by a Varying Force

Consider an object being displaced along the x axis under the action of a varying force, as shown in Figure 5.19a. Such a situation would occur if a person attempted to pull an object at a constant speed along a surface that varied in roughness from point to point. The magnitude of the required force would change as the roughness increased or decreased. In such a situation, we cannot use $W = Fs \cos \theta$ to calculate the work done by the force because *this relationship applies only when the force is constant in magnitude and direction.*

However, we can calculate the work done as the object undergoes a very small displacement, Δx, as shown in Figure 5.19a. During this small displacement the force is approximately constant, and so we can express the work done by the force as

$$\Delta W = F_x \, \Delta x$$

Note that this is just the area of the blue-shaded rectangle in Figure 5.19a. Now, if we imagine that the force-displacement curve is divided into a large number of such intervals, as in Figure 5.19a, then the total work done for the displacement from x_i to x_f is approximately equal to the sum of a large number of such terms.

$$W \approx \sum F_x \, \Delta x \qquad\qquad (5.20)$$

The right side of this equation is the sum of the area of all the rectangles pictured, but if the rectangles are made very narrow by decreasing Δx, the right side of the equation turns out to be just the total area under the curve, which is shown in Figure 5.19. (Note that the word "area" here refers to the region under a curve of F vs. x, not a geometric area.) Thus, we can say that

the total work done by a varying force is equal to the area under the force-displacement curve.

Example 5.13

A force acting on an object varies with x as shown in Figure 5.20. (a) Calculate the work done by the force as the object moves from $x = 0$ to $x = 6$ m.

Solution
The work done by the force is equal to the total area under the curve from $x = 0$ to $x = 6$ m. This area is equal to the area of the rectangular section between $x = 0$ to $x = 4$ m plus the area of the triangular section between $x = 4$ m to $x = 6$ m. The area of the rectangle is (4×5) N $\cdot$ m $= 20$ J, and the area of the triangle is $\frac{1}{2}(2 \times 5)$N $\cdot$ m $= 5$ J. Therefore, the total work done is 25 J.

(b) Following the motion indicated in part (a), the force immediately drops to a constant value of 3 N in the negative direction for the next 2 m, as shown. Find the work done by this force during this interval, and (c) the net work done for the entire 8 m of motion.

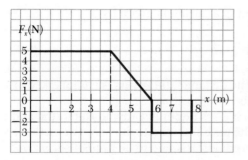

Figure 5.20 (Example 5.13) The force is constant for the first 4 m of motion and then decreases linearly with x from $x = 4$ m to $x = 6$ m. Finally, the force becomes -3 N from $x = 6$ m to $x = 8$ m. The net work done by this force is the area under this curve.

Solution

(b) The work done is the area under the rectangular curve between 6 and 8 m, which is

$$W = Fs = (-3 \text{ N})(2 \text{ m}) = -6 \text{ J}$$

Why is W negative in this case?

(c) The net work done during the entire interval is

$$W_{\text{net}} = 25 \text{ J} - 6 \text{ J} = \boxed{19 \text{ J}}$$

Summary

The **work** done by a *constant* force **F** acting on an object is defined as the product of the component of the force in the direction of the displacement and the magnitude of the displacement. If the force makes an angle θ with the displacement **s**, the work done by **F** is

$$W \equiv (F \cos \theta)s \qquad\qquad (5.1)$$

The **kinetic energy** of an object of mass m moving with a speed v is defined as

$$KE \equiv \frac{1}{2} mv^2 \qquad\qquad (5.5)$$

The **work-energy theorem** states that the net work done on an object by external forces equals the change in kinetic energy of the object:

$$W_{\text{net}} = KE_{\text{f}} - KE_{\text{i}} \qquad\qquad (5.6)$$

The **gravitational potential energy** of an object of mass m that is elevated a distance y above the Earth's surface is given by

$$PE \equiv mgy \qquad\qquad (5.7)$$

The elastic potential energy stored in a spring is given by

$$PE_{\text{s}} = \tfrac{1}{2}kx^2 \qquad\qquad (5.9)$$

where k is the spring constant and x is the distance the spring is compressed or extended from its unstretched (equilibrium) position.

A force is **conservative** if the work it does depends only on the initial and final positions of an object and not on the path selected to move between these points. A force that does not follow this criterion is said to be **nonconservative**.

The **law of conservation of mechanical energy** states that if the only force acting on a system is conservative, total mechanical energy is conserved:

$$KE_i + PE_i = KE_f + PE_f \qquad \text{(5.11)}$$

In the important case in which the only force doing any work on a system is the gravitational force, the law of conservation of mechanical energy reduces to

$$\frac{1}{2} mv_i{}^2 + mgy_i = \frac{1}{2} mv_f{}^2 + mgy_f \qquad \text{(5.12)}$$

The **work-energy theorem** states that the work done by all nonconservative forces acting on a system equals the change in the total mechanical energy of the system:

$$W_{nc} = (KE_f - KE_i) + (PE_f - PE_i) \qquad \text{(5.14)}$$

Average power is defined as the ratio of work done to the time interval during which the work is done:

$$\bar{P} = \frac{W}{\Delta t} \qquad \text{(5.16)}$$

If a force **F** acts on an object moving with an average velocity $\bar{v}$, the average power delivered to the object is given by

$$\bar{P} = F\bar{v} \qquad \text{(5.17)}$$

The total work done by a varying force is equal to the area under the force-displacement curve.

Additional Reading

Freeman J. Dyson, "Energy in the Universe," *Sci. American,* September 1971, p. 50.

A. Einstein and L. Infeld, *Evolution of Physics,* New York, Simon and Schuster, 1938.

"Energy and Power," *Sci. American,* September 1971. This entire issue is devoted to energy-related topics.

R. H. Romer, *Energy: An Introduction to Physics,* San Francisco, Freeman, 1976. This text contains many interesting energy applications.

G. H. Schurr, "Energy," *Sci. American,* September 1963, p. 110.

G. Waring, "Energy and the Automobile," *The Physics Teacher, 18,* 1980, p. 494. An informative article on fuel consumption by automobiles.

R. Wilson and W. J. Jones, *Energy, Ecology, and the Environment,* New York, Academic Press, 1974.

Questions

1. Explain why the work done by the force of sliding friction is negative when an object undergoes a displacement on a rough surface.

2. When a punter kicks a football, is he doing any work on the ball while his toe is in contact with the ball? Is he doing any work on the ball after it loses contact with his toe? Are there any forces doing work on the ball while it is in flight?

3. Discuss whether any physical work is being done by each of the following agents and, if so, whether the work done is positive or negative. (a) a chicken scratching the ground, (b) a person studying, (c) a crane lifting a bucket of concrete, (d) the force of gravity on the bucket in Part c, (e) the leg muscles of a person in the act of sitting down.

4. Does the gravitational force exerted on the Moon by the Earth do any work on the Moon? Explain.

5. As a pendulum swings back and forth, does the tension in the string do any work on the pendulum bob? Does the force of gravity do any work?

6. A child riding in the back seat of a car exerts a constant force on the front seat in an effort to make the car go faster. The child exerts a force on the car and the car is in motion. Does the child do any work?

7. If no force is acting on a body, can it have any energy? Explain.

8. Discuss the work done during "isometric exercise," that is, exercise in which the muscles are flexed with no external bodily movement.

9. Can the kinetic energy of an object have a negative value?

10. Does kinetic energy depend on the frame of reference, or coordinate system, in which it is measured?

11. What can be said about the speed of an object if the net work done on that object is zero?

12. A snowball is thrown against a wall with a velocity great enough that the ball melts on impact. Examine the energy changes in this collision.

13. An investigation at the scene of a collision between two cars shows that one car left skid marks twice the length of those left by the other car. A bystander who viewed the collision claims that the cars applied their brakes at the same time. What conclusions can you draw?

14. Two identical objects move with speeds of 5 m/s and 20 m/s. What is the ratio of their kinetic energies?

15. A ball is thrown straight up into the air. At what position is its kinetic energy a maximum? At what position is its gravitational potential energy a maximum?

16. Can the gravitational potential energy of an object ever have a negative value? Explain.

17. A pile driver is a device used to drive objects into the Earth by dropping a heavy weight on them. By how much does the energy of a pile driver increase when the weight it drops is doubled? (Assume the weight is dropped from the same height each time.)

18. A bowling ball is suspended from the ceiling of a lecture hall by a strong cord. The bowling ball is drawn from its equilibrium position and released from rest at the tip of the demonstrator's nose. If the demonstrator remains stationary, explain why she will not be struck by the ball on its return swing. Would this demonstrator be safe if the ball were given a slight push from this position?

19. A ball is dropped by a person from the top of a building, while another person at the bottom observes its motion. Will these two people always agree on the value of the ball's potential energy? On the change in potential energy of the ball? On the kinetic energy of the ball?

20. How can the work-energy theorem explain why the force of sliding friction always has the effect of reducing the kinetic energy of a particle?

21. Discuss the production and dissipation of mechanical energy in (a) lifting a weight, (b) holding the weight up, (c) lowering the weight slowly. Include the muscles in your discussion.

22. Advertisements for "superballs" once stated that the ball would rebound to a height greater than the height from which it was dropped. Is this possible?

23. During a heartbeat the heart does work (a) to move blood and (b) to inflate the arteries. Discuss this statement with regard to mechanical energy.

24. Discuss the work done by a pitcher when he throws a baseball. What is the approximate distance through which the force acts as the ball is thrown?

25. Discuss the kinetic and potential energy of various parts of the body when jumping off the ground.

26. Discuss the energy transformations that occur during a pole vault event.

27. Discuss the role of kinetic and potential energy in the following sports: (a) baseball, (b) football, (c) tennis, (d) basketball, (e) track.

28. During a stress test of the cardiovascular system, a patient walks and runs on a treadmill. (a) Is the energy expended by the patient equivalent to the energy of walking and running on the ground? Explain your answer. (b) If the treadmill is tilted upward, what effect does this have? Discuss.

29. Many mountain roads are built so that they spiral around the mountain rather than go straight up toward the peak. Discuss this design from the viewpoint of energy and power.

30. Estimate the time it takes you to climb a flight of stairs. Then approximate the power required to perform this task. Express your value in horsepower.

Problems

Section 5.2   Work

1. A tugboat exerts a constant force of 5000 N on a ship moving at constant speed through a harbor. How much work does the tugboat do on the ship if it moves a distance of 3 km?

2. A weight lifter lifts a 350-N set of weights from ground level to a position over his head, a vertical distance of 2 m. How much work does the weight-lifter do, assuming he moves the weights at constant speed?

3. If a man lifts a 20-kg bucket from a well and does 6 kJ of work, how deep is the well? Assume the speed of the bucket remains constant as it is lifted.

4. A scraper is drawn over a tooth 20 times, each time moving it a distance of 0.75 cm. The scraper is held against the tooth with a normal force of 5.0 N. Assuming a coefficient of kinetic friction of 0.9 between the scraper and the tooth, determine the work done to clean the tooth.

5. A shopper in a supermarket pushes a cart with a force of 35 N directed at an angle of 25° downward from the horizontal. Find the work done by the shopper as she moves down a 50-m length of aisle.

6. A man moves a broom horizontally by pushing on it with a force of 80 N directed at 50° below the horizontal. When he stopped moving, he had done 450 J of work. How far did he push the broom?

7. Batman, whose mass is 80 kg, is holding onto the free end of a 12-m rope, the other end of which is fixed to a tree limb above. He is able to get the rope in motion as only Batman knows how, eventually getting it to swing enough that he can reach a ledge when the rope makes a 60° angle with the downward vertical. How much work was done against the force of gravity in this maneuver?

8. A 70-kg baserunner begins his slide into second base when moving at a speed of 4 m/s. The coefficient of friction between his clothes and Earth is 0.7. He slides such that his speed has been reduced to zero just as he reaches the base. (a) How much work was done by friction acting on the runner? (b) How far did he slide?

9. A horizontal force of 150 N is used to push a 40-kg packing crate on a rough horizontal surface through a distance of 6 m. If the crate moves at constant speed, find (a) the work done by the 150-N force, (b) the work done by friction, and (c) the coefficient of kinetic friction.

10. A stewardess pulls her 70-N flight bag a distance of 200 m along an airport floor at constant speed. The force she exerts is 40 N at an angle of 50° above the horizontal. Find (a) the work she does, (b) the work done by the force of friction, and (c) the coefficient of kinetic friction between her flight bag and the floor.

Section 5.3   Work and Kinetic Energy

11. What is the kinetic energy of a 3000-kg car moving at 55 mi/h? How much heat, in joules, will be lost through friction in the brake linings when the car is brought to a stop?

12. A very good major league pitcher can throw a baseball at 100 mi/h, while an average pitcher throws the ball at a speed of about 80 mi/h. If a baseball has a mass of 0.15 kg, find the kinetic energy of the ball in each case.

13. A 0.6-kg particle has a speed of 2 m/s at point A and kinetic energy of 7.5 J at point B. What is (a) its kinetic energy at point A? (b) its velocity at point B? (c) the total work done on the particle as it moves from A to B?

14. A person doing a chin-up weighs 700 N exclusive of the arms. During the first 25.0 cm of the lift, each arm exerts an upward force of 355 N on the torso. If the upward movement starts from rest, what is the person's velocity at this point?

15. A gas molecule in the room you are sitting in may have a speed of about 500 m/s. If its mass is equal to 5×10^{-26} kg, what is its kinetic energy?

16. A 2-g bullet leaves the barrel of a gun at a speed of 300 m/s. (a) Find its kinetic energy. (b) If the length of the barrel is 50 cm, find the average force exerted on the bullet by the expanding gases as it moves the length of the barrel.

17. A 1200-kg vehicle is moving along a horizontal surface with a speed of 20 m/s. What work must be done by the brakes to bring the vehicle to rest in 20 s?

18. A mechanic pushes a 2500-kg car from rest to a speed v doing 5000 J of work in the process. During this time, the car moves 25 m. Neglecting friction between the car and the road, find (a) the final speed, v, of the car and (b) the horizontal force exerted on the car.

19. A crate of mass 10 kg is pulled up a rough incline with an initial speed of 1.5 m/s. The pulling force is 100 N parallel to the incline that makes an angle of 20° with the horizontal. If the coefficient of kinetic friction is 0.4 and the crate is pulled a distance of 5 m, (a) how much work is done by the gravitational force? (b) How much work is done by the friction force? (c) How much work is done by the 100-N force? (d) What is the change in kinetic energy of the crate? (e) What is the speed of the crate after being pulled 5 m?

20. A 2000-kg car moves down a level highway under the action of two forces. One is a forward force on the car of 1000 N from the motor; the second is a 950-N resistive force. Use the work-energy theorem to find the speed of the car after it has moved a distance of 20 m, assuming it starts from rest.

21. A 10-kg sled is given a kick on a frozen pond, imparting to it an initial speed $v_0 = 2$ m/s. The coefficient of kinetic friction between the sled and ice is $\mu_k = 0.1$. Use the work-energy theorem to find the distance the sled moves before coming to rest.

Section 5.4   Gravitational Potential Energy
Section 5.5   Conservative and Nonconservative Forces

22. What is the gravitational potential energy relative to the ground of a 0.15-kg baseball at the top of a 100-m tall building?

23. A 2-kg ball hangs at the end of a string 1 m in length from the ceiling of a room. The height of the room is 3 m. What is the gravitational potential energy of the ball relative to (a) the ceiling, (b) the floor, and (c) a point at the same elevation as the ball?

24. A 1000-kg roller coaster is initially at the top of a rise, point A. It then moves 135 ft at an angle of 40° below the horizontal to a lower point B. (a) Choose point B to be the zero level for gravitational potential energy, and find the potential energy of the roller coaster at points A and B, and the difference in potential energy between these points. (b) Repeat part (a), choosing the zero reference level to be at point A.

25. Find the net force along the slope acting on the roller coaster of Problem 24 and use this to calculate the work done as it moves from A to B. (Ignore any losses due to friction.) Compare your answer to the differences in potential energy found in Problem 24.

26. A 40-N child is in a swing attached to ropes 2 m long. Find the gravitational potential energy of the child relative to the child's lowest position when (a) the ropes are horizontal, (b) the ropes make a 30° angle with the vertical, and (c) at the bottom of the circular arc.

27. A 3-kg particle moves from the origin to the position having coordinates $x = 5$ m and $y = 5$ m under the influence of gravity acting in the negative y direction (Fig. 5.21.) Using Equation 5.1, calculate the work done by gravity in going from O to C along the following paths. (a) OAC, (b) OBC, (c) OC. Your results should all be identical. Why?

28. A particle moves in the xy plane in Figure 5.21 under the influence of a frictional force that opposes its dis-

Figure 5.21
Problems 27 and 28

placement. If the frictional force has a magnitude of 3 N, calculate the total work done by friction along the following *closed* paths: (a) the path OA followed by the return path AO, (b) the path OA followed by AC and the return path CO, and (c) the path OC followed by the return path CO. (d) Your results for the three closed paths should all be different and nonzero. What is the significance of this?

Section 5.6   Conservation of Mechanical Energy

29. A baseball player decides to show his skill by catching a 0.15-kg baseball dropped from the top of a 100-m tall building. What is the speed of the baseball just before it strikes his glove?

30. A child and sled with a combined mass of 50 kg slide down a frictionless hill. If the sled starts from rest and has acquired a speed of 3 m/s by the time it reaches the bottom of the hill, what is the height of the hill?

31. A 70-kg high-jumper leaves the ground with a speed of 6 m/s. How high can he leap?

32. A baseball player throws a baseball straight up into the air with an initial speed of 20 m/s. Find (a) the maximum height to which the ball rises, and (b) the speed of the ball when it is halfway up to its maximum height.

33. Tarzan swings on a vine 30 m long initially inclined at an angle of 37° to the vertical. (a) What is his speed at the bottom of the swing if he starts from rest? (b) If he pushes off with a speed of 4 m/s, what is his speed at the bottom of the swing?

34. The masses of the javelin, discus, and shot are 0.8 kg, 2.0 kg, and 7.2 kg, respectively, and record throws in the track events using these objects are about 89 m, 69 m, and 21 m, respectively. Neglecting air resistance, (a) calculate the minimum initial kinetic energies that would produce these throws, and (b) estimate the average force exerted on each object during the throw, assuming the force acts over a distance of 2 m. (c) Do your results suggest that air resistance is an important factor?

Section 5.7 Nonconservative Forces and the Work-Energy Theorem

35. A child starts from rest at the top of a slide of height 4 m. (a) What is her speed at the bottom if the incline is frictionless? (b) If she reaches the bottom with a speed of 6 m/s, what percentage of her total energy at the top of the slide is lost as a result of friction?

36. A 2000-kg car starts from rest at the top of a driveway 5 m long which is sloped at an angle of 20° with the horizontal. If an average friction force of 4000 N impedes the motion of the car, find the speed of the car at the bottom of the driveway.

37. A softball pitcher hurls a ball of mass 0.25 kg around a vertical circular path of radius 60.0 cm before releasing it. The pitcher maintains a component of force of constant magnitude 30.0 N on the ball *in the direction* of motion around the complete path. The speed of the ball at the top of the circle is 15.0 m/s. If the ball is released at the bottom of the circle, what is its speed upon release?

38. A 70-kg diver steps off a 10-meter tower and drops straight down into the water. If he comes to rest 5 meters beneath the surface of the water, determine the average resistance force exerted on the diver by the water.

39. A 25-kg child on a swing 2 m long is released from rest when the swing supports make an angle of 30° with the vertical. (a) Neglecting friction, find the child's speed at the lowest position. (b) If the speed of the child at the lowest position is 2 m/s, what is the energy lost due to friction?

40. An airplane of mass 1.5×10^4 kg is moving at a speed of 60 m/s. The pilot then increases the speed of the engine such that the forward force of the propeller becomes 7.5×10^4 N. The force of air resistance on the airplane is 4×10^4 N. The pilot then goes into a dive at an angle of 30° with the horizontal. Find the speed of the plane after it has traveled 500 m.

41. In a circus performance, a monkey on a sled is given an initial speed of 4 m/s up a 20° incline. The combined mass of the monkey and sled is 20 kg, and the coefficient of kinetic friction between sled and incline is 0.2. How far will the sled move up the incline?

42. A 2-kg block is projected up a 30° incline with an initial speed of 3 m/s at the bottom. The coefficient of kinetic friction between the block and the incline is 0.7. Find (a) the distance the block will travel up the incline before coming to rest, (b) the total work done by friction while the block is in motion, and (c) the change in potential energy and change in kinetic energy when the block has traveled 0.3 m up the incline.

43. A woman weighing 500 N glides across some ice, starting her glide with a speed of 4.0 m/s. If the coefficient of friction between the skates and the ice is 0.115, how far does she go before coming to rest? Solve this problem by using the method of work-energy.

44. A skier starts from rest at the top of a hill inclined at an angle of 10.5° to the horizontal. The hill is 200 m long, and there is a coefficient of friction 0.075 between the snow and the skis. At the bottom of the hill the snow is level and the coefficient of friction is unchanged. How far does the skier move along the horizontal portion of the snow before coming to rest?

Section 5.9 Power

45. A 1500-kg car accelerates uniformly from rest to a speed of 10 m/s in 3 s. Find (a) the work done on the car in this time, (b) the average power delivered by the engine in the first 3 s, and (c) the instantaneous power delivered by the engine at $t = 2$ s.

46. A 1500-kg car starts from rest and accelerates to a final speed of 18 m/s in a time of 12 s. Assume that the force of air resistance remains constant at 400 N during this time. Find (a) the average power developed by the engine of the car, and (b) the instantaneous power output of the engine at $t = 12$ s.

47. What average power is developed by an 800-N professor while running at constant speed up a flight of stairs rising 6 m if he takes 8 s to complete the climb?

48. A motor is used to pull an 80-kg skier along a horizontal surface at a constant speed of 2 m/s to enable him to learn to keep his balance. If the coefficient of kinetic friction between the skis and the surface is 0.15, what horsepower motor is required?

49. A machine lifts a 300-kg crate at constant speed through a height of 5 m in 8 s. Calculate the power output of the machine.

50. A skier of mass 70 kg is pulled up a slope by a motor-driven cable. (a) How much work is required to pull him a distance of 60 m up a 30° slope (assumed frictionless) at a constant speed of 2 m/s? (b) What horsepower motor is required to perform this task?

51. A 50-kg student climbs a rope 5 m in length and stops at the top. (a) What must her average speed have been in order to match the power output of a 200-W lightbulb? (b) How much work did she do?

52. You have just found the "Lost Dutchman" goldmine, 80 ft below ground level. However, the mine is flooded, and you discover that in order to keep it dry enough to mine, you must pump water out at the rate of 60 lb every second. What is the minimum horsepower motor that can be used to perform this task?

53. An elevator having a mass of 650 kg starts upward from rest. It moves upward for 3.0 s with constant acceleration until reaching its cruising speed of 1.75 m/s. (a) What is the average power of the elevator motor during this period? (b) How does this compare with the power while cruising upward with constant speed?

54. A speedboat requires 130 hp to move at a constant speed of 15 m/s ($\approx$ 33 mi/h). Calculate the resistive force due to the water at that speed.

Section 5.10   Work Done by a Varying Force

55. A 1-kg object starts from rest, but then moves through a distance of 3 m when a nonconstant force begins to act on it. The force varies with position as shown in Figure 5.22. (a) How much work is done on the object by this force and (b) what is its speed at $x = 3$ m?

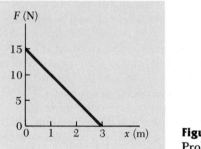

Figure 5.22
Problem 55

56. A body is subjected to a force that varies with position as shown in Figure 5.23. Find the work done by the force on the object as it moves (a) from $x = 0$ to $x = 5$ m, (b) from $x = 5$ m to $x = 10$ m, and (c) from $x = 10$ m to $x = 15$ m. (d) What is the total work done by the force over the distance $x = 0$ to $x = 15$ m?

57. If the object of Problem 56 had a mass of 3 kg and a speed of 0.5 m/s at $x = 0$ find its speed at (a) $x = 5$ m, (b) $x = 10$ m, and (c) $x = 15$ m.

Figure 5.23
Problem 56

Additional Problems

58. Find the height from which you would have to drop this textbook so that it would have a speed of 10 m/s just before it hits the pavement.

59. Figure 5.24 shows the top surface of a rough table. A textbook is moved from position A, the origin of a coordinate system, to position B, which has coordinates (2, 0) where the units are meters. If the magnitude of the frictional force is 4 N, find the work done against friction as the book is moved from A to B along (a) the circular path, (b) the triangular path, and (c) the straight-line path.

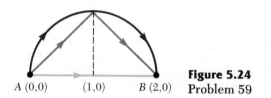

A (0,0)   (1,0) B (2,0)
Figure 5.24
Problem 59

60. The plane shown in Figure 5.25 is designed for vertical takeoff and landing. If the plane has a mass of 8000 kg when fueled, find the net work done on the plane as it accelerates upward at 1 m/s² for a distance of 30 m starting from rest.

Figure 5.25
Problem 60
(Courtesy of NASA)

61. A 0.4-kg bead slides on a curved wire, starting from rest at point A in Figure 5.26. If the wire is frictionless, find the speed of the bead (a) at B and (b) at C.

Figure 5.26
Problems 61 and 65

62. The engine of a jet aircraft develops a forward thrust of 4000 lbs. What horsepower is it developing when flying at its cruising speed of 500 mi/h?

63. A 200-kg crate is pulled parallel to a level surface by an engine. The coefficient of kinetic friction between crate and surface is 0.4. (a) How much power must the engine deliver to move the crate with a constant speed of 5 m/s? (b) How much work is done by the engine in 3 min?

64. A 90-horsepower engine is moving a car at a speed of 60 mi/h. What forward force is the engine exerting on the car?

65. If the wire in Problem 61 (Fig. 5.26) is frictionless between points A and B and rough between points B and C, and the bead starts from rest at point A, (a) find the speed of the bead at B. (b) If the 0.4-kg bead comes to rest at C, find the total work done by friction as the bead goes from B to C.

66. A catcher "gives" with the ball when he catches a 0.15-kg baseball moving at 25 m/s. (a) If he moves his catching glove through a distance of 2 cm, what is the average force acting on his hand? (b) Repeat for the case in which his hand moves 10 cm.

67. A grocery cart weighing 98 N is pushed 12 m across the floor by a shopper who exerts a constant horizontal force of 40 N. If all frictional forces are neglected, what is the final speed of the cart if it starts from rest?

68. If the speed of the cart of Problem 67 is half that found based on the prediction of no friction, find the coefficient of friction between cart and floor surface.

69. (a) A 75-kg man jumps from a window 1 m above a sidewalk. What is his speed just before his feet strike the pavement? (b) If the man jumps with his knees and ankles locked, the only cushion for his fall is an approximate 0.5-cm give in the pads of his feet. Calculate the average force exerted on him by the ground in this situation. This average force is sufficient to cause cartilage damage in the joints or to break bones.

70. Tarzan and Jane, whose total mass is 130 kg, start their swing on a vine 5 m long such that the vine is initially at an angle of 30° with the horizontal. At the bottom of the arc, Jane, whose mass is 50 kg, steps off. What is the maximum height of the tree branch that Tarzan can land on as his swing continues?

71. A 6-kg object is lifted vertically through a distance of 5 m by a light string under a tension of 80 N. Find (a) the work done by the force of the tension, (b) the work done by gravity, and (c) the final speed of the object if it starts from rest.

72. A 0.4-mg flea can jump to a maximum height of about 2.5 cm. Find the speed with which he takes off.

73. A 65-kg athlete runs 600 m at constant speed up a mountain inclined at 20° to the horizontal. He performs this feat in 150 s. Assuming that air resistance is negligible, (a) how much work does he perform and (b) what is his power output during the run?

74. A 3-kg mass is moving such that it has a component of velocity along the x direction of 5 m/s and a component of velocity along the y direction of −3 m/s. (a) What is the kinetic energy at this time? (b) Find the change in kinetic energy if the velocity changes such that the new x component of velocity is 8 m/s and its new y component of velocity is 4 m/s.

75. A projectile of mass m is shot horizontally with an initial speed v_0 from a height h above a flat desert surface. At the instant before the projectile hits the desert surface, find (a) the work done on the projectile by gravity, (b) the change in kinetic energy since the projectile was fired, and (c) the final kinetic energy of the projectile.

76. Two blocks A and B (with mass 50 kg and 100 kg, respectively) are connected by a string as shown in Figure 5.27. The pulley is frictionless and of negligible mass. The coefficient of kinetic friction between block A and the incline is $\mu_k = 0.25$. Determine the change in the kinetic energy of block A as it moves from C to D, a distance of 20 m up the incline.

Figure 5.27
Problem 76

77. A passenger car of mass 1500 kg accelerates from rest to 97 km/h in 10 s. (a) Find the acceleration of the car. (b) Show that the coefficient of friction between the rear tires and road must be at least 0.55. (c) Determine the average power delivered by the engine. (Assume that the normal force on each tire is $\frac{1}{4}$ mg.)

78. A 3-kg block is moved up a 37° incline under the action of a constant horizontal force of 40 N. The coefficient of kinetic friction is 0.1, and the block is pulled 2 m up the incline. Calculate (a) the work done by the 40 N force, (b) the work done by gravity, (c) the work done by friction, and (d) the change in kinetic energy of the block. (Note: the applied force is not parallel to the incline.)

79. A 0.4-kg mass slides on a horizontal circular track of radius 1.5 m. The mass is given an initial speed of 8 m/s. After one revolution, its speed drops to 6 m/s because of friction. Find (a) the work done by the force of friction in one revolution and (b) the coefficient of kinetic friction. (c) How many revolutions does the mass make before coming to rest?

80. A 200-g particle is released from rest at point A along the diameter of the inside of a smooth hemispherical bowl of radius $R = 30$ cm (Fig. 5.28). Calculate (a) its gravitational potential energy at point A relative to point B, (b) its kinetic energy at point B, (c) its speed at point B, and (d) its kinetic energy and potential energy at point C.

81. The particle described in Problem 80 (Fig. 5.26) is released from point A at rest. The speed of the particle at point B is 1.5 m/s. (a) What is its kinetic energy at B? (b) How much energy is lost as a result of friction as the particle goes from A to B? (c) Is it possible to determine μ from these results in any simple manner? Explain.

Figure 5.28
Problems 80 and 81

82. A spring has a force constant of 500 N/m. Show that (a) the energy stored in the spring is 0.400 J when the spring is stretched 4 cm from equilibrium, (b) the stored energy is 0.225 J when the spring is compressed 3 cm from equilibrium, and (c) no energy is stored when the spring is unstretched.

83. A block of mass 0.25 kg is placed on a vertical spring of constant $k = 5000$ N/m and is pushed downward, compressing the spring a distance of 0.1 m. As the block is released, it leaves the spring and continues to travel upward. Show that the block will rise to a height of 10.2 m above the point of release.

84. A block of mass 2 kg is kept at rest by compressing a horizontal massless spring having a spring constant $k = 100$ N/m by 10 cm. As the block is released, it travels on a rough horizontal surface a distance of 0.25 m before it stops. Show that the coefficient of kinetic friction between the horizontal surface and the block is 0.102.

CHAPTER 6

Momentum and Collisions

What would William Tell think? An apple being pierced by a 30-caliber bullet traveling at a supersonic speed of 900 m/s. This collision was photographed with a microflash stroboscope using an exposure time of 0.33 μs. Shortly after the photograph was taken, the apple disintegrated completely. Note that the points of both entry and exit of the bullet are visually explosive. (*Shooting the Apple, 1964. © Harold E. Edgerton. Courtesy of Palm Press, Inc.*)

Consider what happens when a golf ball is struck by a club. The ball is given a very large initial velocity as a result of the collision and consequently is able to travel several hundred feet through the air. The ball experiences a large change in velocity and a correspondingly large acceleration. Furthermore, because the ball experiences this acceleration over a very short time interval, the average force on the ball during collision with the club is very large. By Newton's third law, the club experiences a reaction force that is equal to and opposite the force on the ball. This reaction force produces a change in the velocity of the club. Since the club is much more massive than the ball, however, its change in velocity is much less than the change in velocity of the ball.

One of the main objectives of this chapter is to understand and analyze such events. As a first step, we shall introduce the concept of momentum, a term often used in describing objects in motion. For example, a very massive football player is often said to have a great deal of momentum as he runs down the field. A much less massive player, such as a halfback, can have equal or greater momentum if he moves with a higher velocity. This follows from the fact that momentum is defined as the product of mass and velocity.

Additionally, we shall find that the concept of momentum will lead us to a second conservation law, that of conservation of momentum. This law is especially useful for treating problems that involve collisions between objects.

6.1   Momentum and Impulse

The momentum of an object of mass m moving with a velocity $\mathbf{v}$ is defined as the product of the mass and the velocity:

Momentum

$$\mathbf{p} \equiv m\mathbf{v} \tag{6.1}$$

As its definition shows, momentum is a vector quantity, with its direction the same as that of the velocity. Momentum has dimensions of [M][L]/[T], and its SI units are kilogram-meters per second (kg·m/s).

Often we shall find it advantageous to work with the components of momentum. For two-dimensional motion, these are given by

$$p_x = mv_x \qquad p_y = mv_y$$

where p_x represents the momentum of an object in the x direction and p_y its momentum in the y direction.

The definition of momentum in Equation 6.1 coincides with our customary usage of the word in our daily affairs. When we think of a massive object moving with a high velocity, we often say that this object has a large momentum, in accordance with Equation 6.1. Likewise, a small object moving slowly is said to have a small momentum. On the other hand, a small object moving with a high velocity can have a large momentum.

When Newton first expressed the second law in mathematical form, he did not write it as $\mathbf{F} = m\mathbf{a}$; instead he wrote it as

$$\mathbf{F} = \frac{\text{change in momentum}}{\text{time interval}} = \frac{\Delta\mathbf{p}}{\Delta t} \tag{6.2}$$

where Δt is the time interval during which the momentum changes by $\Delta\mathbf{p}$. This equation says that the rate of change of momentum of an object is equal to the resultant force acting on the object. To see that this is equivalent to $\mathbf{F} = m\mathbf{a}$ for an object of constant mass, consider a constant force $\mathbf{F}$ acting on an object and producing a constant acceleration. We can write Equation 6.2 as

$$\mathbf{F} = \frac{\Delta\mathbf{p}}{\Delta t} = \frac{m\mathbf{v}_f - m\mathbf{v}_i}{\Delta t} = \frac{m(\mathbf{v}_f - \mathbf{v}_i)}{\Delta t} \tag{6.3}$$

Now recall that the velocity of an object moving with constant acceleration varies with time as

$$\mathbf{v}_f = \mathbf{v}_i + \mathbf{a}t$$

If we take $\Delta t = t$ and substitute for v_f in Equation 6.3, we see that $\mathbf{F}$ reduces to the familiar equation

$$\mathbf{F} = m\mathbf{a}$$

Note from Equation 6.2 that

if the resultant force $\mathbf{F}$ is zero, the momentum of the object does not change. In other words, the linear momentum and velocity of an object are conserved when $\mathbf{F} = 0$.

This feature, or property, of momentum will be important to us in analyzing collisions, a subject we shall take up in a later section.

Equation 6.2 can be written as

$$\mathbf{F}\,\Delta t = \Delta \mathbf{p}$$

or

$$\mathbf{F}\,\Delta t = \Delta \mathbf{p} = m\mathbf{v_f} - m\mathbf{v_i} \qquad (6.4)$$

This result is often called the **impulse-momentum theorem.** The term on the left side of the equation, $\mathbf{F}\,\Delta t$, is called the **impulse** of the force $\mathbf{F}$ for the time interval Δt. According to this result, *the impulse of the force acting on an object equals the change in momentum of that object.*

This equation tells us that if we exert a force on an object for a time interval Δt, the effect of this force is to change the momentum of the object from some initial value $m\mathbf{v_i}$ to some final value $m\mathbf{v_f}$. For example, suppose a pitcher throws a baseball with a velocity $\mathbf{v_i}$ and a batter hits the ball head-on so as to reverse the direction of its velocity. The force $\mathbf{F}$ that the bat exerts on the ball can change both the direction and the magnitude of the initial velocity to a higher value $\mathbf{v_f}$.

It is necessary to add a word of caution here. If you were to try to solve a problem such as this using Newton's second law, you would encounter some difficulty in choosing a value for $\mathbf{F}$ because the force exerted on the ball is not constant. Instead, it might be represented by a curve like that of Figure 6.1a. The force starts out small as the bat comes in contact with the ball, rises to a maximum value when they are firmly in contact, and then drops off as the ball leaves the bat. In such instances, it is necessary to define an **average force** $\overline{\mathbf{F}}$, shown as the dashed line in Figure 6.1b. This average force can be thought of as the constant force that gives the same impulse to the object in the time interval Δt as the actual time-varying force gives over this same interval. The short collision between a bullet and a king of diamonds is illustrated in Figure 6.2.

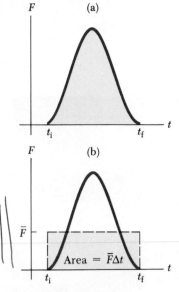

Figure 6.1
(a) A force acting on an object may vary in time. The impulse is the area under the force-time curve. (b) The average force (horizontal dashed line) gives the same impulse to the object in the time interval Δt as the real time-varying force described in (a).

Figure 6.2
Cutting the card quickly: A bullet traveling at about 900 m/s collides with the king of diamonds. The photograph was taken using a microflash stroboscope at an exposure time of less than 1 μs. You should be able to show that the bullet moved less than 1 mm during this time interval. (This explains why the action is "frozen.") (© *Harold E. Edgerton. Courtesy of Palm Press, Inc.*)

The impulse imparted by a force during a time interval Δt is equal to the area under the force-time graph from the beginning to the end of the time interval.

Example 6.1   Teeing Off

A golf ball of mass 50 g is struck with a club, as in Figure 6.3. The force on the ball varies from zero when contact is made up to some maximum value (where the ball is deformed) and then back to zero when the ball leaves the club. Thus the force-time graph is somewhat like that shown in Figure 6.1. Assume that the ball leaves the club face with a velocity of $+44$ m/s, and (a) estimate the impulse due to the collision.

Solution
Before the club hit the ball, the ball was at rest on the tee, and as a result its initial momentum was zero. The momentum immediately after the collision is

$$p_f = mv_f = (50 \times 10^{-3} \text{ kg})(44 \text{ m/s}) = +2.20 \text{ kg·m/s}$$

and thus the impulse imparted to the ball, which equals its change in momentum, is

$$\Delta p = mv_f - mv_i = \quad +2.20 \text{ kg·m/s}$$

(b) Estimate the length of time of the collision and the average force on the ball.

Solution
From Figure 6.3, it appears that a reasonable estimate of the distance the ball travels while in contact with the club is the radius of the ball, about 2.0 cm. The time it takes the club to move this distance (the contact time) is then

$$\Delta t = \frac{\Delta x}{v_i} = \frac{2 \times 10^{-2} \text{ m}}{44 \text{ m/s}} = 4.5 \times 10^{-4} \text{ s}$$

Finally, the magnitude of the average force is estimated to be

$$\overline{F} = \frac{\Delta p}{\Delta t} = \frac{2.2 \text{ kg·m/s}}{4.5 \times 10^{-4} \text{ s}} = \quad 4.9 \times 10^3 \text{ N}$$

Figure 6.3   (Example 6.1) A golf ball being struck by a club. *(Courtesy of Dr. Harold E. Edgerton, MIT)*

$h = 2.0$ m

$h' = 1.5$ m

v_i

v_f

(a)   (b)

Figure 6.4   (Example 6.2) (a) A ball is dropped from a height h and reaches the floor with a velocity v_i. (b) The ball rebounds from the floor with a velocity v_f and reaches a height h'.

Example 6.2   Follow the Bouncing Ball

A ball of mass 100 g is dropped from a height $h = 2$ m above the floor (Fig. 6.4). It rebounds vertically to a height $h' = 1.5$ m after colliding with the floor. (a) Find the momentum of the ball immediately before it collides with the floor, and immediately after it rebounds.

Solution
We can find the velocity of the ball just before it strikes the floor by using the principle of conservation of mechanical energy. Equating the initial potential energy to the final kinetic energy, with the floor as the reference level, gives

$$mgh = \frac{1}{2} mv_i^2$$

$$v_i = \sqrt{2gh} = \sqrt{2(9.80 \text{ m/s}^2)(2 \text{ m})} = -6.26 \text{ m/s}$$

where the minus sign means that the initial velocity is in the negative y direction, as in Figure 6.4.

Likewise, v_f, the ball's velocity after colliding with the floor, is obtained from the energy expression

$$\frac{1}{2} mv_f^2 = mgh'$$

$$v_f = \sqrt{2gh'} = \sqrt{2(9.80 \text{ m/s}^2)(1.5 \text{ m})} = +5.42 \text{ m/s}$$

Because $m = 0.1$ kg, the initial and final momenta are

$$p_i = mv_i = (0.1 \text{ kg})(-6.26 \text{ m/s}) = \boxed{-0.626 \text{ kg·m/s}}$$

and

$$p_f = mv_f = (0.1 \text{ kg})(5.42 \text{ m/s}) = \boxed{+0.542 \text{ kg·m/s}}$$

The initial momentum is negative because the velocity, and hence the momentum, are directed downward, in the negative direction.

(b) Determine the average force exerted by the floor on the ball. Assume the time interval of the collision is 10^{-2} s (a typical value).

Solution
From the impulse-momentum theorem, we find

$$\overline{F} \Delta t = mv_f - mv_i$$

$$\overline{F} = \frac{[0.542 - (-0.626)] \text{ kg·m/s}}{10^{-2} \text{ s}} = \boxed{+1.17 \times 10^2 \text{ N}}$$

Example 6.3   How Good are the Bumpers?

In a particular crash test, an automobile of mass 1500 kg collides with a wall as in Figure 6.5. The initial and final velocities of the automobile are $v_i = -15.0$ m/s and $v_f = +2.6$ m/s, respectively. If the collision lasts for 0.150 s, find the impulse due to the collision and the average force exerted on the automobile.

Solution
The initial and final momenta of the automobile are given by

$$p_i = mv_i$$

$$= (1500 \text{ kg})(-15.0 \text{ m/s}) = -2.25 \times 10^4 \text{ kg·m/s}$$

$$p_f = mv_f = (1500 \text{ kg})(2.6 \text{ m/s}) = +0.39 \times 10^4 \text{ kg·m/s}$$

Figure 6.5   (Example 6.3)

−15.0 m/s   Before   2.6 m/s   After

B has KE
- in coming to rest when in
contact with A through a distance Δs
$\Delta KE = W = F \Delta s$
- F is force on A, work done on A
is $W = F \Delta s \to W = ma \Delta s$
- a can give v and other values
of kinematics

(a)

(b)

Figure 6.6
(a) The collision between two objects as the result of direct contact. (b) The collision between two charged objects.

Hence, the impulse, which equals the change in momentum, is

$$I = \Delta p = p_f - p_i$$
$$= 0.39 \times 10^4 \ \text{kg} \cdot \text{m/s} - (-2.25 \times 10^4 \ \text{kg} \cdot \text{m/s})$$
$$\boxed{I = \ +2.64 \times 10^4 \ \text{kg} \cdot \text{m/s}}$$

The average force exerted on the automobile is

$$\overline{F} = \frac{\Delta p}{\Delta t} = \frac{2.64 \times 10^4 \ \text{kg} \cdot \text{m/s}}{0.150 \ \text{s}} = \boxed{+1.76 \times 10^5 \ \text{N}}$$

6.2   Conservation of Momentum

In this section we shall describe what happens when two objects collide with one another. The collision may be the result of physical contact between two objects, as described in Figure 6.6a. This is a common observation for large-scale objects, such as two billiard balls or a baseball and bat. However, the notion of what we mean by a collision must be generalized because "contact" in the atomic world is different from what we think of as contact in our everyday experiences. For example, Figure 6.6b shows the "collision" between a proton and an alpha particle (the nucleus of a helium atom). Because the two atomic particles are positively charged, they repel each other because of the electrostatic forces between them. Although the two particles never "touch" each other, the results of our analysis here will still apply to this situation.

Figure 6.7 shows a before-and-after picture of two objects colliding. Before the collision, the velocity of object A is $\mathbf{v}_{1i}$ and that of B is $\mathbf{v}_{2i}$; after the collision, the velocities are $\mathbf{v}_{1f}$ and $\mathbf{v}_{2f}$. The impulse-momentum theorem applied to object A becomes

$$\overline{\mathbf{F}}_1 \, \Delta t = m_1 \mathbf{v}_{1f} - m_1 \mathbf{v}_{1i}$$

Likewise, for object B we have

$$\overline{\mathbf{F}}_2 \, \Delta t = m_2 \mathbf{v}_{2f} - m_2 \mathbf{v}_{2i}$$

where $\overline{\mathbf{F}}_1$ is the force exerted on A by B during the collision and $\overline{\mathbf{F}}_2$ is the force exerted on B by A during the collision (Fig. 6.8).

We are using average values for $\overline{\mathbf{F}}_1$ and $\overline{\mathbf{F}}_2$ even though the actual forces may vary in time in a complicated way, such as described in Figure 6.9. However, regardless of how complicated the forces between the objects may be, Newton's third law states that at all times these two forces are equal in magnitude and opposite in direction ($\mathbf{F}_1 = -\mathbf{F}_2$), as shown in Figure 6.9. Additionally, both forces act for the same time interval. Thus,

$$\overline{\mathbf{F}}_1 \, \Delta t = -\overline{\mathbf{F}}_2 \, \Delta t$$

or

$$m_1 \mathbf{v}_{1f} - m_1 \mathbf{v}_{1i} = -(m_2 \mathbf{v}_{2f} - m_2 \mathbf{v}_{2i})$$

from which we find

Figure 6.7
Before and after the head-on collision between two objects. The momentum of each object changes as a result of the collision, but the total momentum of the system remains constant.

$$m_1\mathbf{v}_{1i} + m_2\mathbf{v}_{2i} = m_1\mathbf{v}_{1f} + m_2\mathbf{v}_{2f} \qquad \textbf{(6.5)}$$

Conservation of momentum

This result, known as the **law of conservation of momentum**, indicates that, when no external forces act on a system consisting of two objects, the total momentum of the system before the collision is equal to the total momentum of the system after the collision.

In the statement of conservation of momentum, note that momentum is said to be conserved for a *system* of objects. In the example used to derive Equation 6.5, the system was taken to be two colliding objects. More generally, a system includes all the objects that are interacting with one another. Additionally, we have assumed that the only forces acting during the collision were internal forces, meaning those forces which arise between the interacting objects of the system. For example, in our collision of two objects shown in Figure 6.8, the internal forces are the forces $\mathbf{F}_1$ and $\mathbf{F}_2$. If a third object outside our system (consisting of m_1 and m_2) were to exert a force on either m_1 or m_2 (or both objects) during the collision, momentum would not be conserved for the system. In all the example problems that we shall consider, the system will be assumed to be *isolated*. An isolated system is a system that has no external forces exerted on it, or one in which the external forces are small enough to be considered negligible.

Our derivation has assumed only two objects interacting, but the result remains valid regardless of the number involved. In its most general form, we can state the law of conservation of momentum as follows:

The total momentum of an isolated system of objects is conserved regardless of the nature of the forces between the objects.

To understand what this statement means, consider the following circumstance. Imagine that you are initially standing at rest and then jump upward, leaving the ground with a velocity **v**. Obviously, your momentum is not conserved because your momentum before the jump was zero and became $m\mathbf{v}$ as you began to rise. However, the total momentum of the system is conserved if the system selected includes all objects that exert forces on one another. In the present case, you must include the Earth as part of the system because you exert a downward force on the Earth when you jump. The Earth in turn exerts on you an upward force of the same magnitude, as required by Newton's third law. Momentum is conserved for the system consisting of you and the Earth. Thus, as you move upward with some momentum $m\mathbf{v}$, the Earth moves downward with equal but opposite momentum. The recoil velocity of the Earth due to this event will be imperceptibly small, of course, because the Earth is so massive, but its momentum does exist.

Figure 6.8
When two objects collide, the force $\mathbf{F}_1$ exerted on object **A** is equal to magnitude and opposite in direction to the force $\mathbf{F}_2$ exerted on object **B**.

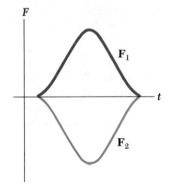

Figure 6.9
The force as a function of time for the two colliding particles described in Figure 6.8. Note that $\mathbf{F}_1 = -\mathbf{F}_2$.

Example 6.4 The Recoiling Pitching Machine

A baseball player attempts to use a pitching machine to help him improve his batting average. To do so, he places the 50-kg machine on a frozen pond as in Figure 6.10. The machine fires a 0.15-kg baseball with a speed of 36 m/s in the horizontal direction. What is the recoil velocity of the machine?

Figure 6.10 (Example 6.4) When the baseball is fired to the right, the pitching machine recoils to the left.

Solution

In this example, the system consists of the baseball and the pitching machine. The system is not really isolated because of the force of gravity. However, this external force acts in the vertical direction while the motion of the system is in the horizontal direction. Therefore, the momentum of the system in the x direction is conserved because there are no external forces in this direction (assuming the surface is frictionless).

The total momentum of the system before firing the baseball is zero. Therefore, the total momentum after firing must also be zero, or

$$m_1 \mathbf{v}_1 + m_2 \mathbf{v}_2 = 0$$

With $m_1 = 0.15$ kg, $\mathbf{v}_1 = +36$ m/s, and $m_2 = 50$ kg, we find the recoil velocity of the pitching machine to be

$$\mathbf{v}_2 = -\frac{m_1}{m_2} \mathbf{v}_1 = -\left(\frac{0.15 \text{ kg}}{50 \text{ kg}}\right)(36 \text{ m/s}) = \boxed{-0.11 \text{ m/s}}$$

The negative sign for $\mathbf{v}_2$ indicates that the pitching machine is moving to the left after firing, in the direction *opposite* the motion of the baseball.

The force from a nitrogen-propelled, hand-controlled device allows an astronaut to move about freely in space without restrictive tethers. *(Courtesy of NASA)*

Inelastic collision

6.3 Collisions

We have seen that, for any type of collision, the total momentum of the system just before collision equals the total momentum just after collision. We can say that the total momentum is always conserved for any type of collision. However, the total kinetic energy is generally not conserved when a collision occurs because some of the kinetic energy is converted to heat or some other form of energy when the objects are deformed during the collision.

We define an **inelastic collision** as a collision in which momentum is conserved but kinetic energy is not.

For a general inelastic collision, we can apply the law of conservation of momentum in the form given by Equation 6.5. The collision of a rubber ball with a hard surface is inelastic because some of the kinetic energy is lost when the ball is deformed while it is in contact with the surface. When two objects collide and stick together, the collision is called **perfectly inelastic.** This is an extreme case of an inelastic collision. For example, if two pieces of putty collide, they stick together and move with some common velocity after the collision. If a meteorite collides head-on with the Earth, it becomes buried in the Earth and the collision is considered perfectly inelastic. However, you should note that not all of the initial kinetic energy is necessarily lost in a perfectly inelastic collision.

PHYSICS IN ACTION

Collisions

In the photograph on the left, the car makes an inelastic collision with a concrete wall. Although the car's momentum changes abruptly, the total momentum of the system (car + wall + Earth) remains constant. However, mechanical energy is *not* conserved in this inelastic collision. The initial kinetic energy of the car is transformed into the energy required to crumple its bumper and fenders, heat energy, energy in the form of sound waves, etc. You should try to identify other forms of energy transfers as a result of such a collision. Why does a passenger in the car tend to be thrown forward in the collision? How do seat belts or air bags help prevent serious injuries?

On the right is a multiflash photograph of an apparatus called the ballistic pendulum, commonly used in the introductory lab (see Example 6.7). A metal ball is fired from a spring-loaded gun into a pendulum that contains a cup which "catches" the ball. The ball makes a totally inelastic collision with the pendulum. If one measures the mass of the ball, the mass of the pendulum, and the vertical height that the system (ball + pendulum) rises, one can calculate the velocity of the ball before the collision. It is important to recognize that only momentum is conserved in this collision. That is, the initial momentum of the ball equals the momentum of the ball and pendulum right after the collision. Although mechanical energy is *not* conserved, we can equate the kinetic energy of the system right *after* the collision to the gravitational potential energy of the system when it reaches its highest position. Why is it incorrect to equate the initial kinetic energy of the ball to the final potential energy of the system?

(Courtesy of CALSPAN Corporation)

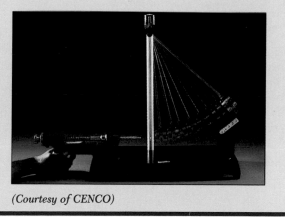

(Courtesy of CENCO)

An **elastic collision** is defined as a collision in which both momentum and kinetic energy are conserved.

Elastic collision

Billiard ball collisions and the collisions of air molecules with the walls of a container at ordinary temperatures are almost elastic. In reality, collisions in the macroscopic world, such as those between billiard balls, can be only approximately elastic because in such collisions there is always some deformation of the objects; hence some kinetic energy is lost in the forms of heat and sound. On the other hand, truly elastic collisions do occur between atomic and subatomic particles.

We summarize the various types of collisions as follows:

1. An elastic collision is one in which both momentum and kinetic energy are conserved.
2. An inelastic collision is one in which momentum is conserved but kinetic energy is not.
3. A perfectly inelastic collision between two objects is an inelastic collision in which the two objects stick together after the collision, so that their final velocities are the same.

In the remainder of this section, we shall treat collisions in one dimension and consider two extreme types: perfectly inelastic and elastic.

Perfectly Inelastic Collisions

Before collision

(a)

After collision

(b)

Figure 6.11
A perfectly inelastic head-on collision between two objects: (a) before the collision and (b) after the collision.

Consider two objects of masses m_1 and m_2 moving with initial velocities v_{1i} and v_{2i} along a straight line, as in Figure 6.11. We shall assume that the objects collide head-on so that they will be moving along the same line of motion after the collision. If the two objects stick together and move with some common velocity v_f after the collision, then only the momentum of the system is conserved. Therefore, we can say that the total momentum before the collision equals the total momentum after the collision:

$$m_1 v_{1i} + m_2 v_{2i} = (m_1 + m_2) v_f \qquad (6.6)$$

It is important to note that v_{1i}, v_{2i}, etc. represent the x components of the vectors $\mathbf{v}_{1i}$, $\mathbf{v}_{2i}$, etc., so one must be careful with signs. For example, in Figure 6.11, v_{1i} would have a positive value (m_1 moving to the right), while v_{2i} would have a negative value (m_2 moving to the left).

In a typical inelastic collision problem, only one quantity in this equation will be unknown. As a result, conservation of momentum is sufficient to tell us what we need to know.

Example 6.5   The Cadillac Versus the "Beetle"

A large luxury car with a mass of 1800 kg stopped at a traffic light is struck from the rear by a compact car with a mass of 900 kg. The two cars become entangled as a result of the collision. If the compact car was moving at 20 m/s before the collision, what is the velocity of the entangled mass after the collision?

Solution
The momentum before the collision is that of the compact car alone because the large car was initially at rest. Thus, we have for the momentum before the collision

$$p_i = m_1 v_i = (900 \text{ kg})(20 \text{ m/s}) = 1.80 \times 10^4 \text{ kg} \cdot \text{m/s}$$

After the collision, the mass that moves is the sum of the masses of the large car plus that of the compact car, and the momentum of the combination is

$$p_f = (m_1 + m_2) v_f = (2700 \text{ kg})(v_f)$$

Equating the momentum before to the momentum after and solving for v_f, the velocity of the wreckage, we have

$$v_f = \frac{p_i}{m_1 + m_2} = \frac{1.80 \times 10^4 \text{ kg} \cdot \text{m/s}}{2700 \text{ kg}} = \boxed{6.67 \text{ m/s}}$$

Example 6.6   Here's Mud in Your Eye

Two balls of mud collide head-on in a perfectly inelastic collision, as in Figure 6.11. Suppose $m_1 = 0.5$ kg, $m_2 = 0.25$ kg, $v_{1i} = +4$ m/s, and $v_{2i} = -3$ m/s. (a) Find the velocity of the composite ball of mud after the collision. (b) How much kinetic energy is lost in the collision?

Solution
(a) Writing Equation 6.6 for conservation of momentum with the positive direction for velocity to the right, we are able to find the velocity of the combined mass after the collision:

$$m_1 v_{1i} + m_2 v_{2i} = (m_1 + m_2)v_f$$

$$(0.5 \text{ kg})(4 \text{ m/s}) + (0.25 \text{ kg})(-3 \text{ m/s}) = (0.75 \text{ kg})(v_f)$$

$$v_f = \boxed{+1.67 \text{ m/s}}$$

(b) The kinetic energy before the collision is

$$KE_i = KE_1 + KE_2 = \frac{1}{2} m_1 v_{1i}^2 + \frac{1}{2} m_2 v_{2i}^2$$

$$= \frac{1}{2} (0.5 \text{ kg})(4 \text{ m/s})^2 + \frac{1}{2} (0.25 \text{ kg})(-3 \text{ m/s})^2 = 5.13 \text{ J}$$

The kinetic energy after the collision is

$$KE_f = \frac{1}{2} (m_1 + m_2)v_f^2 = \frac{1}{2} (0.75 \text{ kg})(1.67 \text{ m/s})^2 = 1.05 \text{ J}$$

Hence, the loss in kinetic energy is

$$KE_i - KE_f = \boxed{4.08 \text{ J}}$$

Most of this lost energy is converted to heat as the objects collide and distort, and a small fraction is converted to energy in the form of sound waves.

Example 6.7   The Ballistic Pendulum

The ballistic pendulum (Fig. 6.12 and the photograph in "Physics in Action" on p. 149) is a device used to measure the velocity of a fast-moving projectile, such as a bullet. The bullet is fired into a large block of wood suspended from some light wires. The bullet is stopped by the block, and the entire system swings through a height h. It is possible to obtain the initial velocity of the bullet by measuring h and the two masses. As an example of the technique, assume that the mass of the bullet, m_1, is 5 g, the mass of the pendulum, m_2, is 1 kg, and h is 5 cm. Find the initial velocity of the bullet, v_{1i}.

Figure 6.12 (Example 6.7) Diagram of a ballistic pendulum. Note that $\mathbf{v_f}$ is the velocity of the system right after the perfectly inelastic collision.

Solution

The collision between the bullet and the block is perfectly inelastic. Writing the conservation of momentum for the collision in the form of Equation 6.6, we have

$$(1) \qquad (5 \times 10^{-3}\ \text{kg})(v_{1i}) = (1.005\ \text{kg})(v_f)$$

There are two unknowns in this equation, v_{1i} and v_f, where the latter is the velocity of the block plus embedded bullet *immediately after the collision*. We must look for additional information if we are to complete the problem. Kinetic energy is not conserved during an inelastic collision. However, mechanical energy is conserved after the collision, and so the kinetic energy of the system at the bottom is transformed into the potential energy of the bullet plus block at the height h:

$$\frac{1}{2}(m_1 + m_2)v_f^2 = (m_1 + m_2)gh$$

$$\frac{1}{2}(1.005\ \text{kg})(v_f^2) = (1.005\ \text{kg})(9.80\ \text{m/s}^2)(5 \times 10^{-2}\ \text{m})$$

giving

$$v_f = 0.990\ \text{m/s}$$

With v_f now known, (1) yields v_{1i}:

$$v_{1i} = \frac{(1.005\ \text{kg})(0.990\ \text{m/s})}{5 \times 10^{-3}\ \text{kg}} = \boxed{199\ \text{m/s}}$$

Exercise   Explain why it would be incorrect to equate the initial kinetic energy of the incoming bullet to the final gravitational potential energy of the bullet-block combination.

Before collision

(a)

After collision

$\mathbf{v_{1f}}$ ◯ ◯ $\mathbf{v_{2f}}$

(b)

Figure 6.13
An elastic head-on collision between two hard spheres: (a) before the collision and (b) after the collision.

Elastic Collisions

Now consider two objects that undergo an elastic head-on collision (Fig. 6.13). In this situation, *both momentum and kinetic energy are conserved*; therefore we can write these conditions as

$$m_1 v_{1i} + m_2 v_{2i} = m_1 v_{1f} + m_2 v_{2f} \qquad \textbf{(6.7)}$$

$$\frac{1}{2} m_1 v_{1i}^2 + \frac{1}{2} m_2 v_{2i}^2 = \frac{1}{2} m_1 v_{1f}^2 + \frac{1}{2} m_2 v_{2f}^2 \qquad \textbf{(6.8)}$$

In a typical problem involving elastic collisions, there will be two unknown quantities, and Equations 6.7 and 6.8 can be solved simultaneously to find these. However, an alternative approach, one that involves a little mathematical manipulation of Equation 6.8, often simplifies this process for head-on collisions. To see this, let's cancel the factor of $\frac{1}{2}$ in Equation 6.8 and rewrite it as

$$m_1(v_{1i}{}^2 - v_{1f}{}^2) = m_2(v_{2f}{}^2 - v_{2i}{}^2)$$

Here we have moved the terms containing m_1 to one side of the equation and those containing m_2 to the other. Next, let us factor both sides of the equation:

$$m_1(v_{1i} - v_{1f})(v_{1i} + v_{1f}) = m_2(v_{2f} - v_{2i})(v_{2f} + v_{2i}) \tag{6.9}$$

We now separate the terms containing m_1 and m_2 in the equation for the conservation of momentum (Eq. 6.7) to get

$$m_1(v_{1i} - v_{1f}) = m_2(v_{2f} - v_{2i}) \tag{6.10}$$

Our final result is obtained by dividing Equation 6.9 by Equation 6.10 to get

$$v_{1i} + v_{1f} = v_{2f} + v_{2i}$$

or

$$v_{1i} - v_{2i} = -(v_{1f} - v_{2f}) \tag{6.11}$$

This equation, in combination with the equation for conservation of momentum, will be used to solve problems dealing with perfectly elastic, head-on collisions. Note that Equation 6.11 says that the relative velocity of the two objects before the collision, $v_{1i} - v_{2i}$, equals the negative of the relative velocity of the two objects after the collision, $-(v_{1f} - v_{2f})$.

Problem-Solving Strategy: Conservation of Momentum

The following procedure is recommended when dealing with problems involving collisions between two objects:

1. Set up a coordinate system and define your velocities with respect to that system. That is, objects moving in the direction selected as the positive direction of the x axis are considered as having a positive velocity and negative if moving in the negative x direction. It is convenient to have the x axis coincide with one of the initial velocities.

2. In your sketch of the coordinate system, draw all velocity vectors with labels and include all the given information.

3. Write expressions for the momentum of each object before and after the collision. (In two-dimensional collision problems, write expressions for the x and y components of momentum before and after the collision.) Remember to include the appropriate signs for the velocity vectors.

4. Now write expressions for the *total* momentum *before* and *after* the collision and equate the two. (For two-dimensional collisions, this

expression should be written for the momentum in both the x and y directions.) It is important to emphasize that it is the momentum of the *system* (the two colliding objects) that is conserved, *not* the momentum of the individual objects.

5. If the collision is **inelastic,** you should then proceed to solve the momentum equations for the unknown quantities.

6. If the collision is **elastic,** kinetic energy is also conserved, so you can equate the total kinetic energy before the collision to the total kinetic energy after the collision. This gives an additional relationship between the various velocities. The conservation of kinetic energy for elastic collisions leads to the expression $v_{1i} - v_{2i} = -(v_{1f} - v_{2f})$, which is often easier to use in solving elastic collision problems than is an expression for conservation of kinetic energy.

Example 6.8   Let's Play Pool

Two billiard balls move toward one another as shown in Figure 6.13. The balls have identical masses, and the collision between them is perfectly elastic. If the initial velocities of the balls are $+30$ cm/s and -20 cm/s, what is the velocity of each ball after the collision?

Solution
We turn first to Equation 6.7. The masses cancel on each side, and after substituting the appropriate values for the initial velocities, we have

$$30 \text{ cm/s} + (-20 \text{ cm/s}) = v_{1f} + v_{2f}$$

$$(1) \qquad 10 \text{ cm/s} = v_{1f} + v_{2f}$$

Since kinetic energy is also conserved, we can apply Equation 6.11, which gives

$$30 \text{ cm/s} - (-20 \text{ cm/s}) = v_{2f} - v_{1f}$$

$$(2) \qquad 50 \text{ cm/s} = v_{2f} - v_{1f}$$

Solving (1) and (2) simultaneously, we find

$$v_{1f} = -20 \text{ cm/s} \qquad v_{2f} = +30 \text{ cm/s}$$

That is, the balls *exchange velocities!* This is always the case when two objects of equal mass collide head-on in an elastic collision.

Exercise   Find the final velocity of the two balls if the ball with initial velocity -20 cm/s has a mass equal to half that of the ball with initial velocity $+30$ cm/s.

Answer $v_{1f} = -3$ cm/s; $v_{2f} = +47$ cm/s.

Example 6.9   Slowing Down Neutrons by Collisions

In a nuclear reactor, neutrons are produced when the isotope $^{235}_{92}U$ undergoes fission. These neutrons are moving at high speeds (typically 10^7 m/s) and must be slowed to about 10^3 m/s. Once the neutrons have slowed down, they have a high probability of producing another fission event and hence a sustained chain reaction, as discussed in Chapter 33. The high-speed neutrons can be slowed down by passing them through a solid or liquid material called a moderator. The neutrons are slowed as they make many elastic collisions with the nuclei of the moderator material. Let us show that a neutron can lose most of its kinetic energy if it collides elastically with a moderator containing light nuclei, such as deuterium and carbon. Hence, the moderator material is usually heavy water (D_2O) or graphite (which contains carbon nuclei).

Assume a neutron has a speed of 10^7 m/s and that it collides head-on with a nucleus of the moderator, assumed to be at rest. We shall consider a collision in which the moderator nucleus is deuterium in D_2O, where deuterium has a mass twice that of the neutron. (a) Find the speed of the neutron after one collision, and (b) the fraction of the kinetic energy lost by the neutron in the collision.

Solution

Equation 6.7, the conservation of momentum, applied to this situation gives

$$m_1 v_{1i} + m_2 v_{2i} = m_1 v_{1f} + m_2 v_{2f}$$

$$m_1(10^7 \text{ m/s}) + 0 = m_1 v_{1f} + 2 m_1 v_{2f}$$

or

$$(1) \quad 10^7 = v_{1f} + 2 v_{2f}$$

Because this is an elastic collision, we also can use Equation 6.11. We find

$$(2) \quad 10^7 = v_{2f} - v_{1f}$$

Equations (1) and (2) can be solved simultaneously to give

$$v_{1f} = -\frac{1}{3} \times 10^7 \text{ m/s}$$

(b) The ratio of the kinetic energy of the neutron after the collision to the kinetic energy before the collision is

$$f = \frac{\frac{1}{2} m_1 (\frac{1}{3} \times 10^7 \text{ m/s})^2}{\frac{1}{2} m_1 (10^7 \text{ m/s})^2} = \frac{1}{9}$$

This says that 89% of the neutron's kinetic energy is transferred to the deuterium nucleus. In practice, the moderator efficiency is reduced because head-on collisions are very unlikely to occur. How would the result differ if graphite were used as the moderator?

The collision problems we have considered up until now have been head-on collisions. That is, the incident mass strikes a second mass head-on and both rebound along a straight-line path that coincides with the line of motion of the incident mass. Anyone who has ever played billiards knows that such collisions are the exception rather than the rule. A more common type of collision is a *glancing collision*, in which the colliding masses rebound at some angle relative to the line of motion of the incident mass. Figure 6.14a shows a blue ball, traveling with an initial speed v_{1i}, striking a red ball obliquely (off-center). After the collision, the blue ball caroms off at an angle θ relative to its incident line of motion and the red ball rebounds at an angle ϕ.

As we emphasized earlier, *momentum is conserved in all collisions*, and glancing collisions are no exception. Since momentum is a vector quantity, the conservation of momentum principle must be written as $\mathbf{p_i} = \mathbf{p_f}$. That is, the

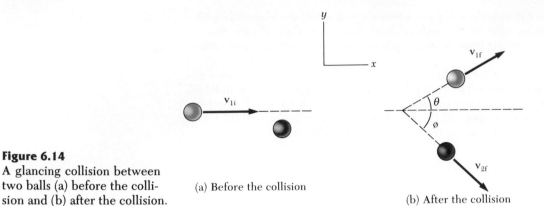

(a) Before the collision

(b) After the collision

Figure 6.14
A glancing collision between two balls (a) before the collision and (b) after the collision.

total initial momentum of the system (the two balls) must equal the total final momentum of the system. For a collision in two dimensions, as in Figure 6.14, this implies that the total momentum is conserved along the x direction *and* along the y direction. We can state this in equation form as

$$\sum p_{ix} = \sum p_{fx} \tag{6.12}$$

and

$$\sum p_{iy} = \sum p_{fy} \tag{6.13}$$

The following example illustrates how to use the principle of conservation of momentum to treat glancing collisions.

Example 6.10   Collision at an Intersection

A 1500-kg car traveling east with a speed of 25 m/s collides at an intersection with a 2500-kg van traveling north at a speed of 20 m/s, as shown in Figure 6.15. Find the direction and magnitude of the velocity of the wreckage after the collision, assuming that the vehicles undergo a perfectly inelastic collision (that is, they stick together).

Solution
Let us choose east to be along the positive x direction and north to be along the positive y direction, as in Figure 6.15. Before the collision, the only object having momentum in the x direction is the car. Thus, the total initial momentum of the system (car plus van) in the x direction is

$$\sum p_{ix} = (1500 \text{ kg})(25 \text{ m/s}) = 37{,}500 \text{ kg} \cdot \text{m/s}$$

Now let us assume that the wreckage moves at an angle θ and speed v after the collision, as in Figure 6.15. The total momentum in the x direction after the collision is

$$\sum p_{fx} = (4000 \text{ kg})(v \cos \theta)$$

Because momentum is conserved in the x direction, we have

$$\sum p_{ix} = \sum p_{fx}$$

$$(1) \qquad 37{,}500 \text{ kg} \cdot \text{m/s} = (4000 \text{ kg})(v \cos \theta)$$

Similarly, the total initial momentum of the system in the y direction is that of the van, which equals (2500 kg)(20 m/s). Applying conservation of momentum to the y direction, we have

Figure 6.15   (Example 6.10) Top view of a perfectly inelastic collision between a car and a van.

$$\sum p_{iy} = \sum p_{fy}$$

$$(2500 \text{ kg})(20 \text{ m/s}) = (4000 \text{ kg})(v \sin \theta)$$

$$(2) \qquad 50{,}000 \text{ kg} \cdot \text{m/s} = (4000 \text{ kg})(v \sin \theta)$$

If we divide (2) by (1), we get

$$\tan \theta = \frac{50{,}000}{37{,}500} = 1.33$$

$$\theta = \boxed{53°}$$

When this angle is substituted into (2)—or alternatively into (1)—the value of v is

$$v = \frac{50{,}000 \text{ kg} \cdot \text{m/s}}{(4000 \text{ kg})(\sin 53°)} = \boxed{15.6 \text{ m/s}}$$

6.5   The Center of Gravity and the Center of Mass

In virtually all the problems that we have worked in which forces acting on an object were considered, we have had to include the weight of the object being analyzed. In this section, we will find that there is a special point in a body, called the **center of gravity,** at which all its weight can be considered to be concentrated. For uniform, symmetrical objects, the location of the center of gravity presents no difficulty. For example, a spherical object, such as a basketball, has its center of gravity at its geometric center. The same statement applies to uniform boxes, cylinders, and so on. Sometimes, however, the center of gravity lies outside the object, as in the case of a coat hanger or a doughnut.

The center of mass of this gymnast follows a parabolic path. (© J. Zimmerman/FPG International)

The motion of the center of mass of ballet dancers as they leap through the air is also parabolic. (© John Terence Turner/FPG International)

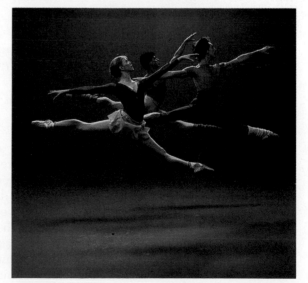

When a system consists of several particles located at different points, the x coordinate of the center of gravity of the system can be found from the equation

$$x_{cg} = \frac{\sum m_i x_i}{\sum m_i}$$ (6.14)

where m_i is the mass of each particle and x_i is its x coordinate. Similarly, the y coordinate of the center of gravity of the system of particles can be found from

$$y_{cg} = \frac{\sum m_i y_i}{\sum m_i}$$ (6.15)

We shall withhold the derivation of these two equations until the problem set of Chapter 8. There, we will give you several hints to guide you toward proving the validity of these equations.

Example 6.11   Where is the Center of Gravity?

Three particles are located in a coordinate system as shown in Figure 6.16. Find the location of the center of gravity.

Solution
The y coordinate of the center of gravity is zero because all particles are located on the x axis. To find the x coordinate of the center of gravity, we use Equation 6.14.

$$x_{cg} = \frac{\sum m_i x_i}{\sum m_i}$$

For the numerator, we find

$$\begin{aligned}
\sum m_i x_i &= m_1 x_1 + m_2 x_2 + m_3 x_3 \\
&= (5 \text{ kg})(-0.5 \text{ m}) + (2 \text{ kg})(0 \text{ m}) + (4 \text{ kg})(1 \text{ m}) \\
&= 1.5 \text{ kg} \cdot \text{m}
\end{aligned}$$

The denominator is $\sum m_i = 11$ kg, therefore

$$x_{cg} = \frac{1.5 \text{ kg} \cdot \text{m}}{11 \text{ kg}} = \boxed{0.136 \text{ m}}$$

Exercise   If a fourth particle of mass 2 kg is placed at the position $x = 0$ and $y = 0.25$ m, find the x and y coordinates of the center of gravity for this system of four particles.

Answer $x_{cg} = 0.115$ m, $y_{cg} = 0.038$ m.

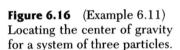

Figure 6.16   (Example 6.11) Locating the center of gravity for a system of three particles.

In the discussion above, we found that there is a special point for any object called the center of gravity and that we can consider the weight of the object as acting at its center. A related point is the *center of mass* of a system. There is a theoretical distinction between the two, but as long as the acceleration due to gravity is constant over the mass distribution, then the center of gravity coincides with the center of mass. Because those circumstances will always prevail in the situations we shall examine, we shall always assume that the center of mass coincides with the center of gravity. That is, in order to locate the center of mass of a system, we shall use the equations

$$x_c = \frac{\sum m_i x_i}{\sum m_i} \qquad y_c = \frac{\sum m_i y_i}{\sum m_i} \tag{6.16}$$

The physical significance and utility of the center of mass arise because *the net external force acting on a system equals the total mass of the system multiplied by the acceleration of the center of mass:*

$$\mathbf{F}_{ext} = M\mathbf{a}_c \tag{6.17}$$

where M is the total mass of the system, given by $M = \Sigma m_i$. This result shows that the center of mass moves like a particle of mass M under the influence of the resultant external force on the system.

From Equation 6.17, we see that, *if the resultant external force on a system of objects is zero, then* $\mathbf{a}_c$ *is also zero.* If the acceleration of the center of mass is zero, the velocity of the center of mass does not change and hence the momentum of the center of mass remains constant.

Suppose an isolated system consisting of two or more objects is at rest. *The center of mass of such a system will remain at rest unless acted upon by an external force.* For example, consider a system made up of a swimmer and a raft, with the system initially at rest. When the swimmer dives off the raft, the center of mass of the system will remain at rest (if we neglect friction between raft and water). Furthermore, the momentum of the diver will be equal in magnitude to the momentum of the raft, but opposite in direction. Next, imagine a pair of figure skaters initially at rest on ice. When the man pushes his female partner away from him, he will recoil with a momentum opposite that of the woman and the center of mass will remain at rest.

Example 6.12 An Exploding Projectile

A projectile is fired into the air and suddenly explodes into several fragments (Fig. 6.17). What can be said about the motion of the fragments after the explosion? (Neglect air resistance.)

Solution

The only external force on the projectile is the force of gravity. Thus, the projectile follows a path like that shown in Figure 6.17. If the projectile did not explode, it would continue to move along the path indicated by the broken line. Because the forces due to the explosion are internal, they do not affect the motion of the center of mass. Thus, after the explosion, the fragments will fly away such that the motion of their center of mass follows the same trajectory the projectile would have followed if there had been no explosion. (Neglecting air resistance, the path is parabolic.)

Motion of center of mass

Figure 6.17 (Example 6.12) When a projectile explodes into several fragments, the center of mass of the fragments follows the same parabolic path as the projectile would have taken had it not exploded.

▶ SPECIAL TOPIC
Rocket Propulsion

When ordinary vehicles, such as automobiles, boats, and locomotives, are propelled, the driving force for the motion is one of friction. In the case of the automobile, the driving force is the force of friction between road and car. A locomotive "pushes"

Additional Reading

A. Einstein and L. Infeld, *Evolution of Physics,* New York, Simon and Schuster, 1938.

H. W. Lewis, "Ballistocardiography," *Sci. American,* February, 1958, p. 89.

F. Ordway, "Principles of Rocket Engines," *Sky and Telescope, 14,* 1954, p. 48. An introduction to the principles of rocket propulsion.

Questions

1. If the kinetic energy of a particle is zero, what is its momentum? If the total energy of a particle is zero, is its momentum necessarily zero? Explain.

2. Consider a field of insects all of essentially equal mass. If the total momentum of the insects is zero, what does this imply about the motion of the insects? If the total kinetic energy of the insects is zero, what does this imply?

3. If the velocity of a particle is doubled, by what factor is its momentum changed?

4. Does a large force always produce a larger impulse on a body than a smaller force? Explain.

5. If two objects collide and one is initially at rest, is it possible for both to be at rest after the collision? Is it possible for one to be at rest after the collision? Explain.

6. Is it possible to have a collision in which all of the kinetic energy is lost? If so, cite an example.

7. If the forward momentum of a bullet is the same as the backward momentum of the gun, why isn't it as dangerous to be hit by the gun as by the bullet?

8. A piece of clay is thrown against a brick wall and sticks to the wall. What happens to the momentum of the clay? Is momentum conserved? Explain.

9. As a ball falls towards the Earth, its momentum increases. How would you reconcile this fact with the law of conservation of momentum?

10. A box slides across the frictionless icy surface of a frozen lake. What happens to the speed of the box as water collects in it from a rainshower? Explain.

11. A skater is standing still on a frictionless ice rink. Her friend throws a Frisbee straight at her. In which of the following cases will the largest momentum be transferred to the skater? (a) The skater catches the Frisbee and holds on to it. (b) The skater catches the Frisbee momentarily, but then drops it vertically downward. (c) The skater catches the Frisbee and holds it momentarily and then throws it back to her friend.

12. If two particles have equal kinetic energies, are their momenta necessarily equal? Explain.

13. In a perfectly elastic collision between two objects, do both objects have the same kinetic energy after the collision? Explain.

14. If two automobiles collide, they usually do not stick together. Does this mean the collision is perfectly elastic? Explain why a head-on collision is likely to be more dangerous than other types of collisions.

15. Early in this century, Robert Goddard proposed sending a rocket to the Moon. Critics took the position that in a vacuum, such as exists between the Earth and the Moon, the gases emitted by the rocket would have nothing to push against to propel the rocket. According to *Scientific American* (January 1975), Goddard placed a gun in a vacuum and fired a blank cartridge from it. (A blank cartridge fires only the hot gases of the burning gunpowder.) What happened when the gun was fired?

16. An astronaut walking in space accidentally severs the safety cord attaching him to the spacecraft. If he happens to have with him a can of aerosol spray deodorant, how could he use this to return safely to his ship?

17. Construct a balloon "rocket." Explain to a friend why the flight of the balloon is analogous to the flight of a NASA rocket, despite the fact that there is no flame belching forth from its "engine."

18. Suppose you are out sailing when the wind calms. If you have a large electric fan on board, can you move the boat by directing the wind from the fan toward the sail? If you remove the sail, will the running fan cause the boat to move? Explain.

19. A boxer wisely moves his head backward just before receiving a punch to his nose. How does that help protect him from the punch?

20. Gymnasts always perform with padded mats. Use the impulse-momentum theorem to discuss how these mats protect the athletes.

21. An open bed sheet is loosely held at its sides by two students to form a "catching net." The instructor asks a third student to throw a raw egg into the sheet as hard as possible, being careful not to miss the sheet. Why is it that the egg does not break?

22. How do car bumpers that collapse on impact help to protect a driver?

23. A pole-vaulter falls from a height of 15 ft onto a foam rubber pad. Could you calculate his velocity just before he reaches the pad? Would you be able to calculate the force exerted on the pole-vaulter due to the collision? Explain.

24. A toy gun shoots rubber bullets at a target. Compare the impulse delivered to the target when the bullets embed in the target to the case in which they strike the target and bounce off.

25. A ball is dropped from a tall building in New York. Identify the system whose linear momentum is conserved.

26. A magician places several dishes, cups, and silverware on a table containing a tablecloth. The magician proceeds to rapidly remove the tablecloth without disturbing the dishes, etc. Explain how this trick is possible on the basis of what you have learned in this chapter.

27. A man is at rest in a boat in the middle of a lake. If he walks to the opposite end of the boat towards the east, why does the boat move towards the west? What can you say about the center of mass of the system consisting of the boat and the man?

28. Does the center of mass of a rocket in free space accelerate? Explain. Can the speed of a rocket exceed the exhaust velocity of the fuel? Explain.

29. A meter stick is balanced in a horizontal position with the index fingers of the right and left hands. If the two fingers are brought together, the stick remains balanced and the two fingers always meet at the 50-cm mark regardless of their original positions (try it!). Carefully explain this observation.

30. Use conservation of momentum to discuss automobile safety, including the roles of seat belts and whiplash from rear-end collisions.

31. Locate the center of gravity for the following uniform objects: (a) a sphere, (b) a cube, (c) a cylinder, (d) a doughnut.

32. The center of gravity of an object may be located outside the object. Give a few examples for which this is the case.

33. Explain why a picture will hang crookedly if the hook is not centered on the frame properly.

34. In order for a chair to be balanced on one leg, where must the center of gravity of the chair be located?

35. A tall crate and a short crate of equal mass are placed side-by-side on an incline (without touching each other). As the incline angle is increased, which crate will tip first? Explain.

36. A male and a female student are asked to do the following task. Face a wall, step three foot lengths away from the wall, and then lean over and touch the wall with your nose, keeping your hands behind your back. The male usually fails, while the female succeeds. How would you explain this?

37. When lifting a heavy object, why is it recommended to straighten your back, as much as possible, rather than bend over and lift mainly with the arms?

38. How many different ways must you support an object to locate its center of gravity? Explain.

39. Why do you tend to lean backward when carrying a box of books in your arms but forward when the same books are carried in a backpack?

40. "During a high jump the person's center of gravity passes under the bar." Can this statement be true? Discuss.

41. Why are sports cars usually built low to the ground and wide-tracked?

Problems

Section 6.1 Momentum and Impulse

1. Calculate the magnitude of the linear momentum for the following cases: (a) a proton moving with a speed of 5×10^6 m/s and whose mass is 1.67×10^{-27} kg, (b) a 15-g bullet moving with a speed of 300 m/s, (c) a 75-kg sprinter running with a speed of 10 m/s, and (d) the Earth (mass = 5.98×10^{24} kg) moving with an orbital speed equal to 2.98×10^4 m/s.

2. A 0.15-kg baseball is thrown with a speed of 10 m/s. (a) How much more momentum does it have if its speed is doubled? (b) How much more kinetic energy will it have?

3. The momentum of a 1500-kg car is equal to the momentum of a 5000-kg truck traveling with a speed of 15 m/s. What is the speed of the car?

4. A 0.10-kg ball is thrown straight up into the air with an initial speed of 15 m/s. Find the momentum of the ball (a) at its maximum height, and (b) halfway up to its maximum height.

5. A 0.10-kg ball is thrown with a speed of 20 m/s at an angle of 30° with the horizontal. Find the momentum of the ball (a) at its maximum height, and (b) just before it strikes the ground.

6. A 1500-kg car moving with a speed of 15 m/s collides with a utility pole and is brought to rest in 0.3 s. Find

the average force exerted on the car during the collision.

7. An 18,000-kg van is moving with a speed of 15 m/s. If Superman is to stop the van in 0.5 s, what average force must he exert on the van?

8. A 0.5-kg football is thrown with a speed of 15 m/s. A stationary receiver catches the ball and brings it to rest in 0.02 s. (a) What is the impulse delivered to the ball? (b) What is the average force exerted on the receiver?

9. An estimated force-time curve for a baseball struck by a bat is shown in Figure 6.20. From this curve, determine (a) the impulse delivered to the ball, and (b) the average force exerted on the ball.

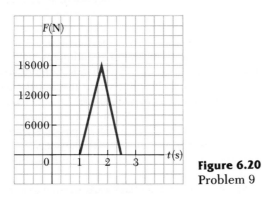

Figure 6.20
Problem 9

10. The force $\mathbf{F}_x$ acting on a 2-kg particle varies in time as shown in Figure 6.21. Find (a) the impulse of the force, (b) the final velocity of the particle if it is initially at rest, and (c) the final velocity of the particle if it is initially moving along the x axis with a velocity of -2 m/s.

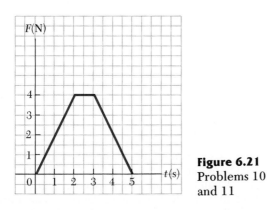

Figure 6.21
Problems 10 and 11

11. Find the average force exerted on the particle described in Figure 16.21 for the time interval $t_i = 0$ to $t_f = 5$ s.

12. A 0.15-kg baseball is thrown with a speed of 20 m/s. It is hit straight back at the pitcher with a final speed of 22 m/s. (a) What is the impulse delivered to the ball? (b) Find the average force exerted by the bat on the ball if the two are in contact for 2×10^{-3} s.

13. An 0.5-kg object is at rest at the origin of a coordinate system. A 3-N force in the $+x$ direction acts on the object for 1.5 s. (a) What is the velocity at the end of this time? (b) At the end of this time, a constant force of 4 N is applied in the $-x$ direction for 3 s. What is the velocity at the end of this time?

14. A prankster drops a plastic bag filled with 2 kg of water from the roof of a dormitory 25 m above street level. Find the force exerted on the head of someone walking along the street by assuming that the bag is spherical, falls freely, and stops in a distance equal to its diameter, which is 15 cm.

Section 6.2   Conservation of Momentum

15. A 40-kg child standing on a frozen pond throws a 0.5-kg stone to the east with a speed of 5 m/s. Neglecting friction between child and ice, find the recoil velocity of the child.

16. A rifle with a weight of 30 N fires a 5-g bullet with a speed of 300 m/s. (a) Find the recoil speed of the rifle. (b) If a 700-N man holds the rifle firmly against his shoulder, find the recoil speed of man and rifle.

17. A popcorn kernel initially at rest explodes into two pieces, one piece having twice the mass of the other. What is the ratio of the velocities of the fragments?

18. The bird perched on the swing in Figure 6.22 has a mass of 52 g and the base of the swing has a mass of 153 g. Assume that the swing and bird are originally at rest and that the bird then takes off in a horizontal direction at 2 m/s. If the base can swing freely (i.e., without friction) around the pivot, how high will the base of the swing rise above its original level?

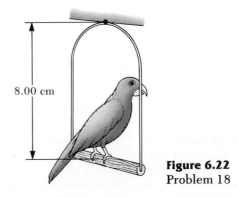

Figure 6.22
Problem 18

19. A 65-kg boy and a 40-kg girl, both wearing skates, face each other at rest on a skating rink. The boy

pushes the girl, sending her eastward with a speed of 4 m/s. Describe the subsequent motion of the boy. (Neglect friction.)

20. A 730-N man stands in the middle of a frozen pond of radius 5 m. He is unable to get to the other side because of a lack of friction between his shoes and the ice. To overcome his difficulty, he throws his 1.2-kg physics textbook horizontally at a speed of 5 m/s toward the north shore. How long does it take him to reach the south shore?

21. A 7-kg bowling ball collides head-on with a 2-kg bowling pin. The pin flies forward with a speed of 3 m/s. If the ball continues forward with a speed of 1.8 m/s, what was the initial speed of the ball?

22. A person with a mass of 65.0 kg throws a 0.045-kg snowball forward with a ground speed of 30.0 m/s. A second person with a mass of 60.0 kg catches the snowball. Both persons are on skates. The first person is initially moving forward with a speed of 2.5 m/s, and the second person is initially at rest. What are the velocities of the two people after the snowball is exchanged? Disregard the friction between the skates and the ice.

Section 6.3   Collisions
Section 6.4   Glancing Collisions

23. A car moving at a speed of 5 m/s crashes into an identical car stopped at a light. What is the velocity of the wreckage immediately after the collision, assuming the cars stick together?

24. A bumper car at an amusement park ride traveling at 0.8 m/s collides with an identical car at rest. This second car moves away with a speed of 0.5 m/s. What is the velocity of the first car after the collision?

25. A 1.2-kg skateboard is coasting along the pavement at a speed of 5 m/s, when a 0.8-kg cat drops from a tree vertically down on the skateboard. What is the speed of the skateboard-cat combination?

26. A railroad car of mass 2×10^4 kg moving with a speed of 3 m/s collides and couples with two other coupled railroad cars each of the same mass as the single car and moving in the same direction at a speed of 1.2 m/s. (a) What is the speed of the three coupled cars after the collision? (b) How much energy is lost in the collision?

27. A 3-kg sphere makes a perfectly inelastic collision with a second sphere initially at rest. The composite system moves with a speed equal to one third the original speed of the 3-kg sphere. What is the mass of the second sphere?

28. A 5-g object moving to the right with a speed of 20 cm/s makes an elastic head-on collision with a 10-g object initially at rest. Find (a) the velocity of each object after the collision and (b) the fraction of the energy transferred to the 10-g object.

29. A 10-g object moving to the right with a speed of 20 cm/s makes an elastic head-on collision with a 15-g object moving in the opposite direction with a speed of 30 cm/s. Find the velocity of each object after the collision.

30. A 25-g object moving to the right with a speed of 20 cm/s overtakes and collides elastically with a 10-g object moving in the same direction with a speed of 15 cm/s. Find the velocity of each object after the collision.

31. An alpha particle of mass 4 u moving to the right with a speed of 10^6 m/s collides with a proton of mass 1 u that is at rest before the collision. Find (a) the speed of each particle after the collision, assuming a perfectly elastic collision, and (b) the kinetic energy of each particle before and after the collision (1 u = 1 atomic mass unit = 1.67×10^{-27} kg).

32. An 8-g bullet is fired into a 2.5-kg ballistic pendulum and becomes embedded in it. If the pendulum rises a vertical distance of 6 cm, calculate the initial speed of the bullet.

33. A 50-kg baseball pitching machine is placed on a frozen pond. It fires a 0.15-kg baseball at an angle of 30° with the vertical at a speed of 20 m/s. What is the recoil velocity of the machine?

34. A 90-kg fullback moving east with a speed of 5 m/s is tackled by a 95-kg opponent running north with a speed of 3 m/s. If the collision is perfectly inelastic, calculate (a) the velocity of the players just after the tackle and (b) the energy lost as a result of the collision. Can you account for the missing energy?

35. An eastern kingbird and a bee are approaching each other at right angles. The bird has a mass of 0.125 kg and a speed of 0.6 m/s, and the bee has a mass of 5.0×10^{-3} kg and a speed of 15.0 m/s. If the bird catches the bee, what is the new speed of the bird?

36. The mass of the blue puck in Figure 6.23 is 20% larger than the mass of the green one. Before colliding, the pucks approach each other with equal and opposite momentum, and the green puck has an initial speed of 10.0 m/s. Find the speeds of the pucks after the collision if half the kinetic energy is lost during the collision.

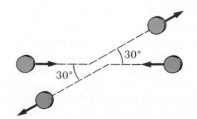

Figure 6.23
Problem 36

37. A 1500-kg car traveling east with a speed of 20 m/s collides at an intersection with a 2500-kg van traveling south at a speed of 15 m/s. The vehicles undergo a perfectly inelastic collision, and the wreckage slides 6 m before coming to rest. Find the magnitude and direction of the constant force that has decelerated them.

38. A neutron in a reactor makes an elastic head-on collision with a carbon atom initially at rest. (The mass of the carbon nucleus is about 12 times the mass of the neutron.) (a) What fraction of the neutron's kinetic energy is transferred to the carbon nucleus? (b) If the initial kinetic energy of the neutron is 1.6×10^{-13} J, find its final kinetic energy and the kinetic energy of the carbon nucleus after the collision.

39. Consider the ballistic pendulum described in Example 6.7 and shown in Figure 6.12. If the mass of the bullet is 8 g and the mass of the pendulum is 2 kg, find the ratio of the kinetic energy after the collision to the kinetic energy before collision. What accounts for this missing energy?

40. A cue ball traveling at a speed of 4 m/s makes a glancing, elastic collision with a target ball of equal mass initially at rest. It is found that the cue ball is deflected such that it makes an angle of 30° with respect to its original direction of travel. Find (a) the angle the second ball will move and (b) the speed of each ball after the collision.

Section 6.5   The Center of Gravity and The Center of Mass

41. Four objects are located along the y axis as follows. A 2-kg object is at +3 m, a 3-kg object is at +2.5 m, a 2.5-kg object is at the origin, and a 4-kg object is at −0.5 m. Where is the center of gravity of these objects located?

42. Where would a fifth object of mass 5 kg have to be placed along the y axis so that the center of gravity of the objects in Problem 41 is at the origin?

43. Three masses are located in a rectangular coordinate system as follows: a 2-kg mass is at $(3, -2)$, a 3-kg mass is at $(-2, 4)$, and a 1-kg mass is at $(2, 2)$, where all distances are in meters. Find the coordinates of the center of gravity for the three masses.

44. A uniform T square used by a draftsman has dimensions as shown in Figure 6.24. Locate the center of gravity with respect to point O. (*Hint:* Note that the mass of each rectangular part is proportional to its area.)

45. A uniform carpenter's square has the shape of an L, as in Figure 6.25. Locate the center of gravity relative to an origin at the lower left corner. (Note the hint for Problem 44.)

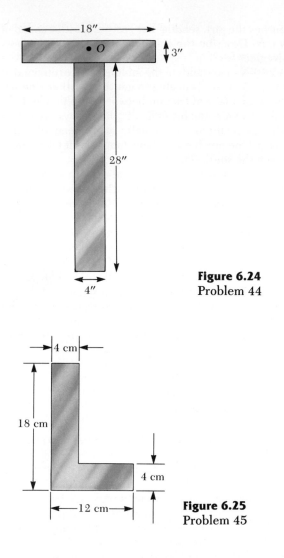

Figure 6.24
Problem 44

Figure 6.25
Problem 45

46. The mass of the Earth is 5.98×10^{24} kg and the mass of the Moon is 7.36×10^{22} kg. The distance of separation measured from their centers is 3.84×10^{8} m. Determine the location of the center of mass of the Earth-Moon system as measured from the center of the Earth.

47. The nucleus of a hydrogen atom is a single proton of mass 1.67×10^{-27} kg. The mass of the electron which circles the proton at a radius of about 0.51×10^{-10} m is 9.1×10^{-31} kg. Find the center of mass of the proton-electron system as measured from the proton.

48. Three masses located in the xy plane have the following coordinates: a 2-kg mass has coordinates $(3, -2)$ m, a 3-kg mass has coordinates $(-2, 4)$ m, and a 1-kg mass has coordinates $(2, 2)$ m. Find the coordinates of the center of mass of the system.

49. An 80.0-kg astronaut is on a space "walk" when the engine on his back fails. He is at rest with respect to

the rocket and 100 m from the rocket. He carries a 100-m rope that is essentially massless and a 20.0-kg engine pack. He ties the engine pack to the end of the rope and throws it toward the rocket, holding on to the other end. (a) How close to the rocket does the engine pack get? (b) What should the astronaut do to get back to the rocket?

Additional Problems

50. A soccer ball of mass 0.4 kg is approaching a player with a speed of 15 m/s in horizontal flight. The player illegally strikes the ball with her hand and causes it to move in the opposite direction with a speed of 22 m/s. What impulse was delivered to the ball by the player?

51. A 4-g marble has a head-on collision with a 2-g marble at rest on a playing surface. The speed of the 4-g marble is reduced from 0.8 m/s to 0.5 m/s in the collision. What is the speed of the 2-g marble after the collision?

52. A swimmer of mass 80 kg stands on the stern of a boat of mass 240 kg at rest on a lake. The swimmer dives off the boat with a velocity that has a horizonal component of 0.8 m/s in the westerly direction. Determine the magnitude and direction of the resulting velocity of the boat. (Neglect any frictional effects of the water.)

53. A 2-g particle moving at a speed of 8 m/s makes a perfectly elastic head-on collision with a resting object of mass 1 g. (a) Find the speed of each after the collision. (b) If the rest particle has a mass of 10 g, find the speed of each after the collision. (c) Find the final kinetic energy of the incident 2-g particle for the situations described in (a) and (b). In which case does the incident particle lose the most kinetic energy?

54. A diver of mass 70 kg drops from a 3 m diving board. (a) What is his momentum just before he hits the water? (b) What is his momentum halfway down?

55. An 80-kg man drops from a 3-m diving board. In 2.0 s after reaching the water, the man comes to rest. What average force did the water exert on him?

56. Two pieces of "Silly Putty" are thrown toward one another and stick together after the collision. If one piece has a mass of 2 g and is traveling with a speed of 4 m/s, and the second has a mass of 3 g and a speed of 3 m/s, (a) what is the speed of the combination after the collision? (b) What speed could the 3-g mass have such that the combination is brought to rest in the collision?

57. A pool ball rolling across a table at 1.5 m/s makes a head-on collision with an identical ball. Assume the collision is elastic and find the speed of each ball after the collision (a) when the second ball is initially at rest, (b) when the second ball is in motion toward the first at a speed of 1 m/s, and (c) when the second ball is in motion away from the first at a speed of 1 m/s.

58. A 0.03-kg bullet is fired vertically at a speed of 200 m/s into a 0.15-kg baseball initially at rest. How high will the combination rise after the collision, assuming the bullet embeds in the ball?

59. An 80-kg astronaut is working on the engines of his ship, which is drifting through space with a constant velocity. The astronaut, wishing to get a better view of the Universe, pushes against the ship and later finds himself 30 m behind the ship. Without a thruster, the only way to return to the ship is to throw his 0.5-kg wrench directly away from the ship. If he throws the wrench with a speed of 20 m/s, how long does it take the astronaut to reach the ship?

60. A 12-g bullet is fired horizontally into a 100-g wooden block initially at rest on a horizontal surface. After impact, the bullet remains embedded in the block and the combined system slides 7.5 m before coming to rest. If the coefficient of friction between the block and the surface is 0.65, what was the speed of the bullet immediately before impact?

61. A 2000-kg car moving eastward at a speed of 10 m/s collides with a 3000-kg car moving northward. The cars stick together and move as a unit after the collision at an angle of 40° north of east at a speed of 5.22 m/s. Find the velocity of the 3000-kg car before the collision.

62. A 0.4-kg bead slides on a curved frictionless wire, starting from rest at point A in Figure 6.26. At point B, the bead collides elastically with a 0.6-kg ball at rest. Find the height that the ball moves up the wire.

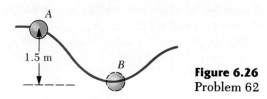

Figure 6.26
Problem 62

63. Tarzan, whose mass is 80 kg, swings from a 3-m vine which is horizontal when he starts. At the bottom of his arc, he picks up 60-kg Jane, in an inelastic collision. What maximum height tree limb can they reach on their upward swing?

64. A truck that weighs 2.5×10^4 N is coasting at 7 m/s along a level roadway. A pallet of construction materials falls onto the truck from an overhead bridge and causes the speed of the truck to decrease by 10%. What is the weight of the pallet of materials?

65. The force shown in the force-time diagram in Figure 6.27 acts on a 1.5-kg mass. Find (a) the impulse of the force, (b) the final velocity of the mass if it is initially at rest, and (c) the final velocity of the mass if it is initially moving along the x axis with a velocity of -2 m/s.

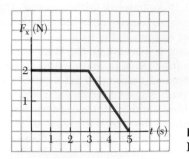

Figure 6.27
Problem 65

66. A 1-kg mass moving with an initial speed of 5 m/s collides with and sticks to a 6-kg mass initially at rest. The combined mass then proceeds to collide with and stick to a 2-kg mass also at rest initially. If the collisions are all head-on, find (a) the final speed of the system and (b) the amount of kinetic energy lost.

67. A 0.3-kg puck, initially at rest on a horizontal, frictionless surface, is struck by a 0.2-kg puck moving initially along the x axis with a velocity of 2 m/s. After the collision, the 0.2-kg puck has a speed of 1 m/s at an angle of $\theta = 53°$ to the positive x axis. (a) Determine the velocity of the 0.3-kg puck after the collision. (b) Find the fraction of kinetic energy lost in the collision.

68. The forces shown in the force-time diagram in Figure 6.28 act on a 1.5-kg mass. Find (a) the impulse for the interval $t = 0$ to $t = 3$ s, and (b) the impulse for the interval $t = 0$ to $t = 5$ s. (c) If the forces act on a particle of mass 1.5 kg which was initially at rest, find the speeds at $t = 3$ s and at $t = 5$ s.

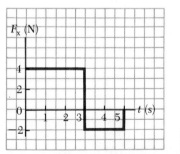

Figure 6.28
Problem 68

69. A tin can with a mass of 0.11 kg is resting on top of a fence post 1.7 m high. A bullet of mass 0.002 kg is fired horizontally at the can. It strikes the can with a speed of 900 m/s, passes through it, and emerges with a speed of 720 m/s. When the can hits the ground, how far is it from the fence post? Disregard friction while the can is in contact with the post.

70. A 6-g bullet is fired into a 200-g block initially at rest at the edge of a table of height 1 m (Fig. 6.29). The bullet remains in the block and after the impact the block lands 2 m from the bottom of the table. Determine the initial speed of the bullet.

Figure 6.29
Problem 70

71. A 7-g bullet is fired into a 1.5-kg ballistic pendulum. The bullet emerges from the block with a speed of 200 m/s, and the block rises to a maximum height of 12 cm. Find the initial speed of the bullet.

72. An unstable nucleus of mass 17×10^{-27} kg initially at rest at the origin of a coordinate system disintegrates into three particles. One particle, having a mass of $m_1 = 5.0 \times 10^{-27}$ kg, moves along the y axis with a velocity $v_1 = 6 \times 10^6$ m/s. Another particle, of mass $m_2 = 8.4 \times 10^{-27}$ kg, moves along the x axis with a velocity of $v_2 = 4 \times 10^6$ m/s. Find the magnitude and direction of the velocity of the third particle.

73. Consider a machine gun mounted on a cart on wheels. The gun fires N bullets of mass m each second. (a) Show that the average force acting on the gun is $Nm\mathbf{v}$. (b) If the gun-cart combination is free to move, find the average acceleration during an interval of 5 s in which the gun fires 50 bullets. Assume the mass of the cart is 70 kg, the mass of each bullet is 5 g, and the muzzle velocity of each bullet is 300 m/s.

74. A candy store clerk drops 1-g pieces of candy from a height of 20 cm onto a scale at a rate of 4 per second. What is the scale reading after 10 s?

75. A 40-kg child stands at one end of a 70-kg boat that is 4 m in length (Fig. 6.30). The boat is initially 3 m from the pier. The child notices a turtle on a rock at the far end of the boat and proceeds to walk to that end to catch the turtle. Neglecting friction between the boat and water, (a) describe the subsequent mo-

Figure 6.30
Problem 75

tion of the system (child + boat). (b) Where will the child be relative to the pier when he reaches the far end of the boat? (c) Will he catch the turtle? (Assume he can reach out 1 m from the end of the boat.)

76. A block of mass M is given an initial velocity v_0 on a rough, horizontal surface. After traveling a distance d, it makes a head-on elastic collision with a block of mass $2M$. How far will the second block move before coming to rest? (Assume the coefficient of friction is the same for each block.)

77. A projectile of mass m collides with a flat surface. The angle between the incident path and the surface equals the angle θ between the reflected path and the surface. The speed of the projectile is unchanged by the collision. The time the projectile is in contact with the surface is t. Find (a) the change in momentum of the projectile in the x direction, (b) the change in momentum in the y direction, and (c) the average force exerted on the surface by the projectile during the collision.

CHAPTER 7

Circular Motion and the Law of Gravity

A view of the Earth as seen from the surface of the Moon. The prominent colors are the blues of the oceans and the clouds, which are seen as swirls of white.
(© Index Stock International, Inc.)

In this chapter we shall look at circular motion, a specific type of two-dimensional motion. We shall encounter such terms as *angular velocity*, *angular acceleration*, *centripetal acceleration*, and *centripetal force*. The results we derive here will enable us to understand the motion of a diverse range of objects in our environment, from a car moving around a circular racetrack to clusters of galaxies orbiting a common center.

We shall also introduce Newton's universal law of gravitation, one of the fundamental laws in nature, and show how this law, together with Newton's laws of motion, enables us to understand a variety of familiar phenomena. Finally, we shall discuss Kepler's laws of planetary motion.

7.1 Angular Velocity

We began our study of linear motion by defining the terms *displacement*, *velocity*, and *acceleration*. This same basic approach will be taken now as we move to a study of rotational motion. To begin, consider Figure 7.1a, which shows a top view of a phonograph record rotating on a turntable. The axis of rotation is at the center of the record at O. A point P on the record is at a distance r from the origin and rotates about O in a circle of radius r. In fact, every point on the record undergoes circular motion about O. To analyze such motion, it is convenient to set up a *fixed* reference line, as shown in Figure

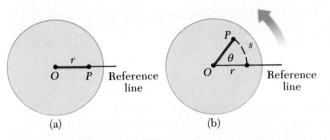

Figure 7.1
(a) The point P on a rotating record at $t = 0$.
(b) As the record rotates, the point P rotates through an arc length s.

7.1a. Let us assume that at time $t = 0$, the point P is on the reference line as in Figure 7.1a and a line is drawn on the record from the origin out to P. After a time Δt has elapsed, P has advanced to a new position, shown in Figure 7.1b. In this time interval, the line OP has rotated through an angle θ with respect to the reference line. Likewise, P has moved through a distance s measured along the circumference of the circle; the distance s is called an *arc length*.

In situations we have encountered thus far, angles have been measured in degrees. However, in scientific work we often measure angles in *radians* rather than degrees because it makes certain equations take on a simpler form. In fact, almost all of the equations we shall derive in this chapter and the next require that we measure angles in radians. With reference to Figure 7.1b, when the arc length s is equal to the radius r, the angle θ swept out by r is equal to one radian. In general, any angle θ, measured in radians, is defined by the relation

$$\theta \equiv s/r \qquad (7.1)$$

It is important to note that the radian is a pure number with no dimensions. This can be seen from Equation 7.1, since θ is the ratio of an arc length (a distance) and the radius of the circle (also a distance).

To find a conversion factor between degrees and radians, note that when point P in Figure 7.1 moves through an angle of $360°$ (one revolution), the arc length s is equal to the circumference of the circle, $2\pi r$. From Equation 7.1, we see that the corresponding angle in radians is $2\pi r/r = 2\pi$ rad. Hence,

$$1 \text{ rad} \equiv 360°/2\pi \approx 57.3°$$

From this definition, it follows that any angle in degrees can be converted to an angle in radians with the expression

$$\theta(\text{rad}) = \frac{\pi}{180°}\, \theta(\text{deg}) \qquad (7.2)$$

For example, $60°$ equals $\pi/3$ rad and $45°$ equals $\pi/4$ rad.

Returning to our phonograph record, we see from Figure 7.2 that, as the record rotates and our point moves from P to Q in a time Δt, the angle through which the record rotates is $\Delta\theta = \theta_2 - \theta_1$. We define $\Delta\theta$ as the **angular displacement**. The *average angular velocity,* $\overline{\omega}$ (ω is the Greek letter omega), of a rotating rigid object is the ratio of the angular displacement, $\Delta\theta$, to the time interval Δt it takes the object to rotate through the angle $\Delta\theta$:

Figure 7.2
As a point on the rotating record moves from P to Q, the record rotates through the angle $\Delta\theta = \theta_2 - \theta_1$.

$$\overline{\omega} \equiv \frac{\theta_2 - \theta_1}{t_2 - t_1} = \frac{\Delta\theta}{\Delta t} \qquad (7.3)$$

Average angular velocity

In analogy to linear velocity, the **instantaneous angular velocity,** ω, is defined as *the value of the angular velocity at any instant of time*. It is important to recognize that points at different radial distances on a rotating rigid body have *different* linear speeds along their circular paths. However, *every point on a rotating rigid body has the same value of angular velocity*. This fact demonstrates the convenience of defining and using angular velocity.

Angular velocity has dimensions of radians per second. You should note that the units for angular velocity are unusual in that the radian is a dimensionless unit. This means that we shall not always be able to cancel the radian unit in our equations as we do with those such as of length, mass, and time. The radian can be viewed as a bookkeeping convenience in our equations. Evidence of this difference will appear soon.

Example 7.1   The 45 rpm Record

Express the angular velocity of a 45 rpm (revolutions per minute) record in units of radians per second. (In this text, we shall sometimes use the abbreviation rpm, but in most cases we shall use the abbreviation rev/min.)

Solution
First we use the fact that 1 rev corresponds to 2π rad and calculate the angular displacement.

$$\Delta\theta = (45 \text{ rev})\left(2\pi \frac{\text{rad}}{\text{rev}}\right) = 90\pi \text{ rad}$$

The time required for this rotation is 60 s, and therefore the angular velocity is

$$\overline{\omega} = \frac{\Delta\theta}{\Delta t} = \frac{90\pi \text{ rad}}{60 \text{ s}} = \boxed{1.5\pi \text{ rad/s}}$$

You could also solve this problem by using the following conversion:

$$1 \frac{\text{rev}}{\text{min}} = \left(1 \frac{\text{rev}}{\text{min}}\right)\left(2\pi \frac{\text{rad}}{\text{rev}}\right)\left(\frac{1 \text{ min}}{60 \text{ s}}\right) = \frac{\pi}{30} \text{ rad/s}$$

Therefore

$$45 \frac{\text{rev}}{\text{min}} = 45\left(\frac{\pi}{30}\right)\frac{\text{rad}}{\text{s}} = 1.5\pi \text{ rad/s}$$

7.2   Angular Acceleration

Figure 7.3 shows a bicycle turned upside down so that a repairperson can work on the rear wheel. The bicycle pedals are turned such that at time t_1 the wheel has an angular velocity ω_1 (Fig. 7.3a), and at a later time t_2 it has an angular velocity ω_2 (Fig. 7.3b).

We define the **average angular acceleration** $\overline{\alpha}$ (α is the Greek letter alpha) of such an object as the ratio of the change in the angular velocity to the time interval Δt it takes the object to undergo the change:

ω_1 ω_2

(a)   (b)

Figure 7.3
An accelerating bicycle wheel rotates with (a) an angular velocity ω_1 at time t_1 and (b) an angular velocity ω_2 at time t_2.

$$\bar{\alpha} \equiv \frac{\text{change in angular velocity}}{\text{time interval}} = \frac{\omega_2 - \omega_1}{t_2 - t_1} = \frac{\Delta\omega}{\Delta t} \qquad \textbf{(7.4)}$$

The **instantaneous angular acceleration** is the value of the angular acceleration at a specific instant of time. Angular acceleration has units of radians per second per second (rad/s^2). Note that

Instantaneous angular acceleration

> when a rigid object rotates about a fixed axis, as does the bicycle wheel, every portion of the object has the same angular velocity and the same angular acceleration.

This, in fact, is precisely what makes these variables so useful in describing rotational motion.

The following argument should convince you that ω and α are the same for every point on the wheel. If a point on the rim of the wheel had a larger angular velocity than a point nearer the center, the shape of the wheel would be changing. The wheel remains circular (symmetrically distributed about the axle) only if all points have the same angular velocity and the same angular acceleration.

7.3   Rotational Motion Under Constant Angular Acceleration

Let us pause for a moment to consider some similarities between the equations we have found thus far for rotational motion and those we found in earlier chapters for linear motion. For example, compare the defining equation for average angular velocity,

$$\bar{\omega} \equiv \frac{\theta_f - \theta_i}{t_f - t_i} = \frac{\Delta\theta}{\Delta t}$$

with the defining equation for linear velocity,

$$\bar{v} \equiv \frac{x_f - x_i}{t_f - t_i} = \frac{\Delta x}{\Delta t}$$

A ten-hour exposure showing circular star trails around the south celestial pole. What would the exposure show if it were taken at the equator? *(© Anglo-Australian Telescope Board 1980)*

The equations are similar in the sense that θ replaces x and ω replaces v. You should take more than a casual note of such similarities as you proceed through the study of rotational motion because virtually every linear quantity that we have seen thus far has a corresponding "twin" in rotational motion. Once you are adept at recognizing such analogies, you will find it unnecessary to memorize many of the equations in this chapter. Additionally, the techniques for solving rotational motion problems are quite similar to those you have already learned for linear motion. For example, problems concerned with objects that rotate with a constant angular acceleration can be solved in much the same manner as those dealing with linear motion under constant acceleration. Therefore, if you have a good understanding of problems involving objects moving with a constant linear acceleration, these new exercises should be little more than a review for you.

An additional analogy between linear motion and rotational motion arises when we compare the defining equation for average angular acceleration,

$$\overline{\alpha} \equiv \frac{\omega_f - \omega_i}{t_f - t_i} = \frac{\Delta\omega}{\Delta t}$$

with the defining equation for average linear acceleration,

$$\overline{a} \equiv \frac{v_f - v_i}{t_f - t_i} = \frac{\Delta v}{\Delta t}$$

Now that we have pointed out the analogies between variables in linear motion and those in rotational motion, it should not surprise you that the equations of rotational motion involve the variables θ, ω, and α. In Section 2.6, we developed a set of kinematic equations for linear motion under constant acceleration. The same procedure can be used to derive a similar set of equations for rotational motion under constant angular acceleration. The resulting equations of rotational kinematics, along with the corresponding equations for linear motion under constant acceleration, are listed in the table below:

Rotational Motion About a Fixed Axis with α Constant (Variables: θ and ω)	Linear Motion with a Constant (Variables: x and v)	
$\omega = \omega_0 + \alpha t$	$v = v_0 + at$	(7.5)
$\theta = \omega_0 t + \frac{1}{2}\alpha t^2$	$x = v_0 t + \frac{1}{2}at^2$	(7.6)
$\omega^2 = \omega_0^2 + 2\alpha\theta$	$v^2 = v_0^2 + 2ax$	(7.7)
$\theta = \frac{1}{2}(\omega_0 + \omega)t$	$x = \frac{1}{2}(v_0 + v)t$	

Again, note the one-to-one correspondence between the rotational equations involving the angular variables θ, ω, and α and the equations of linear motion involving the variables x, v, and a_0.

Example 7.2   The Rotating Wheel

The bicycle wheel of Figure 7.3 rotates with a constant angular acceleration of 3.5 rad/s². If the initial velocity of the wheel is 2.0 rad/s at $t_0 = 0$, (a) through what angle does the wheel rotate in 2 s?

Solution

Since we are given that $\omega_0 = 2.0$ rad/s and $\alpha = 3.5$ rad/s^2, we use

$$\theta = \omega_0 t + \frac{1}{2}\alpha t^2$$

$$\theta = (2.0 \text{ rad/s})(2 \text{ s}) + \frac{1}{2}(3.5 \text{ rad/s}^2)(2 \text{ s})^2$$

$$= 11 \text{ rad} = \boxed{630°}$$

(b) What is the angular velocity at $t = 2$ s?

Solution

If we make use of Equation 7.5, we find

$$\omega = \omega_0 + \alpha t = 2.0 \text{ rad/s} + (3.5 \text{ rad/s}^2)(2 \text{ s}) = \boxed{9.0 \text{ rad/s}}$$

Exercise   Find the angular velocity of the wheel at $t = 2$ s by making use of Equation 7.7 and the results to Part (a).

7.4   Relations Between Angular and Linear Quantities

In this section we shall derive some useful relations between the angular velocity and acceleration of a rotating object and the linear velocity and acceleration of an arbitrary point in the object. In order to do so, you should be aware of the fact that, when a rigid object rotates about a fixed axis, every point in the object moves in a circle whose center is along the axis of rotation.

Consider the arbitrarily shaped object shown in Figure 7.4 rotating about the z axis through the point O. Assume the object rotates through the angle $\Delta\theta$, and hence through the arc length Δs, in a time Δt. We know from the defining equation for radian measure that

$$\Delta\theta = \frac{\Delta s}{r}$$

Let us now divide both sides of this equation by Δt, the time during which the rotation occurred:

$$\frac{\Delta\theta}{\Delta t} = \frac{1}{r}\frac{\Delta s}{\Delta t}$$

If the time interval Δt is very small, then the angle $\Delta\theta$ through which the object rotates is small and the ratio $\Delta\theta/\Delta t$ is the instantaneous angular velocity, ω. Also, Δs is very small when Δt is very small, and the ratio $\Delta s/\Delta t$ equals the instantaneous linear speed, v. Hence, this equation is equivalent to

$$\omega = \frac{v}{r}$$

Figure 7.4 allows us to interpret this equation. The distance Δs is traversed along an arc of the circular path followed by the point P as it rotates during the time Δt. Thus, v must be the linear velocity of a point lying along

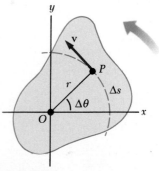

Figure 7.4
Rotation of an object about an axis through O that is perpendicular to the plane of the figure (the z axis). Note that a point P on the object rotates in a circle of radius r centered at O.

this arc, a velocity that is *tangent to the circular path*. For this reason, we often refer to this linear velocity as the **tangential velocity** of a particle moving in a circular path and write

Tangential velocity

$$v_t = r\omega \tag{7.8}$$

That is, the **tangential velocity** of a point on a rotating object equals the distance of that point from the axis of rotation multiplied by the angular velocity.

Note that although every point on the rotating object has the same angular velocity, not every point has the same linear, or tangential, velocity. In fact, Equation 7.8 shows that the linear velocity of a point on the rotating object increases as one moves outward from the center of rotation toward the rim, as you would intuitively expect.

A bit of caution should be exercised when using Equation 7.8. It has been derived using the defining equation for radian measure; hence, the equation is valid only when ω is measured in radians per unit time. Other measures of angular speed, such as degrees per second or revolutions per second, are not to be used in Equation 7.8.

To find a second equation relating linear and angular quantities, imagine that an object rotating about a fixed axis (Fig. 7.4) changes its angular velocity by $\Delta\omega$ in a time Δt. At the end of this time, the velocity of a point on the object, such as P, has changed by an amount Δv_t. From Equation 7.8, we have

$$\Delta v_t = r\,\Delta\omega$$

Dividing by Δt gives

$$\frac{\Delta v_t}{\Delta t} = r\,\frac{\Delta\omega}{\Delta t}$$

If the time interval Δt is very small, then the ratio $\Delta v_t/\Delta t$ is the tangential acceleration of that point and $\Delta\omega/\Delta t$ is the angular acceleration. Therefore, we see that

Tangential acceleration

$$a_t = r\alpha \tag{7.9}$$

That is, the **tangential acceleration** of a point on a rotating object equals the distance of that point from the axis of rotation multiplied by the angular acceleration.

Again, radian measure must be used for the angular acceleration term in this equation.

There is one more equation that relates linear quantities to angular quantities, but we shall withhold the derivation of this relationship until the next section.

Example 7.3   Going for a Spin

The turntable of a record player rotates initially at a rate of 33 rev/min and takes 20 s to come to rest. (a) What is the angular acceleration of the turntable, assuming the acceleration is uniform?

Solution

Recalling that 1 rev = 2π rad, we see that the initial angular velocity is

$$\omega_0 = \left(33 \, \frac{\text{rev}}{\text{min}}\right)\left(2\pi \, \frac{\text{rad}}{\text{rev}}\right)\left(\frac{1 \, \text{min}}{60 \, \text{s}}\right) = 3.46 \, \text{rad/s}$$

If we use $\omega = \omega_0 + \alpha t$ and the fact that $\omega = 0$ at $t = 20$ s, we get

$$\alpha = -\frac{\omega_0}{t} = -\frac{3.46 \, \text{rad/s}}{20 \, \text{s}} = \boxed{-0.173 \, \text{rad/s}^2}$$

where the minus sign for the angular acceleration indicates that ω is decreasing.

(b) How many rotations does the turntable make before coming to rest?

Solution

Equation 7.6 enables us to find the angular displacement in 20 s:

$$\theta = \omega_0 t + \frac{1}{2}\alpha t^2$$

$$= (3.46 \, \text{rad/s})(20 \, \text{s}) + \frac{1}{2}(-0.173 \, \text{rad/s}^2)(20 \, \text{s})^2 = \boxed{34.6 \, \text{rad}}$$

This corresponds to $34.6/2\pi$ rev = 5.51 rev.

(c) If the radius of the turntable is 0.14 m, what is the initial linear speed of a bug riding on the rim?

Solution

The relation $v_t = r\omega$ and the value $\omega_0 = 3.46$ rad/s give

$$v_t = r\omega_0 = (0.14 \, \text{m})(3.46 \, \text{rad/s}) = \boxed{0.484 \, \text{m/s}}$$

(d) What is the magnitude of the tangential acceleration of the bug at time $t = 0$?

Solution

We can use $a_t = r\alpha$, which gives

$$a_t = r\alpha = (0.14 \, \text{m})(0.173 \, \text{rad/s}^2) = \boxed{0.0242 \, \text{m/s}^2}$$

7.5 Centripetal Acceleration

Figure 7.5a shows a car moving in a circular path with *constant linear speed v.* It is often surprising to students to find that *even though the car moves at a constant speed, it still has an acceleration.* To see why this occurs, consider the defining equation for acceleration as

$$\mathbf{a} = \frac{\mathbf{v}_f - \mathbf{v}_i}{t_f - t_i} \tag{7.10}$$

Note that the acceleration depends on *the change in the velocity vector.* Because velocity is a vector, there are two ways in which an acceleration can be produced: by a change in the *magnitude* of the velocity and by a change in the

(a)

(b)

Figure 7.5
(a) Circular motion of a car moving with a constant speed. (b) As the car moves along the circular path from A to B, the direction of its velocity vector changes so the car undergoes a centripetal acceleration.

direction of the velocity. It is the latter situation that is occurring for the car moving in a circular path with constant speed (see Fig. 7.5b). We shall show that the acceleration vector in this case is perpendicular to the path and always points toward the center of the circle. An acceleration of this nature is called a **centripetal acceleration** (center-seeking) and its magnitude is given by

$$a_c = \frac{v^2}{r} \tag{7.11}$$

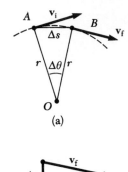

To derive Equation 7.11, consider Figure 7.6a. Here an object is seen first at point A with velocity $\mathbf{v_i}$ at time t_i and then at point B with velocity $\mathbf{v_f}$ at a later time t_f. Let us also assume here that $\mathbf{v_i}$ and $\mathbf{v_f}$ differ only in direction; their magnitudes are the same (that is, $v_i = v_f = v$). In order to calculate the acceleration, let us begin with Equation 7.10, which indicates that we must vectorially subtract $\mathbf{v_i}$ from $\mathbf{v_f}$:

$$\mathbf{a} = \frac{\mathbf{v_f} - \mathbf{v_i}}{t_f - t_i} = \frac{\Delta \mathbf{v}}{\Delta t} \tag{7.12}$$

where $\Delta \mathbf{v} = \mathbf{v_f} - \mathbf{v_i}$ is the change in the velocity. That is, $\Delta \mathbf{v}$ is obtained by adding to $\mathbf{v_f}$ the vector $-\mathbf{v_i}$. This can be accomplished graphically as shown by the vector triangle in Figure 7.6b. Note that when Δt is very small, Δs and $\Delta \theta$ are also very small. In this case, $\mathbf{v_f}$ will be almost parallel to $\mathbf{v_i}$ and the vector $\Delta \mathbf{v}$ will be approximately perpendicular to them, pointing toward the center of the circle. When one considers the limiting case where Δt becomes vanishingly small, $\Delta \mathbf{v}$ will point exactly toward the center of the circle. Furthermore, in this limiting case, the acceleration is also directed toward the center of the circle because it is in the direction of $\Delta \mathbf{v}$.

Now consider the triangle in Figure 7.6a, which has sides Δs and r. This triangle and the one with sides Δv and v in Figure 7.6b are similar. (Two triangles are similar if the angle between any two sides is the same for both triangles and if the ratio of lengths of these sides is the same.) This enables us to write a relationship between the lengths of the sides:

$$\frac{\Delta v}{v} = \frac{\Delta s}{r}$$

This equation can be solved for Δv and the expression so obtained can be substituted into Equation 7.12 to give $a \, \Delta t = v \, \Delta s / r$, or

$$a = \frac{v}{r} \frac{\Delta s}{\Delta t} \tag{7.13}$$

In this situation, Δs is a small distance measured along the arc of the circle (a tangential distance), so $v = \Delta s / \Delta t$, where v is the tangential velocity. Therefore, Equation 7.13 reduces to Equation 7.11:

$$a_c = \frac{v^2}{r} \tag{7.11}$$

Figure 7.6
(a) As the particle moves from A to B, the direction of its velocity vector changes from $\mathbf{v_i}$ to $\mathbf{v_f}$. (b) The construction for determining the direction of the change in velocity, $\Delta \mathbf{v}$, which is toward the center of the circle.

Centripetal acceleration

Since the tangential velocity is related to the angular velocity through the relation $v_t = r\omega$ (Eq. 7.8), an alternative form of Equation 7.11 is

$$a_c = \frac{r^2\omega^2}{r} = r\omega^2 \qquad\qquad \textbf{(7.14)}$$

Thus we conclude that in circular motion the **centripetal acceleration** is directed inward toward the center of the circle and has a magnitude given either by v^2/r or $r\omega^2$.

You should show that the dimensions of a_c are $[L]/[T^2]$, as required.

In order to clear up any misconceptions that might exist concerning centripetal and tangential acceleration, let us consider a car moving around a circular racetrack. Since the car is moving in a circular path, it always has a centripetal component of acceleration because the direction of travel of the car, and hence the direction of its velocity, is continuously changing. If the speed of the car is increasing or decreasing, it also has a tangential component of acceleration. To summarize, the tangential component of acceleration arises when the speed of the car is altered; the centripetal component of acceleration arises when the direction of travel is changed.

When both components of acceleration exist simultaneously, the tangential acceleration is tangent to the circular path and the centripetal acceleration points toward the center of the circular path. Because these components of acceleration are perpendicular to each other, we can find the magnitude of the **total acceleration** using the Pythagorean theorem:

$$a = \sqrt{a_t{}^2 + a_c{}^2} \qquad\qquad \textbf{(7.15)} \qquad \text{Total acceleration}$$

Example 7.4 Let's Go For a Spin

A test car moves at a constant speed of 10 m/s around a circular road of radius 50 m. Find (a) the centripetal acceleration of the car, and (b) the angular velocity of the car.

Solution
(a) The magnitude of the centripetal acceleration of the car is found from Equation 7.11 to be

$$a_c = \frac{v^2}{r} = \frac{(10\ \text{m/s})^2}{50\ \text{m}} = \boxed{2\ \text{m/s}^2}$$

By definition, the direction of $\mathbf{a_c}$ is towards the center of curvature of the road.

(b) The angular velocity of the car can be found using the expression $v_t = r\omega$, which gives

$$\omega = \frac{v_t}{r} = \frac{10\ \text{m/s}}{50\ \text{m}} = \boxed{0.2\ \text{rad/s}}$$

Note that we can also find the centripetal acceleration by using the value of ω found above. This is an alternative method to that Part (a). From Equation 7.12, we have

$$a_c = r\omega^2 = (50\ \text{m})(0.2\ \text{rad/s})^2 = 2\ \text{m/s}^2$$

As expected, both methods for finding the centripetal acceleration give the same answer.

Exercise Find the tangential acceleration and total acceleration of the car.

Answer   The tangential acceleration of the car is zero because the speed of the car remains constant. Since $a_t = 0$, the total acceleration equals the centripetal acceleration, 2 m/s², found in Part (a).

7.6   Centripetal Force

Consider a ball of mass m tied to a string of length r and being whirled in a horizontal circular path, as in Figure 7.7. Let us assume that the ball moves with a constant speed. Because the velocity vector, **v**, changes its direction continuously during the motion, the ball experiences a centripetal acceleration directed toward the center of motion, as described in Section 7.5. This centripetal acceleration has a magnitude

$$a_c = \frac{v^2}{r}$$

Figure 7.7
A ball attached to a string of length r, rotating in a circular path at constant speed.

The inertia of the ball tends to maintain motion in a straight-line path; however, the string prevents this by exerting a force on the ball to make it follow its circular path. This force (equal to the force of tension) is directed along the length of the string toward the center of the circle, as shown in Figure 7.7, and is an example of a class of forces called **centripetal forces**. Thus, the equation for Newton's second law along the radial direction is

Centripetal force

$$F_c = ma_c = m\,\frac{v^2}{r} \qquad (7.16)$$

All centripetal forces act toward the center of the circular path along which the object moves.

Because they act toward the center of rotation, centripetal forces cause a change in the direction of the velocity. Beyond this, they are no different from any of the other forces that we have studied. For example, friction between the tires and the road provides the centripetal force that enables a race car to travel in a circular path on a flat road, and the gravitational force exerted on the Moon by the Earth provides the centripetal force necessary to keep the Moon in its orbit. We shall discuss the gravitational force in more detail in Section 7.8.

Regardless of the example used, if the centripetal force acting on an object moving initially in a circle should vanish, the object would no longer continue to move in its circular path; instead, it would move along a straight-line path tangent to the circle. To illustrate this point, consider a ball attached to a string being whirled in a vertical circle as in Figure 7.8a. If the string happens to break when the ball is at position A, the centripetal force (the tension in the string) would vanish and the ball would move vertically upward. Its subsequent motion would be that of a freely falling body. If the string happens to break when the ball is at the top of its circular path, shown as point B in Figure 7.8b, the ball would initially fly off horizontally in a direction tangent to the path, and would then move in a parabolic path according to the equations of a projectile.

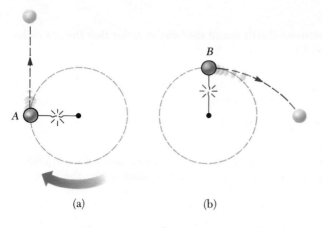

Figure 7.8
(a) When the string breaks at position *A*, the ball moves vertically upwards in free fall. (b) When the string breaks at position *B*, the ball moves along a **parabolic path.**

Problem-Solving Strategy:
Centripetal Forces

The following features involving centripetal forces and centripetal accelerations should be kept in mind:

1. Draw a free-body diagram of the object(s) under consideration, showing all forces that act on it.
2. Choose a coordinate system with one axis perpendicular to the circular path followed by the object and with one axis tangent to the circular path.
3. Find the net force toward the center of the circular path. This is the centripetal force.
4. From this point onward, the steps are virtually identical to those encountered when solving Newton's second law problems with $\Sigma \mathbf{F} = m\mathbf{a}$. Also, you should note that the magnitude of the centripetal acceleration can always be written as $a_c = v^2/r$.

Example 7.5   Buckle Up For Safety

A car travels at a constant speed of 30 mi/h (13.4 m/s) on a level circular turn of radius 50 m as shown in the birds-eye view in Figure 7.9a. What is the minimum

Figure 7.9   (Example 7.5) (a) Top view of a car on a curved path. (b) Free-body diagram for the car showing an end view.

coefficient of static friction between the tires and roadway in order that the car make the circular turn without sliding?

Solution

The centripetal force acting on the car is the force of static friction directed toward the center of the circular path as shown in Figure 7.9a. Thus, Equation 7.16 becomes

$$(1) \qquad f = m\frac{v^2}{r}$$

All the forces acting on the car are shown in Figure 7.9b. Because we have equilibrium in the vertical direction, the normal force upward is balanced by the force of gravity downward, so that

$$n = mg$$

From this expression, we can find the minimum force of friction as follows:

$$(2) \qquad f = \mu_s n = \mu_s mg$$

By setting the right-hand sides of (1) and (2) equal to each other, we find

$$\mu_s mg = m\frac{v^2}{r}$$

or

$$(3) \qquad \mu_s = \frac{v^2}{rg}$$

Therefore, the minimum coefficient of static friction required for the car to make the turn without sliding outwards is

$$\mu_s = \frac{v^2}{rg} = \frac{(13.4 \text{ m/s})^2}{(50 \text{ m})(9.80 \text{ m/s}^2)} = \boxed{0.366}$$

The value of μ_s for rubber on dry concrete is very close to 1. Thus, the car will be able to negotiate the curve with ease. However, if the road is wet or icy, the value for μ_s can be 0.2 or lower. Under this condition, the centripetal force provided by static friction would not be great enough to enable the car to follow the circular path, and it would slide off the roadway.

Example 7.6   Having Fun with a Yo-Yo

A child swings a yo-yo of weight mg in a horizontal circle such that the cord makes an angle of 30° with the vertical, as in Figure 7.10a. Find the centripetal acceleration of the yo-yo.

Solution

Two forces act on the yo-yo: its weight and the tension in the cord, **T**, as shown in Figure 7.10b. Because the yo-yo is in equilibrium in the y direction, we can apply $\Sigma F_y = 0$ in order to find T:

$$\sum F_y = T \cos 30° - mg = 0$$

$$T = \frac{mg}{\cos 30°}$$

The net force acting toward the center of the circular path is the centripetal force, which for this case is the *horizontal component* of **T**.

$$F_c = T \sin 30° = \frac{mg \sin 30°}{\cos 30°} = mg \tan 30°$$

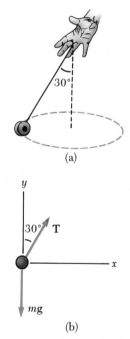

(a)

(b)

Figure 7.10   (Example 7.6) (a) A yo-yo swinging in a horizontal circle such that the cord makes a constant angle of 30° with the vertical. (b) Free-body diagram for the yo-yo.

The centripetal acceleration can now be found by applying Newton's second law along the horizontal direction:

$$F_c = ma_c$$

$$a_c = \frac{F_c}{m} = \frac{mg \tan 30°}{m} = g \tan 30° = (9.80 \text{ m/s}^2)(\tan 30°) = \boxed{5.66 \text{ m/s}^2}$$

7.7 Describing Motion of a Rotating System

We have seen that an object moving in a circle of radius r with constant speed v has a centripetal acceleration whose magnitude is v^2/r and whose direction is toward the center of rotation. The force necessary to maintain this centripetal acceleration, called a centripetal force, must also act toward the center of rotation. In the case of a ball rotating at the end of a string, the force exerted on the ball by the string (equal to the tension in the string) is the centripetal force. For a satellite in a circular orbit around the Earth, the force of gravity is the centripetal force. The centripetal force acting on a car rounding a curve on a level road is the force of friction between the tires and the pavement, and so forth. Centripetal forces are no different from any other forces we have encountered. The term *centripetal* is used simply to indicate that *the force is directed toward the center of a circle*.

In order to better understand the motion of a rotating system, consider a car traveling along a highway at a high speed and approaching a curved exit ramp, as in Figure 7.11. As the car takes the sharp left turn onto the ramp, a person sitting in the passenger seat slides to the right across the seat and hits the door. At that point, the force of the door keeps him from being ejected from the car. What causes the passenger to move toward the door? A popular, but *improper*, explanation is that some mysterious force pushes him outward. (This is sometimes called the "centrifugal" force, but we shall not use this term since it always creates confusion.)

The phenomenon is correctly explained as follows. Before the car enters the ramp, the passenger is moving in a straight-line path. As the car enters the ramp and travels a curved path, the passenger, because of inertia, tends to move along the original straight-line path. This is in accordance with Newton's first law: the natural tendency of a body is to continue moving in a straight line. However, if a sufficiently large centripetal force (toward the center of curvature) acts on the passenger, he will move in a curved path along with the car. The origin of this centripetal force is the force of friction between the passenger and the car seat. If this frictional force is not large enough, the passenger will slide across the seat as the car turns under him. Because of the passenger's inertia, he or she will continue to move in a straight-line path. Eventually, the passenger encounters the door, which provides a large enough centripetal force to enable the passenger to follow the same curved path as the car. The passenger slides toward the door not because of some mysterious outward force but because *there is no centripetal force large enough to allow him to travel along the circular path followed by the car*.

As a second example, consider what happens when you run clothes through the rinse cycle of a washing machine. In the last phase of this cycle, the

Figure 7.11
A car approaching a curved exit ramp.

Passengers on a roller coaster at the Texas State Fair experience the thrill of circular motion. Can you identify the origin of the centripetal force (or forces) acting on a passenger during a loop-the-loop maneuver? (© Zerschling, Photo Researchers)

drum spins rapidly to remove water from the clothes. Why is the water thrown off? An *improper* explanation is that the rotating system creates some mysterious outward force on each drop of water and this force causes the water to be hurled to the outer drum of the machine. The correct explanation goes as follows. When the clothes are at rest in the machine, water is held to them by molecular forces between the water and the fabric. During the spin cycle, the clothes rotate and the molecular forces are not large enough to provide the necessary *centripetal* force to keep the water molecules moving in a circular path along with the clothes. Hence, the water drops, because of their inertia, move in a straight-line path until they encounter the sides of the spinning drum.

In summary, one must be very careful to distinguish real forces from fictitious ones in describing motion in an accelerating frame. An observer in a car rounding a curve is in an accelerating frame and invents a fictitious outward force to explain why he or she is thrown outward. A stationary observer outside the car, however, considers only real forces on the passenger. To this observer, the mysterious outward force *does not exist!* The only real external force on the passenger is the centripetal (inward) force due to friction or the contact force of the door.

7.8   Newton's Law of Gravitation

Prior to 1686, a great mass of data had been collected on the motion of the Moon and the planets but a clear understanding of the forces that caused these celestial bodies to move the way they did was not available. In that year, however, Isaac Newton provided the key that unlocked the secrets of the heavens. He knew, from the first law, that a net force had to be acting on the Moon. If not, it would move in a straight-line path rather than in its almost circular orbit. Newton reasoned that this force arose as a result of an attractive field force between Moon and Earth, which we call the force of gravity. He also concluded that there could be nothing special about the Earth-Moon system or the Sun and its planets that would cause gravitational forces to act on them alone. In other words, he saw that the same force of attraction that causes the Moon to follow its path also causes an apple to fall to Earth from a tree. He wrote, "I deduced that the forces which keep the planets in their orbs must be reciprocally as the squares of their distances from the centers about which they revolve; and thereby compared the force requisite to keep the Moon in her orb with force of gravity at the surface of the Earth; and found them answer pretty nearly."

In 1687 Newton published his work on the **universal law of gravitation.** This law states that

> every particle in the universe attracts every other particle with a force that is directly proportional to the product of their masses and inversely proportional to the square of the distance between them.

If the particles have masses m_1 and m_2 and are separated by a distance r, the magnitude of this gravitational force is

$$F = G\frac{m_1 m_2}{r^2} \qquad \text{(7.17)}$$

where G is a universal constant called the **gravitational constant**, which has been measured experimentally. Its value in SI units is

$$G = 6.673 \times 10^{-11} \frac{\text{N} \cdot \text{m}^2}{\text{kg}^2} \qquad \text{(7.18)}$$

This force law is an example of an **inverse-square law,** in that it varies as the inverse square of the separation. The force acts such that the objects are always attracted to one another. From Newton's third law, we also know that the force on m_2 due to m_1, designated $\mathbf{F}_{21}$ in Figure 7.12, is equal in magnitude to the force on m_1 due to m_2, $\mathbf{F}_{12}$, and in the opposite direction. That is, these forces form an action-reaction pair.

There are several features of the universal law of gravity that deserve some attention:

1. The gravitational force is a field force that always exists between two particles regardless of the medium that separates them.
2. The force varies as the inverse square of the distance between the particles and therefore decreases rapidly with increasing separation.
3. The force is proportional to the product of their masses.

Newton's law of gravitation as given in Equation 7.17 applies to only two cases. It can be used to find the gravitational force between tiny objects (particles) or to find the gravitational force between a sphere and a particle located external to the sphere. In the latter case, the gravitational effect is as though the mass of the sphere were concentrated as its center. For example, the force on a particle of mass m at the Earth's surface has the magnitude

$$F = G\frac{M_e m}{R_e{}^2}$$

where M_e is the Earth's mass and R_e is its radius. This force is directed toward the center of the Earth.

Measurement of the Gravitational Constant

The gravitational constant was first measured in an important experiment by Henry Cavendish in 1798. The apparatus he used consisted of two small spheres, each of mass m, fixed to the ends of a light horizontal rod suspended by a thin metal wire, as in Figure 7.13. Two large spheres, each of mass M, were placed near the smaller spheres. The attractive force between the smaller and larger spheres caused the rod to rotate and the wire to twist. The angle through which the suspended rod rotated was measured by the deflection of a light beam reflected from a mirror attached to the vertical suspension. (Such a moving spot of light is an effective technique for amplifying the motion.) The experiment was carefully repeated with different masses at various separations. In addition to providing a value for G, the results showed that the force is attractive, proportional to the product mM, and inversely proportional to the square of the distance r.

Figure 7.12
The gravitational force between two particles is attractive. Note that according to Newton's third law,
$\mathbf{F}_{12} = -\mathbf{F}_{21}$.

Figure 7.13
Schematic diagram of the Cavendish apparatus for measuring the constant G. The smaller spheres of mass m are attracted to the large spheres of mass M, and the bar rotates through a small angle. A light beam reflected from a mirror attached to the vertical suspension measures the angle of rotation.

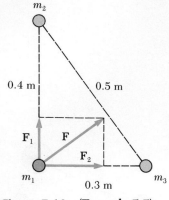

Figure 7.14   (Example 7.7)

Example 7.7   Billiards, Anyone?

Three billiard balls each of mass 0.3 kg are placed on a table at the corners of a right triangle as shown in the overhead view of Figure 7.14. Find the net gravitational force on the ball denoted as m_1 due to the forces exerted on it by the other two balls.

Solution
To find the net gravitational force on m_1, we first calculate the force exerted on m_1 due to m_2. Then, we find the force on m_1 due to m_3. Finally, we must add these two forces *vectorially* to obtain the net force on m_1.

The force exerted on m_1 due to m_2, denoted by $\mathbf{F}_1$ in Figure 7.14, is upwards. The magnitude of this force is calculated using Equation 7.17:

$$F_1 = G\frac{m_1 m_2}{r^2} = (6.67 \times 10^{-11}\ \text{N·m}^2/\text{kg}^2)\frac{(0.3\ \text{kg})(0.3\ \text{kg})}{(0.4\ \text{m})^2}$$

$$= 3.75 \times 10^{-11}\ \text{N}$$

This result shows that gravitational forces between common objects that surround us have extremely small magnitudes.

Now, let us calculate the gravitational force exerted on m_1 due to m_3. This force, denoted by $\mathbf{F}_2$ in Figure 7.14, is directed to the right, and its magnitude is

$$F_2 = G\frac{m_1 m_3}{r^2} = (6.67 \times 10^{-11}\ \text{N·m}^2/\text{kg}^2)\frac{(0.3\ \text{kg})(0.3\ \text{kg})}{(0.3\ \text{m})^2}$$

$$= 6.67 \times 10^{-11}\ \text{N}$$

The net gravitational force exerted on m_1 is found by adding $\mathbf{F}_1$ and $\mathbf{F}_2$ as *vectors*. The magnitude of this net force is given by

$$F = \sqrt{F_1{}^2 + F_2{}^2} = \sqrt{(3.75)^2 + (6.67)^2} \times 10^{-11}\ \text{N} = \boxed{7.65 \times 10^{-11}\ \text{N}}$$

Exercise   Find the direction of the resultant force on m_1.

Answer   The vector $\mathbf{F}$ makes an angle of 29.3° with respect to the line joining m_1 and m_3.

Example 7.8   The Mass of the Earth

Use the gravitational force law to find an approximate value for the mass of the Earth.

Solution
Figure 7.15, obviously not to scale, shows a baseball falling toward the Earth at a location where the acceleration due to gravity is g. We know that the gravitational force exerted on the baseball by the Earth is the same as the weight of the ball and that this is given by $w = m_b g$. Since the force in the gravitational law is the weight of the ball, we find

$$m_b g = G\frac{M_e m_b}{R_e{}^2}$$

We can divide each side of this equation by m_b and solve for the mass of the Earth, M_e:

$$M_e = \frac{g R_e{}^2}{G}$$

Figure 7.15   (Example 7.8) A baseball falling toward the Earth (not drawn to scale).

The falling baseball is close enough to the Earth so that the distance of separation between the center of the ball and the center of the Earth can be taken as the radius of the Earth, 6.38×10^6 m. Thus, the mass of the Earth is

$$M_e = \frac{(9.80 \text{ m/s}^2)(6.38 \times 10^6 \text{ m})^2}{6.67 \times 10^{-11} \text{ N} \cdot \text{m}^2/\text{kg}^2} = \boxed{5.98 \times 10^{24} \text{ kg}}$$

Example 7.9 Gravity and Altitude

Derive an expression that shows how the acceleration due to gravity varies with distance from the center of the Earth at an exterior point.

Solution
The falling baseball of Example 7.8 can be used here also. Now, however, assume that the ball is located at some arbitrary distance r from the Earth's center. The first equation in Example 7.8, with r replacing R_e and m_b removed from both sides, becomes

$$g = G\frac{M_e}{r^2}$$

This indicates that the acceleration due to gravity at an exterior point decreases as the inverse square of the distance from the center of the Earth. Our assumption of Chapter 2 that objects fall with a constant acceleration is obviously incorrect in light of our present example. For small distances of fall, however, this change in g is so small that neglecting the variation does not introduce a significant effort in our results.

Because the true weight of an object is mg, we see that a change in the value of g produces a change in the weight of an object. For example, if you weigh 800 N at the surface of the Earth, you will weigh only 200 N at a height above the Earth equal to the radius of the Earth. Also, we see that if the distance of an object from the Earth becomes infinitely large, the true weight approaches zero. Values of g at various altitudes are given in Table 7.1.

Exercise If an object weighs 270 N at the Earth's surface, what will it weigh at an altitude equal to twice the radius of the Earth?

Answer 30 N.

**Table 7.1
Acceleration Due to
Gravity, g, at Various
Altitudes**

Altitude (km)[a]	g(m/s²)
1000	7.33
2000	5.68
3000	4.53
4000	3.70
5000	3.08
6000	2.60
7000	2.23
8000	1.93
9000	1.69
10,000	1.49
50,000	0.13

[a] All values are distances above the Earth's surface.

7.9 Gravitational Potential Energy Revisited

In Chapter 5 we introduced the concept of gravitational potential energy and found that the potential energy of an object could be calculated using the equation $PE = mgh$, where h is the height of the object above or below some reference level. This equation is actually valid only when the object is near the Earth's surface. For objects high above the Earth's surface, such as a satellite, an alternative expression must be used to compute the gravitational potential energy. The general expression for the gravitational potential energy for an object of mass m at a distance *r from the center of the Earth* can be shown (using integral calculus) to be

$$PE = -G\frac{M_e m}{r} \qquad \textbf{(7.19)}$$

Gravitational potential energy

where M_e is the mass of the Earth.

This equation assumes that the zero level for potential energy is chosen to be at an infinite distance from the center of the Earth. This point is a convenient one because the gravitational force goes to zero when r is set equal to infinity.

Example 7.10   Does the Potential Energy Reduce to *mgh*?

(a) Find expressions for the gravitational potential energy of an object at the surface of the Earth and for the same object when at a height h above the surface of the Earth (Fig. 7.16). (b) From the answers to Part (a), show that the difference in potential energy between these two points reduces to the familiar expression $PE = mgh$ when h is small compared to the Earth's radius.

Solution
Equation 7.19 gives the potential energy at the surface of the Earth as

$$PE_1 = -G\frac{M_e m}{R_e}$$

We can also use Equation 7.19 to find the potential energy of the object at the height h above the surface of the Earth. Taking $r = R_e + h$, we find

$$PE_2 = -G\frac{M_e m}{(R_e + h)}$$

(b) The difference in potential energy between these two points is found as follows:

$$PE_2 - PE_1 = -G\frac{M_e m}{(R_e + h)} - \left(-G\frac{M_e m}{R_e}\right)$$

$$= -GM_e m\left[\frac{1}{(R_e + h)} - \frac{1}{R_e}\right]$$

After finding a common denominator and applying some algebra to the equation above, we find

$$PE_2 - PE_1 = \frac{GM_e mh}{R_e(R_e + h)}$$

When the height h is very small compared to R_e, the denominator in the expression above is approximately equal to $R_e{}^2$. Thus, we have

$$PE_2 - PE_1 \approx \frac{GM_e}{R_e{}^2}mh$$

Now note that the acceleration due to gravity at the surface of the Earth is given by $g = GM_e/R_e{}^2$ (see Example 7.9). Thus,

$$PE_2 - PE_1 \approx mgh$$

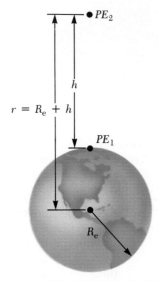

$r = R_e + h$

PE_2

h

PE_1

R_e

Figure 7.16   (Example 7.10)

*7.10   Escape Velocity

Anyone who has ever watched a football game knows what happens to a football after the kickoff. It first goes up and then comes down, just as all common objects will do when projected in the air. The total time the ball remains in the air increases as its initial velocity increases. With this idea, one can imagine a situation in which the ball is kicked with such a large initial velocity that it would simply not come down. This particular velocity is called

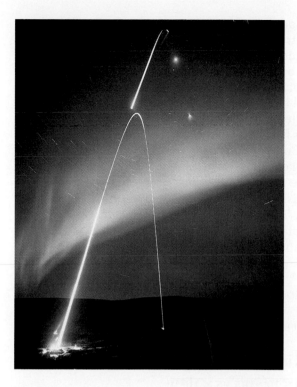

This time-exposure photograph shows the launch of a three-stage Black Brant sounding rocket against the backdrop of the glowing Northern Lights. The rocket was launched from the Poker Flats Research Range near Fairbanks, Alaska, in April 1984. It carried and released two cannisters of barium approximately 890 km above the Earth's surface, resulting in the two bursts seen below the very short streaks of the third stage. *(Photograph by R. Sluder and G. Boquist for Technology International)*

the escape velocity of an object from the Earth. At this velocity, any object will escape from the Earth and soar forever in space until it is captured by the gravitational attraction of some other planet or celestial body. We shall now proceed to derive an expression which will enable us to calculate the escape velocity.

Suppose an object of mass m is projected vertically upward from the Earth's surface with an initial speed v_i as in Figure 7.17. The initial mechanical energy (kinetic plus potential energy) is given by

$$KE_i + PE_i = \frac{1}{2} mv_i{}^2 - G \frac{M_e m}{R_e}$$

We neglect air resistance and assume that the initial velocity is just large enough to allow the object to reach infinity with a velocity of zero. We call this value of v_i the escape velocity v_e. When the object is at an infinite distance from the Earth, its kinetic energy is zero, because $v_f = 0$, and the gravitational potential energy is also zero because our zero level of potential energy was selected at $r = \infty$. Hence, the total mechanical energy is zero, and the law of conservation of energy gives

$$\frac{1}{2} mv_e{}^2 - G \frac{M_e m}{R_e} = 0$$

and

$$v_e = \sqrt{\frac{2GM_e}{R_e}}$$

(7.20) Escape velocity

Figure 7.17
An object of mass m projected upward from the Earth's surface with an initial speed v_i reaches a maximum altitude h (where $M_e \gg m$).

Table 7.2 Escape Velocities for the Planets and the Moon	
Planet	v_e(km/s)
Mercury	4.3
Venus	10.3
Earth	11.2
Moon	2.3
Mars	5.0
Jupiter	60
Saturn	36
Uranus	22
Neptune	24

From this equation, one finds that the escape velocity for Earth is about 11.2 km/s, which corresponds to about 25,000 mi/h (see Example 7.11). Note that this expression for v_e is independent of the mass of the object projected from the Earth. For example, a spacecraft has the same escape velocity as a molecule. A list of escape velocities for the planets and the moon is given in Table 7.2. The data presented in this table help in understanding why some planets have atmospheres while others do not. For example, on the very hot planet Mercury, gas molecules have an average speed that is greater than the escape velocity of 4.3 km/s. Consequently, any gases that might have been present on the surface of the planet at its formation have long since wandered off into space. Likewise, in the atmosphere of our own Earth, hydrogen and helium molecules through collisions often gain speeds greater than 11.2 km/s, the escape velocity on Earth. Therefore, the gases are not retained in the Earth's atmosphere. On the other hand, heavier gases such as oxygen and nitrogen in the Earth's atmosphere have average speeds less than 11.2 km/s, and do not escape.

Example 7.11   Escape Velocity of a Rocket

Calculate the escape velocity from the Earth for a 5000-kg spacecraft.

Solution
Using Equation 7.20 with $M_e = 5.98 \times 10^{24}$ kg and $R_e = 6.38 \times 10^6$ m gives

$$v_e = \sqrt{\frac{2GM_e}{R_e}} = \sqrt{\frac{2(6.68 \times 10^{-11})(5.98 \times 10^{24})}{6.37 \times 10^6}} = \boxed{1.12 \times 10^4 \text{ m/s}}$$

This corresponds to about 25,000 mi/h or about 7 mi/s. Note that the mass of the spacecraft was not required for this calculation.

"The greater the velocity . . . with which [a stone] is projected, the farther it goes before it falls to the Earth. We may therefore suppose the velocity to be so increased, that it would describe an arc of 1, 2, 5, 10, 100, 1000 miles before it arrived at the Earth, till at last, exceeding the limits of the Earth, it should pass into space without touching."—Newton, *System of the World.*

▶SPECIAL TOPIC
Gravity in the Extreme, Black Holes

The maternity wards of space are large clouds of gas and dust called nebulae. The mutual gravitational attraction between particles in the cloud pulls the particles together until eventually a huge collapsing spherical ball is formed. As the separation between the particles decreases, the gravitational force on each increases; the particles become part of a collapsing sphere, and fall together towards the center of the sphere at an increasing speed. Finally, their speed becomes great enough near the center of the collapsing sphere that nuclear reactions begin to occur. (We shall discuss these reactions more completely in Chapter 32.) One manifestation of these nuclear reactions is the continuous release of tremendous amounts of energy that prevents further gravitational collapse of the star. The end result is a star living out its active life in equilibrium between the inward pull of gravitational forces tending to collapse the star, and the outward radiation and thermal forces tending to expand the star.

During the first stages of nuclear burning at the center of a star, hydrogen nuclei are fused together to produce helium nuclei. When the supply of hydrogen is exhausted, the fusion process ceases. At this point, the balance of forces is destroyed since no energy is being released from fusion processes to prevent the collapse by gravitational forces. As the star collapses during its death struggle, the temperature inside the star increases further, and other nuclear reactions can occur involving elements heavier than hydrogen. The most spectacular deaths occur for stars that are originally very massive in nature. At some point in its flirtation with death, such a massive star may undergo a catastrophic explosion, hurling debris into space and leaving behind a central core. Such an exploding star is called a supernova. Supernovae are very rare events, and the only ones that have been observed since the invention of the telescope in 1609 have been in distant galaxies. One exception to this is the one found in 1987 in the large Magellanic Cloud, an irregular galaxy close to our Milky Way galaxy. In A.D. 1054, Chinese observers recorded a "guest star" in the constellation Taurus, in our galaxy, which was so bright that it was visible in daylight. In this same region of space, we now find a nebula, called the Crab nebula, which is believed to be the remnants thrown off by the 1054 supernova.

As we mentioned earlier, when a star explodes as a supernova, most of the star's material is ejected, but a central core is left behind that continues to collapse. What will be the ultimate fate of the collapsing core? The answer to this intriguing question depends on the mass of the core. If the core has a mass less than about 1.4 times the mass of our Sun, it will gradually cool down and end its life as a black dwarf star. However, if the core's mass is greater than about 1.4 solar masses, the star may collapse due to gravity such that the electrons and protons combine to form neutrons. Such a star, in which all the remaining matter is in the form of neutrons, is called a neutron star. The contraction of a neutron star eventually ceases when the neutron gas is compressed to a radius of about 10 km (about 6 miles). A teaspoonful of this material on Earth would weigh about 5 billion tons!

An even more unusual death of a star may occur when the core has a mass greater than about 3 solar masses. No known forces in nature are strong enough to prevent the collapse of such a star. In fact, the collapse of the star may continue until the star becomes a mere point in space called a singularity. Such a singularity is commonly referred to as a **black hole**. *In effect, black holes are remains of stars that have collapsed under their own weight.*

Objects such as spaceships, atoms, and molecules can enter a black hole, but can never escape once in its grips. As soon as they enter a black hole, they are pulled deeper and deeper toward its center and are trapped forever. In fact, even light cannot escape the tremendous gravitational forces set up inside a black hole; hence, the origin of the terminology "black hole."

Using the concept of escape velocity developed in Section 7.10, we can obtain some understanding of how particles become trapped by black holes. (A proper derivation of the results below requires some ideas and equations from relativity. These will be developed in Chapter 28.) Recall that the escape velocity from a spherical body of mass M and radius R is given by

$$v_e = \sqrt{\frac{2GM}{R}}$$

If the escape velocity exceeds the speed of light, $c = 3 \times 10^8$ m/s, light within the body will not be able to escape and the body will appear to be black. The critical radius for which this occurs is called the **Schwarzschild radius** after Karl Schwarzschild, who predicted the existence of black holes in 1916 (see Fig. 7.18). Taking $v_e = c$ in the expression above, and solving for R, we find

$$R = \frac{2GM}{c^2} \tag{7.21}$$

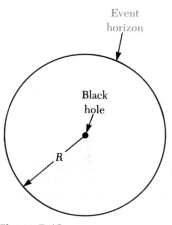

Figure 7.18
The event horizon surrounding a black hole. The distance R equals the Schwarzschild radius.

 Escape velocity

For example, the value of R for a black hole whose mass is equal to that of the Sun is calculated to be 3.0 km (about 2 mi); a black hole whose mass equals that of the Earth would have a radius of about 9 mm (about the size of a one-cent piece). The boundary of the region of radius R surrounding a black hole that permits entrance but prevents escape is called the **event horizon**. Any event occurring inside this horizon is invisible to an outside observer.

years and provided the data on which our currently accepted model of the Solar System was first established. It is interesting to note that these precise observations, made on the planets and 777 stars visible to the naked eye, were carried out with a large sextant and compass because the telescope had not yet been invented.

Kepler, who was Brahe's student, acquired Brahe's astronomical data and spent about 16 years trying to deduce a mathematical model for the motion of the planets. After many laborious calculations, he found that Brahe's precise data on the revolution of Mars about the Sun provided the answer. Such data are difficult to sort out because the Earth is also in motion about the Sun. Kepler's analysis first showed that the concept of circular orbits about the Sun had to be abandoned. He eventually discovered that the orbit of Mars could be accurately described by an ellipse with the Sun at one focus. He then generalized this analysis to include the motion of all planets. The complete analysis is summarized in three statements known as **Kepler's laws**. These laws applied to the Solar System are:

1. All planets move in elliptical orbits with the Sun at one of the focal points.
2. A line drawn from the Sun to any planet sweeps out equal areas in equal time intervals.
3. The square of the orbital period of any planet is proportional to the cube of the average distance from the planet to the Sun.

About 100 years after Kepler's analysis, Newton demonstrated that these laws were the consequence of the gravitational force existing between any two masses.

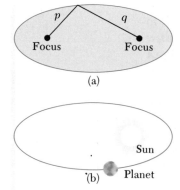

Figure 7.19
(a) The sum $p + q$ is the same for every point on the ellipse. (b) In the Solar System, the Sun is located at one focus of the elliptical orbit of each planet and the other focus is empty.

Kepler's First Law

We shall not attempt to derive Kepler's first law here. Instead, we simply state that it can be shown that the first law arises as a natural consequence of the inverse-square nature of Newton's law of gravitation. That is, any object bound to another by a force that varies as $1/r^2$ will move in an elliptical orbit. As shown in Figure 7.19a, an ellipse is a curve drawn such that the sum of the distances from any point on the curve to two internal points called focal points or foci (singular, focus) is always the same. For the Sun-planet configuration, shown in Figure 7.19b, the Sun is at one focus and the other focus is empty. Because the orbit is an ellipse, the distance from the Sun to the planet continuously changes.

Kepler's Second Law

Kepler's second law states that a line drawn from the Sun to any planet sweeps out equal areas in equal time intervals. Consider a planet in an elliptical orbit about the Sun, as in Figure 7.20. Imagine that at some particular instant of time, we draw a line from the Sun to a planet, tracing out the line AS where S represents the position of the Sun. Exactly 30 days later, we repeat the process and draw the line BS. The area swept out is shown in color in the diagram. When the planet is at point C, we begin the process anew, drawing the line CS from the planet to the Sun. We now wait 30 days, exactly the same time

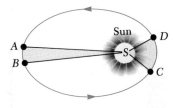

Figure 7.20
The two areas swept out by the planet in its elliptical orbit about the Sun are equal if the time interval between points A and B is equal to the time interval between points C and D.

interval we waited before, and draw the line *DS*. The area swept out in this interval is also shown in color in the figure. These two areas are equal.

Kepler's Third Law

Because of the simplicity of the derivation, we shall derive Kepler's third law here. Consider a planet of mass M_p moving about the Sun of mass M_s in a circular orbit, as in Figure 7.21. (The assumption of a circular orbit rather than an elliptical orbit will not introduce serious error into our approach because the orbits of all planets except Mercury and Pluto are very close to being circular.) Because the gravitational force on the planet is equal to the centripetal force needed to keep it moving in a circle,

$$\frac{GM_sM_p}{r^2} = \frac{M_pv^2}{r}$$

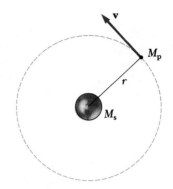

Figure 7.21
A planet of mass M_p moving in a circular orbit about the Sun. The orbits of all planets except Mercury and Pluto are nearly circular.

The speed v of the planet in its orbit is equal to the circumference of the orbit divided by the time for one revolution, T, called the **period** of the planet. That is, $v = 2\pi r/T$ and the above expression becomes

$$\frac{GM_s}{r^2} = \frac{(2\pi r/T)^2}{r}$$

$$T^2 = \left(\frac{4\pi^2}{GM_s}\right)r^3 = K_sr^3 \qquad \textbf{(7.22)} \quad \text{Kepler's third law}$$

where K_s is a constant given by

$$K_s = \frac{4\pi^2}{GM_s} = 2.97 \times 10^{-19} \text{ s}^2/\text{m}^3$$

This is Kepler's third law. It is valid for elliptical orbits if we replace r by a length equal to half the long axis of the ellipse. Note that K_s is independent of the mass of the planet. Therefore Equation 7.22 is valid for any planet. If we consider the orbit of a satellite about the Earth, such as the Moon, then the constant has a different value, with the mass of the Sun replaced by the mass of the Earth. In this case, K_e would equal $4\pi^2/GM_e$.

A collection of useful planetary data is given in Table 7.3. The last column of this table verifies that T^2/r^3 is a constant given by $K_s = 2.97 \times 10^{-19}$ s^2/m^3.

Example 7.12 An Earth Satellite

A satellite of mass m moves in a circular orbit about the Earth with a constant speed v and a height $h = 1000$ km above the Earth's surface, as in Figure 7.22. (For clarity, this figure is not drawn to scale.) Find the orbital speed of the satellite. The radius of the Earth is 6.38×10^6 m, and its mass is 5.98×10^{24} kg.

Solution
The only external force on the satellite is that of the gravitational attraction exerted by the Earth. This force is directed toward the center of the satellite's circular path and is the centripetal force acting on the satellite. Since the force of gravity is GM_em/r^2, we find

Figure 7.22 (Example 7.12)
A satellite of mass m moving in a circular orbit of radius r and with constant speed v around the Earth. The centripetal force is provided by the gravitational force acting on the satellite (not drawn to scale).

Table 7.3
Useful Planetary Data[a]

Body	Mass (kg)	Mean Radius (m)	Period (s)	Mean Distance From Sun (m)	$\dfrac{T^2}{r^3}\left[10^{-19}\left(\dfrac{s^2}{m^3}\right)\right]$
Mercury	3.18×10^{23}	2.43×10^6	7.60×10^6	5.79×10^{10}	2.97
Venus	4.88×10^{24}	6.06×10^6	1.94×10^7	1.08×10^{11}	2.99
Earth	5.98×10^{24}	6.38×10^6	3.156×10^7	1.496×10^{11}	2.97
Mars	6.42×10^{23}	3.37×10^6	5.94×10^7	2.28×10^{11}	2.98
Jupiter	1.90×10^{27}	6.99×10^7	3.74×10^8	7.78×10^{11}	2.97
Saturn	5.68×10^{26}	5.85×10^7	9.35×10^8	1.43×10^{12}	2.99
Uranus	8.68×10^{25}	2.33×10^7	2.64×10^9	2.87×10^{12}	2.95
Neptune	1.03×10^{26}	2.21×10^7	5.22×10^9	4.50×10^{12}	2.99
Pluto	$\approx 1 \times 10^{23}$	$\approx 3 \times 10^6$	7.82×10^9	5.91×10^{12}	2.96
Moon	7.36×10^{22}	1.74×10^6	—	—	—
Sun	1.991×10^{30}	6.96×10^8	—	—	—

[a] For a more complete set of data, see *Handbook of Chemistry and Physics,* Cleveland, Chemical Rubber Publishing Co.

$$F_c = G\frac{M_e m}{r^2} = m\frac{v^2}{r}$$

$$v^2 = \frac{GM_e}{r}$$

In this expression, the distance r is the Earth's radius plus the height of the satellite, that is, $r = R_e + h = 7.38 \times 10^6$ m, so that

$$v^2 = \frac{(6.67 \times 10^{-11}\ \text{N·m}^2/\text{kg}^2)(5.98 \times 10^{24}\ \text{kg})}{7.38 \times 10^6\ \text{m}} = 5.40 \times 10^7\ \text{m}^2/\text{s}^2$$

Therefore,

$$v = 7.35 \times 10^3\ \text{m/s} \approx 16{,}400\ \text{mi/h}$$

Note that *v is independent of the mass of the satellite!*

Exercise   Calculate the period of revolution, *T*, of the satellite (where *T* is the time for one revolution about the Earth).

Answer   105 min.

Summary

The **average angular velocity,** $\overline{\omega}$, of a rigid object is defined as the ratio of the angular displacement $\Delta\theta$, to the time interval Δt:

$$\overline{\omega} \equiv \frac{\theta_2 - \theta_1}{t_2 - t_1} = \frac{\Delta\theta}{\Delta t} \tag{7.3}$$

where $\overline{\omega}$ is in radians per second (rad/s).

The **average angular acceleration**, $\bar{\alpha}$, of a rotating object is defined as the ratio of the change in angular velocity, $\Delta\omega$, to the time interval Δt.

$$\bar{\alpha} \equiv \frac{\omega_2 - \omega_1}{t_2 - t_1} = \frac{\Delta\omega}{\Delta t} \qquad \text{(7.4)}$$

where $\bar{\alpha}$ is in radians per second per second (rad/s^2).

If an object undergoes rotational motion about a fixed axis under constant angular acceleration α, one can describe its motion by the following set of equations:

$$\omega = \omega_0 + \alpha t \qquad \text{(7.5)}$$

$$\theta = \omega_0 t + \frac{1}{2}\alpha t^2 \qquad \text{(7.6)}$$

$$\omega^2 = \omega_0{}^2 + 2\alpha\theta \qquad \text{(7.7)}$$

$$\theta = \frac{1}{2}(\omega_0 + \omega)t$$

When an object rotates about a fixed axis, the angular velocity and angular acceleration are related to the tangential velocity and tangential acceleration through the relationships

$$v_t = r\omega \qquad \text{(7.8)}$$

$$a_t = r\alpha \qquad \text{(7.9)}$$

Any object moving in a circular path has an acceleration directed toward the center of the circular path. This acceleration is called a **centripetal acceleration** and its magnitude is given by

$$a_c = \frac{v^2}{r} = r\omega^2 \qquad \text{(7.11) and (7.14)}$$

Any object moving in a circular path must have a net force exerted on it that is directed toward the center of the circular path. This net force is called a **centripetal force**. From Newton's second law, the centripetal force and centripetal acceleration are related as

$$F_c = m\frac{v^2}{r} \qquad \text{(7.16)}$$

Some examples of centripetal forces are the force of gravity (as in the motion of a satellite), and the force of tension in a string.

Newton's law of gravity states that every particle in the universe attracts every other particle with a force that is directly proportional to the product of their masses and inversely proportional to the square of the distance r between them:

$$F = G\frac{m_1 m_2}{r^2} \qquad (7.17)$$

where $G = 6.673 \times 10^{-11}$ N·m²/kg² is the **gravitational constant.**

The general expression for the gravitational potential energy for an object of mass m at a distance r from the center of the Earth of mass M_e is given by

$$PE = -G\frac{M_e m}{r} \qquad (7.19)$$

where the potential energy is chosen to be zero at $r = \infty$.

The escape velocity for any object at the Earth's surface is independent of the mass of the object and is given by

$$v_e = \sqrt{\frac{2GM_e}{R_e}} \qquad (7.20)$$

where R_e is the radius of the Earth. This expression can also be used to calculate the escape velocity for any planet of mass M and radius R.

Kepler's laws of planetary motion state that

1. All planets move in elliptical orbits with the Sun at one of the focal points.
2. A line drawn from the Sun to any planet sweeps out equal areas in equal time intervals.
3. The square of the orbital period of a planet is proportional to the cube of the mean distance from the planet to the Sun:

$$T^2 = \left(\frac{4\pi^2}{GM_s}\right) r^3 \qquad (7.22)$$

Additional Reading

J. Beams, "Ultra-High-Speed Rotation," *Sci. American,* April 1961, p. 134.

I. B. Cohen, "Newton's Discovery of Gravity," *Sci. American,* March 1981, p. 166.

R. Feynman, *The Character of Physical Law,* Cambridge, MA, MIT Press, 1965, Chaps. 1, 2, and 3.

G. Gamow, *Gravity,* Science Study Series, Garden City, NY, Doubleday, 1962.

D. Layzer, *Constructing the Universe,* Scientific American Library, New York, NY, W. H. Freeman and Co., 1984.

F. Pipkin, "Gravity Up in the Air," *The Sciences,* July/August 1984, p. 24.

R. Smoluchowski, *The Solar System,* Scientific American Library, New York, NY, W. H. Freeman and Co., 1983.

G. Spetz, "Detection of Gravity Waves," *The Physics Teacher,* May 1984, p. 282.

C. Wilson, "How Did Kepler Discover His First Two Laws?" *Sci. American,* March 1972, p. 92.

Questions

1. What is the magnitude of the angular velocity, ω, of the second hand of a clock? What is the angular acceleration, α, of the second hand?
2. When a wheel of radius R rotates about a fixed axis, do all points on the wheel have the same angular velocity? Do they all have the same linear velocity? If the angular velocity is constant and equal to ω_0, describe the linear velocities and linear accelerations of the points at $r = 0$, $r = R/2$, and $r = R$.
3. If a car's wheels are replaced with wheels of larger diameter, will the reading of the speedometer change? Explain.
4. Are the kinematic expressions for θ, ω, and α valid when the angular displacement is measured in degrees instead of in radians?
5. Cite an example of a situation in which an automobile driver can have a centripetal acceleration but no tangential acceleration.
6. Is it possible for a car to move in a circular path in such a way that it has a tangential acceleration but no centripetal acceleration?
7. Cite an example in which you could use Equation 7.6 to describe the motion of a record accelerating on a turntable and a situation in which this equation would not be valid.
8. Correct the following statement: "The racing car rounds the turn at a constant velocity of 90 miles per hour."
9. An object moves in a circular path with constant speed v. (a) Is the velocity of the object constant? (b) Is its acceleration constant? Explain.
10. Describe the path of a moving body in the event that the acceleration is constant in magnitude at all times and (a) perpendicular to the velocity; (b) parallel to the velocity.
11. Imagine that you attach a heavy object to one end of a spring and then whirl the spring and object in a horizontal circle (by holding the other end of the spring). Does the spring stretch? If so, why? Discuss in terms of centripetal force.
12. Why does mud fly off a rapidly turning wheel?
13. A pail of water can be whirled in a vertical path such that none is spilled. Why does the water remain in the pail, even when the pail is above your head?
14. It has been suggested that rotating cylinders about 10 miles in length and 5 miles in diameter be placed in space for colonies. The purpose of the rotations is to simulate gravity for the inhabitants. Explain this concept for producing an effective gravity.
15. Because of the Earth's rotation about its axis, you will weigh slightly less at the equator than at the poles. Why is this?
16. Why is it that an astronaut in a space capsule orbiting the Earth experiences a feeling of weightlessness?
17. Why does a pilot tend to black out when he pulls out of a steep dive?
18. Centrifuges are often used in dairies to separate the cream from the milk. Which remains on the bottom?
19. Use Kepler's second law to convince yourself that the Earth must move faster in its orbit during the winter when it is closest to the Sun than it does during the summer when it is at its greatest distance from the Sun.
20. Estimate the gravitational force between you and a person 2 m away from you.
21. Does the escape velocity of a rocket depend on its mass? Explain.
22. Explain why it takes more fuel for a spacecraft to travel from the Earth to the Moon than for the return trip. Estimate the difference.
23. We often think of the brakes and the gas pedal on a car as the instruments which accelerate and decelerate the car. Could a steering wheel also fall into this category? Explain.
24. Explain the difference between centripetal and angular acceleration.
25. If someone told you that astronauts are weightless in orbit because they are beyond the pull of gravity, would you accept this statement? Explain.

Problems

Section 7.1   Angular Velocity

Section 7.2   Angular Acceleration

1. Convert the following angles in degrees to radians: $30°$, $45°$, $60°$, $90°$, $180°$, $270°$, and $360°$.
2. Convert the following angles in radians to degrees: $\pi/3$, 1.2π, 3π.
3. A wheel has a radius of 4.1 m. How far (path length) does a point on the circumference travel if the wheel is rotated through angles of $30°$, 30 rad, and 30 rev, respectively?
4. Find the angular velocity of the Earth about its axis in rad/s.
5. Find the angular velocity of the Earth about the Sun in units of rad/s and degrees/day.

6. A record has an angular speed of 33 rev/min. (a) What is its angular speed in radians per second? (b) Through what angle in radians does it rotate in 1.5 s?

7. A potter's wheel moves from rest to an angular speed of 0.2 rev/s in a time of 30 s. Find its angular acceleration in rad/s^2.

8. A particle moves in a circle 1.5 m in radius. Through what angle in radians does it rotate if it moves through an arc length of 2.5 m? What is this angle in degrees?

Section 7.3   Rotational Motion Under Constant Angular Acceleration

9. A dentist's drill starts from rest. After 3.2 s of constant angular acceleration it turns at a rate of 2.51×10^4 rev/min. (a) Find the drill's angular acceleration. (b) Determine the angle (in radians) through which the drill rotates during this period.

10. An electric motor rotating a workshop grinding wheel at a rate of 100 rev/min is switched off. Assuming constant negative acceleration of magnitude 2 rad/s^2 (a) how long will it take for the grinding wheel to stop? (b) Through how many radians has the wheel turned during the time found in (a)?

11. The turntable of a record player rotates at $33\frac{1}{3}$ rev/min and takes 20 s to reach this speed from rest. Calculate (a) its angular acceleration and (b) the number of revolutions it makes before reaching its final speed.

12. A tire placed on a machine to balance it in a service station starts from rest and turns through 4.7 revolutions in 1.2 s before reaching its final angular speed. Calculate its angular acceleration.

13. The tub of a washer goes into its spin-dry cycle, starting from rest and reaching an angular speed of 5.0 rev/s in 8.0 s. At this point the person doing the laundry opens the lid, and a safety switch turns off the washer. The tub slows to rest in 12.0 s. Through how many revolutions does the tub turn? Assume constant angular acceleration while starting and stopping.

14. A grinding wheel, initially at rest, is rotated with constant angular acceleration $\alpha = 5$ rad/s^2 for 8 s. The wheel is then brought to rest with uniform negative acceleration in 10 revolutions. Determine the negative acceleration required and the time needed to bring the wheel to rest.

15. A centrifuge in a medical laboratory is rotating at an angular speed of 3600 rev/min. When switched off, it rotates 50 times before coming to rest. Find the constant angular deceleration of the centrifuge.

16. A machine part is rotating at an angular velocity of 0.6 rad/s, and this speed is to be increased to 2.2 rad/s at an angular acceleration of 0.7 rad/s^2. Find the angle through which the part turns before reaching this final speed.

17. An airliner arrives at the terminal, and the engines are shut off. The rotor of one of the engines has an initial clockwise angular velocity of 2000 rad/s. The engine's rotation slows with an angular acceleration of magnitude 80.0 rad/s^2. (a) Determine the angular velocity after 10.0 s. (b) How long does it take the rotor to come to rest?

Section 7.4   Relations Between Angular and Linear Quantities

Section 7.5   Centripetal Acceleration

18. A coin with a diameter of 2.4 cm is dropped onto a horizontal surface. The coin starts out with an initial angular speed of 18.0 rad/s and rolls in a straight line without slipping. If the rotation slows with an angular acceleration of magnitude 1.9 rad/s^2, how far does the coin roll before coming to rest?

19. A racecar travels in a circular track of radius 200 m. If the car moves with a constant speed of 80 m/s, find (a) its angular velocity and (b) its tangential acceleration.

20. The racecar of Problem 19 increases its speed at a constant linear acceleration from 80 m/s to 95 m/s in 10 s. (a) Find the constant angular acceleration and (b) the angle the car moves through in this time.

21. A car is traveling with a speed of 17.0 m/s down a straight horizontal highway, which we assume extends from left to right. The wheels of the car have a radius of 48.0 cm. If the car speeds up with an acceleration of 2.0 m/s^2 for 5.0 s, find the number of revolutions of the wheels during this period.

22. The car in Problem 21 runs out of gas and the rotation of the wheels slows with an angular acceleration of magnitude 1.35 rad/s^2. If the car has a speed of 24.0 m/s when this happens, through how many revolutions do the wheels turn as the car comes to rest?

23. Young David who slew Goliath experimented with slings before tackling the giant. He found that with a sling of length 0.6 m, he could develop a speed of 8 rev/s in his weapon. If he increased the length to 0.9 m, he could revolve the sling only 6 times per second. (a) Which angular speed gives the larger linear speed? (b) What is the centripetal acceleration at 8 rev/s? (c) What is the centripetal acceleration at 6 rev/s?

24. A tire 0.5 m in radius rotates at a constant rate of 200 revolutions per minute. Find the speed and acceleration of a small stone lodged in the tread of the tire (on its outer edge).

25. Find the centripetal acceleration of (a) a point on the equator of the Earth and (b) a point at the north pole of the Earth.

26. It has been suggested that rotating cylinders about 10 mi in length and 5 mi in diameter be placed in space and used as colonies. What angular velocity must such a cylinder have so that the centripetal acceleration at the surface of the cylinder would equal Earth's gravity?

27. (a) What is the tangential acceleration of a bug on the rim of a 78-rpm record 10 in. in diameter if the record moves from rest to its final angular velocity in 3 s? (b) When the record is at its final speed, what is the tangential velocity of the bug? (c) One second after starting from rest, what is the bug's tangential acceleration, the bug's radial acceleration, and the bug's total acceleration?

28. A wheel 60 cm in diameter rotates with a constant angular acceleration of 4 rad/s². The wheel starts at rest at $t = 0$, and a chalk line drawn to a point P on the rim of the wheel makes an angle of 57.3° with the horizontal at this time. At t 2 s, find (a) the angular speed of the wheel, (b) the linear velocity and tangential acceleration of P, and (c) the position of P.

Section 7.6 Centripetal Force

29. A person is flying a 80.0-g model airplane in a horizontal circular path on the end of a string 10.0 m long. If the string is horizontal and exerts a force of 2.65 N on the hand of the person holding it, what is the speed of the plane in meters per second?

30. Tarzan ($m = 85$ kg) tries to cross a river by swinging from a vine. The vine is 10 m long, and his speed at the bottom of the swing (as he just clears the water) is 8 m/s. Tarzan doesn't know that the vine has a breaking strength of 1000 N. Does he make it safely across the river? Justify your answer.

31. A car of mass 2000 kg rounds a circular turn of radius 20 m. If the road is flat and the coefficient of friction is 0.7 between tires and road, how fast can the car go without skidding?

32. A rotating wheel at an amusement park shown in Figure 7.23 is 90 ft in diameter and makes one revolution in 6 s. Each of the 300-kg cars are held in place by two bolts. Find the force each bolt exerts on the car when the wheel is rotating in a horizontal plane.

33. A roller-coaster vehicle has a mass of 500 kg when fully loaded with passengers (Fig. 7.24). (a) If the vehicle has a speed of 20 m/s at point A, what is the force of the track on the vehicle at this point? (b) What is the maximum speed the vehicle can have at B in order that it remain on the track?

Figure 7.23   *(Courtesy Kings Island)*

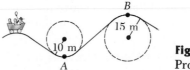

Figure 7.24
Problem 33

34. In a hydrogen atom, the electron in orbit about the proton feels an attraction of about 8.20×10^{-8} N. If the radius of the orbit is 5.3×10^{-11} m, what is the frequency in revolutions per second? See the front cover for additional data.

35. An airplane is flying in a horizontal circle at a speed of 100 m/s. The 80-kg pilot does not want his radial acceleration to exceed $7g$. (a) What is the minimum radius of the circular path? (b) At this radius, what is the *net* centripetal force exerted on the pilot by the seat belts, friction between him and the seat, and so forth?

36. A child stands at the rim of a rotating merry-go-round of radius 2 m. The child has a mass of 50 kg, and the carousel turns with an angular velocity of 3 rad/s. (a) What is the child's centripetal acceleration? (b) What is the minimum force required between her feet and the floor of the carousel to keep her in the circular path? (c) What minimum coefficient of static friction is required? Is the answer found reasonable? In other words, is she likely to be able to stay on the merry-go-round?

37. In a swinging mishap, Tarzan, whose mass is 80 kg, finds himself circling around a tree in a horizontal circle at the end of a vine 3 m long that makes an angle of 5° with the vertical. Find the centripetal force exerted on him by the vine and his centripetal acceleration.

38. An engineer wishes to design a curved exit ramp for a tollroad in such a way that a car will not have to rely on friction to round the curve without skidding. He does so by banking the road in such a way that the centripe-

tal force needed will be supplied by the component of the normal force toward the center of the circular path. (a) Show that for a given speed v and radius r, the curve must be banked at an angle given by $\tan \theta = v^2/rg$. (b) Find the angle at which the curve should be banked if a typical car rounds at a 50 m radius and at a speed of 13.4 m/s.

39. A pail of water is rotated in a vertical circle of radius 1 m (the approximate length of a person's arm). What must be the minimum speed of the pail at the top of the circle if no water is to spill out?

40. A stuntman swings from the end of a 4-m-long rope along the arc of a vertical circle. If his mass is 70 kg, find the tension in the rope required to make him follow his circular path, assuming he starts from rest when the rope is horizontal, (a) at the beginning of his motion, (b) at a height of 1.5 m above the bottom of the circular arc, and (c) at the bottom of his arc.

Section 7.8   Newton's Law of Gravitation

Section 7.11   Kepler's Laws

41. Use the data in Table 7.3 to find the gravitational force that the Sun exerts on the Earth.

42. A 30,000-kg spaceship is halfway between the Earth and the Moon. Use the data in Table 7.3 to find the net gravitational force exerted on the ship by the Earth and the Moon.

43. Two students sitting in adjacent seats in a lecture room have weights of 600 N and 700 N. Assume that Newton's law of gravitation can be applied to these students and find the gravitational force that one student exerts on the other when they are separated by 0.5 m.

44. A coordinate system is constructed on the surface of a pool table and three masses are placed on the coordinate system as follows: a 2-kg mass is at the origin, a 3-kg mass is at (0, 2), and a 4-kg mass is at (4, 0) where the coordinates are in meters. Find the resultant gravitational force exerted on the mass at the origin by the other two masses.

45. Use the data in Table 7.3 and the approach of Example 7.9 to find the acceleration due to gravity at the surface of Venus.

46. Given that the Moon's period about the Earth is 27.32 days and the Earth-Moon distance is 3.84×10^8 m, estimate the mass of the Earth. Assume the orbit is circular. Why do you suppose your estimate is high?

47. A satellite is in a circular orbit just above the surface of the Moon. (The radius of the Moon is 1738 km.) (a) What is the acceleration of the satellite? (b) What is the speed of the satellite? (c) What is the period of the satellite orbit?

48. Communication satellites are placed at an altitude of about 25,000 mi above the surface of the Earth. Find the acceleration due to gravity at this altitude.

49. Use the data of Table 7.3 to verify Kepler's third law for Venus, Mars, and Neptune.

50. A satellite of mass 600 kg is in a circular orbit about the Earth at a height above the Earth equal to the Earth's mean radius. Find (a) the satellite's orbital speed, (b) the period of its revolution, and (c) the gravitational force acting on it.

51. Use the data of Table 7.3 to find the point between the Earth and the Sun that an object could be placed such that the net gravitational force exerted on it by these two objects is zero.

52. Io, a small moon of the giant planet Jupiter, has an orbital period of 1.77 days and an orbital radius of 4.22×10^5 km. From this data, determine the mass of Jupiter.

Section 7.9   Gravitational Potential Energy Revisited

Section 7.10   Escape Velocity

53. A satellite of mass 200 kg is placed in Earth orbit at a height 200 km above the surface. (a) Assuming a circular orbit, how long does the satellite take to complete one orbit? (b) What is the satellite's speed? (c) What is the minimum energy necessary to place this satellite in orbit (assuming no air friction)?

54. Sketch a graph of the gravitational potential energy, PE, versus r for a particle above the Earth's surface. Your graph should extend from the Earth's surface to infinity.

55. A satellite of the Earth has a mass of 100 kg and is at an altitude of 2×10^6 m. (a) What is the potential energy of the satellite at this location? (b) What is the magnitude of the gravitational force on the satellite?

56. Use the data of Table 7.3 to find the escape velocity from (a) the Moon, (b) Mercury, and (c) Jupiter.

57. (a) Use the approach of Example 7.9 and the data in Table 7.3 to find the acceleration due to gravity at the surface of a neutron star of mass 1.5 solar masses having a radius of 10 km. (b) Find the weight of a 0.12-kg baseball on this star. (c) Assume the equation $PE = mgh$ applies and calculate the energy that a 70-kg person would expend climbing a 1-cm-tall mountain on this star.

Additional Problems

58. An athlete swings a ball of mass 5 kg horizontally on the end of a rope. The ball moves in a circle of radius

0.8 m at an angular speed of 0.5 rev/s. What is (a) the tangential velocity of the ball and (b) its centripetal acceleration? (c) If the maximum tension the rope can withstand before breaking is 100 N, what is the maximum tangential velocity the ball can have?

59. A satellite of Mars has a period of 459 min. The mass of Mars is 6.42×10^{23} kg. From this information determine the radius of the satellite's orbit.

60. The Solar Maximum Mission Satellite was placed in a circular orbit of about 150 mi above the Earth. Determine (a) the orbital speed of the satellite and (b) the time for one complete revolution.

61. The angular velocity of a rotating bicycle wheel is 3 rad/s at some instant of time. It then is given an angular acceleration of 1.5 rad/s². A chalk line drawn on the wheel is horizontal at $t = 0$. (a) What angle does this line make with its original direction at $t = 2$ s? (b) What is the angular velocity of the wheel at $t = 2$ s?

62. A floppy disk drive on a computer has a period of 200.2 ms. (a) What is its angular speed? (b) If the floppy disk has a diameter of 5.25 in., what is the linear speed of a point on its rim?

63. A high-speed sander has a disk 6 cm in radius that rotates at a constant rate of 1200 rev/min about its axis. Determine (a) the angular speed of the disk in rad/s, (b) the linear speed of a point 2 cm from its center, (c) the centripetal acceleration of a point on the rim, and (d) the total distance a point on the rim moves in 2 s.

64. Three masses are aligned along the x axis of a rectangular coordinate system such that a 2-kg mass is at the origin, a 3-kg mass is at (2, 0), and a 4-kg mass is at (4, 0) m. Find (a) the gravitational force exerted on the 4-kg mass by the other two masses and (b) the magnitude and direction of the gravitational force exerted on the 3-kg mass by the other two.

65. The downward motion of an elevator is controlled by a cable that unwinds from a cylinder of radius 0.2 m. What is the angular velocity of the cylinder when the downward speed of the elevator is 1.2 m/s?

66. A highway curve has a radius of 150 m and is designed for a traffic speed of 40 mi/h (17.9 m/s). If the curve is not banked, determine the minimum coefficient of friction between car and road needed to keep cars from skidding.

67. At what angle should the curve of Problem 66 be banked if friction is not to be a factor in causing the car to follow its circular path?

68. Suppose the rod in the system shown in Figure 7.25 is made vertical and rotates about this axis. If the mass rotates at a constant speed of 6 m/s in a horizontal plane, determine the tensions in the upper and lower strings.

Figure 7.25
Problems 68 and 69

69. A 4-kg mass is attached to a horizontal rod by two strings as in Figure 7.25. The strings are under tension when the rod rotates about its axis. If the speed of the mass is 4 m/s when observed at the following positions, find the tension in the string when the mass is (a) at its lowest point and (b) in the horizontal position.

70. A car moves across a bridge made in the shape of a circular arc of radius 30 m at a speed of 10 m/s. (a) Find the normal force acting on the car when it is at the top of the arc. (b) At what speed will the normal force go to zero? (Hint: The normal force becomes zero when the car loses contact with the road.)

71. Because of the Earth's rotation about its axis, a point on the equator experiences a centripetal acceleration of 0.034 m/s², while a point at the poles experiences no centripetal acceleration. (a) Show that at the equator the gravitational force on an object (the true weight) must exceed the object's apparent weight. (b) What is the apparent weight at the equator and at the poles of a person having a mass of 75 kg? (Assume the Earth is a uniform sphere and take $g = 9.800$ m/s².)

72. A skier starts at rest from the top of a large hill that is shaped like a hemisphere (Fig 7.26). Neglecting friction, show that the skier will leave the hill and become "air-borne" at a distance $h = R/3$ below the top of the hill. (Hint: At this point, the normal force goes to zero.)

Figure 7.26   Problem 72

73. A frictionless roller coaster is given an initial velocity v_0 at a height h, as in Figure 7.27. The radius of curvature of the track at point A is R. (a) Find the maximum value of v_0 necessary in order that the roller coaster not leave the track at A. (b) Using the value of v_0

calculated in (a), determine the value of h' necessary if the roller coaster is to just make it to point B.

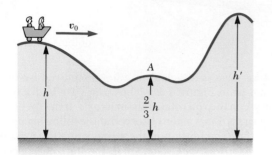

Figure 7.27   Problem 73

74. A car rounds a banked curve. The radius of curvature of the road is R, the banking angle is θ, and the coefficient of static friction is μ. (a) Determine the range of speeds the car can have without slipping up or down the road. (b) What is the range of speeds possible if $R = 100$ m, $\theta = 10°$, and $\mu = 0.1$ (slippery conditions).

75. In a popular amusement park ride, a rotating cylinder of radius 3 m is set into rotation at an angular velocity of 5 rad/s, as in Figure 7.28. The floor then drops away, leaving the riders suspended against the wall in a vertical position. What minimum coefficient of friction between a rider's clothing and the wall is needed to keep the rider from slipping? (*Hint:* Recall that the maximum force of static friction is equal to μn, where n is the normal force, which in this case is the centripetal force.)

Figure 7.28
Problem 75

"I'll be working on the largest and smallest objects in the Universe—superclusters and neutrinos. I'd like you to handle everything in between."

The Big Bang

Sidney C. Wolff

*National Optical
Astronomy Observatory*

Newton's universal law of gravitation — today it is taught in every introductory physics course and is so familiar that it is easy to forget how profoundly this simple equation changed the nature of science. Newton not only offered a precise mathematical description of the motions of material objects, he also showed that this description applied far beyond the boundaries of the Earth. The same force that causes apples to fall to the ground also keeps the Moon revolving about the Earth and holds the planets in their orbits around the Sun. The heavens were transformed; no longer mysterious and unknowable, the Moon, the planets, and potentially even the stars, all subject to the same physical laws that are valid on Earth, became legitimate objects of scientific inquiry.

But is Newton's law of gravitation truly universal? Does it apply to the stars? Newton assumed that it did, but measurements available during his lifetime were not precise enough to offer proof. The influence of gravity is detected by analyzing the motions of objects, and apart from their rising and setting as a consequence of the Earth's rotation, stars did not appear to the astronomers of Newton's time to have motions of their own. The fact that there are clusters or groupings of stars, like the Pleiades and Hyades, suggested that attractive forces do influence stars, but it was meticulous measurements by Sir William Herschel, best known for his discovery of the planet Uranus, that offered the first concrete evidence.

In January 1782, Herschel published a catalog of 269 pairs of stars that appear to be very close together in the sky. At that time, he did not know whether these double stars form physical systems or simply happen to appear in very nearly the same direction but at very different distances from the Earth. About 20 years later, Herschel repeated the observations of the double stars that he had found, and discovered that some pairs had changed position relative to each other in a way that could only be explained if they were revolving around each other. Several more decades of measurements by other astronomers proved that these stars do indeed follow orbits that can be described by the inverse square force of Newton's law of gravitation.

The conclusion that the motions of stars are controlled by gravity raises a problem that was recognized by Newton. Up until the beginning of the 20th century it was assumed that the Universe is static, that is, that it is neither expanding nor contracting. If the theory of gravitation applies to all of the objects in a static Universe, and if the Universe is finite in size, then it is difficult to see why the Universe has not already collapsed to a single mass as all of the objects are attracted to one another. Newton thought that an infinite Universe could perhaps be stable, since gravitational forces of all the stars spread uniformly throughout space might possibly cancel their mutual gravitational attractions.

The next major advance in gravitational theory is embodied in Einstein's general theory of relativity. In 1917, Einstein applied his new theory to the Universe as a whole and found that there were no solutions to his equations that gave a homogeneous static Universe, even if the Universe was assumed to be infinite. Einstein consequently introduced a new term, called the cosmological constant, which represents a repulsion that can balance gravitational attraction over large distances and permit a static Universe. There is, however, no evidence for such a repulsion in nature; the cosmological constant was introduced in order to make general relativity conform to the preconceived notion that the universe was static.

Subsequent theoretical work by the Russian mathematician Aleksandr Friedmann showed that general relativity is compatible with an infinite, homogeneous expanding Universe for all values of the cosmological constant, including zero. At about the same time, observations proved that the Universe is not static but is expanding. American physicist George Gamow reports that Einstein, upon learning that his relativity theory is compatible with an expanding Universe and that

Figure 1
The spiral galaxy M83, the Southern Pinwheel, is about 10 million light years away and has a diameter of 30,000 light years. A galaxy like M83 contains billions of individual stars. *(National Optical Astronomy Observatories)*

the expansion had been directly observed, said that the introduction of the cosmological constant was "the biggest blunder of my life."

The Universe is expanding. This fundamental observation underlies all of modern cosmological thought. In a classic paper published in 1931, Edwin Hubble and Milton Humason compared distances and velocities of remote galaxies, which are giant, gravitationally bound systems of billions of individual stars (Fig. 1). They established that these galaxies are moving away from us with velocities that are proportional to their distances from us. This result, which has been substantiated and extended to still more distant galaxies by subsequent work, demonstrates that the Universe is expanding.

This concept can be easily understood through use of an analogy. Suppose a cook is making raisin bread and suppose that after yeast has been mixed into the dough and the bread is set aside to rise, it doubles in size during the next hour, as in Figure 2. All of the raisins move further apart, and in fact the distance between each pair of raisins also doubles. Since each distance doubled during the hour, if one raisin is selected as the origin of a coordinate system, every other raisin must move away from the one selected at a speed proportional to its distance. The same is true, of course, no matter which raisin you select.

From this analogy, it should be clear that if the Universe is uniformly expanding, all observers, regardless of their location, must see all other galaxies moving away from them at speeds that are greater in proportion to their distances. As Hubble and Humason showed, that is precisely what we observe on Earth.

As the Universe expands, galaxies separate from each other. We can also, however, imagine what must have happened in the past. If we extrapolate backward in time, we find all of the galaxies coming together, until some time in the distant past when all matter was crowded to an extreme density. This is a condition that marks a unique beginning of the Universe, or at least of that Universe we can know about. At that beginning, the Universe suddenly began its expansion with a phenomenon called the "big bang."

Figure 2
Expanding raisin bread. (From George O. Abell, David Morrison, and Sydney C. Wolff, *Exploration of the Universe*, 5th ed., Philadelphia, Saunders College Publishing, 1987; p. 659.)

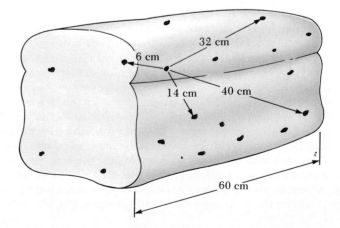

With the data now available, we can even estimate when our Universe began. The total amount of matter and energy of the Universe creates gravitation, whereby all objects pull on all other objects. This mutual attraction must slow the expansion, which means that in the past the expansion must have proceeded at a greater rate than it is today. How much greater depends on the importance of gravitation in decelerating the expansion. At the extreme, if the total mass-energy density is low enough that gravitation is ineffective (an essentially "empty" Universe), the deceleration would be zero, and in that case the Universe would always have been expanding at the present rate.

Clearly that extreme of an empty Universe corresponds to the greatest age of the Universe (since the big bang), because if the expansion had been faster in the past, galaxies would have reached their present separation in a smaller time. Consequently, we can obtain an estimate of the upper limit to the age of the Universe by asking how long it would take for distant galaxies, always moving away from us at their present rates, to have reached those distances. Call that maximum possible age T_0. According to the work of Hubble and Humason, the velocity of a galaxy is proportional to its distance from us, a result that is expressed by the equation:

$$v = Hr$$

where H is the constant of proportionality between velocity and distance (the so-called Hubble constant). But since velocity is just distance divided by time, we can express v as $v = r/T_0$. Hence,

$$\frac{r}{T_0} = Hr$$

or

$$T_0 = \frac{1}{H}$$

We see, then, that the maximum age of the Universe is just the reciprocal of the Hubble constant. The value of the Hubble constant is uncertain, primarily because of the uncertainties in deriving the distances to remote galaxies. The best estimate is that the big bang occurred 10 to 15 billion years ago. A plot of the velocity versus time relation adapted from a 1931 publication by Hubble and Humason is shown in Figure 3.

If our Universe had a beginning, will it also have an end? Will the expansion go on forever? Or will gravity be strong enough to slow the expansion to a halt and force the galaxies to fall together again in a "big crunch?" A major focus of

Figure 3
Hubble and Humason's velocity-distance relation, adapted from their 1931 paper in the *Astrophysical Journal*. Distance is measured in parsecs; one parsec is the distance that light travels in 3.26 years. (From Abell, Morrison, and Wolff, *Exploration of the Universe*, 5th ed., p. 658.)

astronomical research at the present time is the effort to answer these questions. One approach relies on the application of the law of gravitation plus the cosmological principle. The cosmological principle asserts that on a large scale, that is, apart from local irregularities, the Universe at any given time is the same everywhere. If this principle is valid, then we can reach conclusions about the Universe as a whole from measurements of a small portion of it that is close enough to study in detail.

Let us consider a spherical region of the Universe that contains a large number of galaxies. The rest of the Universe can be regarded as a hollow spherical shell, since according to the cosmological principle, the matter in this shell is distributed in the same way in all directions. Newton showed that the gravitational forces exerted by a spherical shell on an object in its hollow interior all cancel exactly, so that the net force is zero. We can then consider the sphere as if it were isolated and uninfluenced by any gravitational forces from the rest of the Universe. The cosmological principle also assures us that any conclusion that we reach about this sphere must apply to the rest of the Universe. If the sphere chosen is a small one so that the velocities of the galaxies within it are small relative to the speed of light, then Newton's law of gravitation can be applied, although it has been shown that general relativity yields the same answer.

Consider then a galaxy located at the surface of the spherical region. It is subject to the gravitational attraction of all of the galaxies within the sphere and is decelerated. The situation is exactly analogous to an object launched upward from the surface of the Earth. If the velocity of the object exceeds the escape velocity, the object will never return to the Earth's surface. If the velocity is less than the escape velocity, the object will fall back to the ground. Similarly, if the galaxy has a velocity that is high enough to overcome the attractive force of all the other galaxies within the sphere, it will continue to move away from them, and the expansion of the Universe will continue. If the velocity is too low, however, gravity will decelerate the galaxy until its velocity becomes zero, and the galaxy will then reverse direction and begin to move closer to the other galaxies within the sphere.

For an object launched from the Earth, the escape velocity is determined by the mass of the Earth. The escape velocity of a galaxy at the surface of the spherical region of galaxies is determined by the total mass within the sphere or by the average density of the Universe (since by the cosmological principle, the density of the Universe is everywhere the same). If the density of the Universe is larger than some critical value, gravity will be strong enough to halt the expansion. If the density is smaller than this critical value, the expansion will go on forever.

The density of the Universe can be estimated by counting up the mass associated with the galaxies within our imaginary sphere. Most galaxies occur in groups or clusters and can be "weighed" by measuring the gravitational influence that they exert on other nearby galaxies (Fig. 4). For example, if two galaxies are in orbit around each other, and if their separations and velocities are known, their combined mass can be estimated from Kepler's third law. There are mathematical procedures that are not much more complex for estimating the total mass of a cluster containing not just two but many galaxies.

The best available observational estimates of the average density of the Universe suggest that the amount of matter that it contains is no more than 10 to 20 percent of the value required to halt the expansion. There are some theoretical arguments for thinking that the density may be even higher and may just exactly equal the critical density required to slow the expansion to a halt, but the deceleration is so gradual in this case that an infinite amount of time is required to reduce the expansion velocity to zero. Therefore, both observation and theory suggest that galaxies will never reverse direction and move closer together.

208

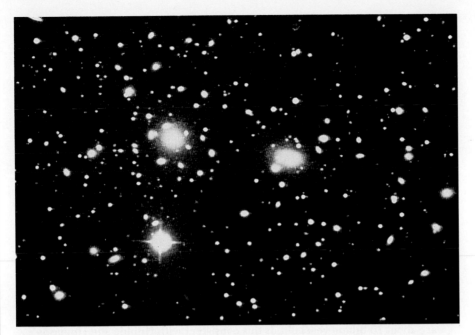

Figure 4
The Coma Berenices cluster contains more than one thousand galaxies. The mass of this cluster can be estimated from measurements of the velocities of the individual member galaxies. The masses of clusters of galaxies are used to estimate the average density of matter in the Universe. *(National Optical Astronomy Observatories)*

If this picture is right, there can be no rejuvenation of the Universe. At the time of the big bang, the Universe was very dense and hot. As it expanded, it cooled, and hydrogen and helium nuclei and then atoms formed. Somehow — theory is not yet very good at explaining this point — hydrogen and helium atoms came together to form concentrations of matter that would become galaxies. Within these protogalaxies stars formed, and in many galaxies star formation has continued to the present. Each star, some in a million years and others only after billions of years, exhausts its sources of nuclear energy. Some stars die gently, others explosively, hurling a portion of their mass into space to be incorporated into a new generation of stars. As they die, most stars also leave behind a portion of their mass in the form of a dense core that becomes a black dwarf, a neutron star, or a black hole. This material is forever lost to the Universe; it radiates no energy and cannot, under any conditions that we now believe likely to occur, be broken apart into gas and dust particles that would provide the raw material for new stars. Ultimately, after the gas and dust are used up, star formation must cease, a final generation of stars will die, and the galaxies will fade into blackness.

Astronomy and physics as we understand them today offer no alternative to this outcome. Today's knowledge, however, may not be — almost certainly is not — complete. At the end of the 19th century, physics was thought to be totally understood. Theories of gravity, heat, electricity, and magnetism all seemed adequate to explain experimental data. Physicists expected to make no major new breakthroughs but simply to refine and improve the existing mathematical descriptions of physical phenomena. The revolution came in 1900 when Max Planck proposed the quantum principle.

With time, the picture presented here of the origin, evolution, and ultimate fate of the material Universe may undergo a comparably radical transformation. Indeed, astronomers are attracted to this area of research in substantial numbers precisely because the subject is not fully understood and seems likely to yield its secrets to a systematic scientific attack. To learn something new, something never known before, is the ultimate challenge and satisfaction of scientific research.

Suggested Reading

T. Ferris, *The Red Limit*, 2nd ed., New York, Morrow, Quill, 1983.

E. R. Harrison, *Cosmology*, Cambridge, England, Cambridge University Press, 1981.

R. Wagoner and D. Goldsmith, *Cosmic Horizons: Understanding the Universe*, New York, W. H. Freeman, 1983.

Essay Questions

1. When Hubble and Humason first reported their results for the relation between distance and velocity for galaxies, they thought that the constant of proportionality H was equal to about 200 km/s per million lightyears. How did their estimate for the age of the Universe differ from the modern value based on $H = 20$ km/s per million lightyears? The oldest stars in the Universe are about 15 billion years old. Is this consistent with Hubble and Humason's value for the Hubble constant H?

2. In the text, it was explained that it was possible to estimate the masses of two galaxies orbiting about each other by measuring their separations and velocities and applying Kepler's third law. Could this be done by measuring the orbital period directly? Why or why not? (Remember that galaxies contain billions of stars and have diameters of about 100,000 lightyears.)

3. Explain why the observation that all galaxies are moving away from us does not mean that we are at the center of the Universe.

Essay Problems

1. Plot the "velocity-distance relation" for the raisins-in-the-bread analogy from the data given in Figure 3. Repeat but use some other raisin than A for a reference. Is your new plot the same as the last one?

2. Current estimates for the Hubble constant H range from 15 to 30 km/s per million lightyears, where one lightyear is the distance that light travels in one year. What is the corresponding range in the estimated age of the universe?

3. Derive an expression for escape velocity in terms of mean density. Now apply this result to two galaxies, separated by a distance R and moving away from each other at just the escape velocity and derive an expression for the critical density at which the escape and expansion velocities are equal. Assume that $H = 20$ km/s per million lightyears. How does your value for the critical density compare with the observed density, which is currently estimated to be less than about 10^{-30} g/cm^3. What does your calculation indicate about the ultimate fate of the Universe?

CHAPTER 8

Objects at Rest and Rotational Dynamics

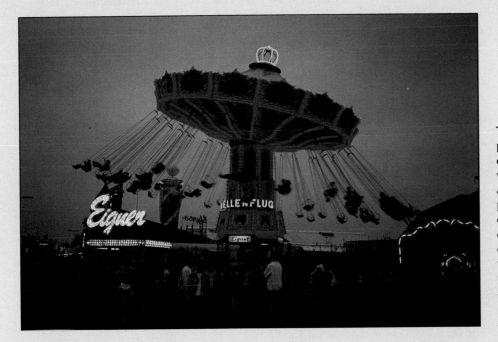

The people in this colorful photograph are set into circular motion about the vertical axis in an exciting amusement park ride. The passengers are seated in chairs connected to long chains. The angle that the chains make with the vertical axis increases as the angular velocity of the apparatus increases. *(Robert Harding Picture Library)*

We begin this chapter by studying objects that are either at rest or moving with constant velocity. Such objects are said to be in equilibrium, and an understanding of the conditions that prevail when an object is in equilibrium is important in a variety of fields. For example, students of architecture or industrial technology will benefit from an understanding of the forces that act on buildings or large machines, and biology students should understand the forces at work in muscles and bones in the human body. Newton's first law as discussed in Chapter 4 will form the basis for much of our work in this chapter. However, in order to fully understand objects in equilibrium, we must also consider the concept of torque. This concept will also play a key role in our description of rotational motion.

In this chapter, we will also complete our study of rotational motion. We shall build on the definitions of angular velocity and angular acceleration encountered in Chapter 7 by examining the relationship between these concepts and the forces that produce rotational motion. Specifically, we shall find the rotational analog of Newton's second law and define a new term that needs to be added into our equation for conservation of mechanical energy: rotational kinetic energy. One of the central points of this chapter will be to develop the concept of angular momentum, a quantity that plays a key role in rotational motion. Finally, just as we found that linear momentum is conserved, we shall also find that the angular momentum of any isolated system is always conserved.

Figure 8.1
A bird's-eye view of a door hinged at O, with a force applied perpendicular to the door.

8.1 Torque

Consider Figure 8.1, which represents a view looking down on a door hinged at point O. From this viewpoint, the door is free to rotate about a line perpendicular to the page and passing through O. When the force **F** is applied at the outer edge as shown, the door can be easily caused to rotate in a counterclockwise rotation, that is, the rotational effect of the force is quite large. On the other hand, the same force applied at a point nearer the hinges produces a smaller rotational effect on the door.

The ability of a force to rotate a body about some axis is measured by a quantity called the **torque**, τ. The torque due to a force **F** has a magnitude given by the equation

$$\tau = Fd \tag{8.1}$$

In this equation, τ (the Greek letter tau) is the torque and the distance d is called the **lever arm** (or moment arm) of the force **F**.

The lever arm is the perpendicular distance from the axis of rotation to a line drawn along the direction of the force.

Figure 8.2
A force **F** acting at an angle ϕ to the horizontal produces a torque of magnitude Fd about the pivot O.

As an example, consider the wrench pivoted about the axis O in Figure 8.2. In this case, the applied force **F** acts at an angle ϕ to the horizontal. If you examine the definition of lever arm given above, you will see that in this case the lever arm is the distance d shown in the figure and not L, the length of the wrench. That is, d is the perpendicular distance from the axis of rotation to the line along which the applied force acts. The distance d is related to L in this case by the expression $d = L \sin \phi$.

If there are two or more forces acting on an object, as in Figure 8.3, then each has a tendency to produce a rotation about the pivot O. For example, $\mathbf{F_2}$ has a tendency to rotate the object clockwise and $\mathbf{F_1}$ has a tendency to rotate the object counterclockwise. We shall use the convention that the sign of the torque resulting from a force is positive if its turning tendency is counterclockwise and negative if its turning tendency is clockwise. In Figure 8.3, then, the torque resulting from $\mathbf{F_1}$, which has a moment arm d_1, is positive and equal to $F_1 d_1$; the torque associated with $\mathbf{F_2}$ is negative and equal to $-F_2 d_2$. The *net torque* acting on the object about O is found by summing the torques:

$$\sum \tau = \tau_1 + \tau_2 = F_1 d_1 - F_2 d_2$$

Notice that the units of torque are units of force times length, such as newton-meter (N·m) or pound-foot (lb·ft).

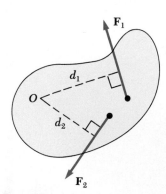

Figure 8.3
The force $\mathbf{F_1}$ tends to rotate the body counterclockwise about O, while $\mathbf{F_2}$ tends to rotate the body clockwise.

Example 8.1   The Spinning Safe

Figure 8.4 shows a top view of a safe being pushed by two equal but opposite forces acting as shown. Find the net torque exerted on the safe if its width is 1 m. Assume an axis of rotation through the center of the safe.

Figure 8.4 (Example 8.1) Top view of a safe being pushed with equal but opposite forces.

Solution

The torque produced by $\mathbf{F}_1$ is

$$\tau_1 = F_1 d_1 = -(500 \text{ N})(0.50 \text{ m}) = -250 \text{ N}\cdot\text{m}$$

and the torque produced by $\mathbf{F}_2$ is

$$\tau_2 = F_2 d_2 = -(500 \text{ N})(0.50 \text{ m}) = -250 \text{ N}\cdot\text{m}$$

Each force produces clockwise rotation, and hence the torques are both negative. Thus, the net torque is $-500 \text{ N}\cdot\text{m}$.

Example 8.2 The Swinging Door

Find the torque produced by the 300-N force applied at an angle of 60° to the door of Figure 8.5a.

Solution

The problem can be solved by proceeding along the lines indicated in Figure 8.5b. The lever arm d can be found from the dashed triangle shown in the figure. Since $\sin 60° = d/(2 \text{ m})$, we see that

$$d = (2 \text{ m})(\sin 60°) = 1.73 \text{ m}$$

Hence, the torque about an axis through the pivot at the left is

$$\tau = (300 \text{ N})(1.73 \text{ m}) = \boxed{520 \text{ N}\cdot\text{m}}$$

Can you explain why the torque is positive in this case?

It is often simpler to solve a problem such as this by first resolving the applied force into components, as shown in Figure 8.5c. The horizontal component of the 300-N force is 150 N, and the vertical component is 260 N.

In this case, the 150-N force produces zero torque about the axis of rotation because the line along which the force acts passes through the axis of rotation and hence the lever arm is zero. The 260-N force has a lever arm of 2 m and hence produces a torque of

$$\tau = (260 \text{ N})(2 \text{ m}) = 520 \text{ N}\cdot\text{m}$$

This agrees with our previous calculation, as it must.

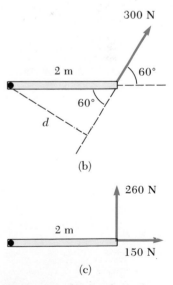

Figure 8.5 (Example 8.2) (a) Top view of a door being pulled by a 300-N force. (b) The moment arm of the 300-N force is d. (c) The components of the 300-N force.

8.2   Torque and the Second Condition for Equilibrium

In Chapter 4 we examined some situations, under the heading of the first condition of equilibrium, in which an object has no net external force acting on it, that is, $\Sigma F_x = 0$ and $\Sigma F_y = 0$. We found there that such an object will either remain at rest if it is at rest, or it will move with a constant velocity if it is in motion. However, the first condition is not sufficient to ensure that an object is in complete equilibrium. This can be understood by considering the situation illustrated in Figure 8.6. Here we see a bird's-eye view of two elephants pushing on a large log with forces that are equal in magnitude but opposite in direction. The first condition for equilibrium is satisfied because the two equal-magnitude and oppositely directed external forces balance each other, and yet the log can still move. It will rotate clockwise because the forces do not act through a common point. This example should point out that, if one is to understand fully the effect of a force or group of forces on an object, not only must one know the magnitude and direction of the force(s), but the point of application of the force(s) must be considered as well. That is, we must consider the net torque which acts on an object.

The **second condition for equilibrium** says that if an object is in rotational equilibrium, the net torque acting on it about any axis must be zero. That is,

$$\sum \tau = 0 \tag{8.2}$$

Thus we see that a body in static equilibrium must satisfy two conditions:

1. The resultant external force must be zero.

$$\sum \mathbf{F} = 0$$

2. The resultant external torque must be zero.

$$\sum \tau = 0$$

The first condition is a statement of translational equilibrium, and the second is a statement of rotational equilibrium. We shall provide a more detailed discussion of torque and its relation to rotational motion later in the chapter.

Position of the Axis of Rotation

In the examples we have been using, an axis of rotation for calculating torques was selected without explanation. Often the nature of a problem will suggest a convenient location for this axis, but just as often no single location will stand out as being preferable. You might ask, "What axis should I choose in calculating the net torque?" The answer is that

it does not matter where you pick the axis of rotation for calculating the net torque if the object is in equilibrium; since the object is not rotating, the location of the axis is completely arbitrary.

Figure 8.6
Top view of two elephants pushing a log with forces of equal magnitude but opposite in direction.

The two conditions for static equilibrium

To see that the location of the axis of rotation is completely arbitrary for an object in equilibrium, consider the object shown in Figure 8.7. Since the object is in equilibrium, the magnitudes of $\mathbf{F}_1$ and $\mathbf{F}_2$ must be equal. Furthermore, both forces have the same lever arm d relative to the arbitrary axis through O. Therefore, the sum of the counterclockwise torque $F_1 d$ and the clockwise torque $-F_2 d$ is equal to zero.

8.3 Examples of Objects in Equilibrium

In Chapter 4 we discussed some techniques that are helpful in solving problems dealing with objects in equilibrium. Recall that when the objects were treated as geometric points, it was sufficient to simply apply the condition that the net force on the object must be zero. In this chapter we have shown that when dealing with objects of finite dimensions, a second condition for equilibrium must be satisfied, namely that the net torque on the object must also be zero. The following general procedure is recommended when solving problems involving objects in equilibrium.

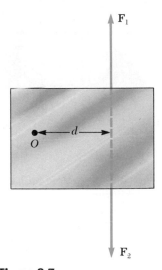

Figure 8.7
In order for the object to be in equilibrium, the magnitudes of $\mathbf{F}_1$ and $\mathbf{F}_2$ must be equal. The axis for calculating the torque is completely arbitrary.

Problem-Solving Strategy:
Objects in Equilibrium

1. Draw a simple, neat diagram of the system.
2. Isolate the object of interest being analyzed. Draw a free-body diagram for this object showing all external forces acting on the object. For systems containing more than one object, draw *separate* diagrams for each object. Do not include forces that the object exerts on its surroundings.
3. Establish convenient coordinate axes for each body and find the components of the forces along these axes. Now apply the first condition of equilibrium for each object under consideration, namely that the net force on the object in the x and y directions must be zero.
4. Choose a convenient origin for calculating the net torque on the object. Now apply the second condition of equilibrium that says that the net torque on the object about any origin must be zero. Remember that the choice of the origin for the torque equation is arbitrary; therefore, choose an origin that will simplify your calculation as much as possible. Note that a force that acts along a line passing through the point chosen as the axis of rotation gives zero contribution to the torque.
5. The first and second conditions for equilibrium will give a set of simultaneous equations with several unknowns. To complete your solution, all that is left is to solve for the unknowns in terms of the known quantities.

The Golden Gate Bridge in San Francisco. Some members of this structure are under tension, while others are under compression. *(© Joachim Messerschmidt/FPG International)*

Figure 8.8   (Example 8.3) Two children balanced on a seesaw.

Example 8.3   The Seesaw

A uniform board of weight 40 N supports two children weighing 500 N and 350 N, as shown in Figure 8.8. If the support (often called the *fulcrum*) is under the center of gravity of the board and if the 500-N child is 1.5 m from the center, (a) determine the upward force **n** exerted on the board by the support.

Solution
First note that, in addition to **n**, the external forces acting on the board are the weights of the children and the weight of the board, all of which act downward. We can assume that the center of gravity of the board is at its geometric center because we were told that the board is uniform. Since the system is in equilibrium, the upward force **n** must balance all the downward forces. From $\Sigma F_y = 0$, we have

$$n - 500 \text{ N} - 350 \text{ N} - 40 \text{ N} = 0 \qquad \text{or} \qquad n = \boxed{890 \text{ N}}$$

It should be pointed out here that the equation $\Sigma F_x = 0$ also applies to this situation, but it is unnecessary to consider this equation because we have no forces acting horizontally on the board.

(b) Determine where the 350-N child should sit to balance the system.

Solution
To find this position, we must invoke the second condition for equilibrium. Taking the center of gravity of the board as the axis for our torque equation, we see from $\Sigma \tau = 0$ that

$$(500 \text{ N})(1.5 \text{ m}) - (350 \text{ N})(x) = 0$$

$$x = \boxed{2.14 \text{ m}}$$

(a)

(b)

Figure 8.9   (Example 8.4) (a) A weight held with the forearm in the horizontal position. (b) The mechanical model for the system.

Example 8.4   A Weighted Forearm

A 50-N weight is held in the hand with the forearm in the horizontal position, as in Figure 8.9a. The biceps muscle is attached 0.03 m from the joint, and the weight is 0.35 m from the joint. Find the upward force that the biceps exerts on the forearm (the ulna) and the downward force on the upper arm (the humerus) acting at the joint. Neglect the weight of the forearm.

Solution
The forces acting on the forearm are equivalent to those acting on a bar of length 0.35 m, as shown in Figure 8.9b, where **F** is the upward force of the biceps and **R** is the downward force at the joint. From the first condition for equilibrium, we have

$$(1) \qquad \sum F_y = F - R - 50 \text{ N} = 0$$

From the second condition for equilibrium, we know that the sum of the torques about any point must be zero. With the joint O as the axis, we have

$$F(0.03 \text{ m}) - (50 \text{ N})(0.35 \text{ m}) = 0$$

$$F = \boxed{583 \text{ N}}$$

This value for F can be substituted into (1) to give $R = 533$ N. These values correspond to $F = 131$ lb and $R = 119$ lb. Hence, the forces at joints and in muscles can be extremely large.

Example 8.5   Walking a Horizontal Beam

A uniform horizontal beam of length 5 m and weight 300 N is attached to a wall by a pin connection that allows the beam to rotate. Its far end is supported by a cable that makes an angle of 53° with the horizontal (Fig. 8.10a). If a 600-N person stands 1.5 m from the wall, find the tension in the cable and the force exerted on the beam by the wall.

Solution

First, we must identify all the external forces acting on the beam and sketch them on a free-body diagram. This is shown in Figure 8.10b. The forces on the beam consist of the weight of the beam, 300 N, acting downward; the downward force on the beam by the man, which is equal in magnitude to his weight, 600 N; the tension force **T** in the cable; and the force of the wall on the beam, **R**. We now resolve the forces **T** and **R** into their horizontal and vertical components as shown in Figure 8.10c. Note that the x component of the tension force ($T \cos 53°$) is to the left, while the y component ($T \sin 53°$) is upward. The horizontal and vertical components of **R** are denoted by R_x and R_y, respectively. Applying the first condition of equilibrium to the beam gives

$$(1) \qquad \sum F_x = R_x - T \cos 53° = 0$$

$$(2) \qquad \sum F_y = R_y + T \sin 53° - 600 \text{ N} - 300 \text{ N} = 0$$

Our unknowns are R_x, R_y, and T. Because there are three unknowns and only two equations, we are unable to find a solution from just the first condition of equilibrium.

Now let us use the second condition of equilibrium. A convenient axis to choose for our torque equation is the one that passes through the pivot at the wall. The feature that makes this point so convenient is that the forces R_x, R_y, and $T \cos 53°$ all have a lever arm of zero, and hence have zero torque about this pivot. Recalling our sign convention for the torque about an axis and noting that the lever arms of the 600-N, 300-N, and $T \sin 53°$ forces are 1.5 m, 2.5 m, and 5 m, respectively, we get

$$(3) \qquad \sum \tau_O = (T \sin 53°)(5 \text{ m}) - (300 \text{ N})(2.5 \text{ m}) - (600 \text{ N})(1.5 \text{ m}) = 0$$

$$T = \boxed{413 \text{ N}}$$

Thus, the torque equation using this axis gives us one of the unknowns immediately! This value for T is then substituted into (1) and (2) to give

$$R_x = \boxed{249 \text{ N}}$$

$$R_y = \boxed{570 \text{ N}}$$

If we had selected some other axis for the torque equation, the solution would have been the same. For example, if the axis were to pass through the center of gravity of the beam, the torque equation would involve both T and R_y. However, this equation, together with (1) and (2), could still be solved for the unknowns. Try it!

Exercise   Repeat this problem, but with the direction of R_x opposite to that shown in Figure 8.10c. What answers do you get for T, R_x, and R_y?

Answer $T = 413$ N, $R_x = -249$ N, and $R_y = 570$ N. The negative sign for R_x means its direction was chosen incorrectly. The direction of R_x must be to the right as shown in Figure 8.10c.

Figure 8.10   (Example 8.5) (a) A uniform beam attached to a wall and supported by a cable. (b) Free-body diagram for the beam. (c) Component form of the free-body diagram.

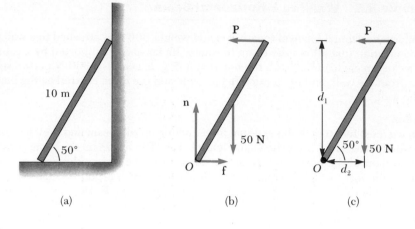

Figure 8.11 (Example 8.6)
(a) A ladder leaning against a
frictionless wall. (b) Free-
body diagram for the ladder.
(c) Lever arms for the forces
w and **P**.

(a)   (b)   (c)

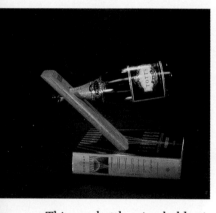

This one-bottle wine holder is
an interesting example of a
balanced mechanical system,
which seems to defy gravity.
The system (wine holder +
bottle) is balanced when its
center of gravity is directly
over the lowest support point
so that the net torque acting
on the system is zero. (*Courtesy
of Henry Leap and Jim Lehman*)

Example 8.6   Don't Climb the Ladder

A uniform ladder of length 10 m and weight 50 N rests against a smooth vertical wall
as in Figure 8.11a. If the ladder is just on the verge of slipping when the angle the
ladder makes with the ground is 50°, find the coefficient of static friction between
the ladder and ground.

Solution
The free-body diagram for the ladder showing all external forces acting on it is illustrated in Figure 8.11b. At the base of the ladder, the Earth exerts an upward normal
force, **n**, and a force of static friction, **f**, acting to the right. The wall exerts a force
P to the left. Note that the force **P** is horizontal because the wall is smooth. (If the
wall were rough, there would be an upward frictional force exerted on the ladder.)
From the first condition of equilibrium applied to the ladder, we have

$$(1) \qquad \sum F_x = f - P = 0$$

$$(2) \qquad \sum F_y = n - 50 = 0$$

From (2) we see that $n = 50$ N. Furthermore, when the ladder is on the verge of
slipping, the force of static friction must be a maximum and given by the relation
$f_{s,max} = \mu_s n = \mu_s (50 \text{ N})$. Thus, (1) reduces to

$$(3) \qquad \mu_s(50 \text{ N}) = P$$

Let us now apply the second condition of equilibrium and take the torques about the
axis O located at the bottom of the ladder as in Figure 8.11c. The force **P** and the
weight of the ladder are the only forces that contribute to the torque about this axis,
and their lever arms are shown in Figure 8.11c. Since the length of the ladder is
10 m, you should note that the lever arm for **P** is $d_1 = 10$ m sin 50°. Likewise, the
lever arm for the 50-N weight force is $d_2 = 5$ m cos 50°, where the weight vector
acts through the center because the ladder is uniform. Thus, we find

$$\sum \tau_O = P(10 \text{ m sin } 50°) - (50 \text{ N})(5 \text{ m cos } 50°) = 0$$

$$P = 21 \text{ N}$$

Now that P is known, we can substitute its value into (3) to find μ_s, giving

$$\mu_s = \frac{21 \text{ N}}{50 \text{ N}} = \boxed{0.42}$$

8.4 Relationship Between Torque and Angular Acceleration

Earlier in this chapter we considered the situation in which both the net force and the net torque acting on an object are equal to zero. Such objects are said to be in static equilibrium. We shall now describe the behavior of an object when the net torque acting on it is not zero. As we shall see, when there is a net torque acting on a rigid object, it will undergo an angular acceleration. Furthermore, we will find that the net torque is directly proportional to the angular acceleration. The end result of our investigation will be an expression that is analogous to $\mathbf{F} = m\mathbf{a}$ in linear motion.

Let us begin by considering the system shown in Figure 8.12, which consists of a mass m connected to a very light rod of length r. The rod is pivoted at the point O and its movement is confined to rotation on a *horizontal* frictionless table. Now let us assume there is a force $\mathbf{F_t}$ acting on m perpendicular to the rod and hence tangent to the circular orbit. Since there is no force to oppose this tangential force, the mass undergoes a tangential acceleration according to Newton's second law:

$$F_t = ma_t$$

Multiplying the left and right sides of this equation by r gives

$$F_t r = mra_t$$

In Chapter 7 we found that the tangential acceleration and angular acceleration for a particle rotating in a circular path are related by the expression

$$a_t = r\alpha$$

so now we find that

$$F_t r = mr^2\alpha \tag{8.3}$$

The left side of Equation 8.3, which should be familiar to you, is the torque acting on the mass about its axis of rotation. That is, the torque is equal in magnitude to the force on m multiplied by the perpendicular distance from the pivot to the line of action of the force, or $\tau = F_t r$. Hence, we can write Equation 8.3 as

$$\tau = mr^2\alpha \tag{8.4}$$

Equation 8.4 shows that the torque on the system is proportional to angular acceleration, where the constant of proportionality mr^2 is called the **moment of inertia** of the mass m. (Because the rod is very light, its moment of inertia can be neglected.)

Torque on a Rotating Object

Now consider a solid disk rotating about its axis, as in Figure 8.13a. The disk consists of many particles located at various distances from the axis of rotation, as in Figure 8.13b. The torque on one of these particles is given by Equation

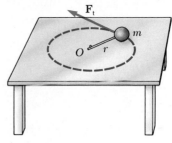

Figure 8.12
A mass m attached to a light rod of length r moves in a circular path on a horizontal frictionless surface while a tangential force $\mathbf{F_t}$ acts on it.

(a)

(b)

Figure 8.13
(a) A solid disk rotating about its axis. (b) The disk consists of many particles, and all the particles have the same angular acceleration.

8.4. The *total* torque on the disk is given by the sum of the individual torques on all the particles:

$$\sum \tau = \left(\sum mr^2 \right) \alpha \tag{8.5}$$

Note that, because the disk is rigid, all particles have the *same* angular acceleration so α is not involved in the sum. If the masses and distances of the particles are labeled with subscripts as in Figure 8.13b, then

$$\sum mr^2 = m_1 r_1^2 + m_2 r_2^2 + m_3 r_3^2 + \ldots$$

This quantity is called the **moment of inertia** of the whole body and is given the symbol I:

Moment of inertia

$$I \equiv \sum mr^2 \tag{8.6}$$

The moment of inertia has units of kg · m² in SI units. Using this result in Equation 8.5, we see that the total torque on a rigid body rotating about a fixed axis is given by

Net torque

$$\sum \tau = I\alpha \tag{8.7}$$

The angular acceleration of an object is proportional to the net torque acting on it. The proportionality constant, I, between the net torque and angular acceleration is the moment of inertia.

It is important to note that the equation $\Sigma \tau = I\alpha$ (Eq. 8.7) is the rotational counterpart to Newton's second law, $F = ma$. Thus, the correspondence between rotational motion and linear motion continues. Recall that in Chapter 7 we found that the linear variables x, v, and a are replaced in rotational motion by the variables θ, ω, and α. Likewise, we now see that *the force and mass in linear motion correspond to torque and moment of inertia in rotational motion.* In this chapter, we shall develop other equations for the rotational kinetic energy and the angular momentum of a body rotating about a fixed axis. Based on the analogies already presented, you should be able to predict the form of these equations.

More on the Moment of Inertia

The moment of inertia of a body as defined by $I = \Sigma mr^2$ (Eq. 8.6) will be used throughout the remainder of this chapter. Thus it will be useful to examine this quantity in more detail before discussing other aspects of rotational motion.

As we saw earlier, a small object (or particle) orbiting about some axis has a moment of inertia equal to mr^2. Now consider a somewhat more complicated system, the baton being twirled by a majorette in Figure 8.14. Let us assume that the baton can be modeled as a very light rod of length 2ℓ and with a heavy mass at each end. (The rod of a real baton has significant mass relative to its ends. Our model neglects the mass of the rod.) Since we are neglecting the mass of the rod, the moment of inertia of the baton about an axis through its center and perpendicular to its length is given by Equation 8.6.

Figure 8.14
A baton of length 2ℓ and mass $2m$ (the mass of the connecting rod is neglected). The moment of inertia about the axis through the baton's center and perpendicular to its length is $2m\ell^2$.

$$I = \sum mr^2$$

Since in this system there are two equal masses equidistant from the axis of rotation, we see that $r = \ell$ for each mass and the sum is

$$I = \sum mr^2 = m\ell^2 + m\ell^2 = 2m\ell^2$$

We pointed out earlier that I is the rotational counterpart of m. However, there are some important distinctions between the two. For example, mass is an intrinsic property of an object that does not change. On the other hand, *the moment of inertia of a system depends upon the axis of rotation and upon the manner in which the mass is distributed. Examples 8.7 and 8.8 illustrate this point.*

Example 8.7 The Baton Twirler

In an effort to be the star of the half-time show, a majorette twirls a highly unusual baton made up of four masses fastened to the ends of light rods as shown in Figure 8.15. Each rod is 1 m long. Find the moment of inertia of the system about an axis perpendicular to the page and through the point where the rods cross.

Solution
Applying Equation 8.6, we get

$$\begin{aligned}
I = \sum mr^2 &= m_1 r_1{}^2 + m_2 r_2{}^2 + m_3 r_3{}^2 + m_4 r_4{}^2 \\
&= (0.2 \text{ kg})(0.5 \text{ m})^2 + (0.3 \text{ kg})(0.5 \text{ m})^2 + (0.2 \text{ kg})(0.5 \text{ m})^2 + (0.3 \text{ kg})(0.5 \text{ m})^2 \\
&= \boxed{0.25 \text{ kg} \cdot \text{m}^2}
\end{aligned}$$

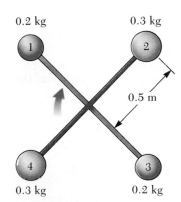

Figure 8.15 (Example 8.7) Four masses connected to light rods rotating in the plane of the paper.

Example 8.8 The Baton Twirler — Second Act

Not satisfied with the crowd reaction from the baton twirling of Example 8.7, the majorette tries spinning her strange baton about the axis OO', as shown in Figure 8.16. Calculate the moment of inertia about this axis.

Solution
Again applying $I = \Sigma mr^2$, we have

$$I = (0.2 \text{ kg})(0)^2 + (0.3 \text{ kg})(0.5 \text{ m})^2 + (0.2 \text{ kg})(0)^2 + (0.3 \text{ kg})(0.5 \text{ m})^2$$

$$= \boxed{0.15 \text{ kg} \cdot \text{m}^2}$$

Calculation of Moments of Inertia for Extended Objects

The method used for calculating moments of inertia in Examples 8.7 and 8.8 is simple enough when you have only a few small masses rotating about an axis. However, the situation becomes much more complex when the object is an extended mass, such as a sphere, a cylinder, or a cone. One extended object that is amenable to a simple solution is a hoop rotating about an axis perpendicular to the plane of the hoop and passing through its center, as shown in Figure 8.17. For example, the object might be a bicycle tire.

Figure 8.16 (Example 8.8) The double baton rotating about the axis OO'.

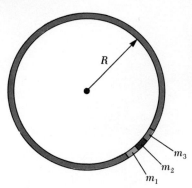

Figure 8.17
The uniform hoop can be divided into a large number of small segments that are equidistant from the center of the hoop.

To evaluate the moment of inertia for the hoop, we can still use the equation $I = \Sigma mr^2$ (Eq. 8.6) and imagine that the hoop is divided into a number of small segments having masses $m_1, m_2, m_3, \ldots$, as in Figure 8.17. This approach is just an extension of the baton problem in our previous examples, except that now we have a large number of small masses in rotation instead of only four.

We can express the sum for I as

$$I = \sum mr^2 = m_1 r_1{}^2 + m_2 r_2{}^2 + m_3 r_3{}^2 + \cdots$$

All of the segments around the hoop are at the *same distance* R from the axis of rotation; thus, we can drop the subscripts on the distances and factor out the common term R:

$$I = (m_1 + m_2 + m_3 + \cdots)R^2$$

We know, however, that the sum of the masses of all the segments must equal the total mass of the hoop, M:

$$M = m_1 + m_2 + m_3 + \cdots$$

and so we can express I as

$$I = Mr^2 \qquad \textbf{(8.8)}$$

This expression can be used for the moment of inertia of any ring-shaped object in any situation in which the ring rotates about an axis through its center and perpendicular to its plane. Note that the result is strictly valid only if the thickness of the ring is small relative to its inner radius.

The hoop we selected as an example is unique in that we were able to find an expression for its moment of inertia by using only simple algebra. Unfortunately, most extended objects are more difficult to work with, and the methods of integral calculus are required. Because calculus techniques are beyond the scope of this text, the moments of inertia for some common shapes are given without proof in Table 8.1. When the need arises, you should refer to this table for an appropriate expression for the moment of inertia.

Example 8.9   Warming Up

A baseball player loosening up his arm before a game tosses a 0.3-kg baseball using only the rotation of his forearm to accelerate the ball, as in Figure 8.18. The ball starts at rest and is released with a speed of 15 m/s in 0.3 s. (a) Find the constant angular acceleration of the arm and ball.

Solution
During its acceleration, the ball moves through an arc of a circle having a radius of 0.35 m. We can determine the angular acceleration using $\omega = \omega_0 + \alpha t$. Since $\omega_0 = 0$, however, $\omega = \alpha t$, or

$$\alpha = \frac{\omega}{t}$$

Figure 8.18   (Example 8.9) A ball being tossed by a pitcher. The forearm is being used to accelerate the ball.

35 cm

We also know that $v = r\omega$, so that we get

$$\alpha = \frac{\omega}{t} = \frac{v}{rt} = \frac{15 \text{ m/s}}{(0.35 \text{ m})(0.3 \text{ s})} = \boxed{143 \text{ rad/s}^2}$$

Table 8.1
Moments of Inertia for Various Rigid Bodies of Uniform Composition

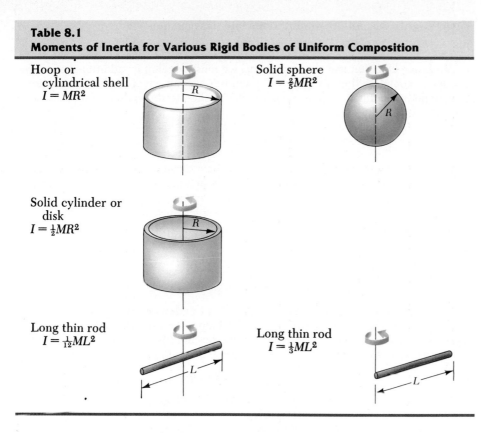

Hoop or cylindrical shell
$I = MR^2$

Solid sphere
$I = \frac{2}{5}MR^2$

Solid cylinder or disk
$I = \frac{1}{2}MR^2$

Long thin rod
$I = \frac{1}{12}ML^2$

Long thin rod
$I = \frac{1}{3}ML^2$

(b) Find the torque exerted on the ball to give it this angular acceleration.

Solution
The moment of inertia of the ball about an axis through the elbow and perpendicular to the arm is

$$I = mr^2 = (0.3 \text{ kg})(0.35 \text{ m})^2 = 3.68 \times 10^{-2} \text{ kg·m}^2$$

Thus, the torque required is

$$\tau = I\alpha = (3.68 \times 10^{-2} \text{ kg·m}^2)(143 \text{ rad/s}^2) = \boxed{5.26 \text{ N·m}}$$

Example 8.10 The Falling Bucket

A frictionless pulley in the shape of a solid cylinder of mass $M = 3$ kg and radius $R = 0.4$ m is used to draw water from a well (Fig. 8.19a). A bucket of mass $m = 2$ kg is attached to a cord wrapped around the cylinder. If the bucket starts from rest at the top of the well and falls for 3 s before hitting the water, how far does it fall?

Solution
Figure 8.19b shows the two forces on the bucket as it falls: **T** is the tension in the cord, and **mg** is the weight of the bucket. We shall choose the positive direction downward and write Newton's second law for the bucket as

$$mg - T = ma$$

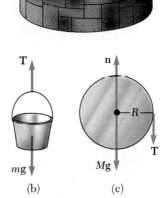

(a)

(b)　　　　(c)

Figure 8.19 (Example 8.10) (a) A water bucket attached to a rope passing over a frictionless pulley. (b) Free-body diagram for the bucket. (c) The tension produces a torque on the cylinder about its axis of rotation.

When the given quantities are substituted into this equation, we have

$$(1) \qquad (2 \text{ kg})(9.80 \text{ m/s}^2) - T = (2 \text{ kg})(a)$$

With one equation and two unknowns, we must develop an additional equation in order to complete the problem. To obtain this second equation, let us consider the cylinder's rotational motion. Equation 8.7 applied to the cylinder gives the necessary expression:

$$\tau = I\alpha = \frac{1}{2} MR^2 \alpha$$

Figure 8.19c shows that the only force producing a torque on the cylinder as it rotates about an axis through its center is **T**, the tension in the cord. There are actually two other forces acting on the cylinder, its weight and an upward force that the axle exerts on the cylinder, but we do not have to consider these here because their lever arm about the axis of rotation is zero. Thus, we have

$$T(0.4 \text{ m}) = \frac{1}{2} (3 \text{ kg})(0.4 \text{ m})^2 (\alpha)$$

$$(2) \qquad T = (0.6 \text{ kg} \cdot \text{m})(\alpha)$$

At this point, it is important to recognize that the downward acceleration of the bucket is equal to the tangential acceleration of a point on the rim of the cylinder. Therefore, the angular acceleration of the cylinder and the linear acceleration of the bucket are related by $a_t = r\alpha$. When this relation is used in (2), we get

$$(3) \qquad T = (1.5 \text{ kg})(a_t)$$

Equations (1) and (3) can now be solved simultaneously to find a_t and T. This procedure gives

$$a_t = 5.60 \text{ m/s}^2 \qquad T = 8.40 \text{ N}$$

Finally, we turn to the equations for motion with constant linear acceleration to find the distance d that the bucket falls in 3 s. Since $v_0 = 0$, we get:

$$d = v_0 t + \frac{1}{2} at^2 = \frac{1}{2} (5.60 \text{ m/s}^2)(3 \text{ s})^2 = \boxed{25.2 \text{ m}}$$

8.5   Rotational Kinetic Energy

In Chapter 5 we defined the kinetic energy of a particle moving through space with a speed v as the quantity $\frac{1}{2}mv^2$. In analogy with this,

> a body rotating about some axis with an angular velocity ω is said to have rotational kinetic energy given by $\frac{1}{2}I\omega^2$.

To prove that this is true, consider a rigid plane body rotating about some axis perpendicular to its plane surface, as in Figure 8.20. The body consists of many small particles, each of mass m. All these particles rotate in a circular path about the axis. If r is the distance of one of the particles from the axis of rotation, the velocity of this particle is $v = r\omega$. Since the *total* kinetic energy of the body is the sum of all the kinetic energies associated with the particles, we have

$$KE_r = \sum \left(\frac{1}{2} mv^2 \right) = \sum \left(\frac{1}{2} mr^2\omega^2 \right) = \frac{1}{2} \left(\sum mr^2 \right) \omega^2$$

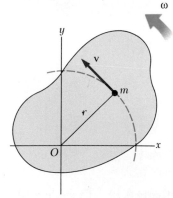

Figure 8.20
A rigid plane body rotating about the z axis with angular velocity ω. The kinetic energy of a particle of mass m is $\frac{1}{2}mv^2$. The total kinetic energy of the body is $\frac{1}{2}I\omega^2$.

$$KE_r \equiv \frac{1}{2}I\omega^2 \qquad \textbf{(8.9)}$$

Rotational kinetic energy

where $I = \Sigma mr^2$ is the moment of inertia of the body. Note that we were able to factor out the ω^2 term since it is the same for every particle.

In linear motion, we found the energy concept to be extremely useful in describing the motion of a system. The energy concept can be equally useful in simplifying the analysis of rotational motion. We now have developed expressions for three types of energy: *gravitational potential energy, PE_g, translational kinetic energy, KE_t,* and *rotational kinetic energy, KE_r.* We must include all these forms of energy in our equation for conservation of mechanical energy:

$$(KE_t + KE_r + PE_g)_i = (KE_t + KE_r + PE_g)_f \qquad \textbf{(8.10)}$$

where "i" and "f" refer to initial and final values, respectively.

Example 8.11 A Ball Rolling Down an Incline

A ball of mass M and radius R starts from rest at a height of 2 m and rolls down a 30° slope, as shown in Figure 8.21. What is the linear velocity of the ball when it leaves the incline? Assume the ball rolls without slipping.

Solution
The initial energy of the ball is gravitational potential energy, and when it reaches the bottom of the ramp, this potential energy has been converted to translational and rotational kinetic energy. The conservation of mechanical energy equation becomes

$$(PE_g)_i = (KE_t + KE_r)_f$$

or

$$Mgh = \frac{1}{2}Mv^2 + \frac{1}{2}\left(\frac{2}{5}MR^2\right)\omega^2$$

where we have used $I = \frac{2}{5}MR^2$ from Table 8.1 as the moment of inertia of the ball. If the ball rolls without slipping, a point on its surface must have the same instantaneous speed as its center of gravity.[1] Thus, we can relate the linear velocity of the ball to its rotational speed:

$$v = R\omega$$

This expression can be used to eliminate ω from the equation for conservation of mechanical energy:

$$Mgh = \frac{1}{2}Mv^2 + \frac{1}{5}Mv^2$$

or, solving for v,

$$v = \sqrt{\frac{10gh}{7}} = \sqrt{\frac{10(9.80 \text{ m/s}^2)(2 \text{ m})}{7}} = \boxed{5.29 \text{ m/s}}$$

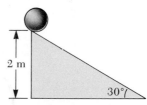

Figure 8.21 (Example 8.11) A ball starts from rest at the top of an incline and rolls to the bottom without slipping.

[1] Note that a point on the surface of the ball travels a distance $2\pi R$ during one rotation. However, if no slipping occurs, this is also the distance that the center of gravity travels in the same time interval. Thus, the center of gravity has a speed equal to the speed of a point on the surface of the ball.

Figure 8.22
An object of mass m rotating in a circular path under the action of a constant torque.

Exercise   Repeat this example for a solid cylinder of the same mass and radius as the ball and released at the same height. In a race between the two objects on the incline, which one would win?

Answer $v = \sqrt{4gh/3} = 5.11$ m/s. The ball would win.

8.6   Angular Momentum

In Figure 8.22, an object of mass m is located a distance r away from a center of rotation. Under the action of a constant torque on the object, its angular velocity will increase from a value ω_0 to a value ω in a time Δt. Thus, we can write

$$\tau = I\alpha = I\left(\frac{\omega - \omega_0}{\Delta t}\right) = \frac{I\omega - I\omega_0}{\Delta t}$$

If we define the product

Angular momentum

$$L \equiv I\omega \qquad\qquad (8.11)$$

as the angular momentum of the object, then we can write

$$\tau = \frac{\text{change in angular momentum}}{\text{time interval}} = \frac{\Delta L}{\Delta t} \qquad (8.12)$$

Equation 8.12 is the rotational analog of Newton's second law, $F = \Delta p/\Delta t$, and states that

> the torque acting on an object is equal to the time rate of change of its angular momentum.

When the net external torque acting on a system is zero, we see from Equation 8.12 that $\Delta L/\Delta t = 0$. In this case, the rate of change of the angular momentum of the system is zero. Therefore, *the product $I\omega$ remains constant in time.* That is, $L_i = L_f$, or

$$I_i\omega_i = I_f\omega_f \qquad \text{if } \sum \tau = 0 \qquad (8.13)$$

> The angular momentum of a system is conserved when the net external torque acting on the system is zero. That is, when $\Sigma\tau = 0$, the initial angular momentum equals the final angular momentum.

Conservation of angular momentum

In Equation 8.13, we have a third conservation law to add to our list: **the conservation of angular momentum.** We now can state that

> the energy, linear momentum, and angular momentum of an isolated system all remain constant.

There are many examples that demonstrate conservation of angular momentum, some of which should be familiar to you. You may have observed a skater spinning in the finale of an act. The angular velocity of the skater

Figure 8.23
(a) The angular velocity of the skater increases when she pulls her arms in close to her body, demonstrating that angular momentum is conserved. (b) Photograph of Elizabeth Manley, winner of a silver medal in the 1988 Olympics. *(© duomo/Paul J. Sutton 1988)*

increases upon pulling her hands and feet close to the trunk of her body, as in Figure 8.23. That is, $\omega_2 > \omega_1$. Neglecting friction between skater and ice, we see that there are no external torques on the skater. The moment of inertia of her body decreases as her hands and feet are brought in. The resulting change in angular velocity is accounted for as follows. Since angular momentum must be conserved, the product $I\omega$ remains constant and a decrease of the moment of inertia of the skater causes a corresponding increase in the angular velocity.

Similarly, when divers or acrobats wish to make several somersaults, they pull their hands and feet close to the trunk of their bodies in order to rotate at a greater angular velocity. In this case, the external force due to gravity acts through the center of gravity and hence exerts no torque about the axis of rotation. Therefore the angular momentum about the center of gravity is conserved. For example, when divers wish to double their angular velocity, they must reduce their moment of inertia to half its initial value.

**Problem-Solving Strategy:
Rotational Motion**

The following facts and procedures should be kept in mind when solving rotational motion problems.

1. There are actually very few new techniques that need to be learned in order to solve rotational motion problems. For example, problems involving the equation $\Sigma\tau = I\alpha$ are very similar to those encountered in Newton's second law problems, $\Sigma\mathbf{F} = m\mathbf{a}$. Note the correspondences between linear and rotational quantities in that $\mathbf{F}$ is replaced by τ, m by I, and $\mathbf{a}$ by α.

2. Other analogs between rotational quantities and linear quantities include x is replaced by θ, and v by ω. These are helpful as memory devices for such rotational motion quantities as rotational kinetic energy, $KE_r = \frac{1}{2}I\omega^2$, and angular momentum, $L = I\omega$.

Hurricane Elena photographed from space. This spiral-shaped, nonrigid mass of air undergoes rotation and has angular momentum. *(Courtesy of NASA)*

3. With the analogies mentioned in step 2 above, conservation of energy techniques remain the same as those examined in Chapter 5, except for the fact that the object's rotational kinetic energy must be included in the expression for the conservation of energy.

4. Likewise, the techniques for solving conservation of angular momentum problems are essentially the same as those used in solving conservation of linear momentum problems, except you are equating total angular momentum before to total angular momentum after as $I_i\omega_i = I_f\omega_f$.

Example 8.12   The Spinning Stool

A student sits on a pivoted stool while holding a pair of weights, as in Figure 8.24. The stool is free to rotate about a vertical axis with negligible friction. The moment of inertia of student, weights, and stool is 2.25 kg·m². The student is set in rotation with an initial angular velocity of 5 rad/s with weights outstretched. As he rotates, he pulls the weights inward so that the new moment of inertia of the system (student, weight, and stool) becomes 1.80 kg·m². What is the new angular velocity of the system?

Solution

The initial angular momentum of the system is

$$L_i = I_i\omega_i = (2.25 \text{ kg·m}^2)(5 \text{ rad/s}) = 11.3 \text{ kg·m}^2/\text{s}$$

When the weights are pulled in, they are closer to the axis of rotation, and as a result the moment of inertia of the system is reduced. The new angular momentum is

$$L_f = I_f\omega_f = (1.80 \text{ kg·m}^2)(\omega_f)$$

Because the net external torque on the system is zero, angular momentum is conserved. Thus, we find

$$(11.3 \text{ kg·m}^2/\text{s}) = (1.80 \text{ kg·m}^2)(\omega_f)$$

$$\omega_f = \boxed{6.28 \text{ rad/s}}$$

Figure 8.24   (Example 8.12) (a) The student is given an initial angular velocity while holding two weights as shown. (b) When the weights are pulled in close to the body, the angular velocity of the system increases. Why?

(a)   (b)

Example 8.13   The Merry-go-round

A student is standing at the edge of a circular platform that rotates in a horizontal plane about a frictionless vertical axle (Fig. 8.25). The platform has a mass $M = 100$ kg and a radius $R = 2$ m. The student, whose mass is $m = 60$ kg, walks slowly from the rim of the disk toward the center. If the angular velocity of the system is 2 rad/s when the student is at the rim, calculate the angular velocity when the student has reached a point 0.5 m from the center.

Solution

The moment of inertia of the platform, I_p, is

$$I_p = \frac{1}{2}MR^2 = \frac{1}{2}(100 \text{ kg})(2 \text{ m})^2 = 200 \text{ kg} \cdot \text{m}^2$$

Treating the student as a point mass, his initial moment of inertia is

$$I_s = mR^2 = (60 \text{ kg})(2 \text{ m})^2 = 240 \text{ kg} \cdot \text{m}^2$$

Thus, the initial angular momentum of the platform plus student is

$$L_i = (I_p + I_s)(\omega_i) = (200 \text{ kg} \cdot \text{m}^2 + 240 \text{ kg} \cdot \text{m}^2)(2 \text{ rad/s}) = 880 \text{ kg} \cdot \text{m}^2/\text{s}$$

When the student has walked to the position 0.5 m from the center, his moment of inertia is

$$I_s' = mr_f{}^2 = (60 \text{ kg})(0.5 \text{ m})^2 = 15 \text{ kg} \cdot \text{m}^2$$

There is no change in the moment of inertia of the platform. Because there are no external torques on the *system* (student plus platform) about the axis of rotation, we can apply the law of conservation of angular momentum:

$$L_i = L_f$$

$$880 \text{ kg} \cdot \text{m}^2/\text{s} = 200\omega_f + 15\omega_f = 215\omega_f$$

$$\omega_f = \boxed{4.09 \text{ rad/s}}$$

Exercise   Calculate the change in kinetic energy of the system (student plus platform) for this situation. What accounts for this change in energy?

This student is holding the axle of a spinning bicycle wheel while seated on a pivoted stool. The student and stool are initially at rest while the wheel is spinning in a horizontal plane. When the wheel is inverted about its center by 180°, the student and stool begin to rotate. This is because the total angular momentum of the system (wheel + student + stool) must be conserved. *(Courtesy of CENCO)*

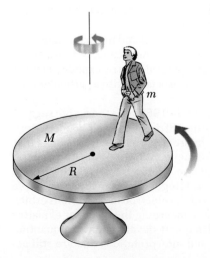

Figure 8.25   (Example 8.13) As the student walks toward the center of the rotating platform, the angular velocity of the system increases because the angular momentum of the system (student + platform) must remain constant.

Answer $KE_f - KE_i = 918$ J. The student must perform positive work in order to walk toward the center of the platform.

►SPECIAL TOPIC
Angular Momentum as a Fundamental Quantity

We have seen that the concept of angular momentum is very useful for describing the motion of macroscopic objects. However, the concept is also valid on a submicroscopic scale and has been used extensively in the development of modern theories of atomic, molecular, and nuclear physics. In these developments, it was found that the angular momentum of a system is a *fundamental* quantity. The word *fundamental* in this context implies that angular momentum is an inherent property of atoms, molecules, and their constituents.

In order to explain the results of a variety of experiments on atomic and molecular systems, it is necessary to assign discrete values to the angular momentum. These discrete values are some multiple of a fundamental unit of angular momentum, which equals $\hbar = h/2\pi$, where h is Planck's constant.

Fundamental unit of angular momentum

$$\hbar = 1.054 \times 10^{-34} \frac{\text{kg} \cdot \text{m}^2}{\text{s}^2}$$

Let us accept this postulate for the time being and show how it can be used to estimate the rotational frequency of a diatomic molecule. Consider the O_2 molecule as a rigid rotor, that is, two atoms separated by a fixed distance d and rotating about the center of mass (Fig. 8.26). Equating the rotational angular momentum to the fundamental unit $\hbar$, we can estimate the lowest rotational frequency:

$$I_c\omega \approx \hbar \qquad \text{or} \qquad \omega \approx \frac{\hbar}{I_c}$$

The moment of inertia of the O_2 molecule about this axis of rotation is equal to 2.03×10^{-46} kg·m². Therefore,

$$\omega \approx \frac{\hbar}{I_c} = \frac{1.054 \times 10^{-34} \text{ kg} \cdot \text{m}^2/\text{s}}{2.03 \times 10^{-46} \text{ kg} \cdot \text{m}^2} = 5.19 \times 10^{11} \text{ rad/s}$$

This result is in good agreement with measured rotational frequencies. Furthermore, the rotational frequencies are much lower than the vibrational frequencies of the molecule, which are typically of the order of 10^{13} Hz.

This simple example shows that certain classical concepts and mechanical models might be useful in describing some features of atomic and molecular systems. However, a wide variety of phenomena on the submicroscopic scale can be explained only if one assumes discrete values of the angular momentum associated with a particular type of motion.

Historically, the Danish physicist Niels Bohr (1885–1962) was the first to suggest this radical idea in his theory of the hydrogen atom. Strictly classical models were unsuccessful in describing many properties of the hydrogen atom, such as the fact that the atom absorbs and emits radiation at discrete frequencies. Bohr postulated that the electron could only occupy circular orbits about the proton for which the orbital angular momentum was equal to $n\hbar$, where n is an integer. From this rather simple model, one can estimate the rotational frequencies of the electron in the various orbits.

Although Bohr's theory provided some insight concerning the behavior of matter at the atomic level, it is basically incorrect. Subsequent developments in quantum mechanics from 1924 to 1930 provided models and interpretations that are still accepted.

Figure 8.26
The rigid-rotor model of the diatomic molecule. The rotation occurs about the center of mass in the plane of the diagram.

Later developments in atomic physics indicated that the electron also possesses another kind of angular momentum, called *spin*, which is also an inherent property of the electron. The spin angular momentum is also restricted to discrete values. We shall return to this important property in Chapter 30 and discuss its great impact on modern physical science.

Special Topic Problems

1. The diatomic oxygen molecule consists of two oxygen atoms separated by a distance of 1.21×10^{-10} m. The mass of each oxygen atom is 2.77×10^{-26} kg. Assume the molecule is in rotation in the xy plane about a z axis that passes through the center of the molecule and is perpendicular to its length. (a) Find the moment of inertia of this molecule about this axis. (b) If its angular momentum is $2\hbar$, find the rotational frequency. (c) How many fundamental units of angular momentum must the oxygen molecule have in order for its rotational frequency to be equal to the typical vibration frequency of this molecule, which is of the order of 10^{13} vibrations per second?

2. In the Bohr model of the hydrogen atom, the electron moves in a circular orbit of radius 0.529×10^{-10} m around the proton. Assuming the orbital angular momentum of the electron is equal to $\hbar$, calculate (a) the orbital speed of the electron, (b) the kinetic energy of the electron, and (c) the angular velocity of the electron. ◀

Summary

The ability of a force to rotate an object about some axis is measured by a quantity called the **torque,** τ. The magnitude of the torque is given by

$$\tau = Fd \tag{8.1}$$

In this equation, d is the **lever arm,** which is the perpendicular distance from the axis of rotation to a line drawn along the direction of the force. The sign of the torque is negative if the turning tendency of the corresponding force is clockwise and positive if its turning tendency is counterclockwise.

An object is in **equilibrium** when the following conditions are satisfied: (1) the resultant external force must be zero and (2) the resultant external torque must be zero about any origin. That is,

$$\sum \mathbf{F} = 0$$
$$\sum \tau = 0$$

The first equation is called the **first condition for equilibrium.** When this equation holds, an object will either be at rest or moving with a constant velocity. The second equation is called the **second condition for equilibrium.** When it holds, an object is said to be in rotational equilibrium.

The **moment of inertia** of a group of particles is

$$I \equiv \sum mr^2 \tag{8.6}$$

If a rigid body free to rotate about a fixed axis has a net external torque acting on it, the body will undergo an angular acceleration α, where

$$\sum \tau = I\alpha \tag{8.7}$$

If a rigid object rotates about a fixed axis with angular velocity ω, its **rotational kinetic energy** is

$$KE_r \equiv \frac{1}{2} I\omega^2 \tag{8.9}$$

where I is the moment of inertia about the axis of rotation.

The **angular momentum** of a rotating object is

$$L \equiv I\omega \tag{8.11}$$

If the net external torque acting on a system is zero, the total angular momentum of the system is constant. Applying the law of conservation of angular momentum to an object whose moment of inertia changes with time gives

$$I_i\omega_i = I_f\omega_f \tag{8.13}$$

Additional Reading

H. Brody, "The Moment of Inertia of a Tennis Racket," *The Physics Teacher*, April 1985, p. 213.

C. Frohlich, "Somersaulting and Twisting," *Sci. American*, March 1980, p. 154.

D. F. Griffing, *The Dynamics of Sports: Why That's the Way the Ball Bounces*, Mohican Publishing Co., 1982.

J. G. Kreifeldt and M. Chuang, "Moment of Inertia: Psychophysical Study of an Overlooked Sensation," *Science*, 1979, p. 588.

K. Laws, "The Physics of Dance," *Physics Today*, February 1985, p. 25.

R. F. Post and S. F. Post, "Flywheels," *Sci. American*, December 1973, p. 17.

Questions

1. If the net force acting on a system is zero, then is it necessarily true that the net torque on it is also zero?
2. Is it possible to calculate the torque acting on a rigid body without specifying the origin? Is the torque independent of the location of the origin?
3. Under what conditions would the tension in the string attached to the bucket in Example 8.10 be equal to the weight of the bucket?
4. Explain why changing the axis of rotation of an object should change its moment of inertia.

5. Two cylinders having the same dimensions are set into rotation about their axes with the same angular velocity. One is hollow, while the other is filled with water. Which cylinder would be easier to stop rotating?

6. Is it possible to change the translational kinetic energy of an object without changing its rotational kinetic energy?

7. Why do tightrope walkers carry a long pole to help balance themselves while walking a tightrope?

8. It is more difficult to do a sit-up with your hands behind your head than it is when you stretch your arms out in front of you. Why?

9. In order for a helicopter to be stable as it flies, it must have two propellers. Why?

10. Three homogeneous rigid bodies — a solid sphere, a solid cylinder, and a hollow cylinder — are placed at the top of an incline (Fig. 8.27). If they all are released from rest at the same elevation and roll without slipping, which reaches the bottom first? Which reaches last? You should try this at home and note that the result is *independent* of the masses and radii.

Figure 8.27
(Question 10)
Which object wins the race?

11. Two spheres look identical and weigh the same. One sphere is hollow, the other solid. Discuss how you might determine which sphere is which.

12. Suppose you remove two eggs from the refrigerator, one that had been hard-boiled and the other uncooked. You wish to determine which is the hard-boiled egg without breaking the eggs. This is possible by spinning the two eggs on the floor and comparing the rotational motions. Which egg spins the fastest? Which will rotate more uniformly? Explain. (This is a neat and helpful trick for Mom.)

13. Stars originate as large bodies of slowly rotating gas. Because of gravity, these clumps of gas slowly decrease in size. What happens to the angular velocity of a star as it shrinks? Explain.

14. As a tetherball winds around a pole, what happens to its angular velocity? Explain.

15. Often when a high diver wants to turn a flip in midair, she will draw her legs up against her chest. Why does this make her rotate faster? What should she do when she wants to come out of her flip?

16. Space colonies have been proposed that would consist of large cylinders placed in space. Gravity would be simulated in these cylinders by setting them into rotation. Discuss the difficulties that would be encountered in attempting to set the cylinder into rotation. (*Hint:* Consider problems presented by the conservation of angular momentum.)

17. Why is it easier to keep your balance on a moving bicycle than on a bicycle at rest?

18. Suppose the planets of the Solar System evolved from the condensation of a primordial gas. From the observation that all the planets revolve around the Sun in the same direction, what would you infer about the state of the gas prior to its condensation?

19. A student sits on a stool that is free to rotate about its vertical axis. The student and stool are set into rotation while the student, with outstretched arms, holds a pair of weights. If she suddenly drops the weights to the floor, what happens to her angular velocity? Explain.

20. A mouse is initially at rest on a horizontal turntable mounted on a frictionless vertical axle. If the mouse begins to walk around the perimeter, what happens to the turntable? Explain.

21. If angular momentum is conserved for a propeller-driven airplane, it seems that if the propeller is turning clockwise, the airplane should be turning counterclockwise. Why doesn't this occur?

22. It is said that if a cat falls out of a tree, it always lands on its feet. Comment on this. See "The Physics of Somersaulting and Twisting" by Cliff Frohlich, *Scientific American*, 242 (March 1980), 155. This article contains an interesting discussion of conservation of angular momentum in diving and somersaulting.

23. A cat usually lands on its feet regardless of the position from which it is dropped. A slow motion film of a cat falling shows that the upper half of its body twists in one direction while the lower half twists in the opposite direction. Why does this type of rotation occur?

24. A ladder is resting inclined against a wall. Would you feel safer climbing up the ladder if you were told that the floor is frictionless but the wall is rough or that the wall is smooth but the floor is rough? Justify your answer.

25. Give an example in which the net force acting on an object is zero and yet the net torque is nonzero.

26. Give an example in which the net torque acting on an object is zero and yet the net force is nonzero.

27. Can an object be in equilibrium if the only torques acting on it tend to produce clockwise rotation?

28. If you measure the net torque and the net force on a system to be zero, (a) could the system still be rotating with respect to you? (b) Could it be translating with respect to you?

Problems

Section 8.1   Torque

1. If the torque required to loosen a nut holding a flat tire in place on a car has a magnitude of 40 N·m, what *minimum* force must be exerted by a garage worker at the end of a 30-cm lug wrench to accomplish the task?
2. Find the torque exerted on the fishing pole by a fish pulling with a force of 100 N as shown in Figure 8.28. Choose the axis of rotation of the pole to be at the hand and perpendicular to the page.

Figure 8.28   Problem 2

3. If the fishing pole of Problem 2 is inclined to the horizontal at an angle of 20°, as shown in Figure 8.29, what is the torque exerted by the fish about an axis perpendicular to the page and passing through the hand of the person holding the pole?

Figure 8.29   Problem 3

4. A 48-kg diver stands at the end of a 3-m-long diving board. What torque does the weight of the diver produce about an axis perpendicular to and in the plane of the diving board through its midpoint?
5. A steel band exerts a horizontal force of 80.0 N on a tooth at the point *B* in Figure 8.30. What is the torque on the root of the tooth about point *A*?

Figure 8.30
Problem 5

6. The person in Figure 8.31 weighs 800 N. The forces F_1 and F_2 have magnitudes of 100 N and 900 N, respectively. Assume the force of gravity acts downward through point *A* as shown. Determine the net torque on the person about axes through points *A*, *B*, and *C* perpendicular to the plane of the paper. (*Hint:* Replace F_1 and F_2 by their horizontal and vertical components.)

Figure 8.31
Problem 6

Section 8.2   Torque and the Second Condition for Equilibrium

7. Two children sit on a seesaw such that a 400-N child is 2 m from the support (the fulcrum). Where should a second child of weight 475 N sit in order to balance the system if the support is at the center of the plank?
8. The 400-N child of Problem 7 decides that she would like to seesaw alone. To do so, she moves the board such that its weight is no longer directly over the fulcrum. She finds that she will be balanced when she is 1.5 m to the left of the fulcrum and the center of the plank is 0.5 m to the right of the fulcrum. What is the weight of the plank?

9. A window washer is standing on a scaffold supported by vertical ropes at each end. The scaffold weighs 200 N and is 3 m long. What is the tension in each rope when the 700-N worker stands 1 m from one end?

10. A 200-N floodlight in a park is supported at the end of a beam of negligible mass hinged to a pole as shown in Figure 8.32. A cable at an angle of 30° with the beam helps to support the light. Find (a) the tension in the cable, and (b) the horizontal and vertical forces exerted on the beam by the pole.

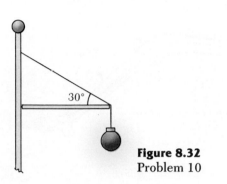

Figure 8.32
Problem 10

11. A hungry bear weighing 160 lb walks out on a beam in an attempt to retrieve some "goodies" hanging at the end of the beam (Fig. 8.33). The beam is uniform, weighs 50 lb, and is 20 ft long; the goodies weigh 20 lb. (a) Draw a free-body diagram for the beam. (b) When the bear is at $x = 3$ ft, find the tension in the wire and the components of the reaction force at the hinge. (c) If the wire can withstand a maximum tension of 200 lb, what is the maximum distance the bear can walk before the wire breaks?

Figure 8.33
Problem 11

12. A uniform plank of length 2 m and mass 30 kg is supported by three ropes as in Figure 8.34. Find the tension in each rope when a 700-N person is 0.5 m from the left end.

Figure 8.34   Problem 12

13. The arm in Figure 8.35 weighs 41.5 N. The weight of the arm acts through point A. Determine the magnitudes of the tension force $\mathbf{F}_t$ in the deltoid muscle and the force $\mathbf{F}_s$ of the shoulder on the humerus (upper-arm bone) to hold the arm in the position shown.

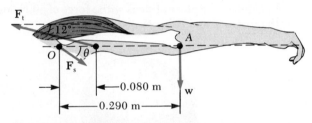

Figure 8.35   Problem 13

14. The force of gravitational attraction on an object is exerted not on the object as a whole but on each molecule of the object. The resultant of all these forces is the weight of the object directed downward. In Figure 8.36 we show an object divided into many very small particles, such as the molecules mentioned above. The masses of these tiny parts are labeled m_1, m_2, m_3, . . . and have coordinates (x_1, y_1), (x_2, y_2),

Figure 8.36
Problem 14

(x_3, y_3) . . . . Also pictured are the vectors indicating the weight of each particle. (a) Find the torque on each of these particles about the origin and add these torques together. (b) The torque due to the total weight, $\mathbf{w} = M\mathbf{g}$, of the object is located at the point (x_{cg}, y_{cg}). Find the torque of the weight about the origin. (c) The net torque due to the individual particles and the torque due to the weight considered to be acting at the center of gravity must be the same. Equate the expressions found in part (a) and part (b) to show that $x_{cg} = \Sigma m_i x_i / \Sigma m_i$.

15. The muscle used for chewing, the masseter, is one of the strongest in the human body. It is attached to the mandible, or lower jawbone, as shown in Figure 8.37a. The jawbone is pivoted about a socket just in front of the auditory canal. The forces acting on the jawbone are equivalent to those acting on a curved bar as shown in Figure 8.37b. The force **C** is the force exerted against the jawbone by the food being chewed, **T** is the tension in the masseter, and **R** is the force exerted on the mandible by the socket. If you bite down on a piece of steak with a force of 50 N, find the forces **T** and **R**.

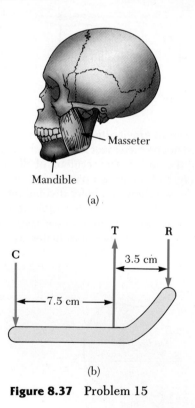

(a)

Figure 8.37   Problem 15

16. The large quadriceps muscle in the upper leg terminates at its lower end in a tendon attached to the upper end of the tibia as shown in Figure 8.38a. The forces on the lower leg when the leg is extended are modeled as in Figure 8.38b, where **T** is the tension in the tendon, **C** is the weight of the lower leg, and **F** is the weight of the foot. Find the tension **T** when the tendon is at an angle of 25° with the tibia, assuming $C = 30$ N, $F = 12.5$ N, and the leg is extended at an angle of 40° with respect to the vertical ($\theta = 40°$). Assume the center of gravity of the lower leg is at its center, and that the tendon attaches to the lower leg at a point one fifth of the way down the leg.

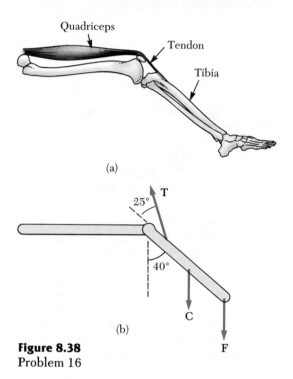

Figure 8.38
Problem 16

17. A 1200-N uniform boom is supported by a cable as in Figure 8.39. The boom is pivoted at the bottom, and a 2000-N weight hangs from its top. Find the tension in the supporting cable and the components of the reaction force on the boom at the hinge.

Figure 8.39
Problem 17

18. A hemispherical sign 1 m in diameter and of uniform mass density is supported by two wires as shown in Figure 8.40. What fraction of the sign's weight is supported by each wire?

Figure 8.40
Problem 18

19. A 15-m uniform ladder weighing 500 N rests against a frictionless wall. The ladder makes an angle of 60° with the horizontal. (a) Find the horizontal and vertical forces that the Earth exerts on the base of the ladder when an 800-N firefighter is 4 m from the bottom. (b) If the ladder is just on the verge of slipping when the firefighter is 9 m up, what is the coefficient of static friction between ladder and ground?

20. An 8-m uniform ladder weighing 200 N rests against a smooth wall. The coefficient of static friction between the ladder and the ground is 0.6, and the ladder makes a 50° angle with the ground. How far up the ladder can an 800-N person climb before the ladder begins to slip?

21. A dormitory door 2.5 m high and 1 m wide weighs 250 N, and its center of gravity is at its geometric center. The door is supported by hinges 0.25 m from top and bottom, each hinge carrying half the weight. Determine the horizontal component of the forces exerted by each hinge on the door.

22. A racing car has a mass of 1600 kg. The distance between the front and rear axles is 3 m. If the center of gravity of the car is 2 m from the rear axle, what is the normal force on each tire?

23. An iron trapdoor 1.25 m wide and 2 m long weighs 360 N and is hinged along the short dimension. Its center of gravity is at its geometric center. What force applied at right angles to the door is required to lift it (a) when it is horizontal and (b) when it has been opened so that it makes an angle of 30° with the horizontal? (Assume that the force is applied at the edge of the door opposite the hinges.)

Section 8.4 Relationships Between Torque and Angular Acceleration

24. A 4.5-kg ball on the end of a chain is whirled in a horizontal circle by an athlete. If the distance of the ball from the axis of rotation is 2.5 m, find the moment of inertia of the ball, assuming it can be considered as a point object.

25. What torque must the track star exert on the ball of Problem 24 to give it an angular acceleration of 2 rad/s²?

26. Four masses are held in position at the corners of a rectangle by light rods as shown in Figure 8.41. Find the moment of inertia of the system about (a) the x axis, (b) the y axis, and (c) an axis through O and perpendicular to the page.

Figure 8.41
Problems 26, 27, 28 and 43

27. If the system shown in Figure 8.41 is set into rotation about each of the axes mentioned in Problem 26, find the torque which will produce an angular acceleration of 1.5 rad/s² in each case.

28. If a constant torque of 20 N·m acts on the system shown in Figure 8.41, causing it to rotate about the y axis, find its angular velocity after 3 s, assuming the system starts from rest.

29. (a) Find the moment of inertia of a solid cylinder of mass 1.5 kg and radius 30 cm about an axis through its center. (b) Repeat for a solid sphere of the same mass and radius about an axis through its center.

30. The cylinder of Problem 29(a) is rotating at an angular velocity of 2 rev/s. What torque is required to stop it in 15 s?

31. A cylindrical fishing reel has a mass of 0.85 kg and a radius of 4.0 cm. A friction clutch in the reel exerts a restraining torque of 1.3 N·m if a fish pulls on the line. The fisherman gets a bite, and the reel begins to spin with an angular acceleration of 66.0 rad/s². (a) What is the force of the fish on the line? (b) How much line unwinds in 0.5 s?

32. A potter's wheel, a thick stone disk of radius 0.5 m and mass 100 kg, is freely rotating at 50 rev/min. The potter can stop the wheel in 6 s by pressing a wet rag against the rim and exerting a radially inward force of 70 N. Find the effective coefficient of kinetic friction between the wheel and the wet rag.

33. A merry-go-round in the shape of a horizontal disk of mass 150 kg and radius 1.5 m is set into motion by wrapping a rope about the rim of the disk and pulling on the rope. What constant force would have to be exerted on the rope to bring it from rest to an angular speed of 0.5 rev/s in 2 s?

34. A pulley in the shape of a cylinder of mass 5 kg and radius 0.6 m is used to lower a bucket of mass 3 kg into a well, as shown in Figure 8.42. The bucket starts from rest and falls for 4 s. (a) What is the linear acceleration of the falling bucket? (b) How far does it drop? (c) What is the angular acceleration of the cylinder?

Figure 8.42   Problems 34 and 53

35. A cable passes over a pulley. Because of friction between the pulley and the cable, the tension in the cable is not the same on opposite sides of the pulley. The force on one side is 120 N, and the force on the other side is 100 N. Assuming the pulley is a uniform disk of mass 2.1 kg and radius 0.811 m, determine the angular acceleration of the pulley.

36. When the motor in Figure 8.43 raises the 1000-kg mass, it produces a tension of 1.14×10^4 N in the cable on the right side of the pulley. The pulley has a moment of inertia of 79.8 kg·m² and a radius of 0.762 m. The cable rides over the pulley without slipping. Determine the acceleration of the 1000-kg mass. (*Hint:* Draw free-body diagrams of the 1000-kg mass and the pulley. Do not assume the tension in the cable is the same on both sides of the pulley.)

1000 kg

Motor

Figure 8.43
Problem 36

37. A string is wrapped around a uniform 2-kg cylinder of radius 15 cm as shown in Figure 8.44. One end of the string is attached to the ceiling and the cylinder is allowed to fall from rest. (a) Write down Newton's second law for the cylinder. (b) Find the net torque about the center of the cylinder and equate this to $I\alpha$. (c) Use the equations found in (a) and (b) along with $a = r\alpha$ to show that the linear acceleration of the cylinder is given by $a = (2/3)g$.

Figure 8.44
Problem 37

38. A grindstone used to sharpen tools is rotating at an angular speed of 8 rad/s when an ax is pressed against it. Because of the torque produced on the grindstone by the force of friction between it and the ax, it comes to rest in four rotations. (a) Find the frictional force produced, assuming the grindstone is a cylinder of mass 5 kg and radius 0.4 m. (b) If the coefficient of friction between ax and grindstone is 0.4, find the normal force with which the ax was pressed against the grindstone.

39. An airliner lands with a speed of 50.0 m/s. Each wheel of the plane has a radius of 1.25 m and a moment of inertia of 110 kg·m². At touchdown the wheels begin to spin under the action of friction. Each wheel supports a weight of 1.4×10^4 N, and the wheels get up to the angular speed of rolling without slipping in 0.48 s. What is the coefficient of kinetic friction between the wheels and the runway? Assume the speed of the plane is constant.

Section 8.5   Rotational Kinetic Energy

40. An automobile tire, considered as a solid disk, has a radius of 35 cm and a mass of 6 kg. Find its rotational kinetic energy when rotating about an axis through its center at an angular velocity of 2 rev/s.

41. An automobile engine part is in the shape of a thin rod of mass 100 g and length 5 cm. When the rod is rotating at an angular velocity of 3 rad/s, find its kinetic energy when (a) rotating about an axis through a point 2.5 cm from each end. (b) Repeat when it is rotating about an axis through one end.

42. Three particles are connected by rigid rods of negligible mass lying along the y axis (Fig. 8.45). If the system rotates about the x axis with an angular speed of 2 rad/s, find (a) the moment of inertia about the x axis and the total kinetic energy evaluated from $\frac{1}{2}I\omega^2$ and (b) the linear speed of each particle and the total kinetic energy evaluated from $\Sigma\frac{1}{2}m_iv_i^2$.

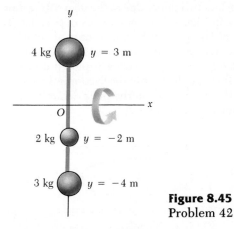

Figure 8.45
Problem 42

43. If the system of masses shown in Figure 8.41 is set into rotation about the x axis with an angular velocity of 2.5 rad/s, (a) find the kinetic energy of the system. (b) Repeat the calculation for the system in rotation at the same speed about the y axis.

44. A horizontal 800-N merry-go-round of radius 1.5 m is started from rest by a constant horizontal force of 50 N applied tangentially to the cylinder. Find the kinetic energy of the solid cylinder after 3 s.

45. What angular velocity must a cylindrical wheel of radius 0.4 m and mass 3 kg have in order to have the same kinetic energy as a sphere of mass 3 kg and radius 0.4 m that is rotating at an angular velocity of 9 rad/s?

46. A cylinder of mass 10 kg rolls without slipping on a rough surface. At the instant its center of mass has a speed of 10 m/s, determine (a) the translational kinetic energy of its center of mass, (b) the rotational kinetic energy about its center of mass, and (c) its total kinetic energy.

47. A solid sphere has a radius of 0.2 m and a mass of 150 kg. How much work is required to get the sphere rolling with an angular speed of 50 rad/s on a horizontal surface? (Assume the sphere starts from rest and rolls without slipping.)

48. A 2-kg solid ball of radius 0.5 m starts at a height of 3 m above the surface of the Earth and *rolls* down a 20° slope. A solid disk and a ring start at the same time and at the same height. Both ring and disk have the same mass and radius as the ball. Which of the three wins the race to the bottom if they all roll without slipping?

49. A car is designed that gets its energy from a rotating flywheel. To use these cars, the flywheel is attached to an electric motor before leaving on a trip, which brings its rotational speed up to 5000 rev/min. The cylinder has a radius of 2 m and a mass of 500 kg. (a) Find the kinetic energy stored in the flywheel. (b) If the flywheel is to supply energy to the car as would a 10-hp motor, find the time the car could run before the flywheel had to be brought back up to speed.

50. The top in Figure 8.46 has a moment of inertia of 4.0×10^{-4} kg · m² and is initially at rest. It is free to rotate about a stationary axis *AA′*. A string, wrapped around a peg along the axis of the top, is pulled in such a manner as to maintain a constant tension of 5.57 N in the string. If the string does not slip while wound around the peg, what is the angular speed of the top after 80 cm of string has been pulled off of the peg? (*Hint:* Consider the work done.)

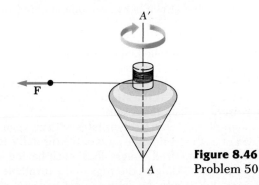

Figure 8.46
Problem 50

51. Assume that the Earth is a homogeneous sphere, and calculate the kinetic energy of rotation about its axis. (Use the table of physical constants inside the front cover.) For how long could this energy be tapped to supply the world's energy needs (assumed constant at 2×10^{20} J/yr) if, in the process, the length of the day is not to be increased by more than 1 min?

52. In a circus performance, a large hoop of mass 5 kg and radius 3 m rolls without slipping. If the hoop is given an angular velocity of 3 rad/s while rolling on the horizontal and allowed to roll up a ramp inclined at 20° with the horizontal, how far (measured along the incline) will the hoop roll?

53. A bucket is attached to a pulley as shown in Figure 8.42. The cylinder has a mass of 5 kg and a radius equal to 0.6 m, while the bucket has a mass of 3 kg. Starting from rest, the bucket falls for 3 m. At this point the rope comes free of the pulley. What is the kinetic energy of the pulley at this time?

Section 8.6 Angular Momentum

54. (a) Calculate the angular momentum of the Earth arising from its spinning motion on its axis and (b) the angular momentum of the Earth arising from its orbital motion about the Sun.

55. The movement of the Moon about the Earth is such that it revolves once in about every 28 days and it rotates on its axis once in this same time period. The net result of this is that the Moon always keeps the same face pointed toward the Earth. (a) Find the angular momentum of the Moon in its orbit, and (b) its angular momentum due to rotation on its axis.

56. If the size of the Earth should increase so that its radius is doubled but the mass is held constant, what would happen to the length of a day?

57. A figure skater rotates about a vertical axis through her center with both arms and one leg extended. Her rotation rate is 0.5 rev/s. She then pulls her arms and leg in closer to her body and begins to spin at 1.7 rev/s. What is the ratio of her new moment of inertia to her old?

58. A skater spins with an angular speed of 12.0 rad/s with her arms outstretched. She lowers her arms, decreasing her moment of inertia by 10%. Ignoring friction on the skates, determine the percentage change in her kinetic energy. Is the kinetic energy increased or decreased? Explain the result.

59. A phonograph turntable freely rotates with an angular speed of $33\frac{1}{3}$ rev/min. The mass of the turntable is 0.225 kg. Two hundred hungry flies, each having a mass of 2.0 g, are riding on the edge of the turntable. All at once they run to the center to get to a spot of jelly. What is the new angular speed of the turntable?

60. A student sits on a rotating stool holding two weights, each of mass 3 kg. When his arms are extended horizontally, the weights are 1 m from the axis of rotation and he rotates with an angular speed of 0.75 rad/s. The moment of inertia of the student plus stool is 3 kg · m² and is assumed to be constant. If the student

pulls the weights horizontally to 0.3 m from the rotation axis, (a) find the new angular speed of the student. (b) Find the kinetic energy of the student before and after the weights are pulled in.

61. A 45-rpm record in the shape of a solid disk 25 cm in diameter and mass 0.1 kg rotates freely about a vertical axis through its center. A 15-g spider drops onto the record at a point 2 cm from the center of the disk. What is the new angular velocity of the record?

62. A merry-go-round rotates at the rate of 0.2 rev/s with an 80-kg man standing at a point 2 m from the axis of rotation. (a) What is the new angular velocity when the man walks to a point 1 m from the center? Assume the merry-go-round is a solid cylinder of mass 25 kg and radius 2 m. (b) Calculate the change in kinetic energy due to this movement. How do you account for this change in kinetic energy?

63. The merry-go-round of Problem 62 rotates at the rate of 0.2 rev/s. (a) What is its new angular velocity if a 75-kg person sits on the outer edge? (b) Calculate the change in kinetic energy due to the man sitting down.

64. The puck in Figure 8.47 has a mass of 0.12 kg. The distance of the puck from the center of rotation is originally 40.0 cm, and the puck is moving with a speed of 80.0 cm/s. The string is pulled downward 15.0 cm through the hole in the frictionless table. Determine the work done on the puck. (*Hint:* Consider the change of kinetic energy.)

Figure 8.47 Problems 64 and 86

65. A woman whose mass is 60 kg stands at the rim of a horizontal turntable having a moment of inertia of 500 kg · m² and a radius of 2 m. The system is initially at rest, and the turntable is free to rotate about a frictionless, vertical axle through its center. The woman then starts walking around the rim in a clockwise direction (looking downward) at a constant speed of 1.5 m/s relative to the Earth. (a) In what direction and with what angular speed does the turntable rotate? (b) How much work does the woman do to set the system into motion?

Additional Problems

66. In a race, a solid sphere, a solid disk, and a cylindrical shell are each set into rotation by a constant torque of 100 N · m. If all objects have a mass of 2 kg and a radius of 0.3 m, which reaches an angular speed of 2 rad/s first? Assume all objects start from rest.

67. A solid sphere of mass 2 kg and radius 0.5 m rotates about an axis through its center. (a) What is its moment of inertia? (b) At what distance, k, from this same axis could a point mass equal to the mass of the sphere be placed so that its moment of inertia is the same as that of the sphere? This distance k is called the *radius of gyration*.

68. A phonograph turntable consists of a solid circular plate 30 cm in diameter that has a mass of 3.5 kg. The turntable is rotating at $33\frac{1}{3}$ rpm when the motor is switched off. A time of 105 s is required for the turntable to come to rest. (a) What is the (constant) angular acceleration? (b) How many revolutions does the turntable make after the motor is turned off? (c) What is the effective decelerating torque?

69. A cylinder with moment of inertia I_1 rotates with angular velocity ω_0 about a vertical, frictionless axle. A second cylinder, with moment of inertia I_2 initially not rotating, drops onto the first cylinder (Fig. 8.48). Since the surfaces are rough, the two eventually reach the same angular velocity ω. (a) Calculate ω. (b) Show that energy is lost in this situation and calculate the ratio of the final to the initial kinetic energy.

Before   After

Figure 8.48   Problem 69

70. A diver springs from a board with arms and legs extended such that her moment of inertia is 1.5 kg·m². She rotates at a rate of 0.1 rev/s. She then pulls her legs up against her chest, reducing her moment of inertia to 0.9 kg·m². What is her new angular velocity?

71. A seesaw is designed such that it can rotate in a horizontal plane. Two men each of mass 80 kg face one another on the seesaw, which has a moment of inertia of 3 kg·m². The men are 1.5 m from the axis of rotation and they are rotating at a speed of 2 rad/s. They then begin to move toward one another. How far apart will they be when the speed has increased to 2.5 rad/s?

72. A sphere 0.2 m in radius and weighing 240 N rolls 6 m down a ramp inclined 37° to the horizontal. What is the angular velocity of the sphere at the bottom of the hill if it starts from rest?

73. A person bends over and lifts a 200-N weight as in Figure 8.49a, with the back in the horizontal position. The back muscle attached at a point two thirds up the spine maintains the position of the back, where the angle between the spine and this muscle is 12°. Using the mechanical model shown in Figure 8.49b and taking the weight of the upper body to be 350 N, find the tension in the back muscle and the compressional force in the spine.

(a)   (b)

Figure 8.49   Problem 73

74. A uniform picture frame weighing 10 N is supported as shown in Figure 8.50. Find the tension in the cords and the magnitude of the horizontal force at P required to hold the frame in the position shown.

Figure 8.50   Problem 74

75. A hinged uniform flagpole weighing 500 N is supported in the horizontal position by a cable, as shown

in Figure 8.51. Find the tension in the cable and the horizontal and vertical forces at the hinge.

Figure 8.51
Problems 75 and 76

76. If the maximum tension that the cable in Figure 8.51 can support without breaking is 2000 N, what is the maximum weight flagpole that can be supported?

77. One end of a uniform, 4-m-long rod of weight W is supported by a string at one end. The other end is resting against the wall where it is held by friction (see Fig. 8.52). The coefficient of static friction between the wall and the rod is $\mu_s = 0.5$. Determine the distance of the nearest point from A from which an additional weight W can be hung without the rod slipping from A.

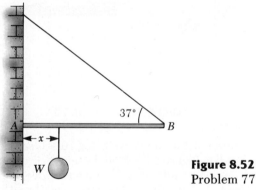

Figure 8.52
Problem 77

78. A crane of mass 3000 kg supports a load of 10,000 kg as in Figure 8.53. The crane is pivoted with a smooth pin at A and rests against a smooth support at B. Find the reaction forces at A and B.

Figure 8.53
Problem 78

79. Two cylinders start from rest at the top of a plane 6 m long inclined at an angle of 30° with the horizontal. Both cylinders have a mass of 0.5 kg and a radius of 10 cm, but one of the cylinders is solid and the other is a cylindrical shell. When the first one reaches the bottom of the incline, how far down the incline has the other one moved?

80. The system of point masses shown in Figure 8.54 is rotating at an angular velocity of 2 rev/s. The masses are connected by light flexible spokes that can be lengthened or shortened. What is the new angular velocity if the spokes are shortened to 0.5 m? (An effect similar to that illustrated in this problem occurred in the early stages of the formation of our Galaxy. As the massive cloud of dust and gas that was the source of the stars and planets contracted, an initially small rotation increased with time.)

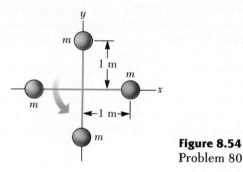

Figure 8.54
Problem 80

81. As a planet moves about the Sun in an elliptical orbit its angular momentum is a constant. Use this fact to show that the angular speed of the planet is greatest when it is closest to the Sun.

82. A spaceship of mass 5000 kg is fired vertically upward from the surface of the Earth. Use conservation of angular momentum to show that the spaceship will not be directly overhead when it has risen to its orbit height of 200 km.

83. A 12-kg mass is attached to a cord that is wrapped around a wheel of radius $r = 10$ cm (Fig. 8.55). The acceleration of the mass down the frictionless incline is measured to be 2.0 m/s². Assuming the axle of the

Figure 8.55
Problem 83

wheel to be frictionless, determine (a) the tension in the rope, (b) the moment of inertia of the wheel, and (c) the angular speed of the wheel 2 s after it begins rotating, starting from rest.

84. A 3-m length of light nylon cord is wound around a uniform cylindrical spool of radius 0.6 m and 1-kg mass. The spool is mounted on a frictionless axle and is initially at rest. The cord is pulled from the spool with a constant acceleration of 2.5 m/s². (a) How much work has been done on the spool when it reaches an angular speed of $\omega = 6$ rad/s? (b) Assuming there is enough cord on the spool, how long will it take the spool to reach an angular speed of 6 rad/s? (c) Is there enough cord on the spool to enable the spool to reach the angular speed of 6 rad/s?

85. A uniform solid sphere of radius r is placed on the inside surface of a hemispherical bowl of radius R. The sphere is released from rest at an angle θ to the vertical and rolls without slipping (Fig. 8.56). Determine the angular speed of the sphere when it reaches the bottom of the bowl.

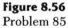

Figure 8.56
Problem 85

86. A 50-g mass is attached to a cord passing through a small hole in a frictionless, horizontal surface (Fig. 8.47). The mass is initially orbiting in a circle of radius 0.3 m with velocity 1.5 m/s. The cord is then slowly pulled from below, decreasing the radius of the circle to 0.1 m. (a) What is the velocity of the mass when the radius is 0.1 m? (b) Find the tension in the cord when $r = 0.1$ m.

Figure 8.57
Problem 87

87. A string is wrapped around a cylinder of mass 0.5 kg and radius 0.1 m. The string is pulled upward, as shown in Figure 8.57, so that the center of mass of the

cylinder does not move. (a) What force must be exerted on the string to cause this? (b) Assuming the cylinder is initially at rest, what angular velocity will it have after 0.2 s?

88. A large printing-press roller in the shape of a cylinder 0.7 m in radius and weighing 1800 N is brought uniformly from rest to an angular speed of 1400 rev/min in 150 s. What average power must be provided by the drive belt during this time interval?

89. A 50-kg child stands on the rim of a merry-go-round of radius 2 m and a mass of 150 kg, shaped as a solid cylinder. The merry-go-round is at rest when the child begins to run at a speed of 2 m/s relative to the Earth clockwise around the rim of the disk. In what direction and with what angular speed does the turntable begin to rotate?

90. A 4-kg mass is connected by a light cord to a 3-kg mass on a smooth surface (Fig. 8.58). The pulley rotates about a frictionless axle and has a moment of inertia of 0.5 kg·m² and radius 0.3 m. Assuming the cord does not slip on the pulley, find (a) the acceleration of the two masses and (b) the tensions T_1 and T_2.

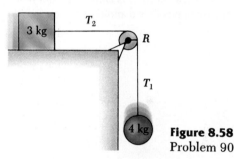

Figure 8.58
Problem 90

91. The pulley in Figure 8.59 has a moment of inertia of 5 kg·m² and radius 0.5 m. If the cord supporting the masses m_1 and m_2 does not slip and the axle is frictionless, find (a) the acceleration of the masses when

Figure 8.59
Problem 91

$m_1 = 2$ kg and $m_2 = 5$ kg. Also, find (b) the tension in the cable supporting m_1 and the tension in the cable supporting m_2 (note that they are different).

92. A small solid sphere of mass m and radius r rolls without slipping along the track shown in Figure 8.60. If it starts from rest at the top of the track, what is the minimum value of h (in terms of the radius of the loop R) such that the sphere completes the loop?

Figure 8.60
Problem 92

Figure 8.61   Problems 93 and 94

93. A uniform solid cylinder of mass M and radius R rotates on a horizontal, frictionless axle (Fig. 8.61). Two equal masses hang from light cords wrapped around the cylinder. If the system is released from rest, find (a) the tension in each cord and (b) the acceleration of each mass after the masses have descended a distance h.

94. Suppose the pulley of Problem 93 has a moment of inertia $I = 5$ kg · m^2 and radius $R = 0.5$ m. If the cord supporting the masses $m_1 = 2$ kg and $m_2 = 5$ kg does not slip and the axle is frictionless, find (a) the acceleration of the masses, and (b) the tension in each cable.

Of all structures built for various utilitarian purposes, a bridge and its structural components are the most visible. The load-carrying tasks of the principal structural components can be comprehended easily; the supporting cables of a suspension bridge are under tension induced by the weight and loads on the bridge.

The arch is another type of structure whose shape indicates that the loads are carried by compression. The arch can be visualized as an up-side-down suspension cable.

The stone arch is one of the oldest existing structures found in buildings, walls, and bridges. Other materials, such as timber, may have been used prior to stone, but nothing of these remains today most likely because of fires, warfare, and the decay processes of nature. Although stone arches were constructed prior to the Roman Empire, the Romans constructed some of the largest and most enduring stone arches.

Before the development of the arch, the principal method of spanning a space was the simple post-and-beam construction (Fig. 1a), in which a horizontal beam is supported by two columns. This type of construction was used to build the great Greek temples. The columns of these temples are closely spaced because of the limited length of available stones. Much larger spans can now be achieved using steel beams, but the spans are limited because the beams tend to sag under heavy loads.

The corbeled arch (or false arch) shown in Figure 1b is another primitive structure; it is only a slight improvement over post-and-beam construction. The stability of this false arch depends upon the horizonal projection of one stone over another and the downward weight of stones from above.

The semicircular arch (Fig. 2a) developed by the Romans was a great technological achievement in architectural design. The stability of this true (or voussoir) arch depends on the compression between its wedge-shaped stones. (That is, the stones are forced to squeeze against each other.) This results in horizontal outward forces at the springing of the arch (where it starts curving), which must be supported by the foundation (abutments) on the stone wall shown on the sides of the arch (Fig. 2a). It is common to use very heavy walls (buttresses) on either side of the arch to provide the horizontal stability. If the foundation of the arch should move, the compressive forces between the wedge-shaped stones may decrease to the extent that the arch collapses. The surfaces of the stones used in the semicircular arches constructed by the Romans were cut, or "dressed," to make a very tight joint; it is interesting to note that mortar was usually not used in these joints. The resistance to slipping between stones was provided by the compression force and the friction between the stone faces.

Another important architectural innovation was the pointed Gothic arch shown in Figure 2b. This type of structure was first used in Europe beginning in

Arch Structures

Gordon Batson

Clarkson University, Potsdam N.Y.

Post-and-beam
(a)

Corbeled (false) arch
(b)

Figure 1
Some methods of spanning space: (a) simple post-and-beam structure and (b) corbeled, or false, arch.

Figure 2
(a) The semicircular arch developed by the Romans. (b) Gothic arch with flying buttresses to provide lateral support. (Typical cross-section of a church or cathedral.) The buttresses transfer the speading forces of the arch by vertical loads to the foundation of the structure.

Semicircular arch (Roman)
(a)

Gothic Arch

Flying buttress

Flying buttress

Pointed arch (Gothic)
(b)

245

Figure 9.1
A model of a solid. The atoms (spheres) are imagined as being attached to each other by springs, which represent the elastic nature of the interatomic forces.

Natural quartz (SiO_2) crystals, one of the most common minerals on earth. Quartz crystals are used to make special lenses and prisms and in certain electronic applications. *(Courtesy of Ward's Natural Science)*

electrically charged particles: the negatively charged electrons and the positively charged ions. Such a highly ionized gas with equal amounts of positive and negative charges is called a **plasma.** The plasma state exists inside stars, for example. If we were to take a grand tour of our Universe, we would find that there is far more matter in the plasma state than in the more familiar forms of solid, liquid, and gas because there are far more stars around than any other form of celestial matter. However, in this chapter we shall ignore this plasma state and concentrate instead on the more familiar solid, liquid, and gaseous forms that make up the environment on our planet.

Everyday experience tells us that a solid has a definite volume and shape. A brick maintains its familiar shape and size day in and day out. We also know that a liquid has a definite volume but no definite shape. For example, when you fill the tank on a lawn mower, the gasoline changes its shape from that of the original container to that of the tank on the mower, but if you have a gallon of gasoline before you pour, you will have a gallon after. Finally, a gas has neither definite volume nor definite shape. These definitions help us to picture the states of matter, but they are somewhat qualitative. For example, asphalt and plastics are normally considered solids, but over long periods of time they tend to flow like liquids. Likewise, water can be a solid, liquid, or gas (or combinations of these) depending on the temperature and pressure.

All matter consists of some distribution of atoms and molecules. The atoms in a solid are held at specific positions with respect to one another by forces that are mainly electrical in origin. The atoms of a solid vibrate about these equilibrium positions because of thermal agitation. However, at low temperatures, this vibrating motion is slight and the atoms can be considered to be almost fixed. As thermal energy (heat) is added to the material, the amplitude of these vibrations increases. The vibrating motion of the atom can be viewed as that which would occur if the atom were bound in its equilibrium position by springs attached to neighboring atoms. Such a vibrating collection of atoms and imaginary springs is shown in Figure 9.1. If a solid is compressed by external forces, we can picture these external forces as compressing these tiny internal springs. When the external forces are removed, the solid tends to return to its original shape and size. For this reason, a solid is said to have elasticity.

Solids can be classified as being either crystalline or amorphous. **A crystalline solid** is one in which the atoms have an ordered, periodic structure. For example, in the sodium chloride crystal (common table salt), sodium and chlorine atoms occupy alternate corners of a cube, as in Figure 9.2a. In an **amorphous solid,** such as glass, the atoms are arranged in a disordered, or random, fashion, as in Figure 9.2b.

In any given substance, the liquid state exists at a higher temperature than the solid state. Thermal agitation is greater in the liquid state than in the solid state. As a result, the molecular forces in a liquid are not strong enough to keep the molecules in fixed positions, and they wander through the liquid in a random fashion (Fig. 9.2c). Solids and liquids have the following property in common. When one tries to compress a liquid or a solid, strong repulsive atomic forces act internally to resist the deformation.

In the gaseous state, the molecules are in constant random motion and exert only weak forces on each other. The average separation distances between the molecules of a gas are quite large compared with the size of the

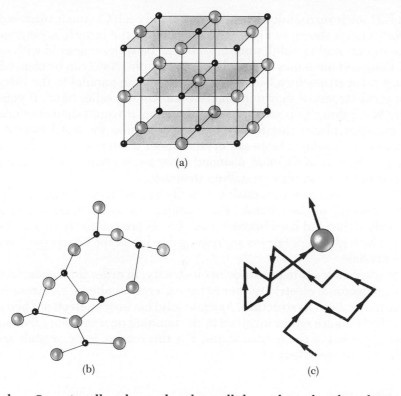

Figure 9.2
(a) The NaCl structure, with the Na$^+$ and Cl$^-$ ions at alternate corners of a cube. The smaller spheres represent Na$^+$ ions, and the larger spheres represent Cl$^-$ ions. (b) In an amorphous solid, the atoms are arranged in a random fashion. (c) Erratic motion of a molecule in a liquid.

molecules. Occasionally, the molecules collide with each other; however, most of the time they move as nearly free, noninteracting particles. We shall have more to say about the properties of gases in subsequent chapters.

9.2 Bonding in Solids

In this section we shall briefly describe how atoms or molecules combine to form solids. As more is learned about electrical forces and quantum mechanics in future chapters, we will fill in many details that must be omitted from our present discussion. For the present we shall focus on three important mechanisms by which atoms are bound together to form solids. These bonding mechanisms, in turn, will enable you to better understand certain characteristics of solids such as their melting point and their electrical properties.

Ionic Bonds

Ionic bonds are fundamentally due to an attractive electrical force between oppositely charged ions. An ion is any atom that has excess electrons or is deficient in electrons. Common table salt, NaCl, is an excellent example of a solid held together by ionic bonds. When mixed together to form the compound NaCl, a sodium atom gives up its single outermost electron to a chlorine atom. As a result, the sodium ion is left with a single positive charge, a state designated as Na$^+$, and the chlorine ion is left with a single negative charge, represented as Cl$^-$. It is essentially the electrical force of attraction between these unlike charges that creates a stable solid structure. In the solid state,

Figure 9.3
In the NaCl structure, each positive sodium ion is surrounded by six negative chlorine ions, while each chlorine ion is surrounded by six sodium ions.

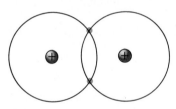

Figure 9.4
The covalent bond formed by the two electrons of the H_2 molecule.

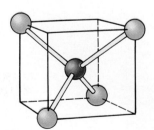

Figure 9.5
Each carbon atom in the diamond structure is covalently bonded to four other carbon atoms.

each Na^+ ion is surrounded by six Cl^- ions, and each Cl^- ion is surrounded by six Na^+ ions, as shown in Figure 9.3. Because of the orderly arrangement of atoms in a crystalline solid, some crystals can be split or cleaved with relative ease along certain planes in the solid. For example, NaCl can be cleaved (with a sharp razor struck by a hammer) along any plane parallel to the sides of its basic cubic structure shown in Figure 9.2a. On the other hand, if you try to cleave NaCl along a diagonal plane of the cube, it would shatter or crumble. The cleavage planes that form the sides of a cube for NaCl can be easily observed by viewing a table salt crystal under a microscope. Other crystals such as calcite ($CaCO_3$) and diamond have more complex cleavage planes because of their complex crystalline structure.

Certain properties of crystals can easily be understood solely on the basis of the strength of their bonds. For example, ionically bonded crystals are relatively stable and hard because ionic bonds are relatively strong. This explains why high temperatures are required to break the bonds apart and melt these crystals.

As we shall learn in our study of electricity, in order for a material to be a good conductor of electricity, it must have a large number of electrons that are free to move about its structure. An ionic solid has no free electrons because all of its electrons are either involved in the bonding process or are contained in the stable states of the normal atoms. For this reason, ionic crystals are poor conductors of electricity.

Covalent Bonds

A covalent bond between two atoms can be visualized as the sharing of electrons supplied by one or both atoms that are bonded together. In many gases, atoms are bonded to other atoms by covalent bonds to form molecules. For example, the hydrogen molecule, H_2, is formed as shown in Figure 9.4. The constituent atoms that form this molecule are two neutral hydrogen atoms, each containing one electron. When the H_2 molecule is formed, the two electrons circle both atomic cores. (This is in contrast to the free atom in which the electron circles only its own nucleus.) Thus, in essence the covalent bond is the result of electron sharing. The actual mechanism that explains the attraction of two hydrogen atoms to form an H_2 molecule involves a quantum-mechanical effect.

The covalent bond is also very strong, comparable to that of the ionic bond. Solid carbon, in the form of diamond, is an example of a crystal whose atoms are covalently bonded. Other examples of covalently bonded solids are silicon and germanium. In general, covalently bonded solids are very hard, have high melting points, and are poor electrical conductors.

In the diamond structure, each carbon atom is covalently bonded to four other carbon atoms located at four corners of a cube (to form a tetrahedron) as in Figure 9.5. The crystalline structure of diamond is shown in Figure 9.6a. Each carbon atom has four valence electrons that are used in creating the covalent bonds. Each bond consists of a pair of electrons between an atom and each of its nearest neighbors. Likewise, each nearest neighbor of a given atom contributes an electron to each bond. Since all four valence electrons of a given atom are used in establishing the four covalent bonds, each atom is strongly bound to its nearest neighbors.

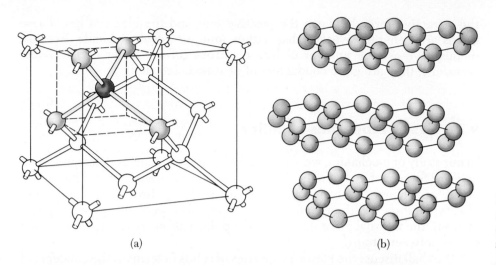

(a)   (b)

Figure 9.6
(a) The crystal structure of diamond. (b) The layered structure of graphite.

The cleavage planes of diamond are very complex as you might expect in view of its crystal structure shown in Figure 9.6a. For this reason, diamond cutters must be very careful and skilled when attempting to cut a typical diamond. One mistake could result in many small diamonds whose fate might be the grit on an industrial grinding tool. Graphite, another form of pure carbon, has a crystal structure which is considerably different than that of diamond. The carbon atoms in the graphite structure are arranged in layers, as shown in Figure 9.6b. Covalent bonding occurs between carbon atoms in a given plane; however, the bonding between atoms in adjacent layers is very weak, so the layers can readily slide past each other. This explains why graphite is very soft and slippery and can be used in powdered form as a lubricant.

Metallic Bonding

The atoms in a metal are bonded in a manner quite different from the bonding of atoms in a molecule or a covalently bonded solid, where electrons are either exchanged or shared. Instead, in a metal, some of the outer electrons are freed from individual atoms and drift through the entire solid much like a gas. In effect, the so-called "free electrons" are shared by many atoms and the bonding is distributed throughout the metal. The positive ions left behind are held in position like buoys in a sea of electrons (Fig. 9.7). The binding mechanism is

A man-made cylinder of nearly pure crystalline silicon (Si), approximately 10 in. long. Such crystals are cut into wafer form and processed to make various semiconductor devices such as integrated circuits and solar cells. (© *Charles D. Winters*)

Metal ion

Electron gas

Figure 9.7
Schematic diagram of a metal. The colored area represents the electron gas, while the spheres represent the positive metal ion cores.

the attractive force between the positive ions and the electron gas. These bonds are generally weaker than either ionic or covalent bonds. Because metals have a large number of free electrons (of the order of 10^{28} electrons/m^3), they are good conductors of heat and electricity.

9.3   Elastic Properties of Solids

In our study of mechanics, we assumed that objects remain undeformed when external forces act on them. In reality, all objects are deformable. That is, it is possible to change the shape or size of a body (or both) through the application of external forces. Although these changes are observed as large-scale deformations, the internal forces that resist the deformation are due to short-range forces between atoms.

We shall discuss the elastic properties of solids in terms of the concepts of stress and strain. **Stress** is a quantity that is related to the force causing a deformation; **strain** is a measure of the degree of deformation. It is found that, for sufficiently small stresses, the stress is proportional to the strain and the constant of proportionality depends on the material being deformed and on the nature of the deformation. We call this proportionality constant the **elastic modulus.** The elastic modulus is therefore the ratio of stress to strain:

$$\text{Elastic modulus} \equiv \frac{\text{stress}}{\text{strain}} \tag{9.1}$$

We shall consider three types of deformation and define an elastic modulus for each:

1. **Young's modulus,** which measures the resistance of a solid to a change in its length
2. **Shear modulus,** which measures the resistance to displacement of the planes of a solid sliding past each other
3. **Bulk modulus,** which measures the resistance that solids or liquids offer to changes in their volume

Figure 9.8
A long bar clamped at one end is stretched by an amount ΔL under the action of a force **F**.

Young's Modulus: Elasticity in Length

Consider a long bar of cross-sectional area A and length L_0 that is clamped at one end (Fig. 9.8). When an external force **F** is applied along the bar and perpendicular to the cross section, internal forces in the bar resist distortion ("stretching"), but the bar attains an equilibrium in which its length is greater and in which the external force is exactly balanced by internal forces. In such a situation, the bar is said to be stressed. We define the **tensile stress** as the ratio of the magnitude of the external force F to the cross-sectional area A. The SI unit of stress is the newton per square meter. This unit is also given a special name, the **pascal** (Pa).

The pascal

$$1 \text{ Pa} \equiv 1 \text{ N/m}^2 \tag{9.2}$$

Table 9.1
Typical Values for Elastic Modulus

Substance	Young's Modulus (Pa)	Shear Modulus (Pa)	Bulk Modulus (Pa)
Aluminum	7.0×10^{10}	2.5×10^{10}	7.0×10^{10}
Brass	9.1×10^{10}	3.5×10^{10}	6.1×10^{10}
Copper	11×10^{10}	4.2×10^{10}	14×10^{10}
Steel	20×10^{10}	8.4×10^{10}	16×10^{10}
Tungsten	35×10^{10}	14×10^{10}	20×10^{10}
Glass	$6.5-7.8 \times 10^{10}$	$2.6-3.2 \times 10^{10}$	$5.0-5.5 \times 10^{10}$
Quartz	5.6×10^{10}	2.6×10^{10}	2.7×10^{10}
Water	—	—	0.21×10^{10}
Mercury	—	—	2.8×10^{10}

The **tensile strain** in this case is defined as the ratio of the change in length, ΔL, to the original length, L_0, and is therefore a dimensionless quantity. Thus, we can use Equation 9.1 to define **Young's modulus**, Y:

$$Y \equiv \frac{\text{tensile stress}}{\text{tensile strain}} = \frac{F/A}{\Delta L/L_0} = \frac{FL_0}{A \, \Delta L} \qquad \textbf{(9.3)}$$

This quantity is typically used to characterize a rod or wire stressed under *either tension or compression.* Note that because the strain is a dimensionless quantity, Y has units of Pa. Typical values are given in Table 9.1. Experiments show that (a) the change in length for a fixed applied force is proportional to the original length and (b) the force necessary to produce a given strain is proportional to the cross-sectional area. Both of these observations are in accord with Equation 9.3.

It is possible to exceed the *elastic limit* of a substance by applying a sufficiently large stress (Fig. 9.9). At the *elastic limit,* the stress-strain curve departs from a straight line. A material subjected to a stress beyond this point will ordinarily not return to its original length when the external force is removed. As the stress is increased even further, the material will ultimately break.

Shear Modulus: Elasticity of Shape

Another type of deformation occurs when a body is subjected to a force **F** tangential to one of its faces while the opposite face is held in a fixed position (Fig. 9.10a). If the object is originally a rectangular block, a shear stress results in a shape whose cross-section is a parallelogram. For this situation, the stress is called a shear stress. A book pushed sideways as in Figure 9.10b is an example of an object under a shear stress. There is no change in volume under this deformation. We define the **shear stress** as F/A, the ratio of the tangential force to the area, A, of the face being sheared. The **shear strain** is defined as the ratio $\Delta x/h$, where Δx is the horizontal distance the sheared face moves and h is the height of the object. In terms of these quantities, the **shear modulus**, S, is

Figure 9.9
Stress-strain curve for an elastic solid.

Figure 9.10
(a) A shear deformation in which a rectangular block is distorted by a force applied tangent to one of its faces.
(b) A book under shear stress.

Figure 9.11
When a solid is under uniform pressure, it undergoes a change in volume but no change in shape. This cube is compressed on all sides by forces normal to its six faces.

Bulk modulus

$$S \equiv \frac{\text{shear stress}}{\text{shear strain}} = \frac{F/A}{\Delta x/h} \qquad (9.4)$$

Values of the shear modulus for some representative materials are given in Table 9.1. Note that the units of shear modulus are force per unit area (Pa).

Bulk Modulus: Volume Elasticity

Finally, we define the bulk modulus of a substance, which characterizes the response of the substance to uniform squeezing. Suppose that the external forces acting on an object are at right angles to all of its faces (Fig. 9.11) and distributed uniformly over all the faces. As we shall see later, this occurs when an object is immersed in a fluid. A body subject to this type of deformation undergoes a change in volume but no change in shape. The **volume stress,** ΔP, is defined as the ratio of the magnitude of the normal force, F, to the area, A. When dealing with fluids, we shall refer to this quantity $\Delta P = F/A$ as the **pressure.** The volume strain is equal to the change in volume, ΔV, divided by the original volume, V. Thus, from Equation 9.1 we can characterize a volume compression in terms of the **bulk modulus,** B, defined as

$$B \equiv \frac{\text{volume stress}}{\text{volume strain}} = -\frac{F/A}{\Delta V/V} = -\frac{\Delta P}{\Delta V/V} \qquad (9.5)$$

Note that a negative sign is included in this defining equation so that B will always be a positive number. This is because an increase in pressure (positive ΔP) causes a decrease in volume (negative ΔV) and vice versa.

Table 9.1 lists bulk modulus values for some materials. If you look up such values in a different source, you will often find that the reciprocal of the bulk modulus is listed. The reciprocal of the bulk modulus is called the **compressibility** of the material. You should note from Table 9.1 that both solids and liquids have a bulk modulus. However, there is no shear modulus and no Young's modulus for liquids because a liquid will not sustain a shearing stress or a tensile stress (it will flow instead).

Example 9.1   Built to Last

A vertical steel beam in a building supports a load of 7×10^4 N. If the length of the beam is 4 m and its cross-sectional area is 8×10^{-3} m^2, find the distance that the beam is compressed along its length.

Solution
Since the beam is under compression, we can use Equation 9.3. Taking Young's modulus, $Y = 20 \times 10^{10}$ Pa, for steel, we have

$$Y = \frac{F/A}{\Delta L/L_0}$$

or

$$\Delta L = \frac{FL_0}{YA} = \frac{(7 \times 10^4 \text{ N})(4 \text{ m})}{(20 \times 10^{10} \text{ Pa})(8 \times 10^{-3} \text{ m}^2)} = \boxed{1.75 \times 10^{-4} \text{ m}}$$

Example 9.2 Squeezing a Lead Sphere

A solid lead sphere of volume 0.5 m³ is dropped in the ocean to a depth where the water pressure is 2×10^7 Pa. Lead has a bulk modulus of 7.7×10^9 Pa. What is the change in volume of the sphere?

Solution
From the definition of bulk modulus, we have

$$B = -\frac{\Delta P}{\Delta V/V}$$

or

$$\Delta V = -\frac{V \Delta P}{B}$$

In this case, the change in pressure, ΔP, has the value 2×10^7 Pa. (This is large relative to atmospheric pressure, 1.01×10^5 Pa.) Taking $V = 0.5$ m³ and the bulk modulus $B = 7.7 \times 10^9$ Pa, we get

$$\Delta V = -\frac{(0.5 \text{ m}^3)(2 \times 10^7 \text{ Pa})}{7.7 \times 10^9 \text{ Pa}} = -1.3 \times 10^{-3} \text{ m}^3$$

The negative sign indicates a *decrease* in volume.

9.4 Density and Pressure

The **density** of a substance of uniform composition is defined as *its mass per unit volume.*

In symbolic form, a substance of mass M and volume V has a density, ρ (Greek rho), given by

$$\rho \equiv \frac{M}{V} \qquad\qquad \textbf{(9.6)} \qquad \text{Density}$$

The units of density are kilograms per cubic meter in the SI system and grams per cubic centimeter in the cgs system. Table 9.2 lists the densities of

Table 9.2
Density of Some Common Substances

Substance	ρ (kg/m³)[a]	Substance	ρ (kg/m³)[a]
Ice	0.917×10^3	Water	1.00×10^3
Aluminum	2.70×10^3	Glycerin	1.26×10^3
Iron	7.86×10^3	Ethyl alcohol	0.806×10^3
Copper	8.92×10^3	Benzene	0.879×10^3
Silver	10.5×10^3	Mercury	13.6×10^3
Lead	11.3×10^3	Air	1.29
Gold	19.3×10^3	Oxygen	1.43
Platinum	21.4×10^3	Hydrogen	8.99×10^{-2}
Uranium	18.7×10^3	Helium	1.79×10^{-1}

[a] All values are at standard atmospheric pressure and temperature (STP). To convert to grams per cubic centimeter, multiply by 10^{-3}.

Figure 9.12
The force of the fluid on a submerged object at any point is perpendicular to the surface of the object. The force of the fluid on the walls of the container is perpendicular to the walls at all points.

various substances. These values vary slightly with temperature because the volume of a substance depends on the temperature (as we shall see in a later chapter). Note that under normal conditions the densities of solids and liquids are about 1000 times greater than the densities of gases. This implies that the average spacing between molecules in a gas under these conditions is about ten times greater than in a solid or liquid.

The **specific gravity** of a substance is defined as the ratio of its density to the density of water at 4°C, which is 1.0×10^3 kg/m³. By definition, specific gravity is a dimensionless quantity. For example, if the specific gravity of a substance is 3, its density is $3(1.0 \times 10^3$ kg/m³) $= 3.0 \times 10^3$ kg/m³.

We have seen that fluids do not sustain shearing stresses, and thus the only stress that can exist on an object submerged in a fluid is one that tends to compress the object. The force exerted by the fluid on the object is always perpendicular to the surfaces of the object, as shown in Figure 9.12.

The pressure at a specific point in a fluid can be measured with the device pictured in Figure 9.13. The device consists of an evacuated cylinder enclosing a light piston connected to a spring. As the device is submerged in a fluid, the fluid presses down on the top of the piston and compresses the spring until the inward force of the fluid is balanced by the outward force of the spring. The fluid pressure can be measured directly if the spring is calibrated in advance. This is accomplished by applying a known force to the spring to compress it a given distance.

If F is the magnitude of the force on the piston and A is the area of the piston, then the **average pressure,** P, of the fluid at the level to which the device has been submerged is defined as the ratio of force to area:

$$P \equiv \frac{F}{A} \tag{9.7}$$

Figure 9.13
A simple device for measuring pressure in a fluid.

The technique just described represents a method for measuring average pressure. It is necessary to call this *average* pressure because the pressure in a fluid varies with depth. That is, the pressure at one point on the piston will be different from that at another point on the piston if the two points are at different depths in the fluid. However, if the measuring device is extremely small, its surface is also small and the variation in pressure over such small distances can be neglected. Thus, in this limit of an extremely small piston, one can assume that the device gives the value of the pressure at a specific point.

Since pressure is defined as force per unit area, it has units of Pa (N/m²).

Example 9.3   The Water Bed

A water bed is 2 m on a side and 30 cm deep. (a) Find its weight.

Solution
Since the density of water is 1000 kg/m³, and the bed's volume is equal to $(2 \times 2 \times 0.30)$ m² $= 1.20$ m³, the mass of the bed is

$$M = \rho V = (1000 \text{ kg/m}^3)(1.20 \text{ m}^3) = 1.20 \times 10^3 \text{ kg}$$

and its weight is

$$w = Mg = (1.20 \times 10^3 \text{ kg})(9.80 \text{ m/s}^2) = \boxed{1.18 \times 10^4 \text{ N}}$$

This is equivalent to approximately 2640 lb. In order to support such a heavy load, you would be well advised to keep your water bed in the basement or on a sturdy, well-supported floor.

(b) Find the pressure that the water bed exerts on the floor when the bed rests in its normal position. Assume that the entire lower surface of the bed makes contact with the floor.

Solution
The weight of the water bed is 1.18×10^4 N. The cross-sectional area is 4 m² when the bed is in its normal position. This gives a pressure exerted on the floor of

$$P = \frac{1.18 \times 10^4 \text{ N}}{4 \text{ m}^2} = \boxed{2.95 \times 10^3 \text{ Pa}}$$

Exercise Calculate the pressure exerted by the bed on the floor if the bed rests on its side.

Answer Since the area of its side is 0.6 m², the pressure is 1.96×10^4 Pa.

9.5 Variation of Pressure with Depth

If a fluid is at rest in a container, *all* portions of the fluid must be in static equilibrium. Furthermore, *all points at the same depth must be at the same pressure.* If this were not the case, a given portion of the fluid would not be in equilibrium. For example, consider the small block of fluid shown in Figure 9.14a. If the pressure were greater on the left side of the block of fluid than on the right, $\mathbf{F}_1$ would be greater than $\mathbf{F}_2$ and the block would accelerate and move and therefore would not be in equilibrium.

Now let us examine the portion of the fluid contained within the volume shown in the darker region in Figure 9.14b. This region has a cross-sectional area A and depth h below the surface of the water. Three external forces act on this volume of fluid: the weight Mg of the fluid, the upward force PA exerted by the fluid below the darker, shaded volume, and a downward force P_aA exerted by the atmosphere, where P_a is atmospheric pressure. Since this block of fluid is in equilibrium, these forces must add to zero and so we get

$$PA - Mg - P_aA = 0 \qquad \textbf{(9.8)}$$

From the relation $M = \rho V = \rho A h$, the weight of the fluid in the volume is found to be

$$w = Mg = \rho g A h \qquad \textbf{(9.9)}$$

When Equation 9.9 is substituted into Equation 9.8, we get

$$\boxed{P = P_a + \rho g h} \qquad \textbf{(9.10)}$$

where normal atmospheric pressure is given by $P_a = 1.01 \times 10^5$ Pa (equivalent to 14.7 lb/in.² in the English system of units). According to Equation 9.10,

the absolute pressure, P, at a depth h below the surface of a liquid open to the atmosphere is greater than atmospheric pressure by an amount $\rho g h$. Moreover, the pressure is not affected by the shape of the vessel.

(a)

(b)

Figure 9.14
(a) If the block of fluid is to be in equilibrium, the force $\mathbf{F}_1$ must balance the force $\mathbf{F}_2$.
(b) The net force on the volume of water within the darker, shaded region must be zero.

This photograph illustrates that the pressure in a liquid is the same at all points having the same elevation. Note that the shape of the vessel does not affect the pressure. *(Courtesy of CENCO)*

In view of the fact that the pressure in a given fluid depends only upon depth, any increase in pressure at the surface must be transmitted to every point in the fluid. This was first recognized by the French scientist **Blaise Pascal** (1623–1662) and is called **Pascal's principle:**

Pressure applied to an enclosed fluid is transmitted undiminished to every point of the fluid and to the walls of the containing vessel.

An important application of Pascal's principle is the hydraulic press, illustrated by Figure 9.15. A downward force $\mathbf{F}_1$ is applied to a small piston of area A_1. The pressure is transmitted through a fluid to a larger piston of area A_2. Because the pressure is the same on both sides, we see that $P = F_1/A_1 = F_2/A_2$. Therefore, the magnitude of the force $\mathbf{F}_2$ is larger than the magnitude of $\mathbf{F}_1$ by the factor A_2/A_1. For this reason, a large load, such as a car, can be supported on the large piston by a much smaller force on the smaller piston. Hydraulic brakes, car lifts, hydraulic jacks, fork lifts, and so on make use of this principle.

Example 9.4   The Car Lift

In a car lift used in a garage, compressed air exerts a force on a small piston having a radius of 5 cm. This pressure is transmitted to a second piston of radius 15 cm. What force must the compressed air exert in order to lift a car weighing 13,300 N? What air pressure will produce this force?

Solution
Because the pressure exerted by the compressed air is transmitted undiminished throughout the fluid, we have

$$F_1 = \left(\frac{A_1}{A_2}\right) F_2 = \frac{\pi (5 \times 10^{-2}\ \text{m})^2}{\pi (15 \times 10^{-2}\ \text{m})^2}\,(13{,}300\ \text{N}) = \boxed{1.48 \times 10^3\ \text{N}}$$

The air pressure that will produce this force is given by

$$P = \frac{F_1}{A_1} = \frac{1.48 \times 10^3\ \text{N}}{\pi (5 \times 10^{-2}\ \text{m})^2} = \boxed{1.88 \times 10^5\ \text{Pa}}$$

This pressure is approximately twice atmospheric pressure.

Figure 9.15
Diagram of a hydraulic press. Since the pressure is the same at the left and right sides, a small force $\mathbf{F}_1$ at the left produces a much larger force $\mathbf{F}_2$ at the right.

Example 9.5   Pressure in the Ocean

Calculate the pressure at an ocean depth of 1000 m. Assume the density of water is $1.0 \times 10^3\ \text{kg/m}^3$ and $P_a = 1.01 \times 10^5\ \text{Pa}$.

Solution

$$\begin{aligned} P &= P_a + \rho g h \\ &= 1.01 \times 10^5\ \text{Pa} + (1.0 \times 10^3\ \text{kg/m}^3)(9.80\ \text{m/s}^2)(10^3\ \text{m}) \end{aligned}$$

$$P \approx \boxed{9.9 \times 10^6\ \text{Pa}}$$

This is approximately 100 times greater than atmospheric pressure! Obviously, the design and construction of vessels that will withstand such enormous pressures are not trivial matters.

Exercise Calculate the total force exerted on the outside of a circular submarine window of diameter 30 cm at this depth.

Answer 7.0×10^5 N.

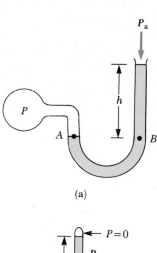

(a)

9.6 Pressure Measurements

One simple device for measuring pressure is the open-tube manometer illustrated in Figure 9.16a. One end of a U-shaped tube containing a liquid is open to the atmosphere, and the other end is connected to a system of unknown pressure P. The pressure at point B equals $P_a + \rho gh$, where ρ is the density of the fluid. The pressure at B, however, equals the pressure at A, which is also the unknown pressure P. Therefore, we conclude that

$$P = P_a + \rho gh$$

The pressure P is called the *absolute pressure*, and $P - P_a$ is called the *gauge pressure*. Thus, if the pressure P in the system is greater than atmospheric pressure, h is positive. If P is less than atmospheric pressure (a partial vacuum), h is negative.

Another instrument used to measure pressure is the *barometer* (Fig. 9.16b) invented by Evangelista Torricelli (1608–1647). A long tube closed at one end is filled with mercury and then inverted into a dish of mercury. The closed end of the tube is nearly a vacuum, and so its pressure can be taken as zero. Therefore, it follows that $P_a = \rho gh$, where ρ is the density of the mercury and h is the height of the mercury column. One atmosphere of pressure is defined to be the pressure equivalent of a column of mercury that is exactly 0.76 m in height at 0°C, with $g = 9.80665$ m/s². At this temperature, mercury has a density of 13.595×10^3 kg/m³; therefore

$$\begin{aligned} P_a = \rho gh &= (13.595 \times 10^3 \text{ kg/m}^3)(9.80665 \text{ m/s}^2)(0.7600 \text{ m}) \\ &= 1.013 \times 10^5 \text{ Pa} \end{aligned}$$

It is interesting to note that the force of the atmosphere on our bodies is extremely large, on the order of 30,000 lb! (Assuming a body area of 2000 in.²) A natural question to raise is, how can we exist under such large forces attempting to collapse our bodies? The answer is that our body cavities and tissues are permeated with fluids and gases pushing outward with this same atmospheric pressure. Consequently, our bodies are in equilibrium under the force of the atmosphere pushing in and an equal internal force pushing out.

Figure 9.16
Two devices for measuring pressure: (a) the open-tube manometer; (b) the mercury barometer.

9.7 Buoyant Forces and Archimedes' Principle

A fundamental principle dealing with objects submerged in fluids was discovered by the Greek mathematician Archimedes. **Archimedes' principle** can be stated as follows:

> Any body completely or partially submerged in a fluid is buoyed up by a force equal to the weight of the fluid displaced by the body.

Archimedes' principle

Figure 9.17
The external forces on the cube of water are its weight **w** and the buoyancy force **B**. Under equilibrium conditions, $B = W$.

Hot air balloons. These colorful balloons rise as the air inside them is heated by a large gas burner. Because warm air is less dense than cooler air, the buoyant force upward can exceed the total downward force, causing the balloons to rise. The vertical motion can also be controlled by releasing weights from the balloon. (© Merrill Peterson/FPG International)

Everyone has experienced Archimedes' principle. As an example of a common experience, recall that it is relatively easy to lift someone if you both are standing in a swimming pool, whereas lifting that same individual on dry land may be a very difficult task. Evidently, water provides partial support to any object placed in it. We say that an object placed in a fluid is buoyed up by the fluid, and we call this upward force the **buoyant force.** According to Archimedes' principle, *the magnitude of the buoyant force always equals the weight of the fluid displaced by the object.* The buoyant force acts vertically upward through what was the center of gravity of the fluid before the fluid was displaced.

Archimedes' principle can be verified in the following manner. Suppose we focus our attention on the indicated cube of water in the container of Figure 9.17. This cube of water is in equilibrium under the action of the forces on it. One of these forces is the weight of the cube of water. What cancels this downward force? Apparently, the rest of the water inside the container is buoying up the cube and holding it in equilibrium. Thus, the buoyant force, **B**, on the cube of water is exactly equal in magnitude to the weight of the water inside the cube:

$$B = w$$

Now, imagine that the cube of water is replaced by a cube of steel of the same dimensions. What is the buoyant force on the steel? The water surrounding a cube will behave in the same way whether a cube of water or a cube of steel is being buoyed up; therefore, *the buoyant force acting on the steel is the same as the buoyant force acting on a cube of water of the same dimensions.* This result applies for a totally submerged object of any shape, size, or density.

Let us show explicitly that the buoyant force is equal in magnitude to the weight of the displaced fluid. The pressure at the bottom of the cube in Figure 9.17 is greater than the pressure at the top by an amount $\rho_f g h$, where ρ_f is the density of the fluid and h is the height of the cube. Since the pressure difference, ΔP, is equal to the buoyant force per unit area, or, $\Delta P = B/A$, we see that $B = (\Delta P)(A) = (\rho_f g h)(A) = \rho_f g V$, where V is the volume of the cube. Since the mass of the water in the cube is $M = \rho_f V$, we see that

$$B = w = \rho_f V g = M g \qquad (9.11)$$

$$B = w_f \text{ displaced}$$

where w is the weight of the displaced fluid.

Note that the weight of the submerged object is $\rho_o V g$, where ρ_o is the density of the object. Therefore, if the density of the object is greater than the density of the fluid, the unsupported object will sink. If the density of the object is less than that of the fluid, the unsupported submerged object will accelerate upward and will ultimately float. When a *floating* object is in equilibrium, part of it is submerged. In this case, *the buoyant force equals the weight of the object.*

It is instructive to compare the forces on a totally submerged object with those acting on a floating object.

Case I: A Totally Submerged Object When an object is *totally* submerged in a fluid of density ρ_f, the upward buoyant force has a magnitude given by $B = \rho_f V_o g$, where V_o is the volume of the object. If the object has a density ρ_o,

Archimedes, a Greek mathematician, physicist, and engineer, was perhaps the greatest scientist of antiquity. He was the first to compute accurately the ratio of a circle's circumference to its diameter and also showed how to calculate the volume and surface area of spheres, cylinders, and other geometric shapes. He is well known for discovering the nature of the buoyant force acting on objects and was also a gifted inventor. One of his practical inventions, still in use today, is Archimedes' screw, an inclined rotating coiled tube used originally to lift water from the holds of ships. He also invented the catapult and devised systems of levers, pulleys, and weights for raising heavy loads. Such inventions were successfully used by the soldiers to defend his native city, Syracuse, during a two-year siege by the Romans.

According to legend, Archimedes was asked by King Hieron to determine whether the king's crown was made of pure gold or had been alloyed with some other metal. The task was to be performed without damaging the crown. Archimedes presumably arrived at a solution while taking a bath, noting a partial loss of weight after submerging his arms and legs in the water. As the story goes, he was so excited about his great discovery that he ran through the streets of Syracuse naked shouting, "Eureka!" which is Greek for "I have found it."

$$M_g \quad (+ \uparrow, - \downarrow)$$

its weight is equal to $w = Mg = \rho_o V_o g$, and the net force on it is given by $B - w = (\rho_f - \rho_o)V_o g$. Hence, if the density of the object is *less* than the density of the fluid as in Figure 9.18a, the unsupported object will accelerate upward. If the density of the object is *greater* than the density of the fluid as in Figure 9.18b, the unsupported object will sink.

Case II: A Floating Object Now consider an object in static equilibrium *floating* on a fluid; that is, one which is *partially* submerged. In this case, the upward buoyant force is balanced by the downward weight of the object. If V is the volume of the fluid displaced by the object (which corresponds to that volume of the object which is beneath the fluid level), then the buoyant force has a magnitude given by $B = \rho_f V g$. Since the weight of the object is given by $w = Mg = \rho_o V_o g$, and $w = B$, we see that $\rho_f V g = \rho_o V_o g$, or

$$\frac{\rho_o}{\rho_f} = \frac{V}{V_o} \tag{9.12}$$

Under normal conditions, the average density of a fish is slightly greater than the density of water. This being the case, a fish would sink if it did not have some mechanism for adjusting its density. The fish accomplishes this by internally regulating the size of its swim bladder. In this manner, fish are able to maintain neutral buoyancy as they swim to various depths.

The human brain is immersed in a fluid of density 1007 kg/m^3, which is slightly less than the average density of the brain, 1040 kg/m^3. Consequently, most of the weight of the brain is supported by the buoyant force of the fluid surrounding the brain. In some clinical procedures it is necessary to remove a portion of this fluid for diagnostic purposes. During such procedures, the nerves and blood vessels in the brain are placed under great strain, which in turn can cause extreme discomfort and pain. Great care must be exercised with such patients until the brain fluid has been restored by the body.

(a)

(b)

Figure 9.18
(a) A totally submerged object with a density *less* than the density of water will experience a net upward force. (b) If a totally submerged object has a density greater than the density of the fluid, it will sink.

Mercury has a density about 13.6 times that of water. As a result, this steel ball is able to float in a pool of mercury. *(Courtesy of Henry Leap and Jim Lehman)*

Example 9.6   A Red Tag Special on Crowns

A bargain hunter purchases a "gold" crown at a flea market. After she gets home, the bargain hunter (with some knowledge of Archimedes' principle) hangs the crown from a scale and finds its weight to be 7.84 N as shown in Figure 9.19a. She then weighs the crown while it is immersed in water of density 1000 kg/m³ as in Figure 9.19b, and finds that the scale reads 6.86 N. Is the crown made of gold?

Solution
When the crown is suspended in air as in Figure 9.19a, the scale reads the true weight, w (neglecting the buoyancy of air). However, when the crown is immersed in water as in Figure 9.19b, the buoyant force **B** reduces the scale reading to $T_2 = w - B$. Hence, the buoyant force on the crown is the difference between its weight in air and its weight in water. That is,

$$B = w - T_2 = 7.84 \text{ N} - 6.86 \text{ N} = 0.98 \text{ N}$$

Because the buoyant force is equal in magnitude to the weight of the displaced fluid, w_f, we have

$$w_f = \rho_f g V_f = 0.98 \text{ N}$$

where V_f is the volume of the displaced fluid, and ρ_f is the density of the displaced fluid (1000 kg/m³). Since the volume of the crown, V_c, is equal to the volume of the displaced fluid (for a completely submerged object), we find

$$V_c = V_f = \frac{0.98 \text{ N}}{g\rho_f} = \frac{0.98 \text{ N}}{(9.80 \text{ m/s}^2)(1000 \text{ kg/m}^3)} = 10^{-4} \text{ m}^3$$

Finally, the density of the crown is given by

$$\rho_c = \frac{m_c}{V_c} = \frac{w_c}{gV_c} = \frac{7.84 \text{ N}}{(9.80 \text{ m/s}^2)(10^{-4} \text{ m}^3)} = 8 \times 10^3 \text{ kg/m}^3$$

From Table 9.2 we see that the density of gold is 19.3×10^3 kg/m³. Thus, the crown is either hollow or is made of an alloy. This was not a good day for the bargain hunter!

Figure 9.19   (Example 9.6)
(a) When the crown is suspended in air, the scale reads the true weight, **w**. (b) When the crown is immersed in water, the buoyant force, **B**, reduces the scale reading to $T_2 = w - B$.

(a)   (b)

Figure 9.20   (Example 9.7)
A raft partially submerged in water.

Example 9.7   Floating Down the River

A raft is constructed of wood having a density of 600 kg/m³. Its surface area is 5.9 m², and its volume is 0.6 m³. When the raft is placed in fresh water of density 1000 kg/m³ as in Figure 9.20, how much of the raft is below water level?

Solution
The weight of the raft is calculated to be

$$w_r = \rho_r g V_r = (600 \text{ kg/m}^3)(9.80 \text{ m/s}^2)(0.6 \text{ m}^3) = 3530 \text{ N}$$

According to Archimedes' principle, the upward buoyant force acting on the raft equals (in magnitude) the weight of the displaced water, which, in turn, must equal the weight of the raft. Thus, we have

$$B = \rho_w g V_w = \rho_w g A h = 3530 \text{ N}$$

Since the area A and density, ρ_w, are known, we can use this result to find the depth, h, the raft sinks into the water:

$$h = \frac{B}{\rho_w g A} = \frac{3530 \text{ N}}{(1000 \text{ kg/m}^3)(9.80 \text{ m/s}^2)(5.9 \text{ m}^2)} = \boxed{0.06 \text{ m}}$$

*9.8   Surface Tension

If you look closely at a dewdrop sparkling in the morning sunlight, you will find that the drop is spherical. The drop takes this shape because of a property of liquid surfaces called **surface tension.** In order to understand the origin of surface tension, consider a molecule at point A in a container of water, as in Figure 9.21. Although nearby molecules exert forces on this molecule, the net force on it is zero because it is completely surrounded by other molecules and hence is attracted equally in all directions. The molecule at B, however, is not attracted equally in all directions. Since there are no molecules above it to exert upward forces, the molecule is pulled toward the interior of the liquid. The contraction at the surface of the liquid ceases when the inward pull exerted on the surface molecules is balanced by the outward repulsive forces that arise from collisions with molecules in the interior of the liquid. *The net effect of this pull on all the surface molecules is to make the surface of the liquid contract and consequently to make the surface area of the liquid as small as possible.* Drops of water take on a spherical shape because a sphere has the smallest surface area for a given volume.

Figure 9.21
The net force on a molecule at A is zero because such a molecule is completely surrounded by other molecules. The net force on a surface molecule at B is downward because it is not completely surrounded by other molecules.

PHYSICS IN ACTION

Surface Tension and Buoyant Forces

On the left is a close-up photograph of water (dew) drops on a blade of grass. The small drops are nearly spherical in shape, since the surface tension is large compared with the weight of the drops. The middle photograph shows droplets of mercury lying on a glass surface. Mercury is the only metal that is a liquid at room temperature. A careful inspection of the different size droplets shows that the smaller droplets are almost spherical, while the larger droplets are flattened. This variation in droplet shape shows that surface tension has more influence on the shape of the small (lighter) droplets. The photograph on the right shows a glass cylinder containing four fluids of different densities. From top to bottom, the fluids are oil (orange), water (yellow), salt water (green), and mercury (silver). The cylinder also contains four objects, which from top to bottom are a Ping-Pong ball, a piece of wood, an egg, and a steel ball. Which of these fluids has the lowest density, and which has the greatest? What can you conclude about the density of each object in the cylinder?

(© Charles Steele)

(Courtesy of Henry Leap and Jim Lehman)

(© Hans Pfletschinger, Peter Arnold, Inc.)

If you place a sewing needle very carefully on the surface of a bowl of water, you will find that the needle floats even though the density of steel is about eight times that of water. This also can be explained by surface tension. A close examination of the needle shows that it actually rests in a depression in the liquid surface, as shown in Figure 9.22. The water surface acts like an elastic membrane under tension. The weight of the needle produces a depression, thus increasing the surface area of the film. Molecular forces now act at all points along the depression in an attempt to restore the surface to its original horizontal position. The vertical components of these forces act to balance **w**, the weight of the needle.

Table 9.3
Surface Tensions for Various Liquids

Liquid	T (°C)	Surface Tension (N/m)
Ethyl alcohol	20	0.022
Mercury	20	0.465
Soapy water	20	0.025
Water	20	0.073
Water	100	0.059

The **surface tension**, γ, in a film of liquid is defined as the ratio of the magnitude of the surface tension force, F, to the length along which the force acts:

$$\gamma \equiv \frac{F}{L} \qquad (9.13)$$

The SI units of surface tension are newtons per meter, and values for a few representative materials are given in Table 9.3.

The concept of surface tension can be thought of as the energy content of the fluid at its surface per unit surface area. To see that this is reasonable, we can manipulate the units of surface tension as

$$[\gamma] = \frac{N}{m} = \frac{N \cdot m}{m^2} = \frac{J}{m^2}$$

In general, *any equilibrium configuration of an object is one in which the energy is a minimum.* Consequently, a fluid will take on a shape such that its surface area is as small as possible. For a given volume, a spherical shape is the one that has the smallest surface area. Therefore, a drop of water takes on a spherical shape, as shown in the left photograph on p. 264.

An apparatus used to measure the surface tension of liquids is shown in Figure 9.23. A circular wire with a circumference ℓ is lifted from a body of liquid. The surface film clings to the inside and outside edges of the wire. This tends to hold back the wire, causing the spring to stretch. If the spring is calibrated, one can measure the force required to overcome the surface tension of the liquid. In this case, the surface tension is given by

$$\gamma = \frac{F}{2\ell}$$

We must use 2ℓ for the length because the surface film exerts forces on the inside and outside of the ring.

The surface tension of liquids decreases with increasing temperature. This occurs because the faster moving molecules of a hot liquid are not bound together as strongly as are those in a cooler liquid. Furthermore, certain ingredients added to liquids decrease surface tension. For example, soap or detergent decreases the surface tension of water. This reduction in surface tension makes it easier for soapy water to penetrate the cracks and crevices of your clothes to clean them better than plain water. An effect similar to this

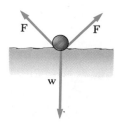

Figure 9.22
End view of a needle resting on the surface of water. The components of surface tension balance the weight force.

Figure 9.23
An apparatus for measuring the surface tension of liquids. The force on the wire ring is measured just before it breaks free of the liquid.

A razor blade floats on water because of surface tension.
(Courtesy of Henry Leap and Jim Lehman)

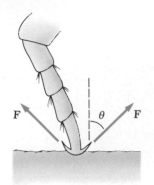

Figure 9.24   (Example 9.8) One foot of an insect resting on the surface of water.

occurs in the lungs. The surface tissue of the air sacs in the lungs contains a fluid that has a surface tension of about 0.050 N/m. A liquid with a surface tension this high would make it very difficult for the lungs to expand as one inhales. However, as the area of the lungs increases with inhalation, the body secretes into the tissue a substance that gradually reduces the surface tension of the liquid. At full expansion, the surface tension of the lung fluid can drop to as low as 0.005 N/m.

Example 9.8   Walking on Water

In this example, we shall illustrate how an insect is supported on the surface of water by surface tension. Let us assume that the insect's "foot" is spherical. When the insect steps onto the water with all six legs, a depression is formed in the water around each foot, as shown in Figure 9.24. The surface tension of the water produces upward forces on the water which tend to restore the water surface to its normally flat shape. If the insect has a mass of 2×10^{-5} kg and if the radius of each foot is 1.5×10^{-4} m, find the angle θ.

Solution

From the definition of surface tension, we can find the net force F directed tangential to the depressed part of the water surface:

$$F = \gamma L$$

The length L along which this force acts is equal to the distance around the insect's foot, $2\pi r$. (It is assumed that the insect depresses the water surface such that the radius of the depression is equal to the radius of the foot.) Thus,

$$F = \gamma \, 2\pi r$$

and the net vertical force is

$$F_v = \gamma \, 2\pi r \cos \theta$$

Since the insect has six legs, this upward force must equal one sixth the weight of the insect, assuming its weight is equally distributed on all six feet. Thus,

$$\gamma \, 2\pi r \cos \theta = \frac{1}{6} w$$

$$(1) \qquad \cos \theta = \frac{w}{12\pi r \gamma} = \frac{(2 \times 10^{-5} \text{ kg})(9.80 \text{ m/s}^2)}{12\pi \, (1.5 \times 10^{-4} \text{ m})(0.073 \text{ N/m})}$$

$$\theta = \boxed{62°}$$

Note that if the weight of the insect were great enough to make the right side of (1) greater than unity, a solution for θ would be impossible because the cosine can never be greater than unity. Under these conditions, the insect would sink.

This water strider resting on the surface of a lake is able to remain on the surface, rather than sink, because an upward surface tension force acts on each leg, which balances the weight of the insect. The interesting pattern around the legs is due to the optical effect of the polarized light in the vicinity of the distorted water surface.

*9.9   The Surface of Liquids

If you have ever closely examined the surface of water in a glass container, you may have noticed that the surface of the liquid near the walls of the glass curves upward, as shown in Figure 9.25a. However, if mercury is placed in a glass container, the mercury surface curves downward, as in Figure 9.25b.

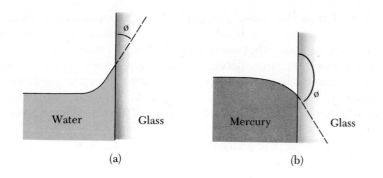

(a)   (b)

Figure 9.25
A liquid in contact with a solid surface. (a) For water, the adhesive force is greater than the cohesive force. (b) For mercury, the adhesive force is less than the cohesive force.

These surface effects can be explained by considering the forces between molecules. In particular, we must consider the forces that the molecules of the liquid exert on one another and the forces that the molecules of the glass surface exert on those of the liquid. In general terms, forces between like molecules, such as the forces between water molecules, are called *cohesive forces* and forces between unlike molecules, such as those of glass on water, are called *adhesive forces*.

Water tends to cling to the walls of the glass because the adhesive forces between the liquid molecules and the glass molecules are *greater* than the cohesive forces between the liquid molecules. In effect, the liquid molecules cling to the surface of the glass rather than fall back into the bulk of the liquid. When this condition prevails, the liquid is said to wet the glass surface. The surface of the mercury curves downward near the walls of the container because the cohesive forces between the mercury atoms are greater than the adhesive forces between mercury and glass. That is, a mercury atom near the surface is pulled more strongly toward other mercury atoms than toward the glass surface; hence mercury does not wet the glass surface.

The angle ϕ between the solid surface and a line drawn tangent to the liquid at the surface is called *the angle of contact* (Fig. 9.25a and 9.25b). Note that ϕ is less than 90° for any substance in which adhesive forces are stronger than cohesive forces and greater than 90° if cohesive forces predominate. For example, if a drop of water is placed on paraffin, the contact angle is approximately 107° (Fig. 9.26a). If certain chemicals, called wetting agents or deter-

Figure 9.26
(a) The contact angle between water and paraffin is about 107°. In this case, the cohesive force is greater than the adhesive force. (b) When a chemical called a wetting agent is added to the water, it wets the paraffin surface, and $\phi < 90°$. In this case, the adhesive force is greater than the cohesive force.

(a)   (b)

gents, are added to the water, the contact angle becomes less than 90°, as shown in Figure 9.26b. The addition of such substances to water is of value when one wants to ensure that water makes intimate contact with a surface and penetrates it. For this reason, detergents are added to water to wash clothes or dishes. On the other hand, it is often necessary to keep water from making intimate contact with a surface, as in waterproofing clothing, where a situation somewhat the reverse of that shown in Figure 9.26 is called for. The clothing is sprayed with a waterproofing agent, which changes ϕ from less than 90° to greater than 90°. Thus, the water beads up on the surface and does not easily penetrate the clothing.

*9.10   Capillarity

Capillary tubes are tubes in which the diameter of the opening is very small. In fact, the word *capillary* means "hair-like." If such a tube is inserted into a fluid for which adhesive forces dominate over cohesive forces, the liquid will rise into the tube, as shown in Figure 9.27. The rising of the liquid in the tube can be explained in terms of the shape of the surface of the liquid and in terms of the surface tension effects in the liquid. At the point of contact between liquid and solid, the upward force of surface tension is directed as shown in Figure 9.27. From Equation 9.13, the magnitude of this force is

$$F = \gamma L = \gamma \, (2\pi r)$$

We use $L = 2\pi r$ here because the liquid is in contact with the surface of the tube at all points around its circumference. The vertical component of this force due to surface tension is

$$F_v = \gamma(2\pi r)(\cos \phi) \tag{9.14}$$

In order for the liquid in the capillary tube to be in equilibrium, this upward force must be equal to the weight of the cylinder of water of height h inside the capillary tube. The weight of this water is

$$w = Mg = \rho Vg = \rho g\pi r^2 h \tag{9.15}$$

Equating F_v in Equation 9.14 to w in Equation 9.15, we have

$$\gamma(2\pi r)(\cos \phi) = \rho g\pi r^2 h$$

Thus, the height to which water is drawn into the tube is

$$h = \frac{2\gamma}{\rho g r} \cos \phi \tag{9.16}$$

If a capillary tube is inserted into a liquid in which cohesive forces dominate over adhesive forces, the level of the liquid in the capillary tube will be below the surface of the surrounding fluid, as shown in Figure 9.28. An analysis similar to that done above would show that the distance h the surface is depressed is given by Equation 9.16.

Capillary tubes are often used to draw small samples of blood from a needle prick in the skin. Capillary action must also be considered in the construction of concrete-block buildings because water seepage through cap-

Figure 9.27
A liquid rises in a narrow tube because of capillary action, a result of surface tension and adhesive forces.

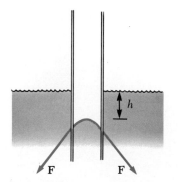

Figure 9.28
When cohesive forces between molecules of the liquid exceed adhesive forces, the liquid level in the capillary tube is below the surface of the surrounding fluid.

illary pores in the blocks or the mortar may cause damage to the inside of the building. To prevent this, the blocks are usually coated with a waterproofing agent either outside or inside the building. Water seepage through a wall is an undesirable effect of capillary action, but paper towels use capillary action in a useful manner to absorb spilled fluids.

Example 9.9 How High Does the Water Rise?

Find the height to which water would rise in a capillary tube with a radius equal to 5×10^{-5} m. Assume that the angle of contact between the water and the material of the tube is small enough to be considered zero.

Solution
The surface tension of water is 0.073 N/m. For a contact angle of $0°$, we have $\cos \phi = \cos 0° = 1$, so that Equation 9.16 gives

$$h = \frac{2\gamma}{\rho g r} = \frac{2(0.073 \text{ N/m})}{(10^3 \text{ kg/m}^3)(9.80 \text{ m/s}^2)(5 \times 10^{-5} \text{ m})} = \boxed{0.29 \text{ m}}$$

Summary

Matter is normally classified as being in one of three states: solid, liquid, or gaseous.

Three primary bonding mechanisms that hold constituent atoms together in a solid are ionic bonds, covalent bonds, and metallic bonds.

The elastic properties of a solid can be described using the concepts of stress and strain. **Stress** is a quantity related to the force producing a deformation; **strain** is a measure of the degree of deformation. Stress is proportional to strain, and the constant of proportionality is the **elastic modulus:**

$$\text{Elastic modulus} \equiv \frac{\text{stress}}{\text{strain}} \tag{9.1}$$

Three common types of deformation are: (1) the resistance of a solid to elongation or compression under a load, characterized by **Young's modulus, Y;** (2) the resistance of a solid to the planes in the solid sliding past each other, characterized by the **shear modulus, S;** (3) the resistance of a solid (or a liquid) to a volume change, characterized by the **bulk modulus, B.**

The **density, ρ,** of a substance of uniform composition is defined as its mass per unit volume and has units of kilograms per cubic meter (kg/m^3) in the SI system:

$$\rho \equiv \frac{M}{V} \tag{9.6}$$

The **pressure, P,** in a fluid is the force per unit area that the fluid exerts on an object immersed in the fluid:

$$P \equiv \frac{F}{A} \tag{9.7}$$

In the SI system, pressure has units of newtons per square meter, and 1 $N/m^2 = 1$ pascal (Pa).

The pressure in a fluid varies with depth h according to the expression

$$P = P_a + \rho g h \tag{9.10}$$

where P_a is atmospheric pressure (1.01×10^5 Pa) and ρ is the density of the fluid.

Pascal's principle states that, when pressure is applied to an enclosed fluid, the pressure is transmitted undiminished to every point of the fluid and to the walls of the containing vessel.

When an object is partially or fully submerged in a fluid, the fluid exerts an upward force on the object called the **buoyant force.** According to **Archimedes' principle,** the buoyant force is equal to the weight of the fluid displaced by the object.

The **surface tension,** γ, in a film of liquid is defined as the ratio of the surface force to the length along which the force acts:

$$\gamma \equiv \frac{F}{L} \tag{9.13}$$

The height to which a fluid will rise (or fall) in a capillary tube is given by

$$h = \frac{2\gamma}{\rho g r} \cos \phi \tag{9.16}$$

where ϕ is the **contact angle,** ρ is the density of the fluid, and r is the radius of the capillary tube.

Additional Reading

F. J. Almgren and J. E. Taylor, "The Geometry of Soap Films and Soap Bubbles," *Sci. American,* July 1976, p. 82.

E. Denton, "The Buoyancy of Marine Animals," *Sci. American,* July 1960, p. 118.

J. J. Gilman, "Fracture in Solids," *Sci. American,* February 1960, p. 94.

V. A. Greenlach, "The Rise of Water in Plants," *Sci. American,* October 1952, p. 78.

R. Hagen and L. Finger, "Crystals at High Pressures," *Sci. American,* May 1985, p. 110.

A. Holden and P. Singer, *Crystals and Crystal Growing,* Doubleday Anchor, 1960.

E. Rogers, *Physics for the Inquiring Mind,* Princeton, NJ, Princeton University Press, 1960. Chapter 6 treats surface tension in detail.

A. Stanger, "Reflections on the Electron Theory of the Chemical Bond: 1900–1925," *The Physics Teacher,* December 1981.

M. H. Zimmerman, "How Sap Moves in Trees," *Sci. American,* March 1963, p. 132.

Questions

1. What kind of deformation does a cube of Jello exhibit when it jiggles?

2. Explain the difference between bending a nail and straightening it again, and bending the nail elastically.

3. A young tree branch bends in the wind. An old tree branch breaks off under the same conditions. Discuss the differences in the physical properties of the two branches.

4. How much force does the atmosphere exert on 1 mi² of land?

5. Indian fakirs do stretch out for a nap on a bed of nails. How is this possible?

6. A woman wearing high-heeled shoes is invited into a home in which the kitchen has a newly installed vinyl floor covering. Why should the homeowner be concerned?

7. Why do your ears pop as you drive up a mountain rapidly?

8. Explain the action of a syringe when used to "draw" blood.

9. A small amount of water is placed at the bottom of a flexible container and brought to a boil. The container is then capped and placed under cool water. The can collapses inward. Why?

10. Explain the need for an extra air hole in the top of a can of fluid if the fluid is to be poured smoothly out of the can.

11. It is dangerous to swim in a pool that is being drained. Why?

12. When you drink a liquid through a straw, you reduce the pressure in your mouth and let the atmosphere move the liquid. Explain how this works. Could you use a straw to sip a drink on the Moon?

13. Two glass tumblers with different shapes and cross-sectional areas are filled to the same level with water. According to the expression $P = P_a + \rho g h$, the pressure is the same at the bottom of both tumblers. In view of this, why does one tumbler weigh more than the other?

14. A typical silo on a farm has many bands wrapped around its perimeter, as shown in the photograph at the right. Why is the spacing between successive bands smaller at the lower regions of the silo?

15. Pascal used a barometer with water as the working fluid. Why is it impractical to use water for a typical barometer?

16. During inhalation, normal pressure in the lungs is slightly less than external pressure, and the muscles controlling exhalation are relaxed. When under water, the body equalizes internal and external pressure. Discuss the condition of the muscles if a person under water is breathing through a snorkel. Would a snorkel work in deep water?

17. If an inflated beachball is placed beneath the surface of a pool of water and released, it will shoot upward out of the water. Use Archimedes' principle to explain why this occurs.

18. A swim bladder alters the average density of a fish. What effect does this have on the fish? Discuss.

19. Will an ice cube float higher in water or in an alcoholic beverage?

20. A person sitting in a boat floating in a small pond throws a heavy anchor overboard. Does the level of the pond rise, fall, or remain the same?

21. Steel is much more dense than water. How, then, do boats made of steel float?

22. A helium-filled balloon will rise until its density becomes the same as that of the air. If a sealed submarine begins to sink, will it go all the way to the bottom of the ocean or will it stop when its density becomes the same as that of the surrounding water?

23. A fish rests on the bottom of a bucket of water while the bucket is being weighed. When it begins to swim around, does the weight change?

24. Will a ship ride higher in the water of an inland lake or in the ocean? Why?

25. If 1,000,000 N were placed on the deck of the World War II battleship North Carolina, it would sink only 2.5 cm lower in the water. What is the cross-sectional area of the ship at water level?

26. Lead has a greater density than iron, and both are denser than water. Is the buoyant force on a lead object greater than, less than, or equal to the buoyant force on an iron object of the same dimensions?

Question 14 (*Photographed by Henry Leap*)

27. An ice cube is placed in a glass of water. What happens to the level of the water as the ice cube melts?

28. A simple device that is used to test the strength of antifreeze in cars consists of a small tube that has several small colored spheres inside it. When antifreeze is admitted into the tube, some of the spheres will float, giving an indication of the strength of the antifreeze. How does this device work?

29. A small piece of steel is tied to a block of wood. When the wood is placed in a tub of water with the steel on top, half of the block is submerged. If the block is inverted so that the steel is under water, will the amount of the block submerged increase, decrease, or remain the same? What happens to the water level in the tub when the block is inverted?

30. In a popular TV advertisement, it is claimed that a certain paper towel absorbs a spilled liquid more readily than its competitors. What is the mechanism responsible for the absorption?

Problems

Section 9.3   Elastic Properties of Solids

1. The heels on a pair of women's shoes have a radius of 0.5 cm at the bottom. If 30% of the weight of a woman weighing 480 N is supported by each heel, find the stress on each heel.

2. A cylindrical aluminum pillar 7.0 m high has a radius of 30.0 cm. If a 2000-kg sculpture is placed on top of the pillar, by how much is the pillar compressed?

3. A mass of 2 kg is supported by a copper wire of length 4 m and diameter 4 mm. Determine (a) the stress in the wire and (b) the elongation of the wire.

4. A 6-m steel beam 1.5 cm in radius is one element used to support an overhead walkway. The beam is designed to not stretch more than 5×10^{-5} m. What is the maximum load in newtons that it can withstand?

5. A child slides across a floor in a pair of rubber-soled shoes. The friction force acting on each foot is 20 N, the cross-sectional area of one foot is 14 cm², and the height of the soles is 5 mm. Find the horizontal distance the sheared face of the sole moves. The shear modulus of the rubber is 3×10^6 Pa.

6. The elastic limit of a material is defined as the maximum stress that can be applied to the material before it becomes permanently deformed. If the elastic limit of steel is 5×10^8 Pa, determine the minimum diameter a steel wire can have if it is to support a 70-kg circus performer without its elastic limit being exceeded.

7. The total cross-sectional area of the load-bearing calcified portion of the two forearm bones (radius and ulna) is approximately 2.5 cm². During a car crash, the forearm is used as a brace against the dashboard. The arm comes to rest from an initial speed of 80 km/h in 5.0 ms. If the arm has an effective mass of 3.0 kg and bone material can normally withstand a maximum compressional stress of 16×10^7 Pa, is the arm likely to withstand the crash?

8. A walkway is supported by pairs of hollow carbon steel pipes, each of which has an outer diameter of 1.8 cm and walls that are 2.0 mm thick. Each pair of pipes supports a section of walkway weighing 4000 N. If the elastic limit (or yield point) for carbon steel is 29×10^7 Pa, determine the number of people that can safely stand on a single section of the walkway. (Assume an average weight of 670 N/person.)

9. A uniform pressure of 5×10^4 Pa is exerted on a copper block having a volume of 10^{-3} m³. What is the change in volume of the block?

10. What increase of pressure would be required to change the volume of a sample of water by 1%?

11. During a tensile strength experiment, a small-diameter fiber is elongated by a force of 0.025 N. The fiber has an initial length of 0.2 m and stretches a distance of 4×10^{-4} m. Young's modulus for the material being tested is equal to 7.6×10^{10} Pa. Determine the diameter of the fiber.

12. If the shear stress in steel exceeds about 4.0×10^8 Pa, the steel ruptures. Determine the shear force necessary to punch a 1-cm-diameter hole in a steel plate that is 0.5 cm thick.

13. A stainless-steel orthodontic wire is applied to a tooth, as in Figure 9.29. The wire has an unstretched length of 3.1 cm and a diameter of 0.22 mm. If the wire is stretched 0.1 mm, find the magnitude and direction of the force on the tooth. Disregard the width of the tooth and assume Young's modulus for stainless steel $= 18 \times 10^{10}$ Pa.

Figure 9.29   Problem 13

14. Determine the elongation of the rod in Figure 9.30 if it is under a tension of 5.8×10^3 N.

0.200 cm

Aluminum   Copper

1.30 m   2.60 m

Figure 9.30 Problem 14

Section 9.4 Density and Pressure

15. A solid cube 5.0 cm on an edge has a mass of 1.31 kg. What is the cube made of, assuming it consists of only one element? (*Hint:* Consult Table 9.2.)
16. A king orders a gold crown having a mass of 0.5 kg. When it arrives from the metalsmith, the volume of the crown is found to be 185 cm³. Is the crown made of gold?
17. A 70-kg man in a 5-kg chair tilts back such that all the weight is balanced on two legs of the chair. Assume each leg of the chair makes contact with the floor over a circular area with radius 1 cm and find the pressure exerted by each leg on the floor.
18. A pipe contains water at a pressure 5×10^5 Pa, above atmospheric. If the only material you have available to patch a 4-mm-diameter hole in the pipe is a piece of bubble gum, how much force must the gum be able to withstand?

Section 9.5 Variation of Pressure with Depth
Section 9.6 Pressure Measurements

19. Determine the absolute pressure at the bottom of a lake that is 30 m deep.
20. Water is to be pumped to the top of the Empire State building, which is 1200 ft high. What gauge pressure is needed in the water line at the base of the building to reach this height?
21. Engineers have developed a bathyscaph that can reach ocean depths of 7 mi. (a) What is the absolute pressure at this depth? (b) If the inside of the vessel is maintained at atmospheric pressure, what is the net force on a porthole of diameter 15 cm?
22. The pressure above the fluid in Figure 9.31 is 1.1 atm. The fluid is ethanol. Determine the pressure in the bulb suspended in the fluid.

Air

40.0 cm   Air

Figure 9.31
Problem 22

23. A collapsible plastic bag (Fig. 9.32) contains a glucose solution. If the average gauge pressure in the artery is 1.33×10^4 Pa, what must be the minimum height h of the bag in order to infuse glucose into the artery? Assume the specific gravity of the solution is 1.02.

Glucose
solution

h

Figure 9.32 Problem 23

24. If the column of mercury in a barometer stands at a height of 74 cm, what is the atmospheric pressure in Pa?
25. To what height in meters would the water in a water barometer stand on a day when a mercury barometer stands at 77 cm?
26. The pressure in a sealed water pipe is 2.5×10^5 Pa above atmospheric on the first floor of an apartment building. In comparison to this value, the pressure on the top floor is only 1.5×10^5 Pa. How tall is the building?
27. A popular television show has a feature in which various objects are compressed by a 90-ton hydraulic press, which has a cross-sectional area of about 400 cm². This force is generated by oil pushing against a smaller piston of area 2 cm². What force must the oil exert on this smaller piston?
28. Figure 9.33 shows the essential parts of a hydraulic brake system. The area of the piston in the master cylinder is 6.4 cm², and the area of the piston in the brake cylinder is 1.75 cm². The coefficient of friction

Wheel
drum

Pedal

Shoe

Master
cylinder   Brake
cylinder

Figure 9.33
Problem 28

between the brake shoe and the wheel is 0.5. If the wheel has a radius of 34.0 cm, determine the frictional torque about the axle of the wheel when a force of 44.0 N is exerted on the brake pedal.

29. Piston 1 in Figure 9.34 has a diameter of 0.25 in; piston 2 has a diameter of 1.5 in. In the absence of friction determine the force **F** necessary to support the 500-lb weight on the jack.

Figure 9.34
Problem 29

30. The U-shaped tube in Figure 9.16 contains mercury. (a) What is the absolute pressure on the left if the column height $h = 20$ cm? (b) What is the gauge pressure?

31. A container is filled to a depth of 20 cm with water. On top of the water floats a 30-cm thickness of oil, with specific gravity 0.7. What is the absolute pressure at the bottom of the container?

Section 9.7   Buoyant Forces and Archimedes' Principle

32. (a) Calculate the buoyant force on a solid object made of copper and having a volume of 0.2 m³ if it is submerged in water. (b) What is the result if the object is made of steel?

33. A frog in a hemispherical pod finds that he just floats without sinking in a fluid of density 1.35 g/cm³. If the pod has a radius of 6 cm and has negligible mass, what is the mass of the frog? (See Fig. 9.35.)

Figure 9.35
Problem 33

34. An empty rubber balloon weighs 0.012 kg. The balloon is filled with helium at a density of 0.181 kg/m³. At this density the balloon has a radius of 0.5 m. If the filled balloon is fastened to a vertical line, what is the tension in the line?

35. A solid object has a weight in air of 5.0 N. When it is suspended from a spring scale and submerged in water, the scale reads 3.5 N. What is the density of the object?

36. A small boat weighing 1000 N has a surface area at water level of 3 m². It floats with only 5 cm above water level when in a fresh-water lake. How high out of the water will it ride in a salt-water lake? Assume the surface area of the boat does not change as it rises. (Salt water has a density of about 1.03×10^3 kg/m³.)

37. Careful analyses of photographs of a "sea monster" show that its volume is 52 m³. Assume that the "sea monster" is weightless when immersed in seawater (density 1030 kg/m³), and calculate its mass in metric tons (1 metric ton equals 1000 kg).

38. A ferry boat is 4 m wide and 6 m long. When a large loaded truck pulls onto the boat it sinks 4 cm lower in the water. What is the weight of the truck?

39. An object is found to have a weight in air of 300 N. It is then tied to a string connected to a balance and immersed in water. The scale now reads 265 N. Finally, the object is immersed in oil and is found to have a weight of 275 N. (a) Find the density of the object and (b) the density of the oil.

40. A rectangular air mattress has a length of 2.0 m, a width of 0.50 m, and a thickness of 0.08 m. If the mass of the material is 2.3 kg, what mass can just be supported by the mattress in water?

41. A hollow brass tube (diam. = 4.0 cm) is sealed at one end and filled with lead shot to give a total mass of 0.20 kg. When the tube is floated in pure water, what will be the depth z of the bottom of the tube?

42. A bar of soap 2 cm thick is floating on the surface of a container of water such that 1.5 cm of the soap is under water. Bath oil of specific gravity 0.6 is poured into the water and floats on top of the water. What is the depth of the oil layer when the bar of soap is just level with the liquid surface?

43. An object of volume 450 cm³ hangs from the end of a copper wire of cross-sectional area 1.5×10^{-5} m². When the object is immersed in water the length of the wire decreases by 0.035 mm. What is the length of the wire?

Section 9.8   Surface Tension

44. A vertical force of 1.61×10^{-2} N is required to lift a wire ring of radius 1.75 cm from the surface of a container of blood plasma. Calculate the surface tension of blood plasma from this information.

45. Each of the six legs of an insect on a water surface makes a depression 0.25 cm in radius with a contact angle of 45°. Calculate the mass of the insect.

46. The surface tension of ethanol is 0.0227 N/m, and the surface tension of tissue fluid is 0.050 N/m. A force of 7.13×10^{-3} N is required to lift a 5-cm diameter wire ring vertically from the surface of ethanol. What should the diameter of a ring be so that this same force would lift it from the tissue fluid?

47. A square metal sheet 5 cm on a side and of negligible thickness is attached to a balance and inserted into a container of fluid. The contact angle is found to be zero, as shown in Figure 9.36a and the balance to which the metal sheet is attached reads 0.40 N. A thin veneer of oil is then spread over the metal sheet and the contact angle becomes 180°, as shown in Figure 9.36b. The balance now reads 0.39 N. What is the surface tension of the fluid?

(a)   (b)

Figure 9.36   Problem 47

Section 9.10   Capillarity

48. A certain fluid has a density of 1080 kg/m³ and is observed to rise to a height of 2.1 cm in a 1-mm-diameter tube. The contact angle between the wall and the fluid is zero. Calculate the surface tension of the fluid.

49. Whole blood has a surface tension of 0.058 N/m and a density of 1050 kg/m³. To what height can whole blood rise in a capillary blood vessel that has a radius of 2×10^{-6} m if the contact angle is zero?

50. Use density and surface tension values from Tables 9.2 and 9.3 to calculate the height to which water will rise in a capillary of diameter 10^{-4} m. Assume a contact angle of zero and a temperature of 20°C.

51. A staining solution used in a microbiology laboratory has a surface tension of 0.088 N/m and a density 1.035 times the density of water. What must the diam-

eter of a capillary tube be so that this solution will rise to a height of 5 cm? (Assume a zero contact angle.)

52. A capillary tube 1 mm in radius is immersed in a beaker of mercury. The mercury level inside the tube is found to be 0.536 cm *below* the level of the reservoir. Use the surface tension for mercury from Table 9.3 and the density from Table 9.2 to determine the contact angle between mercury and glass.

Additional Problems

53. Bone has a Young's modulus of about 14.5×10^9 Pa. Under compression, a bone can withstand a stress of about 160×10^6 Pa before breaking. Estimate the length of your femur (thigh bone) and calculate the amount this bone can be compressed before breaking.

54. The tip of a phonograph needle has a diameter of about 2.5×10^{-4} m. The arm of the phonograph is balanced so that a force of about 0.1 N is exerted downward on the needle. Find the pressure exerted on the record.

55. A particular braking system is designed such that a piston is forced into a tube of diameter 1 cm with a force of about 80 N. This pressure is communicated to a piston 5 cm in diameter at each of the four wheels of a small car. What is the net braking force on the car?

56. Blood of density 1050 kg/m³ is to be administered to a patient. To do so the blood is raised to a height of about 1 m higher than the level of the patient's arm. How much greater is the pressure of the blood at this level than if the container were at the same level as the arm?

57. A person rides up a mountain in a lift, but his ears fail to "pop," that is, the pressure of the inner ear does not equalize with the outside atmosphere. The radius of each eardrum is 0.4 cm. On the way up, the pressure of the atmosphere drops from 1.010 bar to 0.998 bar. (1 bar = 10^5 Pa.) (a) What is the gauge pressure of the inner ear at the top of the mountain? (b) What is the net force on each eardrum?

58. A car weighing 1.2×10^4 N rests on four tires. If the gauge pressure in each tire is 200 kPa, determine the area of each tire in contact with the road surface.

59. The density of ice is 920 kg/m³ and that of seawater is 1030 kg/m³. What fraction of the total volume of an iceberg is exposed?

60. A sample of an unknown material has a weight equal to 300 N in air and 200 N when immersed in alcohol of specific gravity 0.7. (a) What is the volume of the material and (b) what is its density?

61. The distortion of the Earth's crustal plates is an example of shear on a large scale. A particular kind of crustal rock is determined to have a shear modulus equal to 1.5×10^{10} Pa. What shear stress is involved when a 10-km layer of rock is sheared through a distance of 5 m?

62. A particular metal has a weight in air of 50 N. When placed in water it has a weight of 36 N and when placed in oil it has a weight of 41 N. (a) Find the density of the metal and (b) the density of the oil.

63. A U-tube open at both ends contains water to a level of 20 cm above the bottom of the U. A 20-cm column of ethyl alcohol (density 789.9 kg/m³) is added to the right arm of the tube. What is the water level in each arm after equilibrium has been established?

64. A weather balloon has a mass of 600 kg and is designed to lift an experiment package whose mass is equal to 4000 kg. What should the volume of the balloon be after being inflated with helium (density 0.178 kg/m³) in order that the total load can be lifted? Use 1.29 kg/m³ as the density of air.

65. One method of measuring the density of a liquid is illustrated in Fig. 9.37. One side of the U-shaped tube is in the liquid being tested; the other side is in water of density ρ_w. When the air is partially removed at the upper part of the tube, show that the density of the liquid on the left is given by $\rho = (h_w/h)\rho_w$.

Figure 9.37   Problem 65

66. A circular swimming pool has a flat bottom and a 6-m diameter. It is filled with water to a depth of 1.50 m. There is 1 atm of pressure on the top surface. (a) What is the absolute pressure at the bottom of the pool? (b) Two people with a combined mass of 150 kg get into the pool and float there quietly. What is the resulting increase in the absolute pressure at the bottom of the pool?

67. A small ball of specific gravity 0.1 and volume 2 cm³ is pulled below the surface of a pool of water. Find the acceleration of the ball at the instant it is released.

68. A block of wood has a weight of 50 N when weighed in air. A sinker is attached to the block and the weight of the wood-sinker combination is 200 N when the sinker alone is immersed in water. Finally, the wood-

sinker combination is completely immersed and the weight is 140 N. Find the density of the block.

69. A high-speed lifting mechanism supports an 800-kg mass by a steel cable 25 m in length and 4 cm² in cross-sectional area. (a) Determine the elongation of the cable. (b) By what additional amount does the cable increase in length if the mass is accelerated upward at a rate of 3 m/s²? (c) What is the greatest mass that could be accelerated upward at the rate found in part (b) if the stress in the cable is not to exceed the elastic limit of the cable, 2.2×10^8 Pa?

70. A small sphere of density 0.6 times the density of water is dropped from a height of 10 m above the surface of a smooth lake. Determine the maximum depth to which the sphere will sink. Neglect any energy transferred to the water during impact and sinking.

71. In 1657 Otto von Guericke, inventor of the air pump, evacuated a sphere made of two brass hemispheres (Fig. 9.38). Two teams of eight horses each could pull the hemispheres apart only on some trials and then with the greatest difficulty. (a) Show that the force required to pull the evacuated hemispheres apart is $\pi R^2(P_a - P)$, where R is the radius of the hemispheres and P is the pressure inside the hemispheres, which is much less than atmospheric pressure P_a. (b) Determine the force if $P = 0.1\,P_a$ and $R = 0.3$ m.

Figure 9.38   Problem 71. (Photograph by Henry Leap and Jim Lehman.)

72. Oil having a density of 930 kg/m³ floats on the surface of water. A rectangular block of wood 4.0 cm high and with a density of 960 kg/m³ floats on the interface between the two liquids. The oil completely covers the block. How deep is the bottom of the block below the interface?

73. A balloon filled with helium at atmospheric pressure is designed to support a mass M (payload + empty balloon). Show that the volume of the balloon must be at least $V = M/(\rho_a - \rho_{He})$, where ρ_a is the density of air and ρ_{He} is the density of helium. (Ignore the volume of the payload.)

74. A 5.0-g olive 1.0 cm in radius is added to a martini, which is essentially a mixture of 30% ethanol and 70% water *by weight*. Disregarding the loss in volume caused by mixing, determine the apparent weight of the olive.

75. A hollow object with an average density of 900 kg/m³ floats in a pan containing 500 cm³ of water. Ethanol is added to the water until the object is just on the verge of sinking. What volume of ethanol has been added to the water? Disregard the loss of volume caused by mixing.

76. Consider a composite "raft" consisting of two square slabs each of side s, attached face to face. One slab has a density ρ_1 and thickness h_1, while the other has thickness h_2 and density ρ_2 greater than ρ_1. (a) Find the average density of the raft. (b) Assume that the average density is less than that of water, so the raft floats. The raft is placed in water with the denser slab on the bottom. Find d, the depth of the bottom surface of the raft. (c) If the raft is placed in water with the denser slab at the top, find d, the depth of the bottom surface of the raft. Comment on your answer.

77. The *spirit-in-glass thermometer*, invented in Florence, Italy around 1654, consists of a tube of liquid (the spirit) containing a number of submerged glass spheres with slightly different masses as in Figure 9.39. All spheres float at sufficiently low temperatures, but as the temperature rises, the spheres sink to the bottom of the tube one after another. The device is a crude (but rather interesting) tool for measuring temperatures. Suppose the tube is filled with ethyl alcohol whose density is 0.78945 g/cm³ at 20°C and decreases to 0.78097 g/cm³ at 30°C. (a) If one of the spheres in the tube has a radius of 1.000 cm and is in equilibrium halfway up the tube at 20°C, determine its mass. (b) When the temperature increases to 30°C, what mass must a second sphere of radius 1.000 cm have in order to be in equilibrium at the halfway point? (c) At 30°C, the first sphere has fallen to the bottom of the tube. What upward force does the bottom of the tube exert on this sphere?

Figure 9.39 Problem 77. (Photograph and idea courtesy of Prof. Ray Guenther, University of Nebraska at Omaha)

Jupiter's Great Red Spot. The Great Red Spot, a large anticyclonic storm in the southern hemisphere of Jupiter, rotates with a period of about six days and is many times larger than the Earth. *(NASA)*

$\Delta x_1 = v_1 \, \Delta t$, where v_1 is the speed of the fluid at this location. If A_1 is the cross-sectional area in this region, then the mass contained in the bottom blue region is $\Delta M_1 = \rho_1 A_1 \, \Delta x_1 = \rho_1 A_1 v_1 \, \Delta t$, where ρ is the density of the fluid. Similarly, the fluid that moves out of the upper end of the pipe in the same time Δt has a mass $\Delta M_2 = \rho_2 A_2 v_2 \, \Delta t$. However, because mass is conserved and because the flow is steady, the mass that flows into the bottom of the pipe through A_1 in a time Δt must equal the mass that flows out through A_2 in the same time. Therefore, $\Delta M_1 = \Delta M_2$, or

$$\rho_1 A_1 v_1 = \rho_2 A_2 v_2 \qquad \text{(10.1)}$$

This expression is called the **equation of continuity.**

Since ρ is constant for an incompressible fluid, Equation 10.1 reduces to

$$A_1 v_1 = A_2 v_2 \qquad \text{(10.2)}$$

That is, the product of the cross-sectional area of the pipe and the fluid speed at any point along the pipe is a constant.

Therefore, as one would expect, the speed is high where the tube is constricted and low where the tube has a larger diameter. The product Av, which has the dimensions of volume per unit time, is called the **flow rate.**

The condition $Av =$ constant is equivalent to the fact that the amount of fluid that enters one end of the tube in a given time interval equals the amount of fluid leaving the tube in the same time interval, assuming no leaks.

Example 10.1 Filling a Water Bucket

A water hose 2 cm in diameter is used to fill a 20-liter bucket. If it takes 1 min to fill the bucket, what is the speed v at which the water leaves the hose? (You should note that 1 liter = 10^3 cm³.)

Solution
The cross-sectional area of the hose is

$$A = \frac{\pi d^2}{4} = \frac{\pi(2 \text{ cm})^2}{4} = \pi \text{ cm}^2$$

The flow rate of 20 liters/min is equal to the product Av. Thus,

$$Av = 20 \frac{\text{liters}}{\text{min}} = \frac{20 \times 10^3 \text{ cm}^3}{60 \text{ s}}$$

$$v = \frac{20 \times 10^3 \text{ cm}^3}{(\pi \text{ cm}^2)(60 \text{ s})} = \boxed{106 \text{ cm/s}}$$

Exercise If the diameter of the hose is reduced to 1 cm, what will the speed of the water be as it leaves the hose, assuming the same flow rate?

Answer 424 cm/s.

Bernoulli's Equation

As a fluid moves through a pipe of varying cross section and elevation, the pressure will change along the pipe. In 1738 the Swiss physicist Daniel Bernoulli (1700–1782) derived a fundamental expression that relates pressure to fluid speed and elevation. Bernoulli's equation is not a free-standing law of physics. It is, instead, a consequence of energy conservation as applied to our ideal fluid.

In deriving Bernoulli's equation, we shall again assume that the fluid is incompressible and nonviscous and flows in a nonturbulent, steady-state manner. Consider the flow through a nonuniform pipe in a time Δt, as illustrated in Figure 10.3. The force on the lower end of the fluid is $P_1 A_1$, where P_1 is the pressure at the lower end. The work done on the lower end of the fluid by the fluid behind it is

$$W_1 = F_1 \Delta x_1 = P_1 A_1 \Delta x_1 = P_1 V$$

where V is the volume of the lower green region in Figure 10.3. In a similar manner, the work done on the fluid on the upper portion in the time Δt is

$$W_2 = -P_2 A_2 \Delta x_2 = -P_2 V$$

(Note that the volume of fluid that passes through A_1 in a time Δt equals the volume that passes through A_2 in the same time interval.) The work W_2 is negative because the force on the fluid at the top is opposite its displacement. Thus the net work done by these forces in the time Δt is

$$W = P_1 V - P_2 V$$

Part of this work goes into changing the kinetic energy of the fluid, and part goes into changing its gravitational potential energy. If m is the mass passing

Figure 10.3
A fluid flowing through a constricted pipe with streamline flow. The fluid in the section of length Δx_1 moves to the section of length Δx_2. The volumes of fluid in the two sections are equal.

Daniel Bernoulli was a Swiss physicist and mathematician who made important discoveries in hydrodynamics. Born into a family of mathematicians on February 8, 1700, he was the only member of the family to make a mark in physics. He was educated and received his doctorate in Basle, Switzerland.

Bernoulli's most famous work, *Hydrodynamica*, was published in 1738; it is both a theoretical and a practical study of equilibrium, pressure, and velocity of fluids. He showed that as the velocity of fluid flow increases, its pressure decreases. Referred to as "Bernoulli's principle," his work is used to produce a vacuum in chemical laboratories by connecting a vessel to a tube through which water is running rapidly. Bernoulli's principle is an early formulation of the idea of conservation of energy.

Bernoulli's *Hydrodynamica* also attempted the first explanation of the behavior of gases with changing pressure and temperature; this was the beginning of the kinetic theory of gases.

Daniel Bernoulli
(1700–1782)

(Courtesy of The Bettmann Archive)

through the pipe in the time interval Δt, then the change in kinetic energy of the volume of fluid is

$$\Delta KE = \frac{1}{2}\, mv_2{}^2 - \frac{1}{2}\, mv_1{}^2$$

The change in its potential energy is

$$\Delta PE = mgy_2 - mgy_1$$

We can apply the work-energy theorem in the form $W = \Delta KE + \Delta PE$ (Chapter 5) to this volume of fluid to give

$$P_1 V - P_2 V = \frac{1}{2}\, mv_2{}^2 - \frac{1}{2}\, mv_1{}^2 + mgy_2 - mgy_1$$

If we divide each term by V and recall that $\rho = m/V$, this expression reduces to

$$P_1 - P_2 = \frac{1}{2}\, \rho v_2{}^2 - \frac{1}{2}\, \rho v_1{}^2 + \rho gy_2 - \rho gy_1$$

Let us move those terms that refer to point 1 to one side of the equation and those that refer to point 2 to the other side:

$$P_1 + \frac{1}{2}\, \rho v_1{}^2 + \rho gy_1 = P_2 + \frac{1}{2}\, \rho v_2{}^2 + \rho gy_2 \qquad \textbf{(10.3)}$$

This is **Bernoulli's equation.** It is often expressed as

Bernoulli's equation

$$P + \frac{1}{2}\, \rho v^2 + \rho gy = \text{constant} \qquad \textbf{(10.4)}$$

Bernoulli's equation says that the sum of the pressure (P), the kinetic energy per unit volume ($\frac{1}{2}\rho v^2$), and potential energy per unit volume (ρgy), has the same value at all points along a streamline.

Figure 10.4
(a) The pressure P_1 is greater than the pressure P_2, since $v_1 < v_2$. This device can be used to measure the speed of fluid flow. (b) Photograph of a Venturi tube. (*Courtesy of CENCO*)

An important consequence of Bernoulli's law can be demonstrated by considering Figure 10.4a. This figure shows water flowing through a horizontal constricted pipe from a region of large cross-sectional area into a region of smaller cross-sectional area. This device, called a *Venturi tube,* can be used to measure the speed of fluid flow. Let us compare the pressure at point 1 to the pressure at point 2. Because the pipe is horizontal, $y_1 = y_2$ and Equation 10.3 applied to points 1 and 2 gives

$$P_1 + \frac{1}{2}\rho v_1^2 = P_2 + \frac{1}{2}\rho v_2^2$$

The velocity of the liquid in the constriction, v_2, must be greater than the velocity v_1 if the water is not to back up in the pipe. This means that P_2 must be less than P_1. This result is often expressed by the statement that

swiftly moving fluids exert less pressure than do slowly moving fluids.

As we shall see in the next section, this important result will enable us to understand a wide range of everyday phenomena.

Example 10.2   The Leaky Tank

A tank containing a liquid of density ρ has a small hole in its side at a distance y_1 from the bottom (Fig. 10.5). Assume the top of the tank is open to the atmosphere and determine the speed at which the fluid leaves the hole when the liquid level is a distance h above the hole.

Solution
If we assume that the cross-sectional area of the tank is large relative to that of the hole ($A_2 \gg A_1$), then the fluid level will drop very slowly and we can assume that $v_2 \approx 0$. Let us apply Bernoulli's equation to points 1 and 2. If we note that $P_1 = P_a$ at the hole, we get

$$P_a + \frac{1}{2}\rho v_1^2 + \rho g y_1 = P_a + \rho g y_2$$

$$v_1 = \sqrt{2g(y_2 - y_1)} = \boxed{\sqrt{2gh}}$$

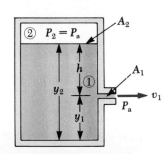

Figure 10.5   (Example 10.2) The water speed v_1, from the hole in the side of the container, is given by $v_1 = \sqrt{2gh}$.

This says that the speed of the water emerging from the hole is equal to the speed acquired by a body falling freely through a vertical distance h. This is known as **Torricelli's law.** If the height h is 50 cm, for example, the velocity of the stream is

$$v = \sqrt{2(9.80 \text{ m/s}^2)(0.5 \text{ m})} = 3.13 \text{ m/s}$$

10.3   Other Applications of Bernoulli's Equation

In this section we shall give a qualitative description of some common phenomena that can be explained, at least in part, by Bernoulli's equation.

First, consider the circulation of air around a thrown baseball. If the ball is not spinning, as in Figure 10.6a, the motion of the airstream past the ball is nearly streamline. In this figure, the ball is moving from right to left. Hence, from the point of view of the baseball, the airstream is moving from left to right. A symmetric region of turbulence occurs behind the ball as shown. When the ball is spinning counterclockwise, as in Figure 10.6b, layers of air near the surface of the ball are carried in the direction of spin because of viscosity, which will be discussed in Section 10.4. The combined effect of the steady flow of air and the air dragged along due to the spinning motion produces the streamlines shown in 10.6b and a corresponding turbulence pattern. The velocity of the air below the ball is greater than the velocity above the ball. Thus, from Bernoulli's equation, the air pressure above the ball is greater than

PHYSICS IN ACTION

Bernoulli's Equation

In the photograph on the left of a large-scale Venturi tube, air is being passed through a constricted open tube. The colored Ping-Pong balls are able to float above the jets of air from the vertical tubes. The pressure below the balls is greater than the pressure above them due to the difference in air speeds. Hence, there is an upward force on each ball which balances the weight of the ball. The different equilibrium heights of the suspended balls can be explained as follows. The pressure is lowest at points where the tube is constricted and the air velocity is largest. Thus, the pressure is least at the center where the constriction is greatest, so the center ball is at the lowest elevation. Note that the outer balls are *not* at the same height, even though the tube diameter is the same at these points. This indicates that the air has slowed down as it passes from right to left. Furthermore, some air has escaped the two vertical tubes on the right before reaching the vertical tube on the left.

In the photograph on the right a ski-jumper in "flight" bends his body forward and keeps his arms close to his body to reach long distances. In this configuration, the jumper's body is similar to the shape of an airfoil. The air

(Courtesy of Henry Leap and Jim Lehman)

(© Thomas Zimmerman/FPG International)

(a)

(b)

Figure 10.6
(a) The airstream around a nonrotating baseball moving from right to left. The streamlines represent the flow relative to the baseball. Note the symmetric region of turbulence behind the ball. (b) The airstream around a spinning baseball. The ball experiences a deflecting force partly because of the Bernoulli principle.

the air pressure below the ball, and the ball experiences a sideways deflecting force. This deflecting force is often called the *dynamic lift*. When a pitcher wishes to throw a curve ball that deflects sideways, the spin axis should be vertical (perpendicular to the page in Fig. 10.6b). On the other hand, if he wishes to throw a "sinker," the spin axis should be horizontal. Tennis balls and golf balls with spin also exhibit dynamic lift.

moving over the top of the jumper's body moves faster than the air moving beneath his body, which results in a pressure difference and a net upward force (lift). If the jumper extends his arms to maintain balance during flight, the drag force will increase, resulting in a shorter jump. The distance of the jump depends on other factors, such as the jumper's speed and his timing of the leap at the end of the ramp before becoming airborne.

As shown in the photograph below, animals such as woodchucks and gophers live in underground burrows. In order to avoid suffocation, they construct their burrows with at least two entrances. In this manner, air is able to circulate through the burrows. Some holes are often higher in elevation than others for a good reason. Wind speed tends to increase with height, which results in a pressure difference between two holes of a burrow, and a corresponding flow of air. Air will flow even if the holes are at the same elevation, since there will be some pressure difference between two holes because of slight air speed variations.

(© Sonja Bullaty, The IMAGE Bank)

Figure 10.7
A stream of air passing over a tube dipped in a liquid will cause the liquid to rise in the tube as shown. This effect is used in perfume bottles and paint sprayers.

Figure 10.8
A flow of blood through a constricted artery.

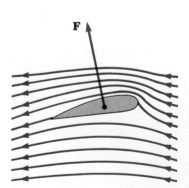

Figure 10.9
Streamline flow around an airplane wing. The pressure above is less than the pressure below, and there is a dynamic lift force upward.

A number of devices operate in the manner described in Figure 10.7. A stream of air passing over an open tube reduces the pressure above the tube. This reduction in pressure causes the liquid to rise into the air stream. The liquid is then dispersed into a fine spray of droplets. You might recognize that this so-called atomizer is used in perfume bottles and paint sprayers. The same principle is used in the carburetor of a gasoline engine. In this case, the low-pressure region in the carburetor is produced by air drawn in by the piston through the air filter. The gasoline vaporizes, mixes with the air, and enters the cylinder of the engine for combustion.

If a person has advanced arteriosclerosis, the Bernoulli effect produces a symptom called vascular flutter. In this situation, the artery is constricted as a result of an accumulation of plaque on its inner walls, as in Figure 10.8. In order to maintain a constant flow rate through such a constricted artery, the driving pressure must increase. Such an increase in pressure requires a greater demand on the heart muscle. If the blood velocity is sufficiently high in the constricted region, the artery may collapse under external pressure, causing a momentary interruption in blood flow. At this point, there is no Bernoulli effect and the vessel reopens under arterial pressure. As the blood rushes through the constricted artery, the internal pressure drops and again the artery closes. Such variations in blood flow can be heard with a stethoscope. If the plaque becomes dislodged and ends up in a smaller vessel that delivers blood to the heart, the person can suffer a heart attack.

The lift on an aircraft wing can also be explained, in part, by the Bernoulli effect. Airplane wings are designed such that the air speed above the wing is greater than the air speed below the wing. As a result, the air pressure above the wing is less than the air pressure below the wing and there is a net upward force on the wing, called the "lift." Another factor influencing the lift on a wing is shown in Figure 10.9. The wing has a slight upward tilt that causes air molecules striking the bottom to be deflected downward. The air molecules bouncing off the wing at the bottom produce an upward force on the wing and a significant lift on the aircraft. Finally, turbulence also produces an effect. If the wing is tilted too much, the flow of air across the upper surface of the wing becomes turbulent, and the pressure difference across the wing is not as great as that predicted by Bernoulli's equation. In an extreme case, this turbulence may cause the aircraft to stall.

10.4   Viscosity

It is considerably easier to pour water out of a container than to pour syrup. This is because syrup has a higher viscosity than water. In a general sense, *viscosity refers to the internal friction of a fluid.* It is very difficult for layers of a viscous fluid to slide past one another. Likewise, it is very difficult for one solid surface to slide past another if there is a highly viscous fluid, such as soft tar, between them.

To better understand the concept of viscosity, consider a liquid layer placed between two solid surfaces, as in Figure 10.10. The lower surface is fixed in position, and the top surface moves to the right with a velocity v under the action of an external force F. Because of this motion, a portion of the liquid is distorted from its original shape, *ABCD*, at one instant to the shape *AEFD* a moment later. With reference to Section 9.3, you will recognize that the liquid has undergone a constantly increasing shear strain. Previous sections in this chapter have assumed ideal fluids that have no internal friction forces between adjacent layers. Such fluids could not have their shape distorted. However, in viscous fluids, there are cohesive forces between molecules in the various layers that can lead to strains which change with time. By definition, the shear stress on the liquid is

$$\text{Shear stress} \equiv \frac{F}{A}$$

where A is the area of the top plate. Furthermore, the shear strain is defined as

$$\text{Shear strain} \equiv \frac{\Delta x}{l}$$

The velocity of the fluid changes from zero at the lower plate to **v** at the upper. Thus, in a time Δt, the fluid at the upper plate moves a distance $\Delta x = v\,\Delta t$. Therefore,

$$\frac{\text{Shear strain}}{\Delta t} = \frac{\Delta x/l}{\Delta t} = \frac{v}{l}$$

This equation states that the rate of change of the shearing strain is v/l.

The **coefficient of viscosity,** η, for the fluid is defined as the ratio of the shearing stress to the rate of change of the shear strain:

$$\eta \equiv \frac{Fl}{Av} \qquad\qquad \textbf{(10.5)}$$

Coefficient of viscosity

The SI units of viscosity are $N \cdot s/m^2$. You should note that the units of viscosity in many reference sources are often expressed in $dyne \cdot s/cm^2$, called 1 **poise** in honor of the French scientist Poiseuille. The relationship between the SI unit of viscosity and the poise is

$$1 \text{ poise} = 10^{-1} \text{ N} \cdot s/m^2 \qquad\qquad \textbf{(10.6)}$$

Small viscosities are often expressed in centipoise (cp), where $1 \text{ cp} = 10^{-2}$ poise. The coefficients of viscosity for some common substances are listed in Table 10.1.

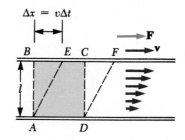

Figure 10.10
A layer of liquid between two solid surfaces in which the lower surface is fixed and the upper surface moves to the right with a velocity **v**.

Table 10.1
The Viscosities of Various Fluids

Fluid	T(°C)	Viscosity $\eta(N \cdot s/m^2)$
Water	20	1.0×10^{-3}
Water	100	0.3×10^{-3}
Whole blood	37	2.7×10^{-3}
Glycerine	20	1500×10^{-3}
10 wt motor oil	30	250×10^{-3}

*10.5   Poiseuille's Law

Figure 10.11 shows a section of a tube containing a fluid under a pressure P_1 at the left end and a pressure P_2 at the right. Because of this pressure difference, the fluid will flow through the tube. The rate of flow (volume per unit time) depends on the pressure difference $(P_1 - P_2)$, the dimensions of the tube, and the viscosity of the fluid. The relationship between these quantities was derived by a French scientist, J. L. Poiseuille (1799–1869), who assumed that the flow was streamline. His result, known as **Poiseuille's law**, is

$$\text{Rate of flow} = \frac{\Delta V}{\Delta t} = \frac{(P_1 - P_2)(\pi R^4)}{8L\eta} \qquad (10.7)$$

where R is the radius of the tube, L is its length, and η is the coefficient of viscosity. We shall not attempt to derive this equation here because the methods of integral calculus are required. However, you should note that the equation does agree with common sense. That is, it is reasonable that the rate of flow should increase if the pressure difference across the tube or the tube radius increases. Likewise, the flow rate should decrease if the viscosity of the fluid or the length of the tube increases. Thus, the presence of R and the pressure difference in the numerator of Equation 10.7 and of L and η in the denominator makes sense.

From Poiseuille's law, we see that in order to maintain a constant flow rate, the pressure difference across the tube has to increase if the viscosity of the fluid increases. This is important when one considers the flow of blood through the circulatory system. The viscosity of blood increases as the number of red blood cells rises. Thus, blood with a high concentration of red blood cells requires greater pumping pressure from the heart to keep it circulating than does blood of lower red blood cell concentration.

Note that the flow rate varies as the radius of the tube raised to the fourth power. Consequently, if a constriction occurs in a vein or artery, the heart will have to work considerably harder in order to produce a higher pressure drop and hence to maintain the required flow rate.

Figure 10.11
Velocity profile of a fluid flowing through a uniform pipe of circular cross-section. The rate of flow is given by Poiseuille's law. Note that the fluid velocity is greatest at the middle of the pipe.

Example 10.3   A Blood Transfusion

A patient receives a blood transfusion through a needle of radius 0.2 mm and length 2 cm. The density of blood is 1050 kg/m³. The bottle supplying the blood is 0.5 m above the patient's arm. What is the rate of flow through the needle?

Solution
The pressure differential between the level of the blood and the patient's arm is

$$P_1 - P_2 = \rho g h = (1050 \text{ kg/m}^3)(9.80 \text{ m/s}^2)(0.5 \text{ m}) = 5.15 \times 10^3 \text{ Pa}$$

Thus, the rate of flow, from Poiseuille's law, is

$$\frac{\Delta V}{\Delta t} = \frac{(P_1 - P_2)(\pi R^4)}{8L\eta} = \frac{(5.15 \times 10^3 \text{ Pa})(\pi)(2 \times 10^{-4} \text{ m})^4}{8(2 \times 10^{-2} \text{ m})(2.7 \times 10^{-3} \text{ N} \cdot \text{s/m}^2)} = \boxed{5.98 \times 10^{-8} \text{ m}^3/\text{s}}$$

Exercise   How long will it take to inject 1 pint (500 cm³) of blood into the patient?

Answer   139 min.

*Reynolds Number

As we mentioned earlier, at sufficiently high velocities, fluid flow changes from simple streamline flow to turbulent flow, that is, flow characterized by a highly irregular motion of the fluid. Experimentally it is found that the onset of turbulence in a tube is determined by a dimensionless factor called the **Reynolds number,** given by

$$RN = \frac{\rho v d}{\eta}$$

(10.8)   Reynolds number

where ρ is the density of the fluid, v is the average velocity of the fluid along the direction of flow, d is the diameter of the tube, and η is the viscosity of the fluid. If RN is below about 2000, the flow of fluid through a tube is streamline; turbulence occurs if RN is above 3000. In the region between 2000 and 3000, the flow is unstable, meaning that the fluid can move in streamline flow but any small disturbance will cause its motion to change to turbulent flow.

Example 10.4   Turbulent Flow of Blood

Determine the velocity at which blood flowing through an artery of diameter 0.2 cm would become turbulent. Assume that the density of blood is 1.05×10^3 kg/m^3 and that its viscosity is 2.7×10^{-3} N·s/m^2.

Solution
At the onset of turbulence, the Reynolds number is 3000. Thus, the velocity of the blood would have to be

$$v = \frac{\eta(RN)}{\rho d} = \frac{(2.7 \times 10^{-3} \text{ N·s/m}^2)(3000)}{(1.05 \times 10^3 \text{ kg/m}^3)(0.2 \times 10^{-2} \text{ m})} = \boxed{3.86 \text{ m/s}}$$

*10.6   Transport Phenomena

When a fluid flows through a tube, the basic mechanism that produces the flow is a difference in pressure across the ends of the tube. This pressure difference is responsible for the transport of a mass of fluid from one location to another. The fluid may also move from place to place because of a second mechanism, one that depends on a concentration difference between two points in the fluid, as opposed to a pressure difference. When the concentration (the number of molecules per unit volume) is higher at one location than at another, molecules will flow from the point where the concentration is high to the point where it is lower. The two fundamental processes involved in fluid transport resulting from concentration differences are called *diffusion* and *osmosis*. This section and the Special Topic section that follows examine the nature and importance of these processes.

*Diffusion

You can imagine what happens when someone wearing a strong shaving lotion or perfume strolls into a crowded room. All eyes turn to seek out the source of the delightful smell. The aroma spreads through the room by a process called diffusion.

In a **diffusion** process, molecules move from a region where their concentration is high to a region where their concentration is lower.

That is, the molecules of the lotion or perfume move from the source (near the person's face), where there are many molecules per unit volume, throughout the room, to regions where the concentration of these molecules is lower. Although the example used here is one of diffusion in air, the process also occurs in liquids and, to a lesser extent, in solids. For example, if a drop of food coloring is placed in a glass of water, the coloring soon spreads throughout the liquid by diffusion. In either case, diffusion ceases when there is a uniform concentration at all locations in the fluid.

To understand why diffusion occurs, consider Figure 10.12, which represents a container in which a high concentration of molecules has been introduced into the left side. For example, this could be accomplished by releasing a few drops of perfume into the left side of the container. The dashed line in Figure 10.12 represents an imaginary barrier separating the region of high concentration from the region of lower concentration. Because the molecules are moving with high speeds in random directions, many of them will cross the imaginary barrier moving from left to right. Very few molecules of perfume will pass through this area moving from right to left simply because there are very few of them on the right side of the container at any instant. Thus, there will always be a *net* movement from the region where there are many molecules to the region where there are fewer molecules. For this reason, the concentration on the left side of the container will decrease in time and that on the right side will increase. There will be no *net* movement across the cross-sectional area once a concentration equilibrium has been reached. That is, when the concentration is the same on both sides, the number of molecules diffusing from right to left in a given time interval will equal the number moving from left to right in the same time interval.

The basic equation for diffusion is **Fick's law,** which in equation form is

Figure 10.12
When the concentration of gas molecules on the left side of the container exceeds the concentration on the right side, there will be a net motion (diffusion) of molecules from left to right.

Fick's law

$$\text{Diffusion rate} = \frac{\text{mass}}{\text{time}} = \frac{\Delta M}{\Delta t} = DA\left(\frac{C_2 - C_1}{L}\right) \tag{10.9}$$

where D is a constant of proportionality. The left side of this equation is called the diffusion rate and is a measure of the mass being transported per unit time. This equation says that

the rate of diffusion is proportional to the cross-sectional area A and to the change in concentration per unit distance, $(C_2 - C_1)/L$, which is called the concentration gradient.

Table 10.2
Diffusion Coefficients for Various Substances at 20°C

Substance	D (m²/s)
Oxygen through air	6.4×10^{-5}
Oxygen through tissue	1×10^{-11}
Oxygen through water	1×10^{-9}
Sucrose through water	5×10^{-10}
Hemoglobin through water	76×10^{-11}

The concentrations C_1 and C_2 are measured in kilograms per cubic meter. The proportionality constant D is called the **diffusion coefficient** and has units of square meters per second. Table 10.2 lists diffusion coefficients for a few substances.

▶SPECIAL TOPIC
The Size of Cells and Osmosis

Diffusion through cell membranes is extremely vital in carrying oxygen to the cells of the body and in removing carbon dioxide and other waste products from them. Oxygen is required by the cells for those metabolic processes in which substances are either synthesized or broken down. In such metabolic processes, the cell uses up oxygen and produces carbon dioxide as a by-product. A fresh supply of oxygen diffuses from the blood, where its concentration is high, into the cell, where its concentration is low. Likewise, carbon dioxide diffuses from the cell into the blood, where it is in lower concentration. Water, ions, and other nutrients also pass into and out of cells by diffusion.

A common characteristic of cells in all plants and animals is their extremely small size. The adult human body contains literally trillions of cells. In order to understand why cells are so small, we must consider the relationship between the surface area of an object and its volume.

Let us consider a cube 2 cm on a side. The area of one of its faces is 2 cm $\times$ 2 cm = 4 cm^2, and because a cube has six sides, the total surface area is 24 cm^2. Its volume is 2 cm $\times$ 2 cm $\times$ 2 cm = 8 cm^3. Hence, the ratio of surface area to volume is 24/8 = 3. Now consider a larger cube, one measuring 3 cm on a side. Repeating the calculations gives us a surface area of 54 cm^2 and a volume of 27 cm^3. In this case, the ratio of surface area to volume is 54/27 = 2. Thus, we see that as the size of an object decreases, the ratio of its surface area to its volume increases. This, of course, says that a small cell has a larger surface-area-to-volume ratio than a large cell. But how does this pertain to the operation of a cell?

A cell can function properly only if it can (a) rapidly receive vital substances such as oxygen and (b) rapidly eliminate waste products. If such substances are to readily move into and out of cells, the cells should have a large surface area. However, if the volume of the cell is too large, it could take a considerable period of time for the nutrients to diffuse into the interior of the cell where they are needed. Under optimum conditions, the surface area of the cell should be large enough so that the exposed membrane area can exchange materials effectively while at the same time the volume should be small enough so that materials can reach or leave particular locations rapidly. To reach these optimum conditions, a small cell with its high surface-area-to-volume ratio is necessary.

Osmosis

As we have seen, the movement of material through cell membranes is necessary for the efficient functioning of cells. The diffusion of material through a membrane is partially determined by the size of the pores (holes) in the membrane wall. That is, small molecules, such as water, may pass through the pores easily while larger molecules, such as sugar, may pass through only with difficulty or not at all. A membrane that allows passage of some molecules but not others is called a selectively permeable membrane.

Osmosis is defined as the movement of water from a region where its concentration is high, across a selectively permeable membrane, into a region where its concentration is lower.

Figure 10.13
(a) The structure of a plant cell. (b) As water accumulates in its interior, the cell expands under turgor pressure.
(c) When water in the cell's interior is depleted, the walls of the cell collapse.

As in the case of diffusion, osmosis continues until the concentrations on the two sides of the membrane are equal. Osmosis is often described simply as the diffusion of water across a membrane.

To understand the effect of osmosis on living cells, let us consider a particular cell in the body that contains a sugar concentration of 1%. (That is, 1 g of sugar is dissolved in enough water to make 100 ml of solution.) Now assume that this cell is immersed in a 5% sugar solution (5 g of sugar dissolved in enough water to make 100 ml). In such a situation, water would diffuse from inside the cell, where its concentration is higher, across the cell wall membrane, to the outside solution, where the concentration of water is lower. This loss of water from the cell would cause it to shrink and perhaps become damaged through dehydration. If the concentrations were reversed, water would diffuse into the cell, causing it to swell and perhaps burst. It should be obvious from this description that normal osmotic relationships must be maintained in the body. If solutions are introduced into the body intravenously, care must be taken to ensure that these solutions do not disturb the osmotic balance of the body because such a disturbance could lead to cell damage. For example, if 9% saline solution surrounds a red blood cell, the cell will shrink. On the other hand, if the saline solution is about 1%, the cell will eventually burst.

Under normal circumstances, the cells of our bodies are in an environment such that there is no net movement of water into or out of them. However, certain one-celled organisms, such as protozoa, do not enjoy this osmotic equilibrium. These organisms usually live in fresh water, which obviously has a higher concentration of water than the solution inside the cell. To prevent an inflow of water to the point of bursting, these organisms possess an organelle (a tiny organ) that acts as a pump and continually forces water out of the cell.

Most plant cells are contained within a rigid wall, as shown in Figure 10.13a. If water accumulates in the cell, it expands (Fig. 10.13b) and exerts a pressure, called **turgor pressure,** against the rigid wall. The rigidity of the wall prevents the cell from bursting. If water within the cell is depleted, the rigid wall collapses inward slightly, as in Figure 10.13c. This causes the plant to wilt. ◀

*10.7   Motion Through a Viscous Medium

When an object falls through air, its motion is impeded by the force of air resistance. In general, this force is dependent on the shape of the falling object and on its velocity. This viscous drag acts on all falling objects, but the exact details of the motion can be calculated only for a few cases in which the object has a simple shape, such as a sphere. In this section, we shall examine the motion of a tiny spherical object falling slowly through a viscous medium.

In 1845 a scientist named George Stokes found that the resistive force on a very small spherical object of radius r falling slowly through a fluid of viscosity η is given by

Stokes's law

$$\mathbf{F_r} = 6\pi\eta r v \tag{10.10}$$

This equation, called **Stokes's law,** has many important applications. For example, it describes the sedimentation of particulate matter in blood samples. It was used by Robert Millikan (1886–1953) to calculate the radius of charged oil droplets falling through air. From this, Millikan was ultimately able to determine the smallest known unit of electric charge. Millikan was awarded the Nobel prize in 1923 for this pioneering work on elemental charge.

As a sphere falls through a viscous medium, three forces act on it, as shown in Figure 10.14: F_r is the force of frictional resistance, **B** is the buoyant force of the fluid, and **w** is the weight of the sphere, whose magnitude is given by

$$w = \rho g V = \rho g \left(\frac{4}{3}\, \pi r^3 \right)$$

where ρ is the density of the sphere and $\frac{4}{3}\pi r^3$ is its volume. According to Archimedes' principle, the magnitude of the buoyant force is equal to the weight of the fluid displaced by the sphere:

$$B = \rho_f g V = \rho_f g \left(\frac{4}{3}\, \pi r^3 \right)$$

where ρ_f is the density of the fluid.

At the instant the sphere begins to fall, the force of frictional resistance is zero because the velocity of the sphere is zero. As it accelerates, the velocity increases and so does F_r. Finally, at a velocity called the **terminal velocity** v_t, *the resultant force goes to zero.* This occurs when the net upward force balances the downward weight force. Hence, the sphere reaches terminal velocity when

$$F_r + B = w$$

or

$$6\pi \eta r v_t + \rho_f g \left(\frac{4}{3}\, \pi r^3 \right) = \rho g \left(\frac{4}{3}\, \pi r^3 \right)$$

When this is solved for v_t, we get

$$v_t = \frac{2r^2 g}{9\eta}\, (\rho - \rho_f) \tag{10.11}$$

Example 10.5 A Falling Pearl

A pearl of density 2×10^3 kg/m³ and radius 2 mm falls through a liquid shampoo of density 1.4×10^3 km/m³ and viscosity 500×10^{-3} N·s/m². Find the terminal speed of the pearl.

Solution
Substituting the values gives into Equation 10.11, we have

$$v_t = \frac{2r^2 g}{9\eta}\, (\rho - \rho_f)$$

$$= \frac{2(2 \times 10^{-3}\ \text{m})^2 (9.80\ \text{m/s}^2)}{9(500 \times 10^{-3}\ \text{N·s/m}^2)}\, (2 \times 10^3\ \text{kg/m}^3 - 1.4 \times 10^3\ \text{kg/m}^3)$$

$$= \ 1.05 \times 10^{-2}\ \text{m/s}$$

Figure 10.14
A sphere falling through a viscous medium. The forces acting on the sphere are the resistive frictional force F_r, the buoyant force **B**, and the weight of the sphere **w**.

By spreading their arms and legs out from the sides of their bodies and by keeping the plane of their bodies parallel to the ground, skydivers experience maximum air drag resulting in a terminal speed of about 60 m/s. *(U.S. Air Force Academy photo by SSgt. West C. Jacobs, 94ATS)*

Sedimentation and Centrifugation

If an object is not spherical, we can still use the basic approach just described to determine its terminal velocity. The only difference will be that we shall not be able to use Stokes's law for the resistive force. Instead, let us assume that the resistive force has a magnitude given by $F_r = kv$, where k is a coefficient of frictional resistance that must be determined experimentally. As we discussed above, the object reaches its terminal velocity when the weight downward is balanced by the net upward force, or

$$w = B + F_r \tag{10.12}$$

where B is the buoyant force, given by $B = \rho_f g V$.

We can use the fact that the volume, V, of the displaced fluid is related to the density of the falling object, ρ, by $V = m/\rho$. Hence, we can express the buoyant force as

$$B = \frac{\rho_f}{\rho} mg$$

Let us substitute this expression for B and $F_r = kv_t$ into Equation 10.12 (terminal velocity condition):

$$mg = \frac{\rho_f}{\rho} mg + kv_t$$

or

$$v_t = \frac{mg}{k}\left(1 - \frac{\rho_f}{\rho}\right) \tag{10.13}$$

Figure 10.15
Simplified diagram of a centrifuge (top view).

The terminal velocity for particles in biological samples is usually quite small. For example, the terminal velocity for blood cells falling through plasma is about 5 cm/h in the gravitational field of the Earth. The terminal velocities for the molecules that make up a cell are many orders of magnitude smaller than this because of their much smaller mass. The velocity at which materials fall through a fluid is called the *sedimentation rate*. This number is often important in clinical analysis.

It is often desired to increase the sedimentation rate in a fluid. A common method used to accomplish this is to increase the effective acceleration g that appears in Equation 10.13. A fluid containing various biological molecules is placed in a centrifuge and whirled at very high angular velocities (Fig. 10.15). Under these conditions, the particles experience a large radial acceleration, $a_c = v^2/r = \omega^2 r$, which is much greater than the acceleration due to gravity, and so we can replace g in Equation 10.13 by $\omega^2 r$:

$$v_t = \frac{m\omega^2 r}{k}\left(1 - \frac{\rho_f}{\rho}\right) \tag{10.14}$$

This equation indicates that those particles having the greatest mass will have the largest terminal velocity. Therefore, the most massive particles will settle out on the bottom of a test tube first.

Example 10.6   The Spinning Test Tube

A centrifuge rotates at 50,000 rev/min, which corresponds to an angular frequency of 5240 rad/s (a typical speed). A test tube placed in this device has its top 5 cm from the axis of rotation and its bottom 13 cm from this axis. Find the effective value of g at the midpoint of the test tube, which corresponds to a distance 9 cm from the axis of rotation.

Solution

The acceleration experienced by the particles of the tube at a distance $r = 9$ cm from the axis of rotation is given by

$$a_c = \omega^2 r = \left(5240\,\frac{\text{rad}}{\text{s}}\right)^2 (9 \times 10^{-2}\,\text{m}) = \boxed{2.47 \times 10^6\,\text{m/s}^2}$$

Exercise If the mass of the contents of the test tube is 15 g, find the centripetal force that the bottom of the tube must exert on the contents of the tube. Assume a centripetal acceleration equal to that found at the midpoint of the tube.

Answer 3.71×10^4 N, or about 8000 lb! (Because of such large forces, the base of the tube in a centrifuge must be rigidly supported to keep the glass from shattering.)

Summary

Various aspects of a fluid in motion can be understood by assuming that the fluid is nonviscous and incompressible and that its motion is in a steady state with no turbulence.

Under these assumptions, two important results regarding fluid flow through a pipe of nonuniform size can be obtained:

1. The flow rate through the pipe is a constant, which is equivalent to stating that the product of the cross-sectional area, A, and the speed, v, at any point is constant:

$$A_1 v_1 = A_2 v_2 \tag{10.2}$$

   This relation is referred to as the **equation of continuity.**
2. The sum of the pressure, the kinetic energy per unit volume, and the potential energy per unit volume has the same value at all points along a streamline:

$$P + \frac{1}{2}\rho v^2 + \rho g y = \text{constant} \tag{10.4}$$

This is known as **Bernoulli's equation.**

Viscosity is a term used to describe the internal friction of a fluid. The **coefficient of viscosity,** η, for a fluid is defined as the ratio of the shearing stress to the rate of change of the shearing strain:

$$\eta \equiv \frac{Fl}{Av} \qquad \text{(10.5)}$$

The SI units of viscosity are $N \cdot s/m^2$.

The rate of flow of a fluid through a tube is given by **Poiseuille's law:**

$$\text{Rate of flow} = \frac{\Delta V}{\Delta t} = \frac{(P_1 - P_2)(\pi R^4)}{8L\eta} \qquad \text{(10.7)}$$

The onset of turbulence in a fluid is determined by a factor called the **Reynolds number,** given by

$$RN = \frac{\rho vd}{\eta} \qquad \text{(10.8)}$$

The movement of atoms or molecules from a region where their concentration is high to a region where their concentration is lower is called **diffusion.** The diffusion rate, which is a measure of the mass being transported per unit time, is given by

$$\text{Diffusion rate} = \frac{\text{mass}}{\text{time}} = DA\left(\frac{C_2 - C_1}{L}\right) \qquad \text{(10.9)}$$

Osmosis is the movement of water from a region where its concentration is high, across a selectively permeable membrane, into a region where its concentration is lower.

Additional Reading

M. Greenwood, C. Hanna, and Rev. J. Melton, "Air Resistance Acting on a Sphere," *The Physics Teacher*, March 1986, p. 153.

D. C. Hazen and R. F. Lehnert, "Low-Speed Flight," *Sci. American*, April 1956, p. 46.

F. Hess, "The Aerodynamics of Boomerangs," *Sci. American*, November 1968, p. 124.

J. A. Lock, "The Physics of Air Resistance," *The Physics Teacher*, March 1982, p. 158.

N. Smith, "Bernoulli and Newton in Fluid Mechanics," *Physics Teacher, 10*, 1972, p. 451.

Questions

1. The water supply for a city is often provided from reservoirs built on high ground. Water flows from the reservoir, through pipes, and into your home when you turn the tap on your faucet. Why is the water flow more rapid out of a faucet on the first floor of a building than in an apartment on a higher floor?

2. Smoke rises in a chimney faster when a breeze is blowing. Use Bernoulli's principle to explain this phenomenon.

3. Why do many trailer trucks use wind deflectors on the top of their cabs? (See photograph.) How do such devices reduce fuel consumption?

The high cost of fuel has prompted many truck owners to install wind deflectors on their cabs to reduce air drag. *(Photo by Henry Leap.)*

4. Consider the cross-section of the wing on an airplane. The wing is designed such that the air travels faster over the top than under the bottom. Explain why there is a net upward force (lift) on the wing due to the Bernoulli effect.

5. When a fast-moving train passes a train at rest, the two tend to be drawn together. How does the Bernoulli effect explain this phenomenon?

6. Prairie dogs ventilate their burrows by building a mound over one of their entrances, which is open to a stream of air. A second entrance at ground level is open to almost stagnant air. How does this construction create an air flow through the burrow?

7. The device in Figure 10.16 is called a water aspirator. It is used for suction. Explain how it works. What is meant by ''suction''?

Figure 10.16
Question 7

Problems

Section 10.2   Fluid Dynamics

Section 10.3   Other Applications of Bernoulli's Equation

1. Each second, 0.25 liter of liquid flows through a tube 1 cm in diameter. (a) What is the flow velocity along the tube? (b) If the tube in part (a) is connected to a

8. If you hold a sheet of paper and blow across the top surface, the paper rises. Explain.

9. If you suddenly turn on your shower water at full speed, why is the shower curtain pushed inward?

10. A tornado or hurricane will often lift the roof of a house. Use the Bernoulli effect to explain why this occurs. Why should you keep your windows open during these conditions?

11. If air from a hair dryer is blown over the top of a ping-pong ball, the ball can be suspended in air. Explain how the ball can remain in equilibrium.

12. A baseball moves past an observer from left to right spinning counterclockwise. In which direction will the ball tend to deflect?

13. How should a pitcher throw a baseball so that it rises as it reaches the batter?

14. Should an airplane take off into the wind or with the wind? Explain.

15. The device shown in Figure 10.17 is a novice's attempt to design an atomizer. Will this design work? If not, what must be done to fix it? (*Hint:* Consider the Bernoulli effect.)

Figure 10.17
Question 15

16. Two ships passing near each other in a harbor tend to be drawn together and run the risk of a sideways collision. How does the Bernoulli effect explain this?

17. How does a bumblebee stay aloft? A helicopter?

18. SAE 10 motor oil has less viscosity at a given temperature than does SAE 40 oil. Which should you use in your car during the winter months? (Actually, a multi-grade oil is best.)

tube 0.75 cm in diameter, what will the flow velocity be in the smaller tube?

2. A cowboy at a dude ranch fills a horse trough that is 1.5 m long, 60 cm wide, and 40 cm tall. He uses a 2-cm diameter hose from which water emerges at 1.5 m/s. How long will it take him to fill the trough?

3. The approximate inside diameter of the aorta is 0.5 cm, that of a capillary is 10.0 μm. The approxi-

mate average flow speed is 1.0 m/s in the aorta and 1.0 cm/s in the capillaries. If all the blood in the aorta eventually flows through the capillaries, estimate the number of capillaries in the circulatory system.

4. A garden hose has an inner radius of 1.0 cm. The hose has a nozzle with an inside radius of 1.5 mm. Water flows through the hose with an average speed of 25.0 cm/s. (a) What is the speed of the water as it emerges from the nozzle? (b) How long does it take to spray 4.0×10^{-3} m^3 of water from the hose?

5. The velocity of flow of a liquid through a water pipe is 0.636 m/s. Determine the flow rate if the radius of the pipe is 1.5 cm.

6. A water tank open to the atmosphere contains water at a depth of 4 m. If a puncture occurs at the base, with what velocity does the stream of water leave the opening?

7. A liquid ($\rho = 1.65$ g/cm^3) flows through two horizontal sections of tubing joined end to end. In the first section the cross-sectional area is 10 cm^2, the flow velocity is 275 cm/s, and the pressure is 1.2×10^5 Pa. The liquid then enters the second section, where the cross-sectional area is 2.5 cm^2. Calculate (a) the flow velocity and (b) the pressure in the section of smaller diameter.

8. While a person inhales, air moves down the bronchus (windpipe) with a speed of 15.0 cm/s. The average flow speed of the air doubles through a constriction in the bronchus. Assuming incompressible flow, determine the pressure drop in the constriction.

9. A hypodermic syringe contains a medicine with the density of water (Fig. 10.18). The barrel of the syringe has a cross-sectional area of 2.5×10^{-5} m^2. The cross-sectional area of the needle is 1.0×10^{-8} m^2. In the absence of a force on the plunger the pressure everywhere is 1.0 atm. A force **F** of magnitude 2.0 N is exerted on the plunger, making medicine squirt from the needle. Determine the flow speed of the medicine emerging from the needle. Assume the pressure in the needle remains equal to 1.0 atm, that the syringe is in a horizontal position, and that the speed of the emerging fluid is the same as the speed of the fluid in the needle.

Figure 10.18   Problem 9

10. (a) If wind blows at a speed of 30 m/s over the roof of your house, what is the pressure difference on your roof between the air inside and the air outside?

(b) What net force does this produce on a roof having an area of 175 m^2? (Use 1.3 kg/m^3 as the density of air.)

11. What is the net upward force on an airplane wing of area 20 m^2 if the velocity of flow across the top of the wing is 300 m/s and the velocity across the bottom of the wing is 280 m/s.

12. A liquid ($\rho = 1.05 \times 10^3$ kg/m^3) flows at constant speed through a plastic tube that forms a vertical loop 1.8 m in diameter. The pressure in the liquid at a point 0.5 m below the top of the loop is 1.3×10^4 Pa. Calculate the pressure (a) at the top of the loop and (b) at the bottom of the loop.

13. Water flows through a pipe of radius 30 cm at the rate of 0.2 m^3/s. The pressure in the pipe is atmospheric. The pipe slants downhill and feeds into a second pipe of radius 15 cm which is located 60 cm lower. What is the gauge pressure in this lower pipe?

14. (a) Calculate the flow rate (in grams per second) of blood ($\rho = 1$ g/cm^3) in an aorta with a cross-sectional area of 2 cm^2 if the flow velocity is 40 cm/s. (b) Assume that the aorta branches to form a large number of capillaries with a combined cross-sectional area of 3×10^3 cm^2. What is the velocity of flow in the capillaries?

15. You are visiting "fish world" while on vacation when a break occurs near the bottom of one of the tanks. If the break is 1.5 m below the surface of the water and 1 m above the floor, how far from the base of the tank will the water strike the floor?

16. A jet of water squirts out horizontally from a leak in the bottom of the tank in Figure 10.19. If the hole has a diameter of 3.5 mm, what is the height h of the water level in the tank?

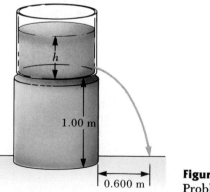

Figure 10.19
Problem 16

17. The inside diameter of the horizontal pipe in Figure 10.20 is 2.5 cm. Water flows to the right at a rate of 1.8×10^{-4} m^3/s. Determine the inside diameter of the constriction.

Figure 10.20 Problem 17

18. The water supply of a building is fed through a main entrance pipe 6 cm in diameter. A 2-cm diameter faucet tap located 2 m above the main pipe is observed to fill a 25-liter container in 30 s. (a) What is the speed at which the water leaves the faucet? (b) What is the gauge pressure in the main pipe? (Assume the faucet is the only outlet in the system.)

19. What is the gauge pressure in a nozzle if the water leaving it sprays the water vertically upward to a height of 4 m?

20. A large storage tank open to the atmosphere at the top and filled with water develops a small hole in the side at a point 16 m below the water level. If the rate of flow from the leak is 2.5×10^{-3} m³/min, determine (a) the speed at which the water leaves the hole and (b) the diameter of the hole.

21. A water tank open to the atmosphere at the top has two holes punched in its side one above the other. The holes are 5 cm and 12 cm above the floor. How high does water stand in the tank if the two streams of water hit the floor at the same place?

22. A horizontal tube of radius 1 cm is joined to a second horizontal tube of radius 0.5 cm. There is a pressure difference of 6660 Pa between the two tubes. (a) Which tube has the higher water pressure? (b) What volume of water flows through the tubes per second?

Section 10.4 Viscosity
Section 10.5 Poiseuille's Law

23. The block of ice (temperature 0° C) shown in Figure 10.21 is drawn over a level surface lubricated by a layer of water 0.1 mm thick. Determine the magni-

Figure 10.21 Problem 23

tude of force **F** needed to pull the block with a constant speed of 0.5 m/s. At 0° C, the viscosity of water is $\eta = 1.79 \times 10^{-3}$ N · s/m².

24. A metal block is pulled over a horizontal surface that has been coated with a layer of lubricant 1.0 mm thick. The face of the block in contact with the surface has dimensions 0.40 m by 0.12 m. A force of 1.9 N is required to move the block at a constant speed of 0.5 m/s. Calculate the coefficient of viscosity of the lubricant.

25. A thin 1.5-mm coating of glycerine has been placed between two microscope slides of width 1 cm and length 4 cm. Find the force required to pull one of the microscope slides at a speed of 0.3 m/s relative to the other.

26. A straight horizontal pipe with a diameter of 1 cm and a length of 50 m carries oil with a coefficient of viscosity of 0.12 Pa · s. At the output of the pipe, the flow rate is 8.6×10^{-5} m³/s and the pressure is 1 atm. Find the gauge pressure at the pipe input.

27. A hypodermic needle is 3 cm in length and 0.3 mm in diameter. What excess pressure is required along the needle so that the flow rate of water through it will be 1 g/s? (Use 1×10^{-3} Pa · s as the viscosity of water.)

28. A needle of radius 0.3 mm and length 3 cm is used to give a patient a blood transfusion. Assume the pressure differential across the needle is achieved by elevating the blood 1 m above the patient's arm. (a) What is the rate of flow of blood through the needle? (b) At this rate of flow, how long will it take to inject 1 pint (approximately 500 cm³) of blood into the patient? The density of blood is 1050 kg/m³, and its coefficient of viscosity is 4×10^{-3} N ·s/m².

29. The water pressure in a horizontal pipe decreases 0.5 atm per 100 m when the flow rate in the pipe is 30.0 liter/min. Determine the radius of the pipe. (Assume a temperature of 20° C.)

30. What diameter needle should be used to inject a volume of 500 cm³ of a solution into a patient in 30 min? Assume that the needle length is 2.5 cm and that the solution is elevated 1 m. Furthermore, assume the viscosity and density of the solution are those of pure water.

31. The pulmonary artery, which connects the heart to the lungs, has an inner radius of 2.6 mm and is 8.4 cm long. If the pressure drop between the heart and lungs is 400 Pa, what is the average speed of blood in the pulmonary artery? (See Problem 28 for the physical properties of blood.)

32. A pipe carrying 20° C water has a diameter of 2.5 cm. Estimate the maximum flow speed if the flow is to be laminar.

33. What is the Reynolds number for the flow of liquid in the 1.2-m-diameter Alaska pipeline? The density of

crude oil is 850 kg/m^3, its velocity is 3 m/s, and its viscosity is 0.3 Pa · s. Is the flow laminar or turbulent?

34. Determine the velocity at which the flow of water through a 0.5-cm-diameter pipe will become turbulent ($RN \geqslant 3000$).

35. Assume a value of 980 for the Reynolds number for blood in an artery and a viscosity of 4×10^{-3} N·s/m^2 for whole blood. If the density of whole blood is equal to 1.05×10^3 kg/m^3, at what velocity does blood flow through an artery 0.45 cm in diameter?

36. The aorta in humans has a diameter of about 2 cm and, at certain times, the blood velocity through it is about 55 cm/s. Is the blood flow turbulent? (Values for several blood properties that you may need are given in Problem 28.)

Section 10.6   Transport Phenomena

37. Sucrose is allowed to diffuse along a 10-cm length of tubing filled with water. The tube is 6 cm^2 in cross-sectional area. The diffusion coefficient is 5×10^{-10} m^2/s, and 8×10^{-14} kg is transported along the tube in 15 s. What is the difference in the concentration levels of sucrose at the two ends of the tube?

38. In a diffusion experiment, it is found that, in 60 s, 3×10^{-13} kg of sucrose will diffuse along a horizontal pipe of cross-sectional area 1 cm^2. If the diffusion coefficient is 5×10^{-10} m^2/s, what is the concentration gradient (that is, the change in concentration per unit length along the path)?

39. Glycerine in water diffuses along a horizontal column that has a cross-sectional area of 2 cm^2. The concentration gradient is 3×10^{-2} kg/m^4, and the diffusion rate is found to be 5.7×10^{-15} kg/s. Determine the diffusion coefficient.

40. Use the data for sucrose given in Problem 37 to calculate how much sucrose will diffuse down a horizontal pipe of cross-sectional area 4 cm^2 in 10 s if the concentration gradient is 0.2 kg/m^4.

Section 10.7   Motion Through a Viscous Medium

41. Small spheres of diameter 1 mm fall through water with a terminal velocity of 1.1 cm/s. Calculate the density of the spheres.

42. A test tube 10 cm tall is filled with water that contains spherical particles of density 1.8 g/cm^3 in suspension. If the radius of the particle is 2×10^{-4} cm, find the time it takes a particle near the top of the tube to reach the bottom.

43. Calculate the viscous force on a spherical oil droplet that is 2×10^{-5} cm in diameter and falling with

a speed of 0.4 mm/s in air. (Use $\eta = 1.8 \times 10^{-5}$ N·s/m^2 as the coefficient of viscosity of air.)

44. The viscous force on an oil drop is measured to be 3×10^{-13} N when the drop is falling through air with a speed of 4.5×10^{-4} m/s. If the radius of the drop is equal to 2.5×10^{-6} m, what is the viscosity of air?

45. An oil drop ($\rho = 800$ kg/m^3) is falling with a terminal velocity of 0.04 mm/s in air of density 1.29 kg/m^3. Calculate the radius of the oil drop. (The coefficient of viscosity of air is $\eta = 1.8 \times 10^{-5}$ N·s/m^2.)

46. Spherical particles of a protein of density 1.8 g/cm^3 are shaken up in a solution of water. The solution is allowed to stand for 1 h. If the depth of water in the tube is 5 cm, find the radius of the largest particles still in solution at the end of the hour.

Additional Problems

47. Water flows through a 2-cm-diameter hose with a speed of 0.5 m/s. A person using the hose pinches the end such that the diameter becomes 0.25 cm. At what speed does the water now leave the hose?

48. Water at a pressure of 3×10^5 Pa flows through a horizontal pipe at a speed of 1 m/s. If the pipe narrows to one-fourth its original diameter, (a) find the speed of flow in the narrow section and (b) the pressure in the narrow section.

49. Use dimensional analysis to convince yourself that the power required to drive a fluid through a pipe is given by the volume rate of flow, Av, times the pressure difference across the pipe, $P_1 - P_2$.

50. The water jet from an amusement park fountain reaches a height of 10 m. If the fountain nozzle has an opening 10 cm in diameter, what power is required to operate the fountain? (Assume that the length of the nozzle is negligible and that the velocity of the fluid in the supply line to the nozzle is also negligible.)

51. One step in an industrial process requires the separation of very small spherical aluminum particles from a suspension of glycerine. If this separation is accomplished by natural sedimentation, calculate the terminal velocity for a particle diameter of 1×10^{-5} m. For glycerine use $\eta = 1500 \times 10^{-3}$ N·s/m^2 and take its density to be $\rho = 900$ kg/m^3.

52. A glass sphere of diameter 1.75 cm and density 2.5 g/cm^3 falls through water with a terminal velocity of 0.3 m/s. A wooden sphere of the same diameter and density 0.85 g/cm^3 is released from the bottom of a tank of water. Determine the terminal velocity with which the wooden sphere rises.

53. Ethanol ($\rho = 791$ kg/m^3) flows through a ground-level horizontal pipe, 30 cm in diameter, at a gauge pressure of 2.5×10^4 Pa and a flow rate of 0.21 m^3/s.

In order to cross a highway, the pipe has a U-shaped section that reaches 2 m below the ground-level section. What is the maximum diameter of the below-ground section such that the gauge pressure at the lowest point will not be greater than 3×10^4 Pa?

54. As part of an industrial process, a liquid for which $\rho = 7.5 \times 10^2$ kg/m^3 and $\eta = 0.45$ N·s/m^2 is pumped under pressure through a horizontal pipe 500 m in length. A flow rate of 0.15 m^3/s is required, and the Reynolds number is not to exceed 1000. Determine (a) the minimum pipe diameter and (b) the pressure difference across the ends of the pipe.

55. A siphon is a device that allows a fluid to seemingly defy gravity (Fig. 10.22). The flow must be initiated by a partial vacuum in the tube, as in a drinking straw.

Figure 10.22   Problem 55

(a) Show that the speed at which the water emerges from the siphon is given by $v = \sqrt{2gh}$. (b) For what values of y will the siphon work?

The Cardiovascular System

The cardiovascular system includes the heart to pump the blood; arteries to carry the blood to the organs, muscle, and skin; and veins to return the blood to the heart (Fig. 3). The blood is pumped into the aorta by the left ventricle. The aorta branches to form smaller arteries, which in turn branch down to even smaller arteries, until finally the blood reaches the very small capillaries of the vascular bed. These capillaries are so small that the red blood cells must pass single file through them. After passing through the capillaries, where materials being carried by the blood are exchanged with the surrounding tissues, the blood flows to the veins and is returned to the heart.

The flow rate of the blood changes as it goes through this system. The cross-sectional area of the vascular bed, which is the product of the cross-sectional area and the number of capillaries, is much greater than the cross-sectional area of the aorta. Because the volume of the blood passing through a cross-sectional area per unit of time is Av, where v is the magnitude of the velocity of the blood, we may express the volume flow rate of the blood as

$$\text{Flow rate} = A_{\text{aorta}} v_{\text{aorta}}$$

Furthermore, because the total average flow rate through the aorta and the capillaries must be the same, we have

$$\text{Flow rate} = A_{\text{aorta}} v_{\text{aorta}} = A_{\text{cap}} v_{\text{cap}}$$

Example 10E.1   Flow of Blood in the Aorta and Capillaries

The velocity of blood in the aorta is 50 cm/s, and it has a radius of 1 cm. (a) What is the rate of flow of blood through this aorta? (b) If the capillaries have a total cross-sectional area of 3000 cm², what is the speed of the blood in the capillaries?

Solution
(a) The area of the aorta is

$$A = \pi r^2 = \pi (1 \text{ cm})^2 = \pi \text{ cm}^2$$

$$\text{Flow rate} = Av = (\pi \text{ cm}^2)(50 \text{ cm/s}) = 50\pi \text{ cm}^3/\text{s}$$

(b) The flow rate in the capillaries $= 50\pi \text{ cm}^3/\text{s} = A_c v_c$,

$$v_c = \frac{\text{Flow rate}}{A_c} = \frac{50\pi \text{ cm}^3/\text{s}}{3000 \text{ cm}^2} = \boxed{0.05 \text{ cm/s}}$$

This low blood velocity in the capillaries is necessary to enable the blood to exchange oxygen, carbon dioxide, and other nutrients with the surrounding tissues.

Essay Questions
1. Explain why some individuals tend to black out when they stand up rapidly.
2. If a person is standing and is at rest, what will be the relation between the blood pressure in the left arm and the left leg? What is the relation if the person is in a horizontal position?
3. At what upward acceleration would you expect the blood pressure in the brain to be zero? (Assume that no body mechanisms are operating to compensate for this condition.)
4. When using a sphygmomanometer to measure blood pressure, will the blood pressure readings depend upon the atmospheric pressure? If the atmospheric pressure decreases rapidly, how will this effect the blood pressure readings?

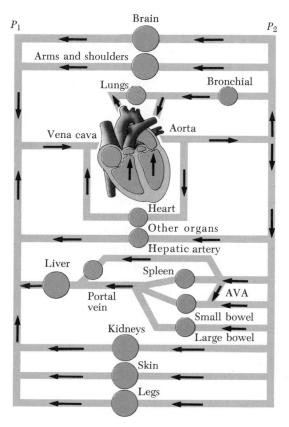

Figure 3
A diagram of the mammalian circulatory system. Pressure P_2 is that in the arterial system, and P_1 is that in the venous system. Arrows indicate the direction of blood flow.

Labels in figure: Brain, Arms and shoulders, Lungs, Bronchial, Vena cava, Aorta, Heart, Other organs, Hepatic artery, Liver, Spleen, AVA, Portal vein, Small bowel, Kidneys, Large bowel, Skin, Legs, P_1, P_2

5. Why is it impractical to measure the pulse rate using a vein?
6. Assuming that an artery is clogged such that the effective radius is one-half its normal radius compared to a normal artery, by what factor must the pressure differential be increased to obtain the normal flow rate through the clogged artery?

Essay Problems

1. Determine the average velocity of the blood in the aorta if it has a radius of 1.2 cm, and the flow rate is 20 liters/min.
2. If the mean blood pressure in the aorta is 100 mm Hg, (a) determine the blood pressure in the artery located 2 ft above the heart. (b) One cannot apply, without significant error, Bernoulli's principle in the smaller arteries and the capillaries. Why not?
3. When the flow rate is 5 liters/min, the velocity in the capillaries is 0.33 mm/s. Assuming the average diameter of a capillary to be 0.008 mm, calculate the number of capillaries in the circulatory system.
4. An artery has a length of 20 cm, and a radius of 0.5 cm, and blood is flowing at a rate of 6 liters/min. What is the difference in the pressure between the ends of the artery?
5. Assuming the internal pressure in the left ventricle is 100 mm Hg and the left ventricle of the heart has an effective radius of 3 cm, calculate the tension in the wall of the left ventricle. (Assume that the external pressure is zero gauge pressure.)
6. If the blood pressure is 10 mm Hg in a capillary which has a radius of 0.005 mm, determine the tension in the wall of the capillary.

then placed in thermal contact with B, and its reading is recorded after thermal equilibrium is reached. If the two readings are the same, then A and B would be in thermal equilibrium with each other. We can summarize these results in a statement known as the **zeroth law of thermodynamics** (the law of equilibrium):

> If bodies A and B are separately in thermal equilibrium with a third body, C, then A and B will be in thermal equilibrium with each other if placed in thermal contact.

This statement, insignificant and obvious as it may seem, is easily proved experimentally and is very important because it can be used to define temperature. We can think of temperature as the property that determines whether or not an object is in thermal equilibrium with other objects. *Two objects in thermal equilibrium with each other are at the same temperature.* Conversely, if two objects have different temperatures, they cannot be in thermal equilibrium with each other.

11.2   Thermometers and Temperature Scales

Thermometers are devices used to define and to measure the temperature of a system. All thermometers make use of a change in some physical property with temperature. Some of these physical properties are (1) the change in volume of a liquid, (2) the change in length of a solid, (3) the change in pressure of a gas held at constant volume, (4) the change in volume of a gas held at constant pressure, (5) the change in electric resistance of a conductor, and (6) the change in color of a very hot object. A temperature scale can be established for a given substance using any one of these physical quantities.

The most common thermometer in everyday use consists of a mass of mercury that expands into a glass capillary tube when heated (Fig. 11.1). Thus, the physical property in this case is the change in volume of a liquid. One can define any temperature change to be proportional to the change in length of the mercury column. The thermometer can be calibrated by placing it in thermal contact with some natural systems that remain at constant temperature. One such system is a mixture of water and ice in thermal equilibrium at atmospheric pressure, which is defined to have a temperature of zero degrees Celsius, written 0°C. (This temperature is called the ice point of water.) Another convenient system for calibrating a thermometer is a mixture of water and steam in thermal equilibrium at atmospheric pressure. The temperature of this system is 100°C (the steam point of water). Once the mercury levels have been established at these two points, the column is divided into 100 equal segments, each denoting a change in temperature of one Celsius degree. (The Celsius temperature scale was once called the centigrade temperature scale.)

Thermometers calibrated in this way do present problems, however, when extremely accurate readings are needed. For instance, an alcohol thermometer calibrated at the ice and steam points of water might agree with a mercury thermometer only at the calibration points. Because mercury and alcohol have different thermal expansion properties, when one reads a temperature of 50°C, say, the other may indicate a slightly different value. The

Figure 11.1
Schematic diagram of a mercury thermometer. Because of thermal expansion, the level of the mercury rises as the mercury is heated from 0°C (the ice point) to 100°C (the steam point).

discrepancies between thermometers are especially large when the temperatures to be measured are far from the calibration points. An additional practical problem of any thermometer is its limited temperature range. A mercury thermometer, for example, cannot be used below the freezing point of mercury, which is $-39°C$. What we need is a universal thermometer whose readings are independent of the substance used. The gas thermometer meets this requirement.

The Constant-Volume Gas Thermometer and the Kelvin Scale

In a **gas thermometer,** the temperature readings are nearly independent of the substance used in the thermometer. One type of gas thermometer is the constant-volume unit shown in Figure 11.2. The physical property used in this device is the pressure variation with temperature of a fixed volume of gas. When this thermometer was first developed, it was calibrated using the ice and steam points of water as follows. (A different calibration procedure, to be discussed shortly, is now used.) The gas flask is inserted into an ice bath, and the column of mercury raised or lowered until the volume of the gas is at some particular value. (Hence, we have a constant-volume device.) The height h of the mercury column indicates the pressure in the flask at $0°C$. Then the flask is inserted into water at the steam point and the mercury level is readjusted until the gas volume is the same as it was in the ice bath. This gives a value for the pressure at $100°C$. These values of pressure and temperatures are then plotted on a graph, as in Figure 11.3. The line connecting these points serves as a calibration curve for measuring unknown temperatures. If we wanted to measure the temperature of a substance, then, we would place the gas flask in thermal contact with the substance and adjust the column of mercury until the gas takes on its specified volume. The height of the mercury column tells us the pressure of the gas, and, since we know the pressure, we can find the temperature of the substance from the graph.

Now suppose that temperatures are measured with various gas thermometers containing different gases. Experiments show that the thermometer readings are nearly independent of the type of gas used, so long as the gas pressure is low and the temperature is well above the point at which the gas liquifies (Fig. 11.4). The agreement among thermometers using various gases improves as the pressure is reduced.

If you extend the curves in Figure 11.4 back toward negative temperatures, you find that the pressure is zero when the temperature is $-273.15°C$. This significant temperature is used as the basis for the Kelvin temperature scale. This scale sets $-273.15°C$ as its zero point, 0 K. The size of a degree on the Kelvin scale is identical to the size of a degree on the Celsius scale. Thus, the relationship that enables one to convert between these temperatures is

$$T_C = T - 273.15 \qquad (11.1)$$

where T_C is the **Celsius temperature** and T is the **Kelvin temperature** (which is sometimes called the **absolute temperature**).

Early gas thermometers made use of the ice point and the steam point as standard temperatures following the procedure just described. However,

Figure 11.2
A constant-volume gas thermometer measures the pressure of the gas contained in the flask on the left. The volume of gas in the flask is kept constant by raising or lowering the column of mercury such that the mercury level remains constant.

$$P_1 + \rho g y_1 = P_a + \rho g y_2$$
$$P_1 \propto y_2$$

Figure 11.3
A typical graph of pressure versus temperature taken with a constant-volume gas thermometer. The dots represent known reference temperatures (the ice point and the steam point).

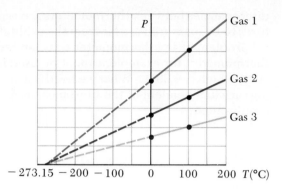

Figure 11.4
Pressure versus temperature for dilute gases. Note that, for all gases, the pressure extrapolates to zero at the unique temperature of $-273.15°C$.

The kelvin

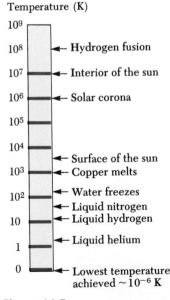

Figure 11.5
Absolute temperature at which various selected physical processes take place.

these points are experimentally difficult to duplicate because they are sensitive to dissolved impurities in the water. For this reason, a new procedure based on a single fixed point was adopted in 1954 by the International Committee on Weights and Measures. The **triple point of water,** which corresponds to *the single temperature and pressure at which water, water vapor, and ice can coexist in equilibrium,* was chosen as a convenient and reproducible reference temperature for the Kelvin scale. The triple point of water occurs at a temperature of $0.01°C$ and a pressure of 4.58 mm of mercury. The temperature at the triple point of water on the Kelvin scale was arbitrarily set at 273.16 kelvins, abbreviated 273.16 K. The SI unit of temperature, the **kelvin,** is defined as *1/273.16 of the temperature of the triple point of water.*

Figure 11.5 shows the Kelvin temperatures for various physical processes and structures. The temperature of 0 K is often referred to as **absolute zero,** and as Figure 11.5 shows, this temperature has never been achieved, although laboratory experiments have come close. In general, experimental attempts to approach absolute zero are based on the principle that when a substance is magnetized it heats up, and when it is demagnetized it cools. In one experiment, copper was magnetized and then placed in liquid helium, thus dropping its temperature to about $-272°C$. When the copper was demagnetized while in the helium, its temperature dropped to a point only 10^{-6} degrees above absolute zero. But what would happen to a substance if its temperature could reach 0 K? As the graph of Figure 11.4 indicates, the pressure that it exerts on the walls of its container would be zero. In Section 11.6, we shall show that the pressure of a gas is proportional to the kinetic energy of the molecules of that gas. Thus, according to classical physics the kinetic energy of the gas would go to zero, and there would be no motion at all of the individual components of the gas, hence the molecules would settle out on the bottom of the container. Quantum theory, to be discussed in Chapter 29, modifies this statement to indicate that there would be some residual energy, called the zero-point energy, at this low temperature.

The Celsius, Fahrenheit, and Kelvin Temperatures Scales

Equation 11.1 shows that the Celsius temperature, T_C, is shifted from the absolute (or Kelvin) temperature, T, by 273.15. From this we see that the size of a degree on the Kelvin scale is the same as on the Celsius scale. In other words, a temperature difference of 5 Celsius degrees, written 5 C°, is equal to a temperature difference of 5 kelvins. The two scales differ only in the choice of the zero point. Furthermore, the ice point (273.15 K) corresponds to 0.00°C, and the steam point (373.15 K) is equivalent to 100.00°C.

The most common temperature scale in everyday use in the United States is the **Fahrenheit scale.** This scale sets the temperature of the ice point at 32°F and the temperature of the steam point at 212°F. The relationship between the Celsius and Fahrenheit temperature scales is

$$T_{\text{F}} = \frac{9}{5} T_{\text{C}} + 32 \qquad \textbf{(11.2)}$$

Equation 11.2 can easily be used to find a relationship between changes in temperature on the Celsius and Fahrenheit scales. It will be left as a problem for you to show that if the Celsius temperature changes by ΔT_{C}, the Fahrenheit temperature changes by an amount ΔT_{F} given by

$$\Delta T_{\text{F}} = \frac{9}{5} \Delta T_{\text{C}} \qquad \textbf{(11.3)}$$

Example 11.1   Converting Temperatures

On a day when the temperature reaches 50°F, what is the temperature in degrees Celsius and in kelvins?

Solution
Let us solve Equation 11.2 for T_{C} and substitute $T_{\text{F}} = 50$°F:

$$T_{\text{C}} = \frac{5}{9}(T_{\text{F}} - 32) = \frac{5}{9}(50 - 32) = \boxed{10\,^{\circ}\text{C}}$$

From Equation 11.1, we find that

$$T = T_{\text{C}} + 273.15 = \boxed{283.15\ \text{K}}$$

Example 11.2   Heating a Pan of Water

A pan of water is heated from 25°C to 80°C. What is the change in its temperature on the Kelvin scale and on the Fahrenheit scale?

Solution
From Equation 11.1, we see that the change in temperature on the Celsius scale equals the change on the Kelvin scale. Therefore,

$$\Delta T = \Delta T_{\text{C}} = 80 - 25 = 55\ \text{C}^{\circ} = \boxed{55\ \text{K}}$$

From Equation 11.3, we find that the change in temperature on the Fahrenheit scale is greater than the change on the Celsius scale by the factor $\frac{9}{5}$. That is,

$$\Delta T_{\text{F}} = \frac{9}{5} \Delta T_{\text{C}} = \frac{9}{5}(80 - 25) = \boxed{99\ \text{F}^{\circ}}$$

In other words, 55 C° = 99 F°, where the notations C° and F° refer to temperature *differences,* and are not to be confused with actual temperatures, which are written °C and °F.

Expansion joints allow bridges to expand and contract safely.

11.3   Thermal Expansion of Solids and Liquids

Our discussion of the mercury thermometer made use of one of the most well-known changes that occurs in a substance: as its temperature increases, its volume increases. (As we shall see shortly, in some materials the volume decreases when the temperature increases.) This phenomenon, known as **thermal expansion,** plays an important role in numerous applications. For example, thermal expansion joints must be included in buildings, concrete highways, and bridges to compensate for changes in dimensions with temperature variations.

The overall thermal expansion of an object is a consequence of the change in the average separation between its constituent atoms or molecules. To understand this, consider a crystalline solid, which consists of a regular array of atoms held together by electric forces. At ordinary temperatures, the atoms vibrate about their equilibrium positions with an amplitude of about 10^{-11} m and a frequency of about 10^{13} cycles/s. The average spacing between the atoms is of the order of 10^{-10} m. As the temperature of the solid increases, the atoms vibrate with larger amplitudes and the average separation between them increases. Consequently, the solid as a whole expands with increasing temperature. If the expansion of an object is sufficiently small compared with its initial dimensions, then the change in any dimension (length, width, or thickness) is, to a good approximation, dependent on the first power of the temperature change.

Suppose an object has an initial length L_0 along some direction at some temperature. The length increases by an amount ΔL for a change in temperature ΔT. Experiments show that, when ΔT is small enough, the change in length is proportional to the temperature change and to the original length. Thus, the basic equation for the expansion of a solid is

$$\Delta L = \alpha L_0 \, \Delta T \tag{11.4a}$$

or

$$L - L_0 = \alpha L_0 (T - T_0) \tag{11.4b}$$

where the constant α is called the **average coefficient of linear expansion** for a given material and has units of $(\text{C}°)^{-1}$.

It may be helpful to think of a thermal expansion as a magnification or as a photographic enlargement. For example, as a metal washer is heated (Fig. 11.6), all dimensions, including the radius of the hole, increase according to Equation 11.4. Table 11.1 lists the average coefficient of linear expansion for various materials. Note that for these materials α is positive, indicating an increase in length with increasing temperature. This is not always the case. For example, some substances, such as calcite ($CaCO_3$), expand along one dimension (positive α) and contract along another (negative α) with increasing temperature.

Because the linear dimensions of an object change with temperature, it follows that the area and volume also change with temperature. Consider a square having an initial length L_0 on a side and therefore an initial area $L_0{}^2$. As the temperature of the square is increased, the length of each side increases to

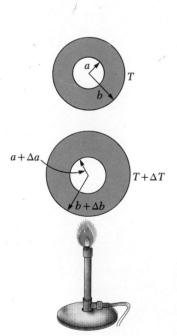

Figure 11.6
Thermal expansion of a homogeneous metal washer. As the washer is heated, all dimensions increase. (Note that the expansion is exaggerated in this figure.)

Table 11.1
Average Coefficients of Linear Expansion for Some Materials Near Room Temperature

Material	Average Coefficient of Linear Expansion $[(C°)^{-1}]$	Material	Average Coefficient of Volume Expansion $[(C°)^{-1}]$
Aluminum	24×10^{-6}	Ethyl alcohol	1.12×10^{-4}
Brass and bronze	19×10^{-6}	Benzene	1.24×10^{-4}
Copper	17×10^{-6}	Acetone	1.5×10^{-4}
Glass (ordinary)	9×10^{-6}	Glycerin	4.85×10^{-4}
Glass (Pyrex)	3.2×10^{-6}	Mercury	1.82×10^{-4}
Lead	29×10^{-6}	Turpentine	9.0×10^{-4}
Steel	11×10^{-6}	Gasoline	9.6×10^{-4}
Invar (Ni-Fe alloy)	0.9×10^{-6}	Air	3.67×10^{-3}
Concrete	12×10^{-6}	Helium	3.665×10^{-3}

$$L = L_0 + \alpha L_0 \, \Delta T$$

The new area $A = L^2$ is

$$L^2 = (L_0 + \alpha L_0 \, \Delta T)(L_0 + \alpha L_0 \, \Delta T) = L_0{}^2 + 2\alpha L_0{}^2 \, \Delta T + \alpha^2 L_0{}^2 \, \Delta T^2$$

The last term in this expression contains the quantity $\alpha \, \Delta T$ raised to the second power. Because $\alpha \, \Delta T$ is much less than unity, squaring it makes it even smaller. Therefore, we can neglect this term to get a simpler expression:

$$A = L^2 = L_0{}^2 + 2\alpha L_0{}^2 \, \Delta T$$

$$A = A_0 + 2\alpha A_0 \, \Delta T$$

or

$$\Delta A = A - A_0 = \gamma A_0 \, \Delta T \qquad \textbf{(11.5)}$$

where $\gamma = 2\alpha$. The quantity γ (Greek letter gamma) is called the **average coefficient of area expansion.**

By a similar procedure, we can show that the *increase in volume* of an object accompanying a change in temperature is given by

$$\Delta V = \beta V_0 \, \Delta T \qquad \textbf{(11.6)}$$

where β, the **average coefficient of volume expansion**, is given by $\beta = 3\alpha$.

Example 11.3 Expansion of a Railroad Track

A steel railroad track has a length of 30 m when the temperature is 0°C. (a) What is its length on a hot day when the temperature is 40°C?

Solution
If we make use of Table 11.1 and note that the change in temperature is 40 C°, we find that the *increase* in length is

$$\Delta L = \alpha L_0 \, \Delta T = [11 \times 10^{-6} (C°)^{-1}](30 \text{ m})(40 \text{ C°}) = 0.013 \text{ m}$$

Therefore, its length at 40°C is 30.013 m.

Thermal expansion: the extreme heat of a July day in Asbury Park, New Jersey, caused these railroad tracks to buckle. *(Wide World Photos)*

(b) Suppose the ends of the rail are rigidly clamped at 0°C so as to prevent expansion. Calculate the thermal stress set up in the rail if its temperature is raised to 40°C. Assume the rail has a cross-sectional area of 30 cm² (0.003 m²).

Solution

From the definition of Young's modulus for a solid (Chapter 9), we have

$$\text{Tensile stress} = \frac{F}{A} = Y\frac{\Delta L}{L_0}$$

Since Y for steel is 20×10^{10} Pa, we have

$$\frac{F}{A} = (20 \times 10^{10} \text{ Pa})\left(\frac{0.013 \text{ m}}{30 \text{ m}}\right) = 8.7 \times 10^7 \text{ Pa}$$

Therefore the force of compression in the rail is

$$F = (8.7 \times 10^7 \text{ Pa})(0.003 \text{ m}^2) = \boxed{2.6 \times 10^5 \text{ N} \approx 59{,}000 \text{ lb}}$$

Obviously, it would be difficult to keep the rail clamped in place under such large forces. For this reason, expansion joints are used to allow rails to expand without bowing.

As Table 11.1 indicates, each substance has its own characteristic coefficient of expansion. For example, a brass rod will expand more than a steel rod of equal length if their temperatures are raised by the same amount from some common initial value because brass has a larger coefficient of expansion than steel. A simple device called a bimetallic strip utilizes this principle for practical applications. Such a strip is made by securely bonding two different metals together, as shown in Figure 11.7a. As the temperature of the strip increases, the two metals expand by different amounts and the strip bends. Figure 11.7b illustrates how a bimetallic strip is used in a thermostat.

Figure 11.7

(a) A bimetallic strip bends as the temperature changes because the two metals have different expansion coefficients. (b) A bimetallic strip used in a thermostat to break or make electrical contact. (c) The blade being heated in the photograph is a bimetallic strip that was straight before being heated. Note that the blade bends when cooled. Which way would it bend if it were heated? *(Courtesy of CENCO)*

Example 11.4

A hole of cross-sectional area 100 cm² is cut in a piece of steel at 20°C. What is the area of the hole if the steel is heated from 20°C to 100°C?

Solution

A hole in a substance expands in exactly the same way as would a piece of the substance having the same shape as the hole. The change in the area of the hole can be found by using Equation 11.5:

$$\Delta A = \gamma A_0 \, \Delta T = [22 \times 10^{-6}(°C)^{-1}](100 \text{ cm}^2)(80 \text{ C}°) = 0.18 \text{ cm}^2$$

Therefore, the area of the hole at 100°C is

$$A = A_0 + \Delta A = \boxed{100.18 \text{ cm}^2}$$

The Unusual Behavior of Water

Liquids generally increase in volume with increasing temperature and have volume expansion coefficients about ten times greater than those of solids. Water is an exception to this rule, as we can see from its density versus temperature curve, in Figure 11.8. As the temperature increases from 0°C to 4°C, water contracts and thus its density increases. Above 4°C, water expands with increasing temperature. The density of water reaches a maximum value of 1000 kg/m³ at 4°C.

We can explain why a pond or lake freezes at the surface from this unusual thermal expansion behavior of water. When the atmospheric temperature drops from, say, 7°C to 6°C, the water at the surface of the pool also cools and decreases in volume. This means that the surface water is denser than the water below it, which has not cooled and decreased in volume. As a result, the surface water sinks and warmer water from below is forced to the surface to be cooled. Between 4°C and 0°C, however, the surface water expands as it cools, becoming less dense than the water below it. The mixing process stops, and eventually the surface water freezes. As the water freezes, the ice remains on the surface because ice is less dense than water. The ice continues to build up

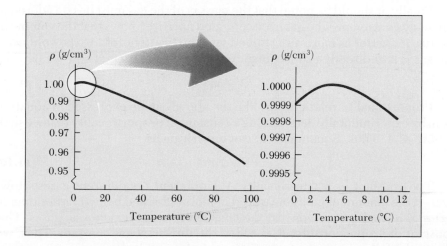

Figure 11.8
The density of water as a function of temperature. The inset at the right shows that the maximum density of water occurs at 4°C.

Figure 11.9
A gas confined to a cylinder whose volume can be varied with a movable piston.

Boyle's law

Charles–Gay-Lussac law

on the surface, while water near the bottom of the pool remains at 4°C. If this did not happen, fish and other forms of marine life would not survive. In fact, if it were not for this peculiarity of water, life as we now know it would not exist.

11.4   Ideal Gases

In this section we shall be concerned with the properties of a gas of mass m confined to a container of volume V at a pressure P and the temperature T. It would be useful to know how these quantities are related. If the gas is maintained at a very low pressure (or low density) and if the conditions of pressure and temperature are such that the gas does not condense into a liquid, the relationship is quite simple. Such a gas is commonly referred to as an *ideal gas.* (A more precise definition of an ideal gas will be given in a later section.) Simple substances that are gases at room temperature and atmospheric pressure behave as if they were ideal gases.

To determine how P, V, and T are related for an ideal gas, consider a gas confined to a cylindrical container whose volume can be varied by means of a movable piston, as in Figure 11.9. We shall assume that the cylinder does not leak and hence that the mass of the gas remains constant. For such a system, experiments provide the following information. First, when the gas is kept at constant temperature, its pressure is inversely proportional to its volume. This experimental fact is called **Boyle's law** and can be expressed as

$$P \propto \frac{1}{V} \tag{11.7}$$

Second, when the pressure of the gas is kept constant, its volume is directly proportional to its temperature. This is called the **Charles–Gay-Lussac law** and can be expressed as

$$V \propto T \tag{11.8}$$

The relationships given in Equations 11.7 and 11.8 can be expressed in a single proportionality as

$$PV \propto T$$

Finally, it should be noted that the mass m of the gas must be considered in our analysis. Anyone who has every pumped up a tire knows that the volume of the tire increases as more air is forced into the tire. Therefore, the volume of the gas is also directly proportional to the mass of the gas. Thus, we have

$$PV \propto mT \tag{11.9}$$

Proportionality relationships like the above can be expressed in equation form by experimentally determining a constant of proportionality that we shall denote as A. Thus, Equation 11.9 can be written as

$$PV = AmT \tag{11.10}$$

The constant of proportionality A is different for different gases. However, if the amount of gas is measured in moles (defined below) rather than in terms of its mass, one finds that the constant A is the *same* for all gases. This constant of proportionality, R, is called the **universal gas constant**. Thus, we obtain the following important relationship for an ideal gas:

P H Y S I C S   I N   A C T I O N

This series of photographs shows what happens when balloons filled with air are cooled in liquid nitrogen, which has a boiling point of −195°C or 77 K. In (a) and (b), the balloons are inserted in the nitrogen, and after a few minutes the balloons collapse to a very small volume. At this point, the air inside the balloons has condensed into a liquid. When the balloons are removed from the nitrogen and begin to warm up, the liquid air returns to a gas and the balloons expand as in (c) and (d). After several minutes, their volumes and shapes are restored to their original values. This demonstration shows that the volume of the gas in the balloons is proportional to the temperature of the gas (Charles–Gay-Lussac law). As T is lowered, the volume occupied by the gas decreases, and the balloon collapses. When the balloons return to room temperature, their original configuration is restored. What do you suppose would happen to the balloons if they were heated to 350 K?

Ideal Gases

(a)

(b)

(c)

(d)

(Photographs courtesy of Henry Leap and Jim Lehman)

Ideal gas law

$$PV = nRT \qquad \qquad \textbf{(11.11)}$$

In this equation, n is the number of moles of the gas. The number of moles of a substance is related to its mass m through the expression

$$n = \frac{m}{M} \qquad \qquad \textbf{(11.12)}$$

where M is a quantity called the **molecular mass** or **atomic mass** of the substance, usually expressed in grams per mole. From Equation 11.12, we see that *1 mole (abbreviated mol) of a pure substance contains a mass in grams numerically equal to the molecular mass of the substance.* For a better understanding of M, you should examine Appendix B, and in particular the column labeled "Atomic Mass." Note that aluminum has an atomic mass of 27 g/mol; thus, 27 g of aluminum constitutes 1 mol of aluminum. The same table shows that 1 mol of lead has a mass of 207 g.

Because we are primarily interested in gases in this section, consider oxygen, which has an atomic mass of approximately 16. In the gaseous state, oxygen consists of two oxygen atoms bound together to form a molecule, O_2. Therefore, the molecular mass of oxygen is 32 g/mol, and the mass of 1 mol of oxygen is 32 g.

In SI units, where pressure is expressed in Pa (N/m^2) and volume in cubic meters (m^3), the product PV has units of newton-meters, or joules, and R has the value

$$R = 8.31 \text{ J/mol} \cdot \text{K}$$

If the pressure is expressed in atmospheres and the volume in liters (1 liter = 10^3 cm^3 = 10^{-3} m^3), then R has the value

$$R = 0.0821 \text{ liter} \cdot \text{atm/mol} \cdot \text{K}$$

It is important to note that when using Equation 11.11, the temperature must always be expressed in kelvins.

Scientists have defined an ideal gas as one that obeys the ideal gas law $PV = nRT$ under all conditions. In reality, no gas is ideal. However, the concept is very useful in view of the fact that real gases behave as if they were ideal gases at and below atmospheric pressure.

Example 11.5   How Many Moles in the Container?

An ideal gas occupies a volume of 100 cm^3 at 20°C and 10^{-3} atm. Determine the number of moles of gas in the container.

Solution
The quantities given are volume, pressure, and temperature: that is, $V = 100$ cm^3 = 0.1 liter, $P = 10^{-3}$ atm, and $T = 20$°C = 293 K. Using Equation 11.11, we get

$$n = \frac{PV}{RT} = \frac{(10^{-3} \text{ atm})(0.1 \text{ liter})}{(0.0821 \text{ liter} \cdot \text{atm/mol} \cdot \text{K})(293 \text{ K})} = 4.16 \times 10^{-6} \text{ mol}$$

Again, note that you must express T as an absolute temperature when using the ideal gas law.

Example 11.6 Squeezing a Tank of Gas

Pure helium gas is admitted into a cylinder containing a movable piston. The initial volume, pressure, and temperature of the gas are 15 liters, 2 atm, and 300 K. If the volume is decreased to 12 liters and the pressure increased to 3.5 atm, find the final temperature of the gas. Assume it behaves like an ideal gas.

Solution

If no gas escapes from the cylinder, the number of moles remains constant; therefore using $PV = nRT$ at the initial and final points gives

$$\frac{P_iV_i}{T_i} = \frac{P_fV_f}{T_f}$$

where "i" and "f" refer to the initial and final values. Solving for T_f, we get

$$T_f = \left(\frac{P_fV_f}{P_iV_i}\right)(T_i) = \frac{(3.5\ \text{atm})(12\ \text{liters})}{(2\ \text{atm})(15\ \text{liters})}(300\ \text{K}) = \boxed{420\ \text{K}}$$

Example 11.7 Heating a Bottle of Air

A sealed glass bottle at 27°C contains air at atmospheric pressure and has a volume of 30 cm^3. It is then tossed into an open fire. When the temperature of the air in the bottle reaches 200°C, what is the pressure inside the bottle? Assume any volume changes of the bottle are small enough to be negligible.

Solution

This example is approached in the same fashion as that used in Example 11.6. We start with the expression

$$\frac{P_iV_i}{T_i} = \frac{P_fV_f}{T_f}$$

Since the initial and final volumes of the gas are assumed equal, this expression reduces to

$$\frac{P_i}{T_i} = \frac{P_f}{T_f}$$

This gives

$$P_f = \left(\frac{T_f}{T_i}\right)(P_i) = \left(\frac{473\ \text{K}}{300\ \text{K}}\right)(1\ \text{atm}) = \boxed{1.58\ \text{atm}}$$

Obviously, the higher the temperature, the higher the pressure exerted by the trapped air. Of course, if the pressure rises high enough, the bottle will shatter.

Exercise In this example, we neglected the change in volume of the bottle. If the coefficient of volume expansion for glass is 27×10^{-6} (C°)$^{-1}$, find the magnitude of this volume change.

Answer 0.14 cm^3.

Example 11.8   The Volume of 1 Mol of Gas

Verify that one mol of oxygen occupies a volume of 22.4 liters at 1 atm and 0°C.

Solution
Let us solve the ideal gas equation for V:

$$V = \frac{nRT}{P}$$

In our problem, the mass of the gas, m, is assumed to be one mol, M. Thus

$$n = \frac{m}{M} = 1 \text{ mol}$$

Let us convert the temperature to the Kelvin scale and substitute into the ideal gas equation:

$$V = \frac{nRT}{P} = \frac{(1 \text{ mol})(0.0821 \text{ liter} \cdot \text{atm/mol} \cdot \text{K})(273 \text{ K})}{1 \text{ atm}} = \boxed{22.4 \text{ liters}}$$

You should note that the answer above has general validity. *One mole of any gas at standard temperature and pressure (STP) occupies a volume of 22.4 liters.*

Exercise   Repeat the calculations in this example problem for hydrogen gas to show that it also occupies a volume of 22.4 liters at STP.

11.5   Avogadro's Number and the Ideal Gas Law

In the early 1800s an important field of experimental investigation was performed in order to determine the relative masses of molecules. The equipment used in one such investigation is shown in Figure 11.10. The lower section of the glass vessel in this figure contains two electrodes that are connected to a battery so that an electric current can be passed through lightly salted water. (The purpose of the salt is to improve the electrical conductivity of the water.) Bubbles of gas are produced at each electrode and become trapped in the column above it. An analytical examination of these gases reveals that the column above the positive electrode contains oxygen gas while the column above the negative electrode contains hydrogen gas. Obviously, the current decomposes the water into its constituent parts. Experimentally, it is found that the volume of hydrogen gas collected is always exactly twice the volume of oxygen gas collected, as you would expect in light of the chemical composition of water, H_2O. Additionally, it is found that if 9 g of water are decomposed, 8 g of oxygen and 1 g of hydrogen are collected. From this information, it is possible to determine the relative masses of oxygen and hydrogen molecules. Following such an experimental investigation, Amedeo Avogadro in 1811 made the following assumption:

> Equal volumes of gas at the same temperature and pressure contain the same number of molecules.

With this assumption, and the fact that the ratio of hydrogen to oxygen molecules collected in the experiment is 2 : 1, we conclude that the mass of an oxygen molecule must be sixteen times that of a hydrogen molecule.

Figure 11.10
When water is decomposed by an electric current, the volume of hydrogen gas collected is twice that of oxygen.

In Example 11.8, we noted that one mole of any gas occupies the same volume at 1 atm and at 0°C. Thus, a corollary to Avogadro's hypothesis is

one mole of any gas at standard temperature and pressure contains the same number of molecules.

Specifically, the number of molecules contained in one mole of any gas is 6.02×10^{23} molecules/mol. This number is known as **Avogadro's number,** and is given the symbol N_A where

$$N_A = 6.02 \times 10^{23} \text{ molecules/mol} \qquad \textbf{(11.13)}$$

Avogadro's number

It is of historical interest to note that Avogadro never knew the precise value for this number. An alternative method for calculating the number of moles of any gas in a container is to divide the total number of molecules present, N, by Avogadro's number, or

$$n = \frac{N}{N_A} \qquad \textbf{(11.14)}$$

With this expression, we are able to rewrite the ideal gas law in the following alternative form:

$$PV = nRT = \frac{N}{N_A} RT$$

or

$$PV = NkT \qquad \textbf{(11.15)}$$

Ideal gas law

where k is a constant called **Boltzmann's constant** and has the value

$$k = \frac{R}{N_A} = 1.38 \times 10^{-23} \text{ J/K} \qquad \textbf{(11.16)}$$

Boltzmann's constant

Example 11.9 What is Avogadro's Number?

One mole of hydrogen has a mass of 1.0078 g. In the early 20th century, it was found that the mass of a hydrogen atom was approximately 1.673×10^{-24} g. Use these values to find Avogadro's number.

Solution

The number of molecules in 1 mol of hydrogen can be found by dividing the mass of 1 mol by the mass per atom. Hence, we find

$$N_A = \frac{1.0078 \text{ g/mol}}{1.673 \times 10^{-24} \text{ g/atm}} = \boxed{6.02 \times 10^{23} \text{ atoms/mol}}$$

11.6 The Kinetic Theory of Gases

In Section 11.4 we discussed the properties of an ideal gas using such quantities as pressure, volume, number of moles, and temperature. In this section we shall show that large-scale properties of a gas such as pressure and temperature can be understood on the basis of what is happening on the atomic scale. In addition, we shall re-examine the ideal gas law in terms of the behavior of individual molecules which compose the gas.

The glass vessel contains dry ice (solid carbon dioxide). The white cloud is carbon dioxide vapor, which is denser than air and hence falls from the vessel as shown. (© R. Folwell/Science Photo Library)

Because the interactions between the molecules of a gas are much weaker than in solids and liquids, our present discussion will be restricted to the molecular behavior of gases. In this view of gas behavior, called the kinetic theory, it is assumed that gas molecules move about in a random fashion, colliding with the walls of the container and with each other. Each molecule will be considered to be a hard sphere and each collision is assumed to be perfectly elastic. In practice, the assumptions of this section work well only for monatomic gases (those in which each particle is a single atom) because the energy of the atoms in a monatomic gas is solely translational kinetic energy. One must modify the theory for molecules consisting of two or more atoms bound together, such as O_2 and CO_2, to include the energy associated with any rotation or vibration of the molecules.

Molecular Model for the Pressure of an Ideal Gas

Assumptions of kinetic theory for an ideal gas

We shall first use the kinetic theory of gases to show that the pressure that a gas exerts on the walls of its container is a consequence of the collisions of the gas molecules with the walls. The following assumptions will be made:

1. *The number of molecules is large, and the average separation between them is large compared with their dimensions.* This means that the molecules occupy a negligible volume compared with the volume of the container.
2. *The molecules obey Newton's laws of motion, but as a whole they move in a random fashion.* By random fashion, we mean that any one molecule can move in any direction with any speed. We also assume that the distribution of velocities does not change in time, despite the collisions between molecules. That is, a certain percentage of molecules move at high speeds, a certain percentage move at low speeds, and so on.
3. *The molecules undergo elastic collisions with each other and with the walls of the container.* Thus, in the collisions *both kinetic energy and momentum are conserved.*
4. *The forces between molecules are negligible except during a collision.* The forces between molecules are short-range, so that the molecules interact with each other only during collisions.
5. *The gas under consideration is a pure substance,* that is, *all molecules are identical.*

Figure 11.11
A cubical box with sides of length d containing an ideal gas. The molecule shown moves with velocity **v**.

Now let us derive an expression for the pressure of N molecules of an ideal gas in a container of volume V. The container is assumed to be in the shape of a cube with edges of length d (Fig. 11.11). We shall focus our attention on one of these molecules, of mass m_0 and assumed to be moving such that its component of velocity in the x direction is v_x (Fig. 11.12). As the molecule collides elastically with any wall, its velocity is reversed. Since the momentum, p, of the molecule is $m_0 v_x$ before the collision and $-m_0 v_x$ after the collision, the *change in momentum of the molecule is*

$$\Delta p_x = mv_f - mv_i = -m_0 v_x - m_0 v_x = -2m_0 v_x$$

Because the momentum of the system consisting of the wall and the molecule must be conserved, we see that, if the change in momentum of the molecule is $-2m_0 v_x$, then the change in momentum of the wall must be $2m_0 v_x$. Applying the impulse-momentum theorem to the wall gives

$$\overline{F}\,\Delta t = \Delta p = 2m_0 v_x$$

$$F_{\Delta}t = \Delta p$$

Before collision

In order for the molecule to make two collisions with the same wall, it must travel a distance $2d$ along the x direction. Therefore, the time between two collisions with the same wall is

$$\Delta t = \frac{2d}{v_x}$$

The substitution of this result into the impulse-momentum equation enables us to express the average force of a molecule on the wall:

$$\overline{F} = \frac{2m_0 v_x}{\Delta t} = \frac{2m_0 v_x^2}{2d} = \frac{m_0 v_x^2}{d}$$

The net average force exerted on the wall by all the molecules is found by adding the forces exerted by the individual molecules:

$$\overline{F}_{\text{net}} = \frac{m_0}{d}\left(v_{x1}^2 + v_{x2}^2 + \cdots + v_{xN}^2\right)$$

In this equation, v_{x1} is the x component of velocity of molecule 1, v_{x2} is the x component of velocity of molecule 2, and so on. The summation terminates when we reach molecule N because there are N molecules in the container.

Figure 11.12
A molecule makes an elastic collision with the wall of the container. Its x component of momentum is reversed, and momentum is imparted to the wall.

To proceed further, note that the average value of the square of the velocity in the x direction for the N molecules is

$$\overline{v_x^2} = \frac{v_{x1}^2 + v_{x2}^2 + \cdots + v_{xN}^2}{N}$$

Thus, the net average force can be written as

$$\overline{F}_{\text{net}} = \frac{m_0}{d}\,N\overline{v_x^2}$$

Now let us focus our attention on one individual molecule in the container and say that this molecule has velocity components v_x, v_y, and v_z. The Pythagorean theorem relates the square of the velocity to the square of these components:

$$v^2 = v_x^2 + v_y^2 + v_z^2$$

Hence, the average value of v^2 for all the molecules in the container is related to the average values of v_x^2, v_y^2, and v_z^2 according to the expression

$$\overline{v^2} = \overline{v_x^2} + \overline{v_y^2} + \overline{v_z^2}$$

However, the average velocity is the same in any direction because the motion is completely random. Therefore,

$$\overline{v_x^2} = \overline{v_y^2} = \overline{v_z^2}$$

and we have

$$\overline{v^2} = 3\overline{v_x^2}$$

Thus, the net average force on the wall is given by

$$\overline{F}_{\text{net}} = \frac{N}{3}\left(\frac{m_0 \overline{v^2}}{d}\right)$$

This expression allows us to find the pressure exerted on the wall:

$$P = \frac{\overline{F}_{net}}{A} = \frac{\overline{F}_{net}}{d^2} = \frac{1}{3}\left(\frac{N}{d^3}\, m_0\overline{v^2}\right) = \frac{1}{3}\left(\frac{N}{V}\right)(m_0\overline{v^2})$$

Pressure of an ideal gas

$$P = \frac{2}{3}\left(\frac{N}{V}\right)\left(\frac{1}{2}\, m_0\overline{v^2}\right) \qquad (11.17)$$

The cork on the champagne bottle is blown off as a result of the large buildup of carbon dioxide gas pressure when the bottle is shaken. *(© Steve Niedorf, The IMAGE Bank)*

This result shows that *the pressure is proportional to the number of molecules per unit volume and to the average translational kinetic energy of the molecules.* With this simplified model of an ideal gas, we have arrived at an important result that relates the large-scale quantity of pressure to an atomic quantity, the average molecular speed. Thus, we have a key link between the atomic world and the large-scale world.

You should note that Equation 11.17 verifies some features of pressure that are probably familiar to you. One way to increase the pressure inside a container is to increase the number of molecules per unit volume in the container. For example, you do this when you add air to a tire. The pressure in the tire can also be increased by increasing the average translational kinetic energy of the molecules in the tire. As we shall see shortly, this can be accomplished by increasing the temperature of the gas inside the tire. This is why the pressure inside a tire increases as the tire heats up during long trips. The continuous flexing of the tires as they move along the road surface generates heat that is transferred to the air inside the tires. This added heat increases the temperature of the air, which in turn produces an increase in pressure in the tires.

Molecular Interpretation of Temperature

We can obtain some insight into the meaning of temperature by first writing Equation 11.17 in the more familiar form

$$PV = \frac{2}{3}\, N\left(\frac{1}{2}\, m_0\overline{v^2}\right)$$

Now let us compare this with the ideal gas equation (Eq. 11.15),

$$PV = NkT$$

Equating the right-hand sides of these two expressions and rearranging terms, we have

$$\frac{2}{3}\, N\left(\frac{1}{2}\, m_0\overline{v^2}\right) = NkT$$

Average kinetic energy

$$\frac{1}{2}\, m_0\overline{v^2} = \frac{3}{2}\, kT \qquad (11.18)$$

This important result says that the average kinetic energy of gas molecules is directly proportional to the absolute temperature of the gas. As the temperature increases, the molecules move with higher average speeds. Conversely, temperature is a direct measure of the average molecular kinetic energy of an ideal gas.

The square root of $\overline{v^2}$ is called the **root mean square** (rms) speed of the molecules. From Equation 11.18, we find this to be

$$v_{rms} = \sqrt{\overline{v^2}} = \sqrt{\frac{3kT}{m_0}}$$ **(11.19)**   Root mean square speed

We can use the fact that $k = R/N_A$ and that $m_0 N_A = M$, where M is the molecular weight, to rewrite Equation 11.19 as

$$v_{rms} = \sqrt{\frac{3kT}{m_0}} = \sqrt{\frac{3RT}{N_A m_0}} = \sqrt{\frac{3RT}{M}}$$ **(11.20)**

This expression for the rms speed shows that at a given temperature lighter molecules move faster, on the average, than heavier molecules. For example, hydrogen, with a molecular mass of 2 g/mol, moves four times as fast as oxygen, whose molecular mass is 32 g/mol if they are at the same temperature. Note that the rms speed is not the speed at which a gas molecule moves across a room because a moving molecule undergoes several billion collisions per second with other molecules under standard conditions. Table 11.2 lists the rms speeds for various molecules at 20°C.

Example 11.10   A Tank of Helium

A tank of volume 0.3 m³ contains 2 moles of helium gas at 20°C. Since the helium behaves like an ideal gas, (a) find the average kinetic energy per atom.

Solution
From Equation 11.18, we see that the average kinetic energy per atom is

$$\frac{1}{2} m_0 \overline{v^2} = \frac{3}{2} kT = \frac{3}{2} (1.38 \times 10^{-23} \text{ J/K})(293 \text{ K}) = \boxed{6.07 \times 10^{-21} \text{ J}}$$

Table 11.2
Some rms Speeds

Gas	Molecular Mass (g/mol)	v_{rms} at 20°C (m/s)[a]
H_2	2.02	1902
He	4.0	1352
H_2O	18	637
Ne	20.1	603
N_2 and CO	28	511
NO	30	494
CO_2	44	408
SO_2	48	390

[a] All values were calculated using Equation 11.19.

(b) Determine the rms speed of the atoms.

Solution
Because the molecular mass of helium is 4 g/mol = 4×10^{-3} kg/mol, the rms speed is

$$v_{\text{rms}} = \sqrt{\frac{3RT}{M}} = \sqrt{\frac{3(8.317 \text{ J/mol} \cdot \text{K})(293 \text{ K})}{4 \times 10^{-3} \text{ kg/mol}}} = \boxed{1.35 \times 10^3 \text{ m/s}}$$

The same result is obtained using $v_{\text{rms}} = \sqrt{3kT/m_0}$. Remember that m_0, the mass of one molecule, is obtained by dividing the molecular mass by the number of particles in 1 mole of the gas (Avogadro's number). Try it!

▶ SPECIAL TOPIC
Evaporation, Humidity, and Boiling

If a container of water is left open, the water slowly evaporates. That is, it spontaneously goes from the liquid state to the gaseous state. The kinetic theory of gases enables us to explain why this phenomenon occurs.

The molecules of a liquid are in continuous random motion. Occasionally, the upward component of velocity of a molecule near the surface of the liquid will be sufficiently large to enable the molecule to escape from the liquid and become a gaseous molecule.

Equation 11.18 shows that the temperature of an ideal gas is a measure of the average kinetic energy of its molecules. This concept can be extended to liquids as well. The molecules that evaporate from the liquid are those that have sufficient energy (greater than the average kinetic energy) to break free of their neighbors on the surface. Since the evaporating molecules are very energetic, the molecules remaining in the liquid must have lower-than-average kinetic energy. For this reason, the liquid becomes cooler as a result of evaporation. You may have noticed this cooling due to evaporation after a swim or shower. Your body feels cooler when you are wet than when you are dry because the water is evaporating. As the rate of evaporation increases, so does the rate of cooling. In another familiar example, when a doctor swabs your arm with alcohol (which evaporates faster than water) before giving you an injection, the alcohol serves two purposes. It sterilizes the area to be injected, and it slightly numbs the area as a result of its rapid evaporation and subsequent cooling of the skin.

Figure 11.13
The evaporation rate increases as air moves over the surface of the liquid.

Humidity

The rate of evaporation is enhanced if air is moved rapidly over the surface of the liquid. This occurs because the liquid molecules that vaporize are removed by the moving air before they have time to undergo collisions with air molecules near the surface that could deflect them back into the liquid (Fig. 11.13). For the same reason, fanning yourself on a hot day helps you cool down.

Let us now look at the exact opposite of fanning. Let us assume that water is placed in a sealed container such that the gaseous molecules cannot escape, as in Figure 11.14. In this case, the air above the liquid will absorb water molecules until it can hold no more. At this point, the air is said to be **saturated** with water vapor. Saturation occurs when the number of water molecules returning to the liquid in some time interval is equal to the number leaving the liquid in the same time interval. The amount of water vapor that air can hold increases as the temperature of the air increases.

Figure 11.14
A liquid placed in a sealed container. The air above the liquid becomes saturated when the number of molecules leaving the liquid equals the number returning to the liquid in some time interval.

The mass of water vapor in the air per unit volume is called the **absolute humidity.** Thus, absolute humidity is a measure of the dampness of the air. If there is a large amount of water vapor in the air, the air feels damp, or humid.

When weather forecasters refer to humidity, they are talking not about absolute humidity but about relative humidity. In order to understand the concept of *relative humidity,* we must first discuss partial pressures and vapor pressure.

We have seen that atmospheric pressure is about 760 mm Hg. The total atmospheric pressure at any location is the sum of the pressures exerted by all the gases that make up the atmosphere. That is, the oxygen molecules in the atmosphere exert a pressure on an object, as do the nitrogen molecules in the atmosphere, and so on for all the molecules contained in our atmosphere. The pressure exerted by each constituent gas is called the **partial pressure** of that gas. According to the **law of partial pressures**

the pressure exerted by each component of a mixture of gases is the same as the pressure that component would exert if it were alone in the volume occupied by the mixture. This means that each component acts virtually independently of the rest of the mixture.

Vapor pressure is *the pressure exerted by a gas when it is in equilibrium with its liquid form.* That is, in the case of the closed container of water discussed earlier, the partial pressure exerted by the gaseous water when saturation exists is equal to its vapor pressure. It should be evident that the partial pressure of a gas at a given temperature may equal but never exceed its vapor pressure. The vapor pressure of water at various temperatures is given in Table 11.3.

The **relative humidity,** *RH,* at a given temperature is defined as

$$RH \equiv \frac{\text{partial pressure of water vapor}}{\text{vapor pressure of water}} \times 100$$

If air saturated with water vapor is cooled, some of the water vapor must condense (liquify) in order to reduce the partial pressure of the water vapor. For example, from Table 11.3, we see that if the temperature of saturated air is reduced from 20°C to 10°C, enough water vapor must condense out to lower the partial pressure from 17.5 mm Hg to 8.9 mm Hg. This is the reason for the formation of fog, clouds, rain, dew, and frost.

Relative humidity can be measured by determining what is called the **dew point,** which is *the temperature at which the air becomes saturated with water vapor.* This temperature can be found by slowly adding ice to a metal container filled with water. (The outside surface of the container should be clean, and the liquid should be stirred as the ice melts.) The temperature at which moisture starts condensing on the outside surface of the container is the dew point.

Table 11.3
Vapor Pressure of Water at Various Temperatures

Temperature (°C)	Vapor Pressure (mm Hg)
0	4.6
10	8.9
20	17.5
100	760.0

Boiling

Boiling can be thought of as evaporation taking place throughout the volume of a liquid. *Boiling begins when the vapor pressure is equal to atmospheric pressure.* From Table 11.3, we see that this occurs for water at 100°C. In order to understand the mechanism of boiling, consider what happens when a bubble of gas forms in the interior of a volume of liquid. If the vapor pressure inside the bubble is less than the pressure exerted on it by the liquid, the bubble collapses. If the pressure inside the bubble is equal to atmospheric pressure, however, the bubble is sustained and rises to the surface and the vapor escapes.

It should be obvious from this discussion that the boiling point of a liquid is influenced by the external pressure exerted on it. In general, as the pressure on the liquid is increased, the boiling point is also increased. That is, in order to reach the boiling point, the temperature of the liquid must be increased until the vapor pressure of the gas equals the external pressure.

Just as an increase in pressure increases the boiling point, a reduction of pressure reduces it. For example, at an altitude of 12 mi above sea level, atmospheric pressure is so low that blood boils at body temperature. Since you would not want to verify this experimentally, let us consider another way the effect of pressure on boiling can be observed. Imagine that water is placed in a container from which the air can be gradually pumped. Furthermore, assume that the water is at room temperature when placed in the container and hence is far below its boiling point. As the air is pumped out of the container, the pressure on the water is reduced, as is its boiling point. Eventually, when the pressure is reduced to about 17.5 mm Hg (Table 11.3), the boiling point is reduced to room temperature and the water boils. Because boiling is merely rapid evaporation (which is a cooling process), the water in the container is being cooled as it boils. If you reduce the pressure even further, the water will boil more rapidly. Finally, the temperature of the liquid will reach the freezing point even though it is boiling. What would the pressure inside the container have to be for this to occur? Refer to Table 11.3 for your answer. ◀

Summary

The **zeroth law of thermodynamics** states that if two objects A and B are separately in thermal equilibrium with a third object, then A and B are in thermal equilibrium with each other.

The relationship between T_C, the *Celsius temperature*, and T, the *Kelvin temperature*, is

$$T_C = T - 273.15 \qquad \qquad (11.1)$$

The relationship between the *Fahrenheit* and *Celsius* temperatures is

$$T_F = \frac{9}{5} T_C + 32 \qquad \qquad (11.2)$$

When a substance is heated, it generally expands. If an object has an initial length L_0 at some temperature, and undergoes a change in temperature ΔT, its length changes by an amount ΔL that is proportional to its initial length and the temperature change:

$$\Delta L = \alpha L_0 \, \Delta T \qquad \text{(11.4)}$$

The parameter α is called the **average coefficient of linear expansion**.
   The change in area of a substance is given by

$$\Delta A = \gamma A_0 \, \Delta T \qquad \text{(11.5)}$$

where γ is the **average coefficient of area expansion** and is equal to 2α.
   Likewise, the change in volume of most substances is proportional to the initial volume, V_0, and the temperature change ΔT:

$$\Delta V = \beta V_0 \, \Delta T \qquad \text{(11.6)}$$

where β is the **average coefficient of volume expansion**, and is equal to 3α.
   An **ideal gas** is one that obeys the equation

$$PV = nRT \qquad \text{(11.11)}$$

where P is the pressure of the gas, V is its volume, n is the number of moles of gas, R is the universal gas constant (8.31 J/mole·K), and T is the absolute temperature in kelvins. A real gas at very low pressures behaves approximately as an ideal gas.
   The **pressure** of N molecules of an ideal gas contained in a volume V is given by

$$P = \frac{2}{3}\left(\frac{N}{V}\right)\left(\frac{1}{2}\,m_0 \overline{v^2}\right) \qquad \text{(11.17)}$$

where $\frac{1}{2}\,m_0\overline{v^2}$ is the **average kinetic energy per molecule**.
   The average kinetic energy of the molecules of a gas is directly proportional to the absolute temperature of the gas:

$$\frac{1}{2}\,m_0\overline{v^2} = \frac{3}{2}\,kT \qquad \text{(11.18)}$$

The **root mean square** (rms) speed of the molecules of gas is

$$v_{\text{rms}} = \sqrt{\frac{3kT}{m_0}} = \sqrt{\frac{3RT}{M}} \qquad \text{(11.19) and (11.20)}$$

Additional Reading

T. B. Greenslade, "The Maximum Density of Water," *The Physics Teacher*, November 1985, p. 474.

M. B. Hall, "Robert Boyle," *Sci. American*, August 1967, p. 84.

E. Jones and R. Childers, "Observational Evidence for Atoms," *The Physics Teacher*, October 1984, p. 354.

F. Jones, "Fahrenheit and Celsius, A History," *Physics Today*, 18, 1980, p. 594.

R. H. Romer, "Temperature Scales: Celsius, Fahrenheit, Kelvin, Reaumur, and Romer," *The Physics Teacher*, October 1982, p. 450.

A. F. Scott, "The Invention of the Balloon and the Birth of Chemistry," *Sci. American*, January 1984, p. 126.

R. E. Wilson, "Standards of Temperature," *Physics Today*, January 1953, p. 10.

Questions

1. Is it possible for two objects to be in thermal equilibrium if they are not in thermal contact with each other? Explain.

2. A piece of copper is dropped into a beaker of water. If the water's temperature rises, what happens to the temperature of the copper? When will the water and copper be in thermal equilibrium?

3. Explain why a column of mercury in a thermometer first descends slightly and then rises when the thermometer is placed in hot water.

4. What would happen if the glass of a thermometer expanded more upon heating than did the liquid inside?

5. "If the pressure above a liquid substance is reduced to the pressure of the triple point, the substance will boil and freeze at the same temperature." Is this statement true?

6. A steel wheel bearing is 1 mm smaller in diameter than an axle. How can the bearing be fit onto the axle without removing any material?

7. Markings to indicate length are placed on a steel tape in a room that has a temperature of $22°C$. Are measurements made with the tape on a day when the temperature is $27°C$ too long, too short, or accurate? Defend your answer.

8. Why is a power line more likely to break in winter rather than the summer even if it is loaded with the same weight?

9. On a hot afternoon you often hear creaking noises in the attic of a house. These noises are also frequently heard at night. Why?

10. If a jar lid is screwed on too tightly to remove, it is sometimes possible to loosen the lid by holding it under hot water. Explain.

11. When drinking glasses become stuck together, an old trick is to fill the inner glass with water and to run water of a different temperature over the sides of the outer glass. Which water should be hot and which cold? Why?

12. Chimneys are never used as a weight-bearing part of the structure of a building. Why is that the case?

13. Determine the number of grams in 1 mole of (a) hydrogen, (b) helium, and (c) carbon monoxide.

14. Two identical cylinders at the same temperature each contain the same kind of gas. If cylinder **A** contains three times more gas than cylinder **B**, what can you say about the relative pressures in the cylinders?

15. Why do vapor bubbles in a pot of boiling water get larger as they approach the surface?

16. A microwave oven is used to heat food in a sealed pouch. Why should the pouch be pricked with a fork before heating?

17. What does the ideal gas law predict about the volume of a gas at absolute zero? Why is this result incorrect?

18. The gas from a carbon-dioxide fire extinguisher emerges so rapidly that no immediate heat transfer occurs between the gas and the atmosphere. Use this information to explain why the gas is very cold.

19. A bottle containing a soft drink is brought inside after being left outside in the cold. The outside temperature is $-10°C$. The soft drink is liquid when it is brought inside, but when it is opened, the contents immediately freeze. Explain this phenomenon.

20. During normal breathing the lungs expand upon inhalation and contract upon exhalation. Does the ideal gas law apply to the volume of the lungs under these circumstances? Explain.

21. If a helium-filled balloon is placed in a freezer, will its volume increase, decrease, or remain the same?

22. What happens to a helium-filled balloon released into the air? Will it expand or contract? Will it stop rising at some height?

23. Milk is a colloidal suspension, which means it contains tiny particles of insoluble matter that float about in the fluid (whey) without settling. Under a high-power microscope the particles are found to be darting around at random (referred to as Brownian motion).

How does the molecular model account for this motion?

24. After cooking food in a pressure cooker, why is it very important to cool the container with cold water before attempting to remove the lid?
25. At room temperature, the average speed of an air molecule is several hundred meters per second. A molecule traveling at this speed should travel across a room in a small fraction of a second. In view of this, why does it take the odor of perfume (or other smells) several seconds to travel across the room?
26. Although the average speed of gas molecules in thermal equilibrium at some temperature is greater than zero, the average velocity is zero. Explain.
27. One container is filled with helium gas and another with argon gas. If both containers are at the same

temperature, which molecules have the higher rms speed?

28. As you go higher in the atmosphere, should you expect to find the percentage of nitrogen molecules to oxygen molecules increasing or decreasing? Explain.
29. A liquid partially fills a container. Explain why the temperature of the liquid decreases when the container is partially evacuated. (Using this technique, it is possible to freeze water at temperatures above 0°C.)
30. In view of the fact that hot air rises, why does it generally become cooler as you climb a mountain? (Note that air is a poor conductor of heat.)

Problems

Section 11.2   Thermometers and Temperature Scales

1. The pressure in a constant volume gas thermometer is 0.700 atm at 100°C and 0.512 atm at 0°C. (a) What is the temperature when the pressure is 0.0400 atm? (b) What is the pressure at 450°C?
2. A constant volume gas thermometer is calibrated in dry ice (−80°C) and in boiling ethyl alcohol (78°C). The two pressures are 0.900 atm and 1.635 atm. (a) What value of absolute zero does the calibration yield? (b) What pressures would be found at the freezing and boiling points of water?
3. Convert the following temperatures to Celsius and kelvins: (a) The normal human body temperature of 98.6°F and (b) the temperature of a cold day of −5°F.
4. Convert the following temperatures to Fahrenheit and kelvins: (a) the boiling point of liquid hydrogen −252.87°C and (b) the temperature of a room at 20°C.
5. The temperature in the interior of some stars is approximately 2.00×10^7 K. Express this temperature in (a) degrees Fahrenheit and (b) degrees Celsius.
6. Recently the temperature on a certain day changed by 54 F°. Express this change in temperature on (a) the Celsius scale and (b) the Kelvin scale.
7. The highest recorded temperature on Earth was 136°F, at Azizia, Libya, in 1922. The lowest recorded temperature was −127°F, at Vostok Station, Antarctica, in 1960. Express these temperature extremes in degrees Celsius.
8. The boiling point of sulfur is 444.60°C. The melting point is 586.1 F° below the boiling point. (a) Deter-

mine the melting point in degrees Celsius. (b) Find the melting and boiling points in degrees Fahrenheit.

9. Show that the temperature −40° is unique in that it has the same numerical value on the Celsius and Fahrenheit scales.
10. Show that if the temperature on the Celsius scale changes by ΔT_C, the Fahrenheit temperature changes by an amount ΔT_F given by $\Delta T_F = (9/5) \Delta T_C$.

Section 11.3   Thermal Expansion of Solids and Liquids

11. The New River Gorge bridge in West Virginia is a steel arch bridge 518 m in length. How much will its length change between temperature extremes of −20°C and 35°C?
12. A copper steam pipe is 2 m long and is installed in a basement when the temperature is 20°C. What is the length of the pipe when it carries steam at 120°C?
13. A structural steel beam is 20 m long when installed at 20°C. How much will its length change over the temperature extremes of −6°C to 30°C?
14. A grandfather clock is controlled by a swinging brass pendulum of length 1.3 m at a temperature of 20°C. (a) What is the length of the pendulum rod when the temperature drops to 0°C? (b) If the period of a pendulum is given by $T = 2\pi\sqrt{L/g}$, where L is its length, does the change in length of the rod cause the clock to run fast or slow?
15. The concrete sections of a certain highway are designed to have a length of 30 m. The sections are poured and cured at 10°C. What minimum spacing should the engineer leave between the sections to

Additional Problems

58. How many kilograms of nitrogen are contained in a tank whose volume is 0.75 m³ when the gauge pressure is 100 atm and the temperature is 27°C?

59. The active element of a certain laser is an ordinary glass rod 20 cm long and 1 cm in diameter. If the temperature of the rod increases by 75°C, find the increase in (a) its length, (b) its diameter, and (c) its volume.

60. Start with the relation $L - L_0 = \alpha L_0 (T - T_0)$ to show that $\Delta V = \beta V_0 \, \Delta T$, where $\beta = 3\alpha$.

61. Show that the total pressure P in a container filled with a mixture of several different ideal gases is given by $P = P_1 + P_2 + P_3 + \ldots$, where P_1, P_2, etc., are the pressures that each gas would exert if it alone filled the container (or the partial pressures of the respective gases). This is known as Dalton's law of partial pressures.

62. A patient in an oxygen tent breathes a mixture of 50% oxygen (O_2) and 50% nitrogen (N_2) *by mass*. If the total pressure is 1.0 atm, what is the partial pressure of each of the two gases in the mixture?

63. A cylinder originally contains 0.30 g of argon (Ar). Nitrogen (N_2) is then added, keeping the total pressure and temperature constant, until the volume is increased by 40%. Determine the mass of nitrogen added.

64. An ideal gas is confined to a container at a pressure of 1.07×10^5 Pa and a temperature of 27°C. What is the temperature of the gas if the pressure is decreased to 8×10^4 Pa at constant volume?

65. If 2 moles of a gas are confined to a 5-liter vessel at a pressure of 8 atm, what is the average kinetic energy of a gas molecule?

66. Calculate the density of helium stored at 300 K under a pressure of 1 atm.

67. Before beginning a long trip on a hot day, a driver inflates an automobile tire to a gauge pressure of 1.8 atm at 300 K. At the end of the trip the pressure in the tire has increased to 2.2 atm. (a) Assuming the volume has remained constant, what is the temperature of the air inside the tire? (b) What volume of air measured at atmospheric pressure should be released from the tire so that the pressure returns to the initial value? Assume that the air is released during a short time interval during which the temperature remains at the value found in Part (a).

68. Calculate the number of molecules per cubic centimeter in an ideal gas at 500 K and pressure at 3 atm.

69. A liquid with a coefficient of volume expansion β just fills a spherical shell of volume V at a temperature T (Fig. 11.16). The shell is made of a material that has a coefficient of linear expansion of α. The liquid is free to expand into a capillary of cross-sectional area A at the top. (a) If the temperature increases by ΔT, show that the liquid rises in the capillary by an amount Δh given by $\Delta h = \dfrac{V}{A}(\beta - 3\alpha) \, \Delta T$. (b) For a typical system, such as a mercury thermometer, why is it a good approximation to neglect the expansion of the shell?

Figure 11.16   Problems 69 and 70

70. A mercury thermometer is constructed as in Figure 11.16. The capillary tube has a diameter of 0.005 cm, and the bulb has a diameter of 0.30 cm. Neglecting the expansion of the glass, find the change in height of the mercury column for a temperature change of 25 C°.

71. A particular gas sample at 1 atm has a density of 3.75×10^{-3} g/cm³. Determine the rms speed of the gas molecules.

72. A glass tube 1.5 m in length and closed at one end is weighted and lowered to the bottom of a freshwater lake. When the tube is recovered, a marking indicator shows that water rose to within 0.4 m of the closed end. Determine the depth of the lake. Assume constant temperature.

73. A hollow aluminum cylinder is to be fitted over a steel piston. At 20°C the inside diameter of the cylinder is 99% of the outside diameter of the piston. To what common temperature should the two pieces be heated in order that the cylinder just fits over the piston?

74. Two small containers of equal volume 100 cm³ contain helium gas at 0°C and 1 atm pressure. The two containers are joined by a small open tube of negligible volume. What is the pressure of the gas in each container if the temperature of one of the containers is raised to 100°C while the other is kept at 0°C?

75. At $T = 0$°C, each one of three metal bars (two of aluminum and one of invar) is drilled with two holes a distance d apart. Pins are put through the holes to create an equilateral triangle. If the bars are then heated to 100°C, what will be the angle between the two aluminum bars?

76. When an ideal gas is held at constant pressure, a small change in volume, ΔV, associated with a small change

in temperature, ΔT, is given by $P(\Delta V) = nR(\Delta T)$.
(a) Show that the volume coefficient of thermal expansion for an ideal gas at constant pressure is given by $\beta = 1/T$, where T is the Kelvin temperature.
(b) What value does this equation predict for β at $0°C$?

77. A vertical cylinder of cross-sectional area 0.05 m² is fitted with a tight-fitting, frictionless piston of mass 5 kg (Fig. 11.17). If there are 3 moles of an ideal gas in the cylinder at 500 K, determine the height h at which the piston will be in equilibrium under its own weight.

Figure 11.17
Problem 77

78. An air bubble originating from a deep-sea diver has a radius of 2 mm at some depth h. When the bubble reaches the surface of the water, it has a radius of 3 mm. Assuming the temperature of the air in the bubble remains constant, determine (a) the depth h of the diver and (b) the absolute pressure at this depth.

"Let's go over to Celsius' place. I hear it's only 36° over there."

If the student is in good shape and lifts the weight, say, once every 5 s, it will take him about 12 h to perform this feat. Clearly, it is much easier to lose weight by dieting.

12.2   Specific Heat

The quantity of heat energy required to raise the temperature of a given mass of a substance by some amount varies from one substance to another. For example, the heat required to raise the temperature of 1 kg of water by 1 C° is 4186 J, but the heat required to raise the temperature of 1 kg of copper by 1 C° is only 387 J. Every substance has a unique value for the amount of heat required to change the temperature of 1 kg of it by 1 C°, and this number is referred to as the specific heat of the substance.

Suppose that Q units of heat are added to m kg of a substance, thereby changing its temperature by ΔT. The **specific heat**, c, of the substance is defined as

Specific heat

$$c \equiv \frac{Q}{m \, \Delta T} \tag{12.2}$$

Table 12.1 lists specific heats for a few substances.

From the definition of specific heat, we can express the heat energy, Q, transferred between a system of mass m and its surroundings for a temperature change ΔT as

$$Q = mc \, \Delta T \tag{12.3}$$

For example, the heat energy required to raise the temperature of 0.5 kg of water by 3 C° is equal to (0.5 kg)(4186 J/kg·C°)(3 C°) = 6280 J. Note that

Table 12.1
Specific Heat of Some Materials at Atmospheric Pressure

Substance	J/kg·C°	cal/g·C°
Aluminum	900	0.215
Beryllium	1820	0.436
Cadmium	230	0.055
Copper	387	0.0924
Germanium	322	0.077
Glass	837	0.200
Gold	129	0.0308
Ice	2090	0.50
Iron	448	0.107
Lead	128	0.0305
Mercury	138	0.033
Silicon	703	0.168
Silver	234	0.056
Steam	2010	0.480
Water	4186	1.00

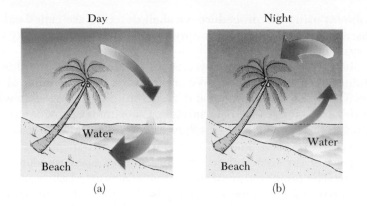

Day   Night

Water   Water

Beach   Beach

(a)   (b)

Figure 12.2
Circulation of air near a large body of water. (a) On a hot day, the air above the warm land warms faster than the air above the cooler water. The cooler air over the water moving toward the beach displaces the rising warmer air. (b) At night, the land cools more rapidly than the water and hence the air currents reverse their direction.

when the temperature increases, ΔT and Q are taken to be *positive*, corresponding to heat flowing *into* the system. Likewise, when the temperature decreases, ΔT and Q are *negative* and heat flows *out* of the system.

Note from Table 12.1 that water has the highest specific heat of any substance we are likely to come in contact with on a routine basis. The high specific heat of water is responsible for the moderate temperatures found in regions near large bodies of water. As the temperature of a body of water decreases during the winter, the water gives off heat to the air, which carries the heat landward when prevailing winds are favorable. For example, the prevailing winds off the western coast of the United States are toward the land, and the heat liberated by the Pacific Ocean as it cools keeps coastal areas much warmer than they would be otherwise. This explains why the western coastal states generally have more favorable winter weather than the eastern coastal states, where the winds do not carry the heat toward land.

That the specific heat of water is higher than that of land is responsible for the pattern of air flow at a beach. During the day, the sun adds roughly equal amounts of heat to beach and water, but the lower specific heat of sand causes the beach to reach a higher temperature than the water. Because of this, the air above the land reaches a higher temperature than that over the water. Consequently, cooler air from above the water is drawn in to displace this rising hot air, resulting in a breeze from ocean to land during the day. The hot air gradually cools as it rises and thus sinks, setting up the circulating pattern shown in Figure 12.2a. During the night, the land cools more quickly than the water and the circulating pattern reverses itself because the hotter air is now over the water (Fig. 12.2b). You should not expect to observe this pattern every time you go to the beach because prevailing winds caused by other factors often obscure it.

12.3   Conservation of Energy: Calorimetry

The principle of conservation of mechanical energy must now be expanded to include energy in the form of heat. Situations in which mechanical energy, such as gravitational energy or kinetic energy, is converted to heat occur frequently. We shall look at some of these in the examples following this section and in the problems at the end of the chapter, but most of our attention here will be directed toward a particular kind of conservation-of-energy situation. Problems using the procedure we shall describe are called *calorimetry* problems, and in them we consider only the transfer of heat energy inside an isolated system.

If a substance in a closed insulated container loses heat, something else in the container must gain this heat. We are ignoring the possibility of any heat transferred between the container and its surroundings. Devices in which this heat transfer occurs are called **calorimeters.** Calorimetry provides a technique for measuring the specific heat of solids and liquids. A known mass of a substance is heated to some known temperature and then placed in a vessel containing water (or some other liquid) of known mass, specific heat, and temperature. The temperature of the system is measured after equilibrium is reached. For example, suppose that m_x is the mass of a substance whose specific heat we wish to determine, c_x its specific heat, and T_x its initial temperature. Likewise, let m_w, c_w, and T_w represent the corresponding values for the water. If T is the final equilibrium temperature after everything is mixed, then from Equation 12.2, we find that the amount of heat gained by the water is $m_w c_w (T - T_w)$ and the amount of heat lost by the substance of unknown specific heat is $m_x c_x (T_x - T)$. Since energy is conserved, we can equate the heat lost to the heat gained:

$$m_w c_w (T - T_w) = m_x c_x (T_x - T)$$

When we solve this equation for c_x, we find

$$c_x = \frac{m_w c_w (T - T_w)}{m_x (T_x - T)} \tag{12.4}$$

You should not attempt to memorize this equation. Instead, you should always start from first principles in solving calorimetry problems. That is, determine which substances lose heat and which substances gain heat, and then equate heat loss to heat gain.

A word about sign conventions: In a later chapter we shall find it necessary to use a sign convention in which a positive sign for Q indicates heat gained by a substance and a negative sign indicates heat lost. However, for calorimetry problems it is less confusing if you ignore the positive and negative signs and instead equate heat loss to heat gain. That is, you should always write ΔT as a positive quantity. For example, if T_f is greater than T_i, let $\Delta T = T_f - T_i$. If T_f is less than T_i, let $\Delta T = T_i - T_f$.

Example 12.2   Cooling a Hot Ingot

A 0.05-kg ingot of metal is heated to 200°C and then dropped into a beaker containing 0.4 kg of water initially at 20°C. If the final equilibrium temperature of the mixed system is 22.4°C, find the specific heat of the metal.

Solution

Because the heat lost by the ingot equals the heat gained by the water, we can write

$$m_x c_x (T_i - T_f) = m_w c_w (T_f - T_i)$$

$$(0.05 \text{ kg})(c_x)(200°\text{C} - 22.4°\text{C}) = (0.4 \text{ kg})(4186 \text{ J/kg·C°})(22.4°\text{C} - 20°\text{C})$$

from which we find that

$$c_x = \boxed{453 \text{ J/kg·C°}}$$

The ingot is most likely iron, as can be seen by comparing this result with the data in Table 12.1

Example 12.3   Fun Time for a Cowboy

A cowboy fires a silver bullet of mass 2 g with a muzzle velocity of 200 m/s into the pine wall of a saloon. Assume that all the internal energy generated by the impact remains with the bullet. What is the temperature change of the bullet?

Solution
The kinetic energy of the bullet is

$$\frac{1}{2}\, mv^2 = \frac{1}{2}\,(2 \times 10^{-3}\ \text{kg})(200\ \text{m/s})^2 = 40\ \text{J}$$

All of this kinetic energy is transformed into internal energy, Q, as the bullet stops in the wall. This has the same effect as adding 40 J of heat to the bullet. Thus,

$$Q = mc\, \Delta T$$

Since the specific heat of silver is 234 J/kg·C° (Table 12.1), we get

$$\Delta T = \frac{Q}{mc} = \frac{40\ \text{J}}{(2 \times 10^{-3}\ \text{kg})(234\ \text{J/kg·C°})} = \boxed{85.5\ \text{C°}}$$

Exercise   Suppose the cowboy runs out of silver bullets and fires a lead bullet of the same mass and velocity into the wall. What is the temperature change of the bullet?

Answer   157 C°.

12.4   Latent Heat

A substance usually undergoes a change in temperature when heat is added to it or extracted from it. There are situations, however, where the flow of heat does not result in a change in temperature. This occurs whenever a substance changes from one physical state to another. Such a change is referred to as a **phase change.** Two common phase changes are solid to liquid (melting) and liquid to gas (boiling).

Consider, for example, the heat required to convert a 1-g block of ice at $-30°$C to steam (water vapor) at $120°$C. Figure 12.3 indicates the experimental results obtained when heat is gradually added to the ice. Let us examine each portion of the curve separately.

Part A   During this portion of the curve, we are changing the temperature of the ice from $-30°$C to $0°$C. Since the specific heat of ice is 2090 J/kg · C°, we can calculate the amount of heat added as follows:

$$Q = m_i c_i\, \Delta T = (10^{-3}\ \text{kg})(2090\ \text{J/kg·C°})(30\ \text{C°}) = 62.7\ \text{J}$$

Part B   When the ice reaches $0°$C, it remains at this temperature—even though heat is being added—until all the ice melts. The heat required to change the phase of a given mass m of a pure substance is

Figure 12.3
A plot of temperature versus
heat added when 1 g of ice
initially at −30°C is converted to steam.

$$Q = mL \qquad\qquad (12.5)$$

where L is the **latent heat** (hidden heat) of the substance. The value of L
depends on the nature of the phase change as well as on the properties of the
substance. The **latent heat of fusion**, L_f, is used when the phase change is from
a solid to a liquid. From Table 12.2, which lists the latent heat of fusion for a
few substances, we see that the latent heat of fusion for water at atmospheric
pressure is 3.34×10^5 J/kg. Thus, the heat required to melt 1 g of ice at 0°C is

$$Q = mL_f = (10^{-3} \text{ kg})(3.34 \times 10^5 \text{ J/kg}) = 334 \text{ J}$$

Part C Between 0°C and 100°C, nothing surprising happens. No phase
change occurs in this region. The heat added to the water is being used to
increase its temperature. The amount of heat necessary to increase the temperature from 0°C to 100°C is

$$Q = m_w c_w \, \Delta T = (10^{-3} \text{ kg})(4.19 \times 10^3 \text{ J/kg} \cdot \text{C}°)(100 \text{ C}°) = 4.19 \times 10^2 \text{ J}$$

Table 12.2
Latent Heats of Fusion and Vaporization

Substance	Melting Point (°C)	Latent Heat of Fusion		Boiling Point (°C)	Latent Heat of Vaporization	
		J/kg	(cal/g)		J/kg	(cal/g)
Helium	−269.65	5.23×10^3	(1.25)	−268.93	2.09×10^4	(4.99)
Nitrogen	−209.97	2.55×10^4	(6.09)	−195.81	2.01×10^5	(48.0)
Oxygen	−218.79	1.38×10^4	(3.30)	−182.97	2.13×10^5	(50.9)
Ethyl alcohol	−114	1.04×10^5	(24.9)	78	8.54×10^5	(204)
Water	0.00	3.34×10^5	(79.7)	100.00	2.26×10^6	(540)
Sulfur	119	3.81×10^4	(9.10)	444.60	3.26×10^5	(77.9)
Lead	327.3	2.45×10^4	(5.85)	1750	8.70×10^5	(208)
Aluminum	660	3.97×10^5	(94.8)	2450	1.14×10^7	(2720)
Silver	960.80	8.82×10^4	(21.1)	2193	2.33×10^6	(558)
Gold	1063.00	6.44×10^4	(15.4)	2660	1.58×10^6	(377)
Copper	1083	1.34×10^5	(32.0)	1187	5.06×10^6	(1210)

Part D At 100°C, another phase change occurs as the water changes from water at 100°C to steam at 100°C. We can find the amount of heat required to produce this phase change by using Equation 12.5. In this case, we must set $L = L_v$, the **latent heat of vaporization.** The latent heat of vaporization for some substances is given in Table 12.2. Since the latent heat of vaporization for water is 2.26×10^6 J/kg, the amount of heat we must add to convert 1 g of water to steam at 100°C is

$$Q = mL_v = (10^{-3} \text{ kg})(2.26 \times 10^6 \text{ J/kg}) = 2.26 \times 10^3 \text{ J}$$

Part E On this portion of the curve, heat is being added to the steam with no phase change occurring. Using 2.01×10^3 J/kg·C° for the specific heat of steam (see Table 12.1), we find that the heat we must add to raise the temperature of the steam to 120°C is

$$Q = m_s c_s \, \Delta T = (10^{-3} \text{ kg})(2.01 \times 10^3 \text{ J/kg·C°})(20 \text{ C°}) = 40.2 \text{ J}$$

The *total amount of heat* that must be added to change one gram of ice at −30°C to steam at 120°C is about 3.11×10^3 J. It should be noted that this process is *reversible.* That is, if we cool steam at 120°C down to the point at which we have ice at −30°C, we must remove 3.11×10^3 J of heat.

Phase changes can be described in terms of a rearrangement of molecules when heat is added or removed from a substance. Consider first the liquid-to-gas phase change. The molecules in a liquid are close together, and the forces between them are stronger than in a gas, where the molecules are far apart. Therefore, work must be done on the liquid against these attractive molecular forces in order to separate the molecules. The latent heat of vaporization is the amount of energy that must be added to the liquid to accomplish this.

Similarly, at the melting point of a solid, we imagine that the amplitude of vibration of the atoms about their equilibrium position becomes large enough to overcome the attractive forces binding the atoms into their fixed positions. The heat energy required to totally melt a given mass of solid is equal to the work required to break these bonds and to transform the mass from the ordered solid phase to the disordered liquid phase.

Because the average distance between atoms in the gas phase is much larger than in either the liquid or the solid phase, we could expect that more work is required to vaporize a given mass of a substance than to melt it. Therefore, it is not surprising that the latent heat of vaporization is much larger than the latent heat of fusion for a given substance (Table 12.2).

**Problem-Solving Strategy:
Calorimetry Problems**

If you are having difficulty with calorimetry problems, one or more of the following factors should be considered.

1. Be sure your units are consistent throughout. That is, if you are using specific heats measured in units of cal/g · C°, be sure that masses are in grams throughout and that temperatures are in Celsius units.

2. Heat losses or gains are found by using $Q = mc \, \Delta T$ only for those intervals in which no phase changes are occurring. Likewise, the equations $Q = mL_f$ and $Q = mL_v$ are only to be used when phase changes *are* taking place.
3. Often sign errors occur in heat loss = heat gain equations. One way to check to see if your equation is correct is to examine the sign of all ΔTs that appear in your equation. Every ΔT that appears should be a positive number.

Example 12.4   Cooling the Steam

What mass of steam initially at 130°C is needed to warm 200 g of water in a 100-g glass container from 20° to 50°C?

Solution
This is a heat transfer problem in which we must equate the heat lost by the steam to the heat gained by the water and glass container. There are three stages as the steam loses heat. In the first stage, the steam is cooled to 100°C. The heat liberated in the process is

$$Q_1 = m_x c_s \, \Delta T = m_x (2.01 \times 10^3 \text{ J/kg} \cdot \text{C°})(30 \text{ C°}) = m_x(6.03 \times 10^4 \text{ J/kg})$$

In the second stage, the steam is converted to water. In this case, to find the heat removed, we use the heat of vaporization and $Q = mL_v$:

$$Q_2 = m_x(2.26 \times 10^6 \text{ J/kg})$$

In the last stage, the temperature of the water is reduced to 50°C. This liberates an amount of heat

$$Q_3 = m_x c_w \, \Delta T = m_x(4.19 \times 10^3 \text{ J/kg} \cdot \text{C°})(50 \text{ C°}) = m_x(2.09 \times 10^5 \text{ J/kg})$$

If we equate the heat lost by the steam to the heat gained by the water and glass and use the given information, we find

$$m_x(6.03 \times 10^4 \text{ J/kg}) + m_x(2.26 \times 10^6 \text{ J/kg}) + m_x(2.09 \times 10^5 \text{ J/kg}) =$$
$$(0.2 \text{ kg})(4.19 \times 10^3 \text{ J/kg} \cdot \text{C°})(30 \text{ C°}) + (0.1 \text{ kg})(837 \text{ J/kg} \cdot \text{C°})(30 \text{ C°})$$

$$m_x = 1.09 \times 10^{-2} \text{ kg} = \boxed{10.9 \text{ g}}$$

Example 12.5   Boiling Liquid Helium

Liquid helium has a very low boiling point, 4.2 K, and a very low heat of vaporization, 2.09×10^4 J/kg (Table 12.2). A constant power of 10 W (1 W = 1 J/s) is transferred to a container of liquid helium from an immersed electric heater. At this rate, how long does it take to boil away 1 kg of liquid helium? Liquid helium has a density of 0.125 g/cm³, so that 1 kg corresponds to 8×10^3 cm³ = 8 liters of liquid.

Solution
Since $L_v = 2.09 \times 10^4$ J/kg for liquid helium, we must supply an energy equal to 2.09×10^4 J to boil away 1 kg. The power supplied to the helium is 10 W = 10 J/s. That is, in 1 s, 10 J of energy is transferred to the helium. Therefore, the time it takes to transfer 2.09×10^4 J is

$$t = \frac{2.09 \times 10^4 \text{ J}}{10 \text{ J/s}} = 2.09 \times 10^3 \text{ s} \approx \boxed{35 \text{ min}}$$

Since 1 kg of helium corresponds to 8 liters, this corresponds to a boil-off rate of about 0.23 liter/min. In contrast, 1 kg of liquid nitrogen would boil away in about 3.4 h at the rate of 10 J/s!

Exercise   If 10 W of power is supplied to 1 kg of water at 100°C, how long will it take for the water to boil away completely?

Answer   62.4 h.

12.5   Heat Transfer by Conduction

There are three ways heat energy can be transferred from one location to another: conduction, convection, and radiation. Regardless of the process, however, there will be no net heat transfer between a system and its surroundings when the two are at the same temperature. In this section, we shall discuss heat transfer by conduction. Convection and radiation will be discussed in Sections 12.6 and 12.7.

Each of the methods of heat transfer can be examined by considering the various ways in which you can warm your hands over an open fire. If you insert a copper rod into the flame, as in Figure 12.4, you will find that the temperature of the metal in your hand increases rapidly. The heat reaches your hand through conduction. The manner in which heat is transferred from the flame, through the copper rod, and to your hand can be understood by examining what is happening to the atoms of the metal. Initially, before the rod is inserted into the flame, the copper atoms are vibrating about their equilibrium positions. As the flame heats the rod, those copper atoms near the flame begin to vibrate with larger and larger amplitudes. These wildly vibrating atoms collide with their neighbors and transfer some of their energy in the collisions. Slowly, copper atoms farther down the rod increase their amplitude of vibration, until those at the end being held are reached. The effect of this increased vibration results in an increase in temperature of the metal, and possibly a burned hand.

Figure 12.4
Heat reaches the hand by conduction through the copper rod.

Although the transfer of heat through a metal can be partially explained by atomic vibrations, the rate of heat conduction also depends on the properties of the substance being heated. For example, it is possible to hold a piece of asbestos in a flame indefinitely. This implies that very little heat is being conducted through the asbestos. In general, metals are good conductors of heat and materials such as asbestos, cork, paper, and fiber glass are poor conductors. Gases also are poor heat conductors because of their dilute nature. Metals are good conductors of heat because they contain large numbers of electrons that are relatively free to move through the metal and transport energy from one region to another. Thus, in a good conductor, such as copper, heat conduction takes place via the vibration of atoms and via the motion of free electrons.

In this section, we shall consider the rate at which heat is transferred from one location to another. If Q is the amount of heat transferred from one location on an object to another in the time Δt, the **heat transfer rate**, H (sometimes called the heat current), is defined as

The ceramic material being heated by a flame is a poor conductor of heat but is able to withstand a large temperature gradient; its left side is buried in ice (0°C) while its right side is extremely hot. *(Courtesy of Corning Glass Works)*

Figure 12.5
Heat transfer through a conducting slab of cross-sectional area A and thickness L. The opposite faces are at different temperatures T_1 and T_2.

$$H \equiv \frac{Q}{\Delta t} \qquad (12.6)$$

Note that H has units of watts when Q is in joules and Δt is in seconds ($1\ W = 1\ J/s$).

The conduction of heat occurs only if there is a difference in temperature between two parts of the conducting medium. Consider a slab of material of thickness L and cross-sectional area A, as in Figure 12.5. Suppose that one face is maintained at a temperature T_2 and the other face is held at a lower temperature T_1. Experimentally, one finds that the rate of heat flow — that is, the heat flow Q per unit time, Δt — is proportional to the difference in temperature $T_2 - T_1$ and the area A, and inversely proportional to the thickness of the slab. Specifically, the rate of flow of heat (or heat current) is given by

$$H = \frac{Q}{\Delta t} = kA \left(\frac{T_2 - T_1}{L} \right) \qquad (12.7)$$

where k is a constant called the **thermal conductivity** of the material. This constant is a property of the material. Table 12.3 lists some values of k for metals, gases, and nonmetals. Note that k is large for metals, which are *good heat conductors,* and small for gases and nonmetals, which are poor heat conductors (good *insulators*).

The fact that different materials have different k values should help you understand the following phenomenon, first mentioned in Chapter 11. If you remove a metal ice tray and a package of frozen food from the freezer, which feels colder? Experience tells you that the metal tray feels colder even though it is at the same initial temperature as the cardboard package. This is explained by noting that, since metal has a much higher thermal conductivity than cardboard, it conducts heat more rapidly and hence removes heat from your hand at a higher rate. Hence, the metal tray *feels* colder than the carton even though it isn't. By use of a similar argument, you should be able to explain why a tile floor feels colder to a bare foot than a carpeted floor.

Table 12.3
Thermal Conductivities

Substance	Thermal Conductivity $(J/s \cdot m \cdot C°)$
Metals (at 25°C)	
Aluminum	238
Copper	397
Gold	314
Iron	79.5
Lead	34.7
Silver	427
Gases (at 20°C)	
Air	0.0234
Helium	0.138
Hydrogen	0.172
Nitrogen	0.0234
Oxygen	0.0238
Nonmetals	
Asbestos	0.08
Concrete	0.8
Glass	0.8
Ice	2
Rubber	0.2
Water	0.6
Wood	0.08

Example 12.6   Heat Transfer Through a Concrete Wall

Find the amount of heat transferred by conduction in 1 h through a concrete wall 2 m high, 3.65 m long, and 0.2 m thick if one side of the wall is held at 20°C and the other side is at 5°C.

Solution
Equation 12.7 gives the rate of heat transfer in joules per second. To find the amount of heat transferred in 1 h, we rewrite the equation as

$$Q = kA\,\Delta t \left(\frac{T_2 - T_1}{L} \right)$$

If we substitute the values given and consult Table 12.3, we find

$$Q = (0.8\ J/s \cdot m \cdot C°)(7.3\ m^2)(3600\ s) \left(\frac{15\ C°}{0.2\ m} \right) = \boxed{1.6 \times 10^6\ J}$$

Early houses were insulated by constructing the walls of thick masonry blocks. This restricted the heat lost by conduction because k is relatively low for masonry. Furthermore, because the thickness, L, is large, the heat loss is also decreased, as shown by Equation 12.7.

Melted snow pattern on a parking lot indicates the presence of underground steam pipes used to aid snow removal. Heat from the steam is conducted to the pavement from the pipes, causing the snow to melt. *(Courtesy of Dr. Albert A. Bartlett, University of Colorado, Boulder)*

*Home Insulation

If you would like to do some calculating to determine whether or not to add insulation to a ceiling or to some other portion of a building, what you have just learned about conduction needs to be modified slightly, for two reasons. (1) The insulating properties of materials used in buildings are usually expressed in engineering rather than SI units. For example, measurements stamped on a package of fiber glass insulating board will be in units such as British Thermal Units, feet, and degrees Fahrenheit. (2) In dealing with the insulation of a building, we must consider heat conduction through a compound slab, with each portion of the slab having a different thickness and a different thermal conductivity. For example, a typical wall in a house will consist of an array of materials, such as wood paneling, dry wall, insulation, sheathing, and wood siding.

It is found that the rate of heat transfer through a compound slab is given by

$$\frac{Q}{\Delta t} = \frac{A(T_2 - T_1)}{\sum_i L_i/k_i} \qquad (12.8)$$

where T_1 and T_2 are the temperatures of the *outer extremities* of the slab and the summation is over all portions of the slab. For example, if the slab consists of three different materials, the denominator will consist of the sum of three terms. In engineering practice, the term L/k for a particular substance is referred to as the **R value** of the material. Thus, Equation 12.8 reduces to

$$\frac{Q}{\Delta t} = \frac{A(T_2 - T_1)}{\sum_i R_i} \qquad (12.9)$$

The R values for a few common building materials are given in Table 12.4 (note the units).

Also, it should be noted that near any vertical surface there is a very thin, stagnant layer of air that must be considered when finding the total R value for a wall. The thickness of this stagnant layer on an outside wall depends on the velocity of the wind. As a result, heat loss from a house on a day when the wind is blowing hard is greater than heat loss on a day when the wind velocity is zero. A representative R value for this stagnant layer of air is given in Table 12.4.

Example 12.7   The R Value of a Typical Wall

Calculate the total R value for a wall constructed as shown in Figure 12.6a. Starting outside the house (to the left in Fig. 12.6a) and moving inward, the wall consists of brick, 0.5 in. of sheathing, a vertical air space 3.5 in. thick, and 0.5 in. of dry wall. Do not forget the dead-air layers inside and outside the house.

Table 12.4
R Values for Some Common Building Materials

Material	R value $(\text{ft}^2 \cdot \text{F}° \cdot \text{h/BTU})$
Hardwood siding (1 in. thick)	0.91
Wood shingles (lapped)	0.87
Brick (4 in. thick)	4.00
Concrete block (filled cores)	1.93
Styrofoam (1-in. thick)	5.0
Fiber glass batting (3.5 in. thick)	10.90
Fiber glass batting (6 in. thick)	18.80
Fiber glass board (1 in. thick)	4.35
Cellulose fiber (1 in. thick)	3.70
Flat glass (0.125 in. thick)	0.89
Insulating glass (0.25-in. space)	1.54
Vertical air space (3.5 in. thick)	1.01
Air film	0.17
Dry wall (0.5 in. thick)	0.45
Sheathing (0.5 in. thick)	1.32

Figure 12.6 (Example 12.7) Cross-sectional view of an exterior wall containing (a) an air space and (b) insulation.

Figure 12.7 Heating your hand by convection.

Solution

Referring to Table 12.4, we find the total R value for the wall as follows:

$$R_1 \text{ (outside air film)} = 0.17 \text{ ft}^2 \cdot \text{F}° \cdot \text{h/BTU}$$
$$R_2 \text{ (brick)} = 4.00$$
$$R_3 \text{ (sheathing)} = 1.32$$
$$R_4 \text{ (air space)} = 1.01$$
$$R_5 \text{ (dry wall)} = 0.45$$
$$R_6 \text{ (inside air film)} = 0.17$$
$$\overline{R_{\text{total}} \qquad\qquad\quad = 7.12 \text{ ft}^2 \cdot \text{F}° \cdot \text{h/BTU}}$$

Exercise If a layer of fiber glass insulation 3.5 in. thick is placed inside the wall to replace the air space as in Figure 12.6b, what is the total R value of the wall? By what factor is the heat loss reduced?

Answer $R = 17 \text{ ft}^2 \cdot \text{F}° \cdot \text{h/BTU}$; a factor of 2.5.

12.6 Convection

At one time or another you probably have warmed your hands by holding them over an open flame, as illustrated in Figure 12.7. In this situation, the air directly above the flame is heated and expands. As a result, the density of the air decreases and the air rises. This warmed mass of air heats your hands as it flows by. *Heat transferred by the movement of a heated substance is said to have been transferred by convection.* When the movement results from differences in density, as in the example of air around a fire, it is referrred to as *natural convection.* When the heated substance is forced to move by a fan or pump, as in some hot-air and hot-water heating systems, the process is called *forced convection*.

The circulating pattern of air flow at a beach pictured in Figure 12.2 is an example of convection. Likewise, the mixing that occurs as water is cooled and eventually freezes at its surface (Chapter 11) is an example of convection in nature. Recall that the mixing by convection currents ceases when the water temperature reaches 4°C. Since the water in the pool cannot be cooled by convection below 4°C, and because water is a relatively poor conductor of heat (see Table 12.3), the water near the bottom remains near 4°C for a long time. As a result, fish have a comfortable temperature in which to live even in periods of prolonged cold weather.

If it were not for convection currents, it would be very difficult to boil water. As water is heated in a teakettle, the lower layers are warmed first. These heated regions expand and rise to the top because their density is lowered. At the same time, the denser cool water replaces the warm water at the bottom of the kettle so that it can be heated.

The same process occurs when a room is heated by a radiator. The hot radiator warms the air in the lower regions of the room. The warm air expands and rises to the ceiling because of its lower density. The denser regions of cooler air from above replace the warm air, setting up the continuous air current pattern shown in Figure 12.8.

Figure 12.8
Convection currents are set up in a room heated by a radiator.

The rate at which heat is transferred by a fluid to a surface of area A is given by the equation

$$\frac{Q}{\Delta t} = hA\,\Delta T \qquad (12.10)$$

where ΔT is the difference in temperature between the fluid and the surface, and h is a constant called the **convection coefficient.** It is found that h depends on several factors such as the shape of the surface and whether the surface is horizontal or vertical. For heated air in contact with a vertical surface, the convection coefficient is given by

$$h = 1.77(\Delta T)^{1/4}\ \text{J/s}\cdot\text{m}^2\cdot\text{C}° \qquad (12.11)$$

Example 12.8 Heat Loss Through Windows

The temperature of the air inside a room is held at 20°C on a day when the outside air temperature is −10°C. Find the rate at which heat is transferred to a single-pane window by convection. Assume the window has a cross-sectional area of 0.5 m² and a thickness of 0.25 cm.

Solution
On a cold day, the temperature of a glass window pane is much lower than the temperature of the air inside the room. In fact, the temperature of the glass is approximately equal to the average of the inside and outside air temperatures. In this example, this average temperature is 5°C. Thus, the temperature difference between the surface and inside air is 15°C, so the convection coefficient can be found from Equation 12.11 as

$$h = 1.77(\Delta T)^{1/4} = 1.77(15°C)^{1/4} = (1.77)(1.968) = 3.48\ \text{J/s}\cdot\text{m}^2\cdot\text{C}°$$

Example 12.10   Who Turned Down the Thermostat?

A student trying to decide what to wear is staying in a room that is at 20°C. If the skin temperature of the unclothed student is 37°C, how much heat is lost from his body in 10 min? Assume that the emissivity of skin is 0.90 and that the surface area of the student is 1.5 m².

Solution
Using Equation 12.13, the rate of heat loss from the skin is

$$P_{net} = \sigma A e(T^4 - T_0^4)$$
$$= (5.67 \times 10^{-8} \text{ W/m}^2 \cdot \text{K}^4)(1.5 \text{ m}^2)(0.90)[(310 \text{ K})^4 - (293 \text{ K})^4] = 143 \text{ J/s}$$

Note that it was necessary to change the temperature to kelvins. At this rate of heat loss, the total heat lost by the skin in 10 min is

$$Q = P_{net} \times \text{time} = (143 \text{ J/s})(600 \text{ s}) = \boxed{8.58 \times 10^4 \text{ J}}$$

*12.8   Hindering Heat Transfer

Figure 12.10
Cross-sectional view of a Dewar vessel used to store hot or cold liquids.

The Thermos bottle, called a *Dewar flask* in scientific applications (after its inventor), is designed to minimize heat transfer by conduction, convection, and radiation. Such a container is used to store either cold or hot liquids for long periods of time. The standard construction (Fig. 12.10) consists of a double-walled Pyrex vessel with silvered inner walls. The space between the walls is evacuated to minimize heat transfer by conduction and convection. The silvered surfaces minimize heat transfer by radiation by reflecting most of the radiant heat. Very little heat is lost through the neck of the flask because Pyrex glass is a poor conductor. A further reduction in heat loss is obtained by reducing the size of the neck. A common usage of Dewar flasks in scientific applications is to store liquid nitrogen (boiling point 77 K) and liquid oxygen (boiling point 90 K). For substances that have a very low specific heat, such as liquid helium (boiling point 4.2 K), it is often necessary to use a double Dewar system in which the Dewar flask containing the liquid is surrounded by a second Dewar flask. The space between the two flasks is filled with liquid nitrogen.

Another application of the principles of minimizing heat transfer is the insulating of buildings, as discussed in Section 12.5. To achieve this, insulating materials are placed in the walls and above the ceilings. These materials, such as fiber glass, are poor conductors of heat. Furthermore, since they are loosely packed, there is air trapped in them in small cavities. Air is a poor heat conductor and cannot carry much heat by convection when it is trapped.

Wool sweaters and down jackets keep us warm by trapping the warmer air in regions close to our bodies and hence reducing heat loss by convection and conduction. In other words, what keeps us warm is the trapped air in the clothing, not the clothing itself.

▶ SPECIAL TOPIC
The Greenhouse Effect

Many of the principles of heat transfer, and the prevention of it, can be understood by considering a florist's greenhouse. One of the basic ideas underlying the operation of these buildings is the fact that glass will allow visible light to pass through but not infrared radiation. During the day, sunlight passes into the greenhouse and is absorbed by the walls, earth, plants, and so on. This absorbed visible light is subsequently re-radiated as infrared radiation, which cannot escape the enclosure. The increasing amount of trapped infrared radiation causes the temperature of the interior to rise.

Convection currents are also inhibited in a greenhouse. This prevents heated air from rapidly passing over those surfaces of the greenhouse exposed to the outside air, which would cause a heat loss through those surfaces. Many experts consider this to be an even more important effect in the operation of a greenhouse than the effect of trapped infrared radiation. In fact, experiments have shown that when the glass over a greenhouse is replaced by a special glass that transmits infrared light, the greenhouse temperature is only slightly lowered. Based on this evidence, it would have to be concluded that the absorption of infrared radiation is not the primary mechanism that heats a greenhouse. It is the inhibition of air flow that occurs with any roof (such as in an attic) that is the main factor in warming a greenhouse.

A phenomenon known as the greenhouse effect can also play a major role in determining the Earth's temperature. This can be understood by first noting that the Earth's atmosphere is a good transmitter (hence a poor absorber) of visible radiation and a good absorber of infrared radiation. Carbon dioxide (CO_2) in the Earth's atmosphere acts somewhat like the glass in a greenhouse in that it allows incoming visible radiation from the Sun to pass through easily but it does not allow infrared radiation to pass through as readily. The visible light that reaches the Earth's surface is absorbed and re-radiated as infrared light, which in turn is absorbed (trapped) by the Earth's atmosphere. An extreme case is the warmest planet, Venus, which has a carbon-dioxide-rich atmosphere and temperatures approaching 850°F.

As fossil fuels (coal, oil, and natural gas) are burned, large amounts of carbon dioxide are released into the atmosphere, causing it to retain more heat. This is of great concern to scientists and governments throughout the world. Many scientists are convinced that the 10% increase in the amount of atmospheric carbon dioxide in the last 30 years could lead to drastic changes in world climate. According to one estimate, doubling the carbon dioxide content in the atmosphere will cause temperatures to increase by 2 C°! In temperate regions, such as Europe and the United States, billions of dollars per year would be saved in fuel costs. Unfortunately, the temperature increase would melt polar ice caps, which could cause flooding and destroy many coastal areas, and increase the frequency of droughts and consequently decrease already low crop yields in tropical and subtropical countries.

While this scenario is rather alarming, other scientists theorize that an atmosphere rich in carbon dioxide could bring on another Ice Age! According to this model, a heavy blanket of carbon dioxide would screen out a large percentage of the Sun's incident rays, causing the Earth's temperature to decrease in time. Another factor adding to the problem is the effect of large quantities of small particles released into the atmosphere by industries. Such particles produce a hazy atmosphere (smog) that reflects some of the incoming radiation back into space, producing a cooling effect.

Obviously, the problem is very complex, and the various models that have been offered are open to question and further study. Nonetheless, it is an important problem that all nations must address. ◀

Summary

Heat flow is an energy transfer that takes place as a consequence of a temperature difference only.

The **calorie** is the amount of heat necessary to raise the temperature of 1 g of water from 14.5°C to 15.5°C.

The **mechanical equivalent of heat** is defined to be

$$1 \text{ cal} \equiv 4.186 \text{ J} \qquad (12.1)$$

The **specific heat** of a substance is defined as

$$c \equiv \frac{Q}{m \, \Delta T} \qquad (12.2)$$

The heat required to change the phase of a given mass m of a pure substance is given by

$$Q = mL \qquad (12.5)$$

where L is the **latent heat** of the substance and depends on the nature of the phase change and the properties of the substance. The **latent heat of fusion**, L_f, is used when the phase change is from a solid to a liquid. The **latent heat of vaporization**, L_v, is used when the phase change is from a liquid to a gas.

Heat may be transferred by three fundamentally distinct processes: **conduction, convection,** and **radiation. Conduction** can be viewed as an exchange of kinetic energy between colliding molecules or by the motion of electrons. The rate at which heat flows by conduction through a slab of area A and thickness L is

$$H = \frac{Q}{\Delta t} = kA \left(\frac{T_2 - T_1}{L} \right) \qquad (12.7)$$

where k is the **thermal conductivity** of the material. Heat transferred by the movement of a heated substance is said to have been transferred by **convection.**

All objects radiate and absorb energy in the form of electromagnetic waves. An object that is hotter than its surroundings radiates more energy than it absorbs, whereas an object that is cooler than its surroundings absorbs more energy than it radiates. The rate at which an object emits radiant energy is given by **Stefan's law:**

$$P = \sigma A e T^4 \qquad (12.12)$$

where σ is a constant equal to 5.6696×10^{-8} W/m² · K⁴ and e is a constant called the **emissivity.**

Additional Readings

P. B. Allen, "Conduction of Heat," *The Physics Teacher,* December 1983, p. 582.

E. Barr, "James Prescott Joule and the Quiet Revolution," *Physics Teacher,* April 1969, p. 199.

B. Chalmers, "How Water Freezes," *Sci. American,* February 1959, p. 144.

J. Dyson, "What Is Heat?" *Sci. American,* September 1954, p. 58.

J. Kelley, "Heat, Cold and Clothing," *Sci. American,* February 1956, p. 194.

R. Revelle, "Carbon Dioxide and World Climate," *Sci. American,* January 1982, p. 35.

M. G. Velarde, "Convection," *Sci. American,* January 1980, p. 92.

M. Wilson, "Count Rumford," *Sci. American,* October 1960, p. 158.

Questions

1. Ethyl alcohol has about one-half the specific heat of water. If equal masses of alcohol and water in separate beakers are supplied with the same amount of heat, compare the temperature increases of the two liquids.
2. A small crucible is taken from a 200°C oven and immersed in a tub of water at room temperature (often referred to as quenching). What is the approximate final equilibrium temperature?
3. The U.S. penny is now made of copper-coated zinc. Can a calorimetric experiment be devised to test for the metal content in a collection of pennies? If so, describe the procedure you would use.
4. What is wrong with the statement: "Given any two bodies, the one with the higher temperature contains more heat."
5. Concrete has a higher specific heat than does soil. Use this fact to explain (partially) why cities have a higher average temperature than the surrounding countryside. If a city is hotter than the surrounding countryside, would you expect breezes to blow from city to country or from country to city? Explain.
6. Pioneers stored fruits and vegetables in underground cellars. Discuss as fully as possible this choice for a storage site.
7. In winter, the pioneers mentioned in Question 6 would store an open barrel of water alongside their produce. Why?
8. What is the major problem that arises in measuring specific heats if a sample with a temperature above 100°C is placed in water?
9. In a daring lecture demonstration, an instructor dips his wetted fingers into molten lead (327°C) and withdraws them quickly, without getting burned. How is this possible? (This is a dangerous experiment, which you should not attempt.)
10. Dogs pant when they are hot. How does this help them stay cool?
11. Why can you get a more severe burn from steam at 100°C than from water at 100°C?
12. During a cold spell, Florida orange growers often spray a mist of water over their trees during the night. Why does this help?
13. Why is it possible to hold a lighted match even when it is burned to within a few millimeters of your fingertips?
14. A piece of paper is wrapped around a rod made half of wood and half of copper. When held over a flame, the paper in contact with the wood burns but the half in contact with the metal does not. Explain.
15. If water is a poor conductor of heat, why can it be heated quickly when placed over a flame?
16. Why does a piece of metal feel colder than a piece of wood when they are at the same temperature?
17. Why can potatoes be baked more quickly when a piece of metal has been inserted through them?
18. A tile floor in a bathroom may feel uncomfortably cold to your bare feet, but a carpeted floor in an adjoining room at the same temperature will feel warm. Why?
19. If your water pipes freeze during the winter, it is almost always the hot water pipes that freeze first. Why? It is often said that if you keep a slow trickle of water running from your pipes they will not freeze. Is this true?
20. If you hold water in a paper cup over a flame, you can bring the water to a boil without burning the cup. How is this possible?
21. A warning sign often seen on highways just before a bridge is "Caution—Bridge surface freezes before road surface." Which of the three heat transfer processes is most important in causing a bridge surface to freeze before a road surface on very cold days?
22. Updrafts of air are familiar to all pilots. What causes these currents?
23. Thermopane windows are usually made with two layers of glass separated by an air space. One manufacturer claims that his panes are farther apart than those of his competitors, which gives them a better insulating factor. Give an argument that would falsify his claim.

24. On a very hot day it is possible to cook an egg on the hood of a car. Would you select a black car or a white car to cook your egg? Why?

25. Two neighboring houses are identical in all respects, save one. One house has a black roof; the other has a white roof. Which house is likely to be a little cooler in the summertime? Explain.

26. Suppose you pour hot coffee for your guests, and one of them chooses to drink the coffee after it has been in the cup for several minutes. In order to have the warmest coffee, should the person add the cream just after the coffee is poured or just before drinking? Explain.

27. Two identical cups both at room temperature are filled with the same amount of hot coffee. One cup contains a metal spoon, while the other does not. If you wait for several minutes, which of the two will have the warmer coffee? Which heat transfer process explains your answer?

28. Most of the heat from the water cooling the engine of your car is transferred to the atmosphere after passing the water through the radiator. Does the radiator transfer the heat by radiation as its name implies?

29. When insulating a wood-frame house, is it better to place the insulation against the cooler outside wall or against the warmer inside wall? (In either case, there is an air barrier to consider.)

30. In the winter you might notice that some roofs are uniformly covered with snow while others have areas where the snow has melted, as shown in the photograph. Which houses would you say are better insulated? Why are there regular spacings between the melted areas in the photograph?

(Question 30) Alternating pattern of snow-covered and exposed roof. *(Courtesy of Dr. Albert A. Bartlett, University of Colorado, Boulder)*

Problems

Section 12.2   Specific Heat

1. A 50-g sample of copper is at 25°C. If 1200 J of heat energy is added to the copper, what is its final temperature?

2. How many joules of energy are required to raise the temperature of 100 g of gold from 20°C to 100°C?

3. A 50-g piece of cadmium is at 20°C. If 400 cal of heat is added to the cadmium, what is its final temperature?

4. A 5-g lead bullet traveling with a speed of 300 m/s is stopped by a large tree. If half the kinetic energy of the bullet is transformed into heat energy and remains with the bullet while the other half is transmitted to the tree, what is the increase in temperature of the bullet?

5. A 75-kg weight-watcher wishes to climb a mountain to work off the equivalent of a large piece of chocolate cake rated at 500 (food) Calories. How high must the person climb?

6. As a part of your exercise routine you climb a 10-m rope. How many food calories do you expend in a single climb up the rope?

7. Consider Joule's apparatus described in Figure 12.1. The two masses are 1.5 kg each, and the tank is filled with 200 g of water. What is the increase in the temperature of the water after the masses fall through a distance of 3 m?

8. A 3-kg rock is initially at rest at the top of a cliff. If the energy equivalent of the rock is sufficient to raise the temperature of 1 kg of water by 0.10°C, how high is the cliff?

9. Water at the top of Niagara Falls has a temperature of 10°C. If it falls through a distance of 50 m and all of its potential energy goes into heating the water, calculate the temperature of the water at the bottom of the falls.

10. Show that 1 BTU = 1054 J.

11. A 1.5-kg copper block is given an initial speed of 3 m/s on a rough horizontal surface. Because of friction, the block finally comes to rest. (a) If 85% of its initial kinetic energy is absorbed by the block in the form of heat, calculate the increase in temperature of the block. (b) What happens to the remaining energy?

12. An aluminum cup of mass 200 g contains 800 g of water in thermal equilibrium at 80°C. The combina-

tion of cup and water is cooled uniformly so that the temperature decreases by 1.5 C° per minute. At what rate is heat energy being removed? Express your answer in watts.

Section 12.3 Conservation of Energy: Calorimetry

13. A 0.4-kg iron horseshoe initially at 500°C is dropped into a bucket containing 20 kg of water at 22°C. What is the final equilibrium temperature? Neglect any heat transfer to or from the surroundings.

14. An unknown liquid of mass 400 g at a temperature of 80°C is poured into 400 g of water at 40°C. The final equilibrium temperature of the mixture is 49°C. What is the specific heat of the unknown liquid?

15. What mass of water at 25°C must be allowed to come to thermal equilibrium with a 3-kg gold bar at 100°C in order to lower the temperature of the bar to 50°C?

16. A block of copper of mass 200 g at a temperature of 90°C is dropped into 400 g of water at 27°C. The water is contained in a 300-g glass container. What is the final temperature of the mixture?

17. An aluminum cup contains 225 g of water at 27°C. A 400-g sample of silver at an initial temperature of 87°C is placed in the water. A 40-g copper stirrer is used to stir the mixture until it reaches its final equilibrium temperature of 32°C. Calculate the mass of the aluminum cup.

18. In a foundry operation, molten lead of mass 90 g at 327.3°C is poured into an iron casting block that has a mass of 300 g and is initially at 20°C. What is the equilibrium temperature of the system? Assume no heat losses to the surroundings.

19. Lead pellets, each of mass 1 g, are heated to 200°C. How many pellets must be added to 500 g of water initially at 20°C to make the equilibrium temperature 25°C? Neglect any heat transfer to or from the container.

20. If 200 g of water is contained in a 300-g aluminum vessel at 10°C and an additional 100 g of water at 100°C is poured into the container, what is the final equilibrium temperature of the mixture?

21. An aluminum calorimeter of mass 100 g contains 250 g of water. They are in thermal equilibrium at 10°C. Two metallic blocks are placed in the water. One is a 50-g piece of copper at 80°C. The other sample has a mass of 70 g and is originally at a temperature of 100°C. The entire system stabilizes at a final temperature of 20°C. Determine the specific heat of the unknown second sample.

22. A Styrofoam cup contains 200 g of mercury at 0°C. To this is added 50 g of ethyl alcohol at 50°C and 100 g of water at 100°C. (a) What is the final temperature of the mixture? (b) How much heat was gained or lost by (i) the mercury, (ii) the alcohol, and (iii) the water? (The specific heat of ethyl alcohol is 2430 J/kg · C° and the heat capacity of the Styrofoam cup is negligible.)

Section 12.4 Latent Heat

23. How much heat is required to change a 40-g ice cube from ice at −10°C to steam at 110°C?

24. A large block of ice at 0°C has a hole chipped into it and 400 g of aluminum pellets at a temperature of 30°C are poured into the hole. How much of the ice melts?

25. A 50-g ice cube at 0°C is heated until 45 g has become water at 100°C and 5 g has become steam at 100°C. How much heat was added to accomplish this?

26. What mass of steam initially at 120°C is needed to warm 350 g of water in a 300-g aluminum container from 20°C to 50°C?

27. Assuming no heat loss, determine the mass of water that boils away when 1.94 kg of mercury at a temperature of 200°C are added to 0.05 kg of water at a temperature of 80.0°C.

28. Determine the final state when 20 g of 0°C ice and 10 g of 100°C steam are mixed together in an insulated container.

29. A 100-g ice cube at 0°C is placed in 650 g of water at 25°C. What is the final temperature of the mixture?

30. A 300-g ice cube at 0°C is placed in 200 g of water at 10°C. The water is in a 100-g aluminum container. How much ice melts?

31. Steam at 100°C is added to ice at 0°C. (a) Find the amount of ice melted and the final temperature when the mass of steam is 10 g and the mass of ice is 50 g. (b) Repeat when the mass of steam is 1 g and the mass of ice is 50 g.

32. A 40-g block of ice is cooled to −78°C. It is added to 560 g of water in an 80-g copper calorimeter at a temperature of 25°C. Determine the final temperature. (If all the ice does not melt, determine how much ice is left.) Remember that the ice must first warm to 0°C, melt, and then continue warming as water. The specific heat of ice is 0.500 cal/g · C°.

33. A beaker of water sits in the sun until it reaches an equilibrium temperature of 30°C. The beaker is made of 100 g of aluminum and contains 180 g of water. In an attempt to cool this system down, 100 g of ice at 0°C is added to the water. (a) Determine the final temperature. If $T = 0$°C, determine how much ice remains. (b) Repeat this for the case when 50 g of ice is used.

34. A 75-kg cross-country skier moves across snow such that the coefficient of friction between skis and snow is 0.2. Assume all the snow beneath his skis is at 0°C and that all the heat generated by friction is added to snow which sticks to his skis until melted. How far would he have to ski to melt 1 kg of snow?

Section 12.5   Heat Transfer by Conduction
Section 12.6   Convection
Section 12.7   Radiation

35. (a) Find the rate of heat flow through a copper block of cross-sectional area 15 cm² and length 8 cm when a temperature difference of 30°C is established across the block. Repeat the calculation assuming the material is (b) a block of air having the same dimensions and (c) a block of wood of these dimensions.

36. A window has a glass surface of 1600 cm² and a thickness of 3 mm. (a) Find the heat transferred by conduction through this pane when the temperature of the inside surface of the glass is 70°F and the outside temperature is 90°F. (b) Repeat for the same inside temperature and an outside temperature of 0°F.

37. Determine the R value for a wall constructed as follows: The outside of the house is covered by lapped wood shingles placed over 0.5 in. thick sheathing, followed by 3 in. of cellulose fiber, and finally 0.5 in of dry wall.

38. A glass window pane has an area of 3 m² and a thickness of 0.6 cm. If the temperature difference between its faces is 25°C, how much heat flows through the window per hour?

39. The average thermal conductivity of the walls (including windows) and roof of the house in Figure 12.11 is 4.8×10^{-4} kW/m · C° and their average thickness is 21.0 cm. The house is heated with natural gas, with a heat of combustion (heat given off per cubic meter of gas burned) of 9300 kcal/m³. How many cubic meters of gas must be burned each day to maintain an inside temperature of 25.0°C if the outside temperature is 0.0°C? Disregard radiation and heat loss through the ground.

40. An iron rod has a length of 60 cm and a cross-sectional area of 2 cm². One end is maintained at 80°C and the other end is at 20°C. At steady state, find (a) the temperature gradient (i.e., the change in temperature per unit length along the rod), (b) the rate of heat transfer through the rod, and (c) the temperature in the rod 20 cm from the hot end.

41. A Thermopane window consists of two glass panes, each 0.5 cm thick, with a sealed layer of air 1.0 cm thick in between. If the inside temperature is 23.0°C and the outside temperature is 0.0°C, determine the rate of heat transfer through 1.0 m² of the window. Compare with the rate of heat transfer through 1.0 m² of a single pane of glass 1.0 cm thick.

42. (a) Find out the construction details of a typical outside wall in your home and estimate its R value. (b) Repeat your estimate for a typical dormitory room on your campus.

43. The temperature of the air inside a room is held at 23°C on a day when the outside temperature is 5°C. Find the rate at which heat is transferred to the window by convection. Assume the window has a cross-sectional area of 0.6 m² and a thickness of 0.125 cm.

44. In Problem 43, find the actual inside and outside temperatures of the glass.

45. Two identical objects are in the same surroundings at 0°C. One is at a temperature of 1200 K, and the other is at 1100 K. Find the ratio of the power emitted by the hotter object to the power emitted by the cooler object.

46. A sphere that is to be considered as a perfect black body radiator has a radius of 0.06 m and is at 200°C in a room where the temperature is 22°C. Calculate the rate at which the sphere radiates energy.

47. A Styrofoam container is in the shape of a box of surface area 0.8 m² with a thickness of 2 cm. The inside temperature is 5°C, and that outside is 25°C. If it takes 8 h for 5 kg of ice to melt in the container, determine the thermal conductivity of the Styrofoam.

48. Calculate the temperature at which a tungsten filament that has an emissivity of 0.25 and a surface area of 2.5×10^{-5} m² will radiate energy at the rate of 25 W in a room where the temperature is 22°C.

49. A copper rod and an aluminum rod of equal diameter are joined end-to-end in good thermal contact. The temperature of the free end of the copper rod is held constant at 100°C and that of the far end of the aluminum rod is held at 0°C. If the copper rod is 0.15 m in length, what must be the length of the aluminum rod so that the temperature at the junction is 50°C?

Figure 12.11   Problem 39

Additional Problems

50. In a showdown on the streets of Laredo, the good guy drops a 5-g silver bullet, at a temperature of 20°C, into a 100-cm³ cup of water at 90°C. Simultaneously, the bad guy drops a 5-g copper bullet, at the same initial temperature, into an identical cup of water. Which one ends the showdown with the coolest cup of water in the west? Neglect any heat transfer into or away from the container.

51. A solar collector 1 m² in area collects radiation from the Sun and focuses it on 250 g of water initially at 23°C. The average thermal energy arriving from the Sun at the surface of the Earth at this location is 550 W/m², and we assume that this is collected with 100% efficiency. Find the time required for the collector to raise the temperature of the water to 100°C.

52. A passive solar home uses a wall of concrete to collect solar energy. The wall has an absorbing area of 20.0 m² and a mass of 1.0×10^4 kg. During a "solar day" the wall is illuminated for 6.0 h, with an average power per unit area of 400 W/m². The wall stores 30% of the energy. (a) If concrete has a specific heat of 920 J/kg · C°, what is the heat capacity of the wall? (Heat capacity is defined as the mass of an object times its specific heat.) (b) What is the energy stored by the wall during a "solar day"? (c) If the initial temperature of the wall is 15°C, what is its temperature at the end of the "solar day"?

53. A solar collector has an effective collecting area of 12.0 m². The collector is thermally insulated, so that conduction is negligible in comparison with radiation. On a cold but sunny winter's day the temperature outside is −20.0°C, and the sun irradiates the collector with a power per unit area of 300 W/m². Treat the collector as a black body (i.e., emissivity = 1.0) and determine the temperature inside the collector after the collector has achieved a steady-state condition (radiating energy as fast as it is received).

54. The bottom of a copper kettle has a 10-cm radius and is 2 mm thick. The temperature of the outside surface is 102°C, and the water inside the kettle is boiling at 1 atm of pressure. Find the rate at which heat is being transferred through the bottom of the kettle.

55. A brass statue (60.0% copper, 40.0 zinc) has a mass of 50.0 kg. If the temperature of the statue increases by 20.0°C, what is the change of internal energy of the statue? (Specific heat of brass = 380 J/kg·C° = 0.092 cal/g·C°.)

56. A 40-g ice cube floats in 200 g of water in a 100-g copper cup all at a temperature of 0°C. A piece of lead at a temperature of 98°C is dropped into the cup, and the final equilibrium temperature is 12°C. What is the mass of the lead that was transferred?

57. An ice cube of mass 10 g is in a copper cup of mass 200 g. Initially the ice and cup are at a temperature of −20°C. (a) What happens to the mixture when 500 cal of heat are added? (b) Repeat when 5000 cal are added.

58. Calculate the temperature increase in 2 kg of water when it is heated with an 800-W immersion heater for a period of 5 min.

59. A class of 10 students taking an exam has a power output per student of about 200 W. Assume the initial temperature of the room is 20°C and that its dimensions are 6 m by 15 m by 3 m. What will be the temperature of the room at the end of 1 h if all the heat remains in the air in the room and none is added by an outside source? The specific heat of air is 837 J/kg·C°, and its density is about 1.3×10^{-3} g/cm³.

60. A 60.0-kg runner dissipates 300 W of power while running a marathon. Assuming that 10.0% of the runner's energy is dissipated in the muscle tissue and that the excess heat is primarily removed from the body by sweating, determine the volume of bodily fluid (assume it is water) lost per hour. (At 37.0°C the latent heat of vaporization of water is 575 kcal/kg.)

61. An iron plate is held against an iron wheel so that there is a sliding frictional force of 50 N acting between the two pieces of metal. The relative speed at which the two surfaces slide over each other is 40 m/s. (a) Calculate the rate at which mechanical energy is converted to heat. (b) The plate and the wheel have a mass of 5 kg each, and each receives 50% of the frictional heat. If the system is run as described for 10 s and each object is then allowed to reach a uniform internal temperature, what is the resultant temperature increase?

62. An automobile has a mass of 1500 kg, and its aluminum brakes have an overall mass of 60 kg. (a) Assuming that all of the frictional heat produced when the car stops is deposited in the brakes, and neglecting heat transfer, how many times could the car be braked to rest starting from 25 m/s (56 mph) before the brakes would begin to melt? (Assume an initial temperature of 20°C.) (b) Identify some effects that are neglected in Part (a) but are likely to be important in a more realistic assessment of the heating of brakes.

63. An aluminum rod, 1 m in length and of cross-sectional area 2 cm², is inserted vertically into a thermally insulated vessel containing liquid helium at 4.2 K. The rod is initially at 300 K. If one half of the rod is inserted into the helium, how many liters of helium boil off by the time the inserted half cools to 4.2 K?

64. A "solar cooker" consists of a curved reflecting mirror that focuses sunlight onto the object to be heated (Fig. 12.12). The solar power per unit area reaching the Earth at some location is 600 W/m², and

whatever interacts with the system. In the large-scale approach to thermodynamics, we describe the state of a system in terms of such quantities as pressure, volume, temperature, and internal energy. The number of such quantities needed to tell us what we need to know about the system depends on the nature of the system. For a simple system, such as a gas containing only one type of molecule, usually only two quantities are needed, such as pressure and volume. In this case, we assume that the gas is in thermal equilibrium internally. That is, we assume that every part of the container is at the same pressure and temperature.

Before we discuss the laws of thermodynamics, we must be sure that we understand three fundamental terms we shall encounter as we proceed: internal energy, work, and heat.

13.2 Internal Energy

The **internal energy** of a substance is the total energy of all the atoms or molecules that make up the substance. For the case of an ideal gas, the internal energy is equal to the random translational kinetic energy of the constituent gas atoms. However, in a more complicated gas, the rotational and vibrational motions of the molecules also contribute to the internal energy of the gas (Fig. 13.1). Atoms or molecules in a solid or a liquid have forces exerted on them by other nearby atoms that are similar to those that would be exerted by small "springs" between the particles. The stretching or compressing of these springs can be characterized by a potential energy that contributes to the internal energy of the system.

As noted above and discussed in Chapter 11, an ideal gas can be considered as a collection of very small particles that do not interact with each other. As a result, the internal energy of an ideal gas arises only from the translational kinetic energy of its constituent particles. Because temperature is related to the average kinetic energy of the particles of an ideal gas, we see that the greater the temperature of a gas, the greater is its internal energy.

It is important to note that energy can be added to a system even when no heat transfer takes place. For example, when two objects are rubbed together, their internal energy increases because mechanical work is done on them. The increase in temperature of the objects is an indicator of this increase in internal energy. Likewise, when an object slides across a surface and comes to rest because of frictional forces, the mechanical work done on the system is partially converted to internal energy, as indicated by the fact that the object and the surface both heat up.

13.3 Work

Consider a system made up of a gas contained in a cylinder fitted with a movable piston (Fig. 13.2a). In equilibrium, the gas occupies a volume V and exerts a pressure P on the cylinder walls and piston. If the piston has a cross-sectional area A, the force exerted by the gas on the piston is $F = PA$. Now let us assume that the gas expands very slowly so that it always remains in equilibrium. As the piston moves up a distance Δy, the work done by the gas is

$$W = F \Delta y = PA \Delta y$$

Rotation

Vibration

Figure 13.1
The rotational and vibrational energy of a diatomic gas contributes to the internal energy of the gas.

(a)   (b)

Figure 13.2
(a) A gas in a cylinder occupying a volume V and at a pressure P. (b) As the gas expands at constant pressure and the volume increases by an amount ΔV, the work done by the gas is $P \Delta V$.

Because $A \, \Delta y$ is the increase in volume of the gas (that is, $\Delta V = A \, \Delta y$), we can express the work done as

$$W = P \, \Delta V \qquad \textbf{(13.1)}$$ **Work**

If the gas expands, as in Figure 13.2b, ΔV is positive and the work done by the gas is positive. If the gas is compressed, ΔV is negative and the work done by the gas is negative. In the latter case, negative work can be interpreted as work being done *on* the system. Clearly, the work done by (or on) the system is zero when the volume remains constant.

You should be aware that Equation 13.1 can be used to calculate the work done on or by the system only when the pressure of the gas remains constant during the expansion or compression. If the pressure changes, the methods of calculus are required to calculate the work done. We shall not attempt such calculations here.

Consider the process represented by the pressure-volume diagram in Figure 13.3. We see here that the gas has expanded from an initial volume V_i to a final volume V_f at a constant pressure P. From Equation 13.1 we see that the work done by the gas is $P(V_f - V_i)$. Note that this is just the area under the pressure-volume curve.

In general, the work done in an expansion from some initial state to a final state is the area under the curve on a *PV* diagram.

This statement is true whether or not the pressure remains constant during the process.

Figure 13.3
The *PV* diagram for a gas expanding at constant pressure. The work done by the gas equals the shaded area.

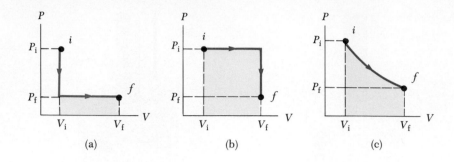

Figure 13.4
The work done by a gas as it is taken from an initial state to a final state depends on the path.

As can be seen from Figure 13.4, the amount of work done in the expansion of a gas from an initial state i to a final state f depends on how the gas is allowed to expand. In the process described in Figure 13.4a, the pressure on the gas is first reduced from P_i to P_f by cooling at constant volume V_i and the gas then expands from V_i to V_f at constant pressure P_f. The work done along this path is $P_f(V_f - V_i)$. In Figure 13.4b, the gas first expands from V_i to V_f at constant pressure P_i and then its pressure is reduced to P_f at constant volume V_f. The work done along this path is $P_i(V_f - V_i)$, which is greater than that for the process described in Figure 13.4a. Finally, for the process described in Figure 13.4c, where both P and V change continuously, the work done has some value intermediate between the values obtained in the first two processes. Therefore, we see that

> the amount of work done by a system depends on how the system goes from the initial to the final state. In other words, the work depends on the initial, final, and intermediate states of the system.

Example 13.1   Work Done by an Expanding Gas

In the system shown in Figure 13.2, the gas in the cylinder is at a pressure of 8000 Pa and the piston has an area of 0.1 m². As heat is slowly added to the gas, the piston is pushed up a distance of 4 cm. Calculate the work done on the surroundings by the expanding gas. Assume the pressure remains constant.

Solution
The change in volume of the gas is

$$\Delta V = A \, \Delta y = (0.1 \text{ m}^2)(4 \times 10^{-2} \text{ m}) = 4 \times 10^{-3} \text{ m}^3$$

and from Equation 13.1, the work done is

$$W = P \, \Delta V = (8000 \text{ Pa})(4 \times 10^{-3} \text{ m}^3) = \boxed{32 \text{ J}}$$

Example 13.2   Work Done Equals the Area Under the *PV* Curve

Find the work done during the thermodynamic process shown on the *PV* diagram of Figure 13.5.

Solution
The work done during this process is equal to the area under the *PV* curve. This can be found by breaking the total area into two parts, one the rectangle outlined by

$P(\times 10^4$ Pa)

$V(\times 10^{-3}$ m³)

Figure 13.5 (Example 13.2)
The work done equals the area under the PV curve.

the light blue area and the second the triangle indicated by the darker blue region.

The area of the rectangle (light area) is

$$A_r = (1 \times 10^4 \text{ N/m}^2)[(10 \times 10^{-3} \text{ m}^3) - (5 \times 10^{-3} \text{ m}^3)] = 50.0 \text{ J}$$

and the area of the triangle (darker area) is

$$A_t = \frac{1}{2}(\text{base} \times \text{height})$$

$$= \frac{1}{2}[(10 \times 10^{-3} \text{ m}^3) - (7.5 \times 10^{-3} \text{ m}^3)][(1.5 \times 10^4 \text{ Pa}) - (1 \times 10^4 \text{ Pa})]$$

$$= 6.3 \text{ J}$$

Thus the total work done, which is the area under the curve, is

$$W = 50.0 \text{ J} + 6.3 \text{ J} = \boxed{56.3 \text{ J}}$$

13.4 Heat

We found in Section 13.3 that the amount of work done on or by a system depends on the sequence used to take the system from the initial to the final state. In a similar manner, the heat transferred into or out of a system is also dependent on the sequence used. This can be demonstrated by considering the situations described in Figure 13.6. In each case, the gas has the same initial volume, temperature, and pressure and is assumed to be ideal. In Figure 13.6a, the gas is in thermal contact with a **heat reservoir,** which is *a source that is assumed to be capable of supplying large amounts of energy to a system without undergoing a temperature change.* For example, if an ice cube is dropped into the ocean, the ocean can be considered a heat reservoir. The change in temperature of the ocean as it melts the cube is zero for all practical purposes since it has such an enormous capacity.

If the pressure of the gas in Figure 13.6a is slightly greater than atmospheric pressure, the gas will expand and cause the piston to rise. During this expansion to some final volume V_f, sufficient heat to maintain a constant temperature T_i will be transferred from the reservoir to the gas.

Now consider the thermally insulated system shown in Figure 13.6b. When the membrane is broken, the gas expands rapidly into the vacuum until it occupies a volume V_f. In this case, the gas does no work because there is no movable piston. Furthermore, no heat is transferred through the thermally

Figure 13.6
(a) A gas at temperature T_i expands slowly by absorbing heat from a reservoir at the same temperature. (b) A gas expands rapidly into an evacuated region after breaking a membrane separating the two regions.

insulated wall, which we call an *adiabatic* wall. (Any process in which no heat enters or leaves the system is called an **adiabatic process**.) The process illustrated in Figure 13.6b is referred to as an adiabatic free expansion, or simply a **free expansion**.

The important point to note here is that the initial values of the pressure, volume, and temperature are the same in both cases; likewise, the final values of the pressure, volume, and temperature are the same in both cases. However, in the first case heat was added to the system, while in the second case there was no heat transferred. Therefore, we conclude that

the amount of heat added or lost by a system depends on the initial, final, and intermediate states of the system.

13.5   The First Law of Thermodynamics

When the law of conservation of energy was first introduced in Chapter 5, it was stated that the mechanical energy of a system is conserved in the absence of nonconservative forces, such as friction. Changes in the internal energy of the system were not included in our model at that time.

The first law of thermodynamics is a generalization of the law of conservation of mechanical energy that includes possible changes in internal energy. It is a universally valid law that can be applied to all kinds of processes.

We have seen that energy can be transferred between a system and its surroundings in two ways. One is *work done* by (or on) the system; the other is *heat transfer*. When energy has been transferred by either of these methods, we say that the system has undergone a **change in internal energy.**

Before we attempt to put this idea on a mathematical basis, let us explain the sign conventions for heat and work. We use the convention that Q is considered to be *positive* if heat is *added* to a system and *negative* if the system *loses* heat. For work, the work done is *positive* if the system does work *on its surroundings* and *negative* if work is done *on the system*.

Consider a gas that undergoes a pressure and volume change from P_i, V_i to P_f, V_f. *Regardless of the process* by which the system changes from the initial to the final state, it is found that the quantity $Q - W$ is *always the same*. Thus, we conclude that the quantity $Q - W$ is determined completely by the initial and final states of the system.

The quantity $Q - W$ is defined as **the change in the internal energy of the system.**

This definition has been substantiated experimentally. It is very important to note that, although both Q and W depend on the sequence used to change from the initial state to the final state, $Q - W$, the change in internal energy, does not. If we represent the internal energy by the letter U, then the change in internal energy, $\Delta U = U_f - U_i$, can be expressed as

The first law of
thermodynamics

$$\Delta U = U_f - U_i = Q - W \qquad \text{(13.2)}$$

where all the quantities must have the same units of energy. Equation 13.2 is known as the **first law of thermodynamics.** Again, note that this law is a generalization of the law of conservation of energy.

On the atomic level, the internal energy of a system includes the kinetic and potential energies of the molecules making up the system. In thermodynamics, we do not concern ourselves with the specific form of the internal energy. We simply use Equation 13.2 as a definition of the change in internal energy. One can make an analogy here between the potential energy associated with an object moving under the influence of gravity without friction. The potential energy is independent of the path, and it is only its change that is of concern. Likewise, the change in internal energy of a thermodynamic system is what matters because only differences in internal energy are important.

The following examples illustrate two important thermodynamic processes: isobaric (constant pressure) and isovolumetric (constant volume).

Example 13.3   An Isobaric Process

A gas is enclosed in a container fitted with a piston of cross-sectional area 0.1 m^2. When the pressure of the gas is maintained at 8000 Pa while heat is slowly added, the piston is pushed up a distance of 4 cm. If 42 J of heat is added to the system during the expansion, what is the internal energy change of the system?

Solution

Any process in which the pressure remains constant is called an **isobaric process.** The work done by the gas is

$$W = P \, \Delta V = (8000 \text{ Pa})(0.1 \text{ m}^2)(4 \times 10^{-2} \text{ m}) = 32 \text{ N} \cdot \text{m} = 32 \text{ J}$$

The change in internal energy is found from the first law:

$$\Delta U = Q - W = 42 \text{ J} - 32 \text{ J} = \boxed{10 \text{ J}}$$

Note that *in an isobaric process the work done and the heat transferred are both nonzero.*

Exercise   If 42 J of heat is added to the system with the piston clamped in a *fixed* position, what is the work done by the gas? What is the change in its internal energy?

Answer   No work is done; $\Delta U = 42 \text{ J}$.

Example 13.4   An Isovolumetric Process

Water of mass 2 kg is held at constant volume in a container while 10,000 J of heat is slowly added by a flame. The container is not well insulated, and as a result 2000 J of heat leaks out to the surroundings. What is the temperature increase of the water?

Solution

A process that takes place at constant volume is called an **isovolumetric process.** In such a process the work is clearly equal to zero. Thus, the first law reduces to

$$\Delta U = Q$$

This indicates that the net heat added to the water goes into increasing the internal energy of the water. The net heat added to the water is

$$Q = 10,000 \text{ J} - 2000 \text{ J} = 8000 \text{ J}$$

The temperature increase of the water is

$$Q = mc \, \Delta T$$

$$\Delta T = \frac{Q}{mc} = \frac{8000 \text{ J}}{(2 \text{ kg})(4.186 \times 10^3 \text{ J/kg} \cdot \text{C}°)} = \boxed{0.956 \text{ C}°}$$

Example 13.5   Boiling Water

One gram of water occupies a volume of 1 cm³ at atmospheric pressure (which is equal to 1.013×10^5 Pa). When this water is boiled, it becomes 1671 cm³ of steam. Calculate the change in internal energy for this process.

Solution
Since the heat of vaporization of water is 2.26×10^6 J/kg at atmospheric pressure, the heat required to boil 1 g is

$$Q = mL_v = (10^{-3} \text{ kg})(2.26 \times 10^6 \text{ J/kg}) = 2260 \text{ J}$$

The work done by the system is positive and equal to

$$W = P(V_{\text{steam}} - V_{\text{water}})$$
$$= (1.013 \times 10^5 \text{ Pa})[(1671 - 1) \times 10^{-6} \text{ m}^3] = 169 \text{ J}$$

Hence, the change in internal energy is

$$\Delta U = Q - W = 2260 \text{ J} - 169 \text{ J} = \boxed{2090 \text{ J}}$$

The positive ΔU tells us that the internal energy of the system has increased. We see that most of the heat (93%) transferred to the liquid goes into increasing the internal energy. Only a small fraction (7%) goes into external work.

Example 13.6   Heat Transferred to a Solid

The internal energy of a solid also increases when heat is transferred to it from its surroundings. To illustrate this point, suppose a 1-kg bar of copper is heated at atmospheric pressure. If its temperature increases from 20°C to 50°C, (a) find the work done by the copper.

Solution
The change in volume of the copper can be calculated using Equation 11.6 and the volume expansion coefficient for copper taken from Table 11.1 (remembering that $\beta = 3\alpha$):

$$\Delta V = \beta V \, \Delta T = [5.1 \times 10^{-5} \text{ (C}°)^{-1}](50°\text{C} - 20°\text{C})V = (1.5 \times 10^{-3})V$$

The volume is equal to m/ρ, and the density of copper is 8.92×10^3 kg/m³. Hence,

$$\Delta V = (1.5 \times 10^{-3}) \left(\frac{m}{\rho} \right) = (1.5 \times 10^{-3}) \left(\frac{1 \text{ kg}}{8.92 \times 10^3 \text{ kg/m}^3} \right)$$
$$= 1.7 \times 10^{-7} \text{ m}^3$$

Since the expansion takes place at constant pressure, the work done is

$$W = P \, \Delta V = (1.013 \times 10^5 \text{ Pa})(1.7 \times 10^{-7} \text{ m}^3) = \boxed{1.9 \times 10^{-2} \text{ J}}$$

(b) What quantity of heat is transferred to the copper?

Solution
The specific heat of copper is given in Table 12.1, and from Equation 12.3 we find that the heat transferred is

$$Q = mc \, \Delta T = (1 \text{ kg})(387 \text{ J/kg} \cdot \text{C}°)(30 \text{ C}°) = \boxed{1.16 \times 10^4 \text{ J}}$$

Exercise What is the increase in internal energy of the copper?

Answer $\Delta U = 1.16 \times 10^4$ J.

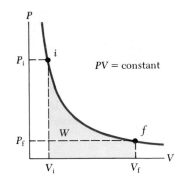

Figure 13.7
The PV diagram for an isothermal expansion of an ideal gas from an initial state to a final state. The curve is a hyperbola.

13.6 Isothermal Expansion of an Ideal Gas

A process in which an ideal gas expands (or is compressed) at a constant temperature is of such importance that we shall consider it separately from the processes described earlier. An expansion or compression of a substance at constant temperature is referred to as an **isothermal process**. The isothermal expansion of a gas can be achieved by placing the gas in thermal contact with a heat reservoir at the same temperature, as in Figure 13.6a.

The plot of the pressure of the gas versus its volume for an isothermal process is illustrated by the PV diagram shown in Figure 13.7. Since T is constant, the curve is a hyperbola described by the equation $PV = \text{constant}$. The work done during this isothermal expansion cannot easily be calculated because the pressure does not remain constant during the process. In fact, the methods of integral calculus must be used to calculate the work done in this situation. The result, which we state without proof, is given by the expression

$$W = nRT \ln \left(\frac{V_f}{V_i} \right) \tag{13.3}$$

where n is the number of moles of gas, R is the ideal gas constant, T is the Kelvin temperature, and ln is the abbreviation for the natural logarithm function. Numerically, this work equals the shaded area under the PV curve in Figure 13.7. If the gas *expands* isothermally, then $V_f > V_i$ and the work calculated using Equation 13.3 is positive, as we would expect based on our previous sign convention. If the gas is *compressed* isothermally, then $V_f < V_i$ and the work is found to be negative.

As noted earlier, the internal energy of an ideal gas depends only on temperature. Hence, for an isothermal process $\Delta U = 0$. Thus, from the first law we have

$$\Delta U = 0 = Q - W$$

or

$$Q = W \quad \text{(Isothermal process)}$$

According to this result, we conclude that the heat that must be added to the gas to allow it to expand at constant temperature is equal to the work done by the gas.

A model steam engine equipped with a built-in horizontal boiler. The water is heated electrically, which generates steam that is used to power the electric generator at the left. *(Courtesy of CENCO)*

We can think of the efficiency as the ratio of what you get (mechanical work) to what you pay for (energy). Equation 13.5 shows that a heat engine has 100% efficiency (*Eff* = 1) only if $Q_c = 0$, that is, if no heat is expelled to the cold reservoir. In other words, a heat engine with perfect efficiency would have to convert all of the absorbed heat energy to mechanical work. One of the consequences of the second law of thermodynamics is that this is impossible. The **second law of thermodynamics** can be stated as follows:

> It is impossible to construct a heat engine that, operating in a cycle, produces no other effect than the absorption of heat from a reservoir and the performance of an equal amount of work.

This form of the second law is useful in understanding the operation of heat engines. With reference to Equation 13.5, the second law says that, during the operation of a heat engine, *W* can never be equal to Q_h or, alternatively, that some heat, Q_c, must be rejected to the environment. As a result, it is theoretically impossible to construct an engine that will work with 100% efficiency. The net result of this assessment is that the first law says *we cannot get more work out of a process than the amount of energy we put in* and the second law says that *we cannot break even!*

13.8   The Gasoline Engine

In this section we shall discuss the four-stroke gasoline engine. Four successive processes occur in each cycle, as illustrated in Figure 13.10. During the intake stroke of the piston shown in Figure 13.10a, the valve on the left opens and air that has been mixed with gasoline vapor in the carburetor is drawn into the cylinder. During the compression stroke (Fig. 13.10b) both valves are closed and the air-fuel mixture is compressed approximately adiabatically. At

Figure 13.10
The four-stroke cycle of a conventional gasoline engine. (a) In the intake stroke, air is mixed with fuel. (b) The intake valve is then closed, and the air-fuel mixture is compressed by the piston. (c) The mixture is ignited by the spark plug, raising it to a higher temperature. (d) In the power stroke, the gas expands against the piston. (e) Finally, the residual gases are expelled. Then the cycle repeats.

Intake	Compression	Spark	Power	Exhaust
(a)	(b)	(c)	(d)	(e)

this point a spark from the sparkplug ignites the air-fuel mixture (Fig. 13.10c), causing a rapid increase in pressure and temperature at nearly constant volume. The burning gases expand approximately adiabatically and force the piston back, which produces the power stroke (Fig. 13.10d). Finally, during the exhaust stroke (Fig. 13.10e), the exhaust valve on the right is opened and the rising piston forces most of the remaining gas out of the cylinder. The cycle is repeated after the exhaust valve is closed and the intake valve is opened.

These processes can be approximated by the **Otto cycle**, a *PV* diagram of which is illustrated in Figure 13.11.

1. During the intake stroke $0 \rightarrow A$ (the horizontal line in Fig. 13.11), air is drawn into the cylinder at atmospheric pressure and the volume increases from V_2 to V_1.
2. In the process $A \rightarrow B$ (compression stroke), the air-fuel mixture is compressed adiabatically from volume V_1 to volume V_2, and the temperature increases from T_A to T_B. The work done on the gas is the area under the curve AB.
3. In the process $B \rightarrow C$, combustion occurs and heat Q_h is added to the gas. This is not an inflow of heat but rather a release of heat from the combustion process. During this time the pressure and temperature rise rapidly, but the volume remains approximately constant. No work is done on the gas.
4. In the process $C \rightarrow D$ (power stroke), the gas expands adiabatically from V_2 to V_1, causing the temperature to drop from T_C to T_D. The work done by the gas equals the area under the curve CD.
5. In the process $D \rightarrow A$, heat Q_c is extracted from the gas as its pressure decreases at constant volume. (Hot gas is replaced by cool gas.) No work is done during this process.
6. In the final process of the exhaust stroke $A \rightarrow 0$ (the horizontal line in Fig. 13.11), the residual gases are exhausted at atmospheric pressure, and the volume decreases from V_1 to V_2. The cycle then repeats itself.

Figure 13.11
The *PV* diagram for the Otto cycle, which approximately represents the processes in the internal combustion engine. No heat is transferred during the adiabatic processes $A \rightarrow B$ and $C \rightarrow D$.

Example 13.8 The Efficiency of an Engine

Find the efficiency of an engine that introduces 2000 J of heat during the combustion phase and loses 1500 J at exhaust and through friction.

Solution
The efficiency of the engine is given by Equation 13.5 as

$$Eff = 1 - \frac{Q_c}{Q_h} = 1 - \frac{1500 \text{ J}}{2000 \text{ J}} = \boxed{0.25, \text{ or } 25\%}$$

Exercise If an engine has an efficiency of 20% and loses 3000 J through friction, how much work is done by the engine?

Answer 750 J.

Example 13.9 The Efficiency of a Gasoline Engine

A gasoline engine is presented by the Otto cycle of Figure 13.12. If the input heat energy is 3000 J, estimate the efficiency of the engine.

Lord Kelvin, British physicist and mathematician (1824–1907). Born William Thomson in Belfast, Kelvin was the first to propose the use of an absolute scale of temperature. Kelvin's study of Carnot's theory led to the idea that heat cannot pass spontaneously from a colder body to a hotter body; this is known as the second law of thermodynamics. *(The Bettmann Archive)*

Figure 13.12   (Example 13.9)
The PV diagram for a gasoline engine.

Solution

If we knew the work output of the engine, we could calculate its efficiency from $Eff = W/Q_h$. We can estimate the work output by recalling that the work done on or by a system is equal to the area under a PV curve. For the process pictured in Figure 13.12, work is done by the air-fuel mixture between points C and D and work is done on the mixture between points A and B. Thus, the net work is the area between the two curves, shown in tan in the diagram. This is an irregular curve, and no simple equations enable us to calculate the area. We can estimate it, however, by finding the area of the rectangle shown in the figure:

$$A_r = [(30 \times 10^5) - (1 \times 10^5)][(800 \times 10^{-6}) - (100 \times 10^{-6})] \ \text{N} \cdot \text{m}$$
$$= 2030 \ \text{J}$$

In this example, one's judgment in estimation comes into play. Let us estimate that one half of the rectangular area is occupied by the Otto cycle. This estimate gives us the work done during the Otto cycle as

$$W = \frac{1}{2} A_r = \frac{1}{2} (2030 \ \text{J}) = 1015 \ \text{J}$$

and the efficiency is

$$Eff = \frac{W}{Q_h} = \frac{1015 \ \text{J}}{3000 \ \text{J}} = \boxed{0.34, \text{ or } 34\%}$$

13.9   Reversible and Irreversible Processes

In the next section we shall discuss a theoretical engine that is the most efficient engine possible. In order to understand the nature of that engine, we must first examine the meaning of reversible and irreversible processes.

> A **reversible** process is one that can be performed such that, at its conclusion, both the system and its surroundings have been returned to their exact initial conditions. A process that does not satisfy these requirements is **irreversible**.

All natural processes are known to be irreversible. From the endless number of examples that could be selected, let us examine the free expansion of a gas, discussed earlier in Section 13.4, and show that it cannot be reversible. A gas is contained in an insulated container, as shown in Figure 13.13, with a membrane separating the gas from a vacuum. If the membrane is punctured,

Figure 13.13
Free expansion of a gas.

the gas expands freely into the vacuum. Because the gas does not exert a force on anything, no work is done by the gas during the free expansion. In addition, no heat is transferred to or from the gas since the container is insulated from its surroundings. Thus, in this process, the system has changed, but the surroundings have not. Now imagine that we try to reverse the process by first compressing the gas to its original volume. To do this, let us imagine an engine being used to force the piston inward. Note, however, that in performing this task we are changing both the system and surroundings. The surroundings are changing because work is being done by an outside agent on the system, and the system is changing because the compression is increasing the temperature of the gas. We can lower the temperature of the gas by allowing it to come into contact with an external reservoir. Although this second procedure would return the gas to its original state, the surroundings would again be affected because heat would be added to the surroundings. If this heat could somehow be used to drive the engine and compress the gas, the system and its surroundings could be returned to their initial state. However, our statement of the second law says that this extracted heat cannot be completely converted into work. We must conclude that a reversible process has not occurred.

Although real processes are always irreversible, there are some processes that are almost reversible. If a real process occurs very slowly so that the system is virtually always in equilibrium, the process can be considered reversible. For example, imagine compressing a gas very slowly by dropping some grains of sand onto a frictionless piston as in Figure 13.14. The compression process is made isothermal by placing the gas in thermal contact with a heat reservoir. The pressure, volume, and temperature of the gas are well defined during this isothermal compression. Each added grain of sand represents a change to a new equilibrium state. The process can be reversed by slowly removing grains of sand from the piston.

A general characteristic of a reversible process is that there can be no dissipative effects present such as turbulence or friction that produce heat. In reality, such effects are impossible to eliminate completely, hence it is not surprising that real processes in nature are irreversible.

Heat reservoir

Figure 13.14
A gas in thermal contact with a heat reservoir is compressed slowly by dropping grains of sand onto the piston. The compression is isothermal and reversible.

13.10 The Carnot Engine

In 1824 a French engineer named Sadi Carnot (1796–1832) described a theoretical engine, now called a *Carnot engine*, that is of great importance from both a practical and a theoretical viewpoint. He showed that a heat engine operating in an ideal, reversible cycle, called a Carnot cycle, between two heat reservoirs is the most efficient engine possible. Such an ideal engine establishes an upper limit on the efficiencies of all engines. *That is, the net work done by a working substance taken through the Carnot cycle is the largest amount of work possible for a given amount of heat supplied to the substance.*

To describe the Carnot cycle, we shall assume that the substance working between temperatures T_c and T_h is an ideal gas contained in a cylinder with a movable piston at one end. The cylinder walls and the piston are thermally nonconducting. Four stages of the Carnot cycle are shown in Figure 13.15, and the *PV* diagram for the cycle is shown in Figure 13.16. The Carnot cycle consists of two adiabatic and two isothermal processes, all reversible.

Figure 13.15
The Carnot cycle. In process $A \rightarrow B$, the gas expands isothermally while in contact with a reservoir at T_h. In process $B \rightarrow C$, the gas expands adiabatically ($Q = 0$). In process $C \rightarrow D$, the gas is compressed isothermally while in contact with a reservoir at $T_c < T_h$. In process $D \rightarrow A$, the gas is compressed adiabatically. The upward arrows on the piston indicate sand being removed during the expansions, and the downward arrows indicate the addition of sand during the compressions.

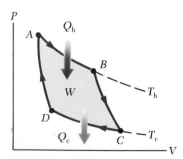

Figure 13.16
The PV diagram for the Carnot cycle.

1. The process $A \rightarrow B$ is an isothermal expansion at temperature T_h, in which the gas is placed in thermal contact with a heat reservoir at temperature T_h (Fig. 13.15a). During the process, the gas absorbs heat Q_h from the reservoir and does work W_{AB} in raising the piston.
2. In the process $B \rightarrow C$, the base of the cylinder is replaced by a thermally nonconducting wall and the gas expands adiabatically, that is, no heat enters or leaves the system (Fig. 13.15b). During the process, the temperature falls from T_h to T_c and the gas does work W_{BC} in raising the piston.
3. In the process $C \rightarrow D$, the gas is placed in thermal contact with a heat reservoir at temperature T_c (Fig. 13.15c) and is compressed isothermally at temperature T_c. During this time, the gas expels heat Q_c to the reservoir and the work done on the gas is W_{CD}.

4. In the final stage, $D \rightarrow A$, the base of the cylinder is replaced by a nonconducting wall (Fig. 13.15d) and the gas is compressed adiabatically. The temperature of the gas increases to T_h and the work done on the gas is W_{DA}.

Carnot showed that the thermal efficiency of a Carnot engine is given by

$$Eff_C = \frac{T_h - T_c}{T_h} = 1 - \frac{T_c}{T_h}$$ **(13.6)**   Carnot efficiency

It is important to note that in this equation, T must be in kelvins. From this result, we see that

> all Carnot engines operating between the same two temperatures have the same efficiency. Furthermore, the efficiency of a Carnot engine operating between two temperatures is greater than the efficiency of any real engine operating between the same two temperatures.

Equation 13.6 can be applied to any working substance operating in a Carnot cycle between two heat reservoirs. According to this result, the efficiency is zero if $T_c = T_h$, as one would expect. The efficiency increases as T_c is lowered and as T_h is increased. However, the efficiency can be unity (100%) only if $T_c = 0$ K. Such reservoirs are not available, and so the maximum efficiency is always less than unity. In most practical cases, the cold reservoir is near room temperature, about 300 K. Therefore, one usually strives to increase the efficiency by raising the temperature of the hot reservoir. *All real engines are less efficient than the Carnot engine because they are all subject to such practical difficulties as friction and heat losses by conduction.*

Model of a Stirling engine. The Stirling engine has an efficiency nearly equal to that of the Carnot engine. *(Courtesy of CENCO)*

Example 13.10   The Steam Engine

A steam engine has a boiler that operates at 500 K. The heat changes water to steam, which drives the piston. The exhaust temperature is that of the outside air, which is about 300 K. What is the maximum thermal efficiency of this steam engine?

Solution
From the expression for the efficiency of a Carnot engine, we find the maximum thermal efficiency for any engine operating between these temperatures:

$$Eff_C = \frac{T_h - T_c}{T_h} = \frac{500 \text{ K} - 300 \text{ K}}{500 \text{ K}} = \boxed{0.4, \text{ or } 40\%}$$

You should note that this is the highest theoretical efficiency of the engine. In practice, the efficiency will be considerably lower.

Example 13.11   The Carnot Efficiency

The highest theoretical efficiency of a gasoline engine, based on the Carnot cycle, is 30%. If this engine expels its gases into the atmosphere, which has a temperature of 300 K, what is the temperature in the cylinder immediately after combustion?

Solution

The Carnot efficiency is used to find T_h:

$$Eff_C = \frac{T_h - T_c}{T_h} = 1 - \frac{T_c}{T_h}$$

$$T_h = \frac{T_c}{1 - Eff_C} = \frac{300 \text{ K}}{1 - 0.3} = \boxed{429 \text{ K}}$$

Exercise   If the heat engine absorbs 837 J of heat from the hot reservoir during each cycle, how much work can it perform in each cycle?

Answer   251 J.

---

*13.11   Heat Pumps

A **heat pump** is a mechanical device that is finding increased acceptance for use in heating and cooling homes and buildings. In the heating mode, a circulating fluid, such as Freon, absorbs heat from the outside and releases it to the interior of the structure. The Freon is usually in the form of a low pressure vapor when in the coils of a unit exterior to the structure, where it absorbs heat either from the air or the ground. This gas is then compressed and enters the structure as a hot, high-pressure vapor. In an interior unit, the gas condenses to a liquid and releases its stored heat. When the heat pump is used as an air conditioner, the cycle above is reversed.

Figure 13.17 is a schematic representation of a heat pump used to heat a structure. The outside temperature is T_c, the inside temperature is T_h, and the heat absorbed by the circulating fluid is Q_c. The compressor does work W on the fluid, and the heat transferred from the pump into the structure is Q_h.

The effectiveness of a heat pump is described in terms of a number called the **coefficient of performance, COP**. This is defined as the ratio of the heat transferred into the hot reservoir and the work required to transfer that heat:

$$\text{COP (heat pump)} \equiv \frac{\text{heat transferred}}{\text{work done by pump}} = \frac{Q_h}{W} \tag{13.7}$$

If the outside temperature is 25°F or higher, the COP for a heat pump is about 4. That is, the heat transferred into the house is about four times greater than the work done by the compressor in the heat pump. However, as the outside temperature decreases, it becomes more difficult for the heat pump to extract sufficient heat from the air and the COP drops. In fact, the COP can fall below unity for temperatures below the midteens.

Although heat pumps used in buildings are relatively new products in the heating and air conditioning field, the refrigerator has been a standard appliance in homes for years. The refrigerator works much like a heat pump, except that it cools its interior by pumping heat from the food storage compartments into the warmer air outside. During its operation, a refrigerator removes a quantity of heat Q_c from the interior of the refrigerator, and in the process its motor does work W. The coefficient of performance of a refrigerator or for a heat pump used in its cooling cycle is given by

$$\text{COP(refrigerator)} = \frac{Q_c}{W}$$

Figure 13.17
Schematic diagram of a heat pump, which absorbs heat Q_c from the cold reservoir and expels heat Q_h to the hot reservoir. The work done on the heat pump is W.

Sadi Carnot, a French physicist, was the first to show the quantitative relationship between work and heat. Carnot was born in Paris on June 1, 1796 and was educated at the École Polytechnique in Paris and at the École Genie in Metz. Carnot's many interests include a wide range of study and research in mathematics, tax reform, industrial development, and the fine arts.

In 1824 he published his only work—*Reflections on the Motive Power of Heat*—which reviewed the industrial, political, and economic importance of the steam engine. In it he defined work as "weight lifted through a height." Carnot began to study the physical properties of gases in 1831, particularly the relationship between temperature and pressure.

On August 24, 1832, Sadi Carnot died suddenly of cholera. In accordance with the custom of his time, all of his personal effects were burned. Some of his notes that fortunately escaped destruction indicate that Carnot had arrived at the idea that heat is essentially work, or work that has changed its form. For this reason, he is considered to be the founder of the science of thermodynamics, from which we learn that energy can never disappear; it can only be altered into other forms of energy. Carnot's notes led Lord Kelvin to confirm and extend the science of thermodynamics in 1850.

Sadi Carnot
(1796–1832)

13.12 An Alternative Statement of the Second Law

The first law of thermodynamics says that energy is conserved in every process. There is more to the story than that, however, and the second law adds the finishing touches. To illustrate what we mean, suppose you wish to cool off a hot piece of pizza by placing it on a block of ice. You will certainly be successful because in every situation like this we have encountered, heat transfer always takes place from a hot object to a cooler one. If you wait a short while, 500 J of heat will be removed from the pizza and added to the ice. Energy is conserved and the first law is satisfied. Yet, there is nothing inherent in the first law that says that this heat transfer could not proceed in the opposite direction. Imagine your astonishment if some day you place a piece of hot pizza on ice and 500 J of heat moves from the cold ice to the pizza. Although this event will most likely never occur, there is nothing in the first law that says it could not. The second law of thermodynamics determines the direction of such natural phenomena.

In effect, the second law governs the order of the sequence of events that occur during any process. An analogy can be made with the impossible sequence of events seen in a movie film running backwards. Such events might be a person diving out of a swimming pool, an apple rising from the ground and latching onto the branch of a tree, or a pot of hot water becoming colder as it rests over an open flame. Such events running backwards in time are impossible because they violate the second law of thermodynamics. In other words, *real processes have a preferred direction of time*, often called the *arrow of time*.

There are many different statements of the second law, but all can be shown to be equivalent. The particular form of the second law you use depends on the application you have in mind. For example, if you were concerned about the heat transfer between pizza and ice, you might choose to concentrate on the second law in this form:

Heat will not flow spontaneously from a cold object to a hot object.

The second law of
thermodynamics

At first glance, this statement of the second law seems to be radically different from the statement given in Section 13.7. They are, in fact, equivalent in all respects. Although we shall not prove it here, it can be shown that if either statement is false, so is the other.

13.13   Entropy

Thus far we have presented two different statements of the second law of thermodynamics. In this section we shall present yet a third statement of this law in terms of an abstract quantity called **entropy,** denoted by the symbol S.

> The **change in entropy** of a system, ΔS, is equal to the heat, ΔQ, flowing into the system as the system changes from one state to another divided by the absolute temperature:

$$\Delta S \equiv \frac{\Delta Q}{T} \tag{13.8}$$

Rudolph Clausius (1822–1888). "I propose . . . to call S the entropy of a body, after the Greek word 'transformation.' I have designedly coined the word 'entropy' to be similar to energy, for these two quantities are so analogous in their physical significance, that an analogy of denominations seems to be helpful." *(AIP Niels Bohr Library, Lande Collection)*

This definition assumes that the process occurring is reversible and isothermal.

The concept of entropy was introduced into the study of thermodynamics by Rudolph Clausius in 1865. One reason it became useful and gained wide acceptance is because it provides another variable to describe the state of a system to go along with pressure, volume, and temperature. The concept of entropy reached a position of even more significance when it was found that

the entropy of the Universe increases in all natural processes.

This statement is yet another way of stating the second law of thermodynamics.

The statement of the second law given above must be interpreted with care. Although the statement says that the entropy of the Universe always increases in all natural processes, there are processes for which the entropy decreases. That is, there are situations in which the entropy of some system may decrease but that decrease takes place at the expense of a net increase in entropy of some other system. For example, we may find a process for which the entropy of system **A** is decreasing. However, in all cases we would find that the process produces an increase in entropy of some other system (system **B**) such that the change in entropy of system **B** is greater than the change in entropy of system **A**.

The concept of entropy is satisfying because it enables us to present the second law of thermodynamics in the form of a mathematical statement. In the next section, we will find that entropy can also be interpreted in terms of probabilities, a relationship which has profound implications for our world.

Example 13.12   Melting a Piece of Lead

Calculate the change in entropy when 300 g of lead melts at 327°C (600 K). Lead has a latent heat of fusion of 2.45×10^4 J/kg.

Solution

The amount of heat added to the lead to melt it is

$$Q = mL_f = (0.3 \text{ kg})(2.45 \times 10^4 \text{ J/kg}) = 7.35 \times 10^3 \text{ J}$$

The entropy change of the lead, from Equation 13.8, is

$$\Delta S = \frac{Q}{T} = \frac{7.35 \times 10^3 \text{ J}}{600 \text{ K}} = \boxed{12.3 \text{ J/K}}$$

Example 13.13   Which Way Does the Heat Flow?

A large cold object is at 273 K, and a large hot object is at 373 K. Show that it is impossible for a small amount of heat energy, say 8 J, to be transferred from the cold object to the hot object without decreasing the entropy of the Universe and hence violating the second law.

Solution

We assume here that during the heat transfer the two systems do not undergo a temperature change. This is not a necessary assumption; it is used to avoid using the techniques of integral calculus. The entropy change of the hot object is

$$\Delta S_h = \frac{Q}{T_h} = \frac{8 \text{ J}}{373 \text{ K}} = 0.0214 \text{ J/K}$$

The cold reservoir loses heat, and its entropy change is

$$\Delta S_c = \frac{Q}{T_c} = \frac{-8 \text{ J}}{273 \text{ K}} = -0.0293 \text{ J/K}$$

The net entropy change of the Universe is

$$\Delta S_U = \Delta S_c + \Delta S_h = -0.0079 \text{ J/K}$$

This is in violation of the concept that the entropy of the Universe always increases in natural processes. That is, *the spontaneous transfer of heat from a cold to a hot object cannot occur.*

Exercise   In the above example, suppose that 8 J of heat is transferred from the hot to the cold object. What would be the net change of the Universe?

Answer   +0.0079 J/K.

13.14   Entropy and Disorder

As you look around at the beauties of nature, it is easy to recognize that the events of natural processes have in them a large element of chance. For example, the spacing between trees in a natural forest is quite random. On the other hand, if you were to discover a forest where all the trees are equally spaced, you would conclude that this was a planted forest. Likewise, leaves fall to the ground with random arrangements. It would be highly unlikely to find the leaves laid out in perfectly straight rows. We can express the results of such observations by saying that **a disorderly arrangement is much more probable than an orderly one if the laws of nature are allowed to act without interference.**

Entropy originally found its place in thermodynamics, but its importance grew tremendously as the field of statistical mechanics developed. This analytical approach provides an alternative interpretation of entropy. In statistical mechanics, the behavior of a substance is described in terms of the statistical behavior of the atoms and molecules contained in the substance. One of the main results of this treatment is that

isolated systems tend toward disorder and entropy is a measure of that disorder.

In light of this new view of entropy, Boltzmann found that an alternative method for calculating entropy is through use of the important relation

$$S = k \ln W$$

(13.9)

where k is Boltzmann's constant ($k = 1.38 \times 10^{-23}$ J/K) and W is a number proportional to the probability of the occurrence of a particular event.

Let us explore the meaning of this equation by presenting a specific example. Imagine that you have a bag of 100 marbles, where 50 are red and 50 are green. You are allowed to draw four marbles from the bag according to the following rules. Draw one marble, record its color, return it to the bag and draw again. Continue this process until four marbles have been drawn. Note that because each marble is returned to the bag before the next one is drawn, the probability of drawing a red marble is always the same as drawing a green one. The results of all the possible drawing sequences that could occur are shown in Table 13.1. For example, the result RRGR means that you drew a red marble on the first draw, a red one on the second, a green one on the third, and a red one on the fourth. This table indicates that there is only one possible way to draw four red marbles. However, there are four possible sequences that could give one green and three red marbles, six sequences that could give two green and two red, four sequences that produce three green and one red, and one sequence that gives all green. Thus, the most likely occurrence is that you will draw two red and two green marbles, which corresponds to the disordered state. There is a much lower probability that you will draw four red or four green marbles, these being the most ordered states. From Equation 13.9, we see that the state with the greatest disorder has the highest entropy because it is the most probable state. On the other hand, the most ordered states (all red or all green) are least likely to occur, and are states of lowest entropy.

Table 13.1
Possible Results of Drawing Four Marbles from a Bag

End Result	Possible Draws	Total Number of Same Results
All R	RRRR	1
1G, 3R	RRRG, RRGR, RGRR, GRRR	4
2G, 2R	RRGG, RGRG, GRRG, RGGR, GRGR, GGRR	6
3G, 1R	GGGR, GGRG, GRGG, RGGG	4
All G	GGGG	1

In summary of this investigation, we see that the outcome of the draw can range from a highly ordered state (say all red marbles), which has the lowest entropy, to a highly disordered state (two green and two red marbles), which has the highest entropy. Thus, entropy can be regarded as an index of how far the system has progressed from an ordered to a disordered state.

It should be noted here that *the second law of thermodynamics is really a statement of what is most probable rather than of what must be.* Let us return to the example of placing an ice cube in contact with a hot piece of pizza. There is nothing in nature that absolutely forbids the transfer of heat from the ice to the much warmer pizza. Statistically, it is possible for a slow-moving molecule in the ice to collide with a faster moving molecule in the pizza such that the slow one transfers some of its energy to the faster one. However, when the large number of molecules present in the ice and pizza are considered, the odds are overwhelmingly in favor of the transfer of energy from the faster-moving molecules to the slower-moving molecules. Furthermore, this example demonstrates that the system naturally tends to move from a state of order to a state of disorder. The initial state in which all the pizza molecules have a high kinetic energy and all the ice molecules have a lower kinetic energy is much more ordered than the final state after heat transfer has taken place and the ice has melted.

As another example, suppose you were able to measure the velocity of all the air molecules in a room at some instant. It is very unlikely that you would find all molecules moving in the same direction with the same speed. This would, indeed, be a highly ordered state. The most probable situation you would find is a system of molecules moving haphazardly in all directions with a distribution of speeds. This is a highly disordered state and also the most likely. Let us compare this example to that of drawing marbles from a bag. Consider a container of gas consisting of 10^{23} molecules. If all of them were found moving in the same direction with the same speed at some instant, the outcome would be similar to drawing marbles from the bag 10^{23} times and finding a red marble on every draw. This is clearly an unlikely set of events.

The tendency of nature to move toward a state of disorder affects the ability of a system to do work. Consider a ball thrown toward a wall. The ball has kinetic energy, and its state is an ordered one. That is, all of the atoms and molecules of the ball move in unison at the same speed and in the same direction (apart from their random thermal motions). When the ball hits the wall, however, part of this ordered energy is transformed to disordered energy. The temperature of the ball and the wall both increase slightly as part of the ball's kinetic energy is transformed into the random, disordered, thermal motion of the molecules in the ball and the wall. Before the collision, the ball is capable of doing work. It could drive a nail into the wall, for example. When part of the ordered energy is transformed to disordered thermal energy, this capability of doing work is reduced. That is, the ball rebounds with less kinetic energy than it had originally because the collision is inelastic.

Various forms of energy can be converted to thermal energy, as in the collision between the ball and the wall, but the reverse transformation is never complete. In general, if two kinds of energy, A and B, can be completely interconverted, we say that they are the *same grade.* However, if form A can be completely converted to form B and the reverse is never complete, then form A is a *higher grade* of energy than form B. In the case of a ball hitting a wall, the

A full house is a very good hand in the game of poker. Can you calculate the probability of being dealt this hand in a standard deck of 52 cards? *(Tom Mareschel, The IMAGE Bank)*

An illustration from Flammarion's novel *La Fin du Monde*, depicting the heat death of the Universe.

kinetic energy of the ball is of a higher grade than the thermal energy contained in the ball and the wall after the collision. Therefore, when high-grade energy is converted to thermal energy, it can never be fully recovered as high-grade energy.

This conversion of high-grade energy to thermal energy is referred to as the **degradation of energy.** *The energy is said to be degraded because it takes on a form that is less useful for doing work.* In other words, *in all real processes where heat transfer occurs, the energy available for doing work decreases.*

Finally, it should be noted once again that the statement that entropy must increase in all natural processes is true only when an isolated system is considered. There are instances in which the entropy of some system may decrease but that decrease takes place at the expense of a net increase in entropy for some other system. When all systems are taken together as the Universe, *the entropy of the Universe always increases.*

Ultimately, the entropy of the Universe should reach a maximum. At this point, the Universe will be in a state of uniform temperature and density. All physical, chemical, and biological processes will have ceased because a state of perfect disorder implies no energy available for doing work. This gloomy state of affairs is sometimes referred to as an ultimate "heat death" of the Universe.

▶SPECIAL TOPIC
Energy Conversion and Thermal Pollution

In recent years, many thoughtful people have been concerned with the release of thermal energy into our environment. This **thermal pollution** is important to the welfare of our planet, and it is important that we understand its sources and develop methods to control it. There are many sources of thermal pollution, one of the primary ones being the waste heat from electric power plants. In this section, we shall focus on the reasons for the intrinsically low efficiency of power plants and on acceptable methods currently being used to dispose of thermal energy from these plants.

In the United States, about 85% of the electric power is produced by steam engines, which burn either fossil fuel (coal, oil, or natural gas) or use nuclear fuel (uranium-235). The remaining 15% is generated by water in hydroelectric plants. The overall thermal efficiency of a modern fossil-fuel plant is about 40%. The actual efficiency of any power plant must be lower than the theoretical efficiency derived from the second law of thermodynamics. There are two reasons that the highest efficiency possible is always sought: First, higher efficiency results in lower fuel costs; second, thermal pollution of the environment is reduced because there is less waste energy in a highly efficient power plant.

The burning of fossil fuels in a electric power plant involves three energy-conversion processes: (1) chemical to thermal energy, (2) thermal to mechanical energy, and (3) mechanical to electrical energy. These are indicated schematically in Figure 13.18.

During the first step, heat energy is transferred from the burning fuel to water, which is converted to steam. In this process about 12% of the available energy is lost up the chimney. In the second step, thermal energy in the form of steam at high pressure and temperature passes through a turbine and is converted to mechanical energy. A well-designed turbine has an efficiency of about 47%. The steam, which leaves the turbine at a lower pressure, is then condensed into water and gives up heat in the process. Finally, in the third step, the turbine drives an electric generator of very high efficiency, typically 99%. Hence, the overall efficiency is the product of the efficiencies of each step, which for the figures given is $(0.88)(0.47)(0.99) = 0.41$ or 41%.

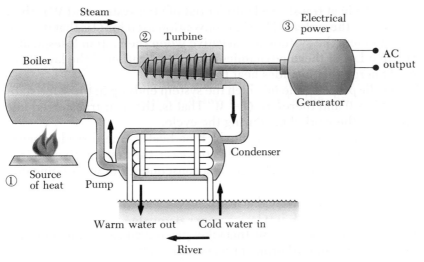

Figure 13.18
Schematic diagram of an electric power plant.

Figure 13.19
A cooling tower at a nuclear reactor site in Oregon. *(Courtesy of U.S. Department of Energy)*

Therefore, if one accounts for the 12% of energy lost to the atmosphere, we conclude that the thermal energy transferred to the cooling water amounts to about 47% of the energy theoretically available from the fuel.

In nuclear power plants, the steam generated by the nuclear reactor is at a lower temperature than in a fossil-fuel plant. This is due primarily to material limitations in the reactor. Typical nuclear power plants have an overall efficiency of about 34%.

Cooling towers (Fig. 13.19) are commonly used to dispose of waste heat. These towers usually use the heat to evaporate water, which is then released to the atmosphere. Cooling towers also present environmental problems since evaporated water can cause increased precipitation, fog, and ice. Another type of tower is the dry cooling tower (nonevaporative), which transfers heat to the atmosphere by conduction. However, this type is more expensive and cannot cool to as low a temperature as an evaporative tower. ◀

Summary

The **work done** as a gas expands or contracts under a constant pressure is

$$W = P\,\Delta V \qquad \textbf{(13.1)}$$

The work done is negative if the gas is compressed and positive if the gas expands. In general, the work done in an expansion from some initial state to a final state is the area under the curve on a PV diagram.

From the first law of thermodynamics, we see that when a system undergoes a change from one state to another, the **change in its internal energy**, ΔU, is

$$\Delta U = U_f - U_i = Q - W \qquad \textbf{(13.2)}$$

9. In practical heat engines, which do we have more control of, T_c or T_h? Explain.

10. A steam-driven turbine is one major component of an electric power plant. Why is it advantageous to increase the temperature of the steam as much as possible?

11. Is it possible to construct a heat engine that creates no thermal pollution?

12. "An adiabatic process can also be called an isentropic process ($\Delta S = 0$)." Is this statement true? Discuss.

13. Electrical energy can be converted to heat energy with an efficiency of 100%. Why is this number misleading with regard to heating a home? That is, what other factors must be considered in comparing the cost of electric heating with the cost of hot air or hot water heating?

14. Discuss three common examples of natural processes that involve an increase in entropy. Be sure to account for all parts of each system under consideration.

15. Give an example of a process in nature that is nearly reversible.

16. Discuss the change in entropy of a gas that expands (a) at constant temperature and (b) adiabatically.

17. Suppose the waste heat at a power plant is exhausted to a pond of water. Could the efficiency of the plant be increased by refrigerating the water?

18. A heat pump is to be installed in a region where the average outdoor temperature in the winter months is $-20°$C. In view of this, why would it be advisable to place the outdoor compressor unit deep into the ground? Why are heat pumps not commonly used for heating in cold climates?

19. Can a heat pump have a coefficient of performance less than one? Explain your answer.

20. A designer of an electric heating unit describes his product as being 100% efficient. Is his claim correct?

21. Energy is conserved; therefore, we should be able to recycle it. Define or rebut this statement.

22. Does water have "more entropy" when frozen or when liquid?

23. If you shake a jar full of jelly beans of different sizes, the larger jelly beans tend to appear near the top, while the smaller ones tend to fall to the bottom. Why does this occur? Does the process violate the second law of thermodynamics?

24. It is often said that the conception and birth of a human child is an example of a situation in which entropy decreases. If so, how can this process be consistent with the second law?

Problems

Section 13.2   Internal Energy
Section 13.3   Work
Section 13.4   Heat

1. Steam moves into the cylinder of a steam engine at a constant pressure of 2×10^5 Pa. The diameter of the piston is 16 cm, and the piston travels 20 cm in one stroke. How much work is done during one stroke?

2. A container of volume 0.4 m³ contains 3 moles of argon gas at 30°C. Assume that argon behaves like an ideal gas and find the total internal energy of the system. (*Hint:* You will have to review the discussion of ideal gases in Chapter 11 to find an expression for the energy of an ideal gas.)

3. Sketch a PV diagram of the following processes. (a) A gas expands at constant pressure P_1 from a volume V_1 to a volume V_2. It is then kept at constant volume while the pressure is reduced to P_2. (b) A gas is reduced in pressure from P_1 to P_2 while its volume is held constant at V_1. It is then expanded at constant pressure P_2 to a final volume V_2. (c) In which of the processes is more work done? Why?

4. A gas expands from I to F along the three paths indicated in Figure 13.20. Calculate the work done by the gas along paths (a) IAF, (b) IF, and (c) IBF.

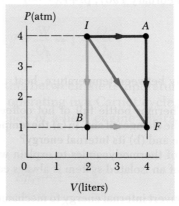

Figure 13.20
Problems 4 and 12

5. Sketch a PV diagram and find the work done by the gas during the following stages. (a) A gas is expanded from a volume of 1 liter to 3 liters at a constant pressure of 3 atm. (b) The gas is then cooled at constant volume until the pressure is reduced to 2 atm. (c) The

gas is then compressed at a constant pressure of 2 atm from a volume of 3 liters to a volume of 1 liter. (*Note:* Be careful of signs.) (d) The gas is heated until its pressure increases from 2 atm to 3 atm at a constant volume. (e) Find the net work done during the complete cycle.

6. Gas in a container is at a pressure of 1.5 atm and a volume of 4 m³. What is the work done by the gas if (a) it expands at constant pressure to twice its initial volume and (b) it is compressed at constant pressure to one-quarter its initial volume?

7. An ideal gas is enclosed in a cylinder. There is a movable piston on top of the cylinder. The piston has a mass of 8 kg, an area of 5 cm², and is free to slide up and down, keeping the pressure of the gas constant. How much work is done as the temperature of 0.2 moles of the gas is raised from 20°C to 300°C?

8. A sample of gas is compressed to one-half its initial volume at a constant pressure of 1.25×10^5 Pa. During the compression, 100 J of work is done on the gas. Determine the final volume of the gas.

9. A cylinder of gas is compressed by a piston from an initial volume of 125 liters to a final volume of 90 liters. The compression occurs at constant pressure, and the work done on the gas by the piston is 10^4 J. What is the gas pressure during the compression?

10. The density of water is 999.17 kg/m³ at 14.5°C and 999.02 kg/m³ at 15.5°C. Calculate the work done against the surrounding atmosphere when 2 kg of water expands as its temperature is increased from 14.5°C to 15.5°C.

Section 13.5 The First Law of Thermodynamics
Section 13.6 Isothermal Expansion of an Ideal Gas

11. A gas is compressed at a constant pressure of 0.3 atm from a volume of 8 liters to a volume of 3 liters. In the process, 400 J of heat energy flows out of the gas. (a) What is the work done by the gas? (b) What is the change in its internal energy?

12. A gas expands from *I* to *F* in Figure 13.20. The heat added to the gas is 418 J when the gas goes from *I* to *F* along the diagonal path. (a) What is the change in internal energy of the gas? (b) How much heat must be added to the gas for the indirect path *IAF* to give the same change in internal energy?

13. A thermodynamic system undergoes a process in which its internal energy decreases by 500 J. If at the same time, 220 J of work is done on the system, find the heat transferred to or from the system.

14. A person takes in a breath at 0.0°C and holds it until the air warms to 37.0°C. The air taken into the lungs

has an initial volume of 0.6 L and a mass equal to 7.7×10^{-4} kg. (a) Determine the work done on the lungs by the air if the pressure remains constant at 1.0 atm. (b) Determine the heat added to the air. (c) Determine the change of internal energy of the air. (Use c = 1010 J/kg·C° for air.)

15. A gas is enclosed in a container fitted with a piston of cross-sectional area 0.15 m². The pressure of the gas is maintained at 6000 Pa as the piston moves inward 20 cm. (a) Calculate the amount of work done by the gas. (b) If the internal energy of the gas decreases by 8 J, find the amount of heat removed from the system during the compression.

16. Two cubic centimeters of water are boiled at atmospheric pressure to become 3342 cubic centimeters of steam also at atmospheric pressure. (a) Calculate the work done by the gas during this process. (b) Find the amount of heat added to the water to accomplish this process. (c) From (a) and (b), find the change in internal energy.

17. Two moles of an ideal gas expands isothermally at 27°C to three times its initial volume. Find (a) the work done by the gas and (b) the heat flow into the system.

18. Consider the cyclic process described by Figure 13.21. If *Q* is negative for the process *BC* and Δ*U* is negative for the process *CA*, determine the signs of *Q*, *W*, and Δ*U* associated with each process.

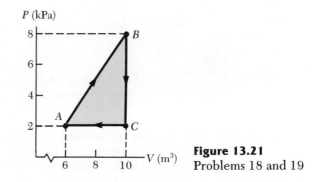

Figure 13.21
Problems 18 and 19

19. A gas is taken through the cyclic process described in Figure 13.21. (a) Find the net heat transferred to the system during one complete cycle. (b) If the cycle is reversed, that is, the process goes along *ACBA*, what is the net heat transferred per cycle?

20. Two moles of an ideal gas are compressed from a volume of 10 liters to a volume of 3 liters at a constant temperature of 27°C. Find (a) the work done by the gas during this process, (b) the change in internal energy of the gas, and (c) the amount of heat that must be added to or removed from the gas.

21. A 100-kg steel support rod in a building has a length of 2 m at a temperature of 20°C. The rod supports a

load of 6000 kg. Find (a) the work done by the rod as the temperature increases to 40°C, (b) the heat added to the rod (assume the specific heat of steel is the same as that for iron), and (c) the change in internal energy of the rod.

22. A 1-kg block of aluminum is heated at atmospheric pressure such that its temperature increases from 22°C to 40°C. Find (a) the work done by the aluminum, (b) the heat added to the aluminum, and (c) the change in internal energy of the aluminum.

23. One mole of helium gas, considered to be an ideal gas, is initially at a temperature of 300 K and pressure 0.2 atm. It is compressed isothermally to a pressure of 0.8 atm. (a) Use the ideal gas law to find the final volume of the gas. (b) Find the work done by the gas, and (c) the heat transferred to the gas.

24. One mole of an ideal gas does 800 J of work during an isothermal expansion at a temperature of 300 K. If its initial volume is 2 liters, find the final volume.

25. One mole of an ideal monatomic gas is carried by a quasi-static isothermal process (at 400 K) to twice its original volume (Fig. 13.22). (a) How much work W_{ab} was done by the gas? (b) How much heat Q_{ab} is supplied to the gas? (c) What is the pressure ratio, P_b/P_a?

Figure 13.22
Problem 25

26. One mole of an ideal gas does 3000 J of work on the surroundings as it expands isothermally to a final pressure of 1 atm and volume of 25 L. Determine (a) the initial volume and (b) the temperature of the gas.

27. A container is placed in a water bath and held at constant volume as a mixture of fuel and oxygen is burned inside it. The temperature of the water is observed to rise during the burning and it also is held at constant volume. (a) Consider the burning mixture to be the system. What are the signs of Q, ΔU, and W? (b) What are the signs of these quantities if the water bath is considered to be the system?

28. One mole of gas is initially at a pressure of 2 atm and a volume of 0.3 liter, and has an internal energy equal to 91 J. In its final state, the pressure is 1.5 atm, the volume is 0.8 liter, and the internal energy equals 182 J. For the paths IAF, IBF, and IF in Figure 13.23, (a) calculate the work done by the gas and (b) the net heat transferred to the gas in the process.

Figure 13.23
Problem 28

Section 13.7   Heat Engines and the Second Law of Thermodynamics
Section 13.8   The Gasoline Engine

29. In each cycle, a heat engine absorbs 375 J of heat and performs 25 J of work. Find (a) the efficiency of the engine and (b) the heat expelled in each cycle.

30. An engine absorbs 1700 J from a hot reservoir and expels 1200 J to a cold reservoir in each cycle. (a) What is the efficiency of the engine? (b) How much work is done in each cycle? (c) What is the power output of the engine if each cycle lasts for 0.3 s?

31. A heat engine performs 200 J of work in each cycle and has an efficiency of 30%. For each cycle of operation, (a) how much heat is absorbed and (b) how much heat is expelled?

32. The heat absorbed by an engine is three times greater than the work it performs. (a) What is its thermal efficiency? (b) What fraction of the heat absorbed is expelled to the cold reservoir?

33. A particular engine has a power output of 5 kW and an efficiency of 25%. If the engine expels 8000 J of heat in each cycle, find (a) the heat absorbed in each cycle and (b) the time for each cycle.

Section 13.10   The Carnot Engine

34. The efficiency of a Carnot engine is 30%. The engine absorbs 800 J of heat per cycle from a hot reservoir at 500 K. Determine (a) the heat expelled per cycle and (b) the temperature of the cold reservoir.

35. A heat engine operates between two reservoirs at temperatures of 20°C and 300°C. What is the maximum efficiency possible for this engine?

36. A steam engine has a boiler that operates at 300°F and the temperature of the exhaust is 150°F. Find the maximum efficiency of this engine.

37. The exhaust temperature of a Carnot heat engine is 300°C. What is the intake temperature if the efficiency of the engine is 30%?

38. In one cycle, a heat engine absorbs 500 J from the high-temperature reservoir and expels 300 J to a low-temperature reservoir. If the efficiency of this engine is 60% of the efficiency of a Carnot engine, what is the ratio of the low temperature to the high temperature in the Carnot engine?

39. A heat engine operates in a Carnot cycle between 80°C and 350°C. It absorbs 21,000 J of heat per cycle from the hot reservoir. The duration of each cycle is 1 s. (a) What is the maximum power output of this engine? (b) How much heat does it expel in each cycle?

40. Water is used to cool the output of the engine of Problem 39. If the temperature of the water is not allowed to increase more than 2°C, find how much water must pass through the engine in one hour.

41. A power plant that uses the temperature gradient in the ocean has been proposed. The system is to operate between 20°C (surface water temperature) and 5°C (water temperature at a depth of about 1 km). (a) What is the maximum efficiency of such a system? (b) If the power output of the plant is 75 MW, how much thermal energy is absorbed per hour? (c) In view of your results to (a) do you think such a system is worthwhile?

42. A heat pump, shown in Figure 13.24, is essentially an air conditioner run backward. It extracts heat from colder air outside and deposits it into a warmer room. Typically, the ratio of the actual heat entering the room to the work done by the device's motor is 10% of the theoretical maximum. Determine the heat entering the room per joule of work done by the motor if the inside temperature is 20.0°C and the outside temperature is −5.0°C.

Figure 13.24
Problem 42

43. Suppose a heat engine is connected to two heat reservoirs, one a pool of molten aluminum (660°C), the other a block of solid mercury (−38.9°C). The engine runs by freezing 1.0 g of aluminum and melting 15.0 g of mercury during each cycle. The latent heat of fusion of aluminum is 3.97×10^5 J/kg and for mercury, it is 1.18×10^4 J/kg. (a) What is the efficiency of this engine? (b) How does the efficiency compare with a Carnot engine?

Section 13.13 Entropy
Section 13.14 Entropy and Disorder

44. What is the change in entropy of 1 kg of water at 100°C as it changes to steam at 100°C?

45. A 70-kg log falls from a height of 25 m into a lake. If the log, the lake, and the air are all at 300 K, find the change in entropy of the Universe for this process.

46. Two 2000-kg cars, both traveling at 20 m/s, undergo a head-on collision and stick together. Find the entropy change of the Universe during the collision if the temperature is 23°C.

47. An ice tray contains 500 g of water at 0°C. Calculate the entropy change of the water as it freezes completely and slowly at 0°C.

48. Prepare a table like Table 13.1 for the following occurrence. You toss four coins into the air simultaneously. Record the results of your tosses in terms of the numbers of heads and tails that result. For example, HHTH and HTHH are two possible ways in which three heads and one tail can be achieved. (a) On the basis of your table, what is the most probable result of a toss? (b) In terms of entropy what is the most ordered state, and (c) what is the most disordered?

49. Repeat the procedure used to construct Table 13.1 (a) for the case in which you draw three marbles from your bag rather than four and (b) for the case in which you draw five rather than four.

50. In tossing two dice what is the total number of ways in which you can obtain (a) a 12 and (b) a 7?

Additional Problems

51. A nuclear power plant has a power output of 1000 MW and operates with an efficiency of 33%. If excess heat is carried away from the plant by a river having a flow rate of 1.0×10^6 kg/s, what is the rise in temperature of the river?

52. One mole of an ideal gas is compressed isothermally at 300 K until its volume is halved. Find the work done on the gas.

53. An engineer claims to have developed an engine that takes in 70,000 J of heat at 500 K and expels 20,000 J at 300 K, with 10,000 J of work being done. Would this be worth investing in? Why or why not?

54. A gas follows the path 123 on the *PV* diagram in Figure 13.25 and 418 J of heat flows into the system. Also, 167 J of work is done. (a) What is the internal energy change of the system? (b) How much heat flows into the system if the process follows the path 143? The work done by the gas along this path is 63 J. What net work would be done on or by the system if the system followed (c) the path 12341 and (d) the path 14321? (e) What is the change in internal energy of the system in the processes described in parts (c) and (d)?

Figure 13.25
Problem 54

55. One mole of hydrogen gas is heated at constant pressure from 300 K to 420 K. Calculate (a) the heat energy transferred to the gas, (b) the change in the internal energy of the gas, and (c) the work done by the gas. Note that hydrogen has a specific heat $c = 28.74$ J/mol·K.

56. A 5-kg block of aluminum is heated at atmospheric pressure such that its temperature increases from 20°C to 90°C. (a) Find the work done by the aluminum, (b) the amount of heat transferred to it, and (c) the increase in its internal energy.

57. One mole of an ideal gas at 1 atm pressure and at 300°C expands isothermally until its volume is tripled. It then is compressed to its original volume at constant pressure. Sketch these two processes on a *PV* diagram.

58. Refer to Problem 57. (a) Calculate the work done by the gas during the isothermal expansion. (b) Find the work done on the gas during the second change.

59. What is the minimum amount of work that must be done to extract 400 J of heat from a massive object at 0°C while rejecting heat to a hot reservoir at a temperature of 20°C?

60. The interior of a refrigerator has a surface area of 4.0 m². It is insulated by material that is 3.0 cm thick and has a thermal conductivity of 2.1×10^{-2} J/(m·s·C°). The ratio of the heat extracted from the interior to the work done by the motor is 7.5% of the theoretical maximum. The temperature of the room is

24.0°C, and the temperature inside the refrigerator is 0.0°C. Determine the power required to run the compressor.

61. A refrigerator has a coefficient of performance of 3. The ice tray compartment is at -20°C, and the room temperature is 22°C. The refrigerator can convert 30 g of water at 22°C to 30 g of ice at -20°C each minute. What input power is required? Ignore any cooling by the refrigerator and give your answer in watts.

62. During each cycle, a particular refrigerator absorbs 25 cal from the cold reservoir and expels 32 cal to the high-temperature reservoir. (a) If the refrigerator completes 60 cycles per second, what power is required? (b) What is the COP for this unit?

63. A 1500-kW heat engine operates at 25% efficiency. The heat energy expelled at the low temperature is absorbed by a stream of water that enters the cooling coils at 20°C. If 60 liters per second flows across the coils, determine the increase in temperature of the water.

64. One mole of an ideal gas is taken through the cycle shown in Figure 13.26. The process *AB* is an isothermal expansion. Calculate the net work done by the gas.

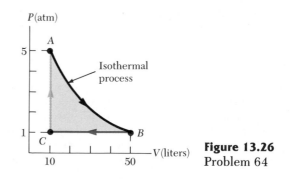

Figure 13.26
Problem 64

65. An ideal refrigerator (or heat pump) is equivalent to a Carnot engine running in reverse. That is, heat Q_c is absorbed from a cold reservoir and heat Q_h is rejected to a hot reservoir. Show that the work that must be supplied to run the refrigerator is given by

$$W = \frac{T_h - T_c}{T_c} Q_c$$

66. One mole of an ideal gas is taken through the reversible cycle shown in Figure 13.27. At point *A*, the pressure, volume, and temperature are P_0, V_0, and T_0. In terms of R and T_0, find (a) the total heat entering the system per cycle, (b) the total heat leaving the system per cycle, (c) the efficiency of an engine operating in this reversible cycle, and (d) the efficiency of an engine operating in a Carnot cycle between the

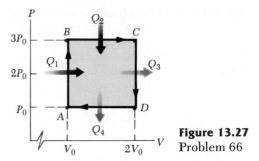

Figure 13.27
Problem 66

temperature extremes for this process. (*Hint:* Recall that work equals the area under a *PV* curve.)

67. An ideal gas initially at pressure P_0, volume V_0, and temperature T_0 is taken through the cycle described in Figure 13.28. (a) Find the net work done by the gas

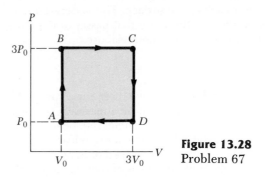

Figure 13.28
Problem 67

per cycle in terms of P_0 and V_0. (b) What is the net heat added to the system per cycle? (c) Obtain a numerical value for the net work done per cycle for 1 mole of gas initially at 0°C. (See the hint for Problem 66.)

68. One mole of an ideal gas at an initial pressure of 4 atm and temperature 300 K is carried through the following reversible cycle: (1) It expands isothermally until its volume is doubled; (2) it is compressed to its original volume at constant pressure; (3) it is compressed isothermally to a pressure of 4 atm; and (4) it expands at constant pressure to its original volume. (a) Make a

plot of the cyclic process on a *PV* diagram. (b) Calculate the work done by the gas per cycle.

69. An electrical power plant has an overall efficiency of 15%. The plant is to deliver 150 MW of power to a city, and its turbines use coal as the fuel. The burning coal produces steam at 190°C, which drives the turbines. This steam is then condensed into water at 25°C by passing it through coils in contact with river water. (a) How many metric tons of coal does the plant consume each day (metric ton = 10^3 kg)? (b) What is the total cost of the fuel per year if the delivered price is \$8/metric ton? (c) If the river water is delivered at 20°C, at what minimum rate must it flow over the cooling coils in order that its temperature not exceed 25°C? (*Note:* The heat of combustion of coal is 7.8×10^3 cal/g.)

"He was working on a theory of entropy, and developed a severe case of it himself."

Essay

Alternative Sources of Energy

Laurent Hodges

Professor of Physics, Iowa State University

The world's major current energy resources are fossil fuels (coal, petroleum, and natural gas) and uranium. However, their disadvantages—their rising costs, their growing scarcity, their vulnerability to political developments around the world, and their contributions to air pollution, acid rain, and global warming—have led society to investigate alternatives such as geothermal energy, fusion energy, and solar energy.

Geothermal Energy

Geothermal energy makes use of the heat in underground rocks or water. It exists everywhere on Earth but is being used only in places with natural reservoirs of geothermal steam or hot water. It is being used for home heating in Iceland; Boise, Idaho, and Klamath Falls, Oregon, and for generating electricity in Larderello, Italy, and Geysers, California.

Geothermal energy is not limited to regions with steam or hot water, but in principle it can be extracted almost everywhere in the world from hot dry rocks deep underground by digging two wells, one to bring cold water down from the surface and the other to bring hot water back to the surface. The underground rocks are fractured with pressurized water, and the water that passes through the fractures is heated. This method is not currently in commercial use.

Fusion Energy

Nuclear power plants in use around the world today derive their energy from the fissioning of heavy nuclei, such as uranium and plutonium. An alternative is fusion energy, the production of energy from the fusing of light nuclei into heavier nuclei of lower total mass. The decrease in mass Δm leads to energy release $(\Delta m)c^2$ in accordance with Einstein's relationship, where c is the speed of light. Fusion is the energy source of sunlight and most starlight, as well as of thermonuclear or "hydrogen" bombs.

There is an abundance on Earth of naturally occurring isotopes suitable for use in fusion energy reactions, notably deuterium (H^2) and lithium. However, controlled fusion has not yet been demonstrated in the laboratory, much less been proved economical. The difficulty is in creating conditions suitable for fusion. In particular, the temperatures must be very high so the kinetic energy of the posi-

An electrical power plant in Geysers, California, whose energy source is geothermal steam. *(Copyright © 1978 KaiDib Films International)*

A portion of a tokamak laboratory fusion reactor at Princeton. The vacuum of the vessel is being tested for leaks.

tively charged nuclei can overcome their electrical repulsion, allowing the strong nuclear interaction to fuse the nuclei. Approaches that are being studied involve confinement of the highly ionized nuclei using magnetic fields or by bombardment with laser radiation.

Solar Energy

The major alternative energy source is the Sun, whose energy comes to the Earth in the form of electromagnetic radiation, mainly in the visible and near infrared parts of the spectrum. Solar energy, broadly defined, occurs in a great variety of forms. In addition to the direct solar radiation, which can be used to heat or cool buildings or to heat domestic water or to produce electricity, there are indirect forms of solar energy, such as water, wind, and photosynthetic power. These are properly classified as solar energy because it is solar radiation that drives the hydrological cycle from which water power is derived, that drives the circulation in the atmosphere from which wind power is derived, and that provides the energy for photosynthesis.

An old windmill in Denmark. This is one example of an indirect source of solar energy. *(Copyright © 1978 KaiDib Films International)*

   Solar energy is often regarded as the energy source of the future, but in fact it was the major energy source for humanity throughout almost all of recorded and unrecorded history. The great explorers all used wind energy to travel the world. Agriculture used beasts of burden, whose energy derived from their food. The U.S. industrial revolution began with solar power in the form of mechanical water power, mechanical wind power, and the combustion of wood fuel.

   Solar energy currently provides approximately 10% of U.S. energy use, outranked only by fossil fuels. The major solar contributions are from hydroelectricity, passive solar energy, and photosynthetic materials (particularly wood).

Hydroelectricity

Hydroelectricity accounts for about 12% of the electricity generated in the United States. The energy source for hydroelectricity is the kinetic energy of water, originally evaporated from the oceans, precipitated in the mountains, and then flowing back to the oceans.

   Where available—as in the Pacific Northwest and the region served by the Tennessee Valley Authority—hydroelectric power is the cheapest form of elec-

An aerial view of Hoover Dam on the Colorado River. This dam has a height of 221 m and a hydroelectric power capacity of 1.354 MW. *(Copyright © 1978 KaiDib Films International)*

Some of the turbines installed in the Hoover Dam hydroelectric power plant. *(Copyright © 1978 KaiDib Films International)*

tric power. But it has major environmental impacts, since its use requires the damming of a river and often the creation of a large reservoir. The many remaining good hydroelectric sites are unlikely to be developed because vast ecosystems would be adversely affected.

Solar Heating of Buildings

The most cost-effective use of solar energy is for winter space heating, the main use of energy in American homes. The space heating requirements can be drastically reduced by energy conservation methods such as good wall and roof insulation, multi-pane windows, and air infiltration control by good caulking and weatherstripping. Solar energy can be used to meet a substantial portion of the remaining heating requirements.

One approach—known as *active solar heating*—uses mechanical means (such as fans and ducts in air systems or pumps and pipes in liquid systems) to distribute the solar energy. Active space heating systems collect solar energy through a flat-plate collector of one or two panes of glass, behind which is a black absorber. In air systems, solar heat is usually stored in a rockbed (a mass of uniformly-sized rocks about the size of a small room), while liquid systems usually store the heat in a large water tank. Active systems were installed on many homes during the 1970s but were never very popular. They tended to be large, complex, and expensive, often costing $10,000 or more. They required expert design and had many parts requiring constant maintenance. They also tended to be very conspicuous and to make a home look far from conventional.

A much more successful method is *passive solar heating*, in which the energy flows are by natural means: natural conduction, natural convection, and natural radiation. Passive systems typically collect solar heat through a large expanse of south-facing windows and store it in masonry such as concrete, stone, brick, or adobe. The masonry absorbs heat when it is in excess and releases it at night. The masonry can be located in such places as walls, floors, partitions, or fireplaces.

Passive solar homes, which cost very little more than conventional homes, work very well in most parts of the United States, particularly in areas that have cold but sunny winters. It is not uncommon for well-insulated passive solar homes in cold northern climates to be heated for a whole winter for only $100 to $200 worth of natural gas heat.

A house with solar collectors on its roof. The solar energy is used to heat this home. *(© James L. Ruhle & Associates)*

A series of solar collectors installed on the roof of a house. Notice the black absorbing surface and the panes of glass on the flat-plate collectors. *(© James L. Ruhle & Associates)*

Photosynthetic Energy

Photosynthetic energy is a major source of energy in many homes, those that use wood for space heating. Wood and agricultural wastes are also being used to generate electricity or industrial process heat, usually by small industries. "Gasohol" is 10% ethanol, which can be produced from grain. Wood was the main energy source in the United States during its first century. Not until the 1880s did coal supplant wood as the leading energy source, and wood remained a significant source through the rest of the 19th century.

Photovoltaic Cells

Photovoltaic or "solar" cells are made of semiconductor materials that develop a voltage when sunlight strikes them. Originally developed for the space program, to enable satellites to have a source of power while in orbit, they are now widely available on small electronic devices such as "solar" calculators, radios, and battery chargers, but they are also sometimes used to power remote installations or electric fences.

A home with photovoltaic cells covering a south-facing roof would be able to collect enough electric energy to meet the home's needs, but the home would need a battery system in order to store the energy for use at night. Such a system would currently cost over $10,000, but by the early 21st century it might be cheap enough to be economical for homeowners. Research is under way, particularly in Europe and Japan, to produce better photovoltaic cells, mainly cells that are higher in efficiency or cheaper to manufacture.

Ocean Thermal Energy Conversion

Ocean thermal energy conversion (OTEC) is the production of electricity using the small temperature difference between warm ocean surface water and cold deep ocean water. An OTEC plant would need to use a working substance other

Silicon solar cells (foreground). These cells are fabricated from large manmade single crystals of silicon (background). *(Copyright © KaiDib Films International)*

A solar power system that uses a line of photovoltaic cells located along the focal point of a parabolic reflector. *(© James L. Ruhle & Associates)*

than steam; ammonia is one possibility. Although the temperature differences are only a few degrees and the plant efficiency would be very low, very large quantities of water and heat are available.

OTEC plants have been tested on and off for many decades, including in the Atlantic Ocean off Florida and the Pacific Ocean near Hawaii. However, OTEC plants are currently far from being technologically and economically feasible, and they may never actually prove commercially successful.

Power Towers

The solar power tower is an electric power plant at the top of a large tower, with its heat source being solar radiation reflected from a large field of mirrors surrounding the tower. The mirrors need to be computer controlled so they can accurately reflect the sunlight to the proper place and yet be defocused quickly wherever an emergency arises.

This concept has been thoroughly researched, and in 1982 an experimental version known as "Solar One" went into operation at Barstow, California. Solar One is a small (10 megawatt) solar electric plant with a steam boiler situated in a field of 1818 helistats. The total land area covered is 130 acres.

It is hoped that solar power towers will prove to be economical, but this is by no means certain. Their best chances to do so in the United States would be in the southern states.

Wind Power

Wind power has long served humanity for transportation, grinding of grain, and other purposes, and it is currently an alternative source of electricity. Many wind generators were installed on midwestern farms in the 1920s and 1930s, and a

A vertical axis wind-powered generator.

surprising number of them are still in operation. More recently, several large "wind farms" have been built to serve electric utilities in the western United States. Currently they produce electricity at a cost higher than the current price of coal-generated electricity, but it is less than the cost of some nuclear-generated electricity.

Essay Questions

1. What enables glass to act as a solar collector? Would transparent plastic materials such as Plexiglas work as well?
2. Passive solar homes normally use double- or triple-pane windows. Why not single-pane windows?
3. Does a solar collector have to face exactly south?

Essay Problems

1. The solar radiation reaching the Earth, above the atmosphere, averages about 1350 W/m^2. Using the facts that the Sun is about 150 million kilometers away and the Earth has a diameter of 12.7 thousand kilometers, find: (a) The total power (in watts) of the solar radiation reaching the Earth, (b) the total power (in watts) of the solar radiation emitted by the Sun, assuming it is emitted equally in all directions, (c) the rate (in kilograms per second) at which the Sun is losing mass, making use of Einstein's relation $E = mc^2$, and (d) the fraction of the Sun's total mass of 2.0×10^{30} kg being lost each year.
2. At a midlatitude in the United States, the average amount of solar radiation falling on a tilted south-facing roof is about 200 W/m^2, averaging over the whole year, day and night, sunny weather and cloudy. Assuming that a homeowner uses solar cells that cover 80% of the roof and have an efficiency of 10% at converting solar radiation into electricity, how large an area (in square meters) would be needed to produce a total of 6000 kWh of electric energy? Is this a reasonable area to find on a real roof?
3. A passive solar home in Iowa has 40 m^2 of south-facing double-pane windows and 100 metric tons (100,000 kg) of concrete heat storage. It has a backup furnace whose heat output is 40 MJ/h. Concrete has a heat capacity of about 800 J/C$^\circ \cdot$kg (about one-fifth that of water). Suppose that on a sunny winter day each square meter of windows admits 16 MJ of solar energy into the home. (a) How many hours would the furnace have to run to provide the home with the same amount of heat as was admitted by the windows? (b) How much would the thermal storage warm up if 40% of the solar heat were lost through the windows, walls, and roof of the house and the other 60% went to heat the storage?

The Riddle of the Third Law

Hans Christian von Baeyer

The College of William and Mary

Invisible barriers in the physical sciences never fail to astonish us. The edge of the detectable Universe—the cosmological horizon—is one invisible barrier, and the relativistic speed limit, beyond which no object can accelerate, is another. The consternation they cause stems from the fact that the extremely large and the extraordinarily fast lie outside the realm of human experience. A third such barrier is the absolute zero of temperature, the state beyond which matter cannot be cooled, in which all molecular motion ceases. Like an invisible wall, this concept jolts us out of a belief that we understand heat and temperature, and it demands deeper reflection.

Absolute zero, as a theoretical concept, was foreshadowed as early as 1699, in the work of the French physicist Guillaume Amontons, who conducted experiments in temperature and its measurement, but a century later it was still being dismissed by the majority of scientists. In 1804, the American adventurer-turned-scientist, Benjamin Thompson, told the Royal Society of London that "all attempts to determine the place of absolute cold, on the scale of a thermometer, must be nugatory." And his French colleague Joseph Louis Gay-Lussac called the notion "altogether chimerical." Not until the late 1800s was the idea revived, and today, when the finitude of the detectable Universe and the speed limit of elementary particles are taken for granted, if not fully comprehended, the quest for the meaning of absolute zero, and the prospect of its experimental achievement, are at the forefront of experimental research.

The difficulty that the concept of absolute zero posed for nineteenth-century physicists stands in contrast to the ease with which its numerical value can be estimated. When a quantity of air is kept at the same pressure and cooled from the temperature of boiling water ($100°C$) to that of freezing water ($0°C$), it loses about a quarter of its volume. If this trend were to continue, the volume would decrease to half by the time it reached $-100°C$, to a quarter by approximately $-200°C$, and to zero at about $-300°C$. Clearly, its temperature cannot fall below that level. Although, in actual practice, air liquefies before it reaches absolute zero, using the relationship between the contraction of gases and decreasing temperature is a valid mathematical exercise that yields the precise figure of $-273.15°C$, or $-459.67°F$.

The first significant step in the direction of achieving absolute zero was taken in the 1870s, when a French mining engineer named Louis-Paul Cailetet liquefied oxygen, at about $-200°F$. Shortly afterward, he achieved the same result with nitrogen, which liquefies at $-319°F$. Each of these advances provided low-temperature physicists with a new liquid coolant that could, in turn, be used to liquefy other gases. The challenge lay in devising bigger and better machines for maintaining sufficiently large and stable quantities of the critical liquids. By the time the Dutch physicist H. Kamerlingh Onnes liquefied helium, in 1908, at $-457°F$, cryogenics laboratories had taken on the appearance of industrial plants, crammed with pumps and compressors, tubes, and containers. Today, when temperatures of less than a degree above absolute zero are achievable, this trend continues at the University of Florida, in Gainesville, where a new cryogenics laboratory opened in 1989.

From cylindrical pits in the laboratory's basement, three 40-ft refrigerators rise like guided missiles protruding from silos through the floors to the upper stories. Their surfaces are highly polished to reflect light and, thereby, insulate their contents. Separating their double walls is a vacuum, which, lacking a medium with which to conduct heat, provides additional thermal protection. To minimize motion (which is synonymous with heat), 18-ft pneumatic shock absorbers, anchored in 5-ton concrete blocks, cradle the refrigerators. The experimental area is surrounded by steel-and-copper shields—to screen out radio waves that generate minute electrical currents, which, in turn, create heat. The rooms that house the refrigerators are soundproof. No one is allowed in the vicinity during experiments (the vibrations from a single footstep could raise the temperature significantly). All operations are handled by remote control.

The architects of this elaborate apparatus intend to cool a piece of copper to a millionth of a kelvin above absolute zero (a kelvin being the scientific unit of temperature and equal to nine-fifths of a degree Fahrenheit). Then they hope to reach the Holy Grail of cryogenics—zero itself. If the Gainesville group does, in fact, fulfill the first of its aims, it will have exceeded the record, established in 1987 at the University of Bayreuth, in West Germany, by 11 μK—a considerable achievement in the cryogenics world. But if they complete the next step, they will have accomplished the impossible, for the third law of thermodynamics expressly forbids an object to reach the state of absolute zero.

Thermodynamic principles are commonly stated in terms of restrictions, which are, for all intents and purposes, ironclad. The first law (also known as the law of conservation of energy) holds that energy can be neither created nor destroyed. The second law, which governs the direction of the flow of heat, dictates that, during every transformation of energy, a certain amount is dissipated and, so, is unavailable for doing work. Anyone who claims to have circumvented either of these laws (by inventing, in the first instance, a machine that creates more energy than it consumes or, in the second instance, a device that is 100% energy-efficient) is considered a fool or, worse, a fraud. It seems, then, that in trying to circumvent the third article of the thermodynamic canon, the Gainesville physicists are embarking on a monumental misadventure.

But maybe not. The third law is not as well understood as the first two; nor is it as firmly established. Indeed, in the light of modern developments in low-temperature physics, its dominion is more questionable than ever. And it is precisely the uncertainty that surrounds the third law—its potential for violation—that makes it provocative to scientists such as those at the University of Florida. The first and second laws, by virtue of their intellectual elegance and explanatory strength, dominate physics, but they are virtually barren as stimuli for experimental and theoretical research, whereas the prohibition against absolute zero raises fruitful questions about the ultimate nature of matter. In effect, then, the third law may turn out to be the Cinderella of thermodynamics, outshining her older, more illustrious stepsisters.

The First and Second Laws

The first law, formulated in the 1840s by the German physicist Hermann von Helmholtz, is the easiest of the three to accept—in large part because it is the easiest to understand. When home appliances convert electricity into motion, heat, and light, energy neither increases nor decreases; it simply changes form. The same is true of conversions that, on first glance, seem imbalanced—producing fire by rubbing together two sticks, for example (mechanical energy is transformed into heat, which, if vigorous enough, breaks atomic bonds in the wood, releasing chemical energy that assumes the form of fire). With every advance in our understanding of energy (including the discovery, in the late 1800s, of one of its most powerful forms, nuclear radiation, and, shortly thereafter, of its fundamental interchangeability with matter), the first law of thermodynamics gained strength. On no scale, from the subnuclear to the cosmic, has there been even a hint of a violation.

The second law of thermodynamics, in its modern form at least, is much more abstract than the first, but it is based on a commonplace observation: When two bodies—say, a pair of copper blocks—with unequal temperatures come into contact, heat always flows from the hot one to the cold one. In fact, the second law was conceived in 1850 by the German physicist Rudolf J. Clausius in just such operational terms, without a real understanding of the underlying mechanism. Nothing in the first law precludes thermal energy's traveling from a cold body to a hot one, making it hotter still; energy is conserved regardless of the direction in which it flows. The second law, therefore, stands on its own as a separate thermodynamic principle.

The meaning of this law becomes apparent when the movement of thermal energy is seen in terms of the atomic activity that makes it possible. In 1872, the Austrian physicist Ludwig Boltzmann identified atomic disorder as an essential thermodynamic quantity and restated the second law in its modern form: In any physical system, disorder increases naturally, and work always is required to reverse this trend. Imagine an orderly room in which toys are stacked neatly in a closet. A child enters and begins playing, scattering the toys with the greatest of ease. As parents well know, restoring the room's order will require greater effort than that expended to make the mess. The basis for this discrepancy is contained in the statistical nature of order: there are many more ways the toys can be spread around the room than ways they can be arranged in the closet. Considering the total number of possible configurations, disorder simply is more probable than order. The same is true of the molecules in, say, a cloud or even something as apparently stable as a block of copper. Indeed, the odds against molecular order in gases, liquids, and solids are so tremendously high that, as time passes, increased disorder is virtually certain.

Boltzmann's achievement was showing how disorder, as expressed on the atomic level, relates to heat flow. He proved that the total amount of disorder in two bodies increases if heat flows from hot to cold. The essence of this demonstration lay in the realization that temperature is a measure of the mean kinetic energy of molecules. The molecules of a hot object jiggle about rapidly, carrying a lot of energy of motion, whereas those of a cold object move at a relatively sluggish pace. Think of a cold copper block placed on top of a hot one. At the outset, the jiggling molecules constitute two orderly piles—low speeds on top, high speeds below. Over the course of an hour, the positions of the blocks' molecules do not change appreciably, but their speeds do. Within the blocks and, especially, along the margin where they meet, warm, fast particles bump into cool, slow ones, transferring their energy. And after several hours, high and low velocities are mixed together helter-skelter, top and bottom. Overall, disorder has increased.

The second law does not rule out the possibility of pushing heat uphill, as it were, from a cold object to a hot one, or of creating order out of disorder. It merely states that such a reversal of the natural flow requires an influx of energy—for instance, a steady supply of electricity. Proof is no farther away than the kitchen, where a refrigerator draws heat from its interior and dumps it, via a metal grid affixed to the back of the machine, into the surrounding air.

The Third Law

Kitchen appliances cannot be used to demonstrate the third law of thermodynamics, because it governs a much rarer province—the realm of the very cold—than the first and second laws, which hold dominion over the entire span of temperatures. Doubtless, the exotic nature of this realm is the reason the third law was the last of the three to be proposed, in 1906, by Walther Nernst, of the University of Berlin (for which he won the 1920 Nobel Prize in chemistry).

Physicists previously had assumed that, when cooled, molecules and atoms gradually slow down until, at absolute zero, they come to rest. But the quantum-mechanical theory of matter, which was emerging at the turn of the century, does not allow that. The lowest energy available to an atom, called its ground state energy, is not zero; there always remains a small, irreducible quivering that can be theoretically described and experimentally measured. The energy associated with this state cannot be shared with other objects (it cannot "flow") or be interpreted as heat, so it does not count as disorderliness. (Think of the child's room again, this time imagining that the toys consist solely of watches of various sizes and shapes. The turning of the watches' gears—their internal quivering—has no bearing on the room's order, on whether the toys are stacked in the closet or scattered across the floor.) Reasoning in this manner, Nernst arrived at a new understanding of absolute zero: rather than the absence of motion, it signifies the absence of disorder, or a state of perfect order—and this became the initial version of the third law of thermodynamics.

Six years later, while surveying the specific heat—the amount of heat lost when temperature drops by a degree—of chemical elements, Nernst discovered a remarkable fact: During the approach toward absolute zero, or perfect order, each step is more difficult than the preceding one. Temperature decreases less with each successive removal of heat. He speculated that this relationship is not merely an accidental property of particular substances but an attribute of all matter. Ascribing the phenomenon to the difficulty of corralling huge numbers of unruly molecules into a single state of perfect order (the chances are astronomical that at least a few will elude capture), Nernst proposed a stronger form of the third law: Absolute zero is unattainable.

Compare a similar injunction, in a different context: Einstein's claim that objects cannot attain the speed of light. When first proposed, this law seemed strange, because the phenomenon occurred in the absence of a specific, countervailing force. But then Einstein showed that, as an object accelerates, its mass increases (as summarized in the formula $E = mc^2$). From this he deduced that, as an object approaches the speed of light, the energy required to further accelerate it approaches infinity. A change of variables, from speed to energy, made an incomprehensible limit appear comprehensible. In a roughly similar way, Nernst's switch from the absence of motion to the absence of disorder as the determinant of absolute zero made his decree appear plausible. The implication is that, as an object approaches a state of perfect order, the effort required to remove the remaining disorder approaches infinity.

About the reality of this barrier there is no doubt; it has been confirmed repeatedly in the laboratory. Whether, however, in all cases, an infinite amount of effort is necessary to reach absolute zero is less certain. Is the third law everywhere and always binding? That is the question lingering in the wings of the drama about to begin in Gainesville.

Can the Third Law Be Broken?

That the question can even be raised goes back to the statistical nature of order. In principle, there is a small but finite probability that an object will reach a temperature of a billionth of a kelvin, or even zero. For macroscopic systems, such as blocks of copper, consisting as they do of trillions of atoms, this probability is vanishingly small. A single atom, on the other hand, could readily jump into its ground state and, hence, into a state of absolute zero, if such a concept were applicable. But it is not. Temperature is a measure of the random motion—the thermal energy—of a group of particles and is meaningless with a single atom.

So whatever doubts might be raised by the third law concern matter that lies between the macroscopic and the microscopic scales, in particular, matter that approaches the small end of the spectrum, where the statistics are the most favorable (fewer atoms means fewer ways of being disordered). In recent years, objects of this kind, called mesoscopic because of their intermediate status, have captured the attention of physicists. Containing a small number of atoms and, therefore, being visible only under powerful microscopes, they occupy a no-man's-land where electrical resistance, normally a fixed property of materials, shifts wildly; where a disturbance in one side of the object can affect the other; where, in short, the ordinary laws of physics do not seem to hold. Chances are, mesoscopic objects are the correct size to bridge the gap between a microkelvin and zero (if the child in the nursery has only two toys to play with, she might put them on the right shelf in the closet by pure chance). It is on this slim statistical possibility that the Gainesville physicists pin their hopes.

Their refrigeration process, called magnetic cooling, is divided into two stages. During the first stage, copper samples will be placed between the poles of a powerful electromagnet and chilled to the lowest temperature attainable by conventional means—being bathed in liquid helium. In the crucial second stage, the magnet will be switched off, and the temperature of the copper will spontaneously fall, within two or three days, to an even lower level.

Magnetic cooling is based on the principle that heat represents the random motion of an object's constituent particles. Since atomic nuclei behave like tiny magnets, they respond to an external magnetic field by lining up in parallel orientation—north to north, south to south. When the field is removed, the nuclei are jostled back into random orientations by collisions with their neighbors. Such collisions slow down atoms, so the temperature is reduced. (Magnetic cooling is reminiscent of the cooling of a rubber band which is stretched and held that way for a few seconds in order to come to room temperature. When the tension is suddenly relaxed, the band feels noticeably colder to the upper lip.)

The process can be repeated again and again, but the technical difficulties mount with every step. A particularly insidious problem at temperatures approaching absolute zero is the minute amount of heat contributed by the residual radioactivity in construction materials. Fallout from, among other things, the atmospheric nuclear tests of the 1950s and, presumably, from the Chernobyl acci-

Magnetic cooling. At high temperature (a) atomic nuclei oscillate rapidly about their equilibrium positions. At lower temperature in a magnetic field (b), they oscillate less and line up in an orderly pattern. When the field is removed (c), they lose their orientation, but their thermal oscillation is also diminished.

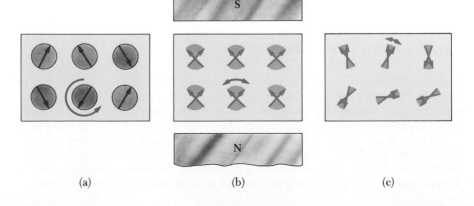

(a)   (b)   (c)

dent, which occurred in 1986, now salts the entire surface of the Earth. This makes it difficult to find metals that are free of radioactive debris or that do not become contaminated during the fabrication of machine parts when, in liquid form, they are exposed to the surfaces of their containers, as well as to the atmosphere. The Gainesville scientists hope this problem will be too small to affect their refrigerators.

Even under the best of conditions, the road to absolute zero will be neither short nor smooth. On the experimental side, the refrigerators will become larger, more complex, and more expensive, as the samples become smaller. Insulative measures will become heroic as the heat leaks grow more difficult to perceive. On the theoretical front, the third law of thermodynamics must, at some point, be reconciled with the quantum-mechanical behavior of mesoscopic systems. The resultant formulation may lack the concision of Nernst's statement that absolute zero is unattainable, but it will touch on philosophical questions about the meaning of quantum mechanics (What is the relationship between randomness in thermodynamics and randomness in quantum theory?) and thermodynamics (Will what we learn about the attainability of absolute zero have any impact on our view of the second law, which also is based on a statistical understanding of atomic behavior?) and about the process of measurement (What sort of thermometer will detect extraordinarily low temperatures without itself introducing heat?).

Work along these lines is just beginning. The third law could go down in the history books as a mistake. It could be dismissed as a trivial practical observation similar to the statement that no object can have infinite energy, because there isn't enough available in the Universe. A new phenomenon could be discovered in the submicrokelvin range that would render the whole discussion obsolete. Or, on the contrary, Nernst's proposition could, by virtue of its intrinsically quantum-mechanical nature, emerge as the most significant of all thermodynamic principles. Faced with so many possibilities, one is reminded of a comment made by Niels Bohr, the father of quantum mechanics:

The opposite of a correct statement is a false statement. But the opposite of a profound truth may well be another profound truth.

Unlike the first and second laws of thermodynamics, the third law fascinates precisely to the extent that it might yet provoke an opposing truth.

Conversation with
Freeman J. Dyson

Freeman Dyson was born in Crowthorne, Berkshire, England, in 1923. He received a B.A. in Mathematics from the University of Cambridge in 1945, studied bombing operations as a professional strategist for the British Air Force from 1943 to 1945, and then taught mathematics at Imperial College in London for a year. Arriving at Cornell University in 1947 with a Commonwealth Fellowship, Freeman Dyson studied theoretical physics with two notable masters of the field, Hans Bethe and Richard Feynman. In the summer of 1948, Dyson achieved recognition for demonstrating the connection between what appeared to be two radically different approaches (one by Julian Schwinger, the other by Feynman) to explaining the breakthrough experiment on hydrogen atoms performed at Columbia University that year. Dyson joined Princeton's Institute for Advanced Study as a visiting member, returned to Cornell as a physics professor, then in 1953 accepted the position of professor of physics at Princeton's Institute for Advanced Study, where he remains today. He has received nu-

Interview conducted by Janice K. Mandel.

416

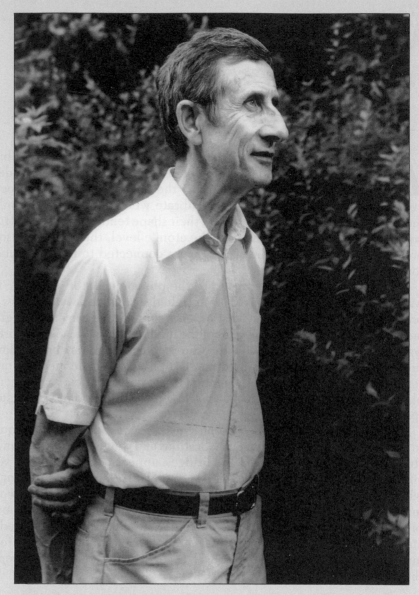

(Courtesy of Mia Dyson)

merous awards and honorary degrees from around the world in recognition of his contributions to his field.

How did you become interested in science?

Sir Frank Dyson was a leader in international astronomy at the height of his career in the 1930s. He not only held the position of Astronomer Royal but was also President of the International Astronomical Union, and just by pure coincidence he happened to have the same name that I do. He was no relation. He came from the same village as my father, and I heard all about him when I was a kid, so he was a role model for me. I don't know whether that was the only reason, but it certainly was one of the reasons I got involved in science. I never met Sir Frank.

Can you describe the scientific events taking place as you grew up?

Einstein discovered general relativity, which was his grand new theory of gravitation, in the middle of World War I. One of the main predictions of his theory was that the light from distant stars would be bent as it came around the Sun by a certain angle. It happened that there was a beautiful solar eclipse in 1919, so immediately after the war was over Sir Frank Dyson organized two expeditions to observe the Sun during the total eclipse, which was very important because this was the only time you could do it. Both expeditions were successful. They brought back very good pictures of the stars around the Sun while the eclipse was going on, and Einstein's theory was confirmed. This was an enormously impressive event for the public, and the whole world made a hero of Einstein and, in addition, Sir Frank got a good deal of glory as well.

You recently found a fiction piece you wrote at age 9 among your mother's papers. How does this reflect the times and the impression they made on you?

That is the only document that happens to have survived from my childhood. It was a fragment of a science fiction story about the collision of the asteroid Eros (which comes fairly close to Earth in its orbit every 37 years)

and the Moon, which of course was a completely fictional event. Eros did happen to pass by the Earth in 1931, however, which was an important event for professional astronomers—it gave them the most accurate measurement of the distance from the Earth to the Sun, which sets the scale for everything else in the Universe. Sir Frank was organizing the observations by professional astronomers and the analysis of the data. This was in the newspapers and I, no doubt, heard about all this while it was going on. I was 7 years old. The characters in my story were borrowed from Jules Verne, but the leading character, Sir Philip, was no doubt modeled after Sir Frank. Sir Philip was the one who calculated the orbit of Eros and predicted its collision with the Moon.

What was it like to come across this story nearly half a century later?

I was amused to see how little I had changed. I was a writer before I was a scientist, and I was always in love with spaceships. What is striking in this little fragment that I wrote is that I'm describing professional astronomers in the story and they never look at the sky. They're always either out fundraising or sitting at their desks doing calculations. That strikes me 50 years later as remarkably realistic. That was exactly as it was and still is.

How did you choose an area of science for concentration?

I had been interested in physics when I was a kid in high school, and I got sidetracked into mathematics because during the war all

the physicists were away working on radar and other such interesting things. So there was no physics being taught, and then I became a pure mathematician more or less because there were still some good pure mathematicians around.

What is a pure mathematician?

A person doing mathematics for its own sake. It's an art form. It is like painting or music: you do it because it's beautiful. It happened that my mathematics teacher in Cambridge was a man called Hardy who was also a rather gifted writer who wrote a famous little book about this subject called *A Mathematician's Apology*, which explains what mathematics means to him. I think it's the best statement about mathematics that has ever been written. He describes what real mathematics is about as opposed to applied mathematics.

What made you decide to turn to physics?

The atomic bomb had been exploded out of the blue sky in August 1945—all the work on nuclear energy had been completely secret. I had been interested in nuclear energy as a possibility, but I'd always thought about it as remote and distant. Smythe, at Princeton at the time, wrote the Smythe Report published on the day that the bombs were announced in 1945. I found it admirable that the United States government published this book explaining roughly what had been done and how and why. To me, that book was a great inspiration. It made it clear that this was a great piece of work and something I should love to be involved with. It wasn't so much that I was interested in building bombs, but interested in understanding how the thing worked. It was obvious that these were really first-rate people who were working on it. So it was reading the Smythe Report that immediately made me switch to physics. Of course, I've always been switching back and forth; I never really gave up mathematics.

You arrived at Cornell in 1947 for a year's study sponsored by a Commonwealth Fellowship and have said that "It was the best possible time for me to be there. Great things were going on." What was happening?

When I came to Cornell in the fall of 1947 there was one tremendous puzzle which everybody was excited about. Some new experiments on hydrogen atoms had been done at Columbia which could not be explained by the existing theories, so it was a real challenge to everybody to try to understand these experiments. So that's what we were working on, particularly Hans Bethe, who was my supervisor at Cornell. He's sort of my spiritual father as far as physics goes. He's a great man and he had produced a very good approximation to a theory of these Columbia experiments, but there were still all sorts of problems with it and so that was the job he gave me to explore.

Can you describe this puzzle in simple terms?

The problem was this: We had a theory of radiation which was developed mostly in Germany around 1930 called Quantum

Electrodynamics. That was the standard theory of atoms and radiation, which gave a generally correct description of atoms emitting and absorbing light. It was generally believed to be a correct theory, but it had all kinds of pathologies, what are called *divergences,* which meant if you tried to calculate things accurately it would give you the answer infinity. Nobody understood why the theory seemed to be to first approximation always good, but if you tried to go beyond the first approximation it seemed to be nonsense.

Then came the Columbia experiments led by Willis E. Lamb in 1946 or 1947. These marvelous experiments forced you to go to the next approximation because they were a thousand times more accurate than the prewar experiments. So we had a thousand times more precise experiments of hydrogen spectrum, and so you had to have a better theory in order to explain these experiments.

While searching for a solution, you enjoyed some fame yourself. Would you explain?

During that year at Cornell there were two completely separate new ideas on the problem: one from Schwinger, one from Feynman. They were professors, each only about age 30 at the time. Feynman was at Cornell; Schwinger was at Harvard. Both of them were getting exciting re-

---

"Science is better done without a plan. It's all just exploring."

---

sults and better agreement with the Columbia experiments, but what was strange was that these theories were so different. It was very hard to understand either of them precisely, and there was also a problem because neither of them was published and so you had to learn about them by talking to people. So, I was in a very happy position of being friendly with Feynman and also, in the summer following the year at Cornell, I went to the University of Michigan in Ann Arbor and had Schwinger for a six-week lecture series. I was just very lucky to have both of them thrown into my hands.

By August 1948, I'd had enough physics, so I went off on holiday in a Greyhound bus and spent a couple of weeks in Berkeley, mostly just reading in the library and wandering around. After two weeks, in the beginning of September, I took the Greyhound back east, and while I was sitting there on the bus in the middle of the night somewhere in Kansas, suddenly the whole thing came together and Schwinger and Feynman suddenly both made sense and fit together beautifully. That was the big moment I'd been waiting for. As usual, you only get some kind of revelation of this kind if you stop working for a few weeks. After that I became famous very rapidly, because that was the latest fad at the time and people tend to run after the latest fad, so I was in the happy position of being sought after for a year or two.

Are moments of revelation like this one of the rewards of science?

It is something that you can't either work for or manage; you're lucky when it happens. I think the main lesson is that idleness is important. They said Shakespeare was always idle between plays. I think that any sort of really creative work needs idleness. If you're always sitting at the computer terminal, things like this don't happen.

You can't teach talent, but what can you teach others about the creative aspects of discovery?

You've got to have both discipline and freedom. I spent a full year of real hard slogging studying Feynman and Schwinger, which meant sitting at my desk for days and days calculating, making sure I understood the details. That's the discipline you've got to master. But then, you must goof off for a few weeks and let the thing settle down in your subconscious and give it a chance to come up with something new. That seems to be how the thing works: you have to do the hard work first, and then be idle for awhile to give it a chance to turn into something new.

Do you need a particular talent to do theoretical physics? Does it require a background in both math and physics?

You certainly need the talent, but even more than that I suppose you need the drive or whatever it is that forces people to spend long hours doing calculations. You have to enjoy doing calculations, not just be good at it.

How does a theory start? Do you start with an impression, an image, an inspiration? How do you work the process?

For me it comes from writing down equations and trying to solve them. It's not really concerned with images at all. There are a great many different kinds of scientists with different ways of thinking. My way of thinking is mostly just in terms of mathematical symbols. When I'm building structures, they are really systems of equations.

Do you have some end in mind when you begin?

I start with a problem, which means I'm trying to calculate something. For example, at the moment I'm interested in the problem of taking a square array of points, like a chess board, throwing down a circle at random and calculating how many points will lie inside the circle. If things go well, out of it there might come a new way of doing the calculation or there might come a new understanding of the problem; anyway, we'll see. Science is better done without a plan. It's all just exploring. So you get a small step ahead and then you look around and see where you might go next.

You knew Richard Feynman as a young professor at Cornell long before he won a Nobel Prize. Can you describe your association?

I found him absolutely fascinating because he was completely different from everybody else, and he was interested in everything. Sometimes he would be furiously busy and he would just say "Go

away!" and not allow you into the room. Then, if he felt like it, he'd go out for a long walk and invite you to come along. So we would go and walk through the hills around Ithaca, and he would tell me the story of his life. I just got to admire him enormously. He struck me as being so unusually sane and, at the same time, in a pleasant way, crazy. It was a very good combination.

Can you tell me about some of those crazier times?

Dick invited me along on a trip to Albuquerque. For me it was just a joy ride. Dick was going there to make up his mind and find out if he was going to marry a girlfriend he had there. His first wife had died of tuberculosis about two years before, and he always said "If you've been happily married once, you can't stay single." I spent three days driving across the country in a car with Feynman, and that was the time I got to know him best. We got caught in a flood, and we had some wonderful nights in sleazy motels where we had the chance to talk about all kinds of things.

Among the things you learned from Feynman, what stands out in your mind?

He had been deeply involved in the Manhattan Project in Los Alamos in 1945 and had an extremely realistic view of the atomic bomb, which I found very helpful. He considered the thing to be unavoidable, a trap we'd fallen into. But he didn't see any good coming of it at all, and as far as he was concerned, he was not going to have anything more to do with it. So for the rest of his life he kept out of it. He didn't regret having been there, but he certainly didn't want to do it again. His view was "You should say yes to everything once and then find out if it's any good. If it turns out not to be any good then say no."

What is your view of science as it relates to weapons?

My viewpoints come from the work I did for the British Bomber Command, which left me deeply scarred. It was not nuclear, but it was a bombing campaign of a very bloody kind. It was indiscriminate bombing of cities on a large scale. I worked on the operations trying to understand how to use the bomber force effectively and particularly how to reduce the heavy bomber losses we were suffering because the German defenses were very good. I got a grandstand view of that whole campaign, and that gave me a burning compulsion for the rest of my life to try to fight the strategic bomber people. To me, it doesn't matter whether bombs are nuclear or not, it's an evil thing and it has a bad effect on people.

> "We tinker around and build toys and suddenly they go boom and the whole universe is disturbed. And that's the story of my life in a way . . ."

What is your view of the Strategic Defense Initiative Program (SDI)?

On the whole, I'm in favor of it because I think saving lives by shooting down missiles is a good idea. Unfortunately, the SDI program as it has been run has been a very corrupt kind of an enterprise. I'm not against it on any political grounds, but on technical grounds it doesn't do what it promises. What the Patriot System does very nicely, as we've just seen in the Middle East, is to shoot down missiles when they're coming to a well-defined small area, and not too many at a time. That's a technical problem that we know how to solve. Whereas what the SDI was advertised to do is to defend a huge area against a huge threat, and it never really had the hope of doing that to the extent that was promised. So the program has led to very little that's valuable.

Which is unfortunate because I think from the long-range point of view it was not such a bad idea. I would say the best thing to do would be to shut it down and start afresh. Typical SDI activities have been really public relations jobs more than science.

Would you like to discuss a few thoughts on *Disturbing the Universe,* your book that was published in 1979?

The Sloane Foundation, a general charitable trust that supports educational activities, decided to commission a series of autobiographies of scientists. The idea was to give people a sense of what being a scientist is about. Essentially, it is an autobiography but of a very sketchy kind. "Disturbing the Universe" is a phrase taken out of a poem by T. S. Eliot, "The Love Song of J. Alfred Prufrock," where a man is going up the stairs, feeling very shy and wondering whether he should knock on the door where his girlfriend is living, and he comes up the stairs and says "Do I dare, do I dare, do I dare disturb the Universe?" So, that's the theme of the book, that's what we're doing as scientists. It's sort of crazy, but that's the way it works. We tinker around and build toys and suddenly they go boom and the whole Universe is disturbed. And that's the story of my life in a way, at least that's the metaphor.

CHAPTER 14

Vibrations and Waves

A surfer "riding the pipe" on an ocean wave.
(© J. Divine/FPG International)

This chapter marks a brief return to the subject of mechanics as we examine various forms of periodic motion. We shall concentrate on a very special type of motion, that which occurs when the force on an object is proportional to the displacement of the object from its equilibrium position. When this force acts only toward the equilibrium position, the result is a back-and-forth motion called simple harmonic motion. An object moving in this manner oscillates, or vibrates, between two extreme positions for an indefinite period of time with no loss of energy. The terms *harmonic motion* and *periodic motion* will be used interchangeably throughout this chapter. Both refer to some form of back-and-forth motion. You are most likely familiar with several types of periodic motion, such as the oscillations of a mass on a spring, the motion of a pendulum, and the vibrations of a stringed musical instrument.

Since vibrations can move through a medium, we shall also study wave motion in this chapter. There are many kinds of waves in nature, such as sound waves, seismic waves, and electromagnetic waves. We shall end this chapter with a brief discussion of some terms and concepts that are common to all types of waves, and in later chapters we shall focus our attention on specific categories of waves.

14.1 Hooke's Law

One of the simplest types of vibrational motion is that of a mass attached to a spring, as in Figure 14.1. Let us assume that the mass moves on a horizontal smooth surface. If the spring is stretched or compressed a small distance x from its unstretched, or equilibrium, position and then released, it will exert a force on the mass. This spring force is found from experiment to obey the equation

$$F_s = -kx \qquad \text{(14.1)}$$

where x is the displacement of the mass from its unstretched ($x = 0$) position and k is a positive constant called the **spring constant.** This force law for springs was discovered by Robert Hooke in 1678 and is known as **Hooke's law.** The value of k is a measure of the stiffness of the spring. Stiff springs have large k values, and soft springs have small k values.

The negative sign in Equation 14.1 signifies that the force exerted by the spring is always directed *opposite* the displacement of the mass. When the mass is to the right of the equilibrium position, as in Figure 14.1a, x is positive and F_s is negative. This means that the force is in the negative direction, to the left. When the mass is to the left of the equilibrium position, as in Figure 14.1c, x is negative and F_s is positive, indicating that the direction of the force is to the right. Of course, when $x = 0$, as in Figure 14.1b, the spring is unstretched and $F_s = 0$. Because the spring force always acts toward the equilibrium position, it is sometimes called a restoring force. *The direction of the restoring force is such that the mass is being either pulled or pushed toward the equilibrium position.*

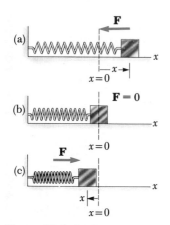

Figure 14.1
The force of a spring on a mass varies with the displacement of the mass from the equilibrium position $x = 0$.
(a) When x is positive (stretched spring), the spring force is to the left. (b) When x is zero (unstretched spring), the spring force is zero.
(c) When x is negative (compressed spring), the spring force is to the right.

Let us examine the motion of the mass if it is initially pulled a distance A to the right and released from rest. The force exerted on the mass by the spring pulls the mass back toward the equilibrium position. As the object moves toward $x = 0$, the magnitude of the force decreases (because x decreases) and reaches zero at $x = 0$. However, the speed of the mass increases as it moves toward the equilibrium position. In fact, the speed reaches its maximum value when $x = 0$. The momentum achieved by the mass causes it to overshoot the equilibrium position and to compress the spring. As the mass moves to the left of the equilibrium position (negative x values), the force acting on it begins to increase to the right and the speed of the mass begins to decrease. The mass finally comes to a stop at $x = -A$. The process is then repeated, and the mass continues to oscillate back and forth over the same path. This type of motion is referred to as **simple harmonic motion.** You should note that

> simple harmonic motion occurs when the net force along the direction of motion is a Hooke's law type of force, that is, when the net force is proportional to the displacement and in the opposite direction.

It should be emphasized here that not all repetitive motion over the same path can be classified as simple harmonic motion. For example, a car could start at one end of a city block, accelerate to the other end of the block, stop, back up to its original position, and then repeat the process. The car is moving back and forth over the same path, but the motion is not simple harmonic. Unless the force acting on an object along the direction of motion is of the form of Equation 14.1, the object does not exhibit simple harmonic motion.

The motion of a mass suspended from a vertical spring is also simple harmonic. In this case, the force of gravity acting on the attached mass stretches the spring until equilibrium is reached (where the mass is suspended and at rest). This position of equilibrium establishes the position of the mass for which $x = 0$. When the mass is stretched a distance x beyond this equilibrium position and then released, a net force acts (in the form of Hooke's law) toward the equilibrium position. Because the net force is proportional to x, the motion is simple harmonic.

Before we can discuss simple harmonic motion in more detail, it is necessary to define a few terms relative to such motion.

1. The **amplitude**, A, is the *maximum distance that an object moves away from its equilibrium position*. In the absence of friction, an object will continue in simple harmonic motion and reach a maximum displacement equal to the amplitude on each side of the equilibrium position during each cycle.
2. The **period**, T, is *the time it takes the object to execute one complete cycle of the motion*.
3. The **frequency**, f, is *the number of cycles or vibrations per unit of time*.

Example 14.1   Measuring the Spring Constant

A common technique used to evaluate the spring constant is illustrated in Figure 14.2. The spring is hung vertically as in Figure 14.2a, and a body of mass m is attached to the lower end of the spring as in Figure 14.2b. The spring stretches a distance d from its initial position under the action of the "load" mg. Since the spring force is upward, it must balance the weight mg downward when the system is at rest. In this case, we can apply Hooke's law to give

$$F_s = kd = mg$$

$$k = \frac{mg}{d}$$

For example, if a spring is stretched 2.0 cm by a mass of 0.55 kg, the force constant is

$$k = \frac{mg}{d} = \frac{(0.55 \text{ kg})(9.80 \text{ m/s}^2)}{2.0 \times 10^{-2} \text{ m}} = 2.7 \times 10^2 \text{ N/m}$$

Figure 14.2 (Example 14.1) Determining the spring constant. The elongation, d, of the spring is due to the suspended weight, mg. Since the spring force upward balances the weight, it follows that $k = mg/d$.

Example 14.2   Simple Harmonic Motion on a Slick Surface

A 0.35-kg mass attached to a spring of spring constant 130 N/m is free to move on a horizontal frictionless surface, as in Figure 14.1. If the mass is released from rest at $x = 0.1$ m, find the force on it and its acceleration at (a) $x = 0.1$ m, (b) $x = 0.05$ m, (c) $x = 0$ m, and (d) $x = -0.05$ m.

Solution
(a) The point at which the mass is released defines the amplitude of the motion. In this case, $A = 0.1$ m. The mass moves continuously between the limits of 0.1 m and -0.1 m. When x is a maximum $(x = A)$, the force on the mass is a maximum and is calculated as follows:

$$F = -kx$$

$$F_{max} = -kA = -(130 \text{ N/m})(0.1 \text{ m}) = \boxed{-13.0 \text{ N}}$$

The negative sign indicates that the force acts to the left, in the negative x direction. We can use Newton's second law to calculate the acceleration at this position:

$$F_{max} = ma_{max}$$

$$a_{max} = \frac{F_{max}}{m} = -\frac{13.0 \text{ N}}{0.35 \text{ kg}} = \boxed{-37.1 \text{ m/s}^2}$$

Again, the negative sign indicates that the acceleration is to the left.

(b) We can use the same approach to find the force and acceleration at other positions. At $x = 0.05$ m, we have

$$F = -kx = -(130 \text{ N/m})(0.05 \text{ m}) = \boxed{-6.50 \text{ N}}$$

$$a = \frac{F}{m} = -\frac{6.50 \text{ N}}{0.35 \text{ kg}} = \boxed{-18.5 \text{ m/s}^2}$$

Note that the acceleration of an object moving with simple harmonic motion *is not constant* since F is not constant.

(c) At $x = 0$, the spring force is zero (since $F = -kx = 0$) and the acceleration is zero. In other words, when the spring is unstretched, it exerts no force on the mass attached to it.

(d) At $x = -0.05$ m, we have

$$F = -kx = -(130 \text{ N/m})(-0.05 \text{ m}) = \boxed{6.50 \text{ N}}$$

$$a = \frac{F}{m} = \frac{6.50 \text{ N}}{0.35 \text{ kg}} = \boxed{18.5 \text{ m/s}^2}$$

This result shows that the force and acceleration are positive when the mass is on the negative side of the equilibrium position. This indicates that the force of the spring on the mass is acting to the right as the spring is being compressed. At the same time, the mass is slowing down as it moves from $x = 0$ to $x = -0.05$ m.

Exercise   Find the force and acceleration when $x = -0.1$ m.

Answer   13.0 N; 37.1 m/s².

As indicated in Example 14.2, the acceleration of an object moving with simple harmonic motion can be found by using Hooke's law as the force in the equation for Newton's second law, $F = ma$. This gives

$$-kx = ma$$

$$\boxed{a = -\frac{k}{m} x} \tag{14.2}$$

Since the maximum value of x is defined to be the amplitude, A, we see that the acceleration ranges over the values $-kA/m$ to $+kA/m$. Equation 14.2 enables us to find the acceleration of the object as a function of its position. In subsequent sections, we shall find equations for the velocity as a function of position and the position as a function of time.

Earlier we stated that an object will move with simple harmonic motion when the net force acting on it is proportional to its displacement from equilibrium. Equation 14.2 provides an alternative definition of simple harmonic motion. An object moves with simple harmonic motion if its acceleration is proportional to its displacement.

14.2   Elastic Potential Energy

So far, for the most part, we have worked with three types of mechanical energy: gravitational potential energy, translational kinetic energy, and rotational kinetic energy. In Chapter 5 we briefly discussed a fourth type of mechanical energy, **elastic potential energy,** and we shall now consider this form of energy in more detail.

An object has potential energy by virtue of its shape or position. As we learned in Chapter 5, an object of mass m at a distance h above the ground has gravitational potential energy equal to mgh. This means that the object can do work after it is released. Likewise, a compressed spring has potential energy by virtue of its shape. In this case, the compressed spring, when allowed to expand, can move an object and thus do work on the object. As an example, Figure 14.3 shows a ball being projected from a spring-loaded toy gun, where the spring is compressed by a distance, x. As the gun is fired, the compressed spring does work on the ball and imparts kinetic energy to it.

The energy stored in a stretched or compressed spring or other elastic material is called **elastic potential energy.**

We defined the **elastic potential energy,** PE_s, as

$$PE_s \equiv \frac{1}{2}kx^2$$

(14.3)   Elastic potential energy

Note that the elastic potential energy stored in a spring is zero when the spring is unstretched or uncompressed ($x = 0$). *Energy is stored in a spring only when it is either stretched or compressed.* Furthermore, *the elastic potential energy is a maximum when a spring has reached its maximum compression or extension.* Finally, the potential energy in a spring is always positive because it

Energy = elastic PE

x

Energy = KE

Figure 14.3
A ball projected from a spring-loaded gun. The elastic potential energy stored in the spring is transferred to the kinetic energy of the ball.

is proportional to x^2. We now include this new form of energy in our equation for conservation of mechanical energy:

$$(KE + PE_g + PE_s)_i = (KE + PE_g + PE_s)_f \qquad \textbf{(14.4)}$$

If there are nonconservative forces, such as friction, present, then the final mechanical energy will not equal the initial mechanical energy. In this case, the difference in the two energies must equal the work done by the nonconservative force, W_{nc}. According to the work-energy theorem, we have

$$W_{nc} = (KE + PE_g + PE_s)_f - (KE + PE_g + PE_s)_i \qquad \textbf{(14.5)}$$

As an example of the energy conversions that take place when a spring is included in the system, consider Figure 14.4. A block of mass m slides on a frictionless horizontal surface with constant velocity $\mathbf{v}_i$ and collides with a coiled spring. The description that follows is greatly simplified by assuming that the spring is very light and therefore has negligible kinetic energy. As the spring is compressed, it exerts a force to the left on the block. At maximum compression, the block momentarily comes to rest (Fig. 14.4c). The initial total energy in the system before the collision (block plus spring) is the kinetic energy of the block. After the block collides with the spring and the spring is partially compressed, as in Figure 14.4b, the block has kinetic energy $\frac{1}{2}mv^2$ (where $v < v_i$) and the spring has potential energy $\frac{1}{2}kx^2$. When the block comes momentarily to rest after colliding with the spring, the kinetic energy is zero. Since the spring force is conservative and since there are no external forces that can do work on the system, *the total mechanical energy of the system consisting of the block and spring remains constant*. Thus, there is a transfer of energy from kinetic energy of the block to potential energy stored in the spring. As the spring expands, the block moves in the opposite direction and regains all of its initial kinetic energy, as in Figure 14.4d.

Figure 14.4
A block sliding on a frictionless horizontal surface collides with a light spring. (a) Initially, the mechanical energy is entirely the kinetic energy of the block. (b) The mechanical energy at some arbitrary position is the sum of the kinetic energy of the block and the elastic potential energy stored in the spring. (c) When the block comes to rest, the mechanical energy is entirely elastic potential energy stored in the compressed spring. (d) When the block leaves the spring, the mechanical energy is equal to its kinetic energy. The total energy remains constant.

Figure 14.5 (Example 14.3) A car starts from rest on a hill at the position shown. When it reaches the bottom of the hill, the car collides with a spring-loaded guard rail.

Example 14.3 Stop That Car!

A 13,000-N car starts at rest and rolls down a hill from a height of 10 m (Fig. 14.5). It then moves across a level surface and collides with a light spring-loaded guard rail. Neglecting any losses due to friction, find the maximum distance the spring is compressed. Assume the spring constant is 10^6 N/m.

Solution

The initial potential energy of the car is completely converted to elastic potential energy in the spring at the end of the trip. (This assumes we are neglecting any energy losses due to friction during the collision.) Thus, conservation of energy gives

$$mgh = \frac{1}{2} kx^2$$

Solving for x gives

$$x = \sqrt{\frac{2mgh}{k}} = \sqrt{\frac{2(13,000 \text{ N})(10 \text{ m})}{10^6 \text{ N/m}}} = \boxed{0.5 \text{ m}}$$

Note that it was not necessary to calculate the velocity of the car at any point to obtain a solution. This demonstrates the power of the principle of conservation of energy. One has to work with the initial and final energy values only, without having to consider all the details in between.

Exercise What is the speed of the car just before it collides with the guard rail?

Answer 14 m/s.

Example 14.4 Motion With and Without Friction

A block of mass 1.6 kg is attached to a spring with a spring constant of 10^3 N/m, as in Figure 14.1. The spring is compressed a distance of 2.0 cm, and the block is released from rest. (a) Calculate the velocity of the block as it passes through the equilibrium position, $x = 0$, if the surface is frictionless.

Solution

By using Equation 14.3, we can find the initial elastic potential energy of the spring when $x = -2.0$ cm $= -2 \times 10^{-2}$ m:

$$PE_s = \frac{1}{2} kx_i^2 = \frac{1}{2} (10^3 \text{ N/m})(-2 \times 10^{-2} \text{ m})^2 = 0.20 \text{ J}$$

Since the block is always at the same height above the Earth's surface, its gravitational potential energy remains constant. Hence, the initial potential energy stored in the spring is converted to kinetic energy at $x = 0$. That is,

$$\frac{1}{2} kx_i^2 = \frac{1}{2} mv_f^2$$

$$0.2 \text{ J} = \frac{1}{2}(1.6 \text{ kg})(v_f^2)$$

$$v_f = \boxed{0.5 \text{ m/s}}$$

(b) Calculate the velocity of the block as it passes through the equilibrium position if a constant frictional force of 4.0 N retards its motion.

Solution

Since sliding friction is present in this situation, we know that the final mechanical energy will be less than the initial mechanical energy. The work done by the frictional force for a displacement of 2×10^{-2} m is

$$W_{nc} = W_f = -fs = -(4.0 \text{ N})(2 \times 10^{-2} \text{ m}) = -0.08 \text{ J}$$

Applying Equation 14.5 to this situation gives

$$-0.08 \text{ J} = \frac{1}{2}(1.6 \text{ kg})(v_f^2) - 0.20 \text{ J}$$

$$v_f = \boxed{0.39 \text{ m/s}}$$

Note that this value for v_f is less than that obtained in the frictionless case. Does the result make sense?

Exercise   How far does the block travel before coming to rest? Assume a constant friction force of 4.0 N and the same initial conditions as stated above.

Answer   5 cm.

Figure 14.6
(a) A mass attached to a spring on a frictionless surface is released from rest with the spring extended a distance A. Just before release, the total energy is elastic potential energy, $kA^2/2$. (b) When the mass reaches position x, it has kinetic energy $mv^2/2$ and the elastic potential energy has decreased to $kx^2/2$.

14.3   Velocity as a Function of Position

Conservation of energy provides us with a simple method for deriving an expression that gives the velocity of a mass undergoing periodic motion as a function of position. The mass in question is initially located at its maximum extension A, as in Figure 14.6a, and is then released from rest. In this situation, the initial energy of the system is entirely elastic potential energy stored in the spring, $\frac{1}{2}kA^2$. As the mass moves toward the origin to some new position x (Fig. 14.6b), part of this energy is transformed to kinetic energy and the potential energy stored in the spring is reduced to $\frac{1}{2}kx^2$. Since the total energy of the system is equal to $\frac{1}{2}kA^2$ (the initial energy stored in the spring), we can equate this to the sum of the kinetic and potential energies at the final position:

$$\frac{1}{2} kA^2 = \frac{1}{2} mv^2 + \frac{1}{2} kx^2$$

When this is solved for v, we get

$$v = \pm \sqrt{\frac{k}{m}(A^2 - x^2)} \qquad \text{(14.6)}$$

This expression shows us that the speed is a maximum at $x = 0$ and zero at the extreme positions, $x = \pm A$.

The right side of Equation 14.6 is preceded by the $\pm$ sign because the square root of a number can be either positive or negative. The sign of v that should be selected depends on the circumstances of the motion. If the mass in Figure 14.6 is moving to the right, v must be positive; if the mass is moving to the left, v must be negative.

Example 14.5   The Mass-Spring System Revisited

A mass of 0.5 kg connected to a light spring of spring constant 20 N/m oscillates on a horizontal frictionless surface. (a) Calculate the total energy of the system and the maximum speed of the mass if the amplitude of the motion is 3 cm.

Solution
From Equation 14.3, we have

$$E = PE_s = \frac{1}{2} kA^2 = \frac{1}{2} (20 \text{ N/m})(3 \times 10^{-2} \text{ m})^2 = \boxed{9.0 \times 10^{-3} \text{ J}}$$

When the mass is at $x = 0$, $PE_s = 0$ and $E = \frac{1}{2} mv_{max}^2$; therefore

$$\frac{1}{2} mv_{max}^2 = 9 \times 10^{-3} \text{ J}$$

$$v_{max} = \sqrt{\frac{18 \times 10^{-3} \text{ J}}{0.5 \text{ kg}}} = \boxed{0.19 \text{ m/s}}$$

(b) What is the velocity of the mass when the displacement is 2 cm?

Solution
We can apply Equation 14.6 directly:

$$v = \pm \sqrt{\frac{k}{m} (A^2 - x^2)} = \pm \sqrt{\frac{20}{0.5} (3^2 - 2^2) \times 10^{-4}} = \boxed{\pm 0.14 \text{ m/s}}$$

The $\pm$ sign indicates that the mass could be moving to the right or to the left at this instant.

(c) Compute the kinetic and potential energies of the system when the displacement equals 2 cm.

Solution
The results to (b) can be used to give

$$KE = \frac{1}{2} mv^2 = \frac{1}{2} (0.5 \text{ kg})(0.14 \text{ m/s})^2 = \boxed{5.0 \times 10^{-3} \text{ J}}$$

$$PE_s = \frac{1}{2} kx^2 = \frac{1}{2} (20 \text{ N/m})(2 \times 10^{-2} \text{ m})^2 = \boxed{4.0 \times 10^{-3} \text{ J}}$$

Note that the sum $KE + PE_s$ equals the total energy, E, found in Part (a).

Exercise   For what values of x does the speed of the mass equal 0.10 m/s?

Answer ± 2.55 cm.

Figure 14.7
Experimental setup for demonstrating the connection between simple harmonic motion and uniform circular motion. As the ball rotates on the turntable with constant angular velocity, its shadow on the screen moves back and forth with simple harmonic motion.

14.4   Comparing Simple Harmonic Motion with Uniform Circular Motion

We can better understand and visualize many aspects of simple harmonic motion along a straight line by looking at their relationships to uniform circular motion. Figure 14.7 shows an experimental arrangement useful for developing this concept. This figure represents a top view of a ball attached to the rim of a phonograph turntable of radius A, illuminated by a lamp. Rather than concentrating on the ball, let us focus our attention on the shadow that the ball casts on the screen. We find that *as the turntable rotates with constant angular velocity, the shadow of the ball moves back and forth with simple harmonic motion.*

In order to understand why the shadow moves with simple harmonic motion, let us examine Equation 14.6 more closely. This equation says that the velocity of an object moving with simple harmonic motion is related to the displacement as

$$v = C\sqrt{A^2 - x^2}$$

where C is a constant. To see that the shadow also obeys this relation, consider Figure 14.8, which shows the ball moving with a constant speed, v_0, in a direction tangent to the circular path. At this instant, the velocity of the ball in the x direction is given by $v = v_0 \sin \theta$, or

$$\sin \theta = \frac{v}{v_0}$$

Likewise, the larger triangle in Figure 14.8 that contains the angle θ enables us to write

$$\sin \theta = \frac{\sqrt{A^2 - x^2}}{A}$$

Equating the right-hand sides of the last two expressions, we see that v is related to the displacement x as

$$\frac{v}{v_0} = \frac{\sqrt{A^2 - x^2}}{A}$$

or

$$v = \frac{v_0}{A} \sqrt{A^2 - x^2} = C\sqrt{A^2 - x^2}$$

Thus, the velocity of the ball along the x direction is related to the displacement x in exactly the same manner as that of an object undergoing simple harmonic motion. Hence, the shadow moves with simple harmonic motion.

Period and Frequency

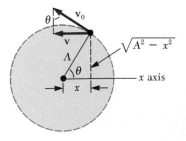

Figure 14.8
The ball rotates with constant speed v_0. The x component of the velocity of the ball equals the projection of $\mathbf{v_0}$ on the x axis.

Note that the period, T, of the shadow, which equals the time for one complete trip back and forth, is also equal to the time it takes the ball to make one complete circular trip on the turntable. Since the ball moves through a distance $2\pi A$ (the circumference of the circle) in the time T, the speed v_0 of the ball around the circular path is

$$v_0 = \frac{2\pi A}{T}$$

$$T = \frac{2\pi A}{v_0}$$

However, for our purposes, let us consider only a fraction of the complete trip. Imagine that the ball moves from P to Q in Figure 14.7, which represents one quarter of a revolution. This requires a time equal to one fourth of the period, and the distance traveled by the ball is $2\pi A/4$. Therefore, we see that

$$\frac{T}{4} = \frac{2\pi A}{4v_0} \qquad\qquad \textbf{(14.7)}$$

Now imagine that the motion of the shadow is equivalent to the motion of a mass on the end of a spring moving in the horizontal direction. During this quarter of a cycle, the shadow moves from a point where its energy is solely elastic potential energy to a point where its energy is solely kinetic energy. That is,

$$\frac{1}{2}\,kA^2 = \frac{1}{2}\,mv_0{}^2$$

$$\frac{A}{v_0} = \sqrt{\frac{m}{k}}$$

Substituting for A/v_0 in Equation 14.7, we find that the **period** is

$$T = 2\pi\,\sqrt{\frac{m}{k}} \qquad\qquad \textbf{(14.8)}$$

The period of a mass-spring system moving with simple harmonic motion

This expression gives the time required for an object to make a complete cycle of its motion. Now recall that the definition of frequency, f, is the number of cycles per unit of time. The symmetry in the units of period and frequency should lead you to see that the two must be related inversely as

$$f = \frac{1}{T} \qquad\qquad \textbf{(14.9)}$$

Therefore, the **frequency** of the periodic motion is

$$f = \frac{1}{2\pi}\sqrt{\frac{k}{m}} \qquad\qquad \textbf{(14.10)} \quad \text{Frequency}$$

The units of frequency are s^{-1}, or hertz (Hz).

Example 14.6   That Car Needs a New Set of Shocks!

A car of mass 1300 kg is constructed using a frame supported by four springs. Each spring has a spring constant of 20,000 N/m. If two people riding in the car have a combined mass of 160 kg, find the frequency of vibration of the car when it is driven over a pot hole in the road.

Solution
We shall assume that the weight is evenly distributed. Thus, each spring supports one fourth of the load. The total mass supported by the springs is 1460 kg, and therefore each spring supports 365 kg. Hence, the frequency of vibration is

$$f = \frac{1}{2\pi}\sqrt{\frac{k}{m}} = \frac{1}{2\pi}\sqrt{\frac{20{,}000 \text{ N/m}}{365 \text{ kg}}} = \boxed{1.18 \text{ Hz}}$$

Exercise   How long does it take the car to execute three complete vibrations?

Answer   2.54 s.

14.5   Position as a Function of Time

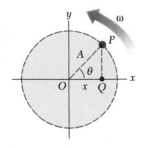

Figure 14.9
The reference circle. As the ball at P rotates in a circle with uniform angular velocity, its projection Q along the x axis moves with simple harmonic motion.

We can obtain an expression for the position of an object moving with simple harmonic motion as a function of time by returning to the relationship between simple harmonic motion and uniform circular motion. Again, consider a ball on the rim of a rotating turntable of radius A, as in Figure 14.9. Let us assume that the turntable rotates at a constant angular velocity ω. We shall refer to this circle as the *reference circle* for the motion. As the ball rotates on the reference circle, the angle θ that the line OP makes with the x axis changes with time. Furthermore, as the ball rotates, the projection of P on the x axis, labeled point Q, moves back and forth along the axis with simple harmonic motion.

From the right triangle, OPQ, we see that $\cos \theta = x/A$. Therefore, the x coordinate of the ball is

$$x = A \cos \theta$$

Since the ball rotates with constant angular velocity, it follows that $\theta = \omega t$ (see Chapter 7). Therefore,

$$x = A \cos(\omega t) \tag{14.11}$$

In one complete revolution, the ball rotates through an angle of 2π rad in a time equal to a period T. In other words, the motion repeats itself every T seconds. Therefore,

$$\omega = \frac{\Delta \theta}{\Delta t} = \frac{2\pi}{T} = 2\pi f \tag{14.12}$$

where f is the frequency of the motion. Consequently, Equation 14.11 can be written

$$x = A \cos(2\pi f t) \tag{14.13}$$

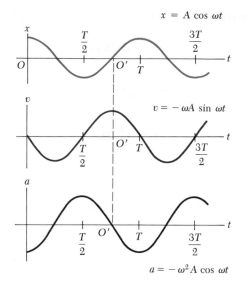

$$x = A \cos \omega t$$

$$v = -\omega A \sin \omega t$$

$$a = -\omega^2 A \cos \omega t$$

Figure 14.10
(a) Displacement, (b) velocity, and (c) acceleration versus time for an object moving with simple harmonic motion under the initial conditions that $x_0 = A$ and $v_0 = 0$ at $t = 0$.

This equation represents the position of an object moving with simple harmonic motion as a function of time. A graph of this equation is shown in Figure 14.10a. The curve should be familiar to you from trigonometry. Note that x varies between A and $-A$ since the cosine function varies between 1 and -1.

Figure 14.10b and 14.10c represent curves for the velocity and acceleration as a function of time. Although we shall not prove it here, it can be shown that the velocity and acceleration are also sinusoidal functions of time. Note that when x is a maximum or minimum, the velocity is zero, and when x is zero, the magnitude of the velocity is a maximum. Furthermore, when x has its maximum positive value, the acceleration is a maximum but in the negative x direction, and when x is at its maximum negative position, the acceleration has its maximum value in the positive direction. These curves are consistent with what we discussed earlier concerning the points at which v and a reach their maximum, minimum, and zero values.

One experimental arrangement that demonstrates simple harmonic motion is illustrated in Figure 14.11. A mass connected to a spring has a marking pen attached to it. While the mass vibrates vertically, a sheet of paper is moved horizontally with constant speed, and the pen traces out a sinusoidal pattern.

Motion of paper

Figure 14.11
An experimental apparatus for demonstrating simple harmonic motion. A pen attached to the oscillating mass traces out a sinusoidal wave on the moving chart paper.

Example 14.7   The Vibrating Mass-Spring System

Find the amplitude, frequency, and period of motion for an object vibrating at the end of a spring if the equation for its position as a function of time is

$$x = (0.25 \text{ m}) \cos\left(\frac{\pi}{8} t\right)$$

Solution
We can find two of our unknowns by comparing this equation with the general equation for such motion:

$$x = A \cos(2\pi ft)$$

A disturbance traveling from right to left on a stretched spring.

What Is a Wave?

As we said before, when you drop a pebble into a pool of water, the disturbance produced by the pebble excites water waves, which move away from the point at which the pebble entered the water. If you were to examine carefully the motion of a leaf floating near the disturbance, you would see that it moves up and down and back and forth about its original position but does not undergo any net displacement attributable to the disturbance. That is, the water wave (or disturbance) moves from one place to another *but the water is not carried with it.*

Einstein and Infeld[1] make these remarks about wave phenomena:

A bit of gossip starting in Washington reaches New York very quickly, even though not a single individual who takes part in spreading it travels between these two cities. There are two quite different motions involved, that of the rumor, Washington to New York, and that of the persons who spread the rumor. The wind, passing over a field of grain, sets up a wave which spreads out across the whole field. Here again we must distinguish between the motion of the wave and the motion of the separate plants, which undergo only small oscillations. . . . The particles constituting the medium perform only small vibrations, but the whole motion is that of a progressive wave. The essentially new thing here is that for the first time we consider the motion of something which is not matter, but energy propagated through matter.

When we observe what we call a water wave, what we see is a rearrangement of the water's surface. Without the water there would be no wave. A wave traveling on a string would not exist without the string. Sound waves travel through air as a result of pressure variations from point to point. (We shall discuss sound waves in Chapter 15.) The wave motion we shall consider in this chapter corresponds to the disturbance of a body or medium. Therefore, we can consider a wave *to be the motion of a disturbance.* In a later chapter, we shall discuss electromagnetic waves, which do not require a medium.

The mechanical waves discussed in this chapter require (1) some source of disturbance, (2) a medium that can be disturbed, and (3) some physical connection or mechanism through which adjacent portions of the medium can influence each other. All waves carry energy and momentum. The amount of energy transmitted through a medium and the mechanism responsible for the transport of energy will differ from case to case. For instance, the energy carried by ocean waves during a storm is much greater than that carried by a sound wave generated by a single human voice.

14.9   Types of Waves

One of the simplest ways to demonstrate wave motion is to flip one end of a long rope that is under tension and has its opposite end fixed, as in Figure 14.17. The bump (called a pulse) travels to the right with a definite speed. A

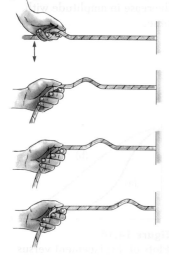

Figure 14.17
A wave pulse traveling down a stretched rope.

[1] Albert Einstein and Leopold Infeld, *The Evolution of Physics,* New York, Simon and Schuster, 1961.

disturbance of this type is called a **traveling wave.** Figure 14.7 shows three consecutive "snapshots" of the traveling wave.

An important point about a traveling wave is that, as the wave pulse travels along the rope, *each segment of the rope that is disturbed moves in a direction perpendicular to the wave motion.* Figure 14.18 illustrates this point for one particular segment, labeled *P*. Never does the rope move in the direction of the wave.

A traveling wave such as this, in which the particles of the disturbed medium move perpendicular to the wave velocity, is called a **transverse wave.**

Other examples of transverse waves are electromagnetic waves, such as light, radio, and television waves.

In another class of waves, called **longitudinal waves,** the particles of the medium undergo displacement in a direction parallel to the direction of travel of the wave.

Sound waves are longitudinal waves. The disturbance in a sound wave corresponds to a series of high- and low-pressure regions that travel through a material. A longitudinal pulse can be easily produced in a stretched spring, as in Figure 14.19. The left end of the spring is given a sudden jerk (consisting of a brief push to the right and an equally brief pull to the left) along the length of the spring; this creates a sudden compression of the coils. The compressed region *C* travels along the spring, and so we see that the disturbance is parallel to the wave motion. Region *C* is followed by region *R*, where the coils are spread farther apart than normal.

Picture of a Wave

The curve in Figure 14.20 shows the shape of a vibrating string. This particular pattern, sometimes called a waveform, should be familiar to you from our study of simple harmonic motion; it is a sinusoidal curve. The red curve can be thought of as a snapshot of a traveling wave taken at some instant of time, say $t = 0$; the blue curve represents a snapshot of the same traveling wave at a later time. It is not difficult to imagine that the picture of the wave on a string in Figure 14.20 can as easily be used to represent a wave on water. In such a case, point *A* would correspond to the *crest* of the wave and point *B* would correspond to the low point, or *trough,* of the wave.

Figure 14.18
A pulse traveling on a stretched rope is a transverse wave. Any element *P* on the rope moves in a direction *perpendicular* to the motion of the pulse.

Figure 14.19
A longitudinal pulse moving along a stretched spring. The displacement of the coils is in the direction of the wave motion. For the starting motion described in the text, the compressed region *C* is followed by an extended region *R*.

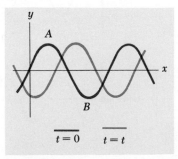

Figure 14.20
A one-dimensional harmonic wave traveling to the right with a speed v. The red curve represents a snapshot of the wave at $t = 0$, and the blue curve at some later time t.

Figure 14.21
(a) A longitudinal wave on a spring. (b) The crests of the waveform correspond to positions on the spring where the coils are close together, and the troughs of the wave correspond to positions of loosely spaced coils. (c) The displacement wave.

You may be surprised to find out that this same waveform can be used to describe a longitudinal wave. To see how this is done, consider a longitudinal wave traveling on a spring. Figure 14.21a represents a snapshot of the longitudinal wave at some instant, and Figure 14.21b shows the sinusoidal curve that represents the wave. Points where the coils of the spring are packed tightly together correspond to the crests of the waveform, and loosely spaced coils correspond to troughs.

The wave represented by the curve in Figure 14.21b is often referred to as a density or pressure wave. The reason for this terminology is that the crests of the curve indicate locations where the spring coils are close together (high density) and the troughs represent locations where the coils are far apart (low density).

An alternative method for representing wave motion along a spring is through the concept of a displacement wave, shown in Figure 14.21c. In this representation, coils that are displaced the greatest distance from equilibrium in one direction are indicated by crests, and coils that are displaced the greatest distance from equilibrium in the opposite direction are represented by troughs. Both of these wave representations will be used in future sections.

14.10 Frequency, Amplitude, and Wavelength

One method of producing a wave on a very long string is shown in Figure 14.22. One end of the string is connected to a blade that is set into vibration. As the blade oscillates vertically with simple harmonic motion, a traveling wave moving to the right is set up on the string. Figure 14.22 represents snapshots of the wave at intervals of one quarter of a period. Note that *each particle of the string, such as P, oscillates vertically in the y direction with simple harmonic motion.* This must be the case because each particle follows the simple harmonic motion of the blade. Therefore, every segment of the string can be treated as a simple harmonic oscillator vibrating with a frequency equal to the frequency of vibration of the blade that drives the string.

Figure 14.22
One method for producing traveling waves on a continuous string. The left end of the string is connected to a blade that is set into vibration. Note that every segment, such as *P*, oscillates with simple harmonic motion in the vertical direction.

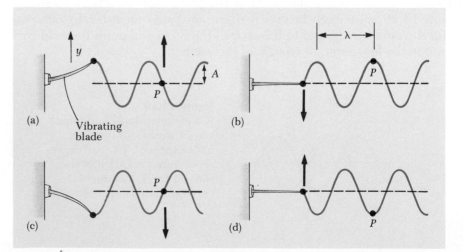

The frequencies of the waves we shall study will range from rather low values for waves on a string and waves on water to values between 20 and 20,000 Hz (recall that 1 Hz = 1 s^{-1}) for sound waves and to much higher frequencies for electromagnetic waves.

The horizontal dashed line in Figure 14.22 represents the shape of the string if no wave was present. The maximum distance that the string is raised above this equilibrium value is called the **amplitude, A,** of the wave. The amplitude can also be designated as the maximum distance that the string falls below the equilibrium value. Note that, for the waves we shall be working with, the amplitudes at the crest and the trough are identical.

Another characteristic of a wave is shown in Figure 14.22b. The horizontal arrows show the distance between two successive points that behave identically. This distance is called the **wavelength,** λ (Greek letter lambda).

We can use these definitions to derive an expression for the velocity of a wave. We start with the defining equation for velocity:

$$v = \frac{\Delta x}{\Delta t}$$

A little reflection should convince you that a wave will advance a distance equal to one wavelength in a time equal to one period of the vibration. Substitution in the velocity equation gives

$$v = \frac{\lambda}{T}$$

Since the frequency is equal to the reciprocal of the period, we have

$$v = \lambda f \qquad\qquad \textbf{(14.15)} \quad \text{Wave velocity}$$

We shall apply this equation to many types of waves. For example, we shall use it often in our study of sound and electromagnetic waves.

Example 14.9 A Traveling Wave

A wave traveling in the positive x direction is pictured in Figure 14.23. Find the amplitude, wavelength, period, and speed of the wave if it has a frequency of 8 Hz.

Solution
The amplitude and wavelength can be read directly off the figure:

$$A = \boxed{15 \text{ cm}} \qquad \lambda = 40 \text{ cm} = \boxed{0.40 \text{ m}}$$

The period of the wave is

$$T = \frac{1}{f} = \frac{1}{8} \text{ s} = \boxed{0.125 \text{ s}}$$

and the speed is

$$v = \lambda f = (0.40 \text{ m})(8 \text{ s}^{-1}) = \boxed{3.2 \text{ m/s}}$$

Figure 14.23 (Example 14.9) A harmonic wave of wavelength $\lambda = 40$ cm and amplitude $A = 15$ cm.

Example 14.10   Give Me a "C" Note

The note middle C on a piano has a frequency of approximately 264 Hz and a wavelength in air of 1.31 m. Find the speed of sound in air.

Solution
By direct substitution into Equation 14.15, we find

$$v = \lambda f = (1.31 \text{ m})(264 \text{ s}^{-1}) = \boxed{346 \text{ m/s}}$$

Example 14.11   The Speed of Radio Waves

An FM station broadcasts at a frequency of 100 MHz (M = mega = 10^6) with a radio wave having a wavelength of 3 m. Find the speed of the radio wave.

Solution
As in the last example, we use Equation 14.15:

$$v = \lambda f = (3 \text{ m})(100 \times 10^6 \text{ s}^{-1}) = \boxed{3 \times 10^8 \text{ m/s}}$$

This, in fact, is the speed of *all* electromagnetic waves traveling through empty space.

Exercise   Find the wavelength of an electromagnetic wave whose frequency is 9 GHz = 9×10^9 Hz, which is the microwave range.

Answer   0.033 m.

14.11   The Velocity of Waves on Strings

In this section, we shall focus our attention on the speed of a wave on a stretched string. Rather than deriving the equation, we shall use dimensional analysis to verify that the expression can be valid.

It is easy to understand why the wave velocity depends on the tension in the string. If a string under tension is pulled sideways and released, the tension in the string is responsible for accelerating a particular segment back toward its equilibrium position. The acceleration and wave velocity increase with increasing tension in the string. Likewise, the wave velocity is inversely proportional to the mass per unit length of the string. This is because it is more difficult to accelerate (and impart a large wave velocity) to a massive string compared to a light string. Thus, the wave velocity is directly proportional to the tension and is inversely proportional to the mass per unit length of the string. The exact relationship between the wave velocity v, the tension F, and the mass per length μ, is given by

$$v = \sqrt{\frac{F}{\mu}} \tag{14.16}$$

From this we see that the velocity of mechanical waves, such as a wave on a string, depends only on the properties of the medium through which the disturbance travels.

Now, let us verify that this expression is dimensionally correct. The dimensions of F are $[M][L]/[T^2]$, and the dimensions of μ are $[M]/[L]$. Therefore, the dimensions of F/μ are $[L^2]/[T^2]$ and those of $\sqrt{F/\mu}$ are $[L]/[T]$, which are indeed the dimensions of velocity. No other combination of F and μ is dimensionally correct, assuming these are the only variables relevant to the situation.

Equation 14.16 indicates that we can increase the velocity of a wave on a stretched string by increasing the tension in the string. It also shows that if we wrap a string with a metallic winding, as on the bass strings of a piano or a guitar, we decrease the velocity of a transmitted wave because the mass per unit length is increased.

Example 14.12 A Pulse Traveling on a String

A uniform string has a mass, M, of 0.3 kg and a length, L, of 6 m. Tension is maintained in the string by suspending a 2-kg block from one end (Fig. 14.24). Find the speed of a pulse on this string.

Solution
The tension, F, in the string is equal to the mass, m, of the block multiplied by the gravitational acceleration:

$$F = mg = (2 \text{ kg})(9.80 \text{ m/s}^2) = 19.6 \text{ N}$$

For the string, the mass per unit length, μ, is

$$\mu = \frac{M}{L} = \frac{0.3 \text{ kg}}{6 \text{ m}} = 0.05 \text{ kg/m}$$

Therefore, the wave speed is

$$v = \sqrt{\frac{F}{\mu}} = \sqrt{\frac{19.6 \text{ N}}{0.05 \text{ kg/m}}} = \boxed{19.8 \text{ m/s}}$$

Exercise Find the time it takes the pulse to travel from the wall to the pulley.

Answer 0.253 s.

Figure 14.24 (Example 14.12) The tension, F, in the string is maintained by the suspended block. The wave speed is given by the expression $v = \sqrt{F/\mu}$.

14.12 Superposition and Interference of Waves

Many interesting wave phenomena in nature cannot be described by a single moving wave. Instead, one must analyze what happens when two or more waves attempt to pass through the same region of space. To analyze such wave combinations, one can make use of the **superposition principle:**

> If two or more traveling waves are moving through a medium, the resultant wave is found by adding the displacements of the individual waves together point by point.

Experiments show that the superposition principle is valid only when the individual waves have small amplitudes of displacement. This small amplitude approximation will be assumed in all our examples.

One consequence of the superposition principle is that *two traveling waves can pass through each other without being destroyed or even altered.* For

(a)

(b)

(c)

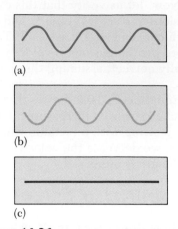

(a)

(b)

(c)

Figure 14.25
Constructive interference. If two waves having the same frequency and amplitude are in phase as in (a) and (b), the resultant wave (c) when they combine has the same frequency as the individual waves but twice their amplitude.

Figure 14.26
Destructive interference. When two waves having the same frequency and amplitude are 180° out of phase as in (a) and (b), the result when they combine is complete cancellation, as in (c).

Interference patterns produced by outward spreading waves from two drops of liquid dropped into a body of water. (*Martin Dohrn/Science Photo Library*)

instance, when two pebbles are thrown into a pond, the expanding circular waves do not destroy each other. In fact, the ripples pass through each other. Likewise, when sound waves from two sources move through air, they also pass through each other. The sound one hears at a given location is the result of both disturbances.

In Figures 14.25a and 14.25b, two waves of the same amplitude and frequency are shown. If at some instant of time, these two waves attempt to travel through the same region of space, the resultant wave at that instant would look like the wave shown in Figure 14.25c. For example, suppose these are water waves of amplitude 1 m. At the instant they overlap such that crest meets crest and trough meets trough, the resultant wave has an amplitude of 2 m. Waves coming together like this are said to be *in phase* and undergo **constructive interference.**

In Figures 14.26a and 14.26b, two similar waves are shown. In this case, however, the crest of one coincides with the trough of the other, that is, one wave is *inverted* relative to the other. The resultant wave, shown in Figure 14.26c, is seen to be a state of complete cancellation. If these were water waves coming together, one of the waves would be trying to pull an individual drop of water upward at the same instant that the second wave was trying to pull the same drop downward. The result is no motion of the water at all. In this situation, the two waves are said to be 180° out of phase and undergo **destructive interference.** Figure 14.27 illustrates the interference of water waves produced in a ripple tank.

Figure 14.28 shows constructive interference in two pulses moving toward each other along a stretched spring. Figure 14.29 shows destructive

Figure 14.27
Interference of water waves produced in ripple tank. The sources of the waves are two objects that vibrate perpendicular to the surface of the tank. *(Courtesy of CENCO)*

Figure 14.28
Constructive interference. Superposition of two equal and symmetric pulses traveling in opposite directions on a stretched spring. *(Courtesy of Education Development Center, Newton, Mass.)*

Figure 14.29
Destructive interference. Superposition of two symmetric pulses traveling in opposite directions, where one is inverted relative to the other. *(Courtesy of Education Development Center, Newton, Mass.)*

Figure 14.30
The reflection of a traveling wave at the fixed end of a stretched string. Note that the reflected pulse is inverted, but its shape remains the same.

interference in two pulses. You should note from these figures that, when the two pulses separate, their shapes are unchanged, as if they had never met!

14.13   Reflection of Waves

In our discussion so far, we have assumed that the waves we are analyzing can travel indefinitely without striking anything. Obviously, such conditions are not realized in practice. Whenever a traveling wave reaches a boundary, part or all of the wave is reflected. For example, consider a pulse traveling on a string fixed at one end (Fig. 14.30). When the pulse reaches the wall, it will be reflected.

Note that *the reflected pulse is inverted.* This can be explained as follows. When the pulse meets the wall, the string exerts an upward force on the wall. By Newton's third law, the wall must exert an equal and opposite (downward) reaction force on the string. This downward force causes the pulse to invert upon reflection.

Now consider another case, one where the pulse arrives at the end of the string shown in Figure 14.31, where the end of the string is attached to a ring of negligible mass that is free to slide along the post without friction. Again, the pulse will be reflected, but this time it is not inverted. Upon reaching the post, the pulse exerts a force on the ring, causing it to accelerate upward. The ring is then returned to its original position by the downward component of the tension.

An alternative method of showing that a pulse is reflected without inversion when it strikes a free end of a string is to send a pulse down a string hanging vertically. When the pulse hits the free end, it is reflected without inversion similar to the situation shown in Figure 14.31.

Figure 14.31
The reflection of a traveling wave at the free end of a stretched string. In this case, the reflected pulse is not inverted.

Summary

Simple harmonic motion occurs when the net force along the direction of motion is a **Hooke's law** type of force, that is, when the net force is proportional to the displacement and in the opposite direction:

$$F_s = -kx \tag{14.1}$$

The time for one complete vibration is called the **period** of the motion. The inverse of the period is the **frequency** of the motion, which equals the number of oscillations per second.

When an object is moving with simple harmonic motion, its **acceleration** as a function of location is

$$a = -\frac{k}{m}x \tag{14.2}$$

The energy stored in a stretched or compressed spring or other elastic material is called **elastic potential energy**:

$$PE_s \equiv \frac{1}{2} kx^2 \qquad (14.3)$$

The **velocity** of an object as a function of position when the object is moving with simple harmonic motion is

$$v = \pm \sqrt{\frac{k}{m}(A^2 - x^2)} \qquad (14.6)$$

The **period** of an object of mass m moving with simple harmonic motion while attached to a spring of spring constant k is

$$T = 2\pi \sqrt{\frac{m}{k}} \qquad (14.8)$$

Because $f = 1/T$, the **frequency** of a mass-spring system is

$$f = \frac{1}{2\pi} \sqrt{\frac{k}{m}} \qquad (14.10)$$

The **position** of an object as a function of time when the object is moving with simple harmonic motion is

$$x = A \cos(2\pi ft) \qquad (14.13)$$

A **simple pendulum** of length L moves with simple harmonic motion for small angular displacements from the vertical, with a period given by

$$T = 2\pi \sqrt{\frac{L}{g}} \qquad (14.14)$$

The period is independent of the suspended mass.

A **transverse wave** is a wave in which the particles of the medium move in a direction perpendicular to the direction of the wave velocity. An example is a wave on a stretched string.

A **longitudinal wave** is a wave in which the particles of the medium move in a direction parallel to the direction of the wave velocity. An example is a sound wave.

The relationship between the velocity, wavelength, and frequency of a wave is

$$v = \lambda f \qquad (14.15)$$

The speed of a wave traveling on a stretched string of a mass per unit length μ and under tension F is

$$v = \sqrt{\frac{F}{\mu}} \qquad \text{(14.16)}$$

The **superposition principle** states that, if two or more traveling waves are moving through a medium, the resultant wave is found by adding the individual waves together point by point. When waves meet crest to crest and trough to trough, they undergo **constructive interference.** If crest meets trough, the waves undergo complete **destructive interference.**

When a wave pulse reflects from a rigid boundary, the pulse is inverted. If the boundary is free, the reflected pulse is not inverted.

Additional Reading

W. Bascom, "Ocean Waves," *Sci. American*, August 1959, p. 74.

W. Bascom, *Waves and Beaches: The Dynamics of the Ocean Surface*, New York, Doubleday Anchor Books, 1980.

A. Einstein and L. Infeld, *The Evolution of Physics*, New York, Simon and Schuster, 1961.

B. Gilbert and P. Glanz, "Springs: Distorted and Combined," *The Physics Teacher*, October 1983, p. 430.

S. D. Kelby and R. P. Middleton, "The Vibrations of Hand Bells," *Physics Education*, Vol. 15, 1980, p. 320.

J. Oliver, "Long Earthquake Waves," *Sci. American*, March 1959, p. 14.

T. D. Rossing, "The Physics of Kettledrums," *Sci. American*, November 1982, p. 172.

R. A. Waldron, *Waves and Oscillations*, Momentum Series, Princeton, NJ, Van Nostrand, 1964.

Questions

1. What is the total distance traveled by an object moving with simple harmonic motion in a time equal to its period if its amplitude is A?

2. Does the acceleration of a simple harmonic oscillator remain constant during its motion? Is the acceleration ever zero? Explain.

3. Explain why the kinetic and potential energies of a mass-spring system can never be negative.

4. Given a spring, a clock, and a single known mass, devise an experiment to measure the force constant of the spring.

5. A mass-spring system undergoes simple harmonic motion with an amplitude A. Does the total energy change if the mass is doubled but the amplitude is not changed? Do the kinetic and potential energies depend on the mass? Explain.

6. If a mass-spring system is hung vertically and set into oscillation, why does the motion eventually stop?

7. What happens to the period of a simple pendulum if its length is doubled? What happens if the suspended mass is doubled?

8. If a pendulum clock keeps perfect time at the base of a mountain, will it also keep perfect time when moved to the top of the mountain? Explain.

9. If a grandfather clock were running slow, how could we adjust the "length" of the pendulum to correct the time?

10. A pendulum bob is made with a ball filled with water. What happens to the frequency of vibration of this pendulum if there is a hole in the ball that allows water to slowly leak out?

11. Fill your bathroom sink or tub with water. Touch the surface gently with one fingertip and observe the wave pulses. What happens when the wave fronts reach the water's edge? Try touching the surface with the long edge of a ruler. Make a sketch of the resulting wavefronts.

12. If a duck lands in a lake, it bobs up and down but essentially remains in one place as waves pass by. Explain why the duck is not carried along by the wave motion.

13. A flag waves in the breeze. The waves on the flag consist of straight sinusoidal fronts. Describe the motion of various points on the flag. What type of waves are they?

14. How do transverse waves differ from longitudinal waves?

15. A solid may transport a longitudinal wave as well as a transverse wave, but a fluid may only transport a longitudinal wave. Why?

16. How would you create a longitudinal wave in a stretched spring? Would it be possible to create a transverse wave in a spring?

17. If you were to periodically shake the end of a stretched rope up and down three times each second, what would be the period of the waves set up in the rope?

18. What happens to the wavelength of a wave on a string when the frequency is doubled? Assume the tension in the string remains the same.

19. What happens to the velocity of a wave on a string when the frequency is doubled? Assume the tension in the string remains the same.

20. Consider a wave traveling on a stretched rope. What is the difference, if any, between the speed of the wave and the speed of a small section of the rope?

21. By what factor would you have to increase the tension in a stretched string in order to double the wave speed?

22. If a long rope is hung from a ceiling and waves are sent up the rope from its lower end, the waves do not ascend with constant speed. Explain.

23. When all the strings on a guitar are stretched to the same tension, will the velocity of a wave along the more massive bass strings be faster or slower than the velocity of a wave on the lighter strings?

24. Suppose two pulses are moving toward one another on a string. How could you tell whether such pulses reflect off or pass through one another?

25. When two waves interfere, can the resultant wave be larger than either of the two original waves? Under what conditions?

Problems

Section 14.1   Hooke's Law

1. (a) A mass of 400 g is suspended from a spring hanging vertically, and the spring is found to stretch 8 cm. Find the spring constant. (b) How much will the spring stretch if the suspended mass is 575 g?

2. A 0.4-kg mass is attached to a spring of spring constant 160 N/m such that the mass is allowed to move on a horizontal frictionless surface. The mass is released from rest when the spring is compressed 0.15 m. Find (a) the force on the mass and its acceleration at this instant and (b) the maximum values of the force and acceleration.

3. A load of 50 N attached to a spring hanging vertically will stretch the spring 5 cm. The spring is now placed horizontally on a table and stretched 11 cm. (a) What force is required to stretch the spring by this amount? (b) Plot a graph of force (on the y axis) versus spring displacement from the equilibrium position along the x axis.

4. A 3-kg mass is attached to a spring and pulled out horizontally to a maximum displacement from equilibrium of 0.5 m. What spring constant must the spring have if the mass is to achieve an acceleration equal to that of gravity?

Section 14.2   Elastic Potential Energy

5. The mat of a trampoline is held by 32 springs, each having a spring constant of 5000 N/m. A person with a mass of 40.0 kg jumps from a platform 1.93 m high onto the trampoline. Determine the stretch of each of the springs. Assume the springs were initially unstretched and that they stretch equally.

6. An archer pulls his bow string back 0.4 m by exerting a force that increases uniformly from zero to 230 N. (a) What is the equivalent spring constant of the bow? (b) How much work is done in pulling the bow?

7. In an arcade game a 0.1-kg disk is shot across a horizontal frictionless surface by compressing it against a spring and releasing it. If the spring has a spring constant of 200 N/m and is compressed from its equilibrium position by 6 cm, find the speed with which the disk slides across the surface.

8. The disk of Problem 7 is projected up a 10° frictionless incline by the spring. Find (a) the distance the disk moves up the incline before coming to rest and (b) its speed when it is halfway to the point where it comes to rest.

9. A spring having a spring constant of 19.6 N/m is stretched 0.4 m. Find the ratio of the energy stored in

the spring at this extension to the energy stored when the extension is 0.2 m.

10. When a 4-kg mass is hung vertically on a certain light spring that obeys Hooke's law, the spring stretches 2.5 cm. If the 4-kg mass is removed, (a) how far will the spring stretch if a 1.5-kg mass is hung on it? (b) How much work must an external agent do to stretch the same spring 4.0 cm from its unstretched position?

11. A simple harmonic oscillator has a total energy *E*. (a) Determine the kinetic and potential energies when the displacement equals one-half the amplitude. (b) For what value of the displacement does the kinetic energy equal the potential energy?

12. A block of mass 1.5 kg is attached to a spring with spring constant of 2000 N/m. The spring is then stretched a distance of 0.3 cm and the block is released from rest. (a) Calculate the speed of the block as it passes through the equilibrium position if no friction is present. (b) Calculate the speed of the block as it passes through the equilibrium position if a constant frictional force of 2.0 N retards its motion. (c) What would be the strength of the frictional force if the block reaches the equilibrium position the first time with zero velocity?

13. A bullet of mass 10 g is fired into and embeds in a 2-kg block attached to a spring of spring constant 19.6 N/m. How far will the spring be compressed if the speed of the bullet just before striking the block is 300 m/s and the block slides on a frictionless surface? (*Hint:* You must use conservation of momentum in this problem. Why?)

14. The spring constant of the spring in Figure 14.32 is 19.6 N/m and the mass of the object is 1.5 kg. The spring is unstretched and the surface is frictionless. A constant 20-N force is applied to the object horizontally as shown. Find the speed of the object after it has moved a distance of 0.3 m.

Figure 14.32
Problem 14

Section 14.3   Velocity as a Function of Position

15. A 50-g mass is attached to a horizontal spring of spring constant 10 N/m and released from rest with an amplitude of 25 cm. What is the velocity of the mass when it is halfway to the equilibrium position if the surface is frictionless?

16. A mass of 0.4 kg connected to a light spring of spring constant 19.6 N/m oscillates on a horizontal frictionless surface. If the spring is compressed 4 cm and released from rest, determine (a) the maximum speed of the mass, (b) the speed of the mass when the spring is compressed 1.5 cm, and (c) the speed of the mass when the spring is stretched 1.5 cm. (d) For what value of *x* does the speed equal one-half the maximum speed?

17. A spring of spring constant 19.6 N/m is compressed 5 cm. A mass of 0.3 kg is attached to the spring and released from rest. Find (a) the maximum elastic potential energy stored in the spring and (b) the maximum speed of the mass.

18. At an outdoor market, a bunch of bananas is set into oscillatory motion with an amplitude of 20 cm on a spring with a spring constant of 16 N/m. It is observed that the maximum speed of the bunch of bananas is 40 cm/s. What is the weight of the bananas in newtons?

19. A mass-spring system oscillates with an amplitude of 3.5 cm. If the spring constant is 250 N/m and the mass is 0.5 kg, determine (a) the mechanical energy of the system, (b) the maximum speed of the mass, and (c) the maximum acceleration.

Section 14.4   Comparing Simple Harmonic Motion with Uniform Circular Motion

20. A 200-g mass is attached to a spring and executes simple harmonic motion with a period of 0.25 s. If the total energy of the system is 2 J, find (a) the force constant of the spring and (b) the amplitude of the motion.

21. A ball moves with constant speed of 5 m/s in a circular path of radius 0.4 m as shown in Figure 14.8. Find the *x* component of the velocity of the ball when θ equals (a) 0°, (b) 60°, (c) 90°, (d) 180°, (e) 270°.

22. (a) Find the frequency and (b) the period of the ball described in Problem 21.

23. The frequency of vibration of a mass-spring system is 5 Hz when a 4-g mass is attached to the spring. What is the spring constant of the spring?

24. A ball rotates in a circle of radius 20 cm, making one complete revolution every 2 s. What is (a) the speed of the ball, (b) the frequency of motion in Hz, and (c) the angular velocity of the ball?

25. The amplitude of a system moving with simple harmonic motion is doubled. Determine the change in (a) the total energy, (b) the maximum velocity, (c) the maximum acceleration, and (d) the period.

26. A spring stretches by 3.9 cm when a 10-g mass is hung from it. If a total mass of 25 g attached to this spring oscillates in simple harmonic motion, calculate the period of motion.
27. The frequency of vibration of a mass-spring system is 5 Hz when a 4-g mass is attached to the spring. What is the force constant of the spring?
28. When four people having a combined mass of 320 kg sit down in a car they find that the car moves 0.8 cm lower on its springs. They then get out of the car and bounce it up and down. What is the frequency of its vibration if the mass of the car when empty is 2000 kg?

Section 14.5 Position as a Function of Time

29. An object attached to the end of a spring vibrates with an amplitude of 20 cm according to Equation 14.13. Find the position of the object at these times: 0, $T/8$, $T/4$, $3T/8$, $T/2$, $5T/8$, $3T/4$, $7T/8$, and T, where T is the period of vibration. Plot your results (position along the vertical axis and time along the horizontal axis).
30. The motion of an object is described by the equation

$$x = (0.3 \text{ m}) \cos (\pi t/3).$$

Find (a) the position of the object at $t = 0$ and $t = 0.2$ s, (b) the amplitude of the motion, (c) the frequency of the motion, and (d) the period of the motion.
31. A 0.5-kg object is attached to a spring of spring constant 30 N/m and released from rest at $t = 0$ from a position 0.25 m from equilibrium. Find the equation that describes its x location as a function of time.
32. An object on a spring vibrates with a period of 3 s. The motion is initiated by releasing the object from its point of maximum displacement $x = A$. After what time will the object first be at (a) $x = A/2$, (b) $x = -A/2$, and (c) $x = 0$?
33. A spring stretches 3.0 cm from its relaxed length when a force of 7.5 N is applied. A particle with a mass of 0.5 kg rests on a frictionless horizontal surface and is attached to the free end of the spring. The particle is pulled horizontally so that it stretches the spring 5.0 cm and is then released from rest at $t = 0$. (a) What is the force constant of the spring? (b) What are the period, frequency, f, and the angular frequency, ω, of the motion? (c) What is the total energy of the system? (d) What is the amplitude of the motion? (e) What are the maximum velocity and the maximum acceleration of the particle? (f) Determine the displacement x of the particle from the equilibrium position at $t = 0.5$ s.

Section 14.6 The Pendulum

34. A simple pendulum 2.00 m in length oscillates in a location where $g = 9.80$ m/s². How many complete oscillations will it make in 5 min?
35. The period of a 50-cm-long pendulum is found to be 1.4 s. What is the acceleration due to gravity at this location?
36. A pendulum is to be used in a clock. What length should the pendulum be so that its period of vibration is 1 s?
37. (a) Find the ratio of the period of a pendulum on Earth to the period of an identical pendulum on the Moon, where the acceleration due to gravity is one-sixth that on Earth. (b) If the period of the pendulum is 2.5 s on Earth, what will be its period on the Moon?
38. A "seconds" pendulum is one that moves through its equilibrium position once each second. (The period of the pendulum is 2 s.) The length of a seconds pendulum is 0.9927 m at Tokyo and 0.9942 m at Cambridge, England. What is the ratio of the acceleration due to gravity at these two locations?
39. If the length of a simple pendulum is quadrupled, what happens to (a) its frequency and (b) its period?

Section 14.10 Frequency, Amplitude, and Wavelength

40. A cork resting on the surface of a pond bobs up and down two times per second on some ripples having a wavelength of 8.5 cm. If the cork is 10.0 m from shore, how long does it take a ripple passing the cork to reach the shore?
41. If the frequency of oscillation of the wave emitted by an FM radio station is 88 MHz, determine (a) the period of vibration of the wave and (b) the wavelength of the wave. (Radio waves travel at the speed of light.)
42. When a particular wire is vibrating with a frequency of 4 Hz, a transverse wave of wavelength 60 cm is produced. Determine the speed of wave pulses along the wire.
43. A piano emits frequencies that range from a low of about 28 Hz to a high of about 4200 Hz. Find the range of wavelengths spanned by this instrument. The speed of sound in air is approximately 343 m/s.
44. A harmonic wave is traveling along a rope. It is observed that the oscillator that generates the wave completes 40 vibrations in 30 s. Also, a given maximum travels 425 cm along the rope in 10 s. What is the wavelength?

Figure 14.33
Problem 45

45. A wave traveling in the positive x direction is pictured in Figure 14.33. Find (a) the amplitude, (b) wavelength, (c) period, and (d) speed of the wave if it has a frequency of 25 Hz.

46. For a certain transverse wave, it is observed that the distance between two successive maxima is 1.2 m. It is also noted that eight crests, or maxima, pass a given point along the direction of travel every 12 s. Calculate the wave speed.

47. (a) How long does it take light to reach us from the Sun, 93,000,000 mi away? (The speed of light is equal to 3×10^8 m/s.) (b) An astronaut communicates with Earth from the Moon (3.84×10^8 m away). How long does it take his signal to reach us?

48. How far in meters does light travel in one year? This distance is called a light-year.

49. A sound wave, traveling at 343 m/s, is emitted by the foghorn of a tugboat. An echo is heard 2.6 s later. How far away is the reflecting object?

Section 14.11   The Velocity of Waves on Strings

50. A circus performer stretches a tightrope between two towers. He strikes one end of the rope and sends a wave along it toward the other tower. He notes that it takes the wave 0.8 s to reach the opposite tower 20 m away. If one meter of the wire has a mass of 0.35 kg, find the tension in the tightrope.

51. A phone cord is 4.0 m long. The cord has a mass of 0.2 kg. If a transverse wave pulse travels from the receiver to the phone box in 0.1 s, what is the tension in the cord?

52. One end of a string 3 m long is attached to a wall while the other end hangs over a pulley and is attached to a hanging 2.0-kg mass. The speed of a pulse on the string is observed to be 15 m/s. What is the mass of the string?

53. Transverse waves with a speed of 50 m/s are to be produced on a stretched string. A 5-m length of string with a total mass of 0.06 kg is used. (a) What is the required tension in the string? (b) Calculate the wave speed in the string if the tension is 8 N.

54. The elastic limit of a piece of steel wire is equal to 2.7×10^9 Pa. What is the maximum velocity at which transverse wave pulses can propagate along this wire without exceeding this stress? (The density of the steel is 7.86 g/cm³.)

55. Transverse waves travel with a speed of 20 m/s on a string under a tension of 6 N. What tension is required for a wave speed of 30 m/s in the same string?

56. When a steel wire stretched between two clamps is plucked, waves travel along the wire with a speed of 80 m/s. A second wire of the same material but of twice the length and twice the radius of the first is stretched between two points under the same tension as the first. At what speed will transverse waves travel along the second wire?

Section 14.12   Superposition and Interference of Waves
Section 14.13   Reflection of Waves

57. A series of pulses of amplitude 0.15 m are sent down a string attached to a post at one end. The pulses are reflected at the post and travel back along the string without loss of amplitude. What is the amplitude at a point on the string where two pulses are crossing (a) if the string is rigidly attached to the post and (b) if the end at which reflection occurs is free to slide up and down?

58. A wave of amplitude 0.3 m interferes with a second wave of amplitude 0.2 m. What is (a) the largest and (b) the smallest value of the absolute value of the resultant displacement that can occur and what will be the conditions in which these maxima and minima occur?

Additional Problems

59. A string under a tension of 100 N has a wave on it traveling at 90 m/s. What is the speed of a wave on this string if the tension is doubled?

60. The position of a 0.3-kg object attached to a spring is described by

$$x = (0.25 \text{ m}) \cos (0.4\pi t)$$

Find (a) the amplitude of the motion, (b) the spring constant, (c) the position at $t = 0.3$ s, and (d) its velocity at $t = 0.3$ s.

61. A mass of 1.5 kg is attached to a spring of spring constant 1000 N/m. Find the force on the mass and its acceleration at the positions (a) 0.5 m, (b) 0.1 m, (c) 0 m, (d) -0.20 m, and (e) -0.5 m. (The positions are measured from equilibrium, $x = 0$.)

62. A string is 50 cm long and has a mass of 3 g. A wave travels with a speed of 5 m/s along this string. A second string has the same length but a mass half that of

the first. If both strings are under the same tension, what is the speed of a wave along the second string?

63. Tension is maintained in a string as shown in Figure 14.34. The observed wave speed is 24 m/s when the suspended mass is 3 kg. (a) What is the mass per unit length of the string? (b) What will be the wave speed when the suspended mass is 2 kg?

Figure 14.34
Problem 63

3 kg

64. A spring having a spring constant of 30 N/m is stretched to a distance of 0.2 m from its equilibrium position. How much work must be done to stretch it an additional 0.1 m?

65. A block of mass 500 g is released from rest and slides down a frictionless track of height 2 m above the horizontal, as shown in Figure 14.35. At the bottom of the track, where the surface is horizontal, the block strikes and sticks to a light spring of spring constant 20 N/m. Find the maximum distance the spring is compressed.

500 g

2 m

k

Figure 14.35
Problem 65

66. One end of a hacksaw blade is clamped in a vise and the other end is set into vibration. The free end makes 5 complete vibrations each second with an amplitude of 0.4 cm. Find the speed of the end of the hacksaw blade when the displacement from the equilibrium position is 0.1 cm.

67. A spring in a toy gun has a spring constant of 9.8 N/m and can be compressed 20 cm beyond the equilibrium position. A small pellet having a mass of 1 g resting against the spring is propelled forward when the spring is released. (a) Find the muzzle velocity of the pellet. (b) If the pellet is fired horizontally from a height of 1 m above the floor, what is its range?

68. When an object attached to the end of a metal rod or wire is suspended and then released, the object is found to vibrate with simple harmonic motion. Show that the effective spring constant for this wire is given by $k = YA/L$, where Y is Young's modulus, A is the cross-sectional area of the wire, and L is its unstretched length.

69. A mass of 5.5 g is suspended from a cylindrical sample of collagen 3.5 cm long and 2.0 mm in diameter. If the mass vibrates up and down with a frequency of 36.0 Hz, what is the Young's modulus of the collagen?

70. Figure 14.36 shows a crude model of an insect wing. The mass m represents the entire mass of the wing, which pivots about the fulcrum F. The spring represents the surrounding connective tissue. Motion of the wing corresponds to vibration of the spring. Suppose the mass of the wing is 0.3 g and the effective spring constant of the tissue is 4.7×10^{-4} N/m. If the mass m moves up and down a distance of 2.0 mm from its position of equilibrium, what is the maximum speed of the outer tip of the wing?

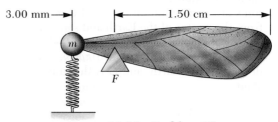

3.00 mm — 1.50 cm

m

F

Figure 14.36 Problem 70

71. A 5-g bullet moving with an initial speed of 400 m/s is fired into and passes through a 1-kg block, as in Figure 14.37. The block, initially at rest on a frictionless, horizontal surface, is connected to a spring of spring constant 900 N/m. If the block moves a distance of

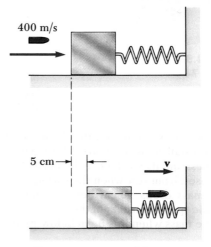

400 m/s

5 cm

v

Figure 14.37 Problem 71

types of whistles produce ultrasonic waves. Even though the human ear is insensitive to the waves emitted by these whistles, certain animals, such as dogs, are able to hear frequencies in this region.

15.3   The Speed of Sound

The speed of a sound wave depends on the compressibility and inertia of the medium through which the wave is traveling. If the medium has a bulk modulus B and an equilibrium density ρ, the **speed of sound** in that medium is

$$v = \sqrt{\frac{B}{\rho}} \tag{15.1}$$

Recall from Chapter 9 that bulk modulus is defined as the ratio of the change in pressure, ΔP, to the resulting fractional change in volume, $\Delta V/V$:

$$B = -\frac{\Delta P}{\Delta V/V} \tag{15.2}$$

Note that B is always positive because an increase in pressure (positive ΔP) results in a decrease in volume. Hence the ratio $\Delta P/\Delta V$ is always negative.

It is interesting to compare Equation 15.1 with the expression for the speed of transverse waves on a string, $v = \sqrt{F/\mu}$, discussed in Chapter 14. In both cases, the wave speed depends on an elastic property of the medium (B or F) and on an inertial property of the medium (ρ or μ). In fact, the speed of all mechanical waves follows an expression of the general form

$$v = \sqrt{\frac{\text{elastic property}}{\text{inertial property}}}$$

Another example of this general form is the **speed of a longitudinal wave in a solid**, which is given by the expression

$$v = \sqrt{\frac{Y}{\rho}} \tag{15.3}$$

where Y is the Young's modulus of the solid, defined as the longitudinal stress divided by the longitudinal strain (Chapter 9), and ρ is its density.

Table 15.1 lists the speed of sound in various media. As you can see, the speed of sound is much higher in solids than in gases. This makes sense because the molecules in a solid are closer together than in a gas and hence respond more rapidly to a disturbance. In general, sound travels more slowly in liquids than in solids because liquids are more compressible and hence have a smaller bulk modulus.

The velocity of sound also depends on the temperature of the medium. For sound traveling through air, the relationship between the speed of sound and temperature is

Table 15.1
Speed of Sound in Various Media

Medium	v(m/s)
Gases	
Air (0°C)	331
Air (100°C)	366
Hydrogen (0°C)	1290
Oxygen (0°C)	317
Helium (0°C)	972
Liquids at 25°C	
Water	1490
Methyl alcohol	1140
Seawater	1530
Solids	
Aluminum	5100
Copper	3560
Iron	5130
Lead	1320
Vulcanized rubber	54

$$v = (331 \text{ m/s}) \sqrt{1 + \frac{T}{273}} \tag{15.4}$$

where 331 m/s is the speed of sound in air at $0°C$ and T is the temperature in degrees Celsius. Using the equation above, you should show that at $20°C$, the speed of sound in air is approximately 343 m/s.

Example 15.1   Sound Waves in a Solid Bar

If a solid bar is struck at one end with a hammer, a longitudinal pulse will propagate down the bar. Find the speed of sound in a bar of aluminum, which has a Young's modulus of 7.0×10^{10} Pa and a density of 2.7×10^3 kg/m³.

Solution
From Equation 15.3 we find that

$$v_{Al} = \sqrt{\frac{Y}{\rho}} = \sqrt{\frac{7.0 \times 10^{10} \text{ Pa}}{2.7 \times 10^3 \text{ kg/m}^3}} \approx 5100 \text{ m/s}$$

This is a typical value for the speed of sound in solids (see Table 15.1).

Example 15.2   Speed of Sound in a Liquid

Find the speed of sound in water, which has a bulk modulus of about 2.1×10^9 Pa and a density of about 10^3 kg/m³.

Solution
From Equation 15.1, we find

$$v_{water} = \sqrt{\frac{B}{\rho}} = \sqrt{\frac{2.1 \times 10^9 \text{ Pa}}{1 \times 10^3 \text{ kg/m}^3}} \approx 1500 \text{ m/s}$$

15.4   Energy and Intensity of Sound Waves

As the tines of a tuning fork move back and forth through the air, they exert a force on a layer of air and cause it to move. In other words, tines do work on the layer of air. The fact that the fork pours energy into the air as sound energy is one of the reasons that the vibration of the fork slowly dies out. (Other factors, such as the energy lost to friction as the tines bend, also are responsible for the diminution of movement.)

We define the **intensity I of a wave** to be the rate at which energy flows through a unit area A perpendicular to the direction of travel of the wave.

In equation form this is

$$I \equiv \frac{1}{A}\frac{\Delta E}{\Delta t} \tag{15.5}$$

Example 15.4 The Noisy Typewriter

A rather noisy typewriter produces a sound intensity of 10^{-5} W/m². Find the decibel level of this machine and calculate the new decibel level when a second identical typewriter is added to the office.

Solution

The decibel level of the single typewriter is

$$\beta = 10 \log\left(\frac{10^{-5}\ \text{W/m}^2}{10^{-12}\ \text{W/m}^2}\right) = 10 \log(10^7) = \boxed{70\ \text{dB}}$$

Adding the second typewriter doubles the energy input into sound and hence doubles the intensity. The new decibel level is

$$\beta = 10 \log\left(\frac{2 \times 10^{-5}\ \text{W/m}^2}{10^{-12}\ \text{W/m}^2}\right) = \boxed{73\ \text{dB}}$$

Federal regulations now demand that no office or factory worker can be exposed to noise levels that average more than 90 dB over an 8-h day. The results in this example read like one of the old jokes that start "There is some good news and some bad news." First the good news. Imagine that you are a manager analyzing the noise conditions in your office. One typewriter in the office produces a noise level of 70 dB. When you add a second typewriter, the noise level increases by only 3 dB. Because of the logarithmic nature of decibel levels, doubling the intensity does not double the decibel level; in fact, it alters it by a surprisingly small amount. This means that additional equipment can be added to an office or factory without appreciably altering the decibel level of an environment.

Now the bad news. The results also work in reverse. As you remove noise machinery, the decibel level is not lowered appreciably. For example, consider an office with 60 typewriters producing a noise level of 93 dB, which is 3 dB above the maximum allowed. In order to reduce the noise level by 3 db, half the machines would have to be removed! That is, you would have to remove 30 typewriters to reduce the noise level to 90 dB. To reduce the level another 3 db, you would have to remove half of the remaining machines, and so on.

15.5 Spherical and Plane Waves

If a small spherical object oscillates such that its radius changes periodically with time, a spherical sound wave will be produced (Fig. 15.3). The wave moves outward from the source at a constant speed.

Because all points on the vibrating sphere behave in the same way, we conclude that the energy in a spherical wave propagates equally in all directions. That is, no one direction is preferred over any other. If P_{av} is the average power emitted by the source, then at any distance r from the source, this power must be distributed over a spherical surface of area $4\pi r^2$ assuming no absorption in the medium. (Recall that $4\pi r^2$ is the surface area of a sphere.) Hence, the **intensity** of the sound at a distance r from the source is

$$I = \frac{\text{average power}}{\text{area}} = \frac{P_{av}}{A} = \frac{P_{av}}{4\pi r^2} \tag{15.8}$$

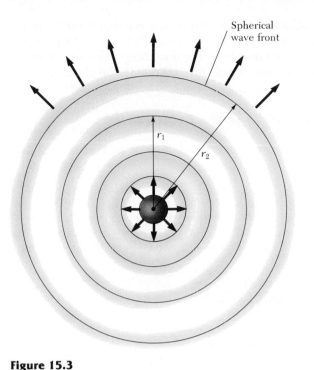

Figure 15.3
A spherical wave propagating radially outward from an oscillating sphere. The intensity of the spherical wave varies as $1/r^2$.

Figure 15.4
Spherical waves emitted by a point source. The circular arcs represent the spherical wavefronts concentric with the source. The rays are radial lines pointing outward from the source, perpendicular to the wavefronts.

This shows that the intensity of a wave decreases with increasing distance from its source, as you might expect. The fact that I varies as $1/r^2$ is a result of the assumption that the small source (sometimes called a **point source**) emits a spherical wave. Since the average power is the same through any spherical surface centered at the source, we see that the intensities at distance r_1 and r_2 (Fig. 15.3) from the center of the source are

$$I_1 = \frac{P_{av}}{4\pi r_1{}^2} \qquad I_2 = \frac{P_{av}}{4\pi r_2{}^2}$$

Therefore, the ratio of intensities at these two spherical surfaces is

$$\frac{I_1}{I_2} = \frac{r_2{}^2}{r_1{}^2}$$

It is useful to represent spherical waves graphically by a series of circular arcs concentric with the source representing part of a spherical surface, as in Figure 15.4. We call such a surface a **wavefront**. The distance between adjacent wavefronts equals the wavelength, λ. The radial lines pointing outward from the source and cutting the arcs perpendicularly are called **rays**.

Now consider a small portion of the wavefront that is at a *large* distance (large relative to λ) from the source, as in Figure 15.5. In this case, the rays are nearly parallel to each other and the wavefronts are very close to being planes. Therefore, at distances from the source that are large relative to the wavelength, we can approximate the wavefront by parallel planes. We call such

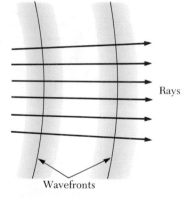

Figure 15.5
Far away from a point source, the wavefronts are nearly parallel planes and the rays are nearly parallel lines perpendicular to the planes. Hence, a small segment of a spherical wavefront is approximately a plane wave.

Figure 15.6
Representation of a plane wave moving in the positive x direction with a speed v. The wavefronts are planes parallel to the yz plane.

waves plane waves. Any small portion of a spherical wave that is far from the source can be considered a **plane wave.** Figure 15.6 illustrates a plane wave propagating along the x axis. If x is taken to be the direction of the wave motion (or rays) in Figure 15.6, then the wavefronts are parallel to the plane containing the y and z axes.

Example 15.5   Intensity Variations of a Point Source

A small source emits sound waves with a power output of 80 W. (a) Find the intensity 3 m from the source.

Solution
A small source emits energy in the form of spherical waves (see Fig. 15.3). Let P_{av} be the average power output of the source. At a distance r from the source, the power is distributed over the surface area of a sphere, $4\pi r^2$. Therefore, the intensity at a distance r from the source is given by Equation 15.8. Since $P_{av} = 80$ W and $r = 3$ m, we find that

$$I = \frac{P_{av}}{4\pi r^2} = \frac{80 \text{ W}}{4\pi (3 \text{ m})^2} = \boxed{0.71 \text{ W/m}^2}$$

which is close to the threshold of pain.

(b) Find the distance at which the sound level is 40 dB.

Solution
We can find the intensity at the 40-dB intensity level by using Equation 15.7 with $I_0 = 10^{-12}$ W/m^2:

$$40 = 10 \log(I/I_0)$$
$$4 = \log(I/I_0)$$
$$I = 10^4 I_0 = 10^{-8} \text{ W/m}^2$$

When this value for I is used in Equation 15.8, solving for r gives

$$r = \left(\frac{P_{av}}{4\pi I}\right)^{1/2} = \left(\frac{80 \text{ W}}{4\pi \times 10^{-8} \text{ W/m}^2}\right)^{1/2} = \boxed{2.5 \times 10^4 \text{ m}}$$

which is approximately 15 mi!

15.6   The Doppler Effect

When a car or truck is moving while its horn is blowing, the frequency of the sound you hear is higher as the vehicle approaches you and lower as it moves away from you. This is one example of the **Doppler effect,** named for the Austrian physicist Christian Doppler (1803–1853), who discovered the effect.

In general, a **Doppler effect** is experienced whenever there is relative motion between source and observer. When the source and observer are moving toward each other, the frequency heard by the observer is higher than the frequency of the source. When the source and observer are moving away from each other, the observer hears a frequency lower than the source frequency.

Although the Doppler effect is most commonly experienced with sound waves, it is a phenomenon common to all waves. For example, the frequencies of light waves are also shifted by relative motion.

First let us consider the case in which the observer is moving and the sound source is stationary. For simplicity, we shall assume that the air is also stationary and all velocity measurements are made relative to this stationary medium. Figure 15.7 describes the situation when the observer is moving with a speed v_o toward the source (considered a point source), which is at rest ($v_s = 0$).

We shall take the frequency of the source to be f, the wavelength to be λ, and the speed of sound in air to be v. Clearly, if both observer and source were stationary, the observer would detect f wavefronts per second. (That is, when $v_o = 0$ and $v_s = 0$, the observed frequency equals the source frequency.) When the observer is moving toward the source, he or she moves a distance $v_o t$ in a time of t seconds. In this time, *the observer detects an additional number of wavefronts*. The number of extra wavefronts detected will be equal to the distance traveled, $v_o t$, divided by the wavelength λ. Thus, we have

$$\text{Additional wavefronts detected} = \frac{v_o t}{\lambda}$$

The number of additional wavefronts detected *per second* is v_o/λ. Hence, the frequency f' heard by the observer is *increased* and given by

$$f' = f + \frac{v_o}{\lambda}$$

Using the fact that $\lambda = v/f$, we see that $v_o/\lambda = (v_o/v)f$. Hence f' can be expressed as

$$f' = f\left(\frac{v + v_o}{v}\right) \tag{15.9}$$

An observer traveling *away* from the source, as in Figure 15.8, *detects fewer wavefronts per second*. Thus, from Equation 15.9, it follows that the frequency heard by the observer in this case is *lowered* and given by

$$f' = f\left(\frac{v - v_o}{v}\right) \tag{15.10}$$

We can incorporate these two equations into one:

$$f' = f\left(\frac{v \pm v_o}{v}\right) \tag{15.11}$$

This general equation applies when an observer is moving with a speed v_o relative to a stationary source. *The positive sign is used when the observer is moving toward the source and the negative sign is used when the observer is moving away from the source.*

Now consider the situation in which the source is in motion and the observer is at rest. If the source is moving directly toward observer A in Figure 15.9a, the wavefronts heard by A are closer together because the source is moving in the direction of the outgoing wave. As a result, the wavelength λ'

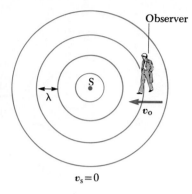

Figure 15.7
An observer moving with a speed v_0 *toward* a stationary point source S hears a frequency f' that is *greater* than the source frequency f.

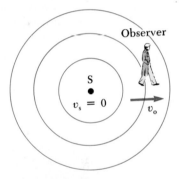

Figure 15.8
An observer moving with a speed v_o *away* from a stationary source hears a frequency f' that is *lower* than the source frequency.

Observed frequency—
Observer in motion

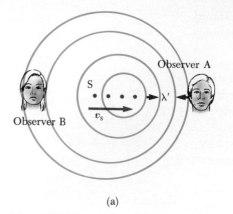

(a)

Figure 15.9
(a) A source S moving with a speed v_s toward a stationary observer A and away from a stationary observer B. Observer A hears an *increased* frequency, and observer B hears a *decreased* frequency. (b) The Doppler effect in water observed in a ripple tank. The source producing the water waves is moving to the right. *(Courtesy Educational Development Center, Newton, Mass.)*

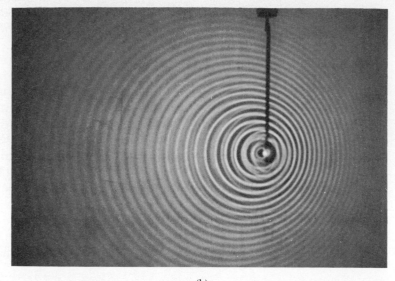

(b)

measured by observer A is shorter than the true wavelength, λ, of the source. During each vibration, which lasts for a time T (the period), the source moves a distance $v_s T = v_s/f$ and *the wavelength is shortened by this amount.* Therefore, the observed wavelength, λ', is given by

$$\lambda' = \lambda - \frac{v_s}{f}$$

Since $\lambda = v/f$, the frequency heard by observer A is

$$f' = \frac{v}{\lambda'} = \frac{v}{\lambda - \dfrac{v_s}{f}} = \frac{v}{\dfrac{v}{f} - \dfrac{v_s}{f}} = f\left(\frac{v}{v - v_s}\right) \tag{15.12}$$

That is, *the observed frequency is increased when the source is moving toward the observer.*

In Figure 15.9a the source is moving away from observer B, who is at rest to the left of the source. Thus observer B measures a wavelength that is *greater* than λ and hears a *decreased* frequency given by

$$f' = f\left(\frac{v}{v + v_s}\right) \tag{15.13}$$

Combining Equations 15.12 and 15.13, we can express the general relationship for the observed frequency when the source is moving and the observer is at rest:

Observed frequency — source in motion

$$f' = f\left(\frac{v}{v \mp v_s}\right) \tag{15.14}$$

The *negative* sign is used when the source is moving *away* from the observer, and the *positive* sign is used when the source is moving *toward* the observer.

Finally, if both the source and the observer are in motion, one finds the following general relationship for the observed frequency:

$$f' = f\left(\frac{v \pm v_o}{v \mp v_s}\right)$$ **(15.15)**   Observed frequency —
                                                                            Observer and source in motion

where the sign conventions for v_o and v_s are specified immediately following Equations 15.11 and 15.14.

Example 15.6   Listen, but Don't Stand on the Track

A train moving at a speed of 40 m/s sounds its whistle, which has a frequency of 500 Hz. Determine the frequency heard by a stationary observer as the train approaches the observer.

Solution
We can use Equation 15.12 to get the apparent frequency as the train approaches the observer. Taking $v = 345$ m/s as the speed of sound in air, we have

$$f' = f\left(\frac{v}{v - v_s}\right) = (500 \text{ Hz})\left(\frac{345 \text{ m/s}}{345 \text{ m/s} - 40 \text{ m/s}}\right) = 566 \text{ Hz}$$

Exercise   Determine the frequency heard by the stationary observer as the train recedes from the observer.

Answer   448 Hz.

Example 15.7   The Noisy Siren

An ambulance travels down a highway at a speed of 75 mi/h, its siren emitting sound at a frequency of 400 Hz. What is the frequency heard by a passenger in a car traveling at 55 mi/h in the opposite direction as the car (a) approaches and (b) moves away from the ambulance?

Solution
Let us take the velocity of sound in air to be $v = 345$ m/s and use the conversion 1 mi/h = 0.447 m/s. Therefore, $v_s = 75$ mi/h = 33.5 m/s and $v_o = 55$ mi/h = 24.6 m/s. We can use Equation 15.15 in both cases.

(a) As the ambulance and car approach each other, the observed frequency is

$$f' = f\left(\frac{v + v_o}{v - v_s}\right) = (400 \text{ Hz})\left(\frac{345 \text{ m/s} + 24.6 \text{ m/s}}{345 \text{ m/s} - 33.5 \text{ m/s}}\right) = \boxed{475 \text{ Hz}}$$

(b) As the two vehicles recede from each other, the passenger in the car hears a frequency

$$f' = f\left(\frac{v - v_o}{v + v_s}\right) = (400 \text{ Hz})\left(\frac{345 \text{ m/s} - 24.6 \text{ m/s}}{345 \text{ m/s} + 33.5 \text{ m/s}}\right) = \boxed{339 \text{ Hz}}$$

(a)   (b)

Figure 15.10
(a) Representation of a shock wave produced when a source moves from S_0 to S_n with a speed v_s that is *greater* than the wave speed, v, in that medium. The envelope of the wavefronts forms a cone whose half-angle is given by $\sin \theta = v/v_s$. (b) Stroboscopic photograph of a bullet moving at supersonic speed through the hot air above a candle. *(Harold Edgerton, Courtesy of Palm Press, Inc.)*

The V-shaped bow wave of a boat is formed because the boat travels at a speed greater than the speed of water waves. This is analogous to a shock wave formed by an airplane traveling at a speed greater than the speed of sound. *(© Mike Sheldon/FPG International)*

Shock Waves

Now let us consider what happens when the source velocity, v_s, *exceeds* the wave velocity, v. This situation is described graphically in Figure 15.10a. The circles represent spherical wavefronts emitted by the source at various times during its motion. At $t = 0$, the source is at point S_0, and at some later time t, the source is at point S_n. In the time t, the wavefront centered at S_0 reaches a radius of vt. In this same interval, the source travels a distance $v_s t$ to S_n. At the instant the source is at S_n, the waves just beginning to be generated at this point have wavefronts of zero radius. The line drawn from S_n to the wavefront centered on S_0 is tangent to all other wavefronts generated at intermediate times. All such tangent lines lie on the surface of a cone. The angle θ between one of these tangent lines and the direction of travel is given by

$$\sin \theta = \frac{v}{v_s}$$

The ratio v_s/v is referred to as the **Mach number**. The conical wavefront produced when $v_s > v$ (supersonic speeds) is known as a **shock wave**. Figure 15.10b is a photograph of a bullet traveling at supersonic speed through the hot air rising above a candle. Note the shock waves in the vicinity of the bullet. Another interesting example of shock waves is the V-shaped wavefronts produced by a boat (the bow wave) when the boat's speed exceeds the speed of the water waves (see photograph at the left).

Jet airplanes traveling at supersonic speeds produce shock waves, which are responsible for the loud explosion, or sonic boom, one hears. The shock wave carries a great deal of energy concentrated on the surface of the cone, with correspondingly large pressure variations. Such shock waves are unpleasant to hear and can damage buildings when aircraft fly supersonically at low altitudes. In fact, an airplane flying at supersonic speeds produces a double boom because two shock waves are formed, one from the nose of the plane and one from the tail (Fig. 15.11).

15.7 Interference in Sound Waves

The fact that sound waves can be made to interfere with each other can be demonstrated with the device shown in Figure 15.12. Sound from a loudspeaker at S is sent into a tube at P, where there is a T-shaped junction. Half the sound intensity travels in one direction and half in the opposite direction. Thus, the sound waves that reach the receiver at R travel along two different paths. If the two paths are of the same length, a crest of the wave that enters the junction will separate into two halves, travel the two paths, and then combine again at the receiver. The upper and lower waves reunite at the receiver such that constructive interference takes place, and thus a loud sound is heard at the detector.

However, suppose one of the path lengths is adjusted by sliding the upper U-shaped tube upward so that the upper path is half a wavelength *longer* than the lower path. In this case, an entering sound wave splits and travels the two paths as before but now the wave along the upper path must travel a distance equivalent to half a wavelength farther than the wave traveling along the lower path. As a result, the crest of one wave meets the trough of the other when they merge at the receiver. Since this is the condition for destructive interference, no sound is detected at the receiver.

You should be able to predict what will be heard if the upper path is adjusted to one full wavelength longer than the lower path. In this case, constructive interference of the two waves occurs and a loud sound is detected at the receiver.

There are many other examples of interference phenomena in nature. In a later chapter, we shall describe several interesting interference effects involving light waves.

Figure 15.11
The two shock waves produced by the nose and tail of a jet airplane traveling at supersonic speed.

Figure 15.12
An acoustical system for demonstrating interference of sound waves. Sound from the speaker enters the tube and splits into two parts at P. The two waves combine at the opposite side and are detected at R. The upper path length is varied by the sliding section.

Figure 15.13   (Example 15.8) Two loudspeakers driven by the same source can produce interference.

Example 15.8   Interference from Two Loudspeakers

Two loudspeakers are placed as in Figure 15.13 and driven by the same source at a frequency of 2000 Hz. The top speaker is then moved to the left to position A. At this location, an observer at a large distance from the speakers and directly in front of them notices that the intensity of the sound from the two sources has decreased to a minimum. What minimum distance has the top speaker been moved? Assume that the speed of sound in air is 345 m/s.

Solution

Initially, both speakers are at the same distance from the observer. Hence the sound from each speaker must travel the same distance and the observer hears a loud sound corresponding to constructive interference. When the top speaker is at position A, its sound must travel farther to reach the observer. Since the observer notices a minimum in the sound level when the top speaker is at A, destructive interference is taking place between the two separate sound signals. This means that the top speaker has been moved half a wavelength. With the speed of sound in air equal to 345 m/s, we can calculate the wavelength:

$$\lambda = \frac{v}{f} = \frac{345 \text{ m/s}}{2000 \text{ s}^{-1}} = 0.173 \text{ m}$$

Therefore, the distance moved is half this value, or 0.0865 m.

15.8   Standing Waves

If a stretched string is clamped at both ends, traveling waves will reflect from the fixed ends, creating waves traveling in both directions. The incident and reflected waves will combine according to the superposition principle. For example, if the string is vibrated at exactly the right frequency, a crest moving toward one end and a reflected trough will meet at some point along the string. The two waves cancel at this point, which is called a **node.** The resulting pattern on the string is one in which the wave appears to stand still, and we have what is called a **standing wave** on the string. There is no motion in the string at the nodes, but at a distance equal to $\lambda/4$ away from each node, the string vibrates with a large amplitude. These points are called **antinodes.** The distance between successive nodes, or between successive antinodes, is equal to $\lambda/2$.

The oscillation of a standing wave during one half of a cycle is shown in Figure 15.14. *Notice that all points on the string oscillate vertically with the same frequency except for the node, which is stationary.* (The points at which the string is attached to the wall are also nodes, which are labeled N in Figure 15.14a.) Furthermore, the various points have different amplitudes of motion.

Consider a string of length L that is fixed at both ends, as in Figure 15.15. The string has a number of natural patterns of vibration, called normal modes. Three of these are pictured in Figures 15.15b, 15.15c, and 15.15d. Each of these has a characteristic frequency, which we shall now calculate.

First, note that *the ends of the string must be nodes because these points are fixed.* If string is displaced at its midpoint and released, the vibration shown in Figure 15.15b can be produced, in which case the center of the string is an antinode. For this normal mode, the length of the string equals $\lambda/2$ (the distance between nodes). Thus,

Figure 15.14
A standing wave pattern in a stretched string showing snapshots of the string during one half of a cycle.

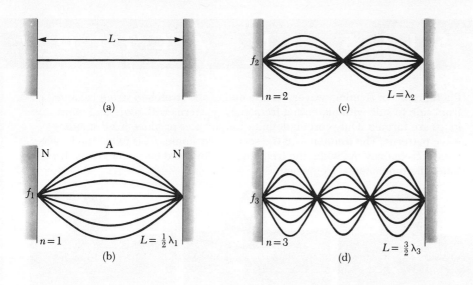

Figure 15.15
(a) Standing waves in a stretched string of length L fixed at both ends. The normal frequencies of vibration form a harmonic series: (b) the fundamental frequency, or first harmonic, (c) the second harmonic, and (d) the third harmonic.

$$L = \frac{\lambda_1}{2} \quad \text{or} \quad \lambda_1 = 2L$$

and the frequency of this vibration is

$$f_1 = \frac{v}{\lambda_1} = \frac{v}{2L} \tag{15.16}$$

In Chapter 14, the speed of a wave on a string was given as $v = \sqrt{F/\mu}$, where F is the tension in the string and μ is its mass per unit length. Thus, we can express Equation 15.16 as

$$f_1 = \frac{1}{2L} \sqrt{\frac{F}{\mu}} \tag{15.17}$$

Fundamental frequency

This lowest frequency of vibration is called the **fundamental frequency.**

The next normal mode, of wavelength λ_2 (Fig. 15.15c), occurs when the length of the string equals one wavelength, that is when $\lambda_2 = L$. Hence

$$f_2 = \frac{v}{L} = \frac{2v}{2L} = 2f_1 \tag{15.18}$$

Note that this frequency is equal to *twice* the fundamental frequency. You should convince yourself that the next highest frequency of vibration, shown in Figure 15.15d, is given by

$$f_3 = \frac{3v}{2L} = 3f_1 \tag{15.19}$$

In general, the characteristic frequencies are given by

$$f_n = nf_1 = \frac{n}{2L} \sqrt{\frac{F}{\mu}} \tag{15.20}$$

PHYSICS IN ACTION

Standing Waves

Photographs of standing waves. As one end of the stretched string is moved from side to side with increasing frequency, patterns with more and more loops are formed. Only certain definite frequencies produce fixed standing wave patterns. The frequency of the second harmonic, f_2, is twice the fundamental frequency, f_1, while the frequency of the third harmonic, f_3, is three times the fundamental frequency.

(Courtesy Educational Development Center, Newton, MA)

where $n = 1, 2, 3, \ldots$. In other words, the frequencies are integral multiples of the fundamental frequency. The set of frequencies f_1, $2f_1$, $3f_1$, and so on form a **harmonic series.** The fundamental f_1 corresponds to the **first harmonic;** the frequency $f_2 = 2f_1$ corresponds to the **second harmonic,** and so on.

When a stretched string is distorted to a shape that corresponds to any one of its harmonics, after being released it will only vibrate at the frequency of that harmonic. If the string is struck or bowed, however, the resulting vibration will include frequencies of various harmonics, including the fundamental. Waves not in the harmonic series are quickly damped out on a string fixed at both ends. In effect, the string "selects" the normal-mode frequencies when disturbed. As we shall see later, the presence of several harmonics on a string gives the stringed instruments their characteristic sound, which enables us to distinguish one from another even when they are sounding identical fundamental frequencies.

The frequency of a string on a musical instrument can be changed either by varying the tension or by changing the length. For example, the tension in guitar and violin strings is varied by turning pegs on the neck of the instrument. As the tension is increased, the frequency of the normal modes increases according to Equation 15.20. Once the instrument is tuned, the musician varies the frequency by moving his or her fingers along the neck, thereby changing the length of the vibrating portion of the string. As the length is shortened, the frequency increases, as Equation 15.20 indicates.

Finally, Equation 15.20 shows that if a string of fixed length can be made to vibrate at a lower fundamental frequency by increasing its mass per unit length. This is achieved in the bass strings of guitars and pianos by wrapping the strings with a metal winding.

Concert style harp. *(Photo Lyon & Healy Harps, Chicago)*

Example 15.9 Harmonics of a Stretched String

Find the first four harmonics of a string 1 m long if the string has a mass per unit length of 2×10^{-3} kg/m and is under a tension of 80 N.

Solution
The speed of wave on the string is

$$v = \sqrt{\frac{F}{\mu}} = \sqrt{\frac{80 \text{ N}}{2 \times 10^{-3} \text{ kg/m}}} = 200 \text{ m/s}$$

The fundamental frequency can be found by using Equation 15.16:

$$f_1 = \frac{v}{2L} = \frac{200 \text{ m/s}}{2(1 \text{ m})} = \boxed{100 \text{ Hz}}$$

The frequencies of the next three modes are $f_2 = 2f_1$, $f_3 = 3f_1$, and $f_4 = 4f_1$. Thus, $f_2 = 200$ Hz, $f_3 = 300$ Hz, and $f_4 = 400$ Hz.

Exercise Find the tension in the string if the fundamental frequency is increased to 120 Hz.

Answer 115 N.

15.9 Forced Vibrations and Resonance

In Chapter 14 we learned that the energy of a damped oscillator decreases in time because of friction. It is possible to compensate for this energy loss by applying an external force that does positive work on the system.

For example, suppose a mass-spring system having some natural frequency of vibration f_0 is pushed back and forth with a periodic force whose frequency is f. The system will vibrate at the frequency f of the driving force. This type of motion is referred to as a **forced vibration**. The amplitude of the motion reaches a maximum when the frequency of the driving force equals the natural frequency of the system, f_0, called the **resonant frequency** of the system. Under this condition, the system is said to be in **resonance**.

In Section 15.8 we learned that a stretched string is able to vibrate in one or more of its natural modes. Here again, if a periodic force is applied to the

Top view of a concert grand piano. Why do you suppose the sounding board has this unusual shape?

Figure 15.16
Resonance: If pendulum A is set into oscillation, only pendulum C, whose length is the same as that of A, will eventually oscillate with large amplitude, or resonate. The arrows indicate motion perpendicular to the page.

Vibrating
blade

Figure 15.17
Standing waves can be set up in a stretched string by connecting one end of the string to a vibrating blade. When the blade vibrates at one of the natural frequencies of the string, large amplitude standing waves are created.

string, the amplitude of vibration will increase as the frequency of the applied force approaches one of the natural frequencies of vibration.

Resonance vibrations occur in a wide variety of circumstances. One experiment that demonstrates a resonance condition is illustrated in Figure 15.16. Several pendulums of different lengths are suspended from a flexible beam. If one of them, such as A, is set into motion, the others will begin to oscillate because they are coupled by vibrations in the flexible beam. Pendulum C, whose length is the same as that of A, will oscillate with the greatest amplitude since its natural frequency matches that of pendulum A (the driving force).

Another apparatus that can easily be set up in the laboratory is a stretched string fixed at one end and connected at the opposite end to a vibrating blade, as in Figure 15.17. As the blade oscillates, transverse waves sent down the string are reflected at the fixed end. As we found in Section 15.8, the string has natural frequencies that are determined by its mass per unit length and the tension (Eq. 15.20). When the frequency of the vibrating blade equals one of the natural frequencies of the string, standing waves will be produced and the string will vibrate with a large amplitude.

Another simple example of resonance is a child being pushed on a swing, which is essentially a pendulum with a natural frequency that depends on the length. The swing is kept in motion by a series of appropriately timed pushes. In order to increase the height of the swing, you must push it each time it returns to your hands. This corresponds to a frequency equal to the natural frequency of the swing. If the energy you put into the system per cycle of motion exactly equals the energy lost due to friction, the amplitude remains constant.

Opera singers have been known to set crystal goblets into audible vibration with their powerful voices, as shown in the photograph below. This is yet another example of resonance: the sound waves emitted by the singer are able to set up large-amplitude vibrations in the glass. If a highly amplified sound wave has the right frequency, the amplitude of forced vibrations in the glass will increase to the point where the glass becomes heavily strained and shatters. In this case, resonance occurs when the wavelength of the emitted sound wave equals the circumference of the glass.

A wine glass shattered by the amplified sound of a human voice. Resonant vibrations are set up in the wine glass when the frequency of the singer's sound wave matches one of the natural frequencies of vibration of the wine glass. *(Ben Rose/The IMAGE Bank)*

A striking example of structural resonance occurred in 1940, when the Tacoma Narrows bridge in the state of Washington was set into oscillation by the wind (Fig. 15.18). The amplitude of these oscillations increased steadily until the bridge was ultimately destroyed.

15.10   Standing Waves in Air Columns

Standing longitudinal waves can be set up in a tube of air, such as an organ pipe, as the result of interference between sound waves traveling in opposite directions. The relationship between the incident wave and the reflected wave depends on whether the reflecting end of the tube is open or closed. A portion of the sound wave is reflected back into the tube at the open end. *If the reflecting end is closed, a node must exist at this end because the movement of air molecules is restricted. If the end is open, the air molecules have complete freedom of motion and an antinode exists.*

The first three modes of vibration of a pipe open at both ends are shown in Figure 15.19a. When air is directed against an edge at the left, longitudinal standing waves are formed and the pipe vibrates at its natural frequencies. Note that for the fundamental frequency the wavelength is twice the length of the pipe and hence $f_1 = v/2L$. Similarly, one finds that the frequencies of the second and third harmonics are $2f_1$, $3f_1$, ... Thus,

> in a pipe open at both ends, the natural frequencies of vibration form a series in which all **harmonics are present** and are equal to integral multiples of the fundamental.

(a)

(b)

Figure 15.18
The collapse of the Tacoma Narrows suspension bridge in 1940 is a vivid demonstration of mechanical resonance. High winds set up standing waves in the bridge, causing it to oscillate at one of its natural frequencies. Once established, this resonance condition led to the bridge's collapse. *(United Press International Photo)*

$\lambda_1 = 2L$
$f_1 = \dfrac{v}{\lambda_1} = \dfrac{v}{2L}$ First harmonic

$\lambda_2 = L$
$f_2 = \dfrac{v}{L} = 2f_1$ Second harmonic

$\lambda_3 = \dfrac{2}{3}L$
$f_3 = \dfrac{3v}{2L} = 3f_1$ Third harmonic

(a)

$\lambda_1 = 4L$
$f_1 = \dfrac{v}{\lambda_1} = \dfrac{v}{4L}$ First harmonic

$\lambda_3 = \dfrac{4}{3}L$
$f_3 = \dfrac{3v}{4L} = 3f_1$ Third harmonic

$\lambda_5 = \dfrac{4}{5}L$
$f_5 = \dfrac{5v}{4L} = 5f_1$ Fifth harmonic

(b)

Figure 15.19
(a) Standing longitudinal waves in an organ pipe open at both ends. The natural frequencies f_1, $2f_1$, $3f_1$ . . . form a harmonic series. (b) Standing longitudinal waves in an organ pipe closed at one end. Only *odd* harmonics are present, and the natural frequencies are f_1, $3f_1$, $5f_1$, etc.

We can express this harmonic series as

Pipe open at both ends,
all harmonics present

$$f_n = n \frac{v}{2L} \qquad n = 1, 2, 3, \ldots \qquad \text{(15.21)}$$

where v is the speed of sound in air.

If a pipe is closed at one end and open at the other, the closed end is a node (Fig. 15.19b). In this case, the wavelength for the fundamental mode is four times the length of the tube. Hence, $f_1 = v/4L$ and the frequencies of the third and fifth harmonics are $3f_1$, $5f_1$, ... That is,

in a pipe closed at one end and open at the other, **only odd harmonics are present,**

and these are given by

Pipe closed at one end,
odd harmonics present

$$f_n = n \frac{v}{4L} \qquad n = 1, 3, 5, \ldots \qquad \text{(15.22)}$$

Example 15.10   Harmonics of a Pipe

A pipe has a length of 2.46 m. (a) Determine the frequencies of the first three harmonics if the pipe is open at each end. Take 345 m/s as the speed of sound in air.

Solution
The fundamental frequency of a pipe open at both ends can be found from Equation 15.21, with $n = 1$.

$$f_1 = \frac{v}{2L} = \frac{345 \text{ m/s}}{2(2.46 \text{ m})} = \boxed{70 \text{ Hz}}$$

Since all harmonics are present in a pipe open at both ends, the second and third harmonics have frequencies given by $f_2 = 2f_1 = 140$ Hz and $f_3 = 3f_1 = 210$ Hz.

(b) What are the lowest possible frequencies if the pipe is closed at one end and open at the other?

Solution
The fundamental frequency of a pipe closed at one end can be found from Equation 15.22, with $n = 1$.

$$f_1 = \frac{v}{4L} = \frac{345 \text{ m/s}}{4(2.46 \text{ m})} = \boxed{35 \text{ Hz}}$$

In this case, only odd harmonics are present, and so the third and fifth harmonics have frequencies given by $f_3 = 3f_1 = 105$ Hz and $f_5 = 5f_1 = 175$ Hz.

Exercise   If the pipe is open at one end, how many harmonics are possible in the normal hearing range, 20 Hz to 20,000 Hz?

Answer   286.

Example 15.11 Resonance in a Tube of Variable Length

A simple apparatus for demonstrating resonance in a tube is described in Figure 15.20a. A long tube open at both ends is partially submerged in a beaker of water, and a vibrating tuning fork of unknown frequency is placed near the top. The length of the air column, L, is adjusted by moving the tube vertically. The sound waves generated by the fork are reinforced when the length of the air column corresponds to one of the resonant frequencies of the tube. The smallest value of L for which a peak occurs in the sound intensity is 9 cm. For this measurement, determine the frequency of the tuning fork and the value of L for the next two resonant vibrations.

Figure 15.20 (Example 15.11)
(a) Apparatus for demonstrating the resonance of sound waves in a tube closed at one end. The length, L, of the air column is varied by moving the tube vertically while it is partially submerged in water. (b) The first three normal frequencies of vibration for the system.

Solution
Once the tube is in the water, this setup represents a pipe closed at one end, and the fundamental has a frequency of $v/4L$ (Fig. 15.20b). If we take $v = 345$ m/s for the speed of sound in air and $L = 0.09$ m, we get

$$f_1 = \frac{v}{4L} = \frac{345 \text{ m/s}}{4(0.09 \text{ m})} = \boxed{958 \text{ Hz}}$$

The fundamental wavelength of the pipe is given by $\lambda = 4L = 0.36$ m. Since the frequency of the source is constant, we see that the next resonance positions (Fig. 15.20b) correspond to lengths of $3\lambda/4 = 0.27$ m and $5\lambda/4 = 0.45$ m.

This arrangement is often used to measure the speed of sound, in which case the frequency of the tuning fork and the lengths at which resonance occurs must be known.

*15.11 Beats

The interference phenomena we have been dealing with up until now involve the superposition of two or more waves with the same frequency traveling in opposite directions. Let us now consider another type of interference effect which results from the superposition of two waves with slightly different frequencies. In this situation, the waves at some fixed point are periodically in and out of phase corresponding to an alternation in time between constructive and destructive interference. In order to understand this phenomenon, con-

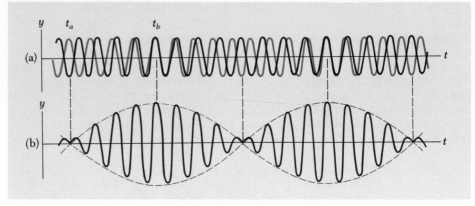

Figure 15.21
Beats are formed by the combination of two waves of slightly different frequencies traveling in the same direction. (a) The individual waves. (b) The combined wave has an amplitude (broken line) that oscillates in time.

sider Figure 15.21. The two waves shown in Figure 15.21a represent the waves emitted by two tuning forks having slightly different frequencies, while Figure 15.21b shows the superposition of these two waves. At some time indicated as t_a in Figure 15.21a, the two waves are out of phase, and destructive interference occurs as demonstrated by the resultant curve in Figure 15.21b. At some later time, however, the vibrations of the two forks move into step with one another. At the time indicated as t_b in Figure 15.21a, the two forks simultaneously emit compressions, and constructive interference results as demonstrated by the curve in Figure 15.21b. As time continues, the vibrations of the two forks will continue to move back out of phase, then into phase, and so on. Consequently, a listener at some fixed point will hear an alternation in loudness, known as **beats.** The number of beats per second, or beat frequency, equals the difference in frequency between the two sources. One can use beats to tune a stringed instrument, such as a piano, by beating a note on the instrument against a note of known frequency. The string can then be tuned to the desired frequency by adjusting the tension until no beats are heard.

Example 15.12   Sour Notes

A particular piano string is supposed to vibrate at a frequency of 440 Hz. In order to check its frequency, a tuning fork known to vibrate at a frequency of 440 Hz is sounded at the same time the piano key is struck, and a beat frequency of 4 beats per second is heard. Find the possible frequencies at which the string could be vibrating.

Solution
The number of beats per second is equal to the difference in frequency between the two sound sources. In this case, since one of the source frequencies is 440 Hz, 4 beats per second would be heard if the frequency of the string (the second source) were either 444 Hz or 436 Hz.

*15.12   Quality of Sound

The sound wave patterns produced by most musical instruments are very complex. Some characteristic waveforms produced by a tuning fork, a flute, and a clarinet, each playing the same note, are shown in Figure 15.22. Although each instrument has its own characteristic pattern, Figure 15.22 shows that each of the waveforms is periodic. Note that the tuning fork produces only one harmonic (the fundamental frequency) but the two instruments emit a mixture of harmonics. Figure 15.23 shows the harmonics of the waveforms of Figure 15.22. In Figure 15.23b, we see that, when this note is played on the flute, part of the sound consists of a vibration at the fundamental frequency, an even higher intensity is contributed by the second harmonic, the fourth harmonic produces about the same intensity as the fundamental, and so on. These sounds add together according to the principle of superposition to give the complex waveform for the flute shown in Figure 15.22b. Figure 15.23c shows that the clarinet emits a certain intensity at a frequency of the first harmonic, about half as much intensity at the frequency of the second harmonic, and so forth. The resultant superposition of these frequencies produces the pattern for the clarinet shown in Figure 15.22c. Figure 15.23a shows that the tuning fork emits sound intensity only at the frequency of the first harmonic.

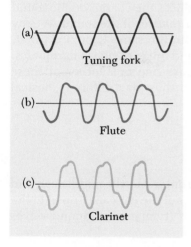

Figure 15.22
Waveforms produced by (a) a tuning fork, (b) a flute, and (c) a clarinet, each playing the same note. *(Adapted from C. A. Culver, Musical Acoustics, 4th ed., New York, McGraw-Hill, 1956)*

Figure 15.23
Harmonics of the waveforms shown in Figure 15.22. Note the variations in intensities of the various harmonics. *(Adapted from C. A. Culver, Musical Acoustics, 4th ed., New York, McGraw-Hill, 1956)*

Each musical instrument has its own characteristic sound and mixture of harmonics. Instruments shown are (a) the violin, (b) the saxophone, and (c) the trumpet. *(Photographs courtesy of (a) © 1989 Gary Buss/FPG International), (b) © 1989 Michael Simpson/FPG International), and (c) © 1989 Richard Laird/FPG International)*

In music, the mixture of harmonics that produces the characteristic sound of any instrument is referred to as the *quality,* or *timbre,* of the sound. We say that the note C on a flute differs in quality from the same C on a clarinet. Instruments such as the bugle, trumpet, violin, and tuba are rich in harmonics. A musician playing a wind instrument can cause one or another of these harmonics to be emphasized by lip control and thus play different musical notes with the same valve openings.

*15.13   The Ear

The human ear is divided into three regions; the outer ear, the middle ear, and the inner ear (Fig. 15.24a). The outer ear consists of the ear canal (open to the atmosphere), which terminates on the eardrum (tympanum). Sound waves

Figure 15.24
(a) Structure of the human ear. (b) The three tiny bones (ossicles) that connect the eardrum to the window of the cochlea act as a double lever system to decrease the amplitude of vibration and hence increase the pressure on the fluid in the cochlea.

travel down the ear canal to the eardrum, which vibrates in and out in phase with the pushes and pulls caused by the alternating high and low pressures of the sound wave. Behind the eardrum are three small bones of the middle ear, called the hammer, the anvil, and the stirrup because of their shapes (Fig. 15.24b). These bones transmit the vibration to the inner ear, which contains the cochlea, a snail-shaped tube about 2 cm long. The cochlea makes contact with the stirrup of the oval window and is divided along its length by the basilar membrane, which consists of small hairs and nerve fibers. This membrane varies in mass per unit length and tension along its length, and different portions of it resonate at different frequencies. (Recall that the natural frequency of a string depends on its mass per unit length and on the tension on it.) Along the basilar membrane are numerous nerve endings, which sense the vibration of the membrane and in turn transmit impulses to the brain. The brain interprets the impulses as sounds of varying frequency, depending on the location along the basilar membrane of the impulse-transmitting nerves and on the rate at which the impulses are transmitted.

Figure 15.25 is a frequency response curve for an average human ear for sounds of equal loudness, ranging from 0 dB to 120 dB. It shows that the decibel level at which pain is experienced does not vary greatly from 120 dB regardless of the frequency of the tone. However, the threshold of hearing shown by the lowest curve is very strongly dependent on frequency. The easiest frequencies to hear are around 3300 Hz, whereas frequencies about 12,000 Hz or below about 50 Hz must be relatively intense to be heard. A person with very good hearing (about 1% of the population) would have a threshold of hearing corresponding to the lowest curve. About 50% of the population has a threshold of hearing corresponding to the second lowest curve.

Figure 15.25
Curves of intensity level versus frequency for sounds that are perceived to be of equal loudness. Note that the ear is most sensitive at a frequency of about 3300 Hz. The lowest curve corresponds to the threshold of hearing for only about 1% of the population.

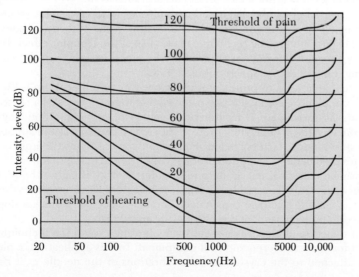

The exact mechanism by which sound waves are amplified and detected by the ear are rather complex and not fully understood. However, we can give a qualitative description of the amplification mechanisms. The small bones in the middle ear represent an intricate lever system that increases the force on the oval window over a given force on the eardrum. The pressure is greatly magnified because the surface area of the eardrum is about 20 times that of the oval window (in analogy with the hydraulic press). The middle ear, together with the eardrum and oval window, in effect acts as a matching network between the air in the outer ear and the liquid in the inner ear. The overall energy transfer between the outer ear and inner ear is highly efficient, with pressure amplification factors of several thousand. In other words, pressure variations in the inner ear are much greater than those in the outer ear.

The ear has its own built-in protection against loud sounds. The muscles connecting the three bones to the walls of the middle ear control the volume of the sound by changing the tension on the bones as sound builds up, thus hindering their ability to transmit vibrations. In addition, the eardrum becomes stiffer. These two occurrences cause the ear to be less sensitive to loud sounds. There is a time delay between the onset of loud sound and the ear's protective reaction, however, so that a very sudden loud sound can still damage the ear.

▶SPECIAL TOPIC
Ultrasound and Its Applications

Ultrasonic waves are sound waves whose frequencies are in the range of 20 kHz to 100 kHz, which is beyond the audible range. Because of their high frequency, and corresponding short wavelengths, ultrasonic waves can be used to produce images of small objects and are currently in wide use in medical applications, both as a diagnostic tool and in certain treatments. Various internal organs in the body can be examined through the images produced by the reflection and absorption of ultrasonic waves. Although ultrasonic waves are far safer than x-rays, their images do not always provide as much detail. On the other hand, certain organs, such as the liver and the spleen, are invisible to x-rays but can be diagnosed with ultrasonic waves.

It is possible to measure the speed of blood flow in the body using a device called an ultrasonic flow meter. The technique makes use of the Doppler effect. By comparing the frequency of the waves scattered by the blood vessels with the incident frequency, one can obtain the speed of blood flow.

The technique used to produce ultrasonic waves for clinical use is illustrated in Figure 15.26. Electrical contacts are made to the opposite faces of a crystal, such as quartz or strontium titanate. If an alternating voltage of very high frequency is applied to these contacts, the crystal will vibrate at the same frequency as the applied voltage. As the crystal vibrates, it emits a beam of ultrasonic waves. At one time, almost all of the headphones used in radio reception produced their sound in this manner. This method of transforming electrical energy into mechanical energy is called the **piezoelectric effect.** This effect is also reversible. That is, if some external source causes the crystal to vibrate, an alternating voltage is produced across the crystal. Hence, a single crystal can be used to both transmit and receive ultrasonic waves.

The production of electric voltages by a vibrating crystal is a technique that has been used for years in stereo and hi-fi equipment. In this application, a phonograph needle is attached to the crystal, and the vibrations of the needle as it rides in the

Direction of vibration

Electrical connections

Crystal

Figure 15.26
An alternating voltage applied to the faces of a piezoelectric crystal causes the crystal to vibrate.

groove of the record are translated by the crystal into an alternating voltage. This voltage is then amplified and used to drive the system's speakers.

The primary physical principle that makes ultrasound imaging possible is the fact that a sound wave is partially reflected whenever it is incident on a boundary between two materials having different densities. It is found that, if a sound wave is traveling in a material of density ρ_i and strikes a material of density ρ_t, the percentage of the incident sound wave reflected, PR, is given by

$$PR = \left(\frac{\rho_i - \rho_t}{\rho_i + \rho_t}\right)^2 \times 100 \qquad (15.23)$$

This equation assumes that the incident sound wave travels perpendicular to the boundary and that the speed of sound is approximately the same in both materials. This latter assumption holds very well for the human body since the speed of sound does not vary much in the various organs of the body.

Physicians commonly use ultrasonic waves to observe a fetus. This technique offers far less risk than x-rays, which can be genetically dangerous to the fetus and can produce birth defects. First the abdomen of the mother is coated with a liquid, such as mineral oil. If this is not done, most of the incident ultrasonic waves from the piezo-electric source will be reflected at the boundary between the air and the skin of the mother. Mineral oil has a density similar to that of skin, and as Equation 15.23 indicates, a very small fraction of the incident ultrasonic wave is reflected when $\rho_i \approx \rho_t$. The ultrasound energy is emitted as pulses rather than as a continuous wave so that the same crystal can be used as a detector as well as a transmitter. The source-receiver is then passed over a particular line along the mother's abdomen. The reflected sound waves picked up by the receiver are converted to an electric signal, which forms an image along a line on a fluorescent screen. The sound source is then moved a few centimeters on the mother's body, and the process is repeated. The reflected signal produces a second line on the fluorescent screen. In this fashion a complete scan of the fetus can be made. Difficulties with the pregnancy, such as the likelihood of abortion or of breech birth, are easily detected with this technique. Also, such fetal abnormalities as spina bifida and water on the brain are readily observable.

Another interesting application of ultrasound is the ultrasonic ranging unit designed by the Polaroid Corporation. This device is used in some of their cameras to provide an almost instantaneous measurement of the distance between the camera and object to be photographed. The principal component of this device is a crystal that acts as both a loudspeaker and a microphone. A pulse of ultrasonic waves is transmitted from the transducer to the object to be photographed. The object reflects part of the signal, producing an echo that is detected by the device. The time interval between the outgoing pulse and the detected echo is then electronically converted to a distance value, since the speed of sound is a known quantity. ◀

An ultrasound image of the heart. *(Schleichkorn, Custom Medical Stock Photo)*

just barely audible to a person with a normal threshold of hearing? Disregard absorption.

16. A skyrocket explodes 100 m above the ground, as shown in Figure 15.27. Three observers are spaced 100 m apart, and the first (A) is directly under the point of the explosion. What is the ratio of sound intensities heard by observers A and B, and A and C?

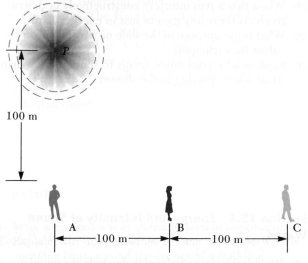

Figure 15.27   Problem 16

17. A stereo speaker that we shall consider as a small source emits sound waves with a power output of 100 W. (a) Find the intensity at a distance of 10 m from the source. (b) Find the intensity level in dB at this distance. (c) At what distance would you have to be such that the sound was at the threshold of pain, 120 dB?

18. A man shouting loudly produces a dB level of about 70 dB at a distance of 5 m. How many watts of power does the man emit? (Treat the man as a point source.)

19. The sound level 3 m from a point source is 120 dB. At what distance will the sound level be 100 dB?

20. The dB level at a distance of 2 m from a sound source is 100 dB. What is the dB level at distances of (a) 4 m, (b) 6 m, (c) 8 m?

Section 15.6   The Doppler Effect

21. A train at rest emits a sound at a frequency of 1000 Hz. An observer in a car travels away from the sound at a speed of 30 m/s. What is the frequency heard by the observer?

22. The observer of Problem 21 stops the car. Determine the frequency he hears when (a) the train approaches him at a speed of 30 m/s, (b) when the train recedes from him at a speed of 30 m/s.

23. At rest a car's horn sounds the note A (440). While the car is moving down the street, the horn is sounded. A bicyclist moving in the same direction with one-third the car's speed hears the note A♭ (415). How fast is the car moving? Is the cyclist ahead of or behind the car?

24. A train passenger hears a frequency of 520 Hz as the train approaches a bell on a trackside safety gate; the bell is actually emitting a signal of 500 Hz. What frequency will the passenger hear just after passing the bell?

25. An ambulance siren emits a note of 500 Hz when the ambulance is at rest. When the ambulance is in motion at a speed of 10 m/s, what is the frequency heard by an observer traveling at (a) 15 m/s toward the source, and (b) 15 m/s after the observer has passed the ambulance and recedes from it?

26. An alert physics student stands beside the tracks as a train rolls slowly past. He notes that the frequency of the train whistle is 442 Hz when the train is approaching him and is 441 Hz when the train is receding from him. From this he is able to find the speed of the train. What value does he find?

27. Two trains on separate tracks move toward one another. Train 1 has a speed of 130 km/h and train 2 a speed of 90.0 km/h. Train 2 blows its horn, emitting a frequency of 500 Hz. What is the frequency heard by the engineer on train 1?

28. A horn emits a frequency of 1000 Hz. A 14.0-m/s wind is blowing toward a listener. What is the frequency of the sound heard by the listener? (*Hint:* Consider a frame of reference in which the air is at rest. What can you conclude from the result of this problem?)

29. At what speed should a supersonic aircraft fly so that the angle θ shown in Figure 15.10a will be (a) 40°, (b) 30°?

30. The Concorde flies at mach 1.5, which means the speed of the plane is 1.5 times the speed of sound in air. What is the angle between the direction of propagation of the shock wave and the direction of the plane's velocity?

Section 15.7   Interference in Sound Waves

31. Two loudspeakers are placed above and below one another as in Figure 15.13 and driven by the same source at a frequency of 500 Hz. (a) What minimum distance should the top speaker be moved back in order to have destructive interference between the two speakers? (b) If the top speaker is moved back twice the distance calculated in Part (a), will constructive or destructive interference occur?

32. A source of unknown frequency is attached to the two loudspeakers of Problem 31. It is found that when the

upper loudspeaker is moved back 0.5 m, *constructive* interference occurs between the two. What is the frequency of the sound generator?

33. The sound interferometer shown in Figure 15.12 is driven by a speaker emitting a 400-Hz note. If *destructive* interference occurs at a particular instant, how much must the path length in the U-shaped tube be increased in order to hear (a) constructive interference, (b) destructive interference once again?

34. The two speakers of Figure 15.13 are 0.6 m apart, and both emit a frequency of 1000 Hz. An observer on the center line between the speakers begins to walk along a line parallel to a line joining the speakers. When he is 5 m from the closest speaker, he hears destructive interference. How far is he from the most distant speaker?

35. Two speakers are driven by a common oscillator at 800 Hz and face each other at a distance of 1.25 m. Locate the points along a line joining the two speakers where relative minima would be expected. (Use $v = 343$ m/s.)

36. The ship in Figure 15.28 travels along a straight line parallel to the shore and 600 m from the shore. Its radio receives simultaneous signals of the same frequency from antennas A and B. The signals interfere constructively at point C, which is equidistant from A and B. The signal goes through the first minimum at point D. Determine the wavelength of the radio waves.

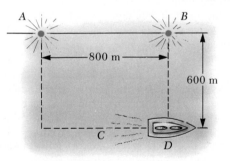

Figure 15.28 Problem 36

Section 15.8 Standing Waves

37. A string 50 cm long has a mass per unit length equal to 20×10^{-5} kg/m. To what tension should this string be stretched if its fundamental frequency is to be (a) 20 Hz and (b) 4500 Hz?

38. A steel wire in a piano has a length of 0.70 m and a mass of 4.3×10^{-3} kg. To what tension must this wire be stretched in order that the fundamental vibration correspond to middle C ($f_C = 261.6$ Hz on the chromatic musical scale)?

39. A stretched string is 160 cm long and has a linear density of 0.015 g/cm. What tension in the string will result in a second harmonic of 460 Hz?

40. A wire of mass 0.3 g is stretched between two points 70 cm apart. If the tension in the wire is 600 N, find the first, second, and third harmonics of this string.

41. (a) To what tension would the wire of Problem 40 have to be stretched in order for the frequency of the fundamental to be 20,000 Hz? (b) Does it seem likely that this could be accomplished?

42. A stretched string fixed at each end has a mass of 40 g and a length of 8 m. The tension in the string is 49 N. (a) Determine the position of the nodes and antinodes for the third harmonic. (b) What is the vibration frequency for this harmonic?

43. Two pieces of steel wire having identical cross sections have lengths of L and $2L$. The wires are each fixed at both ends and stretched such that the tension in the longer wire is four times greater than that in the shorter wire. If the fundamental frequency in the shorter wire is 60 Hz, what is the frequency of the second harmonic in the longer wire?

44. A stretched string of length L is observed to vibrate in five equal segments when driven by a 630-Hz oscillator. What oscillator frequency will set up a standing wave such that the string vibrates in three segments?

Section 15.9 Forced Vibrations and Resonance

45. A child's swing has a length of 2 m. List three frequencies at which this swing could be pushed to set up resonant vibrations in the swing.

46. If the circumference of the mouth of a crystal goblet is 10 cm, at what fundamental frequency would an opera singer have to sing in order to set the glass into a resonant vibration?

Section 15.10 Standing Waves in Air Columns

47. The fundamental frequency of an open organ pipe corresponds to middle C (261.6 Hz on the chromatic musical scale). The third resonance of a closed organ pipe has the same frequency. What are the lengths of the two pipes?

48. An organ pipe open at both ends is vibrating in its third harmonic with a frequency of 748 Hz. The length of the pipe is 0.7 m. Determine the speed of sound in air in the pipe.

49. Middle C has a frequency of 261.6 Hz. (a) What is the length of the shortest pipe open at both ends which would produce a sound of this frequency? (b) What would be the required length if the pipe is closed at one end?

50. A closed organ pipe is 3 m long. At what frequencies between 20 Hz and 20,000 Hz will this pipe resonate?

51. A pipe open at both ends has a fundamental frequency of 300 Hz when the temperature is 0°C. (a) What is the length of the pipe? (b) What is the fundamental frequency at a temperature of 30°C?

52. A tuning fork is sounded above a resonating tube as in Figure 15.20. The first resonant point is 0.08 cm from the top of the tube, and the second is at 0.24 m. (a) Where would the third resonant point be located? (b) What is the frequency of the tuning fork?

53. An air column 2 m in length is open at both ends. The frequency of a certain harmonic is 410 Hz, and the frequency of the next higher harmonic is 492 Hz. Determine the speed of sound in the air column.

54. The range of a certain pipe organ is from 8 Hz to 30,000 Hz. What range of pipe lengths is necessary if they are (a) open at both ends and (b) closed at one end? (Assume that the speed of sound is 343 m/s.)

Section 15.11   Beats

55. A piano tuner strikes a 440-Hz tuning fork at the instant she strikes a piano key that should emit a tone of 440 Hz and hears a beat frequency of 2 Hz. What are the possible frequencies the piano key could be emitting?

56. A flute is designed such that it plays a frequency of 261.6 Hz, middle C, when all the holes are covered and the temperature is 20°C. (a) Consider the flute as a pipe open at both ends and find the length of the flute, assuming that the frequency above is the fundamental. (b) A second player nearby in a colder room also attempts to play middle C on an identical flute. A beat frequency of 3 Hz is heard. What is the temperature of the room?

57. The G string on a violin has a fundamental frequency of 196 Hz. It is 30 cm long and has a mass of 0.5 g. A nearby violinist fingers her identical violin until a beat frequency of 2 Hz is heard between the two. What is the length of her violin string?

Section 15.13   The Ear

58. If the ear canal of a person can be considered as acting like an organ pipe closed at one end that resonates at a fundamental frequency of 3000 Hz, what is the length of the canal? Use normal body temperature for your determination of the speed of sound in the canal.

59. Some studies indicate that the upper limit of hearing is determined by the diameter of the eardrum. The wavelength of the sound wave and the diameter of the eardrum are approximately equal at this upper limit of frequency. If this is so, what is the diameter of the eardrum of a person capable of hearing 20,000 Hz? (Assume a body temperature of 37°C.)

Additional Problems

60. What is the lowest frequency of the standing wave of sound that can be set up between two walls that are 8 m apart if the temperature is 22°C?

61. A commuter train passes a passenger platform at a constant speed of 40 m/s. The train horn is sounded at a frequency of 320 Hz when the train is at rest. (a) What is the frequency observed by a person on the platform as the train approaches and (b) as the train recedes from him? (c) What wavelength does the observer find in each case?

62. Two train whistles have identical frequencies of 180 Hz. When one train is at rest in the station sounding its whistle, a beat frequency of 2 Hz is heard from a moving train. What two possible speeds and directions can the moving train have?

63. A pipe open at each end has a fundamental frequency of 300 Hz when the speed of sound in air is 333 m/s. (a) What is the length of the pipe? (b) What is the frequency of the second harmonic when the temperature of the air is increased so that the speed of sound in the pipe is 344 m/s?

64. A standing wave is established in a string that is 240 cm long and fixed at both ends. The string vibrates in four segments when driven at 120 Hz. (a) Determine the wavelength. (b) What is the fundamental frequency?

65. A string of length L, mass per unit length μ, and tension F is vibrating at its fundamental frequency. What effect will the following have on the fundamental frequency? (a) The length of the string is doubled with all other factors held constant. (b) The mass per unit length is doubled with all other factors held constant. (c) The tension is doubled with all other factors held constant.

66. Two point sound sources have measured intensities of $I_1 = 100$ W/m² and $I_2 = 200$ W/m². By how many decibels is the level of source 1 lower than the level of source 2? (Assume that the observer is the same distance from both sources.)

67. A string of 8-g mass and 5-m length has one end attached to a wall while the other end is draped over a pulley and attached to a hanging mass of 4 kg. If this string is plucked, what is the fundamental frequency of vibration?

68. A typical sound level for a buzzing mosquito is 40 dB, and normal conversation is approximately 50 dB. How many buzzing mosquitoes will produce a sound intensity equal to that of normal conversation?

69. When high-energy charged particles move through a transparent medium with a velocity greater than the velocity of light in that medium, a shock wave, or bow wave, of light is produced. This phenomenon is called the *Cerenkov effect* and can be observed in the vicinity of the core of a swimming pool nuclear reactor due to high-speed electrons moving through the water. In a particular case, the Cerenkov radiation produces a wavefront with a cone angle of 53°. Calculate the velocity of the electrons in the water. (Take the speed of light in water to be 2.25×10^8 m/s.)

70. Refer to Table 15.2, which gives dB values for representative sounds from various sources. Determine the resultant intensity in dB when a vacuum cleaner and a power mower are operated against a background of busy traffic.

71. A variable-length air column is placed just below a vibrating wire fixed at both ends. The length of the air column is gradually increased from zero until the first position of resonance is observed at $L = 34$ cm. The wire is 120 cm in length and is vibrating in its third harmonic. If the speed of sound in air is 340 m/s, what is the speed of transverse waves in the wire?

72. Two pipes are each open at one end and have equal lengths. Each has a fundamental frequency of 480 Hz at 300 K. The air temperature is increased in one pipe to 305 K. If the two pipes are sounded together, what beat frequency will result?

73. Two trains when at rest have sirens that emit a frequency of 300 Hz. The two trains travel toward one another and toward an observer stationed between them. One of the trains moves at 30 m/s, and the observer hears a beat frequency of 3 beats per second. What is the velocity of the second train, which travels faster than the first?

74. By proper excitation, it is possible to produce both longitudinal and transverse waves in a long metal rod. In one particular case, the rod is 150 cm in length and 0.2 cm in radius and has a mass of 50.9 g. Young's modulus for the material is 6.8×10^{11} dynes/cm². Determine the required tension in the rod so that the ratio of the speed of longitudinal waves to the speed of transverse waves is 8.

75. A speaker at the front of a room and an identical speaker at the rear of the room are being driven by the same sound source at 456 Hz. A student walks at a uniform rate of 1.5 m/s along the length of the room. How many beats does the student hear per second?

76. A student located several meters in front of a smooth reflecting wall is holding a board on which a wire is fixed at each end. The wire, vibrating in its third harmonic, is 75 cm long, has a mass of 2.25 g, and is under a tension of 400 N. A second student is moving toward the wall and hears 8.3 beats per second. At what speed does the moving student approach the wall? Use 340 m/s as the speed of sound in air.

77. Two ships are moving along a line due east. The trailing vessel has a speed relative to a land-based observation point of 64 km/h, and the leading ship has a speed of 45 km/h relative to that station. The two ships are in a region of the ocean where the current is moving uniformly due west at 10 km/h. The trailing ship transmits a sonar signal at a frequency of 1200 Hz. What frequency is monitored by the leading ship? (Use 1520 m/s as the speed of sound in ocean water.)

78. Three metal rods are located relative to each other as shown in Figure 15.29, where $L_1 + L_2 = L_3$. Values of the density and Young's modulus for the three materials are $\rho_1 = 2.7 \times 10^3$ kg/m³, $Y_1 = 7 \times 10^{10}$ Pa, $\rho_2 = 11.3 \times 10^3$ kg/m³, $Y_2 = 1.6 \times 10^{10}$ Pa, $\rho_3 = 8.8 \times 10^3$ kg/m³, $Y_3 = 11 \times 10^{10}$ Pa. (a) If $L_3 = 1.5$ m, what must the ratio L_1/L_2 be if a sound wave is to travel the length of rods 1 and 2 in the same time required to travel the length of rod 3?

Figure 15.29
Problem 78

79. A duck swims in a pond. Water waves move outward from the duck with a speed of 0.5 m/s. If the wavelength of the waves behind the duck is 50.0% longer than the wavelength of the waves ahead of the duck, how fast is the duck moving?

"I love hearing that lonesome wail of the train whistle as the magnitude of the frequency of the wave changes due to the Doppler effect."

Consider a point charge q located a distance r from a test charge q_0. According to Coulomb's law, the *magnitude* of the force on the test charge is

$$F = k \frac{|q||q_0|}{r^2}$$

Because the magnitude of the electric field at the position of the test charge is defined as $E = F/q_0$, we see that the *magnitude* of the electric field due to the charge q at the position of $|q_0|$ is

Electric field due to a charge q

$$E = k \frac{|q|}{r^2} \qquad \text{(16.5)}$$

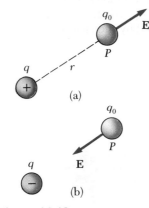

(a)

(b)

Figure 16.12
A test charge q_0 at P is a distance r from a point charge q. (a) If q is positive, the electric field at P points radially *outward* from q. (b) If q is negative, the electric field at P points radially *inward* toward q.

If q is *positive*, as in Figure 16.12a, the field at P due to this charge is *radially outward* from q. If q is *negative*, as in Figure 16.12b, the field at P is directed *toward* q. Equation 16.5 points out an important property of electric fields that makes them such useful quantities in describing electrical phenomena. As the equation indicates, an electric field at a given point depends only on the charge q on the object setting up the field and the distance r from that object to a specific point in space. As a result, we can say that an electric field exists at point P in Figure 16.12 whether or not there is a charge located at P.

The principle of superposition holds when calculating the electric field due to a group of point charges. We first use Equation 16.5 to calculate the electric field produced by each charge individually at a point and then add these electric fields together as vectors.

Example 16.4   Electric Force on a Proton

Find the electric force on a proton placed in an electric field of 2×10^4 N/C that is directed along the positive x axis.

Solution
Because the charge on a proton is $+e = +1.6 \times 10^{-19}$ C, the electric force acting on the proton is

$$F = eE = (1.6 \times 10^{-19} \text{ C})(2 \times 10^4 \text{ N/C}) = \boxed{3.2 \times 10^{-15} \text{ N}}$$

where the force is in the positive x direction. The weight of the proton has the value $mg = (1.67 \times 10^{-27} \text{ kg})(9.80 \text{ m/s}^2) = 1.64 \times 10^{-26}$ N. Hence, we see that the magnitude of the gravitational force is negligible compared with that of the electric force.

Example 16.5   Electric Field Due to Two Point Charges

A charge $q_1 = 7 \ \mu$C is located at the origin, and a charge $q_2 = -5 \ \mu$C is located on the x axis 0.3 m from the origin (Fig. 16.13). Find the electric field at the point P, which has coordinates (0, 0.4) m.

Solution
First let us find the magnitudes of the electric fields due to each charge. The fields at P, $\mathbf{E}_1$ due to the 7-μC charge and $\mathbf{E}_2$ due to the -5-μC charge, are shown in Figure 16.13. Their magnitudes are

$$E_1 = k \frac{|q_1|}{r_1{}^2} = (9.0 \times 10^9 \text{ N} \cdot \text{m}^2/\text{C}^2) \frac{(7 \times 10^{-6} \text{ C})}{(0.4 \text{ m})^2} = 3.94 \times 10^5 \text{ N/C}$$

$$E_2 = k \frac{|q_2|}{r_2{}^2} = (9.0 \times 10^9 \text{ N} \cdot \text{m}^2/\text{C}^2) \frac{(5 \times 10^{-6} \text{ C})}{(0.5 \text{ m})^2} = 1.80 \times 10^5 \text{ N/C}$$

The vector $\mathbf{E_1}$ has an x component of zero. The vector $\mathbf{E_2}$ has an x component given by $E_2 \cos \theta = E_2(0.3/0.5) = \frac{3}{5}E_2 = 1.08 \times 10^5$ N/C and a negative y component given by $-E_2 \sin \theta = -E_2(0.4/0.5) = -\frac{4}{5}E_2 = -1.44 \times 10^5$ N/C. Hence, the resultant component in the x direction is

$$E_x = 1.08 \times 10^5 \text{ N/C}$$

and the resultant component in the y direction is

$$E_y = E_{y1} + E_{y2} = 3.94 \times 10^5 \text{ N/C} - 1.44 \times 10^5 \text{ N/C} = 2.50 \times 10^5 \text{ N/C}$$

From the Pythagorean theorem ($E = \sqrt{E_x{}^2 + E_y{}^2}$), we find that $\mathbf{E}$ has a magnitude of 2.72×10^5 N/C and makes an angle ϕ of 64.4° with the positive x axis.

Exercise Find the force on a positive test charge of 2×10^{-8} C placed at P.

Answer 5.44×10^{-3} N in the same direction as $\mathbf{E}$.

Figure 16.13 (Example 16.5) The total electric field $\mathbf{E}$ at P equals the vector sum $\mathbf{E_1} + \mathbf{E_2}$, where $\mathbf{E_1}$ is the field due to the positive charge q_1 and $\mathbf{E_2}$ is the field due to the negative charge q_2.

16.6 Electric Field Lines

A convenient aid for visualizing electric field patterns is to draw lines pointing in the direction of the electric field vector at any point. These lines, called **electric field lines**, are related to the electric field in any region of space in the following manner: (1) the electric field vector $\mathbf{E}$ is tangent to the electric field lines at each point and (2) the number of lines per unit area through a surface perpendicular to the lines is proportional to the strength of the electric field in a given region; this means that $\mathbf{E}$ is large when the field lines are close together and small when they are far apart.

Some representative electric field lines for a single positive point charge are shown in Figure 16.14a. Note that in this two-dimensional drawing we

Figure 16.14
The electric field lines for a point charge. (a) For a positive point charge, the lines are radially outward. (b) For a negative point charge, the lines are radially inward. Note that the figures show only those field lines that lie in the plane containing the charge. (c) The dark areas are small pieces of thread suspended in oil, which align with the electric field produced by a small charged conductor at the center. *(Photo courtesy of Harold M. Waage, Princeton University)*

(a)   (b)   (c)

start at ⊕ end at ⊖

(a)

(b)

Figure 16.15
(a) The electric field lines for two equal and opposite point charges (an electric dipole). Note that the number of lines leaving the positive charge equals the number terminating at the negative charge.
(b) The photograph was taken using small pieces of thread suspended in oil, which align with the electric field produced by two charged conductors. *(Photo courtesy of Harold M. Waage, Princeton University)*

show only the field lines that lie in the plane containing the point charge. The lines are actually directed radially outward from the charge in *all* directions, somewhat like the needles of a porcupine. Since a positive test charge placed in this field would be repelled by the charge q, the lines are directed radially away from the positive charge. Similarly, the electric field lines for a single negative point charge are directed toward the charge (Fig. 16.14b). In either case, the lines are along the radial direction and extend all the way to infinity. Note that the lines are closer together as they get near the charge, indicating that the strength of the field is increasing. Equation 16.5 verifies that this should indeed be the case.

The rules for drawing electric field lines for any charge distribution are as follows:

1. The lines must begin on positive charges and terminate on negative charges, or at infinity in the case of a lone charge.
2. The number of lines drawn leaving a positive charge or approaching a negative charge is proportional to the magnitude of the charge.
3. No two field lines can cross.

The electric field lines for two point charges of equal magnitude but of opposite sign are shown in Figure 16.15. This charge configuration is called an **electric dipole.** In this case the number of lines that begin at the positive charge must equal the number that terminate at the negative charge. At points very near the charges, the lines are nearly radial. The close spacing of lines between the charges indicates a strong electric field in this region.

Figure 16.16 shows the electric field lines in the vicinity of two equal positive charges. Again, the lines are nearly radial at points close to either charge. The same number of lines are shown emerging from each charge because the charges are equal in magnitude. The lines are not close together between the charges, indicating that this is a region of low electric field strength.

Finally, in Figure 16.17 we sketch the electric field lines associated with a positive charge $+2q$ and a negative charge $-q$. In this case, the number of lines leaving the charge $+2q$ is twice the number entering the charge $-q$. Hence, only half the lines that leave the positive charge end on the negative charge. The remaining half terminate at infinity.

Figure 16.16
(a) The electric field lines for two positive point charges. (b) The photograph was taken using small pieces of thread suspended in oil, which align with the electric field produced by two charged conductors. *(Photo courtesy of Harold M. Waage, Princeton University)*

(a)

(b)

514

16.7 Conductors in Electrostatic Equilibrium

A good electric conductor, such as copper, contains charges (electrons) that are not bound to any atom and are free to move about within the material but cannot easily leave. When there is no net motion of charge within a conductor or on its surface, the conductor is said to be in **electrostatic equilibrium.** The electric field in and around a conductor in electrostatic equilibrium has some important characteristics, which we shall now consider. As we shall see, a conductor in *electrostatic equilibrium* has the following properties:

1. The electric field is zero everywhere inside the conductor.
2. Any excess charge on an isolated conductor resides entirely on its surface.
3. The electric field just outside a charged conductor is perpendicular to the conductor's surface.
4. On an irregularly shaped conductor, the charge per unit area is greatest at locations where the curvature of the surface is greatest, that is, at sharp points.

The first property can be understood by examining what would happen if it were *not* true. If there were an electric field inside a conductor, the free charge there would move and a flow of charge, or current, would be created. However, if there were a net movement of charge, we would no longer have electrostatic equilibrium.

Property 2 is a direct result of the repulsion between like charges, described by Coulomb's law. If by some means an excess of charge is placed inside a conductor, the repulsive forces arising between the charges would force them as far apart as possible, causing them to quickly migrate to the surface.

Property 3 can be understood by again considering what would happen if it were not true. In Figure 16.18a we assume that the electric field is not perpendicular to the surface. In this case, the electric field would have a component along the surface, and this component would cause the free charges of the conductor to move (to the left in the figure). If the charges

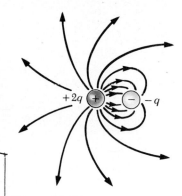

Figure 16.17
The electric field lines for a point charge $+2q$ and a second point charge $-q$. Note that two lines leave the charge $+2q$ for every one that terminates on the charge $-q$.

(a)

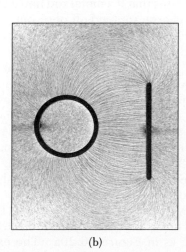

(b)

Figure 16.18
(a) Negative charges at the surface of a conductor. If the electric field were at an angle to the surface, as shown, an electric force would be exerted on the charges along the surface and they would move to the left. Since the conductor is assumed to be in electrostatic equilibrium, **E** cannot have a component along the surface and hence must be perpendicular to it. (b) Electric field pattern of a charged conducting plate near an oppositely charged conducting cylinder. Small pieces of thread suspended in oil align with the electric field lines. Note that (1) the electric field lines are perpendicular to the conductors and (2) there are no lines inside the cylinder ($\mathbf{E} = 0$). *(Courtesy of Harold M. Waage, Princeton University)*

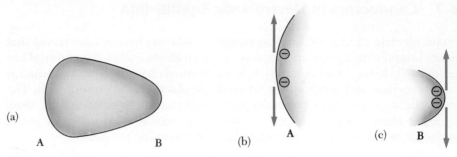

(a)

A   B   (b)   A   (c)   B

Figure 16.19
(a) The conductor shaped as shown has a flatter end **A** and a sharp surface **B**. Excess charge placed on a conductor resides entirely at its surface and is distributed such that (b) there is less charge per unit area on the flatter surface and (c) there is a large charge per unit area on the sharp surface.

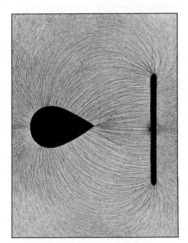

Electric field pattern of a charged conducting plate near an oppositely charged pointed conductor. Small pieces of thread suspended in oil align with the electric field lines. Note that the electric field is most intense near the pointed part of the conductor where the radius of curvature is the smallest. Also, the lines are perpendicular to the conductors. *(Courtesy of Harold M. Waage, Princeton University)*

move, however, a current is created and we no longer have electrostatic equilibrium. Hence, **E** must be perpendicular to the surface.

To see why Property 4 must be true, consider Figure 16.19a, which shows a conductor that is relatively flat at one end and sharply pointed at the other. Any excess charge placed on the object moves to its surface. Figure 16.19b shows the forces between two of these charges at the flatter end of the object. These forces are predominantly directed parallel to the surface. Thus, the charges move apart until repulsive forces from other nearby charges create an equilibrium situation. At the sharp end, however, the forces of repulsion between two charges are directed predominantly away from the surface, as in Figure 16.19c. As a result, there is less tendency for the charges to move apart along the surface here, and the amount of charge per unit area is greater than at the flat end. The cumulative effect of many such outward forces by nearby charges at the sharp end produces a large force directed away from the surface that could be great enough to cause charges to leap from the surface into the surrounding air.

Property 4 indicates that if a metal rod having sharp points is attached to a house, most of the charge sprayed off of or onto the house will pass through these points, thus eliminating the induced charge on the house produced by storm clouds. In addition, a lightning discharge striking the house can pass through the metal rod and be safely carried to the ground through wires leading from the rod to the Earth. Lightning rods using this principle were first developed by Benjamin Franklin. It is an interesting sidelight to American history to note that some foreign countries could not accept the fact that such a worthwhile idea could originate in this country. As a result, they "improved" the design by eliminating the sharp points. This modification in design drastically reduced the efficiency of the lightning rod.

Many experiments have been performed to show that the net charge on a conductor resides on its surface. The experiment we shall describe here was first performed by Michael Faraday. A metal ball having a positive charge was lowered at the end of a silk thread into an uncharged hollow conductor insulated from ground, as in Figure 16.20a. (The experiment is referred to as

Faraday's ice-pail experiment because he used a metal ice pail as the hollow conductor.) As the ball was lowered into the pail, the needle on an electrometer attached to the outer surface of the pail was observed to deflect. (An electrometer is a device used to measure charge.) The needle deflected because the charged ball induced a negative charge on the inner wall of the pail, which left an equal positive charge on the outer wall, as indicated in Figure 16.20b.

Faraday also noted that the needle deflection did not change when the ball touched the inner surface of the pail (Fig. 16.20c). After the ball had touched the inside of the pail and then been removed, the electrometer reading still remained unchanged (Fig. 16.20d). Furthermore, Faraday found that the ball was now uncharged. Apparently, when the ball touched the inside of the pail, the excess positive charge on the ball was neutralized by the induced negative charge on the inner surface of the pail.

Faraday concluded that since the electrometer deflection did not change when the charged ball touched the inside of the pail, the negative charge induced on the inside surface of the pail was just enough to neutralize the positive charge on the ball. As a result of these investigations, he concluded that a charged object suspended inside a metal container will cause a rearrangement of charge on the container in such a manner that the sign of the charge on the inside surface of the container is *opposite* the sign of the charge on the suspended object. This produces a charge on the outside surface of the container of the same sign as that on the suspended object.

Faraday also found that if the electrometer was connected to the inside surface of the pail after the experiment had been run, the needle showed no deflection. Thus, the *excess* charge acquired by the pail when contact was made between ball and pail appeared on the outer surface of the pail.

▶SPECIAL TOPIC
The Millikan Oil-Drop Experiment

During the period 1909 to 1913, Robert Andrews Millikan (1868–1953) performed a brilliant set of experiments at the University of Chicago in which he measured the elementary charge on an electron, e, and demonstrated the quantized nature of the electronic charge. The apparatus used by Millikan, diagrammed in Figure 16.21, contains two parallel metal plates. Oil droplets charged by friction in an atomizer are allowed to pass through a small hole in the upper plate. A light beam directed horizontally is used to illuminate the oil droplets, which are viewed by a telescope whose axis is at right angles to the beam. When the droplets are viewed in this manner, they appear as shining stars against a dark background, and the rate of fall of individual drops may be determined.

Let us assume a single drop having a mass m and carrying a charge q is being viewed, and that its charge is negative. If there is no electric field present between the plates, the two forces acting on the charge are its weight, mg, acting downward, and an upward viscous drag force **D**, as indicated in Figure 16.22a. The drag force is proportional to the speed of the drop. When the drop reaches its terminal speed v, the two forces balance each other ($mg = D$).

Now suppose that an electric field is set up between the plates by connecting a battery such that the upper plate is positively charged. In this case, a third force $q\mathbf{E}$ acts on the charged drop. Since q is negative and **E** is downward, the electric force is

Figure 16.20
An experiment showing that any charge transferred to a conductor resides on its surface in electrostatic equilibrium. The hollow conductor is insulated from ground, and the small metal ball is supported by an insulating thread.

Figure 16.21
A schematic view of the Millikan oil-drop apparatus.

upward as in Figure 16.22b. If this force is large enough, the drop will move upward and the drag force **D′** will act downward. When the upward electric force, $q\mathbf{E}$, balances the sum of the weight and the drag force both acting downward, the drop reaches a new terminal speed v'.

With the field turned on, a drop moves slowly upward, typically at rates of *hundredths* of a centimeter per second. The rate of fall in the absence of a field is comparable. Hence, a single droplet with constant mass and radius may be followed for hours, alternately rising and falling, by simply turning the electric field on and off.

After making measurements on thousands of droplets, Millikan and his coworkers found that all drops, to within about 1% precision, had a charge equal to some integer multiple of the elementary charge e. That is,

$$q = ne \qquad n = 0, \pm 1, \pm 2, \pm 3, \ldots \qquad \textbf{(16.6)}$$

where $e = 1.60 \times 10^{-19}$ C. Millikan's experiment is conclusive evidence that charge is quantized. He was awarded the Nobel Prize in physics in 1923 for this work. ◄

Figure 16.22
The forces on a charged oil droplet in the Millikan experiment.

(a) Field off   (b) Field on

*16.8   The Van de Graaff Generator

In 1929 Robert J. Van de Graaff designed and built an electrostatic generator that is used extensively in nuclear physics research. The principles of its operation can be understood with the knowledge of electric fields and charges we have developed in this chapter. The basic construction details of this device are shown in Figure 16.23. A motor-driven pulley, *P*, moves a belt past positively charged comb-like metallic needles located at *A*. Negative charges are attracted to these needles from the belt, leaving the left side of the belt with a net positive charge. These positive charges attract electrons onto the belt as it moves past a second comb of needles at *B*, increasing the excess positive charge on the dome. Since the electric field inside the metal dome is negligible, the positive charge on it can easily be increased regardless of how much charge is already present. The result is that the dome is left with a large amount of positive charge.

This accumulation of charge on the dome cannot continue indefinitely, because eventually an electric discharge through the air will take place. To understand why this discharge occurs, consider that as more and more charge appears on the surface of the dome, the magnitude of the electric field at the surface of the dome is also increasing. Finally, the strength of the field becomes great enough to partially ionize the air near the surface, thus making the air partially conducting. Charges on the dome now have a pathway to leak off into the air, which can produce some spectacular "lightning bolts" as the discharge occurs. As noted earlier, charges find it easier to leap off a surface at points where the curvature is great. As a result, one way to inhibit the electric discharge, and to increase the amount of charge that can be stored on the dome, is to increase its radius. Another method for inhibiting discharge is to place the entire system in a container filled with a high-pressure gas. Such a gas is significantly more difficult to ionize than air at atmospheric pressure.

If protons (or other charged particles) are introduced into a tube attached to the dome, the large electric field of the dome exerts a repulsive force on the protons, causing them to accelerate to energies high enough to initiate nuclear reactions between the protons and various target nuclei.

*16.9   The Oscilloscope

The oscilloscope is an electronic instrument widely used in making electrical measurements. The main component of the oscilloscope is the cathode ray tube (CRT), shown in Figure 16.24a. This tube is commonly used to obtain a visual display of electronic information for other applications, including radar systems, television receivers, and computers. The CRT is essentially a vacuum tube in which electrons are accelerated and deflected under the influence of electric fields.

The electron beam is produced by an assembly called an *electron gun,* located in the neck of the tube. The assembly shown in Figure 16.24a consists of a heater (H), a cathode (C), and a positively charged anode (A). An electric current through the heater causes its temperature to rise, which in turn heats the cathode. The cathode reaches temperatures high enough to cause electrons to be "boiled off." Although they are not shown in the figure, the

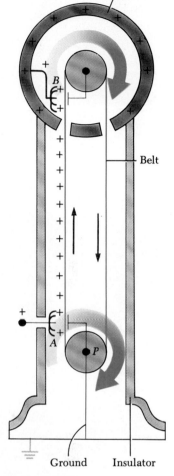

Figure 16.23
Diagram of a Van de Graaff generator. Charge is transferred to the dome by means of a rotating belt. The charge is deposited on the belt at point *A* and transferred to the dome at point *B*.

(a)   (b)

Figure 16.24
(a) Diagram of a cathode ray tube. Electrons leaving the hot cathode C are accelerated to the anode A. The electron gun is also used to focus the beam, and the plates
deflect the beam. (b) Photograph of a "Maltese Cross" tube showing the shadow of
a beam of cathode rays falling on the tube's luminescent screen. The hot filament
also produces a beam of light and a second shadow of the cross. *(Courtesy of CENCO)*

electron gun also includes an element that focuses the electron beam and one
that controls the number of electrons reaching the anode (that is, a brightness
control). The anode has a hole in its center that allows the electrons to pass
through without striking the anode. These electrons, if left undisturbed, travel
in a straight-line path until they strike the face of the CRT. The screen at the
front of the tube is coated with a fluorescent material that emits visible light
when bombarded with electrons. This results in a visible spot on the screen of
the CRT.

The electrons are moved in various directions by two sets of deflection
plates placed at right angles to each other in the neck of the tube. In order to
understand how the deflection plates operate, first consider the horizontal
deflection plates in Figure 16.24a. External electric circuits can change the
amount of charge present on these plates, with positive charge being placed on
one plate and negative on the other. (In Chapter 17 we shall see that this can be
accomplished by applying a voltage across the plates.) This increasing charge
creates an increasing electric field between the plates, which causes the electron beam to be deflected from its straight-line path. The tube face is slightly
phosphorescent and therefore glows briefly after the electron beam moves
from one point to another on the screen. Slowly increasing the charge on the
horizontal plates causes the electron beam to move gradually from the center
toward the side of the screen. Because of the phosphorescence, however, one
sees a horizontal line extending across the screen instead of the simple movement of the dot. The horizontal line can be maintained on the screen by rapid,
repetitive tracing.

The vertical deflection plates act in exactly the same way as the horizontal
plates, except that changing the charge on them causes a vertical line on the
tube face. In practice, the horizontal and vertical deflection plates are used
simultaneously. To see how the oscilloscope can display visual information, let
us examine how we could observe the sound wave from a tuning fork on the
screen. For this purpose, the charge on the horizontal plates changes in such a

manner that the beam sweeps across the face of the tube at a constant rate. The tuning fork is then sounded into a microphone, which changes the sound signal to an electric signal that is applied to the vertical plates. The combined effect of the horizontal and vertical plates causes the beam to sweep horizontally and up and down at the same time, with the vertical motion corresponding to the tuning fork signal. (A pattern like that shown in Figure 16.25 is swept out on the screen.)

*16.10 Electric Flux and Gauss' Law

This section describes an elegant technique for calculating electric fields that was developed by Karl Friedrich Gauss (1777–1855). Even though the procedure developed by Gauss is applicable only to situations in which the charge distribution is highly symmetric, it serves as a guide for understanding more complicated problems. In order to develop the law, we must first understand the concept of electric flux.

Electric Flux

Many qualitative features of electric field lines have been described in this chapter. We shall now use the concept of electric flux to put these ideas on a quantitative basis. First, consider an electric field that is uniform in both magnitude and direction as in Figure 16.26. The electric field lines penetrate a surface of area A, which is perpendicular to the field. The technique used for drawing a figure such as Figure 16.26 is that the number of lines per unit area, Φ, is proportional to the magnitude of the electric field. Therefore,

$$\Phi = EA \tag{16.6}$$

where Φ is the electric flux, which has units of $N \cdot m^2/C$ in SI units. Thus, if the area shown in Figure 16.26 is 1 m², the magnitude of the electric field is 14 N/C. (Count the lines in Fig. 16.26.) If the surface under consideration is not perpendicular to the field, the expression for the electric flux is given by

$$\Phi = EA \cos \theta \tag{16.7}$$

Equation 16.7 can be easily understood by considering Figure 16.27, where the area A is at an angle θ with respect to the field. Note that the number of lines that cross this area is equal to the number that cross the projected area A', which is perpendicular to the field. We see that the two areas are related by $A' = A \cos \theta$. From Equation 16.7 we see that the flux through a surface of

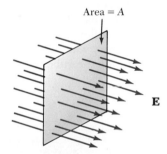

Figure 16.25
A sinusoidal wave produced by a wave generator and displayed on the oscilloscope. *(Courtesy of Henry Leap and Jim Lehman)*

Area = A

E

Figure 16.26
Field lines of a uniform electric field penetrating a plane of area A perpendicular to the field. The electric flux, Φ, through this area is equal to EA.

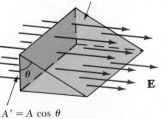

A

θ

E

$A' = A \cos \theta$

Figure 16.27
Field lines for a uniform electric field through an area A that is at an angle θ to the field. Since the number of lines that go through the shaded area A' is the same as the number that go through A, we conclude that the flux through A' is equal to the flux through A and is given by $\Phi = EA \cos \theta$.

fixed area has the maximum value, EA, when the surface is perpendicular to the field (when $\theta = 0°$) and that the flux is zero when the surface is parallel to the field (when $\theta = 90°$).

> When the area is constructed such that a closed surface is formed, we shall adopt the convention that flux lines passing into the interior of the volume are negative and those passing out of the interior of the volume are positive.

Example 16.6   Flux Through a Cube

Consider a uniform electric field oriented in the x direction. Find the net electric flux through the surface of a cube of edges L oriented as shown in Figure 16.28.

Solution

The net flux can be evaluated by summing up the fluxes through each face of the cube. First, note that the flux through four of the faces is zero, since **E** is parallel to the area on these surfaces. These surfaces are those that are parallel to the xy and xz planes. For these surfaces, $\theta = 90°$, so $\Phi = EA \cos 90° = 0$. For surface 1 that lies in the yz plane in Figure 16.28, the flux lines pass into the interior of the cube, and the flux is taken to be negative. We have

$$\Phi_1 = -EA = -EL^2$$

For surface 2 at the right face of the cube, the flux is positive and given by

$$\Phi_2 = EA = EL^2$$

The net flux through the surface of the cube is

$$\Phi_{net} = \Sigma EA = \Phi_1 + \Phi_2 = -EL^2 + EL^2 = 0$$

Figure 16.28   (Example 16.6) A hypothetical surface in the shape of a cube in a uniform electric field parallel to the x axis. The net flux through the surface is zero when the net charge inside the cube is zero.

Gauss' Law

We shall now describe a general relation between the net electric flux through a closed surface (often called a *gaussian surface*) and the charge *enclosed* by the surface. In our discussion of electric field lines we saw that the number of field lines originating on a positive charge or terminating on a negative charge is proportional to the magnitude of the charge. That is, the net flux passing through a closed surface surrounding a charge Q is proportional to the magnitude of Q, or

$$\Phi_{net} = \Sigma EA \cos \theta \propto Q$$

In free space, the constant of proportionality is $1/\epsilon_0$ where ϵ_0 is called the permittivity of free space and has the value

$$\epsilon_0 = \frac{1}{4\pi k} = \frac{1}{4\pi(9 \times 10^9 \text{ N} \cdot \text{m}^2/\text{C}^2)} = 8.85 \times 10^{-12} \text{ C}^2/\text{N} \cdot \text{m}^2$$

We shall examine the physical significance of ϵ_0 more closely in our discussion of capacitors in the next chapter. Thus, the net electric flux through a closed surface is

Gauss' law

$$\Sigma EA \cos \theta = \frac{Q}{\epsilon_0} \tag{16.8}$$

This result, known as **Gauss' law,** states that

the net electric flux through any closed gaussian surface is equal to the net charge inside the surface divided by ϵ_0.

In principle, Gauss' law can always be used to calculate the electric field of a system of charges or a continuous distribution of charge. However, in practice, the technique is useful only in a limited number of situations where there is a high degree of symmetry.

To indicate the procedure for applying Gauss' law, consider the problem of calculating the electric field at a distance r from a positive point charge q. The first step is to construct a gaussian surface. To construct this surface, we must (1) consider the symmetry of the charge distribution, and (2) allow the gaussian surface to pass through the point at which we want to calculate the electric field. These two considerations lead us to the construction of a gaussian surface, which is a sphere with q at its center and having a radius of r as shown in Figure 16.29.

Since there is only one surface to be considered, there is only one term in the sum on the left side of Gauss' law, and the left side of Equation 16.8 can be expressed as $EA \cos \theta$. Furthermore, the electric field is perpendicular to the area A at all points, so $\cos \theta = \cos 0° = 1$ at all points on the gaussian surface. Thus, the left side of Equation 16.8 reduces to

$$EA = E4\pi r^2$$

where $4\pi r^2$ is the surface area of the gaussian surface. Recall that the Q in Gauss' law is equal to the *net* charge enclosed by the gaussian surface, so the right side of Equation 16.8 reduces to

$$\frac{Q}{\epsilon_0} = \frac{q}{\epsilon_0}$$

and we have

$$E4\pi r^2 = \frac{q}{\epsilon_0}$$

or

$$E = \frac{q}{4\pi r^2 \epsilon_0} = k\frac{q}{r^2},$$

which is the equation used to calculate the magnitude of an electric field set up by a point charge q used earlier in this chapter.

The following examples should clarify the procedure for applying Gauss' law to some simple situations.

Gaussian surface

Figure 16.29
A spherical surface of radius r surrounding a point charge q. When the charge is at the center of the sphere, the electric field is normal to the surface and constant in magnitude everywhere on the surface.

Example 16.7   The Electric Field of a Charged Thin Spherical Shell

A thin spherical shell of radius a has a total charge q distributed uniformly over its surface (Fig. 16.30a). Find the electric field at points (a) outside and (b) inside the shell.

Solution
(a) The calculation of the field outside the shell is identical to that carried out in the body of the text for a point charge. Since the charge distribution is spherically symmetric, we select a spherical gaussian surface of radius r, concentric with the shell,

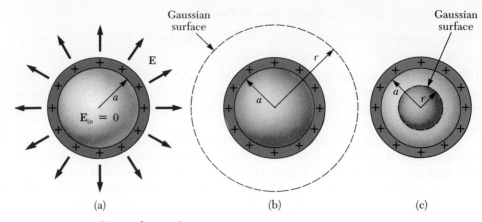

Figure 16.30   (Example 16.7)
(a) The electric field inside a uniformly charged spherical shell is *zero*. The field outside is the same as that of a point charge having a total charge Q located at the center of the shell. (b) The construction of a gaussian surface for calculating the electric field *outside* a spherical shell. (c) The construction of a gaussian surface for calculating the electric field *inside* a spherical shell.

as in Figure 16.30b. Following the same line of reasoning as that for a point charge, the left side of Gauss' law reduces to

$$\Sigma EA \cos \theta = E4\pi r^2$$

and the right side becomes

$$\frac{Q}{\epsilon_0} = \frac{q}{\epsilon_0}$$

Thus, we have

$$E4\pi r^2 = \frac{q}{\epsilon_0}$$

or

$$E = \frac{q}{4\pi r^2\epsilon_0} = k\frac{q}{r^2}$$

Therefore, the field at a point outside the shell is equivalent to that of a point charge located at the center of the shell.

   (b) *The electric field inside the spherical shell is zero.* This follows from Gauss' law applied to a spherical gaussian surface of radius r placed inside the shell, as shown in Figure 16.30c. Since the net charge inside the surface is zero, $Q = 0$, we see that $\mathbf{E} = 0$ in the region inside the shell.

Example 16.8   A Nonconducting Plane Sheet of Charge

Find the electric field due to a nonconducting infinite plane sheet of charge with uniform charge per unit area σ.

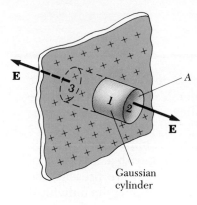

Figure 16.31   (Example 16.8)
A cylindrical gaussian surface penetrating an infinite sheet of charge. The flux through each end of the gaussian surface is *EA*. There is no flux through the cylinder's surface.

Solution

The symmetry of the situation shows that the electric field must be perpendicular to the plane and that the direction of the field on one side of the plane must be opposite its direction on the other side, as shown in Figure 16.31. It is convenient to choose for our gaussian surface a small cylinder whose axis is perpendicular to the plane and whose ends each have an area *A*. The left side of Gauss' law, $\Sigma EA \cos \theta$, now is the sum of three terms, for each of the three surfaces labeled (1), (2), and (3) in Figure 16.31. Thus,

$$\Sigma EA \cos \theta = (EA \cos \theta)_1 + (EA \cos \theta)_2 + (EA \cos \theta)_3$$

But, we see that since **E** is parallel to the cylindrical surface, surface 1, $\cos \theta = \cos 90° = 0$, so $(EA \cos \theta)_1 = 0$.

Now consider surface (2). For this surface, we see that $\cos \theta = \cos 0° = 1$, and

$$(EA \cos \theta)_2 = EA$$

By the same reasoning, for surface (3),

$$(EA \cos \theta)_3 = EA$$

Therefore, the left side of Gauss' law becomes

$$(EA \cos \theta)_2 + (EA \cos \theta)_3 = 2EA$$

Since the total charge *inside* the gaussian surface is σA, the right side of Gauss' law becomes

$$\frac{Q}{\epsilon_0} = \frac{\sigma A}{\epsilon_0}$$

Applying Gauss' law to this situation gives

$$2EA = \frac{\sigma A}{\epsilon_0}$$

or,

$$E = \frac{\sigma}{2\epsilon_0}$$

Since the distance of the surfaces from the plane does not appear in our result, we conclude that $E = \sigma/\epsilon_0$ at *any* distance from the plane. That is, the electric field is uniform everywhere.

Problem-Solving Strategy: Electric Forces and Fields

1. **Units:** When performing calculations that involve the use of the Coulomb constant k that appears in Coulomb's law, charges must be in coulombs and distances in meters. If they are given in other units, you must convert them to SI.

2. **Applying Coulomb's law to point charges:** It is important to use the superposition principle properly when dealing with a collection of interacting point charges. When several charges are present, the resultant force on any one of them is found by finding the individual force that every other charge exerts on it and then finding the vector sum of all these forces. The magnitude of the force that any charged object exerts on another is given by Coulomb's law, and the direction of the force is found by noting that the forces are repulsive between like charges and attractive between unlike charges.

3. **Calculating the electric field of point charges:** Remember that the superposition principle can also be applied to electric fields, which are also vector quantities. To find the total electric field at a given point, first calculate the electric field at the point due to each individual charge. The resultant field at the point is the vector sum of the fields due to the individual charges.

Summary

Electric charges have the following important properties:

1. Unlike charges attract one another, and like charges repel one another.
2. Electric charge is always conserved. That is, charge is never created or destroyed.
3. The smallest unit of negative charge is the charge on an electron, which has the value -1.60×10^{-19} C. The charge on the proton is identical in magnitude, but it is positive ($+1.60 \times 10^{-19}$ C).

Conductors are materials through which charges can move freely. Some examples of good electric conductors are copper, aluminum, and silver. **Insulators** are materials that do not readily allow charge to move through them. Some examples of good electric insulators are glass, rubber, and wood.

Coulomb's law states that the electric force between two stationary charged particles separated by a distance r has a magnitude given by

$$F = k \frac{|q_1||q_2|}{r^2} \tag{16.1}$$

where $|q_1|$ and $|q_2|$ are the magnitudes of the charges on the particles in coulombs and k is the **Coulomb constant,** which has the value

$$k \approx 9.0 \times 10^9 \ \text{N} \cdot \text{m}^2/\text{C}^2 \qquad \text{(16.3)}$$

The magnitude of the **electric field E** at some point in space is defined as the magnitude of the electric force that acts on a small positive charge placed at that point divided by the magnitude of its charge, $|q_0|$.

$$E \equiv \frac{|\mathbf{F}|}{|q_0|} \qquad \text{(16.4)}$$

The direction of the electric field at a point in space is defined to be in the direction of the electric force that would be exerted on a small positive charge placed at that point.

The electric field due to a *point charge q*, at a distance r from the point charge, has a magnitude

$$E = k \frac{|q|}{r^2} \qquad \text{(16.5)}$$

Electric field lines are useful for describing the electric field in any region of space. These lines are related to the electric field as follows: (1) the electric field is tangent to the electric field lines at every point and (2) the number of electric field lines per unit area through a surface perpendicular to the lines is proportional to the strength of the electric field in that region.

A conductor in electrostatic equilibrium has the following properties:

1. The electric field is zero everywhere inside the conductor.
2. Any excess charge on an isolated conductor must reside entirely on its surface.
3. The electric field just outside a charged conductor is perpendicular to the conductor's surface.
4. On an irregularly shaped conductor, charge tends to accumulate where the curvature of the surface is greatest, that is, at sharp points.

Gauss' Law states that the net electric flux through any closed surface is equal to the net charge inside the surface divided by ϵ_0, where ϵ_0 is the permittivity of free space.

Additional Reading

W. F. Magie, *Source Book in Physics,* Cambridge, MA, Harvard University Press, 1963. This includes extracts from the works of Coulomb and others.

H. W. Meyer, *History of Electricity and Magnetism,* Cambridge, MA, MIT Press, 1971.

A. D. Moore, "Electrostatics," *Sci. American,* March 1972, p. 46.

A. F. Moyer, "Benjamin Franklin: Let the Experiment Be Made," *Am. J. Phys.* 44:536, 1976.

R. W. Purcell, "Bolts From the Blue," *The Sciences,* July/August 1986, p. 48.

D. Roller and D. H. D. Roller, *The Development of the Concept of Electric Charge,* Cambridge, MA, Harvard University Press, 1954.

Questions

1. Explain from an atomic viewpoint why charge is usually transferred by electrons.
2. If a metal object receives a positive charge, does its mass increase, decrease, or stay the same? What happens to the mass if the object is given a negative charge?
3. When the air is very dry, sparks are often observed (or heard) when clothes are removed in the dark. Explain.
4. Because of a higher moisture content, air is a better conductor of electricity in the summer than in the winter. Would you expect the shocks from static electricity to be more severe in summer or winter? Discuss.
5. A balloon is negatively charged by rubbing and then clings to a wall. Does this mean that the wall is positively charged? Why does the balloon eventually fall?
6. A charged comb will often attract small bits of dry paper that fly away when they touch the comb. Explain.
7. If a suspended object A is attracted to object B, which is charged, can we conclude that object A is charged? Explain.
8. A light piece of aluminum foil is draped over a wooden rod. When a rod with a positive charge is brought close to the foil, the foil leaves stand apart. Why? What kind of charge is on the foil?
9. What is the difference between charging an object by induction and charging by conduction?
10. Personnel must wear special conducting shoes while working around oxygen in an operating room. Why? Contrast with what might happen when personnel wear rubber shoes.
11. A student stands on a piece of insulating material, places his hand on top of a Van de Graaff generator, and then turns on the generator. Will he be shocked? Why or why not? He will find that his hair stands on end as in the photograph. Why does this occur? Why

does the second student, who is also on an insulated stand, find that his hair stands up when he touches the first student?
12. It has been reported that in some instances people near where a lightning bolt strikes the Earth have had their clothes thrown off. Explain why this might happen.
13. Would life be different if the electron were positively charged and the proton were negatively charged? Could we identify a difference? Would any physical laws be different?
14. Compare and contrast Newton's law of universal gravitation with Coulomb's force law.
15. Assume that someone proposes a theory that says people are bound to the Earth by electric forces rather than by gravity. How could you prove this theory wrong?
16. When defining the electric field, why is it necessary to specify that the magnitude of the test charge be very small (i.e., take the limit as $q \rightarrow 0$)?
17. Is it possible for an electric field to exist in empty space? Explain.
18. Explain why electric field lines never cross. (*Hint:* **E** must have a unique direction at all points.)
19. A "free" electron and "free" proton are placed in an identical electric field. Compare the electric forces on each particle. Compare their accelerations.
20. In Figure 16.17, where do the extra lines leaving the charge $+2q$ end?
21. Are the occupants of a steel-frame building safer than those in a wood-frame house during an electrical storm or vice versa? Explain.
22. Suppose someone tells you that it is the rubber tires on a car that make a car safe during a lightning storm. What would be your response? Do the relatively new steel-belted tires on a car affect the safety of an occupant in a car?
23. Why should a ground wire be connected to the metal support rod for a television antenna?
24. Why should you get out of a swimming pool during a lightning storm?
25. Why is it not a good idea to seek shelter under a tree during a lightning storm?

Question 11. (© *L. V. Bergman/Fran Heyl Associates*)

Problems

Section 16.3   Coulomb's Law

1. Calculate the net charge on a substance consisting of (a) 5×10^{14} electrons and (b) a combination of 7×10^{13} protons and 4×10^{13} electrons.

2. Two point charges of magnitude 3×10^{-9} C and 6×10^{-9} C are separated by a distance of 0.3 m. Find the electric force of repulsion between them.

3. A 4.5×10^{-9} C charge is located 3.2 m from a -2.8×10^{-9} C charge. Find the electrostatic force exerted by one charge on the other.

4. Two identical conducting spheres are placed with a distance of 0.3 m between their centers. One is given a charge of 12×10^{-9} C and the other a charge of -18×10^{-9} C. (a) Find the electrostatic force exerted on one sphere by the other. (b) The spheres are connected by a conducting wire. After equilibrium has occurred, find the electrostatic force between the two.

5. Suppose that 1 g of hydrogen is separated into electrons and protons. Suppose also that the protons are placed at the Earth's north pole and the electrons are placed at the south pole. What is the resulting compressional force on the Earth?

6. Determine what the mass of a pair of protons would be if the gravitational and electrical forces between them were equal in magnitude.

7. The Moon and Earth are bound together by gravity. If, instead, the force of attraction was the result of each having a charge of the same magnitude, yet opposite in sign, find the quantity of charge that would have to be placed on each to produce the required force.

8. An electron is released above the surface of the Earth. A second electron directly below it exerts an electrostatic force on the first just great enough to cancel the gravitational force on it. How far below the first electron is the second?

9. A 2.2×10^{-9} C charge is located on the x axis at $x = -1.5$ m, a 5.4×10^{-9} C charge is located on the x axis at $x = 2$ m, and a 3.5×10^{-9} C charge is located at the origin. Find the net force on the 3.5×10^{-9} C charge.

10. Calculate the magnitude and direction of the Coulomb force on each of the three charges in Figure 16.32.

Figure 16.32   Problems 10 and 27

11. A molecule of DNA (deoxyribonucleic acid) is 2.17 μm long. The ends of the molecule become singly ionized—negative on one end, positive on the other. The helical molecule acts like a spring and compresses 1.0% upon becoming charged. Determine the effective spring constant of the molecule.

12. Three charges are arranged as shown in Figure 16.33. Find the magnitude and direction of the electrostatic force on the charge at the origin.

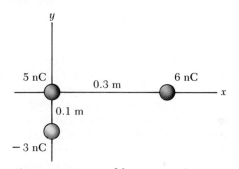

Figure 16.33   Problems 12 and 36

13. Three charges are arranged as shown in Figure 16.34. Find the magnitude and direction of the electrostatic force on the 6-nC charge.

Figure 16.34
Problem 13

14. In the Bohr theory of the hydrogen atom, an electron moves in a circular orbit about a proton, where the radius of the orbit is 0.51×10^{-10} m. (a) Find the electrostatic force between the two. (b) If this force serves as the centripetal force on the electron, what is the speed of the electron?

15. Two point charges of equal magnitude repel each other with a force of 2 N when separated by 5 cm. Find the magnitude of the charge on each.

16. Find (a) the total number of electrons in one gram of copper and (b) the total charge of all these electrons.

17. A charge of 2×10^{-9} C is placed at the origin, and a charge of 4×10^{-9} C is placed at $x = 1.5$ m. Find the location of the point between the two charges where a charge of 3×10^{-9} C should be placed so that the net electric force on it is zero.

18. Two small metallic spheres each of mass 0.2 g are suspended by light strings as pendulums from a common point. Each is given the same electric charge, and it is found that the two come to equilibrium when each string is at an angle of 5° with the vertical. If the length of each string is 30 cm, what is the magnitude of the charge on each sphere?

19. A charge of 6×10^{-9} C and a charge of -3×10^{-9} C are separated by a distance of 60 cm. Find the position at which a third charge of 12×10^{-9} C can be placed so that the net electrostatic force on it is zero.

20. The total charge on two spheres is 600 μC. The two spheres are placed 0.9 m apart and the force of repulsion between the two is 30 N. What is the charge on each sphere?

Section 16.5   The Electric Field

21. The electric force on a point charge of 5×10^{-9} C at some point is 3.8×10^{-3} N in the positive x direction. What is the magnitude of the electric field at this location?

22. The magnitude of the electric field at a certain location is 500 N/C and the field is directed east to west. (a) Find the magnitude and direction of the force acting on a proton placed at this point. (b) Repeat for an electron located at the point.

23. Find the magnitude and direction of the electric field at a distance of 10 cm from an electron.

24. The electric field at a distance of 0.8 m from a certain charge is found to have a magnitude of 200 N/C. What is the magnitude of the charge?

25. What are the magnitude and direction of the electric field set up by the proton at the position of the electron in the hydrogen atom? (See Problem 14.)

26. Find the electric field at a point midway between two charges of (a) $+30 \times 10^{-9}$ C and $+60 \times 10^{-9}$ C separated by 30 cm. (b) Repeat when the magnitude of the second charge is -60×10^{-9} C.

27. (a) Determine the electric field strength at a point 1.00 cm to the left of the middle charge shown in Figure 16.32. (b) If a charge of -2.00 μC is placed at this point, what are the magnitude and direction of the force on it?

28. What are the magnitude and direction of the electric field that will balance the weight of (a) an electron and (b) a proton? (Use the data in Table 16.1.)

29. A piece of aluminum foil of mass 5×10^{-2} kg is suspended by a string in an electric field directed vertically upward. If the charge on the foil is 3 μC, find the strength of the field that will reduce the tension in the string to zero.

30. An object having a net charge of 24 μC is placed in a uniform electric field of 610 N/C directed vertically. What is the mass of this object if it "floats" in this electric field?

31. An electron is accelerated by a constant electric field of magnitude 300 N/C. (a) Find the acceleration of the electron. (b) Use the equations of motion with constant acceleration to find its speed after 10^{-8} s, assuming it starts from rest.

32. A constant electric field directed along the positive x axis has a strength of 2000 N/C. Find (a) the force exerted on the proton by the field, (b) the acceleration of the proton, and (c) the time required for the proton to reach a speed of 10^6 m/s, assuming it starts from rest.

33. A proton accelerates from rest in a uniform electric field of 640 N/C. At some later time, its speed is 1.20×10^6 m/s. (a) Find the acceleration of the proton. (b) How long does it take the proton to reach this velocity? (c) How far has it moved in this time? (d) What is its kinetic energy at this time?

34. Consider an electron which is released from rest in a uniform electric field. (a) If the electron is accelerated to 1% of the speed of light after traveling 2 mm, what is the strength of the electric field? (b) What speed will the electron have after traveling 4 mm from rest?

35. The electrons in a particle beam each have a kinetic energy of 1.6×10^{-17} J. What are the magnitude and direction of the electric field that will stop these electrons in a distance of 10 cm?

36. Find the net electric field exerted on the charge at the origin in Figure 16.33.

37. Positive charges are located at three corners of a rectangle as shown in Figure 16.35. Find the electric field at the fourth corner.

6 nC

0.2 m

0.6 m

3 nC   5 nC

Figure 16.35
Problem 37

38. Repeat Problem 37 if the 5-nC charge is replaced by a negative charge of the same magnitude.

39. Three identical charges ($q = -5$ μC) are located along a circle of 2-m radius at angles of 30°, 150°, and 270° as shown in Figure 16.36. What is the resultant electric field at the center of the circle?

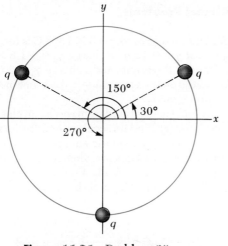

Figure 16.36   Problem 39

40. Three identical point charges $(q = +2.7\,\mu C)$ are placed on the corners of an equilateral triangle whose sides have a length $a = 35$ cm as shown in Figure 16.37. (a) At what point in the plane of the charges (other than ∞) is the electric field zero? (b) What are the magnitude and direction of the electric field at the point P *due to the two charges at the base of the triangle?*

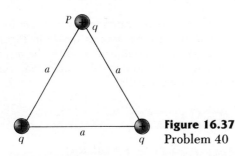

Figure 16.37
Problem 40

41. Three charges are at the corners of an equilateral triangle as shown in Figure 16.38. Calculate the electric field at a point midway between the two charges on the x axis.

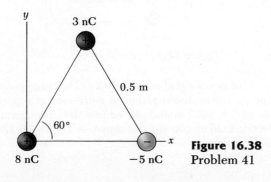

Figure 16.38
Problem 41

42. What is the strength of the electric field that would give a proton an acceleration equal to that of gravity?

43. In Figure 16.39, determine the point (other than infinity) at which the total electric field is zero.

Figure 16.39   Problem 43

44. Repeat Problem 43 for the case in which the 6-μC charge is replaced by a charge of $-6\,\mu C$.

Section 16.6   Electric Field Lines
Section 16.7   Conductors in Electrostatic Equilibrium

45. (a) Sketch the electric field lines around an isolated 1-μC positive point charge. (b) Sketch the electric field pattern around an isolated negative point charge of magnitude $-2\,\mu C$.

46. (a) Sketch the electric field pattern around two positive point charges of magnitude 1 μC placed close together. (b) Sketch the electric field pattern around two negative point charges of magnitude $-2\,\mu C$ placed close together. (c) Sketch the pattern around a 1-μC positive point charge and a -2-μC charge placed close together.

47. Figure 16.40 shows the electric field lines for two point charges separated by a small distance. (a) Determine the ratio q_1/q_2. (b) What are the signs of q_1 and q_2?

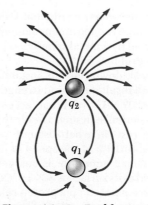

Figure 16.40   Problem 47

48. (a) Sketch the electric field pattern set up by a positively charged hollow sphere. Include a sketch of the lines both inside and outside the sphere. (b) A conducting cube is given a positive charge. Sketch the electric field pattern both inside and outside the cube.

49. Consider a rod of finite length having a uniform negative charge. Sketch the pattern of the electric field lines in a plane containing the rod.

50. Two point charges are a small distance apart. (a) Sketch the electric field lines for the two if one has a charge four times that of the other and both charges are positive. (b) Repeat if both charges are negative.

51. Refer to Figure 16.20. The charge lowered into the center of the hollow conductor has a magnitude of $5 \mu C$. Find the magnitude and sign of the charge on the inside and outside of the hollow conductor when the charge is as shown in (a) Figure 16.20a, (b) Figure 16.20b, (c) Figure 16.20c, and (d) Figure 16.20d.

52. Repeat Problem 51 if the charge has a magnitude of $-5 \mu C$.

Section 16.8   The Van de Graaff Generator

53. The dome of a Van de Graaff generator receives a charge of 2×10^{-4} C. Find the strength of the electric field (a) inside the dome, (b) at the surface of the dome, assuming it has a radius of 1 m, and (c) 4 m from the center of the dome. (*Hint:* Refer to Section 16.7 to review properties of conductors in electrostatic equilibrium. Also use the fact that the points on the surface are outside a spherically symmetric charge distribution; the total charge may be considered as being located at the center of the sphere.)

54. If the electric field strength in air exceeds 3×10^6 N/C, the air becomes a conductor. Using this information, determine the maximum amount of charge that can be carried by a metal sphere 2.0 m in radius. (See the hint in Problem 53.)

55. Air will break down (lose its insulating quality) and sparking will result if the field strength is increased to about 3×10^6 N/C. (a) What acceleration will an electron experience in such a field? (b) If the electron starts from rest, in what distance will it acquire a speed equal to 10% of the speed of light?

56. A Van de Graaff generator is charged such that the electric field at its surface is 3×10^4 N/C. Find (a) the electric force exerted on a proton released at its surface and (b) the acceleration of the proton at this instant of time.

Additional Problems

57. A point charge $q_1 = -3.5 \times 10^{-9}$ C is located on the y axis at $y = 0.12$ m, a charge $q_2 = -1.8 \times 10^{-9}$ C is located at the origin, and a charge $q_3 = 2.6 \times 10^{-9}$ C is located on the x axis at $x = -0.12$ m. Find the resultant force on q_3.

58. A proton is accelerated from rest by an electric field of 400 N/C. (a) Find its acceleration, (b) its speed after 10^{-8} s, and (c) its kinetic energy at this time.

59. Three point charges are aligned along the x axis as shown in Figure 16.41. Find the electric field at (a) the position (2, 0) and (b) the position (0, 2).

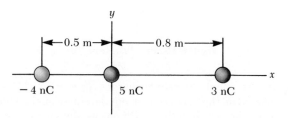

Figure 16.41   Problem 59

60. Three point charges lie along the y axis. A charge $q_1 = -9 \mu C$ is at $y = 6.0$ m, and a charge $q_2 = -8 \mu C$ is at $y = -4.0$ m. Where must a third positive charge, q_3, be placed such that the resultant force on it is zero?

61. A small 2-g plastic ball is suspended by a 20-cm-long string in a uniform electric field as shown in Figure 16.42. If the ball is in equilibrium when the string makes a 15° angle with the vertical as indicated, what is the net charge on the ball?

Figure 16.42   Problem 61

62. Two charges of equal magnitude 2 nC but of opposite sign are connected together at a distance of separation of 3×10^{-8} m and placed in a uniform upward electric field of 200 N/C, as shown in Figure 16.43a.

(a) Find the net torque on the charges. (b) Repeat when the charges are aligned with respect to the field as shown in Figure 16.43b.

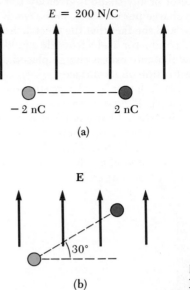

$E = 200$ N/C

-2 nC 2 nC

(a)

E

30°

Figure 16.43
(b)   Problem 62

63. Three point charges lie along the y axis. A charge $q_1 = -3$ μC is at $y = 5.0$ m, and a charge $q_2 = 8$ μC is at $y = 2.0$ m. Where must a positive charge q_3 be placed such that the resultant force on it is zero?

64. A proton is shot vertically upward with a speed of 2×10^5 m/s in a downward-directed electric field of 500 N/C. How high will it rise?

65. Two small spheres each of mass 2 g are suspended by light strings 10 cm in length (Fig. 16.44). A uniform electric field is applied in the x direction. If the spheres have charges of -5×10^{-8} C and $+5 \times 10^{-8}$ C, determine the electric field intensity that enables the spheres to be in equilibrium at $\theta = 10°$.

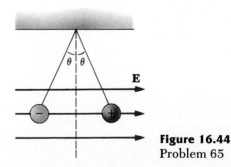

θ θ

E

Figure 16.44
Problem 65

66. Four point charges are situated at the corners of a square of sides a as in Figure 16.45. Find the expression for the resultant force on the positive charge q.

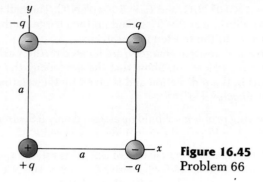

Figure 16.45
Problem 66

67. An electron traveling with an initial speed equal to 4.0×10^6 m/s enters a region of a uniform electric field of magnitude 2.5×10^4 N/C. The direction of travel of the electron is in the same direction as the field. (a) Find the acceleration of the electron. (b) Determine the time it takes for the electron to come to rest after it enters the field. (c) How far does the electron move in the electric field before coming to rest?

68. Two point charges like those shown in Figure 16.46 are called an electric dipole. Show that the electric field at a distant point along the x axis is given by the expression $E_x = 4kqa/x^3$.

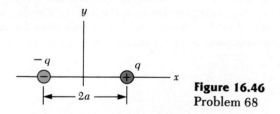

Figure 16.46
Problem 68

69. A charged cork ball of mass 1 g is suspended on a light string in the presence of a uniform electric field as in Figure 16.47. The electric field has components,

Figure 16.47
Problem 69

$E_x = 3 \times 10^5$ N/C, and $E_y = 5 \times 10^5$ N/C. The ball is in equilibrium at $\theta = 37°$. Find (a) the charge on the ball and (b) the tension in the string.

70. Two equal positive charges q are located on the x axis at $x = a$ and $x = -a$. Show that the field along the y axis is in the y direction and is given by the relation $E_y = 2kqy(y^2 + a^2)^{-3/2}$.

The following problems should be assigned only if Section 16.10, Electric Flux and Gauss' Law, is covered.

71. A flat surface having an area of 3.2 m² is rotated in a uniform electric field of intensity $E = 6.2 \times 10^5$ N/C. (a) Show that the electric flux through this area equals 1.98×10^6 N·m²/C when the electric field is perpendicular to the surface and (b) is equal to zero when the electric field is parallel to the surface.

72. A 40-cm diameter loop is rotated in a uniform electric field until the position of maximum electric flux is found. The flux in this position is measured to be 5.2×10^5 N·m²/C. Show that the electric field strength is 4.14×10^6 N/C.

73. A point charge of $+5\ \mu$C is located at the center of a sphere with a radius of 12 cm. Show that the electric flux through the surface of the sphere is 5.65×10^5 N·m²/C.

74. A point charge of magnitude q is located at the center of a spherical shell of radius a, which has a charge uniformly distributed on its surface of magnitude $-q$. Show that the electric field (a) equals zero for all points outside the spherical shell and (b) has the value

$E = kq/r^2$ for a point inside the shell a distance r from the center.

75. Show that the electric field just outside the surface of a good conductor of any shape is given by $E = \sigma/\epsilon_0$, where σ is the charge per unit area on the conductor.

76. Use Gauss' law and the fact that the electric field inside any closed conductor in electrostatic equilibrium is zero to show that any excess charge placed on the conductor must reside on its surface.

77. An infinite plane conductor has charge spread out on its surface as shown in Figure 16.48. Use Gauss' law to show that the electric field at any point outside the conductor is given by $E = \sigma/\epsilon_0$, where σ is the charge per unit area on the conductor. (*Hint:* Choose a gaussian surface in the shape of a cylinder with one end inside the conductor and one end outside the conductor.)

Figure 16.48
Problem 77

CHAPTER 17

Electrical Energy and Capacitance

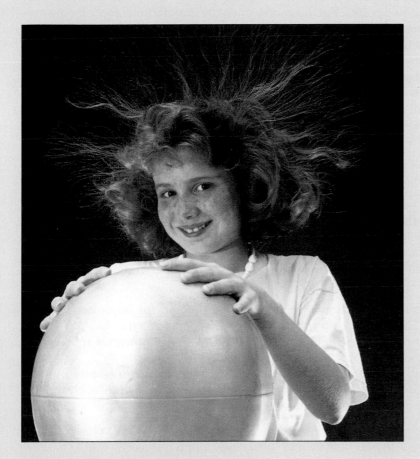

Jennifer is holding onto a charged sphere that reaches a potential of about 100,000 volts. The device that generates this high potential is called a Van de Graaff generator. Why do you suppose Jennifer's hair stands on end like the needles of a porcupine? Why is it important that she stand on a pedestal insulated from ground?
(Courtesy of Henry Leap and Jim Lehman)

The concept of potential energy was first introduced in Chapter 5. A potential energy function can be defined for all conservative forces, such as the force of gravity. By using the law of conservation of energy, we were often able to avoid working directly with forces when solving various problems. In this chapter we shall see that the energy concept is also useful in the study of electricity. Because the Coulomb force is conservative, we shall be able to define an electrical potential energy corresponding to the Coulomb force. This concept of potential energy will be of value to us, but perhaps of even more value will be a quantity called electric potential, defined as the potential energy per unit charge.

Electric potential is of great practical value in dealing with electric circuits. For example, when speaking of a voltage applied between two points, we are actually referring to an electric potential difference between those points. We shall take our first steps toward an investigation of circuits with a discussion of electric potential, and we shall carry this forward with an investigation of a common circuit element called a capacitor.

17.1   Potential Difference and Electric Potential

In Chapter 5 we showed that the gravitational force is a conservative force. As you may recall, this means that the work this force does on an object depends only on the initial and final positions of the object and not on the path followed between the two points. Furthermore, since the gravitational force is conservative, it is possible to define a potential energy function, which we call gravitational potential energy. Because the Coulomb force is of the same form as the universal law of gravity, it follows that *the electrostatic force is also conservative.* Therefore, it is possible to define an electrical potential energy function associated with this force.

We shall consider potential energy from the point of view of the particular situation shown in Figure 17.1. Imagine a small positive charge placed at point A in a uniform electric field of magnitude E. As the charge moves from point A to point B under the influence of the electric force, qE, exerted on it, the work done on the charge by the electric force is

Figure 17.1
When a charge q moves in a uniform electric field **E** from point A to point B, the work done on the charge by the electric force is qEd.

$$W = Fd = qEd$$

where d is the distance between A and B.

By definition, *the work done by a conservative force equals the negative of the change in potential energy,* ΔPE. Therefore we find the change in electrical potential energy to be

Change in potential energy

$$\Delta PE = -W = -qEd \qquad \textbf{(17.1)}$$

Note that although potential energy can be defined for any electric field, *Equation 17.1 is valid only for the case of a uniform electric field.* We shall examine situations in which the electric field is not uniform in subsequent sections.

We shall have occasion to use electrical potential energy often in the coming pages, but of even more practical importance in the study of electricity is the concept of electric potential.

> The potential difference between points A and B, $V_B - V_A$, is defined as the change in potential energy (final value minus initial value) of a charge q moved from A to B, divided by the charge.

Potential difference between two points

$$\Delta V \equiv V_B - V_A = \frac{\Delta PE}{q} \qquad \textbf{(17.2)}$$

Potential difference should not be confused with potential energy. The change in potential between two points is proportional to the change in potential energy of a charge as it moves between the points, and we see from Equation 17.2 that the two are related as $\Delta PE = q\,\Delta V$. Because potential energy is a scalar quantity, *electric potential is also a scalar quantity.* From Equation 17.2 we see that potential difference is a measure of energy per unit charge, or, as an alternative viewpoint, electrical potential difference is the work done to move a charge from a point A to a point B divided by the magnitude of the change. Thus, the SI units of potential are joules per coulomb, defined to be equal to a unit called the volt (V)[1]:

[1] Note that the symbol V (italic letter) represents potential, while V (Roman letter) is the symbol for the units of this quantity (volts). The two symbols should not be confused with each other.

$$1\ V \equiv 1\ J/C \qquad\qquad \textbf{(17.3)}$$

This says that 1 J of work must be done to move a 1-C charge between two points that are at a potential difference of 1 V. In the process, the 1-C charge gains (or loses) 1 J of energy as it moves through a potential difference of 1 V. Dividing Equation 17.1 by q gives

$$\frac{\Delta PE}{q} = V_B - V_A = -Ed \qquad\qquad \textbf{(17.4)}$$

This equation shows that potential difference also has units of electric field times distance. From this, it follows that the SI unit of electric field (N/C) can also be expressed as volts per meter:

$$1\ N/C = 1\ V/m$$

Since Equation 17.4 is directly related to Equation 17.1, it, too, is valid only for the case of a uniform field.

Let us examine the changes in energy associated with movements in the electric field pictured in Figure 17.2. Because the positive charge q tends to move in the direction of the electric field, we would have to apply an upward external force on the charge to move it from A to B. Since work is done on the charge, this means that *a positive charge will gain electrical potential energy when it is moved in a direction opposite the electric field.* This is analogous to a mass gaining gravitational potential energy when it rises to higher elevations in the presence of gravity. If a positive charge is released from rest at point A, it experiences a force qE in the direction of the field (downward in Figure 17.2). Therefore, it accelerates downward, gaining kinetic energy. *As it gains kinetic energy, it loses an equal amount of potential energy.*

On the other hand, if the test charge q is negative, the situation is reversed. *A negative charge loses electrical potential energy when it moves in the direction opposite the electric field.* That is, a negative charge released from rest in the field **E** accelerates in a direction *opposite* the field.

Let us pause briefly to discuss a situation that illustrates the concept of potential difference. Consider the common 12-V automobile battery. Such a battery maintains a potential difference across its terminals, where the positive terminal is 12 V higher in potential than the negative terminal. In practice, the negative terminal is usually connected to the metal body of the car, which can be considered at a potential of zero volts. The battery becomes a useful device when connected by conducting wires to such things as light bulbs, a radio, powered windows, and so forth. Now consider a charge of + 1 C which is to be moved around a circuit which contains the battery connected to some external devices. As the charge is moved inside the battery from the negative terminal (at 0 V) to the positive terminal (at 12 V), the work done on the charge by the battery is 12 J. Thus, every coulomb of positive charge that leaves the positive terminal of the battery carries an energy of 12 J. As this charge moves through the external circuit toward the negative terminal, it gives up its 12 J of electrical energy to the external devices. When the charge reaches the negative terminal, its electrical energy is zero. At this point, the battery takes over and again replenishes 12 J of energy to the charge as it is moved from the negative to the positive terminal, enabling it to make another transit of the circuit. The actual amount of charge that leaves the battery

Figure 17.2
If we wish to move a positive charge q from A to B, against the external electric field **E**, we have to apply an upward force on the charge. The electric potential energy of the positive charge increases as it is moved in the direction opposite the electric field.

depends on the properties of the external devices as we shall see in the next chapter.

Example 17.1   The Field Between Two Parallel Plates of Opposite Charge

Figure 17.3 (Example 17.1) A 12-V battery connected to two parallel plates. The electric field **E** between the plates has a magnitude given by the potential difference divided by the plate separation, d.

Figure 17.3 illustrates a situation in which a constant electric field can be set up. In this situation, a 12-V battery is connected between two parallel metal plates separated by 0.3 cm. Find the strength of the electric field.

Solution
The electric field is uniform (except near the edges of the metal plates), and thus the relationship between potential difference and the magnitude of the field is given by Equation 17.4:

$$V_B - V_A = -Ed$$

As noted above, chemical forces inside a battery maintain one electrode, called the positive terminal, at a higher potential than a second electrode, the negative terminal. Thus, in Figure 17.3, plate B, which is connected to the negative terminal, must be at a lower potential than plate A, which is connected to the positive terminal. That is,

$$V_B - V_A = -12 \text{ V}$$

This gives a value for E of

$$E = -\frac{(V_B - V_A)}{d} = -\frac{(-12 \text{ V})}{0.3 \times 10^{-2} \text{ m}} = \boxed{4 \times 10^3 \text{ V/m}}$$

The direction of this field is from the positive plate to the negative plate. A device consisting of two plates separated by a small distance is called a *parallel-plate capacitor*. We shall examine these devices later in this chapter.

Example 17.2   Motion of a Proton in a Uniform Electric Field

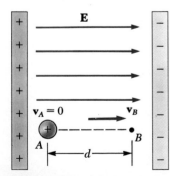

Figure 17.4 (Example 17.2) A proton accelerates from A to B in the direction of the uniform electric field.

A proton is released from *rest* in a uniform electric field of magnitude 8×10^4 V/m directed along the positive x axis (Fig. 17.4). The proton undergoes a displacement of 0.5 m in the direction of the field. (a) Find the *change* in the electric potential of the proton as a result of this displacement.

Solution
From Equation 17.4, we have

$$\Delta V = V_B - V_A = -Ed = -(8 \times 10^4 \text{ V/m})(0.5 \text{ m}) = \boxed{-4 \times 10^4 \text{ V}}$$

Thus, the electric potential of the proton *decreases* as it moves from A to B.

(b) Find the change in potential energy of the proton for this displacement.

Solution

$$\Delta PE = q \, \Delta V = e \, \Delta V = (1.6 \times 10^{-19} \text{ C})(-4 \times 10^4 \text{ V}) = \boxed{-6.4 \times 10^{-15} \text{ J}}$$

The negative sign here means that the potential energy of the proton decreases as it moves in the direction of the electric field. This makes sense since, as the proton *ac-*

celerates in the direction of the field, it gains kinetic energy and at the same time loses potential energy (mechanical energy is conserved).

(c) Find the speed of the proton after it has been moved 0.5 m, starting from rest.

Solution
If there are no forces acting on the proton other than the conservative electrical force, we can apply the principle of conservation of mechanical energy in the form

$$KE_i + PE_i = KE_f + PE_f$$

In our case, $KE_i = 0$, hence the expression above gives

$$KE_f = PE_i - PE_f = -\Delta PE$$

With this equation and the results of part (b), we find

$$\frac{1}{2} m v_f^2 = 6.4 \times 10^{-15} \text{ J}$$

and

$$v_f^2 = \frac{2(6.4 \times 10^{-15} \text{ J})}{1.67 \times 10^{-27} \text{ kg}} = 7.66 \times 10^{12} \text{ m}^2/\text{s}^2$$

$$v_f = \boxed{2.77 \times 10^6 \text{ m/s}}$$

17.2   Electric Potential and Potential Energy Due to Point Charges

In electric circuits, a point of zero potential is often defined by grounding (connecting to Earth) some point in the circuit. For example, if the negative plate in Example 17.1 had been grounded, its potential would be considered to have a value of zero and the positive plate to have a value of 12 V. It is also possible to define the potential at a point in space due to a point charge. In this case, the point of zero potential is taken to be at an infinite distance from the charge. With this choice, the methods of calculus can be used to show that *the electric potential due to a point charge q at any distance r from the charge is given by*

$$V = k \frac{q}{r} \qquad \qquad \textbf{(17.5)}$$

Electric potential due to a point charge

Equation 17.5 points out a significant property of electric potential that makes it an important quantity in the study of electricity. As the equation indicates, the potential at a given point depends only on the charge q on the object setting up the potential and the distance r from that object to a specific point in space. As a result, we can say that a potential exists at some point in space whether or not there is a charge located at that point.

The electric potential of two or more charges is obtained by applying the **superposition principle.** That is, *the total potential at some point P due to several point charges is the algebraic sum of the potentials due to the individual charges.* This is similar to the method used in Chapter 16 to find the resultant

Superposition principle

Figure 17.5
If two point charges q_1 and q_2 are separated by a distance r_{12}, the potential energy of the pair of charges is kq_1q_2/r_{12}.

Potential energy of a pair of charges

electric field at a point in space. However, note that in the case of potentials, one must evaluate an *algebraic sum* of individual potentials to obtain the total potential since *potentials are scalar quantities*. Thus, it is much easier to evaluate the electric potential at some point due to several charges than to evaluate the electric field, which is a vector quantity.

We now consider the potential energy of interaction of a system of two charged particles. If V_1 is the electric potential due to charge q_1 at a point P, then the work required to bring a charge q_2 from infinity to the point P without acceleration is q_2V_1. By definition, this work equals the potential energy, PE, of the two-particle system when the particles are separated by a distance r_{12} (Fig. 17.5).

Therefore, we can express the potential energy of the *pair* of charges as

$$PE = q_2V_1 = k\frac{q_1q_2}{r_{12}} \qquad (17.6)$$

Note that if the charges are of the *same* sign, PE is positive. This is consistent with the fact that like charges repel, and so positive work must be done on the system to bring two charges near one another. Conversely, if the charges are of *opposite* sign, the force is attractive and PE is negative. This means that negative work must be done to bring unlike charges close together.

Problem-Solving Strategy:
Electric Potential

1. When working problems involving electric potential, remember that potential is a *scalar quantity* (rather than a vector quantity like the electric field), so there are no components to worry about. Therefore, when using the superposition principle to evaluate the electric potential at a point due to a system of point charges, you simply take the algebraic sum of the potentials due to each charge. However, you must keep track of signs. The potential due to each positive charge is positive, while the potential due to each negative charge is negative. The basic equation to use is $V = kq/r$.
2. As in mechanics, only changes in potential are significant, hence the point where you choose the potential to be zero is arbitrary.

Example 17.3

A 5-μC point charge is located at the origin, and a point charge of -2 μC is located on the x axis at the position (3, 0) m, as in Figure 17.6. (a) If the potential is taken to be zero at infinity, find the total electric potential due to these charges at point P, whose coordinates are (0, 4) m.

Solution
The total potential at P due to the two charges is

$$V_P = k\left(\frac{q_1}{r_1} + \frac{q_2}{r_2}\right)$$

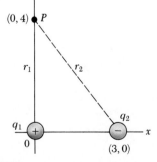

Figure 17.6   (Example 17.3) The electric potential at point P due to the point charges q_1 and q_2 is the algebraic sum of the potentials due to the individual charges.

From the data given, we find

$$V_P = \left(9 \times 10^9 \ \frac{N \cdot m^2}{C^2}\right)\left(\frac{5 \times 10^{-6} \ C}{4 \ m} - \frac{2 \times 10^{-6} \ C}{5 \ m}\right) = \boxed{7.65 \times 10^3 \ V}$$

(b) How much work is required to bring a third point charge of 4 μC from infinity to P?

Solution

$$W = q_3 V_P = (4 \times 10^{-6} \ C)(7.65 \times 10^3 \ V)$$

Since 1 V = 1 J/C, W reduces to

$$W = \boxed{3.06 \times 10^{-2} \ J}$$

Exercise Find the magnitude and direction of the electric field at the point P.

Answer 2.28×10^3 N/C at an angle of 79.1° with the x axis.

17.3 Potentials and Charged Conductors

In order to examine the potential at all points on a charged conductor, let us combine Equations 17.1 and 17.2. From Equation 17.1, we see that the work done on a charge by electric forces is related to the change in electrical potential energy of the charge as

$$W = -\Delta PE$$

Furthermore, from Equation 17.2 we see that the change in potential energy between two points A and B is related to the potential difference between these points as

$$\Delta PE = q(V_B - V_A)$$

Combining these two equations, we find

$$\boxed{W = -q(V_B - V_A)} \qquad \textbf{(17.7)}$$

As we see from this result, no work is required to move a charge between two points that are at the same potential. That is, $W = 0$ when $V_B = V_A$.

In Chapter 16 we found that, when a conductor is in electrostatic equilibrium, a net charge placed on it resides entirely on its surface. Furthermore, we showed that the electric field just outside the surface of a charged conductor in electrostatic equilibrium is perpendicular to the surface and that the field inside the conductor is zero. We shall now show that *every point on the surface of a charged conductor in electrostatic equilibrium is at the same potential.*

Consider a surface path connecting any points A and B on a charged conductor, as in Figure 17.7. The electric field, **E**, is always perpendicular to the displacement along this path; therefore no work is done if a charge is moved between these points. From Equation 17.7, we see that if the work done is zero, the difference in potential, $V_B - V_A$, is also zero. Therefore,

Figure 17.7
An arbitrarily shaped conductor with an excess positive change. When the conductor is in electrostatic equilibrium, all of the charge resides at the surface, **E** = 0 inside the conductor, and the electric field just outside the conductor is perpendicular to the surface. The potential is constant inside the conductor and is equal to the potential at the surface.

the electric potential is a constant everywhere on the surface of a charged conductor in equilibrium.

Properties of a charged conductor in equilibrium

Furthermore, because the electric field is zero inside a conductor, no work is required to move a charge between two points inside a conductor. Again, Equation 17.7 shows that if the work done is zero, the difference in potential between any two points inside a conductor must also be zero. Thus, we conclude that the electric potential is constant everywhere inside a conductor.

Finally, since one of the points could be arbitrarily close to the surface of the conductor, we conclude that

the electric potential is constant everywhere inside a conductor and equal to its value at the surface.

Consequently, no work is required to move a charge from the interior of a charged conductor to its surface. (Note that the potential is not zero inside a conductor even though the electric field is zero in its interior.)

The Electron Volt

A unit of energy commonly used in atomic and nuclear physics is the **electron volt (eV)**.

Definition of the electron volt

The **electron volt** is defined as the energy that an electron (or proton) gains when accelerated through a potential difference of 1 V.

Since $1 \text{ V} = 1 \text{ J/C}$ and since the magnitude of charge on the electron or proton is 1.6×10^{-19} C, we see that the electron volt is related to the joule through the relation

$$1 \text{ eV} = 1.6 \times 10^{-19} \text{ C} \cdot \text{V} = 1.6 \times 10^{-19} \text{ J} \qquad \textbf{(17.8)}$$

Example 17.4   An Electron in a Television Tube

An electron in the beam of a typical television picture tube is accelerated through a potential difference of 20,000 V before it strikes the face of the tube. What is the energy of this electron in electron volts, and what is its speed when it strikes the screen?

Solution
The initial potential energy of the electron in joules is

$$PE_i = qV = (1.60 \times 10^{-19} \text{ C})(20 \times 10^3 \text{ V}) = 3.20 \times 10^{-15} \text{ J}$$

Converting this to electron volts, we have

$$PE_i = (3.20 \times 10^{-15} \text{ J}) \left(\frac{1 \text{ eV}}{1.6 \times 10^{-19} \text{ J}} \right) = \boxed{20,000 \text{ eV}}$$

An inspection of the definition of the electron volt shows that this answer could have been obtained without any calculations.

The speed of the electron is found from the law of conservation of energy by equating the initial electrical potential energy of 3.20×10^{-15} J to the final kinetic energy, where the mass of the electron is 9.11×10^{-31} kg:

$$PE_i = KE_f$$

$$3.20 \times 10^{-15} \text{ J} = \frac{1}{2}(9.11 \times 10^{-31} \text{ kg})v^2$$

$$v = \boxed{8.38 \times 10^7 \text{ m/s}}$$

This is about 28% of the speed of light, whose value is 3×10^8 m/s.

---

*17.4   Equipotential Surfaces

If all points on a surface are at the same potential, the surface is called an **equipotential surface.** The potential difference between any two points on an equipotential surface is zero. Hence, *no work is required to move a charge at constant speed on an equipotential surface.* Equipotential surfaces are related to the electric field at any point in a very simple manner. *An equipotential surface associated with any charge distribution is always perpendicular to the electric field at any point.* If the electric field had a component parallel to the surface, this component of **E** would produce an electric force on a charge placed on the surface. This force would do work on the charge as it moved from one point to another, in contradiction to the definition of an equipotential surface.

It is convenient to represent equipotential surfaces on a diagram by drawing **equipotential lines,** which are two-dimensional views of the intersection of the surfaces with the plane of the drawing. Figure 17.8a shows the equipotential lines associated with a positive point charge. Note that the equipotential lines (in blue) are perpendicular to the electric field lines (in red) at all points. Recall that the potential due to a point charge q is given by $V = kq/r$. This relation shows that the potential is constant on any surface in which r is constant. Therefore, the equipotential surfaces of a point charge are a family of spheres centered on the point charge. Figure 17.8b shows the equipotential lines associated with two equal but opposite charged particles.

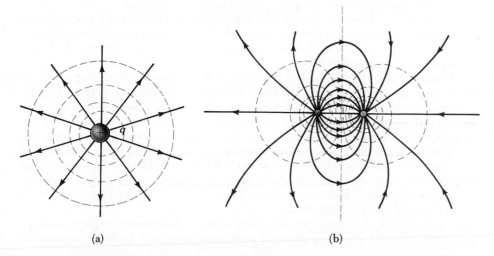

(a)   (b)

Figure 17.8
Equipotential surfaces (dashed blue lines) and electric field lines (red lines) for (a) a positive point charge and (b) two equal but opposite point charges. In all cases, the equipotential surfaces are *perpendicular* to the electric field lines at every point.

Figure 17.9
Schematic diagram of an electrostatic precipitator. The high voltage maintained on the central wires creates an electric discharge in the vicinity of the wire.

▶ S P E C I A L   T O P I C
The Electrostatic Precipitator

One important application of electric discharge in gases is a device called an *electrostatic precipitator*. This device is used to remove particulate matter from combustion gases, thereby reducing air pollution. It is especially useful in coal-burning power plants and in industrial operations that generate large quantities of smoke. Current systems are able to eliminate approximately 90% (by mass) of the ash and dust from the smoke. Unfortunately, a very high percentage of the lighter particles still escape, and these contribute significantly to smog, haze, and so forth.

Figure 17.9 shows the basic idea of the electrostatic precipitator. A high voltage (typically 40 kV to 100 kV) is maintained between a wire running down the center of a duct and the outer wall, which is grounded. The wire is maintained at a negative potential with respect to the wall, and so the electric field is directed towards the wire. The electric field near the wire reaches high enough values to cause a discharge around the wire and the formation of positive ions, electrons, and negative ions, such as O_2^-. As the electrons and negative ions are accelerated toward the outer wall by the nonuniform electric field, the dirt particles in the streaming gas become charged by collisions and ion capture. Since most of the charged dirt particles are negative, they are also drawn to the outer wall by the electric field. When the duct is shaken, the particles fall loose and are collected at the bottom.

In addition to reducing the amount of harmful gases and particulate matter in the atmosphere, the electrostatic precipitator also recovers valuable materials from the stack in the form of metal oxides. ◀

▶ S P E C I A L   T O P I C
Xerography and Laser Printers

The process of xerography is widely used for making photocopies of letters, documents, and other printed materials. The basic idea for the process was developed by Chester Carlson, who was granted a patent for his invention in 1940. In 1947, the Xerox Corporation launched a full-scale program to develop automated duplicating machines using this process. The huge success of this development is quite evident; today, practically all offices and libraries have one or more duplicating machines, and the capabilities of modern machines are on the increase.

Some features of the xerographic process involve simple concepts from electrostatics and optics. However, the one idea that makes the process unique is the use of photoconductive material to form an image. (A photoconductor is a material that is a poor conductor of electricity in the dark but becomes a reasonably good electric conductor when exposed to light.)

The sequence of steps used in the xerographic process is illustrated in Figure 17.10. First, the surface of a plate or drum is coated with a thin film of the photoconductive material (usually selenium or some compound of selenium) and the photoconductive surface is given a positive electrostatic charge in the dark (Fig. 17.10a). The page to be copied is then projected onto the charged surface (Fig. 17.10b). The photoconducting surface becomes conducting only in areas where light strikes. In these areas, the light produces charge carriers in the photoconductor, which neutralize the positively charged surface. The charges remain on those areas of the photoconductor not exposed to light, however, leaving a hidden image of the object in the form of a positive surface charge distribution.

Next, a negatively charged powder called a *toner* is dusted onto the photoconducting surface (Fig. 17.10c). The charged powder adheres only to those areas of the

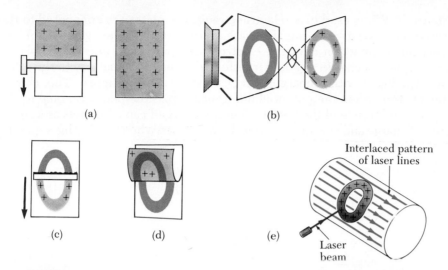

(a) (b)

(c) (d) (e)

Interlaced pattern of laser lines

Laser beam

Figure 17.10
The xerographic process. (a) The photoconductive surface is positively charged. (b) Through the use of a light source and lens, a hidden image is formed on the charged surface in the form of positive charges. (c) The surface containing the image is covered with a negatively charged powder, which adheres only to the image area. (d) A piece of paper is placed over the surface and given a charge. This transfers the image to the paper, which is then heated to "fix" the powder to the paper. (e) The image on the drum of a laser printer is produced by turning a laser beam on and off as it sweeps across the selenium-coated drum.

surface that contain the positively charged image. At this point, the image becomes visible. The image is then transferred to the surface of a sheet of positively charged paper.

Finally, the toner material is "fixed" to the surface of the paper through the application of heat (Fig. 17.10d). This results in a permanent copy of the original.

The steps for producing a document on a laser printer are similar to those used in describing a Xerox machine, in that parts (a), (c), and (d) of Figure 17.10 remain essentially the same. The difference in the two techniques lies in the way that the image is formed on the selenium-coated drum. In a laser printer, the command to print, say, the letter O is sent to a laser from the memory of a computer. To prepare a document, a rotating mirror inside the laser printer causes the beam of a laser to sweep across the selenium-coated drum in an interlaced pattern shown in Figure 17.10e. Electrical signals generated by the printer turn the laser beam on and off in a pattern that traces out the letter O in the form of positive charges on the selenium. Toner is then applied to the drum and the transfer to paper is accomplished as in a Xerox machine. ◀

17.5   The Definition of Capacitance

A **capacitor** is a device used in a variety of electric circuits. For example, capacitors are used (1) to tune the frequency of radio receivers, (2) to eliminate sparking in automobile ignition systems, and (3) as short-term energy-storing devices in electronic flash units. A typical design for a capacitor is shown in Figure 17.11. It consists of two parallel metal plates separated by a

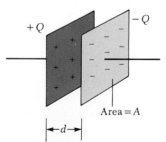

$+Q$ $-Q$

Area $= A$

d

Figure 17.11
A parallel-plate capacitor consists of two parallel plates, each of area A, separated by a distance d. The plates carry equal and opposite charge.

distance d. When used in an electric circuit, the plates are connected to the positive and negative terminals of a battery or some other voltage source. When this connection is made, electrons are pulled off one of the plates, leaving it with a charge $+Q$, and transferred through the battery to the other plate, leaving it with a charge of $-Q$, as shown in the figure. This charge transfer stops when the potential difference across the plates is equal to the potential difference of the battery. Thus, a charged capacitor acts as a storehouse of charge and energy which can be reclaimed when needed for a specific application.

Capacitance of a pair of conductors

The capacitance, C, of a capacitor is defined as the ratio of the magnitude of the charge on either conductor to the magnitude of the potential difference between them:

$$C \equiv \frac{Q}{V} \tag{17.9}$$

From this equation, we see that in order to store a large amount of charge for a given applied voltage, a large capacitance is needed. Also, we see that capacitance has SI units of coulombs per volt. The SI unit of capacitance is called the **farad** (F), in honor of Michael Faraday. That is

$$1 \text{ F} \equiv 1 \text{ C/V}$$

The farad is a very large unit of capacitance. In practice, typical devices have capacitances ranging from microfarads ($1 \ \mu\text{F} = 10^{-6} \text{ F}$) to picofarads ($1 \text{ pF} = 10^{-12} \text{ F}$).

Example 17.5   The Charge on the Plates of a Capacitor

A 3-μF capacitor is connected to a 12-V battery. What is the magnitude of the charge on each plate of the capacitor?

Solution
The definition of capacitance, Equation 17.9, gives us

$$Q = CV = (3 \times 10^{-6} \text{ F})(12 \text{ V}) = \boxed{36 \ \mu\text{C}}$$

17.6   The Parallel-Plate Capacitor

The capacitance of a device depends on the physical characteristics and geometric arrangement of the conductors. For example, the capacitance of a parallel-plate capacitor whose plates are separated by air (see Fig. 17.11) is

Capacitance of a parallel-plate capacitor

$$C = \epsilon_0 \frac{A}{d} \tag{17.10}$$

where A is the area of one of the plates, d is the distance of separation of the plates, and ϵ_0 (Greek letter epsilon) is a constant called the **permittivity of free space,** which has the value

$$\epsilon_0 = 8.85 \times 10^{-12} \; C^2/N \cdot m^2$$

The permittivity of free space is related to the Coulomb constant, k, as

$$k = \frac{1}{4\pi\epsilon_0}$$

Although we shall not derive Equation 17.10, we shall attempt to make it seem plausible. As you can see from the definition of capacitance, $C \equiv Q/V$, the amount of charge a given capacitor is able to store for a given potential difference across its plates increases as the capacitance increases. Therefore, it seems reasonable that a capacitor constructed from plates having a large area should be able to store a large charge. Furthermore, if the oppositely charged plates are close together, the attractive force between them will be large. Therefore, for a given potential difference, the charge on the plates increases with decreasing plate separation.

The Inkjet Printer

Capacitors are used in many practical applications including all types of electronic circuits. One rather interesting application is the inkjet printer, whose basic parts are shown in Figure 17.12. Droplets of ink are ejected from a nozzle at speeds of approximately 50 mi/h and directed toward a sheet of paper. Before reaching the paper, the droplets pass between the plates of two capacitors. The first capacitor is used to charge the droplets while the second capacitor either deflects them to a gutter or allows them to pass. The voltage across the charging capacitor is large enough to charge the droplets. As these charged droplets move between the plates of the deflecting capacitor, they respond to the applied voltage. If the voltage is large enough, a charged droplet can be deflected down to the gutter. However, when the voltage is zero, the droplet moves in a straight line path to the paper, producing one dot in the final image. This particular procedure would require many sweeps of the printhead across the face of the paper. In an alternative procedure, the voltage on the deflecting capacitor is varied such that one droplet passes straight through, the next one is deflected downward slightly, the next slightly more, and so forth. However, none of these droplets are deflected enough to end up in the gutter. In this manner, a single letter in the alphabet can be obtained with only one pass of the printhead.

Figure 17.12
Ink droplets are sent at high speeds between the plates of two capacitors and toward a sheet of paper. The capacitor on the left charges the droplets, which are then deflected by the second capacitor to the gutter or allowed to pass on through and strike the paper.

Example 17.6   Calculating C for a Parallel-Plate Capacitor

A parallel-plate capacitor has an area of $A = 2$ cm^2 $= 2 \times 10^{-4}$ m^2 and a plate separation of $d = 1$ mm $= 10^{-3}$ m. Find its capacitance.

Solution
From $C = \epsilon_0 A/d$ we find

$$C = \epsilon_0 \frac{A}{d} = (8.85 \times 10^{-12} \text{ C}^2/\text{N}\cdot\text{m}^2) \left(\frac{2 \times 10^{-4} \text{ m}^2}{1 \times 10^{-3} \text{ m}} \right)$$

$$= 1.77 \times 10^{-10} \text{ F} = \boxed{177 \text{ pF}}$$

Exercise   Show that 1 C^2/N$\cdot$m equals 1 F.

17.7   Combinations of Capacitors

Two or more capacitors are often combined in circuits in several ways. The equivalent capacitance of certain combinations can be calculated using methods described in this section. The symbol that is commonly used to represent a capacitor in a circuit is ——||——, or sometimes ——)|——. The circuit symbol =|‖+— is used to designate a battery (or any other direct current source). The positive terminal of the battery is at the higher potential and is represented by the longer vertical line in the battery symbol. In the next chapter, we shall discuss another circuit element, called a resistor, which will be represented by the symbol ∿∿. Those wires in a circuit that do not have appreciable resistance compared to other elements in the circuit will be represented as straight lines.

Electric field pattern of two oppositely charged conducting parallel plates. Note the nonuniform nature of the electric field at the ends of the plates. *(Courtesy of Harold M. Waage, Princeton University)*

Parallel Combination

Two capacitors connected as shown in Figure 17.13a are known as a *parallel combination* of capacitors. The left plate of each capacitor is connected by a conducting wire to the positive terminal of the battery and the left plates are therefore at the same potential. Likewise, the right plates are connected to the negative terminal of the battery. When the capacitors are first connected in the circuit, electrons are transferred from the left plates through the battery to the right plates, leaving the left plates positively charged and the right plates negatively charged. The energy source for this charge transfer is the internal chemical energy stored in the battery, which is converted to electrical energy. The flow of charge ceases when the voltage across the capacitors is equal to that of the battery. The capacitors reach their maximum charge when the flow of charge ceases. Let us call the maximum charges on the two capacitors Q_1 and Q_2. Then the *total charge*, Q, stored by the two capacitors is

$$Q = Q_1 + Q_2 \tag{17.11}$$

Suppose we wish to replace these two capacitors by one equivalent capacitor having a capacitance C_{eq}. This equivalent capacitor must have exactly the same external effect on the circuit as the original two. That is, it must store Q units of charge. We also see from Figure 17.13b that

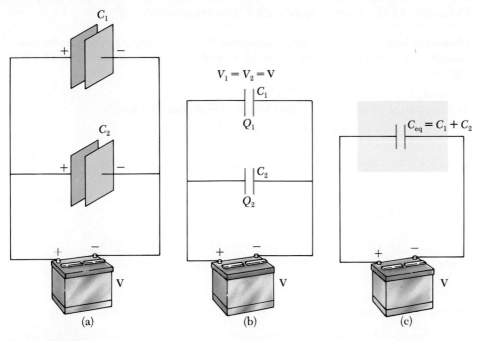

Figure 17.13
(a) A parallel connection of two capacitors. (b) The circuit diagram for the parallel combination. (c) The potential difference is the same across each capacitor, and the equivalent capacitance is $C_{eq} = C_1 + C_2$.

the potential difference across each capacitor in the parallel circuit is the same and is equal to the voltage of the battery, V.

V is the same across capacitors connected in parallel

From Figure 17.12c, we see that the voltage across the equivalent capacitor is also V. Thus, we have

$$Q_1 = C_1V \qquad Q_2 = C_2V$$

and, for the equivalent capacitor,

$$Q = C_{eq}V$$

Substituting these relations into Equation 17.11 gives

$$C_{eq}V = C_1V + C_2V$$

or

$$C_{eq} = C_1 + C_2 \qquad \binom{\text{parallel}}{\text{combination}} \qquad \textbf{(17.12)}$$

If we extend this treatment to three or more capacitors connected in parallel, the equivalent capacitance is found to be

$$C_{eq} = C_1 + C_2 + C_3 + \ldots \qquad \binom{\text{parallel}}{\text{combination}} \qquad \textbf{(17.13)}$$

Thus we see that *the equivalent capacitance of a parallel combination of capacitors is larger than any of the individual capacitances.*

Figure 7.14 (Example 17.7) Four capacitors connected in parallel.

Example 17.7   Four Capacitors Connected in Parallel

Obtain the capacitance of the single capacitor that is equivalent to the parallel combination of capacitors shown in Figure 17.14 and find the charge on the 12-μF capacitor.

Solution

The equivalent capacitance is found by use of Equation 17.13:

$$C_{eq} = C_1 + C_2 + C_3 + C_4$$

$$= 3\ \mu F + 6\ \mu F + 12\ \mu F + 24\ \mu F = \boxed{45\ \mu F}$$

The potential difference across the 12-μF capacitor (and all other capacitors in this case) is equal to the voltage of the battery, and so

$$Q = CV = (12 \times 10^{-6}\ F)(18\ V) = 216 \times 10^{-6}\ C = \boxed{216\ \mu C}$$

Series Combination

Now consider two capacitors connected in *series,* as illustrated in Figure 17.15a.

Q is the same for all capacitors connected in series

For this series combination of capacitors, the magnitude of the charge must be the same on all the plates.

To see why this must be true, let us consider the charge transfer process in some detail. We start with uncharged capacitors and follow what happens just after a battery is connected to the circuit. When the battery is connected, electrons are transferred from the left plate of C_1 to the right plate of C_2

Figure 17.15
A series combination of two capacitors. The charge on each capacitor is the same, and the equivalent capacitance can be calculated from the reciprocal relationship $1/C_{eq} = (1/C_1) + (1/C_2)$.

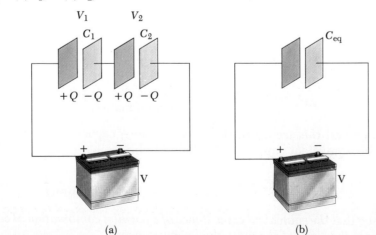

(a)   (b)

through the battery. As this negative charge accumulates on the right plate of C_2, an equivalent amount of negative charge is removed from the left plate of C_2, leaving it with an excess positive charge. The negative charge leaving the left plate of C_2 accumulates on the right plate of C_1, where again an equivalent amount of negative charge is removed from the left plate. The result of this is that *all of the right plates gain a charge of* $-Q$ *while all the left plates have a charge of* $+Q$. (This is a consequence of the conservation of charge.)

Suppose an equivalent capacitor performs the same function as the series combination. After it is fully charged, *the equivalent capacitor must end up with a charge of* $-Q$ *on its right plate and* $+Q$ *on its left plate.* By applying the definition of capacitance to the circuit shown in Figure 17.15b, we have

$$V = \frac{Q}{C_{eq}}$$

where V is the potential difference between the terminals of the battery and C_{eq} is the equivalent capacitance. From Figure 17.15a, we see that

$$V = V_1 + V_2 \tag{17.14}$$

where V_1 and V_2 are the potential differences across capacitors C_1 and C_2. (This is a consequence of the conservation of energy.)

In general, the potential difference across any number of capacitors (or other circuit elements) in series is equal to the sum of the potential differences across the individual capacitors.

Since $Q = CV$ can be applied to each capacitor, the potential difference across each is given by

$$V_1 = \frac{Q}{C_1} \qquad V_2 = \frac{Q}{C_2}$$

Substituting these expressions into Equation 17.14, and noting that $V = Q/C_{eq}$, we have

$$\frac{Q}{C_{eq}} = \frac{Q}{C_1} + \frac{Q}{C_2}$$

Cancelling Q, we arrive at the relationship

$$\frac{1}{C_{eq}} = \frac{1}{C_1} + \frac{1}{C_2} \qquad \left(\begin{array}{l}\text{series} \\ \text{combination}\end{array}\right) \tag{17.15}$$

If this analysis is applied to three or more capacitors connected in series, the equivalent capacitance is found to be

$$\frac{1}{C_{eq}} = \frac{1}{C_1} + \frac{1}{C_2} + \frac{1}{C_3} + \ldots \qquad \left(\begin{array}{l}\text{series} \\ \text{combination}\end{array}\right) \tag{17.16}$$

As we shall see in the following example, this shows that *the equivalent capacitance of a series combination is always less than any individual capacitance in the combination.*

Problem-Solving Strategy: Capacitors

1. Be careful with your choice of units. To calculate the capacitance of a device in farads, make sure that distances are in meters and use the SI value of ϵ_0.
2. When two or more unequal capacitors are connected in *series*, they carry the same charge, but the potential differences across them are not the same. Their capacitances add as reciprocals, and the equivalent capacitance of the combination is always *less* than the smallest individual capacitor.
3. When two or more capacitors are connected in *parallel*, the potential difference across each is the same. The charge on each capacitor is proportional to its capacitance; hence, the capacitances add directly to give the equivalent capacitance of the parallel combination.
4. A complicated circuit consisting of capacitors can often be reduced to a simple circuit containing only one capacitor. To do so, examine your initial circuit and replace any capacitors in series or any in parallel using the rules of Steps 2 and 3 above. Draw a sketch of your new circuit after these changes have been made. Examine this new circuit and replace any series or parallel combinations. Continue this process until a single, equivalent capacitor is found.
5. If the charge on or the potential difference across one of the capacitors in the complicated circuit is to be found, start with the final circuit found in Step 4 and gradually work your way back through the circuits using $C = Q/V$ and the rules given in Steps 2 and 3 above.

Example 17.8   Four Capacitors Connected in Series

Figure 17.16   (Example 17.8) Four capacitors connected in series.

Four capacitors are connected in series with a battery, as in Figure 17.16. Find (a) the capacitance of the equivalent capacitor and (b) the charge on the 12-μF capacitor.

Solution
(a) The equivalent capacitance is found from Equation 17.16:

$$\frac{1}{C_{eq}} = \frac{1}{3\ \mu F} + \frac{1}{6\ \mu F} + \frac{1}{12\ \mu F} + \frac{1}{24\ \mu F}$$

$$C_{eq} = \boxed{1.6\ \mu F}$$

Note that the equivalent capacitance is less than the capacitance of any of the individual capacitors in the combination.

(b) To find the charge on the 12-μF capacitor, we can find the charge on the equivalent capacitor:

$$Q = C_{eq}V = (1.6 \times 10^{-6}\ F)(18\ V) = \boxed{28.8\ \mu C}$$

This is also the charge on each of the capacitors it replaced. Thus, the charge on the 12-μF capacitor in the original circuit is 28.8 μC.

Example 17.9   Equivalent Capacitance

Find the equivalent capacitance between a and b for the combination of capacitors shown in Figure 17.17a. All capacitances are in μF.

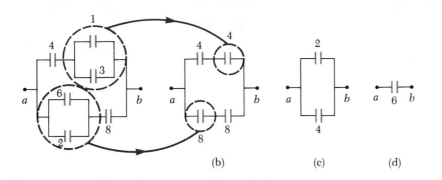

(b)   (c)   (d)

Figure 17.17 (Example 17.9) To find the equivalent capacitance of the circuit in (a), the circuit is reduced in steps as indicated in (b), (c), and (d), using the series and parallel rules described in the text.

Solution

Using Equations 17.13 and 17.16, we reduce the combination step by step as indicated in the figure. The 1-μF and 3-μF capacitors are in *parallel* and combine according to $C_{eq} = C_1 + C_2$. Their equivalent capacitance is 4 μF. Likewise, the 2-μF and 6-μF capacitors are also in *parallel* and have an equivalent capacitance of 8 μF. The upper branch in Figure 17.17b now consists of two 4-μF capacitors in *series*, which combine according to

$$\frac{1}{C_{eq}} = \frac{1}{C_1} + \frac{1}{C_2} = \frac{1}{4\ \mu F} + \frac{1}{4\ \mu F} = \frac{1}{2\ \mu F}$$

$$C_{eq} = 2\ \mu F$$

Likewise, the lower branch in Figure 17.17b consists of two 8-μF capacitors in *series*, which give an equivalent capacitance of 4 μF. Finally, the 2-μF and 4-μF capacitors in Figure 17.17c are in *parallel* and have an equivalent capacitance of 6 μF. Hence, the equivalent capacitance of the circuit is 6 μF.

17.8   Energy Stored in a Charged Capacitor

Almost everyone that works with electronic equipment has at some time verified that a capacitor is able to store energy. If the plates of a charged capacitor are connected together by a conductor, such as a wire, charge will transfer from one plate to the other until the two are uncharged. The discharge can often be observed as a visible spark. If you should accidentally touch the opposite plates of a charged capacitor, your fingers would act as a pathway by which the capacitor can discharge, which results in an electric shock. The degree of shock you would receive depends on the capacitance and voltage applied to the capacitor. Such a shock could be fatal, where high voltages and large quantities of charge are present, such as in the power supply of a television set.

If a capacitor is initially uncharged (both plates neutral), so that the plates are at the same potential, almost no work is required to transfer a small amount of charge ΔQ from one plate to the other. However, once this charge has been transferred, a small potential difference $\Delta V = \Delta Q/C$ will appear between the

Very large capacitors are used as components in high-voltage power generators. In this photograph, a technician installs a safety switch in the "oil section" that contains the generators seen at the right of the picture. This is the main high-voltage mechanical switch linking the generators to a particle accelerator. *(Courtesy of Sandia National Laboratories)*

plates. Therefore, work must be done to transfer additional charge through this potential difference. As more charge is transferred from one plate to the other, the potential difference increases in proportion. If the potential difference at any instant during the charging process is V, the work required to move more charge ΔQ through this potential difference is $V\,\Delta Q$, that is,

$$\Delta W = V\,\Delta Q$$

Figure 17.18
A plot of voltage versus charge for a capacitor is a straight line whose slope is $1/C$. The work required to move a charge ΔQ through a potential difference V across the capacitor plates is $\Delta W = V\,\Delta Q$, which equals the area of the colored rectangle. The *total work* required to charge the capacitor to a final charge Q is the area under the straight line, which equals $QV/2$.

We know that $V = Q/C$ for a capacitor that has a total charge Q. Therefore, a plot of voltage versus charge gives a straight line whose slope is $1/C$, as shown in Figure 17.18. Since the work ΔW is the area of the shaded rectangle, the total work done in charging the capacitor to a final voltage V is the area under the voltage-charge curve, which in this case equals the area under the straight line. Since the area under this line is the area of a triangle (which is one half the product of the base and height), the total work done is

$$W = \frac{1}{2}\,QV \qquad\qquad \textbf{(17.17)}$$

Note that this is also the energy stored in the capacitor since the work required to charge the capacitor will equal the energy stored in the capacitor after it is charged. From the definition of capacitance, we find $Q = CV$, hence we can express the energy stored as

$$\text{Energy stored} = \frac{1}{2}\,QV = \frac{1}{2}\,CV^2 = \frac{Q^2}{2C} \qquad\qquad \textbf{(17.18)}$$

This result applies to any capacitor. In practice, there is a limit to the maximum energy (or charge) that can be stored because electrical breakdown will ultimately occur between the plates of the capacitor at a sufficiently large value of V. For this reason, capacitors are usually labeled with a maximum operating voltage.

Example 17.10   Energy Stored in a Charged Capacitor

Find the amount of energy stored in a 5-μF capacitor when it is connected across a 120-V line.

Solution
Using Equation 17.18, we have

$$\text{Energy stored} = \frac{1}{2}\,CV^2 = \frac{1}{2}\,(5 \times 10^{-6}\text{ F})(120\text{ V})^2 = \;\boxed{3.6 \times 10^{-2}\text{ J}}$$

*17.9   Capacitors with Dielectrics

A **dielectric** is an insulating material, such as rubber, glass, or waxed paper. When a dielectric is inserted between the plates of a capacitor, the capacitance increases. If the dielectric completely fills the space between the plates, the capacitance increases by a factor κ, called the **dielectric constant**.

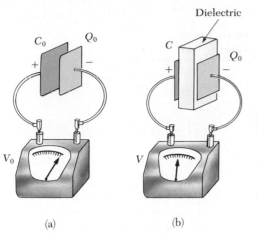

(a)   (b)

Figure 17.19
(a) With air between the plates, the voltage across the capacitor is V_0, the capacitance is C_0, and the charge is Q_0. (b) With a dielectric between the plates, the charge remains at Q_0, but the voltage and capacitance both change.

The following experiment can be performed to illustrate the effect of a dielectric in a capacitor. Consider a parallel-plate capacitor of charge Q_0 and capacitance C_0 in the absence of a dielectric. The potential difference across the capacitor plates can be measured, and is found to be given by $V_0 = Q_0/C_0$ (Fig. 17.19a). Because the capacitor is not connected to an external circuit, there is no pathway for charge to leave or to be added to the plates. If a dielectric is now inserted between the plates as in Figure 17.19b, it is found that the voltage across the plates *decreases* by a factor κ to a value V, where

$$V = \frac{V_0}{\kappa}$$

Since V is less than V_0, we see that κ must be greater than 1. Because the charge Q_0 on the capacitor does not change, we conclude that the capacitance in the presence of the dielectric, C, must change to the value

$$C = \frac{Q_0}{V} = \frac{Q_0}{V_0/\kappa} = \frac{\kappa Q_0}{V_0}$$

or

$$C = \kappa C_0 \qquad \textbf{(17.19)}$$

According to this result, the capacitance *increases* by a factor κ when the dielectric completely fills the region between the plates. For a parallel-plate capacitor, where the capacitance in the absence of a dielectric is $C_0 = \epsilon_0 A/d$, we can express the capacitance in the presence of a dielectric as

$$C = \kappa \epsilon_0 \frac{A}{d} \qquad \textbf{(17.20)}$$

From this result, it would appear that the capacitance could be made very large by decreasing d, the distance between the plates. In practice, the lowest

This photograph illustrates dielectric breakdown in air. Sparks are produced when a large alternating voltage is applied across the electrodes using a high-voltage induction coil power supply. *(Courtesy of CENCO)*

Table 17.1
Dielectric Constants and Dielectric Strengths of Various Materials at Room Temperature

Material	Dielectric Constant κ	Dielectric Strength (V/m)
Vacuum	1.00000	—
Air	1.00059	3×10^6
Bakelite	4.9	24×10^6
Fused quartz	3.78	8×10^6
Pyrex glass	5.6	14×10^6
Polystyrene	2.56	24×10^6
Teflon	2.1	60×10^6
Neoprene rubber	6.7	12×10^6
Nylon	3.4	14×10^6
Paper	3.7	16×10^6
Strontium titanate	233	8×10^6
Water	80	—
Silicone oil	2.5	15×10^6

value of d is limited by the electric discharge that could occur through the dielectric material separating the plates. For any given plate separation, there is a maximum electric field that could be produced in the dielectric before it breaks down and begins to conduct. This maximum electric field is called the **dielectric strength,** and for air its value is about 3×10^6 V/m. Most insulating materials have dielectric strengths greater than that of air, as indicated by the values in Table 17.1.

Commercial capacitors are often made using metal foil interlaced with thin sheets of paraffin-impregnated paper or mylar, which serves as the dielectric material. These alternate layers of metal foil and dielectric are then rolled into the shape of a cylinder to form a small package (Fig. 17.20a).

Figure 17.20
Three commercial capacitor designs: (a) a tubular capacitor whose plates are separated by paper and then rolled into a cylinder, (b) a high-voltage capacitor consisting of many parallel plates separated by oil, and (c) an electrolytic capacitor.

Metal foil

Paper
(a)

Plates

Oil
(b)

Case

Electrolyte

Contacts

Metallic foil + oxide layer
(c)

High-voltage capacitors commonly consist of a number of interwoven metal plates immersed in silicone oil (Fig. 17.20b). Small capacitors are often constructed from ceramic materials. Variable capacitors (typically 10 to 500 pF) usually consist of two interwoven sets of metal plates, one fixed and the other movable, with air as the dielectric.

An electrolytic capacitor is often used to store large amounts of charge at relatively low voltages. This device, shown in Figure 17.19c, consists of a metal foil in contact with an electrolyte — a solution that conducts electricity by virtue of the motion of ions contained in the solution. When a voltage is applied between the foil and the electrolyte, a thin layer of metal oxide (an insulator) is formed on the foil, and this layer serves as the dielectric. Very large values of capacitance can be obtained because the dielectric layer is very thin.

When electrolytic capacitors are used in circuits, the polarity (the plus and minus signs on the device) must be installed properly. If the polarity of the applied voltage is opposite what is intended, the oxide layer will be removed and the capacitor will conduct electricity rather than store charge. Furthermore, reversing the polarity could result in such a large current that the capacitor could either burn or produce steam and explode.

Collection of capacitors used in various applications. *(Courtesy of Henry Leap and Jim Lehman)*

The Stud Finder

If you have ever tried to hang a picture on a wall, you know that it can sometimes be difficult to locate a wooden stud in the wall into which you can securely anchor your nail. The principles discussed in this section can help to solve the problem of finding the hidden stud via a device called a stud finder, shown in Figure 17.21. The primary element in this device is a capacitor with its plates arranged side by side instead of facing one another. As the detector is moved across a wall, the capacitance of the device changes when it passes across a stud because the dielectric constant of the material between the plates has changed, and this change in capacitance can be used to cause a light to come on to indicate the presence of the stud.

Figure 17.21
(a) The materials between the plates of the capacitor are the drywall and the air behind it. (b) The materials become drywall and wood when the detector moves across a stud in the wall. The change in the dielectric constant causes a light to come on, indicating the presence of the stud.

Figure 17.22
The water molecule, H_2O, has a permanent polarization resulting from its bent geometry.

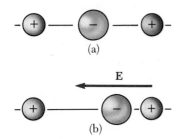

Figure 17.23
(a) A symmetrical molecule has no permanent polarization. (b) An external electric field induces a polarization in the molecule.

Figure 17.24
When a dielectric is placed between the plates of a charged parallel-plate capacitor, the dielectric becomes polarized. This creates a net positive induced charge on the left side of the dielectric and a net negative induced charge on the right side. As a result, the capacitance of the device increases by the factor κ.

Example 17.11   A Paper-Filled Capacitor

A parallel-plate capacitor has plates of dimensions 2 cm by 3 cm. The plates are separated by a 1-mm thickness of paper. (a) Find the capacitance of this device, and (b) find the maximum charge that can be placed on the capacitor.

Solution
(a) Since $\kappa = 3.7$ for paper (Table 17.1), we get

$$C = \kappa\epsilon_0 \frac{A}{d} = 3.7 \left(8.85 \times 10^{-12} \frac{C^2}{N \cdot m^2}\right)\left(\frac{6 \times 10^{-4} \ m^2}{1 \times 10^{-3} \ m}\right)$$

$$= 19.6 \times 10^{-12} \ F = \boxed{19.6 \ pF}$$

(b) From Table 17.1, we see that the dielectric strength of paper is 16×10^6 V/m. Because the paper thickness is 1 mm, the maximum voltage that can be applied before electrical breakdown occurs is

$$V_{max} = E_{max}d = (16 \times 10^6 \ V/m)(1 \times 10^{-3} \ m) = 16 \times 10^3 \ V$$

Hence, the maximum charge that can be placed on the capacitor is

$$Q_{max} = CV_{max} = (19.6 \times 10^{-12} \ F)(16 \times 10^3 \ V) = \boxed{0.31 \ \mu C}$$

An Atomic Description of Dielectrics

Our explanation of why a dielectric increases the capacitance of a capacitor is based on an atomic description of the material. In order to do this, we shall first have to explain a property of some molecules called **polarization.** Molecules are said to be polarized when there is a separation between the "center of gravity" of the negative charges and that of the positive charges that make up the molecule. In some molecules, such as water, this condition is always present. To see why this is the case, consider the geometry of a water molecule (Fig. 17.22). The molecule is arranged so that the negative oxygen atom is bonded to the positively charged hydrogen atoms with a 105° angle between the two bonds. The center of negative charge is at the oxygen atom, and the center of positive charge lies at a point midway along the line joining the hydrogen atoms (point x in the diagram). Materials composed of molecules that are permanently polarized in this fashion have large dielectric constants, and, indeed, Table 17.1 shows that the dielectric constant of water is quite large ($\kappa = 80$).

A symmetrical molecule (Fig. 17.23a) must have no permanent polarization, but a polarization can be induced by an external electric field. A field directed to the left, as in Figure 17.23b, would cause the center of positive charge to shift to the left from its initial position and the center of negative charge to shift to the right. This *induced polarization* is the effect that predominates in most materials used as dielectrics in capacitors.

To understand why the polarization of a dielectric can affect the capacitance, consider Figure 17.24, which shows a slab of dielectric placed between the plates of a parallel-plate capacitor. The dielectric becomes polarized as shown because it is in the electric field that exists between the metal plates.

Notice that a net positive charge appears on the surface of the dielectric that is adjacent to the negatively charged metal plate. The presence of the positive charge on the dielectric effectively cancels some of the negative charge on the metal. Therefore, more negative charge can be stored on the capacitor plates for a given applied voltage. From the definition of capacitance, $C = Q/V$, we see that the capacitance must increase since the plates are able to store more charge for a given voltage.

▶S P E C I A L   T O P I C
Living Cells as Capacitors

Although we shall examine the subject of electrical phenomena in the human body in detail in Chapter 19, for now let us consider a feature of living cells that gives them capacitor-like characteristics.

As shown in Figure 17.25, the presence of charged ions in a cell and in the fluid surrounding the cell sets up a charge distribution across the membrane wall. From this charge distribution, we see that the cell is equivalent to a small capacitor separated by a dielectric, with the membrane wall acting as the dielectric. The potential difference across this "capacitor" can be measured using the technique described in Figure 17.26. A tiny probe is forced through the cell wall, and a second probe is placed in the extracellular fluid. With this technique, typical potential differences across the cell wall are found to be of the order of 100 mV.

The charge distribution of a cell can be understood on the basis of the theory of transport through selectively permeable membranes, discussed in Chapter 10. The primary ionic constituents of a cell are potassium ions (K^+) and chloride ions (Cl^-). The cell wall is highly permeable to K^+ ions but only moderately permeable to Cl^- ions. Since the K^+ ions can cross the cell wall with ease, the charge distribution on a cell wall is determined primarily by the movement of these ions.

In order to explain the process, let us assume that the cellular fluid is electrically neutral at some point. Under normal circumstances, the concentration of K^+ ions outside the cell is much smaller than inside the cell. Therefore, K^+ ions will diffuse out of the cell, leaving behind a net negative charge. As diffusion commences, electrostatic forces of attraction across the membrane wall produce a layer of positive charge on the exterior surface of the wall and a layer of negative charge on the interior of the wall. At the same time, the electric field set up by these charges *impedes* the flow of additional K^+ ions diffusing out of the cell. Because of this impeding effect to diffusion, an equilibrium is finally established such that there is no net movement of K^+ ions through the cell wall. When this equilibrium situation is reached, the potential difference across the cell wall is given by the **Nernst potential,**

$$V_N = \frac{kT}{q} \ln\left(\frac{c_i}{c_o}\right) \qquad (17.21)$$

where k is Boltzmann's constant (1.38×10^{-23} J/K), T is the Kelvin temperature, q is 1.6×10^{-19} C, and c_i and c_o are the concentrations of K^+ ions inside and outside the cell. At normal body temperatures ($T = 310$ K), this equation becomes

$$V_N = (26.7 \text{ mV}) \ln\left(\frac{c_i}{c_o}\right) \qquad (17.22)$$

Figure 17.25
A living cell is equivalent to a small capacitor separated by a dielectric, which is the membrane wall of the cell.

Figure 17.26
An experimental technique for measuring the potential difference across the walls of a living cell.

The electrical balance achieved in the cell is necessary in order for it to function properly. A threat to this balance occurs when the extracellular fluid contains a higher concentration of sodium ions (Na^+) than is normally found inside a cell. Although the cell wall is not highly permeable to these ions, a fraction are able to penetrate the wall. This encroachment by Na^+ ions inside the cell will gradually deplete the negative charge on the interior of the cell wall. If this occurs, more K^+ ions will escape, thus endangering the cell. Fortunately, the cell is able to prevent this by effectively "pumping" Na^+ ions out of the cell while "pumping" K^+ ions into the cell. The mechanism for these processes is not fully understood.

Example   Concentration Ratio

Find the ratio of the concentration of K^+ ions inside a cell to the concentration outside the cell if the Nernst potential is measured to be 90 mV.

Solution
Solving Equation 17.22 for $\ln(c_i/c_o)$, we obtain

$$\ln\left(\frac{c_i}{c_o}\right) = \frac{V_N}{26.7 \times 10^{-3}} = \frac{90 \times 10^{-3}\ V}{26.7 \times 10^{-3}\ V} = 3.37$$

The number whose natural logarithm is 3.37 is 29.1. (Use your calculator to verify this.) Thus, the required concentration ratio is

$$\frac{c_i}{c_o} = \boxed{29.1}$$

◀

Summary

The **difference in potential** between two points A and B is defined as

$$V_B - V_A \equiv \frac{\Delta PE}{q} \tag{17.2}$$

where ΔPE is the *change* in electrical potential energy experienced by a charge q as it moves between A and B. The units of potential difference are joules per coulomb, defined to be equal to a unit called the **volt**; that is, $1\ J/C = 1\ V$.

The **potential difference** between two points A and B in a *uniform electric field* **E** is

$$V_B - V_A = -Ed \tag{17.4}$$

where d is the distance between A and B, and E is the strength of the electric field in that region.

The **electric potential** due to a point charge q, at a distance r from the point charge, is

$$V = k \frac{q}{r} \tag{17.5}$$

The **electrical potential energy** of a pair of point charges separated by a distance r_{12} is

$$PE = k \frac{q_1 q_2}{r_{12}} \tag{17.6}$$

Every point on the surface of a charged conductor in electrostatic equilibrium is at the same potential. Furthermore, the potential is constant everywhere inside the conductor and is equal to its value on the surface.

The **electron volt** is defined as the energy that an electron (or proton) gains when accelerated through a potential difference of 1 V. The conversion between electron volts and joules is

$$1 \text{ eV} = 1.6 \times 10^{-19} \text{ J} \tag{17.8}$$

A **capacitor** consists of two metal plates whose charges are equal in magnitude but opposite in sign. The **capacitance** (C) of any capacitor is defined to be the ratio of the magnitude of the charge Q on either plate to the potential difference V between them:

$$C \equiv \frac{Q}{V} \tag{17.9}$$

Capacitance has units of coulombs per volt, which is defined to be equal to the **farad:** 1 C/V $\equiv$ 1 F.

The capacitance of two *parallel metal plates* of area A separated by a distance d is

$$C = \epsilon_0 \frac{A}{d} \tag{17.10}$$

where ϵ_0 is a constant called the **permittivity of free space** and has the value

$$\epsilon_0 = 8.85 \times 10^{-12} \text{ C}^2/\text{N} \cdot \text{m}^2$$

The **equivalent capacitance** of a **parallel** combination of capacitors is

$$C_{eq} = C_1 + C_2 + C_3 + \ldots \tag{17.13}$$

If two or more capacitors are connected in **series**, the **equivalent capacitance** of the series combination is

$$\frac{1}{C_{eq}} = \frac{1}{C_1} + \frac{1}{C_2} + \frac{1}{C_3} + \dots \qquad (17.16)$$

Three equivalent expressions for calculating the **energy stored** in a charged capacitor are

$$\text{Energy stored} = \frac{1}{2}QV = \frac{1}{2}CV^2 = \frac{Q^2}{2C} \qquad (17.18)$$

When a nonconducting material, called a **dielectric,** is placed between the plates of a capacitor, the capacitance increases by a factor κ, which is called the **dielectric constant** and is a property of the dielectric material. The capacitance of a parallel-plate capacitor filled with a dielectric is

$$C = \kappa \epsilon_0 \frac{A}{d} \qquad (17.20)$$

Additional Reading

Alan H. Cromer, *Physics for the Life Sciences*, New York, McGraw-Hill, 1974, Chap. 17.

A. Einstein and L. Infeld, *The Evolution of Physics*, New York, Simon and Schuster, 1938.

A. L. Stanford Jr., *Foundations of Biophysics*, New York, Academic Press, 1975. Electric dipoles.

Questions

1. A constant electric field is parallel to the x axis. In what direction can a charge be displaced in this field without any external work being done on the charge?
2. If the potential is constant in a certain region, what is the electric field in that region?
3. Explain why equipotential surfaces are always perpendicular to the electric field lines. Do equipotential surfaces ever intersect?
4. If a proton is released from rest in a uniform electric field, does its electric potential increase or decrease? What about its potential energy?
5. If an electron is released from rest in a uniform electric field, does its electric potential increase or decrease? What about its potential energy?

6. In your own words, distinguish between electric potential and electrical potential energy.
7. Two charges have the same electric potential energy. Do they also have the same electric potential?
8. If the electric potential at some point is zero, can you conclude that there are no charges in the vicinity of that point? Explain.
9. Give a physical explanation of the fact that the potential energy of a pair of like charges is positive whereas the potential energy of a pair of unlike charges is negative.
10. Explain why, under static conditions, all points in a conductor must be at the same electric potential.
11. Why is it important to avoid sharp edges, or points, on conductors used in high-voltage equipment?

12. Two capacitors have the same potential difference across them, but one has a large capacitance and one a small capacitance. When you touch the leads to the capacitors, you get a greater shock when you touch the one with high capacitance. From this information, what causes the sensation of shock?

13. What happens to the charge on a capacitor if the potential difference between the conductors is doubled?

14. Why is it dangerous to touch the terminals of a high-voltage capacitor even after the applied voltage has been turned off? What could be done to make the capacitor safe to handle after the voltage source has been removed?

15. The plates of a capacitor are connected to a battery. What happens to the charge on the plates if the connecting wires are removed from the battery? What happens to the charge if the wires are removed from the battery and connected to each other?

16. A capacitor is designed such that one plate is large while the other is small. Do each of the plates have the same charge when connected to a battery?

17. Since the net charge in a capacitor is always zero, what does a capacitor store?

18. Since the charges on the plates of a parallel-plate capacitor are equal and opposite, they attract each other. Hence, it would take positive work to increase the plate separation. What happens to the external work done in this process?

19. If the potential difference across a capacitor is doubled, by what factor does the energy stored change?

20. Two parallel plates are uncharged. Does the set of plates have a capacitance? Explain.

21. If you were asked to design a capacitor where small size and large capacitance were required, what factors would be important in your design?

22. Explain why a dielectric increases the maximum operating voltage of a capacitor although the physical size of the capacitor does not change.

23. A pair of capacitors is connected in parallel while an identical pair is connected in series. Which pair would be more dangerous to handle after being connected to the same voltage source?

24. What advantage might there be in using two identical capacitors in parallel connected in series with another identical parallel pair, rather than using a single capacitor by itself?

25. Is it always possible to reduce a combination of capacitors to one equivalent capacitor with the rules we have developed? Explain your answer.

Problems

Section 17.1   Potential Difference and Electric Potential

1. (a) How much work is done on a proton by a uniform electric field of 200 N/C as the charge moves a distance of 2 cm in the field? (b) What is the difference in potential energy between these two points?

2. Show that the units of N/C and V/m are equivalent.

3. A uniform electric field of magnitude 250 V/m is directed in the positive x direction. A $+12$-μC charge moves from the origin to the point $(x, y) = (20$ cm, 50 cm). (a) What was the change in the potential energy of this charge? (b) Through what potential difference did the charge move?

4. The difference in potential between the accelerating plates of a TV set is about 25,000 V. If the distance between these plates is 1.5 cm, find the magnitude of the uniform electric field in this region.

5. A capacitor consists of two parallel plates separated by a distance of 0.3 mm. If a 20-V potential difference is maintained between those plates, calculate the electric field strength in the region between the plates.

6. A pair of oppositely charged, parallel plates are separated by 5.33 mm. There is a potential difference of 600 V between the plates. (a) What is the magnitude of the electric field strength between the plates? (b) What is the magnitude of the force on an electron between the plates? Make a sketch showing the direction of the force on the electron. (c) How much work must be done on the electron to move it to the negative plate if it is initially positioned 2.9 mm from the positive plate?

7. How much work is done (by a battery, generator, or some other source of electrical energy) in moving Avogadro's number of electrons from a point where the electric potential is 6 V to a point where the electric potential is -10 V?

8. A spherical cell is 3.6 μm in diameter, and its outer membrane is 0.11 μm thick. The potential difference between the inner and outer surfaces of the cell is 90 mV, and the inner surface is negative. How much work is required to eject a positive sodium ion (Na$^+$) from the interior of the cell?

9. An ion, after being accelerated through a potential difference of 60 V, experiences an increase of poten-

tial energy of 1.92×10^{-17} J. Calculate the charge on the ion.

10. A proton is between two plates, separated by a distance of 2 cm, across which there is a potential difference of 5000 V. (a) Find the electrical force exerted on the proton, and (b) its acceleration.

11. (a) Calculate the speed of a proton that is accelerated from rest through a potential difference of 120 V. (b) Calculate the speed of an electron that is accelerated through the same potential difference.

12. What potential difference is needed to stop an electron with an initial speed of 4.2×10^5 m/s?

13. (a) Through what potential difference would an electron need to accelerate in order for it to achieve a speed of 60% of the speed of light, starting from rest? The speed of light is 3×10^8 m/s. (b) Repeat your calculation for a proton. (Do not consider relativistic effects.)

14. An electron moves from one plate to another across which there is a potential difference of 2000 V. (a) Find the velocity with which it strikes the positive plate. (b) Repeat part (a) for a proton moving from the positive toward the negative plate.

Section 17.2   Electric Potential and Potential Energy Due to Point Charges
Section 17.3   Potentials and Charged Conductors

15. At what distance from a point charge of 6 μC would the potential equal 2.7×10^4 V?

16. (a) Find the potential at a distance of 1 cm from a proton. (b) What is the potential difference between two points 1 cm and 2 cm from a proton? (c) Repeat Parts (a) and (b) for an electron.

17. In the Bohr model of the hydrogen atom an electron circles a proton in an orbit of radius 0.51×10^{-10} m. Find the potential at the position of the electron.

18. Two point charges are located along the y axis. One charge of magnitude 3×10^{-9} C is at the origin and a second of magnitude 6×10^{-9} C is at the point $y = 30$ cm. (a) Calculate the potential at $y = 60$ cm, and at (b) $y = -60$ cm. (c) Repeat this problem assuming the 6×10^{-9} C charge is replaced by a -6×10^{-9} C charge.

19. A 6×10^{-9} C charge is located at the point (0, 0.15) where the coordinates are in meters, and a second charge, -4×10^{-9} C, is at $(-0.45, 0)$. Find the potential at (a) $(0, -0.45)$ and (b) (0.15, 0).

20. A charge of -3×10^{-9} C is located at the origin of a coordinate system while a charge of 8×10^{-9} C is on the x axis at $x = 2$ m. At what two location(s) on the x axis is the electric potential equal to zero?

21. A point charge of 9×10^{-9} C is located at the origin. How much work is required to bring a positive charge of 3×10^{-9} C from infinity to the location $x = 30$ cm?

22. Two point charges, $Q_1 = +5$ nC and $Q_2 = -3$ nC, are separated by 35 cm. (a) What is the electric potential at a point midway between the charges? (b) What is the potential energy of the pair of charges? What is the significance of the algebraic sign of your answer?

23. Three charges are located at corners of a rectangle as in Figure 17.27. How much energy would be expended in moving the 8 μC to infinity?

8 μC   6 cm

3 cm

2 μC   4 μC   **Figure 17.27**   Problem 23

24. How much work is required to move an electron from a point 50 cm distant from a proton to a point 100 cm away from the proton?

25. Calculate the speed of (a) an electron that has a kinetic energy of 1 eV and (b) a proton that has a kinetic energy of 1 eV.

26. A proton is accelerated from rest through a potential difference of 25,000 V. (a) What is the kinetic energy of the proton in eV? (b) What is the speed of the proton after this acceleration?

27. How much energy is gained by a charge of 75 μC moving through a potential difference of 90 V? Express your answer in joules and electron volts.

Section 17.5   The Definition of Capacitance
Section 17.6   The Parallel-Plate Capacitor

28. An air-filled parallel-plate capacitor is to have a capacitance of 1 F. If the distance between the plates is 1 mm, calculate the required surface area of each plate. Convert your answer to square miles.

29. (a) How much charge is on each plate of a 4-μF capacitor when it is connected to a 12-V battery? (b) If this same capacitor is connected to a 1.5-V battery, what charge is stored?

30. To what difference in potential would a 2-μF capacitor have to be connected in order to store 98 μC of charge?

31. The plates of a parallel-plate capacitor are separated by 0.1 mm. If the material between the plates is air, what plate area is required to provide a capacitance of 2 pF?

32. The potential difference between a pair of oppositely charged parallel plates is 400 V. (a) If the spacing between the plates is doubled without altering the charge on the plates, what is the new potential difference between the plates? (b) If the plate spacing is doubled while keeping the potential difference between the plates constant, what is the ratio of the final charge on one of the plates to the original charge?

33. A parallel-plate capacitor has an area of 2 cm² and a plate separation of 2 mm. How much charge does this capacitor store when connected to a 6-V battery? (There is air between the plates.)

34. If the area of the plates of a parallel plate capacitor is doubled while the spacing between the plates is halved, how is the capacitance affected?

35. A parallel-plate capacitor has an area of 5 cm² and the plates are separated by 1 mm with air between the plates. It stores a charge of 400 pC. (a) What is the potential difference across the plates of the capacitor? (b) What is the magnitude of the uniform electric field in the region between the plates?

36. Consider the Earth and a cloud layer 800 m above the Earth as the plates of a parallel-plate capacitor. (a) If the cloud layer has an area of 1 km² = 10^6 m², what is the capacitance? (b) If an electric field strength larger than 2.0×10^6 N/C causes the air to break down and conduct electricity (lightning), what is the maximum charge the cloud can hold?

Section 17.7 Combinations of Capacitors

37. Two capacitors, $C_1 = 5\ \mu F$ and $C_2 = 12\ \mu F$, are connected in parallel, and the resulting combination is connected to a 9-V battery. What is the value of the equivalent capacitance of the combination?

38. (a) What is the potential difference across each capacitor in Problem 37? (b) What is the charge stored on each of these capacitors?

39. The two capacitors of Problem 37 are now connected in series and to a 9-V battery. Find (a) the value of the equivalent capacitance of the combination and (b) the charge on and the voltage across each.

40. A series circuit consists of a 0.05-μF capacitor, a 0.10-μF capacitor, and a 400-V battery. Find the charge (a) on each of the capacitors and (b) on each of the capacitors if they are reconnected in parallel across the battery.

41. Three capacitors, $C_1 = 5\ \mu F$, $C_2 = 4\ \mu F$, and $C_3 = 9\ \mu F$, are connected together. (a) Find the effective capacitance of the group if they are all in parallel. (b) Find the effective capacitance of the group if they are all in series.

42. The combination of capacitors in Problem 41 is connected to a 12-V battery. Find the charge on and the potential difference across each of the capacitors (a) when they are connected in parallel and (b) when they are connected in series.

43. Find the charge on each of the capacitors in Figure 17.28.

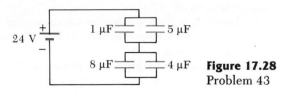

Figure 17.28
Problem 43

44. (a) Find the equivalent capacitance of the group of capacitors shown in Figure 17.29. (b) Find the charge on and the potential difference across each.

Figure 17.29
Problems 44 and 56

45. (a) Determine the equivalent capacitance for the capacitor network shown in Figure 17.30. (b) If the network is connected to a 12-V battery, calculate the potential difference across each capacitor and the charge on each capacitor.

Figure 17.30
Problem 45

46. Evaluate the effective capacitance of the configuration shown in Figure 17.31. Each of the capacitors is identical and has capacitance C.

Figure 17.31
Problem 46

47. A purchasing director gets a quantity discount for buying identical capacitors in large numbers. How many different values of effective capacitance can a technician produce using all combinations of one to three separate capacitors? Evaluate the capacitance, in terms of the capacitance C of a single capacitor, for each combination.

48. How should four 2-μF capacitors be connected to have a total capacitance of (a) 8 μF, (b) 2 μF, (c) 1.5 μF, and (d) 0.5 μF?

49. Consider the group of capacitors shown in Figure 17.32. Find (a) the equivalent capacitance of the group and (b) the charge and potential difference across each capacitor.

Figure 17.32
Problem 49

50. Consider the combination of capacitors shown in Figure 17.33. (a) What is the equivalent capacitance of the group? (b) Determine the charge on each capacitor.

Figure 17.33
Problem 50

51. A 1.0-μF capacitor is first charged by connecting it across a 10.0-V battery. The capacitor is then disconnected from the battery and connected across an uncharged 2.0-μF capacitor. Determine the charge on each of the capacitors.

Section 17.8 Energy Stored in a Charged Capacitor

52. (a) A 3-μF capacitor is connected to a 12-V battery. How much energy is stored in the capacitor? (b) If the capacitor had been connected to a 6-V battery how much energy would have been stored?

53. A capacitor is connected to a 120-V source and holds a charge of 36 μC. (a) What is the capacitance of the capacitor? (b) Find the energy stored by the capacitor by the use of at least two different equations.

54. A parallel-plate capacitor is charged and then disconnected from a battery. By what fraction does the stored energy change (increase or decrease) when the plate separation is doubled?

55. A parallel-plate capacitor has plates of area 2 cm², a distance of separation of 5 mm, and air between the plates. If a 12-V battery is connected to this capacitor, how much energy does it store?

56. Find the energy stored by each capacitor in Figure 17.29.

Section 17.9 Capacitors with Dielectrics

57. Find the capacitance of a parallel-plate capacitor that uses Bakelite as a dielectric if the plates each have an area of 5 cm² and the plate separation is 2 mm.

58. A capacitor that has air between its plates is connected across a potential difference of 12 V and stores 48 μC of charge. It is then disconnected from the source while still charged. (a) Find the capacitance of the capacitor. (b) If a piece of Teflon is now inserted between the plates, find the voltage and charge on the capacitor, and (c) its new capacitance.

59. A 3-μF capacitor is connected to a 12-V battery. If the material between the plates is initially air, by what factor will the energy stored change if a sheet of nylon is placed between the plates?

60. A capacitor with air between the plates is charged to 100 V and then disconnected from the battery. When a piece of glass is placed between the plates, the voltage across the capacitor drops to 25 V. What is the dielectric constant of the glass? Assume the glass completely fills the space between the plates.

61. Determine (a) the capacitance and (b) the maximum voltage which can be applied to a Teflon-filled parallel-plate capacitor having a plate area of 175 cm² and insulation thickness of 0.04 mm.

62. (a) How much charge can be placed on a capacitor with air between the plates before it breaks down if the area of the plates is 5 cm²? (b) If polystyrene is used between the plates instead of air, find the maximum charge.

Additional Problems

63. A proton moving in a uniform electric field experiences an increase in kinetic energy of 9×10^{-18} J after being displaced 1 cm in a direction parallel to the field. What is the magnitude of the electric field?

64. At a distance r away from a point charge q, the electric potential is $V = 600$ V and the magnitude of the electric field is $E = 200$ N/C. Determine the values of q and r.

65. Find the potential at point P for the rectangular grouping of charges shown in Figure 17.34.

Figure 17.34
Problem 65

66. The three charges shown in Figure 17.35 are at the vertices of an isosceles triangle. Calculate the electric potential at the midpoint of the base, taking the charge $q = 5 \times 10^{-9}$ C.

Figure 17.35
Problem 66

67. Find the equivalent capacitance of the group of capacitors shown in Figure 17.36.

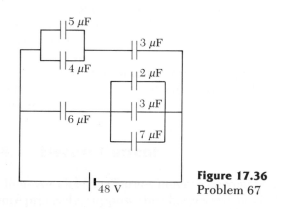

Figure 17.36
Problem 67

68. Find the equivalent capacitance of the group of capacitors shown in Figure 17.37.

Figure 17.37
Problem 68

69. Find the equivalent capacitance of the group of capacitors shown in Figure 17.38.

Figure 17.38
Problems 69 and 70

70. Find the total energy stored by the group of capacitors in Figure 17.38.

71. Suppose that the extracellular fluid of a cell has a potassium ion concentration of 0.0045 mole/liter and the intracellular fluid has a potassium ion concentration of 0.138 mole/liter. Calculate the potential difference across the cell membrane. Neglect the diffusion of any other ions across the membrane and assume a temperature of 310 K.

72. A commercial capacitor is made by rolling two strips of aluminum separated by two strips of paraffin-coated paper into a cylinder. Each strip of foil and paper is 5 cm wide. The foil is 0.0005 cm thick, and the paper is 0.002 cm thick and has a dielectric constant of 3. What length should the strips be if a capacitance of 2×10^{-8} F is desired? (Use the parallel-plate equation.)

73. When a certain air-filled parallel-plate capacitor is connected across a battery, it acquires a charge (on each plate) of 150 μC. While the battery connection is maintained, a dielectric slab is inserted into and fills the region between the plates. This results in the accumulation of an additional charge of 200 μC on each plate. What is the dielectric constant of the dielectric slab?

Figure 18.1
Charges in motion through an area A. The time rate of flow of charge through the area is defined as the current I. The direction of the current is in the direction of flow of positive charges.

If ΔQ is the amount of charge that passes through this area in a time interval Δt, the current, I, is equal to the ratio of the charge to the time interval:

$$I \equiv \frac{\Delta Q}{\Delta t} \qquad (18.1)$$

The SI unit of current is the **ampere** (A), where

$$1 \text{ A} = 1 \text{ C/s} \qquad (18.2)$$

Thus, 1 A of current is equivalent to 1 C of charge passing through the cross-sectional area in a time interval of 1 s. In practice, smaller units of current are often used, such as the milliampere ($1 \text{ mA} = 10^{-3}$ A) and the microampere ($1 \ \mu\text{A} = 10^{-6}$ A).

The charges flowing through a surface as in Figure 18.1 can be positive, negative, or both. It is *conventional to choose the direction of the current to be in the direction of flow of positive charge.* In a metallic conductor such as copper, the current is due to the motion of the negatively charged electrons. Therefore, when we speak of current in such a conductor, *the direction of the current will be opposite the flow of electrons.*

Students studying physics for the first time often wonder why the direction of current is defined in this manner. The primary reason for this choice of direction lies in an historical lack of understanding of what actually moves when there is a current in a wire. In the early study of electric circuits, it was not known what type of charge was moving, and the direction of current was arbitrarily taken to be that of the motion of positive charges. Even though we now know that it is negative charges (electrons) that move in a metallic conductor, the historical legacy of a current being in the direction of motion of positive charges remains with us.

If one considers a beam of positively charged protons in an accelerator, like the Van de Graaff accelerator discussed in Chapter 16, the current is in the direction of motion of the protons. In some cases, the current is the result of the flow of both positive and negative charges. This occurs, for example, in semiconductors, which we shall discuss in Chapter 31. It is common to refer to a moving charge (whether it is positive or negative) as a *mobile charge carrier.* For example, the mobile charge carriers in a metal are electrons.

Example 18.1 The Current in a Wire

A charge of 6×10^{-3} C passes through a cross-sectional area of a wire in 2 s. What is the current in the wire?

Solution
From Equation 18.1, we have

$$I = \frac{\Delta Q}{\Delta t} = \frac{6 \times 10^{-3} \text{ C}}{2 \text{ s}} = 3 \times 10^{-3} \text{ A} = \boxed{3 \text{ mA}}$$

Exercise If a wire carries a current of 5 A, how much charge moves through a cross section in 4 s?

Answer 20 C.

18.2 Current and Drift Velocity

It is instructive to relate current to the motion of the charged particles. To illustrate this point, consider the current in a conductor of cross-sectional area A (Fig. 18.2). We shall focus our attention on a small segment of the conductor of length x; the volume of this segment is Ax. If n represents the number of mobile charge carriers per unit volume, then the total number of mobile charge carriers in this volume element is nAx. Therefore, the total charge ΔQ that is free to move in this volume is

$$\Delta Q = \text{number of charges} \times \text{charge per particle} = (nAx)(q)$$

where q is the charge on each particle. Let us choose a time interval Δt such that all the charges in the segment of length x will move through the cross-sectional area A. If the charge carriers move with a speed v_d, the distance they move in a time Δt is $x = v_d\Delta t$. Therefore, we can write ΔQ in the form

$$\Delta Q = (nAv_d\Delta t)(q)$$

If we divide both sides of this equation by Δt, we see that the current in the conductor is given by

$$I = \frac{\Delta Q}{\Delta t} = nqv_dA \qquad (18.3)$$

The velocity of the charge carriers, v_d, is actually an average velocity and is called the **drift velocity.** To understand the meaning of drift velocity, consider a conductor in which the charge carriers are free electrons. In such a conductor, the electrons undergo random motion similar to that of gas molecules. When a battery is connected across the ends of the conductor, an electric field is set up in the conductor, which exerts an electric force on the electrons. This electric force causes the electrons to drift in a direction opposite that of the electric field (the drift is opposite the field because the electrons are negatively charged). This motion of charge constitutes a current. In reality, the electrons do not simply move in straight lines along the conductor. Instead, they undergo repeated collisions with the metal atoms, which results in a complicated zigzag motion (Fig. 18.3). Despite the collisions, the electrons do move slowly (drift) along the conductor, and their average velocity is what is called the drift velocity. However, if the voltage across the conductor is zero, the drift velocity is zero.

Before we end our discussion of current, it is instructive to clarify what might appear to be a discrepancy regarding a property of electric fields in conductors. We stated above that an electric field is set up in a conductor when a voltage is applied across its ends, but in Chapter 16 we stated that the electric field is zero inside a conductor. There is no conflict between these statements because an electric field *can* exist in a conductor when we are dealing with charges in motion. This is in contrast to the discussion of Chapter 16, where electrostatic equilibrium conditions were assumed (the charges are at rest). In the latter case, the electric field inside the conductor must indeed be zero.

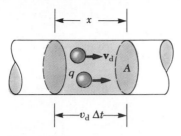

Figure 18.2
A section of a uniform conductor of cross-sectional area A. The charge carriers move with a speed v_d, and the distance they travel in a time Δt is given by $x = v_d\,\Delta t$. The number of mobile charge carriers in the section of length x is given by $nAv_d\,\Delta t$, where n is the number of mobile charge carriers per unit volume.

Current is proportional to the drift velocity

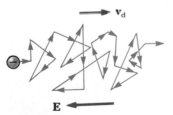

Figure 18.3
A schematic representation of the zigzag motion of a charge carrier in a conductor. The changes in direction are due to collisions with atoms in the conductor. Note that the net motion of electrons is opposite the direction of the electric field.

Example 18.2   The Drift Velocity in a Copper Wire

A copper wire of cross-sectional area 3×10^{-6} m² carries a current of 10 A. Assuming that each copper atom contributes one free electron to the metal, find the drift velocity of the electrons in this wire. The density of copper is 8.95 g/cm³.

Solution
From the periodic table of the elements, we find that the atomic weight of copper is 63.5 g/mole. Recall that one atomic mass of any substance contains Avogadro's number of atoms, 6.02×10^{23} atoms. Knowing the density of copper enables us to calculate the volume occupied by 63.5 g of copper:

$$V = \frac{m}{\rho} = \frac{63.5 \text{ g}}{8.95 \text{ g/cm}^3} = 7.09 \text{ cm}^3$$

If we now assume that each copper atom contributes one free electron to the body of the material, we have

$$n = \frac{6.02 \times 10^{23} \text{ electrons}}{7.09 \text{ cm}^3} = 8.48 \times 10^{22} \text{ electrons/cm}^3$$

$$= \left(8.48 \times 10^{22} \, \frac{\text{electrons}}{\text{cm}^3}\right)\left(10^6 \, \frac{\text{cm}^3}{\text{m}^3}\right)$$

$$= 8.48 \times 10^{28} \text{ electrons/m}^3$$

From Equation 18.3, we find that the drift velocity is

$$v_d = \frac{I}{nqA} = \frac{10 \text{ C/s}}{(8.48 \times 10^{28} \text{ electrons/m}^3)(1.6 \times 10^{-19} \text{ C})(3 \times 10^{-6} \text{ m}^2)}$$

$$= \quad 2.46 \times 10^{-4} \text{ m/s}$$

Example 18.2 shows that typical drift velocities are very small. In fact, the drift velocity is much smaller than the average velocity between collisions. For instance, electrons traveling with this velocity would take about 68 min to travel 1 m! In view of this low speed, you might wonder why a light turns on almost instantaneously when a switch is thrown. This can be explained by considering the flow of water through a pipe. If a drop of water is forced in one end of a pipe that is already filled with water, a drop must be pushed out the other end of the pipe. While it may take individual drops of water a long time to make it through the pipe, a flow initiated at one end produces a similar flow at the other end very quickly. In a conductor, the electric field that drives the free electrons travels through the conductor with a speed close to that of light. Thus, when you flip a light switch, the message for the electrons to start moving through the wire (the electric field) reaches them at a speed of the order of 10^8 m/s.

18.3   Ohm's Law

For many conductors, the current is proportional to the potential difference, V, across the conductor, that is, $I \propto V$. This says that if one wire is connected to a 3-V battery and an identical wire to a 1.5-V battery, the current in the first wire will be twice the current in the second wire.

An additional factor that affects the flow of charges through a wire is the **resistance, R,** of the wire. The current in a wire is inversely proportional to the resistance of the wire, that is, $I \propto 1/R$. According to this relationship, if we double the resistance of a conductor without changing the applied voltage, the current is reduced to one-half its original value.

George Simon Ohm (1787–1854) was the first to note that the current in a conductor is directly proportional to the applied voltage as

$$I \propto V$$

Later, a proportionality constant R was introduced to provide the following useful equation:

$$V = IR \qquad\qquad \textbf{(18.4)} \qquad \text{Ohm's law}$$

This equation is called **Ohm's law.** The letter R is called the **resistance** of the conductor and has the units of **ohms.** The symbol Ω (Greek letter omega) is usually used to represent ohms. For example, the filament of a lightbulb might have a resistance of 100 Ω when hot.

Ohm's law is not a fundamental law of nature. Rather, *it is based on experimental observation and is valid only for certain materials.* Materials that obey Ohm's law are said to be *ohmic,* and those that do not are called *non-ohmic.*

From Equation 18.4, we see that resistance has the units of volts per ampere. One volt per ampere is defined to be one ohm:

$$1\ \Omega \equiv 1\ \text{V/A} \qquad\qquad \text{Definition of the ohm unit}$$

That is, if a potential difference of 1 V across a conductor produces a current of 1 A, the resistance of the conductor is equal to 1 Ω.

Example 18.3   The Resistance of a Steam Iron

All electric devices are required to have an identifying plate that specifies their electrical characteristics. For example, the plate on a certain steam iron states that the iron carries a current of 6 A when connected to a 120-V source. What is the resistance of the steam iron?

Solution
From Ohm's law, we find the resistance to be

$$R = \frac{V}{I} = \frac{120\ \text{V}}{6\ \text{A}} = \boxed{20\ \Omega}$$

Exercise   The resistance of a hot plate is equal to 40 Ω. How much current does the plate carry when connected to a 120-V source?

Answer   3 A.

18.4   Resistivity and Resistance

Figure 18.4
A conductor of length L and cross-sectional area A.

In an earlier section we pointed out that electrons do not move in straight-line paths through a conductor. Instead, they undergo repeated collisions with the metal atoms. The state of affairs can be pictured as follows. Consider a conductor with a voltage applied between its ends. An electron gains speed as the electric force associated with the internal electric field accelerates the electron, giving it a velocity in the direction opposite the direction of the electric field. A collision with an atom randomizes the electron's velocity, thus reducing its velocity in the direction opposite the field. The process then repeats itself. The net effect these collisions have on the electron is somewhat like a force of internal friction. This is the origin of the resistance of a material. If a conductor has a length L and cross-sectional area A, as shown in Figure 18.4, its resistance is given by

$$R = \rho \frac{L}{A} \qquad\qquad \textbf{(18.5)}$$

where ρ is an intrinsic property of the conductor called the **resistivity**.[1] Every material has a characteristic resistivity, which depends on the electronic structure of the material and on temperature. Good electric conductors have very low resistivities, and good insulators have very high resistivities. Table 18.1 gives the resistivities of a variety of materials at 20°C. Because resistance values are in ohms, resistivity values must be in ohm-meters.

[1] The symbol ρ used for resistivity should not be confused with the same symbol used earlier in the text for density. Very often, the same symbol is used to represent different quantities.

An assortment of resistors used for various applications in electronic circuits. *(Courtesy of Henry Leap and Jim Lehman)*

Table 18.1
Resistivities and Temperature Coefficients of Resistivity for Various Materials[a]

Material	Resistivity $(\Omega \cdot m)$	Temperature Coefficient of Resistivity $[(C°)^{-1}]$
Silver	1.59×10^{-8}	3.8×10^{-3}
Copper	1.7×10^{-8}	3.9×10^{-3}
Gold	2.44×10^{-8}	3.4×10^{-3}
Aluminum	2.82×10^{-8}	3.9×10^{-3}
Tungsten	5.6×10^{-8}	4.5×10^{-3}
Iron	10×10^{-8}	5.0×10^{-3}
Platinum	11×10^{-8}	3.92×10^{-3}
Lead	22×10^{-8}	3.9×10^{-3}
Nichrome[b]	150×10^{-8}	0.4×10^{-3}
Carbon	3.5×10^{5}	-0.5×10^{-3}
Germanium	0.46	-48×10^{-3}
Silicon	640	-75×10^{-3}
Glass	$10^{10} - 10^{14}$	
Hard rubber	$10^{13} - 10^{6}$	
Sulfur	10^{15}	
Quartz (fused)	75×10^{16}	

[a] All values at 20°C.

[b] A nickel-chromium alloy commonly used in heating elements.

Although the analogy is not perfect, the relationship of the variables in Equation 18.5 can be understood by comparison to a common experience, that of trying to walk through a crowded hallway without bumping into anyone. It should seem reasonable to you that you are more likely to traverse the hallway without many collisions if the hallway is short or if the hallway is wide. Likewise, the electrons in a wire are more likely to travel through a piece of wire with fewer randomizing collisions if the wire is short or if its cross-sectional area is large, in agreement with Equation 18.5.

Example 18.4   The Resistance of a Conductor

Calculate the resistance of a piece of aluminum that is 10 cm long and has a cross-sectional area of 10^{-4} m². Repeat the calculation for a piece of hard rubber of resistivity 10^{13} $\Omega \cdot$m.

Solution

From Equation 18.5 and Table 18.1, we can calculate the resistance of the aluminum bar:

$$R = \rho \frac{L}{A} = (2.82 \times 10^{-8}\ \Omega \cdot m) \left(\frac{0.1\ m}{10^{-4}\ m^2} \right) = \boxed{2.82 \times 10^{-5}\ \Omega}$$

Similarly, for rubber we find

$$R = \rho \frac{L}{A} = (10^{13}\ \Omega \cdot m) \left(\frac{0.1\ m}{10^{-4}\ m^2} \right) = \boxed{10^{16}\ \Omega}$$

As you might expect, aluminum has a much lower resistance than rubber. This is the reason that aluminum is a good conductor and rubber is a poor conductor.

Example 18.5   The Resistance of Nichrome Wire

(a) Calculate the resistance per unit length of a 22-gauge nichrome wire of radius 0.321 mm.

Solution

The cross-sectional area of this wire is

$$A = \pi r^2 = \pi (0.321 \times 10^{-3}\ m)^2 = 3.24 \times 10^{-7}\ m^2$$

The resistivity of nichrome is 1.5×10^{-6} $\Omega \cdot$m (Table 18.1). Thus, we can use Equation 18.5 to find the resistance per unit length:

$$\frac{R}{L} = \frac{\rho}{A} = \frac{1.5 \times 10^{-6}\ \Omega \cdot m}{3.24 \times 10^{-7}\ m^2} = \boxed{4.6\ \Omega/m}$$

(b) If a potential difference of 10 V is maintained across a 1-m length of the nichrome wire, what is the current in the wire?

Solution

Since a 1-m length of this wire has a resistance of 4.6 Ω, Ohm's law gives

$$I = \frac{V}{R} = \frac{10\ V}{4.6\ \Omega} = \boxed{2.2\ A}$$

The resistivity of nichrome wire is about 100 times larger than that of copper wire. Specifically, a copper wire of the same gauge would have a resistance per unit length of only 0.052 Ω/m. A 1-m length of 22-gauge copper wire would carry the same current (2.2 A) with an applied voltage of only 0.11 V.

Photograph of an old-fashioned carbon filament incandescent lamp. The resistance of such a lamp is typically 10 Ω, but it changes with temperature. Does the resistance increase or decrease as the filament is heated?
(Courtesy of CENCO)

Because of its high resistivity, nichrome is often used for heating elements in toasters, irons, and electric heaters.

Exercise   What is the resistance of a 6-m length of 22-gauge nichrome wire? How much current does it carry when connected to a 120-V source?

Answer   28 Ω, 4.3 A.

18.5   Temperature Variation of Resistance

The resistivity, and hence the resistance, of a conductor depends on a number of factors. One of the most important of these factors is the temperature of the metal. For most metals, resistivity increases with increasing temperature. This increase can be understood as follows. As the temperature of the material increases, its constituent atoms vibrate with larger and larger amplitudes. Just as it is more difficult to weave one's way through a crowded room when the people are in motion than it is when they are standing still, so do the electrons find it more difficult to pass the atoms moving with large amplitudes.

The resistivity of a conductor varies with temperature over a limited range according to the expression

$$\rho = \rho_0[1 + \alpha(T - T_0)] \tag{18.6}$$

where ρ is the resistivity at some temperature T (in °C), ρ_0 is the resistivity at some reference temperature T_0 (usually taken to be 20°C), and α is a parameter called the **temperature coefficient of resistivity.** The temperature coefficients for various materials are given in Table 18.1.

Since the resistance of a conductor with uniform cross section is proportional to the resistivity according to Equation 18.5 ($R = \rho L/A$), the temperature variation of resistance can be written

$$R = R_0[1 + \alpha(T - T_0)] \tag{18.7}$$

Precise temperature measurements are often made using this property, as shown in the following example.

Example 18.6   A Platinum Resistance Thermometer

A resistance thermometer made from platinum has a resistance of 50.0 Ω at 20°C. When the device is immersed in a vessel containing melting indium, its resistance increases to 76.8 Ω. From this information, find the melting point of indium. Note that $\alpha = 3.92 \times 10^{-3}$ (C°)$^{-1}$ for platinum.

Solution
If we solve Equation 18.7 for $T - T_0$, we obtain

$$T - T_0 = \frac{R - R_0}{\alpha R_0} = \frac{76.8\ \Omega - 50.0\ \Omega}{[3.92 \times 10^{-3}\ (\text{C}°)^{-1}][50.0\ \Omega]} = 137\ \text{C}°$$

Since $T_0 = 20°\text{C}$, we find that the melting point of indium is $T =$ 157°C

Carbon
granules

Steel
diaphragm

Transformer

Figure 18.5
Diagram of a carbon microphone.

*The Carbon Microphone

The carbon microphone is commonly used in the mouthpiece of a telephone. The construction of this device is illustrated in Figure 18.5. A flexible steel diaphragm is placed in contact with carbon granules inside a container. The carbon granules serve as the primary resistance medium in a circuit containing a source of current (shown here as a battery) and a transformer, whose operation will be described in a later chapter.

The magnitude of the current in the circuit changes when a sound wave strikes the diaphragm. When a compression arrives at the microphone, the diaphragm flexes inward, causing the carbon granules to press together into a smaller-than-normal volume, corresponding to a decrease in the length of the resistance medium. This results in a lower circuit resistance and hence an increase in current in the circuit. When a rarefaction arrives at the microphone, the diaphragm relaxes, and the carbon granules become more loosely packed. This results in an increase in the circuit resistance and a corresponding decrease in current. These variations in current, which follow the changes of the sound wave, are sent through the transformer into the transmission line of the telephone company. A speaker in the listener's earpiece then converts these electric signals back to a sound wave.

The carbon microphone has a very poor frequency response. It reproduces frequencies below 4000 Hz adequately and is therefore suitable for speech transmission, since the critical frequencies in normal conversation are usually below this value. However, its capabilities fall off dramatically at higher frequencies, rendering it useless for high-fidelity purposes, which require reliable sound reproduction at all frequencies between 20 Hz and 20,000 Hz.

►SPECIAL TOPIC
Superconductivity

There is a class of metals and compounds whose resistance goes virtually to *zero* below a certain temperature, T_c, called the *critical temperature*. These materials are known as **superconductors**. The resistance-temperature graph for a superconductor follows that of a normal metal at temperatures above T_c (Fig. 18.6). When the temperature is at or below T_c, the resistance drops suddenly to zero. This phenomenon was discovered in 1911 by the Dutch physicist H. Kamerlingh-Onnes when he and a graduate student were working with mercury, which is a superconductor below 4.2 K. Recent measure-

Figure 18.6
Resistance versus temperature for a sample of mercury. The graph follows that of a normal metal above the critical temperature T_c. The resistance drops to zero at the critical temperature, which is 4.2 K for mercury, and remains at zero for lower temperatures.

**Table 18.2
Critical Temperatures
for Various
Superconductors**

Material	T_c (K)
Nb_3Ge	23.2
Nb_3Sn	18.05
Nb	9.46
Pb	7.18
Hg	4.15
Sn	3.72
Al	1.19
Zn	0.88
$YBa_2Cu_3O_7$	90
Bi-Sr-Ca-Cu-O	105
Tl-Ba-Ca-Cu-O	125

ments have shown that the resistivities of superconductors below T_c are less than 4×10^{-25} $\Omega \cdot m$, which is around 10^{17} times smaller than the resistivity of copper and considered to be *zero* in practice.

Today there are thousands of known superconductors, including such common metals as aluminum, tin, lead, zinc, and indium. Table 18.2 lists the critical temperatures of several superconductors. The value of T_c is sensitive to chemical composition, pressure, and crystalline structure. It is interesting to note that copper, silver, and gold, which are excellent conductors, do not exhibit superconductivity.

One of the truly remarkable features of superconductors is the fact that, once a current is set up in them, it will persist *without any applied voltage* (since $R = 0$). In fact, steady currents have been observed to persist in superconducting loops for many years with no apparent decay!

One of the most important recent developments in physics that has created much excitement in the scientific community has been the discovery of high temperature copper-oxide-based superconductors. The excitement began with a 1986 publication by J. Georg Bednorz and K. Alex Müller, two scientists working at the IBM Zurich Research Laboratory in Switzerland, who reported evidence for superconductivity in the 30 K range in an oxide of barium, lanthanum, and copper. Bednorz and Müller were awarded the Nobel Prize in 1987 for their incredible and important discovery. Shortly thereafter, a new family of compounds was open for investigation, and research activity in the field of superconductivity proceeded vigorously. In early 1987, groups at the University of Alabama at Huntsville and the University of Houston announced the discovery of superconductivity at about 92 K in an oxide of yttrium, barium, and copper ($YBa_2Cu_3O_7$). Late in 1987, teams of scientists from Japan and the United States reported superconductivity at 105 K in an oxide of bismuth, strontium, calcium, and copper. More recently, scientists have reported superconductivity as high as 125 K in an oxide containing thallium. At this point, one cannot rule out the possibility of room temperature superconductivity, and the search for novel superconducting materials continues. These developments are very exciting and important both for scientific reasons and because practical applications become more probable and widespread as the critical temperature is raised. The interesting essay at the end of this chapter discusses superconductivity and the new generation of superconductors in more detail.

An important and useful application of superconductivity has been the construction of superconducting magnets in which the magnetic field intensities are about ten times greater than those of the best normal electromagnets. Such superconducting magnets are being considered as a means of storing energy. The idea of using superconducting power lines for transmitting power efficiently is also receiving some consideration. Modern superconducting electronic devices consisting of two thin-film superconductors separated by a thin insulator have been constructed. These devices include magnetometers (a magnetic-field measuring device) and various microwave devices. ◄

18.6   Electrical Energy and Power

If a battery is used to establish an electric current in a conductor, there is a continuous transformation of the chemical energy stored in the battery into potential energy of the charge carriers. This extra energy added to the charge carriers is lost when they collide with the atoms of the conductor, resulting in a temperature increase in the conductor. In other words, the chemical energy stored in the battery is continuously transformed into thermal energy.

Consider a simple circuit consisting of a battery whose terminals are connected to a resistor, indicated by the symbol 〰, as shown in Figure 18.7. Let

Figure 18.7
A circuit consisting of a battery and resistance R. Positive charge flows in the clockwise direction, from the positive to the negative terminal of the battery. Point A is grounded.

us now follow a positive quantity of charge ΔQ as it moves around the circuit from point A through the battery and resistor and back to A. Point A is a reference point that is grounded (ground symbol is ⏚), and its potential is zero. As the charge moves from A to B through the battery, its electrical potential energy increases by an amount $V \Delta Q$ (where V is the potential at B) while the chemical potential energy in the battery decreases by the same amount. However, as the charge moves from C to D through the resistor, it loses this electrical potential energy as it undergoes collisions with the atoms of the resistor. The electrical potential energy is converted to thermal energy. Note that there is no loss in energy for the paths BC and DA because the connecting wires are assumed to have negligible resistance in comparison to other elements in the circuit. (In all future circuit diagrams, straight lines will indicate connecting wires with zero resistance.) When the charge returns to the negative terminal of the battery, *it must have the same potential energy (zero) as it had at the start.* If this were not the case, the charge would gain energy during each trip around the circuit, which would be a violation of the law of conservation of energy.

The electrical potential energy that is transformed into heat as the charge ΔQ flows through the resistor is $V \Delta Q$. Thus, the rate of conversions of electrical energy to heat is

$$\frac{\Delta Q}{\Delta t} V = IV$$

where I is the current in the circuit. Of course, the charge regains this energy when it passes through the battery. Since the rate at which the charge loses energy equals the **power**, P, lost in the resistor, we have

$$P = IV \tag{18.8}$$

High-voltage transmission lines for delivering electrical power. *(Hank Delespinasse/The IMAGE Bank)*

Power

In this case, the power is supplied to a resistor by a battery. However, Equation 18.8, often referred to as **Joule's law,** can be used to calculate the power transferred to any device carrying a current I and having a potential difference V between its terminals.

By combining Equation 18.8 and the voltage-current relationship for a resistor, $V = IR$, we can express the power lost in the resistor in either of two alternative forms:

$$P = I^2 R = \frac{V^2}{R} \tag{18.9}$$

Power lost in a resistor

When I is in amperes, V in volts, and R in ohms, the SI unit of power is the **watt** (W). (Recall that $1 \text{ W} = 1 \text{ J/s}$.)

Example 18.7 The Power Consumed by an Electric Heater

An electric heater is constructed by applying a potential difference of 50 V to a nichrome wire of total resistance 8 Ω. Find the current carried by the wire and the power rating of the heater.

Solution

Since $V = IR$, we have

$$I = \frac{V}{R} = \frac{50 \text{ V}}{8 \text{ }\Omega} = \boxed{6.25 \text{ A}}$$

We can find the power rating using $P = I^2R$:

$$P = I^2R = (6.25 \text{ A})^2(8 \text{ }\Omega) = \boxed{313 \text{ W}}$$

Exercise   If we were to double the applied voltage to the heater, what would happen to the current and power?

Answer   The current would double, and the power would quadruple.

Example 18.8   Electrical Rating of a Lightbulb

A lightbulb is rated at 120 V and 75 W. That is, its operating voltage is 120 V and it has a power rating of 75 W. The bulb is powered by a 120-V direct current power supply. Find the current in the bulb and its resistance.

Solution

Since we know that the power rating of the bulb is 75 W and the operating voltage is 120 V, we can use $P = IV$ to find the current:

$$I = \frac{P}{V} = \frac{75 \text{ W}}{120 \text{ V}} = \boxed{0.625 \text{ A}}$$

Using Ohm's law, $V = IR$, the resistance is calculated to be

$$R = \frac{V}{I} = \frac{120 \text{ V}}{0.625 \text{ A}} = \boxed{192 \text{ }\Omega}$$

Exercise   What would the resistance be in a lamp rated at 120 V and 100 W?

Answer   144 Ω.

(Left) Photograph of a small permanent magnet levitated above a disk of the superconductor $YBa_2Cu_3O_{7-\delta}$ which is at 77 K. *(Courtesy of IBM Research) (Right)* This prototype train, constructed in Japan, has superconducting magnets built into its base. A powerful magnetic field both levitates the train a few inches above the track, and propels it smoothly at speeds of 300 miles per hour or more. *(© T. Matsumoto/SYGMA)*

*18.7   Energy Conversion in Household Circuits

The heat generated when current flows through a resistive material is used in many common devices. A cross-sectional view of the spiral heating element of an electric range is shown in Figure 18.8a. The material through which the current passes is surrounded by an insulating substance in order to prevent the current from flowing through the cook to the Earth when he or she touches the pan. A material that is a good conductor of heat surrounds the insulator.

Figure 18.8b shows a common hair dryer, in which a fan blows air past heating coils. In this case the warm air can be used to dry hair, but on a broader scale this same principle is used to dry clothes and to heat buildings.

A final example of a household appliance that uses the heating effect of electric currents is the steam iron shown in Figure 18.8c. A heating coil warms the bottom of the iron and simultaneously turns water to steam, which is sprayed from jets located in the bottom of the iron.

Regardless of the way in which you use electrical energy in your home, you ultimately must pay for it or risk having your power turned off. The unit of energy the electric company uses to calculate consumption, the **kilowatt-hour,** is defined in terms of the unit of power. One kilowatt-hour (kWh) is the energy converted or consumed in 1 h at the constant rate of 1 kW. The numerical value of 1 kWh is

$$1 \text{ kWh} = (10^3 \text{ W})(3600 \text{ s}) = 3.6 \times 10^6 \text{ J} \qquad \textbf{(18.10)}$$

The kilowatt-hour is a unit of energy

On your electric bill, the amount of electricity used is usually stated in multiples of kWh.

Example 18.9   The Cost of Operating a Lightbulb

How much does it cost to burn a 100-W lightbulb for 24 h if electricity costs eight cents per kilowatt-hour?

Figure 18.8
(a) The cross-section of a heating element used in an electric range. (b) In a hair dryer, warm air is produced by blowing air from a fan past the heating coils. (c) In a steam iron, water is turned into steam by heat from a heating coil.

(a)   (b)   (c)

Solution

A 100-W lightbulb is equivalent to a 0.1-kW bulb. Since the energy consumed equals power × time, the amount of energy you must pay for, expressed in kWh, is

$$\text{Energy} = (0.10 \text{ kW})(24 \text{ h}) = 2.4 \text{ kWh}$$

If energy is purchased at eight cents per kWh, the cost is

$$\text{Cost} = (2.4 \text{ kWh})(\$0.08/\text{kWh}) = \boxed{\$0.19}$$

That is, it will cost 19 cents to operate the light bulb for one day. This is a small amount, but when larger and more complex devices are being used, the costs go up rapidly.

Demands on our energy supplies have made it necessary to be aware of the energy requirements of our electric devices. This is true not only because they are becoming more expensive to operate but also because, with the dwindling of the coal and oil resources that ultimately supply us with electrical energy, increased awareness of conservation becomes necessary. On every electric appliance is a label that contains the information you need to calculate the power requirements of the appliance. The power consumption in watts is often stated directly, as on a lightbulb. In other cases, the amount of current used by the device and the voltage at which it operates are given. This information and Equation 18.8 are sufficient to calculate the operating cost of any electric device.

Exercise   If electricity costs eight cents per kilowatt-hour, what does it cost to operate an electric oven, which operates at 20 A and 220 V, for 5 h?

Answer $1.76.

Summary

The **electric current,** I, in a conductor is defined as

$$I \equiv \frac{\Delta Q}{\Delta t} \tag{18.1}$$

where ΔQ is the charge that passes through a cross section of the conductor in a time Δt. The SI unit of current is the **ampere** (A), where $1 \text{ A} = 1 \text{ C/s}$. By convention, the direction of current is in the direction of flow of positive charge.

The current in a conductor is related to the motion of the charge carriers through the relationship

$$I = nqv_{d}A \tag{18.3}$$

where n is the number of mobile charge carriers per unit volume, q is the charge on each carrier, v_{d} is the drift velocity of the charges, and A is the cross-sectional area of the conductor.

Ohm's law provides a relationship between the voltage drop across a conductor (V), the current through it (I), and the resistance of the conductor (R):

$$V = IR \qquad \textbf{(18.4)}$$

If a conductor has a length L and cross-sectional area A, its **resistance** is

$$R = \rho \frac{L}{A} \qquad \textbf{(18.5)}$$

where ρ is an intrinsic property of the conductor called the **electrical resistivity**. The SI unit of resistance is the **ohm** (Ω), and the SI unit of resistivity is the **ohm-meter** ($\Omega \cdot m$).

The resistivity of a conductor varies with temperature over a limited temperature range according to the expression

$$\rho = \rho_0[1 + \alpha(T - T_0)] \qquad \textbf{(18.6)}$$

where α is the **temperature coefficient of resistivity** and ρ_0 is the resistivity at some reference temperature T_0 (usually taken to be $20°C$).

The resistance of a conductor varies with temperature according to the expression

$$R = R_0[1 + \alpha(T - T_0)] \qquad \textbf{(18.7)}$$

If a potential difference V is maintained across a conductor, the **power**, or rate at which energy is supplied to the conductor, is given by

$$P = IV \qquad \textbf{(18.8)}$$

Using the fact that $V = IR$ for a resistor, the power lost in a resistor can be expressed as

$$P = I^2R = \frac{V^2}{R} \qquad \textbf{(18.9)}$$

A **kilowatt-hour** is the amount of energy converted or consumed in one hour by a device being supplied with power at the rate of 1 kW. This is equivalent to

$$1 \text{ kWh} = 3.6 \times 10^6 \text{ J} \qquad \textbf{(18.10)}$$

Additional Reading

M. Azbcl et al., "Conduction Electrons in Metals," *Sci. American*, January 1973, p. 88.

H. Ehrenreich, "The Electrical Properties of Materials," *Sci. American*, September 1967, p. 194.

T. Geballe, "Superconductors in Electric Power Technology," *Sci. American*, November 1980, p. 138.

R. M. Hazen, *The Breakthrough; The Race for the Super-conductor*, New York, Summit Books, 1988. An account of recent discoveries in high-temperature superconductivity.

R. T. Matthias, "Superconductivity," *Sci. American*, November 1957, p. 92.

D. Snowden, "Superconductors in Power Transmission," *Sci. American*, April 1972, p. 84.

Questions

1. In an analogy between traffic flow and electrical current, what would correspond to the charge Q? What would correspond to the current I?

2. Suppose you charge a pocket comb by friction and then walk around the room carrying the comb. Are you producing an electric current?

3. We have seen that an electric field must exist inside a conductor that carries a current. How is this possible in view of the fact that, in our study of electrostatics, we concluded that the electric field must be zero inside a conductor?

4. What is the difference between the drift velocity of an electron and its velocity?

5. What would happen to the drift velocity of the electrons in a wire and to the current in the wire if the electrons could move freely without resistance through the wire?

6. If charges flow very slowly through a metal, why does it not require several hours for a light to come on when you throw a switch?

7. What factors affect the resistance of a conductor?

8. What is the difference between resistance and resistivity?

9. Two wires, A and B, are made of the same metal and have equal lengths, but the resistance of wire A is three times greater than that of wire B. What is the ratio of their cross-sectional areas? How do their radii compare?

10. Use the atomic theory of matter to explain why the resistance of a material should increase as its temperature increases.

11. How does the resistance change with temperature for copper and silicon? Why are they different?

12. To reduce static electricity, a certain material is to be sprayed on a nylon rug. Describe the necessary electrical properties of this material.

13. Edison's original lightbulb contained a carbon filament. How should carbon be described electrically?

14. When the voltage across a certain conductor is doubled, the current is observed to increase by a factor of 3. What can you conclude about the conductor?

15. When incandescent lamps burn out, they usually do so just after they are switched on. Why?

16. Some homes have light dimmers that you operate by rotating a knob. What are you changing in the electric circuit when you rotate the knob?

17. In the water analogy of an electric circuit, what corresponds to the power supply, resistor, charge, and potential difference?

18. What single experimental requirement makes superconducting devices expensive to operate? In principle, can this limitation be overcome?

19. If materials could be produced that would be superconducting at room temperature, list some ways that such materials could benefit mankind.

20. Two lightbulbs operate from 120 V, but one has a power rating of 25 W and the other a power rating of 100 W. Which bulb has the higher resistance? Which carries the greater current?

21. Two conductors of the same length and radius are connected across the same potential difference. One conductor has twice the resistance of the other. Which conductor will dissipate more power?

22. A hot pad for sore, aching muscles carries a label indicating the proper voltage and current. How are these specifications related to the amount of heat produced by the hot pad? What would happen if the hot pad were accidentally plugged into the wrong voltage? Consider cases in which the voltage is greater and less than the proper voltage.

23. If you were to design an electric heater using nichrome wire as the heating element, what parameters of the wire could you vary to meet a specific power output, such as 1000 W?

24. If electrical power is transmitted over long distances, the resistance of the wires becomes significant. Why? Which mode of transmission would result in less energy loss—high current and low voltage or low current and high voltage? Discuss.

25. A typical monthly utility rate structure might be: $1.60 for the first 16 kWh, 7.05 cents/kWh for the next 34 kWh used, 5.02 cents/kWh for the next

50 kWh, 3.25 cents/kWh for the next 100 kWh, 2.95 cents/kWh for the next 200 kWh, 2.35 cents/kWh for all in excess of 400 kWh. Based on these rates, what would be the charge for 527 kWh? From the standpoint of encouraging conservation of energy, what is wrong with this pricing method?

Problems

Section 18.1   Electric Current
Section 18.2   Current and Drift Velocity

1. The compressor on an air conditioner draws 90 A when it starts up. If the start-up time is about 1/2 s, how much charge passes a cross-sectional area of the circuit in this time?

2. How many electrons pass through the cross-sectional area in Problem 1 during the 1/2-s interval?

3. If Avogadro's number of electrons pass by a given cross-sectional area in one hour, find the current in the conductor.

4. A single positive point charge of 1.67 μC travels with a speed of 80.0 km/s around a circular orbit of radius 2.15 m. Determine the time-averaged current corresponding to the circulating charge.

5. In a particular television picture tube, the measured beam current is 60 μA. How many electrons strike the screen every second?

6. If a current of 80.0 mA exists in a metal wire, how many electrons flow past a given cross section of the wire in 10.0 min? Sketch the direction of current and the direction of the electrons' motion.

7. If 3.25×10^{-3} kg of gold is deposited on the negative electrode of an electrolytic cell in a period of 2.78 h, what is the current through the cell in this period? Assume the gold ions carry one elementary unit of positive charge.

8. A teapot with a surface area of 700 cm² is to be silver plated. It is attached to the negative electrode of an electrolytic cell containing silver nitrate ($Ag^+NO_3^-$). If the cell is powered by a 12.0-V battery and has a resistance of 1.8 Ω, how long does it take to build up a 0.133-mm layer of silver on the teapot? (Density of silver = 10.5×10^3 kg/m³.)

9. Calculate the number of free electrons per cubic meter for gold, assuming one free electron per atom. (Density of gold = 19.3×10^3 kg/m³.)

10. An aluminum wire having a cross-sectional area of 4×10^{-6} m² carries a current of 5 A. Find the drift velocity of the electrons in the wire. The density of aluminum is 2.7 g/cm³. (Assume one electron is supplied by each atom.)

11. A metal wire 1.0 mm in diameter contains 2.50×10^{22} free electrons per cubic centimeter. If the electrons travel through the wire with an average drift speed of 0.5 mm/s, what is the current in the wire?

12. A typical aluminum wire has about 1.5×10^{20} free electrons in every centimeter of its length, and the drift speed of these electrons is about 0.03 cm/s. (a) How many electrons pass through a cross-sectional area of the wire every second? (b) What is the current in the wire?

Section 18.3   Ohm's Law
Section 18.4   Resistivity and Resistance

13. Find the resistance of a piece of silver 1 m long if its cross-sectional area is 10^{-5} m².

14. An electrician finds that the resistance of a 1-m length of wire is 0.20 Ω. If he plans to use 120 m of this wire in wiring a house, what is the total resistance of all the wiring used?

15. Calculate the diameter of a 2-cm length of tungsten filament in a small lightbulb if its resistance is 0.05 Ω.

16. Eighteen-gauge wire has a diameter of 1.024 mm. Calculate the resistance of 15 m of 18-gauge copper wire at 20°C.

17. A person notices a mild shock if the current along a path through the thumb and index finger exceeds 80 μA. Compare the maximum allowable voltage without shock across the thumb and index finger with a dry-skin resistance of 4.0×10^5 Ω and a wet-skin resistance of 2000 Ω.

18. A lightbulb has a resistance of 240 Ω when operating at a voltage of 120 V. What is the current through the lightbulb?

19. A typical color television draws about 2.5 A when connected to a 120-V source. What is the effective resistance of the TV set?

20. When operating at 120 V, a resistor carries a current of 0.5 A. What current would be carried if (a) the operating voltage is lowered to 90 V, (b) the voltage is raised to 130 V?

21. A sales representative promoting energy-efficient products advertises that a certain product has a resistance of 20 Ω and carries a current of only 0.6 A when connected to a 120-V source. Would you buy this product? Why or why not?

22. (a) Calculate the resistance of a piece of nichrome wire 200 cm long with a cross-sectional area of 0.02 cm². (b) What current would the wire carry if connected to a 3-V flashlight battery?

23. Suppose that a uniform wire of resistance R is stretched uniformly to three times its original length. What will be its new resistance, assuming its density and resistivity remain constant?

24. The heating element of a toaster is made of nichrome wire 1.0 mm in diameter. If the toaster has a resistance of 8.5 Ω, how long is the wire?

25. A potential difference of 12 V is found to produce a current of 0.4 A in a 3.2-m length of wire having a uniform radius of 0.4 cm. (a) Calculate the resistivity of the wire. (b) What is the resistance of the wire?

26. A 2-m piece of iron wire carries a current of 0.25 A when connected to a 6-V battery. What length of gold wire would carry the same current when connected to the battery if the gold wire has the same radius as the iron?

27. A rectangular block of copper has sides of length 10 cm, 20 cm, and 40 cm. If the block is connected to a 6-V source across opposite faces of the rectangular block, what is (a) the maximum current that can be carried and (b) the minimum current?

28. A battery establishes a uniform electric field of 100 V/m in a wire of radius 0.5 cm. If the current in the wire is 5 A, what is the resistivity of the material?

Section 18.5   Temperature Variation of Resistance

29. If a silver wire has a resistance of 10 Ω at 20°C, what resistance will it have at 40°C? Neglect any change in length or cross-sectional area due to the change in temperature.

30. The copper wire used in a house has a cross-sectional area of 3 mm². If 10 m of this wire is used to wire a circuit in the house at 20°C, find the resistance of the wire at a temperature of (a) 30°C and (b) 10°C.

31. At 40°C, the resistance of a segment of gold wire is 100 Ω. When the wire is placed in a liquid bath, the resistance decreases to 97 Ω. What is the temperature of the bath?

32. At 20°C the resistance of the carbon resistor in an electric circuit is 200 Ω and it is connected to a 5-V battery. What is the current in the circuit when the temperature of the carbon rises to 80°C?

33. A wire 3 m in length and 0.45 mm² in cross-sectional area has a resistance of 41 Ω at 20°C. If the resistance of the wire increases to 41.4 Ω at 29°C, what is the temperature coefficient of resistivity?

34. Calculate the percentage change in the resistance of a carbon filament when it is heated from 20°C to 160°C.

35. How much would the temperature of a copper wire have to be increased to raise its resistance by 20% over the value it had at 20°C?

36. A certain lightbulb has a tungsten filament with a resistance of 19 Ω when cold and 140 Ω when hot. Assume that Equation 18.7 can be used over the large temperature range involved here and find the temperature of the filament when hot. Assume an initial temperature of 20°C.

37. A copper wire of length 100 cm and radius 0.5 cm has a potential difference across it sufficient to produce a current of 3 A at 20°C. (a) What is the potential difference? (b) If the temperature of the wire is increased to 200°C, what potential difference is now required to produce a current of 3 A?

38. At what temperature will tungsten have a resistivity four times the resistivity of copper at room temperature?

Section 18.6   Electrical Energy and Power

39. A toaster is rated at 600 W when connected to a 120-V source. What current does the toaster carry and what is its resistance?

40. The power supplied to a typical black-and-white television set is 90 W when the set is connected to 120 V. (a) How much energy does this set consume in one hour? (b) A color television set draws about 2.5 A when connected to 120 V. How long is required for it to consume the same energy as does the black-and-white model in one hour?

41. How much current is being supplied by a 240-V generator delivering 120 kW of power?

42. The sticker on a compact disk player says that it draws 300 mA of current at 9 V. What power does it dissipate?

43. The tungsten heating element in a 1500-W heater is 3 m long, and the resistor is to be connected to a 120-V source. What is the cross-sectional area of the wire?

44. Determine the loss of electrical power per meter of a copper wire 2.0 mm in diameter if a current of 40.0 A exists in the wire.

45. Suppose that a voltage surge produces 140 V for a moment. By what percentage will the output of a 120-V, 100-W lightbulb increase, assuming its resistance does not change?

46. How many 100-W lightbulbs can you use in a 120-V circuit without tripping a 15-A circuit breaker? (*Note:* The bulbs are connected in parallel.)

47. A 0.25-hp motor is connected to a 120-V line. What current is carried by the motor?

48. What is the required resistance of an immersion heater that will increase the temperature of 1.5 kg of water from 10°C to 50°C in 10 min while operating at 120 V?

49. In a hydroelectric installation, a turbine delivers 2000 hp to a generator, which in turn converts 90%

of the mechanical energy to electrical energy. Under these conditions, what current will the generator deliver at a potential difference of 3000 V?

50. Two conductors made of the same material are connected across a common potential difference. Conductor A has twice the diameter and twice the length of conductor B. What is the ratio of the power delivered to the two conductors?

Section 18.7   Energy Conversion in Household Circuits

51. How much does it cost to watch a complete 21-hour-long World Series on a 90-W black-and-white television set? Assume electricity costs 7 cents per kWh.

52. A certain household uses an average of 2500 kWh of energy per month. If coal supplies 7×10^6 cal/kg, how much coal must be burned in a 40% efficient generator to supply this much energy?

53. If electricity costs 8 cents per kWh, estimate how much it costs a person to dry his hair with a 1500-W blow dryer during a year's time.

54. A small motor draws a current of 1.75 A from a 120-V line. The output power of the motor is 0.2 hp. (a) At a rate of $0.06/kWh, what is the cost of operating the motor for 4 h? (b) What is the efficiency of the motor?

55. An electric resistance heater is to deliver 1500 kcal/h to a room using 110-V electricity. If fuses come in 10-A, 20-A, and 30-A sizes, what is the smallest fuse we can safely use in the heater circuit?

56. An advertisement for a vaporizer that operates on 110 V states that it converts 400 cm³ of water to steam in an hour. Ignoring heat losses, determine the resistance of the vaporizer.

57. A 110-V motor produces 2.5 hp of mechanical power. If this motor is 90% efficient in converting electrical power into mechanical power, find (a) the current drawn by the motor and (b) the total electrical energy used by this motor after running for one hour. (c) If electrical energy costs 8 cents/kWh, how much did it cost to run the motor for this hour?

58. If an electric stove is capable of heating 500 cm³ of water from 20°C to boiling in 10 minutes, what is the power rating of the stove? Assume 50% efficiency.

Additional Problems

59. Storage batteries are often rated in terms of the amount of charge they can deliver. How much charge can a 90 amp-hour battery deliver?

60. A steam iron draws 6 A from a 120-V line. How many joules of heat are produced in 20 min?

61. How much does it cost at $0.08/kWh to run the steam iron of Problem 60 for 20 minutes?

62. The headlights on a car are rated at 80 W. If these are connected to a fully charged 90 amp·h, 12-V battery, how long will it take the battery to completely discharge?

63. A particular wire has a resistivity of 3×10^{-8} Ω·m and a cross-sectional area of 4×10^{-6} m². A length of this wire is to be used as a resistor that will develop 48 W of power when connected across a 20-V battery. What length of wire is required?

64. Birds resting comfortably on a high-voltage power line are a common sight. The copper wire on which a bird stands is 2.2 cm in radius and carries a current of 50 A. If the bird's feet are 4 cm apart, calculate the potential difference across his body.

65. A length of metal wire has a radius of 5×10^{-3} m and a resistance of 0.1 Ω. When the potential difference across the wire is 15 V, the electron drift velocity is found to be 3.17×10^{-4} m/s. Based on these data, calculate the density of free electrons in the wire.

66. Two conductors of the same length and same material are connected in series (one after the other) across a difference in potential. The wires have different cross-sectional areas. (a) When in series, the current is the same in both wires. Does the wire of larger cross-sectional area dissipate more or less power than the smaller wire? (b) Repeat the problem if the wires are connected in parallel (so that the voltage is the same across each).

67. A small sphere that carries a charge of 8 nC is whirled in a circle at the end of an insulating string. The rotation frequency is 100π rad/s. What average current does this rotating charge represent?

68. The current through a resistor decreases by 3 A when the voltage applied across the resistor decreases from 12 V to 6 V. Find the resistance of the resistor.

69. (a) A 115-g mass of aluminum is formed into a right circular cylinder shaped so that the diameter of the cylinder equals its height. Calculate the resistance between the top and bottom faces of the cylinder at 20°C. (b) Calculate the resistance between opposite faces if the same mass of aluminum is formed into the cube.

70. A resistor is constructed by forming a material of resistance 3.5×10^5 Ω·m into the shape of a hollow cylinder of length 4 cm and inner and outer radii of 0.5 cm and 1.2 cm, respectively. In use, a potential difference is applied between the ends of the cylinder, producing a current parallel to the length of the cylinder. Find the resistance of the cylinder.

71. The current in a conductor varies in time as shown in Figure 18.9. (a) How many coulombs of charge pass through a cross section of the conductor in the interval $t = 0$ to $t = 5$ s? (b) What constant current would

transport the same total charge during the 5-s interval as does the actual current?

Figure 18.9
Problem 71

72. A wire of initial length L_0 and radius r_0 has a measured resistance of $1\ \Omega$. The wire is drawn under tensile stress to a new uniform radius of $r = 0.25r_0$. What is the new resistance of the wire?

73. A 50-g sample of a conducting material is all that is available. The resistivity of the material is measured to be $11 \times 10^{-8}\ \Omega\cdot m$, and the density is $7.86\ g/cm^3$. It is desired to shape the material into a wire that has a total resistance of $1.5\ \Omega$. (a) What length is required and (b) what must be the diameter of the wire?

74. (a) A sheet of copper ($\rho = 1.7 \times 10^{-8}\ \Omega\cdot m$) is 2 mm thick and has surface dimensions of 8 cm $\times$ 24 cm. If the long edges are joined to form a hollow tube 24 cm in length, what is the resistance between the ends? (b) What mass of copper would be required to manufacture a spool of copper cable 1500 m in length and having a total resistance of $4.5\ \Omega$?

75. An engineer is in need of a resistor that is to have zero overall temperature coefficient of resistance at 20°C. The design is a composite of right circular cylinders of two materials, as in Figure 18.10. The ratio of the resistivities of the two materials is $\rho_1/\rho_2 = 3.2$, and the ratio of the lengths of the sections is $\ell_1/\ell_2 = 2.6$. The radius is uniform throughout. Assuming that the temperature of the two sections remains equal, calculate α_1/α_2, the required ratio of temperature coefficients of resistivity of the two materials.

Figure 18.10   Problem 75

"I'm afraid Mr. Edison is too busy to discuss your idea with you at this time. Please call again, Mr. Fluorescent."

Super-conductivity

Introduction

At a temperature typically between 1 and 10 K, many metals and alloys undergo a change of phase. At higher T they exhibit ordinary resistance, R, to current (Ohm's law), but at low T, the resistance vanishes. The high T phase is called the normal (or N) state of the electrons, and the low T phase is called the superconducting (or S) state. When you plot R vs. T, you find a sharp transition at a "critical temperature," called T_c. The purer the material, the sharper the transition. The critical temperature for pure mercury is 4.2 K (see Fig. 1).

The discovery of this phenomenon came in 1911 at the laboratory of Heike Kamerlingh-Onnes in Leiden, Holland. Several aspects of this amazing behavior were realized early on. First, there is a close analogy to the well-studied phase transition of water to ice, commonly called freezing. This too takes place at a specific temperature, 0°C for pure water. Impure water is liquid to a lower temperature. The clearest property showing the sharp transition for water is density, since water abruptly expands upon freezing. The higher T phase has the higher entropy (liquid water), and lower T has lower entropy (solid ice). The same idea applies to normal conduction electrons when compared with superconducting electrons.

Second, any unusual electrical property is accompanied by an equally unusual magnetic property, because the same mobile charges are responsible for both. In the 1930s it was established that at T below T_c, a moderate magnetic flux is expelled from the interior of a superconductor. A pure superconductor expels 100% of the flux, except for a narrow surface layer. However, a large magnetic flux destroys superconductivity, and the metal reverts to the normal state even at T below T_c. The large flux penetrates the metal in the normal way.

By the 1950s numerous electromagnetic and thermal properties of the S state were known. They were explained by the theory of superconductivity due to John Bardeen, Leon Cooper, and Robert Schrieffer ("the BCS theory") at the University of Illinois and by mathematical and conceptual extensions due to many others.

Explanation

The BCS theory treats the conduction electrons as an "electron fluid." This fluid is capable of freezing at a low enough T. But the freezing process is not the same as what water molecules undergo. The orderly arrangement of molecules in ice takes place in what we call "position space." This is the space we live in and look at. We place axes labeled x, y, z in this space, and use them to designate the position of any particle. A regular arrangement of positions is a low entropy state.

An alternative transition for a fluid takes place in an abstract space called "velocity space." This is the space that is conventionally used to describe motions of particles (or motions of the centers of mass of bodies). We place axes labeled v_x, v_y, v_z in this space and use them to designate the velocity of any particle. A traffic engineer could study the distribution of automobile velocities on a highway. This would show the fraction of drivers obeying the posted speed limits.

Automobiles travel at a variety of speeds. In my state 55 mph is the limit, and 40 mph is the minimum on an interstate highway. Some cars go much faster than that. It is inconceivable that all cars would go at 55, even tolerating slight variations and accepting 54 or 56 mph. If they did (and stayed in their lanes), the number of accidents would be greatly reduced.

What is inconceivable for automobiles is achievable by electrons that conspire to form a superconducting phase. This is the state of the electron fluid in which all electrons do exactly the same thing. What they do is form bound pairs,

David Markowitz
University of Connecticut

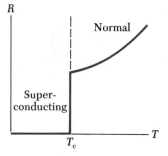

Figure 1
Resistance versus temperature for a pure superconductor, such as mercury, lead, or aluminum. The metal behaves normally down to the critical temperature, T_c, below which the resistance drops to zero.

and each pair follows the identical motion. They tolerate no variation at all, and they never have an accident. The process of forming this state from previously unpaired electrons is "freezing in velocity space." The regular arrangement of velocities is a low entropy (hence, low T) state.

Some questions arise. Why do electrons want to form identical pairs? Answer: Pairing lowers their potential energy of interaction. They pair for the same reason a released ball falls to the Earth. Why are electrons able to form identical pairs? Answer: Only the fact that modern physics governs electrons allows them to do that. In classical physics properties of motion are continuous, but in modern physics they are quantized. An analogy may help. Throw a number of darts at a target. They will not hit the same point, because there are so many (an infinite number) nearby points. But they may all hit the same ring of the target. The rings are quantized; neighboring rings are clearly separated.

Present Situation

There are different classifications of superconductors. They are all electron systems, but they are not all regarded as fluids. Still, it appears that they all rely on the pairing mechanism, and it is pairs that carry current without resistance.

The simplest class of superconductors is elemental metals, and the next simplest is alloys formed from them. Apparently, any structure is permissible to a metallic superconductor, as long as the material satisfies two criteria: large enough number of valence electrons per atom, and strong enough interaction between electrons and the polarizable lattice. Since these properties are widespread among elements in the periodic table, these classes contain large numbers of members. Lead and its alloys are superconductors; copper and brass are not.

Some materials turn superconducting when they are amorphous but not when they are crystalline. Bismuth acts this way. Others are semiconducting in one crystalline form, metallic superconductors in another. Tin fits this description. This suggests that structure influences electronic behavior. Then too, some materials have strikingly different superconducting properties when they are fashioned into thin films, rather than bars or wires. A film that is thin enough acts as if it constrains electrons to move in two dimensions instead of the usual three, implying that dimensionality also affects electronic behavior.

These features reach their peak of subtlety, surprise, and importance in the newest class of superconductors to be discovered: ceramics that possess the highest values of T_c known. Credit for this discovery belongs to J. Georg Bednorz and Karl Alex Müller, physicists at the **IBM** Research Laboratory in Zurich, Switzer-

J. Georg Bednorz and Karl Alex Müller, discoverers of the first high-temperature superconductor. *(Courtesy of IBM Research)*

A small permanent magnet levitated above a disk of the superconductor $YBa_2Cu_3O_7$ cooled to liquid nitrogen temperature. *(Courtesy of IBM Research)*

land. While metallic superconductors are generally restricted to the range 1 – 20 K, the new class has T_c up to temperatures above 130 K. Higher critical temperatures are claimed but not verified. These ceramics possess a special type of structure, called "layered." It is as if sheets of strongly bound atoms were laid on top of each other with weaker bonds. Above T_c they are electrical insulators (non-fluid electrically); below they are superconductors. They are known as high temperature superconductors.

There are high hopes for these materials. One hope is that much higher T_c will be found so as to achieve a room temperature superconductor. Because of the repulsion between a superconductor and a magnet, engineers have designed, built, and tested an invention called a "Mag Lev," meaning a magnetically levitated vehicle. This vehicle would be quieter and smoother in performance, and more energy-efficient than any vehicle now in use or seriously contemplated. Other large-scale uses include electrical energy transport over long distances. However, it appears that for now, small-scale precision devices are the best bets for economical technological uses. These include magnetometers and computer memory elements.

Computers form one of the largest industries in the world. In a computer, photons and electrons can play the same roles. For example, the storage of one bit of information can be a tiny electric current or a tiny ray of light. As another example, the switching of information from one value to another can be a change in current or in illumination. It is not known which means will eventually be more reliable and economical. In this rivalry between electronics and photonics, high T_c superconductors may play the decisive role.

Figure 19.2
Series connection of two resistors, R_1 and R_2. The current in each resistor is the same, and the equivalent resistance of the combination is given by $R_{eq} = R_1 + R_2$.

A series connection of three lamps, all rated at 120 V, with power ratings of 60 W, 75 W, and 150 W. Why are the intensities of the lamps different? Which lamp has the greatest resistance? How would their relative intensities differ if they were connected in parallel? *(Courtesy of Henry Leap and Jim Lehman)*

Figure 19.2 shows two resistors, R_1 and R_2, connected to a battery in a circuit called a series circuit. In such a circuit, there is only one pathway for the current. Charges must pass through both resistors and the battery as they traverse the circuit. Hence, all charges in a series circuit must follow the same conducting path.

Note that the current is the same through each resistor in a series circuit because any charge that flows through R_1 must also flow through R_2.

This is analogous to water flowing through a pipe with two constrictions, corresponding to R_1 and R_2. Whatever volume of water flows in one end in a given time interval must exit the opposite end.

Since the potential drop from a to b in Figure 19.2b equals IR_1 and the potential drop from b to c equals IR_2, the potential drop from a to c is

$$V = IR_1 + IR_2$$

Figure 19.2c shows an equivalent resistor, R_{eq}, that can replace the two resistors of the original circuit. Applying Ohm's law to this resistor, we have

$$V = IR_{eq}$$

Equating these two expressions, we have

$$IR_{eq} = IR_1 + IR_2$$

or

$$R_{eq} = R_1 + R_2 \qquad \text{(series combination)} \qquad \textbf{(19.1)}$$

The resistance R_{eq} is equivalent to the series combination $R_1 + R_2$ in the sense that the circuit current is unchanged when R_{eq} replaces R_1 and R_2.

An extension of this analysis shows that the equivalent resistance of three or more resistors connected in series is

$$R_{eq} = R_1 + R_2 + R_3 + \ldots \qquad \text{(series combination)} \qquad \textbf{(19.2)}$$

From this, we see that *the equivalent resistance of a series combination of resistors is always greater than any individual resistance.* As a somewhat crude analogy here, placing several resistors end to end in a circuit is somewhat like creating a longer resistor. Based on the equation $R = \rho L/A$, show why we would expect the resistance of two or more resistors in series to be greater than any individual resistance.

Note that if the filament of one lightbulb in the series circuit shown in Figure 19.2a were to break, or burn out, the circuit would no longer be complete and the second bulb would also go out. Some Christmas tree light sets are connected in this way, and the agonizing experience of determining which bulb in a set is the culprit that has burned out is a familiar one. Frustrating experiences such as this illustrate how inconvenient it would be to have all appliances in a house connected in series.

Example 19.1   Four Resistors in Series

Four resistors are arranged as shown in Figure 19.3a. Find (a) the equivalent resistance and (b) the current in the circuit if the emf of the battery is 6 V.

Solution
(a) The equivalent resistance is found from Equation 19.2 to be

$$R_{eq} = R_1 + R_2 + R_3 + R_4 = 2\,\Omega + 4\,\Omega + 5\,\Omega + 7\,\Omega = \boxed{18\,\Omega}$$

(b) If we apply Ohm's law to the equivalent resistor in Figure 19.3b, we find the current in the circuit to be

$$I = \frac{V}{R_{eq}} = \frac{6\ \text{V}}{18\ \Omega} = \boxed{\frac{1}{3}\ \text{A}}$$

Exercise   If $\frac{1}{3}$ A flows through the equivalent resistor, this must also be the current in each resistor of the original circuit. Find the voltage drop across each resistor.

Answer $V_{2\Omega} = \frac{2}{3}$ V, $V_{4\Omega} = \frac{4}{3}$ V, $V_{5\Omega} = \frac{5}{3}$ V, $V_{7\Omega} = \frac{7}{3}$ V.

Figure 19.3   (Example 19.1) (a) Four resistors connected in series. (b) The equivalent resistance of the circuit in (a).

Georg Simon Ohm (1787–1854), German physicist. *(Courtesy of AIP Niels Bohr Library, E. Scott Barr Collection)*

19.3   Resistors in Parallel

Now consider two resistors connected in parallel, as shown in Figure 19.4.

> When resistors are connected in parallel, the potential difference is the same across each resistor.

This must be true because the left side of each resistor is connected to a common point, the positive side of the battery, point a in Figure 19.4b, and the right side of each resistor is connected to a common point, the negative terminal of the battery in Figure 19.4b.

However, the current in each resistor is in general not the same. The currents will be the same only if the resistors have the same resistance. For example, three identical bulbs connected in parallel with a battery would each draw the same current. When the current, I, reaches point a (called a junction) in Figure 19.4b, it splits into two parts, I_1 going through R_1 and I_2 going through R_2. If R_1 is greater than R_2, then I_1 will be less than I_2. That is, *the charge will tend to follow the path of least resistance. Since charge is conserved, the current I that enters point a must equal the total current leaving this point, $I_1 + I_2$.* That is,

$$I = I_1 + I_2$$

This is a special case of one of two Kirchhoff rules discussed in the next section. The potential drop must be the same across each resistor and must also equal the potential difference across the battery. Ohm's law applied to each resistor gives

$$I_1 = \frac{V}{R_1} \qquad I_2 = \frac{V}{R_2}$$

If we apply Ohm's law to the equivalent resistor in Figure 19.4c, we have

$$I = \frac{V}{R_{eq}}$$

(a)

Figure 19.4
(a) Parallel connection of two resistors. (b) Circuit diagram for the parallel combination. (c) The voltage across each resistor is the same, and the equivalent resistance of the combination is given by the reciprocal relationship $1/R_{eq} = 1/R_1 + 1/R_2$.

(b)

(c)

When these expressions for the current are substituted into the equation $I = I_1 + I_2$ and V is cancelled, we obtain

$$\frac{1}{R_{eq}} = \frac{1}{R_1} + \frac{1}{R_2}$$ (parallel combination)   **(19.3)**

An extension of this analysis to three or more resistors in parallel gives the following general expression for the equivalent resistance:

$$\frac{1}{R_{eq}} = \frac{1}{R_1} + \frac{1}{R_2} + \frac{1}{R_3} + \dots$$ (parallel combination)   **(19.4)**

Three incandescent lamps with power ratings of 25 W, 75 W, and 150 W, connected in parallel to a voltage source of about 100 V. All lamps are rated at the same voltage. Why do the intensities of the lamps differ? Which lamp draws the most current? Which has the least resistance? *(Courtesy of Henry Leap and Jim Lehman)*

From this result, you should be able to show that *the equivalent resistance of two or more resistors connected in parallel is always less than the smallest resistance in the group.* As a crude analogy to help you understand why this statement is true, you can consider the process of connecting resistors in parallel as like that of creating a "fatter" resistor. As an exercise, you should determine why this analogy is appropriate and, based on the use of $R = \rho L / A$, show why this would lead to the equivalent resistance of a parallel combination being less than the smallest resistance in the combination.

Household circuits are always wired such that the lights and appliances are connected in parallel. In this manner, each device operates independently of the others, so that if one is switched off, the others remain on. Equally important, each device operates on the same voltage.

Problem-Solving Strategy: Resistors

1. Be careful with your choice of units. To calculate the resistance of a device in ohms, make sure that distances are in meters and use the SI value of ρ.
2. When two or more unequal resistors are connected in *series*, they carry the same current, but the potential differences across them are not the same. The resistors add directly to give the equivalent resistance of the series combination.
3. When two or more unequal resistors are connected in *parallel*, the potential difference across each is the same. Since the current is inversely proportional to the resistance, the current will not be the same through each. The equivalent resistance of a parallel combination of resistors is found through reciprocal addition, and the equivalent resistor is always *less* than the smallest individual resistor.
4. A complicated circuit consisting of resistors can often be reduced to a simple circuit containing only one resistor. To do so, examine the initial circuit and replace any resistors in series or any in parallel using the steps outlined in 2 and 3 above. Draw a sketch of the new circuit after these changes have been made. Examine this new circuit and replace any series or parallel combinations. Continue this process until a single, equivalent resistance is found.

5. If the current through or the potential difference across a resistor in the complicated circuit is to be found, start with the final circuit found above and gradually work your way back through the circuits using $V = IR$ and the rules of Steps 2 and 3 above.

Example 19.2   Three Resistors in Parallel

Three resistors are connected in parallel as in Figure 19.5. A potential difference of 18 V is maintained between points a and b. (a) Find the current in each resistor.

Solution
The resistors are in parallel, and so the potential difference across each is 18 V. Let us apply $V = IR$ to find the current in each resistor:

$$I_1 = \frac{V}{R_1} = \frac{18\text{ V}}{3\ \Omega} = \boxed{6\text{ A}}$$

$$I_2 = \frac{V}{R_2} = \frac{18\text{ V}}{6\ \Omega} = \boxed{3\text{ A}}$$

$$I_3 = \frac{V}{R_3} = \frac{18\text{ V}}{9\ \Omega} = \boxed{2\text{ A}}$$

Figure 19.5 (Example 19.2) Three resistors connected in parallel. The voltage across each resistor is 18 V.

(b) Calculate the power dissipated by each resistor and the total power dissipated by the three resistors.

Solution
Applying $P = I^2R$ to each resistor gives

$$3\ \Omega: \quad P_1 = I_1{}^2R_1 = (6\text{ A})^2(3\ \Omega) = \boxed{108\text{ W}}$$

$$6\ \Omega: \quad P_2 = I_2{}^2R_2 = (3\text{ A})^2(6\ \Omega) = \boxed{54\text{ W}}$$

$$9\ \Omega: \quad P_3 = I_3{}^2R_3 = (2\text{ A})^2(9\ \Omega) = \boxed{36\text{ W}}$$

This shows that the smallest resistance dissipates the most power since it carries the most current. (Note that you can also use $P = V^2/R$ to find the power dissipated by each resistor.) Summing the three quantities gives a total power of 198 W.

(c) Calculate the equivalent resistance of the three resistors.

Solution
We can use Equation 19.4 to find R_{eq}:

$$\frac{1}{R_{eq}} = \left(\frac{1}{3} + \frac{1}{6} + \frac{1}{9}\right)\Omega^{-1}$$

$$R_{eq} = \boxed{\frac{18}{11}\ \Omega}$$

Exercise   Use the result for R_{eq} to calculate the total power dissipated in the circuit.

Answer   198 W.

Example 19.3   Find the Equivalent Resistance

Four resistors are connected as shown in Figure 19.6a. (a) Find the equivalent resistance between points a and c.

Solution
The circuit can be reduced in steps, as shown in Figure 19.6b. The 8-Ω and 4-Ω resistors are in series, and so the equivalent resistance between a and b is 12 Ω (Eq. 19.2). The 6-Ω and 3-Ω resistors are in parallel, and so from Equation 19.4 we find that the equivalent resistance from b to c is 2 Ω. Hence, the equivalent resistance from a to c is 14 Ω.

(b) What is the current in each resistor if a 42-V battery is placed between a and c?

Solution
The current I in the 8-Ω and 4-Ω resistors is the same since they are in series. Let us use Ohm's law and the results from (a) to get

$$I = \frac{V_{ac}}{R_{eq}} = \frac{42 \text{ V}}{14 \text{ } \Omega} = 3 \text{ A}$$

When this current enters the junction at b, it splits; part of the current goes through the 6-Ω resistor (I_1), and part goes through the 3-Ω resistor (I_2). Since the potential difference across these resistors, V_{bc}, is the *same* (they are in parallel), we see that $(6 \text{ } \Omega)(I_1) = (3 \text{ } \Omega)(I_2)$, or $I_2 = 2I_1$. Using this result and the fact that $I_1 + I_2 = 3$ A, we find that $I_1 = 1$ A and $I_2 = 2$ A. We could have guessed this from the start by noting that the current through the 3-Ω resistor has to be twice the current through the 6-Ω resistor in view of their relative resistances and the fact that the same voltage is applied to both.

As a final check, note that $V_{bc} = 6I_1 = 3I_2 = 6$ V and $V_{ab} = 12I = 36$ V; therefore, $V_{ac} = V_{ab} + V_{bc} = 42$ V, as it must.

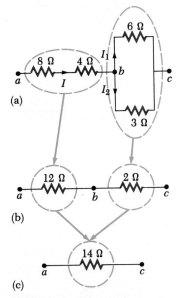

Figure 19.6 (Example 19.3) The four resistors shown in (a) can be reduced in steps to an equivalent 14-Ω resistor.

19.4   Kirchhoff's Rules

As Section 19.3 indicates, we can analyze simple circuits using Ohm's law and the rules for series and parallel combinations of resistors. However, there are many ways in which resistors can be connected such that the circuits formed cannot be reduced to a single equivalent resistor. The procedure for analyzing more complex circuits is greatly simplified by the use of two simple rules called **Kirchhoff's rules:**

1. The sum of the currents entering any junction must equal the sum of the currents leaving that junction. (A junction is any point in a circuit where the current can split.) This rule is often referred to as the **junction rule.**
2. The sum of the potential differences across each element around any closed circuit loop must be zero. This rule is usually called the **loop rule.**

The first rule is a statement of *conservation of charge.* Whatever current enters a given point in a circuit must leave that point because charge cannot build up or disappear at a point. If we apply this rule to the junction shown in Figure 19.7a, we get

$$I_1 = I_2 + I_3$$

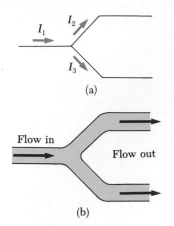

I_1 I_2 I_3

(a)

Flow in

Flow out

(b)

Figure 19.7
(a) A schematic diagram illustrating Kirchhoff's junction rule. Conservation of charge requires that whatever current enters a junction must leave that junction. Therefore, in this case, $I_1 = I_2 + I_3$. (b) A mechanical analog of the junction rule: the flow out must equal the flow in.

Figure 19.7b represents a mechanical analog to this situation in which water flows through a branched pipe with no leaks. The flow rate into the pipe equals the total flow rate out of the two branches.

The second rule is equivalent to the law of *conservation of energy.* Any charge that moves around any closed loop in a circuit (it starts and ends at the same point) must gain as much energy as it loses. Its energy may decrease in the form of a potential drop, $-IR$, across a resistor or as a result of flowing backward through a source of emf, that is, from the positive to the negative terminal inside the battery. In the latter case, electrical energy is converted to chemical energy as the battery is charged.

When applying Kirchhoff's rules, there are two decisions that you must make at the beginning of the problem.

1. You must assign symbols and directions to the currents in the various branches of the circuit. If you should happen to guess the wrong direction for a current, the end result will be a negative answer for that current, but its magnitude will be correct.
2. When applying the loop rule, you must choose a direction (clockwise or counterclockwise) for moving around the loop. As you traverse the loop, record voltage drops and increases according to the rules given below.

As an aid in applying the loop rule, the following points should be noted. These points are summarized in Figure 19.8, where it is assumed we are moving from point a toward point b:

1. If a resistor is traversed in the direction of the current, the change in potential across the resistor is $-IR$ (Fig. 19.8a).
2. If a resistor is traversed in the direction opposite the current, the change in potential across the resistor is $+IR$ (Fig. 19.8b).
3. If a source of emf is traversed in the direction of the emf (from $-$ to $+$ on the terminals), the change in potential is $+\mathcal{E}$ (Fig. 19.8c).
4. If a source of emf is traversed in the direction opposite the emf (from $+$ to $-$ on the terminals), the change in potential is $-\mathcal{E}$ (Fig. 19.8d).

There are limitations on the number of times you can use the junction rule and the loop rule. The junction rule can be used as often as needed so long as each time you write an equation, you include in it a current that has not been used in a previous junction rule equation. In general, the number of times the junction rule can be used is one fewer than the number of junction points in the circuit. The loop rule can be used as often as needed so long as a new circuit element (resistor or battery) or a new current appears in each new equation. In general, the number of independent equations you need must at least equal the number of unknowns in order to solve a particular circuit problem.

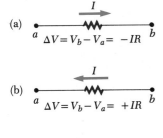

(a) I

a $\Delta V = V_b - V_a = -IR$ b

(b) I

a $\Delta V = V_b - V_a = +IR$ b

(c) $\mathcal{E}$

a $\Delta V = V_b - V_a = +\mathcal{E}$ b

(d) $\mathcal{E}$

a $\Delta V = V_b - V_a = -\mathcal{E}$ b

Figure 19.8
Rules for determining the potential changes across a resistor and a battery.

Problem-Solving Strategy:
Kirchhoff's Rules

1. First, draw the circuit diagram and assign labels and symbols to all the known and unknown quantities. You must assign a *direction* to the currents in each part of the circuit. Do not be alarmed if you guess

the direction of a current incorrectly; the result will have a negative value, but *its magnitude will be correct*. Although the assignment of current directions is arbitrary, you must stick with your guess throughout as you apply Kirchhoff's rules.

2. Apply the junction rule to any junction in the circuit. The junction rule may be applied as many times as a new current (one not used in a previous application) appears in the resulting equation.

3. Now apply Kirchhoff's loop rule to as many loops in the circuit as are needed to solve for the unknowns. In order to apply this rule, you must correctly identify the change in potential as you cross each element in traversing the closed loop. Watch out for signs!

4. Finally, you must solve the equations simultaneously for the unknown quantities. Be careful in your algebraic steps, and check your numerical answers for consistency.

Example 19.4 Applying Kirchhoff's Rules

Find the currents in the circuit shown in Figure 19.9.

Solution
Although this circuit could be handled much more easily by Ohm's law and the rules for series and parallel circuits, we shall use Kirchhoff's rules to demonstrate the techniques. Also, because students often find the application of Kirchhoff's rules confusing, we shall go into considerable detail in this example.

The first step is to assign currents to each branch of the circuit; these are our unknowns and are labeled I_1, I_2, and I_3 in Figure 19.9. Furthermore, it is necessary to guess a direction for these currents. The experience that you have with circuits like this should tell you that the directions of all three have been chosen correctly. However, recall that if a current direction were chosen incorrectly, the numerical answer will turn out negative, but the magnitude will be correct. This point will be demonstrated in the next example.

We now apply Kirchhoff's rules. First we can apply the junction rule using either point c or point d, the only two junctions in the circuit. Let us choose point c. The net current into this junction is I_1, and the net current leaving the junction is $I_2 + I_3$. Thus, the junction rule applied to point c gives

$$I_1 = I_2 + I_3$$

Recall that you can apply the junction rule over and over until you reach a situation in which no new currents appear in an equation. We have reached that point with one application in this example. If we apply the junction rule at point d, we find that $I_1 = I_2 + I_3$, exactly the same equation.

We have three unknowns in our problem, I_1, I_2, and I_3; thus, we need two more independent equations before we can find a solution. These are obtained by applying the loop rule to the two loops indicated in the figure. Note that there are actually three loops in the circuit, but these two are sufficient for us to complete the problem. (Where is the loop that we do not use?)

When applying the loop rule, we must first choose the loops to be traversed, and then the direction in which to traverse them. We have selected the two loops indicated in the figure, and have decided to traverse both of them in the clockwise direction. Although other choices could have been selected, the final result would be the same.

Figure 19.9 (Example 19.4) A multiloop circuit.

Starting at point a and moving clockwise around the large loop, we encounter the following voltage changes (see Figure 19.8 for the basic rules):

1. Going from a to b, we encounter a voltage rise of 6 V.
2. Going from b to c through the 4-Ω resistor results in a voltage drop of $-4I_1$.
3. Going from c to d through the 9-Ω resistor results in a voltage drop of $-9I_3$.

No voltage change occurs going from d back to a. Now that we have made a complete traversal of the loop, we can equate the sum of the voltage changes to zero:

$$6 - 4I_1 - 9I_3 = 0$$

Moving clockwise around the small loop from point c, we have the following:

1. Going from c to d through the 5-Ω resistor results in a voltage drop of $-5I_2$.
2. Going from d to c through the 9-Ω resistor results in a voltage rise of $+9I_3$.

We find

$$-5I_2 + 9I_3 = 0$$

Thus we have the following three equations to be solved for the three unknowns:

$$I_1 = I_2 + I_3$$
$$6 - 4I_1 - 9I_3 = 0$$
$$-5I_2 + 9I_3 = 0$$

If you need help in solving three equations with three unknowns, see Example 19.5. You should be able to show that the following answers are obtained:

$$I_1 = \boxed{0.83 \text{ A}} \qquad I_2 = \boxed{0.53 \text{ A}} \qquad I_3 = \boxed{0.30 \text{ A}}$$

You should now solve this problem using the methods for series and parallel circuits to show that the answers given can be found by this alternate method. However, the next example can be solved only by applying Kirchhoff's rules.

Example 19.5

Find the currents I_1, I_2, and I_3 in the circuit shown in Figure 19.10.

Solution

We shall choose the directions of the currents as shown in Figure 19.10. Kirchhoff's junction rule applied at c gives

$$(1) \qquad I_1 + I_2 = I_3$$

There are *three* loops in the circuit, *abcda*, *befcb*, and *aefda* (the outer loop), but we need only two loop equations to determine the unknown currents. The third loop equation would give no new information. When Kirchhoff's second rule is applied to loops *abcda* and *befcb* traversed in the clockwise direction, we obtain the following expressions:

$$(2) \qquad \text{Loop } abcda: \quad 10 - 6I_1 - 2I_3 = 0$$

$$(3) \qquad \text{Loop } befcb: \quad -14 - 10 + 6I_1 - 4I_2 = 0$$

Note that, in loop *befcb*, a positive sign is obtained when traversing the 6-Ω resistor since the current, I_1, is in the opposite direction. A loop equation for *aefda* gives $-14 = 2I_3 + 4I_2$, which is just the sum of (2) and (3).

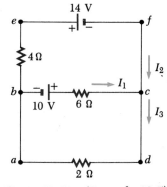

Figure 19.10 (Example 19.5) A circuit containing three loops.

In (1), (2), and (3) we now have three independent expressions with three unknowns. We can solve the problem as follows. If (1) is substituted into (2), we have

$$10 - 6I_1 - 2(I_1 + I_2) = 0$$

$$(4) \quad 10 = 8I_1 + 2I_2$$

Let us divide each term in (3) by 2 and rearrange the equation to obtain

$$(5) \quad -12 = -3I_1 + 2I_2$$

Subtracting (5) from (4) eliminates I_2, giving

$$22 = 11I_1$$

$$I_1 = \boxed{2 \text{ A}}$$

This value of I_1 can be inserted into (5) to give a value for I_2:

$$2I_2 = 3I_1 - 12 = 3(2) - 12 = -6$$

$$I_2 = \boxed{-3 \text{ A}}$$

The fact that I_2 is negative indicates only that we chose the *wrong* direction for this current.

Exercise Calculate the value of I_3.

Answer -1 A.

19.5 *RC* Circuits

So far we have been concerned with circuits with constant currents. We shall now consider direct current circuits containing capacitors, in which the current varies with time. Consider the series circuit shown in Figure 19.11. Let us assume that the capacitor is initially uncharged. There is no current when the switch is open (Fig. 19.11b). If the switch is closed at $t = 0$, a current will pass through the resistor and the capacitor will begin to charge (Fig. 19.11c). Note that, during charging, charges do not jump across the plates of the capacitor because the gap between the plates represents an open circuit.

Figure 19.11
(a) A capacitor in series with a resistor, battery, and switch.
(b) Circuit diagram representing this system before the switch is closed ($t < 0$).
(c) Circuit diagram after the switch is closed ($t > 0$).

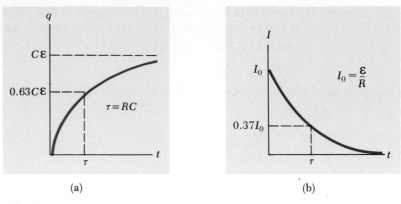

(a)   (b)

Figure 19.12
(a) Plot of the charge on the capacitor versus time for the circuit shown in Figure 19.11. After one time constant, τ, the charge is 63% of the maximum value, $C\mathcal{E}$. The charge approaches its maximum value as t approaches infinity. (b) Plot of the current versus time for the RC circuit shown in (a). At $t = 0$, the current has its maximum value, $I_0 = \mathcal{E}/R$ and decays to zero exponentially as t approaches infinity. After one time constant, τ, the current decreases to 37% of its initial value.

This charge transfer ceases when the voltage drop across the capacitor is equal to the emf of the battery, $\mathcal{E}$. Thus, from the definition of capacitance, the maximum charge on the capacitor is given by $Q = C\mathcal{E}$. Once the capacitor reaches its maximum charge, the current in the circuit is zero.

It is found that the charge on the capacitor increases with time in a manner pictured by Figure 19.12a. Note that the charge is zero at time $t = 0$ and approaches the maximum value of $C\mathcal{E}$ as t approaches infinity. Figure 19.12b shows the current in the circuit as a function of time. Because the capacitor is uncharged at $t = 0$, the only element in the circuit that inhibits the current is the resistor R. Thus, from Ohm's law, the current in the circuit at $t = 0$ is given by $I_0 = \mathcal{E}/R$. The current decreases exponentially to zero as t approaches infinity, as pictured in Figure 19.12b.[1]

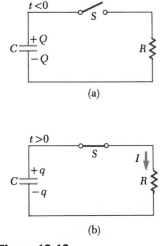

(a)

(b)

Figure 19.13
(a) A charged capacitor connected to a resistor and a switch, which is open at $t < 0$. (b) After the switch is closed, a nonsteady current is set up in the direction shown and the charge on the capacitor decreases exponentially with time.

[1] The methods of calculus can be used to show that the current in an RC circuit as a function of time is given by

$$I = I_0 e^{-t/RC} = \frac{\mathcal{E}}{R}\, e^{-t/RC}$$

where e is the base of the natural logarithm, $e = .2.718$, and $I_0 = \mathcal{E}/R$. The charge on the capacitor as a function of time is given by

$$q = Q(1 - e^{-t/RC}) = C\mathcal{E}\,(1 - e^{-t/RC})$$

where $Q = C\mathcal{E}$ is the maximum charge on the capacitor.

As an example of how to use the two equations above, assume that a circuit like that in Figure 19.11 has $\mathcal{E} = 12$ V, $C = 5$ μF, and $R = 8 \times 10^5$ Ω. Find the current in the circuit and the charge on the capacitor at $t = 2$ s.

Since the time constant is $\tau = RC = (8 \times 10^5 \ \Omega)(5 \times 10^{-6} \ \text{F}) = 4$ s, the current is found as follows:

$$I = \frac{\mathcal{E}}{R}\, e^{-t/RC} = \left(\frac{12 \ \text{V}}{8 \times 10^5 \ \Omega}\right) e^{-(2\,\text{s}/4\,\text{s})}$$

$$= (1.5 \times 10^{-5})e^{-0.5} \ \text{A} = (1.5 \times 10^{-5})(0.607)$$
$$= 9.1 \times 10^{-6} \ \text{A} = 9.1 \ \mu\text{A}$$

The charge on the capacitor is given by

$$q = C\mathcal{E}\,(1 - e^{-t/RC}) = (5 \times 10^{-6} \ \text{F})(12 \ \text{V})(1 - e^{-0.5})$$
$$= (6 \times 10^{-5})(1 - 0.607) \ \text{C} = 2.36 \times 10^{-5} \ \text{C} = 0.236 \ \mu\text{C}$$

It takes an infinite amount of time for a capacitor to become fully charged, but the capacitor has stored a substantial fraction of its final charge in a very short period of time. A quantity called the *time constant, τ* (Greek letter tau), is used to describe the rate at which the charge on the capacitor varies with time. For a circuit containing a capacitor and resistor in series

$$\tau = RC$$

The time constant represents *the time for the charge to increase from zero to 63% of its maximum value.* That is, after one time constant, the charge on the capacitor has risen from zero to $0.63C\mathcal{E}$. Also in one time constant, the current in the circuit has fallen from its initial value of I_0 to $0.37I_0$.

An important feature that you should note here is that the capacitor in a circuit with a large time constant will charge (or discharge) slowly. It can easily be seen that the time constant should be directly dependent on R, the resistance in the circuit. For large values of R, the current in the circuit will be small, and thus, it will take a longer time for charge to move onto the plates of the capacitor. Since $C = Q/V$, we see that a large capacitor will store a large amount of charge. Thus, one would also expect the time constant to depend directly on C, since it will take a longer time for the necessary charge to accumulate.

The following dimensional analysis shows that τ has the unit of time:

$$[\tau] = [RC] = \left[\frac{V}{I} \times \frac{Q}{V}\right] = \left[\frac{Q}{Q/T}\right] = [T]$$

Now consider the circuit in Figure 19.13, consisting of a capacitor with an initial charge Q, a resistor, and a switch. When the switch is open (Fig. 19.13a), there is a potential difference Q/C across the capacitor and zero potential difference across the resistor because $I = 0$. If the switch is closed at $t = 0$, the capacitor begins to discharge through the resistor. The charge does not disappear instantaneously but decays exponentially to zero, as shown in Figure 19.14. Likewise, the current decays to zero exponentially. The time constant, $\tau = RC$, gives the time it takes both current and charge to fall to 37% of their initial values.

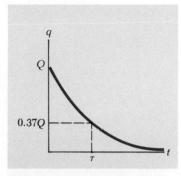

Figure 19.14
The charge on the capacitor of Figure 19.13 as a function of time.

Example 19.6   Charging a Capacitor in an *RC* Circuit

An uncharged capacitor and a resistor are connected in series to a battery, as in Figure 19.15. If $\mathcal{E} = 12$ V, $C = 5\ \mu F$, and $R = 8 \times 10^5\ \Omega$, find the time constant of the circuit, the maximum charge on the capacitor, and the charge on the capacitor after one time constant.

Solution
The time constant of the circuit is

$$\tau = RC = (8 \times 10^5\ \Omega)(5 \times 10^{-6}\ F) = \boxed{4\ s}$$

The maximum charge on the capacitor is

$$Q = C\mathcal{E} = (5 \times 10^{-6}\ F)(12\ V) = \boxed{60\ \mu C}$$

Figure 19.15   (Example 19.6)
The switch of this series *RC* circuit is closed at $t = 0$.

After a time equal to one time constant, the charge on the capacitor is 63% of its maximum value:

$$q = 0.63Q = 0.63(60 \times 10^{-6}\ \text{C}) = \boxed{37.8\ \mu\text{C}}$$

Exercise   Calculate the maximum current in the circuit and the current in the circuit after one time constant has elapsed.

Answer $15\ \mu\text{A}$; $5.6\ \mu\text{A}$.

19.6   Measurement of Resistance

The Voltmeter-Ammeter Method

From Ohm's law, $V = IR$, it is easy to see that the resistance of a conductor can be obtained if the current through the conductor and the potential difference across it are measured simultaneously. A device that measures current is called an **ammeter,** and one that measures potential difference across two terminals is called a **voltmeter.** A multimeter is an instrument that is designed (with proper switch settings) to be used as either an ammeter, a voltmeter, or an ohmeter, a device to measure resistance. Figure 19.16 is a photo of a typical multimeter. We shall describe the construction of these instruments in Chapter 20. For now, we shall describe a few of their characteristics and the procedure for using them.

An ammeter is constructed so that it must be placed in series with the current that it is to measure (Fig. 19.17). When using an ammeter to measure direct currents, one must be sure to connect it such that current enters at the positive terminal of the instrument and exits at the negative terminal. *Ideally, an ammeter should have zero resistance so as not to alter the current being measured.* In the circuit shown in Figure 19.17, this condition requires that the ammeter's resistance be small compared to R_1 and R_2. Since any ammeter always has some resistance, it presence in the circuit will slightly reduce the current from its value when the ammeter is not present. If its resistance is known, one can calibrate the ammeter to give the value of the current that would be present in the absence of the ammeter.

The potential difference across a resistor can be measured with a voltmeter connected in parallel with the resistor, as in Figure 19.18. Again it is

Figure 19.16
Voltages, currents, and resistances are frequently measured by digital multimeters like the one shown in this photograph. *(Courtesy of Henry Leap and Jim Lehman)*

Figure 19.17
The current in a circuit can be measured with an ammeter Ⓐ connected in series with the resistor and battery. An ideal ammeter has zero resistance.

Figure 19.18
The potential difference across a resistor can be measured with a voltmeter Ⓥ connected in parallel with the resistor. An ideal voltmeter has infinite resistance and does not affect the circuit.

necessary to observe the polarity of the instrument. The positive terminal of the voltmeter must be connected to the end of the resistor at the higher potential, and the negative terminal to the low-potential end of the resistor. *An ideal voltmeter has infinite resistance so that no current will pass through it.* In practice, the resistance of a voltmeter should be large compared with the resistance to be measured.

The ammeter-voltmeter method is not recommended for precise measurements because most meters are not reliable to more than three-digit accuracy.

The Wheatstone Bridge

Unknown resistances can be accurately measured using a circuit known as a **Wheatstone bridge.** This circuit consists of the unknown resistance, R_x; a sensitive current detector called a galvanometer; three known resistances, R_1, R_2, and R_3, and a source of emf (Fig. 19.19). When the bridge is being used, R_1 is varied until the galvanometer reading is zero, that is, until there is no current from a to b in Figure 19.19. (Note that the symbol used for R_1 is ⤢, which indicates that it is a variable resistor.) When there is no current through the galvanometer, the bridge is said to be *balanced* and the potential at point a must equal the potential at point b (otherwise there would be a current between these points). This means that the potential difference across R_1 must equal the potential difference across R_2, and the potential difference across R_3 must equal the potential difference across R_x. From these considerations, we see that

$$(1) \qquad I_1 R_1 = I_2 R_2$$

$$(2) \qquad I_1 R_3 = I_2 R_x$$

Dividing (1) by (2) and solving for R_x, we find

$$R_x = \frac{R_2 R_3}{R_1} \qquad\qquad \textbf{(19.5)}$$

Since R_1, R_2, and R_3 are known quantities, R_x can be calculated.

When high resistances are to be measured (above $10^5 \ \Omega$), it becomes difficult to use the Wheatstone bridge. As a result of recent advances in solid-state technology, modern electronic instruments are capable of measuring resistances as high as $10^{12} \ \Omega$. Such instruments are designed to have an extremely high effective resistance (as high as $10^{10} \ \Omega$) between the input terminals.

Example 19.7   Poorly Designed Meters

(a) A poorly designed ammeter has a resistance of 6 Ω and is to be used to measure the current in the circuit shown in Figure 19.20a. Find the current in the absence of the meter and the current that the meter will read when inserted in the circuit.

Solution

In the absence of the ammeter, the current in the circuit is found from Ohm's law as

$$I = \frac{V}{R} = \frac{6 \text{ V}}{12 \ \Omega} = \frac{1}{2} \text{ A}$$

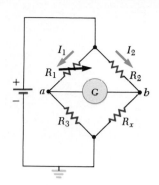

Figure 19.19
Circuit diagram for a Wheatstone bridge. This circuit can be used to measure an unknown resistance R_x in terms of known resistances R_1, R_2, and R_3. When the bridge is balanced, there is no current in the galvanometer.

Figure 19.20   Example 19.7

When the ammeter is inserted in series as in Figure 19.20b, the total resistance in the circuit is 18 Ω, so the current in the circuit is

$$I = \frac{V}{R} = \frac{6\text{ V}}{18\text{ }\Omega} = \boxed{\frac{1}{3}\text{ A}}$$

No measuring instrument should affect the value of the quantity it is designed to measure. In this case, the ammeter has reduced the current from 1/2 A to 1/3 A because its resistance is too high. Typical ammeters have resistances of a fraction of an ohm.

(b) A poorly designed voltmeter whose resistance is 6 Ω is to be used to measure the voltage across one of the 6-Ω resistors in the circuit of Figure 19.20a. Find the voltage across the resistor before and after insertion of the meter.

Solution

In the absence of the voltmeter, the current in the circuit is 1/2 A, and the voltage drop across one of the 6-Ω resistors is found from Ohm's law to be (1/2 A)(6 Ω) = 3 V. When the 6-Ω voltmeter is connected across one of the resistors as in Figure 19.20c, the equivalent resistance of the circuit becomes 9 Ω. (You should use the rules for combining resistors in parallel and series to verify this value.) Thus, the current in the circuit now becomes

$$I = \frac{V}{R} = \frac{6\text{ V}}{9\text{ }\Omega} = \frac{2}{3}\text{ A}$$

and the voltage across the voltmeter is

$$V = IR = \left(\frac{2}{3}\text{ A}\right)(3\text{ }\Omega) = \boxed{2\text{ V}}$$

Again, the meter has affected the quantity it is to measure, which in this case is voltage, because its resistance is too low. Properly designed voltmeters have very high resistances, typically several thousand ohms or higher.

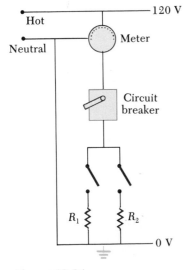

Figure 19.21
Wiring diagram for a household circuit. The resistances R_1 and R_2 represent appliances or other electrical devices which operate with an applied voltage of 120 V.

*19.7   Household Circuits

Household circuits represent a very practical application of some of the ideas we have presented in this chapter. In a typical installation, the utility company distributes electric power to individual houses with a pair of wires, or power lines. A circuit in a house is then connected in parallel to these lines, as shown in Figure 19.21. The potential difference between the two wires is about 120 V. (These are alternating currents and voltages, but for the present discussion we shall assume that they are direct currents and voltages.) One of the wires is connected to ground, and the other wire, sometimes called the "hot" wire, has an average potential of 120 V. A meter and circuit breaker (or in older installations, a fuse) are connected in series with the wire entering the house, as indicated in Figure 19.21. A cutaway view of a fuse is shown in Figure 19.22. The current passing through the fuse flows through a small metallic strip, which melts if the current is large enough. If a circuit does not include a fuse, excessive currents could damage circuit elements, overheat wires, and perhaps cause a fire.

In modern home construction, circuit breakers are used in place of fuses. When the current in a circuit exceeds some value (typically 15 A), the circuit breaker acts as a switch and opens the circuit. One design for a circuit breaker is shown in Figure 19.23. In this design, current passes through a bimetallic strip, the top of which bends to the left when excessive current heats it. If the strip bends far enough to the left, it settles into a groove in the spring-loaded metal bar. When this occurs, the bar drops sufficiently downward to open the circuit at the contact point. The bar also flips the switch to indicate that the circuit breaker is not operational. After the overload is removed, the switch can be flipped back on. Circuit breakers using this design have the disadvantage that some time is required for the heating of the strip. Thus the circuit is not opened immediately when it is overloaded. For this reason, many circuit breakers use electromagnets, which we shall discuss later.

The wire and circuit breaker are carefully selected to meet the current demands of a circuit. If the circuit is to carry currents as large as 30 A, a heavy-duty wire and appropriate circuit breaker must be used. Household circuits normally used to power lamps and small appliances often require only 15 A. Therefore, each circuit has its own circuit breaker to accommodate various load expectations.

As an example, consider a circuit containing a toaster, a microwave oven, and a heater (represented by R_1, R_2, . . . in Fig. 19.21). We can calculate the current through each appliance using the equation $P = IV$. The toaster, rated at 1000 W, would draw a current of 1000/120 = 8.33 A. The microwave oven, rated at 800 W, would draw a current of 6.67 A, and the heater, rated at 1300 W, would draw a current of 10.8 A. If the three appliances are operated simultaneously, they will draw a total current of 25.8 A. Therefore the breaker should be able to handle at least this much current or else it will be tripped. Alternatively, one could operate the toaster and microwave oven on one 15-A circuit and the heater on a separate 15-A circuit.

Many heavy-duty appliances, such as electric ranges and clothes dryers, require 240 V for their operation. The power company supplies this voltage by providing a third wire, also considered live, which is 120 V below ground potential (Fig. 19.24). Therefore, the potential difference between this wire and the other live wire (which is 120 V above ground potential) is 240 V. An appliance that operates from a 240-V line requires half the current of one operating from a 120-V line; therefore smaller wires can be used in the higher-voltage circuit without overheating of the wires becoming a problem.

Figure 19.22
A cutaway view of a fuse.

Figure 19.23
A circuit breaker uses a bimetallic strip for its operation.

Figure 19.24
Power connections for a 240-V appliance.

*19.8   Electrical Safety

A person can be electrocuted by touching a live wire (which commonly happens because of a frayed cord or other exposed conductors) while in contact with ground. The ground contact might be made either by the person touching a water pipe (which is normally at ground potential) or by standing on ground with wet feet, since water is a good electric conductor. Such situations should be avoided at all costs.

Electric shock can result in fatal burns, or it can cause the muscles of vital organs, such as the heart, to malfunction. The degree of damage to the body depends on the magnitude of the current, the length of time it acts, and the part of the body through which it passes. Currents of 5 mA or less can cause a sensation of shock but ordinarily do little or no damage. If the current is larger than about 10 mA, the hand muscles contract and the person may be unable to release the live wire. If a current of about 100 mA passes through the body for only a few seconds, the result could be fatal. Such large currents will paralyze the respiratory muscles and prevent breathing. In some cases, currents of about 1 A through the body may produce serious (and sometimes fatal) burns.

In order to provide an additional safety feature to consumers, electrical equipment manufacturers now use electrical cords that have a third wire, called a case ground. To understand how this works, consider the drill being used in Figure 19.25. Figure 19.25a shows a two-wire device that has one wire, called the "hot" wire, that is connected to the high-potential (120 V) side of the input power line, while the second wire is connected to ground (0 V). Under normal operating conditions, the path of the current through the drill is like that shown in Figure 19.25a. However, the cord inside the device can become frayed and a "short circuit" can occur if the high voltage wire comes into contact with the case of the drill, as shown in Figure 19.25b. In this undesirable circumstance, the pathway for the current is from the high voltage wire through the consumer and to Earth — a pathway that can lead to death for

Figure 19.25
(a) When the drill is operated with two wires, the "hot wire" at 120 V is always fused on this side of the circuit for safety. (b) This illustrates the potentially dangerous situation of the two-wire connection. If the high voltage side comes in contact with the case of the drill, the person holding the drill will get an electrical shock. (c) Protection from shock can be obtained by having a third wire running from the case of the drill to the ground.

the user. Protection is provided by a third wire, which is connected to the case of the drill, as shown in Figure 19.25c. In this case, if a short occurs, the path of least resistance for the current is from the high-voltage wire through the case and back to ground through the third wire. The high current produced can blow a fuse or trip a circuit breaker before the consumer is injured.

Special power outlets, called ground-fault interrupters (GFI's) are now being used in kitchens, bathrooms, basements, and other hazardous areas of new homes. These devices are designed to protect persons from electrical shock by sensing small currents (approximately 5 mA) leaking to ground. When an excessive leakage current is detected, the current is shut off (interrupted) in less than one millisecond.

▶ SPECIAL TOPIC
Voltage Measurements in Medicine

Electrocardiograms

Every action involving the body's muscles is initiated by electrical activity. The voltages produced by muscular action in the heart are particularly important to physicians. Voltage pulses cause the heart to beat, and the waves of electrical excitation associated with the heart beat are conducted through the body via the body fluids. These voltage pulses are large enough to be detected by suitable monitoring equipment attached to the skin. Standard electric devices can be used to record these voltage pulses because the amplitude of a typical pulse associated with heart activity is of the order of 1 mV. These voltage pulses are recorded on an instrument called an **electrocardiograph,** and the pattern recorded by this instrument is called an **electrocardiogram (EKG)**. In order to understand the information contained in an EKG pattern, it is useful to first describe the underlying principles concerning electrical activity in the heart.

The right atrium of the heart contains a specialized set of muscle fibers called the SA (sinoatrial) node, which initiate the heartbeat (Fig. 19.26). Electric impulses that originate in these fibers gradually spread from cell to cell throughout the right and left atrial muscles, causing them to contract. The pulse that passes through the muscle cells is often called a *depolarization wave* because of its effect on individual cells. If an individual muscle cell were examined, an electric charge distribution would be found on its surface, as shown in Figure 19.27a. (See Living Cells as Capacitors in Ch. 17 for an explanation of how this charge distribution arises.) The impulse generated by the SA node momentarily changes the cell's charge distribution to that shown in Figure 19.27b. The positively charged ions on the surface of the cell are temporarily able to

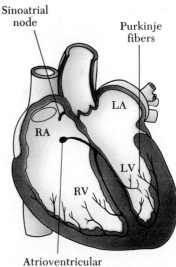

Figure 19.26
The electrical conduction system of the human heart. (RA: right atrium; LA: left atrium; RV: right ventricle; LV: left ventricle.)

Figure 19.27
(a) Charge distribution of a muscle cell in the atrium before the depolarization wave has passed through the cell. (b) Charge distribution as the wave passes.

(a)   (b)

Figure 19.28
An EKG response for a normal heart.

diffuse through the membrane wall such that the cell attains an excess positive charge on its inside surface. As the depolarization wave travels from cell to cell throughout the atria, the cells recover to the charge distribution shown in Figure 19.27a. When the impulse reaches the AV (atrioventricular) node (Fig. 19.26), the muscles of the atria begin to relax, and the pulse is directed by the AV node to the ventricular muscles. The muscles of the ventricles then contract as the depolarization wave spreads through the ventricles along a group of fibers called the *Purkinje fibers*. The ventricles then relax after the pulse has passed through. At this point, the SA node is again triggered and the cycle is repeated.

A sketch of the electrical activity registered on an EKG for one beat of a normal heart is shown in Figure 19.28. The pulse indicated by P occurs just before the atria begin to contract. The QRS pulse occurs in the ventricles just before they contract, and the T pulse occurs when the cells in the ventricles begin to recover. EKGs for an abnormal heart are shown in Figure 19.29. The QRS portion of the pattern shown in Figure 19.29a is wider than normal. This indicates that the patient may have an enlarged heart. Figure 19.29b indicates that there is no relationship between the P pulse and the QRS pulse. This suggests a blockage in the electrical conduction path between the SA and AV nodes. This can occur when the atria and ventricles beat independently. Finally, Figure 19.29c shows a situation in which there is no P pulse and an irregular spacing between the QRS pulses. This is symptomatic of irregular atrial contraction, which is called fibrillation. In this situation, the atrial and ventricular contractions are irregular.

Electroencephalography

The electrical activity of the brain can be measured with an instrument called an **electroencephalograph** in much the same way that an electrocardiograph measures the electrical activity of the heart. The voltage pattern measured by an electroencephalograph is referred to as an **electroencephalogram** (EEG). An EEG pattern is recorded by placing electrodes on the patient's scalp. While an EKG voltage pulse is typically 1 mV, a typical voltage amplitude associated with brain activity is only a few *micro*volts and is therefore more difficult to measure.

Figure 19.29
Abnormal EKGs.

(a)   (b)

(c)

The EEGs in Figure 19.30 represent the brain wave pattern of a patient awake and then in various stages of sleep. In Figure 19.30b the patient begins to fall into a light sleep, and in Figure 19.30d the patient is in a deep sleep. Figure 19.30c is the EEG during a type of sleep called **REM** (rapid eye movement) sleep, which occurs approximately every 2 h. In this stage, the brain activity is quite similar to that of the patient while awake. It is interesting to note that a person in this stage of sleep is extremely difficult to awaken. Apparently, REM sleep is necessary for psychological well-being. Anyone who is deprived of this stage of sleep for an extended period of time becomes extremely fatigued and irritable.

The EEG is an extremely important diagnostic tool for detecting epilepsy, brain tumors, brain hemorrhage, meningitis, and so forth. For example, the EEG pattern of a patient suffering an epileptic seizure would show very little structure, indicating that the brain activity is greatly reduced.

Brain waves are often discussed in terms of their frequencies. Frequencies of about 10 Hz are referred to as *alpha waves*, frequencies between 10 Hz and 60 Hz are called *beta waves*, and those below 10 Hz are called *delta waves*. As a person falls asleep, the frequency of brain wave activity generally decreases. For example, the brain wave frequency for a person in a very deep sleep may drop as low as 1 or 2 Hz. ◄

Figure 19.30
Brain waves from an individual in various stages of sleep.

Summary

A source of electromotive force (emf) is any device that will transform non-electrical energy into electrical energy.

The **equivalent resistance** of a set of resistors connected in **series** is

$$R_{eq} = R_1 + R_2 + R_3 + \ldots \qquad \textbf{(19.2)}$$

The **equivalent resistance** of a set of resistors connected in **parallel** is

$$\frac{1}{R_{eq}} = \frac{1}{R_1} + \frac{1}{R_2} + \frac{1}{R_3} + \ldots \qquad \textbf{(19.4)}$$

Complex circuits are conveniently solved by using two rules called **Kirchhoff's rules:**

1. The sum of the currents entering any junction must equal the sum of the currents leaving that junction.
2. The sum of the potential differences across each element around any closed circuit loop must be zero.

The first rule is a statement of **conservation of charge.** The second is a statement of **conservation of energy.**

If a capacitor is charged by a battery through a resistor, it is found that the current approaches zero as the capacitor is being charged. The **time constant,** $\tau = RC$, represents the time it takes the charge to increase from zero to 63% of its maximum value.

A **Wheatstone bridge** is a circuit that can be used to measure an unknown resistance.

Additional Reading

P. F. Baker, "The Nerve Axon," *Sci. American*, March 1966, p. 74.

P. Davidovits, *Physics in Biology and Medicine*, Englewood Cliffs, NJ, Prentice-Hall, 1977.

B. Katz, "The Nerve Impulse," *Sci. American*, November 1952, p. 55.

K. Kordesch and K. Tomantschger, "Primary Batteries," *The Physics Teacher*, January 1981, p. 12.

T. F. Robinson, S. M. Factor, and E. H. Sonnenblick, "The Heart as a Suction Pump," *Sci. American*, June 1986, p. 84.

A. M. Scher, "The Electrocardiogram," *Sci. American*, November 1961, p. 132.

G. M. Shepherd, "Microcircuits in the Nervous System," *Sci. American*, February 1978, p. 92.

A. K. Solomon, "Pumps in the Living Cell," *Sci. American*, August 1962, p. 100.

Questions

1. Is the direction of current through a battery always from negative to positive on the terminals?

2. Car batteries are often rated in ampere-hours. Does this designate the amount of current, power, energy, or charge that can be drawn from the battery?

3. Two sets of Christmas tree lights are available. In set A, when one bulb is removed or burns out, all the others remain illuminated. In set B, when one bulb is removed, the remaining bulbs will not operate. Explain the difference in the wiring for the two sets.

4. How would you connect resistors in order for the equivalent resistance to be larger than the individual resistances? Give an example.

5. How would you connect resistors in order for the equivalent resistance to be smaller than the individual resistances? Give an example.

6. When resistors are connected in series, which of the following will be the same, for each resistor: potential difference, current, power?

7. When resistors are connected in parallel, which of the following will be the same for each resistor: potential difference, current, power?

8. Given three lightbulbs and a battery, sketch as many different electric circuits as you can.

9. The circuits in a car are "single-wire" circuits. To work properly, a ground wire must be attached from the chassis of the car to one of the battery posts. Why? What happens if the ground wire corrodes?

10. Explain why a bird can roost on a high-voltage power line with no ill effect. Is there any connection between the bird-on-the-wire effect and the admonition given to experimenters to "keep one hand in the pocket" when working around high voltage?

11. Suppose you fall from a building and on the way down grab a high-voltage wire. Assuming that the wire holds you, will you be electrocuted? If the wire then breaks, should you continue to hold onto an end of the wire as you fall?

12. Are the two headlights on a car wired in series or in parallel? How can you tell? Suppose you rewired your headlights so that they were in series. What would be the effect (even before one burns out)?

13. An incandescent lamp connected to a 120-V source with a short extension cord will provide more illumination than the same lamp connected to the same source with a long extension cord. Explain.

14. In the circuits considered in this chapter, the wires connecting the various elements are idealized to be perfect conductors (represented schematically by lines connecting the elements). Discuss the effect of the real wires in a circuit. Consider the resistance of the wires in relation to resistive elements of the circuit.

15. A "short circuit" is a circuit containing a path of very low resistance in parallel with some other part of the circuit. Discuss the effect of a short circuit on the portion of the circuit it is in parallel with. Use a lamp with a frayed line cord as an example.

16. A series circuit consists of three identical lamps connected to a battery as shown in Figure 19.31. When the switch S is closed, (a) what happens to the intensi-

Figure 19.31   Question 16

ties of lamp A and B? (b) What happens to the intensity of lamp C? (c) What happens to the current in the circuit? (d) Does the power dissipated in the circuit increase, decrease, or remain the same?

17. What advantage might there be in using two identical resistors in parallel connected in series with another identical parallel pair, rather than just using a single resistor?

18. Embodied in Kirchhoff's rules are two conservation laws. What are they?

19. When can the potential difference across a resistor be positive?

20. With reference to Figure 19.32, describe what happens to the lightbulb after the switch is closed. Assume the capacitor has a large capacitance and is initially uncharged, and assume that the light will illuminate when connected directly across the battery terminals.

21. Would a fuse work successfully if it were placed in parallel with the device it is supposed to protect?

22. Does it matter whether a fuse is placed in the high-voltage line of a household circuit or in the ground line?

23. What advantage does 120-V operation offer over 240 V? What disadvantages?

24. Suppose you are flying a kite when it strikes a high-voltage wire. What factors determine how great a shock you receive?

25. If it is the current flowing through the body that determines how serious a shock will be, why do we see warnings of high voltage rather than high current near electric equipment?

26. Why is it dangerous to turn on a light when you are in the bathtub?

27. When electricians work with potentially live wires, they often use the backs of their hands or fingers to move the wires. Why do you suppose they use this technique?

28. What procedure would you use to try to save a person who is "frozen" to a live high-voltage wire without endangering your own life?

29. At what levels of current do you experience (a) the sensation of shock, (b) involuntary muscle contractions, (c) paralysis of respiratory muscle, and (d) serious and possibly fatal burns? In practice what current level is regarded as safe?

Figure 19.32
Question 20

Problems

Section 19.2   Resistors in Series
Section 19.3   Resistors in Parallel

1. A 4-Ω, an 8-Ω, and a 12-Ω resistor are connected in series with a 24-V battery. What is (a) the equivalent resistance and (b) the current in each resistor?

2. A 9-Ω and a 6-Ω resistor are connected in series to a power supply. The current in the circuit is found to be 1/3 A. What is the voltage setting of the power supply?

3. The resistors of Problem 1 are connected in parallel across a 24-V battery. Find (a) the equivalent resistance and (b) the current in each resistor.

4. A television repairman needs a 100-Ω resistor to repair a malfunctioning set. He is temporarily out of resistors of this value. All he has in his tool box is a 500-Ω resistor and two 250-Ω resistors. How can the desired resistance be obtained from the resistors on hand?

5. A technician has a box full of resistors, each having the same resistance R. How many different values of effective resistance could the technician make using all combinations of one to three separate resistors? Express the effective resistance of each combination in terms of R.

6. A 9-Ω and a 6-Ω resistor are connected in parallel across a power supply. If the current through the equivalent resistor is 1/3 A, find the setting of the power supply.

7. Three resistors (10 Ω, 20 Ω, 30 Ω) are connected in parallel. The total current through this network is 5 A. (a) What is the voltage drop across the network? (b) What is the current in each resistor?

8. A battery with an internal resistance of 10.0 Ω produces an open-circuit voltage of 12.0 V. A variable load resistance with a range of 0.0–30.0 Ω is connected across the battery. (*Note*: A battery has a resistance to the flow of current which depends on the

condition of the chemicals in it and increases as the battery ages. This so-called internal resistance may be represented in a simple circuit diagram as a resistor in series with the battery.) (a) Plot a graph of the power dissipated in the load resistor as a function of the load resistance. (b) From your graph demonstrate the following important theorem: *The power delivered to a load is maximum if the load resistance is equal to the internal resistance of the source.*

9. Find the equivalent resistance of the circuit shown in Figure 19.33.

Figure 19.33   Problem 9

10. An 18-Ω and a 6-Ω resistor are connected in series across an 18-V battery. (a) Find the current through each resistor and the voltage drop across each resistor. (b) Repeat Part (a) for the situation in which the resistors are connected in parallel.

11. Find the equivalent resistance of the circuit shown in Figure 19.34.

Figure 19.34   Problem 11

12. A 9-Ω and a 6-Ω resistor are connected in series with a power supply. (a) The voltage drop across the 6-Ω resistor is measured and found to be 12 V. Find the voltage output of the power supply. (b) The two resistors are connected in parallel across a power supply and the current through the 9-Ω resistor is found to be 1/4 A. Find the voltage setting of the power supply.

13. Find the equivalent resistance of the circuit shown in Figure 19.35.

14. If the total power supplied to the circuit shown in Figure 19.35 is 4 W, find the emf of the battery.

Figure 19.35   Problems 13 and 14

15. Find the current in the 12-Ω resistor in Figure 19.36.

Figure 19.36   Problem 15

16. Two resistors A and B are connected in series to a 6-V battery. A voltmeter connected across resistor A measures a voltage of 4 V. When the two resistors are connected in parallel across the 6-V battery, the current through B is found to be 2 A. Find the resistance of A and B.

Section 19.4   Kirchhoff's Rules

17. Find the current through each resistor of Figure 19.37 by (a) the rules for resistors in series and parallel and (b) by the use of Kirchhoff's rules.

Figure 19.37   Problem 17

18. Determine the potential on terminals A, B, and C in Figure 19.38. (a) if terminal A is grounded; (b) if terminal B is grounded; (c) if terminal C is grounded.

Figure 19.38 Problem 18

19. A pair of 1.50-V penlight batteries in series power a transistor radio. The batteries hold a total charge of 180 C. How long will they last if the effective resistance of the radio is 300 Ω?

20. Two 1.50-V batteries — with their positive terminals in the same direction — are inserted in series into the barrel of a flashlight. One battery has an internal resistance of 0.255 Ω, the other an internal resistance of 0.153 Ω. When the switch is closed, a current of 0.6 A occurs in the lamp. (See note in Problem 8.) (a) What is the lamp's resistance? (b) What fraction of the power dissipated is dissipated in the batteries?

21. An unmarked battery has an unknown internal resistance. If the battery is connected to a fresh 5.6-V battery (negligible internal resistance) positive to positive and negative to negative, the current through the circuit is 10.0 mA. If we reverse the polarity of the unknown battery, the current increases to 25.0 mA. Determine the source voltage and internal resistance of the unknown battery. Assume that in each case the direction of the current is negative to positive in the 5.6-V battery. (See note in Problem 8.)

22. Determine the current in the circuit of Figure 19.39 and the voltage drop across the 400-Ω resistor.

Figure 19.39 Problem 22

23. Figure 19.40 shows a circuit diagram. Determine (a) the current, (b) the potential of wire A, and (c) the voltage drop across the 1500-Ω resistor.

Figure 19.40 Problem 23

24. Find the current in each resistor in Figure 19.41.

Figure 19.41 Problem 24

25. Find the current in each branch of the circuit in Figure 19.42 and the value of $\mathcal{E}$.

Figure 19.42 Problem 25

26. Determine the potential difference V_{ab} for the circuit shown in Figure 19.43.

Figure 19.43 Problem 26

Figure 19.46 Problem 29

27. The two branch currents in the circuit shown in Figure 19.44 are $I_1 = \frac{1}{3}$A and $I_2 = \frac{1}{2}$A. Determine the emfs, $\mathcal{E}_1$ and $\mathcal{E}_2$.

30. Find the potential difference across each resistor shown in Figure 19.47.

Figure 19.44 Problem 27

Figure 19.47 Problem 30

Section 19.5 *RC* Circuits

28. Calculate each of the unknown currents I_1, I_2, and I_3 for the circuit of Figure 19.45.

31. Consider a series RC circuit for which $C = 6\ \mu$F, $R = 2 \times 10^6\ \Omega$, and $\mathcal{E} = 20$ V. Find (a) the time constant of the circuit, (b) the maximum charge on the capacitor after a switch in the circuit is closed, and (c) the current in the circuit at the instant just after the switch in the circuit is closed.

32. An uncharged capacitor and a resistor are connected in series to a source of emf. If $\mathcal{E} = 9$ V, $C = 20\ \mu$F, and $R = 100\ \Omega$, find (a) the time constant of the circuit, (b) the maximum charge on the capacitor, (c) the maximum current in the circuit, (d) the charge on the capacitor after one time constant, and (e) the current in the circuit after one time constant.

33. Consider an RC circuit in which the capacitor is being charged by a battery connected in the circuit. In two time constants, what percentage of the *final* charge is on the capacitor?

34. Consider a series RC circuit (Fig. 19.11c) for which $R = 1$ MΩ, $C = 5\ \mu$F, and $\mathcal{E} = 30$ V. Find the current in the resistor 10 s after the switch is closed.

Figure 19.45 Problem 28

29. What is the emf $\mathcal{E}_1$ of the battery in the circuit shown in Figure 19.46?

35. An uncharged capacitor, $C = 10 \ \mu F$, and a resistor are connected in series to an emf of 9 V. (a) What should the resistance equal to give a time constant of 5 s? (b) What is the current in this circuit, after two time constants have elapsed, if the resistance has the same value found in (a) above?

36. (This problem assumes coverage of the material in Footnote 1 in Section 19.5.) A series circuit contains an emf of 9 V, a resistor of $3 \times 10^6 \ \Omega$, and a capacitor of $2 \ \mu F$. Find (a) the charge on the capacitor after 10 s and (b) the current in the circuit at this time.

Section 19.6 Measurement of Resistance

37. (a) Calculate the potential difference across the 6-Ω resistor shown in Figure 19.48. (b) If you attempt to measure this potential difference by connecting across it a voltmeter that has a resistance of 1 Ω, what voltage will your meter read? (c) Repeat your calculation in part (b) assuming a voltmeter resistance of $1 \times 10^6 \ \Omega$.

Figure 19.48 Problem 37

38. A voltmeter with an internal resistance of $3.08 \times 10^4 \ \Omega$ measures the voltage across R_1 in Figure 19.49. (a) What is the voltage across R_1 in the absence of the meter? (b) Draw the circuit diagram with the meter in place. Calculate the effective resistance of the circuit. (c) What is the voltage across R_1 with the meter connected across it?

Figure 19.49 Problem 38

39. An ammeter with an internal resistance of 5.0 Ω measures the current in a simple series circuit consisting of a 100-V battery connected to a resistance of 25.0 Ω. (a) What is the current in the circuit in the absence of the meter? (b) What current does the meter read? Determine the percentage error in the measurement. Draw the circuit in each case.

40. A circuit consists of an emf of 12 V connected to a 6-Ω resistor. (a) Calculate the current in the circuit. (b) Assume you attempt to measure this current by using an ammeter having a resistance of 100 Ω. What current will your meter read? (c) Repeat your calculations in part (b) assuming an ammeter resistance of 0.1 Ω.

41. The Wheatstone bridge in Figure 19.19 is balanced when $R_1 = 15 \ \Omega$, $R_2 = 25 \ \Omega$, and $R_3 = 40 \ \Omega$. Calculate the value of R_x.

42. The Wheatstone bridge in Figure 19.19 is modified, as shown in Figure 19.50. The result is that the resistors in one branch of the circuit have been replaced by a 1-m length of nichrome wire. Find an expression for R_x in terms of the resistor R_3, the length of wire between points a and b in Figure 19.50, and the length between points b and c when the bridge is balanced.

Figure 19.50 Problem 42

43. A Wheatstone bridge of the type shown in Figure 19.19 is used to measure an unknown resistance. The resistor shown in the circuit as $R_3 = 1 \ k\Omega$. If the bridge is balanced by adjusting R_1 such that $R_1 = 2.5 R_2$, what is the resistance of the unknown resistor?

Section 19.7 Household Circuits

44. An electric heater is rated at 1300 W, a toaster is rated at 1000 W, and an electric grill is rated at 1500 W. The three appliances are connected in parallel to a common 120-V circuit. (a) How much current does each appliance draw? (b) Is a 30-A circuit sufficient in this situation? Explain.

45. How many 100-W lightbulbs can you connect in a 120-V house circuit without tripping a 20-A circuit breaker?

46. A heating element in a stove is designed to dissipate 3000 W when connected to 240 V. (a) Calculate the current in this element if it is connected to 120 V and (b) calculate the power it dissipates at this voltage.

47. An 8-foot extension cord has two 18-gauge copper wires, each having a diameter of 1.024 mm. (a) How much power does this cord dissipate when carrying a current of 1 A? (b) How much power does this cord dissipate when carrying a current of 10 A?

48. Sometimes aluminum wiring is used instead of copper for economic reasons. According to the National Electrical Code, the maximum allowable current for 12-gauge copper wire with rubber insulation is 20 A. What should be the maximum allowable current in a 12-gauge aluminum wire if it is to dissipate the same power per unit length as the copper wire?

49. A lamp $(R = 150\ \Omega)$, an electric heater $(R = 25\ \Omega)$, and a fan $(R = 50\ \Omega)$ are connected in parallel across a 120-V line. (a) What total current is supplied to the circuit? (b) What is the voltage across the fan? (c) What is the current in the lamp? (d) What power is expended in the heater?

Additional Problems

50. If you have a 2-Ω, a 4-Ω, and a 6-Ω resistor, how many different resistance values can you obtain from the set? Show how you would connect them to get each resistance value.

51. The meter in Figure 19.51 reads 1 mA when a 2-V power supply is connected between points A and B. (a) Find the resistance R_1. (b) The meter continues to read 1 mA if a 20-V power supply is now connected between points A and C. Find R_2. (c) The meter continues at its reading of 1 mA when a 200-V power supply is connected between points A and D. Find R_3.

Figure 19.51   Problem 51

52. In Figure 19.52, $R_1 = 0.1\ \Omega$, $R_2 = 1.0\ \Omega$, and $R_3 = 10.0\ \Omega$. Find the equivalent resistance of the circuit

Figure 19.52   Problem 52

and the current in each resistor when a 5-V power supply is connected between (a) points A and B, (b) points A and C, and (c) points A and D.

53. Find the equivalent resistance of the circuit shown in Figure 19.53.

Figure 19.53   Problem 53

54. For the circuit of Problem 53, find (a) each current in the circuit, (b) the potential difference across each resistor, and (c) the power dissipated by each resistor.

55. If the current in the 5-Ω resistor of Figure 19.54 is 3 A, find the current in all the other resistors and the emf of the battery.

Figure 19.54   Problem 55

56. A series RC circuit has a time constant of 0.96 s. The battery has an emf of 48 V, and the maximum current in the circuit is 500 mA. What is (a) the value of the capacitance and (b) the charge stored in the capacitor 1.92 s after the switch is closed?

57. In the circuit of Figure 19.55, find the power delivered or absorbed by each battery and the power dissipated in each resistor.

Figure 19.55   Problem 57

58. Consider the circuit shown in Figure 19.56. Find the values of I_1, I_2, and I_3.

Figure 19.56 Problem 58

59. In a certain circuit, a current of 1.00 A is drawn from a battery. The current then divides and passes through two resistors in parallel. One of the resistors has a value of 41 Ω, and the current through it is 0.10 A. What is the value of the other resistor?

60. (a) Calculate the value of R for the circuit shown in Figure 19.57. (b) Determine the currents in the 6-Ω and 8-Ω resistors.

Figure 19.57 Problem 60

61. Find the equivalent resistance between points a and b in Figure 19.58.

Figure 19.58 Problem 61

62. Consider the circuit shown in Figure 19.59. Calculate (a) the equivalent resistance of the circuit, and (b) the power dissipated by the entire circuit. Find (c) the current in the 5-Ω resistor.

Figure 19.59
Problem 62

63. (This problem assumes coverage of Footnote 1 in Section 19.5.) An emf of 10 V is connected to a series RC circuit of a resistor of 2×10^6 Ω and a capacitor of 3 μF. Find the time taken for the charge on the capacitor to reach 90% of its final value.

64. Consider the circuit shown in Figure 19.60. In this circuit, the 4-V battery and the 2-V battery have internal resistances of 0.5 Ω and 0.25 Ω, respectively. (*Hint:* See note in Problem 8.) (a) Find the current in the circuit. (b) Find the power dissipated as heat in the circuit. (c) Find the power absorbed by the 2-V battery as it charges, including the power dissipated as heat within the battery.

Figure 19.60
Problem 64

65. When a battery of unknown emf is connected to a 5-Ω resistor, the current in the circuit is 0.3 A. If the battery is now connected to an 8-Ω resistor, the current is 0.2 A. What is the emf of the battery and its internal resistance? (See note in Problem 8.)

66. The current in a circuit that has a resistance of R_1 is 2 A. The current is reduced to 1.6 A when an additional resistor $R_2 = 3$ Ω is added in series with R_1. What is the value of R_1? Assume the voltage applied to the circuit remains constant.

67. Calculate the current in the 6-V battery in Figure 19.61.

Figure 19.61
Problem 67

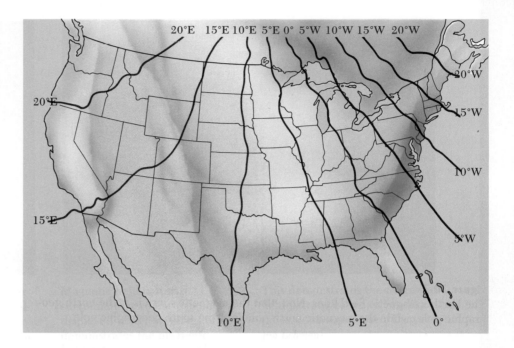

Figure 20.3
A map of the United States showing the declination of a compass from true north.

Although the magnetic field pattern of the Earth is similar to that which would be set up by a bar magnet deep within the Earth, it is easy to understand why the source of the Earth's field cannot be large masses of permanently magnetized material. The Earth does have large deposits of iron ore deep beneath its surface, but the high temperatures in the Earth's core prevent the iron from retaining any permanent magnetization. It is considered more likely that the true source of the Earth's field is charge-carrying convection currents in its core. Charged ions or electrons circling in the liquid interior could produce a magnetic field. There is also some evidence to indicate that the strength of a planet's field is related to the planet's rate of rotation. For example, Jupiter rotates faster than the Earth, and recent space probes indicate that Jupiter's magnetic field is stronger than ours. Venus, on the other hand, rotates more slowly than the Earth, and its magnetic field is found to be weaker. Investigation into the cause of the Earth's magnetism remains open.

There is an interesting sidelight concerning the Earth's magnetic field. It has been found that the direction of the field has been reversed several times during the last million years. Evidence for this is provided by basalt (a type of rock that contains iron) that is spewed forth by volcanic activity on the ocean floor. As the lava cools, it solidifies and retains a picture of the Earth's magnetic field direction. The rocks can be dated by other means to provide the evidence for these periodic reversals of the magnetic field.

20.3   Magnetic Fields

Experiments show that a stationary charged particle does not interact with a static magnetic field. However, *a charged particle experiences a force when moving through a magnetic field*. This force has its maximum value when the charge moves perpendicular to the magnetic field lines, decreases in value at other angles, and becomes zero when the particle moves along the field lines. We shall make use of these observations in describing the characteristics of the magnetic field.

In our discussion of electricity, the electric field at some point in space was defined as the electric force per unit charge acting on some test charge placed at that point. In a similar manner, we shall describe the properties of the magnetic field, **B**, at some point, in terms of the magnetic force exerted on a test charge at that point. Our test object is assumed to be a charge q moving with velocity **v**. It is found that the strength of the magnetic force on the particle is proportional to the magnitude of the charge q, the magnitude of the velocity **v**, the strength of the external magnetic field, **B**, and the sine of the angle θ between the direction of **v** and the direction of **B**. That is, $F \propto qvB \sin \theta$. Taking the constant of proportionality to be one, we obtain the equation

$$F = qvB \sin \theta \qquad \textbf{(20.1)}$$ Magnetic force

This expression is used to define the magnitude of the magnetic field as

$$B \equiv \frac{F}{qv \sin \theta} \qquad \textbf{(20.2)}$$ Magnetic field defined

If F is in newtons, q in coulombs, and v in meters per second, the SI unit of magnetic field is the **tesla** (T), also called the **weber per square meter** (that is, $1 \text{ T} = 1 \text{ Wb/m}^2$). Thus, if a 1-C charge moves through a magnetic field of magnitude 1 T with a velocity of 1 m/s in a direction perpendicular to the field ($\sin \theta = 1$), the magnetic force exerted on the charge is 1 N. We can express the units of **B** as

$$[B] = \text{T} = \frac{\text{Wb}}{\text{m}^2} = \frac{\text{N}}{\text{C} \cdot \text{m/s}} = \frac{\text{N}}{\text{A} \cdot \text{m}} \qquad \textbf{(20.3)}$$

In practice, the cgs unit for magnetic field, the **gauss** (G), is often used. The gauss is related to the tesla through the conversion

$$1 \text{ T} = 10^4 \text{ G}$$

Conventional laboratory magnets can produce magnetic fields as large as about 25,000 G, or 2.5 T. Superconducting magnets that can generate magnetic fields as high as 3×10^5 G, or 30 T, have been constructed. These values can be compared with the value of the Earth's magnetic field near its surface, which is about 0.5 G, or 0.5×10^{-4} T.

From Equation 20.1, we see that the force on a charged particle moving in a magnetic field has its maximum value when the particle moves *perpendicular* to the magnetic field, corresponding to $\theta = 90°$, so that $\sin \theta = 1$. The magnitude of this maximum force has the value

$$F_{max} = qvB \qquad \textbf{(20.4)}$$

Also, note from Equation 20.1 that F is zero when **v** is parallel to **B** (corresponding to $\theta = 0°$ or $180°$). Thus, there is no magnetic force exerted on a charged particle when it moves in the direction of the magnetic field or opposite the field.

Figure 20.6 (Example 20.2) The magnetic force **F** on a proton is in the positive z direction when **v** and **B** lie in the xy plane.

Since the mass of the proton is 1.67×10^{-27} kg, its initial acceleration is

$$a = \frac{F}{m} = \frac{2.77 \times 10^{-12} \text{ N}}{1.67 \times 10^{-27} \text{ kg}} = 1.66 \times 10^{15} \text{ m/s}^2$$

in the positive z direction.

Exercise Calculate the acceleration of an electron that moves through the same magnetic field at the same speed as the proton. The mass of an electron is equal to 9.11×10^{-31} kg.

Answer 3.04×10^{18} m/s^2.

20.4   Magnetic Force on a Current-Carrying Conductor

If a force is exerted on a single charged particle when it moves through a magnetic field, it should not surprise you to find that a current-carrying wire also experiences a force when placed in a magnetic field. This follows from the fact that the current represents a collection of many charged particles in motion; hence, the resultant force on the wire is due to the sum of the individual forces on the charged particles. The force on the particles is transmitted to the "bulk" of the wire through collisions with the atoms making up the wire.

The force on a current-carrying conductor can be demonstrated by hanging a wire between the faces of a magnet as in Figure 20.7. In this figure, the magnetic field is directed into the page and covers the region within the shaded circle. When the current in the wire is zero, the wire remains vertical as in Figure 20.7a. However, when a current is set up in the wire directed upwards as in Figure 20.7b, the wire deflects to the left. If we reverse the current, as in Figure 20.7c, the wire deflects to the right.

Figure 20.7
A flexible vertical wire which is partially stretched between the faces of a magnet with the field (blue crosses) directed into the paper. (a) When there is no current in the wire, it remains vertical. (b) When the current is upwards, the wire deflects to the left. (c) When the current is downwards, the wire deflects to the right.

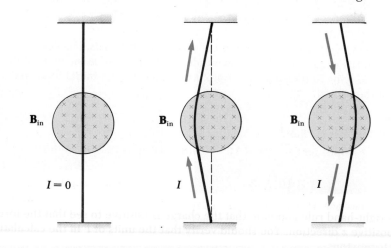

Consider a straight segment of wire of length ℓ and cross-sectional area A, carrying a current I in a uniform external magnetic field **B**, as in Figure 20.8. Let us assume that the magnetic field is perpendicular to the wire and is directed into the page. (Note that the direction of magnetic fields directed into the page is represented by crosses, and fields coming out of the page are represented by dots. The crosses can be thought of as the tail feathers of arrows moving away from you, while the dots represent the points of arrows coming toward you.) Because of the external magnetic field, each charge carrier in the wire experiences a force $F_{max} = q v_d B$, where v_d is the drift velocity of the charge. This force causes the charge carrier to be deflected upward if it is positive and downward if it is negative. This single force would be imperceptible by itself, but every charge carrier in the wire has a similar force on it. Hence, the combined effect of all these forces on the charge carriers (that is, the net magnetic force) can be great enough to move the wire. To find the total force on the wire, we multiply the force on one charge carrier by the number of carriers in the segment. Since the volume of the segment is $A\ell$, the number of carriers is $nA\ell$, where n is the number of carriers per unit volume. Hence, the total magnetic force on the wire of length ℓ is

Total force = (force on each charge carrier)(total number of carriers)

$$F_{max} = (q v_d B)(nA\ell)$$

From Chapter 18, however, we know that the current in the wire is given by $I = nq v_d A$. Therefore, F_{max} can be expressed as

$$F_{max} = BI\ell \qquad (20.5)$$

This equation can be used only when *the current and magnetic field are at right angles to each other.*

If the wire is not perpendicular to the field, but is at some arbitrary angle, as in Figure 20.9, the magnitude of the magnetic force on the wire is

$$F = BI\ell \sin \theta \qquad (20.6)$$

where θ is the angle between **B** and the direction of the current. The direction of this force can be obtained using the right-hand rule. However, in this case, you must place your thumb in the direction of the current rather than in the direction of **v**. In Figure 20.9, the direction of the magnetic force on the wire is out of the page.

Finally, you should note that when the current is either in the direction of the field or opposite the direction of the field, the magnetic force on the wire is zero.

Example 20.3   A Current-Carrying Wire in the Earth's Magnetic Field

A wire carries a current of 20 A from east to west. Assume that the magnetic field of the Earth is horizontal at this location, directed from south to north, and that it has a magnitude of 0.5×10^{-4} T. Find the force on a 30-m length of wire. How is the force changed if the current runs west to east?

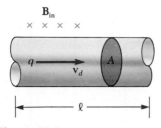

Figure 20.8
A section of a wire containing moving charges in an external magnetic field **B**.

This apparatus demonstrates the force on a current-carrying conductor in an external magnetic field. Why does the bar swing *away* from the magnet after the switch is closed? *(Courtesy of Henry Leap and Jim Lehman)*

Figure 20.9
A wire carrying a current I in the presence of an external magnetic field **B** that makes an angle θ with the wire.

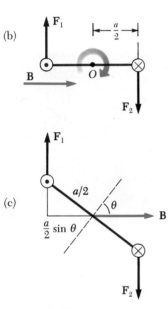

Figure 20.10
(a) A rectangular loop in a uniform magnetic field **B**. There are no magnetic forces on the sides of length a parallel to **B**, but there are forces acting on the sides of length b. (b) An end view of the rectangular loop shows that the forces $\mathbf{F}_1$ and $\mathbf{F}_2$ on the sides of length b create a torque that tends to twist the loop clockwise as shown. (c) If **B** is at an angle θ with respect to a line perpendicular to the plane of the loop, the torque is given by $IAB \sin \theta$.

Solution

Because the direction of the current and the direction of the magnetic field are at right angles, we can use Equation 20.5. The magnitude of the force is

$$F_{max} = BI\ell = (0.5 \times 10^{-4} \text{ T})(20 \text{ A})(30 \text{ m}) = \boxed{3 \times 10^{-2} \text{ N}}$$

The right-hand rule shows that the force on the wire is directed toward the Earth.

If the direction of the current is from west to east, the magnitude of the force remains the same but its direction will be upward, away from the Earth.

Exercise   If the current flows north to south, what is the magnetic force on the wire?

Answer   Zero.

20.5   Torque on a Current Loop

In the previous section we showed how a force is exerted on a current-carrying conductor when the conductor is placed in an external magnetic field. With this as a starting point, we shall show that a torque is exerted on a current loop placed in a magnetic field. The results of this analysis will be of great practical value when we discuss the galvanometer in this chapter and in a discussion of generators in a future chapter.

Consider a rectangular loop carrying a current I in the presence of an external uniform magnetic field directed as shown in Figure 20.10a. The forces on the sides of length a are zero because these wires are parallel to the field. The magnitude of the forces on the sides of length b, however, is

$$F_1 = F_2 = BIb$$

The direction of $\mathbf{F}_1$, the force on the left side of the loop, is out of the paper and that of $\mathbf{F}_2$, the force on the right side of the loop, is into the paper. If we were to view the loop from below the loop, as in Figure 20.10b, we would see the forces directed as shown. If we assume that the loop is pivoted so that it can rotate about point O, we see that these two forces produce a torque about O that rotates the loop clockwise. The magnitude of this torque, τ_{max}, is

$$\tau_{max} = F_1 \frac{a}{2} + F_2 \frac{a}{2} = (BIb)\frac{a}{2} + (BIb)\frac{a}{2} = BIab$$

where the moment arm about O is $a/2$ for both forces. Since the area of the loop is $A = ab$, the torque can be expressed as

$$\boxed{\tau_{max} = BIA} \tag{20.7}$$

Note that this result is valid only when the magnetic field is parallel to the plane of the loop, as shown in the view of the loop pictured in Figure 20.10b. If the field makes an angle θ with respect to a line perpendicular to the plane of the loop, as shown in Figure 20.10c, the moment arm for each force is given by $(a/2) \sin \theta$. An analysis like that just used shows that the magnitude of the torque is given by

$$\tau = BIA \sin \theta \qquad \textbf{(20.8)}$$

This result shows that the torque has the *maximum* value *BIA* when the field is parallel to the plane of the loop ($\theta = 90°$) and is *zero* when the field is perpendicular to the plane of the loop ($\theta = 0$). As we see in Figure 20.10c, the loop tends to rotate to smaller values of θ (that is, such that the normal to the plane of the loop rotates toward the direction of the magnetic field).

Although this analysis is for a rectangular loop, a more general derivation would indicate that Equation 20.8 applies regardless of the shape of the loop. Furthermore, the torque on a loop with *N* turns is equal to

$$\tau = NBIA \sin \theta \qquad \textbf{(20.9)}$$

Example 20.4   The Torque on a Circular Loop in a Magnetic Field

A circular loop of wire of radius 50 cm is oriented at an angle of 30° to a magnetic field of 0.5 T, as shown in an end view in Figure 20.11. The current in the loop is 2 A in the direction shown. Find the magnitude of the torque at this instant.

Solution
Regardless of the shape of the loop, Equation 20.8 is valid:

$$\tau = BIA \sin \theta = (0.5 \text{ T})(2 \text{ A})[\pi(0.5 \text{ m})^2](\sin 30°) = \boxed{0.393 \text{ N·m}}$$

Exercise   Find the torque on the loop if it has three turns rather than one.

Answer   The torque is three times greater than on the one-turn loop, or 1.18 N·m.

Figure 20.11   (Example 20.4) The edge view of a circular current loop in an external magnetic field **B**.

20.6   The Galvanometer and Its Applications

The Galvanometer

The *galvanometer* is a device used in the construction of both ammeters and voltmeters. The basic operation of this instrument makes use of the fact that a torque acts on a current loop in the presence of a magnetic field. The main components of a galvanometer are shown in Figure 20.12a. It consists of a coil

(a)   Spring   Coil

(b)

Figure 20.12
(a) The principal components of a galvanometer. When current passes through the coil, situated in a magnetic field, the magnetic torque causes the coil to twist. The angle through which the coil rotates is proportional to the current through it.
(b) Large-scale demonstration model of a galvanometer movement. Why does the coil rotate about the vertical axis after the switch is closed? *(Courtesy of Jim Lehman)*

of wire mounted such that it is free to rotate on a pivot in a magnetic field provided by a permanent magnet. The torque experienced by the coil is proportional to the current through it. This means that the larger the current, the larger the torque and the more the coil will rotate before the spring tightens enough to stop the movement. Hence, the amount of deflection is proportional to the current. Once the instrument is properly calibrated, it can be used in conjunction with other circuit elements to measure either currents or potential differences. Figure 20.12b is a photograph of a large-scale galvanometer movement.

A Galvanometer Is the Basis of an Ammeter

A typical off-the-shelf galvanometer is usually not suitable for use as an ammeter (a current-measuring device). One of the main reasons for this is that a typical galvanometer has a resistance of about 60 Ω. An ammeter resistance this large could considerably alter the current in the circuit in which it is placed. This can be easily understood by considering the following example. Suppose you were to construct a simple series circuit containing a 3-V battery and a 3-Ω resistor. The current in such a circuit is 1 A. However, if you include a 60-Ω galvanometer in the circuit in an attempt to measure the current, the total resistance of the circuit is now 63 Ω and the current is reduced to 0.048 A.

A second factor that limits the use of a galvanometer as an ammeter is the fact that a typical galvanometer will give a full-scale deflection for very low currents, of the order of 1 mA or less. Consequently, such a galvanometer cannot be used directly to measure currents greater than this.

Now let us assume we wish to convert a 60-Ω, 1-mA galvanometer to an ammeter that deflects full scale when 2 A passes through it. In spite of the factors just described, this can be accomplished by simply placing a resistor, R_p, in *parallel* with the galvanometer, as in Figure 20.13. (The galvanometer and parallel resistor combination is called an ammeter.) The size of the resistor must be selected such that when 2 A passes through the ammeter, only 0.001 A passes through the galvanometer and the remaining 1.999 A passes through the resistor R_p, sometimes called the *shunt resistor*. Because the galvanometer and shunt resistor are in parallel, the potential difference across each is the same. Thus, using Ohm's law we get

$$(0.001 \text{ A})(60 \text{ } \Omega) = (1.999 \text{ A})R_p$$

$$R_p = 0.03 \text{ } \Omega$$

Notice that the shunt resistance, R_p, is extremely small. Thus, we see that the configuration shown in Figure 20.13 solves both problems associated with converting a galvanometer to an ammeter. The ammeter just described is able to measure a large current (2 A), and it has a low resistance, on the order of 0.03 Ω. (Recall that the equivalent resistance of two resistors in parallel is always *less* than the value of the individual resistors.)

A Galvanometer Is the Basis of a Voltmeter

With the proper modification, the basic galvanometer can also be used to measure potential differences in a circuit. In order to understand how this can be accomplished, let us first calculate the largest voltage that can be measured

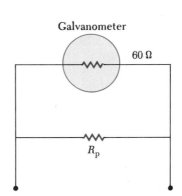

Galvanometer

60 Ω

R_p

Figure 20.13
When a galvanometer is to be used as an ammeter, a resistor R_p is connected in parallel with the galvanometer.

with a galvanometer. If the galvanometer has a resistance of 60 Ω and gives a maximum deflection for a current of 1 mA, the largest voltage it can measure is

$$V_{max} = (0.001 \text{ A})(60 \text{ } \Omega) = 0.06 \text{ V}$$

From this result, we see that some modification is required to enable this device to measure larger voltages. Furthermore, a voltmeter must have a very high resistance in order to insure that it will not disturb the circuit in which it is placed. The basic galvanometer, with a resistance of only 60 Ω, is not acceptable for direct voltage measurements.

The circuit in Figure 20.14 shows the basic modification that must be made to convert a galvanometer to a voltmeter. Let us assume we want to construct a voltmeter capable of measuring a maximum voltage of 100 V. In this situation, a resistor, R_s, is placed in *series* with the galvanometer. The value of R_s is found by noting that a current of 1 mA must pass through the galvanometer when the voltmeter is connected across a potential difference of 100 V. Applying Ohm's law to this circuit gives

$$100 \text{ V} = (0.001 \text{ A})(R_s + 60 \text{ } \Omega)$$

$$R_s = 99,940 \text{ } \Omega$$

This result shows that this voltmeter has a very high resistance.

When a voltmeter is constructed with several available ranges, various values of R_s may be selected by using a switch that can be connected to a preselected set of resistors. The required value of R_s increases as the maximum voltage to be measured increases.

Figure 20.14
When a galvanometer is to be used as a voltmeter, a resistor R_s is connected in series with the galvanometer.

20.7 Motion of a Charged Particle in a Magnetic Field

Consider the case of a positively charged particle moving in a uniform magnetic field such that the direction of the particle's velocity is *perpendicular to the field*, as in Figure 20.15. The label $\mathbf{B}_{in}$ indicates that $\mathbf{B}$ is directed into the page. Application of the right-hand rule at point P shows that the direction of the magnetic force, $\mathbf{F}$, at this location is upward. This causes the particle to alter its direction of travel and to follow a curved path. Application of the right-hand rule at any point shows that *the magnetic force is always toward the center of the circular path;* therefore the magnetic force is effectively a centripetal force that changes only the direction of $\mathbf{v}$ and not its magnitude. Since $\mathbf{F}$ is a centripetal force, we can equate its magnitude, qvB in this case, to the mass of

A magnet placed near the screen of a television tube will distort the picture as shown. This is due to the deflection of the electron beam in the tube in the presence of the magnetic field. *(Courtesy of Henry Leap and Jim Lehman)*

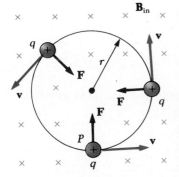

Figure 20.15
When the velocity of a charged particle is perpendicular to a uniform magnetic field, the particle moves in a circle whose plane is perpendicular to $\mathbf{B}$, which is directed into the page. (The crosses represent the tail of the magnetic field vector.) The magnetic force $\mathbf{F}$ on the charge is always directed toward the center of the circle.

Aurora borealis, the Northern Lights, photographed near Fairbanks, Alaska. Auroras occur when cosmic rays—electrically charged particles originating mainly from the Sun—become trapped in the Earth's atmosphere over Earth's magnetic poles and collide with other atoms, resulting in the emission of visible light. *(Jack Finch/Science Photo Library)*

The bending of an electron beam in an external magnetic field. The tube contains gas at very low pressure, and the beam is made visible as the electrons collide with the gas atoms, which in turn emit visible light. The apparatus used to take this photograph is part of a system used to measure the ratio e/m. *(Courtesy of Henry Leap and Jim Lehman)*

the particle multiplied by the centripetal acceleration, v^2/r. From Newton's second law, we find that

$$F = qvB = \frac{mv^2}{r}$$

which gives

$$r = \frac{mv}{qB} \qquad \textbf{(20.10)}$$

This says that the radius of the path is proportional to the momentum, mv, of the particle and is inversely proportional to the magnetic field.

If the initial direction of the velocity of the charged particle is not perpendicular to the magnetic field but instead is directed at an angle to the field, as shown in Figure 20.16, the path followed by the particle is a spiral (called a helix) along the magnetic field lines.

Example 20.5 A Proton Moving Perpendicular to a Uniform Magnetic Field

A proton is moving in a circular orbit of radius 14 cm in a uniform magnetic field of magnitude 0.35 T directed perpendicular to the velocity of the proton. Find the orbital speed of the proton.

Solution
From Equation 20.10, we get

$$v = \frac{qBr}{m} = \frac{(1.6 \times 10^{-19}\ \text{C})(0.35\ \text{T})(14 \times 10^{-2}\ \text{m})}{1.67 \times 10^{-27}\ \text{kg}} = 4.7 \times 10^6\ \text{m/s}$$

Exercise If an electron moves perpendicular to the same magnetic field with this speed, what is the radius of its circular orbit?

Answer 7.6×10^{-5} m.

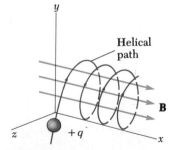

Figure 20.16
A charged particle having a velocity vector that has a component parallel to a uniform magnetic field moves in a helical path.

Example 20.6   The Mass Spectrometer

Two singly ionized atoms move out of a slit at point S in Figure 20.17 and into a magnetic field of 0.1 T. They each have a velocity of 10^6 m/s. The nucleus of the first atom contains one proton and has a mass of 1.68×10^{-27} kg, and the nucleus of the second atom contains a proton and a neutron and has a mass of 3.36×10^{-27} kg. Atoms with the same chemical properties but different masses are called isotopes. The two isotopes here are hydrogen and deuterium. Find their distance of separation when they strike a photographic plate at P.

Solution
The radius of the circular path followed by the lighter isotope, hydrogen, is

$$r_1 = \frac{m_1 v}{qB} = \frac{(1.68 \times 10^{-27} \text{ kg})(10^6 \text{ m/s})}{(1.6 \times 10^{-19} \text{ C})(0.1 \text{ T})} = 0.11 \text{ m}$$

The radius of the path of the heavier isotope, deuterium, is

$$r_2 = \frac{m_2 v}{qB} = \frac{(3.36 \times 10^{-27} \text{ kg})(10^6 \text{ m/s})}{(1.6 \times 10^{-19} \text{ C})(0.1 \text{ T})} = 0.21 \text{ m}$$

The distance of separation is

$$x = 2r_2 - 2r_1 = \boxed{0.20 \text{ m}}$$

The concepts used in this example underlie the operation of a device called a **mass spectrometer**. This instrument is sometimes used to separate isotopes according to their mass-to-charge ratio but it is used more often to measure masses.

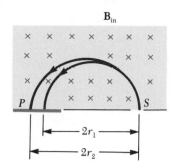

Figure 20.17 (Example 20.6) Two isotopes leave the slit at point S and travel in different circular paths before striking a photographic plate at P.

20.8   Magnetic Field of a Long, Straight Wire and Ampere's Law

The relationship between electricity and magnetism was discovered in 1819 when, during a lecture demonstration, the Danish scientist Hans Oersted (1777–1851) found that an electric current in a wire deflected a nearby compass needle. This discovery, linking a magnetic field with an electric current, was the beginning of our understanding of the origin of magnetism.

Figure 20.18 illustrates an experiment that can be performed to show that a current produces a magnetic field. Several compasses are placed in a horizon-

Hans Christian Oersted (1777–1851), Danish physicist. *(The Bettmann Archive)*

Figure 20.18
(a) When there is no current in the vertical wire, all compass needles point in the same direction. (b) When the wire carries a strong current, the compass needles deflect in a direction tangent to the circle, which is the direction of **B** due to the current.

tal plane near a long vertical wire. The compasses are positioned on a circular path concentric with the wire. When there is no current in the wire, all compasses point in the same direction, as in Figure 20.18a (that of the Earth's field), as one would expect. However, when there is a steady current flowing upward, as in Figure 20.18b, the compasses point as shown. The fact that the compasses point in a direction tangent to the circle indicates that the magnetic field lines of the wire are circles concentric with the wire. If the current in the wire is reversed, the compasses point in the opposite direction, indicating that the direction of the magnetic field is also reversed.

A convenient rule for determining the direction of a magnetic field produced by a current in a wire is the following:

Grasp the wire with the right hand, with the thumb along the direction of the current, as shown in Figure 20.19a. The four fingers wrap around the wire in the direction of the magnetic field lines.

Shortly after Oersted's discovery, scientists were able to arrive at an expression that gives the strength of the magnetic field due to a long, straight wire. The magnetic field strength at a distance r from a wire carrying a current I is

$$B = \frac{\mu_0 I}{2\pi r} \tag{20.11}$$

This result shows that the magnitude of the magnetic field is proportional to the current and decreases as the distance from the wire increases, as one might intuitively expect. The proportionality constant μ_0, called the **permeability of free space,** is defined to have the value

$$\mu_0 \equiv 4\pi \times 10^{-7} \text{ T·m/A} \tag{20.12}$$

Figure 20.19
(a) The right-hand rule for determining the direction of the magnetic field due to a long, straight wire carrying a current. Note that the magnetic field lines form circles around the wire. (b) Circular magnetic field lines surrounding a current-carrying wire, as displayed with iron filings. *(Courtesy of Henry Leap and Jim Lehman)*

(a)   (b)

*Ampère's Law

Equation 20.11 enables us to calculate the magnetic field due to a long, straight wire carrying a current. A generalization of this result was proposed by the French scientist André-Marie Ampère (1775–1836). This scheme provides a relation between the current in an arbitrarily shaped wire and the magnetic field produced by the wire.

Consider a circular path surrounding a current, as in Figure 20.19a. The path can be divided into many short segments, each of length $\Delta\ell$. Let us now multiply one of these lengths by the component of the magnetic field parallel to that segment, where the product is labeled $B_\parallel \Delta\ell$. According to Ampère, the sum of all such products over the closed path is equal to μ_0 times the net current I that passes through the surface bounded by the closed path. This statement, known as **Ampère's circuital law,** can be written

$$\sum B_\parallel \Delta\ell = \mu_0 I \qquad \textbf{(20.13)}$$

Ampère's circuital law

where $\Sigma B_\parallel \Delta\ell$ means that we take the sum over all the products $B_\parallel \Delta\ell$ around the closed path.

We can use Ampère's circuital law to derive the magnetic field due to a long, straight wire carrying a current I. As we discussed earlier, the magnetic field lines of this configuration form circles with the wire at their centers, as shown in Figure 20.19a. The magnetic field is tangent to this circle at every point and has the same value, $B_\parallel$, over the entire circumference of a circle of radius r. We now calculate the sum $\Sigma B_\parallel \Delta\ell$ over a circular path and note that $B_\parallel$ can be removed from the sum (since it has the same value for each element on the circle). Equation 20.13 then gives

$$\sum B_\parallel \Delta\ell = B_\parallel \sum \Delta\ell = B_\parallel (2\pi r) = \mu_0 I$$

If we divide both sides by $2\pi r$, we obtain

$$B = \frac{\mu_0 I}{2\pi r}$$

This is identical to Equation 20.11, which is the magnetic field of a long, straight current.

Ampère's circuital law is extremely important because it provides an elegant and simple method for calculating the magnetic fields of current configurations with high symmetry. However, it cannot be used to calculate magnetic fields for complex current configurations that lack symmetry.

Example 20.7   The Magnetic Field of a Long Wire

A long, straight wire carries a current of 5 A. Find the magnitude of the magnetic field 4 cm from the wire.

Solution
From Equation 20.11, we find

$$B = \frac{\mu_0 I}{2\pi r} = \frac{(4\pi \times 10^{-7}\ \text{T·m/A})(5\ \text{A})}{2\pi(4 \times 10^{-2}\ \text{m})} = \boxed{2.5 \times 10^{-5}\ \text{T}}$$

André-Marie Ampère
(1775–1836)

André-Marie Ampère was a French mathematician, chemist, and philosopher who founded the science of electrodynamics. The unit of measure for electric current was named in his honor.

Ampère's genius, particularly in mathematics, became evident early in his life: he had mastered advanced mathematics by the age of 12. In his first publication, *Considerations on the Mathematical Theory of Games,* an early contribution to the theory of probability, he proposed the inevitability of a player's losing a game of chance to a player with greater financial resources.

Ampère is credited with the discovery of electromagnetism—the relationship between electric current and magnetic fields. His work in this field was influenced by the findings of Danish physicist Hans Christian Oersted. Ampère presented a series of papers expounding the theory and basic laws of electromagnetism, which he called electrodynamics, to differentiate it from the study of stationary electric forces, which he called electrostatics.

The culmination of Ampère's studies came in 1827 when he published his *Mathematical Theory of Electrodynamic Phenomena Deduced Solely from Experiment,* in which he derived precise mathematical formulations of electromagnetism, notably Ampère's law.

Many stories are told of Ampère's absent-mindedness, a trait he shared with Newton. In one instance, he forgot to honor an invitation to dine with the Emperor Napoleon.

Ampère's personal life was filled with tragedy. His father, a wealthy city official, was guillotined during the French Revolution, and his wife's death in 1803 was a major blow. Ampère died at the age of 61 of pneumonia. His judgment of his life is clear from the epitaph he chose for his gravestone: *Tandem felix* (Happy at last).

Exercise   At what distance from the wire does the field have a magnitude equal to 0.5×10^{-5} T?

Answer   20 cm.

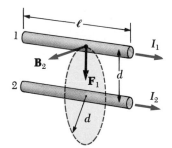

Figure 20.20
Two parallel wires, each carrying a steady current, exert a force on each other. The field $\mathbf{B}_2$ at wire 1 due to wire 2 produces a force on wire 1 given by $F_1 = B_2 I_1 \ell$. The force is attractive if the currents flow in the same direction, as shown, and repulsive if the two currents flow in opposite directions.

20.9   Magnetic Force Between Two Parallel Conductors

As we have seen, a magnetic force acts on a current-carrying conductor when the conductor is placed in an external magnetic field. Since a current in a conductor creates its own magnetic field, it is easy to understand that two current-carrying wires placed close together will exert magnetic forces on each other. Consider two long, straight, parallel wires separated by a distance d and carrying currents I_1 and I_2 in the same direction, as shown in Figure 20.20. Let us determine the force on one wire due to a magnetic field set up by the other wire.

Wire 2, which carries a current I_2, sets up a magnetic field $\mathbf{B}_2$ at wire 1. The direction of $\mathbf{B}_2$ is perpendicular to the wire, as shown in the figure. Using Equation 20.11, we see that the magnitude of this magnetic field is

$$B_2 = \frac{\mu_0 I_2}{2\pi d}$$

According to Equation 20.5, the magnetic force on wire 1 in the presence of the field $\mathbf{B}_2$ due to I_2 is

$$F_1 = B_2 I_1 \ell = \left(\frac{\mu_0 I_2}{2\pi d}\right) I_1 \ell = \frac{\mu_0 I_1 I_2 \ell}{2\pi d}$$

We can rewrite this in terms of the force per unit length as

$$\frac{F_1}{\ell} = \frac{\mu_0 I_1 I_2}{2\pi d} \qquad\qquad \textbf{(20.14)}$$

The direction of $\mathbf{F}_1$ is downward, toward wire 2, as indicated by the right-hand rule. If one considers the field set up at wire 2 due to wire 1, the force $\mathbf{F}_2$ on wire 2 is found to be equal to and opposite $\mathbf{F}_1$. This is what one would expect from Newton's third law of action-reaction.

We have shown above that parallel conductors carrying currents in the same direction *attract* each other. Now you use the approach indicated by Figure 20.20 and the steps leading to Equation 20.14 to show that parallel conductors carrying currents in opposite directions *repel* each other.

The force between two parallel wires carrying a current is used to define the SI unit of current, the **ampere,** as follows:

> If two long, parallel wires 1 m apart carry the same current and the force per unit length on each wire is 2×10^{-7} N/m, then the current is defined to be 1 A.

The SI unit of charge, the **coulomb,** can now be defined in terms of the ampere as follows:

> If a conductor carries a steady current of 1 A, then the quantity of charge that flows through any cross section in 1 s is 1 C.

Example 20.8   Levitating a Wire

Two wires, each having a weight per unit length of 10^{-4} N/m, are strung parallel to one another above the surface of the Earth, with one being placed directly above the other. The wires are aligned in a north-south direction so that the Earth's magnetic field will not affect them. When their distance of separation is 0.1 m, what must the current be in each in order for the lower wire to levitate the upper wire? Assume that each wire carries the same current traveling in opposite directions.

Solution
If the upper wire is to float, it must be in equilibrium under the action of two forces: its weight and magnetic repulsion. In our example, the weight per unit length (10^{-4} N/m) must be equal and opposite the magnetic force per unit length given in Equation 20.14. Since the currents are the same, we have

$$\frac{F_1}{\ell} = \frac{mg}{\ell} = \frac{\mu_0 I^2}{2\pi d}$$

$$10^{-4} \text{ N/m} = \frac{(4\pi \times 10^{-7} \text{ T·m/A})(I^2)}{(2\pi)(0.1 \text{ m})}$$

We solve for the current to find

$$I = \boxed{7.1 \text{ A}}$$

Exercise   If the current in each wire is increased by a factor of 2, what is the equilibrium separation of the two wires?

Answer   0.4 m.

20.10   Magnetic Field of a Current Loop

The strength of the magnetic field set up by a piece of wire carrying a current can be enhanced at a specific location if the wire is formed into a loop. This can be understood by considering the effect of several small segments of the current loop, as in Figure 20.21. The small segment labeled Δx_1 at the top of the loop produces at the loop's center a magnetic field of magnitude B_1 and directed out of the page. The direction of **B** can be verified using the right-hand rule for a long straight wire. Grab the wire with your right hand with your thumb pointing in the direction of the current. Your fingers curl around in the direction of **B**.

A segment at the bottom of the loop, Δx_2, also contributes to the field at the center, thus increasing its strength. The magnitude of the field produced at the center of the current loop by the segment Δx_2 has the same magnitude as B_1 and is also directed out of the page. Similarly, all other such segments of the current loop contribute to the field. The net effect is to produce a magnetic field for the current loop as pictured in Figure 20.22a.

Notice in Figure 20.22a that the magnetic field lines enter at the left side of the current loop and exit at the right. Thus, one side of the loop acts as though it were the north pole of a magnet and the other acts as a south pole. The fact that the field set up by such a current loop bears a striking resemblance to the field of a bar magnet shown in Figure 20.22c will be of more than casual interest to us in a future section.

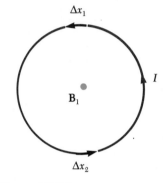

Figure 20.21
All segments of the current loop produce a magnetic field at the center of the loop directed *out of the page*.

Figure 20.22
(a) Magnetic field lines for a current loop. Note that the magnetic field lines of the current loop resemble those of a bar magnet. (b) Field lines of a current loop as displayed by iron filings. (Photo Education Development Center, Newton, Mass.) (c) The magnetic field of a bar magnet is similar to that of a current loop.

(a)

(b)

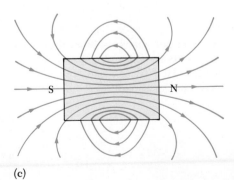

(c)

20.11   Magnetic Field of a Solenoid

If a long, straight wire is wrapped into a coil of several closely spaced loops, the resulting device is a **solenoid** or, as it is often called, an **electromagnet**. This device is important in many applications since it acts like a magnet only when it carries a current. As we shall see, the magnetic field inside a solenoid increases with the current and is proportional to the number of coils per unit length.

Figure 20.23 shows the magnetic field lines of a loosely wound solenoid of length ℓ and total number of turns N. Note that the field lines inside the solenoid are nearly parallel, uniformly spaced, and close together. This indicates that the field inside the solenoid is nearly uniform and strong. The exterior field at the sides of the solenoid is nonuniform and is much weaker than the field inside.

If the turns are closely spaced, the field lines are as shown in Figure 20.24a. In this case, the field lines enter at one end of the solenoid and emerge at the other. This means that one end of the solenoid acts as though it were the north pole of a magnet while the other end acts as though it were the south pole. If the length of the solenoid is much greater than its radius, the field outside the solenoid becomes very weak relative to the field inside. This occurs because the lines that leave the north end of the solenoid spread out over a wide region before returning to enter the south end. Hence, we see that the magnetic field lines outside are widely separated, indicative of a weak field. This is in contrast to a much stronger field inside the solenoid, where the lines are close together. Also, the field inside the solenoid has a constant magnitude at all points far from its ends. The expression for the field inside the solenoid is

$$B = \mu_0 n I \qquad (20.15)$$

The field inside a solenoid

where $n = N/\ell$ is the number of turns per unit length.

Figure 20.23
The magnetic field lines for a loosely wound solenoid. (Adapted from D. Halliday and R. Resnick, *Physics*, New York, Wiley, 1978)

(a)   (b)

Figure 20.24
(a) Magnetic field lines for a tightly wound solenoid of finite length carrying a steady current. The field inside the solenoid is nearly uniform and strong. Note that the field lines resemble those of a bar magnet, so that the solenoid effectively has north and south poles. (b) Magnetic field pattern of a bar magnet, as displayed by small iron filings on a sheet of paper. *(Courtesy of Henry Leap and Jim Lehman)*

Example 20.9   The Magnetic Field Inside a Solenoid

A solenoid consists of 100 turns of wire and has a length of 10 cm. Find the magnetic field inside the solenoid when it carries a current of 0.5 A.

Solution
The number of turns per unit is

$$n = \frac{N}{\ell} = \frac{100 \text{ turns}}{0.1 \text{ m}} = 1000 \text{ turns/m}$$

so

$$B = \mu_0 n I = (4\pi \times 10^{-7} \text{ T·m/A})(1000 \text{ turns/m})(0.5 \text{ A}) = \boxed{6.28 \times 10^{-4} \text{ T}}$$

Exercise   How many turns should the solenoid have (assuming it carries the same current) if the field inside is to be increased by a factor of 5?

Answer   500 turns.

*20.12   Magnetic Domains

The magnetic field produced by a current in a coil of wire gives us a hint as to what might cause certain materials to exhibit strong magnetic properties. A single coil like that of Figure 20.22 has a north and a south pole, but if this is true for a coil of wire, it should also be true for any current confined to a circular path. In particular, an individual atom should act like a magnet because of the motion of the electrons about the nucleus. Each electron, with its charge of 1.6×10^{-19} C, circles the atom once in about 10^{-16} s. If we divide the electronic charge by this time interval, we see that the orbiting electron is equivalent to a current of 1.6×10^{-3} A. Such a current would produce a magnetic field of the order of 20 T at the center of the circular path. From this result, we see that a very strong magnetic field would be produced if several of these atomic magnets could be aligned inside a material. However, this does not occur because the simple model we have described is not the complete story. A thorough analysis of atomic structure shows that the magnetic field produced by one electron in an atom is often canceled by an oppositely revolving electron in the same atom. The net result is that *the magnetic effect produced by the electrons orbiting the nucleus is either zero or very small for most materials.*

The magnetic properties of many materials are explained by the fact that an electron also spins on its axis like a top as it circles in its orbit, as indicated in Figure 20.25. (This classical description should not be taken literally. The property of spin can be understood only by relying on the methods of quantum mechanics, which we shall not discuss here.) The spinning electron represents a charge in motion which produces a magnetic field. The field due to the spinning motion is generally stronger than the field due to the orbital motion. In atoms containing many electrons, the electrons usually pair up with their spins opposite each other, so that their fields cancel each other. Hence most substances are not magnets. However, for certain strongly magnetic materials such as iron, cobalt, and nickel, the magnetic fields produced by the electron

Figure 20.25
Model of a spinning electron.

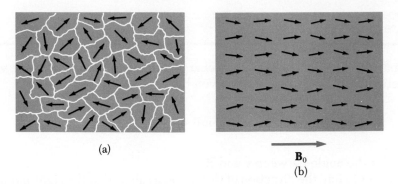

Figure 20.26
(a) Random orientation of domains in an unmagnetized substance. (b) When an external magnetic field **B**$_0$ is applied, the domains tend to align with the magnetic field.

spins do not cancel completely. Such strongly magnetic materials whose spins do not completely cancel are said to be **ferromagnetic.** In these materials there is a strong coupling between neighboring atoms to form large groups of atoms whose spins are aligned. Typically, the size of one of these groups, called a **domain,** ranges from about 10^{-4} cm to 0.1 cm. In an unmagnetized substance, the domains are randomly oriented, as shown in Figure 20.26a. When an external field is applied, as in Figure 20.26b, the magnetic field of each domain tends to align with the external field, resulting in a magnetized substance.

Two effects, both dependent on the domains, are responsible for the material's becoming magnetized. In some cases, the domains rotate slightly to become more closely aligned with the external field. This is pictured in Figure 20.27b. A second effect, which occurs in some substances, is one in which domains that are already aligned with the field tend to grow at the expense of the others, as shown in Figure 20.27c.

In what are called hard magnetic materials, domain alignment persists after the external field is removed, and such materials are referred to as **permanent magnets.** In soft magnetic materials, such as iron, once the external field is removed, thermal agitation produces domain motion and the material quickly returns to an unmagnetized state.

The alignment of domains explains why the strength of an electromagnet is increased dramatically by the insertion of an iron core into the magnet's center. The magnetic field produced by the current in the loops causes alignment of the domains, thus producing a large net external field. The use of iron, a soft magnetic material, as a core is also advantageous since the iron loses its magnetism almost instantaneously after the current in the coils is turned off.

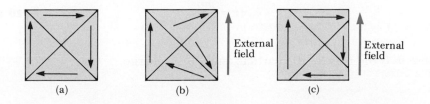

Figure 20.27
(a) An unmagnetized sample of four domains. (b) Magnetization due to rotation of the domains. (c) Magnetization due to a shift in domain boundaries.

Summary

The **magnetic force** that acts on a charge q moving with a velocity $\mathbf{v}$ in a magnetic field $\mathbf{B}$ has a magnitude given by

$$F = qvB \sin \theta \tag{20.1}$$

where θ is the angle between $\mathbf{v}$ and $\mathbf{B}$.

In order to find the direction of this force, a rule called the **right-hand rule** applies: Using your open right hand, place your fingers in the direction of $\mathbf{B}$ and point your thumb along the direction of the velocity $\mathbf{v}$. The force $\mathbf{F}$ on a positive charge is directed out of the palm of your hand.

If the charge is *negative* rather than positive, the force is directed opposite the force given by the right-hand rule.

The SI unit of magnetic field is the **tesla** (T), also called the weber per square meter (Wb/m²). An additional commonly used unit for magnetic field is the **gauss** (G), where $1 \text{ T} = 10^4 \text{ G}$.

If a straight conductor of length ℓ carries a current I, the magnetic force on that conductor when it is placed in a uniform external magnetic field, $\mathbf{B}$, is

$$F = BI\ell \sin \theta \tag{20.6}$$

The right-hand rule also gives the direction of the magnetic force on the conductor. In this case, however, you must place your thumb in the direction of the current rather than in the direction of $\mathbf{v}$.

The torque, τ, on a current-carrying loop of wire in a magnetic field $\mathbf{B}$ has a magnitude given by

$$\tau = BIA \sin \theta \tag{20.8}$$

where I is the current in the loop and A is its cross-sectional area. The angle θ is the angle between $\mathbf{B}$ and a line drawn perpendicular to the plane of the loop.

The galvanometer can be used in the construction of both ammeters and voltmeters.

If a charged particle moves in a uniform magnetic field such that its initial velocity is perpendicular to the field, the particle will move in a circular path whose plane is perpendicular to the magnetic field. The radius r of the circular path is

$$r = \frac{mv}{qB} \tag{20.10}$$

where m is the mass of the particle and q is its charge.

The magnetic field at a distance r from a **long, straight wire** carrying a current I has a magnitude

$$B = \frac{\mu_0 I}{2\pi r} \qquad (20.11)$$

where $\mu_0 = 4\pi \times 10^{-7}$ T·m/A is the **permeability of free space.** The magnetic field lines around a long, straight wire are circles concentric with the wire.

Ampère's law can be used to find the magnetic field around certain simple current-carrying conductors. It is stated as

$$\sum B_\parallel \Delta\ell = \mu_0 I \qquad (20.13)$$

where $B_\parallel$ is the component of **B** tangent to a small current element of length $\Delta\ell$ that is part of a closed path and I is the total current that penetrates the closed path.

The force per unit length between two parallel wires separated by a distance d and carrying currents I_1 and I_2 has a magnitude given by

$$\frac{F}{\ell} = \frac{\mu_0 I_1 I_2}{2\pi d} \qquad (20.14)$$

The force is attractive if the currents are in the same direction and repulsive if they are in opposite directions.

The magnetic field inside a solenoid has a magnitude

$$B = \mu_0 n I \qquad (20.15)$$

where n is the number of turns of wire per unit length, $n = N/\ell$.

Additional Reading

S. Banerjee, "Polar Flip-Flops," *The Sciences*, November/December 1984, p. 24.

J. J. Becker, "Permanent Magnets," *Sci. American*, December 1970, p. 92.

F. Bitter, *Magnets: The Education of a Physicist*, Science Study Series, Garden City, NY, Doubleday, 1959.

C. Carrigan and D. Gubbins, "The Source of the Earth's Magnetic Field," *Sci. American*, February 1979, p. 118.

R. A. Carrigan, Jr. and W. P. Trower, "Superheavy Magnetic Monopoles," *Sci. American*, April 1982, p. 106.

B. Dibner, *"Oersted and the Discovery of Electromagnetism,"* Blaisdell, 1962.

S. Felch, "Searches for Magnetic Monopoles and Fractional Electric Charge," *The Physics Teacher*, March 1984, p. 142.

H. H. Kolm and A. J. Freeman, "Intense Magnetic Fields," *Sci. American*, April 1965, p. 66.

A. Nier, "The Mass Spectrometer," *Sci. American*, March 1953, p. 68.

Questions

1. The north-seeking pole of a magnet is attracted toward the geographic north pole of the Earth. Yet, like poles repel. What is the way out of this dilemma?

2. Which way would a compass point if you were at the north magnetic pole?

3. Two charged particles are projected into a region where there is a magnetic field perpendicular to their velocities. If the particles are deflected in opposite directions, what can you say about them?

4. If a charged particle moves in a straight line through some region of space, can you say that the magnetic field in that region is zero?

5. Suppose an electron is chasing a proton up this page when suddenly a magnetic field is applied perpendicular to the page. What will happen to the particles?

6. Why does the picture on a television screen become distorted when a magnet is brought near the screen?

7. At a given instant, a proton moves in the positive x direction in a region where there is a magnetic field in the negative z direction. What is the direction of the magnetic force? Will the proton continue to move in the positive x direction? Explain.

8. A proton moving horizontally enters a region where there is a uniform magnetic field perpendicular to the proton's velocity, as shown in Figure 20.28. Describe its subsequent motion. How would an electron behave under the same circumstances?

Figure 20.28   Question 8

9. List several similarities and differences in electric and magnetic forces.

10. How can the motion of a charged particle be used to distinguish between a magnetic field and an electric field? Give a specific example to justify your argument.

11. Can a magnetic field set a resting electron into motion? If so, how?

12. Justify the following statement: "It is impossible for a constant (i.e., time independent) magnetic field to alter the speed of a charged particle."

13. A charged particle moves in a circular path because of an applied magnetic field. Does the particle gain energy from the magnetic field?

14. Explain why it is not possible to determine the charge and mass of a charged particle separately by electric and magnetic forces.

15. How can a current loop be used to determine the presence of a magnetic field in a given region of space?

16. Is it possible to orient a current loop in a uniform magnetic field such that the loop will not tend to rotate?

17. If a solenoid were suspended by a string so that it could rotate freely, could it be used as a magnet when it carries a direct current? Could it also be used if the current were alternating in direction?

18. What would be the internal resistance of an ideal ammeter and voltmeter? Why should these be called ideal meters?

19. The *bubble chamber* is a device used for observing tracks of particles that pass through the chamber, which is immersed in a magnetic field. If some of the tracks are spirals and others are straight lines, what can you say about the particles?

20. It is found that charged particles from outer space, called cosmic rays, strike the Earth more frequently at the poles than at the equator. Why does this occur?

21. Explain why two parallel wires carrying currents in opposite directions repel each other.

22. Two wires carrying equal but opposite currents are twisted together in the construction of a circuit. Why does this technique reduce stray magnetic fields?

23. Describe the change in the magnetic field inside a solenoid carrying a steady current if (a) the length of the solenoid is doubled but the number of turns remains the same and (b) the number of turns is doubled but the length remains the same.

24. The electron beam in the photograph below is projected to the right. The beam deflects downward in

Question 24 *(Courtesy of CENCO)*

the presence of a magnetic field produced by a pair of current-carrying coils. (a) What is the direction of the magnetic field? (b) What would happen to the beam if the current in the coils were reversed?

25. A magnet attracts a piece of iron. The iron can then attract another piece of iron. Explain, on the basis of alignment of the domains, what happens in each piece of iron.

26. Will a nail be attracted to either pole of a magnet? Explain what is happening inside the nail.

27. Why will hitting a magnet with a hammer cause its magnetism to be reduced?

28. You are an astronaut stranded on a planet with no test equipment or minerals around. The planet does not even have a magnetic field. You have two iron bars in your possession: one is magnetized, one is not. How could you determine which is magnetized?

29. A Hindu ruler once suggested that he be entombed in a magnetic coffin with the polarity arranged such that he would be forever suspended between heaven and Earth. Is such magnetic levitation possible? Discuss.

Problems

Section 20.3   Magnetic Fields

1. A proton is moving at right angles to a magnetic field of 2 T. What speed does the proton have if the magnetic force on it has a magnitude of 6×10^{-11} N?

2. A proton moves at right angles to a magnetic field of 0.1 T with a speed of 2×10^7 m/s. Find the magnitude of the acceleration of the proton.

3. A proton travels with a speed of 3×10^6 m/s at an angle of $37°$ with respect to the direction of a magnetic field of 0.3 T in the $+y$ direction. What is (a) the magnitude of the magnetic force on the proton and (b) its acceleration?

4. A proton moves with a speed of 2.5×10^6 m/s horizontally at right angles to a magnetic field. (a) What strength field is required to just balance the weight of the proton and keep it moving horizontally? (b) Should the direction of the magnetic field be in a horizontal or a vertical plane?

5. (a) Find the direction of the force on a proton moving as shown through the magnetic fields pictured in Figure 20.29. (b) Repeat part (a) assuming the moving particle is an electron.

6. Consider an electron near the magnetic equator. In which direction will it tend to be deflected if its velocity is directed (a) downward, (b) northward, (c) westward, or (d) southeastward?

7. Find the direction of the magnetic field on the positively charged particle moving in the various situations shown in Figure 20.30 if the direction of the magnetic force acting on it is as indicated.

8. Sodium ions (Na^+) with a speed of 0.851 m/s move through the bloodstream in the arm of a person standing near a large magnet. The magnetic field has a strength of 0.254 T and makes an angle of $51.0°$ with respect to the motion of the sodium ions. The arm contains 100 cm³ of blood with 3.0×10^{20} Na^+ ions

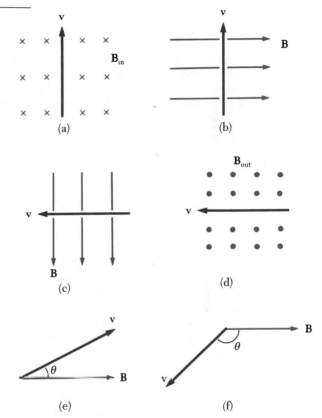

Figure 20.29   Problems 5 and 17.
For problem 17, replace the velocity vector by a current in that direction.

Figure 20.30   Problems 7 and 18.
For Problem 18, replace the velocity vector by a current in that direction.

per cubic centimeter. If there were no other ions present in the arm, what would be the magnetic force on the arm?

9. A proton moving with a speed of 5×10^7 m/s through a magnetic field of 2 T experiences a magnetic force of 3×10^{-12} N. What is the angle between the proton's velocity and the field?

10. An electron is accelerated through 2400 V from rest and then enters a region where there is a uniform 1.7-T magnetic field. What are (a) the maximum and (b) the minimum values of the magnetic force this charge can experience?

11. A proton moves perpendicular to a uniform magnetic field $\mathbf{B}$ with a speed of 10^7 m/s and experiences an acceleration of 2×10^{13} m/s^2 in the $+x$ direction when its velocity is in the $+z$ direction. Determine the magnitude and direction of the field.

12. Show that the work done by the magnetic force on a charged particle moving in a uniform magnetic field is zero for any displacement of the particle.

Section 20.4   Magnetic Force on a Current-Carrying Conductor

13. Calculate the force on a 2-m length of conductor carrying a current of 10 A in a region where a uniform magnetic field has a magnitude of 1.2 T and is directed perpendicular to the conductor.

14. A wire carries a current of 10 A in a direction of 30° with respect to the direction of a magnetic field of strength 0.3 T. Find the magnetic force on a 5-m length of the wire.

15. A vertical wire carries a current of 10 A directed upward at a location where the magnetic field of the Earth is horizontal and has a value of 5×10^{-5} T. If the wire is 20 m long, find the magnitude and direction of the net magnetic force on it.

16. A current $I = 15$ A is directed along the positive x axis and perpendicular to a magnetic field. The conductor experiences a magnetic force per unit length of 0.12 N/m in the negative y direction. Calculate the magnitude and direction of the magnetic field in the region through which the current passes.

17. In Figure 20.29, assume the velocity vector shown in each case is replaced with a wire carrying a current in the direction of the velocity vector. Find the direction of the magnetic force acting on the wire for each case.

18. In Figure 20.30, assume the velocity vector shown in each case is replaced with a wire carrying a current in the direction of the velocity vector. Find the direction of the magnetic field that will produce the magnetic force shown in each case.

19. A wire with a mass of 1 g/cm is placed on a horizontal surface with a coefficient of friction of 0.2. The wire

carries a current of 1.5 A toward the east, and moves horizontally to the north. What are the magnitude and the direction of the *smallest* magnetic field that enables the wire to move in this fashion?

20. A wire with a mass of 0.5 g/m carries a 2-A current horizontally to the south. What are the direction and magnitude of the minimum magnetic field needed to lift this wire vertically upward?

21. A thin copper rod 1 m long has a mass of 50 g. What is the minimum current in the rod that will cause it to float in a magnetic field of 2 T?

22. An unusual message delivery system is pictured in Figure 20.31. A 15-cm length of conductor that is free to move is held in place between two thin conductors. When a 5-A current is directed as shown in the figure, the wire segment moves upward at a constant velocity. If the mass of the wire is 0.15 kg, find the magnitude and direction of the minimum magnetic field required to move the wire. (The wire slides without friction on the two vertical conductors.)

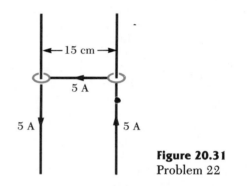

Figure 20.31
Problem 22

23. The Earth has a magnetic field of 0.6×10^{-4} T, pointing 75° below the horizontal in a north-south plane. A 10-m-long straight wire carries a 15-A current. (a) If the current is directed horizontally toward the east, what are the magnitude and direction of the magnetic force on the wire? (b) What are the magnitude and direction of the force if the current is directed vertically upward?

Section 20.5   Torque on a Current Loop

24. A rectangular loop consists of 100 closely wrapped turns and has dimensions 0.4 m by 0.3 m. The loop is hinged along the y axis, and the plane of the coil makes an angle of 30° with the x axis (Fig. 20.32). What is the magnitude of the torque exerted on the loop by a uniform magnetic field of 0.8 T directed along the x axis when the current in the windings has a value of 1.2 A in the direction shown? What is the expected direction of rotation of the loop?

Figure 20.32 Problem 24

Figure 20.33 Problem 31

25. A rectangular coil of 500 turns has dimensions of 4 cm by 5 cm and is suspended in a magnetic field of 0.65 T. What is the current in the coil if the maximum torque exerted on it by the magnetic field is 0.18 N·m?

26. A single circular loop of wire of radius 50 cm and carrying a current of 2 A is in a magnetic field of 0.4 T. (a) Find the maximum torque which acts on this loop. (b) Find the angle that the plane of the loop makes with respect to the field when the torque is one-half the value found in Part (a).

Section 20.6   The Galvanometer and Its Applications

27. A 50-Ω, 10-mA galvanometer is to be converted to an ammeter that reads 3 A at full-scale deflection. What value of R_p should be placed in parallel with the coil?

28. An ammeter reads 10 A at full-scale deflection. The meter was constructed by inserting a resistor in parallel with a 50-Ω galvanometer coil that will deflect full scale when the voltage across it is 50 mV. What is the value of the parallel resistance that should be used?

29. Consider a galvanometer with an internal resistance of 60 Ω. If the galvanometer deflects full scale when it carries a current of 0.5 mA, what value series resistance must be connected to the galvanometer if this combination is to be used as a voltmeter having a full-scale deflection for a potential difference of 1.0 V?

30. A 40-Ω, 2-mA galvanometer is to be converted to a voltmeter that reads 150 V at full-scale deflection. What value of series resistance should be used with the galvanometer coil?

31. The galvanometer of Problem 32 is to be converted to a multirange ammeter using the circuit shown in Figure 20.33. Find the values of R_1, R_2, and R_3 that will give the full-scale readings indicated on the figure.

32. A galvanometer has an internal resistance of 100 Ω and deflects full scale for a current of 100 μA. This galvanometer is to be used to construct a multirange voltmeter as shown in Figure 20.34. Find the value of R_1, R_2, and R_3 that will enable the meter to give the full-scale readings indicated on the figure.

Figure 20.34 Problem 32

Section 20.7   Motion of a Charged Particle in a Magnetic Field

33. A singly charged positive ion of mass 6.68×10^{-27} kg moves clockwise in a circular path of radius 3 cm with a speed of 10^4 m/s. Find the direction and strength of the magnetic field.

34. A proton moves with a velocity of 2×10^5 m/s perpendicular to a uniform magnetic field of strength 0.2 T. What is the radius of the path?

35. A proton moves in a circular orbit perpendicular to a uniform magnetic field of 0.758 T. Find the time it takes the proton to make one pass around the orbit.

36. An electron travels in a circular orbit of radius 1.77 m in a region of uniform magnetic field. If the magnetic field strength doubles, what is the new radius of the orbit?

37. What is the radius of the circular path of an electron with a kinetic energy of 1.6×10^{-19} J moving perpendicular to a 0.2-T magnetic field?

38. Figure 20.35a shows a diagram of a device called a velocity selector. In this device, particles of a specific velocity pass through undeflected while those with greater or smaller velocities are deflected either up-

ward or downward. In this device, an electric field is directed perpendicular to a magnetic field. This produces on the charged particle an electric force and a magnetic force which are equal in magnitude and in opposite directions as in Figure 20.35b, and hence cancel. Show that particles with a speed of $v = E/B$ will pass through undeflected.

(a)   (b)

Figure 20.35   Problem 38

39. A beam of protons is fired through crossed electric and magnetic fields as described in Problem 38. If protons of speed 10^5 m/s are to be selected, find the strength of the required magnetic field when the electric field is 5×10^5 N/C.

40. A singly charged positive ion has a mass equal to 2.5×10^{-26} kg. After being accelerated through a potential difference of 250 V, the ion enters a magnetic field of 0.5 T along a direction perpendicular to the field. Calculate the radius of the path of the ion in the field.

41. Consider the mass spectrometer shown schematically in Figure 20.36. The electric field btween the plates of the velocity selector is 950 V/m, and the magnetic field in both the velocity selector and the deflection chamber has a magnitude of 0.93 T. Calculate the

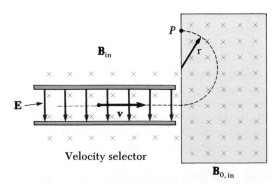

Figure 20.36   Problem 41
A mass spectrometer. Charged particles are first sent through a velocity selector. They then enter a region where the magnetic field B_0 (inward) causes positive ions to move in a semicircular path and strike a photographic film at P.

radius of the path in the system for a singly charged ion with a mass $m = 2.18 \times 10^{-26}$ kg. (*Hint:* see Problem 38.)

42. Two ions of mass 6.64×10^{-27} kg move out of the slit of a mass spectrometer and into a region where the magnetic field is 0.2 T. They each have a speed of 10^6 m/s, but one ion is singly charged while the other is doubly charged. Find (a) the radius of the circular path followed by each in the field and (b) the distance of separation when they have moved through one-half their circular path and strike a piece of photographic paper.

43. A mass spectrometer is used to examine the isotopes of uranium. Ions in the beam emerge from the velocity selector with a speed equal to 3×10^5 m/s and enter a uniform magnetic field of 0.6 T directed perpendicular to the velocity of the ions. What is the distance between the impact points formed on the photographic plate by singly charged ions of ^{235}U and ^{238}U?

44. A singly charged positive ion moving with a speed of 4.6×10^5 m/s leaves a spiral track of radius 7.94 mm in a photograph along a direction perpendicular to the magnetic field of a bubble chamber. The magnetic field applied for the photograph has a magnitude of 1.8 T. Compute the mass of this particle.

Section 20.8   Magnetic Field of a Long, Straight Wire and Ampère's Law

45. Find the magnetic field strength at a distance of (a) 10 cm from a long, straight wire carrying a current of 5 A. Repeat the calculation for distances of (b) 50 cm and (c) 2 m.

46. At what distance from a long, straight wire carrying a current of 5 A is the magnetic field due to the wire equal to the strength of the Earth's field, approximately 5×10^{-5} T?

47. Niobium metal becomes a superconductor (with electrical resistance equal to zero) when cooled below 9 K. If superconductivity is destroyed when the surface magnetic field exceeds 0.1 T, determine the maximum current a 2-mm-diameter niobium wire can carry and remain superconducting.

48. Find the direction of the current in the wire shown in Figure 20.37 that would produce a magnetic field directed as shown in each case.

49. The two wires shown in Figure 20.38 each carry currents of 5 A in opposite directions and are separated by 10 cm. Find (a) the direction and magnitude of the net magnetic field at a point midway between the wires, (b) at a point P_1, that is, 10 cm to the right of the wire on the right, and (c) at a point P_2, that is, 20 cm to the left of the wire on the left.

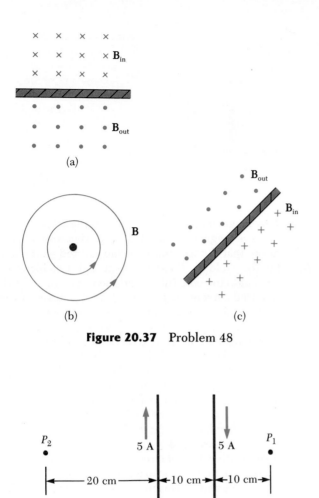

(a)

(b)   (c)

Figure 20.37   Problem 48

Figure 20.38   Problem 49

50. The two wires shown in Figure 20.39 carry currents of 3 A and 5 A in the direction indicated. (a) Find the direction and magnitude of the magnetic field at a point midway between the wires. (b) Find the magnitude and direction of the magnetic field at a point 20 cm above the wire carrying the 5-A current.

Figure 20.39
Problem 50

Section 20.9   Magnetic Force Between Two Parallel Conductors

51. Two parallel conductors are each 0.5 m long and carry 10-A currents in opposite directions. (a) What center-to-center separation must the conductors have if they are to repel each other with a force of 1.0 N? (b) Is this physically possible?
52. Two long, straight wires separated by a distance of 0.3 m carry currents in the same direction. If the current in one wire is 5 A and the current in the other is 8 A, find the magnitude and direction of the force that one exerts on a 2-m length of the other.
53. Two parallel wires are 10 cm apart and each carries a current of 10 A. (a) If the currents are in the same direction, find the force per unit length exerted on one of the wires by the other. Are the wires attracted or repelled? (b) Repeat the problem if the currents are in opposite directions.
54. A wire 2 m long and weighing 0.08 N per meter is suspended directly above a second wire. The top wire carries a current of 30 A and the bottom wire carries a current of 60 A. Find the distance of separation between the wires such that the top wire will be held in place by magnetic repulsion.

Section 20.10   Magnetic Field of a Current Loop
Section 20.11   Magnetic Field of a Solenoid

55. What current in a solenoid 15 cm long wound with 100 turns would produce a magnetic field at the center equal to that of the Earth, which is about 5×10^{-5} T?
56. Consider the solenoid described in Problem 55. If it carries the same current, how many turns must it have to increase the strength of the magnetic field by a factor of 3?
57. A single-turn square loop of wire, 2 cm on a side, carries a current of 0.2 A. The loop is inside a solenoid, with the plane of the loop perpendicular to the magnetic field of the solenoid. The solenoid has 30 turns per centimeter and carries a current of 15 A. Find the force on each side of the loop and the torque acting on it.
58. Use the path shown in Figure 20.40 and Ampère's Law to derive the expression for the magnetic field

Figure 20.40
Problem 58

inside a solenoid of n turns per unit length and current I. (*Hint:* assume the strength of the magnetic field is zero outside the solenoid, a good assumption.)

Additional Problems

59. What current is required in the windings of a solenoid that has 500 turns uniformly distributed over a length of 0.2 m in order to produce a magnetic field of 1.2×10^{-4} T at the center of the solenoid?

60. An electron moves at a speed of 0.6 the speed of light along the $+x$ direction. A magnetic field of strength 0.6 T is in the $-y$ direction. Find the magnitude and direction of the force on the electron.

61. What magnetic field would be required to constrain an electron whose kinetic energy is 400 eV to a circular path of radius 0.8 m?

62. A circular coil consisting of a single loop of wire has a radius of 30 cm and carries a current of 25 A. It is placed in an external magnetic field of 0.3 T. Find the torque on the wire when the plane of the coil makes an angle of 35° with the direction of the field.

63. Two species of singly charged positive ions of mass 20×10^{-27} kg and 23.4×10^{-27} kg enter a magnetic field at the same location with a speed of 10^5 m/s. If the strength of the field is 0.2 T, and they move perpendicular to the field, find their distance of separation after completing one half of their circular path.

64. A conductor suspended by two cords as in Figure 20.41 has a mass per unit length of 0.04 kg/m. What current must exist in the conductor in order for the tension in the supporting wires to be zero, if the magnetic field over the region is 3.6 T and directed into the page? What is the required direction of the current?

Figure 20.41
Problem 64

65. Two long, straight wire cross one another at right angles, as shown in Figure 20.42. (a) Find the direction and magnitude of the magnetic field at point P, which is in the same plane as the two wires. (b) Find the magnetic field at a point 30 cm above the point of intersection. This will be a point 30 cm out of the page toward you.

Figure 20.42
Problem 65

66. A rectangular coil of 150 turns and area 0.12 m² is in a uniform magnetic field of 0.15 T. Measurements indicate that the maximum torque exerted on the loop by the field is 6×10^{-4} N·m. (a) Calculate the current in the coil. (b) Would the value found for the required current be different if the 150 turns of wire were used to form a single-turn coil of larger area? Explain.

67. A rectangular loop of wire carrying a current of 2 A is suspended vertically and attached to the right arm of a balance. After the system is balanced, an external magnetic field, **B**, is introduced. The field threads the lower end of the loop in a direction perpendicular to the wire. If the width of the loop is 20 cm and it takes 13.5 g of added mass on the left arm to rebalance the system, determine **B**.

68. A conducting wire of circular cross section formed of a material that has a mass density of 2.7 g/cm³ is placed in a uniform magnetic field with the axis of the wire perpendicular to the direction of the field. A current density of 2.4×10^6 A/m² is established in the wire and the magnetic field increased until the magnetic force on the wire just balances the gravitational force. Calculate the value of **B** when this condition is met.

69. Two parallel conductors carry currents in opposite directions as shown in Figure 20.43. One conductor

Figure 20.43
Problem 69

carries a current of 10 A. Point *A* is at the midpoint between the wires and point *C* is a distance of 5 cm to the right of the 10-A current. *I* is adjusted so that the magnetic field at *C* is zero. Find (a) the value of the current *I* and (b) the value of the magnetic field at *A*.

70. Four long, parallel conductors all carry a current of 4 A. An end view of the conductors is shown in Figure 20.44. The current direction is out of the page at points *A* and *B* (indicated by the dots) and into the page at *C* and *D* (indicated by the crosses). Calculate the magnitude and direction of the magnetic field at point *P*, located at the center of the square of edge length 0.2 m.

Figure 20.44
Problem 70

71. An electron circles at a speed of 10^4 m/s in a radius of 2 cm in a solenoid. The magnetic field of the solenoid is perpendicular to the plane of the electron's path. Find (a) the strength of the magnetic field inside the solenoid and (b) the current in the solenoid if it has 25 turns per centimeter.

72. Consider a particle of mass *m* and charge *q* moving with a velocity **v**. The particle enters a region perpendicular to a magnetic field **B**. Show that, while in the region of the magnetic field, the kinetic energy on the particle is proportional to the square of the radius of the orbit.

73. For the arrangement shown in Figure 20.45, the current in the straight conductor has the value $I_1 = 5$ A and lies in the plane of the rectangular loop, which carries a current $I_2 = 10$ A. The dimensions are $c = 0.1$ m, $a = 0.15$ m, and $\ell = 0.45$ m. Find the magnitude and direction of the *net force* exerted on the rectangle by the magnetic field of the straight current-carrying conductor.

Figure 20.45
Problem 73

74. A straight wire of mass 10 g and length 5 cm is suspended from two identical springs that, in turn, form a closed circuit (Fig. 20.46). The springs stretch a distance of 0.5 cm under the weight of the wire. The circuit has a total resistance of 12 Ω. When a magnetic field is turned on, directed out of the page (indicated by the dots in Fig. 20.46), the springs are observed to stretch an additional 0.3 cm. What is the strength of the magnetic field? (The upper portion of the circuit is fixed.)

Figure 20.46
Problem 74

75. When the spring on a galvanometer is put under tension, a restoring torque given by $\tau = K\theta$ is set up, where *K* is called the torque constant and θ is the angle through which the end of the spring is turned. A coil is wrapped so that the loop is equivalent to a straight wire whose length perpendicular to the field is 0.7 m. The radius of the coil is 1.5 cm, and the radial magnetic field has a magnitude of 0.5 T. The torque constant for the spring is 2×10^{-7} N·m/rad. What value of current will produce a full-scale deflection of $\pi/4$ rad?

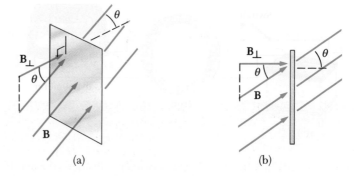

Figure 21.2
(a) A uniform magnetic field **B** making an angle θ with the normal to the plane of a loop of wire of area A. (b) Edge view of the loop.

where $B_\perp$ is the component of **B** perpendicular to the plane of the loop, as in Figure 21.2a, and θ is the angle between **B** and the direction perpendicular to the plane of the loop. An end view of the loop and the penetrating magnetic field lines is shown in Figure 21.2b. When the field is perpendicular to the plane of the loop, as in Figure 21.3a, $\theta = 0$ and Φ has a maximum value given by $\Phi_{max} = BA$. When the plane of the loop is parallel to **B**, as in Figure 21.3b, $\theta = 90°$ and $\Phi = 0$. Since B has the units of T, or Wb/m^2, the units of flux are T·m^2, or Wb.

We can emphasize the significance of Equation 21.1 by first drawing magnetic field lines as in Figure 21.3. The number of lines per unit area increases as the field strength increases. The value of the magnetic flux *is proportional to the total number of lines passing through the loop.* Thus, we see that a maximum number of lines pass through the loop when its plane is perpendicular to the field, as in Figure 21.3a, and so the flux has its maximum value. As Figure 21.3b shows, no lines pass through the loop when its plane is parallel to the field, and so in this case $\Phi = 0$.

Figure 21.3
The edge view of a loop in a uniform magnetic field. (a) When the field lines are perpendicular to the plane of the loop, the magnetic flux through the loop is a maximum, and equal to $\phi = BA$. (a) When the field lines are parallel to the plane of the loop, the magnetic flux through the loop is zero.

Michael Faraday was a British physicist and chemist who is often regarded as the greatest experimental scientist of the 1800s. His many contributions to the study of electricity include the invention of the electric motor, electric generator, and transformer, as well as the discovery of electromagnetic induction, the laws of electrolysis, the discovery of benzene, and the theory that the plane of polarization of light is rotated in an electric field.

Faraday was born in 1791 in rural England, but his family moved to London shortly thereafter. One of ten children and the son of a blacksmith, Faraday received a minimal education and became apprenticed to a bookbinder at age 14. He was fascinated by articles on electricity and chemistry and was fortunate to have an employer who allowed him to read books and attend scientific lectures. He received some education in science from the City Philosophical Society.

When Faraday finished his apprenticeship in 1812, he expected to devote himself to bookbinding rather than to science. That same year, Faraday attended a lecture by Humphry Davy, who made many contributions in the field of heat and thermodynamics. Faraday sent 386 pages of notes, bound in leather, to Davy; Davy was impressed and appointed Faraday his permanent assistant at the Royal Institution. Faraday toured France and Italy from 1813 to 1815 with Davy, visiting leading scientists of the time such as Volta and Vauquelin.

Despite his limited mathematical ability, Faraday succeeded in making the basic discoveries on which virtually all our uses of electricity depend. He conceived the fundamental nature of magnetism and, to a degree, that of electricity and light.

A modest man who was content to serve science as best he could, Faraday declined a knighthood and an offer to become president of the Royal Society. He was also a moral man; he refused to take part in the preparation of poison gas for use in the Crimean War.

Faraday died in 1867. His many achievements are recognized by the use of his name. The Faraday constant is the quantity of electricity required to deliver a standard amount of substance in electrolysis, and the SI unit of capacitance is the farad.

Michael Faraday
(1791 – 1867)

(Courtesy of AIP Niels Bohr Library)

21.2 Faraday's Law of Induction

The usefulness of the concept of magnetic flux will become obvious as we describe another simple experiment that demonstrates the basic idea of electromagnetic induction. Consider a loop of wire connected to a galvanometer, as in Figure 21.4. If a magnet is moved toward the loop, the galvanometer needle will deflect in one direction, as in Figure 21.4a. If the magnet is moved away from the loop, the galvanometer needle will deflect in the opposite direction, as in Figure 21.4b. If the magnet is held stationary and the loop is moved either toward or away from the magnet, the needle will also deflect. From these observations, it can be concluded that *a current is set up in the circuit as long as there is relative motion between the magnet and the loop.*

These results are quite remarkable in view of the fact that *a current is set up in the circuit even though the circuit contains no batteries!* We call such a current an *induced current* because it is produced by an *induced emf.*

Figure 21.4
(a) When a magnet is moved toward a loop of wire connected to a galvanometer, the galvanometer deflects as shown. This shows that a current is induced in the loop. (b) When the magnet is moved away from the loop of wire, the galvanometer deflects in the opposite direction, indicating that the induced current is opposite that shown in (a).

This experiment has something in common with the Faraday experiment discussed in Section 21.1. In both cases, an emf is induced in a current when the magnetic flux through the circuit changes with time. In fact, we can make the following general statement that summarizes such experiments involving induced currents and emfs:

The instantaneous emf induced in a circuit is equal to the rate of change of magnetic flux through the circuit.

If a circuit contains N loops and the flux changes by an amount $\Delta\Phi$ during a time Δt, the average induced emf during the time Δt is

Faraday's law

$$\varepsilon = -N\frac{\Delta\Phi}{\Delta t} \qquad \text{(21.2)}$$

This is a statement of **Faraday's law of induction.** The minus sign is included to indicate the polarity of the induced emf, which can be found by use of **Lenz's law:**

Lenz's law

The polarity of the induced emf is such that it produces a current whose magnetic field opposes the change in magnetic flux through the loop. That is, the induced current tends to maintain the original flux through the circuit.

We shall consider several applications of Lenz's law in Section 21.4.

PHYSICS IN ACTION

Demonstrations of Electromagnetic Induction

(Left) To demonstrate electromagnetic induction, an ac voltage is applied to the lower coil in the apparatus shown in the photograph. A voltage is induced in the upper coil as indicated by the illuminated lamp connected to the upper coil. What do you think happens to the lamp's intensity as the upper coil is moved over the vertical tube? In order to answer this, you should note that the magnetic field associated with the lower coil varies along the axis of the tube. *(Right)* The coil at the bottom of the photograph is connected to a battery and a switch (not shown). The core of the coil is filled with iron, and an aluminum ring whose diameter is larger than that of the coil is slipped over the coil. When the switch in the circuit is closed and current builds up in the coil, the aluminum ring is observed to spring upward as shown. This is explained by the fact that a current is induced in the ring by the changing magnetic field associated with the coil. (The iron core is included to increase the strength of the magnetic field.) The magnetic field associated with the induced current opposes the field due to the coil, thereby causing the ring to be repelled by the coil. What do you suppose would happen if the experiment were repeated with the battery replaced by a source of alternating current? In this case, when the metal ring is held down over the coil, it becomes warmer. Why is this?

(Photographs courtesy of CENCO)

Example 21.1   Application of Faraday's Law

A coil is wrapped with 200 turns of wire on a square frame of sides 18 cm. Each turn has the same area, equal to that of the frame, and the total resistance of the coil is 2 Ω. A uniform magnetic field is applied perpendicular to the plane of the coil. If the field changes uniformly from 0 to 0.5 T in 0.8 s, find the magnitude of the induced emf in the coil while the field is changed.

Solution
The area of the coil is $(0.18 \text{ m})^2 = 0.0324 \text{ m}^2$. The magnetic flux through the coil at $t = 0$ is zero because $B = 0$. At $t = 0.8$ s, the magnetic flux through the coil is

$$\Phi_f = BA = (0.5 \text{ T})(0.0324 \text{ m}^2) = 0.0162 \text{ T·m}^2$$

Therefore, the *change* in flux through the coil during the 0.8-s interval is

$$\Delta\Phi = \Phi_f - \Phi_i = 0.0162 \text{ T·m}^2$$

Faraday's law of induction enables us to find the magnitude of the induced emf:

$$|\mathcal{E}| = N\frac{\Delta\Phi}{\Delta t} = (200 \text{ turns})\left(\frac{0.0162 \text{ T·m}^2}{0.8 \text{ s}}\right) = \boxed{4.05 \text{ V}}$$

(Note that $1 \text{ T·m}^2 = 1 \text{ V·s}$.)

Exercise   Find the magnitude of the induced current in the coil while the field is changing.

Answer   2.03 A.

21.3   Motional emf

Figure 21.5
A straight conductor of length ℓ moving with a velocity **v** through a uniform magnetic field **B** directed perpendicular to **v**. The vector **F** is the force on an electron in the conductor. An emf equal to $B\ell v$ is induced between the ends of the bar.

In Section 21.2, we considered a situation in which an emf is induced in a circuit when the magnetic field changes with time. In this section we describe a particular application of Faraday's law in which a so-called **motional emf** is produced. This is the emf induced in a conductor moving through a magnetic field.

First consider a straight conductor of length ℓ moving with constant velocity through a uniform magnetic field directed into the paper, as in Figure 21.5. For simplicity, we shall assume that the conductor is moving perpendicular to the field. The electrons in the conductor will experience a force directed downward along the conductor and having a magnitude $F = qvB$. Because of this magnetic force, the electrons will move to the lower end and accumulate there, leaving a net positive charge at the upper end. An electric field is therefore produced in the conductor as a result of this charge separation. The charge at the ends builds up until the downward magnetic force, qvB, is balanced by the upward electric force, qE. At this point, charge stops flowing and the condition for equilibrium requires that

$$qE = qvB \qquad \text{or} \qquad E = vB$$

Since the electric field is constant, the electric field produced in the conductor is related to the potential difference across the ends according to the relation $V = E\ell$. Thus,

$$V = E\ell = B\ell v \qquad \textbf{(21.3)}$$

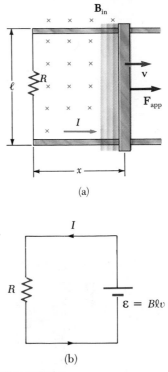

Since there is an excess of positive charge at the upper end of the conductor and an excess of negative charge at the lower end, the upper end is at a higher potential than the lower end. Thus,

a potential difference is maintained across the conductor as long as there is motion through the field. If the motion is reversed, the polarity of the potential difference is also reversed.

A more interesting situation occurs if we now consider what happens when the moving conductor is part of a closed conducting path. This approach is particularly useful for illustrating how a changing magnetic flux induces a current in a closed circuit. Consider a circuit consisting of a conducting bar of length ℓ sliding along two fixed parallel conducting rails, as in Figure 21.6a. For simplicity, we shall assume that the moving bar has zero resistance and that the stationary part of the circuit has a resistance R. A uniform and constant magnetic field $\mathbf{B}$ is applied perpendicular to the plane of the circuit. As the bar is pulled to the right with a velocity $\mathbf{v}$, under the influence of an applied force $\mathbf{F}_{app}$, the free charges in the bar experience a magnetic force along the length of the bar. This force in turn sets up an induced current because the charges are free to move in a closed conducting path. In this case, the changing magnetic flux through the loop and the corresponding induced emf across the moving bar arise from the change in area of the loop as the bar moves through the magnetic field.

Let us assume that the bar moves a distance Δx, as shown in Figure 21.7. The increase in flux, $\Delta\Phi$, through the loop is the amount that now passes through the portion of the circuit that has an area equal to $\ell\Delta x$:

$$\Delta\Phi = BA = B\ell\Delta x$$

Let us use Faraday's law, noting that there is one loop ($N = 1$); we find that the induced emf has a magnitude

Figure 21.6
(a) A conducting bar sliding with a velocity $\mathbf{v}$ along two conducting rails under the action of an applied force $\mathbf{F}_{app}$. A counterclockwise current is induced in the loop. (b) The equivalent circuit of (a).

$$\mathcal{E} = \frac{\Delta\Phi}{\Delta t} = B\ell\,\frac{\Delta x}{\Delta t} = B\ell v \qquad \textbf{(21.4)}$$

This induced emf is often called a *motional emf* because it arises from the motion of a conductor through a magnetic field.

Furthermore, if the resistance of the circuit is R, the magnitude of the induced current is

$$I = \frac{\mathcal{E}}{R} = \frac{B\ell v}{R} \qquad \textbf{(21.5)}$$

Figure 21.7
The magnetic flux through the loop increases as the bar moves to the right.

The equivalent circuit diagram for this example is shown in Figure 21.6b.

Example 21.2   The Electrified Airplane Wing

An airplane with a wing span of 50 m flies horizontal to the Earth at a location where the downward component of the Earth's magnetic field is 0.6×10^{-4} T. Find the difference in potential between the tips of the wing when the speed of the plane is 250 m/s.

Solution

Because the plane is flying horizontally, we do not have to concern ourselves with the horizontal component of the Earth's field. Thus we find

$$\mathcal{E} = B\ell v = (0.6 \times 10^{-4} \text{ T})(50 \text{ m})(250 \text{ m/s}) = \boxed{0.75 \text{ V}}$$

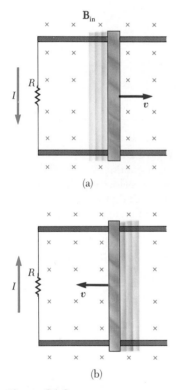

Figure 21.8
(a) As the conducting bar slides on the two fixed conducting rails, the magnetic flux through the loop increases in time. By Lenz's law, the induced current must be *counterclockwise* so as to produce a counteracting flux *out of the paper.* (b) When the bar moves to the left, the induced current must be *clockwise.* Why?

21.4   Lenz's Law Revisited

In order to obtain a better understanding of Lenz's law, let us return to the example of a bar moving to the right on two parallel rails in the presence of a uniform magnetic field directed into the paper (Fig. 21.8a). As the bar moves to the right, the magnetic flux through the circuit increases with time because the area of the loop increases. Lenz's law says that the induced current must be in a direction such that the flux *it* produces opposes the change in the external magnetic flux. Since the flux due to the external field is increasing *into* the paper, the induced current, if it is to oppose the change, must produce a flux *out* of the paper. Hence, the induced current must be counterclockwise when the bar moves to the right to give a counteracting flux out of the paper. (Use the right-hand rule to verify this direction.) On the other hand, if the bar is moving to the left, as in Figure 21.8b, the magnetic flux through the loop decreases with time. Since the flux is into the paper, the induced current has to be clockwise to produce a flux into the paper. In either case, the induced current tends to maintain the original flux through the circuit.

Let us look at this situation from the viewpoint of energy considerations. Suppose that the bar is given a slight push to the right. In the above analysis, we found that this motion leads to a counterclockwise current in the loop. Let us see what happens if we assume that the current is clockwise. For a clockwise current I, the direction of the magnetic force, $BI\ell$, on the sliding bar would be to the right. This force would accelerate the rod and increase its velocity. This, in turn, would cause the area of the loop to increase more rapidly, thus increasing the induced current, which would increase the force, which would increase the current, which would. . . . In effect, the system would acquire energy with zero input energy. This is clearly inconsistent with all experience and with the law of conservation of energy. Thus, we are forced to conclude that the current must be counterclockwise.

Consider another situation, one in which a bar magnet is moved to the right toward a stationary loop of wire, as in Figure 21.9a. As the magnet moves to the right toward the loop, the magnetic flux through the loop increases with time. To counteract this increase in flux to the right, the induced current produces a flux to the left, as in Figure 21.9b; hence the induced current is in the direction shown. Note that the magnetic field lines associated with the

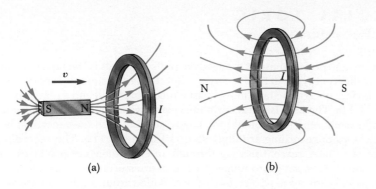

(a) (b)

Figure 21.9
(a) When the magnet is moved toward the stationary conducting loop, a current is induced in the direction shown. (b) This induced current produces its own flux to the left to counteract the increasing external flux to the right.

induced current oppose the motion of the magnet. Therefore, the left face of the current loop is a north pole and the right face is a south pole.

On the other hand, if the magnet were moving to the left, its flux through the loop, which is toward the right, would decrease in time. Under these circumstances, the induced current in the loop would be in a direction such as to set up a field through the loop directed from left to right in an effort to maintain a constant number of flux lines. Hence, the induced current in the loop would be opposite that shown in Figure 21.9b. In this case, the left face of the loop would be a south pole and the right face would be a north pole.

Example 21.3   Application of Lenz's Law

A coil of wire is placed near an electromagnet as shown in Figure 21.10a. Find the direction of the induced current in the coil (a) at the instant the switch is closed, (b) after the switch has been closed for several seconds, and (c) when the switch is opened.

Solution
(a) When the switch is closed, the situation changes from a condition in which no lines of flux pass through the coil to one in which lines of flux pass through in the direction shown in Figure 21.10b. To counteract this change in the number of lines, the coil must set up a field from left to right in the figure. This requires a current directed as shown in Figure 21.10b.

(b) After the switch has been closed for several seconds, there is no change in the number of lines through the loop; hence the induced current is zero.

(c) Opening the switch causes the magnetic field to change from a condition in which flux lines thread through the coil from right to left to a condition of zero flux. The induced current must then be as shown in Figure 21.10c, so as to set up its own field from right to left.

Figure 21.10   (Example 21.3)

ε (a) ε (b) ε (c)

▶**SPECIAL TOPIC**
Tape Recorders

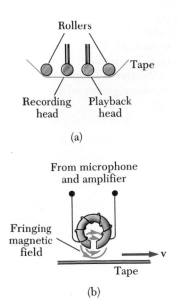

Figure 21.11
(a) The head of a magnetic tape recorder. (b) The fringing magnetic field magnetizes the tape during recording.

One interesting practical use of induced currents and emfs is the tape recorder. There are many different types of tape recorders, but the basic principles are the same for all. In this device, a magnetic tape moves past a recording head and a playback head, as in Figure 21.11a. The tape is a plastic ribbon coated with iron oxide or chromium oxide. We shall examine the function of both heads, starting with the recording head.

The steps of recording are illustrated in Figure 21.11b. The recording process uses the fact that a current passing through an electromagnet produces a magnetic field. A sound wave sent into a microphone is transformed into an electric current, amplified, and allowed to pass through a wire coiled around a doughnut-shaped piece of iron, which functions as the recording head. The iron ring and the wire constitute an electromagnet, which contains the lines of the magnetic field completely inside the iron except at the point where a slot is cut in the ring. At this location the magnetic field fringes out of the iron and magnetizes the small pieces of iron oxide embedded in the tape. Thus, as the tape moves past the slot, it becomes magnetized in a pattern that reproduces both the frequency and the intensity of the sound signal entering the microphone.

To reconstruct the sound signal, the tape is allowed to pass through a recorder with the playback head in operation. This head is very similar to the recording head in that it consists of a wire-wound doughnut-shaped piece of iron with a slot in it. When the tape moves past this head, the varying magnetic fields on the tape produce changing field lines through the wire coil. These changing lines induce a current in the coil that corresponds to the current in the recording head that produced the tape originally. This changing electric current can be amplified and used to drive a speaker. Playback is thus an example of induction of a current by a moving magnet. ◀

21.5   Generators

Generators and motors are important practical devices that operate on the principle of electromagnetic induction. First, let us consider the **alternating current generator** (or ac generator), a device that converts mechanical energy to electrical energy. In its simplest form, the ac generator consists of a loop of wire rotated by some external means in a magnetic field (Fig. 21.12a). In commercial power plants, the energy required to rotate the loop can be derived from a variety of sources. For example, in a hydroelectric plant, falling

Figure 21.12
(a) Schematic diagram of an ac generator. An emf is induced in a coil which rotates by some external means in a magnetic field. (b) The alternating emf induced in the loop plotted versus time.

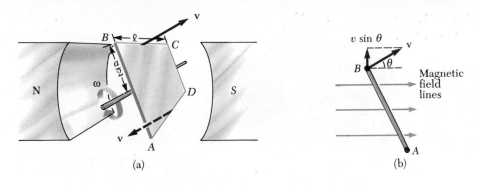

Figure 21.13
(a) A loop rotating at constant angular velocity in an external magnetic field. The emf induced in the loop varies sinusoidally with time. (b) An edge view of the rotating loop.

water directed against the blades of a turbine produces the rotary motion; in a coal-fired plant, the heat produced by burning coal is used to convert water to steam and this steam is directed against the turbine blades. As the loop rotates, the magnetic flux through it changes with time, inducing an emf and a current in an external circuit. The ends of the loop are connected to slip rings that rotate with the loop. Connections to the external circuit are made by stationary brushes in contact with the slip rings.

We can derive an expression for the emf generated in the rotating loop by making use of the expression for motional emf, $\mathcal{E} = B\ell v$. Figure 21.13a shows a loop of wire rotating clockwise in a uniform magnetic field directed to the right. The magnetic force (qvB) on the charges in wires AB and CD is not along the length of the wires. (The force on the electrons in wires AB and CD is perpendicular to the wires, rather than along their length.) Hence, an emf is generated only in wires BC and AD. At any instant, wire BC has a velocity **v** at an angle θ with the magnetic field as shown in Figure 21.13b. (Note that the component of velocity parallel to the field has no effect on the charges in the wire whereas the component of velocity perpendicular to the field produces a magnetic force on the charges that moves electrons from C to B.) The emf generated in wire BC is equal to $B\ell v_\perp$, where ℓ is the length of the wire and $v_\perp$ is the component of velocity perpendicular to the field. An emf of $B\ell v_\perp$ is also generated in wire DA, and the sense of this emf is the same as in wire BC. Since $v_\perp = v \sin \theta$, the total emf is

$$\mathcal{E} = 2B\ell v_\perp = 2B\ell v \sin \theta \tag{21.6}$$

If the loop rotates with a constant angular velocity ω, we can use the relation $\theta = \omega t$ in Equation 21.6. Furthermore, since every point on wires BC and DA rotates in a circle about the axis of rotation with the same angular velocity ω, we have $v = r\omega = (a/2)\omega$, where a is the length of sides AB and CD. Therefore, Equation 21.6 reduces to

$$\mathcal{E} = 2B\ell \left(\frac{a}{2}\right) \omega \sin \omega t = B\ell a \omega \sin \omega t$$

If the loop has N turns, the emf is N times larger because each loop has the same emf induced in it. Furthermore, since the area of the loop is $A = \ell a$, the total emf is

$$\mathcal{E} = NBA\omega \sin \omega t \tag{21.7}$$

Figure 21.14
(a) Schematic diagram of a dc generator. (b) The emf fluctuates in magnitude but always has the same polarity.

This result shows that the emf varies sinusoidally with time, as plotted in Figure 21.12b. Note that the maximum emf has the value

$$\mathcal{E}_{max} = NBA\omega \tag{21.8}$$

which occurs when $\omega t = 90°$ or $270°$. In other words, $\mathcal{E} = \mathcal{E}_{max}$ when the plane of the loop is parallel to the magnetic field. Furthermore, the emf is zero when $\omega t = 0$ or $180°$, that is, when the magnetic field is perpendicular to the plane of the loop. The frequency of rotation for commercial generators in this country and Canada is 60 Hz, whereas in some European countries, 50 Hz is used. (Recall that $\omega = 2\pi f$, where f is the frequency in hertz.)

The **direct current** (dc) **generator** is illustrated in Figure 21.14a. The components are essentially the same as those of the ac generator, except that the contacts to the rotating loop are made by a split ring, or commutator. In this design, the output voltage always has the same polarity and the current is a pulsating direct current, as in Figure 21.14b. The reason for this can be understood by noting that the contacts to the split ring reverse their roles every half cycle. At the same time, the polarity of the induced emf reverses; hence the polarity of the split ring remains the same.

A pulsating dc current is not suitable for most applications. To obtain a more steady dc current, commercial dc generators use many loops and commutators distributed around the axis of rotation so that the sinusoidal pulses from the various loops are out of phase. When these pulses are superimposed, the dc output is almost free of fluctuations.

Example 21.4   Emf Induced in an ac Generator

An ac generator consists of eight turns of wire of area $A = 0.09$ m² and total resistance 12 Ω. The loop rotates in a magnetic field of 0.5 T at a constant frequency of 60 Hz. (a) Find the maximum induced emf.

Solution
First note that $\omega = 2\pi f = 2\pi(60 \text{ Hz}) = 377 \text{ s}^{-1}$. When we substitute the appropriate numerical values into Equation 21.8, we obtain

$$\mathcal{E}_{max} = NAB\omega = 8(0.09 \text{ m}^2)(0.5 \text{ T})(377 \text{ s}^{-1}) = \boxed{136 \text{ V}}$$

(b) What is the maximum induced current?

Solution
From Ohm's law and the results to (a), we find that the maximum induced current is

$$I_{max} = \frac{\mathcal{E}_{max}}{R} = \frac{136 \text{ V}}{12 \text{ } \Omega} = \boxed{11.3 \text{ A}}$$

(c) Determine the time variation of the induced emf.

Solution
We can use Equation 21.7 to obtain the time variation of $\mathcal{E}$:

$$\mathcal{E} = \mathcal{E}_{max} \sin \omega t = \boxed{136 \sin 377t \text{ V}}$$

Exercise Determine the time variation of the induced current.

Answer $I = 11.3 \sin 377t$ A.

Photograph of turbines which generate electrical energy at a hydroelectric power plant.
(Luis Castaneda/The IMAGE Bank)

*Motors and Back emf

Motors are devices that convert electrical energy to mechanical energy. Essentially, *a motor is a generator run in reverse.* Instead of a current being generated by a rotating loop, a current is supplied to the loop by a source of emf and the magnetic force on the current-carrying loop causes it to rotate.

A motor can perform useful mechanical work when a shaft connected to its rotating coil is attached to some external device. As the coil in the motor rotates, however, the changing magnetic flux through it induces an emf that always acts to reduce the current in the coil. If this were not the case, Lenz's law would be violated. The phrase **back emf** is used to indicate an emf that tends to reduce the supplied current. The back emf increases in magnitude as the rotational speed of the coil increases. We can picture the state of affairs in such situations as the equivalent circuit shown in Figure 21.15. For illustrative purposes, we assume in Figure 21.15 that the external power source attempting to drive current through the coil of the motor has a voltage of 120 V, that the coil has a resistance of 10 Ω, and that the back emf induced in the coil at this instant is 70 V. Thus, the voltage available to supply current equals the difference between the supply voltage and the back emf, 50 V in this case. So we see that the current is limited by the back emf.

When a motor is first turned on, there is initially no back emf and the current is very large because it is limited only by the resistance of the coil. As the coil begins to rotate, the induced back emf opposes the applied voltage and the current in the coil is reduced. If the mechanical load increases, the motor will slow down, which causes the back emf to decrease. This reduction in the back emf increases the current in the coil and therefore also increases the power needed from the external voltage source. For this reason, the power requirements are greater for starting a motor and for running it under heavy loads. If the motor is allowed to run under no mechanical load, the back emf reduces the current to a value just large enough to overcome energy losses by heat and friction.

Figure 21.15
A motor can be represented as a resistance plus a back emf.

Figure 21.16
An apparatus that demonstrates the formation of eddy currents in a conductor moving through a magnetic field. As the plate enters or leaves the field, the changing magnetic flux sets up an induced emf, which causes the eddy currents in the plate.

Figure 21.17
As the conducting plate enters the magnetic field in position 1, the eddy currents are counterclockwise. At position 2, however, the currents are clockwise. In either case, the magnetic force retards the motion of the plate.

Example 21.5   The Induced Current in a Motor

Assume that a motor having coils with a resistance of 10 Ω is supplied by a voltage of 120 V. When the motor is running at its maximum speed, the back emf is 70 V. Find the current in the coils (a) when the motor is first turned on and (b) when the motor has reached maximum speed.

Solution
(a) When the motor is first turned on, the back emf is zero. (The coils are motionless.) Thus the current in the coils is a maximum and equal to

$$I = \frac{\mathcal{E}}{R} = \frac{120 \text{ V}}{10 \text{ }\Omega} = \boxed{12 \text{ A}}$$

(b) At the maximum speed, the back emf has its maximum value. Thus, the effective supply voltage is now that of the external source minus the back emf. Hence, the current is reduced to

$$I = \frac{\mathcal{E} - \mathcal{E}_{\text{back}}}{R} = \frac{120 \text{ V} - 70 \text{ V}}{10 \text{ }\Omega} = \frac{50 \text{ V}}{10 \text{ }\Omega} = \boxed{5 \text{ A}}$$

Exercise   If the current in the motor is 8 A at some instant, what is the back emf at this time?

Answer   40 V.

*21.6   Eddy Currents

As we have seen, an emf and a current are induced in a circuit by a changing magnetic flux. In the same manner, circulating currents called **eddy currents** are set up in pieces of metal moving through a magnetic field. This can easily be demonstrated by allowing a flat metal plate at the end of a bar to swing through a magnetic field (Fig. 21.16). The metal should be a nonmagnetic material, such as aluminum or copper. As the plate enters the field, the changing flux creates an induced emf in the plate, which in turn causes the free electrons in the metal to move, producing the swirling eddy currents. According to Lenz's law, the direction of the eddy currents must oppose the change that causes them. For this reason, the eddy currents must produce effective magnetic poles on the plate, which are repelled by the poles of the magnet, thus giving rise to a repulsive force that opposes the swinging motion of the plate.

As indicated in Figure 21.17, when the magnetic field is into the paper, the eddy current is counterclockwise as the swinging plate enters the field at position 1. This is because the external flux into the paper is increasing, and hence by Lenz's law the induced current must provide a flux out of the paper. The opposite is true as the plate leaves the field at position 2, where the current is clockwise. Since the induced eddy current always produces a retarding force when the plate enters or leaves the field, the swinging plate quickly comes to rest.

If slots are cut in the metal plate, as in Figure 21.18, the eddy currents and the corresponding retarding force are greatly reduced. This can be understood by noting that the cuts in the plate are open circuits for any large current loops that might otherwise be formed.

The braking system on many rapid transit cars make use of electromagnetic induction and eddy currents. An electromagnet, which can be energized with a current, is positioned near the steel rails. The braking action occurs when a large current is passed through the electromagnet. The relative motion of the magnet and rails induces eddy currents in the rails, and the direction of these currents produces a drag force on the moving vehicle. The loss in mechanical energy of the vehicle is transformed into heat. Because the eddy currents decrease steadily in magnitude as the vehicle slows, the braking effect is quite smooth.

Eddy currents are undesirable in motors and transformers because they dissipate energy in the form of heat. To reduce this energy loss, moving conducting parts are often laminated, that is, built up in thin layers separated by a nonconducting material, such as lacquer or metal oxide. This layered structure increases the resistance of the possible paths of the eddy currents and effectively confines the currents to individual layers. Such a laminated structure is used in the cores of transformers and motors to minimize eddy currents and thereby increase the efficiency of those devices.

Figure 21.18
When slots are cut in the conducting plate, the eddy currents are reduced and the plate swings more freely through the magnetic field.

21.7 Self-Inductance

Consider a circuit consisting of a switch, a resistor, and a source of emf, as in Figure 21.19. When the switch is closed, the current does not immediately change from zero to its maximum value, $\mathcal{E}/R$. The law of electromagnetic induction, Faraday's law, prevents this from occurring. What happens is the following. As the current increases with time, the magnetic flux through the loop due to this current also increases. This increasing flux induces an emf in the circuit that opposes the change in magnetic flux. By Lenz's law, the induced electric field in the loop must therefore be opposite the direction of flow of current. That is, the induced emf is in the direction indicated by the dashed battery in Figure 21.19. The net potential difference across the resistor is the emf of the battery minus the opposing induced emf. As the magnitude of the current increases, the *rate* of increase becomes smaller and hence the induced emf decreases. This opposing emf results in a gradual increase in the current. For the same reason, when the switch is opened, the current gradually decreases to zero. This effect is called **self-induction** because the changing flux through the circuit arises from the circuit itself. The emf that is set up in this case is called a **self-induced emf.**

As a second example of self-inductance, consider Figure 21.20, which shows a coil wound on a cylindrical form. (A practical device would have several hundred turns.) Assume that the current changes with time. When the

Figure 21.19
After the switch in the circuit is closed, the current produces its own magnetic flux through the loop. As the current increases toward its equilibrium value, the flux changes in time and induces an emf in the loop.

Figure 21.20
(a) A current in the coil produces a magnetic field directed to the left. (b) If the current increases, the coil acts as a source of emf directed as shown by the dashed battery. (c) The emf of the coil changes its polarity if the current decreases.

current is in the direction shown, a magnetic field is set up inside the coil directed from right to left. As a result, some lines of magnetic flux pass through the cross-sectional area of the coil. The flux through the coil changes as the current changes with time, which induces an emf in the coil. Application of Lenz's law shows that the direction of this induced emf is such as to oppose the change in the current. That is, if the current is increasing, the induced emf will be as pictured in Figure 21.20b and if the current is decreasing, the induced emf will be as shown in Figure 21.20c.

To obtain a quantitative description of self-inductance, first note that Faraday's law says that the induced emf is as given by Equation 21.2:

$$\mathcal{E} = -N\frac{\Delta\Phi}{\Delta t}$$

The magnetic flux is proportional to the magnetic field, which is proportional to the current in the circuit. Thus, *the self-induced emf is always proportional to the time rate of change of the current*:

Self-induced emf

$$\mathcal{E} \equiv -L\frac{\Delta I}{\Delta t} \qquad \textbf{(21.9)}$$

where L is a proportionality constant called the **inductance** of the device. The negative sign that appears in this equation indicates that a changing current induces an emf which opposes that change. This means that if the current is increasing (ΔI positive), then the induced emf is negative, indicative of opposition to the increase in current. Likewise, if the current is decreasing (ΔI negative), then the sign of the induced emf is positive to indicate that the emf is acting to oppose the decrease.

The inductance depends on the cross-sectional area of the coils and other quantities, all of which can be grouped under the general heading of geometric factors. The SI unit of inductance is the **henry** (**H**), which from Equation 21.9 is seen to be equal to 1 volt-second per ampere:

$$1\ \text{H} = 1\ \text{V}\cdot\text{s/A}$$

Examples 21.6 and 21.7 discuss some simple situations for which self-inductances are easily evaluated. When evaluating self-inductances, it is often convenient to equate Equations 21.2 and 21.9 to find an expression for L:

$$N\frac{\Delta\Phi}{\Delta t} = L\frac{\Delta I}{\Delta t}$$

Inductance

$$L = N\frac{\Delta\Phi}{\Delta I} = \frac{N\Phi}{I} \qquad \textbf{(21.10)}$$

where we assume $\Phi = 0$ and $I = 0$ at $t = 0$.

Joseph Henry, an American physicist who carried out early experiments in electrical induction, was born in Albany, New York, in 1797. The son of a laborer, Henry had little schooling and was forced to go to work at a very young age. After working his way through Albany Academy to study medicine, then engineering, Henry became professor of mathematics and physics in 1826. He later became professor of natural philosophy at New Jersey College (now Princeton University).

In 1848, Henry became the first director of the Smithsonian Institute, where he introduced a weather-forecasting system based on meteorological information received by the electric telegraph. He was also the first president of The Academy of Natural Science, a position he held until his death in 1878.

Many of Henry's early experiments were with electromagnetism. He improved the electromagnet of William Sturgeon and made one of the first electromagnetic motors. By 1830, Henry had made powerful electromagnets by using many turns of fine insulated wire wound around iron cores. He discovered the phenomenon of self-induction but failed to publish his findings; as a result, credit was given to Michael Faraday.

Henry's contribution to science was ultimately recognized: in 1893 the unit of inductance was named the henry.

Joseph Henry
(1797 – 1878)

(Courtesy of AIP Niels Bohr Library, E. Scott Barr Collection)

Example 21.6 Inductance of a Solenoid

Find the inductance of a uniformly wound solenoid with N turns and length ℓ. Assume that ℓ is large compared with the radius and that the core of the solenoid is air.

Solution
In this case, we can take the interior field to be uniform and given by Equation 20.16:

$$B = \mu_0 nI = \mu_0 \frac{N}{\ell} I$$

where n is the number of turns per unit length, N/ℓ. The flux through each turn is

$$\Phi = BA = \mu_0 \frac{N}{\ell} AI$$

where A is the cross-sectional area of the solenoid. From this expression and Equation 21.10, we find that

$$L = \frac{N\Phi}{I} = \frac{\mu_0 N^2 A}{\ell} \qquad \textbf{(21.11)}$$

This shows that L depends on geometric factors and μ_0 and is proportional to the square of the number of turns. Since $N = n\ell$, we can also express the result in the form

$$L = \mu_0 \frac{(n\ell)^2}{\ell} A = \mu_0 n^2 A\ell = \mu_0 n^2 (\text{volume}) \qquad \textbf{(21.12)}$$

where $A\ell$ is the volume of the solenoid.

Example 21.7   Calculating Inductance and Self-Induced emf

(a) Calculate the inductance of a solenoid containing 300 turns if the length of the solenoid is 25 cm and its cross-sectional area is 4 cm^2 = 4 $\times$ 10^{-4} m^2.

Solution
Using Equation 21.11, we get

$$L = \frac{\mu_0 N^2 A}{\ell} = (4\pi \times 10^{-7}\ \text{T}\cdot\text{m/A})\ \frac{(300)^2 (4 \times 10^{-4}\ \text{m}^2)}{25 \times 10^{-2}\ \text{m}}$$

$$= 1.81 \times 10^{-4}\ \text{T}\cdot\text{m}^2/\text{A} = \boxed{0.181\ \text{mH}}$$

(b) Calculate the self-induced emf in the solenoid described in (a) if the current through it is decreasing at the rate of 50 A/s.

Solution
Equation 21.9 ($\mathcal{E} = -L\,\Delta I/\Delta t$) can be combined with $\Delta I/\Delta t = -50$ A/s to obtain

$$\mathcal{E} = -L\frac{\Delta I}{\Delta t} = -(1.81 \times 10^{-4}\ \text{H})(-50\ \text{A/s}) = \boxed{9.05\ \text{mV}}$$

*21.8   *RL* Circuits

A circuit element that has a large inductance, such as a closely wrapped coil of many turns, is called an **inductor**. The circuit symbol for an inductor is ℓℓℓ. We shall always assume that the self-inductance of the remainder of the circuit is negligible compared with that of the inductor.

In order to gain some insight into the effect of an inductor in a circuit, consider the two circuits shown in Figure 21.21. Figure 21.21a consists of a resistor connected to the terminals of a battery. Applying Ohm's law to this circuit gives

$$\mathcal{E} = RI \tag{21.13}$$

In the past, we have interpreted resistance as a measure of the *opposition to the current*.

Now consider the circuit shown in Figure 21.21b, consisting of an inductor connected to the terminals of a battery. At the instant the switch in this

Figure 21.21
Comparing the effect of a resistor to that of an inductor in a simple circuit.

circuit is closed, the emf of the battery is equal to the back emf generated in the coil. Thus we have

$$\mathcal{E} = -L \frac{\Delta I}{\Delta t}$$ **(21.14)**

By comparing this equation with Equation 21.13, we see that the position of *R* is now occupied by *L* and the position of *I* is now occupied by $\Delta I/\Delta t$. As a result, we can interpret *L* as a measure of the *opposition to the rate of change in current.*

Consider the circuit consisting of a resistor, inductor, and battery shown in Figure 21.22. Suppose the switch is closed at $t = 0$. The current will begin to increase, but the inductor will produce an emf that opposes the increasing current. Thus, the current is unable to change from zero to its maximum value of $\mathcal{E}/R$ instantaneously. This induced emf has its maximum value when the switch is first closed. That is, Equation 21.14 shows that the induced emf is a maximum when the current is changing at its most rapid rate, which occurs when the switch is first closed. As the current approaches its steady-state value, the back emf of the coil falls off because the current is changing more slowly with time. Finally, when the current reaches its steady-state value, the rate of change is zero and the back emf is also zero. Figure 21.23 represents a plot of current in the circuit as a function of time.[1] This plot is very similar to that of the charge on the plate of a capacitor as a function of time discussed in Section 19.5. In that case, we found it convenient to introduce a quantity called the time constant of the circuit, which told us something about the time required for the capacitor to approach its steady-state charge. In the same fashion, time constants are defined for circuits containing resistors and inductors. The **time constant** in an *RL* circuit is defined as

$$\tau \equiv \frac{L}{R}$$ **(21.15)**

The time required for the current in the circuit to reach 63% of its final value, $\mathcal{E}/R$, is the time constant, τ, for the circuit.

Example 21.8   The Time Constant for an *RL* Circuit

The circuit shown in Figure 21.24 consists of a 30-mH inductor, a 6-Ω resistor, and a 12-V battery. The switch is closed at $t = 0$. Find (a) the time constant of the circuit and (b) the current after one time constant has elapsed.

Solution
(a) The time constant is given by Equation 21.15:

$$\tau = \frac{L}{R} = \frac{30 \times 10^{-3}\,\text{H}}{6\Omega} = \boxed{5\ \text{ms}}$$

[1] The equation for the current in the circuit as a function of time is

$$I = \frac{\mathcal{E}}{R}(1 - e^{-Rt/L})$$

Figure 21.22
A series *RL* circuit. As the current increases toward its maximum value, the inductor produces an emf that opposes the increasing current.

Figure 21.23
Plot of the current versus time for the *RL* circuit shown in Figure 21.22. The switch is closed at $t = 0$, and the current increases toward its maximum value, $\mathcal{E}/R$. The time constant, τ, is the time it takes the current to reach 63% of its maximum value.

Figure 21.24   (Example 21.8)

(b) After a time equal to one time constant, the current in the circuit has risen to 63% of its final value. Thus, the current is

$$I = 0.63 \frac{\mathcal{E}}{R} = (0.63)\left(\frac{12 \text{ V}}{6 \, \Omega}\right) = \boxed{1.26 \text{ A}}$$

Exercise   What is the voltage drop across the resistor (a) at $t = 0$ and (b) after one time constant?

Answer   (a) 0 (b) 7.56 V.

21.9   Energy Stored in a Magnetic Field

The induced emf set up by an inductor prevents a battery from establishing a current in a circuit instantaneously. Hence, a battery has to do work to produce a current. We can think of this work needed to produce the current as being energy stored by the inductor in its magnetic field. In a manner quite similar to the one we used in Section 17.7 to find the energy stored by a capacitor, we find that the energy stored by an inductor is

Energy stored in an inductor

$$PE_L = \frac{1}{2} L I^2 \tag{21.16}$$

Note that the result is similar in form to the expression for the energy stored in a capacitor, which is

Energy stored in a capacitor

$$PE_C = \frac{1}{2} CV^2$$

Summary

The **magnetic flux, Φ,** through a closed loop is defined as

$$\Phi \equiv BA \cos\theta \tag{21.1}$$

where B is the strength of the uniform magnetic field, A is the cross-sectional area of the loop, and θ is the angle between **B** and the direction perpendicular to the plane of the loop.

Faraday's law of induction states that the instantaneous emf induced in a circuit is equal to the rate of change of magnetic flux through the circuit:

$$\mathcal{E} = -N \frac{\Delta\Phi}{\Delta t} \tag{21.2}$$

where N is the number of loops in the circuit,

Lenz's law states that the polarity of the induced emf is such that it produces a current whose magnetic field opposes the *change* in magnetic flux through a circuit.

If a conducting bar of length ℓ moves through a magnetic field with a speed v such that **B** is perpendicular to the bar, the emf induced in the bar, often called a **motional emf**, is

$$\mathcal{E} = B\ell v \tag{21.3}$$

When a coil of wire having N turns, each of area A, rotates with constant angular velocity ω in a uniform magnetic field **B**, the emf induced in the coil is

$$\mathcal{E} = NAB\omega \sin \omega t \tag{21.7}$$

When the current in a coil changes with time, an emf is induced in the coil according to Faraday's law. This **self-induced emf** is defined by the expression

$$\mathcal{E} \equiv -L \frac{\Delta I}{\Delta t} \tag{21.9}$$

where L is the **inductance** of the coil. Inductance has the SI unit of henry (H), where $1 \text{ H} = 1 \text{ V} \cdot \text{s/A}$.

The **inductance**, L, of a coil can be found from the expression

$$L = \frac{N\Phi}{I} \tag{21.10}$$

where N is the number of turns on the coil, I is the current in the coil, and Φ is the magnetic flux through the coil.

If a resistor and inductor are connected in series to a battery and a switch is closed at $t = 0$, the current in the circuit does not rise instantly to its maximum value. After one **time constant**, $\tau = L/R$, the current in the circuit is 63% of its final value, $\mathcal{E}/R$.

The **energy stored** in the magnetic field of an inductor carrying a current I is

$$PE_L = \frac{1}{2} LI^2 \tag{21.16}$$

Additional Reading

H. Kondo, "Michael Faraday," *Sci. American,* October 1953, p. 90.

D. K. C. McDonald, *Faraday, Maxwell and Kelvin,* New York, Doubleday Anchor, 1964.

H. L. Sharlin, "From Faraday to the Dynamo," *Sci. American,* May 1961, p. 107.

G. Shiers, "The Induction Coil," *Sci. American,* May 1971, p. 80.

Questions

1. What is the difference between magnetic flux and magnetic field?
2. A loop of wire is placed in a uniform magnetic field. For what orientation of the loop is the magnetic flux a maximum? For what orientation is the flux zero?
3. A circular loop is located in a uniform and constant magnetic field. Describe how an emf can be induced in the loop in this situation.
4. We saw in the last chapter that a steady magnetic field does not change the energy of a charge moving in it. Can a changing magnetic field affect the energy of a charge?
5. Will dropping a magnet down a long copper tube produce a current in the tube? Explain.
6. A spacecraft circling the Earth has a coil of wire in it. An astronaut notes that there is a current in the coil although no battery is connected to it and there are no magnets on the spacecraft. What is causing the current?
7. As the conducting bar in Figure 21.25 moves to the right, an electric field directed downward is set up in the conductor. If the bar were moving to the left, explain why the electric field would be upward.

Figure 21.25
Questions 7 and 8

8. As the bar in Figure 21.25 moves perpendicular to the magnetic field, is an external force required to keep it moving with constant velocity? Explain.
9. Explain why an external force is necessary to keep the bar in Figure 21.26 moving with a constant velocity.
10. The bar in Figure 21.26 moves to the right with a velocity **v**, and a uniform, constant magnetic field is directed *outward* (represented by the dots). Why is the induced current clockwise? If the bar were moving to the left, what would be the direction of the induced current?

Figure 21.26   Questions 9 and 10

11. Magnetic storms on the Sun can cause difficulties with communications here on the Earth. Why do these sunspots affect us in this way?
12. Would you expect the tape from a tape recorder to be attracted to a magnet? (Try it, but not with a recording you wish to save.)
13. Wearing a metal bracelet in a region of strong magnetic field could be hazardous. Discuss.
14. How is electrical energy produced in dams (that is, how is the energy of motion of the water converted into ac electricity)?
15. What happens when the coil of a generator is rotated at a faster rate?
16. Could a current be induced in a coil by rotating a magnet inside the coil? If so, how?
17. "If the shaft of a dc motor is cranked manually, the motor becomes a dc generator." Is this statement true? Discuss.
18. A piece of aluminum is dropped vertically downward between the poles of an electromagnet. Does the magnetic field affect the velocity of the aluminum?
19. A bar magnet is dropped toward a conducting ring laying on the floor. As the magnet falls toward the ring does it move as a freely falling body?
20. In a beam balance scale, an aluminum plate is sometimes used to slow the oscillations of the beam near equilibrium. The plate is mounted at the end of the beam, and moves between the poles of a small horseshoe magnet, attached to the frame. Why are the oscillations of the beam strongly damped near equilibrium?
21. Why is the induced emf that appears in an inductor called a "counter" or "back" emf?
22. A circuit containing a coil, resistor, and battery is in a steady state, that is, the current has reached a constant value. Does the coil have an inductance? Does the coil affect the value of the current in the circuit?
23. Does the inductance of a coil depend on the current in the coil? What parameters affect the inductance of a coil?

24. Suppose the switch in a circuit containing an inductor, a battery, and a resistor has been closed for a long time and is suddenly opened. Does the current instantaneously drop to zero? Why does a spark usually appear at the switch contacts when the switch is opened?

25. If the current in an inductor is doubled, by what factor does the stored energy change?

Problems

Section 21.1   Induced emf and Magnetic Flux

1. A magnetic field of strength 0.25 T is directed perpendicular to a circular loop of wire of radius 30 cm. Find the magnetic flux through the area.

2. Find the flux of the Earth's magnetic field of magnitude 5×10^{-5} T through a square loop of area 20 cm² when the field is (a) perpendicular to the plane of the loop, (b) when the field makes a 30° angle with the normal to the plane of the loop, and (c) when the field makes a 90° angle with the normal to the plane.

3. A square loop 2 m on a side is placed in a magnetic field of strength 0.3 T. If the field makes an angle of 50° with the normal to the plane of the loop, as in Figure 21.2, determine the magnetic flux through the loop.

4. A long, straight wire lies in the plane of a circular coil of radius 0.010 m. The wire carries a current of 2 A and is placed along a diameter of the coil. (a) What is the net flux through the coil? (b) If the wire passes through the center of the coil and is perpendicular to the plane of the coil, find the net flux through the coil.

5. A solenoid 4 cm in diameter and 20 cm in length has 250 turns and carries a current of 15 A. Calculate the magnetic flux through the circular cross-sectional area of the solenoid.

6. The solenoid of Problem 5 is surrounded by a single loop of wire 10 cm in diameter. The loop is positioned perpendicular to and centered on the axis of the solenoid as shown in Figure 21.27. Find the magnetic flux through the loop of wire. (The field outside the solenoid is small enough to be negligible.)

Figure 21.27
Problem 6

7. A cube of edge length $\ell = 2.5$ cm is positioned as shown in Figure 21.28. There is a uniform magnetic field throughout the region with components of $B_x = +5$ T, $B_y = +4$ T, and $B_z = +3$ T. (a) Calculate the flux through the shaded face of the cube. (b) What is the total flux through the six faces of the cube?

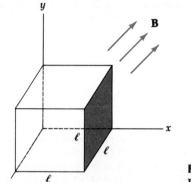

Figure 21.28
Problem 7

8. A circular loop of wire of radius R is placed in a uniform magnetic field **B** and is then spun at constant angular velocity ω about an axis through its diameter. Determine the magnetic flux through the loop as a function of time if the axis of rotation is (a) perpendicular to **B** and (b) parallel to **B**.

Section 21.2   Faraday's Law of Induction

9. The magnetic flux through a loop consisting of two turns of wire is changing at the rate of 3 Wb/s. Find the magnitude of the induced emf in the loop.

10. A coil of radius 20 cm is placed in an external magnetic field of strength 0.20 T such that the plane of the coil is perpendicular to the field. The coil is pulled out of the field in a time of 0.3 s. Find the average induced emf during this time.

11. The bolt of lightning depicted in Figure 21.29 strikes the ground 200 m from a 100-turn coil oriented as

Figure 21.29   Problem 11

shown. If the current in the lightning bolt falls from 6.02×10^6 A to zero in 10.5 μs, what is the average voltage induced in the coil? Assume that the distance to the center of the coil determines the average magnetic induction at the coil's position. Treat the lightning bolt as a vertical wire.

12. A powerful electromagnet has a field of 1.6 T and a cross-sectional area of 0.2 m². If we place a coil of 200 turns with a total resistance of 20 Ω around the electromagnet, and then turn off the power to the electromagnet in 0.02 s, what will be the induced current in the coil?

13. A coil has a radius of 15 cm and is placed in an external magnetic field of strength 0.25 T with the plane of the coil perpendicular to the field direction. The field increases to 0.50 T in a time of 0.7 s. (a) Find the magnitude of the average induced emf during this time. (b) If the average current during this time was 0.8 A, find the resistance of the coil.

14. A loop of wire of radius 0.30 m lies such that an external magnetic field of strength +0.30 T is perpendicular to the loop. The field changes to −0.20 T in a time of 1.5 s. (The plus and minus signs here refer to opposite directions through the coil.) Find the magnitude of the average induced emf in the coil during this time.

15. A 500-turn circular-loop coil 15.0 cm in diameter is initially aligned so that its axis is parallel to the Earth's magnetic field. In 2.77 ms the coil is flipped so that its axis is perpendicular to the Earth's magnetic field. If a voltage of 0.166 V is induced in the coil, what is the value of the Earth's magnetic field?

16. A square, single-turn coil 0.20 m on a side is placed with its plane perpendicular to a constant magnetic field. An emf of 18 mV is induced in the winding when the area of the coil decreases at a rate of 0.1 m²/s. What is the magnitude of the magnetic field?

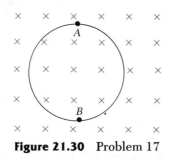

Figure 21.30   Problem 17

17. The flexible loop shown in Figure 21.30 has a radius of 12 cm and is in a magnetic field of strength 0.15 T. The loop is grasped at points A and B and stretched until it closes. If it takes 0.2 s to close the loop, find the magnitude of the average induced emf in it during this time.

18. The plane of a rectangular coil of dimensions 5 cm by 8 cm is perpendicular to the direction of a magnetic field **B**. If the coil has 75 turns and a total resistance of 8 Ω, at what rate must the magnitude of **B** change in order to induce a current of 0.1 A in the windings of the coil?

19. A magnetic field of 0.2 T exists within a solenoid of 500 turns and a diameter of 10 cm. How rapidly (that is, within what period of time) must the field be reduced to zero magnitude if the average magnitude of the induced emf within the coil during this time interval is to be 10 kV?

20. A 300-turn solenoid with a length of 20 cm and a radius of 1.5 cm carries a current of 2 A. A second coil of 4 turns is wrapped tightly about this solenoid such that it can be considered to have the same radius as the solenoid. (a) Find the change in the magnetic flux through the coil and (b) the magnitude of the average induced emf in the coil when the current in the solenoid increases to 5 A in a period of 0.9 s.

21. A square coil of wire having a single turn 1 cm on a side is placed inside a solenoid that has a circular cross section of radius 3 cm, as shown in Figure 21.31. The solenoid is 20 cm long and wound with 100 turns of wire. (a) When the current in the solenoid is 3 A, find the flux through the coil. (b) If the current in the solenoid is reduced to zero in 3 s, find the magnitude of the average induced emf in the coil.

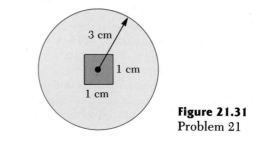

Figure 21.31
Problem 21

22. A coil formed by wrapping 50 turns of wire in the shape of a square is positioned in a magnetic field so that the normal to the plane of the coil makes an angle of 30° with the direction of the field. It is observed that, if the magnitude of the magnetic field is increased uniformly from 200 μT to 600 μT in 0.4 s, an emf of 80 mV is induced in the coil. What is the total length of the wire forming the coil?

Section 21.3   Motional emf

23. A sliding bar moves with a velocity of 20 cm/s toward the right along a U-shaped piece of metal, as shown in Figure 21.7. The magnetic field is out of the page and

has a strength of 0.2 T. Find the induced emf in the circuit if the length of the bar is 30 cm.

24. An airplane with a wingspan of 40 m flies horizontal to the Earth at a location where the downward component of the Earth's magnetic field is 0.6×10^{-4} T. At what speed would the plane have to fly to produce a difference in potential of 1.5 V between the tips of its wings? Is this likely to occur in a typical airplane flight?

25. Over a region where the *vertical* component of the Earth's magnetic field is 40 μT directed downward, a 5-m length of wire is held along an east-west direction and moved horizontally to the north with a speed of 10 m/s. Calculate the potential difference between the ends of the wire and determine which end is positive.

26. Consider the arrangement shown in Figure 21.32. Assume that $R = 6\ \Omega$, $\ell = 1.2$ m, and that a uniform 2.5-T magnetic field is directed *into* the page. At what speed should the bar be moved to produce a current of 0.5 A in the resistor?

Figure 21.32   Problem 26

27. Over a region where the vertical component of the Earth's magnetic field is 40 μT, a 50-cm length of wire is moved with a speed of 5 m/s along U-shaped rails as shown in Figure 21.7. Find (a) the induced emf in the circuit and (b) the induced current in the circuit if there is a resistor of 5 Ω in the U-shaped track. The resistance of the wire and track can be considered negligible.

28. For the circuit described in Problem 27, find (a) the average power supplied to the circuit by the induced emf, (b) the power dissipated in the 5-Ω resistor, (c) the force required to keep the length of wire moving at a constant speed, and (d) the power supplied to the circuit by the agent exerting the force calculated in part (c).

29. A steel beam 12 m in length is accidentally dropped by a construction crane from a height of 9 m. The horizontal component of the Earth's magnetic field over the region is 18 μT. What is the induced emf in the beam just before impact with the Earth assuming its long dimension remains in a horizontal plane and

oriented perpendicular to the horizontal component of the Earth's magnetic field?

Section 21.4   Lenz's Law Revisited

30. A bar magnet is held above a loop of wire in a horizontal plane as shown in Figure 21.33. The south end of the magnet is toward the loop of wire. The magnet is dropped toward the loop. Find the direction of the current through the resistor (a) while the magnet is falling toward the loop and (b) after the magnet has passed through the loop and moves away from it.

Figure 21.33   Problem 30

31. A bar magnet is positioned near a coil of wire as shown in Figure 21.34. What is the direction of the current through the resistor when the magnet is moved (a) to the left and (b) to the right?

Figure 21.34   Problem 31

32. Find the direction of the current through the resistor in Figure 21.35 (a) at the instant the switch is closed, (b) after the switch has been closed for several minutes, and (c) at the instant the switch is opened.

Figure 21.35   Problem 32

33. Find the direction of the current in the resistor R in Figure 21.36 when the following steps are taken in the order given. (a) The switch is closed. (b) The variable resistance in series with the battery is decreased. (c) The circuit containing resistor R is moved to the left. (d) The switch is opened.

Figure 21.36   Problem 33

34. What is the direction of the current induced in the resistor at the instant the switch in the circuit of Figure 21.37 is closed?

Figure 21.37
Problem 34

35. What is the direction of the current induced in the resistor when the current in the long, straight wire shown in Figure 21.38 decreases rapidly to zero?

Figure 21.38
Problem 35

36. A square loop of wire is placed perpendicular to a magnetic field directed as shown in Figure 21.39a.

(a)   (b)

Figure 21.39   Problem 36

Find the *direction* of the induced current through the resistor while the loop is being rotated from the position shown in (a) to that shown in Figure 21.39b.

37. The magnetic field shown in Figure 21.40 has a uniform magnitude of 25.0 mT directed into the paper. The initial diameter of the kink is 2.00 cm. (a) The wire is quickly pulled taut, and the kink shrinks to a diameter of zero in 50.0 ms. Determine the average voltage induced between endpoints A and B. Include the polarity. (b) Suppose the kink is undisturbed, but the magnetic field increases to 100 mT in 4.0×10^{-3} s. Determine the average voltage across terminals A and B, including polarity, during this period.

Figure 21.40   Problem 37

38. A bar magnet is spun at constant angular velocity ω about an axis as shown in Figure 21.41. A flat rectangular conducting loop surrounds the magnet, and at $t = 0$, the magnet is oriented as shown. Sketch the induced current in the loop as a function of time, plotting counterclockwise currents as positive and clockwise currents as negative.

Figure 21.41
Problem 38

Section 21.5   Generators

39. The alternating voltage of a generator is represented by the equation $\mathcal{E} = 240 \sin 500t$, where $\mathcal{E}$ is in volts and t is in seconds. Find the frequency of the voltage and the maximum voltage output of the source.

40. Coils rotating in a magnetic field are often used to measure unknown magnetic fields. As an example, consider a coil of radius 1 cm with 50 turns that is rotated about an axis perpendicular to the field at a rate of 20 Hz. If the maximum induced emf in the coil is 3 V, find the strength of the magnetic field.

41. A square coil (20 cm × 20 cm) that consists of 100 turns of wire rotates about a vertical axis at 1500 rpm, as indicated in Figure 21.42. The horizontal component of the Earth's magnetic field at the location of the loop is 2×10^{-4} T. Calculate the maximum emf induced in the coil by the Earth's field.

Figure 21.42
Problem 41

42. A 500-turn circular coil of radius 20 cm is rotating about an axis perpendicular to a magnetic field of 0.1 T. What angular velocity will produce a maximum induced emf of 2 mV?

43. An ac generator with its terminals shorted together consists of 40 turns of wire of area 0.12 m² and with total resistance 30 Ω. The loop rotates in a magnetic field of 0.1 T at a constant frequency of 60 Hz. (a) Find the maximum induced emf, (b) the maximum induced current, (c) an expression for the time variation of $\mathcal{E}$, and (d) an expression for the time variation of the induced current.

44. A motor has coils with a resistance of 30 Ω and operates from a voltage of 240 V. When the motor is operating at its maximum speed, the back emf is 145 V. Find the current in the coils (a) when the motor is first turned on and (b) when the motor has reached maximum speed. (c) If the current in the motor is 6 A at some instant, what is the back emf at this time?

45. When the coil of a motor is rotating at maximum speed, the current in the windings is 4 A. When the motor is first turned on, the current in the windings is 11 A. If the motor is operated at 120 V, find the back emf in the coil and the resistance of the windings.·

46. What is the maximum torque delivered by an electric motor if it has 80 turns of wire wrapped on a rectangular coil, of dimensions 2.5 cm by 4 cm? Assume that the motor uses 10 A of current and that a uniform 0.8-T magnetic field exists within the motor.

47. In a model ac generator, a 500-turn rectangular coil of dimensions 8 cm by 20 cm rotates at 120 rev/min in a uniform magnetic field of 0.6 T. (a) What is the maximum emf induced in the coil? (b) What is the instantaneous value of the emf in the coil at $t = (\pi/32)$ s? Assume that the emf is zero at $t = 0$. (c) What is the smallest value of t for which the emf will have its maximum value?

Section 21.7   Self-Inductance

48. A coil has an inductance of 3 mH and a current through it changes from 0.2 A to 1.5 A in a time of 0.2 s. Find the magnitude of the average induced emf in the coil during this time.

49. A Slinky toy spring has a radius of 4 cm and an inductance of 275 μH when extended to a length of 1.5 m. What is the total number of turns in the spring?

50. A solenoid of radius 2.5 cm has 400 turns and a length of 20 cm. (a) Find the inductance of the solenoid and (b) the rate at which current must change through it to produce an emf of 75 mV.

51. At the instant the emf is 75 mV for the solenoid of Problem 50, find the rate of change of magnetic flux through the cross-sectional area of the solenoid.

52. An emf of 24 mV is induced in a 500-turn coil at an instant when the current is 4 A and is changing at a rate of 10 A/s. What is the magnetic flux through each turn of the coil?

53. Calculate the magnetic flux through each turn of a 300-turn, 7.2-mH coil when the current in the coil is 10 mA.

Section 21.8   *RL* Circuits

54. Show that the inductive time constant τ has SI units of seconds.

55. A 6-V battery is connected in series with a resistor and an inductor. The series curcuit has a time constant of 600 μs, and the maximum current is 300 mA. What is the value of the inductance?

56. A 25-mH inductor, an 8-Ω resistor, and a 6-V battery are connected in series. The switch is closed at $t = 0$. Find the voltage drop across the resistor (a) at $t = 0$ and (b) after one time constant has passed. Also, find the voltage drop across the inductor (c) at $t = 0$ and (d) after one time constant has elapsed.

57. The switch in a series RL circuit in which $R = 6\ \Omega$, $L = 3$ H, and $\mathcal{E} = 24$ V is closed at $t = 0$. (a) What is the maximum current in the circuit? (b) What is the current when $t = 0.5$ s?

58. This problem assumes coverage of Footnote 1 in Section 21.8. Find the current in the circuit of Problem 57 at $t = 0.7$ s.

Section 21.9   Energy Stored in a Magnetic Field

59. How much energy is stored in a 70-mH inductor at the instant when the current is 2 A?

60. A 24-V battery is connected in series with a resistor and an inductor, where $R = 8\ \Omega$ and $L = 4$ H. What energy is stored in the inductor (a) when the current reaches its maximum value and (b) one time constant after the switch is closed?

Additional Problems

61. A 50-turn rectangular coil of dimensions 0.2 m by 0.3 m is rotated at 90 rad/s in a magnetic field so that the axis of the coil is perpendicular to the direction of the field. The maximum emf induced in the coil is 0.5 V. What is the magnitude of the field?

62. Calculate the energy associated with the magnetic field of a 200-turn solenoid in which a current of 1.75 A produces a flux of 3.7×10^{-4} Wb in each turn.

63. A tightly wound circular coil has 50 turns, each of radius 0.2 m. A uniform magnetic field is introduced perpendicular to the plane of the coil. If the field increases in strength from 0 to 0.3 T in 0.4 s, what average emf is induced in the windings of the coil?

64. Find the self-inductance of a coil if an emf of 100 mV is produced when the current changes from 0 to 1.5 A in a time of 0.3 s.

65. An automobile starter motor draws a current of 3.5 A from a 12-V battery when operating at normal speed. A broken pulley locks the armature in position and the current increases to 18 A. What is the back emf of the motor?

66. A circular coil of radius 15 cm and 5 turns is oriented with the plane of the coil perpendicular to a uniform magnetic field of 0.15 T. This field increases at a uniform rate of 0.20 T in 3 s. If the resistance of the coil is 8 Ω, find the amount of charge that passes through the coil during this 3-s interval.

67. Consider a 100-turn rectangular coil of cross-sectional area 0.06 m^2 rotating with an angular velocity of 20 rad/s about an axis perpendicular to a magnetic field of 2.5 T. Plot the induced voltage as a function of time over one complete period of rotation.

68. A 300-turn inductor has a radius of 5 cm and a length of 20 cm. Find the energy stored in it when the current is 0.5 A.

69. An open hemispherical surface of radius 0.1 m is located in a magnetic field of 0.15 T. The circular cross section of the surface is perpendicular to the direction of the field. Calculate the magnetic flux through the surface.

70. The coil of Problem 68 is connected in a series circuit with a 1000-Ω resistor and a 12-V battery. Find the energy stored in the inductor (a) after one time constant and (b) after two time constants.

71. Shown in Figure 21.43 is a graph of the induced emf versus time for a coil of N turns rotating with angular velocity ω in a uniform magnetic field directed perpendicular to the axis of rotation of the coil. Copy this sketch (to a larger scale), and on the same set of axes show the graph of emf versus t when (a) the number of turns in the coil is doubled, (b) the angular velocity is doubled, and (c) the angular velocity is doubled while the number of turns in the coil is halved.

Figure 21.43
Problem 71

72. Two coil, A and B, are wound using equal lengths of wire. Each coil has the same number of turns per unit length, but coil A has twice as many turns as coil B. What is the ratio of the self-inductance of A to the self-inductance of B? (*Note:* The radii of the two coils are not equal.)

73. A single-turn, circular loop of radius 0.2 m is coaxial with a long solenoid of radius 0.05 m, length 0.8 m, and having 1600 turns, as in Figure 21.44. The vari-

Variable resistor

Figure 21.44
Problem 73

able resistor is changed so that the solenoid current decreases linearly from 6 A to 1.5 A in 0.2 s. Calculate the induced emf in the circular loop. (The field just outside the solenoid is small enough to be negligible.)

74. A series circuit contains a 12-V battery, a 2000-Ω resistor, and a 3-mH inductor. If the switch to the battery is closed at $t = 0$, find the time required for the current in the circuit to reach 90% of its final value. (See Footnote 1, Section 21.8.)

75. A rectangular coil of 50 turns, length 0.15 m, width 0.25 m, and resistance 12 Ω rotates with an angular velocity 20 rad/s about the y axis in a region where a uniform magnetic field of 1.4 T is directed along the x axis. The rotation is initiated when the plane of the coil is perpendicular to the direction of the magnetic field. Calculate (a) the maximum induced emf in the coil, (b) the value of the induced emf at $t = 0.05$ s, and (c) the torque exerted on the coil by the magnetic field at the instant when the emf is a maximum.

76. A horizontal wire is free to slide on the vertical rails of a conducting frame as in Figure 21.45. The wire has a mass m and length ℓ, and the resistance of the circuit is R. If a uniform magnetic field is directed perpendicular to the frame, what is the terminal velocity of the wire as it falls under the force of gravity? (Neglect friction.)

77. A thin metal strip is allowed to slide down parallel frictionless rails of negligible resistance connected at the bottom end and elevated at an angle of 30° above the horizontal as in Figure 21.46. A uniform magnetic field of 2.60 T is directed vertically upward throughout the region. The strip has a mass $m = 40$ g, resistance $R = 30$ Ω, and length between the rails $\ell = 0.2$ m. Calculate the terminal speed achieved by the strip sliding along the incline.

Figure 21.45
Problem 76

Figure 21.46
Problem 77

Exponential Growth

Albert A. Bartlett
University of Colorado

If something is growing in size at a constant rate, such as 5% per year, we refer to the growth as steady growth. It is also referred to as exponential growth because the size, N, of the growing quantity at some time t in the future is related to its present size, N_0 (at time $t = 0$), by the exponential function

$$N = N_0 e^{kt} \qquad \text{(E.1)}$$

where $e = 2.718 \ldots$ is the base of the natural logarithms and k is the annual percent growth rate, P, divided by 100:

$$k = \frac{P}{100} \qquad \text{(E.2)}$$

Note that if k is positive, N increases exponentially with time (exponential growth). If k is negative, N decreases exponentially with time (exponential decay). Some examples of exponential decay are radioactive decay (Section 32.4) and the decay of the quantity of charge on the plates of a capacitor as the capacitor is discharging through a resistor (Section 19.5). This essay describes several examples of exponential growth.

We have steady growth whenever the rate of change of a quantity is proportional to the quantity,

$$\frac{\Delta N}{\Delta t} = kN \qquad \text{(E.3)}$$

Here ΔN is the small change in N during the short time interval Δt. If we can show that a quantity N obeys Equation E.3, then it follows automatically that N will increase in accord with Equation E.1.

If N is the number of dollars in a savings account, then ΔN is the interest added to the account in a time Δt, and k is the interest rate. If I have \$250 in the bank at an annual interest rate of 3% ($k = 0.03$), then the amount added to the account in $\Delta t = 1$ year is

$$\frac{\Delta N}{\Delta t} = 0.03 \times 250 = \$7.50$$

This is in accord with Equation E.3, so the sum in the account will grow according to the relation

$$N = 250 \, e^{0.03t}$$

and when $t = 1$ year this gives

$$N = 250 \times 1.03045 = 257.61$$

Note that if the interest is compounded once a year, at the end of the year the account will contain \$250 + \$7.50 = \$257.50. However, if the interest is "compounded continuously," the amount follows Equation E.1, and the amount at the end of the year is \$257.61. With the advent of computers, banks are able to compound their interest continuously or "daily."

It has been shown that the number of miles of highways in the United States obeys Equation E.3 so that the number of miles of highways will grow exponentially in accord with Equation E.1.[1]

---

[1] A. A. Bartlett, *Civil Engineering.* December 1969, pp. 71–72.

In steady growth, it takes a fixed length of time for a quantity to grow by a fixed fraction such as 5 percent. From this it follows that it takes a fixed longer length of time for that quantity to grow by 100 percent. Let us calculate the time required for the quantity N to double in value, which is called the **doubling time**, T_2. We can obtain an expression for T_2 by writing Equation E.1 as $N/N_0 = e^{kt}$ and taking the natural logarithm of each side:

$$\ln\left(\frac{N}{N_0}\right) = kt$$

If we set $N = 2N_0$ (that is, we double N_0), then T_2 (which is the time t when $N = 2N_0$) is

$$T_2 = \frac{\ln(2N_0/N_0)}{k} = \frac{\ln 2}{k} = \frac{0.693}{k}$$

Since $k = P/100$, this becomes[2]

$$T_2 \approx \frac{70}{P} \tag{E.4}$$

Example E.1 Compound Interest

Suppose you put \$1 in a savings account at 10 percent annual interest to be compounded continuously for 200 years. How much money would be in the account at the end of the period?

Solution
In this case, $N_0 = \$1$, $k = 10\%/100 = 0.10$ per year, and $t = 200$ years. Therefore, from Equation E.1 we have

$$N = \$1 \times e^{(0.10)(200)} = (\$1) \times e^{20} = \$4.85 \times 10^8$$
$$= \$485 \text{ million!}$$

Now you can see why a famous financier once said that he could not name the seven wonders of the world but surely the eighth wonder would have to be compound interest!

Example E.2 The Consequences of Inflation

Let us use the doubling time instead of the quantity e to estimate the consequences of an annual inflation rate of 14 percent that continued for 50 years.

Solution
First, we calculate T_2 from Equation E.3:

$$T_2 = \frac{70}{14} = 5 \text{ years}$$

This inflation rate will cause prices to double every five years!
    In the next step, we calculate the number of doubling times in 50 years:

$$\text{Number of doublings} = \frac{50 \text{ years}}{5 \text{ years/doubling}} = 10$$

---

[2] If you wanted to calculate T_3, the time it takes N to triple, it is $T_3 = (100 \ln 3)/P$.

Finally, we count up the consequence of each doubling by making use of the following table:

Number of Doublings	Price Increase Factor
1	$2 = 2^1$
2	$4 = 2^2$
3	$8 = 2^3$
4	$16 = 2^4$
5	$32 = 2^5$
6	$64 = 2^6$
7	$128 = 2^7$
8	$256 = 2^8$
9	$512 = 2^9$
10	$1024 = 2^{10}$

The table shows us that, in ten doubling times, prices will increase by a factor of 1024, which is approximately 1000. Thus, in 50 years of 14% annual inflation, the cost of a $4 ticket to the movies would increase to roughly $4000!

It is very convenient to remember that 10 doublings give an increase by a factor of 10^3, that 20 doublings give an increase of a factor of 10^6, that 30 doublings give an increase by a factor of 10^9, and so on.

Example E.3   The Increasing Rate of Energy Consumption

For many years before 1975, consumption of electrical energy in the United States grew steadily at a rate of about 7% per year. By what factor would consumption increase if this growth rate continued for 40 years?

Solution
In this case, $P = 7$, and so the doubling time from Equation E.3 is

$$T_2 = \frac{70}{7} = 10 \text{ years}$$

and the number of doublings in 40 years is

$$\text{Number of doublings} = \frac{40 \text{ years}}{10 \text{ years/doubling}} = 4$$

Therefore, in 40 years the amount of power consumed would be $2^4 = 16$ times the amount used today. That is, 40 years from now we would need 16 times as many electric generating plants as we have at the present. Furthermore, if those additional plants are similar to today's, then each day they would consume 16 times as much fuel as is used by our present plants and there would be 16 times as much pollution and waste heat to contend with!

Populations tend to grow steadily. In 1990 the population of the Earth had reached 5.3×10^9 people. The world birth rate was estimated to be 28 per 1000 each year, while the annual death rate was estimated to be 11 per 1000. Thus for every 1000 people, the population increase each year is $28 - 11 = 17$. For this growth rate we find

$$k = \frac{17}{1000} = 0.017 \text{ per year}$$

$$P = 100k = 1.7\% \text{ per year}$$

This growth rate seems so small that many people regard it as trivial and inconsequential. A proper perspective of this rate appears only when we calculate the doubling time:

$$T_2 = \frac{70}{1.7} = 41 \text{ years}$$

This simple calculation indicates that it is most likely that the world population will double within the life expectancy of today's students! At the most elemental level, this means that we have approximately 41 years to double world food production if we want the fraction of the population that is starving to be no larger than it is today. This would be twice as many people starving as are starving today.

Example E.4   Annual Increase in World Population

What is the annual increase in the Earth's population? Since $\Delta N \ll N$ we can get a good answer from Equation E.3.

$$\frac{\Delta N}{\Delta t} = 0.017 \times 5.3 \times 10^9 = 90 \text{ million per year}$$

This annual increase in the world population is roughly one third of the population of the United States.

Some illuminating calculations can be made based on the assumption that this rate of growth has been constant and will remain constant. These calculations will demonstrate that the growth rate has not been constant at this value in the past and cannot remain this high for very long.

Example E.5   When did Adam and Eve Live?

If we use Equation E.1, setting $N = 5.3 \times 10^9$, $N_0 = 2$ (Adam and Eve) and $k = 0.017$, we have

$$5.3 \times 10^9 = 2e^{0.017t}$$

This gives $t = 1276$ years ago, or about A.D. 714! This result proves that through essentially all of human history the population growth rate was very much smaller than it is today. It must have been near zero through most of human history.

Example E.6

The land area of the continents (excluding Antarctica) is 1.24×10^{14} m^2. How long would it take for the population to reach a density of one person per square meter on the continents?

$$1.24 \times 10^{14} = 5.3 \times 10^9 \, e^{0.017t}$$

Solving, we find t is slightly less than 600 years!

Figure 22.1
A series circuit consisting of a resistor R connected to an ac generator, designated by the symbol

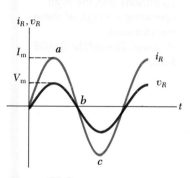

Figure 22.2
Plot of current and voltage across a resistor versus time.

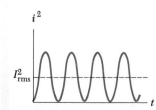

Figure 22.3
A plot of the square of the current in a resistor versus time. The rms current is the square root of the average of the square of the current.

arbitrarily called the positive direction. Between points a and b, the current is decreasing in magnitude but is still in the positive direction. At point b, the current is momentarily zero, and then begins to increase in the opposite (negative) direction between points b and c. At point c, the current has reached its maximum value in the negative direction.

Note that the current and voltage are in step with each other since they vary identically with time.

Since the current and the voltage reach their maximum values at the same time, they are said to be in phase.

Note that *the average value of the current over one cycle is zero.* That is, the current is maintained in one direction (the positive direction) for the same amount of time and at the same magnitude as it is in the opposite direction (the negative direction). However, the direction of the current has no effect on the behavior of the resistor in the circuit. This can be understood by realizing that collisions between electrons and the fixed atoms of the resistor result in an increase in the temperature of the resistor. Although this temperature increase depends on the magnitude of the current, it is independent of its direction.

This discussion can be made quantitative by recalling that the rate at which electrical energy is converted to heat in a resistor, which is the power P, is given by

$$P = i^2 R$$

where i is the instantaneous current in the resistor. Since the heating effect of a current is proportional to the *square* of the current, it makes no difference whether the current is direct or alternating, that is, whether the sign associated with the current is positive or negative. However, the heating effect produced by an alternating current having a maximum value of I_m *is not the same* as that produced by a direct current of the same value. This is explained by the fact that the alternating current is at this maximum value for only a very brief instant of time during a cycle. What is of importance in an ac circuit is an average value of current referred to as the rms current. The term **rms** refers to *root mean square*, which simply means that the square root of the average value of the square of the current is taken. Since i^2 varies as $\sin^2 2\pi ft$, one can show[1] that the average value of i^2 is $\frac{1}{2}I_m^2$ (Fig. 22.3). Therefore, the rms current, I, is related to the maximum value of the alternating current, I_m, as

---

[1] The fact that the square root of the average value of the square of the current is equal to $I_m/\sqrt{2}$ can be shown as follows. The current in the circuit varies with time according to the expression $i = I_m \sin 2\pi ft$, so that $i^2 = I_m^2 \sin^2 2\pi ft$. Therefore we can find the average value of i^2 by calculating the average value of $\sin^2 2\pi ft$. Note that a graph of $\cos^2 2\pi ft$ versus time is identical to a graph of $\sin^2 2\pi ft$ versus time, except that the points are shifted on the time axis. Thus, the time average of $\sin^2 2\pi ft$ is equal to the time average of $\cos^2 2\pi ft$ when taken over one or more cycles. That is,

$$(\sin^2 2\pi ft)_{av} = (\cos^2 2\pi ft)_{av}$$

With this fact and the trigonometric identity $\sin^2 \theta + \cos^2 \theta = 1$, we get

$$(\sin^2 2\pi ft)_{av} + (\cos^2 2\pi ft)_{av} = 2(\sin^2 2\pi ft)_{av} = 1$$

$$(\sin^2 2\pi ft)_{av} = \frac{1}{2}$$

When this result is substituted in the expression $i^2 = I_m^2 \sin^2 2\pi ft$, we get $(i^2)_{av} = I^2 = I_m^2/2$, or $I = I_m/\sqrt{2}$, where I is the rms current.

$$I = \frac{I_m}{\sqrt{2}} = 0.707 I_m \qquad \textbf{(22.2)} \quad \text{rms current}$$

This equation says that an alternating current whose maximum value is 2 A will produce the same heating effect in a resistor as a direct current of $\frac{2}{\sqrt{2}}$ A. Thus, we can say that the average power dissipated in a resistor that carries an alternating current I is $P_{av} = I^2R$, where I is the rms current.

Alternating voltages are also best discussed in terms of rms voltages, and the relationship here is identical to the above, that is, the rms voltage, V, is related to the maximum value of the alternating voltage, V_m, as

$$V = \frac{V_m}{\sqrt{2}} = 0.707 V_m \qquad \textbf{(22.3)} \quad \text{rms voltage}$$

When we speak of measuring an ac voltage of 120 V from an electric outlet, we are really referring to an rms voltage of 120 V. A quick calculation using Equation 22.3 shows that such an ac voltage actually has a peak value of about 170 V. In this chapter we shall use rms values when discussing alternating currents and voltages. One reason for this is that ac ammeters and voltmeters are designed to read rms values. Furthermore, we shall find that if we use rms values, many of the equations we use will have the same form as those used in the study of direct current (dc) circuits. Table 22.1 summarizes the notation that will be used in this chapter.

Consider the series circuit shown in Figure 22.1, consisting of a resistor connected to an ac generator. A resistor limits the current in an ac circuit in the same way it does in a dc circuit. Therefore, Ohm's law is valid for an ac circuit, and we have

$$V_R = IR \qquad \textbf{(22.4)}$$

That is, *the rms voltage across a resistor is equal to the rms current in the circuit times the resistance.* This equation also applies if maximum values of current and voltage are used. That is, the maximum voltage drop across a resistor is equal to the maximum current in the resistor times the resistance.

Table 22.1
Notation Used in This Chapter

	Voltage	Current
Instantaneous value	v	i
Maximum value	V_m	I_m
rms value	V	I

Example 22.1   What is the rms Current?

An ac voltage source has an output given by $v = (200\ \text{V})\sin 2\pi ft$. This source is connected to a 100-Ω resistor as in Figure 22.1. Find the rms current through the resistor.

Solution

Compare the expression for the voltage output given above with the general form, $v = V_m \sin 2\pi ft$. We see that the maximum output voltage of the device is 200 V. Thus, the rms voltage output of the source is

$$V = \frac{V_m}{\sqrt{2}} = \frac{200\ \text{V}}{\sqrt{2}} = 141\ \text{V}$$

Ohm's law can be used in resistive ac circuits as well as in dc circuits. The calculated rms voltage can be used with Ohm's law to find the rms current in the circuit:

$$I = \frac{V}{R} = \frac{141\ \text{V}}{100\ \Omega} = \boxed{1.41\ \text{A}}$$

Exercise   Find the maximum current in the circuit.

Answer   2 A.

22.2   Capacitors in an ac Circuit

In order to consider the effect of a capacitor on the behavior of a circuit containing an ac voltage source, let us first recall what happens when a capacitor is placed in a circuit containing a dc source, such as a battery. At the instant a switch is closed in a series circuit containing a battery, a resistor, and a capacitor, there is no charge on the plates of the capacitor. Therefore, the motion of charge through the circuit is relatively free, and initially there is a large current in the circuit. As more charge accumulates on the capacitor, the voltage across it increases, opposing the current. After some time interval has elapsed, which depends on the time constant, RC, the current approaches zero. From this, we see that a capacitor in a dc circuit will limit, or impede, the current such that it approaches zero after a brief time.

Now consider the simple series circuit shown in Figure 22.4, consisting of a capacitor connected to an ac generator. Let us sketch a curve of current versus time and one of voltage versus time and then attempt to make the graphs seem reasonable. These graphs are given in Figure 22.5. First, note

Figure 22.4
A series circuit consisting of a capacitor C connected to an ac generator.

Figure 22.5
Plots of the current and voltage across a capacitor versus time in an ac circuit. The voltage lags the current by 90°.

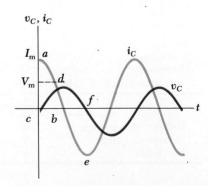

that the segment of the current curve from a to b indicates that the current starts out at a rather large value. This can be understood by recognizing that there is no charge on the capacitor at $t = 0$; consequently, there is nothing in the circuit except the resistance of the wires to hinder the flow of charge at this instant. However, the current decreases as the voltage across the capacitor increases, from c to d on the voltage curve. When the voltage is at point d, the current reverses and begins to increase in the opposite direction (from b to e on the current curve). During this time, the voltage across the capacitor decreases in going from d to f because the plates are now losing the charge that they had accumulated earlier. The remainder of the cycle for both voltage and current is a repeat of what happened during the first half cycle. The current reaches a maximum value in the opposite direction at point e on the current curve and then decreases as the voltage across the capacitor builds up.

Note that the current and voltage are not in step with each other, as they are in a purely resistive circuit. The curves of Figure 22.5 indicate that

> when an alternating voltage is applied across a capacitor, the voltage reaches its maximum value one quarter of a cycle after the current reaches its maximum value. In this situation, it is common to say that the voltage always lags the current through a capacitor by 90°.

The impeding effect of a capacitor to the flow of current in an ac circuit is expressed in terms of a factor called the **capacitive reactance**, X_C, defined as

$$X_C \equiv \frac{1}{2\pi f C}$$

(22.5)   Capacitive reactance

It is left as a problem to show that when C is in farads and f is in hertz, the unit of X_C is the ohm.

Let us examine whether or not Equation 22.5 is reasonable. If we have a dc source (a dc source can be considered as an ac source with zero frequency), X_C is infinitely large. This means that a capacitor will impede the flow of direct current the same way a resistor of infinitely large resistance would. The current in such a circuit is zero. Indeed, we found that to be the case in Chapter 17. On the other hand, Equation 22.5 predicts that as the frequency increases, the capacitive reactance decreases. This means that before the charge on a capacitor has time to build up to the point where the current is zero, the direction of current flow has reversed.

The analogy between capacitive reactance and resistance allows us to write an equation that has the same form as Ohm's law to describe ac circuits containing capacitors. This equation relates the rms voltage and rms current in the circuit to the reactance as

$$V_C = I X_C$$

(22.6)

Example 22.2   A Purely Capacitive ac Circuit

An 8-μF capacitor is connected to the terminals of an ac generator whose rms voltage is 150 V and whose frequency is 60 Hz. Find the capacitive reactance and the rms current in the circuit.

Solution

From Equation 22.5 and the fact that $2\pi f = 337 \text{ s}^{-1}$, we have

$$X_C = \frac{1}{2\pi fC} = \frac{1}{(377 \text{ s}^{-1})(8 \times 10^{-6} \text{ F})} = \boxed{332 \ \Omega}$$

If we substitute this result into Equation 22.6, we find that

$$I = \frac{V_C}{X_C} = \frac{150 \text{ V}}{332 \ \Omega} = \boxed{0.452 \text{ A}}$$

Exercise   If the frequency is doubled, what happens to the capacitive reactance and the current?

Answer X_C is halved, and I is doubled.

22.3   Inductors in an ac Circuit

$$v = V_m \sin 2\pi ft$$

Figure 22.6
A series circuit consisting of an inductor L connected to an ac generator.

Inductive reactance

Now consider an ac circuit consisting only of an inductor connected to the terminals of an ac generator, as in Figure 22.6. (In any real circuit, there is some resistance in the wire forming the inductive coil, but we ignore this for now.) The changing current output of the generator produces a back emf in the coil of magnitude

$$v_L = L \frac{\Delta I}{\Delta t} \tag{22.7}$$

Thus, the current in the circuit is impeded by the back emf of the inductor. The effective resistance of the coil in an ac circuit is measured by a quantity called the **inductive reactance**, X_L, defined by

$$X_L \equiv 2\pi fL \tag{22.8}$$

It is left as a problem to show that when f is in hertz and L is in henries, the unit of X_L is the ohm. Note that the inductive reactance increases with increasing frequency and increasing inductance.

In order to understand the meaning of inductive reactance, let us compare this equation for X_L with Equation 22.7. First, note from Equation 22.8 that the inductive reactance depends on the inductance, L. This seems reasonable because the back emf (Eq. 22.7) is large for large values of L. Second, note that the inductive reactance depends on the frequency, f. This, too, seems reasonable because the back emf depends on $\Delta I/\Delta t$, a quantity that is large when the current is changing at a rapid rate, as it would be for large frequencies.

With inductive reactance defined in this manner, we can write an equation for the voltage across the coil or inductor that has the same form as Ohm's law:

$$V_L = IX_L \tag{22.9}$$

where V_L is the rms voltage drop across the coil and I is the rms current in the coil.

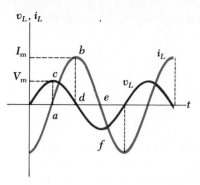

Figure 22.7
Plots of the current and voltage across an inductor versus time in an ac circuit. The voltage leads the current by 90°.

The instantaneous voltage and instantaneous current across the coil as functions of time are shown in Figure 22.7.

> When a sinusoidal voltage is applied across an inductor, the voltage reaches its maximum value one quarter of an oscillation period before the current reaches its maximum value. In this situation, we say that the voltage always leads the current by 90°.

To understand why this phase relationship between voltage and current should exist, let us examine a few points on the curves of Figure 22.7. Note that at point a on the current curve, the current is beginning to increase in the positive direction. At this instant, the rate of change of current is at a maximum, and consequently we see from Equation 22.7 that the voltage across the inductor is also at a maximum at this time. As the current rises, between points a and b on the curve, $\Delta I / \Delta t$ (the slope of the current curve) gradually decreases until it reaches zero at point b. As a result, the voltage across the inductor is decreasing during this same time interval, as the segment between c and d on the voltage curve indicates. Immediately after point b, the current begins to decrease, although it is still in the same direction as during the previous quarter cycle. As the current decreases to zero (from b to e on the curve), a voltage is again induced in the coil (d to f), but the sense of this voltage is opposite the sense of the voltage induced between c and d. This occurs because back emfs are always directed to oppose the change in the current. We could continue to examine other segments of the curves, but no new information would be gained since the current and voltage variations are repetitive.

Example 22.3   A Purely Inductive ac Circuit

In a purely inductive ac circuit (see Fig. 22.6), $L = 25$ mH and the rms voltage is 150 V. Find the inductive reactance and rms current in the circuit if the frequency is 60 Hz.

Solution
First, note that $2\pi f = 2\pi(60) = 377$ s^{-1}. Equation 22.8 then gives

$$X_L = 2\pi f L = (377 \text{ s}^{-1})(25 \times 10^{-3} \text{ H}) = \boxed{9.43 \ \Omega}$$

Substituting this result into Equation 22.9 gives

$$I = \frac{V_L}{X_L} = \frac{150 \text{ V}}{9.43 \ \Omega} = \boxed{15.9 \text{ A}}$$

Figure 22.8
A series circuit consisting of a resistor, an inductor, and a capacitor connected to an ac generator.

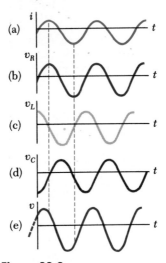

Figure 22.9
Phase relations in the series *RLC* circuit shown in Figure 22.8.

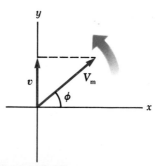

Figure 22.10
Phasor diagram for the voltage in an ac circuit, where ϕ is the phase angle between the voltage and current, and v is the instantaneous voltage.

Exercise   Calculate the inductive reactance and rms current in the circuit if the frequency is 6 Hz.

Answer $X_L = 943\ \Omega$, $I_m = 0.159$ A.

22.4   The *RLC* Series Circuit

In the previous sections, we examined the effects that an inductor, a capacitor, and a resistor have when connected separately across an ac voltage source. We shall now consider what happens when combinations of these devices are used.

Figure 22.8 shows a circuit containing a resistor, an inductor, and a capacitor connected in series across an ac generator. The current in the circuit varies sinusoidally with time, as indicated in Figure 22.9a. Thus,

$$i = I_m \sin 2\pi ft$$

Earlier we learned that the voltage across each element may or may not be in phase with the current. The instantaneous voltages across the three elements, shown in Figure 22.9, have the following phase relations to the instantaneous current:

1. The instantaneous voltage across the resistor, v_R, is *in phase* with the instantaneous current. (See Fig. 22.9b.)
2. The instantaneous voltage across the inductor, v_L, *leads* the current by 90°. (See Fig. 22.9c.)
3. The instantaneous voltage across the capacitor, v_C, lags the current by 90°. (See Fig. 22.9d.)

The net instantaneous voltage, v, across all three elements is the sum of the instantaneous voltages across the separate elements:

$$v = v_R + v_C + v_L \tag{22.10}$$

This net voltage can be obtained graphically as shown in Figure 22.9e. Note that the instantaneous voltage across the resistor (Fig. 22.9b) is in phase with the current, the instantaneous voltage across the inductor (Fig. 22.9c) leads the current by a quarter cycle, and the instantaneous voltage across the capacitor (Fig. 22.9d) lags the current by a quarter cycle. Because both v_L and v_C are out of phase with v_R, the net voltage is a complicated function of the voltage amplitudes across the individual elements.

Rather than following the analytical procedure of adding the individual instantaneous voltages to find the net voltage, it is helpful to use another technique involving vectors. In this technique, we represent the voltage across each element by a rotating vector as in Figure 22.10. Such a diagram is called a **phasor diagram**, and the rotating vectors are referred to as phasors. This particular diagram represents the circuit voltage given by the expression $v = V_m \sin(2\pi ft + \phi)$, where V_m is the maximum voltage (the amplitude of the phasor) and ϕ is the angle between the phasor and the $+x$ axis. The phasor can be viewed as a vector of magnitude V_m rotating at a constant frequency f, such that its projection along the y axis is the instantaneous voltage in the circuit. Since ϕ is the phase angle between the voltage and current in the circuit, the phasor for the current (not shown in Figure 22.10) would lie along the $+x$ axis, and would be expressed by the relation $i = I_m \sin(2\pi ft)$.

The phasor diagrams given in Figure 22.11 are useful for analyzing the *series RLC* circuit. These diagrams are plotted such that voltages in phase with the current are represented by vectors along the $+x$ axis, while voltages out of phase with the current lie along other axes. Thus, V_R is horizontal and to the right because the voltage across the resistor is in phase with the current. Likewise, V_L is represented by a phasor along the $+y$ axis since V_L leads the current by 90°. Finally, V_C is along the $-y$ axis since V_C lags the current by 90°. If the phasors are added as vector quantities, Figure 22.11a shows that the only x component for the voltages is V_R, while the net y component is $V_L - V_C$. It is convenient to now add the phasors vectorially as shown in Figure 22.11b, where V is the total rms voltage in the circuit. The right triangle in Figure 22.11b gives the following equations for the total voltage and phase angle:

$$V = \sqrt{V_R{}^2 + (V_L - V_C)^2} \tag{22.11}$$

$$\tan \phi = \frac{V_L - V_C}{V_R} \tag{22.12}$$

where all voltages are rms values. Note that although we choose to use rms voltages in our analysis, the equations above apply equally well to peak voltages because the two quantities are related to each other by the same factor for all circuit elements. The result for the total voltage V, as given by Equation 22.11, reinforces the fact that *the rms voltages across the resistor, capacitor, and inductor are not in phase, so one cannot simply add the individual voltages across the separate elements to get the voltage across the combination of elements.*

We can write Equation 22.11 in the form of Ohm's law using the relations $V_R = IR$, $V_L = IX_L$, and $V_C = IX_C$, where I is the rms current in the circuit:

$$V = I\sqrt{R^2 + (X_L - X_C)^2} \tag{22.13}$$

It is convenient to define a parameter called the **impedance**, Z, of the circuit as

$$Z \equiv \sqrt{R^2 + (X_L - X_C)^2} \tag{22.14}$$ **Impedance**

so that Equation 22.13 becomes

$$V = IZ \tag{22.15}$$

Note that Equation 22.15 is in the form of Ohm's law, $V = IR$, where R is replaced by the impedance, which has the unit of ohm. Equation 22.15 can be regarded as a generalized form of Ohm's law applied to a series ac circuit. Note that the current in the circuit depends upon the resistance, the inductance, the capacitance, *and* the frequency since the reactances are frequency-dependent.

It is useful to represent the impedance Z by a vector diagram, as shown in Figure 22.11c. A right triangle is constructed whose right side is the quantity

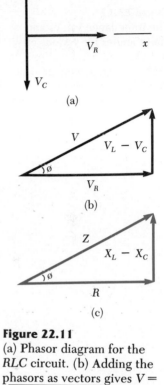

Figure 22.11
(a) Phasor diagram for the *RLC* circuit. (b) Adding the phasors as vectors gives $V = \sqrt{V_R{}^2 + (V_L - V_C)^2}$. (c) The reactance triangle which gives the impedance relation, $Z = \sqrt{R^2 + (X_L - X_C)^2}$.

Figure 22.12
The impedance values and phase angles for various circuit element combinations. In each case, an ac voltage (not shown) is applied across the combination of elements (that is, across the dots).

Circuit Elements	Impedance, Z	Phase angle, ϕ
R	R	$0°$
C	X_C	$-90°$
L	X_L	$+90°$
$R \quad C$	$\sqrt{R^2 + X_C^2}$	Negative, between $-90°$ and $0°$
$R \quad L$	$\sqrt{R^2 + X_L^2}$	Positive, between $0°$ and $90°$
$R \quad L \quad C$	$\sqrt{R^2 + (X_L - X_C)^2}$	Negative if $X_C > X_L$ Positive if $X_C < X_L$

$X_L - X_C$, whose base is R, and whose hypotenuse is Z. Applying the pythagorean theorem to this triangle, we see that

$$Z = \sqrt{R^2 + (X_L - X_C)^2}$$

which is consistent with Equation 22.14. Furthermore, we see from the vector diagram that the phase angle, ϕ, between the current and the voltage is given by

Phase angle ϕ

$$\tan \phi = \frac{X_L - X_C}{R} \tag{22.16}$$

The physical significance of the phase angle will become apparent in Sections 22.5 and 22.6.

Figure 22.12 gives impedance values and phase angles for various series circuits containing different combinations of circuit elements.

There are also many parallel alternating current circuits that are useful in everyday applications. We shall not discuss them here, however, because the analysis of such circuits is beyond the scope of this text.

Problem-Solving Strategy:
Alternating Current

The following procedures are recommended when solving alternating current problems:

1. The first step in analyzing alternating current circuits is to calculate as many of the unknown quantities such as X_L and X_C as possible. (Note that when calculating X_C, the capacitance should be expressed in farads, rather than, say, microfarads.)
2. Apply the equation $V = IZ$ to that portion of the circuit of interest. For example, if you want to know the voltage drop across the combination of an inductor and a resistor, the equation for the voltage drop reduces to $V = I\sqrt{R^2 + X_L^2}$.

Example 22.4   Analyzing a Series *RLC* ac Circuit

Analyze a series *RLC* ac circuit for which $R = 250\ \Omega$, $L = 0.6$ H, $C = 3.5\ \mu$F, $f = 60$ Hz, and $V = 150$ V.

Solution

The reactances are given by $X_L = 2\pi fL = 226\ \Omega$ and $X_C = 1/2\pi fC = 758\ \Omega$. Therefore, the impedance is

$$Z = \sqrt{R^2 + (X_L - X_C)^2} = \sqrt{(250\ \Omega)^2 + (226\ \Omega - 758\ \Omega)^2} = 588\ \Omega$$

The rms current is

$$I = \frac{V}{Z} = \frac{150\ \text{V}}{588\ \Omega} = 0.255\ \text{A}$$

The phase angle between the current and voltage is

$$\phi = \tan^{-1}\left(\frac{X_L - X_C}{R}\right) = \tan^{-1}\left(\frac{226\ \Omega - 758\ \Omega}{250\ \Omega}\right) = -64.8°$$

Since the circuit is more capacitive than inductive (that is, $X_C > X_L$), ϕ is negative. A negative phase angle means that the current leads the applied voltage.

The rms voltages across the elements are

$$V_R = IR = (0.255\ \text{A})(250\ \Omega) = 63.8\ \text{V}$$

$$V_L = IX_L = (0.225\ \text{A})(226\ \Omega) = 57.6\ \text{V}$$

$$V_C = IX_C = (0.225\ \text{A})(758\ \Omega) = 193\ \text{V}$$

Note that the sum of the three rms voltages, $V_R + V_L + V_C$, is 314 V, which is much larger than the rms voltage of the generator, 150 V. The sum of 314 V is a meaningless quantity because, when alternating voltages are added, *both their amplitude and their phase* must be taken into account. That is, the voltages must be added in a way that takes account of the different phases. The relationship between V, V_R, and V_C is given by Equation 22.11. You should use the values found above to verify this equation.

Nikola Tesla (1856–1943) was born in Croatia (now part of Yugoslavia) but spent most of his professional life as an inventor in the United States. He was a key figure in the development of alternating current electricity, high voltage transformers, and the transport of electrical power using ac transmission lines. Tesla's approach was at odds with the ideas of Edison, who committed himself to the use of dc current in power transmission. After much bitterness between the two inventors, Tesla's ac approach for power transmission was the winner. *(Courtesy of Tesla Memorial Society, Inc.)*

22.5   Power in an ac Circuit

As we shall see in this section, *there are no power losses associated with capacitors and pure inductors in an ac circuit.* (A pure inductor is defined as one with no resistance or capacitance.) First, let us analyze the power dissipated in an ac circuit containing only a generator and a capacitor.

When the current begins to increase in one direction in an ac circuit, charge begins to accumulate on the capacitor and a voltage drop appears across it. When the voltage drop across the capacitor reaches its maximum value, the energy stored in the capacitor is

$$PE_C = \tfrac{1}{2}CV_{\text{m}}^2$$

However, this energy storage is only momentary. When the current reverses direction, this charge leaves the capacitor plates. In this discharging process, the charge returns to the voltage source. Thus in one cycle, the capacitor is being charged half of the time and the charge is being returned to the voltage source during the other half of the time. Therefore, *the average power supplied*

These high-voltage transmission lines typically operate at 750 kV. *(Robert Kristofik/The IMAGE Bank)*

by the source is zero. In other words, *a capacitor in an ac circuit does not dissipate energy.*

Similarly, the source must do work against the back emf of an inductor, which carries a current. When the current reaches its maximum value, the energy stored in the inductor is a maximum and is given by

$$PE_L = \tfrac{1}{2}LI_m{}^2$$

When the current begins to decrease in the circuit, this stored energy is returned to the source as the inductor attempts to maintain the current in the circuit.

The only element in an *RLC* circuit that dissipates energy is the resistor. The average power lost in a resistor is

$$P_{av} = I^2R \qquad (22.17)$$

where *I* is the rms current in the circuit. An alternative equation for the average power dissipated in an ac circuit can be found by substituting (from Ohm's law) $R = V_R/I$ into Equation 22.17:

$$P_{av} = IV_R$$

It is convenient to refer to the voltage triangle that shows the relationship between V, V_R, and $V_L - V_C$ as in Figure 22.10b. From this figure, we see that the voltage drop across a resistor can be written in terms of the voltage of the source:

$$V_R = V \cos \phi$$

Thus, the average power dissipated in an ac circuit is

Average power

$$P_{av} = IV \cos \phi \qquad (22.18)$$

where the quantity $\cos \phi$ is called the **power factor**.

Example 22.5   Calculate the Average Power

Calculate the average power delivered to the series *RLC* circuit described in Example 22.4.

Solution
We are given that the rms voltage supplied to the circuit is 150 V, and we have calculated that the rms current in the circuit is 0.255 A and the phase angle ϕ is $-64.8°$. Thus the power factor, $\cos \phi$, is 0.426. From these values, we calculate the average power using Equation 22.18:

$$P_{av} = IV \cos \phi = (0.255 \text{ A})(150 \text{ V})(0.426) = \boxed{16.3 \text{ W}}$$

The same result can be obtained using Equation 22.17.

22.6   Resonance in a Series *RLC* Circuit

In general, the current in a series *RLC* circuit can be written as

$$I = \frac{V}{Z} = \frac{V}{\sqrt{R^2 + (X_L - X_C)^2}} \qquad \textbf{(22.19)}$$

From this, we see that the current has its *maximum* value when the impedance has its *minimum* value. This occurs when $X_L = X_C$. In such a circumstance, the impedance of the circuit reduces to $Z = R$. The frequency, f_0, at which this occurs is called the **resonance frequency** of the circuit. To find f_0, we set $X_L = X_C$, which gives, from Equations 22.5 and 22.8,

$$2\pi f_0 L = \frac{1}{2\pi f_0 C}$$

$$f_0 = \frac{1}{2\pi\sqrt{LC}} \qquad \textbf{(22.20)}$$

Resonance frequency

A plot of current as a function of frequency is shown in Figure 22.13 for a circuit containing a fixed value of capacitance and inductance. From Equation 22.19, it must be concluded that the current would become infinite at resonance when $R = 0$. Although Equation 22.19 predicts this result, real circuits always have some resistance, which limits the value of the current.

The receiving circuit of a radio is an important application of a series resonance circuit. The radio is tuned to a particular station (which transmits a specific radio-frequency signal) by varying a capacitor, which changes the resonance frequency of the receiving circuit. When the resonance frequency of the circuit matches that of the incoming radio wave, the current in the receiving circuit increases. The techniques for removing the information from the radio wave will be discussed in Chapter 23.

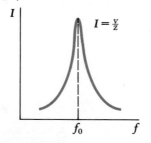

Figure 22.13
Current amplitude in a series *RLC* circuit plotted versus frequency of the generator voltage. Note that the current reaches its maximum value at the resonance frequency, f_0.

Example 22.6   The Capacitance of a Circuit in Resonance

Consider a series *RLC* circuit for which $R = 150\ \Omega$, $L = 20$ mH, $V = 20$ V, and $2\pi f = 5000\ \text{s}^{-1}$. Determine the value of the capacitance for which the rms current is a maximum.

Solution
The current is a maximum at the resonance frequency, f_0, which should be made to match the driving frequency of 5000 s^{-1} in this problem:

$$2\pi f_0 = 5 \times 10^3\ \text{s}^{-1} = \frac{1}{\sqrt{LC}}$$

$$C = \frac{1}{(25 \times 10^6\ \text{s}^{-2})L} = \frac{1}{(25 \times 10^6\ \text{s}^{-2})(20 \times 10^{-3}\ \text{H})} = \boxed{2.0\ \mu\text{F}}$$

Exercise   Calculate the maximum rms current in the circuit.

Answer   0.133 A.

Figure 22.14
An ideal transformer consists of two coils wound on the same soft iron core. An ac voltage V_1 is applied to the primary coil, and the output voltage V_2 is observed across the load resistance R.

*22.7   The Transformer

There are many situations in which it is desirable or necessary to change a small ac voltage to a larger one or vice versa. Before we examine a few cases in which such changes are made, let us consider the device that makes such conversions possible, the **ac transformer.**

In its simplest form, the ac transformer consists of two coils of wire wound around a core of soft iron, as in Figure 22.14. The coil on the left, which is connected to the input ac voltage source and has N_1 turns, is called the *primary* winding (or primary). The coil on the right, which is connected to a resistor R and consists of N_2 turns, is called the *secondary*. The purpose of the common iron core is to increase the magnetic flux and to provide a medium in which nearly all the flux through one coil passes through the other.

When an input ac voltage V_1 is applied to the primary, the induced voltage across it is given by

$$V_1 = -N_1 \frac{\Delta \Phi}{\Delta t} \qquad \textbf{(22.21)}$$

where Φ is the magnetic flux through each turn. If we assume that no flux leaks from the iron core, then the flux through each turn of the primary equals the flux through each turn of the secondary. Hence, the voltage across the secondary coil is

$$V_2 = -N_2 \frac{\Delta \Phi}{\Delta t} \qquad \textbf{(22.22)}$$

The term $\Delta \Phi / \Delta t$ is common to Equations 22.21 and 22.22. Thus, we see that

$$V_2 = \frac{N_2}{N_1} V_1 \qquad \textbf{(22.23)}$$

When N_2 is greater than N_1, and thus V_2 exceeds V_1, the transformer is referred to as a *step-up transformer*. When N_2 is less than N_1, making V_2 less than V_1, we speak of a *step-down transformer*.

It should be clear that a voltage is generated across the secondary only when there is a *change* in the number of flux lines passing through it. Thus, the input current in the primary must change with time, which is what happens when an alternating current is used. However, when the input at the primary is a direct current, there is a voltage output at the secondary only at the instant when a switch in the primary circuit is opened or closed. Once the current in the primary reaches a steady value, the output voltage at the secondary is zero.

At first thought, it seems that a transformer is a device in which it is possible to get something for nothing. For example, a step-up transformer can change an input voltage from, say, 10 V to 100 V. This says that each 1 C of charge leaving the secondary has 100 J of energy while each 1 C of charge entering the primary has only 10 J of energy. However, this is not an example of a breakdown in the law of conservation of energy since *the power input to the primary is equal to the power output at the secondary*, that is,

$$I_1 V_1 = I_2 V_2 \qquad \textbf{(22.24)}$$

Coil

To other
spark plugs

12 V

Breaker
points

Distributor

Spark
plug

Figure 22.15
Circuit diagram of an automobile ignition system.

Thus, we see that if the voltage at the secondary is ten times that at the primary, the current at the secondary is reduced by a factor of 10. Equation 22.24 assumes an **ideal transformer,** in which there are no power losses between the primary and the secondary. Typical transformers have power efficiencies ranging from 90% to 99%. Power losses are due to such factors as eddy currents induced in the iron core of the transformer. These currents dissipate energy in the form of an I^2R loss.

When electric power is transmitted over large distances, it is economical to use a high voltage and low current because the power lost via resistive heating in the transmission lines varies as I^2R. This means that if a utility company can reduce the current by a factor of 10, for example, the power loss will be reduced by a factor of 100. In practice, the voltage is stepped up to around 230,000 V at the generating station, then stepped down to around 20,000 V at a distribution station, and finally stepped down to 120 V at the customer's utility pole.

A situation in which a transformer is used with a dc source is the electrical system of automobiles. In an automobile engine, spark plugs produce a spark to ignite a gasoline-air mixture in the cylinders. In order for a plug to work, there must be a voltage high enough to cause the spark to jump across the gap in the plug. The technique for creating this high voltage is indicated in Figure 22.15. The primary coil of the transformer is connected to the 12-V battery of the car. The secondary is connected to the spark plugs through the distributor. The distributor is a mechanical device that connects, in turn, each spark plug to the secondary coil at the instant the plug is supposed to fire. At this instant, a switch (called the "breaker points" or simply the "points") breaks the circuit between the battery and the primary. This interrupts the current in the primary and causes a voltage to be induced in the secondary. The transformer used is a step-up transformer, and the resulting voltage in the secondary is quite high (about 20,000 V), high enough to cause a spark to jump when the voltage is applied across the gap of the spark plug.

Example 22.7 A Step-Up Transformer

A generator produces 10 A of current at 400 V. The voltage is stepped up to 4500 V by an ideal transformer and transmitted a long distance through a power line of total resistance 30 Ω. (a) Determine the percentage of power lost.

$$\tan \phi = \frac{X_L - X_C}{R} \qquad (22.16)$$

The **average power** delivered by the generator in an *RLC* ac circuit is

$$P_{av} = IV \cos \phi \qquad (22.18)$$

where the constant $\cos \phi$ is called the **power factor**.

Additional Reading

L. Barthold and H. G. Pfeiffer, "High-Voltage Transmission," *Sci. American*, May 1964, p. 38.

T. Geballe and J. K. Hulm, "Superconductors in Electric-Power Technology," *Sci. American*, November 1980, p. 138.

A. L. Hammond, W. D. Metz, and T. H. Maugh, *Energy and the Future,* American Association for the Advancement of Science, 1973.

J. J. O'Neill, *Prodigal Genius, The Life of Nikola Tesla,* Ives Washburn, 1944.

Questions

1. How can the average value of a current be zero and yet the square root of the average value squared not be zero?

2. Do ac ammeters and voltmeters read peak, rms, or average values?

3. Explain why the reactance of a capacitor decreases with increasing frequency whereas the reactance of an inductor increases with increasing frequency.

4. What is meant by the statement, "The voltage across an inductor leads the current by 90°"?

5. A memory device often used with alternating current circuits is ELI the ICE man. ICE means that in a capacitive circuit, the current leads the voltage. What does ELI mean?

6. "At high frequencies a capacitor looks like a short circuit." Is this statement true? Discuss. Explain what is meant by high frequencies.

7. If it were not for the eye's persistence of vision, one could see a lightbulb operated with an ac voltage flicker on and off. Would you see 60 or 120 flickers per second?

8. Would an inductor and a capacitor used together in an ac circuit dissipate any power?

9. If the frequency is doubled in a series *RLC* circuit, what happens to the resistance, the inductive reactance, and the capacitive reactance?

10. Is the voltage applied to a circuit always in phase with the current through a resistor in the circuit?

11. In a series *RLC* circuit, what is the possible range of values for the phase angle?

12. Does the phase angle depend on frequency? What is the phase angle when the inductive reactance equals the capacitive reactance?

13. What is the impedance of an *RLC* circuit at the resonance frequency?

14. Show that the maximum current in an *RLC* circuit occurs when the circuit is in resonance.

15. If you wished to make a lamp dimmer with a coil of a variable inductance, how might you accomplish this? Would the dimmer work with dc?

16. A series RLC circuit models a driven harmonic oscillator. If the inductance represents the mass of the oscillator, what do the capacitance and resistance represent? Discuss.

17. Will a transformer operate if a battery is used for the input voltage across the primary?

18. Why are the primary and secondary coils of a transformer wrapped on an iron core that passes through both coils?

19. The tapped coil in Figure 22.16 acts as an autotransformer. This device steps voltage up or down with a single coil. Explain how this is possible. Does the coil shown step the voltage up or down? What modifications would be needed to do the opposite?

Figure 22.16   Question 19

Figure 22.17   Question 23

20. Why do power lines carry electrical energy at several thousand volts potential, but it is always stepped down to 240 V or 120 V as it enters your home?

21. What determines the peak voltage that can be used on a transmission line?

22. At one time there were debates in this country concerning whether we should use ac or dc for transmitting electrical power. What disadvantage would dc have in the transmission of power over a long distance?

23. With reference to Figure 22.17, explain why the capacitor prevents a dc voltage from passing between A and B, yet allows an ac signal to pass from A to B. (The circuits are said to be capacitively coupled.)

24. With reference to Figure 22.18, one finds that if C is made sufficiently large, an ac signal passes from A to ground rather than into B. Hence, the capacitor acts as a filter. Explain.

Figure 22.18   Question 24

25. A night-watchman is fired by his boss for being wasteful and keeping all the lights on in the building. The night-watchman defends himself by claiming that the building is electrically heated, so his boss's claim is unfounded. Who should win the argument if this were to end up in a court of law?

Problems

Section 22.1   Resistors in an ac Circuit

1. An rms voltage of 100 V is applied to a device that has a resistance of 5 Ω. Find (a) the maximum voltage applied, (b) the maximum current, and (c) the rms current supplied.

2. What are the maximum values of a 6-A rms current and a 240-V rms voltage?

3. If we plug a toaster into a source of alternating voltage with an rms value of 110 V, an rms current of 7.5 A occurs in the heating element. What is (a) the resistance of the toaster's heating element? (b) the average power input to the toaster?

4. (a) What is the resistance of a lightbulb that uses an average power of 75 W when connected to a 60-Hz power source with a rms voltage of 120 V? (b) What is the resistance of a 100-W bulb?

5. The circuit in Figure 22.19 shows three lamps connected to the 110-V ac (rms) household supply voltage. Lamps 1 and 2 have 150-W bulbs; lamp 3 has a 100-W bulb (average power). Find the rms value of the current I and the resistance of each of the bulbs.

6. An audio amplifier represented by the ac source and the resistor R in Figure 22.20 delivers alternating volt-

Figure 22.19   Problem 5

Figure 22.20   Problem 6

ages at audio frequencies to the speaker. If the source puts out an alternating voltage with an amplitude of 15.0 V, the resistance R is 8.2 Ω, and the speaker is equivalent to a resistance of 10.4 Ω, what is the time-averaged power input to the speaker?

7. An ac voltage source has an output given by $v = 150 \sin 377t$. Find (a) the rms voltage output, (b) the frequency of the source, and (c) the voltage at $t = 1/120$ s. (d) Find the maximum current in the circuit when the generator is connected to a 50-Ω resistor.

8. For a particular ac generator, the output voltage is $v = 0.25\ V_m$ when $t = 0.002$ s. When is the operating frequency of the generator?

Section 22.2   Capacitors in an ac Circuit

9. Show that the SI unit of capacitive reactance is the ohm.

10. A certain capacitor in a circuit has a capacitive reactance of 30 Ω when the frequency is 120 Hz. What capacitive reactance does the capacitor have at a frequency of 10,000 Hz?

11. What rms current will be delivered by an ac generator whose rms output voltage is 48 V at a frequency of 90 Hz when connected across a 3.7-μF capacitor?

12. The current in a circuit containing a 4-μF capacitor is 0.3 A when connected to a generator whose rms output is 30 V. What is the frequency of the source?

13. The generator in a purely capacitive ac circuit has an angular frequency of $120\ \pi$ rad/s. If $V_m = 140$ V and $C = 6\ \mu$F, what is the rms current in the circuit?

14. What value of capacitor must be inserted in a 60-Hz circuit in series with a generator of 170 V maximum output voltage to produce an rms current output of 0.75 A?

Section 22.3   Inductors in an ac Circuit

15. Find the inductive reactance of a 3-mH inductor at a frequency of (a) 10 Hz and (b) 10^6 Hz.

16. Show that the inductive reactance, X_L, has the SI unit of ohm.

17. A 2.4-μF capacitor is connected across an alternating voltage with an rms value of 9.0 V. The rms current through the capacitance is 25.0 mA. (a) What is the source frequency? (b) If the capacitor is replaced by an ideal coil with an inductance of 0.160 H, what is the rms current through the coil?

18. An inductor is connected to a 20-Hz power supply that produces a 50-V rms voltage. What inductance is needed to keep the maximum current in the circuit below 80 mA?

19. In a purely inductive circuit, the rms voltage is 120 V. (a) If the rms current is 10 A at a frequency of 60 Hz, calculate the inductance. (b) At what frequency will the rms current be reduced to 5 A?

20. In a purely inductive circuit, $L = 30$ mH, the frequency is 60 Hz, and the maximum voltage is 200 V.

Find (a) the rms current in the circuit. (b) Repeat for a frequency of 6000 Hz.

21. Calculate the inductance of a coil that has an inductive reactance of 40 Ω at an angular frequency of 754 rad/s.

22. An inductor has a 54-Ω reactance at 60 Hz. What will be the peak current if this inductor is connected to a 50-Hz source that produces a 100-V rms voltage?

Section 22.4   The *RLC* Series Circuit

23. An *RLC* circuit consists of a 150-Ω resistor, a 21-μF capacitor, and a 460-mH inductor, connected in series with a 120-V, 60-Hz power supply. (a) What is the phase angle between the current and the applied voltage? (b) Does the current or voltage reach its peak earlier?

24. A 40-μF capacitor is connected to a 50-Ω resistor and a generator whose rms output is 30 V at 60 Hz. What is (a) the rms current in the circuit, (b) the voltage drop across the resistor, (c) the voltage drop across the capacitor, and (d) the phase angle for the circuit? (e) Sketch the phasor diagram for this circuit.

25. A 10-μF capacitor and a 2-H inductor are connected in series with a 60-Hz source whose rms output is 100 V. Find (a) the rms current in the circuit, (b) the voltage drop across the inductor, (c) the voltage drop across the capacitor, and (d) the phase angle for the circuit. (e) Sketch the phasor diagram for this circuit.

26. At what frequency will the inductive reactance of a 57-μH inductance equal the capacitive reactance of a 57-μF capacitor?

27. A coil has an inductance of 320 mH and a resistance of 100 Ω, which can be considered to be in series. At what frequency is the impedance of the coil 141 Ω?

28. A pure 20-mH inductor is connected in series with a 20-Ω resistor and a 60-Hz, 100-V rms source. Find (a) the rms current in the circuit, (b) the voltage drop across the inductor, (c) the voltage drop across the resistor, and (d) the phase angle for this circuit. (e) Sketch the phasor diagram for this circuit.

29. A 60-Hz source with an rms output of 100 V is connected to a series combination of a 30-Ω resistor, a 3-μF capacitor, and a 2-H inductor. Find (a) the impedance of the circuit, (b) the rms current in the circuit, (c) the voltage drop across each element, and (d) the phase angle. (e) Sketch the phasor diagram for this circuit.

30. A 50-Ω resistor, a 0.1-H inductor, and a 10-μF capacitor are connected in series to a 60-Hz source. The rms current in the circuit is 2.75 A. Find the rms voltage across (a) the resistor, (b) the inductor, (c) the capacitor, and (d) the *RLC* combination. (e) Sketch the phasor diagram for this circuit.

31. A 60-Ω resistor, a 3-μF capacitor, and a 0.4-H inductor are connected in series to a 90-V, 60-Hz source. Find (a) the voltage drop across the *LC* combination and (b) the voltage drop across the *RC* combination.

32. A person is working near the secondary of a transformer, as shown in Figure 22.21. The primary voltage is 110 V at 60.0 Hz. The stray capacitance between the person's finger and the secondary winding C_s is 20.0 pF. Assuming the person has a body resistance to ground R_b of 50.0 kΩ, as indicated, determine the rms voltage across the person's body. (*Hint:* Redraw the circuit with the secondary of the transformer as a simple ac source.)

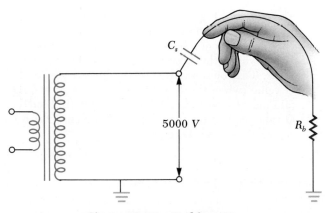

Figure 22.21   Problem 32

33. An ac source with a peak voltage of 150 V and $f = 50$ Hz is connected between points *a* and *d* in Figure 22.22. Calculate the *peak* voltages between points (a) *a* and *b*, (b) *b* and *c*, (c) *c* and *d*, (d) *b* and *d*.

$$a \overset{b}{\underset{40\ \Omega}{-\!\!\bigwedge\!\!\bigwedge\!\!-}} \overset{c}{\underset{185\ \text{mH}}{-\!\!\text{000}\!\!-}} \overset{d}{\underset{65\ \mu\text{F}}{-\!\!|\!|\!\!-}}$$

Figure 22.22   Problem 33

34. A 50-Ω resistor is connected in series with a 15-μF capacitor and a 60-Hz, 120-V source. (a) Find the current in the circuit. (b) What value inductor must be inserted in the circuit to reduce the current to one-half that found in (a)?

35. A 100-Ω resistor and a 3-μF capacitor are connected in series. What is the current in the circuit (a) when a dc source of 10 V is connected to the circuit and (b) when a 10-V, 60-Hz source is connected?

36. A 20-Ω resistor, a 40-μF capacitor, and a 0.2-H inductor are connected in series with a 60-Hz, 70-V source. The inductor is found to have a resistance of 15 Ω. Find (a) the current in the circuit and (b) the voltage drop across the inductor.

37. A resistor ($R = 900\ \Omega$), a capacitor ($C = 0.25\ \mu$F), and an inductor ($L = 2.5$ H) are connected in series across a 240-Hz ac source for which $V_m = 140$ V. Calculate the (a) impedance of the circuit, (b) peak current delivered by the source, and (c) phase angle between the current and voltage. (d) Is the current leading or lagging behind the voltage?

Section 22.5   Power in an ac Circuit

38. Consider the circuit described in Problem 37. (a) What is the power factor of the circuit? (b) What is the rms current in the circuit? (c) What average power is delivered by the source?

39. A series ac circuit contains a 50-Ω resistor, a 15-μF capacitor, a 0.2-H inductor, and a 60-Hz generator that has an rms output of 90 V. Find the average power delivered to the circuit.

40. A 50-Ω resistor is connected to a 30-μF capacitor and to a 60-Hz, 100-V rms source. (a) Find the power factor and the average power delivered to the circuit. (b) Repeat Part (a) when the capacitor is replaced with a 0.3-H inductor.

41. A series *RLC* circuit has a resistance of 80 Ω and an impedance of 180 Ω. What average power will be delivered to the circuit when the rms voltage is 120 V?

42. In a certain *RLC* circuit, the rms current is 6 A, the rms voltage is 240 V, and the current leads the voltage by 53°. (a) What is the total resistance of the circuit? (b) Calculate the total reactance $X_L - X_C$. (c) Find the average power dissipated in the circuit.

43. A series *RLC* circuit takes 108 W from a 110-V line. The applied voltage leads the voltage across the resistor by 37°. (a) What is the value of the current in the circuit? (b) What is the impedance of the circuit?

44. An inductor and a resistor are connected in series. When connected to a 60-Hz, 90-V source the voltage drop across the resistor is found to be 50 V and the power dissipated in the circuit is 14 W. Find (a) the value of the resistance and (b) the value of the inductance.

Section 22.6   Resonance in a Series *RLC* Circuit

45. An *RLC* circuit is used in a radio to tune into an FM station broadcasting at 99.7 MHz. The resistance in the circuit is 12 Ω and the inductance is 1.40 μH. What capacitance should be used?

46. A circuit consists of a source of alternating voltage with a constant amplitude of 5.0 V and a variable frequency in series with a resistance of 100 Ω, a capacitance of 2.25 μF, and an inductance of 0.25 H.

(a) Determine the rms current at frequencies of 10.0 Hz, 100 Hz, 1000 Hz, and 10,000 Hz. (b) Determine the phase of the source voltage relative to the current at these same frequencies. (c) Determine the frequency at which the rms current is at a maximum.

47. A 40-Ω resistor, a 2-μF capacitor, and a 2-mH inductor are connected in series to a source with an rms output of 60 V. Find (a) the frequency that would produce resonance, (b) the current at resonance, and (c) the current in the circuit when the frequency is one-half the resonance frequency.

48. A resonant circuit in a radio receiver is tuned to a certain station when the inductor has a value of 0.2 mH and the capacitor has a value of 30 pF. Find the frequency of the radio station and the wavelength sent out by the station.

49. The AM band extends from approximately 500 kHz to 1600 kHz. If a 2-μH inductor is used in a tuning circuit for a radio, what are the extremes that a capacitor must reach in order to cover the complete band of frequencies?

50. Figure 22.23 shows a tuning circuit for the AM broadcast frequencies, 500–1600 kHz. If the inductance of the coil is 100 μH and the fixed capacitance C_1 is 80.0 pF, what should be the range of the variable capacitance C_2?

Figure 22.23 Problem 50

51. The resonance frequency of a series RLC circuit is found to be 4.2×10^4 Hz. If the capacitance of the circuit is 300 pF, what is the value of the inductance?

52. A series circuit contains a 3-H inductor, a 3-μF capacitor, and a 30-Ω resistor connected to a 120-V rms source of variable frequency. Find the power delivered to the circuit when the frequency of the source is (a) the resonance frequency, (b) one-half the resonance frequency, (c) one-fourth the resonance frequency, (d) two times the resonance frequency, and (e) four times the resonance frequency. From your calculations can you draw a conclusion about the frequency at which the maximum power is delivered to the circuit?

53. The Q value of an RLC circuit is defined as the voltage drop across the inductor (or capacitor) at resonance divided by the voltage drop across the resistor. The larger the Q value, the sharper, or narrower, is the curve of power versus frequency. Figure 22.24 shows such curves for small R and for large R. (a) Show that the Q value is given by the expression $Q = 2\pi f_0 L/R$, where f_0 is the resonance frequency. (b) Calculate the Q value for the circuit of Problem 52.

Figure 22.24 Problems 53 and 54

54. The curves of Figure 22.24 are characterized by the width at half-maximum, Δf. This width is the difference in frequency between the two points on the curve where the power is one-half the maximum value. It can be shown that this width is related to the Q value (when the Q value is large) as $\Delta f = f_0/Q$. (a) Find Δf for the circuit in Problem 52. (b) Find the Q value and Δf for the circuit in Problem 52 if R is replaced by a 300-Ω resistor.

Section 22.7 The Transformer

55. An ideal transformer has 150 turns on the primary winding and 600 turns on the secondary. If the primary is connected across a 110-V rms generator, what is the rms output voltage at the secondary?

56. A transformer at a public utility reduces the voltage on a line from 360,000 V to 3600 V. The primary has 10,000 turns. (a) How many turns are there on the secondary? (b) If the current in the secondary is 600 A, find the current in the primary. Assume an ideal transformer.

57. The primary current of an ideal transformer is 6.5 A when the primary voltage is 96 V. Calculate the voltage across the secondary when a current of 0.8 A is delivered to a resistor.

58. A transformer on a pole near a factory steps the voltage down from 3600 V to 120 V. The transformer is to deliver 1000 kW to the factory at 90% efficiency. Find (a) the power delivered to the primary, (b) the current in the primary, and (c) the current in the secondary.

59. Each inhabitant of a city of 20,000 people turns on a 100-W lightbulb at the same time. Assume no other power in the city is being used. (a) If the utility company were to furnish this total power at 120 V, calculate the current in the power lines from the utility to the city. (b) Calculate this current if the power company first steps up the voltage to 200,000 V. (c) How much heat is lost in each 1-m length of the power lines

if the resistance of the lines is 5×10^{-4} Ω/m? Repeat this calculation for situation (a) and for situation (b). (d) If an individual line from the utility can handle only 100 A, how many lines are required to handle the current for each situation described?

Additional Problems

60. A 300-Ω resistor is in series with a 0.3-H inductor, a 15-μF capacitor, and a 50-Hz, 50-V generator. Find (a) the current in the circuit, (b) the voltage drop across the resistor-capacitor combination, and (c) the power delivered to the circuit.

61. A 250-mH inductor is connected across an ac source of $V_m = 90$ V. (a) At what frequency will the inductive reactance equal 20 Ω? (b) What is the rms value of current in the inductor at the frequency found in part (a)?

62. A 0.25-H inductor is connected to a capacitor and a 30-Ω resistor along with a 60-Hz, 30-V generator. (a) To what value would the capacitor have to be adjusted to produce resonance? (b) At resonance find the voltage drop across the capacitor and inductor combination.

63. A coil with an inductance of 15.3 mH and a resistance of 5 Ω is connected to a variable frequency ac generator. At what frequency would the voltage across the coil lead the current by 60°?

64. The primary coil of a certain transformer has an inductance of 2.5 H and a resistance of 80.0 Ω. (a) If the primary coil is connected to an ac source with a frequency of 60.0 Hz and a voltage of 110 V rms, what is the rms current through the primary? (b) If the primary is connected to 110 V dc, what is the current through the primary? Disregard initial effects. (c) Compare the power dissipated in the resistance in each case.

65. A resistor, inductor, and capacitor are connected in series. The rms voltages are 75 V across the resistor, 75 V across the capacitor, and 150 V across the inductor. Find (a) the voltage of the source and (b) the phase angle.

66. A transformer to be used with a small portable cassette player has a flexible design that enables it to deliver either 3 V, 6 V, or 9 V when connected to a 60-Hz, 120-V rms source. If the primary has 200 turns, (a) find the total number of turns in the secondary and (b) the number of turns that must be tapped to produce each of the voltages above.

67. A particular inductor has appreciable resistance. When the inductor is connected to a 12-V battery, the current through the inductor is 3 A. When it is connected to an ac source whose rms output is 12 V and whose frequency is 60 Hz, the current drops to 2 A.

What is (a) the impedance at 60 Hz and (b) the inductance of the inductor?

68. A 0.7-H inductor is connected in series with a fluorescent lamp to limit the current drawn by the lamp. If the combination is connected to a 60-Hz, 120-V line, and if the voltage across the lamp is to be 40 V, what is the current in the circuit? (The lamp is a pure resistive load.)

69. A small transformer is used to supply an ac voltage of 6 V to a model railroad lighting circuit. The primary has 220 turns and is connected to a standard 110-V, 60-Hz line. Although the resistance of the primary may be neglected, it has an inductance of 150 mH. (a) How many turns are required on the secondary winding? (b) If the transformer is left plugged in, what current will be drawn by the primary when the secondary is open? (c) What power will be drawn by the primary when the secondary is open?

70. A 200-Ω resistor is connected in series with a 5-μF capacitor and a 60-Hz, 120-V rms line. If electrical energy costs $0.08 per kWh, how much does it cost to leave this circuit connected for 24 h?

71. A 100-Ω resistor is connected in a series circuit to a 0.6 H inductor and a 25-μF capacitor. This *RLC* combination is then connected across a voltage source that varies as $v = 80 \sin(1000t/\pi)$ V. Calculate (a) the maximum current, (b) the phase angle, (c) the power factor, (d) the rms voltage across the inductor, and (e) the power delivered to the circuit.

72. The current in a purely resistive circuit equals 70% of the maximum current at $t = 0.003$ s. What is the frequency of the generator?

73. A series *RLC* circuit has a resonance frequency of $2000/\pi$ Hz. When operating at a frequency $\omega > \omega_0$, $X_L = 12$ Ω and $X_C = 8$ Ω. Calculate the values of L and C for the circuit.

74. An inductor is to be made of a 5-m length of copper wire of diameter 1 mm wound on a coil of radius 3 cm. (The coil has an air core.) (a) Estimate the length of the completed solenoid and number of turns on it. Find (b) the inductance of the completed coil, and (c) the resistance of the completed coil. (d) Find the current that would be drawn when this device is connected to a 60-Hz, 20-V rms source.

75. Two connections allow contact with two circuit elements in series inside a box, but it is not known whether the circuit elements are *R*, *L*, or *C*. In an attempt to find what is inside the box, you perform the following set of measurements. When a 3-V dc power supply is connected across the terminals, there is a direct current of 300 mA in the circuit. When a 3-V, 60-Hz source is connected, the current becomes 200 mA. (a) What are the two elements in the box? (b) What are their values of *R*, *L*, or *C*?

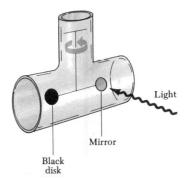

Figure 23.9
An apparatus for measuring the pressure of light. In practice, the system is contained in a high vacuum.

Since $E = cB = B/\sqrt{\mu_0 \epsilon_0}$, this can also be expressed as

$$\text{Average power per unit area} = \frac{E_m^2}{2\mu_0 c} = \frac{c}{2\mu_0} B_m^2 \qquad (23.6)$$

Note that the power per unit area in these expressions represents the *average* power per unit area. Also note that the values to be used for E and B in these equations are the *maximum* values. It is of some interest to note that a detailed analysis would show that the energy carried by an electromagnetic wave is shared equally by the electric and magnetic fields.

Electromagnetic waves transport momentum as well as energy. This is demonstrated by the fact that radiation pressure is exerted on a surface when an electromagnetic wave impinges upon it. Although radiation pressures are very small (about 5×10^{-6} N/m^2 for direct sunlight), they have been measured using a device like the one pictured in Figure 23.9. Light is allowed to strike a mirror and a black disk connected to each other by a horizontal bar suspended from a fine fiber. Light striking the black disk is completely absorbed, and so *all* of the momentum of the light is transferred to the disk. Light striking the mirror head on is totally reflected; hence the momentum transfer to the mirror is *twice* as great as that transmitted to the disk. As a result, the horizontal bar supporting the disks twists counterclockwise as seen from above. The bar comes to equilibrium at some angle under the action of the torques due to the radiation pressure and the torque produced as the fiber twists. The radiation pressure can be determined by measuring the angle at which equilibrium occurs. The apparatus must be placed in a high vacuum to eliminate the effects of air currents.

In summary, electromagnetic waves traveling through free space have the following properties:

1. Electromagnetic waves travel with the speed of light.
2. Electromagnetic waves are transverse waves since the electric and magnetic fields are perpendicular to the direction of propagation of the wave and to each other.
3. The ratio of the electric to the magnetic field in an electromagnetic wave equals the speed of light.
4. Electromagnetic waves carry both energy and momentum, which can be delivered to a surface.

Example 23.1   Solar Energy

The Sun delivers an average power per unit area of about 700 W/m^2 to the Earth's surface. Calculate the total power incident on a roof of dimensions 8 m by 20 m. Assume the radiation is incident *normal* to the roof (Sun directly overhead).

Solution
The power per unit area, or light intensity, is 700 W/m^2. For normal incidence, we get

$$\text{Power} = (700 \text{ W/m}^2)(8 \times 20 \text{ m}^2) = \boxed{1.12 \times 10^5 \text{ W}}$$

Note that if this power could *all* be converted to electric power, it would provide more than enough power for the average home. Unfortunately, solar energy is not

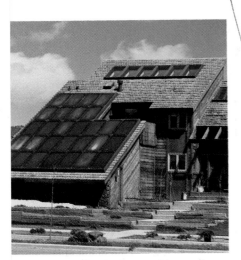

A typical solar home used in the southern and southwestern parts of the United States. The rooftop collectors are heated by the Sun, and this heat in turn is transferred to a circulating fluid that is used to heat the home. (© *Barbara Burnes/Photo Researchers, Inc.*)

easily harnessed, and the prospects for large-scale conversion are not as bright as they may appear from this simple calculation. For example, the conversion efficiency from solar to electrical energy is far less than 100% (typically, it is 10% for photovoltaic cells). Roof systems for converting solar energy to thermal energy have been built with efficiencies of around 50%; however, there are other practical problems with solar energy that must be considered, such as overcast days, geographic location, and energy storage.

Exercise   How much solar energy (in joules) is incident on the roof in 1 h?

Answer 4.03×10^8 J.

23.5   The Spectrum of Electromagnetic Waves

Near-infrared radiation was discovered in 1800 by William Herschel. By the mid-1800's, Maxwell had recognized that both infrared radiation and visible light were electromagnetic waves. Working from Maxwell's prediction of electromagnetic waves, Hertz generated and detected radio waves in 1887. It is now known that there are electromagnetic waves of many other frequencies and wavelengths.

   Because all electromagnetic waves travel through a vacuum with a speed c, their frequency and wavelength are related by the important expression

$$c = f\lambda \tag{23.7}$$

The various types of electromagnetic waves are listed in Figure 23.10. Note the wide range of frequencies and wavelengths. For instance, a radio wave of frequency 5 MHz (a typical value) has a wavelength in vacuum of

$$\lambda = \frac{c}{f} = \frac{3 \times 10^8 \text{ m/s}}{5 \times 10^6 \text{ s}^{-1}} = 60 \text{ m}$$

   In the sections that follow, we shall give a brief description of these various waves in order of decreasing wavelength. There is no sharp dividing point between one kind of wave and the next.

Example 23.2   An Electromagnetic Wave

A plane electromagnetic sinusoidal wave of frequency 40 MHz travels in free space in the x direction, as in Figure 23.7. At some point and at some instant, the electric field has its *maximum* value of 750 N/C and is along the y axis. Determine the wavelength and period of the wave.

Solution
Since $c = \lambda f$ and $f = 40$ MHz $= 4 \times 10^7$ s^{-1}, we get

$$\lambda = \frac{c}{f} = \frac{3 \times 10^8 \text{ m/s}}{4 \times 10^7 \text{ s}^{-1}} = \boxed{7.5 \text{ m}}$$

The period of the wave, T, equals the inverse of the frequency:

$$T = \frac{1}{f} = \frac{1}{4 \times 10^7 \text{ s}^{-1}} = \boxed{2.5 \times 10^{-8} \text{ s}}$$

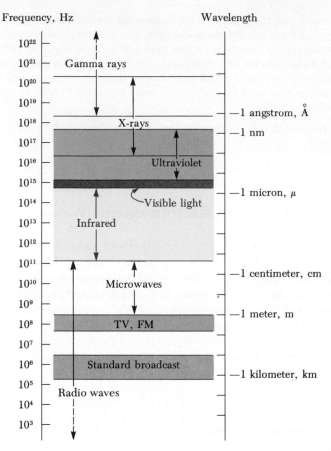

Figure 23.10
The spectrum of electromagnetic radiation.

(b) Calculate the magnitude and direction of the magnetic field when $E = 750$ N/C and is in the y direction.

Solution
From Equation 23.4 we see that

$$B = \frac{E}{c} = \frac{750 \text{ N/C}}{3 \times 10^8 \text{ m/s}} = \boxed{2.5 \times 10^{-6} \text{ T}}$$

Because the magnetic and electric fields must be perpendicular to each other and both must be perpendicular to the direction of wave propagation (x in this case), we conclude that **B** is along the z axis.

Radio Waves

Radio waves are produced by charges accelerating through conducting wires, as discussed in Section 23.3. They get their name because their principal use is as a carrier of information from a radio transmitting station to your home receiver. Because communication via radio signals is so important in today's world, we shall examine the fundamentals of this process.

All domestic radio stations are assigned a *carrier frequency* by the federal government. This assigned frequency is the basic frequency of the electromagnetic wave the station is allowed to broadcast. If the station is an **AM** *(amplitude modulation)* station, its carrier frequency will be between 540 and 1600 kHz, while **FM** *(frequency modulation)* stations operate at frequencies from 88 to 108 MHz.

The term *modulation* means alteration or change. To visualize the process for an AM station, let the carrier wave be represented as in Figure 23.11a. The sound signal to be transmitted by this carrier wave is converted by a microphone to an electric signal and superimposed on the carrier wave. For example, if the audio signal is a 100-Hz tone, the amplitude of the carrier wave would fluctuate at a rate of 100 times per second. Figure 23.11b shows a waveform representing an arbitrary audio signal, and Figure 23.11c represents the carrier wave with the audio signal impressed on it.

The modulated carrier wave travels through space as an unnoticed presence in our environment until it is intercepted by an antenna. There are two common types of antennas used to detect these waves. The one most often used in cars consists of a metal rod extended into the air. In this type of antenna, the changing electric field in the carrier wave causes electrons in the rod to oscillate. The amplitude of these oscillations varies with the strength of the electric field in the modulated carrier wave. That is, the strength of the receiving antenna current varies with a frequency matching that of the audio wave impressed on the carrier wave.

The second type of antenna is a coil of wire built into the back of the radio receiver. The changing magnetic field portion of the carrier wave induces a current in the coil that emulates the variations in amplitude of the carrier wave.

Satellite television antennas are common for receiving television stations from satellites in orbit around the Earth and are most widely used in rural locations. *(© Hank Delespinasse/The IMAGE Bank)*

Figure 23.11
Amplitude modulation.

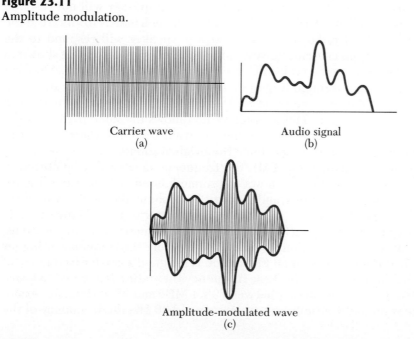

Carrier wave
(a)

Audio signal
(b)

Amplitude-modulated wave
(c)

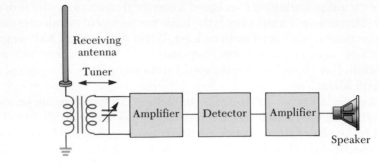

Figure 23.12
The various stages of a radio receiver.

Figure 23.13
(a) The average value of this electrical signal is zero.
(b) The dashed colored line represents the average of the signal shown by the solid line.

Impinging on the receiving antenna are not one but several such electromagnetic waves, one from every radio station within range. Figure 23.12 illustrates one method in which the tuner of the receiver selects the desired signal. The oscillating antenna current induces a current in the tuner coil via the coil shown on the left. As we have seen, a circuit consisting of an inductor and a capacitor can serve as a resonant circuit and enhance a certain frequency while suppressing others. As you tune your radio, you vary the capacitance in the tuner until the circuit is in resonance at a frequency equal to the carrier wave of the desired radio station signal.

The current delivered to the receiver by the antenna is very weak and must be strengthened by an amplifier circuit (Fig. 23.12) before it is delivered to the next stage of the radio receiver, the *detector*, which removes the audio information encoded on the carrier wave. The detector in most radios is either a vacuum tube or a transistor, but we can understand the principle of what occurs by considering our detector to be a diode. A diode is a solid-state device that allows a current in only one direction. If we attempt to feed the high-frequency current from the antenna to a speaker, the speaker will not respond because it is a mechanical device and thus has too much inertia to vibrate at the high frequencies of radio waves. Instead, the speaker will respond to the average value of the current; however, for an input signal like that shown in Figure 23.13a, the average value of the current in the circuit is zero. Allowing the current to pass through a diode clips off half the current, as shown in Figure 23.13b. The average value for this current is represented by the red dashed line in the figure. This average value follows the same pattern as the audio signal on the carrier wave. The speaker is able to follow these lower-frequency variations and thus reproduce the original sound.

In frequency modulation (FM), the frequency is altered, as pictured in Figure 23.14. Figure 23.14a shows the unmodulated carrier wave, Figure 23.14b depicts an audio signal, and Figure 23.14c shows the carrier wave after it has been modulated. The wider-than-normal portions in Figure 23.14c indicate that the carrier has a lower-than-normal frequency at these locations, and the closely spaced portions indicate points where the frequency is higher than normal. For example, if an FM station is assigned a carrier frequency of 98.6 MHz, this represents the basic frequency the station is allowed to transmit, but this frequency can vary between 98.4 MHz and 98.8 MHz. For example, if the station is transmitting a frequency of 100 Hz, the frequency of the carrier wave varies 100 times per second within its allowed range. The function of the radio receiver is, as before, to select the desired station, separate the audio signal from the carrier, and reproduce the audio signal with a speaker.

Example 23.3   Tuning a Radio

The tuner coil in Figure 23.12 has an inductance of 5 μH. If you desire to listen to a radio station that broadcasts at a frequency of 1400 kHz, to what value will you have to adjust the capacitor?

Solution
The frequency at which the LC circuit will resonate is given by Equation 23.1. Solving this equation for C and substituting, we find

$$C = \frac{1}{4\pi^2 f_0^2 L} = \frac{1}{4\pi^2 (1400 \times 10^3 \text{ s}^{-1})^2 (5 \times 10^{-6} \text{ H})}$$

$$= 2.6 \times 10^{-9} \text{ F} = \boxed{2.6 \text{ pF}}$$

Exercise   What is the wavelength of the carrier wave transmitted by this radio station?

Answer   214 m.

Carrier wave
(a)

Audio signal
(b)

Frequency-modulated wave
(c)

Figure 23.14
Frequency modulation.

Microwaves

Microwaves are short-wavelength radio waves that have wavelengths between about 1 mm and 30 cm and are generated by electronic devices. Microwave ovens represent an interesting domestic application of these waves. To understand how these ovens cook food, consider that water and fat molecules have a permanent separation between the center of their positive charge and negative charge (see Chapter 16), and as a result, each of these molecules has an internal electric field that originates at the center of positive charge and terminates at the center of negative charge. Such molecules are said to be electrically *polarized*. The electric field of the microwave radiation interacts with this electric field of the molecule, causing the molecule to rotate. This rotation bends and twists the bonds between the individual molecules, which produces an effect similar to friction, leading to the production of heat. Thus, the energy absorbed from the microwave beam heats the food internally.

Microwaves are also used in communication, particularly in the transmission of long-distance telephone messages. In this process, the information to be transmitted is encoded on microwaves by either amplitude or frequency modulation and carried from point to point via microwave relay towers.

In a recent proposal, it was suggested that solar energy could be harnessed by placing a solar collector in space and then beaming down the collected energy by microwave to a receiving station on Earth. The obvious advantage to this scheme is that the Sun shines on the collector continuously, and there are no problems with cloud cover.

One of the most common applications of microwaves is in the radar (<u>ra</u>dio <u>d</u>etection <u>a</u>nd <u>r</u>anging) systems used in aircraft navigation and police cars. The principles of radar are quite similar to those used in radio broadcasting. As in

These antennas are used to receive cable television signals transmitted as microwaves.

Radar detector on the tower of a ship. (© *John V. A. F. Neal/ Photo Researchers, Inc.*

broadcasting, an antenna is used to transmit an electromagnetic wave. The wave produced is not a continuous wave but consists of short bursts, or pulses. After the emission of a pulse, the transmitting antenna is deactivated as a transmitter momentarily and used as a receiver to detect any reflected pulses. A signal, or "blip," proportional to the distance of the reflecting object can easily be displayed on a fluorescent screen or calculated as outlined in the following example.

Example 23.4   Measuring Distances with Radar

A radar receiver indicates that a transmitted pulse is returned as an echo 20 μs after transmission. How far away is the reflecting object?

Solution
Traveling with a speed of 3×10^8 m/s for 20 μs, the pulse travels a total distance

$$x = ct = (3 \times 10^8 \text{ m/s})(20 \times 10^{-6} \text{ s}) = 6 \times 10^3 \text{ m}$$

Because the pulse must travel to the reflecting object and back, the distance of the object from the transmitter is half this value, or 3×10^3 m.

Infrared Waves

Infrared waves (sometimes called heat waves) have wavelengths ranging from about 1 mm to the longest wavelength of visible light, 7×10^{-7} m. These waves, produced by hot objects, are readily absorbed by most materials. The infrared energy absorbed by a substance appears as heat because the energy agitates the atoms of the substance, increasing their vibrational or translational motion, which is detected as a temperature rise.

Photographic film made with an emulsion sensitive to infrared waves has found increased usage. One advantage is that infrared rays can penetrate fog and haze better than visible light. Additionally, when these films are used in cameras on airplanes or satellites, the developed film provides more information about land areas than photographs taken with visible light. This occurs because there are distinct differences between the amount of radiation emitted by the ground, by water, and by vegetation. It is possible, for example, to distinguish between a field of corn and a field of wheat because the two emit different wavelengths of infrared radiation. Remote disease detection is also feasible since a diseased crop differs from a healthy crop in the amount of infrared radiation emitted.

There are special telescopes that can immediately change an image produced by infrared radiation into a visible image. The visible image produced in this way, however, is considerably different from a normal image. For example, the exposed portions of a person's body would emit more infrared radiation than would the cooler clothes that cover most of the body. This means that the uncovered areas would appear much brighter than the rest of the body. Slightly revised versions of this process enable doctors to diagnose cancerous growths. An infrared scan may reveal hot spots in a patient's body. These hotter-than-normal regions are characteristic of malignant growths.

Visible Light

Visible light, the most familiar form of electromagnetic waves, may be defined as that part of the electromagnetic spectrum that the human eye can detect. Light is produced by the rearrangement of electrons in atoms and molecules, as we shall discuss in Chapters 29 and 30. The various wavelengths of visible light are classified by color, ranging from violet ($\lambda \approx 4 \times 10^{-7}$ m) to red ($\lambda \approx 7 \times 10^{-7}$ m). The eye's sensitivity is a function of wavelength, the sensitivity being a maximum at a wavelength of about 5.6×10^{-7} m (yellow-green).

Light is the basis of the science of optics and optical instruments, which we shall discuss in Chapters 24 through 27. The units often used to designate short wavelengths and distances are abbreviated as follows:

$$1 \text{ micron } (\mu\text{m}) = 10^{-6} \text{ m}$$

$$1 \text{ nanometer } (\text{nm}) = 10^{-9} \text{ m}$$

$$1 \text{ angstrom } (\text{Å}) = 10^{-10} \text{ m}$$

In these units, the wavelengths of visible light range from 0.4 to 0.7 μm, or 400 to 700 nm, or 4000 to 7000 Å.

Ultraviolet Light

Ultraviolet light is the name given to wavelengths ranging from approximately 3.8×10^{-7} m (380 nm) down to 6×10^{-8} m (60 nm). The Sun is an important source of ultraviolet light, which is the main cause of suntans. Most of the ultraviolet light from the Sun is absorbed by atoms in the upper atmosphere, or stratosphere. This is fortunate because ultraviolet light in larger quantities is harmful to humans. One important constituent of the stratosphere is ozone (O_3), which results from reactions between oxygen and ultraviolet radiation. This ozone shield converts lethal high-energy ultraviolet radiation to infrared radiation, which in turn warms the stratosphere. Recently there has been a great deal of controversy concerning the possible depletion of the protective ozone layer as a result of the release into the air of chlorofluorocarbons used in aerosol spray cans and as refrigerants.

The tanning process of the human body occurs as a result of a protective mechanism of the body against the harmful effects of ultraviolet rays. Normal skin contains melanin pigment, which gives the skin a brownish coloration. Excessive ultraviolet light stimulates the body to produce more melanin pigment, resulting in a rosy tan. Ultraviolet light in moderation may give the body a healthy glow, but if carried to extremes, tanning may, over a period of years, lead to unsightly thickening of the skin and even to skin cancer.

Ultraviolet radiation is often used to kill germs. Ultraviolet lamps are used to sterilize hospital operating rooms and surgical instruments. Low-intensity ultraviolet lamps are sometimes placed above grocery meat counters to reduce spoilage.

Ordinary glass provides an effective shield against ultraviolet light. As a result, you cannot get a tan when separated from the Sun by a glass window. Some light sources inherently produce large amounts of ultraviolet radiation; mercury lamps are a case in point. These lamps are enclosed in a glass shield if they are to be used for the visible light they produce. If obtaining the ultraviolet light is desired, quartz housings are used because ultraviolet light can readily pass through quartz.

X-rays

X-rays are electromagnetic waves with wavelengths in the range from about 10^{-8} m (10 nm) down to 10^{-13} m (10^{-4} nm). X-rays are produced when high-energy electrons decelerate after striking a metal target. These rays are used as a treatment for certain forms of cancer, but their most widely known use is as a diagnostic tool in medicine and dentistry. If a beam of x-rays is focused on a hand, for example, the penetrating power of the rays is sufficient to allow the rays to pass right through the hand and expose a photographic film on which the hand rests. Because it is easier for the waves to penetrate fatty tissue than bone, the exposed film shows the bones standing out in vivid detail. Furthermore, since x-rays damage or destroy living tissues and organisms, care must be taken to avoid unnecessary exposure or overexposure. We shall discuss x-rays in more detail in Chapter 30.

Gamma Rays

Gamma rays are electromagnetic waves emitted by radioactive nuclei and during certain nuclear reactions. They have wavelengths ranging from about 10^{-10} m to less than 10^{-14} m. They are highly penetrating and produce serious damage when absorbed by living tissues. Consequently, those working near such dangerous radiation must be protected with heavily absorbing materials, such as thick layers of lead. Under controlled conditions, gamma rays can be used to kill cancerous cells.

Note in Figure 23.10 that the wavelength ranges of x-rays and gamma rays overlap. This means that the two are essentially the same in their physical nature. They are divided according to their origin rather than to some physical quantity, such as wavelength. We shall discuss the origin of gamma rays in greater detail in Chapter 30. Finally, note that the boundary between ultraviolet rays and x-rays is as blurred as that between x-rays and gamma rays. Again, the distinction between the two is based on how they are produced.

▶ S P E C I A L   T O P I C
Television

In this section we shall describe the underlying physical principles of television transmission and reception. As we shall see, the techniques used in this form of communication include a multitude of phenomena from the fields of electricity and optics.

Television signals, like radio signals, are carried from the transmitting station to our homes via electromagnetic carrier waves. However, television signals are sent out on two carrier waves: one carries the audio (sound) signal and the other carries the video (picture) signal. The basic idea of sound transmission by television carrier waves is identical to that for FM radio wave transmission, discussed in Section 23.5. Therefore, let us confine our discussion to the video portion of the television signal.

Video signals displayed on your television screen use two forms of optical deception, one of which is used in motion pictures, the other in photography. The optical deception used in motion pictures involves the fact that a series of still pictures scanned

Figure 23.15
An image orthicon tube, which is the basic element of a television camera. *(Adapted from J. Wilson,* Technical College Physics, *Saunders College Publishing Co., 1982)*

rapidly by the eye is perceived as continuous motion in the scene that was photographed. A theater movie projector flashes 24 different still pictures before your eyes each second. Each picture is only slightly different from the preceding one, and the persistence of eye vision blends them together as continuous action. In the case of television signals, 30 different pictures appear on the television screen each second.

The second optical deception used in video display can be understood by careful examination of a photograph in any newspaper. Close scrutiny of such a photograph with a good magnifying glass will reveal that the photograph is not continuous. Rather, it consists of a large number of closely spaced dots. At large distances from the photograph, the eye is not able to distinguish the individual dots and therefore is tricked into thinking that the photograph is a continuous scene. A television camera reduces an individual picture into a series of tiny dots similar to the scheme used in photographs. The camera produces 30 complete pictures each second, one dot at a time, which are sent out on the carrier wave. Your television set receives these strings of dots and reassembles them into a picture on the screen of the picture tube.

The basic element in a television camera is an image orthicon tube (Fig. 23.15), which generates the video signal. Light from the object being televised is focused by lenses onto a light-sensitive surface that releases electrons. Few electrons are released in regions where the surface is dark, and many are released in highly illuminated regions. Thus, an image of the object appears on the surface, which is in fact an electron image. The surface is then scanned with an electron beam, which "reads" the information by converting the image to electric signals.

The signals produced in this manner are transmitted to the receiver, along with the FM audio signal, via the electromagnetic carrier wave (Fig. 23.16). The picture tube (cathode ray tube) in the home receiver is synchronized with the message sent to it by the carrier wave. The electron beam in the picture tube scans the fluorescent screen, producing hundreds of horizontal lines in a fraction of a second. Signals from the

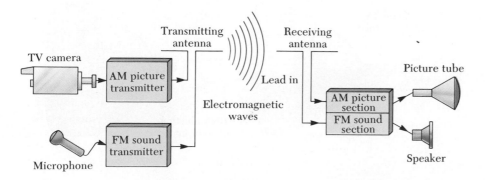

Figure 23.16
Simplified diagram of television transmission and reception. Television combines both a picture AM signal and an audio FM signal. *(Adapted from J. Wilson,* Technical College Physics, *Saunders College Publishing Co., 1982)*

Figure 23.17
In the color camera, the incoming light beam is separated into three light beams of primary colors. In the receiving set, the beams are recombined to produce a colored image. *(Adapted from J. Wilson, Technical College Physics, Saunders College Publishing Co., 1982)*

television camera are used to control the variations in the electron beam striking the screen. In this manner, an image is reproduced on the screen of the picture tube as a mosaic of light and dark dots. The dots are perceived by the eye as a complete black-and-white picture.

Color television cameras and sets operate in basically the same way, except that a color camera contains three image tubes instead of one. A series of filters and mirrors (Fig. 23.17) separates the light from the scene being televised into the three primary colors: red, green, and blue. Information associated with the strength of each color signal is then transmitted to the home receiver. The color picture tube in the receiver has three electron guns (Fig. 23.18a), each of which produces an electron beam whose intensity depends on the intensity of the red, green, or blue signal reaching it. These three electron beams are directed to the fluorescent screen on the tube face (Fig. 23.18b). The signal from the blue electron gun strikes a dot on the screen, which causes it to give off blue light. Likewise, the red and green guns produce their respective colors. Because these dots are very close together, the eye cannot distinguish them individually from a normal viewing distance. Instead, the eye blends the three colors together to produce a color that depends on the intensity of the light being emitted by each dot. If each dot emits the same light intensity, the colors blend together to produce a white spot at that location. Any color can be reproduced at a given screen location by a proper combination of the three primary colors. ◀

Figure 23.18
(a) The color television picture tube has three electron guns. (b) The screen of a color television picture tube contains three sets of colored dots. The size of the dots has been exaggerated for clarity.

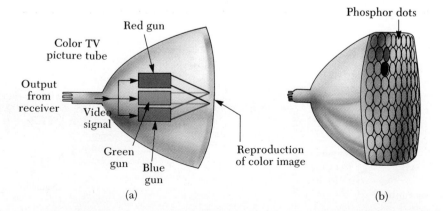

(a)

(b)

Summary

Electromagnetic waves were predicted by James Clerk Maxwell and later generated and detected by Heinrich Hertz. These waves have the following properties:

1. Electromagnetic waves are transverse waves since the electric and magnetic fields are perpendicular to the direction of travel.
2. Electromagnetic waves travel with the speed of light.
3. The ratio of the electric to the magnetic field in an electromagnetic field equals the speed of light, that is

$$\frac{E}{B} = c \qquad\qquad (23.4)$$

4. Electromagnetic waves carry energy as they travel through space. The average power per unit area is

$$\frac{E_m B_m}{2\mu_0} = \frac{E_m{}^2}{2\mu_0 c} = \frac{c}{2\mu_0} B_m{}^2 \qquad (23.5, 23.6)$$

   where E_m and B_m are the maximum values of the electric and magnetic fields.
5. Electromagnetic waves transport momentum as well as energy. The speed, c, frequency, f, and wavelength, λ, of an electromagnetic wave are related as

$$c = f\lambda \qquad\qquad (23.7)$$

The **electromagnetic spectrum** includes waves covering a broad range of frequencies and wavelengths. These waves have a variety of applications and characteristics, depending on their frequency or wavelength.

Additional Reading

A. H. W. Beck, *Words and Waves*, World University Library, New York, McGraw-Hill, 1967.

J. Hertz, *Heinrich Hertz: Memoirs, Letters, Diaries*, San Francisco Press, 1977.

O. Kedem and U. Ganiel, "Solar Energy, How Much Do We Receive?" *The Physics Teacher*, December 1983, p. 573.

D. K. C. McDonald, *Faraday, Maxwell and Kelvin*, Science Study Series, Garden City, NY, Doubleday, 1964.

J. R. Pierce, *Electrons and Waves*, Science Study Series, Garden City, NY, Doubleday, 1964.

Questions

1. Does the wire connected to a battery emit an electromagnetic wave? Explain.
2. If you charge a comb by running it through your hair and then hold the comb next to a bar magnet, do the electric and magnetic fields produced constitute an electromagnetic wave?
3. If the fundamental source of a sound wave is a vibrating object, what is the fundamental source of an electromagnetic wave?
4. When light (or other electromagnetic radiation) travels across a given region, what is it that moves?
5. List as many similarities and differences as you can between sound waves and light waves.
6. What does a radio wave do to the charges in the receiving antenna to provide a signal for your car radio?
7. Explain how the (dipole) VHF antenna of a television set works.
8. Explain how the (loop) UHF antenna of a television set works.
9. Explain why the voltage induced in a UHF antenna depends on the frequency of the signal, while the voltage in a VHF antenna does not.
10. Certain orientations of the receiving antenna of a television set give better reception than others. Furthermore, the best orientation varies from station to station. Explain these observations.
11. Often when you touch the indoor antenna on a television receiver, the reception instantly improves. Why?
12. Radio stations often advertise "instant news." If what they mean is that you hear the news at the instant they speak it, is their claim true? About how long would it take for a message to travel across this country by radio waves, assuming that these waves could travel this great distance and still be detected?
13. For a given incident energy of an electromagnetic wave, why is the radiation pressure on a perfectly reflecting surface twice as large as the pressure on a perfectly absorbing surface?
14. Is there any difference between 1.0-nm ultraviolet light and 1.0-nm x-rays? Discuss.
15. Explain what is meant by monochromatic x-rays.
16. What is meant by the terms "amplitude modulation" and "frequency modulation"?
17. Why would you expect to find less static on the FM radio band than on the AM band?
18. If a high-frequency current is passed through a solenoid containing a metallic core, the core will heat up by induction. This process also cooks foods in microwave ovens. Explain why the materials heat up in these situations.
19. An empty plastic or glass dish removed from a microwave oven is cool to the touch. How can this be possible?
20. Suppose a creature from another planet had eyes that were sensitive to infrared radiation. Describe what he would see if he looked around the room you are now in. That is, what would be bright and what would be dim?
21. Why should an infrared photograph of a person look different from a photograph taken with visible light?
22. A welder must wear protective clothing to prevent sunburn. What does this imply about the light produced by the welding?

Problems

Section 23.2   Hertz's Discoveries

1. If the coil in the resonant circuit of a radio has an inductance of $2 \mu H$, what range of values must the tuning capacitor have in order to cover the complete range of FM frequencies? (See the section on Radio Waves to find the FM range of frequencies.)
2. A standing wave interference pattern is set up by radio waves between two metal sheets located 2 m apart. This is the shortest distance between the plates which will produce a standing wave pattern. What is the fundamental frequency?

Section 23.4   Properties of Electromagnetic Waves

3. Experimenters at the National Bureau of Standards have made precise measurements of the speed of light using the fact that, in vacuum, the velocity of electromagnetic waves is $c = 1/\sqrt{\mu_0 \epsilon_0}$, where $\mu_0 = 4\pi \times 10^{-7}$ $N \cdot s^2/C^2$ and $\epsilon_0 = 8.854 \times 10^{-12}$ $C^2/N \cdot m^2$. What value (to four significant figures) does this give for the speed of light in vacuum?
4. Verify that Equation 23.2 has the proper dimensions.
5. An electromagnetic wave in vacuum has an electric field amplitude of 150 V/m. Calculate the amplitude of the accompanying magnetic field.

6. Find the average power per unit area associated with the wave described in Problem 5.

7. An incandescent lamp is radiating uniformly in all directions at 15 W. Calculate the maximum values of the electric and magnetic fields (a) 1 m and (b) 5 m from the source. (*Hint:* Use the formula for the surface area of a sphere, $A = 4\pi r^2$, to find the area on which the total energy is incident in each case, and then use Equation 23.6.)

8. Assume that the solar radiation incident on the Earth is 1340 W/m² (at the top of the Earth's atmosphere). Calculate the total power radiated by the Sun, taking the average separation between the Earth and the Sun to be 1.49×10^{11} m.

9. The Sun delivers an average power of 1340 W/m² to the top of the Earth's atmosphere. Find the magnitudes of $\mathbf{E_m}$ and $\mathbf{B_m}$ for the electromagnetic waves at the top of the atmosphere.

10. A radio transmitter broadcasts uniformly in all directions with an average power of 15 kW. Compute the maximum value of the electric field in its radio wave at the following distances from the transmitter: (a) 1 km, (b) 10 km, (c) 100 km.

11. A particular electromagnetic wave traveling in vacuum has a magnetic field intensity of 1.5×10^{-7} T. Find (a) the electric field intensity and (b) the average power per unit area associated with the wave.

12. An electromagnetic wave is traveling west with the electric field vibrating at a frequency of 10^{14} Hz perpendicular to the surface of the Earth. What is (a) the frequency of vibration of the magnetic field and (b) the direction of vibration?

13. The light from a 5-mW laser is spread out in a cylindrical beam whose diameter is about 0.5 cm. What are the peak values of $\mathbf{E}$ and $\mathbf{B}$ in this beam?

14. How much solar energy is incident on a collector with dimensions 20 m by 30 m at the top of the Earth's atmosphere in 8 h? (See Problem 9.)

15. A community plans to build a facility to convert solar radiation into electrical power. They require 1 MW of power, and the system to be installed has an efficiency of 30% (that is, 30% of the solar energy incident on the surface is converted to electrical energy). What must be the effective area of a perfectly absorbing surface used in such an installation, assuming a constant power per unit area of 1000 W/m² in the incident solar radiation?

16. A plane electromagnetic wave has an average power per unit area of 300 W/m². A flat, rectangular surface of dimensions 20 cm by 40 cm is placed perpendicular to the direction of the plane wave. If the surface absorbs half the energy and reflects half, calculate the net energy delivered to the surface in 1 min.

Section 23.5 The Spectrum of Electromagnetic Waves

17. What is the frequency of an electromagnetic wave that has a wavelength of (a) 2 m, (b) 20 m, and (c) 200 m?

18. Locate the position in the electromagnetic spectrum of the waves that have the following frequencies, and find the wavelength associated with each of these frequencies: (a) 10^6 Hz, (b) 10^8 Hz, (c) 10^{10} Hz, (d) 10^{13} Hz, (e) 10^{15} Hz, (f) 10^{17} Hz, and (g) 10^{21} Hz. (*Hint:* See Fig. 23.10.)

19. What are the wavelength ranges in the (a) AM radio band (540–1600 kHz) and (b) the FM radio band (88–108 MHz)?

20. The eye is most sensitive to light having a wavelength of 5.5×10^{-7} m, which is in the green-yellow region of the electromagnetic spectrum. What is the frequency of this light?

21. A "rabbit ears" television antenna consists of a pair of adjustable rods. If we set each rod to a quarter-wavelength for channel 3, which has a central frequency of 63.0 MHz, how long are the rods?

22. A wave has a wavelength of 4×10^{-7} m and a speed of 2.5×10^8 m/s in a particular material. (a) What is the frequency of this wave in the material? (b) The frequency of a wave does not change as the wave moves from one material to another. Find the wavelength of this wave in vacuum.

23. A radio receiver is located 200 km from a transmitter. A signal travels to the receiver via a direct line-of-sight path and also via a wave from an ionospheric layer at an altitude of 100 km. Make the approximation of a flat Earth and calculate the time difference between the arrivals of the two signals. (*Hint:* Look ahead at the discussion of the law of reflection in Chapter 24.)

24. A singer's voice is transmitted by a radio wave to a person 100 km away. (a) How much time passes before the distant listener hears the sound? (b) By the time the radio message reaches a listener, how far from the singer has the sound wave moved in the auditorium? Assume that the speed of sound is 345 m/s.

25. The Moon is approximately 250,000 mi from us and the Sun is approximately 93,000,000 mi away. How much time is required for an electromagnetic wave to reach us from (a) the Sun and (b) the Moon?

26. A tuning circuit in a radio receiver is set such that $L = 5.4\ \mu$H and $C = 3.1$ pF. What is the wavelength of the wave being detected?

27. A radar pulse returns to the receiver after a total travel time of 4×10^{-4} s. How far away is the object that reflected the wave?

28. An airplane is traveling at 50 m/s and is 6000 m from a ship. (a) What is the total time of flight of the radar wave from the ship to the plane and back? (b) What distance does the plane move in this time?

Additional Problems

29. At one location on the Earth, the amplitude of the magnetic field due to solar radiation is $2.4 \ \mu$T. From this value calculate the magnitude of the electric field due to solar radiation.
30. Find the frequency of waves having the following wavelengths in vacuum: (a) 1 micron, (b) 1 nanometer, and (c) 1 angstrom.
31. What value of inductance should be used in series with a capacitor of 1.50 pF to form an oscillating circuit that will radiate a wavelength of 5.25 m?
32. Show that the ratio E/B has units of m/s.
33. The wave number, k, of a wave of wavelength λ is defined by the equation $k = 2\pi/\lambda$. What is the wave number corresponding to light whose frequency is 4.8×10^{14} Hz?
34. The nearest star to us is approximately 4 lightyears away, where 1 lightyear is the distance light travels in one year. Find the distance to this star in meters.
35. As the spaceship Voyager neared Jupiter, a radio command from Earth was sent to it to turn on its TV cameras. If Jupiter is 630×10^6 km away, what is the minimum time required for a TV signal to be received on Earth?
36. A source of electromagnetic waves radiates uniformly in all directions with a power of 100 W. At what distance from this source will the electric field amplitude be 15 V/m?
37. What power must be radiated uniformly in all directions by a source if the amplitude of the magnetic field is 7×10^{-8} T at a distance of 2 m?
38. A microwave transmitter emits electromagnetic waves of a single wavelength. The maximum electric field 1 km from the transmitter is 6.0 V/m. Assuming the transmitter is a point source and neglecting waves reflected from the Earth, calculate (a) the maximum magnetic field at this distance and (b) the total power emitted by the transmitter.

39. A dish antenna with a diameter of 20 m receives (at normal incidence) a radio signal from a distant source, as shown in Figure 23.19. The radio signal is a continuous sinusoidal wave with amplitude $E_m = 0.2 \ \mu$V/m. Assume the antenna absorbs all the radiation that falls on the dish. (a) What is the amplitude of the magnetic field in this wave? (b) What is the intensity of the radiation received by this antenna? (c) What is the power received by the antenna?

Figure 23.19   Problem 39

40. Compute the average energy content of a liter of sunlight as it reaches the top of the Earth's atmosphere, where its intensity is 1340 W/m².

PART 5

Light and Optics

Sir Isaac Newton
1642-1727

Scientists have long been intrigued by the nature of light, and philosophers have had endless arguments concerning the proper definition and perception of light. It is important to understand the nature of light because it is one of the basic ingredients of life on Earth. Plants convert light energy from the Sun to chemical energy through photosynthesis. Light is the means by which we are able to transmit and receive information from objects around us and throughout the Universe.

The nature and properties of light have been subjects of great interest and speculation since ancient times. The Greeks believed that light consisted of tiny particles (corpuscles) that were emitted by a light source and then stimulated the perception of vision upon striking the observer's eye. Newton used this corpuscular theory to explain the reflection and refraction of light. In 1670, one of Newton's contemporaries, the Dutch scientist Christian Huygens, was able to explain many properties of light by proposing that light was wave-like in character. In 1801, Thomas Young showed that light beams can interfere with one another, giving strong support to the wave theory. In 1865, Maxwell developed a brilliant theory that electromagnetic waves travel with the speed of light (Chap. 23). By this time, the wave theory of light seemed to be on firm ground.

However, at the beginning of the 20th century, Max Planck and Albert Einstein returned to the corpuscular theory of light in order to explain the radiation emitted by hot objects and the electrons emitted by a metal exposed to light (the photoelectric effect). We shall discuss these and other topics in modern physics in the last part of this text.

Today, scientists view light as having a dual nature. Experiments can be devised which will display either its particle-like or its wave-like nature.

In this part of the book, we shall concentrate on those aspects of light that are best understood through the wave model. First, we shall discuss the reflection of light at the boundary between two media and the refraction (bending) of light as it travels from one medium into another. We shall use these ideas to study the refraction of light as it passes through lenses and the reflection of light from mirrored surfaces. Finally, we shall describe how lenses and mirrors can be used to view objects with such instruments as cameras, telescopes, and microscopes.

"I procured me a Triangular glass-Prisme to try therewith the celebrated Pheaenomena of Colours. . . . I placed my Prisme at his entrance (the sunlight), that it might thereby be refracted to the opposite wall. It was a very pleasing divertisement to view the vivid and intense colours produced thereby; . . . I have often with admiration beheld that all the colours of the Prisme being made to converge, and thereby to be again mixed, as they were in the light before it was incident upon the Prisme, reproduced light, entirely and perfectly white, and not at all sensibly differing from the direct light of the sun. . . ."
ISAAC NEWTON

Conversation with
M. Brian Maple

M. Brian Maple spent his undergraduate years at Pomona College, the University of California at Berkeley, and San Diego State University. He graduated from San Diego State University in 1963 with an A.B. in mathematics and a B.S. in physics. He received his Ph.D. in physics from the University of California at San Diego (UCSD) in 1969 and was appointed to the faculty in the early 1970s, where he remains today teaching physics and conducting research in superconductivity and several other areas of condensed matter physics. He currently holds the Bernd T. Matthias Endowed Chair in Physics at UCSD and participates in an unusually wide variety of national and international advisory committees, including the Scientific and Academic Advisory Committee to the President of the University of California for Lawrence Livermore National Laboratory and Los Alamos National Laboratories, two national laboratories managed by the University of California.

What influenced you to get involved in the area of superconductivity?

In a way, it's hard to trace the reason back because I feel that all areas of science, and probably all areas of anything, are interesting once you get involved in them. So I don't know how conscious my choice was in the beginning. The reason I got in-

Interview conducted by Janice K. Mandel

volved in the area of superconductivity and magnetism is primarily because when I was a graduate student, there was an opportunity to join a research group at UCSD headed by Professor Bernd T. Matthias, one of the pioneers and leading world authorities in superconductivity. In fact, at the moment I hold an endowed chair that was established in his name after he passed away in 1980. In any case, I had the opportunity to get involved, and I knew this was very exciting research.

Why did you believe this was an exciting area of research?

It was an area that seemed to have a promising future, and many advances were being made. The phenomenon of superconductivity itself—the conduction of electrical current with no resistance—is a very exciting phenomenon; it's almost unbelievable. In graduate school at UCSD I had the opportunity to

meet well-known scientists from all over the world who came to give talks, and I soon learned about superconductivity and magnetism and other interesting physical phenomena. That's how I guess I first learned how exciting it was. And then my professor himself had some ideas, particularly about science, that were unconventional. He was outgoing, rather flamboyant and controversial, but very highly respected. That added to the excitement of the research.

What was it like at UCSD in the early '60s?

When UCSD started in the early '60s, the first and only department was the physics department, with a faculty of about 30 to 35 well-known, established physicists together with some promising young physicists. The department started out at Scripps Institution of Oceanogra-

phy, which is down by the ocean; the main campus wasn't constructed at that time. It started off right at the top, yet it was small and informal. There were many visits from world authorities in various fields of physics. In fact, the atmosphere was so informal that each graduate student had a key to the library, and we could go into the library at three o'clock in the morning and sit in there and read by ourselves.

What keeps you drawn to science?

The most attractive thing that I find about science after reaching a certain point is just the activity of research itself, because you really do work at the cutting edge of knowledge and pursue the unknown. It's the closest thing to being an explorer that you can do in these modern times.

Is it hard to get started in physics?

Physics has a bad reputation in the sense that people automatically think it's so hard that they can't learn it. I think that's not the case. It's a very logical, well-organized subject with interesting interrelationships, and after you study it for a while, you see a certain beauty and harmony in it. You have to learn some new ideas and concepts as well as the language of physics, once you do, you can then see things that you probably were prevented from seeing before because you hadn't developed this language. I think it's as exciting as anything you can imagine.

What kinds of research have you worked on?

Probably the common thread that goes through most of my research has been the interrelationship between superconductivity and magnetism. I have worked on that problem in a variety of different contexts throughout my entire career. I work in several different areas in addition to superconductivity, such as magnetism, surface science, and heavy fermions. The latter name refers to some remarkable phenomena that occur when electrons in solids interact very strongly with one another and produce effects that, prior to around 1970, no one ever knew could exist in metals. The potential in these materials is to understand new forms of superconductivity and new forms of magnetism.

Can you define superconductivity simply?

It's pretty mysterious in a way. It turns out that if you take most metals, such as aluminum, lead, or tin, and you cool them down far below room temperature (room temperature is about 300 K on the Kelvin scale, and 0 K is "absolute zero"), they go into what's called a superconducting state. The most dramatic property they have in this state is that they become perfect electrical conductors. And I mean perfect. They conduct electricity without any resistance whatsoever.

Would you expect that to happen?

No. And no one really anticipated it. It was a discovery made in 1911 by a Dutch physicist named Kamerlingh Onnes. A few years prior to that discovery, he had succeeded in liquefying helium gas, which in liquid form will boil at 4.2 K. Then he took the element mercury, which is a liquid at room temperature, and froze it so it would become a metallic solid, and he measured its resistance to the flow of an electrical current at low temperature by cooling it with the liquid helium. Once it was cooled below about 4.2 K, however, it lost all of its resistance to the flow of an electrical current, or in other words, the element mercury became a perfect conductor, a "superconductor"!

Is it practical to use liquid helium for commercial applications?

Yes, but it is rather expensive. Helium is a nonrenewable resource, it is expensive to liquefy, and once it is liquefied, it's difficult to handle because you have to put it into insulated containers to keep it from evaporating too rapidly. So the discovery of superconducting properties in oxide materials, which normally are very good electrical insulators, was very exciting.

Why was the discovery of superconducting properties in oxide materials such an exciting phenomenon?

It was something that the experts thought was completely impossible. The promise of the new oxide materials that have been made metallic by doping them with other elements is that some of them can be cooled into the superconducting state by immersion in liquid nitrogen, which boils at 77 K and is a renewable resource that is much cheaper and more convenient to work with than liquid helium. This means that many known applications of superconductivity, such as a superconducting power transmission line, could become economically more feasible. The reason high-temperature superconductivity wasn't discovered earlier is that it occurs in a class of materials, oxides, which people never believed were promising candidates for high-temperature superconductivity. This is now the class of materials where the highest temperature superconductivity has been found. It's amazing, and nobody really understands it.

What causes electrical resistance in conducting electricity in common metals such as copper?

The flowing electrons that constitute an electrical current in a metal experience collisions with imperfections in the metal, such as missing atoms or impurity atoms. That slows the electrons down. And the other thing that slows them down is that the atoms out of which the metal is made vibrate about their equilibrium positions due to their thermal energy. So as they're sitting there shaking around, the electrons come by and collide with them because they have moved out of position.

Do all metals become superconductors if you cool them enough?

No. Some metals, like gold and copper, have not been observed to become superconducting when cooled to temperatures below 1 mK. Other elements like iron, nickel, and cobalt, never become superconducting be-

cause they become ferromagnetic instead. However, there are certain materials that were discovered, in our laboratory here in the '70s, that do become simultaneously magnetic and superconducting, and that was a big surprise. That's one of the subjects I have worked on for many years.

Did you spend some bleary-eyed nights on this research?

I've seen a lot of sunrises during my career. In fact, recently there were a lot of people throughout the world having sleepless nights because of the extreme competition and excitement that abounded in the field when J. Georg Bednorz and Karl Alex Müller reported their breakthrough in high-temperature superconductivity (for which they received the Nobel Prize in Physics in 1987). When their initial, carefully worded report was confirmed by several other groups, people in laboratories throughout the entire world started working day and night, because nobody ever thought high-temperature superconductivity above about 25 K or 30 K would ever be observed. Now superconducting oxides have been found that become superconducting at temperatures as high as 125 K! It turns out that in 1987 I was the Vice Chairman of the Division of Condensed Matter Physics of the American Physical Society, and at the March meeting of the Society, which was held in New York City, we had convened a special session so that the various researchers could report their findings in this new chapter in the field of superconductivity. That became known as "The Woodstock of Physics."

Would you describe what it was like at "The Woodstock of Physics"?

At the beginning of these developments, it was decided to have this special session at the upcoming annual APS March meeting. There were so many scientists involved that ultimately we allowed anybody to speak for 5 minutes each in this session. In the end, this unique event started at 7:30 in the evening and didn't end until 3:30 in the morning. There were at least 4,000 people there, and we couldn't get a room big enough. So we had closed circuit televisions put in lobbies and other rooms in the hotel so those people who couldn't get into the main hall could see it from other vantage points. I don't think there has ever been a session like this, and I don't know whether there ever will be another one.

Do scientists really understand why superconductors have these properties?

Conventional superconductivity is pretty well understood. But superconductivity that occurs in these new oxide materials that have been discovered since the latter part of 1986 is not very well understood. It is not known whether it's the same kind of superconductivity that we understand in conventional materials, or whether it's something different and brought about by another mechanism. That's a major theoretical question, and there is a lot of work going on to develop a theory that can explain the properties of these superconducting materials. This will allow us to predict how to use them in applications, refine them to make superconducting properties that are better suited to technology, or even to discover superconductors that have properties that make them more useful.

Can you describe one of the more interesting applications of superconductivity?

Magnetically levitated trains are the most romantic application of superconductivity (see Special Topic: Superconductivity in Chapter 18). One scheme being pursued by the Japanese involves the use of on-board superconducting electromagnets, a metallic track, and a train with wheels to ride on when it's at rest or moving slowly. As the train picks up speed, the magnetic field that the magnets generate induces currents to flow in the conducting track, and those currents in turn interact with the magnetic field and repel the magnets and the train to which they're attached. That's what causes the train to levitate. The faster the train moves, the greater the height at which it levitates. The train literally flies through the air supported on a magnetic cushion. Some of the models tested move at speeds in the neighborhood of 300 miles per hour. It's a very smooth ride. In Japan they have prototype trains on Kyushu Island that are capable of carrying passengers.

What other promising applications have emerged affecting other fields?

One application is the magnetic resonance imaging machines that are now used in hospitals throughout the world to map out the body and organs for medical diagnostic purposes. Those machines rely on very large super-

conducting electromagnets made into the form of a coil with a diameter large enough to put a person inside.

Another medical application being developed uses another superconducting phenomenon called the Josephson effect (a prediction for which Brian Josephson received the Nobel Prize in 1973). That effect can be used to make very sensitive magnetic field sensing devices called Superconducting QUantum Interference Devices (SQUIDs). Arrays of SQUIDs are currently being developed to measure the magnetic fields produced by the brain for medical diagnostic purposes. I can imagine the apparatus would resemble a hat that fits over the subject's head.

What interests you now? What do you foresee working on in the future?

Of course, one never knows exactly what one is going to do, but I foresee looking for new oxide materials, new heavy electron materials, and studying their superconducting properties and the properties that they have when they're in the normal state to try to advance our general understanding of these materials. Many of the new superconductors were discovered by trying to take a known superconducting material and alloying different elements into it. What sometimes happens is that you stumble on a new structure, and this new material itself can sometimes have more dramatic properties and give you a new class of materials that you can then study. Materials are so complicated that you need to have some kind of guiding intuition based upon your experience with materials and your understanding of the available theories.

Does the unpredictability of working with new materials make it fun for you?

Absolutely. You never know what's going to happen until you do it, and sometimes remarkable surprises occur and, of course, that's where breakthroughs are often made. When Bednorz and Müller discovered the high-temperature superconductivity that started this recent revolution, I am sure it wasn't because they knew it was going to happen. They just had a working hypothesis and set out to explore this high-risk problem, and they hit pay dirt.

What can students expect if they enter the field today?

Other breakthroughs could occur, you can't rule that out. At this point, I think the experts agree—although they don't know—that room-temperature superconductivity may be possible. This would truly have a profound effect on technology. There are problems that have not been overcome. So, there's a big future in it, and this future is multifaceted. This interdisciplinary subject involves materials scientists, chemists, physicists, and engineers, and it will affect all of them. The future scientists and engineers will be the ones that are going to apply all of this to technology as well as advance the field further. The theory that explains superconductivity in oxide materials has yet to be developed. There are many more superconducting compounds that await to be discovered, and presumably there will be materials with even better superconducting properties. I can't believe that we've reached the end.

CHAPTER 24

Reflection and Refraction of Light

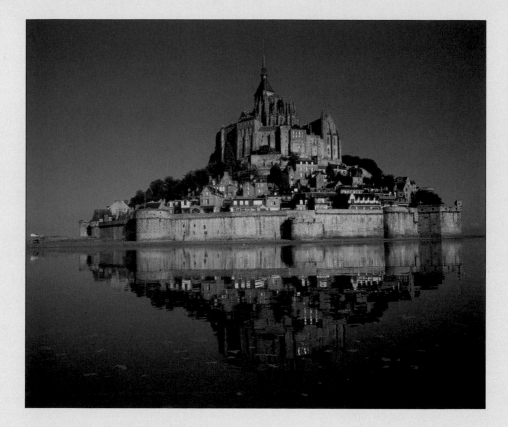

Dramatic photograph of Mont-Saint-Michel in France and its "mirror" reflection in water.
(© Travelpix 1989/FPG International)

24.1 The Nature of Light

Until the beginning of the 19th century, light was considered to be a stream of particles that were emitted by a light source and stimulated the sense of sight upon entering the eye. The chief architect of this particle theory of light was Newton. With this theory, Newton was able to provide a simple explanation of some known experimental facts concerning the nature of light, namely, the laws of reflection and refraction.

Most scientists accepted Newton's particle theory of light. However, during Newton's lifetime another theory was proposed, one that argued that light might be some sort of wave motion. In 1678, a Dutch physicist and astronomer, Christian Huygens (1629–1695), showed that a wave theory of light could explain the laws of reflection and refraction. The wave theory did not receive immediate acceptance for several reasons. All the waves known at the time (sound, water, and so on) traveled through some sort of medium, but light from the Sun could travel to us through space. Furthermore, it was argued that if light were some form of wave motion, the waves could bend around obstacles; hence, we should be able to see around corners. It is now known that light

Christian Huygens was a Dutch physicist and astronomer who is best known for his contributions to the fields of dynamics and optics. As a physicist, his accomplishments include invention of the pendulum clock and the first exposition of a wave theory of light. As an astronomer, Huygens was the first to recognize the rings around Saturn and to discover Titan, a satellite of Saturn.

Huygens was born in 1629 into a prominent family in The Hague. He was the son of Constantin Huygens, one of the most important figures of the Renaissance in Holland. Educated at the University of Leyden, Christian became a close friend of René Descartes, who was a frequent guest at the Huygens home. Huygens published his first paper in 1651 on the subject of the quadrature of various curves.

Huygens' reputation in optics and dynamics spread throughout Europe, and in 1663 he was elected a charter member of the Royal Society. Louis XIV lured Huygens to France in 1666, in accordance with the king's policy of collecting scholars for the glory of his regime. While in France, Huygens became one of the founders of the French Academy of Science.

In 1673, in Paris, Huygens published *Horologium Oscillatorium*. In this work he described a solution to the problem of the compound pendulum, for which he calculated the equivalent simple pendulum length. In the same publication he also derived a formula for computing the period of oscillation of a simple pendulum, and he explained the force governing uniform motion in a circle.

Huygens returned to Holland in 1681, constructed some lenses of large focal lengths, and invented the achromatic eyepiece for telescopes. Shortly after returning from a visit to England, where he met Isaac Newton, Huygens published his treatise on the wave theory of light. To Huygens, light was a vibratory motion in the ether, spreading out and producing the sensation of light when impinging on the eye. On the basis of this theory, he was able to deduce the laws of reflection and refraction and to explain the phenomenon of double refraction.

Huygens, second only to Newton among the greatest scientists in the second half of the 17th century, was the first to proceed in the field of dynamics beyond the point reached by Galileo and Descartes. A solitary man, Huygens did not attract students or disciples and was very slow in publishing his findings. Huygens died in 1695 after a long illness.

Christian Huygens
(1629–1695)

(Photo courtesy of Rijksmuseum voor de Geschiedenis der Natuurwetenschappen. Courtesy AIP Niels Bohr Library)

does indeed bend around the edges of objects. This phenomenon, known as *diffraction,* is not easy to observe because light waves have such short wavelengths. Thus, although experimental evidence for the diffraction of light was discovered by Francesco Grimaldi (1618–1663) around 1660, most scientists rejected the wave theory and adhered to Newton's particle theory for more than a century. This was, for the most part, due to Newton's great reputation as a scientist.

The first clear demonstration of the wave nature of light was provided in 1801 by Thomas Young (1773–1829), who showed that, under appropriate conditions, light exhibits interference behavior. That is, at certain points in the vicinity of two sources, light waves can combine and cancel each other by destructive interference. Such behavior could not be explained at that time by a particle theory because there was no conceivable way by which two or more particles could come together and cancel one another. Several years later, a French physicist, Augustin Fresnel (1788–1829), performed a number of detailed experiments dealing with interference and diffraction phenomena. In 1850, Jean Foucault (1791–1868) provided further evidence of the inadequacy of the particle theory by showing that the speed of light in liquids is less than in air. According to the particle model of light, the speed of light would be higher in glasses and liquids than in air. Additional developments during the 19th century led to the general acceptance of the wave theory of light.

The most important development concerning the theory of light was the work of Maxwell, who in 1865 predicted that light was a form of high-frequency electromagnetic wave (Chap. 23). His theory predicted that these waves should have a speed of about 3×10^8 m/s. Within experimental error, this value is equal to the speed of light. As discussed in Chapter 23, Hertz provided experimental confirmation of Maxwell's theory in 1887 by producing and detecting electromagnetic waves. Furthermore, Hertz and other investigators showed that these *waves exhibited reflection, refraction, and all the other characteristic properties of waves.*

Although the classical theory of electricity and magnetism was able to explain most known properties of light, some subsequent experiments could not be explained by assuming that light was a wave. The most striking of these is the *photoelectric effect*, also discovered by Hertz. The photoelectric effect is the ejection of electrons from a metal whose surface is exposed to light. As one example of the difficulties that arose, experiments showed that the kinetic energy of an ejected electron is *independent* of the light intensity. This was in contradiction of the wave theory, which held that a more intense beam of light should add more energy to the electron. An explanation of this phenomenon was proposed by Einstein in 1905. Einstein's theory used the concept of quantization developed by Max Planck (1858–1947) in 1900. The quantization model assumes that the energy of a light wave is present in bundles of energy called *photons*; hence the energy is said to be *quantized*. (Any quantity that appears in discrete units is said to be quantized. For example, electric charge is quantized because it always appears in multiples of 1.6×10^{-19} C.) Einstein modified and reinterpreted the work of Planck to show that the energy of a photon is proportional to the frequency of the electromagnetic wave:

Energy of a photon

$$E = hf \qquad \qquad \textbf{(24.1)}$$

where $h = 6.63 \times 10^{-34}$ J·s is *Planck's constant*. This theory retains some features of both the wave theory and the particle theory of light. As we shall discuss later, the photoelectric effect is the result of energy transfer from a single photon to an electron in the metal. That is, the electron interacts with one photon of light as if it, the electron, had been struck by a particle. Yet this photon has wave-like characteristics because its energy is determined by the frequency (wave-like quantity).

In view of these developments, light must be regarded as having a *dual nature*. That is, *in some cases light acts like a wave and in others it acts like a particle.* Classical electromagnetic wave theory provides an adequate explanation of light propagation and of the effects of interference, whereas the photoelectric effect and other experiments involving the interaction of light with matter are best explained by assuming that light is a particle. Light is light, to be sure. However, the question, "Is light a wave or a particle?" is an inappro-priate one. Sometimes it acts like a wave, sometimes like a particle. Fortunately, it never acts like both in the same experiment.

24.2   Measurements of the Speed of Light

Light travels at such a high speed ($c \approx 3 \times 10^8$ m/s) that early attempts to measure its speed were unsuccessful. Galileo attempted to measure the speed of light by positioning two observers in towers separated by about 5 miles. Each observer carried a shuttered lantern. One observer would open his lantern first, and then the other would open his lantern at the moment he saw the light from the first lantern. The velocity could then be obtained, in principle, knowing the transit time of the light beams between lanterns. However, at a speed of 186,000 mi/s, light would travel the ten-mile roundtrip in a time of approximately 54 μs. Thus, it is impossible to measure the speed of light in this manner because the transit time is very small compared with the reaction time of the observers.

Roemer's Method

The first successful estimate of the speed of light was made in 1675 by the Danish astronomer Ole Roemer (1644–1710). His technique involved astronomical observations of one of the moons of Jupiter, called Io. At that time only four of Jupiter's moons had been discovered, and the periods of their orbits were known. Io, the innermost moon, has a period of 42.5 h, and its orbit, the orbit of Jupiter, and the orbit of the Earth all lie in approximately the same plane. As a result, Io goes into eclipse behind Jupiter every 42.5 h as seen from Earth. Using the orbital motion of Io as a clock, Roemer expected to find a constant period over long time intervals. Instead, Roemer observed a systematic variation in Io's period. He found that the periods were larger than average when Earth receded from Jupiter and smaller than average when Earth approached Jupiter. If Io had a constant period, Roemer should have been able to observe a particular eclipse and be able to predict when the next eclipse would occur. However, when Roemer checked to see if the second eclipse did occur at the predicted time, he often found that his predictions did not agree with the actual time of the eclipse. For example, consider the situation shown in Figure 24.1. Suppose an eclipse had occurred when the Earth was at position E_1. Knowing the period of Io, Roemer was able to predict when an eclipse should occur three months later when the Earth was at position E_2. However, the eclipse at E_2 actually occurred at a time approximately 600 s later than the predicted time. Roemer attributed this discrepancy to the fact that the distance between the Earth and Jupiter was changing between these observations. In three months (one quarter of the Earth's period), the Earth has moved through one fourth of its orbit from E_1 to E_2 as shown in Figure 24.1. In the same time interval, Jupiter, whose period is about 12 years, has moved a much shorter distance from J_1 to J_2. Therefore as the Earth moves from E_1 to E_2 light from Jupiter has to travel an additional distance equal to the radius of the Earth's orbit.

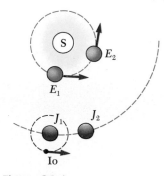

Figure 24.1
As the Earth moves from E_1 to E_2, Jupiter moves only from J_1 to J_2.

Using data available at that time, Roemer would have estimated the speed of light to be about 2.1×10^8 m/s. The large discrepancy between this value and the currently accepted value of 3×10^8 m/s is due to a large error in the assumed radius of the Earth's orbit. This experiment is important historically because it demonstrated that light does have a finite speed and established a rough estimate of the magnitude of this speed.

Fizeau's Technique

The first successful method of measuring the speed of light using purely terrestrial techniques was developed in 1849 by Armand H. L. Fizeau (1819–1896). Figure 24.2 represents a simplified diagram of his apparatus. The basic idea is to measure the total time it takes light to travel from some point to a distant mirror and back. If d is the distance between the light source and the mirror and if the transit time for one round trip is t, then the speed of light is $c = 2d/t$. To measure the transit time, Fizeau used a rotating toothed wheel, which converts an otherwise continuous beam of light into a series of light pulses. Additionally, the rotation of the wheel controls what an observer at the light source sees. For example, if the light passing the opening at point A in Figure 24.2 should return at the instant that tooth B had rotated into position to cover the return path, the light would not reach the observer. At a faster rate of rotation, the opening at point C could move into position to allow the reflected beam to pass and reach the observer. Knowing the distance, d, the number of teeth in the wheel, and the angular velocity of the wheel, Fizeau arrived at a value of $c = 3.1 \times 10^8$ m/s. Similar measurements made by subsequent investigators yielded more accurate values for c, approximately 2.9977×10^8 m/s.

A variety of other, more accurate measurements have been reported for the value of c. A recent value obtained using laser techniques is

$$c = 2.997924574(12) \times 10^8 \text{ m/s}$$

where the (12) indicates the uncertainty in the last two digits. The number of significant figures here is certainly impressive. In fact, the speed of light has been determined with such high accuracy that it is now used to define the SI unit of length, the meter. As noted in Chapter 1, the meter is defined as the distance traveled by light in a vacuum during a time of 1/299,792,458 s.

Figure 24.2
Fizeau's method for measuring the speed of light using a rotating, toothed wheel.

Example 24.1   Measuring the Speed of Light with Fizeau's Toothed Wheel

Assume the toothed wheel of the Fizeau experiment has 360 teeth and is rotating with a speed of 27.5 rev/s when the light from the source is extinguished, that is, when a burst of light passing through opening A in Figure 24.2 is blocked by tooth B on return. If the distance to the mirror is 7500 m, find the speed of light.

Solution
If the wheel has 360 teeth, it will turn through an angle of 1/720 rev in the time that passes while the light makes its round trip. From the definition of angular velocity, we see that the time is

$$t = \frac{\theta}{\omega} = \frac{(1/720) \text{ rev}}{27.5 \text{ rev/s}} = 5.05 \times 10^{-5} \text{ s}$$

Hence, the speed of light is

$$c = \frac{2d}{t} = \frac{2(7500 \text{ m})}{5.05 \times 10^{-5} \text{ s}} = \boxed{2.97 \times 10^8 \text{ m/s}}$$

24.3 Huygens' Principle

We shall develop the laws of reflection and refraction by using a geometric method proposed by Huygens in 1678. Huygens assumed that light is some form of wave motion rather than a stream of particles. He had no knowledge of the nature of light or of its electromagnetic character. Nevertheless, his simplified wave model is adequate for understanding many practical aspects of the propagation of light.

Huygens' principle is a geometric construction for determining the position of a new wavefront at some instant from the knowledge of an earlier wavefront. (A wavefront is a surface passing through those points of a wave that are behaving identically. For instance, a wavefront could be a surface over which the wave is at a crest.) In Huygens' construction,

> all points on a given wavefront are taken as point sources for the production of spherical secondary waves, called wavelets, which propagate outward with speeds characteristic of waves in that medium. After some time has elapsed, the new position of the wavefront is the surface tangent to the wavelets.

Figure 24.3 illustrates two simple examples of Huygens' construction. First, consider a plane wave moving through free space, as in Figure 24.3a. At $t = 0$, the wavefront is indicated by the plane labeled AA'. In Huygens' construction, each point on this wavefront is considered a point source. For clarity, only a few points on AA' are shown. With these points as sources for the wavelets, we draw circles each of radius ct, where c is the speed of light in vacuum and t is the time of propagation from one wavefront to the next. The surface drawn tangent to these wavelets is the plane BB', which is parallel to AA'. In a similar manner, Figure 24.3b shows Huygens' construction for an outgoing spherical wave.

A convincing demonstration of Huygens' principle is obtained with water waves in a shallow tank (called a ripple tank), as in Figure 24.4. Plane waves

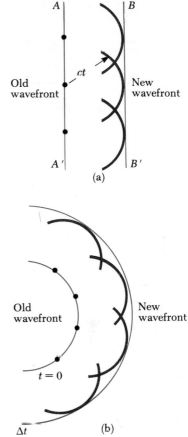

Figure 24.3
Huygens' construction for (a) a plane wave propagating to the right and (b) a spherical wave.

Figure 24.4
Water waves in a ripple tank, which demonstrates Huygens' wavelets. A plane wave is incident on a barrier with a small opening. The opening acts as a source of circular wavelets. *(Photograph courtesy of Education Development Center, Newton, MA)*

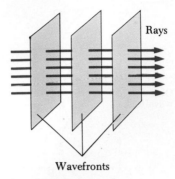

Figure 24.5
A plane wave traveling to the right. Note that the rays, corresponding to the direction of wave motion, are straight lines perpendicular to the wavefronts.

produced below the slit emerge above the slit as two-dimensional circular waves propagating outward.

The Ray Approximation in Geometric Optics

In studying geometric optics here and in Chapter 25, we shall use what is called the *ray approximation*. To understand this approximation, first recall that the direction of energy flow of a wave, corresponding to the direction of wave propagation, is called a ray. The rays of a given wave are straight lines that are perpendicular to the wavefronts, as illustrated in Figure 24.5 for a plane wave. In the ray approximation, we assume that a wave moving through a medium travels in a straight line in the direction of its rays. That is, a ray is a line drawn in the direction in which the light is traveling. For example, a beam of sunlight passing through a darkened room traces out the path of a ray.

24.4   Reflection and Refraction

Reflection of Light

When a light ray traveling in a medium encounters a boundary leading into a second medium, part of the incident ray is reflected back into the first medium. Figure 24.6a shows several rays of a beam of light incident on a smooth, mirror-like, reflecting surface. The reflected rays are parallel to each other, as indicated in the figure. Reflection of light from such a smooth surface is called *specular reflection*. On the other hand, if the reflecting surface is rough, as in Figure 24.6b, the surface will reflect the rays in various directions. Reflection

Figure 24.6
Schematic representation of (a) specular reflection, where the reflected rays are all parallel to each other, and (b) diffuse reflection, where the reflected rays travel in random directions. (c) and (d) Photographs of specular and diffuse reflection using laser light. *(Photographs courtesy of Henry Leap and Jim Lehman)*

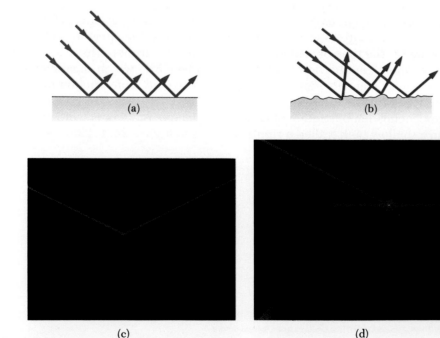

from any rough surface is known as *diffuse reflection.* A surface will behave as a smooth surface as long as the surface variations are small compared with the wavelength of the incident light. Photographs of specular and diffuse reflection using laser light are shown in Figures 24.6c and 24.6d.

For instance, consider the two types of reflection one can observe from a road's surface while driving a car at night. When the road is dry, light from oncoming vehicles is scattered off the road in different directions (diffuse reflection) and the road is quite visible. On a rainy night, when the road is wet, the road irregularities are filled with water. Because the water surface is quite smooth, the light undergoes specular reflection. In this book, we shall concern ourselves only with specular reflection, and we shall use the term *reflection* to mean specular reflection.

Consider a light ray traveling in air and incident at an angle on a flat, smooth surface, as in Figure 24.7. The incident and reflected rays make angles θ_1 and θ_1', respectively, with a line drawn perpendicular to the surface at the point where the incident ray strikes the surface. We shall call this line the *normal* to the surface. Experiments show that *the angle of reflection equals the angle of incidence*, that is,

$$\theta_1' = \theta_1 \tag{24.2}$$

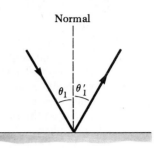

Figure 24.7
According to the law of reflection, $\theta_1 = \theta_1'$.

Example 24.2 The Double-Reflecting Light Ray

Two mirrors make an angle of 120° with each other, as in Figure 24.8. A ray is incident on mirror M_1 at an angle of 65° to the normal. Find the direction of the ray after it is reflected from mirror M_2.

Solution
From the law of reflection, we see that the first reflected ray also makes an angle of 65° with the normal. Thus, it follows that this same ray makes an angle of 90° − 65°, or 25°, with the horizontal. From the triangle made by the first reflected ray and the two mirrors, we see that the first reflected ray makes an angle of 35° with M_2 (since the sum of the interior angles of any triangle is 180°). This means that this ray makes an angle of 55° with the normal of M_2. Hence, from the law of reflection, it follows that the second reflected ray makes an angle of 55° with the normal to M_2.

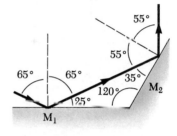

Figure 24.8 (Example 24.2) Mirrors M_1 and M_2 make an angle of 120° with each other.

Refraction of Light

When a ray of light traveling through a transparent medium encounters a boundary leading into another transparent medium, as in Figure 24.9a, part of the ray is reflected and part enters the second medium. The ray that enters the second medium is bent at the boundary and is said to be *refracted.* The incident ray, the reflected ray, and the refracted ray all lie in the same plane. The **angle of refraction,** θ_2 in Figure 24.9a, depends on the properties of the two media and on the angle of incidence through the relationship

$$\frac{\sin \theta_2}{\sin \theta_1} = \frac{v_2}{v_1} = \text{constant} \tag{24.3}$$

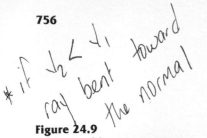

if $v_2 < v_1$ ray bent toward the normal

Figure 24.9
(a) A ray obliquely incident on an air-glass interface. The refracted ray is bent toward the normal since $v_2 < v_1$.
(b) Light incident on the lucite block bends when it enters the block and also when it leaves the block.

(a)

(b)

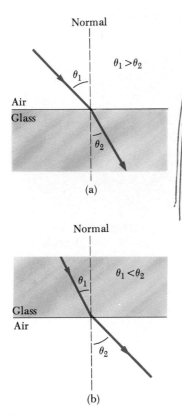

Figure 24.10
(a) When the light beam moves from air into glass, its path is bent toward the normal. (b) When the beam moves from glass into air, its path is bent away from the normal.

where v_1 is the speed of light in medium 1 and v_2 is the speed of light in medium 2. The experimental discovery of this relationship is usually credited to Willebrord Snell (1591–1626) and is therefore known as **Snell's law.** In Section 24.8, we shall verify the laws of reflection and refraction using Huygens' principle.

It is found that *the path of a light ray through a refracting surface is reversible.* For example, the ray in Figure 24.9a travels from point A to point B. If the ray originated at B, it would follow the same path to reach point A. In the latter case, however, the reflected ray would be in the glass.

When light moves from a material in which its speed is high to a material in which its speed is lower, the angle of refraction, θ_2, is less than the angle of incidence, as shown in Figure 24.10a. If the ray moves from a material in which it moves slowly to a material in which it moves more rapidly, it is bent away from the normal, as shown in Figure 24.10b.

The behavior of light as it passes from air into another substance and then re-emerges into air is often a source of confusion to students. Let us take a look at what happens and see why this behavior is so different from other occurrences in our daily lives. The speed of light in air is about 3×10^8 m/s, and its speed is reduced to about 2×10^8 m/s upon entering a block of glass. When the light re-emerges into air, its speed instantaneously increases to its original value of 3×10^8 m/s. This is far different from what happens, for example, when a bullet is fired through a block of wood. In this case, the speed of the bullet is reduced as it moves through the wood because some of its original energy is used to tear apart the fibers of the wood. When the bullet enters the air once again, it emerges at the speed it had just before leaving the block of wood.

In order to see why light behaves as it does, consider Figure 24.11, which represents a beam of light entering a piece of glass from the left. Once inside the glass, the light may encounter an electron bound to an atom, indicated as point A in the figure. Let us assume that light is absorbed by the atom, which causes the electron to oscillate. The oscillating electron then acts as an antenna and radiates the beam of light toward an atom at point B, where the light is again absorbed by an atom at that point. The details of these absorptions and emissions are best explained in terms of quantum mechanics, a subject we shall study in Chapter 29. For now, it is sufficient to think of the process as one in which the light passes from one atom to another through the glass. (The situation is somewhat analogous to a relay race in which a baton is passed between runners on the same team.) Although light travels from one atom to another with a speed of 3×10^8 m/s, the processes of absorption and emission of light by the atoms take time. Enough time is required, in fact, to lower the speed of light in glass to 2×10^8 m/s. Once the light emerges into the air, the absorptions and emissions cease and its speed returns to the original value.

Figure 24.11
Light passing from one atom to another in a material medium.

24.5   The Law of Refraction

When light passes from one medium to another, it is refracted because the speed of light is different in the two media. In general, the speed of light in any material is less than the speed of light in vacuum. In fact, *light travels at its maximum speed in vacuum.* It is convenient to define the **index of refraction**, n, of a medium to be the ratio

$$n \equiv \frac{\text{speed of light in vacuum}}{\text{speed of light in a medium}} = \frac{c}{v} \qquad \textbf{(24.4)}$$

Index of refraction

From this definition, we see that the index of refraction is a dimensionless number greater than unity since v is always less than c. Furthermore, n is equal to unity for a vacuum. The indices of refraction for various substances measured with respect to vacuum are listed in Table 24.1.

As light travels from one medium to another, *the frequency of the light does not change.* To see why this is so, consider Figure 24.12. Wavefronts pass an observer at point A in medium 1 with a certain frequency and are incident on the boundary between medium 1 and medium 2. The frequency with which the wavefronts pass an observer at point B in medium 2 must equal the frequency at which they arrive at point A in medium 1. If this were not the case, wavefronts would either be piling up at the boundary or they would be destroyed or created at the boundary. Since there is no mechanism for this to occur, the frequency must be a constant as a light ray passes from one medium into another.

Figure 24.12
As the wave moves from medium 1 to medium 2, its wavelength changes but its frequency remains constant.

Table 24.1
Index of Refraction for Various Substances Measured with Light of Vacuum Wavelength $\lambda_0 = 589$ nm

Substance	Index of Refraction	Substance	Index of Refraction
Solids at 20° C		Liquids at 20° C	
Diamond (C)	2.419	Benzene	1.501
Fluorite (CaF_2)	1.434	Carbon disulfide	1.628
Fused quartz (SiO_2)	1.458	Carbon tetrachloride	1.461
Glass, crown	1.52	Ethyl alcohol	1.361
Glass, flint	1.66	Glycerine	1.473
Ice (H_2O) (at 0° C)	1.309	Water	1.333
Polystyrene	1.49	Gases at 0° C, 1 atm	
Sodium chloride (NaCl)	1.544	Air	1.000293
Zircon	1.923	Carbon dioxide	1.00045

if $v_2 < v_1$ and
then $n_2 > n_1$,
$\lambda_2 < \lambda_1$

Therefore, because the relation $v = f\lambda$ must be valid in both media and because $f_1 = f_2 = f$, we see that

$$v_1 = f\lambda_1 \qquad \text{and} \qquad v_2 = f\lambda_2$$

where the subscripts refer to the two media. A relationship between index of refraction and wavelength can be obtained by dividing these two equations and making use of the definition of the index of refraction given by Equation 24.4:

$$\frac{\lambda_1}{\lambda_2} = \frac{v_1}{v_2} = \frac{c/n_1}{c/n_2} = \frac{n_2}{n_1} \tag{24.5}$$

which gives

$$\lambda_1 n_1 = \lambda_2 n_2 \tag{24.6}$$

If medium 1 is a vacuum, or for all purposes air, then $n_1 = 1$. Hence, it follows from Equation 24.6 that the index of refraction of any medium can be expressed as the ratio

$n_1 \mid n_2$

$|\!\!\leftarrow\!\!\lambda_1\!\!\rightarrow\!\!|$ $|\!\leftarrow\!\lambda_2\!\rightarrow\!|$

$n_2 > n_1$
$\lambda_2 < \lambda_1$

Figure 24.13
Schematic diagram of the *reduction* in wavelength when light travels from a medium of low index of refraction to one of higher index of refraction.

$$n = \frac{\lambda_0}{\lambda_n} \tag{24.7}$$

where λ_0 is the wavelength of light in vacuum and λ_n is the wavelength in the medium whose index of refraction is n. A schematic representation of this reduction in wavelength is shown in Figure 24.13.

We are now in a position to express Snell's law (Eq. 24.3) in an alternative form. If we substitute Equation 24.5 into Equation 24.3, we get

Snell's law

$$n_1 \sin \theta_1 = n_2 \sin \theta_2 \tag{24.8}$$

This is the most widely used and practical form of **Snell's law.**

Example 24.3   An Index of Refraction Measurement

A beam of light of wavelength 550 nm traveling in air is incident on a slab of transparent material. The incident beam makes an angle of 40° with the normal, and the refracted beam makes an angle of 26° with the normal. Find the index of refraction of the material.

Solution
Snell's law of refraction (Eq. 24.8) together with the given data, $\theta_1 = 40°$, $n_1 = 1.00$ for air, and $\theta_2 = 26°$, gives

$$n_1 \sin \theta_1 = n_2 \sin \theta_2$$

$$n_2 = \frac{n_1 \sin \theta_1}{\sin \theta_2} = (1.00)\frac{\sin 40°}{\sin 26°} = \frac{0.643}{0.440} = \boxed{1.46}$$

If we compare this value with the data in Table 24.1, we see that the material is probably fused quartz.

Exercise   What is the wavelength of light in the material?

Answer   377 nm

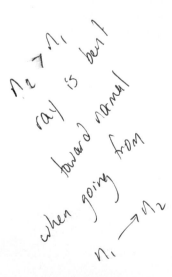

$n_2 > n_1$
ray is bent
toward normal
when going from
$n_1 \to n_2$

Example 24.4 Angle of Refraction for Glass

A light ray of wavelength 589 nm (produced by a sodium lamp) traveling through air is incident on a smooth, flat slab of crown glass at an angle of 30° to the normal, as sketched in Figure 24.14. Find the angle of refraction, θ_2.

Solution
Snell's law given by Equation 24.8 can be rearranged as

$$\sin \theta_2 = \frac{n_1}{n_2} \sin \theta_1$$

From Table 24.1, we find that $n_1 = 1.00$ for air and $n_2 = 1.52$ for crown glass. Therefore, the unknown refraction angle is

$$\sin \theta_2 = \left(\frac{1.00}{1.52}\right) (\sin 30°) = 0.329$$

$$\theta_2 = \sin^{-1} (0.329) = \boxed{19.2°}$$

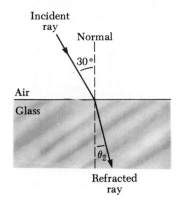

Figure 24.14 (Example 24.4) Refraction of light by glass.

Thus we see that the ray is bent *toward* the normal, as expected.

Exercise If the light ray moves from inside the glass toward the glass-air interface at an angle of 30° to the normal, determine the angle of refraction.

Answer 49.5° *away* from the normal.

Example 24.5 The Speed of Light in Quartz

Light of wavelength 589 nm in vacuum passes through a piece of fused quartz whose index of refraction is ($n = 1.458$). (a) Find the speed of light in quartz.

Solution
The speed of light in quartz can be easily obtained from Equation 24.4:

$$v = \frac{c}{n} = \frac{3 \times 10^8 \text{ m/s}}{1.458} = \boxed{2.058 \times 10^8 \text{ m/s}}$$

It is interesting to note that the speed of light in vacuum, 3×10^8 m/s, is an upper limit on the speed of material objects. In our treatment of relativity in Chapter 28, we shall find that this upper limit is consistent with experimental observations. However, it is possible for a particle moving in a medium to have a speed that exceeds the speed of light in that medium. For example, it is theoretically possible for a particle to travel through fused quartz at a speed greater than 2.06×10^8 m/s, but it must have a speed less than 3×10^8 m/s when moving in a vacuum.

(b) What is the wavelength of this light in quartz?

Solution
We can use $\lambda_n = \lambda_0/n$ (Eq. 24.7) to calculate the wavelength in quartz, noting that we are given $\lambda_0 = 589$ nm $= 589 \times 10^{-9}$ m:

$$\lambda_n = \frac{\lambda}{n} = \frac{589 \text{ nm}}{1.458} = \boxed{404 \text{ nm}}$$

Exercise Find the frequency of the light passing through the quartz.

Answer 5.09×10^{14} Hz.

Figure 24.15 (Example 24.6) When light passes through a flat slab of material, the emerging beam is parallel to the incident beam, and therefore $\theta_1 = \theta_3$.

Example 24.6   Light Passing Through a Slab

A light beam passes from medium 1 to medium 2 through a thick slab of material whose index of refraction is n_2 (Fig. 24.15). Show that the emerging beam is parallel to the incident beam.

Solution

First, let us apply Snell's law to the upper surface:

$$(1) \qquad \sin \theta_2 = \frac{n_1}{n_2} \sin \theta_1$$

Applying Snell's law to the lower surface gives

$$(2) \qquad \sin \theta_3 = \frac{n_2}{n_1} \sin \theta_2$$

Substituting (1) into (2) gives

$$\sin \theta_3 = \frac{n_2}{n_1} \left(\frac{n_1}{n_2} \sin \theta_1 \right) = \sin \theta_1$$

That is, $\theta_3 = \theta_1$, and so the layer does not alter the direction of the beam. It does, however, produce a displacement of the beam. The same result is obtained when light passes through multiple layers of materials.

24.6   Dispersion and Prisms

An important property of the index of refraction is that it is different for different wavelengths of light. A graph of the index of refraction for three materials is shown in Figure 24.16. Since n is a function of wavelength, Snell's law indicates that light of *different wavelengths* will be bent at *different angles* when incident on a refracting material. As we see from Figure 24.16, the index of refraction decreases with increasing wavelength. This means that blue light will bend more than red light when passing into a refracting material.

Any substance in which n varies with wavelength is said to exhibit **dispersion**. To understand the effects that dispersion can have on light, let us con-

Figure 24.16
Variations of index of refraction with wavelength for three materials.

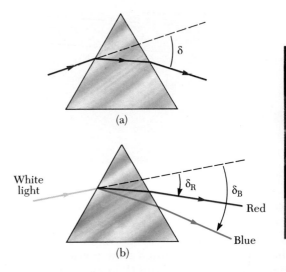

(a)

White light

δ_R δ_B

Red

Blue

(b)

Figure 24.17
(a) A prism refracts a light ray and deviates the light through an angle δ. (b) When light is incident on a prism, the blue light is bent more than the red. (c) Photograph of light being refracted by a prism.

sider what happens when light strikes a prism, as in Figure 24.17a. A single ray of light incident on the prism from the left emerges bent away from its original direction of travel by an angle δ, called the **angle of deviation.** Now suppose a beam of white light (a combination of all visible wavelengths) is incident on a prism, as in Figure 24.17b. The blue component of the incident beam is bent more than the red component, and the rays that emerge from the second face of the prism spread out in a series of colors known as a **spectrum,** as shown in Figure 24.18. These colors, in order of decreasing wavelength, are red, orange, yellow, green, blue, indigo, and violet. Clearly, the angle of deviation, δ, depends on the wavelength of a given color. Violet light deviates the most, red light deviates the least, and the remaining colors in the visible spectrum fall between these extremes. When light is spread out by a substance such as the prism, the light is said to be dispersed into a spectrum.

Figure 24.18
(a) Dispersion of white light by a prism. Since n varies with wavelength, the prism disperses the white light into its various spectral components. (b) The various colors are refracted at different angles because the index of refraction of the glass depends on wavelength. The blue light deviates the most, while red light deviates the least.
(Photograph courtesy of Bausch and Lomb)

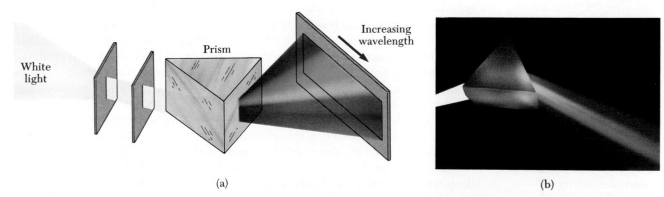

White light

Prism

Increasing wavelength

(a)

(b)

Figure 24.19
(a) Diagram of a prism spectrometer. The various colors in the spectrum are viewed through a telescope. (b) Photograph of a prism spectrometer. *(Courtesy of CENCO)*

(a)   (b)

A prism is often used in an instrument known as a **prism spectrometer,** the essential elements of which are shown in Figure 24.19a. The instrument is commonly used to study the wavelengths emitted by a light source, such as a sodium vapor lamp. Light from the source is sent through a narrow, adjustable slit to produce a parallel, or collimated, beam. The light then passes through the prism and is dispersed into a spectrum. The refracted light is observed through a telescope. The experimenter sees an image of the slit through the eyepiece of the telescope. The telescope can be moved or the prism can be rotated in order to view the various wavelengths, which have different angles of deviation. Figure 24.19b is a photograph of a prism spectrometer used in undergraduate laboratories.

All hot low-pressure gases emit their own characteristic spectrum. Thus, one use of a prism spectrometer is to identify gases. For example, sodium emits only two wavelengths in the visible spectrum; these are two closely spaced yellow lines. Thus, a gas emitting these and only these colors can be identified as sodium. Likewise, mercury vapor has its own characteristic spectrum, consisting of four prominent wavelengths — orange, green, blue, and violet lines — along with some wavelengths of lower intensity. The particular wavelengths emitted by a gas serve as "fingerprints" of that gas.

*24.7 The Rainbow

The dispersion of light into a spectrum is demonstrated most vividly in nature through the formation of a rainbow, often seen by an observer located between the Sun and a rain shower. To understand how a rainbow is formed, consider Figure 24.20. A ray of light passing overhead strikes a drop of water in the atmosphere and is refracted and reflected as follows. It is first refracted at the front surface of the drop, with the violet light deviating the most and the red light the least. At the back surface of the drop, the light is reflected and returns to the front surface, where it again undergoes refraction as it moves from water into air. The rays leave the drop such that the angle between the incident white light and the returning violet ray is 40°, while the angle between the white light and the returning red ray is 42° (Fig. 24.20). This small angular difference between the returning rays causes us to see the bow.

Now consider an observer viewing a rainbow, as in Figure 24.21a. If a raindrop high in the sky is being observed, the red light returning from the drop is able to reach the observer because it is deviated the most, but the violet

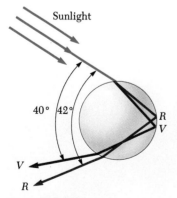

Figure 24.20
Refraction of sunlight by a spherical raindrop.

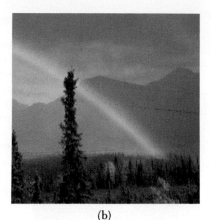

(a)

(b)

Figure 24.21
(a) The formation of a rainbow. (b) A primary rainbow is formed as sunlight undergoes refraction and internal reflection in water droplets in the atmosphere. Under certain conditions, a more faint secondary rainbow can also be seen whose order of colors is reversed from that of the primary rainbow. *(© D. Davis/FPG International)*

light passes over the observer because it is deviated the least. Hence, the observer sees this drop as being red. Similarly, a drop lower in the sky would direct violet light toward the observer and appear to be violet. (The red light from this drop would strike the ground and not be seen.) The remaining colors of the spectrum would reach the observer from raindrops lying between these two extreme positions. Figure 24.21b is a scenic view showing a portion of a rainbow.

▶ SPECIAL TOPIC
Color

In Chapter 23 we found that the wavelengths in the visible portion of the electromagnetic spectrum range from 400 nm to 700 nm, with the shortest wavelength corresponding to violet and the longest corresponding to red. Each wavelength between these two extremes has its own characteristic hue, or shade. The spectrum of colors can be remembered by the boy's name "ROY G BIV," where each letter represents one of the prominent spectral colors: red, orange, yellow, green, blue, indigo, and violet. (Indigo is not really distinct from violet, but it does allow the boy's last name to have a vowel.) If all the colors of the visible spectrum are incident on the eye simultaneously, the mixture is interpreted as white by the eye. On the other hand, black denotes the complete absence of color.

A thorough study of color and of why we see colors as we do would constitute a rather long chapter or perhaps even a textbook to do it justice. In this section, we shall present some concepts that will help you have some understanding of the colors in our world.

The color of an object depends on one or more of the following three processes:

1. The emission of light by the object
2. The reflection of light by the object
3. The transmission of light by the object

The color of flames can sometimes reveal the presence of specific atoms or ions in a sample. *(Top)* A sample containing Na^+ ions placed in a Bunsen burner produces a yellow flame. *(Bottom)* A sample containing Cu^+ ions produces a blue-green color.

Emission of Light

Any object whose atoms have been sufficiently excited by the absorption of energy will emit light. For example, if you heat an iron poker to the point where it glows, the added heat energy causes a rearrangement of the electrons in the atoms of the poker, and light is emitted by the atoms as they return to the state they were in before the rearrangement took place. Hence, the light we see when the poker glows arises from this emission.

Likewise, the light from a neon sign in a store window is emitted by the excited gas atoms contained in the tubes. The colors emitted by a heated gas depend on the characteristics of atoms contained in the tubes.

A more complete discussion of emission requires a knowledge of the concepts of quantum physics, which we shall present in Chapter 29.

Transmission of Light

The color of a piece of transparent material depends, to a large extent, on the wavelengths of light that the material transmits. A piece of colored glass receives its color by absorbing most of the wavelengths of light incident upon it while allowing others to pass through. Those wavelengths that pass through (or are transmitted) give the glass a distinctive color. For example, a piece of red glass in the stained glass window of a church appears red because it absorbs all colors in the visible spectrum except those the eye interprets as red. Because the various wavelengths of light carry energy, a piece of glass is slightly warmed when it absorbs light. A clear piece of glass, such as a window pane, allows all colors in the visible spectrum to pass through and so has no characteristic color.

Pieces of colored glass are often used as optical filters to select a particular wavelength for certain experiments. A color television camera uses three filters to separate the light from a scene into the three primary colors — red, blue, and green. (See the Special Topic in Chapter 23 for a discussion of the television camera.)

Reflection of Light

Most of the objects around us achieve their color by reflection. For example, a red rose neither emits nor transmits visible light. Rather, its color arises because it reflects only those wavelengths in the visible spectrum that are interpreted by the eye as red. Thus, when a red rose is placed in sunlight, it absorbs all the colors in the visible spectrum except red. Now imagine that we take our red rose into a room that is illuminated with only a green light source. There is no red light available from the source, and hence the rose will not reflect any light at all. For this reason, the rose will appear to be black when viewed with green light. On the other hand, the stem and leaves of the rose will appear to have their natural green color when viewed in green light.

Before you read the next few sentences, see if you can guess what the American flag would look like if you observed it in a room illuminated with only a blue light source. The red stripes would appear black because there is no red light for them to reflect. The white stripes and stars would appear blue because all the colors in the spectrum are reflected by a white object. Finally, the blue square in the corner would appear blue because there is blue light in the room for it to reflect.

The Additive Primaries and Color Vision

As we have seen, when all the colors of the visible spectrum are mixed together, the light appears to be white. However, if equal intensities of red, green, and blue light are mixed together, the result is also interpreted by the eye as white. In fact, *all of the colors in the visible spectrum can be generated by mixing together proper proportions of these three colors.* Because of their ability to reproduce the sensation of any color when added to each other in varying amounts, red, green, and blue are called the **additive primaries.**

You can demonstrate some of the properties of the additive primaries by placing a piece of red cellophane over the face of one flashlight, a piece of green cellophane over the face of a second flashlight, and a piece of blue cellophane over the face of a third flashlight. When the three beams are superimposed on a white wall, the illuminated area will appear as in Figure 24.22. The area where the green light overlaps the red light appears yellow, a magenta area appears where the red light overlaps the blue, and a cyan (bluish green) area appears where the green light overlaps the blue. However, the central region, where all the three colors overlap, is white.

The color television set, discussed in Chapter 23, is a good example of how to make practical use of color mixing. The colors observed on the screen of a color television set are produced when the electron beams strike the screen, causing the dots to fluoresce in various colors. If three dots at a particular location on the screen are emitting equal intensities of light, the blue, red, and green light coming from this location appear to be white light. If the blue dot at another location is emitting no light while the red and green dots are emitting light of equal intensity, that spot appears to be yellow. (See the color wheel, Fig. 24.22a.)

The light sensors on the retina of the eye are of two types, rods and cones. The rods are sensitive enough to respond to dim light and to small variations in light intensity. However, they cannot distinguish between the various wavelengths, or colors, of light. The cones are the sensors that enable us to distinguish colors. There are three types of cones, each sensitive to one of the three additive primary colors. The cones sensitive to red light respond when illuminated by red, orange, yellow, and, to a lesser extent, green light. The signal sent to the brain from the red-sensitive cones is the same regardless of which of these colors strikes the cones. Likewise, the green-sensitive cones respond primarily to green and yellow light and, to a lesser extent, blue and orange light. Finally, the blue-sensitive cones are sensitive to blue and violet light.

Now let us assume that light entering the eye stimulates only the red-sensitive cones, which in turn causes a message to be sent to the brain. Based on this information

Figure 24.22
(a) The color wheel. Different colors in the visible spectrum can be produced by an additive or subtractive process. (b) Additive mixing is demonstrated using the three primary colors, red, blue, and green. The overlap areas involving a combination of any two primary colors are yellow (red + green), cyan (green + blue), and magenta (red + blue). The central region where all three primary colors overlap is white.
(Fritz Goro, Life Magazine)

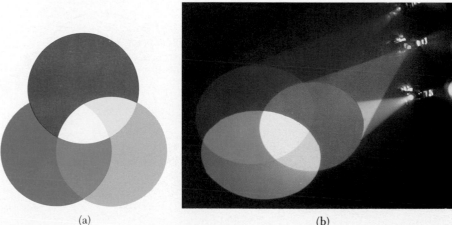

(a)   (b)

alone, the brain would have no way of determining the color of the light. In addition, however, the brain recognizes that no signals are being sent by the green-sensitive cones. Hence, the brain concludes that the light must be red since it receives signals from the red-sensitive cones only. If pure yellow light enters the eye, it stimulates both the red-sensitive and the green-sensitive cones. The brain interprets such stimulation as yellow. It is interesting that light of this single wavelength appears to be yellow, even though we see the same yellow color when a combination of green and red light strikes the retina. Such a combination appears to be yellow even though there is no yellow light present. This is explained by the fact that red and green wavelengths stimulate both the red-sensitive and green-sensitive cones in the same manner as light of a single yellow wavelength.

Most animals are completely color blind. Except for the other primates and a few other species, notably bees, the human being is the only animal capable of seeing the world in full color. Complete lack of color vision is very rare in humans, but partial color blindness is common. About one man in 12 and one woman in 120 have what is known as red-green blindness. The eyes of such individuals respond in the same way to red and green light, and therefore the brain cannot distinguish these colors.

It is in order to remark here that any textbook on art seems to disagree with our earlier statement concerning the three primary colors. Such books claim that if you mix red, green, and blue paint, the resultant color will be black or muddy brown, rather than white as we have stated; it is their assertion that the primary colors are magenta (bluish red), yellow, and cyan (turquoise). Actually, both claims are correct. The art textbook is referring to the colors that are obtained when mixing various *pigments of paint*, whereas we are referring to the mixing of *light of different wavelengths*. The magenta, yellow, and cyan of the artist can combine in various proportions to give all the other colors of the spectrum. However, this phenomenon depends on the ability of the pigments to *absorb* the various wavelengths of white light, and as a result, these colors are called the **subtractive primaries.** Let's see how these colors do their job.

Consider placing a drop of cyan paint on a piece of paper, as shown in Figure 24.23a. When white light falls on the cyan paint, the pigment reflects the colors blue and green while absorbing red (see the color wheel). The two colors blue and green are blended together by the eye, and the drop is interpreted as having its characteristic cyan color. Now here is an exercise for you to work on. What color will be produced when equal proportions of magenta paint and yellow paint are mixed? To answer this, recall that magenta will reflect only red and blue light, while yellow will reflect only red and green (refer to the color wheel). Thus, the combination will absorb every color except red, and the mixture, thus, appears red. Refer to the color wheel and convince yourself that blue will be produced when magenta and cyan paint are mixed in equal

Figure 24.23
(a) Cyan paint reflects only the colors blue and green, which are blended by the eye to produce cyan. (b) Subtractive mixing occurs when white light passes through colored filters (yellow, magenta, and cyan) that allow certain wavelengths to pass. *(Fritz Goro, Life Magazine/© 1944 Time Inc.)*

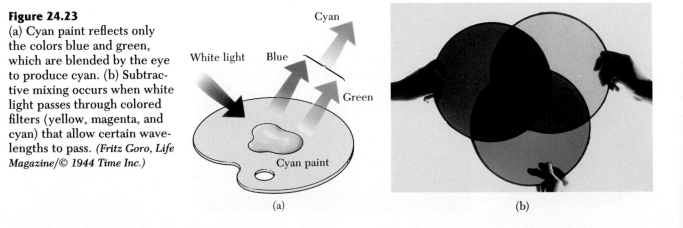

(a)   (b)

amounts. Subtractive mixing can be demonstrated by passing white light through colored filters that allow certain wavelengths to pass as in Figure 24.23b. For example, when white light passes through a combination of yellow and magenta filters, every color is removed except red, so the color we see is red. When the light passes through all three filters, all colors are absorbed, so the region appears black. ◀

24.8   Huygens' Principle Applied to Reflection and Refraction

The laws of reflection and refraction were stated earlier in this chapter without proof. We shall now derive these laws using Huygens' principle. Figure 24.24a will be used in our consideration of the law of reflection. The line AA' represents a wavefront of the incident light. As ray 3 travels from A' to C, ray 1 reflects from A and produces a spherical wavelet of radius AD. (Recall that the radius of a Huygens wavelet is equal to vt.) Since the two wavelets having radii $A'C$ and AD are in the same medium, they have the same velocity, v, and thus $AD = A'C$. Meanwhile, the spherical wavelet centered at B has spread only half as far as the one centered at A since ray 2 strikes the surface later than ray 1.

From Huygens' principle, we find that the reflected wavefront is CD, a line tangent to all the outgoing spherical wavelets. The remainder of the analysis depends upon geometry, as summarized in Figure 24.24b. Note that the right triangles ADC and $AA'C$ are congruent because they have the same hypotenuse, AC, and because $AD = A'C$. From Figure 24.24b we have

$$\sin \theta_1 = \frac{A'C}{AC} \qquad \text{and} \qquad \sin \theta_1' = \frac{AD}{AC}$$

Thus,

$$\sin \theta = \sin \theta_1'$$
$$\theta_1 = \theta_1'$$

which is the law of reflection.

Now let us use Huygens' principle and Figure 24.25a to derive Snell's law of refraction. Note that in the time interval Δt, ray 1 moves from A to B and ray 2 moves from A' to C. The radius of the outgoing spherical wavelet centered at

(a)

(b)

Figure 24.24
(a) Huygens' construction for proving the law of reflection. (b) Triangle ADC is identical to triangle $AA'C$.

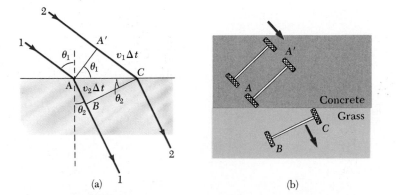

(a)   (b)

Figure 24.25
(a) Huygens' construction for proving the law of refraction. (b) A mechanical analog of refraction.

A is equal to $v_2 \Delta t$. The distance $A'C$ is equal to $v_1 \Delta t$. Geometric considerations show that angle $A'AC$ equals θ_1 and angle ACB equals θ_2. From triangles $AA'C$ and ACB, we find that

$$\sin \theta_1 = \frac{v_1 \Delta t}{AC} \quad \text{and} \quad \sin \theta_2 = \frac{v_2 \Delta t}{AC}$$

If we divide these two equations, we get

$$\frac{\sin \theta_1}{\sin \theta_2} = \frac{v_1}{v_2}$$

But from Equation 24.4 we know that $v_1 = c/n_1$ and $v_2 = c/n_2$. Therefore,

$$\frac{\sin \theta_1}{\sin \theta_2} = \frac{c/n_1}{c/n_2} = \frac{n_2}{n_1}$$

$$n_1 \sin \theta_1 = n_2 \sin \theta_2$$

which is the law of refraction.

A mechanical analog of refraction is shown in Figure 24.25b. The wheels on a device such as a lawnmower change their direction as they move from a concrete surface to a grass surface.

24.9   Total Internal Reflection

An interesting effect called *total internal reflection* can occur when light attempts to move from a medium having a *high* index of refraction to one having a *lower* index of refraction. Consider a light beam traveling in medium 1 and meeting the boundary between medium 1 and medium 2, where n_1 is greater than n_2 (Fig. 24.26). Various possible directions of the beam are indicated by

Figure 24.26
(a) A ray traveling from a medium of index n_1 to a medium of index n_2, where $n_1 > n_2$. As the angle of incidence increases, the angle of refraction increases until θ_2 is 90° (ray 4). For even larger angles of incidence, total internal reflection occurs (ray 5). (b) The angle of incidence producing an angle of refraction equal to 90° is often called the *critical angle*, θ_c.

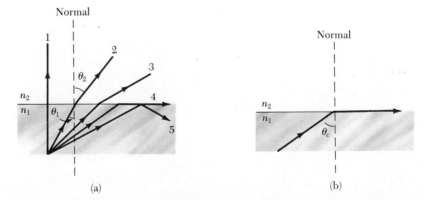

rays 1 through 5. Note that the refracted rays are bent away from the normal because n_1 is greater than n_2. At some particular angle of incidence, θ_c, called the **critical angle,** the refracted light ray will move parallel to the boundary so that $\theta_2 = 90°$ (Fig. 24.26b). *For angles of incidence greater than θ_c, the beam is entirely reflected at the boundary.* Ray 5 in Figure 24.26a shows this occurrence. This ray is reflected at the boundary as though it had struck a perfectly reflecting surface. This ray, and all those like it, obey the law of reflection; that is, the angle of incidence equals the angle of reflection.

We can use Snell's law to find the critical angle. When $\theta_1 = \theta_c$, $\theta_2 = 90°$ and Snell's law (Eq. 24.8) gives

$$n_1 \sin \theta_c = n_2 \sin 90° = n_2$$

$$\sin \theta_c = \frac{n_2}{n_1} \qquad \text{for } n_1 > n_2 \qquad \textbf{(24.9)} \qquad \text{Critical angle}$$

Note that this equation can be used only when n_1 is greater than n_2. That is,

total internal reflection occurs only when light attempts to move from a medium of high index of refraction to a medium of lower index of refraction.

If n_1 were less than n_2, Equation 24.9 would give $\sin \theta_c > 1$, which is an absurd result because the sine of an angle can never be greater than unity.

The critical angle is small for substances with a large index of refraction, such as diamond, where $n = 2.42$ and $\theta_c = 24°$. For crown glass, $n = 1.52$ and $\theta_c = 41°$. In fact, this property combined with proper faceting causes diamonds and crystal glass to sparkle.

The direction of travel of a light beam can be altered by use of a prism and the phenomenon of total internal reflection. Two such possibilities are illustrated in Figure 24.27. In one case the light beam is deflected by 90° (Fig. 24.27a), and in the second case the path of the beam is reversed (Fig. 24.27b). A common application of total internal reflection is in a submarine periscope. In this device, two prisms are arranged as in Figure 24.27c so that an incident beam of light follows the path shown and one is able to "see around corners."

This photograph shows non-parallel light rays entering a glass prism. The bottom two rays undergo total internal reflection at the longest side of the prism. The top three rays are refracted at this side as they leave the prism. *(Courtesy of Henry Leap and Jim Lehman)*

Figure 24.27
Internal reflection in a prism. (a) The ray is deviated by 90°. (b) The direction of the ray is reversed. (c) Two prisms used as a periscope.

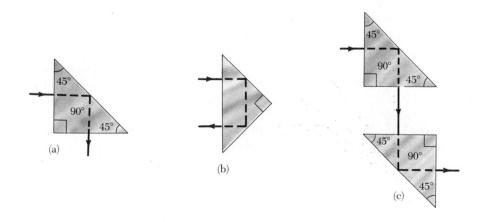

Example 24.7   A View from the Fish's Eye

(a) Find the critical angle for a water-air boundary if the index of refraction of water is 1.33.

Solution

Applying Equation 24.9, we find the critical angle to be

$$\sin \theta_c = \frac{n_2}{n_1} = \frac{1}{1.33} = 0.752$$

$$\theta_c = \boxed{48.8°}$$

(b) Use the results of (a) to predict what a fish will see if it looks upward toward the water surface at an angle of 40°, 48.8°, and 60°.

Solution

Because the path of a light ray is reversible, the fish can see out of the water if it looks toward the surface at an angle less than the critical angle. Thus, at 40°, the fish can see into the air above the water. At an angle of 48.8°, the critical angle for water, the light that reaches the fish has to skim along the water surface before being refracted to the fish's eye. At angles greater than the critical angle, the light reaching the fish comes via internal reflection at the surface. Thus, at 60°, the fish sees a reflection of some object on the bottom of the pool.

Fiber Optics

Figure 24.28
Light travels in a curved, transparent rod by multiple internal reflections.

Another interesting application of total internal reflection is the use of glass or transparent plastic rods to "pipe" light from one place to another. As indicated in Figure 24.28, light is confined to traveling within the rods, even around gentle curves, as the result of successive internal reflections. Such a "light pipe" will be flexible if thin fibers are used rather than thick rods. If a bundle of parallel fibers is used to construct an optical transmission line, images can be transferred from one point to another.

This technique is used in a sizable industry known as *fiber optics*. There is very little light intensity lost in these fibers as a result of reflections on the sides. Any loss in intensity is due essentially to reflections from the two ends and absorption by the fiber material. These devices are particularly useful when one wishes to view an image produced at inaccessible locations. Physicians often use fiber optics cables to aid in the diagnosis and repair of certain medical problems, without the intrusive effects of major surgery. For example, a fiber optics cable can be threaded through the esophagus and into the stomach to look for ulcers. In this application, the cable actually consists of two fiber optics lines, one to transmit a beam of light into the stomach for illumination and the other to allow this light to be transmitted out of the stomach. The resulting image can, in some cases, be viewed directly by the physician but most often is displayed on a television monitor or captured on film. In a similar fashion, the cables can be used to examine the colon or to do repair work without the use of large incisions. Damaged knees or other joints in athletes can sometimes be repaired using a process called arthroscopic surgery. In this technique, a small incision is made into the joint. Repair is accomplished by inserting a small fiber optics cable through the cut to provide illumination and then by trimming cartilage or damaged tissue with a small knife at the end of a second cable. The field of fiber optics is also finding increasing use in telecommunications, since the fibers can carry a much higher volume of telephone calls or other forms of communication than electrical wires.

Summary

Huygens' principle states that all points on a wavefront are point sources for the production of spherical secondary waves called wavelets. These wavelets propagate outward at a speed characteristic of waves in a particular medium. After some time has elapsed, the new position of the wavefront is the surface tangent to the wavelets.

The **index of refraction** of a material, n, is defined as

$$n \equiv \frac{c}{v} \tag{24.4}$$

where c is the speed of light in a vacuum and v is the speed of light in the material. In general, n varies with wavelength as

$$n = \frac{\lambda_0}{\lambda_n} \tag{24.7}$$

where λ_0 is the wavelength of the light in vacuum and λ_n is its wavelength in the material.

The **law of reflection** states that a wave reflects from a surface such that the *angle of reflection*, θ_1', equals the *angle of incidence*, θ_1.

The **law of refraction**, or **Snell's law**, states that

$$n_1 \sin \theta_1 = n_2 \sin \theta_2 \tag{24.8}$$

Total internal reflection can occur when light attempts to move from a material having a high index of refraction to one having a lower index of refraction. The *maximum angle of incidence*, θ_c, for which light can move from a medium of index n_1 into a medium of index of refraction n_2, where n_1 is greater than n_2, is called the **critical angle** and is given by

$$\sin \theta_c = \frac{n_2}{n_1} \qquad \text{for } n_1 > n_2 \tag{24.9}$$

A bundle of quartz fibers acts as "light pipes" to transmit light from one location to another. The light is confined to the fibers through a succession of internal reflections. *(© Michael Keller/FPG International)*

Additional Reading

W. S. Boyle, "Light-Wave Communications," *Sci. American,* August 1977, p. 40.

W. Bragg, *The Universe of Light,* Dover, 1959.

A. B. Fraser and W. H. Mach, "Mirages," *Sci. American,* January 1976, p. 102.

N. S. Kapany, "Fiber Optics," *Sci. American,* November 1960, p. 72.

E. A. Lacy, *Fiber Optics,* Englewood Cliffs, NJ, Prentice-Hall, 1982.

E. H. Land, "The Retinex Theory of Color Vision," *Sci. American,* December 1977, p. 108.

"Light," *Sci. American,* September 1968 (entire issue).

D. F. Mandoli and W. R. Briggs, "Fiber Optics in Plants," *Sci. American,* August 1984, p. 90.

E. W. Stark, "Diffuse Reflection: Uses That Affect Our Lives," *The Physics Teacher,* March 1986, p. 144.

A. C. S. van Heel and C. H. F. Velzel, *What Is Light,* World University Library, New York, McGraw-Hill, 1968, Chap. 1.

S. Williamson and H. Cummins, *Light and Color in Nature and Art,* John Wiley and Son, 1983.

Questions

1. Explain why we see lightning before we hear thunder.
2. Why do astronomers looking at distant galaxies talk about looking backward in time?
3. Would the Huygens' wavelets be spherical in a non-isotropic medium (a medium in which the wave speed depends on the direction of the wave propagation)? Discuss.
4. A solar eclipse occurs when the Moon gets between the Earth and Sun. Use a diagram to show why some areas of the Earth see a total eclipse, other areas see a partial eclipse, and most areas see no eclipse.
5. Some department stores have their windows slanted slightly inward at the bottom. This is to decrease the glare from streetlights or the Sun, which would make it difficult for shoppers to see the display inside. Draw a sketch of a light ray reflecting off such a window to show how this technique works.
6. You can make a corner reflector by placing three plane mirrors in the corner of a room where the ceiling meets the walls. Show that no matter where you are in the room, you can see yourself reflected in the mirrors—upside down.
7. Several corner reflectors were left on the Moon's Sea of Tranquility by the astronauts of Apollo 11. How can scientists utilize a laser beam sent from Earth even today to determine the precise distance from the Earth to the Moon?
8. The rectangular aquarium sketched in Figure 24.29 contains only one goldfish. When the fish is located near a corner of the aquarium and is viewed along a direction which makes an equal angle with two adjacent faces, one observes two fish as shown, which appear as mirror images of each other. Explain this observation.
9. As light travels from one medium to another, does its wavelength change? Does its frequency change? Does its velocity change? Explain.

Figure 24.29   Question 8

10. Does a light ray traveling from one medium into another always bend toward the normal as in Figure 24.14?
11. Explain why an oar in the water appears to be bent.
12. The level of water in a clear, colorless glass is easily observed with the naked eye. The level of liquid helium in a clear glass is extremely difficult to see with the naked eye. Explain.
13. Suppose you are told only that two colors of light (X and Y) are sent through a prism and that X is bent more than Y. Which color travels more slowly in the glass of the prism?
14. Under certain circumstances, sound can be heard over extremely long distances. This frequently happens over a body of water, where the air near the water surface is cooler than the air higher up. Explain how the refraction of sound waves in such a situation could increase the distance over which the sound can be heard.
15. A laser beam passing through a nonhomogeneous sugar solution is observed to follow a curved path. Explain.

16. A laser beam ($\lambda = 632.8$ nm) is incident on a piece of Lucite as in Figure 24.30. Part of the beam is reflected and part is refracted. What information can you get from this photograph?

Figure 24.30   Questions 16 and 17
Light from a helium-neon laser beam ($\lambda = 632.8$ nm) is incident on a block of Lucite. The photograph shows both reflected and refracted rays. Can you identify the incident, reflected, and refracted rays? From this photograph, estimate the index of refraction of Lucite at this wavelength. *(Courtesy of Henry Leap and Jim Lehman)*

17. Suppose blue light were used instead of red light in the experiment shown in Figure 24.30. Would the refracted beam be bent at a larger or smaller angle?
18. What are the conditions for the production of a mirage? On a hot day, what is it that we are seeing when we observe "water on the road"?
19. Sound waves have much in common with light waves, including the properties of reflection and refraction. Give examples of such phenomena for sound waves.
20. Explain why a diamond shows more "sparkle" than a glass crystal of the same shape and size.
21. A color-television image is formed by superimposing three separate images of red, green, and blue phosphors. Discuss the effect of viewing the composite image through a prism.
22. Why does a diamond show flashes of color when observed under ordinary white light?
23. Why does the arc of a rainbow appear with red colors on top and violet hues on the bottom?
24. How is it possible that a complete *circular* rainbow can sometimes be seen from an airplane?
25. Is it possible to have total internal reflection for light incident from air on water? Explain.
26. What would happen to a light pipe if the coating separating individual fibers were to wear off?
27. Redesign the periscope of Figure 24.27c so that it can show you where you have been rather than where you are going.

Problems

Section 24.1   The Nature of Light
Section 24.2   Measurements of the Speed of Light

1. As a result of his observations, Roemer concluded that the time interval between successive eclipses of the moon Io by the planet Jupiter increased by 22 minutes during a six-month period as the Earth moved from a point in its orbit on the side of the Sun nearer Jupiter to a point on the side opposite Jupiter. Using 1.5×10^8 km as the average radius of the Earth's orbit about the Sun, calculate the speed of light from these data.
2. Use Roemer's value of 22 min discussed in Problem 1 and the presently accepted value of c (2.998×10^8 m/s) to find an average value for the distance between the Earth and the Sun.
3. If the Fizeau experiment is performed such that the round-trip distance for the light is 40 m, (a) find the two lowest speeds of rotation that allow the light to pass through the notches. Assume that the wheel has 360 teeth and that the speed of light is 3×10^8 m/s. (b) Repeat for a round-trip distance of 4000 m.
4. Albert A. Michelson very carefully measured the speed of light using an alternative version of the technique developed by Fizeau. Figure 24.31 shows the

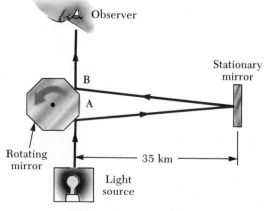

Figure 24.31   Problem 4

approach used. Light was reflected from one face of a rotating eight-sided mirror toward a stationary mirror 35 km away. At certain rates of rotation, the returning beam of light would be directed toward the eye of an observer as shown. (a) What minimum angular velocity must the rotating mirror have in order that side **A** will have rotated to position **B** so that the light will be reflected to the eye? (b) What is the next highest angular velocity that will enable the source of light to be seen?

5. In Galileo's attempt to determine the speed of light, he and his companion were located on hilltops 9 km apart. What time interval between uncovering his lantern and seeing the light from his companion's lantern would Galileo have had to measure in order to obtain a value for the speed of light?

Section 24.4   Reflection and Refraction
Section 24.5   The Law of Refraction

6. You are standing with a mirror at the center of a giant floral clock. Someone at 12 o'clock shines a beam of light toward you, and you want to use the mirror to reflect the beam toward an observer at 5 o'clock. What should be the angle of incidence to achieve this?

7. The angle between the two mirrors in Figure 24.32 is a right angle. The beam of light in the vertical plane P strikes mirror 1 as shown. (a) Determine the distance the reflected light beam travels before striking mirror 2. (b) In what direction does the light beam travel after being reflected from mirror 2?

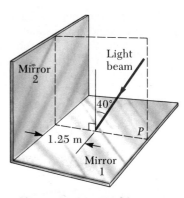

Figure 24.32   Problem 7

8. How many times will the incident beam shown in Figure 24.33 be reflected by each of the parallel mirrors?

Figure 24.33   Problem 8

9. Find the speed of light in benzene.

10. Find the speed of light in (a) flint glass, (b) water, and (c) zircon.

11. The wavelength of sodium light in air is 589 nm. Find its wavelength in ethyl alcohol.

12. Light of wavelength 436 nm in air enters a fishbowl filled with water then exits through the crown glass wall of the container. What is the wavelength of the light (a) while in the water and (b) while in the glass?

13. A light ray in air is incident on a water surface at an angle of 30° with respect to the normal to the surface. What is the angle of the refracted ray relative to the normal to the surface?

14. A beam of light strikes the surface of mineral oil at an angle of 23.1° with respect to the normal to the surface. If the light travels with a speed of 2.17×10^8 m/s through the oil, what is the angle of refraction?

15. A narrow beam of sodium yellow light ($\lambda_0 = 589$ nm) is incident from air on a smooth surface of water at an angle $\theta_1 = 35°$. Determine the angle of refraction θ_2 and the wavelength of the light in water.

16. An underwater swimmer observes that a beam of sunlight in the water makes an angle of 31.0° with the vertical. Assuming the surface of the water is level and horizontal, determine the beam's angle as it enters the water.

17. A narrow beam of ultrasonic waves reflects off the liver tumor in Figure 24.34. If the wave speed is 10.0% less in the liver than the surrounding medium, determine the depth of the tumor.

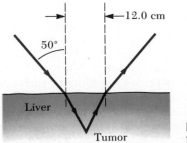

Figure 24.34
Problem 17

18. A ray of light is incident on the surface of a block of clear ice at an angle of 40° with the normal. Part of the light is reflected and part is refracted. Find the angle between the reflected and refracted light.

19. A ray of light strikes a flat block of glass ($n = 1.50$) of thickness 2 cm at an angle of 30° with the normal (Fig. 24.35). Trace the light beam through the glass and find the angles of incidence and refraction at each surface.

Figure 24.35
Problems 19, 20, and 22

20. When the light of Problem 19 passes through the glass block, it is shifted laterally by a distance d (Fig. 24.35). Find the value of d.

21. A light ray initially in water enters a transparent substance at an angle of 37° with respect to the normal, and the transmitted ray is refracted at an angle of 25°. Calculate the speed of light in the transparent substance.

22. Find the time for the light to pass through the glass block described in Problem 19.

23. A ray of light strikes the midpoint of one face of an equiangular glass prism ($n = 1.5$) at an angle of incidence of 30°. (a) Trace the path of the light ray through the glass and find the angles of incidence and refraction at each surface. (b) If a small fraction of light is also reflected at each surface, find the angle of incidence and angle of reflection at these surfaces.

24. The light beam shown in Figure 24.36 makes an angle of 20.0° with the normal line NN' in the linseed

oil. Determine the angles θ and θ'. (Refractive index = 1.48 for linseed oil)

25. Three sheets of plastic have unknown indices of refraction. Sheet 1 is placed on top of sheet 2, and a laser beam is directed onto the sheets from above so that the beam strikes the interface at an angle of 26.5° with respect to the normal. The refracted beam in sheet 2 makes an angle of 31.7° with the normal. The experiment is repeated with sheet 3 on top of sheet 2, and with the same angle of incidence the refracted beam makes an angle of 36.7° with the normal. If the experiment is repeated again with sheet 1 on top of sheet 3, what is the expected angle of refraction in sheet 3? Assume the same angle of incidence.

26. A submarine is 300 m horizontally out from the shore and 100 m beneath the surface of the water. A laser beam is sent from the sub such that it strikes the surface of the water at a point 210 m from the shore. If the beam just strikes the top of a building standing directly at the water's edge, find the height of the building.

27. (a) What is the speed of light in crown glass if the wavelength in vacuum is 589 nm? (b) What thickness of crown glass will equal 100 wavelengths of this light (measured in the glass)? (Make use of the curve for crown glass in Figure 24.16.)

28. Two light pulses are emitted simultaneously from a source. Both pulses travel to a detector, but one passes first through 6.2 m of ice. Determine the difference in the pulses' times of arrival at the detector.

29. How far does a beam of light travel in water in the same time it takes it to travel 10 m in glass of index of refraction 1.5?

30. A drinking glass is 4 cm wide at the bottom, as shown in Figure 24.37. When an observer's eye is placed as shown, the observer sees the edge of the bottom of the glass. When this glass is filled with water, the observer sees the center of the bottom of the glass. Find the height of the glass.

Figure 24.36   Problem 24

Figure 24.37   Problem 30

31. A cylindrical tank with an open top has a diameter of 3 m and is completely filled with water. When the setting Sun reaches an angle of 28° above the horizon, sunlight ceases to illuminate the bottom of the tank. How deep is the tank?

Section 24.6   Dispersion and Prisms

32. A certain kind of glass has an index of refraction of 1.6500 for blue light at wavelength 430 nm and an index of 1.615 for red light of wavelength 680 nm. If a beam containing these two colors is incident at an angle of 30° on a piece of this glass, what is the angle between the two beams inside the glass?

33. Light of wavelength 400 nm is incident at an angle of 45° on acrylic and is refracted as it passes into the material. What wavelength of light incident on fused quartz at an angle of 45° would be refracted at exactly this same angle? (See Fig. 24.16.)

34. Light of wavelength 700 nm is incident on the face of a fused quartz prism at an angle of 75° with respect to the normal to the surface. The apex angle of the prism is 60°. Use the value of n from Figure 24.16 and calculate the angle (a) of refraction at the first surface, (b) of incidence at a second surface, (c) of refraction at the second surface, and (d) between the incident and emerging rays.

35. The glass described in Problem 32 is shaped into an equiangular prism and the red and blue wavelengths of Problem 32 are directed toward one face at an angle of 75° with the normal. (a) Find the angles δ_R and δ_B shown in Figure 24.17. (b) Calculate the dispersion of $\delta_B - \delta_R$ produced.

Section 24.9   Total Internal Reflection

36. As shown in Figure 24.38, a light ray is incident normally on one face of a 30°-60°-90° block of dense flint

glass (a *prism*) that is immersed in water. (a) Determine the exit angle θ_4 of the ray. (b) A substance is dissolved in the water to increase the index of refraction. At what value of n_2 will total internal reflection cease at point P?

37. Calculate the critical angle for the following materials when surrounded by air: (a) zircon, (b) fluorite, (c) ice. Assume that $\lambda = 589$ nm.

38. Repeat Problem 37 when the materials are surrounded by water.

39. A beam of light is incident from air on the surface of a liquid. If the angle of incidence is 30° and the angle of refraction is 22°, find the critical angle for the liquid when surrounded by air.

40. A jewel thief hides a diamond by placing it on the bottom of a public swimming pool. He places a circular piece of metal on the surface of the water directly above and centered on the diamond. If the pool is 2 m deep, find the diameter of the piece of metal such that the diamond cannot be seen.

41. A light ray is incident on the interface between a diamond and air from within the diamond. What is the critical angle for total internal reflection? Use Table 24.1. (The smallness of θ_c for diamond means that light is easily "trapped" within a diamond and eventually emerges from the many cut faces; this makes a diamond more "brilliant" than stones with smaller n and larger θ_c.)

42. The light beam in Figure 24.39 strikes surface 2 at the critical angle. Determine the angle of incidence θ_i.

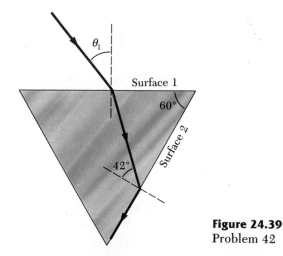

Figure 24.39
Problem 42

43. A light ray is incident perpendicular to the long face (the hypotenuse) of a 45°-45°-90° prism surrounded by air, as shown in Figure 24.27b. Calculate the minimum index of refraction of the prism for which the ray will follow the path shown.

44. Repeat Problem 43 if the prism is surrounded by water.

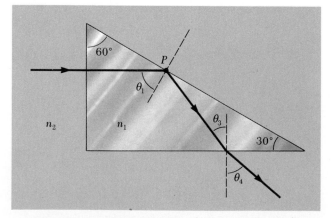

Figure 24.38   Problem 36

45. A large Lucite cube ($n = 1.59$) has a small air bubble (a defect in the casting process) below one surface. When a penny (diameter 1.9 cm) is placed directly over the bubble, it cannot be seen at any angle from the opposite site of the cube. However, when a dime (diameter 1.75 cm) is placed directly over the bubble, it can be seen from the opposite surface of the cube. What is the range of the possible depths of the air bubble beneath the surface?

46. A plastic light pipe has an index of refraction of 1.53. For total internal reflection what is the minimum angle of incidence with respect to the normal to the wall of the pipe if the pipe is in (a) air and (b) water?

47. A light pipe consists of a central strand of material surrounded by an outer coating. The interior portion of the pipe has an index of refraction of 1.60. If all rays striking the interior walls of the pipe with incident angles greater than $59.5°$ are subject to total internal reflection, what is the index of refraction of the coating?

48. Determine the maximum angle θ for which the light rays incident on the end of the pipe in Figure 24.40 are subject to total internal reflection along the walls of the pipe. Assume the pipe has an index of refraction of 1.36 and the outside medium is air.

2.00 μm

θ

Figure 24.40
Problem 48

Additional Problems

49. Light is incident normally on a 1-cm layer of water that lies on top of a flat Lucite plate whose thickness is 0.5 cm. How much longer is required for light to pass through this double layer than is required to traverse the same distance in air? ($n_{\text{Lucite}} = 1.59$)

50. A layer of ice floats on water. If light is incident on the upper surface of the ice at an angle of $30°$, what is the angle of refraction in the water?

51. A beam of light is incident at an angle of $37°$ on the surface of a block of transparent material. The angle of refraction is found to be $25°$. What is the speed of light in the material?

52. Light is incident on the surface of a prism, $n = 1.8$, as shown in Figure 24.27a. If the prism is surrounded by a fluid, what is the maximum index of refraction of the fluid that would still cause total internal reflection?

53. The Sun is $10°$ above the horizon. If you are swimming beneath the surface of a pool of water, at what angle above the horizon does the Sun appear to be?

54. A narrow beam of light is incident from air onto a glass surface of index of refraction 1.56. Find the angle of incidence for which the corresponding angle of refraction will be one-half the angle of incidence. (*Hint:* You might want to use the trigonometric identify $\sin 2\theta = 2 \sin \theta \cos \theta$.)

55. A thick plate of flint glass ($n = 1.66$) rests on top of a thick plate of transparent acrylic ($n = 1.50$). A beam of light is incident on the top surface of the flint glass at an angle θ_i. The beam passes through the glass and the acrylic and emerges from the acrylic at an angle of $40°$ with respect to the normal. Calculate the value of θ_i. A sketch of the light path through the two plates of refracting material would be helpful.

56. One technique to measure the angle of a prism is shown in Figure 24.41. A parallel beam of light is directed on the angle such that the beam reflects from opposite sides. Show that the angular separation of the two beams is given by $B = 2A$.

θ_i

θ_r

A

B

Figure 24.41
Problem 56

57. Repeat Example 24.2 for the case in which the angle between the two mirrors is $60°$.

58. Repeat Example 24.2 for the case in which the angle is $90°$. Show that the ray of light is always reflected from the second mirror so that it travels opposite to the original direction.

59. Light of wavelength λ_0 in vacuum has a wavelength of 438 nm in water and a wavelength of 390 nm in benzene. What is the index of refraction of water relative to benzene at the wavelength λ_0?

60. A light ray of wavelength 589 nm is incident at an angle θ on the top surface of a block of polystyrene, as shown in Fig. 24.42. (a) Find the maximum value of θ

θ

Figure 24.42
Problem 60

for which the refracted ray will undergo total internal reflection at the left vertical face of the block. (b) Repeat the calculation for the case in which the polystyrene block is immersed in water. (c) What happens if the block is immersed in carbon disulfide?

61. For this problem refer to Figure 24.17. For various angles of incidence, it can be shown that the angle δ is a minimum when the ray passes through the glass such that the ray is parallel to the base. A measurement of this minimum angle of deviation enables one to find the index of refraction of the prism material. Show that n is given by the expression

$$n = \frac{\sin\left[\frac{1}{2}(A + \delta_{min})\right]}{\sin\left(\frac{A}{2}\right)}$$

where A is the apex angle of the prism.

62. A light ray is incident on a prism and refracted at the first surface as shown in Figure 24.43. Let ϕ represent the apex angle of the prism and n its index of refraction. Find in terms of n and ϕ the smallest allowed value of the angle of incidence at the first surface for which the refracted ray will not undergo internal reflection at the second surface.

Figure 24.43 Problem 62

"Most mirrors reverse left and right. This one reverses top and bottom."

CHAPTER 25

Mirrors and Lenses

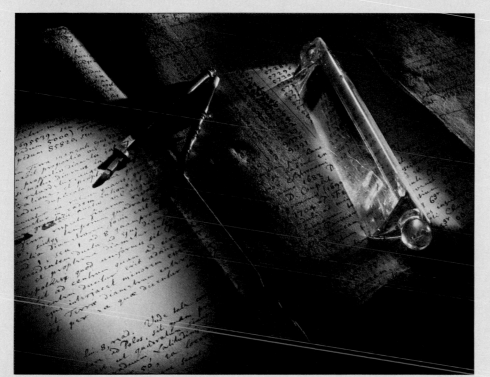

White light directed through a triangular glass prism creates a "Pheae-nomena of Colours" on an old manuscript.
(© *Erich Lessing, Magnum Photos*)

Mirrors and lenses are basic components of such ordinary devices as cameras, telescopes, and microscopes. In Chapter 27, we shall examine the construction and properties of such optical instruments. This chapter is concerned solely with the formation of images when light rays fall on plane and spherical surfaces. These surfaces are either reflecting, as in the case of mirrors, or refracting, as in the case of lenses. Our studies will show that the images formed are the result of reflection or refraction.

25.1 Plane Mirrors

One of the objectives of this chapter will be to discuss the manner in which optical elements such as lenses and mirrors form images. We shall begin this investigation by considering the simplest possible mirror, the plane mirror. Throughout our discussion of both mirrors and lenses, we shall use the ray model of light.

Consider a point source of light placed at O in Figure 25.1, a distance s in front of a plane mirror. The distance s is often referred to as the **object distance.** Light rays leave the source and are reflected from the mirror. After reflection, the rays diverge (spread apart), but they appear to the viewer to come from a point I located behind the mirror. The eye and the brain treat all

Figure 25.1
An image formed by reflection from a plane mirror. The image point, I, is located behind the mirror at a distance s', which is equal to the object distance, s.

779

light rays which enter the eye as though the rays follow straight-line paths. As a result, the eye follows the rays backward along a straight-line path to the point from which they appear to originate, which is point I in Figure 25.1. Point I is called the **image** of the object at O. Regardless of the system under study, images are always formed in the same way. *Images are formed at the point where rays of light actually intersect or at the point from which they appear to originate.* Since the rays in Figure 25.1 appear to originate at I, which is a distance s' behind the mirror, this is the location of the image. The distance s' is often referred to as the **image distance.**

Images are classified as real or virtual. *A real image is one in which light actually intersects, or passes through, the image point; a virtual image is one in which the light does not really pass through the image point but appears to diverge from that point.* The image formed by the plane mirror in Figure 25.1 is a virtual image. The images seen in plane mirrors *are always virtual* for real objects. Real images can always be displayed on a screen (as at a movie), but virtual images cannot be displayed on a screen.

We shall examine some of the properties of the images formed by plane mirrors by using the simple geometric techniques shown in Figure 25.2. In order to find out where an image is formed, it is always necessary to follow at least two rays of light as they reflect from the mirror. One of those rays starts at P, follows a horizontal path to the mirror, and reflects back on itself. The second ray follows the oblique path PR and reflects as shown. An observer to the left of the mirror would trace the two reflected rays back to the point from which they appear to have originated, that is, point P'. A continuation of this process for points on the object other than P would result in a virtual image (drawn as a yellow arrow) to the right of the mirror. Since triangles PQR and $P'QR$ are congruent, $PQ = P'Q$. Hence, we conclude that *the image formed by an object placed in front of a plane mirror is as far behind the mirror as the object is in front of the mirror.* Geometry also shows that the object height, h, equals the image height, h'. Let us define **lateral magnification, M,** as follows:

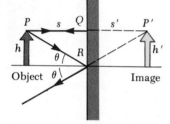

Figure 25.2
Geometric construction used to locate the image of an object placed in front of a plane mirror. Because the triangles PQR and $P'QR$ are congruent, $s = s'$, and $h = h'$.

$$M \equiv \frac{\text{image height}}{\text{object height}} = \frac{h'}{h} \qquad (25.1)$$

Note that $M = 1$ for a plane mirror because $h' = h$ in this case.

The image formed by a plane mirror has one more important property, that of right-left reversal between image and object. This reversal can be seen by standing in front of a mirror and raising your right hand. The image you see raises its left hand. Likewise, your hair appears to be parted on the opposite side and a mole on your right cheek appears to be on your left cheek.

Thus, we conclude that the image formed by a plane mirror has the following properties:

1. The image is as far behind the mirror as the object is in front.
2. The image is unmagnified, virtual, and erect. (By erect we mean that if the object arrow in Figure 25.2 points upward, so does the image arrow.)
3. The image has right-left reversal.

Example 25.1   Multiple Images Formed by Two Mirrors

Two plane mirrors are at right angles to each other, as in Figure 25.3, and an object is placed at point O. In this situation, multiple images are formed. Locate the positions of these images.

Solution

The image of the object is at I_1 in mirror 1 and at I_2 in mirror 2. In addition, a third image is formed at I_3, which will be considered to be the image of I_1 in mirror 2 or, equivalently, the image of I_2 in mirror 1. That is, the image at I_1 (or I_2) serves as the object for I_3. When viewing I_3, note that the rays reflect twice after leaving the object at O.

Exercise   Sketch the rays corresponding to viewing the images at I_1 and I_2 and show that the light is reflected only once in these cases.

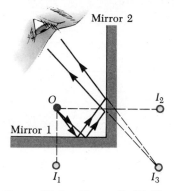

Figure 25.3   (Example 25.1) When an object is placed in front of two perpendicular mirrors as shown, three images are formed.

25.2   Images Formed by Spherical Mirrors

Concave Mirrors

A spherical mirror, as its name implies, has the shape of a segment of a sphere. Figure 25.4 shows a spherical mirror with light reflecting from its silvered inner surface. Such a mirror, in which light is reflected from the inner, concave surface, is called a **concave mirror.** The mirror has a radius of curvature R, and its center of curvature is located at point C. Point V is the center of the spherical segment, and a line drawn from C to V is called the **principal axis** of the mirror.

Now consider a point source of light placed at point O in Figure 25.4, located on the principal axis and outside point C. Several diverging rays originating at O are shown. After reflecting from the mirror, these rays converge and meet at I, called the **image point.** The rays then continue to diverge from I as if there were an object there. As a result, we have a real image formed. *Real images are always formed at a point when reflected light actually passes through that point.*

In what follows, we shall assume that all rays that diverge from the object make a small angle with the principal axis. All such rays reflect through the image point, as in Figure 25.4. Rays that are far from the principal axis, as in Figure 25.5, converge to other points on the principal axis, producing a

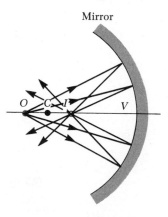

Figure 25.4
A point object placed at O outside the center of curvature of a concave spherical mirror forms a real image at I as shown. If the rays diverge from O at small angles, they all reflect through the same image point.

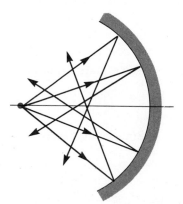

Figure 25.5
Rays at large angles from the horizontal axis reflect from a spherical, concave mirror to intersect the optic axis at different points, resulting in a blurred image. This is called *spherical aberration.*

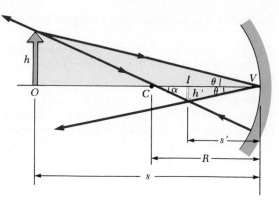

Figure 25.6
Ray diagram for a spherical concave mirror, where the object, at O, lies outside the center of curvature, C.

blurred image. This effect, called **spherical aberration,** is present to some extent for any spherical mirror and will be discussed in Section 25.8.

We can use the geometry shown in Figure 25.6 to calculate the image distance, s', from a knowledge of the object distance, s, and radius of curvature, R. By convention, these distances are measured from point V. Figure 25.6 shows two rays of light leaving the tip of the object. One of these rays passes through the center of curvature, C, of the mirror, hitting the mirror head on (perpendicular to the mirror surface) and reflecting back on itself. The second ray strikes the mirror at the center, point V, and reflects as shown, obeying the law of reflection. The image of the tip of the arrow is located at the point where these two rays intersect. From the largest triangle in Figure 25.6 we see that $\tan \theta = h/s$, while the smallest triangle gives $\tan \theta = -h'/s'$. The negative sign signifies that the image is inverted, and so h' is negative. Thus, from Equation 25.1 and these results, we find that the magnification of the mirror is

Magnification

$$M = \frac{h'}{h} = -\frac{s'}{s} \qquad \text{(25.2)}$$

We also note from two other triangles in the figure that

$$\tan \alpha = \frac{h}{s - R} \qquad \text{and} \qquad \tan \alpha = -\frac{h'}{R - s'}$$

from which we find that

$$\frac{h'}{h} = -\frac{R - s'}{s - R} \qquad \text{(25.3)}$$

We equate Equation 25.2 to Equation 25.3 to give

$$\frac{R - s'}{s - R} = \frac{s'}{s}$$

Simple algebra reduces this to

$$\frac{1}{s} + \frac{1}{s'} = \frac{2}{R} \qquad \text{(25.4)}$$

Figure 25.7
(a) Light rays from a distant object ($s = \infty$) reflect from a concave mirror through the focal point, F. In this case, the image distance $s' = R/2 = f$, where f is the focal length of the mirror. (b) Photograph of the reflection of parallel rays from a concave mirror. *(Courtesy of Jim Lehman, James Madison University)*

This expression is called the **mirror equation.**

If the object is very far from the mirror, that is, if the object distance, s, is large enough compared with R that s can be said to approach infinity, then $1/s \approx 0$, and we see from Equation 25.4 that $s' \approx R/2$. That is, when the object is very far from the mirror, *the image point is halfway between the center of curvature and the center of the mirror,* as in Figure 25.7a. Note that the incoming rays are essentially parallel in this figure because the source is assumed to be very far from the mirror. We call the image point in this special case the **focal point,** F, and the image distance the **focal length,** f, where

$$f = \frac{R}{2}$$

Focal length

The mirror equation can therefore be expressed in terms of the focal length:

$$\frac{1}{s} + \frac{1}{s'} = \frac{1}{f}$$

Mirror equation

Note that objects at infinity are always focused at the focal point.

You should be alerted to the fact that in order to apply the equations of this section to specific problems, you will have to follow a sign convention for the quantities s, s', f, and R. We shall summarize this sign convention, which is applicable to all spherical mirrors, in the next section.

25.3   Convex Mirrors and Sign Conventions

Figure 25.8 shows the formation of an image by a **convex mirror,** that is, one silvered such that light is reflected from the outer, convex surface. This is sometimes called a **diverging mirror** because the rays from any point on the

Figure 25.8
Ray diagram for a spherical convex mirror. Note that the image is virtual and erect.

Front, or
real, side | Back, or
virtual, side

s and s' positive | s and s' negative

Incident light

No light

Reflected light

Convex or
concave mirror

Figure 25.9
Diagram for describing the signs of s and s' for convex or concave mirrors.

object diverge after reflection as though they were coming from some point behind the mirror. Note that the image in Figure 25.8 is virtual rather than real because it lies behind the mirror at the location from which the reflected rays appear to originate. Furthermore, whenever the object is located on the left side of the mirror, the image will always be erect, virtual, and smaller than the object, as shown in the figure. Convex mirrors are commonly used at various vantage points around a place of business in order to spot shoplifters in action. They also are used as rearview mirrors in most new automobiles.

We shall not attempt to derive any equations for convex spherical mirrors. The results of such derivations would show that the equations developed for concave mirrors can be used if we adhere to a sign convention that we shall now summarize.

We can use Equations 25.2, 25.4, and 25.6 for either concave or convex mirrors if we adhere to the following procedure. Let us refer to the region in which light rays move as the *front side* of the mirror and the other side, where virtual images are formed, as the *back side*. For example, in Figures 25.6 and 25.8, the side to the left of the mirrors is the front side and the side to the right of the mirrors is the back side. Figure 25.9 is useful in understanding the rules for object and image distances, and Table 25.1 summarizes the sign conventions for all the necessary quantities.

Table 25.1
Sign Convention for Mirrors

s is $+$ if the object is in front of the mirror (real object).
s is $-$ if the object is in back of the mirror (virtual object).

s' is $+$ if the image is in front of the mirror (real image).
s' is $-$ if the image is in back of the mirror (virtual image).

Both f and R are $+$ if the center of curvature is in front of the mirror (concave mirror).
Both f and R are $-$ if the center of curvature is in back of the mirror (convex mirror).

If M is positive, the image is erect.
If M is negative, the image is inverted.

s = object distance; s' = image distance; f = focal length; R = radius of curvative;
M = lateral magnification.

Ray Diagrams for Mirrors

The position and size of images formed by mirrors can be conveniently deter-mined by using *ray diagrams*. These graphical constructions tell us the total nature of the image and can be used to check parameters calculated from the mirror and magnification equations. In these diagrams, the position of the object and the location of the center of curvature need to be known. In order to locate the image, three rays are then constructed, as shown by the various examples in Figure 25.10. These rays all start from any object point (chosen to be the top) and are drawn as follows:

1. The first, labeled ray 1, is drawn from the top of the object parallel to the optical axis and is reflected back through the focal point, *F*.
2. The second, labeled ray 2, is drawn from the top of the object through the focal point. Thus, it is reflected parallel to the optic axis.
3. The third, labeled ray 3, is drawn from the top of the object through the center of curvature, *C*, which is reflected back on itself.

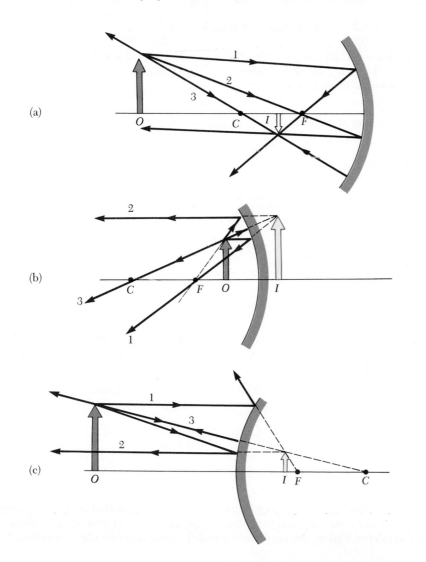

Figure 25.10
Ray diagrams for spherical mirrors. (a) The object is lo-cated outside the center of curvature of a spherical con-cave mirror. (b) The object is located between the spherical concave mirror and the focal point *F*. (c) The object is lo-cated in front of a spherical convex mirror.

The intersection of any *two* of these rays at a point locates the image. The third ray serves as a check on your construction. The image point obtained in this fashion must always agree with the value of s' calculated from the mirror formula.

In the case of a concave mirror, note what happens as the object is moved closer to the mirror. The real, inverted image in Figure 25.10a moves to the left as the object approaches the focal point. When the object is at the focal point, the image is infinitely far to the left. However, when the object lies between the focal point and the vertex, as in Figure 25.10b, the image is virtual and erect. Finally, for the convex mirror shown in Figure 25.10c, the image of a real object is always virtual and erect. In this case, as the object distance increases, the virtual image decreases in size and approaches the focal point as s approaches infinity. You should construct other diagrams to verify the variation of image position with object position.

The following examples should help you become familiar with these sign conventions.

Example 25.2   The Image for a Concave Mirror

Assume that a certain concave spherical mirror has a focal length of 10 cm. Find the location of the image for object distances of (a) 25 cm, (b) 10 cm, and (c) 5 cm. Describe the image in each case.

Solution

(a) For an object distance of 25 cm, we find the image distance using the mirror equation:

$$\frac{1}{s} + \frac{1}{s'} = \frac{1}{f}$$

$$\frac{1}{25 \text{ cm}} + \frac{1}{s'} = \frac{1}{10 \text{ cm}}$$

$$s' = \boxed{16.7 \text{ cm}}$$

The magnification is given by Equation 25.2:

$$M = -\frac{s'}{s} = -\frac{16.7 \text{ cm}}{25 \text{ cm}} = \boxed{-0.67}$$

Thus, the image is smaller than the object. Furthermore, the image is inverted because M is negative. Finally, because s' is positive, the image is located on the front side of the mirror and is real. This situation is pictured in Figure 25.10a.

(b) When the object distance is 10 cm, the object is located at the focal point. Substituting the values $s = 10$ cm and $f = 10$ cm into the mirror equation, we find

$$\frac{1}{10 \text{ cm}} + \frac{1}{s'} = \frac{1}{10 \text{ cm}}$$

$$s' = \boxed{\infty}$$

Thus, we see that rays of light originating from an object located at the focal point of a concave mirror are reflected such that the image is formed at an infinite distance from the mirror; that is, the rays travel parallel to one another after reflection.

(c) When the object is at the position $s = 5$ cm, it is inside the focal point of the mirror. In this case, the mirror equation gives

$$\frac{1}{5 \text{ cm}} + \frac{1}{s'} = \frac{1}{10 \text{ cm}}$$

$$s' = \boxed{-10 \text{ cm}}$$

That is, the image is virtual since it is located behind the mirror. The magnification is

$$M = -\frac{s'}{s} = -\left(\frac{-10 \text{ cm}}{5 \text{ cm}}\right) = \boxed{2}$$

From this, we see that the image is magnified by a factor of 2, and the positive sign indicates that the image is erect (Fig. 25.10b).

Note the characteristics of the images formed by a concave spherical mirror. When the object is outside the focal point, the image is inverted and real; at the focal point, the image is formed at infinity; inside the focal point, the image is erect and virtual.

Exercise If the object distance is 20 cm, find the image distance and the magnification of the mirror.

Answer $s' = 20$ cm, $M = -1$.

Example 25.3 The Image for a Convex Mirror

An object 3 cm high is placed 20 cm from a convex mirror having a focal length of 8 cm. Find (a) the position of the final image and (b) the magnification of the mirror.

Solution
(a) Since the mirror is convex, its focal length is negative. To find the image position, we use the mirror equation:

$$\frac{1}{s} + \frac{1}{s'} = \frac{1}{f}$$

$$\frac{1}{20 \text{ cm}} + \frac{1}{s'} = \frac{1}{-8 \text{ cm}}$$

$$s' = \boxed{-5.71 \text{ cm}}$$

The negative value of s' indicates that the image is virtual, or behind the mirror, as in Figure 25.10c.

(b) The magnification of the mirror is

$$M = -\frac{s'}{s} = -\left(\frac{-5.71 \text{ cm}}{20 \text{ cm}}\right) = \boxed{0.286}$$

The image is erect because M is positive.

Exercise Find the height of the image.

Answer 0.857 cm.

Example 25.4   Who is Fairest of Them All?

When a woman stands 40 cm in front of a cosmetic mirror, the erect image is twice the size of her face. What is the focal length of this mirror?

Solution
The magnification equation gives us a relationship between the object and image distances:

$$M = -\frac{s'}{s} = 2$$

$$s' = -2s = -2(40 \text{ cm}) = -80 \text{ cm}$$

First, note that a virtual image is formed because the woman is able to see her erect image in the mirror. This explains why the image distance is negative. Substitute $s' = -80$ cm into the mirror equation to give

$$\frac{1}{40 \text{ cm}} - \frac{1}{80 \text{ cm}} = \frac{1}{f}$$

$$f = \boxed{80 \text{ cm}}$$

The positive sign for the focal length indicates that the mirror is concave, a fact that we already knew because the mirror magnified the object (a convex mirror would have produced an image of diminished size).

25.4   Images Formed by Refraction

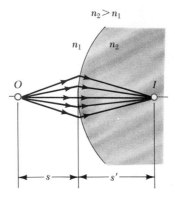

Figure 25.11
Image formed by refraction at a spherical surface. Rays making small angles with the optic axis diverge from a point object at O, and pass through the image point I.

We shall now consider the process by which images are formed by refraction at a spherical surface. Consider two transparent media with indices of refraction n_1 and n_2, separated by a spherical surface of radius R (Fig. 25.11). Let us assume that the medium to the right has a higher index of refraction than the one to the left, that is, $n_2 > n_1$. This would be the situation for light entering a curved piece of glass from air or for light entering the water in a fish bowl from air. Note that the rays originating at the object location, O, are refracted at the spherical surface and then converge to the image point, I. You can begin with Snell's law of refraction and use simple geometric techniques to show that the object distance, image distance, and radius of curvature are related by the equation

$$\frac{n_1}{s} + \frac{n_2}{s'} = \frac{n_2 - n_1}{R} \tag{25.7}$$

Furthermore, the magnification of a refracting surface is given by

$$M = \frac{h'}{h} = -\frac{n_1 s'}{n_2 s} \tag{25.8}$$

As was the case for mirrors, we must use a sign convention if we are to apply these equations to a variety of circumstances. First note that real images

Table 25.2
Sign Convention for Refracting Surfaces

s is $+$ if the object is in front of the surface (real object).
s is $-$ if the object is in back of the surface (virtual object).

s' is $+$ if the image is in back of the surface (real image).
s' is $-$ if the image is in front of the surface (virtual image).

R is $+$ if the center of curvature is in back of the surface.
R is $-$ if the center of curvature is in front of the surface.

s = object distance; s' = image distance; R = radius of curvature.

are formed on the side of the surface that is *opposite* the side from which the light comes, in contrast to mirrors, where real images are formed on the *same* side of the reflecting surface. Therefore, *the sign convention for spherical refracting surfaces is the same as for mirrors, recognizing the change in sides of the surface for real and virtual images.* For example, in Figure 25.11, s, s', and R are all positive.

The sign convention for spherical refracting surfaces is summarized in Table 25.2. The same sign convention will be used for thin lenses, which will be discussed in the next section. As with mirrors, we assume that the front of the refracting surface is the side from which the light approaches the surface.

Plane Refracting Surfaces

If the refracting surface is a plane, then R approaches infinity and Equation 25.7 reduces to

$$\frac{n_1}{s} = -\frac{n_2}{s'}$$

or

$$s' = -\frac{n_2}{n_1} s \qquad (25.9)$$

From this equation we see that the sign of s' is opposite that of s. Thus, *the image formed by a plane refracting surface is on the same side of the surface as the object.* This is illustrated in Figure 25.12 for the situation in which n_1 is greater than n_2, where a virtual image is formed between the object and the surface. If n_1 is less than n_2, the image will still be virtual but will be formed to the left of the object.

Example 25.5 Gaze into the Crystal Ball

A coin 2 cm in diameter is embedded in a solid glass ball of radius 30 cm (Fig. 25.13). The index of refraction of the ball is 1.5, and the coin is 20 cm from the surface. Find the position and height of the image.

Solution
First, note that the rays originating from the object are refracted away from the normal at the surface and diverge outward. Hence, the image is formed in the glass

Figure 25.12
The image formed by a plane refracting surface is virtual; that is, it forms to the left of the refracting surface.

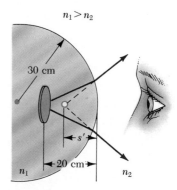

Figure 25.13   (Example 25.5) A coin embedded in a glass ball forms a virtual image between the coin and the glass surface.

and is virtual. Applying Equation 25.7 and taking $n_1 = 1.5$, $n_2 = 1$, $s = 20$ cm, and $R = -30$ cm, we get

$$\frac{n_1}{s} + \frac{n_2}{s'} = \frac{n_2 - n_1}{R}$$

$$\frac{1.5}{20 \text{ cm}} + \frac{1}{s'} = \frac{1 - 1.5}{-30 \text{ cm}}$$

$$s' = \boxed{-17 \text{ cm}}$$

The negative sign indicates that the image is in the same medium as the object (the side of incident light), in agreement with our ray diagram. Since the image is in the same medium as the object, it must be virtual.

To find the image height, we first use Equation 25.8 for the magnification:

$$M = -\frac{n_1 s'}{n_2 s} = -\frac{1.5(-17 \text{ cm})}{1(20 \text{ cm})} = \frac{h'}{h} = 1.28$$

Therefore,

$$h' = 1.28h = (1.28)(2 \text{ cm}) = \boxed{2.56 \text{ cm}}$$

The positive value for M indicates an erect image.

Example 25.6   The One That Got Away

A small fish is swimming at a depth d below the surface of a pond (Fig. 25.14). What is the *apparent depth* of the fish as viewed from directly overhead?

Solution
In this example, the refracting surface is a plane, and so R is infinite. Hence, we can use Equation 25.9 to determine the location of the image. Using the facts that $n_1 = 1.33$ for water and $s = d$ gives

$$s' = -\frac{n_2}{n_1} s = -\frac{1}{1.33} d = \boxed{-0.75d}$$

Again, since s' is negative, the image is virtual, as indicated in Figure 25.14. The apparent depth is three fourths the actual depth. For instance, if $d = 4$ m, $s' = -3$ m.

Exercise   If the fish is 10 cm long, how long is its image?

Answer   10 cm.

$n_2 = 1$

$n_1 = 1.33$

s'

d

Figure 25.14 (Example 25.6) The apparent depth s' of the fish is less than the true depth d.

*25.5   Atmospheric Refraction

Images formed by refraction in our atmosphere lead to some interesting results. In this section, we shall look at two examples. One situation that occurs daily is that we are able to see the Sun at dusk even though it has passed below the horizon. Figure 25.15 shows why this occurs. Rays of light from the Sun strike the Earth's atmosphere, represented by the shaded area around the

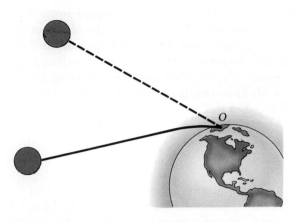

Figure 25.15
Because of refraction, an observer at O sees the Sun even though it has fallen below the horizon.

Earth, and are bent as they pass into a medium having an index of refraction different from that of the almost empty space in which they were traveling. The bending in this situation is somewhat different from those cases previously considered because here there is a gradual, continuous bending as the light moves through the atmosphere toward an observer at point O. This occurs because the light moves through layers of air having a continuously changing index of refraction. When the rays reach the observer, the eye will follow them back along the direction from which they appear to have come; this is indicated by the dashed path in the figure. The end result is that the Sun is seen to be above the horizon even after it has fallen below it.

The **mirage** is another phenomenon of nature produced by refraction in the atmosphere. A mirage can be observed when the ground is so hot that the air directly above it is warmer than the air at higher elevations. The desert is, of course, a region in which such circumstances prevail, but mirages are also seen on heated roadways during the summer. The layers of air at different heights above the Earth have different densities and different refractive indices. The effect that this can cause is pictured in Figure 25.16. In this situation the observer is able to see a tree, for example, in two different ways. One group of light rays reaches the observer by the straight-line path A, and the eye traces these back to see the tree in the normal fashion. In addition, a second group of

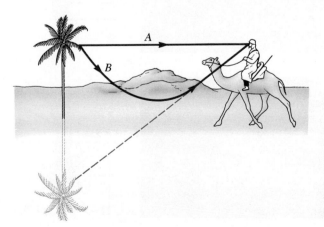

Figure 25.16
A mirage is produced by the bending of light rays in the atmosphere when there are large temperature differences between the ground and the air.

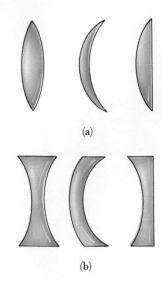

(a)

(b)

Figure 25.17
Various lens shapes: (a) Converging lenses have a positive focal length and are thickest at the middle. (b) Diverging lenses have a negative focal length and are thickest at the edges.

rays travels along the curved path *B*. These rays are directed toward the ground and are bent as a result of refraction. Consequently, the observer also sees an inverted image of the tree as these rays are traced back to the point from which they appear to originate. An erect image and an inverted image are seen when one observes the image of a tree formed in a reflecting pool of water; therefore, the observer calls upon this past experience and concludes that a pool of water must be in front of the tree.

25.6   Thin Lenses

A typical **thin lens** consists of a piece of glass or plastic ground so that its two refracting surfaces are segments of spheres. Lenses are commonly used to form images by refraction in optical instruments, such as cameras, telescopes, and microscopes. The equation that relates object distances and image distances for a lens is virtually identical to the mirror equation derived earlier, and the method used to derive the equation is similar.

Figure 25.17 shows some representative shapes for lenses. Note that we have placed these lenses in two groups. Those in Figure 25.17a are thicker at the center than at the rim, and those in Figure 25.17b are thinner at the center than at the rim. The first group are examples of **converging lenses,** and the second group are called **diverging lenses.** The reason for these names will become apparent shortly.

As was the case for mirrors, it is convenient to define a point called the **focal point** for a lens. For example, in Figure 25.18a a group of rays parallel to the axis pass through the focal point, *F*, after being converged by the lens. The distance from the focal point to the lens is called the **focal length,** *f*. *The focal length is the image distance that corresponds to an infinite object distance.*

Figure 25.18
(Left) Photographs of the effect of converging and diverging lenses on parallel rays. *(Courtesy of Jim Lehman, James Madison University) (Right)* The principal focal points of (a) the biconvex lens and (b) the biconcave lens.

(a)

(b)

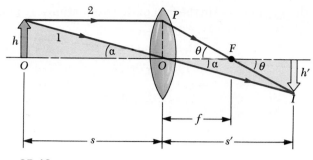

Figure 25.19
Geometrical construction for developing the thin-lens equation.

Recall that we are considering the lens to be very thin. As a result, it makes no difference whether we consider the focal length to be the distance from the focal point to the surface of the lens or the distance from the focal point to the center of the lens because the difference in lengths is negligible.

Rays parallel to the axis diverge after passing through a lens of the shape shown in Figure 25.18b. In this case, the focal point is defined to be the point from which the diverged rays appear to originate, labeled F in the figure. Figures 25.18a and 25.18b indicate why the names *converging* and *diverging* are applied to these lens. A converging lens brings together, or converges, parallel rays of light; a diverging lens spreads apart, or diverges, parallel rays.

Consider a ray of light that passes through the center of a lens, shown as ray 1 in Figure 25.19. If we followed this ray through the lens by applying Snell's law at both surfaces, we would find that this ray is deflected from its original direction of travel by a distance δ, shown in Figure 25.20. We shall make here what is called the *thin-lens approximation*, that is, *the thickness of the lens is assumed to be negligible*, and as a result the distance δ becomes vanishingly small. Thus, this ray will pass through the lens undeflected. Ray 2 in Figure 25.19 is parallel to the principal axis of the lens, and as a result it passes through the focal point, F, after refraction. The point at which these rays intersect is the point at which the image is formed.

We first note that the tangent of the angle α can be found by using the shaded triangles in Figure 25.19:

$$\tan \alpha = \frac{h}{s} \qquad \text{or} \qquad \tan \alpha = -\frac{h'}{s'}$$

from which

$$M = \frac{h'}{h} = -\frac{s'}{s} \qquad\qquad \textbf{(25.10)}$$

Thus, the equation for magnification by a lens is the same as the equation for magnification by a mirror. We also note from Figure 25.19 that the tangent of θ is

$$\tan \theta = \frac{PO}{f} \qquad \text{or} \qquad \tan \theta = -\frac{h'}{s' - f}$$

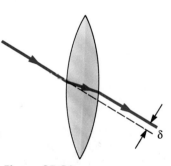

Figure 25.20
A ray passing through the center of the lens is deviated from its path by a distance δ.

Virtual side	Real side
s positive s' negative	s negative s' positive
Incident light	Refracted light

Concave or convex
thin lens or refracting surfaces

Figure 25.21
Diagram for obtaining the signs of s and s' for a thin lens or a refracting surface.

Thin-lens equation

Lens makers' equation

However, the height PO used in the first of these equations is the same as h, the height of the object. Therefore we get

$$\frac{h}{f} = -\frac{h'}{s'-f}$$

$$\frac{h'}{h} = -\frac{s'-f}{f}$$

Using this in combination with Equation 25.10 gives

$$\frac{s'}{s} = \frac{s'-f}{f}$$

which reduces to

$$\frac{1}{s} + \frac{1}{s'} = \frac{1}{f} \qquad \textbf{(25.11)}$$

This equation, called the **thin-lens equation,** can be used with either converging or diverging lenses if we adhere to a set of sign conventions. Figure 25.21 is useful for obtaining the signs of s and s', and the complete sign convention for lenses is given in Table 25.3. Note that a *converging lens has a positive focal length* under this convention, and a *diverging lens has a negative focal length*. Hence the names *positive* and *negative* are often given to these lenses.

The focal length for a lens in air is related to the curvatures of its front and back surfaces and to the index of refraction n of the lens material as

$$\frac{1}{f} = (n-1)\left(\frac{1}{R_1} - \frac{1}{R_2}\right) \qquad \textbf{(25.12)}$$

where R_1 is the radius of curvature of the front surface of the lens and R_2 is the radius of curvature of the back surface. Equation 25.12 enables us to calculate the focal length from the known properties of the lens, and it is called the **lens makers' equation.**

Table 25.3
Sign Convention for Thin Lenses

s is $+$ if the object is in front of the lens.
s is $-$ if the object is in back of the lens.

s' is $+$ if the image is in back of the lens.
s' is $-$ if the image is in front of the lens.

R_1 and R_2 are $+$ if the center of curvature for each surface is in back of the lens.
R_1 and R_2 are $-$ if the center of curvature for each surface is in front of the lens.

f is $+$ for a converging lens.
f is $-$ for a diverging lens.

$s =$ object distance; $s' =$ image distance; $R_1 =$ radius of curvature of front surface; $R_2 =$ radius of curvature of back surface; $f =$ focal length.

Figure 25.22
Ray diagrams for locating the image of an object. (a) The object is located outside the focal point of a converging lens. (b) The object is located inside the focal point of a converging lens. (c) The object is located outside the focal point of a diverging lens.

Ray Diagrams for Thin Lenses

Graphical methods, or ray diagrams, are very convenient for determining the image of a thin lens or a system of lenses. Such constructions should also help clarify the sign conventions that have been discussed. Figure 25.22 illustrates this method for three different single-lens situations. To locate the image of a converging lens (Figs. 25.22a and 25.22b), the following three rays are drawn from the top of the object:

1. The first ray is drawn parallel to the optic axis. After being refracted by the lens, this ray passes through (or appears to come from) one of the focal points.
2. The second ray is drawn through the center of the lens. This ray continues in a straight line.
3. The third ray is drawn through the focal point F, and emerges from the lens parallel to the optic axis.

A similar construction is used to locate the image of a diverging lens, as shown in Figure 25.22c. The point of intersection of *any two* of the rays in these diagrams can be used to locate the image. The third ray serves as a check on your construction.

For the converging lens in Figure 25.22a, where the object is *outside* the front focal point ($s > f$), the image is real and inverted. On the other hand, when the real object is *inside* the front focal point ($s < f$), as in Figure 25.22b, the image is virtual, erect, and enlarged. Finally, for the diverging lens shown in Figure 25.22c, the image is always virtual and erect. These geometric constructions are reasonably accurate only if the distance between the rays and the principal axis is small compared to the radii of the lens surfaces.

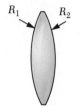

Figure 25.23
This lens has two curved surfaces with radii of curvature of R_1 and R_2. For this converging lens, the center of curvature R_1 lies to the right of the lens, while the center of R_2 lies to the left.

Problem-Solving Strategy:
Lenses and Mirrors

Success or failure in working lens and mirror problems is largely determined by whether or not you make sign errors when substituting into the lens and mirror equations, and the only way to insure that you don't make sign errors is to become adept at using the sign conventions. The best way to develop proficiency in working with the sign conventions is to work a multitude of problems on your own. Watching your instructor or reading the example problems is no substitute for your own practice.

Example 25.7   The Lens Makers' Equation

A lens shaped as shown in Figure 25.23 has an index of refraction of 1.50. the magnitude of the radius of curvature of the front surface is 10 cm, and the radius of curvature of the back surface has a magnitude of 15 cm. Find the focal length of the lens.

Solution
From the sign conventions in Table 25.3 we find that $R_1 = +10$ cm and $R_2 = -15$ cm. Thus, using the lens makers' equation, we have

$$\frac{1}{f} = (n-1)\left(\frac{1}{R_1} - \frac{1}{R_2}\right) = (1.5-1)\left(\frac{1}{10 \text{ cm}} - \frac{1}{-15 \text{ cm}}\right)$$

$$f = \boxed{12 \text{ cm}}$$

Example 25.8   An Image Formed by a Converging Lens

A converging lens of focal length 10 cm forms an image of an object placed (a) 30 cm, (b) 10 cm, and (c) 5 cm from the lens. Find the image distance and describe the image in each case.

Solution
(a) The thin-lens equation, Equation 25.11, can be used to find the image distance:

$$\frac{1}{s} + \frac{1}{s'} = \frac{1}{f}$$

$$\frac{1}{30 \text{ cm}} + \frac{1}{s'} = \frac{1}{10 \text{ cm}}$$

$$s' = \boxed{15 \text{ cm}}$$

The positive sign for the image distance tells us that the image is on the real side of the lens (Fig. 25.21). The magnification of the lens is

$$M = -\frac{s'}{s} = -\frac{15 \text{ cm}}{30 \text{ cm}} = \boxed{-0.50}$$

Thus, the image is reduced in size by one half, and the negative sign for M tells us that the image is inverted. The situation is like that pictured in Figure 25.22a.

(b) No calculation should be necessary for this case because we know that, when the object is placed at the focal point, the image will be formed at infinity. This is readily verified by substituting $s = 10$ cm into the lens equation.

(c) We now move inside the focal point, to an object distance of 5 cm. In this case, the lens equation gives

$$\frac{1}{5 \text{ cm}} + \frac{1}{s'} = \frac{1}{10 \text{ cm}}$$

$$s' = \boxed{-10 \text{ cm}}$$

and

$$M = -\frac{s'}{s} = -\left(\frac{-10 \text{ cm}}{5 \text{ cm}}\right) = \boxed{2}$$

The negative image distance tells us that the image is formed on the side of the lens from which the light is incident, the virtual side (Fig. 25.21). The image is enlarged, and the positive sign for M tells us that the image is erect, as shown in Figure 25.22b.

Example 25.9 The Case of a Diverging Lens

Repeat the problem of Example 25.8 for a *diverging* lens of focal length 10 cm.

Solution

(a) Let us apply the lens equation with an object distance of 30 cm:

$$\frac{1}{s} + \frac{1}{s'} = \frac{1}{f}$$

$$\frac{1}{30 \text{ cm}} + \frac{1}{s'} = -\frac{1}{10 \text{ cm}}$$

$$s' = \boxed{-7.5 \text{ cm}}$$

The magnification is

$$M = -\frac{s'}{s'} = -\left(\frac{-7.5 \text{ cm}}{30 \text{ cm}}\right) = \boxed{0.25}$$

Thus, the image is virtual, smaller than the object, and erect.

(b) When the object is at the focal point, $s = 10$ cm, we have

$$\frac{1}{10 \text{ cm}} + \frac{1}{s'} = -\frac{1}{10 \text{ cm}}$$

$$s' = \boxed{-5 \text{ cm}}$$

and

$$M = -\frac{s'}{s'} = -\left(\frac{-5 \text{ cm}}{10 \text{ cm}}\right) = \boxed{0.50}$$

(c) When the object is inside the focal point, at 5 cm, we have

$$\frac{1}{5 \text{ cm}} + \frac{1}{s'} = -\frac{1}{10 \text{ cm}}$$

$$s' = \boxed{-3.33 \text{ cm}}$$

and

$$M = -\left(\frac{-3.33 \text{ cm}}{5 \text{ cm}}\right) = \boxed{0.66}$$

Again, we have a virtual image that is smaller than the object and erect as in Figure 25.22c.

25.7   Multiple Lens Systems

Many practical optical systems consist of two or more lenses in a row. In order to solve problems involving several lenses, Equation 25.11 can be applied to each lens in turn. For example, if two lenses are used to form an image, the system can be treated by finding the image of the first lens imagining that the second lens was not present. Next, it is important to note that

the image of the first lens is treated as the object of the second lens.

The image of the second lens is the final image for the system. The same procedure can be extended to three or more lenses in a row. *The overall magnification of the system is the product of the magnifications for each individual lens.* The techniques for solving such problems are best shown through the following example problems.

Example 25.10   Two Lenses in a Row

Two converging lenses are placed 20 cm apart as shown in Figure 25.24. If the first lens has a focal length of 10 cm and the second 20 cm, find the location of the final image formed of an object 30 cm in front of the first lens. Also, find the magnification of the system.

Figure 25.24   (Examples 25.10 and 25.11)

PHYSICS IN ACTION

(Left) The "professor in the box" appears to be balancing himself on a few fingers with both of his feet elevated from the floor. Since the professor is able to maintain this position for a long time, he appears to defy gravity. This is one example of an optical illusion used by magicians which makes use of a mirror. Can you explain where the mirror was placed in the box to create this illusion? (The same strange professor and some of his friends from VMI have been known to "fly" at times above a lecture bench.)

    (Right) This photograph shows two converging lenses separated by about 5 cm. Parallel light rays are incident from the left on the larger lens whose focal length was measured by an independent experiment to be 11 cm. The final image formed is to the right of the smaller lens as shown. Using the image of the larger lens as the object for the smaller lens, and the information given in the photograph, you should be able to show that the focal length of the smaller lens is about 5.2 cm.

(Courtesy of Henry Leap and Jim Lehman)

Solution

The location of the image formed by the first lens is found by the thin-lens equation as

$$\frac{1}{30 \text{ cm}} + \frac{1}{s'} = \frac{1}{10 \text{ cm}}$$

$$s' = 15 \text{ cm}$$

and the magnification of this lens is

$$M_1 = -\frac{s'}{s} = -\frac{15 \text{ cm}}{30 \text{ cm}} = -\frac{1}{2}$$

The image formed by this lens becomes the object for the second lens. Thus, the object distance for the second lens is 5 cm. We now apply the thin-lens equation again to find the location of the final image.

$$\frac{1}{5 \text{ cm}} + \frac{1}{s'} = \frac{1}{20 \text{ cm}}$$

$$s' = \boxed{-6.67 \text{ cm}}$$

The magnification of the second lens is

$$M_2 = -\frac{s'}{s} = -\frac{(-6.67 \text{ cm})}{5 \text{ cm}} = 1.33$$

Thus, the final image is located 6.67 cm to the left of the second lens, and the overall magnification of the system is

$$M = M_1 M_2 = \left(-\frac{1}{2}\right)(1.33) = \boxed{-0.667}$$

The negative sign indicates that the final image is inverted with respect to the initial object.

Example 25.11   Move the Lenses Closer Together

Find the location of the final image in Example 25.10 if the two lenses are separated by 10 cm.

Solution

As above, the first lens attempts to form an image 15 cm to the right of the lens. However, before this image is formed, the light rays are intercepted by the second lens. Regardless of this, the second lens treats the image formed (or trying to be formed) by the first lens as its object. Thus, the object position for the second lens is 5 cm to its right. This is a virtual object, and the thin-lens equation becomes

$$\frac{1}{-5 \text{ cm}} + \frac{1}{s'} = \frac{1}{20 \text{ cm}}$$

$$s' = \boxed{4 \text{ cm}}$$

Thus, the final image is formed 4 cm to the right of the second lens.

Exercise   Find the overall magnification of the system.

Answer $M = -0.40$

*25.8   Lens Aberrations

One of the basic problems of lenses and lens systems is the imperfect quality of the images. The simple theory of mirrors and lenses assumes that rays make small angles with the optic axis. In this simple model, all rays leaving a point source focus at a single point, producing a sharp image. Clearly, this is not always true. For those cases where the approximations used in this theory do not hold, imperfect images are formed.

If we wish to perform a precise analysis of image formation, it is necessary to trace each ray using Snell's law at each refracting surface. This procedure shows that the rays from a point object do *not* focus at a single point. This means there is no single point image; instead, the image is *blurred*. The departures of real (imperfect) images from the ideal image predicted by the simple theory are called **aberrations.** Two types of aberrations will now be described.

Spherical Aberrations

Spherical aberrations result from the fact that the focal points of light rays far from the optic axis of a spherical lens (or mirror) are different from the focal points of rays passing through the center. Figure 25.25 illustrates spherical aberration for parallel rays passing through a converging lens. Rays near the middle of the lens have longer focal lengths than rays at the edges. Hence, there is no single focal length for a lens. Many optical devices are equipped with an adjustable aperture to reduce spherical aberration when possible. (An aperture is an opening that only allows light rays to pass that are close to the lens axis.) Sharper images are produced as the aperture size is reduced, since only the central portion of the lens is exposed to the incident light. At the same time, however, less light is imaged. This is precisely the reason that a pinhole camera (one with a very small aperture) is able to produce such sharp images.

In the case of mirrors, spherical aberration can be eliminated, or at least minimized, by using a parabolic surface rather than a spherical surface. Parabolic surfaces are not often used, however, because they are very expensive to make. Parallel light rays incident on such a surface focus at a common point. Parabolic reflecting surfaces are used in astronomical telescopes to enhance the image intensity. They are also used in flashlights, where a parallel light beam is produced from a small lamp placed at the focal point of the surface.

Figure 25.25
Spherical aberration produced by a converging lens. Does a diverging lens produce spherical aberration? (Angles are greatly exaggerated for clarity.)

Chromatic Aberrations

The fact that different wavelengths of light refracted by a lens focus at different points gives rise to chromatic aberrations. In Chapter 24, we described how the index of refraction of a material varies with wavelength. When white light passes through a lens, we find, for example, that violet light rays are refracted more than red light rays (Fig. 25.26). From this we see that the focal length is larger for red light than for violet light. Other wavelengths (not shown in Figure 25.26) would have intermediate focal points. The chromatic aberration for a diverging lens is opposite that for a converging lens. Chromatic aberration can be greatly reduced by using a combination of a converging and diverging lens made from two different types of glass. Photographs of various forms of lens aberrations are shown in Figure 25.27.

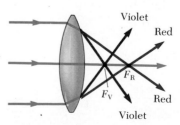

Figure 25.26
Chromatic aberration produced by a converging lens. Rays of different wavelengths focus at different points. (Angles are greatly exaggerated for clarity.)

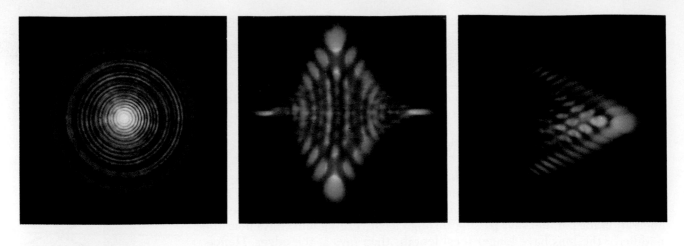

Figure 25.27
Lenses can produce various forms of aberrations, as shown by the blurred images of a point source in these photos. (a) Spherical aberration occurs when light passing through the lens at different distances from the optical axis is focused at different points. (b) Astigmatism is an aberration that occurs for objects that are not located on the optical axis of the lens. (c) Coma. This aberration occurs as light passing through the lens far from the optical axis focuses at a different part of the focal plane from light passing near the center of the lens. (*Photos by Norman Goldberg*)

Summary

Images are formed where rays of light intersect or at the point from which they appear to originate. **A real image** is one in which light intersects, or passes through, an image point. **A virtual image** is one in which the light does not pass through the image point but appears to diverge from that point.

The image formed by a plane mirror has the following properties:

1. The image is as far behind the mirror as the object is in front.
2. The image is unmagnified, virtual, and erect.
3. The image has right-left reversal.

The **magnification**, M, of a mirror is defined as the ratio of **image height**, h', to **object height**, h, which is equal to the negative of the ratio of image distance, s', to object distance, s:

$$M = \frac{h'}{h} = -\frac{s'}{s} \tag{25.2}$$

The **object distance** and **image distance** for a spherical mirror of radius R are related by the **mirror equation:**

$$\frac{1}{s} + \frac{1}{s'} = \frac{1}{f} \tag{25.6}$$

where $f = R/2$ is the **focal length** of the mirror.

An image can be formed by refraction at a spherical surface of radius R. The object and image distances for refraction from such a surface are related as

$$\frac{n_1}{s} + \frac{n_2}{s'} = \frac{n_2 - n_1}{R} \tag{25.7}$$

The **magnification** of a refracting surface is

$$M = \frac{h'}{h} = -\frac{n_1 s'}{n_2 s} \tag{25.8}$$

where the object is located in the medium of index of refraction n_1 and the image is formed in the medium whose index of refraction is n_2.

For a **thin lens** the **magnification** is

$$M = \frac{h'}{h} = -\frac{s'}{s}$$

(25.10)

and the object and image distances are related by the **thin-lens equation:**

$$\frac{1}{s} + \frac{1}{s'} = \frac{1}{f}$$

(25.11)

Aberrations are responsible for the formation of imperfect images by lenses and mirrors. **Spherical aberration** results from the fact that the focal points of light rays far from the optic axis of a spherical lens or mirror are different from those of rays passing through the center. **Chromatic aberration** arises from the fact that light rays of different wavelengths focus at different points when refracted by a lens.

Additional Reading

A. B. Fraser and M. W. Hirsch, *Mirages,* Springer-Verlag, 1975.

T. B. Greenslade, "Multiple Images in Plane Mirrors," *The Physics Teacher,* January 1982, p. 29.

R. C. Jones, "How Images Are Detected," *Sci. American,* September 1968, p. 111.

F. D. Smith, "How Images Are Formed," *Sci. American,* September 1968, p. 97.

W. Tape, "The Topology of Mirages," *Sci. American,* June 1985, p. 120.

D. E. Thomas, "Mirror Images," *Sci. American,* December 1980, p. 206.

P. K. Tien, "Integrated Optics," *Sci. American,* April 1974, p. 28.

A. C. S. van Heel and C. H. F. Velzel, *What is Light?* World University Library, New York, McGraw-Hill, 1968, Chaps. 1 and 2.

Questions

1. Why do some emergency vehicles have the symbol ƎƆИA⅃UᗺMA written on the front?

2. When you look in a mirror, the image of your left and right sides is reversed, yet the image of your head and legs is not reversed. Explain.

3. The rear-view mirror on late model cars warns the user that objects may be closer than they appear. What kind of mirror is being used and why was that type selected?

4. Consider a concave spherical mirror with a real object. Is the image always inverted? Is the image always real? Give conditions for your answers.

5. Repeat the previous question for a convex spherical mirror.

6. Consider a spherical concave mirror, with the object located to the left of the mirror beyond the focal point. Using ray diagrams, show that the image of the object moves to the left as the object approaches the focal point.

7. What is the magnification of a plane mirror? What is its focal length?

8. Why does a clear stream always appear to be shallower than it actually is?

9. A person spearfishing from a boat sees a fish located 3 m from the boat at a depth of 1 m. In order to spear the fish, should the person aim at, above, or below the image of the fish?

10. Explain why a fish in a spherical goldfish bowl appears larger than it really is.

11. A mirage is formed when the air gets gradually cooler as the height above the ground increases. What might happen if the air becomes gradually warmer as the height is increased? This often happens over bodies of water or snow-covered ground: the effect is called looming.

12. If a cylinder of solid glass or clear plastic is placed above the words LEAD OXIDE and the words viewed from above, as in Figure 25.28. LEAD appear inverted but OXIDE does not. Explain.

Figure 25.28   Question 12
(Courtesy of Henry Leap and Jim Lehman)

13. Explain this statement: "The focal point of a lens is the image of a point object at infinity." Discuss the notion of infinity in real terms as it applies to object distances. Based on this statement, can you think of a "quick and dirty" method for determining the focal length of a positive lens?

14. One method for determining the position of an image, either real or virtual, is by means of *parallax*. If a finger or other object is placed at the position of the image, as in Figure 25.29, and the finger and image are viewed simultaneously (the image through the lens if it is virtual), the finger and image will have the same parallax; that is, if it is viewed from different positions, the image will appear to move along with

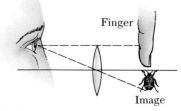

Figure 25.29   Question 14

the finger. Try using this method to locate the image formed by a lens. Explain why the method works.

15. It is well known that distant objects viewed under water with the naked eye appear blurred and out of focus. On the other hand, the use of goggles provides the swimmer with a clear view of objects. Explain this, using the fact that the indices of refraction of the cornea, water, and air are 1.376, 1.333, and 1.00029, respectively.

16. Lenses used in eyeglasses, whether converging or diverging, are always designed such that the middle of the lens curves away from the eye, like the center lenses on Figure 25.17a and 25.17b. Why?

17. Describe lenses that can be used to start a fire.

18. In a Jules Verne novel, a piece of ice is shaped to form a magnifying lens to focus sunlight to start a fire. Is this possible?

19. Consider the image formed by a thin converging lens. Under what conditions will the image be (a) inverted, (b) erect, (c) real, (d) virtual, (e) larger than the object, and (f) smaller than the object?

20. Repeat Question 19 for a thin diverging lens.

21. Discuss the proper position of a slide relative to the lens in a slide projector. What type of lens must the slide projector have?

22. Discuss why the following statements are true or false: (a) A virtual image can be a virtual object. (b) A virtual image can be a real object. (c) A real image can be a virtual object. (d) A real image can be a real object.

23. Explain why a mirror cannot give rise to a chromatic aberration.

24. Discuss the type of aberration involved in the following situations: (a) The edges of the image appear reddish. (b) A clear focus of the image's central portion cannot be obtained. (c) A clear focus of the image's outer portion cannot be obtained. (d) The central portion of the image is enlarged relative to the outer portions.

25. A solar furnace can be constructed by using a concave mirror to reflect and focus sunlight into a furnace enclosure. What factors in the design of the reflecting mirror would guarantee that very high temperatures may be achieved?

Problems

Section 25.1   Plane Mirrors

1. Use Figure 25.2 to give a geometric proof that the virtual image formed by a plane mirror is the same distance behind the mirror as the object is in front of it.

2. (a) What is the minimum length for a plane mirror that allows a person 6 ft tall to see his full height? (b) Does the distance from the mirror affect your answer? (*Hint:* A diagram showing the path followed by

the rays which leave the person would help. Follow those that leave the top of the head and the feet, strike the mirror, and then enter the eye.)

3. A person walks into a room that has two plane mirrors on opposite walls that produce multiple images. When the person is 5 ft from the mirror on the left wall and 10 ft from the mirror on the right wall, find the distance from the person to the first three images seen in the mirror on the left.

4. Consider the case in which a light ray is incident on mirror 1 in Figure 25.3. The reflected ray is incident on mirror 2 and is subsequently reflected. Let the angle of incidence (with respect to the normal) on mirror 1 be 53°. Determine the angle between the ray incident on mirror 1 and the ray reflected from mirror 2.

Section 25.2   Images Formed by Spherical Mirrors
Section 25.3   Convex Mirrors and Sign Conventions

Note that in the problems below, the algebraic signs are not given. We leave it to you to determine the correct algebraic sign to use with each quantity from an analysis of the problem and the sign conventions in Table 25.1.

5. A concave spherical mirror has a focal length of 20 cm. Find the location of the image for object distances of (a) 40 cm, (b) 20 cm, and (c) 10 cm. For each case, state whether the image is real or virtual and erect or inverted. Find the magnification for each case.

6. A concave spherical mirror has a radius of curvature of 20 cm. Find the location of the image for object distances of (a) 40 cm, (b) 20 cm, and (c) 10 cm. For each case, state whether the image is real or virtual and erect or inverted. Find the magnification for each case.

7. Repeat Problem 5 for a convex spherical mirror.

8. Repeat Problem 6 for a convex spherical mirror.

9. A dentist uses a mirror to examine a tooth. The tooth is 1.0 cm in front of the mirror, and the image is formed 10.0 cm behind the mirror. Determine (a) the mirror's radius of curvature and (b) the magnification of the image.

10. A convex mirror monitors the aisles in a store. The mirror has a radius of curvature of 0.55 m. Locate and describe the image of a customer 10.0 m from the mirror. Determine the magnification.

11. A virtual image is formed 20 cm from a concave mirror, having a radius of curvature of 40 cm. Find the position of the object.

12. A convex spherical mirror of focal length 15 cm is to form an image 10 cm from the mirror. Where should the object be placed?

13. A spherical Christmas tree ornament is 6 cm in diameter. What is the magnification of an object placed 10 cm away from the ornament?

14. A convex mirror has a focal length of 20 cm. Determine the object location for which the image will be one half the size of the object.

15. The real image height of a concave mirror is observed to be four times larger than the object height when the object is 30 cm in front of a mirror. What is the radius of curvature of the mirror?

16. A concave make-up mirror is designed so that a person 25 cm in front of the mirror sees an erect image magnified by a factor of two. What is the radius of curvature of the mirror?

17. An object 2 cm high is placed 10 cm in front of a mirror. What type of mirror and what radius of curvature are needed for an upright image that is 4 cm high?

18. A man standing 1.52 m in front of a shaving mirror produces an inverted image 18.0 cm in front of it. How close to the mirror should he stand if he wants to form an upright image of his chin that is twice the actual size of his chin?

19. A convex spherical mirror with radius of curvature of 10 cm produces a virtual image one third the size of the object. Where is the object located?

20. A child holds a candy bar 10 cm in front of a convex mirror and notices that the image is reduced by one half. What is the radius of curvature of the mirror?

21. A concave mirror has a focal length of 40 cm. Determine the object position for which the resulting image will be erect and four times the size of the object.

22. An object 2 cm in height is placed 3 cm in front of a concave mirror. If the image is 5 cm in height and virtual, what is the focal length of the mirror?

23. What type of mirror is required to form an image on a wall 2 m from the mirror of an object placed 10 cm in front of the mirror? What is the magnification of the image?

24. Under certain limiting conditions Equation 25.4 can be used for a plane mirror. (a) What value must be assumed for R if the mirror is plane? (b) What is the relationship between the object and image distance in this limiting case? (c) Does the result of (b) agree with the discussion given previously for the plane mirror?

Section 25.4   Images Formed by Refraction

25. A swimming pool is 2 m deep. How deep does it appear to be (a) when completely filled with water, (b) when filled halfway with water?

26. A cubical block of ice 50 cm on a side is placed on a level floor over a speck of dust. Find the location of the image of the speck if the index of refraction of ice is 1.309.

27. A colored marble is dropped into a large tank filled with benzene ($n = 1.50$). What is the depth of the tank if the apparent depth of the marble when viewed from directly above the tank is 35 cm?

28. A transparent sphere of unknown composition is observed to form an image of the Sun on the surface of the sphere opposite the Sun. What is the refractive index of the sphere material?

29. A goldfish is swimming in water inside a spherical plastic bowl of index of refraction 1.33. If the goldfish is 10 cm from the wall of the 15-cm-radius bowl, where does the goldfish appear to an observer outside the bowl?

30. A paperweight is made of a solid glass hemisphere of index of refraction 1.50. The radius of the circular cross-section is 4 cm. The center of the hemisphere is placed directly over a line drawn to a length of 2.5 mm on a sheet of paper. What is the length of the line as seen looking vertically down on the hemisphere?

31. One end of a long glass rod ($n = 1.50$) is formed into the shape of a convex surface of radius 8 cm. An object is located in air along the axis of the rod. Find the image position corresponding to the following object positions, (a) 20 cm, (b) 8 cm, (c) 4 cm, and (d) 2 cm.

32. Repeat Problem 31 assuming the rod is placed in water instead of air.

33. Calculate the image positions for the object positions of Problem 31 assuming the end of the rod has a concave shape of radius 8 cm.

34. Repeat Problem 33 if the object is in water rather than air.

Section 25.6   Thin Lenses

35. A contact lens is made of plastic with an index of refraction of 1.58. The lens has a focal length of $+25$ cm and its inner surface has a radius of curvature of $+18.0$ mm. What is the radius of curvature of the outer surface?

36. A lens with radii of curvature of 52.5 cm and -61.9 cm has a focal length of $+60$ cm. Find its index of refraction.

37. A converging lens has a focal length of 20 cm. Find the location of the image for object distances of (a) 40 cm, (b) 20 cm, and (c) 10 cm. For each case state whether the image is real or virtual and erect or inverted. Find the magnification for each case.

38. A glass converging lens ($n = 1.50$) is designed to look like the lens shown in Figure 25.22a. The radius of the first surface is 15 cm and the radius of the second surface is 10 cm. (a) Find the focal length of the lens. Determine the position of the image for object distances of (b) infinity, (c) $3f$, (d) f, and (e) $f/2$.

39. A thin lens has a focal length of 25.0 cm. Locate the image when the object is placed (a) 26.0 cm and (b) 24.0 cm in front of the lens. Describe the image in each case.

40. A thin lens has a focal length of -25.0 cm. Locate the image when the object is placed as in Problem 39. Describe the image in each case. Determine the magnification.

41. A diverging lens has a focal length of 20 cm. Find the location of the image for object distances of (a) 40 cm, (b) 20 cm, and (c) 10 cm. For each case state whether the image is real or virtual and erect or inverted. Find the magnification for each case.

42. A diverging lens ($n = 1.50$) is shaped like that shown in Figure 25.22c. The radius of the first surface is 15 cm and the radius of the second surface is 10 cm. (a) Find the focal length of the lens. Determine position of the image for object distances of (b) infinity, (c) $3|f|$, (d) $|f|$, and (e) $|f|/2$.

43. Repeat Problem 42 for the case in which the radius of the first surface is 10 cm and the radius of the second surface is 15 cm.

44. A microscope slide is placed in front of a converging lens with a focal length of 2.44 cm. The lens forms an image of the slide 12.9 cm from the slide. How far is the lens from the slide if the image is (a) real and (b) virtual?

45. The nickel's image in Figure 25.30 has twice the diameter of the nickel and is 2.84 cm from the lens. Determine the focal length of the lens.

Figure 25.30
Problem 45

46. An object is placed 30 cm from a lens and a virtual image is formed 15 cm from the lens. (a) What is the focal length of the lens? (b) Is the lens converging or diverging?

47. A diverging lens is used to form a virtual image of an object. The object is 80 cm to the left of the lens, and the image is 40 cm to the left of the lens. Determine the focal length of the lens.

48. We want to form an image 30.0 cm from a diverging lens with a focal length of −40.0 cm. Where must we place the object? Determine the magnification.

49. An object's distance from a converging lens is 10 times the focal length. How far is the image from the focal point? Express the answer as a fraction of the focal length.

50. A person looks at a gem with a jeweler's microscope —a converging lens with a focal length of 12.5 cm. The microscope forms a virtual image 30.0 cm from the lens. Determine the magnification. Is the image upright or inverted?

51. An object is placed 50 cm from a screen. Where should a converging lens with a 10-cm focal length be placed in order to form an image on the screen? Find the magnification(s).

52. A diverging lens is to be used to produce a virtual image one-third the size of the object. Where should the object be placed?

Section 25.7 Multiple Lens Systems

53. Two converging lenses, each of focal length 15 cm, are placed 40 cm apart, and an object is placed 30 cm in front of the first. Where is the final image formed, and what is the magnification of the system?

54. Repeat Problem 53 for the case in which the two lenses are placed (a) 20 cm apart and (b) 30 cm apart.

55. An object is placed 20 cm to the left of a converging lens of focal length 25 cm. A diverging lens of focal length 10 cm is located 25 cm to the right of the converging lens. Find the position of the final image and its magnification.

56. Repeat Problem 55 for the case in which the light first encounters the diverging lens.

57. Lens L_1 in Figure 25.31 has a focal length of 15.0 cm, whereas lens L_2 has a focal length of 13.3 cm. The

distance d of lens L_2 from the film plane can be varied from 5.0 cm to 10.0 cm. Determine the range of distances for which objects can be focused on the film.

58. An object 1 cm in height is placed 4 cm to the left of a converging lens of focal length 8 cm. A diverging lens of focal length −16 cm is located 6 cm to the right of the converging lens. Find the position and size of the final image. Is the image inverted or erect? Real or virtual?

59. Two converging lenses having focal lengths of 10.0 cm and 20.0 cm are located 50.0 cm apart as shown in Figure 25.32. The final image is to be located between the lenses at the position indicated. (a) How far to the left of the first lens should the object be positioned? (b) What is the overall magnification? (c) Is the final image erect or inverted?

Figure 25.32 Problem 59

Additional Problems

60. The lens and mirror in Figure 25.33 have focal lengths of +80 cm and −50 cm, respectively. An object is placed 1.0 m to the left of the lens as drawn. Locate the final image. State whether the image is erect or inverted and determine the overall magnification.

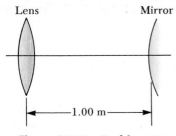

Figure 25.33 Problem 60

61. The object in Figure 25.34 is midway between the lens and the mirror. The mirror's radius of curvature is 20.0 cm, and the lens has a focal length of −16.7 cm. Considering only the light that leaves the

Figure 25.31
Problem 57

object and travels first toward the mirror, locate the final image formed by this system. Is this image real or virtual? Is it erect or inverted? What is the overall magnification?

Figure 25.34   Problem 61

62. A concave mirror has a focal length of 30 cm. (a) What is its radius of curvature? Locate and describe the image when the object is at (b) 100 cm and (c) 10 cm.

63. A planoconvex lens ($n = 1.50$) is flat on one side. What must be the radius of curvature of the curved side to produce a converging lens of focal length 20 cm?

64. Repeat Problem 63 for a planoconcave lens having a focal length of -20 cm.

65. A converging lens has a focal length of 20 cm. Find the position of the image for object distances of (a) 50 cm, (b) 30 cm, and (c) 10 cm. (d) For each object distance, determine the magnification of the lens and whether the image is erect or inverted.

66. An object placed 10 cm from a concave spherical mirror produces a real image 8 cm from the mirror. If the object is moved to a new position 20 cm from the mirror, what is the position of the image? Is the final image real or virtual?

67. A converging lens of focal length 20 cm is separated by 50 cm from a converging lens of focal length 5 cm. (a) Find the final position of the image of an object placed 40 cm in front of the first lens. (b) If the height of the object is 2 cm, what is the height of the final image? Is it real or virtual? (c) Determine the image position of an object placed 5 cm in front of the two lenses in contact.

68. A real object is located at the zero end of a meter stick. A large concave mirror located at the 100-cm end of the meter stick forms an image of the object at the 70-cm position. A small convex mirror placed at the 20-cm position forms a final image at the 10-cm point. What is the radius of curvature of the convex mirror?

69. An object is located 36 cm to the left of a biconvex lens of index of refraction 1.5. The left surface of the lens has a radius of curvature of 20 cm. The right surface of the lens is to be shaped so that a real image will be formed 72 cm to the right of the lens. What is the required radius of curvature of the second surface?

70. An object is placed 12 cm to the left of a diverging lens of focal length -6 cm. A converging lens of focal length 12 cm is placed a distance d to the right of the diverging lens. Find the distance d such that the final image is at infinity.

71. A colored marble is dropped in a large tank filled with benzene ($n = 1.50$). If the marble has a diameter of 1.5 cm, what is its apparent diameter when viewed from directly above, outside the tank? The apparent depth of the marble is 35 cm when viewed from directly above.

72. An object O_1 is located 15 cm to the left of a converging lens of 10-cm focal length. A second lens is positioned 10 cm to the right of the first lens and is observed to form a final image located at the position of the original object O_1. (a) What is the focal length of the second lens? (b) What is the overall magnification of this system? (c) What is the nature (i.e., real or virtual, erect or inverted) of the final image?

73. Use the fact that the image formed by the first surface becomes the object for the second surface to work this problem. Figure 25.35 shows a piece of glass whose index of refraction is 1.50. The ends are hemispheres with radii 2 cm and 4 cm, and the centers of the hemispherical ends are separated by a distance of 8 cm. A point object is located in air 1 cm from the left end of the glass. Find the location of the image of the object due to refraction at the two spherical surfaces.

74. Repeat Problem 73 for the case in which the distance between the centers of the hemispherical ends is 4 cm and the object is placed 28 cm to the left of the first surface.

75. Find the object distances (in terms of f) of a thin converging lens of focal length f if (a) the image is real and the image distance is four times the focal length and (b) the image is virtual and the image distance is three times the focal length. (c) Calculate the magnification of the lens for cases (a) and (b).

Figure 25.35   Problem 73

CHAPTER 26

Wave Optics

Thin film interference. A thin film of oil on water displays interference, as shown by the pattern of colors when white light is incident on the film. The film thickness varies in the vicinity of the blade, thereby producing the interesting color pattern. *(Peter Aprahamian/Science Photo Library)*

Our discussion of light thus far has been concerned with examining what happens when light passes through a lens or reflects from a mirror. This part of optics is often called geometric optics because an explanation of such phenomena relies on a geometric analysis of the light rays. We shall now expand our study of light into an area called wave optics. The three primary topics we shall examine in this chapter are interference, diffraction, and polarization. These phenomena cannot be adequately explained with ray optics, but wave theory leads to a satisfying description.

26.1 Conditions for Interference

In our discussion of wave interference in Section 14.12, we found that two waves could add together constructively or destructively. In constructive interference, the amplitude of the resultant wave is greater than that of either of the individual waves, while in destructive interference, the resultant amplitude is less than that of either of the individual waves. Light waves also undergo interference. Fundamentally, all interference associated with light waves arises as a result of the combining of the electric and magnetic field vectors that constitute the individual waves.

Interference effects in light waves are not easy to observe because of the short wavelengths involved (about 4×10^{-7} m to about 7×10^{-7} m). In order to observe sustained interference between two sources of light, the following conditions must be met:

1. The sources must be **coherent.** (We shall discuss this new term below.)
2. The sources must be **monochromatic,** that is, of a single wavelength.
3. The **superposition principle** must apply.

We shall now describe the characteristics of **coherent sources.** As we have said, two sources (producing two traveling waves) are needed to create interference. However, in order to produce a stable interference pattern, *the individual waves must maintain a constant phase relationship with one another.* A constant phase relationship implies that the two waves move together throughout their oscillations. For example, both waves may always be at a crest, or one wave might always be at a crest when the other is at a trough, or one wave might always be at a crest while the other is at some other point in its oscillation. As an example, the sound waves emitted by two side-by-side loudspeakers driven by a single amplifier can produce interference because the two speakers respond to the amplifier in the same way at the same time. Now, if two separate light sources are placed side by side, no interference effects are observed because in this case the light waves emitted by the two sources undergo independent random frequency and amplitude fluctuations and hence do not maintain a constant phase relationship with each other over the time of observation. For constructive interference to occur, two waves must overlap such that crests meet crests and troughs meet troughs. (A crest for a light wave means that the electric or magnetic field of the wave is a maximum in some direction. A trough occurs when the fields are at a maximum in the opposite direction.) If for some reason one of the waves undergoes some random change, perfect overlap is not maintained. Light from an ordinary light source undergoes such random changes about once every 10^{-8} s. Therefore the conditions for constructive interference, destructive interference, or some intermediate state last for times of the order of 10^{-8} s. The result is that no interference effects are observed since the eye cannot follow such short-term changes. Such light sources are said to be **noncoherent.**

A common method for producing two coherent light sources is to use one monochromatic source to illuminate a screen containing two small openings (usually in the shape of slits). The light emerging from the two slits is coherent because a single source produces the original light beam and the two slits serve only to separate the original beam into two parts (which, after all, is what was done to the sound signal discussed above). A random change in the light emitted by the source will occur in the two separate beams at the same time, and interference effects can be observed. An alternative to the use of conventional light sources is the laser, which will be discussed in more detail in Chapter 30. One of the most interesting and useful properties of this device is the highly coherent nature of the light it produces. In fact, one of the most unique applications of interference phenomena which uses laser light is the production of three-dimensional images called holograms.

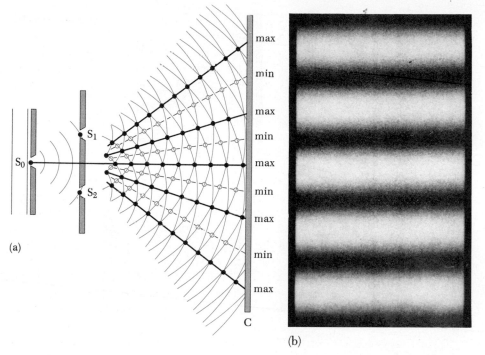

(a)

C

(b)

Figure 26.1
(a) Schematic diagram of Young's double-slit experiment. The narrow slits act as sources of waves. Slits S_1 and S_2 behave as coherent sources which produce an interference pattern on screen C. (Note that this is not to scale.) (b) The fringe pattern formed on screen C could look like this.

26.2 Young's Double-Slit Interference

The phenomenon of interference in light waves from two sources was first demonstrated by Thomas Young in 1801. A schematic diagram of the apparatus used in this experiment is shown in Figure 26.1a. Light is incident on a screen, which is provided with a narrow slit S_0. The waves emerging from this slit arrive at a second screen, which contains two narrow, parallel slits, S_1 and S_2. These two slits serve as a pair of coherent light sources because waves emerging from them originate from the same wavefront and therefore are always in phase. The light from the two slits produces a visible pattern on screen C; the pattern consists of a series of bright and dark parallel bands called **fringes** (Fig. 26.1b). When the light from slits S_1 and S_2 arrives at a point on the screen such that constructive interference occurs at that location, a bright line appears. When the light from the two slits combines destructively at any location on the screen, a dark line results. Figure 26.2 is a photograph of an interference pattern produced by two coherent vibrating sources in a water tank.

Figure 26.3 is a schematic diagram of some of the ways the two waves can combine at the screen. In Figure 26.3a, the two waves, which leave the two slits in phase, strike the screen at the central point P. Since these waves travel an equal distance, they arrive in phase at P, and as a result constructive interference occurs at this location and a bright area is observed. In Figure 26.3b, the two light waves again start in phase, but the upper wave has to travel one wavelength farther to reach point Q on the screen. Since the upper wave falls behind the lower one by exactly one wavelength, they still arrive in phase at Q,

Figure 26.2
Water waves set up by two vibrating sources produce an effect similar to Young's double-slit interference. The waves interfere constructively at X and destructively at Y.

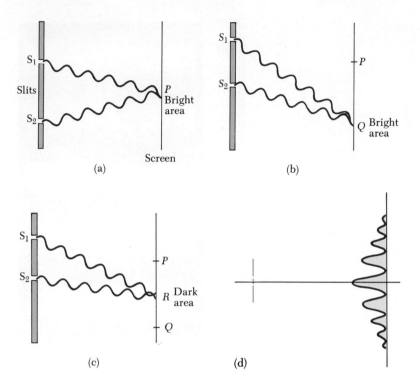

Figure 26.3
(a) Constructive interference occurs at P when the waves combine. (b) Constructive interference also occurs at Q. (c) Destructive interference occurs at R when the wave from the upper slit falls one-half wavelength behind the wave from the lower slit. (d) The intensity of the fringes decreases as one moves to higher orders. (Note that these figures are not drawn to scale.)

and so a second bright light appears at this location. Now consider point R, midway between P and Q in Figure 26.3c. At this location, the upper wave has fallen half a wavelength behind the lower wave. This means that the trough from the bottom wave overlaps the crest from the upper wave, giving rise to destructive interference at R. For this reason, a dark region at this location can be observed. Figure 26.3d shows the intensity distribution on the screen. Notice that the central fringe is most intense and that the intensity decreases for higher order fringes.

We can obtain a quantitative description of Young's experiment with the help of Figure 26.4. Consider a point P on the viewing screen; the screen is located a perpendicular distance L from the screen containing slits S_1 and S_2, which are separated by a distance d. Let us assume that the source is monochromatic. Under these conditions, the waves emerging from S_1 and S_2 have the same frequency and amplitude and are in phase. The light intensity on the screen at P is the resultant of the light coming from both slits. Note that a wave

Figure 26.4
Geometric construction for describing Young's double-slit experiment. The path difference between the two rays is $r_2 - r_1 = d \sin \theta$. (Note that this figure is not to scale.)

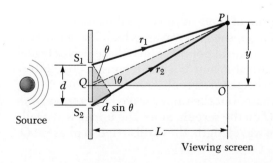

from the lower slit travels farther than a wave from the upper slit by an amount equal to $d \sin \theta$. This distance is called the **path difference**, δ, where

$$\delta = r_2 - r_1 = d \sin \theta \qquad (26.1)$$

Path difference

This equation assumes that r_1 and r_2 are parallel, which is approximately true because L is much greater than d. As noted earlier, the value of this path difference will determine whether or not the two waves are in phase when they arrive at P. If the path difference is either zero or some integral multiple of the wavelength, the two waves are in phase at P and constructive interference results. Therefore, the condition for bright fringes, or **constructive interference,** at P is given by

$$\delta = d \sin \theta = m\lambda \qquad m = 0, \pm 1, \pm 2, \ldots \qquad (26.2)$$

Constructive interference

The number m is called the **order number** of the fringe. The central bright fringe at $\theta = 0$ ($m = 0$) is called the *zeroth-order maximum.* The first maximum on either side, when $m = \pm 1$, is called the *first-order maximum,* and so forth.

Similarly, when the path difference is an odd multiple of $\lambda/2$, the two waves arriving at P will be 180° out of phase and will give rise to destructive interference. Therefore, the condition for dark fringes, or **destructive interference,** at P is given by

$$\delta = d \sin \theta = \left(m + \frac{1}{2}\right) \lambda \qquad m = 0, \pm 1, \pm 2, \ldots \qquad (26.3)$$

Destructive interference

If $m = 0$ in this equation, the path difference is $\delta = \lambda/2$, which is the condition for the location of the first dark line on either side of the central (bright) maximum. Likewise, if $m = 1$, $\delta = 3\lambda/2$, which is the condition for the second dark line on each side, and so forth.

It is useful to obtain expressions for the positions of the bright and dark fringes measured vertically from O to P. We shall assume that $L \gg d$ (Fig. 26.4) and that $d \gg \lambda$. The first assumption says that the distance from the slits to the screen is much larger than the distance between the two slits, while the second says that the distance between the two slits is much larger than the wavelength. This situation can prevail in practice because L is often on the order of 1 m while d is a fraction of a millimeter and λ is less than a micrometer for visible light. Under these conditions θ will be small for the first several orders, and so we can use the approximation $\sin \theta \cong \tan \theta$. From the triangle OPQ in Figure 26.4, we see that

$$\sin \theta \approx \tan \theta = \frac{y}{L} \qquad (26.4)$$

Using this result together with Equation 26.2, we see that the positions of the *bright fringes* measured from O are given by

$$y_{\text{bright}} = \frac{\lambda L}{d} m \qquad (26.5)$$

Reflection, interference, and diffraction can be seen in this aerial photograph of waves in the sea. As the waves pass through a narrow gap, they spread out (diffract), and the interference of two waveforms is manifested in cross-patterned areas. *(Photo by John S. Shelton)*

Similarly, using Equations 26.3 and 26.4, we find that the *dark fringes* are located at

$$y_{\text{dark}} = \frac{\lambda L}{d}\left(m + \frac{1}{2}\right) \tag{26.6}$$

As we shall demonstrate in Example 26.1, Young's double-slit experiment provides a method for measuring the wavelength of light. In fact, Young used this technique to make the first measurement of the wavelength of light. Additionally, the experiment gave the wave model of light a great deal of credibility. It was inconceivable that particles of light coming through the slits could cancel each other in a way that would explain the regions of darkness. Today we still use the phenomenon of interference to explain wave-like behavior in many observations.

Example 26.1   Measuring the Wavelength of a Light Source

A screen is separated from a double-slit source by 1.2 m. The distance between the two slits is 0.03 mm. The second-order bright fringe ($m = 2$) is measured to be 4.5 cm from the center line. Determine (a) the wavelength of the light and (b) the distance between adjacent bright fringes.

Solution
(a) We can use Equation 26.5 with $m = 2$, $y_2 = 4.5 \times 10^{-2}$ m, $L = 1.2$ m, and $d = 3 \times 10^{-5}$ m:

$$\lambda = \frac{y_2 d}{mL} = \frac{(4.5 \times 10^{-2}\text{ m})(3 \times 10^{-5}\text{ m})}{2(1.2\text{ m})} = 5.6 \times 10^{-7}\text{ m} = \boxed{560\text{ nm}}$$

(b) The location of the third-order bright fringe is found by using the results of (a) as follows:

$$y_3 = \frac{\lambda L}{d}\,m = \frac{(5.6 \times 10^{-7}\text{ m})(1.2\text{ m})}{3 \times 10^{-5}\text{ m}}\,(3) = 6.7 \times 10^{-2}\text{ m}$$

Thus, the distance between the second and third bright fringes is

$$y_3 - y_2 = 6.7 \times 10^{-2}\text{ m} - 4.5 \times 10^{-2}\text{ m} = 2.2 \times 10^{-2}\text{ m} = \boxed{2.2\text{ cm}}$$

26.3   Newton's Rings

An interesting interference effect can be observed by placing a converging lens that is flat on one side on a glass surface, as in Figure 26.5a. When monochromatic light falls on the surface from above, a pattern of light and dark rings is seen, as shown in Figure 26.5b. These circular fringes, discovered by Newton, are referred to as **Newton's rings**. Newton's particle model of light could not explain the origin of these rings.

The interference effect is due to the combination of ray 1, reflected from the flat glass plate, with ray 2, reflected from the lower part of the lens (Fig. 26.5a). At first thought, it would seem that the central spot in the pattern

(a)

(b)

Figure 26.5
(a) The combination of rays reflected from the glass plate and the curved surface of the lens gives rise to an interference pattern known as Newton's rings. (b) Photograph of Newton's rings. *(Courtesy of Bausch & Lomb Optical Co.)*

should be bright. At this location, indicated as point O in Figure 26.5a, the extra distance that ray 1 has to travel is extremely small. It is not exactly zero, because there is always a thin layer of air trapped between the two surfaces. Therefore, when rays 1 and 2 recombine, you would think they would interfere constructively, producing a bright spot at the center. The photograph of Newton's rings shown in Figure 26.5b shows that this is not the case. Instead, a dark spot is observed at the center. Early observers of this phenomenon used great care to polish and clean the surfaces in order to bring them closer together, but the more they polished and cleaned, the more evident it became that the central spot was indeed dark. This difficulty was resolved when it was realized that

a light wave undergoes a phase change of 180° upon reflection from a medium having an index of refraction greater than the medium in which the wave is traveling.

In the case of a light wave, a 180° phase shift means that a crest (a maximum in the electric field) incident on a surface is reflected as a trough (a minimum in the electric field).

It is useful to draw an analogy between reflected light waves and the reflection of a transverse wave on a stretched string (Sec. 14.13). The reflected pulse on a string undergoes a phase change of 180° when it is reflected from a rigid support. Figure 26.6a compares this situation with that in which a light wave traveling in a medium of index of refraction n_1 is reflected from a medium of index n_2, where n_1 is less than n_2. (An example of this would be light traveling through air and reflecting off a glass plate.) In this case the light

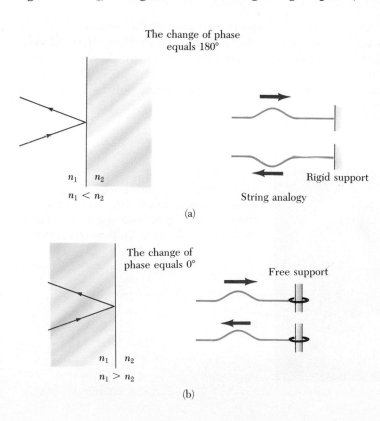

(a)

(b)

Figure 26.6
(a) A ray reflecting from a medium of higher refractive index undergoes a 180° phase change. The right side shows the analogy with a reflected pulse on a string. (b) A ray reflecting from a medium of lower refractive index undergoes *no* phase change.

undergoes a 180° phase change upon reflection. On the other hand, there is no phase change if a pulse on a string reflects from a support that is free to move, as in Figure 26.6b. Likewise, when light traveling in a medium of index of refraction n_1 reflects off a boundary with a lower index of refraction n_2, as in Figure 26.6b, the reflected wave undergoes no phase change.

In view of this information, we are now in a position to explain the pattern observed in Newton's ring experiment. Ray 2 in Figure 26.5a is traveling in glass and reflects off the glass-air interface between the lens and plate. This ray does not undergo a phase shift upon reflection. However, ray 1 travels through the air between the plate and the lens before it reflects off the plate, which has a higher index of refraction than the air. Therefore, ray 1 undergoes a 180° phase shift upon reflection. At point O, ray 1 does not travel appreciably farther than ray 2, but it does undergo a 180° phase shift because of reflection. Consequently when rays 1 and 2 recombine after reflection, they undergo destructive interference.

Let us now consider what happens to other rays of light striking at points away from the center of the convex lens. In this case, the extra distance that ray 1 travels relative to ray 2 becomes important. For example, suppose the extra distance of travel is $\lambda/2$. In this case, ray 1 loses 180° upon reflection, corresponding to half a wavelength, and then loses another half wavelength because of the extra distance it must travel. Rays 1 and 2 therefore combine in phase upon reflection, and a bright spot appears. Actually, constructive interference is occurring at all points P located the same distance r from the center of the lens, as in Figure 26.5. Therefore, what is really observed is a bright ring of light. Under what conditions would the first dark ring be seen? The second bright ring?

One of the important uses of Newton's rings is in the testing of optical lenses. A circular pattern like that pictured in Figure 26.5b is obtained only when the lens is ground to a perfectly symmetric curvature. Variations from such symmetry might produce a pattern like that in Figure 26.7. These variations give an indication of how the lens must be ground and polished in order to remove the imperfections.

Figure 26.7
This asymmetrical interference pattern indicates imperfections in the lens. *(From Physical Science Study Committee,* College Physics, *Lexington, MA, D. C. Heath and Co., 1968)*

The photograph on the left shows interference in soap bubbles. White light incident on soap bubbles (thin films of soap) forms a beautiful pattern of colors as a result of interference in the films. The middle photograph shows interference in a vertical soap film of variable thickness. The top of the film appears darkest where the film is thinnest. In the photograph on the right a thin film of oil on water displays interference, as shown by the pattern of colors when white light is incident on the film. The film thickness varies, thereby producing the interesting color pattern.

Interference

(Peter Aprahamian/Science Photo Library)

(© 1983 Larry Mulvehill/ Photo Researchers, Inc.)

(© Tom Branch 1984/Photo Researchers, Inc.)

26.4   Interference in Thin Films

Everyone has seen the vivid colors produced when light shines on a thin film of oil floating on water. Likewise, the film of a soap bubble can produce striking colors. The explanation of how these colors are produced follows much the same argument used to explain Newton's rings. The interference in this case is caused by the interference of waves reflected from the opposite surfaces of the film.

Consider a thin film of uniform thickness t and index of refraction n, as in Figure 26.8. Let us assume that the light rays traveling in air are nearly normal to the surface of the film. In order to determine whether the reflected rays interfere constructively or destructively, we must first note the following facts:

1. There is a phase change of 180° upon reflection if the reflecting medium has a higher index of refraction than the medium in which the wave is traveling.
2. The wavelength of light λ_n in a medium whose index of refraction is n (Chap. 24) is

$$\lambda_n = \frac{\lambda}{n} \tag{26.7}$$

where λ is the wavelength of light in free space.

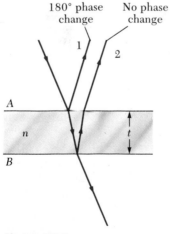

Figure 26.8
Interference observed in light reflected from a thin film is due to a combination of rays reflected from the upper and lower surfaces.

Layer of bubbles on water produced by soap film. The colors, produced just before the bubbles burst, are due to interference between light rays reflected from the front and back of the thin film of water making the bubble. The colors depend on the thickness of the film, ranging from black where the film is at its thinnest, to red as the film gets thicker. *(Dr. Jeremy Burgess/Science Photo Library)*

Let us apply these facts to the film shown in Figure 26.8. We find that ray 1, which is reflected from the upper surface (*A*), undergoes a phase change of 180° with respect to the incident wave. Ray 2, which is reflected from the lower surface (*B*), undergoes no phase change with respect to the incident wave. Therefore rays 1 and 2 are 180° out of phase following reflection. However, we must also consider that ray 2 travels an extra distance equal to $2t$ before the waves recombine. For example, if $2t = \lambda_n/2$, rays 1 and 2 will recombine in phase and constructive interference will result.

There are many permutations and combinations of surfaces which may be encountered in problems dealing with thin film interference. For example, suppose the film shown in Figure 26.8 is water, with air above its upper surface and glass below its lower surface. In this situation, rays 1 and 2 would both undergo a 180° phase change. Because of a wide variety of situations which could occur, it is impossible to derive a single equation that will enable you to solve such problems. Instead, each situation must be examined to determine whether or not phase changes occur upon reflection, and the conditions for constructive or destructive interference.

Problem-Solving Strategy: Thin Film Interference

The following features should be kept in mind while working thin film interference problems:

1. Identify the thin film from which interference effects are being observed.
2. The type of interference that occurs in a specific problem is determined by the phase relationship between that portion of the wave reflected at the upper surface of the film and that portion reflected at the lower surface of the film.
3. Phase differences between the two portions of the wave occur because of differences in the distance traveled by the two portions and by phase changes occurring upon reflection.

   The wave reflected from the lower surface of the film has to travel a distance equal to twice the thickness of the film before it returns to the upper surface of the film where it interferes with that portion of the wave reflected at the upper surface. When *distances alone are considered*, if the film thickness is equal to an integral multiple of $\lambda/2$, the interference will be constructive; if the extra distance equals $\lambda/4$, $3\lambda/4$, and so forth, destructive interference will occur.

   Reflections may change the results above, which are based solely on the distance traveled. When a wave traveling in a particular medium reflects off a surface having a higher index of refraction than the one it is in, a 180° phase shift occurs. This has the same effect as if the wave lost $\lambda/2$. These losses must be considered in addition to those losses that occur because of the extra distance one wave travels over another.

4. When distance and phase changes upon reflection are both taken into account, the interference will be constructive if the waves are out of phase by an integral multiple of λ. Destructive interference will occur when the phase difference is $\lambda/2$, $3\lambda/2$, and so forth.

Example 26.2   Interference in a Soap Film

Calculate the minimum thickness of a soap bubble film ($n = 1.33$) that will result in constructive interference in the reflected light if the film is illuminated with light whose wavelength in air is 600 nm.

Solution
This situation is the same as that shown in Figure 26.8. Because there is air on each side of the soap film, ray 1 undergoes a phase change of 180°, while ray 2 does not undergo a phase change. In order for the two waves to be in phase (the condition for constructive interference), ray 2 must travel a minimum extra distance equal to $\lambda_n/2$. Setting $2t = \lambda_n/2$, and using Equation 26.7, $\lambda_n = \lambda/n$, we get

$$t = \frac{\lambda}{4n} = \frac{600 \text{ nm}}{4(1.33)} = \boxed{113 \text{ nm}}$$

Exercise   What other film thicknesses will produce constructive interference?

Answer $3t$, $5t$, $7t$, and so on.

Example 26.3   Nonreflecting Coatings for Solar Cells

Solar cells are often coated with a transparent thin film, such as silicon monoxide (SiO, $n = 1.45$), in order to minimize reflective losses from the surface (Fig. 26.9). A silicon solar cell ($n = 3.5$) is coated with a thin film of silicon monoxide for this purpose. Determine the minimum thickness of the film that will produce the least reflection of a wavelength of 550 nm, which is the center of the visible spectrum.

Solution
The reflected light is a minimum when rays 1 and 2 meet the condition of destructive interference. Note that *both* rays undergo a 180° phase change upon reflection in this case. Hence, the net change in phase because of reflection is zero and the condition for a reflection *minimum* is $2t = \lambda_n/2$ or, using Equation 26.7

$$t = \frac{\lambda}{4n} = \frac{550 \text{ nm}}{4(1.45)} = 94.8 \text{ nm}$$

Typically, such antireflecting coatings reduce the reflective loss from 30% (with no coatings) to 10% (with coating). Such coatings will therefore increase the cell's efficiency since more light will be available to create charge carriers in the cell. Note that, in reality, the coating will never be perfectly nonreflecting because the required thickness is wavelength-dependent and the incident light covers a wide range of wavelengths.

Glass lenses used in cameras and other optical instruments are usually coated with a transparent thin film, such as magnesium fluoride (MgF_2), to reduce or eliminate unwanted reflection.

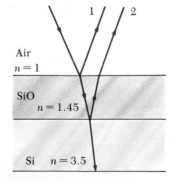

Figure 26.9   (Example 26.3) Reflective losses from a silicon solar cell are minimized by coating it with a thin film of silicon monoxide, SiO.

Figure 26.10 (Example 26.4) Interference bands in reflected light can be observed by illuminating a wedge-shaped film with monochromatic light. The dark areas correspond to positions of destructive interference.

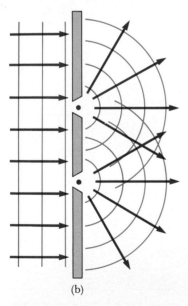

(a)

(b)

Figure 26.11
(a) If light did not spread out after passing through the slits, no interference would occur. (b) The light from the two slits overlaps as it spreads out, producing interference fringes.

Example 26.4   Interference in a Wedge-Shaped Film

A thin, wedge-shaped film of refractive index n is illustrated with monochromatic light of wavelength λ, as illustrated in Figure 26.10. Describe the interference pattern observed for this case.

Solution
The interference pattern is that of a thin film of variable thickness surrounded by air. Hence, the pattern will be a series of alternating bright and dark parallel bands. A dark band corresponding to destructive interference appears at point O, the apex, since the upper reflected ray undergoes a 180° phase change while the lower one does not. Other dark bands appear when $t_1 = \lambda/2n$, $t_2 = \lambda/n$, $t_3 = 3\lambda/2n$, and so on. Similarly, bright bands will be observed for thicknesses of $\lambda/4n$, $3\lambda/4n$, $5\lambda/4n$, and so on. If white light is used, bands of different colors will be observed at different points, corresponding to the different wavelengths of light.

26.5   Diffraction

Suppose a light beam is incident on two slits, as in Young's double-slit experiment. If a light truly traveled in straight-line paths after passing through the slits, as in Figure 26.11a., the waves would not overlap and no interference pattern would be seen. Instead, Huygens' principle requires that the waves spread out from the slits as shown in Figure 26.11b. In other words, the light deviates from a straight-line path and enters the region that would otherwise be shadowed. **Diffraction** is defined as the bending of light when it passes an obstacle or barrier. Note that the phenomenon of diffraction should not be confused with the refraction of light discussed earlier. Refraction is concerned with the bending of light as it passes from one material to another having a different index of refraction. In general, diffraction occurs when waves pass through small openings, around obstacles, or by relatively sharp edges. The diffraction of sound waves is easy to experience. For example, if a radio located in another room is turned on, you can hear the sound waves it emits even if you are not standing in front of your doorway. On the other hand, the diffraction of light is not as easy to detect because the bending is usually very small. The most common way of detecting the diffraction of light is by observing interference patterns in the deviated light.

As an example of diffraction, consider the following. When an opaque object is placed between a point source of light and a screen, as in Figure 26.12a, the boundary between the shadowed and illuminated regions on the

(a)

(b)

Figure 26.12
(a) Light bends around the opaque object. (b) The result is a series of dark and light bands in the area that would be completely in the shadow if light did not bend around the object.

screen is not sharp. A careful inspection of the boundary shows that a small amount of light bends into the shadowed region. The region outside the shadow contains alternating light and dark bands, as in Figure 26.12b. If no diffraction occurred, the region behind the obstacle would be completely shadowed, and the region outside the barrier would be uniformly illuminated. When the effects of diffraction are considered, the first bright band has an intensity greater than the intensity of the region under uniform illumination. Effects of this type were first reported in the 17th century by Francesco Grimaldi.

Figure 26.13 shows the shadow of the diffraction pattern of a penny. There is a bright spot at the center, circular fringes near the shadow's edge, and another set of fringes outside the shadow. This particular type of diffraction pattern was first observed in 1818 by Dominique Arago (1756–1853). The bright spot at the center of the shadow can be explained only through the use of the wave theory of light, which predicts constructive interference at this point. This was certainly a most dramatic experimental proof of the wave nature of light.

Diffraction phenomena are usually classified as being of two types, which are named after the men who first explained them. The first type, called **Fraunhofer diffraction,** occurs when the rays reaching a point are approximately parallel. This can be achieved experimentally either by placing the observing screen far from the opening or by using a converging lens to focus the parallel rays on the screen, as in Figure 26.14a. Note that a bright fringe is

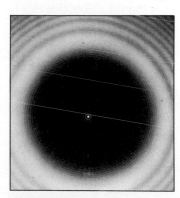

Figure 26.13
Diffraction pattern of a penny, taken with the penny midway between the screen and source. *(Courtesy of P. M. Rinard, from* Am. J. Phys., *44:70, 1976)*

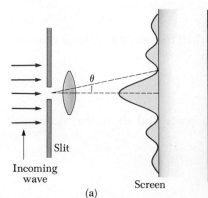

(a)

(b)

Figure 26.14
(a) Fraunhofer diffraction pattern of a single slit. The parallel rays are brought into focus on the screen with a converging lens. The pattern consists of a central bright region flanked by much weaker maxima. (Note that this is not to scale.)
(b) Photograph of a single-slit Fraunhofer diffraction pattern. *(From M. Cagnet, M. Francon, and J. C. Thierr,* Atlas of Optical Phenomena, *Berlin, Springer-Verlag, 1962, plate 18)*

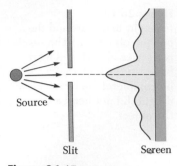

Figure 26.15
A Fresnel diffraction pattern of a single slit is observed when the incident rays are not parallel and the observing screen is at a finite distance from the slit. (Note that this is not to scale.)

Figure 26.16
Fresnel diffraction pattern for a rectangular aperture. *(From M. Cagnet, M. Francon, and J. C. Thierr, Atlas of Optical Phenomena, Berlin, Springer-Verlag, 1962, plate 34)*

observed along the axis at $\theta = 0$, with alternating dark and bright fringes on either side of the central bright fringe. Figure 26.14b is a photograph of a single-slit Fraunhofer diffraction pattern.

When the observing screen is placed a finite distance from the slit and no lens is used to focus parallel rays, the observed pattern is called a **Fresnel diffraction** pattern (Fig. 26.15). The diffraction patterns shown in Figures 26.12b and 26.13 are examples of Fresnel diffraction. Another example, shown in Figure 26.16, is the diffraction pattern of a rectangular slit. Fresnel diffraction is rather complex to treat quantitatively. Therefore, the following discussion will be restricted to Fraunhofer diffraction.

26.6   Single-Slit Diffraction

Let us discuss the nature of the Fraunhofer diffraction pattern produced by a single slit. We can deduce some important features of this problem by examining waves coming from various portions of the slit, as shown in Figure 26.17. According to Huygens' principle, *each portion of the slit acts as a source of waves.* Hence, *light from one portion of the slit can interfere with light from another portion,* and the resultant intensity on the screen will depend on the direction θ.

To analyze the diffraction pattern, it is convenient to divide the slit in two halves, as in Figure 26.17. All the waves that originate from the slit are in phase. Consider waves 1 and 3, which originate from the bottom and center of the slit, respectively. Wave 1 travels farther than wave 3 by an amount equal to the path difference $(a/2) \sin \theta$, where a is the width of the slit. Similarly, the path difference between waves 2 and 4 is also $(a/2) \sin \theta$. If this path difference is exactly one half of a wavelength (corresponding to a phase difference of 180°), the two waves cancel each other and destructive interference results. This is true, in fact, for any two waves that originate at points separated by half the slit width because the phase difference between two such points is 180°. Therefore, waves from the upper half of the slit interfere *destructively* with waves from the lower half of the slit when

$$\frac{a}{2} \sin \theta = \frac{\lambda}{2}$$

or when

$$\sin \theta = \frac{\lambda}{a}$$

If we divide the slit into four parts rather than two and use similar reasoning, we find that the screen is also dark when

$$\sin \theta = \frac{2\lambda}{a}$$

Likewise, we can divide the slit into six parts and show that darkness occurs on the screen when

$$\sin \theta = \frac{3\lambda}{a}$$

Therefore, the general condition for **destructive interference** is

$$\sin \theta = m \frac{\lambda}{a} \qquad m = \pm 1, \pm 2, \pm 3, \ldots \qquad \textbf{(26.8)}$$

Equation 26.8 gives the values of θ for which the diffraction pattern has zero intensity, that is, where a dark fringe is formed. However, it tells us nothing about the variation in intensity along the screen. The general features of the intensity distribution along the screen are shown in Figure 26.18. A broad central bright fringe is observed, flanked by much weaker, alternating bright fringes. The various dark fringes (points of zero intensity) occur at the values of θ that satisfy Equation 26.8. The position of the points of constructive interference lies approximately halfway between the dark fringes. Note that the central bright fringe is twice as wide as the weaker bright fringes.

In this section we referred to the alternating dark and bright bands on the screen as a diffraction pattern, while in our discussion of Young's double-slit experiment we referred to a similar pattern as an interference pattern. Also, you should note that the derivation of the equations associated with interference and diffraction effects are similar in that they consider the addition of waves that are either in or out of phase. The effects of interference and diffraction can be distinguished as follows. When the waves to be added come from two or more openings, as in Young's experiment, the resulting pattern is called an interference pattern. When various portions of a single wave interfere, as in this section, the result is a diffraction pattern.

Figure 26.17
Diffraction of light by a narrow slit of width a. Each portion of the slit acts as a point source of waves. The path difference between rays 1 and 3 or between rays 2 and 4 is equal to $(a/2)$ $\sin \theta$. (Note that this is not to scale, and that the rays are assumed to converge at a distant point.)

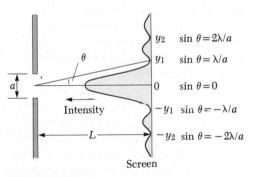

Figure 26.18
Positions of the various minima for the Fraunhofer diffraction pattern of a single slit of width a. (Note that this is not to scale.)

Example 26.5   Where are the Dark Fringes?

Light of wavelength 580 nm is incident on a slit of width 0.30 mm. The observing screen is placed 2 m from the slit. Find the positions of the first dark fringes and the width of the central bright fringe.

Solution
The first dark fringes that flank the central bright fringe correspond to $m = \pm 1$ in Equation 26.8. Hence, we find that

$$\sin \theta = \pm \frac{\lambda}{a} = \pm \frac{5.8 \times 10^{-7} \text{ m}}{0.3 \times 10^{-3} \text{ m}} = \pm 1.93 \times 10^{-3}$$

From the triangle in Figure 26.18, we see that $\tan \theta = y_1 / L$. Since θ is very small, we can use the approximation $\sin \theta \approx \tan \theta$, so that $\sin \theta \approx y_1 / L$. Therefore, the positions of the first minima measured from central axis are

$$y_1 \approx L \sin \theta = \pm L \frac{\lambda}{a} = \boxed{\pm 3.86 \times 10^{-3} \text{ m}}$$

The positive and negative signs correspond to the first dark fringes on either side of the central bright fringe. Hence, the width of the central bright fringe is given by $2|y_1| = 7.72 \times 10^{-3}$ m $= 7.72$ mm. Note that this value is *much larger* than the width of the slit. However, as the width of the slit is *increased*, the diffraction pattern will *narrow*, corresponding to smaller values of θ. In fact, for *large* values of a, the various maxima and minima will be so closely spaced that the only pattern observed will be a large central bright area that resembles the geometric image of the slit.

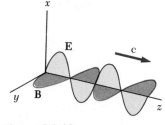

Figure 26.19
Schematic diagram of a polarized electromagnetic wave propagating in the z direction. The electric field vector **E** vibrates in the xz plane, and the magnetic field vector **B** vibrates in the yz plane.

26.7   Polarization of Light Waves

In Chapter 23 we characterized electromagnetic waves as transverse waves because *the oscillating electric and magnetic fields vibrate perpendicular to the direction in which the wave travels.* Figure 26.19 shows a polarized electromagnetic wave moving in the positive z direction. The phenomenon of **polarization,** which will be described in this section, is firm evidence of the transverse nature of electromagnetic waves.

An ordinary beam of light consists of a large number of waves emitted by the atoms or molecules contained in the light source. Each atom produces a wave having a particular orientation of the electric field vector. The result is a huge collection of waves with electric fields oscillating in all conceivable transverse directions. We can picture this state of affairs using Figure 26.20a, which represents an **unpolarized** light wave. In contrast to this situation, a wave is said to be **linearly polarized** if only one of these directions of vibration of the electric field vector exists at a particular point, as in Figure 26.20b. (Sometimes such a wave is described as being plane-polarized or simply polarized.)

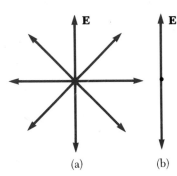

Figure 26.20
(a) An unpolarized light beam viewed along the direction of propagation (perpendicular to the page). The transverse electric field vector can vibrate in any direction with equal probability. (b) A linearly polarized light beam with the electric field vector vibrating in the vertical direction.

It is possible to obtain a linearly polarized beam from an unpolarized beam by removing all waves from the beam except those whose electric field vectors oscillate along a single direction. We shall now discuss four processes for producing polarized light from unpolarized light: (1) selective absorption, (2) reflection, (3) double refraction, and (4) scattering.

PHYSICS IN ACTION

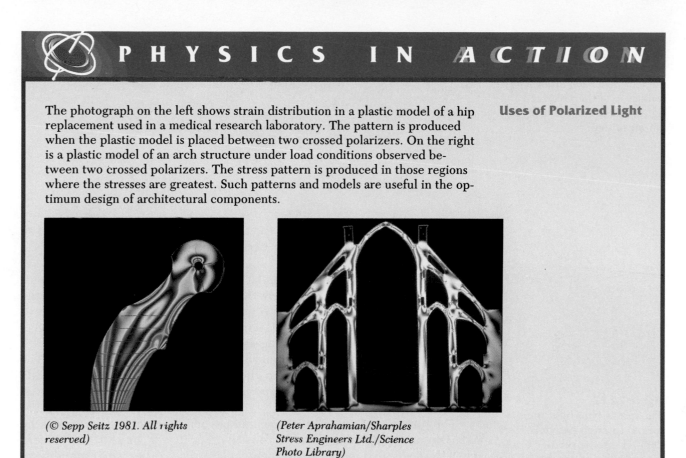

The photograph on the left shows strain distribution in a plastic model of a hip replacement used in a medical research laboratory. The pattern is produced when the plastic model is placed between two crossed polarizers. On the right is a plastic model of an arch structure under load conditions observed between two crossed polarizers. The stress pattern is produced in those regions where the stresses are greatest. Such patterns and models are useful in the optimum design of architectural components.

Uses of Polarized Light

(© Sepp Seitz 1981. All rights reserved)

(Peter Aprahamian/Sharples Stress Engineers Ltd./Science Photo Library)

Polarization by Selective Absorption

The most common technique for obtaining polarized light is to use a material that will transmit only those waves whose electric field vectors vibrate in a plane parallel to a certain direction and will absorb those waves whose electric field vectors vibrate in other directions. Any material that has the property of transmitting light with the electric field vector vibrating in only one direction is called a **dichroic** substance.

In 1938, E. H. Land discovered a material, which he called **Polaroid,** that polarizes light through selective absorption by oriented molecules. This material is fabricated in thin sheets of long-chain hydrocarbons, such as polyvinyl alcohol. The sheets are stretched during manufacture so that the long-chain molecules align. After a sheet is dipped into a solution containing iodine, the molecules can conduct electric charges along their hydrocarbon chains. As a result, the molecules readily *absorb* light whose electric field vector is *parallel* to their length and *transmit* light whose electric field vector is *perpendicular* to their length. It is common to refer to the direction perpendicular to the molecular chains as the *transmission axis.* In an ideal polarizer, all light with

Figure 26.21
Two polarizing sheets whose transmission axes make an angle θ with each other. Only a fraction of the polarized light incident on the analyzer is transmitted.

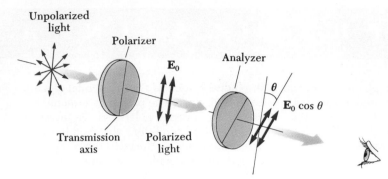

the electric field vector parallel to the transmission axis is transmitted and all light with the electric field vector perpendicular to the transmission axis is absorbed.

Figure 26.21 represents an unpolarized beam of light incident on a polarizing sheet, called a **polarizer,** where the transmission axis is indicated by the straight lines on the polarizer. The light that passes through this sheet is plane-polarized vertically as shown, where the transmitted electric field vector is labeled E_0. A second polarizing sheet, called an **analyzer,** intercepts this beam with its transmission axis at an angle θ to the axis of the polarizer. The component of E_0 *perpendicular* to the axis of the analyzer is completely *absorbed*, and the component of E_0 *parallel* to the axis of the analyzer is *transmitted.*

The transmitted intensity is a maximum when the transmission axes of the polarizer and analyzer are parallel to each other ($\theta = 0°$ or $180°$) and is zero when the transmission axes are perpendicular to each other (corresponding to complete absorption by the analyzer). This variation in transmitted intensity through a pair of polarizing sheets is illustrated in Figure 26.22.

Polarization by Reflection

Another method for obtaining polarized light is by reflection. When an unpolarized light beam is reflected from a surface, the reflected light is completely polarized, partially polarized, or unpolarized, depending on the angle of incidence. If the angle of incidence is either $0°$ or $90°$ (normal or grazing angles), the reflected beam is unpolarized. However, for intermediate angles of inci-

Figure 26.22
The intensity of light transmitted through two polarizers depends on the relative orientation of their transmission axes. (a) The transmitted light has *maximum* intensity when their transmission axes are *aligned* with each other.
(b) The transmitted light intensity diminishes when their transmission axes are at an angle of 45° with each other.
(c) The transmitted light intensity is a *minimum* when their transmission axes are at *right angles* to each other.

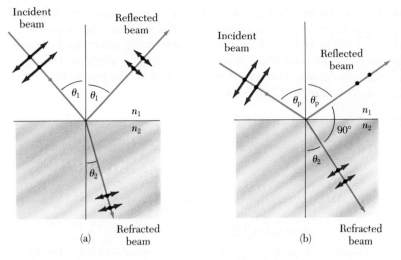

Figure 26.23
(a) When unpolarized light is incident on a reflecting surface, the reflected and refracted beams are partially polarized.
(b) The reflected beam is completely polarized when the angle of incidence equals the polarizing angle, θ_p, which satisfies the equation $n = \tan \theta_p$.

dence, the reflected light is polarized to some extent. In fact, for one particular angle of incidence, the reflected beam is completely polarized.

Suppose an unpolarized light beam is incident on a surface, as in Figure 26.23a. The beam can be described by two electric field components, one parallel to the surface (the dots) and the other perpendicular to the first component and to the direction of propagation (the arrows). It is found that the parallel component reflects more strongly than the other components, and this results in a partially polarized reflected beam. Furthermore, the refracted beam is also partially polarized.

Now suppose that the angle of incidence, θ_1, is varied until the angle between the reflected and refracted beams is 90° (Fig. 26.23b). At this particular angle of incidence, the reflected beam is completely polarized, with its electric field vector parallel to the surface, and the refracted beam is partially polarized. The angle of incidence at which this occurs is called the **polarizing angle, θ_p**.

An expression can be obtained relating the polarizing angle to the index of refraction of the reflecting surface. From Figure 26.23b, we see that, at the polarizing angle, $\theta_p + 90° + \theta_2 = 180°$, so that $\theta_2 = 90° - \theta_p$. Using Snell's law, we have

$$n = \frac{\sin \theta_1}{\sin \theta_2} = \frac{\sin \theta_p}{\sin \theta_2}$$

Because $\sin \theta_2 = \sin(90° - \theta_p) = \cos \theta_p$, the expression for n can be written

$$n = \frac{\sin \theta_p}{\cos \theta_p} = \tan \theta_p \qquad\qquad \textbf{(26.9)} \qquad \text{Brewster's law}$$

This expression is called **Brewster's law,** and the polarizing angle, θ_p, is sometimes called **Brewster's angle** after its discoverer, David Brewster (1781–1868). Because n varies with wavelength for a given substance, Brewster's angle is also a function of wavelength.

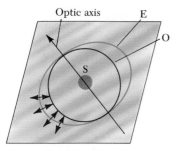

Unpolarized light

Calcite

E ray

O ray

Figure 26.24
When an unpolarized light beam is incident on a calcite crystal, it splits into an ordinary (O) ray and an extraordinary (E) ray. The two rays are polarized in mutually perpendicular directions. (Note that this is not to scale.)

Polarization by reflection is a common phenomenon. Sunlight reflected from water, glass, or snow is partially polarized. If the surface is horizontal, the electric field vector of the reflected light will have a strong horizontal component. Sunglasses made of polarizing material reduce the glare of reflected light. The transmission axes of the lenses are oriented vertically to absorb the strong horizontal component of the reflected light.

Example 26.6   Brewster's Angle for Water

At what angle of incidence will unpolarized light from the Sun become completely polarized upon reflection from a pool of water?

Solution
In this case, $n = 1.33$, and Equation 26.9 gives Brewster's angle as

$$\theta_p = \tan^{-1} 1.33 = \boxed{53°}$$

Polarization by Double Refraction

Optic axis

E

O

S

Figure 26.25
A point source S inside a doubly refracting crystal produces a spherical wavefront corresponding to the O ray and an elliptical wavefront corresponding to the E ray. The two waves propagate with the same velocity along the optic axis.

When light travels through an amorphous material, such as glass, it travels with a speed that is the same in all directions. That is, glass has a single index of refraction. However, in certain crystalline materials, such as calcite and quartz, the speed of light is *not* the same in all directions. Such materials are characterized by two indices of refraction. Hence, they are often referred to as **double-refracting** or **birefringent** materials.

When an unpolarized beam of light enters a calcite crystal, the beam splits into two plane-polarized rays that travel with different velocities, corresponding to two different angles of refraction, as in Figure 26.24. The two rays are polarized in two mutually perpendicular directions, as indicated by the dots and arrows. One ray, called the **ordinary (O) ray,** is characterized by an index of refraction, n_O, that is the *same* in all directions. This means that if one could place a point source of light inside the crystal, as in Figure 26.25, the ordinary waves would spread out from the source as spheres.

The second plane-polarized ray, called the **extraordinary (E) ray,** travels with *different* speeds in different directions and hence is characterized by an index of refraction, n_E, that *varies* with the direction of propagation. A point source of light inside such a crystal would send out an extraordinary wave having wavefronts that are elliptical in cross section (Fig. 26.25). Note from Figure 26.25 that there is one direction, called the **optic axis,** along which the ordinary and extraordinary rays have the *same* velocity, corresponding to the direction for which $n_O = n_E$. The difference in velocity for the two rays is a maximum in the direction perpendicular to the optic axis. For example, in calcite, $n_O = 1.658$ at a wavelength of 589.3 nm and n_E varies from 1.658 along the optic axis to 1.486 perpendicular to the optic axis.

If one places a piece of calcite on a sheet of paper and then looks through the crystal at any writing on the paper, two images of the writing are seen, as shown in Figure 26.26. As can be seen from Figure 26.24, these two images

Figure 26.26
A calcite crystal produces a double image because it is a birefringent (double-refracting) material. *(Courtesy of Henry Leap and Jim Lehman)*

correspond to one formed by the ordinary ray and the second formed by the extraordinary ray. If the two images are viewed through a sheet of rotating polarizing glass, they will alternately appear and disappear because the ordinary and extraordinary rays are plane-polarized along mutually perpendicular directions.

Polarization by Scattering

When light is incident on a system of particles, such as a gas, the electrons in the medium can absorb and reradiate part of the light. The absorption and reradiation of light by the medium, called **scattering,** is what causes sunlight reaching an observer on the Earth from straight overhead to be partially polarized. You can observe this effect by looking directly up through a pair of sunglasses made of polarizing glass. At certain orientations of the lenses, less light passes through than at others.

Figure 26.27 illustrates how the sunlight becomes partially polarized. The left side of the figure shows an incident unpolarized beam of sunlight on the verge of striking an air molecule. When this beam strikes the air molecule, it sets the electrons of the molecule into vibration. These vibrating charges act like the vibrating charges in an antenna, except that these charges are vibrating in a complicated pattern. The horizontal part of the electric field vector in the incident wave causes the charges to vibrate horizontally, and the vertical part of the vector simultaneously causes them to vibrate vertically. A polarized wave is directed to an observer as shown in Figure 26.27.

Uses of Polarized Light

Many important practical applications of polarized light involve the use of certain materials which display the property of **optical activity.** A substance is said to be optically active if it rotates the plane of polarization of transmitted light. In order to understand how this process occurs, suppose unpolarized light is incident on a polarizer from the left, as in Figure 26.28a. The transmit-

Figure 26.27
The scattering of unpolarized sunlight by air molecules. The light observed at right angles is plane polarized because the vibrating molecule has a horizontal component of vibration.

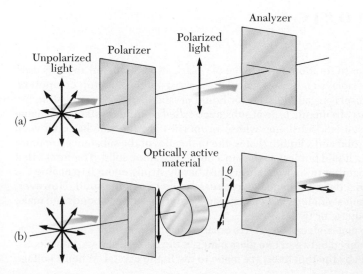

Figure 26.28
(a) When crossed polarizers are used, none of the polarized light can pass through the analyzer. (b) An optically active material rotates the direction of polarization through an angle θ, enabling some of this light to pass through the analyzer.

(Top) The strain distribution in a plastic wrench under stress. The test piece is viewed between two crossed polarizers. *(Courtesy of Henry Leap and Jim Lehman)* *(Bottom)* Stress patterns in several plastic draftsman's tools. *(Peter Aprahamian/Science Photo Library)*

ted light is polarized vertically as shown. If this light is then incident on an analyzer whose axis is perpendicular to that of the polarizer, no light emerges from it. If an optically active material is placed between the polarizer and analyzer as in Figure 26.28b, the material causes the direction of the polarized beam to rotate through an angle θ. As a result, some light is able to pass through the analyzer. The angle through which the light is rotated by the material can be found by rotating the polarizer until the light is again extinguished. It is found that the angle of rotation depends on the length of the sample and on the concentration if the substance is in solution. One optically active material is a solution of common sugar, dextrose. A standard method for determining the concentration of sugar solutions is to measure the rotation produced by a fixed length of the solution.

Optical activity occurs in a material because of an asymmetry in the shape of its constituent molecules. For example, some proteins are optically active because of their spiral shape. Environmental influences such as temperature changes can affect the shape and properties of these molecules. Studies of the changes in optical activity provides a great deal of information on such properties and their relation to the environment.

Some materials, such as glass and plastic, become optically active when placed under stress. If polarized light is passed through an unstressed piece of plastic and then through an analyzer whose axis is parallel to that of the polarizer, all of the polarized light is transmitted. However, if the plastic is placed under stress, the regions of greatest stress produce the largest angles of rotation of polarized light. As a result, a series of light and dark bands in the transmitted light, with the darkest bands corresponding to points of greatest stress, is observed. Engineers often use this technique to assist in the design of various structures ranging from bridges to small tools. A plastic model is built and analyzed under different load conditions to determine positions of potential weakness and failure under stress. If the design is poor, the pattern of light and dark bands will indicate the points of greatest weakness and the design can be corrected at an early stage of development. An example of the strain distribution in a plastic rectangular beam under stress is shown in the photograph on the left.

▶ SPECIAL TOPIC
Liquid Crystals

An effect that is similar to rotation of the plane of polarization discussed above is used to form the familiar displays on pocket calculators, wristwatches, laptop computers, and so forth. These displays, called LCDs for liquid crystal displays, are possible because of the properties of a unique type of substance called a **liquid crystal.** As its name implies, a liquid crystal is a substance whose properties are intermediate between those of a crystalline solid and a liquid; that is, the molecules of the substance are more orderly than those in a liquid but less so than in a pure crystalline solid. The forces that hold the molecules together in such a state are just barely strong enough to enable the substance to maintain a definite shape, so it is reasonable to call it a solid. However, small inputs of mechanical or electrical energy can disrupt these weak bonds and make the substance flow, rotate, or twist.

To see how these materials can be used to create a display, consider Figure 26.29a. The liquid crystal is placed between two glass plates in the pattern shown, and electrical contacts, indicated by the thin lines, are made to the liquid crystal. When a voltage

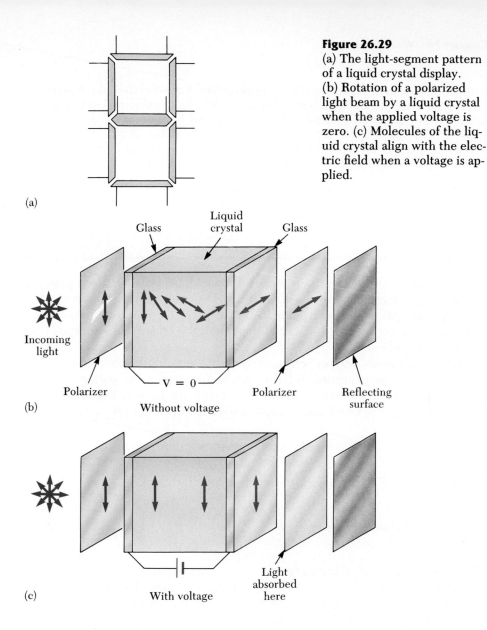

(a)

Glass   Liquid crystal   Glass

Incoming light

Polarizer   V = 0   Polarizer   Reflecting surface

(b)   Without voltage

With voltage

Light absorbed here

(c)

is applied across any segment in the display, that segment turns dark. In this fashion, any number between 0 and 9 can be formed by the pattern, depending on the voltages applied to the eight segments.

To see why a segment can be changed from dark to light by the application of a voltage, consider Figure 26.29b, which shows the basic construction of a portion of the display. The liquid crystal is placed between two glass substrates that are packaged between two pieces of Polaroid material with their transmission axis perpendicular. A reflecting surface is placed behind one of the pieces of Polaroid. First consider what happens when light falls on this package when no voltages are applied to the liquid crystal, as shown in Figure 26.29b. Incoming light is polarized by the first polarizer and then falls on the liquid crystal. The liquid crystal is applied such that its molecules gradually change orientation by 90° from the first piece of glass to the second. The effect of this is to cause a rotation of the plane of polarization of the light as it moves through the crystal. Thus, when light strikes the second polarizer, it has rotated 90° and is allowed to pass through to the reflecting surface. The light reflects and retraces its path, and the segment appears light to a viewer at the left of the figure. When a voltage is applied as in Figure 26.29c, the molecules of the liquid crystal rotate to align with the electric field, that is, parallel to the direction of travel of the light. As a result, no rotation of the plane of polarization occurs, and the light is absorbed by the second polarizer. Thus, there is no light to be reflected, and a viewer at the left sees this segment of the liquid crystal as black. Thus, changing the voltage applied to the crystal in a precise pattern and at precise times can make the pattern tick off the seconds on a watch, display a letter on a computer display, and so forth. ◄

Summary

Interference occurs when two or more light waves overlap at a given point. A sustained interference pattern is observed if (1) the sources are coherent (that is, they maintain a constant phase relationship with one another), (2) the sources are monochromatic (of a single wavelength), and (3) the superposition principle is applicable.

In **Young's double-slit experiment,** two slits separated by a distance d are illuminated by a monochromatic light source. An interference pattern consisting of bright and dark fringes is observed on a screen a distance L from the slits. The condition for **bright fringes** (constructive interference) is

$$d \sin \theta = m\lambda \qquad m = 0, \pm 1, \pm 2, \ldots \qquad \textbf{(26.2)}$$

The condition for **dark fringes** (destructive interference) is

$$d \sin \theta = \left(m + \frac{1}{2} \right) \lambda \qquad m = 0, \pm 1, \pm 2, \ldots \qquad \textbf{(26.3)}$$

The number m is called the **order number** of the fringe.

An electromagnetic wave undergoes a phase change of 180° upon reflection from a medium whose index of refraction is higher than that of the medium in which the wave is traveling.

The wavelength of light, λ_n, in a medium whose refractive index is n is

$$\lambda_n = \frac{\lambda}{n} \qquad \textbf{(26.7)}$$

where λ is the wavelength of the light in free space.

The **diffraction pattern** produced by a single slit on a distant screen consists of a central bright maximum and alternating bright and dark regions of lower intensity. The angles θ at which the diffraction pattern has zero intensity (regions of destructive interference) are given by

$$\sin \theta = \frac{m\lambda}{a} \qquad m = \pm 1, \pm 2, \pm 3, \ldots \qquad \textbf{(26.8)}$$

where a is the width of the slit and λ is the wavelength of the light incident on the slit.

Unpolarized light can be polarized by selective absorption, reflection, double refraction, and scattering.

In general, light reflected from an amorphous material, such as glass, is partially polarized. However, the reflected light is completely polarized when the angle of incidence is such that the angle between the reflected and refracted beams is 90°. This angle of incidence, called the **polarizing angle,** θ_p, satisfies **Brewster's law,** given by

$$n = \tan \theta_p \qquad\qquad (26.9)$$

where n is the index of refraction of the reflecting medium.

Additional Reading

P. Baumeister and G. Pincus, "Optical Interference Coatings," *Sci. American,* December 1970, p. 59.

C. Bohrens and A. Fraser, "Colors of the Sky," *The Physics Teacher,* May 1985, p. 267.

H. M. Nussensvieg, "The Theory of the Rainbow," *Sci. American,* April 1977, p. 116.

P. M. Rinard, "Large Scale Diffraction Patterns from Circular Objects," *American Journal of Physics,* January 1976, p. 70.

Scientific American, September 1968. The entire issue is devoted to light.

W. A. Shurcliffe and S. S. Ballard, *Polarized Light,* Van Nostrand, 1964.

A. C. S. van Heel and C. H. F. Velzel, *What Is Light?* World University Library, New York, McGraw-Hill, 1968, Chaps. 3, 4, and 5.

H. C. Von Baeyer, "Rainbows, Whirlpools, and Clouds," *The Scientist,* July/August 1984, p. 24.

R. Wehner, "Polarized Light Navigation by Insects," *Sci. American.* July 1976, p. 106.

A. Wood and F. Oldham, *Thomas Young, Natural Philosopher, 1773–1829,* Cambridge University Press, 1954.

Questions

1. What is the necessary condition on path length difference between two waves that interfere (a) constructively and (b) destructively?

2. Explain why two flashlights held close together will not produce an interference pattern on a distant screen.

3. A simple way of observing an interference pattern is to look at a distant light source through a stretched handkerchief or an opened umbrella. Explain how this works.

4. Why is it so much easier to perform interference experiments with a laser than an ordinary light source?

5. In Young's double-slit experiment, why do we use monochromatic light? If white light is used, how would the pattern change?

6. If Young's double-slit experiment were performed under water, how would the observed interference pattern be affected?

7. A lens with outer radius of curvature R and index of refraction n rests on a flat glass plate. It is illuminated with white light from above. Is there a dark spot or a light spot at the center of the lens? What does it mean if the observed rings are noncircular?

8. In the process of evaporation, a soap bubble appears black just before it breaks. Explain this phenomenon in terms of the phase changes that occur upon reflection from the two surfaces.

9. If a soap film on a wire loop is held in air, it appears black in the thinnest regions when observed by reflected light and shows a variety of colors in thicker regions, as in the photograph below. Explain.

Question 9

10. If an oil film is observed on water, the film appears brightest at the outer regions, where it is thinnest. From this information, what can you say about the index of refraction of the oil relative to that of water?

11. Lenses with a "nonreflective" coating usually appear to have a purplish cast. Explain the color.

12. Would it be possible to place a nonreflective coating on an airplane to cancel radar waves of wavelength 3 cm?

13. In order to observe interference in a thin film, why must the film thickness be thin compared with the wavelengths of visible light? (*Hint:* How far apart are the two reflected waves when they attempt to interfere if the film is thick?)

14. Washed dishes that are not rinsed well often have colored bands or rings across them. Discuss the interference effect that causes this.

15. What is the difference between interference and diffraction?

16. If you observe a distant streetlight between two fingers pinched together, you will observe a pattern of alternating light and dark bands. Is this an interference or a diffraction pattern?

17. Observe the shadow of your book when it is held a few inches above a table with a lamp several feet above the book. Why is the shadow of the book fuzzy at the edges?

18. Describe the change in width of the central maximum of a single-slit diffraction pattern as the slit is made narrower.

19. A point source of light occurs inside a container that is opaque except for a single hole. Discuss what happens to the image on the screen as the diameter of the hole is reduced (a) in the ray approximation and (b) in an actual situation.

20. Light falls on a slit that is twice as wide as it is high. What kind of pattern will be formed on a screen? In which direction will the beam be the most spread out?

21. Show, by a sketch, that the waves emitted by a radio antenna are polarized.

22. Can a sound wave be polarized? Explain.

23. Certain sunglasses use a polarizing material to reduce the intensity of light reflected from shiny surfaces, such as water or the hood of a car. What orientation of polarization should the material have in order to be most effective?

24. Why is the sky black when viewed from the Moon?

25. Is light from the sky polarized? Why is it that clouds seen through Polaroid glasses stand out in bold contrast to the sky?

Problems

Section 26.2   Young's Double-Slit Interference

1. Monochromatic light falls on a screen 1.5 m from two slits spaced 1.3 mm apart. The first- and second-order bright fringes are found to be 0.5 mm apart. What is the wavelength of the light?

2. If the distance between two slits is 0.05 mm and the distance to a screen is 2.5 m, find the spacing between the first- and second-order bright fringes for yellow light of 600 nm wavelength.

3. A pair of narrow, parallel slits separated by 0.25 mm are illuminated by the green component from a mercury vapor lamp ($\lambda = 546.1$ nm). The interference pattern is observed on a screen 1.2 m from the plane of the parallel slits. Calculate the distance (a) from the central maximum to the first bright region on either side of the central maximum and (b) between the first and second dark bands in the interference pattern.

4. A double slit with a spacing of 0.083 mm between the slits is 2.5 m from a screen. (a) If yellow light of wavelength 570 nm strikes the double slit, what is the separation between the zeroeth- and first-order maxima on the screen? (b) If blue light of wavelength 410 nm strikes the double slit, what is the separation between the second- and fourth-order maxima? (c) Repeat Parts (a) and (b) for the minima.

5. A warehouse on a dock has two open doors, as in Figure 26.30. A boat sounds its horn from across the river. To person A the sound is loud and clear. To person B the sound is barely audible. The principal wavelength of the sound waves is 3.0 m. Assuming person B is at the position of the first minimum, determine the distance between the doors, center to center.

Figure 26.30   Problem 5

6. Light of wavelength 575 nm falls on a double slit and the first bright fringe is seen at an angle of 16.5°. Find the distance between the slits.

7. Light of wavelength 460 nm falls on two slits spaced 0.3 mm apart. What is the required distance from the

slits to a screen if the spacing between the first and second dark fringes is to be 4 mm?

8. White light spans the wavelength range between about 400 nm and 700 nm. If white light passes through two slits 0.30 mm apart and falls on a screen 1.5 m from the slits, find the distance between the first-order violet and red fringes.

9. The yellow component of light from a helium discharge tube ($\lambda = 587.5$ nm) is allowed to fall on a plate containing parallel slits that are 0.2 mm apart. A screen is located such that the second bright fringe in the interference pattern is at a distance equal to 10 slit spacings from the central maximum. What is the distance between the plate and the screen?

10. A Young's interference experiment is performed with blue-green argon laser light. The separation between the slits is 0.50 mm and the interference pattern on a screen 3.3 m away shows the first maximum at a distance of 3.4 mm from the center of the pattern. What is the wavelength of argon laser light?

11. Two radio antennas separated by 300 m as shown in Figure 26.31 simultaneously transmit identical signals (assume waves) on the same wavelength. A radio in a car traveling due north receives the signals. (a) If the car is at the position of the second maximum, what is the wavelength of the signals? (b) How much farther must the car travel to encounter the next minimum in reception? (*Caution:* Avoid small angle approximations in this problem.)

Figure 26.31 Problem 11

12. Waves from a radio station have a wavelength of 300 m. They arrive by two paths at a home receiver 20 km from the transmitter. One path is a direct path, and the second is by reflection from a mountain directly behind the home receiver. What is the minimum distance from the mountain to the receiver such that destructive interference occurs at the receiver? (Assume no phase change occurs on reflection from the mountain.)

13. Waves broadcast by a 1500-kHz radio station arrive at a home receiver by two paths. One is a direct path, and the second is from reflection off an airplane directly above the home receiver. The airplane is approximately 100 m above the home receiver, and the direct distance from the station to the home is 20 km. What is the exact height of the airplane if destructive interference is occurring? (Assume no phase change occurs on reflection from the plane.)

14. The waves from a radio station can reach a home receiver by two different paths. One is a straight line path from the transmitter to the home, a distance of 30 km. The second path is by reflection from the ionosphere (a layer of ionized air molecules near the top of the atmosphere). Assume this reflection takes place at a point midway between receiver and transmitter. If the wavelength broadcast by the radio station is 350 m, find the minimum height of the ionospheric layer that would produce destructive interference between the direct and reflected beams. (Assume no phase changes on reflection.)

15. Radio waves of wavelength 250 m from a star reach a radio telescope by two separate paths. One is a direct path to the receiver, which is situated on the edge of a cliff by the ocean. The second is by reflection off the water. The first minimum of destructive interference occurs when the star is 25° above the horizon. Find the height of the cliff. (Assume no phase change on reflection.)

16. One of the bright bands in Young's interference pattern is located 12 mm from the central maximum. The screen is 119 cm from the pair of slits that serve as sources. The slits are 0.241 mm apart and are illuminated by the blue light from a hydrogen discharge tube ($\lambda = 486$ nm). How many bright lines are observed between the central maximum and the 12-mm position?

Section 26.3 Newton's Rings
Section 26.4 Interference in Thin Films

17. Suppose the film shown in Figure 26.8 has an index of refraction of 1.36 and is surrounded by air on both sides. Find the minimum thickness other than zero that will produce constructive interference in the reflected light when illuminated by light of wavelength 500 nm.

18. Repeat Problem 17 for the case in which the thin film is resting on a glass plate of index 1.50.

19. A thin layer of liquid methylene iodide ($n = 1.756$) is sandwiched between two flat parallel plates of glass. What must be the thickness of the liquid layer if normally incident light with $\lambda = 600$ nm is to be strongly reflected?

20. A thin film of glass ($n = 1.52$) with a thickness of $0.42\ \mu$m is viewed in white light at near-normal incidence. Visible light with what wavelength is most strongly reflected by the film when surrounded by air?

21. Two parallel glass plates are placed in contact and illuminated from above with light of wavelength 580 nm. As the plates are slowly moved apart, darkness occurs at certain separations when the reflected light is viewed. What are the distances (other than, perhaps, zero) of the first three of these separations?

22. Repeat Problem 12 if a phase change of $180°$ occurs on reflection.

23. A beam of light of wavelength 580 nm passes through two closely spaced glass plates, as shown in Figure 26.32. For what minimum nonzero value of the plate separation, d, will the transmitted light be bright?

Figure 26.32   Problem 23

24. A planoconvex (flat on one side, convex on the other) lens rests with its curved side on a flat glass surface and is illuminated from above by light of wavelength 500 nm (see Fig. 26.5). It is observed that there is a dark spot at the center surrounded by 19 concentric dark rings (with bright rings in between). How much thicker is the air wedge at the position of the 19th dark ring than at the center?

25. A planoconvex lens (see Problem 24) with a radius of curvature $R = 3.0$ m is in contact with a flat plate of glass. A light source and the observer's eye are located close to the normal, as shown in Figure 26.5. The radius of the 50th of the bright Newton's rings is found to be 9.8 mm. What is the wavelength of the light produced by the source?

26. A transparent oil with an index of refraction of 1.29 spills on the surface of water (index of refraction, 1.33), producing a maximum of reflection with normally incident orange light (wavelength, 600.0 nm in air). Assuming the maximum occurs in the first order, determine the thickness of the oil slick.

27. An oil film 500 nm thick floats on water. It is illuminated with white light in the direction perpendicular to the film. What wavelengths will be strongly reflected in the range from 300 nm to 700 nm? Take $n = 1.46$ for oil.

28. A thin film of glass ($n = 1.50$) floats on a liquid of $n = 1.35$. When illuminated by light of $\lambda = 580$ nm incident from air above the glass, find the minimum thickness other than zero that will produce destructive interference in the reflected light.

29. A thin film of $MgF_2\ 10^{-5}$ cm thick ($n = 1.38$) is used to coat a camera lens. Will any wavelengths in the visible spectrum be intensified in the reflected light?

30. A coating is applied to a lens to minimize reflections. The index of refraction of the coating is 1.55, and that of the lens is 1.48. If the coating is 177.4 nm thick, which reflected wavelengths are minimally reflected for normal incidence in the lowest order?

31. Determine the minimum thickness of a soap film ($n = 1.33$) that will result in constructive interference of (a) the red H_α line ($\lambda = 656.3$ nm) and (b) the blue H_γ line ($\lambda = 434.0$ nm).

32. A soap bubble of index of refraction 1.4 strongly reflects both red and green colors in white light. What thickness of soap bubble allows this to happen? (In air, λ red $= 700$ nm, λ green $= 500$ nm.)

33. Two rectangular optically flat plates ($n = 1.52$) are in contact along one end and are separated along the other end by a sheet of paper that is 4×10^{-3} cm thick (Fig. 26.33). The top plate is illuminated by monochromatic light of wavelength 546.1 nm. Calculate the number of dark parallel bands crossing the top plate (including the dark band at zero thickness along the edge of contact between the two plates).

Figure 26.33   Problems 33, 34, and 35

34. A hair replaces the sheet of paper in Problem 33, and when the reflected light is viewed, it is found that seven dark bands (including the one along the edge of contact) cross the top plate, with the seventh being directly over the hair. What is the thickness of the hair? The wavelength of light remains the same.

35. An air wedge is formed between two thick glass plates in a manner like that described in Problem 33. Light of wavelength 434 nm is incident vertically on the top plate. In this case, there are 20 bright parallel interference fringes across the top plate. Calculate the thickness of the paper separating the plates.

36. A flat piece of glass is supported horizontally above the flat end of a metal rod of length 10 cm which has its lower end rigidly fixed. The thin film of air between the rod and glass is observed to be bright when illuminated by light of wavelength 500 nm. As the temperature is slowly increased by $25°$ C, the film changes from bright to dark and back to bright 200

times. What is the coefficient of linear expansion of the metal?

37. If the Newton's ring arrangement is such that the radius of curvature of the lens R is much greater than the thickness of the air gap, the radius of the dark rings is given by $r = \sqrt{m\lambda R}$, where m is an integer counting the rings, starting with 0 at the central spot and increasing outward, and λ is the wavelength of the light used. Show that this relationship for r is valid.

Section 26.6 Single-Slit Diffraction

38. Microwaves of wavelength 5.0 cm enter a long narrow window in a building otherwise essentially opaque to the microwaves. If the window is 36.0 cm wide, what is the distance from the central maximum to the first-order minimum along a wall 6.5 m from the window?

39. Infrared radiation of wavelength 12.2 μm passes through a single slit. (a) If the separation between the two first-order minima is 0.4 mm on a screen 84.0 cm away, what is the width of the slit? (b) If the width is reduced by half, what is the new separation between the first-order minima?

40. Light of wavelength 600 nm falls on a slit 0.4 mm wide and forms a diffraction pattern on a screen 1.5 m away. (a) Find the position of the first dark band on each side of the central maximum. (b) Find the width of the central maximum.

41. A screen is placed 50 cm from a single slit, which is illuminated with light of wavelength 680 nm. If the distance between the first and third minima in the diffraction pattern is 3.0 mm, what is the width of the slit?

42. Light of wavelength 550 nm falls on a single slit of width 0.2 mm. Find the angle between the first dark bands on each side of the central maximum.

43. Light of wavelength 587.5 nm illuminates a single slit 0.75 mm in width. (a) At what distance from the slit should a screen be located if the first minimum in the diffraction pattern is to be 0.85 mm from the central maximum? (b) Calculate the width of the central maximum.

44. A slit of width 0.5 mm is illuminated with light of wavelength 500 nm, and a screen is located 120 cm in front of the slit. Find the widths of the first and second maxima on each side of the central maximum.

Section 26.7 Polarization of Light Waves

45. Determine Brewster's angle for light incident on benzene.

46. If the polarizing angle for cubic zirconia (ZrO$_2$) is 65.6°, what is the index of refraction for this material?

47. A light beam is incident on heavy flint glass ($n = 1.65$) at the polarizing angle. Calculate the angle of refraction for the transmitted ray.

48. Equation 26.9 assumes the incident light is in air. If the light is incident from a medium of index n_1 on a medium of index n_2, use the procedure used to derive Equation 26.9 to show that $\tan \theta_p = n_2/n_1$.

49. Use the results of Problem 48 to find Brewster's angle when light is reflected off a piece of glass ($n = 1.50$) submerged in water.

50. At what angle above the horizon is the Sun if light from it is completely polarized upon reflection from water?

51. The angle of incidence of a light beam in air onto a reflecting surface is continuously variable. The reflected ray is found to be completely polarized when the angle of incidence is 48°. (a) What is the index of refraction of the reflecting material? (b) If some of the light passes into the material, what is the angle of refraction?

52. The index of refraction of the ordinary ray in a calcite crystal is 1.66, and the index of the extraordinary ray is 1.49. Find the angle between these two rays inside the crystal if the incident beam strikes the refracting surface at an angle of incidence of 20°.

53. The critical angle for total internal reflection for sapphire surrounded by air is 34.4°. Calculate the polarizing angle for sapphire if the light is incident from the air.

54. For a particular transparent medium surrounded by air, show that the critical angle for internal reflection and the polarizing angle are related by the expression $\cot \theta_p = \sin \theta_c$.

Additional Problems

55. When a monochromatic beam of light is incident from air at an angle of 37° with the normal on the surface of a glass block, it is observed that the refracted ray is directed at 22° with the normal. What angle of incidence from air would result in total polarization of the reflected beam?

56. A thin layer of oil ($n = 1.25$) is floating on water. How thick is the oil in the region that reflects green light ($\lambda = 525$ nm)?

57. Helium-neon laser light ($\lambda = 632.8$ nm) is sent through a 0.3-mm-wide single slit. What is the width of the central maximum on a screen 1 m in back of the slit?

58. The transmitting antenna on a submarine is 5 m above the ocean surface when the ship is surfaced. The captain wishes to transmit a message to a receiver on a 90-m-tall cliff at the ocean shore. If his signal is to be completely polarized by reflection off the ocean surface, how far must the ship be from the shore?

59. Monochromatic light incident on a slit of width 0.05 mm forms a diffraction pattern on a screen 2 m away. The second-order dark fringe is observed at an angle of 1.56° (see Fig. 26.18). Calculate the wavelength of the light.

60. Light of wavelength 546 nm (the intense green line from a mercury source) produces a Young's interference pattern in which the second minimum from the central maximum is along a direction that makes an angle of 18 min of arc relative to the direction to the central maximum. What is the distance between the parallel slits?

61. A light beam containing light of wavelengths λ_1 and λ_2 is incident on a set of parallel slits. In the interference pattern, the fourth bright line of the λ_1 light occurs at the same position as the fifth bright line of the λ_2 light. If λ_1 is known to be 540 nm, what is the value of λ_2?

62. When a beam of light passes through a pair of polarizing disks (polarizer and analyzer), the intensity of the transmitted beam is a maximum when the transmission axes of the polarizer and the analyzer are parallel to each other ($\theta = 0°$ or $180°$) and zero when the axes are perpendicular ($\theta = 90°$). For any value of θ between $0°$ and $90°$, the transmitted intensity is given by **Malus' law**:

$$I = I_i \cos^2 \theta$$

where I_i is the intensity of the light incident on the analyzer after the light has passed through the polarizer. (a) Polarized light is incident on the analyzer such that the direction of polarization makes an angle of 45° with respect to the transmission axes. What is the ratio I/I_i? (b) What should the angle between the transmission axes be to make $I/I_i = 1/3$?

63. Three polarizing disks whose planes are parallel are centered on a common axis. The directions of the transmission axes relative to the common vertical direction are shown in Figure 26.34. A plane-polarized beam of light with the plane of polarization parallel to the vertical reference direction is incident from the left on the first disk with intensity $I_i = 10$ units (arbitrary). Calculate the transmitted intensity I_f when $\theta_1 = 20°$, $\theta_2 = 40°$, and $\theta_3 = 60°$. (*Hint:* Refer to Problem 62.)

64. Suppose that a slit 6 cm wide is placed in front of a microwave source operating at a frequency of 7.5 GHz. Calculate the angle (measured from the central maximum) where the first minimum in the diffraction pattern occurs.

65. A diffraction pattern is produced on a screen 140 cm from a single slit using monochromatic light of wavelength 500 nm. The distance from the center of the central maximum to the first-order maximum is 3 mm. Calculate the slit width. (*Hint:* Assume the first-order maximum is halfway between the first- and second-order minima.)

66. In a double-slit arrangement, $d = 0.15$ mm, $L = 140$ cm, $\lambda = 643$ nm, and $y = 1.8$ cm. (a) What is the path difference δ for the two slits at this y location? (b) Express this path difference in terms of the wavelength. (c) Will the interference occurring correspond to a maximum, a minimum, or an intermediate condition?

67. A glass plate ($n = 1.61$) is covered with a thin uniform layer of oil ($n = 1.2$). A light beam in air of variable wavelength is normally incident on the oil surface. Observation of the reflected beam shows destructive interference at 500 nm and constructive interference at 750 nm. Calculate the thickness of the oil film from this information.

68. A piece of transparent material having an index of refraction n is cut into the shape of a wedge as shown in Figure 26.35. The angle of the wedge is small, and monochromatic light of wavelength λ is normally incident from above. If the height of the wedge is h and the length is ℓ, show that bright fringes occur at the positions

$$x = \frac{\lambda \ell \left(m + \frac{1}{2} \right)}{2hn}$$

and dark fringes occur at the positions $x = \dfrac{\lambda \ell m}{2hn}$ where $m = 0, 1, 2, \ldots$ and x is measured as shown.

Figure 26.34   Problem 63

Figure 26.35   Problem 68

69. The discussion of the techniques for determining constructive and destructive interference by reflection from a thin film in air has been confined to rays striking the film at nearly normal incidence. Assume that a ray is incident at an angle of 30° (relative to the normal) on a film with index of refraction 1.38. Calculate the minimum thickness for constructive interference if the light is sodium light having a wavelength of 590 nm.

70. Figure 26.36 illustrates the formation of an interference pattern by the Lloyd's mirror method. Light from a source S strikes the screen by two different pathways. One is a direct path, and the second is the path followed as light reflects from a horizontal mirror before arriving at the screen. The interference is as though light from two different sources, S and S', had interfered as in the Young's double-slit arrangement. Assume the actual source S and the virtual source S' are in a plane 25 cm to the left of the mirror and the screen is a distance $L = 120$ cm to the right of this plane. The source S is a distance $d = 2.5$ mm above the top surface of the mirror and the light is monochromatic with $\lambda = 620$ nm. Determine the distance of the first bright fringe above the surface of the mirror.

Figure 26.36 Problem 70

CHAPTER 27

Optical Instruments

A microscope is a common device used to view small objects with the assistance of a light source. This simple microscope is accompanied by a laser beam.
(Garry Gay © The IMAGE Bank)

We make use of devices made from lenses, mirrors, or other optical components every time we put on a pair of eyeglasses to read the daily newspaper, take a photograph, look at the heavens through a telescope, and so on. In this chapter we shall examine how these and other optical instruments work. For the most part, our analyses will involve the laws of reflection and refraction and the procedures of geometric optics. However, for some instruments and in some instances we must use the wave nature of light to explain certain phenomena.

27.1 The Camera

The single-lens photographic **camera** is a simple optical instrument whose essential features are shown in Figure 27.1a. It consists of a lighttight box, a converging lens that produces a real image, and a film behind the lens to receive the image. Focusing is accomplished by varying the distance between lens and film with an adjustable bellows in older style cameras or some other

(a)   (b)

Figure 27.1
(a) Cross-sectional view of a simple camera. (b) Cutaway
view of a modern 35-mm camera. *(Courtesy of Minolta
Corporation)*

mechanical arrangement in newer models. For proper focusing, or sharp
images, the lens-to-film distance will depend on the object distance as well as
on the focal length of the lens. The shutter, located behind the lens, is a
mechanical device that is opened for selected time intervals. With this arrangement, we can photograph moving objects by using short exposure times
or dark scenes (low light levels) by using long exposure times. If this arrangement were not available, it would be impossible to take stop-action photographs. For example, a rapidly moving racecar could move enough in the
interval that the shutter was open to produce a blurred image. Typical shutter
speeds are 1/30, 1/60, 1/125, and 1/250 s. A stationary object is normally shot
with a shutter speed of 1/60 s.

More expensive cameras like that shown in Figure 27.1b also have an
aperture of adjustable diameter either behind or in front of the lens to provide
further control of the intensity of the light reaching the film. When small
aperture diameters are used, only light from the central portion of the lens
reaches the film and so the aberration is reduced somewhat. (See Chapter 25
for a discussion of aberration.)

The brightness of the image focused on the film depends on the focal
length and diameter of the lens. Clearly, the amount of light that reaches the
film increases as the size of the lens increases. That is, as the lens diameter
increases, the brightness of the image formed on the film increases. We can see
that the focal length of the lens affects the brightness of the image formed by
considering the magnification equation for a thin lens:

$$M = \frac{h'}{h} = \frac{s'}{s}$$

$$h' = h\frac{s'}{s}$$

where h and h' are the object and image heights, respectively, and s and s' are
the object and image distances. When s is large, s' is approximately equal to
the focal length, f. Thus we have

$$h' = \frac{h}{s}f$$

From this result, we see that a lens having a short focal length produces a small image, corresponding to a small value of h'.

A small image will be brighter than a larger one because all of the incoming light is spread over a much smaller area. Because the brightness of the image depends on f and D, the diameter of the lens, a quantity called the *f*-number is defined as

$$f\text{-number} \equiv \frac{f}{D} \qquad\qquad (27.1)$$

The *f*-number is a measure of the "light-concentrating" power of a lens and determines what is called the speed of the lens. A fast lens has a small *f*-number and is usually one with a small focal length and large diameter. Camera lenses are often marked with various *f*-numbers, such as $f/2.8$, $f/4$, $f/5.6$, $f/8$, $f/11$, $f/16$. These *f*-numbers are obtained by adjusting the aperture, which effectively changes D. When the *f*-number is increased by one position (or one "stop"), the light admitted decreases by a factor of 2. Likewise, the shutter speed is changed in steps whose factor is 2. The smallest *f*-number corresponds to the case where the aperture is wide open and as much as possible of the lens area is in use. Fast lenses, with *f*-numbers as low as 1.2, are more expensive because it is more difficult to keep aberrations acceptably small. Simple cameras for routine snapshots usually have a fixed focal length and fixed aperture size, with *f*-number of about $f/11$.

Example 27.1   Choosing the *f*-Number

Suppose you are using a single-lens 35-mm camera (where 35 mm is the width of the film strip), and the camera has only two *f*-numbers, $f/2.8$ and $f/22$. Which *f*-number would you use on a cloudy day and why?

Solution
Substituting the given *f*-numbers in Equation 27.1, we have

$$2.8 = \frac{f_1}{D_1} \qquad \text{and} \qquad 22 = \frac{f_2}{D_2}$$

The focal length of the camera is fixed ($f_1 = f_2$ in the above), but the diameter of the aperture is not. On a cloudy day, you should make the shutter opening as large as possible. As these equations indicate, the largest value of D produces the smallest *f*-number. Thus, you should use the 2.8 setting.

27.2   The Eye

The eye is an extremely complex part of the body, and because of its complexity, certain defects often arise that can cause the impairment of vision. In these cases, external aids, such as eyeglasses, are often used. In this section we shall describe the parts of the eye, their purpose, and some of the corrections that can be made when the eye does not function properly. You will find that the eye has much in common with the camera. Like the camera, the eye gathers light and produces a sharp image. However, the mechanisms by which the eye controls the amount of light admitted and adjusts itself to produce correctly focused images are far more complex, intricate, and effective than those in the most sophisticated camera. In all respects, the eye is an architectural wonder.

Figure 27.2 shows the essential parts of the eye. The front is covered by a transparent membrane called the *cornea*. This is followed by a clear liquid region (the *aqueous humor*), a variable aperture (the *iris* and *pupil*), and the *crystalline lens*. Most of the refraction occurs in the cornea because the liquid medium surrounding the lens has an average index of refraction close to that of the lens. The iris, which is the colored portion of the eye, is a muscular diaphragm that controls the size of the pupil. The iris regulates the amount of light entering the eye by dilating the pupil in light of low intensity and contracting the pupil in high-intensity light. The *f*-number range of the eye is about $f/2.8$ to $f/16$.

Light entering the eye is focused by the cornea-lens system onto the back surface of the eye, called the *retina*. The surface of the retina consists of millions of sensitive structures called *rods* and *cones*. When stimulated by light, these receptors send impulses via the optic nerve to the brain, where an image is perceived. By this process, a distinct image of an object is observed when the image falls on the retina.

The eye focuses on a given object by varying the shape of the pliable crystalline lens through an amazing process called **accommodation**. An important component in accommodation is the *ciliary muscle*, which is attached to the lens. When the eye is focused on distant objects, the ciliary muscle is relaxed. For an object distance of infinity, the focal length of the eye (the distance between the lens and the retina) is about 1.7 cm. The eye focuses on nearby objects by tensing the ciliary muscle. This action effectively decreases the focal length by slightly decreasing the radius of curvature of the lens, which allows the image to be focused on the retina. This lens adjustment takes place so swiftly that we are not even aware of the change. Again in this respect,

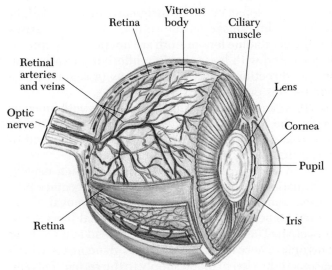

Figure 27.2
Essential parts of the eye. Note the similarity between the eye and the simple camera. Can you correlate the essential parts of the eye with those of the camera?

Close-up photograph of the cornea of the human eye. (© *Lennart Nilsson*)

even the finest electronic camera is a toy compared with the eye. It is evident that there is a limit to accommodation because objects that are very close to the eye produce blurred images.

> The **near point** represents the closest distance for which the lens will produce a sharp image on the retina. This distance usually increases with age and has an average value of around 25 cm.

Typically, at age ten the near point of the eye is about 18 cm. This increases to about 25 cm at age 20, to 50 cm at age 40, and to 500 cm or greater at age 60.

Defects of the Eye

Although the eye is one of the most remarkable organs in the body, it often does not function properly. The eye may have several abnormalities, which can often be corrected with eyeglasses, contact lenses, or surgery.

When the relaxed eye produces an image of a distant object *behind* the retina, as in Figure 27.3a, the abnormality is known as **hyperopia,** and the person is said to be *farsighted*. With this defect, distant objects are seen clearly but near objects are blurred. Either the hyperopic eye is too short or the ciliary muscle is unable to change the shape of the lens enough to properly focus the image. The condition can be corrected with a converging lens, as shown in Figure 27.3b.

Another condition, known as **myopia,** or *nearsightedness*, occurs either when the eye is longer than normal or when the maximum focal length of the lens is insufficient to produce a clearly formed image on the retina. In this case, light from a distant object is focused in front of the retina (Fig. 27.4a). The distinguishing feature of this imperfection is that distant objects are not seen clearly. Nearsightedness can be corrected with a diverging lens, as in Figure 27.4b.

Beginning with middle age, most people lose some of their accommodation power as a result of a weakening of the ciliary muscle and a hardening of the lens. This causes an individual to become farsighted, and the condition can be corrected with converging lenses.

A person may also have an eye defect known as **astigmatism,** in which light from a point source produces a line image on the retina. This defect arises either when the cornea or the crystalline lens or both are not perfectly spherical. Astigmatism can be corrected with lenses having different curvatures in two mutually perpendicular directions. A cylindrical lens (a segment of a cylinder) is typically used for this purpose.

Bifocals contain lenses that can correct two different defects with one pair of eyeglasses. For example, the lower half of the lens could be ground to correct for farsightedness as an aid to reading and the upper half ground to correct a problem such as astigmatism.

The eye is also subject to several diseases. One disease, which usually occurs in old age, is the formation of **cataracts,** where the lens becomes partially or totally opaque. One remedy for cataracts is surgical removal of the lens. Another disease, called **glaucoma,** arises from an abnormal increase in fluid pressure inside the eyeball. The pressure increase can lead to a swelling of the lens and to strong myopia. There is a chronic form of glaucoma in which the pressure increase causes a reduction in blood supply to the retina. This can

(a)

(b)

Figure 27.3
(a) A farsighted eye is slightly shorter than normal, hence the image of a nearby object focuses *behind* the retina.
(b) The condition can be corrected with a converging lens. (The size of the object is assumed to be very small in these figures.)

eventually lead to blindness because the nerve fibers of the retina eventually die. If the disease is discovered early enough, it can be treated with medicine or surgery.

Optometrists and ophthalmologists usually prescribe lenses measured in **diopters.**

> The **power**, P, of a lens in diopters equals the inverse of the focal length in meters, that is $P = 1/f$.

For example, a converging lens whose focal length is $+20$ cm has a power of $+5$ diopters, and a diverging lens whose focal length is -40 cm has a power of -2.5 diopters.

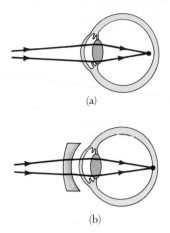

(a)

(b)

Figure 27.4
(a) A nearsighted eye is slightly longer than normal, hence the image of a distant object focuses *in front of* the retina. (b) The condition can be corrected with a diverging lens. (The size of the object is assumed to be very small in these figures.)

Example 27.2   Prescribing a Lens

The near point of an eye is 50 cm. (a) What focal length lens should be used so that the eye can clearly see an object 25 cm away?

Solution
The thin lens equation (Eq. 25.11) enables us to solve this problem. We have placed our object at 25 cm, and we want our lens to form an image at the closest point that the eye can see clearly. This corresponds to the near point, 50 cm. Applying the thin lens equation, we have

$$\frac{1}{25 \text{ cm}} + \frac{1}{(-50 \text{ cm})} = \frac{1}{f}$$

$$f = \boxed{50 \text{ cm}}$$

Why did we use a negative sign for the image distance? Note that the focal length is positive, indicating the need for a converging lens to correct farsightedness problems such as this.

(b) What is the power of this lens?

Solution
The power of the lens is the reciprocal of the focal length in meters:

$$P = \frac{1}{f} = \frac{1}{0.5 \text{ m}} = \boxed{2 \text{ diopters}}$$

Example 27.3   A Case of Nearsightedness

A particular nearsighted person is unable to see objects clearly when they are beyond 50 cm (the far point of the eye). What should the focal length of the lens prescribed to correct this problem be?

Solution
The purpose of the lens in this instance is to "move" an object from infinity to a distance where it can be seen clearly. From the thin lens equation, we have

$$\frac{1}{s} + \frac{1}{s'} = \frac{1}{\infty} + \frac{1}{(-50 \text{ cm})} = \frac{1}{f}$$

$$f = \boxed{-50 \text{ cm}}$$

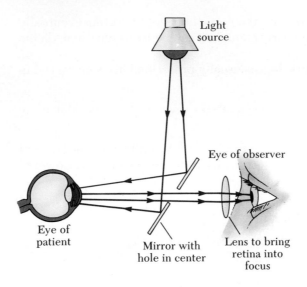

Why did we use a negative sign for the image distance? As you should have suspected, the lens must be a diverging lens (negative focal length) to correct near-sightedness.

Exercise   What is the power of this lens?

Answer -2 diopters.

---

*The Ophthalmoscope

One of the earliest optical instruments to be used for medical purposes is the **ophthalmoscope,** invented almost a century ago. This device, which is simply a mirror with a hole in the middle, is used to examine the retina of the eye, and Figure 27.5 indicates how it works. Light rays from the source are reflected by a small mirror into the eye of the patient. If the eye is relaxed, the light is reflected from the retina and the rays emerge from the eye traveling in parallel lines. This parallel beam of light then passes through a small hole in the center of the mirror and into the eye of the examiner. This instrument can be used to examine the retina for abnormal conditions, such as detachment. It can also be used to give limited information about the refractive properties of the eye lens. If the lens has a problem associated with refraction, a set of lenses on a wheel attached to the mirror of the ophthalmoscope can be used to bring the retina into focus.

·Figure 27.6
The size of the image formed on the retina depends on the angle θ subtended at the eye.

27.3   The Simple Magnifier

The **simple magnifier** is one of the simplest and most basic of all optical instruments because it consists of only a single converging lens. As the name implies, this device is used to increase the apparent size of an object. Suppose an object is viewed at some distance s from the eye, as in Figure 27.6. Clearly, the size of the image formed at the retina depends on the angle θ subtended by

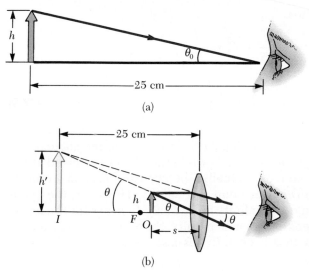

Figure 27.7
(a) An object placed at the near point ($s = 25$ cm) subtends an angle θ_0 at the eye, where $\theta_0 \approx h/25$. (b) An object placed near the focal point of a converging lens produces a magnified image, which subtends an angle $\theta \approx h'/25$ at the eye.

the object at the eye. As the object moves closer to the eye, θ increases and a larger image is observed. However, a normal eye is unable to focus on an object closer than about 25 cm, the near point (Fig. 27.7a). Try it! Therefore, θ is maximum at the near point.

To further increase the apparent angular size of an object, a converging lens can be placed in front of the eye with the object located at point O, just inside the focal point of the lens, as in Figure 27.7b. At this location, the lens forms a virtual, erect, and enlarged image, as shown. Clearly, the lens increases the angular size of the object. We define the **angular magnification,** m, as the ratio of the angle subtended by the object with a lens in use (angle θ in Figure 27.7b) to that subtended by the object when it is placed at the near point with no lens (angle θ_0 in Figure 27.7a):

$$m \equiv \frac{\theta}{\theta_0} \qquad \textbf{(27.2)}$$

Angular magnification

The angular magnification is a maximum when the image is at the near point of the eye, that is, when $s' = -25$ cm (see Fig. 27.7b). The object distance corresponding to this image distance can be calculated from the thin lens formula:

$$\frac{1}{s} + \frac{1}{-25 \text{ cm}} = \frac{1}{f}$$

$$\textbf{(27.3)}$$

$$s = \frac{25f}{25 + f}$$

where f is the focal length in centimeters. Let us now make the small angle approximation as follows:

$$\tan \theta_0 \approx \theta_0 \approx \frac{h}{25} \qquad \text{and} \qquad \tan \theta \approx \theta \approx \frac{h}{s} \qquad \textbf{(27.4)}$$

Thus, Equation 27.2 becomes

$$m = \frac{\theta}{\theta_0} = \frac{h/s}{h/25} = \frac{25}{s} = \frac{25}{25f/(25+f)}$$

$$m = 1 + \frac{25 \text{ cm}}{f} \qquad \textbf{(27.5)}$$

The magnification given by Equation 27.5 is the ratio of the angular size seen with the lens to the angular size seen when the object is viewed at the near point of the eye with no lens. Actually, the eye can focus on an image formed anywhere between the near point and infinity. However, the eye is more relaxed when the image is at infinity (Sec. 27.2). In order for the image formed by the magnifying lens to appear at infinity, the object has to be placed at the focal point of the lens, that is, $s = f$. In this case, Equation 27.4 becomes

$$\theta_0 \approx \frac{h}{25} \qquad \text{and} \qquad \theta \approx \frac{h}{f}$$

and the magnification is

$$m = \frac{\theta}{\theta_0} = \frac{25 \text{ cm}}{f} \qquad \textbf{(27.6)}$$

With a single lens, it is possible to obtain angular magnifications up to about 4 without serious aberrations. Magnifications up to about 20 can be achieved by using a second lens to correct for aberrations.

Example 27.4   Maximum Magnification of a Lens

What is the maximum magnification of a lens having a focal length of 10 cm, and what is the magnification of this lens when the eye is relaxed?

Solution
The maximum magnification occurs when the image formed by the lens is located at the near point of the eye. Under these circumstances. Equation 27.5 gives us the magnification as

$$m = 1 + \frac{25 \text{ cm}}{f} = 1 + \frac{25 \text{ cm}}{10 \text{ cm}} = \boxed{3.5}$$

When the eye is relaxed, the image is at infinity. In this case, we use Equation 27.6:

$$m = \frac{25}{f} = \frac{25 \text{ cm}}{10 \text{ cm}} = \boxed{2.5}$$

27.4   The Compound Microscope

A simple magnifier provides only limited assistance in inspecting the minute details of an object. Greater magnification can be achieved by combining two lenses in a device called a compound microscope, a schematic diagram of

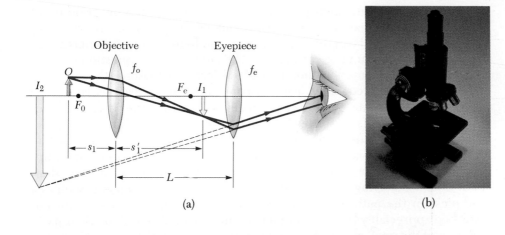

(a)   (b)

Figure 27.8
(a) Diagram of a compound microscope, which consists of an objective lens and an eyepiece, or ocular, lens. (b) An old-fashioned compound microscope. The three-objective turret allows the user to switch to several different powers of magnification. Combinations of oculars with different focal lengths and different objectives can produce a wide range of magnifications. *(Photo by Henry Leap and Jim Lehman)*

which is shown in Figure 27.8a. It consists of an objective lens with a very short focal length, f_o (where $f_o < 1$ cm), and an ocular, or eyepiece lens, having a focal length, f_e, of a few centimeters. The two lenses are separated by a distance L, where L is much greater than either f_o or f_e. The object, O, which is placed just outside the focal length of the objective, forms a real, inverted image at I_1, which is at or just inside the focal point of the eyepiece. The image at I_1 is real and much enlarged. (For clarity, the enlargement of I_1 is not shown in Figure 27.8a.) The eyepiece, which serves as a simple magnifier, uses the image at I_1 as its object, and produces an image at I_2. The image which is seen by the eye at I_2 is virtual, inverted, and very much enlarged.

The lateral magnification, M_1, of the first image is $-s_1'/s_1$. Note in Figure 27.8a that s_1' is approximately equal to L. This occurs because the object is placed close to the focal point of the objective lens, which ensures that the image formed will be far from the objective lens. Furthermore, since the object is very close to the focal point of the objective lens, $s_1 \approx f_o$. This gives a magnification for the objective of

$$M_1 = -\frac{s_1'}{s_1} \approx -\frac{L}{f_o}$$

The angular magnification of the eyepiece for an object (corresponding to the image at I_1) placed at the focal point is found from Equation 27.6 to be

$$m_e = \frac{25 \text{ cm}}{f_e}$$

The overall magnification of the compound microscope is defined as the product of the lateral and angular magnifications:

$$M = M_1 m_e = -\frac{L}{f_o}\left(\frac{25 \text{ cm}}{f_e}\right) \qquad \textbf{(27.7)}$$

The negative sign indicates that the image is inverted with respect to the object.

The microscope has extended our vision into the previously unknown depths of incredibly small objects, and the capabilities of this instrument have

Some common instruments used for navigation. *(Murray Alcosser/The IMAGE Bank)*

steadily increased with improved techniques in the precision grinding of lenses. A question that is often asked about microscopes is, "If you were extremely patient and careful, would it be possible to construct a microscope that would enable you to see an atom?" The answer to this question is no, as long as visible light is used to illuminate the object. The reason is that, in order to be seen, the object under a microscope must be at least as large as a wavelength of light. An atom is many times smaller than the wavelengths of visible light, and so its mysteries have to be probed using other techniques.

The wavelength dependence of the "seeing" ability of a wave can be illustrated by water waves set up in a bathtub in the following manner. Suppose you vibrate your hand in the water until waves having a wavelength of about 6 in. are moving along the surface. If you fix a small object, such as a toothpick, in the path of the waves, you will find that the waves are not disturbed appreciably by the toothpick but instead continue along their path, oblivious of the small object. Now suppose you fix a larger object, such as a toy sailboat, in the path of the waves. In this case, the waves are considerably "disturbed" by the object. In the first case, the toothpick is much smaller than the wavelength of the waves, and as a result the waves do not "see" the toothpick. In the second case, the toy sailboat is about the same size as the wavelength of the waves and hence the sailboat creates a disturbance. Light waves behave in this same general way. The ability of an optical microscope to view an object depends on the size of the object relative to the wavelength of the light used to observe it. Hence, it will never be possible to observe atoms or molecules with such a microscope, since their dimensions are small (≈ 0.1 nm) relative to the wavelength of the light (≈ 500 nm).

Example 27.5   Magnifications of a Microscope

A certain microscope has two possible objectives that can be used. One has a focal length of 20 mm, and the second has a focal length of 2 mm. Also available are two eyepieces of focal lengths 2.5 cm and 5 cm. If the length of the microscope is 18 cm, what magnifications are possible?

Solution

The solution consists of applying Equation 27.7 to four different combinations of lenses. For the combination of the two long focal lengths, we have

$$M = -\frac{L}{f_o}\left(\frac{25 \text{ cm}}{f_e}\right) = -\frac{18}{2}\left(\frac{25}{5}\right) = \boxed{-45}$$

The combination of the 20-mm objective and the 2.5-cm eyepiece gives

$$M = -\frac{18}{2}\left(\frac{25}{2.5}\right) = \boxed{-90}$$

The 2-mm and 5-cm combination produces

$$M = -\frac{18}{0.2}\left(\frac{25}{5}\right) = \boxed{-450}$$

and finally the two short focal lengths give

$$M = -\frac{18}{0.2}\left(\frac{25}{2.5}\right) = \boxed{-900}$$

27.5 The Telescope

There are two fundamentally different types of **telescopes,** both designed to aid in viewing distant objects, such as the planets in our Solar System. The two classifications are (1) the **refracting telescope,** which uses a combination of lenses to form an image, and (2) the **reflecting telescope,** which uses a curved mirror and a lens to form an image.

First, let us consider the refracting telescope. In this device, two lenses are arranged such that the objective forms a real, inverted image of the distant object very near the focal point of the eyepiece (Fig. 27.9a). Furthermore, the image at I_1 is formed at the focal point of the objective because the object is essentially at infinity. Hence, the two lenses are separated by a distance $f_o + f_e$, which corresponds to the length of the telescope's tube. The eyepiece finally forms, at I_2, an enlarged, inverted image of the image at I_1.

The angular magnification of the telescope is given by θ/θ_o, where θ_o is the angle subtended by the object at the objective and θ is the angle subtended by the final image. From the triangles in Figure 27.9a, and for small angles, we have

$$\theta \approx \frac{h'}{f_e} \quad \text{and} \quad \theta_o \approx \frac{h'}{f_o}$$

Hence, the angular magnification of the telescope can be expressed as

$$m = \frac{\theta}{\theta_o} = \frac{h'/f_e}{h'/f_o} = \frac{f_o}{f_e} \tag{27.8}$$

This says that the angular magnification of a telescope equals the ratio of the objective focal length to the eyepiece focal length. Here again, the magnification is the ratio of the angular size seen with the telescope to the angular size seen with the unaided eye.

In some applications, such as observing nearby objects like the Sun, Moon, or planets, magnification is important. However, stars are so far away that they

Figure 27.9
(a) Diagram of a refracting telescope, with the object at infinity. (b) Photograph of a refracting telescope. *(Photo courtesy of Henry Leap and Jim Lehman)*

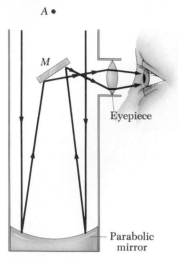

A •

M

Eyepiece

Parabolic
mirror

Figure 27.10
A reflecting telescope with a
Newtonian focus.

always appear as small points of light regardless of how much magnification is used. Large research telescopes used to study very distant objects must have a large diameter in order to gather as much light as possible. It is difficult and expensive to manufacture large lenses for refracting telescopes. Another difficulty with large lenses is that their large weight leads to sagging, which is an additional source of aberration. These problems can be partially overcome by replacing the objective lens with a reflecting, concave mirror. Figure 27.10 shows the design for a typical reflecting telescope. Incoming light rays pass down the barrel of the telescope and are reflected by a parabolic mirror at the base. These rays converge toward point A in the figure, where an image would be formed. However, before this image is formed, a small flat mirror at point M reflects the light toward an opening in the side of the tube that passes into an eyepiece. This particular design is said to have a Newtonian focus because it was Newton who developed it. Note that the light never passes through glass in the reflecting telescope (except through the small eyepiece). As a result, problems associated with chromatic aberration are virtually eliminated. Also, difficulties arising from spherical aberration are reduced because of the parabolic shape of the mirror.

The largest single-mirrored telescope in the world is the 6-m-diameter reflecting telescope on Mount Pastukhov in the Caucasus, Soviet Union. The largest single-mirrored reflecting telescope in this country is the 5-m-diameter instrument on Mount Palomar in California. In contrast, the largest refracting telescope in the world, which is located in the Yerkes Observatory in Williams Bay, Wisconsin, has a diameter of only 1 m.

Example 27.6   Magnification of a Reflecting Telescope

A reflecting telescope has an 8-in.-diameter objective mirror with a focal length of 1500 mm. What is the magnification of this telescope when an eyepiece having an 18-mm focal length is used?

Solution
The equation for finding the magnification of a reflector is the same as that for a refractor. Thus, Equation 27.8 gives

$$m = \frac{f_o}{f_e} = \frac{1500 \text{ mm}}{18 \text{ mm}} = \boxed{83}$$

27.6   Resolution of Single-Slit and Circular Apertures

The ability of optical systems such as microscopes and telescopes to distinguish between closely spaced objects is limited because of the wave nature of light. To understand this difficulty, consider Figure 27.11, which shows two light sources far from a narrow slit of width a. The sources can be considered as two point sources, S_1 and S_2, that are *not* coherent. For example, they could be two distant stars. If no diffraction occurred, one would observe two distinct bright spots (or images) on the screen at the right in the figure. However, because of diffraction, each source is imaged as a bright central region flanked

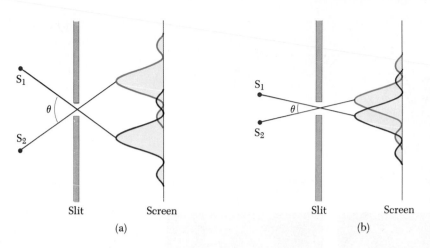

Slit   Screen

(a)

Slit   Screen

(b)

Figure 27.11
Two point sources at some distance from a small aperture each produce a diffraction pattern. (a) The angle subtended by the sources at the aperture is large enough so that the diffraction patterns are distinguishable. (b) The angle subtended by the sources is so small that the diffraction patterns are not distinguishable. (Note that the angles are greatly exaggerated.)

by weaker bright and dark rings. What is observed on the screen is the sum of two diffraction patterns, one from S_1, and the other from S_2.

If the two sources are separated such that their central maxima do not overlap, as in Figure 27.11a, their images can be distinguished and are said to be *resolved*. If the sources are close together, however, as in Figure 27.11b, the two central maxima may overlap and the images are *not resolved*. To decide when two images are resolved, the following condition applied to their diffraction patterns is often used:

> When the central maximum of one image falls on the first minimum of another image, the images are said to be just resolved. This limiting condition of resolution is known as **Rayleigh's criterion**.

Figure 27.12 shows the diffraction patterns for three situations. When the objects are far apart, their images are well resolved (Fig. 27.12a). The images are just resolved when their angular separation satisfies Rayleigh's criterion (Fig. 27.12b). Finally, the images are not resolved in Figure 27.12c.

From Rayleigh's criterion, we can determine the minimum angular separation, θ_m, subtended by the source at the slit such that their images will be just resolved. In Chapter 26, we found that the first minimum in a single-slit diffraction pattern occurs at the angle that satisfies the relationship

$$\sin \theta = \frac{\lambda}{a}$$

where a is the width of the slit. According to Rayleigh's criterion, this expression gives the smallest angular separation for which the two images will be resolved. Because $\lambda \ll a$ in most situations, $\sin \theta$ is small and we can use the approximation $\sin \theta \approx \theta$. Therefore, the limiting angle of resolution for a slit of width a is

$$\theta_m \approx \frac{\lambda}{a} \qquad \textbf{(27.9)}$$

where θ_m is expressed in radians. Hence, the angle subtended by the two sources at the slit must be *greater* than λ/a if the images are to be resolved.

Figure 27.12
The diffraction patterns of two point sources (solid curves) and the resultant pattern
(dashed curve) for various angular separations of the sources. (a) The sources are far
apart, and the patterns are well resolved. (b) The sources are closer together, and
their patterns are just resolved. (c) The sources are so close together that their patterns are not resolved. *(From M. Cagnet, M. Francon, and J. C. Thierr,* Atlas of Optical Phenomena, *Berlin, Springer-Verlag, 1962, plate 16).*

Many optical systems use circular apertures rather than slits. The diffraction pattern of a circular aperture, illustrated in Figure 27.13, consists of a
central circular bright disk surrounded by progressively fainter rings. Analysis
shows that the limiting angle of resolution of the circular aperture is

**Resolution for a circular
aperture**

$$\theta_m = 1.22 \, \frac{\lambda}{D} \qquad (27.10)$$

where D is the diameter of the aperture. Note that Equation 27.10 is similar to
Equation 27.9 except for the factor of 1.22, which arises from a complex
mathematical analysis of diffraction from the circular aperture.

Example 27.7   Limiting Resolution of a Microscope

Sodium light of wavelength 589 nm is used to view an object under a microscope. If
the aperture of the objective is 0.9 cm, (a) find the limiting angle of resolution.
(b) Using visible light of any wavelength you desire, what is the maximum limit of
resolution for this microscope? (c) Suppose water of index of refraction 1.33 fills
the space between the object and the objective. What effect would this have on the
resolving power of the microscope?

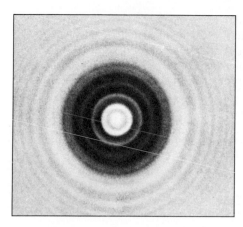

Figure 27.13
The Fresnel diffraction pattern of a circular aperture consists of a central bright disk surrounded by concentric bright and dark rings. *(From M. Cagnet, M. Francon, and J. C. Thierr,* Atlas of Optical Phenomena, *Berlin, Springer-Verlag, 1962, plate 34).*

Solution

(a) From Equation 27.10, we find the limiting angle of resolution to be

$$\theta_m = 1.22\left(\frac{589 \times 10^{-9} \text{ m}}{0.9 \times 10^{-2} \text{ m}}\right) = \boxed{8 \times 10^{-5} \text{ rad}}$$

This means that any two points on the object subtending an angle at the objective less than 8×10^{-5} rad cannot be distinguished in the image.

(b) To obtain the maximum angle of resolution, we have to use the shortest wavelength available in the visible spectrum. Violet light of wavelength 400 nm gives us a limiting angle of resolution of

$$\theta_m = 1.22\left(\frac{400 \times 10^{-9} \text{ m}}{0.9 \times 10^{-2} \text{ m}}\right) = \boxed{5.4 \times 10^{-5} \text{ rad}}$$

(c) In this case, the wavelength of the sodium light in the water is found by $\lambda_w = \lambda_a/n$ (Chap. 24). Thus, we have

$$\lambda_w = \frac{\lambda_a}{n} = \frac{589 \text{ nm}}{1.33} = 443 \text{ nm}$$

The limiting angle of resolution at this wavelength is

$$\theta_m = 1.22\left(\frac{443 \times 10^{-9} \text{ m}}{0.9 \times 10^{-2} \text{ m}}\right) = \boxed{6 \times 10^{-5} \text{ rad}}$$

Example 27.8   Resolution of a Telescope

The Hale telescope at Mount Palomar has a diameter of 200 in. What is its limiting angle of resolution at a wavelength of 600 nm?

Solution
Because $D = 200$ in. $= 5.08$ m and $\lambda = 6 \times 10^{-7}$ m, Equation 27.10 gives

$$\theta_m = 1.22\frac{\lambda}{D} = 1.22\left(\frac{6 \times 10^{-7} \text{ m}}{5.08 \text{ m}}\right) = 1.44 \times 10^{-7} \text{ rad} = \boxed{0.03 \text{ s of arc}}$$

Therefore, any two stars that subtend an angle greater than or equal to this value will be resolved (assuming ideal atmospheric conditions).

A large reflecting telescope used in many observatories. *(Guy Marche/FPG International)*

Photograph of the Hubble space telescope. The telescope, whose mirror diameter is 2.4 m, was supposed to have a resolution limited only by diffraction. *(Hughes Optical Systems)*

It is interesting to compare this value with the resolution of a large radiotelescope, such as the system at Arecibo, Puerto Rico, which has a diameter of 1000 ft (305 m). This telescope detects radio waves at a wavelength of 0.75 m. The corresponding minimum angle of resolution is calculated to be 3×10^{-3} rad (10 min 19 s of arc), which is more than 10,000 times larger than the calculated minimum angle for the Hale telescope.

The Hale telescope can never reach its diffraction limit. Instead, its limiting angle of resolution is always set by atmospheric blurring. This seeing limit is usually about 1 s of arc and is never smaller than about 0.1 s of arc. (This is one of the reasons for the current interest in the Hubble space telescope.)

Example 27.9   Compare the Two Telescopes

Two telescopes have the following properties:

Telescope	Diameter of Objective (in.)	Focal Length of Objective (mm)	Focal Length of Eyepiece (mm)
A	6	1000	6
B	8	1250	25

(a) Which has the better resolving power? (b) Which has the greater light-gathering ability? (c) Which will produce the larger magnification?

Solution

(a) The telescope with the larger objective has the greatest ability to discriminate between nearby objects. Thus, we should choose telescope B.

(b) The telescope with the larger objective is able to collect more light. Hence, B is again the choice.

(c) The magnification of telescope A is

$$m = \frac{f_o}{f_e} = \frac{1000 \text{ mm}}{6 \text{ mm}} = 167$$

and for telescope B, we have

$$m = \frac{1250 \text{ mm}}{25 \text{ mm}} = 50$$

Thus, select telescope A if magnification is the most important consideration.

27.7   The Michelson Interferometer

The camera and telescope are examples of optical instruments that are common in our daily activities. In contrast, the Michelson interferometer is not familiar to most people. It is, however, an optical instrument that has great scientific importance. The interferometer, invented by the American physicist A. A. Michelson (1852–1931), is an ingenious device that splits a light beam into two parts and then recombines them to form an interference pattern. The device is used for obtaining accurate length measurements.

Figure 27.14
Diagram of the Michelson interferometer. A single beam is split into two rays by the half-silvered mirror M. The path difference between the two rays is varied with the adjustable mirror M_1.

A schematic diagram of the interferometer is shown in Figure 27.14. A beam of light provided by a monochromatic source is split into two rays by a partially silvered mirror M inclined at an angle of 45° relative to the incident light beam. One ray is reflected vertically upward to mirror M_1, and the second ray is transmitted horizontally through mirror M to mirror M_2. Hence, the two rays travel separate paths, L_1 and L_2. After reflecting from mirrors M_1 and M_2, the two rays eventually recombine to produce an interference pattern, which can be viewed through a telescope. The glass plate P, equal in thickness to mirror M, is placed in the path of the horizontal ray in order to insure that the two rays travel the same distance through glass.

The interference pattern for the two rays is determined by the difference in their path lengths. When the two rays are viewed as shown, the image of M_2 is at M_2' parallel to M_1. Hence, the space between M_2' and M_1 forms the equivalent of a parallel air film. The effective thickness of the air film is varied by moving mirror M_1 in the direction indicated by the arrows in Figure 27.14 with a finely threaded screw. If one of the mirrors is tipped slightly with respect to the other, the thin film between the two is wedge shaped, and an interference pattern consisting of parallel fringes is set up, as described in Chapter 26, Example 26.4. Now suppose we focus on one of the dark lines with the crosshairs of a telescope. As the mirror M_1 is moved to make the path length L_1 longer, the thickness of the wedge increases. When the thickness increases by $\lambda/4$, the destructive interference which initially produced the dark fringe has changed to constructive interference, and we now observe a bright fringe at the location of the crosshairs. The term *fringe shift* is used to describe the change in a fringe from dark to light or light to dark. Thus, successive light and dark fringes are formed each time M_1 is moved a distance of $\lambda/4$. The wavelength of light can be measured by counting the number of fringe shifts for a measured displacement of M_1. Conversely, if the wavelength is accurately known (as with a laser beam), mirror displacements can be measured to within a fraction of the wavelength. Because the interferometer can measure displacements precisely, it is often used to make highly accurate measurements of the dimensions of mechanical components.

If the mirrors are perfectly aligned, rather than tipped with respect to one another, the path difference is slightly different for different angles of view. This results in an interference pattern that resembles Newton's rings. The pattern can be used in a fashion similar to that described for tipped mirrors. To do so, one concentrates on the center spot in the interference pattern. For example, suppose the spot is initially dark, indicating that destructive interference is occurring. If M_1 is now moved by a distance of $\lambda/4$, this central spot changes to a light region, corresponding to a fringe shift.

27.8   The Diffraction Grating

The diffraction grating, a very useful device for analyzing light sources, consists of a large number of equally spaced parallel slits. A grating can be made by scratching parallel lines on a glass plate with a precision machining technique. The spaces between the scratches are transparent to the light and hence act as separate slits. A typical grating contains several thousand lines per centimeter. For example, a grating ruled with 5000 lines/cm has a slit spacing d equal to the inverse of this number, hence $d = (1/5000)$ cm $= 2 \times 10^{-4}$ cm.

A schematic diagram of a section of a plane diffraction grating is illustrated in Figure 27.15. A plane wave is incident from the left, normal to the plane of the grating. A converging lens can be used to bring the rays together at point P. The intensity of the pattern on the screen is the result of the combined effects

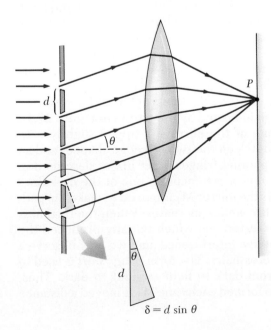

Figure 27.15
Side view of a diffraction grating. The slit separation is d, and the path difference between adjacent slits is $d \sin \theta$.

of interference and diffraction. Each slit produces diffraction, and the diffracted beams in turn interfere with each other to produce the pattern. Moreover, each slit acts as a source of waves, where all waves start at the slits in phase. However, for some arbitrary direction θ measured from the horizontal, the waves must travel *different* path lengths before reaching a particular point P on the screen. From Figure 27.15, note that the path difference between waves from any two adjacent slits is equal to $d \sin \theta$. If this path difference equals one wavelength or some integral multiple of a wavelength, waves from all slits will be in phase at P and a bright line will be observed. Therefore, the condition for **maxima** in the interference pattern at the angle θ is

$$d \sin \theta = m\lambda \qquad m = 0, 1, 2, \ldots \qquad \textbf{(27.11)}$$

This expression can be used to calculate the wavelength from a knowledge of the grating space and the angle of deviation, θ. The integer m represents the **order number** of the diffraction pattern. If the incident radiation contains several wavelengths, each wavelength will deviate through a specific angle which can be found from Equation 27.11. All wavelengths are focused at $\theta = 0$, corresponding to $m = 0$. This is called the *zeroth-order maximum*. The *first-order maximum*, corresponding to $m = 1$, is observed at an angle that satisfies the relationship $\sin \theta = \lambda/d$; the *second-order maximum*, corresponding to $m = 2$, is observed at a larger angle θ, and so on. A sketch of the intensity distribution for the various orders produced by a diffraction grating is shown in Figure 27.16. Let us compare the diffraction pattern of the grating with the pattern produced by two slits (see Young's double-slit experiment discussed in Chapter 26). The two patterns differ as follows: (1) The bright lines of the diffraction grating are sharper (narrower), and the dark areas are broader. (2) The lines of the two-slit pattern are more intense because the light has to travel through two slits rather than many slits.

A simple arrangement that can be used to measure the various orders of a diffraction pattern is shown in Figure 27.17a. This is a form of diffraction grating spectrometer. The light to be analyzed passes through a slit and is formed into a parallel beam by a lens. The light then strikes the grating at a 90° angle. The diffracted light leaves the grating at angles that satisfy Equation 27.11. A telescope is used to view the image of the slit. The wavelength can be determined by measuring the angles at which the images of the slit appear for the various orders. Figure 27.17b shows a diffraction grating dispersing light into its various spectral components.

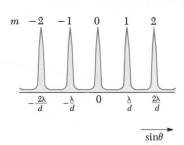

Figure 27.16
Intensity versus $\sin \theta$ for the diffraction grating. The zeroth-, first-, and second-order principal maxima are shown.

(a)

(b)

Figure 27.17
(a) Diagram of a diffraction grating spectrometer. The collimated beam incident on the grating is diffracted into the various orders at the angles θ that satisfy the equation $d \sin \theta = m\lambda$, where $m = 0, 1, 2, \ldots$. (b) White light is dispersed into its various spectral components using a diffraction grating. *(Courtesy of PASCO Scientific)*

Example 27.10   The Orders of a Diffraction Grating

Monochromatic light from a helium-neon laser ($\lambda = 632.8$ nm) is incident normally on a diffraction grating containing 6000 lines/cm. Find the angles at which one would observe the first-order maximum, the second-order maximum, and so forth.

Solution

First we must calculate the slit separation, which is equal to the inverse of the number of lines per centimeter:

$$d = \frac{1}{6000} \text{ cm} = 1.667 \times 10^{-4} \text{ cm} = 1667 \text{ nm}$$

For the first-order maximum ($m = 1$), we get

$$\sin \theta_1 = \frac{\lambda}{d} = \frac{632.8 \text{ nm}}{1667 \text{ nm}} = 0.3796$$

$$\theta_1 = \boxed{22.31°}$$

For $m = 2$ we find that

$$\sin \theta_2 = \frac{2\lambda}{d} = \frac{2(632.8 \text{ nm})}{1667 \text{ nm}} = 0.7592$$

$$\theta_2 = \boxed{49.39°}$$

However, for $m = 3$ we find that $\sin \theta_3 = 1.139$. Since $\sin \theta$ cannot exceed unity, this does not represent a realistic solution. Hence, only zeroth-, first-, and second-order maxima will be observed for this situation.

Resolving Power of the Diffraction Grating

The diffraction grating is most useful for making accurate wavelength measurements. Like the prism, the diffraction grating can be used to disperse a spectrum into its components. Of the two devices, the grating is more precise if one wants to distinguish between two closely spaced wavelengths. We say that the grating spectrometer has a higher resolution than the prism spectrometer. If λ_1 and λ_2 are two nearly equal wavelengths between which the spectrometer can just barely distinguish, the **resolving power**, R, of the grating is defined as

$$R \equiv \frac{\lambda}{\lambda_2 - \lambda_1} = \frac{\lambda}{\Delta \lambda} \qquad \textbf{(27.12)}$$

where $\lambda \approx \lambda_1 \approx \lambda_2$ and $\Delta\lambda = \lambda_2 - \lambda_1$. Thus, we see that a grating with a high resolving power can distinguish small differences in wavelength. Furthermore, if N lines of the grating are illuminated, it can be shown that the resolving power in the mth-order diffraction equals the product Nm:

Resolving power of a grating

$$R = Nm \qquad \textbf{(27.13)}$$

Thus, the resolving power increases with increasing order number. Furthermore, R is large for a grating with a large number of illuminated slits. Note that for $m = 0$, $R = 0$, which signifies that *all wavelengths are indistinguishable* for the zeroth-order maximum (all wavelengths fall at the same point on the screen). However, consider the second-order diffraction pattern of a grating that has 5000 rulings illuminated by the light source. The resolving power of such a grating in second order is $R = 5000 \times 2 = 10,000$. Therefore, the *minimum* wavelength separation between two spectral lines that can be just resolved, assuming a mean wavelength of 600 nm, is calculated from Equation 27.12 to be $\Delta\lambda = \lambda/R = 6 \times 10^{-2}$ nm. For the third-order principal maximum $R = 15,000$ and $\Delta\lambda = 4 \times 10^{-2}$ nm, and so on.

Example 27.11 Resolving the Sodium Spectral Lines

Two strong lines in the spectrum of sodium have wavelengths of 589.00 nm and 589.59 nm. (a) What must the resolving power of a grating be in order to distinguish these wavelengths?

Solution
From Equation 27.12, we find

$$R = \frac{\lambda}{\Delta\lambda} = \frac{589 \text{ nm}}{589.59 \text{ nm} - 589.00 \text{ nm}} = \frac{589}{0.59} = \boxed{998}$$

(b) In order to resolve these lines in the second-order spectrum, how many lines of the grating must be illuminated?

Solution
From Equation 27.13 and the results to (a), we find that

$$N = \frac{R}{m} = \frac{998}{2} = \boxed{499 \text{ lines}}$$

Summary

The light-concentrating power of a lens of focal length f and diameter D is determined by the *f*-number, defined as

$$f\text{-number} \equiv \frac{f}{D} \qquad (27.1)$$

The smaller the *f*-number of a lens, the greater the brightness of the image formed.

Hyperopia (farsightedness) is a defect of the eye that results either when the eyeball is too short or when the ciliary muscle is unable to change the shape of the lens enough to form a properly focused image. **Myopia** (nearsightedness) is caused either when the eye is longer than normal or when the maximum focal length of the lens is insufficient to produce a clearly focused image on the retina.

The **power** of a lens in **diopters** equals the inverse of the focal length in meters.

The **angular magnification of a lens** is defined as

$$m \equiv \frac{\theta}{\theta_0} \qquad \text{(27.2)}$$

where θ is the angle subtended by an object at the eye with a lens in use and θ_0 is the angle subtended by the object when it is placed at the near point of the eye and no lens is used. The **maximum magnification of a lens** is

$$m = 1 + \frac{25 \text{ cm}}{f} \qquad \text{(27.5)}$$

When the eye is relaxed, the magnification is

$$m = \frac{25 \text{ cm}}{f} \qquad \text{(27.6)}$$

The overall **magnification of a compound microscope** of length L is equal to the product of the magnification produced by the objective of focal length f_o and the magnification produced by the eyepiece of focal length f_e:

$$M = -\frac{L}{f_o}\left(\frac{25 \text{ cm}}{f_e}\right) \qquad \text{(27.7)}$$

The **magnification of a telescope** is given by

$$m = \frac{f_o}{f_e} \qquad \text{(27.8)}$$

where f_o is the focal length of the objective and f_e is the focal length of the eyepiece.

Two images are said to be **resolved** when the central maximum of the diffraction pattern for one image falls on the first minimum of the other image. This limiting condition of resolution is known as **Rayleigh's criterion.** The limiting angle of resolution for a **slit** of width a is

$$\theta_m \approx \frac{\lambda}{a} \qquad \text{(27.9)}$$

The limiting angle of resolution of a **circular aperture** is

$$\theta_m = 1.22 \frac{\lambda}{D} \qquad \text{(27.10)}$$

where D is the diameter of the aperture.

A **diffraction grating** consists of a large number of equally spaced, identical slits. The condition for **maximum intensity** in the interference pattern of a diffraction grating is

$$d \sin \theta = m\lambda \qquad m = 0, 1, 2, \ldots \qquad \textbf{(27.11)}$$

where d is the spacing between adjacent slits and m is the order number of the diffraction pattern. The **resolving power** of a diffraction grating in the mth order is

$$R = Nm \qquad \textbf{(27.13)}$$

where N is the number of rulings on the grating.

Additional Reading

J. Bahcall and L. Spitzer, Jr., "The Space Telescope," *Sci. American*, July 1982, p. 40.

E. Bandana, "The Mystery of Myopia," *The Sciences*, Nov./Dec. 1985, p. 46.

D. Marr, *Vision*, W. H. Freeman, 1982.

C. R. Michael, "Retinal Processing of Visual Images," *Sci. American*, May 1969, p. 105.

U. Neisser, "The Processes of Vision," *Sci. American*, September 1968, p. 204.

W. H. Price, "Photographic Lens," *Sci. American*, August 1976, p. 72.

M. Ruiz, "Camera Optics," *The Physics Teacher*, September 1982, p. 372.

R. S. Shankland, "Michelson and His Interferometer," *Physics Today*, April 1976, p. 72.

G. Wald, "Eye and Camera," *Sci. American*, August 1950, p. 32.

T. D. Walker, *Light and Its Uses*, W. H. Freeman, 1980.

Questions

1. A photoflash unit makes a very bright light for a very short time. Discuss differences in the exposure of film with and without a flash unit. Consider the f-stop and shutter speed in your deliberation. Which is important: the intensity of the light, or the product of the intensity and the time of the flash?

2. Estimate the shutter speed and f-number necessary to photograph a center driving for the goal in a well-lighted basketball gymnasium. Repeat for a racecar during the final lap on a cloudy day. Why will the latter probably be blurred anyway?

3. A pinhole camera can be constructed by punching a small hole in one side of a cardboard box. If the opposite side is cut out and replaced with white paper, the image formed by the box can be examined. When the box is pointed toward an open window with the pinhole facing the incoming light from a distant object, you should be able to see an image on the white paper. No lens is involved here. Can you explain why the image is formed? Why is the image sharp?

4. Compare and contrast the eye and a camera. What parts of the camera correspond to the iris, the retina, and the cornea of the eye?

5. A classic science-fiction story, "The Invisible Man," tells of a person who becomes invisible by changing the index of refraction of his body to that of air. This story has been criticized by students who know how the eye works by claiming that the invisible man would be unable to see. On the basis of your knowledge of the eye, could he see or not?

6. In a darkened room we can see reflected images of a luminous object in a person's eye (Fig. 27.18). Upon accommodation, image b shifts and changes size, and the other two images do not. Discuss the origin of the three images.

Figure 27.18 Question 6

7. Suppose we look through a person's eyeglasses, with the lens a few centimeters from our eyes. If we move the glasses from side to side, the image appears to move. Which way does the image move if the person has (a) myopia and (b) hyperopia? Discuss.

8. The optic nerve and the brain invert the image formed on the retina. Why do we not see everything upside down?

9. If you want to use a converging lens to set fire to a piece of paper, why should the light source be farther away than the focal point of the lens?

10. If you want to examine the fine detail in an object, could you hold the object farther away than the focal length of a lens used as a simple magnifier? Why or why not?

11. Explain why you could not use a diverging lens as a simple magnifier.

12. In what respect are the optical system of a microscope and a telescope (a) similar and (b) different?

13. How should a telescope or microscope be altered to photograph the image? Discuss.

14. What difficulty would astronauts encounter in finding their landing site on the Moon if astronomers were not aware of the properties of the telescope?

15. Large telescopes are usually reflecting rather than refracting. List several reasons why this is true.

16. Explain why it is theoretically impossible to see an object as small as an atom regardless of how good a light microscope is used.

17. Assuming that the headlights of a car are point sources, estimate the maximum distance from an observer to the car at which the headlights are distinguishable from each other.

18. The diffraction grating effect is easily observed with everyday equipment. For example, a long-playing record can be held so that light is reflected from it at a small glancing angle. When the record is held this way, various colors in the reflected light can be seen. Explain how this works.

19. A laser beam is incident at a shallow angle on a ma-chinist's ruler that has a finely calibrated scale. The rulings on the scale give rise to a diffraction pattern on a screen. Discuss how you can use this technique to obtain a measure of the wavelength of the laser light.

20. Describe the zeroth-order spectrum of a grating spectroscope. Include the angular location relative to the source and the color(s) of the image in the event that the source emits white light.

21. Short of looking inside a spectroscope, how would we know whether the instrument is a grating or prism type?

22. Can a diffraction grating be used for infrared, ultraviolet, or x-rays? Could a prism be used to examine the spectrum of these parts of the electromagnetic spectrum?

23. Discuss the physical principles involved in each of the following instruments: (a) an endoscope—a device for examining the inside of a person's stomach, (b) a telephoto lens, (c) a spectrum analyzer, (d) a pair of binoculars.

24. The diffraction grating effect is easily observed with everyday equipment. For example, a compact disk can be held so that light is reflected from it at a glancing angle as in Figure 27.19. When held this way, various colors in the reflected light can be seen. Explain how this works.

Figure 27.19   Question 24. Compact disks act as diffraction gratings when observed under white light. *(© Bobbie Kingsley/Photo Researchers, Inc.)*

Problems

Section 27.1   The Camera

1. The focal length of the lens in a simple camera is 10 cm, and the size of the image formed on the film is to be 35 mm high. How far would a person 2 m tall have to stand from the camera so that the person's image will fit on the film?

2. A camera used by a professional photographer to shoot portraits has a focal length of 25 cm. The photographer takes a portrait of a person 1.5 m in front of the camera. Where will the image be formed, and what is the magnification?

3. A photographic image of a building is 0.092 m high. The image was made with a lens with a focal length of 52.0 mm. If the lens was 100 m from the building when the photograph was made, determine the height of the building.

4. The f-numbers of a simple camera are $f/4$, $f/8$, and $f/16$. What are the possible aperture openings for this camera if the focal length is 12 cm?

5. A certain lens forms a real image 15 cm to its right. The object is 20 cm to the left of the lens. If the lens has a diameter of 2 cm, find its f-number.

6. If a certain type of film requires an exposure time of 0.01 s with an *f*-stop of 11.0, and another type of film requires twice the light energy for the same exposure, what *f*-stop is needed with the same exposure time for the second type of film?

7. A camera is found to give proper film exposure when it is set at *f*/16 and the shutter is open for 1/32 s. Determine the correct exposure time if a setting of *f*/8 is used. Assume the lighting conditions are the same.

8. A camera is being used with the correct exposure at *f*/4 and a shutter speed of 1/32 s. In order to "stop" a fast-moving subject, the shutter speed is changed to 1/256 s. Find the new *f*-number setting that should be used to maintain satisfactory exposure, assuming no change in lighting conditions.

Section 27.2   The Eye

9. The near point of an eye is 100 cm. (a) What focal length lens should be used so that the eye can see clearly an object 25 cm in front of it? (b) What is the power of the lens?

10. A certain farsighted person uses lenses with a power of 2.9 diopters. What is her near point?

11. A particular nearsighted person is unable to see objects clearly when they are beyond 100 cm (the far point of the eye). What power lens should be used to correct the eye?

12. A person has far points 8.44 cm from the right eye and 12.2 cm from the left eye. Write a prescription for the power of the corrective lenses.

13. An artificial lens is implanted in a person's eye in place of a diseased lens. The distance between the artificial lens and the retina is 2.80 cm. In the absence of the lens, the image of a distant object falls 2.53 cm behind the retina. The lens is designed to put the image of the distant object on the retina. What is the power of the implanted lens? (*Hint*: Consider the image formed by the eye as a virtual object.)

14. An individual is nearsighted such that his near point is 13 cm and his far point is 20 cm. (a) What lens power is needed to correct his nearsightedness? (b) When the lenses are in use, what is the person's near point?

15. A person required a lens with a power of −3 diopters in order to see distant objects clearly. What is the far point of this person's eye?

16. A person is to be fitted with bifocals. She can see clearly when the object is between 30 cm and 1.5 m from the eye. (a) The upper portion of the bifocals should be designed to enable her to see distant objects clearly. What power should the upper portion have? (b) The lower portion of the bifocals should enable her to see objects comfortably at 25 cm. What power should the lower portion have (Fig. 27.20)?

Figure 27.20   Problems 16 and 17

17. An older person has far points of 1.0 m and near points of 0.667 m in both eyes. Bifocal lenses (Fig. 27.20) are prescribed. Determine the powers of the upper and lower portions of the bifocal lens.

18. A person wearing contact lenses finds that the correction is imperfect. With the lenses in place, the near points are 50.5 cm from the eyes, whereas they are 84.0 cm away without the lenses. Determine the effective focal length of the lenses when they are in place on the eyes.

Section 27.3   The Simple Magnifier

19. A leaf of length *h* is located 71.0 cm in front of a converging lens with a focal length of 39.0 cm. An observer views the image of the leaf from a position 1.26 m behind the lens as shown in Figure 27.21. (a) What is the magnification (ratio of image size to object size) produced by the lens? (b) What angular magnification does the observer obtain by viewing the image of the leaf rather than viewing the leaf directly?

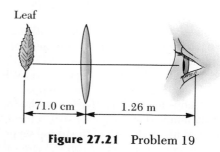

Figure 27.21   Problem 19

20. A lens with focal length 5 cm is used as a magnifying glass. (a) To obtain maximum magnification, where should the object be placed? (b) What is the magnification?

21. A biology student uses a simple magnifier to examine the structural features of the wing of an insect. The lens is mounted in a frame 3.5 cm above the work surface, and the image is formed 25 cm from the eye. What is the focal length of the lens?

22. A stamp collector uses a 7.5-cm focal length lens as a simple magnifier. The virtual image produced is formed at the normal near point (25 cm). (a) How far from the lens should the stamp be placed, and (b) what is the expected magnification?

23. (a) What is the angular magnification of a lens having a focal length of 25 cm? (b) What is the magnification of this lens when the eye is relaxed?

24. A boy scout uses a certain lens to set a fire by using the Sun as an object and placing some kindling 5 cm from the lens. The boy scout has a near point of 15 cm. When used as a simple magnifier, (a) what is the maximum magnification that can be achieved, and (b) what is the magnification when the eye is relaxed? (*Caution:* The equations derived in the text for a simple magnifier assume a "normal" eye.)

Section 27.4   The Compound Microscope
Section 27.5   The Telescope

25. The desired overall magnification of a compound microscope is 140. The objective alone produces a lateral magnification of 12. Determine the required focal length of the eyepiece. Assume the final image will be 25 cm from the eye.

26. The length of microscope tube is 15 cm. The focal length of the objective lens is 1 cm, and the focal length of the eyepiece is 2.5 cm. What is the magnification of the microscope, assuming that it is adjusted so the eye is relaxed?

27. A microscope is to have a magnification of 800. If the tube length is 15 cm and the focal length of the objective is 0.5 cm, what focal length eyepiece should be selected?

28. The objective lens in a microscope with a 20-cm-long tube has a magnification of 50 and the eyepiece has a magnification of 20. What is the focal length (a) of the objective and (b) of the eyepiece? (c) What is the overall magnification of the microscope?

29. A microscope has an objective lens with a focal length of 16.22 mm and an eyepiece with a focal length of 9.5 mm. With the length of the barrel set at 29.0 cm, the diameter of a red blood cell's image subtends an angle of 1.43 mrad with the eye. If the final image distance is 29.0 cm from the eyepiece, what is the actual diameter of the red blood cell?

30. A certain telescope has a 5-inch diameter and an objective lens of focal length 1250 mm. (a) What is the *f*-number of this lens? (b) What is the magnification of this telescope when used with a 25-mm eyepiece?

31. What is the magnification of a telescope that uses a 2-diopter objective and a 30-diopter eyepiece?

32. An astronomical telescope has an objective with a focal length of 75 cm and an eyepiece with a focal length of 4 cm. What is the magnifying power of this instrument?

33. A telescope adjusted for very distant objects has an objective with a focal length of 100 cm and an eyepiece with a focal length of 10 cm. A bee buzzes in front of the telescope at a distance of 150 cm from the objective. (a) Use the thin lens equation twice (once for each lens) to find the location of the final image of the bee. (b) On the basis of your answer to (a), could you see the bee through the telescope? Why or why not?

34. The objective lens of a refracting telescope has a focal length of 1.75 m. The eyepiece has a focal length of −9.73 cm. How much larger does a distant tree appear to be when viewed through the telescope?

35. A certain telescope has an objective of focal length 1500 cm. If the Moon is used as an object, a 1-cm length of the image formed by the objective corresponds to what distance in miles on the Moon? Assume the Earth-Moon distance is 3.8×10^8 m.

36. A person decides to use an old pair of eyeglasses to make some optical instruments. He knows that the near point in his left eye is 50 cm and that the near point in his right eye is 100 cm. (a) What is the maximum magnification of the telescope he produces? (b) If he places the lens 10 cm apart, what is the maximum magnification of the microscope he can produce? (Go back to basics and use the thin-lens equation to solve Part (b).)

Section 27.6   Resolution of Single-Slit and Circular Apertures

37. If the distance from the Earth to the Moon is 250,000 mi, what diameter telescope objective would be required to resolve a Moon crater 300 m in diameter? Assume a wavelength of 500 nm.

38. Two telescopes have the following characteristics. Telescope A has a diameter of 5 in. and a focal length for the objective of 1200 mm and is used with an 18-mm eyepiece. Telescope B has a diameter of 3 in. and a focal length for the objective of 1000 mm, and is used with a 25-mm eyepiece. (a) What is the limiting angle of resolution for each telescope when used with 600-nm light? (b) What is the magnification of each telescope?

39. What is the minimum distance between two points that will permit them to be resolved at 1 km (a) using

a terrestrial telescope with a 6.5-cm diameter objective (assume $\lambda = 550$ nm) and (b) using the unaided eye (assume a pupil diameter of 2.5 mm)?

40. A positive lens with a diameter of 30.0 cm forms an image of a satellite passing overhead. The satellite has two green lights (wavelength 500 nm) spaced 1.0 m apart. If the lights can just be resolved according to the Rayleigh criterion, what is the altitude of the satellite?

41. A spy satellite circles the Earth at an altitude of 200 km and carries out surveillance with a special high-resolution telescopic camera having a lens diameter of 35 cm. If the angular resolution of this camera is limited by diffraction, estimate the separation of two small objects on the Earth's surface that are just resolved in yellow-green light ($\lambda = 5500$ Å).

42. Two stars are 8 lightyears away and can just be resolved by a 20-in. telescope that is equipped with a filter that only allows light of wavelength 500 nm to pass. What is the distance between the two stars?

43. A light source that emits light of wavelength 580 nm is lowered into water ($n = 1.33$). If this source is viewed through a telescope that is in the water and has a 5-cm diameter objective, find the minimum angle of resolution.

44. Two radio telescopes are used in a special technique called long-baseline interferometry, which causes the two to act as a single telescope having an effective diameter of 1000 mi. At what distance from this telescope would two objects 250,000 mi apart be at the limit of resolution? Assume that the wavelength of the radio waves from these objects is about 1 m.

45. (a) Calculate the limiting angle of resolution for the eye, assuming a pupil diameter of 2 mm, a wavelength of 500 nm in air, and an index of refraction for the eye equal to 1.33. (b) What is the maximum distance from the eye at which two points separated by 1 cm could be resolved?

46. Two motorcycles separated laterally by 2 m are approaching an observer who is holding an infrared detector sensitive to light of wavelength 885 nm. What aperture diameter is required in the detector if the two headlights are to be resolved at a distance of 10 km?

47. The photosensitive screen of a television camera is 4 cm wide and 3 cm high. It contains 2.0×10^5 light-sensitive elements (pixels) arranged in a square lattice of 387 rows and 516 columns. The camera lens has a focal length of 15 cm. (a) What is the spacing between centers of adjacent pixels on the screen? (b) What is the minimum diameter of the camera lens if the resolution of the camera is to be limited by the pixel spacing on the screen (assume a 500-nm wavelength)? (c) If the leaves of a tree have a typical linear dimension of the order of 9.0 cm, what is the maxi-

mum distance from the camera at which individual leaves of the tree can be resolved?

48. The separate stars of a double star are just resolved by a lens with a diameter of 10.0 cm and a focal length of 1.4 m. If the image is viewed from a distance of 30.0 cm, what is the angle subtended by the lines of sight of the two stars in the image? [Assume a 500-nm wavelength.]

Section 27.7 The Michelson Interferometer

49. A Michelson interferometer is used to measure an unknown wavelength. The mirror in one arm of the instrument is moved 0.12 mm as 481 dark fringes are counted. Determine the wavelength of the light used.

50. An interferometer is used to measure the length of an insect. The wavelength of the light used is 650 nm. As one arm of the interferometer is moved from one end of the insect to the other, 155 dark fringes are counted. How long is the insect?

51. Light of wavelength 550 nm is used to calibrate a Michelson interferometer. By use of a micrometer screw, the platform on which one mirror is mounted is moved 0.18 mm. How many dark fringe shifts are counted?

52. The Michelson interferometer can be used to measure the index of refraction of a gas by placing an evacuated transparent tube along one arm of the device. Fringe shifts occur as the gas is slowly added to the tube. Assume that the tube is 5 cm long and that 60 fringe shifts occur as the pressure of the gas in the tube increases to atmospheric pressure. What is the index of refraction of the gas? (*Hint:* The fringe shifts occur because the wavelength of the light changes inside the gas-filled tube.)

53. A thin sheet of transparent material has an index of refraction of 1.40 and is 15 μm thick. When it is inserted along one arm of an interferometer in the path of 600-nm light, how many fringe shifts occur in the pattern?

Section 27.8 The Diffraction Grating

54. The second-order image formed by a diffraction grating having 6000 lines/cm is observed at an angle of 29°. What is the wavelength of the light used?

55. A diffraction grating is calibrated by using the 546.1-nm line of mercury vapor. It is found that the first-order line is located at an angle of 21°. Calculate the number of lines per centimeter on this grating.

56. A grating having 1500 slits per centimeter is illuminated with light of wavelength 500 nm. (a) What is the highest order number that can be observed with

this grating? (b) Repeat for a grating of 15,000 slits per centimeter.

57. Three discrete spectral lines occur at angles of 10.09°, 13.71°, and 14.77° in the first-order spectrum of a grating spectroscope. (a) If the grating has 3660 slits/cm, what are the wavelengths of the light? (b) At what angles are these lines found in the second-order spectra?

58. In a spectrum analyzed by a diffraction grating the fifth-order green maximum ($\lambda = 500$ nm) falls atop the fourth-order yellow maximum. What is the wavelength of the yellow light?

59. Light from a hydrogen source is incident on a diffraction grating. The incident light contains four wavelengths: $\lambda_1 = 410.1$ nm, $\lambda_2 = 434.0$ nm, $\lambda_3 = 486.1$ nm, and $\lambda_4 = 656.3$ nm. The diffraction grating has 410 lines/mm. Calculate the angles between (a) λ_1 and λ_4 in the first-order spectrum and (b) λ_1 and λ_3 in the third-order spectrum.

60. Visible light ranging from 400 nm to 700 nm is focused on a diffraction grating. The entire first order is observed, but no second-order spectrum is seen. What is the maximum spacing between lines on this grating?

61. Light containing two different wavelengths passes through a diffraction grating with 1200 slits/cm. On a screen 15.0 cm from the grating the third-order maximum of the shorter wavelength falls on top of the first-order minimum of the longer wavelength. If the neighboring maxima of the longer wavelength are 8.44 mm apart on the screen, what are the wavelengths in the light? (*Hint:* Use small angle approximation.)

62. A light source contains two major spectral lines, an orange line of wavelength 610 nm and a blue-green line of wavelength 480 nm. If the spectrum is resolved by a diffraction grating having 5000 lines/cm and viewed on a screen 2.0 m from the grating, what is the distance between the two spectral lines in the second-order spectrum?

63. A diffraction grating is 3 cm wide and contains lines uniformly spaced at 775 nm. Determine the minimum wavelength difference that can be resolved in first order at 600 nm by this grating. Assume that the full grating width is illuminated.

64. The H_{α} line in hydrogen has a wavelength of 656.2 nm. This line differs in wavelength from the corresponding spectral line in deuterium (the heavy stable isotope of hydrogen) by 0.18 nm. (a) Determine the minimum number of lines a grating must have in order to resolve these two wavelengths in the first order. (b) Repeat part (a) for the second order.

Additional Problems

65. A telescope has an objective of focal length 100 cm and an eyepiece of focal length 1.5 cm. (a) What is the distance between the two lenses when the telescope is used to view a distant object? (b) What is the angular magnification of the telescope?

66. The near point of an eye is 75 cm. (a) What power lens should be prescribed so that the eye can see an object clearly at 25 cm? (b) If the user can actually see an object clearly at 26 cm rather than 25 cm, by how many diopters did the lens grinder miss the prescription?

67. The 546.1-nm line in mercury is measured at an angle of 81° in the third-order spectrum of a diffraction grating. Calculate the number of lines per millimeter for the grating.

68. The far point of an eye is 75 cm. What power lens should be prescribed?

69. A certain lens forms a real image 20 cm from a lens when an object is placed 40 cm from the lens. What is the maximum magnification produced by this lens when used as a simple magnifier?

70. The wavelengths of the sodium doublet are $\lambda_1 = 589.0$ nm and $\lambda_2 = 589.6$ nm. Determine the minimum number of lines in a grating that will allow resolution of the sodium doublet in (a) the first order and (b) the third order.

"Just checking."

71. Light consisting of two wavelength components is incident on a grating. The shorter wavelength component has a wavelength of 440 nm. The third-order image of this component is coincident with the second-order image of the longer wavelength component. Determine the wavelength of the longer wavelength component.

72. A grating is 4 cm wide and is entirely illuminated with monochromatic light of wavelength 577 nm. The second-order maximum is formed at an angle of 41.25°. What is the total number of lines in this grating?

73. If a typical eyeball is 2 cm long and has a pupil opening which can range from about 2 mm to 6 mm, what is (a) the focal length of the eye when focused on objects 1 m away, (b) the smallest f-number of the eye when focused at objects 1 m away, and (c) the largest f-number of the eye when focused at objects 1 m away?

74. The 501.5-nm line in helium is observed at an angle of 30° in the second-order spectrum of a diffraction grating. Calculate the angular deviation of the 667.8-nm line in helium in the first-order spectrum for the same grating.

75. A fringe pattern is established in the field of view of a Michelson interferometer using light of wavelength 580 nm. A parallel-faced sheet of transparent material 2.5 μm thick is placed in front of one of the mirrors perpendicular to the incident and reflected light beams. An observer counts a fringe shift of six dark fringes. What is the index of refraction of the sheet?

76. A person uses a $+2$ diopter lens in order to comfortably see an object placed 25 cm away. By the time he decides to have his prescription changed, he must hold an object 35 cm away to see it. What power lens does he now need?

77. If the aqueous humor of the eye has an index of refraction of 1.34 and the distance from the vertex of the cornea to the retina is 2 cm, what is the radius of curvature of the cornea for which distant objects will be focused on the retina? (Assume, incorrectly, that all refraction occurs in the aqueous humor.)

PART 6
Modern Physics

At the end of the 19th century, scientists believed that they had learned most of what there was to know about physics. Newton's laws of motion and his universal theory of gravitation, Maxwell's theoretical work in unifying electricity and magnetism, and the laws of thermodynamics and kinetic theory were highly successful in explaining a wide variety of phenomena.

However, at the turn of the 20th century, a major revolution shook the world of physics. In 1900 Planck provided the basic ideas that led to the formulation of the quantum theory, and in 1905 Einstein formulated his brilliant special theory of relativity. The excitement of the times is captured in Einstein's own words: "It was a marvelous time to be alive." Both ideas were to have a profound effect on our understanding of nature. Within a few decades, these theories inspired new developments and theories in the fields of atomic physics, nuclear physics, and condensed matter physics.

The discussion of modern physics in this last part of the text will begin with a treatment of the special theory of relativity in Chapter 28. Although the concepts underlying this theory often violate our common sense, the theory provides us with a new and deeper view of physical laws. Next we shall discuss various developments in quantum theory (Chapter 29), which provides us with a successful model for understanding electrons, atoms, and molecules. The last five chapters of the text are concerned with applications of quantum theory. Chapters 30 and 31 discuss the structure and properties of atoms and molecules in terms of concepts from quantum mechanics. Chapter 32 is concerned with the structure and properties of the atomic nucleus, and Chapter 33 discusses many practical applications of nuclear physics, including nuclear reactors, and the interaction of radiation with matter. We conclude the text with a discussion of elementary particles in Chapter 34.

You should keep in mind that, although modern physics has been developed during this century and has led to a multitude of important technological achievements, the story is still incomplete. Discoveries will continue to evolve during our lifetime, many of which will deepen or refine our understanding of nature and the world around us. It is still "a marvelous time to be alive."

"The scientist does not study nature because it is useful; he studies it because he delights in it, and he delights in it because it is beautiful. If nature were not beautiful, it would not be worth knowing, and if nature were not worth knowing, life would not be worth living."
HENRI POINCARÉ

Conversation
with
Hans Bethe

Born in Strasbourg in 1906, Hans Albrecht Bethe studied physics, chemistry, and mathematics at the University of Frankfurt and passed his doctoral exam in physics at the University of Munich in 1928. There he studied with Arnold Sommerfeld, an influential theoretical physicist whose students included Wolfgang Pauli, Werner Heisenberg, Peter Debye, and Max von Laue, all of whom later won the Nobel Prize. He collaborated with Sommerfeld on a long article containing the basis for the modern solid-state theory of metals in 1932, and he was subsequently the author of two other important articles. One was on atomic physics, and the other was the first comprehensive treatment of nuclear physics.

As the Nazi movement grew, Bethe lost his job as an assistant professor at Tübingen because his mother was Jewish; he was given a fellowship by Sommerfeld, who aided many physicists of Jewish ancestry. Bethe spent the years 1930 and 1931 at Cambridge University on a Rockefeller Foundation fellowship and at the University of Rome, where he worked with Enrico Fermi. In 1935 he joined the faculty of Cornell University in Ithaca, New York, where he remains today as professor emeritus.

Interview conducted by Janice K. Mandel

Bethe's many honors include receiving the President's Medal of Merit in 1946 for his work as the head of the Theoretical Division of the Los Alamos atomic bomb project (known as the Manhattan Project), and receiving the 1967 Nobel Prize in Physics for his work on the energy production in stars.

Why did you decide to become a physicist?

I was always interested in numbers as a child. My mother found a little booklet in which I had done all sorts of exercises with numbers like powers of 2, 3, and 5 when I was about 7 years old. Then [numbers] seemed a little too narrow a subject, and so, as I learned later in high school about physics, I saw that the real place I could use my interest in numbers was physics, and that was it.

How did you become involved in astrophysics?

I got into this by good luck. The Carnegie Foundation sponsored a conference in Washington in which astrophysicists were discussing solar energy and reported on their investigations of the interior of stars. They were convinced that the energy in stars comes from nuclear reactions, and so I said to myself, "I know nuclear reactions; I have written a long article about it in *The Review of Modern Physics*, so I probably could answer that question on just what nuclear reactions are responsible."

I didn't want to go to that conference. I was invited by my friend Edward Teller, but it was not a subject along the main line of my interest, so I was very hesitant in accepting and only after some persuasion did I go. When I came home after that conference I set to work. Karl Weizsäcker had previously made the suggestion that the source of stellar energy was the collision between two protons (hydrogen nuclei) making the nucleus of heavy hydrogen. Following this suggestion, a young graduate student in Washington, Critchfield, put some actual numbers into the model. Later, Critchfield and I wrote a paper together on this reaction.

However, I now wanted to know whether this was the only nuclear reaction that can take place in stars, and so I searched through the periodic table of the elements. Nothing worked until I came to carbon, and then it worked beautifully. Carbon adds four protons one after the other; of course, that takes millions of years in the star. Then a beautiful thing happens. The protons combine to make a helium nucleus, which releases a great amount of energy, and the carbon comes back out the same way it got in. Hydrogen is very plentiful in all stars, but carbon is not present in such large amounts, so it is most important that this is a cycle and that you get the carbon back out which you put in and that you use only the abundant hydrogen. This is known as the carbon cycle. I found this out within about two weeks of coming home from the conference. And this is the piece of work that is known about me.

Is this the work for which you received the Nobel Prize in 1967?

Yes, to a large extent.

Before this work, had theoretical physicists and astrophysicists worked together much?

Not very much. The behavior of nuclei was known to theoretical physicists but not to astrophysicists. On the other hand, the behavior of stars was known only to astrophysicists. So it was a very happy marriage.

From your experience, would you say it's a good idea to look to other disciplines for opportunities in your own field?

Yes. It isn't possible today for anyone to know all the sciences, and so you cannot be prepared for all the opportunities that may come along. I would not have been inclined to study astrophysics if it had not been for that one conference.

What was the Manhattan Project in Los Alamos, and why was it organized?

The Manhattan Project was designed to build an atomic weapon based on the fission of uranium and on the chain reaction that can occur when neutrons come in contact with a rare isotope of uranium, called uranium-235. This process had been

> "I believe that in the last month I've finally found out the key to the problem . . . You think of one additional point which you have forgotten until now, and suddenly you have an insight which makes everything fit together."

discovered by two German chemists and interpreted by two German refugee physicists, Lise Meitner and Otto Frisch. And people everywhere in many countries conceived that this might lead to a very powerful explosive. So, the Manhattan Project was meant to exploit this process and to build an atomic bomb.

Why did you participate in making the atomic bomb?

I was fully aware of what a terrible weapon we were fashioning there. But one has to remember that the project started in 1942

when Hitler had conquered many countries in Europe and was well on the way to conquering Russia. Things looked quite desperate for the West and in particular for Western civilization. We all cherished Western civilization and did not want it to end, and this is why many of us joined. There were between 500 and 1,000 actual scientists working on the project, in addition to enormous numbers of technicians. The total staff of the laboratory was probably close to 5,000 employees organized into various divisions by the laboratory Director, Robert Oppenheimer. I was the leader of the Theoretical Division, which had increased to 80 people by the end of the war. I believe the people working on this project had their hearts and souls in it. We were very devoted to the United States and to winning the war.

Why were there so many people involved?

There was so much to do. First we had to find out what are really the properties of uranium and neutrons. We had to determine how big the bomb would have to be. That was important because the director of the project, General Leslie Groves, needed to know how much he had to produce. And we could only conclude this from experiments and theoretical analysis of these experiments, because all we had was a small speck—maybe a few milligrams—of uranium-235, and it took a long time before we got a similar small speck—maybe 10 mg—of plutonium. From experiments on

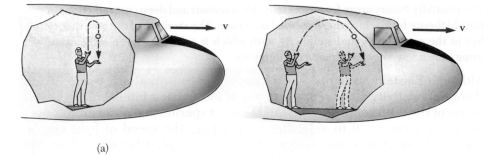

(a)

Figure 28.1
(a) The observer on the airplane sees the ball move in a vertical path when thrown upwards. (b) The Earth observer views the path of the ball to be a parabola.

(b)

moving with a constant velocity as in Figure 28.1a. If a passenger in the airplane throws a ball straight up in the air, the passenger observes that the ball moves in a vertical path. The motion of the ball is precisely the same as if the ball were thrown while at rest on Earth. The law of gravity and the equations of motion under constant acceleration are obeyed whether the airplane is at rest or in uniform motion. Now consider the same experiment when viewed from another observer at rest on the Earth. This stationary observer views the path of the ball to be a parabola as in Figure 28.1b. Furthermore, according to this observer, the ball has a velocity to the right equal to the velocity of the plane. Although the two observers disagree on certain aspects of the experiment, both will agree that the motion of the ball obeys the law of gravity and the laws of motion. Thus, we draw the following important conclusion:

> There is no preferred frame of reference for describing the laws of mechanics.

28.3   The Michelson-Morley Experiment

Many experiments similar to the one described in the previous section show us that the laws of mechanics are the same in all inertial frames of reference. When similar inquiries are made into the laws of other branches of physics, the results are contradictory. In particular, it is found that the laws of electricity and magnetism depend on the frame of reference used. It might be argued that it is these laws that are wrong, but this is difficult to accept because the laws are in total agreement with all known experimental results. The Michelson-Morley experiment was an attempt to find an answer to this dilemma.

The Michelson-Morley experiment was designed because of a misconception early physicists had concerning the manner in which light propagates. The properties of mechanical waves, such as water or sound waves, were well known, and all of these require a medium to support the disturbances. In the 19th century, physicists thought that electromagnetic waves also required a medium through which to propagate. They proposed that such a medium existed, and they named it the **luminiferous ether.** This ether was assumed to

Albert A. Michelson (1852–1931). *(AIP Niels Bohr Library)*

be present everywhere, even in a vacuum, and to have the unusual property of being a massless but rigid medium. Indeed, this is a strange concept. Additionally, it was found that the troublesome laws of electricity and magnetism would take on their simplest form in a frame of reference *at rest* with respect to this ether. This frame was called the **absolute frame.** The laws of electricity and magnetism would be valid but would have to be modified in any reference frame moving with respect to it.

As a result of the importance attached to this absolute frame, it became of considerable interest in physics to prove its existence by experiment. However, all attempts to detect the presence of the ether (and hence the absolute frame) proved futile!

The most famous experiment designed to show the presence of the ether was performed in 1887 by A. A. Michelson (1852–1931) and E. W. Morley (1838–1923). Later in this section we shall discuss the Michelson-Morley experiment in more detail. We should state at the outset that the outcome of the experiment was *negative,* thus contradicting the ether hypothesis. The experiment was designed to determine the velocity of the Earth with respect to the ether, and the experimental tool used was the interferometer discussed in Section 27.7. Suppose one of the arms of the interferometer is aligned along the direction of the motion of the Earth through space. The Earth moving through the ether would be equivalent to the ether flowing past the Earth in the opposite direction. This "ether wind" blowing in the direction opposite the Earth's motion should cause the speed of light as measured in the Earth's frame of reference to be $c - v$ as the light approaches mirror M_2 in Figure 28.2 and $c + v$ after reflection, where c is the speed of light in the ether frame and v is the speed of the Earth through space and hence the speed of the ether wind. The incident and reflected beams of light would recombine, and an interference pattern consisting of alternating dark and bright bands would be formed.

During the experiment, the interference pattern was observed while the interferometer was rotated through an angle of 90°. The idea was that this rotation would change the speed of the ether wind along the direction of the arms of the interferometer, and consequently the fringe pattern would shift slightly but measurably. Measurements failed to show any change in the interference pattern! The Michelson-Morley experiment was repeated by other researchers under various conditions and at different locations, but the results were always the same: *no fringe shift of the magnitude required was ever observed.*

The problem that arises can be understood by considering the following simple situation. Suppose an observer riding in the back of a truck moving at a speed of 50 mi/h throws a baseball in the direction of travel of the truck such that the ball leaves his hand with a speed of 30 mi/h with respect to the truck. What speed will a stationary observer by the side of the road find for the baseball? The answer is quite simple. The stationary observer would add the speed of the truck to the speed of the baseball and conclude that its speed with respect to his frame of reference is 80 mi/h. (If the ball were thrown in the direction opposite that of the truck's motion, the stationary observer would find a speed of 20 mi/h.) This simple calculation is called a *transformation* because we have transformed a measurement made in one reference frame to a measurement made in another. The transformation we have described is in accord with our common sense that a velocity measurement made in one reference frame will be different in another.

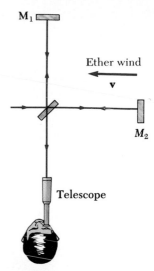

Figure 28.2
According to the ether wind theory, the speed of light should be $c - v$ as the beam approaches mirror M_2 and $c + v$ after reflection.

Now let us consider a transformation involving the measurement of the speed of light. Suppose an astronaut traveling in a rocket ship at a speed of 1×10^8 m/s turns on a flashlight pointed in the direction of motion of the ship. The astronaut measures the speed of the light to be 3×10^8 m/s. If we assume that velocities add vectorially as in the case of the moving baseball, a stationary observer should measure the speed of the light beam to be the sum of the rocket's speed and the speed of light, or 4×10^8 m/s. This result would mean that the two observers in the different reference frames would get different results when attempting to measure the same electromagnetic effect. As we shall see later, the assumption that velocities add vectorially is not valid when the relative velocity between reference frames is very large. Furthermore, Maxwell's equations, which form the basis of all electromagnetic phenomena, take on a different form when transformed from one inertial frame to another. In fact, they take on their simplest form in a reference frame where the speed of light is 3×10^8 m/s. This strange result is at odds with the fact that the laws of physics must be the same in all inertial frames of reference.

The negative result of the Michelson-Morley experiment was in contradiction with the ether hypothesis. This result also meant that it was impossible to measure the absolute (orbital) velocity of the Earth with respect to the ether frame. From a theoretical viewpoint, this meant that it was impossible to find the absolute frame. However, as we shall see in the next section, Einstein offered a postulate that places a different interpretation on these negative results. In later years, when more was known about the nature of light, the idea of an ether that permeated all space was relegated to the ash heap of worn-out concepts. Light is now understood to be *an electromagnetic wave that requires no medium for its propagation.* As a result, the idea of having an ether in which these waves could travel became an unnecessary construct.

*Details of the Michelson-Morley Experiment

As we mentioned earlier, the Michelson-Morley experiment was designed to detect the motion of the Earth with respect to the ether. Before we examine the details of this important, historical experiment, it is instructive to first consider a race between two airplanes as shown in Figure 28.3a. One airplane

Figure 28.3
(a) If an airplane wishes to travel from O to A with a wind blowing to the right, it must head into the wind at some angle. (b) Vector diagram for determining the heading of the airplane for the trip from O to A. (c) Vector diagram for determining the heading for the trip from A to O.

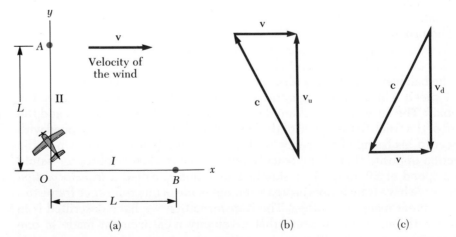

flies from point O to point A perpendicular to the wind, while the second airplane flies from point O to point B parallel to the direction of the wind. We shall assume that they start at O at the same time, travel the same distance L, with the same cruising speed c with respect to the wind, and return to O. Which airplane will win the race? In order to answer this question, we shall first calculate the time of flight for both airplanes.

First, consider the airplane that moves along path I parallel to the wind. As it moves to the right, its speed is enhanced by the wind and its velocity with respect to the Earth is $c + v$. On its return journey as it moves to the left, it must fly opposite the wind, hence its speed with respect to the Earth is $c - v$. Thus, the times of flight to the right and to the left are, respectively,

$$t_R = \frac{L}{c + v} \quad \text{and} \quad t_L = \frac{L}{c - v}$$

and the total time of flight for the airplane moving along path I is

$$t_1 = t_R + t_L = \frac{L}{c + v} + \frac{L}{c - v} = \frac{2Lc}{c^2 - v^2}$$

$$= \frac{2L}{c \left(1 - \dfrac{v^2}{c^2} \right)} \tag{28.1}$$

Now consider the airplane flying along path II. If the pilot points the airplane directly towards point A, the airplane will be blown off course by the wind and will not reach its destination. To compensate for the wind, the pilot must point the airplane into the wind at some angle as shown in Figure 28.3a. This angle must be selected such that the vector sum of $\mathbf{c}$ and $\mathbf{v}$ leads to a velocity vector pointed directly toward A. The resultant vector diagram is shown in Figure 28.3b, where $\mathbf{v}_u$ is the velocity of the airplane with respect to the ground as it moves from O to A. From the Pythagorean theorem, the magnitude of the vector $\mathbf{v}_u$ is

$$v_u = \sqrt{c^2 - v^2} = c \sqrt{1 - \frac{v^2}{c^2}}$$

Likewise, on the return trip from A to O, the pilot must again head into the wind such that the airplane's velocity with respect to the Earth, $\mathbf{v}_d$, will be directed toward O, as shown in Figure 28.3c. From this figure, we see that

$$v_d = \sqrt{c^2 - v^2} = c \sqrt{1 - \frac{v^2}{c^2}}$$

Thus, the total time of flight for the trip along path II is

$$t_2 = \frac{L}{v_u} + \frac{L}{v_d} = \frac{L}{c \sqrt{1 - \dfrac{v^2}{c^2}}} + \frac{L}{c \sqrt{1 - \dfrac{v^2}{c^2}}}$$

$$= \frac{2L}{c \sqrt{1 - \dfrac{v^2}{c^2}}} \tag{28.2}$$

Comparing Equations 28.1 and 28.2, we see that the airplane flying along path II wins the race. The difference in flight times is given by

$$\Delta t = t_1 - t_2 = \frac{2L}{c}\left[\frac{1}{\left(1 - \dfrac{v^2}{c^2}\right)} - \frac{1}{\sqrt{1 - \dfrac{v^2}{c^2}}}\right]$$

This expression can be simplified using the following binomial expansions after dropping all terms higher than second-order:

$$\left(1 - \frac{v^2}{c^2}\right)^{-1} \approx 1 + \frac{v^2}{c^2}$$

and

$$\left(1 - \frac{v^2}{c^2}\right)^{-1/2} \approx 1 + \frac{1}{2}\frac{v^2}{c^2}$$

This gives

$$\Delta t = \frac{Lv^2}{c^3} \tag{28.3}$$

The analogy between this airplane race and the Michelson-Morley experiment is shown in Figure 28.4a. Two beams of light travel along two arms of an interferometer. In this case, the "wind" is the ether blowing across the Earth from left to right as the Earth moves through the ether from right to left. Because the speed of the Earth in its orbital path is approximately equal to 3×10^4 m/s, the speed of the wind should be at least this great. The two light beams start out in phase and return to form an interference pattern. Let us assume that the interferometer is adjusted for parallel fringes and that a telescope is focused on one of these fringes. The time difference between the two light beams gives rise to a phase difference between the beams, producing an interference pattern when they combine at the position of the telescope. The

Figure 28.4
(a) Top view of the Michelson-Morley interferometer, where **v** is the velocity of the ether and L is the length of each arm. (b) When the interferometer is rotated by 90°, the role of each arm is reversed.

(a)

(b)

difference in the pattern is detected by rotating the interferometer through 90° in a horizontal plane, such that the two beams exchange roles (Fig. 28.4b). This results in a net time shift of twice the time difference given by Equation 28.3. Thus, the net time difference is

$$\Delta t_{\text{net}} = 2 \; \Delta t = \frac{2Lv^2}{c^3} \tag{28.4}$$

The path difference which corresponds to this time difference is

$$\Delta d = c \; \Delta t_{\text{net}} = \frac{2Lv^2}{c^2} \tag{28.5}$$

In the first experiments by Michelson and Morley, each light beam was reflected by mirrors many times to give an increased effective path length L of about 11 meters. Using this value and taking v to be equal to 3×10^4 m/s gives a path difference of

$$\Delta d = \frac{2(11 \text{ m})(3 \times 10^4 \text{ m/s})^2}{(3 \times 10^8 \text{ m/s})^2} = 2.2 \times 10^{-7} \text{ m}$$

This extra distance of travel should produce a noticeable shift in the fringe pattern. Specifically, calculations show that if the pattern is viewed while the interferometer is rotated through 90°, a shift of about 0.4 fringes should be observed. The instrument used by Michelson and Morley had the capability of detecting a shift in the fringe pattern as small as 0.01 fringes. However, *they detected no shift in the fringe pattern*. Since then, the experiment has been repeated many times by various scientists under various conditions and no fringe shift has ever been detected. Thus, it was concluded that the motion of the Earth with respect to the ether cannot be detected.

Many efforts were made to explain the null results of the Michelson-Morley experiment. For example, perhaps the Earth drags the ether with it in its motion through space. To test this assumption, interferometer measurements were made at various altitudes, but again no fringe shift was detected. In the 1890's, G. F. Fitzgerald and H. A. Lorentz tried to explain the null results by making the following ad hoc assumption. They proposed that the length of an object moving along the direction of the ether wind would contract by a factor of $\sqrt{1 - v^2/c^2}$. The net result of this contraction would be a change in length of one of the arms of the interferometer such that no path difference would occur as it was rotated.

No experiment in the history of physics has received such valiant efforts to try to explain the absence of an expected result as did the Michelson-Morley experiment. The stage was set for the brilliant Albert Einstein, who solved the problem in 1905 with his special theory of relativity.

28.4   Einstein's Postulates

The special theory of relativity deals with situations involving inertial reference frames, that is, *frames in which Newton's first law is valid*, and is based on two postulates. First, Einstein postulated that

the laws of physics are the same in every inertial frame of reference.   **The first postulate of relativity**

Albert Einstein
(1879–1955)

Albert Einstein, one of the greatest physicists of all times, was born in Ulm, Germany. He showed little intellectual promise as a youngster, and he left the highly disciplined German school system after one teacher stated "You will never amount to anything, Einstein." Following a vacation in Italy, he completed his education at the Swiss Federal Polytechnic School in 1901. Although Einstein attended very few lectures, he was able to pass the courses with the help of excellent lecture notes taken by a friend. Unable to find an academic position, Einstein accepted a position as a junior official in the Swiss Patent Office in Berne. In this setting, and during his "spare time," he continued his independent studies in theoretical physics. In 1905, at the age of 26, he published four scientific papers that revolutionized physics. (In that same year, he earned his Ph.D.) One of these papers, for which he was awarded the 1921 Nobel prize in physics, dealt with the photoelectric effect. Another was concerned with Brownian motion, the irregular motion of small particles suspended in a liquid. The remaining two papers were concerned with what is now considered his most important contribution of all, the special theory of relativity. In 1915, Einstein published his work on the general theory of relativity, which relates gravity to the structure of space and time. The most dramatic prediction of this theory is the degree to which light would be deflected by a gravitational field. Measurements made by astronomers on bright stars in the vicinity of the eclipsed Sun in 1919 confirmed Einstein's prediction, and Einstein suddenly became a world celebrity. This and other predictions of the general theory of relativity are discussed in the delightful essay by Clifford Will that follows this chapter.

In 1913, following academic appointments in Switzerland and Czechoslovakia, Einstein accepted a special position created for him at the Kaiser Wilhelm Institute in Berlin. This made it possible for him to be able to devote all of his time to research, free of financial troubles and routine duties. Einstein left Germany in 1933, which was then under Hitler's power, thereby escaping the fate of millions of other European Jews. In the same year he accepted a special position at the Institute for Advanced Study in Princeton where he remained for the rest of his life. He became an American citizen in 1940. Although he was a pacifist, Einstein was persuaded by Leo Szilard to write a letter to President Franklin D. Roosevelt urging him to initiate a program to develop a nuclear bomb. The result was the successful six-year Manhattan project and two nuclear explosions in Japan which ended World War II in 1945.

Einstein made many important contributions to the development of modern physics, including the concept of the light quantum and the idea of stimulated emission of radiation, which led to the invention of the laser 40 years later. Einstein was deeply disturbed by the development of quantum mechanics in the 1920s despite his own role as a scientific revolutionary. In particular, he could never accept the probabilistic view of events in nature that is a central feature of the highly successful quantum theory. He once said, "God does not play dice with nature." The last few decades of his life were devoted to an unsuccessful search for a unified theory that would combine gravitation and electromagnetism into one picture.

(Photograph Courtesy of AIP Niels Bohr Library)

In Einstein's own words, "The same laws of electrodynamics and optics will be valid for all frames of reference for which the equations of mechanics hold good." This is, in effect, a generalization of Newton's principle of relativity, which applies only to the laws of mechanics.

Figure 28.5
A pulse of light is sent out by
a person in a moving boxcar.
According to Newtonian relativity, the speed of the pulse
should be $c + v$ relative to a
stationary observer.

Now consider Einstein's second postulate, which states that

the speed of light has the same value for all observers, independent of their
motion or of the motion of the light source.

The second postulate of
relativity

Here we are faced with a fundamental problem. We can demonstrate the
nature of the problem by considering a light pulse sent out by an observer in a
boxcar moving with a velocity v (Fig. 28.5). The light pulse has a velocity c
relative to observer S′ in the boxcar. According to the ideas of Newtonian
relativity, the speed of the pulse relative to stationary observer S outside the
boxcar should be $c + v$. This is in obvious contradiction to Einstein's second
postulate, which states that the velocity of the light pulse is the same for all
observers. According to Einstein's theory, the stationary and the moving observer should both measure the same velocity for the light pulse. This conclusion seems strange because it contradicts our intuition, or what we often call
common sense. However, common sense ideas are based on everyday experiences, which do not involve speed-of-light measurements.

Although the Michelson-Morley experiment was performed before Einstein published his work on relativity, it is not clear whether or not Einstein
was aware of the details of the experiment. Nonetheless, the second postulate
explains the null result of the experiment for, in effect, the second postulate
means that the premises of the experiment were incorrect. For example, in
explaining the expected results, we stated that when light traveled against the
ether wind its speed was $c - v$. However, if the state of motion of the observer
or of the source has no influence on the value found for the speed of light, the
value will always be measured to be c. Likewise, after reflection from the
mirror, the light makes the return trip with a speed of c, not $c + v$. Thus, if we
accept Einstein's second postulate, the motion of the Earth should not influence the fringe pattern observed in the Michelson-Morley experiment and a
null result should be expected.

If we accept Einstein's postulates, we must conclude that *relative motion is
unimportant when measuring the speed of light.* In effect, Einstein altered our
concepts of space and time in such a way as to give the same result for the speed
of light measured by any observer located in any inertial frame.

28.5   Consequences of Special Relativity

Almost everyone who has dabbled even superficially in science is aware of
some of the startling predictions that arise because of Einstein's approach to
relative motion. As we examine some of the consequences of relativity in this

section, we shall find that they conflict with some of our basic notions of space and time. We shall restrict our discussion to the concepts of length, time, and simultaneity, which are quite different in relativistic mechanics from what they are in Newtonian mechanics. For example, we shall see that *the distance between two points and the time interval between two events depend on the frame of reference in which they are measured.* That is, *in relativity, there is no such thing as absolute length or absolute time. Furthermore, events at different locations that occur simultaneously in one frame are not simultaneous in another frame.*

> Absolute length and absolute time intervals are meaningless in relativity

Simultaneity and the Relativity of Time

A basic premise of Newtonian mechanics is that there is a universal time scale that is the same for all observers. In fact, Newton wrote, "Absolute, true, and mathematical time, of itself, and from its own nature, flows equably without relation to anything external." In his special theory of relativity, Einstein abandoned this assumption. According to Einstein, *time interval measurements depend on the reference frame in which they are made.*

Einstein devised the following thought experiment to illustrate this point. A boxcar moves with uniform velocity, and two lightning bolts strike its ends, as in Figure 28.6a, leaving marks on the boxcar and the ground. The marks left on the boxcar are labeled A' and B', and those on the ground are labeled A and B. An observer at O' moving with the boxcar is midway between A' and B', and an observer on the ground at O is midway between A and B. The events seen by the observers are the light signals from the lightning bolts.

Let us assume that the two light signals reach the observer at O at the same time, as indicated in Figure 28.6b. This observer realizes that the light signals have traveled at the same speed over distances of equal length. Thus, the observer at O concludes that the events at A and B occurred simultaneously. Now consider the same events as viewed by the observer on the boxcar at O'. By the time the light has reached the observer at O, the observer at O' has moved as indicated in Figure 28.6b. Thus, the light signal from B' has already swept past O', while the light from A' has not yet reached O'. According to Einstein's second postulate, the observer at O' must find that light travels at the same speed as that measured by the observer at O. Therefore, the observer at O' concludes that the light reaches him from the front of the boxcar before it reaches him from the back. This thought experiment clearly demonstrates that the two events which appear to be simultaneous to the observer at O do not appear to be simultaneous to the observer at O'. In other words,

Figure 28.6
Two lightning bolts strike the end of a moving boxcar.
(a) The events appear to be simultaneous to the stationary observer at O, who is midway between A and B. (b) The events do not appear to be simultaneous to the observer at O', who claims that the front of the train is struck *before* the rear.

(a)   (b)

two events that are simultaneous in one reference frame are in general not simultaneous in a second frame moving with respect to the first. That is, simultaneity is not an absolute concept.

At this point, you might wonder which observer is right concerning the two events. The answer is that *both are correct* because the principle of relativity states that *there is no preferred inertial frame of reference.* Although the two observers reach different conclusions, both are correct in their own reference frame because the concept of simultaneity is not absolute.

Simultaneity is not an absolute concept

Time Dilation

Consider a vehicle moving to the right with a speed v, as in Figure 28.7a. A perfectly reflecting mirror is fixed to the ceiling of the vehicle, and an observer at O' at rest in this system holds a flash gun a distance d below the mirror. At some instant, the flash gun goes off and a pulse of light is released. Because the light pulse has a speed c, the time it takes it to travel from the observer to the mirror and back again can be found from the definition of velocity:

$$\Delta t' = \frac{\text{distance traveled}}{\text{velocity}} = \frac{2d}{c} \qquad \textbf{(28.6)}$$

where the prime notation indicates that this is the time measured by the observer in the reference frame of the moving vehicle.

Now consider the same set of events as viewed by an observer at O in a stationary frame (Fig. 28.7b). According to this observer, the mirror and flash gun are moving to the right with a speed v. The sequence of events just described would appear entirely different to this stationary observer. By the time the light from the flash gun reaches the mirror, the mirror will have moved a distance $v\Delta t/2$, where Δt is the time it takes the light pulse to travel from O' to the mirror and back, as measured by the stationary observer. In other words, the stationary observer concludes that, because of the motion of the system, the light, if it is to hit the mirror, will leave the flash gun at an angle with respect to the vertical. Comparing Figures 28.7a and 28.7b, we see that the light must travel farther in the stationary frame than in the moving frame.

Now, according to Einstein's second postulate, the speed of light must be c as measured by both observers. Therefore, it follows that the time interval Δt, measured by the observer in the stationary frame, is *longer* than the time

Figure 28.7
(a) A mirror is fixed to a moving vehicle, and a light pulse leaves O' at rest in the vehicle. (b) Relative to a stationary observer on Earth, the mirror and O' move with a speed v. Note that the distance the pulse travels is greater than $2d$ as measured by the stationary observer. (c) The right triangle for calculating the relationship between Δt and $\Delta t'_{O'}$.

(a)   (b)

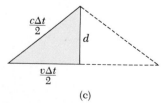

(c)

interval $\Delta t'$, measured by the observer in the moving frame. To obtain a relationship between Δt and $\Delta t'$, it is convenient to use the right triangle shown in Figure 28.7c. The Pythagorean theorem applied to this triangle gives

$$\left(\frac{c\Delta t}{2}\right)^2 = \left(\frac{v\Delta t}{2}\right)^2 + d^2$$

Solving for Δt gives

$$\Delta t = \frac{2d}{\sqrt{c^2 - v^2}} = \frac{2d}{c\sqrt{1 - v^2/c^2}} \qquad \textbf{(28.7)}$$

Because $\Delta t' = 2d/c$, we can express Equation 28.7 as

Time dilation

$$\Delta t = \frac{\Delta t'}{\sqrt{1 - v^2/c^2}} = \gamma\Delta t' \qquad \textbf{(28.8)}$$

where $\gamma = 1/\sqrt{1 - v^2/c^2}$. This result says that the time interval measured by the observer in the stationary frame is *longer* than that measured by the observer in the moving frame (because γ is always greater than unity).

For example, suppose an observer in a moving vehicle has a clock that he uses to measure the time required for the light flash to leave the gun and return. Let us assume that the measured time interval in this frame of reference, $\Delta t'$, is one second. (This would require a very tall vehicle.) Now let us find the time interval as measured by a stationary observer using an identical clock. If the vehicle is traveling at half the speed of light ($v = 0.5c$), then $\gamma = 1.15$, and according to Equation 28.6 $\Delta t = \gamma\Delta t' = 1.15$ (1 s) = 1.15 s. Thus, when the observer on the moving vehicle claims that 1 s has passed, a stationary observer claims that 1.15 s has passed. From this we may conclude that

A moving clock runs slower than an identical stationary clock

according to a stationary observer, a moving clock runs slower than an identical stationary clock by a factor of γ^{-1}. This effect is known as **time dilation.**

The time interval $\Delta t'$ in Equation 28.8 is called the **proper time.** In general, **proper time** is defined as *the time interval between two events as measured by an observer who sees the events occur at the same place.* In our case, the observer at O' measures the proper time. That is, *proper time is always the time measured by an observer moving along with the clock.*

We have seen that moving clocks run slow by a factor of γ^{-1}. This is true for ordinary mechanical clocks as well as for the light clock just described. In fact, we can generalize these results by stating that *all physical processes, including chemical and biological reactions, slow down relative to a stationary clock when they occur in a moving frame.* For example, the heartbeat of an astronaut moving through space has to keep time with a clock inside the spaceship. Both the spaceship clock and the heartbeat are slowed down relative to a stationary clock. The astronaut would not have any sensation of life slowing down in the spaceship.

Time dilation is a very real phenomenon that has been verified by various experiments. For example, muons are unstable elementary particles that have a charge equal to that of the electron and a mass 207 times that of the electron.

Muons can be produced by the absorption of cosmic radiation high in the atmosphere. These unstable particles have a lifetime of only 2.2 μs when measured in a reference frame at rest with respect to them. If we take 2.2 μs as the average lifetime of a muon and assume that their speed is close to the speed of light, we find that these particles can travel only about 600 m before they decay into something else (Fig. 28.8a). Hence, they could never reach the Earth from the upper atmosphere where they are produced. However, experiments show that a large number of muons *do* reach the Earth, and the phenomenon of time dilation explains how. Relative to an observer on Earth, the muons have a lifetime equal to $\gamma\tau$, where $\tau = 2.2$ μs is the lifetime in a frame of reference traveling with the muons. For example, for $v = 0.99c$, $\gamma \approx 7.1$ and $\gamma\tau \approx 16$ μs. Hence, the average distance traveled as measured by an observer on Earth is $\gamma v\tau \approx 4800$ m, as indicated in Figure 28.8b.

In 1976, experiments with muons were conducted at the laboratory of the European Council for Nuclear Research (CERN) in Geneva. Muons were injected into a large storage ring, reaching speeds of about $0.9994c$. Electrons produced by the decaying muons were detected by counters around the ring, enabling scientists to measure the decay rate, and hence the lifetime of the muons. The lifetime of the moving muons was measured to be about 30 times as long as that of the stationary muon, in agreement with the prediction of relativity to within two parts in a thousand.

The results of an experiment reported by Hafele and Keating provided direct evidence for the phenomenon of time dilation.[2] The experiment involved the use of very stable cesium-beam atomic clocks. Time intervals measured with four such clocks in jet flight were compared with time intervals measured by reference atomic clocks at the U.S. Naval Observatory. (Because of the Earth's rotation about its axis, a ground-based clock is not in a true inertial frame.) Time intervals measured with the flying clocks were compared with time intervals measured with the Earth-based reference clocks. In order to compare the results with the theory, many factors had to be considered, including periods of acceleration and deceleration relative to the Earth, variations in direction of travel, and the weaker gravitational field experienced by the flying clocks. Their results were in good agreement with the predictions of the special theory of relativity. In their paper, Hafele and Keating report the following: "Relative to the atomic time scale of the U.S. Naval Observatory, the flying clocks lost 59 ± 10 ns during the eastward trip and gained 273 ± 7 ns during the westward trip. . . . These results provide an unambiguous empirical resolution of the famous clock paradox with macroscopic clocks."

Figure 28.8
(a) The muons travel only about 600 m as measured in the muons' reference frame, where their lifetime is about 2.2 μs. Because of time dilation, the muons' lifetime is longer as measured by the observer on Earth. (b) Muons traveling with a speed $0.99c$ travel a distance of about 4800 m as measured by an observer on Earth.

Example 28.1 What Is the Period of the Pendulum?

The period of a pendulum is measured to be 3 s in the inertial frame of the pendulum. What is the period when measured by an observer moving with a speed of $0.95c$ with respect to the pendulum?

Solution
In this case, the proper time is equal to 3 s. We can use Equation 28.8 to calculate the period measured by the moving observer:

[2] J. C. Hafele and R. E. Keating, "Around the World Atomic Clocks: Relativistic Time Gains Observed," *Science*, July 14, 1972, p. 168.

$$T = \gamma T' = \frac{1}{\sqrt{1 - \dfrac{(0.95c)^2}{c^2}}} \, T' = (3.2)(3 \text{ s}) = \boxed{9.6 \text{ s}}$$

That is, the observer moving with a speed of $0.95c$ observes that the pendulum slows down.

The Twin Paradox

An interesting consequence of time dilation is the so-called twin paradox. Consider a controlled experiment involving 20-year-old twin brothers Speedo and Goslo. Speedo, the more venturesome twin, sets out on a journey toward a star located 30 lightyears from Earth. His spaceship is able to accelerate to a speed close to the speed of light. After reaching the star, Speedo becomes very homesick and immediately returns to Earth at the same high speed. Upon his return, he is shocked to find that many things have changed. Old cities have expanded and new cities have appeared. Lifestyles, people's appearance, and transportation systems have changed dramatically. Speedo's twin brother, Goslo, has aged to about 80 years old and is now wiser, feeble, and somewhat hard of hearing. Speedo, on the other hand, has aged only about ten years. This is because his bodily processes slowed down during his travels in space.

The space traveler ages more slowly than his twin who remains on Earth

It is quite natural to raise the question, "Which twin actually traveled at a speed close to the speed of light, and therefore would be the one who did not age?" Herein lies the paradox: From Goslo's frame of reference, he was at rest while his brother Speedo traveled at a high velocity. On the other hand, according to the space traveler Speedo, it was he who was at rest while his brother zoomed away from him on Earth and then returned. This leads to contradictions as to which twin actually aged.

In order to resolve this paradox, it should be pointed out that the trip is not as symmetrical as we may have led you to believe. Speedo, the space traveler, had to experience a series of accelerations and decelerations during his journey to the star and back home, and therefore is not always in uniform motion. This means that Speedo has been in a noninertial frame during a large part of his trip, so that predictions based on special relativity are not valid in his frame. On the other hand, the brother on Earth has been in an inertial frame, and he can make reliable predictions based on the special theory. The situation is not symmetrical since Speedo experiences forces when his spaceship turns around, whereas Goslo is not subject to such forces. Therefore, the space traveler will indeed be younger upon returning to Earth.

Length Contraction

We have seen that measured time intervals are not absolute, that is, the time interval between two events depends on the frame of reference in which it is measured. Likewise, the measured distance between two points depends on the frame of reference. The **proper length** of an object is defined as *the length of the object measured in the reference frame in which the object is at rest*. The length of an object measured in a reference frame in which the object is

Figure 28.9
A meter stick moves to the right with a speed v. (a) The meter stick as viewed by an observer at rest with respect to the meter stick. (b) The meter stick as seen by an observer moving with a speed v with respect to the meter stick. The moving meter stick is always measured to be *shorter* than in its own rest frame by a factor $\sqrt{1 - v^2/c^2}$.

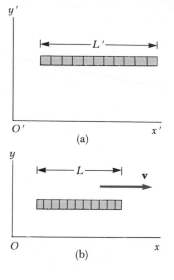
(a)

(b)

moving is always less than the proper length. This effect is known as **length contraction.**

To understand length contraction quantitatively, let us consider a spaceship traveling with a speed v from one star to another, as seen by two observers. An observer at rest on Earth (and also assumed to be at rest with respect to the two stars) measures the distance between the stars to be L' (where L' is the proper length), and according to this observer, it takes a time $\Delta t = L'/v$ for the spaceship to complete the voyage. What does an observer in the spaceship measure? Because of time dilation, the space traveler measures a smaller time of travel: $\Delta t' = \Delta t/\gamma$. The observer in the spaceship claims to be at rest and sees the destination star as moving toward the ship with speed v. Since the space traveler reaches the star in the time Δt, he or she concludes that the distance, L, between the stars is shorter than L'. This distance is given by

$$L = v\,\Delta t' = v\,\frac{\Delta t}{\gamma}$$

Since $L' = v\,\Delta t$, we see that

$$L = \frac{L'}{\gamma}$$

or,

$$L = L'\sqrt{1 - v^2/c^2} \qquad \textbf{(28.9)} \qquad \text{Length contraction}$$

According to this result, which is illustrated in Figure 28.9, if an observer at rest with respect to an object measures its length to be L', an observer moving with a relative speed v with respect to the object will find it to be shorter than its rest length by the factor $\sqrt{1 - v^2/c^2}$.

You should note that *the length contraction takes place only along the direction of motion.* For example, suppose a spaceship as seen in a rest frame is in the shape of a cube 1 km on a side, as in Figure 28.10a. When this ship is in motion with a speed v relative to an observer at rest, measurements made by the observer will indicate that the ship has the shape shown in Figure 28.10b.

(a)   (b) $v = 0.8c$

Figure 28.10
A spaceship in the shape of a cube when viewed in a rest frame as in (a) takes on the shape shown in (b) when in motion to the right.

Example 28.2 The Contraction of a Spaceship

A spaceship is measured to be 100 m long while it is at rest with respect to an observer. If this spaceship now flies by the observer with a speed of $0.99c$, what length will the observer find for the spaceship?

Solution
From Equation 28.9, the length measured by the observer is

$$L = L'\sqrt{1 - v^2/c^2} = (100 \text{ m})\sqrt{1 - \frac{(0.99c)^2}{c^2}} = \boxed{14 \text{ m}}$$

Exercise If the ship moves past the observer with a speed of $0.01c$, what length will the observer measure?

Answer 99.99 m.

Example 28.3 How High is the Spaceship?

An observer on Earth sees a spaceship at an altitude of 435 m moving downward toward the Earth with a speed of $0.97c$. What is the altitude of the spaceship as measured by an observer in the spaceship?

Solution
The moving observer in the ship finds the altitude to be

$$L = L'\sqrt{1 - v^2/c^2} = (435 \text{ m})\sqrt{1 - \frac{(0.97c)^2}{c^2}} = \boxed{106 \text{ m}}$$

Example 28.4 The Triangular Spaceship

A spaceship in the form of a triangle flies by an observer with a speed of $0.95c$. When the ship is at rest (Fig. 28.11a), the distances x and y are found to be 50 m and 25 m, respectively. What is the shape of the ship as seen by an observer at rest when the ship is in motion along the direction shown in Figure 28.11b?

Solution
The observer sees the horizontal length of the ship to be contracted to a length of

$$L = L'\sqrt{1 - v^2/c^2} = (50 \text{ m})\sqrt{1 - \frac{(0.95c)^2}{c^2}} = \boxed{15.6 \text{ m}}$$

The 25-m vertical height is unchanged because it is perpendicular to the direction of relative motion between the observer and the spaceship. Figure 28.11b represents the shape of the spaceship as seen by the observer at rest.

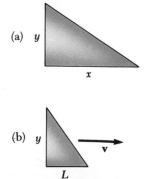

Figure 28.11 (Example 28.4) (a) When the spaceship is at rest, its shape is as shown. (b) The spaceship appears to look like this when it moves to the right with a speed v. Note that only its x dimension is contracted in this case.

28.6 Relativistic Momentum

In order to properly describe the motion of particles within the framework of special relativity, we must generalize Newton's laws and the definitions of momentum and energy. These generalized definitions reduce to the classical (nonrelativistic) definitions when v is much less than c.

First, recall that conservation of momentum states that when two objects collide, the total momentum of the system remains constant, assuming that the objects are isolated (that is, they interact only with each other). If such a collision is analyzed within the framework of Einstein's postulates of relativity, it is found that momentum is not conserved if the classical definition of momentum, $p = m_0 v$, where the subscript on m_0 is used in discussions of relativity to indicate the mass of an object at rest or moving at low speed is used. However, according to the principle of relativity, momentum must be conserved in all systems. In view of this condition, it is necessary to modify the definition of momentum to satisfy the following conditions:

1. The relativistic momentum must be conserved in all collisions.
2. The relativistic momentum must approach the classical value, $m_0 v$, as the quantity v/c approaches zero.

The correct relativistic equation for momentum that satisfies these conditions is

$$p \equiv \frac{m_0 v}{\sqrt{1 - v^2/c^2}} = \gamma m_0 v \qquad \textbf{(28.10)} \qquad \text{Momentum}$$

where m_0 is the mass of the particle and v is its velocity. The theoretical proof of this generalized expression for momentum is beyond the scope of this text. Note that when v is much less than c, the denominator of Equation 28.10 approaches unity and so p approaches $m_0 v$. Therefore, the relativistic equation for momentum reduces to the classical expression when v is small compared with c.

Example 28.5   The Relativistic Momentum of an Electron

An electron, which has a mass of 9.11×10^{-31} kg, moves with a speed of $0.75c$. Find its relativistic momentum and compare this value with the momentum calculated from the classical expression.

Solution
From Equation 28.10, with $v = 0.75c$, we have

$$p = \frac{m_0 v}{\sqrt{1 - v^2/c^2}}$$

$$= \frac{(9.11 \times 10^{-31} \text{ kg})(0.75 \times 3 \times 10^8 \text{ m/s})}{\sqrt{1 - (0.75c)^2/c^2}}$$

$$= 3.10 \times 10^{-22} \text{ kg} \cdot \text{m/s}$$

The classical expression gives

$$\text{Momentum} = m_0 v = 2.05 \times 10^{-22} \text{ kg} \cdot \text{m/s}$$

The (correct) relativistic result is 50% greater than the classical result!

Figure 28.12
Variation of mass with speed. This theoretical curve is in excellent agreement with experimental data on particles such as electrons.

28.7   Mass and the Ultimate Speed

We have seen that the values of length and time intervals depend on the reference frame in which they are measured. On this basis, mass must be expected to be a relative quantity also. In fact, Einstein showed that *the observed mass of an object increases with speed according to the relation*

$$m \equiv \frac{m_0}{\sqrt{1 - v^2/c^2}} \qquad (28.11)$$

where m is the mass of the object as measured by an observer moving with a speed v and m_0 is the mass of the object as measured by an observer at rest with respect to the object. We shall refer to m_0 as the **rest mass** of the object. From this definition, the relativistic momentum given by Equation 28.10 can be interpreted as being the product of the relativistic mass, m, and the velocity of the object; that is, $p = mv$. Equation 28.11 indicates that *the mass of an object increases as its speed increases*, as shown in Figure 28.12. This prediction has been borne out through observations on elementary particles, such as energetic electrons.

It is also possible to see from Equation 28.11 that the ultimate speed of an object is the speed of light. The equation indicates that, as an object is accelerated, its mass increases, and hence the unbalanced force on the object has to increase in proportion in order to maintain a constant acceleration. Finally, when $v = c$, the mass becomes infinite. This means that an infinite amount of energy would be required to accelerate the object to the speed of light. Thus, *the speed of light is the ultimate speed for any material object.*

Example 28.6   The Mass of a Speedy Ball

Superman, who has an exceptionally strong arm, throws a fast ball with a speed of $0.9c$. If the rest mass of the ball is 0.5 kg, what is its mass in flight?

Solution
This is a straightforward application of Equation 28.11. The relativistic mass is

$$m = \frac{m_0}{\sqrt{1 - v^2/c^2}} = \frac{0.5 \text{ kg}}{\sqrt{1 - \frac{(0.9c)^2}{c^2}}} = \boxed{1.15 \text{ kg}}$$

28.8   Relativistic Addition of Velocities

Imagine a motorcycle rider moving with a speed of $0.8c$ past a stationary observer, as shown in Figure 28.13. If the rider tosses a ball in the forward direction with a speed of $0.7c$ relative to himself, what is the speed of the ball as seen by the stationary observer at the side of the road? Common sense, and the ideas of Newtonian relativity, says that the speed should be the sum of the two speeds, or $1.5c$. This answer must be incorrect since it contradicts the assertion in the last section that no material object can travel faster than the speed of light.

Figure 28.13
A motorcycle moves past a stationary observer with a speed $0.8c$; the motorcyclist throws a ball in the direction of motion with a speed $0.7c$ relative to himself.

Einstein resolved this dilemma by deriving an equation for the relativistic addition of velocities. For one-dimensional motion, this equation is

$$u = \frac{v + u'}{1 + \dfrac{vu'}{c^2}}$$

(28.12)   Velocity addition

where u is the velocity of an object as seen in one frame of reference (frame **A**), u' is the velocity of the same object as measured in a different frame of reference (frame **B**), and v is the velocity of the reference frame in which u' is measured (**B**) relative to the other frame (**A**). Let us apply this equation to the case of the speedy motorcycle rider and the stationary observer.

The velocity of the motorcycle with respect to the stationary observer is $0.8c$. Thus, $v = 0.8c$. The velocity of the ball in the frame of the motorcyclist is $0.7c$. That is, $u' = 0.7c$. Therefore, the velocity, u, of the ball relative to the stationary observer is

$$u = \frac{0.8c + 0.7c}{1 + \dfrac{(0.8c)(0.7c)}{c^2}} = 0.96c$$

Example 28.7   Measuring the Speed of a Light Beam

Suppose that the motorcyclist moving with a speed of $0.8c$ turns on a beam of light that moves away from the motorcycle with a speed of c in the same direction as the moving motorcycle. What would the stationary observer measure for the speed of the beam of light?

Solution
In this case, $v = 0.8c$ and $u' = c$. We can use these values in Equation 28.12 to give

$$u = \frac{0.8c + c}{1 + \dfrac{(0.8c)(c)}{c^2}} = \boxed{c}$$

This is consistent with the statement made earlier that *all observers measure the speed of light to be c regardless of the motion of the source of light.*

28.9   Relativistic Energy

We have seen that the definition of momentum required generalization to make it compatible with the principle of relativity. Likewise, the definition of kinetic energy requires modification in relativistic mechanics. In view of the definition of relativistic momentum, we might guess that the classical expression for kinetic energy, $mv^2/2$, could be translated to a relativistic expression by substituting the value for m defined in Equation 28.11. This, however, is not correct. Einstein found that the correct expression for the **kinetic energy** of an object is

Kinetic energy

$$KE = mc^2 - m_0 c^2 \qquad \textbf{(28.13)}$$

The term $m_0 c^2$, which is independent of the speed of the object, is called the **rest energy** of the object. The term mc^2, which depends on the object speed, is therefore the sum of the kinetic and rest energies. We define mc^2 to be the **total energy**, E, that is,

$$E = mc^2 = KE + m_0 c^2 \qquad \textbf{(28.14)}$$

This is Einstein's famous *mass-energy equivalence equation.* If we make use of Equation 28.11, we can express the total energy as[3]

Total energy

$$E = \frac{m_0 c^2}{\sqrt{1 - v^2/c^2}} \qquad \textbf{(28.15)}$$

The relation $E = mc^2$ shows that *mass is a form of energy.* Furthermore, this result shows that *even a small mass corresponds to an enormous amount of energy.* The concept stated in Equation 28.15 has revolutionized the field of nuclear physics. The validity of the relationship between mass and energy has been proved beyond question.

In many situations, the momentum or energy of a particle is known rather than its speed. It is therefore useful to have an expression relating the total

---

[3] At first glance, Equation 28.13 does not look even remotely similar to the classical expression for the kinetic energy. However, we shall show that the two equations are equivalent in the limit of low velocities. First, let us substitute the expression for the total energy from Equation 28.15 into Equation 28.13 to find

$$KE = mc^2 - m_0 c^2 = \frac{m_0 c^2}{\sqrt{1 - \dfrac{v^2}{c^2}}} - m_0 c^2$$

$$= m_0 c^2 \left[\left(1 - \frac{v^2}{c^2} \right)^{-1/2} - 1 \right]$$

In the limit of low velocities, v becomes very small compared to c, hence the first term in the brackets can be approximated by the first two terms in the binomial series expansion for $\left(1 - \dfrac{v^2}{c^2} \right)^{-1/2}$ to give

$$KE \approx m_0 c^2 \left[1 + \frac{1}{2} \frac{v^2}{c^2} - 1 \right] = \frac{1}{2} m_0 v^2$$

energy, E, to the relativistic momentum, p. This is accomplished by using the expressions $E = mc^2$ and $p = mv$, as shown in the footnote. The result is

$$E^2 = p^2c^2 + (m_0c^2)^2 \qquad (28.16)$$

When the particle is at rest, $p = 0$, and so $E = m_0c^2$. That is, by definition, *the total energy of a particle at rest equals its rest energy*. As we shall discuss in Chapter 29, it is well established that there are particles that have zero rest mass, such as photons. If we set m_0 equal to zero in Equation 28.16, we see that the *total energy of a photon* can be expressed as

$$E = pc \qquad (28.17)$$

This equation is an exact expression relating energy and momentum for photons, which always travel at the speed of light.

When dealing with electrons or other subatomic particles, it is convenient to express the energy in electron volts (eV) because the particles are usually given energy by acceleration through a potential difference. Recall that

$$1 \text{ eV} = 1.60 \times 10^{-19} \text{ J}$$

For example, the mass of an electron is 9.11×10^{-31} kg. Hence, its rest energy is

$$m_0c^2 = (9.11 \times 10^{-31} \text{ kg})(3 \times 10^8 \text{ m/s})^2 = 8.2 \times 10^{-14} \text{ J}$$

Converting this to electron volts, we have

$$m_0c^2 = (8.2 \times 10^{-14} \text{ J})\left(\frac{1 \text{ eV}}{1.60 \times 10^{-19} \text{ J}}\right) = 0.511 \text{ MeV}$$

where 1 MeV = 10^6 eV.

Example 28.8 The Energy Contained in a Baseball

If a 0.5-kg baseball could be converted completely to energy, how much energy would be released?

---

[4] The expression relating the total energy, E, to the momentum, p, can be found by evaluating the quantity $m^2c^4 - m_0{}^2c^4$ as

$$m^2c^4 - m_0{}^2c^4 = \frac{m_0{}^2c^4}{1 - v^2/c^2} - m_0{}^2c^4 = m_0{}^2c^4\left(\frac{v^2/c^2}{1 - v^2/c^2}\right) = \frac{m_0{}^2v^2c^2}{1 - v^2/c^2}$$

But,

$$p = \frac{m_0v}{\sqrt{1 - v^2/c^2}}$$

so,

$$m^2c^4 - m_0{}^2c^4 = p^2c^2$$

Thus,

$$m^2c^4 = p^2c^2 + (m_0c^2)^2$$

Since $E^2 = m^2c^4$, we have

$$E^2 = p^2c^2 + (m_0c^2)^2$$

Solution

The energy equivalent of the baseball is found from Equation 28.14 (with $KE = 0$):

$$E = m_0 c^2 = (0.5 \text{ kg})(3 \times 10^8 \text{ m/s})^2 = \boxed{4.5 \times 10^{16} \text{ J}}$$

This is enough energy to keep a 100-W lightbulb burning for approximately ten million years. However, you should be aware that it is impossible to achieve complete conversion from mass to energy in practical situations. For example, mass is converted to energy in nuclear power plants, but only a small fraction of the mass actually undergoes conversion.

Example 28.9   The Energy of a Speedy Electron

An electron moves with a speed of $v = 0.85c$. Find its total energy and kinetic energy in electron volts.

Solution

The fact that the rest energy of an electron is 0.511 MeV, along with Equation 28.15, gives

$$E = \frac{m_0 c^2}{\sqrt{1 - v^2/c^2}} = \frac{0.511 \text{ MeV}}{\sqrt{1 - \dfrac{(0.85c)^2}{c^2}}}$$

$$= 1.90(0.511 \text{ MeV}) = \boxed{0.970 \text{ MeV}}$$

The kinetic energy is obtained by subtracting the rest energy from the total energy:

$$KE = E - m_0 c^2 = 0.970 \text{ MeV} - 0.511 \text{ MeV} = \boxed{0.459 \text{ MeV}}$$

Example 28.10   The Energy of a Speedy Proton

The total energy of a proton is three times its rest energy. (a) Find the proton's rest energy in electron volts.

Solution

$$\text{Rest energy} = m_0 c^2 = (1.67 \times 10^{-27} \text{ kg})(3 \times 10^8 \text{ m/s})^2$$

$$= (1.50 \times 10^{-10} \text{ J}) \left(\frac{1 \text{ eV}}{1.60 \times 10^{-19} \text{ J}} \right) = \boxed{938 \text{ MeV}}$$

(b) With what speed is the proton moving?

Solution

Since the total energy, E, is three times the rest energy, Equation 28.15 gives

$$E = mc^2 = 3m_0 c^2 = \frac{m_0 c^2}{\sqrt{1 - v^2/c^2}}$$

$$3 = \frac{1}{\sqrt{1 - v^2/c^2}}$$

Solving for v gives

$$1 - \frac{v^2}{c^2} = \frac{1}{9}$$

$$\frac{v^2}{c^2} = \frac{8}{9}$$

$$v = \frac{\sqrt{8}}{3}c = \boxed{2.83 \times 10^8 \text{ m/s}}$$

(c) Determine the kinetic energy of the proton in electron volts.

Solution

$$KE = E - m_0c^2 = 3m_0c^2 - m_0c^2 = 2m_0c^2$$

Since $m_0c^2 = 938$ MeV, $KE = \boxed{1876 \text{ MeV}}$

Summary

The two basic postulates of the **special theory of relativity** are

1. The laws of physics are the same in every inertial frame of reference.
2. The speed of light is the same for all inertial observers, independent of their motion or of the motion of the source of light.

Some of the consequences of the special theory of relativity are as follows:

1. Clocks in motion relative to an observer slow down. This is known as **time dilation**. The relationship between time intervals in the moving and at rest systems is

$$\Delta t = \gamma \Delta t' \tag{28.8}$$

where Δt is the time interval in the system in relative motion with respect to the clock, $\gamma = 1/\sqrt{1 - v^2/c^2}$, and $\Delta t'$ is the time interval in the system moving with the clock.
2. The length of an object in motion is *contracted* in the direction of motion. The equation for **length contraction** is

$$L = L'\sqrt{1 - v^2/c^2} \tag{28.9}$$

where L is the length in the system in motion relative to the object, and L' is the length in the system in which the object is at rest.
3. Events that are simultaneous for one observer are not simultaneous for another observer in motion relative to the first.

The relativistic expression for the **momentum** of a particle moving with a velocity v is

$$p \equiv \frac{m_0 v}{\sqrt{1 - v^2/c^2}} = \gamma m_0 v \qquad \textbf{(28.10)}$$

The **mass** of an object increases as its speed increases as

$$m \equiv \frac{m_0}{\sqrt{1 - v^2/c^2}} \qquad \textbf{(28.11)}$$

where m_0 is the rest mass of the object, that is, its mass as measured by an observer at rest with respect to it.

The relativistic expression for the addition of velocities is

$$u = \frac{v + u'}{1 + \dfrac{vu'}{c^2}} \qquad \textbf{(28.12)}$$

where u is the velocity of an object as seen in one frame of reference, u' is the velocity of the same object in a different frame of reference, and v is the relative velocity of the two frames.

The relativistic expression for the **kinetic energy** of an object is

$$KE = mc^2 - m_0 c^2 \qquad \textbf{(28.13)}$$

where $m_0 c^2$ is the **rest energy** of the object.

The **total energy** of a particle is

$$E = \frac{m_0 c^2}{\sqrt{1 - v^2/c^2}} \qquad \textbf{(28.15)}$$

This is Einstein's famous mass-energy equivalence equation.

The relativistic momentum is related to the total energy through the equation

$$E^2 = p^2 c^2 + (m_0 c^2)^2 \qquad \textbf{(28.16)}$$

Additional Reading

H. Bondi, *Relativity and Common Sense,* Science Study Series, Garden City, NY, Doubleday, 1964.

J. Bronowski, "The Clock Paradox," *Sci. American,* February 1963, p. 134.

R. W. Clark, *Einstein: The Life and Times,* New York, World Publishing, 1971.

J. Crelinsten, "Relativity, Einstein, Physicists, and the Public," *The Physics Teacher,* February 1980, p. 115.

J. Crelinsten, "Physicists Receive Relativity: Revolution and Reaction," *The Physics Teacher,* February 1980, p. 187.

A. Einstein, *Ideas and Opinions,* New York, Crown, 1954.

A. Einstein, *Out of My Later Years,* Secaucus, NJ, Citadel Press, 1973.

G. Gamow, *Mr. Tomkins in Wonderland,* New York, Cambridge University Press, 1939.

R. D. Henry, "Special Relativity Made Transparent," *The Physics Teacher,* December 1985, p. 536.

L. Infeld, *Albert Einstein,* New York, Scribner's, 1950.

P. K. MacKeown, "Gravity is Geometry," *The Physics Teacher,* December 1984, p. 557.

J. Schwinger, *Einstein's Legacy,* Scientific American Library, New York, NY, W. H. Freeman and Co., 1985.

R. S. Shankland, "The Michelson-Morley Experiment," *Sci. American,* November 1964, p. 107.

G. J. Whitrow, ed., *Einstein, The Man and His Achievement,* New York, Dover, 1967.

Questions

1. What one measurement will two observers in relative motion always agree upon?

2. It is said that Einstein, in his teenage years, asked the question, "What would I see in a mirror if I carried it in my hands and ran at the speed of light?" How would you answer this question?

3. Some of the distant stars, called quasars, are receding from us at half the speed of light (or greater). What is the speed of light we receive from these quasars?

4. The speed of light in a transparent material medium is less than the speed of light in a vacuum. Is this a violation of relativity? Discuss.

5. High-energy particles sometimes travel through a material medium faster than the speed of light in that medium. Does this violate the principles of relativity?

6. A spaceship in the shape of a sphere moves past an observer on Earth with a speed of $0.5c$. What shape will the observer see as the spaceship moves past?

7. When defining length, explain why it is necessary to specify that the positions of the ends of a rod are to be determined simultaneously.

8. Suppose two observers who are moving relativistically each have meter sticks that they wish to use as standards. How can the observers be sure that their sticks have the same proper length? Discuss.

9. An astronaut moves away from the Earth at a speed close to the speed of light. If an observer on Earth could measure the astronaut's size and pulse rate, what changes (if any) would he or she measure? Would the astronaut measure any changes?

10. Suppose astronauts were paid according to the time spent traveling in space. After a long voyage at a speed near that of light, a crew of astronauts return and open their pay envelopes. What will their reaction be?

11. Assuming that the Universe started with a big bang 10 billion years ago and that evolution proceeded at the same pace everywhere, would it make any sense to try to listen for intelligent life elsewhere in the Universe? Discuss.

12. Causality means that cause precedes effect. What limitations does special relativity place on causality in the event that a cause occurring in one location produces an effect at another location? Discuss.

13. Imagine a rocket with a flasher on top that flashes a sequence of numbers 1, 2, 3, . . . at regular proper time intervals. Discuss the sequence in which the numbers are emitted, the position of the flasher, and the time interval between the emission of each number in a frame of reference in which the rocket is moving with velocity $0.99c$.

14. Give a physical argument showing that it is impossible to accelerate an object of mass m to the speed of light, even with a continuous force acting on it.

15. What happens to the density of an object as its speed increases?

16. Since mass is a form of energy, can we conclude that a compressed spring has more mass than the same spring when it is not compressed? On the basis of your answer, which has more mass, a spinning planet or an otherwise identical but nonspinning planet?

17. Consider the incorrect statement, "Matter can neither be created nor destroyed." How would you correct this statement in view of the special theory of relativity?

18. The expression $E = mc^2$ is often given in popular descriptions of Einstein's work. Is this expression strictly correct? For example, does it accurately take into account the kinetic energy of a moving mass?

19. List some ways in which our day-to-day lives would change if the speed of light were 100 mi/h.

20. Relativistic quantities must make a smooth transition to their Newtonian counterparts as the velocity of a system becomes small with respect to the speed of light. Explain.

Problems

Section 28.3 The Michelson-Morley Experiment

1. Two airplanes fly paths I and II specified in Figure 28.3a. Both planes have air speeds of 100 m/s and fly for a distance of 200 km. The wind blows at 20 m/s in the direction shown in the figure. Find (a) the time of flight to each city, (b) the time to return, and (c) the difference in total flight times.

2. In one version of the Michelson-Morley experiment, the lengths L in Figure 28.4 were 28 m. Take v to be 3×10^4 m/s and find (a) the time difference caused by rotation of the interferometer and (b) the expected fringe shift, assuming that the light used has a wavelength of 550 nm.

Section 28.5   Consequences of Special Relativity

3. An astronaut at rest on Earth has a heartbeat rate of 70 beats/min. When the astronaut is traveling in a space-ship at $0.9c$ what will be this rate as measured (a) by an observer also in the ship and (b) by an observer at rest on the Earth?

4. The period of a pendulum is found to be 2 s when measured in the inertial frame of the pendulum. An observer moves by at a high rate of speed and concludes that the period is 7 s. How fast is he moving?

5. With what speed will a clock have to be moving in order to run at a rate that is one-half the rate of a clock' at rest?

6. A deep-space probe moves away from Earth with a speed of $0.8c$. An antenna on the probe requires 3.0 s, probe time, to rotate through 1.0 rev. How much time is required for 1.0 rev according to an observer on Earth?

7. If astronauts could travel at $v = 0.95c$, we on Earth would say it takes $(4.2/0.95) = 4.42$ years to reach Alpha Centauri, 4.2 lightyears away. The astronauts disagree. (a) How much time passes on the astronaut's clocks? (b) What is the distance to Alpha Centauri, as measured by the astronauts?

8. The average lifetime of a pi meson in its own frame of reference is 2.6×10^{-8} s. (This is the proper lifetime.) If the meson moves with a speed of $0.98c$, what is (a) its mean lifetime as measured by an observer on Earth and (b) the average distance it travels before decaying as measured by an observer on Earth? (c) What distance would it travel if time dilation did not occur?

9. How fast must a meter stick be moving if its length is observed to shrink to 0.5 m?

10. A cube is 2 cm on a side when at rest. (a) What shape does it take on when moving past an observer at a speed of 2.5×10^8 m/s, and (b) what is the length of each side?

11. An unidentified flying object (UFO) flashes across the night sky. A UFO enthusiast on the top of Pike's Peak determines that the length of the UFO is 100 m along the direction of motion. If the UFO is moving with a speed of $0.9c$, how long is the UFO according to its pilot?

12. A rocket sled is designed so that if it encounters a hole in the ice, it will pass safely over the hole as long as the length of the hole is shorter than the length of the sled. Suppose the sled encounters a hole whose proper length is $\frac{2}{3}$ the proper length of the sled. What is the largest safe speed of the sled?

13. The nearest stars to Earth are approximately 4 lightyears away. How fast would you have to travel in a spaceship to cause that distance to shrink to 1.5 lightyears?

14. In a high-energy physics experiment, a certain elementary particle is created that travels a distance of 0.15 m at a speed of 2.2×10^8 m/s before it decays into two other particles. Find the lifetime of this particle in its frame of reference.

15. A friend travels by you in a spaceship at a high speed. He tells you that his ship is 20 m long and that the identical ship you are sitting in is 19 m long. According to your observations, (a) how long is your ship, (b) how long is his ship, and (c) what is the speed of your friend's ship?

16. A moving rod is 2.0 m long, and its length is oriented at an angle of 30° with respect to the direction of motion. The rod has a speed of $0.995c$. (a) What is the proper length of the rod? (b) What is the orientation angle in the proper frame?

17. Observer A measures the length of two rods, one stationary, the other moving with a speed of $0.955c$. She finds that the rods have the same length. A second observer B travels along with the moving rod. What is the ratio of the length of A's rod to the length of B's rod according to observer B?

18. The nearest star to Earth is approximately 4 lightyears away. If you travel at a speed of 2.5×10^8 m/s in a spaceship, how long does it take to get there (a) according to an Earthbound observer and (b) according to an observer on the spaceship?

19. An atomic clock is placed on a jet airplane. The clock measures a time interval of 3600 s when the jet moves with a speed of 400 m/s. What corresponding time interval does an identical clock held by an observer on the ground measure? (*Hint:* For $v/c \ll 1$, you can use the approximation $\gamma \approx 1 + v^2/2c^2$.)

20. A 2-m-long car is traveling at a speed of 65 mi/h. What is the decrease in its length as seen by a stationary observer? (See Hint in Problem 19.)

Section 28.6   Relativistic Momentum

21. Calculate the momentum of an electron moving with a speed of (a) $0.01c$, (b) $0.5c$, (c) $0.9c$.

22. Repeat Problem 21 for a proton.

23. An electron has a momentum that is 90% larger than its classical momentum. (a) Find the speed of the electron. (b) How would your result change if the particle were a proton?

24. An electron has a speed $v = 0.9c$. At what speed will a proton have a momentum equal to that of the electron?

25. An unstable particle at rest breaks up into two fragments of *unequal* mass. The rest mass of the lighter fragment is 2.5×10^{-28} kg, and that of the heavier fragment is 1.67×10^{-27} kg. If the lighter fragment has a speed of $0.893c$ after the breakup, what is the speed of the heavier fragment?

Section 28.7   Mass and the Ultimate Speed

26. What is the speed of an electron that has been accelerated in some device such that its mass is ten times its rest mass?

27. At what speed must a 0.15-kg baseball be thrown to have a mass of 0.4 kg?

28. At what speed is the mass of a particle equal to twice its rest mass? Does your answer depend on the initial mass of the particle?

29. At low speeds, a charged object in a magnetic field moves in a circular path of radius $r = mv/qB$. If an electron moves in an orbit of radius 10 cm with a speed of 10^5 m/s, what will be the radius when its speed is $0.96c$?

30. A synchrocyclotron is a particle accelerator capable of accelerating protons to a speed of $0.66c$. At this speed what is the mass of a proton?

31. A cube of mass 8 kg is 0.5 m on a side. (a) What is its density (mass per unit volume) as seen by an observer when the cube is moving away from the observer with a speed of $0.9c$? (b) What is the density as measured by an observer moving with the cube?

32. An electron is accelerated along a straight path in a linear accelerator. The mass of the electron becomes 10,000 times its rest mass when it reaches its final speed. (a) What is its final speed? (b) If the length of the accelerator is 3500 m, what is its length in the frame of reference of the electron?

Section 28.8   Relativistic Addition of Velocities

33. A space vehicle is moving with a speed of $0.75c$ with respect to an external observer. An atomic particle is projected with a speed of $0.9c$ in the same direction as the ship's velocity with respect to an observer inside the vehicle. What is the speed of the projectile as seen by the external observer?

34. An astronaut in a spaceship travels at a speed of $0.8c$ away from an observer at rest on the Earth. He turns around and shines a beam of light toward the stationary observer. What does the stationary observer measure for the speed of light?

35. Observer A sees an electron moving to the right at $0.8c$. What will be the velocity of the electron as measured by observer B who is moving to the left at $0.95c$ relative to A?

36. A rocket blasts off from a space station with a velocity of $0.92c$ to the left relative to the space station. Determine the velocity of the rocket in a frame of reference in which the space station has a velocity of $0.95c$ to the right.

37. An electron moves to the right with a speed of $0.9c$ relative to the laboratory frame. A proton moves to the right with a speed of $0.7c$ relative to the electron. Find the speed of the proton relative to the laboratory frame.

38. Spaceship R is moving toward the right with a speed of $0.7c$ with respect to the Earth. A second spaceship, L, moves to the left with the same speed with respect to the Earth. What is the speed of L with respect to R?

39. A pulsar is a stellar object that emits light in short bursts. Suppose a pulsar with a speed of $0.95c$ approaches the Earth, and a rocket with speed of $0.995c$ heads toward the pulsar (both speeds measured in the Earth's frame of reference). If the pulsar emits 10.0 pulses per second in its own frame of reference, at what rate are the pulses emitted in the rocket's frame of reference?

40. A rocket moves with a velocity of $0.92c$ to the right with respect to a stationary observer A. An observer B moving relative to observer A finds that the rocket is moving with a velocity of $0.95c$ to the left. What is the velocity of observer B relative to observer A? (*Hint:* Consider observer B's velocity in the frame of reference of the rocket.)

41. According to an observer A, a rocket moves at $0.91c$ to the left. In A's frame of reference the rocket has a mass of 1.2×10^4 kg. According to a second observer B, the rocket has a mass of 2.7×10^4 kg. (a) What is the *speed* of the rocket according to observer B? (b) What is the *velocity* of B relative to observer A? (*Caution:* There are two possible answers to Part (b) of this problem.)

Section 28.9   Relativistic Energy

42. What is the speed of a particle whose kinetic energy is equal to its own rest energy?

43. A proton moves with a speed of $0.95c$. Calculate its (a) rest energy, (b) total energy, and (c) kinetic energy.

44. Find the rest energy of a proton in eV.

45. Find the speed of a particle whose total energy is 50% greater than its rest energy.

46. A mass of 0.5 kg is converted completely into energy. (a) How much energy is produced and (b) how long equivalence principle applied to planetary sized bodies. Other measurements of planetary and lunar orbits have shown that the gravitational constant is a true constant of nature; it does not vary with time as the Universe ages. Ambitious experimenters are currently designing an experiment using supercooled quartz gyroscopes to be placed in Earth orbit, to try to detect an important general relativistic effect called the "dragging of inertial frames," caused by the rotation of the Earth.

General relativity has passed every experimental test to which it has been put, and many alternative theories have fallen by the wayside. Most physicists now take the theory for granted, and look to see how it can be used as a practical tool in physics and astronomy.

Gravitational Radiation

One of these new tools is gravitational radiation, a subject that is almost as old as general relativity itself. By 1916, Einstein had succeeded in showing that the field equations of general relativity admitted wavelike solutions analogous to those of electromagnetic theory. For example, a dumbbell rotating about an axis passing at right angles through its handle will emit gravitational waves that travel at the speed of light. They also carry energy away from the rotating dumbbell, just as electromagnetic waves carry energy away from a light source.

In 1968, Joseph Weber made the stunning announcement that he had detected gravitational radiation of extraterrestrial origin using massive aluminum bars as detectors. A passing gravitational wave acts as an oscillating gravitational force field that alternately compresses and extends the bar in the lengthwise direction. However, subsequent observations by other researchers using bars with sensitivity that was claimed to be better than Weber's failed to confirm Weber's results. His results are now generally regarded as a false alarm, although there is still no good explanation for the "events" that he recorded in his bars.

Nevertheless, Weber's experiments did initiate the program of gravitational-wave detection, and inspired other groups to build better detectors. Currently a dozen laboratories around the world are engaged in building and improving upon the basic "Weber bar" detector, striving to reduce noise from thermal, electrical, and environmental sources as much as possible to have a hope of detecting the very weak oscillations produced by a gravitational wave. For a bar of one meter length, the challenge is to detect a variation in length smaller than 10^{-20} meters, or 10^{-5} of the radius of a proton.

Another important type of detector is the laser interferometer, currently under development at several laboratories. This device operates on the same principle as the interferometer used by Michelson and Morley, but uses a laser as the light source. A passing gravitational wave will change the length of one arm of the apparatus relative to the other, and cause the interference pattern to vary. One ambitious proposal is to build two independent interferometers, each with arm lengths of 4 km, and separated by 5000 km.

It is hoped that these detectors will eventually be sensitive enough to detect gravitational waves from many sources, both in our galaxy and in distant galaxies, such as collapsing stars, double star systems, colliding black holes, and possibly even gravitational waves left over from the big bang. When this happens, a new field of "gravitational-wave astronomy" will be born.

Although gravitational radiation has not been detected directly, we know that it exists, through a remarkable system known as the binary pulsar. Discovered in 1974 by radio astronomers Russell Hulse and Joseph Taylor, it consists of a pulsar (which is a rapidly spinning neutron star) and a companion star in orbit around each other. Although the companion has not been seen directly, it is also believed to be a neutron star.

The pulsar acts as an extremely stable clock, its pulse period of approximately 59 milliseconds drifting by only a quarter of a nanosecond per year. By measuring the arrival times of radio pulses at Earth, observers were able to determine the motion of the pulsar about its companion with amazing accuracy. For example, the accurate value for the orbital period is 27906.980895 seconds, and the orbital eccentricity is 0.617131.

Like a rotating dumbbell, an orbiting binary system should emit gravitational radiation, and in the process lose some of its orbital energy. This energy loss will cause the pulsar and its companion to spiral in toward each other, and the orbital period to shorten. According to general relativity, the predicted decrease in the orbital period is 75.8 microseconds per year. The observed decrease rate is in agreement with the prediction to better than 1%.

This confirms the existence of gravitational radiation and the general relativistic equations that describe it.

Black Holes

One of the most important and exciting aspects of the relativity renaissance is the study of and search for black holes. However, the first glimmerings of the black hole idea date back to the 18th century, in the writings of a British amateur astronomer, the Reverend John Michell. Reasoning on the basis of the corpuscular theory that light would be attracted by gravity in the same way as ordinary matter is attracted, he noted that light emitted from the surface of a body such as the Earth or the Sun would be reduced in velocity by the time it reached great distances (Michell of course did not know special relativity). It would therefore be possible for a body to be sufficiently massive and compact to prevent light from escaping from its surface.

Although the general relativistic solution for a nonrotating black hole was discovered by Karl Schwarzschild in 1916, and a calculation of gravitational collapse to a black hole state was performed by J. Robert Oppenheimer and Hartland Snyder in 1939, black hole physics didn't really begin until the middle 1960s, when astronomers confronted the problem of the energy output of the quasars, and started to take black holes seriously.

A black hole is formed when a star has exhausted the thermonuclear fuel necessary to produce the heat and pressure that support it against gravity. The star begins to collapse, and if it is massive enough, it continues to collapse until the radius of the star approaches a value called the gravitational radius or Schwarzschild radius. In the nonrotating spherical case, this radius has a value given by $2GM/c^2$, where M is the mass of the star. For a body of one solar mass, the gravitational radius is about 3 km; for a body of the mass of the Earth, it is about 9 mm. An observer sitting on the surface of the star sees the collapse continue to smaller and smaller radii, until both star and observer reach the origin $r = 0$, with consequences too horrible to describe in detail. On the other hand, an observer at great distances observes the collapse to slow down as the radius approaches the gravitational radius, a result of the gravitational redshift of the light signals sent outward. The distant observer never sees any signals emitted by the falling observer once the latter is inside the gravitational radius. This is because any signal emitted inside can never escape the sphere bounded by the gravitational radius, called the "event horizon."

In 1974, Stephen Hawking discovered that the laws of quantum mechanics, when applied to the physics outside a black hole, required it to evaporate by the creation of particles with a thermal energy spectrum, and to have an associated temperature and entropy. The temperature of a Schwarzschild black hole is $T = hc^3/8\pi kGM$, where h is Planck's constant, and k is Boltzmann's constant. This discovery demonstrated a remarkable connection between gravity, thermodynamics, and quantum mechanics that helped renew the theoretical quest for a grand synthesis of all the fundamental interactions. On the other hand, for black holes of astronomical masses, the evaporation is completely negligible, since for a solar-mass black hole, $T \approx 10^{-6}$ K.

Although a great deal is known about black holes in theory, rather less is known about them observationally. There are several instances in which the evidence for the existence of black holes is impressive, but in all cases it is indirect. For instance, in the x-ray source Cygnus X1, the source of the x-rays is believed to be a black hole with a mass larger than about 6 solar masses in orbit around a giant star. The x-rays are emitted by matter pulled from the surface of the companion star and sent into a spiralling orbit around the black hole. Similarly, there is evidence in the centers of certain galaxies, such as M87, and possibly even our own, of black holes of between 10^2 and 10^8 solar masses. Accretion of matter onto supermassive rotating black holes may produce the jets of outflowing matter that are observed in many quasars and active galactic nuclei. These and other astrophysical processes that might aid in the detection of black holes are being studied by relativists and astrophysicists.

Cosmology

The other area in which there has been a renaissance for general relativity is cosmology. Although Einstein in 1917 first used general relativity to calculate a model for the Universe as a whole, the subject was not considered a serious branch of physics until the 1960s, when astronomical observations lent firm credence to the idea that the Universe was expanding from a "Big Bang." In 1965 came the discovery of the cosmic background radiation by Arno Penzias and Robert Wilson. This radiation is the remains of the hot electromagnetic blackbody radiation that once dominated the Universe in its earlier phase, now cooled to 3 K by the subsequent expansion of the Universe. Next came calculations of the amount of helium that would be synthesized from hydrogen by thermonuclear fusion in the very early Universe, around 1000 seconds after the Big Bang. The amount, approximately 25% by weight, was in agreement with the abundances of helium observed in stars and in interstellar space. This was an important confirmation of the hot Big-Bang picture, because the amount of helium believed to be produced by fusion in the interiors of stars is woefully inadequate to explain the observed abundances.

Today, the general relativistic hot Big-Bang model of the Universe has broad acceptance, and cosmologists now focus their attention on more detailed issues, such as how galaxies and other large-scale structures formed out of the hot primordial soup, and on what the Universe might have been like earlier than 1000 seconds. One of the leading models for this "early Universe" phase proposes that the laws of elementary-particle physics played a major role in the evolution, leading to a brief period of super-fast expansion known as "inflation."

Acceptance of General Relativity

One of the outgrowths of the renaissance of general relativity that has occurred since 1960 has been a change in attitude about the importance and use of the theory. Its importance as a fundamental theory of the nature of spacetime and gravitation has not been diminished in the least; if anything it has been enhanced by the flowering of research in the subject that has taken place. Its importance as a foundation for other theories of physics has been strengthened by current searches for unified quantum theories of nature that incorporate gravity along with the other interactions.

But the real change in attitude about general relativity has been in its use as a tool in the real world. In astrophysical situations, general relativity plays a central role in the study of neutron stars, black holes, relativistic binary star systems, and the Universe as a whole. Gravitational radiation may one day provide a completely new observational tool for exploring and examining the cosmos.

Relativity even plays a role in everyday life. For example, the gravitational redshift effect on clocks *must* be taken into account in satellite-based navigation systems, such as the U.S. Air Force's Global Positioning System, in order to achieve the required positional accuracy of a few meters or time transfer accuracy of a few nanoseconds.

To general relativists, always eager to find practical consequences of their subject, these have been very welcome developments!

Suggested Reading

P. C. W. Davies, *The Search for Gravity Waves*, Cambridge, England, Cambridge University Press, 1980.

G. Greenstein, *Frozen Star: Of Pulsars, Black Holes and the Fate of Stars*, New York, Freundlich Books, 1984.

S. W. Hawking, *A Brief History of Time*, New York, Bantam Books, 1988.

S. Weinberg, *The First Three Minutes: A Modern View of the Origin of the Universe*, New York, Basic Books, 1977.

C. M. Will, *Was Einstein Right? Putting General Relativity to the Test*, New York, Basic Books, 1986.

Essay Questions

1. Two stars are a certain distance from each other as seen in the night sky. Half a year later, the stars are now overhead during the day, and the Sun is now located midway between the two stars. Because of the deflection of light by the Sun, do the stars now appear closer together or farther apart?

2. When gravitational waves are emitted by the binary pulsar, the double-star system loses energy. As a consequence, its orbital velocity increases, and thus its kinetic energy increases. How is this possible?

3. The temperature of a black hole is inversely proportional to its mass, so that when it gains energy (which is equivalent to mass), its temperature goes down, and when it loses energy, its temperature goes up. What do you conclude about the sign of its heat capacity? Can a black hole ever be in thermal equilibrium with an infinite reservoir at a fixed temperature?

Essay Problems

1. What is the gravitational frequency shift between two atomic clocks separated by 1000 km, both at sea level?

2. A pair of identical twin scientists set out to test the gravitational redshift effect in the following way. One climbs to the top of Mount Baldy and lives there for a year, while the other remains behind in the laboratory at the base of the mountain. When they are reunited after one year, one of the twins is slightly older than the other. Which one is older and why? If Mount Baldy is 5000 m high, what is the difference in age between the twins after one year?

3. Calculate the deflection of light for light rays that pass the Sun at distances of 2 solar radii, 10 solar radii, and 100 solar radii.

4. Gravitational waves travel with the same speed as light, 3×10^{10} cm/s. Calculate the wavelength of gravitational waves emitted by the following

sources: (a) a star collapsing to form a black hole, emitting waves with a frequency around 1000 Hz, and (b) the pulsar, emitting waves with a period of four binary hours (one half the orbital period).

5. What is the fractional decrease in the orbital period of the binary pulsar per year? Estimate how long the binary system will live before gravitational radiation will reduce the orbital period to zero (in other words will bring the stars so close together that they will merge into one object).

CHAPTER 29

Quantum Physics

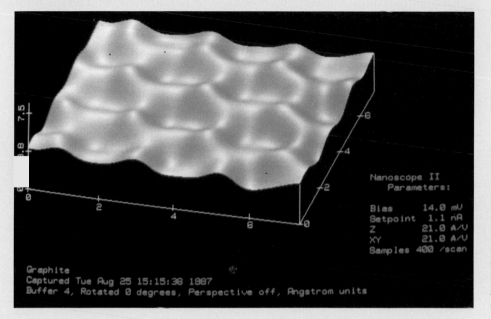

Nanoscope II
Parameters:

Bias   14.0 mV
Setpoint   1.1 nA
Z   21.0 A/V
XY   21.0 A/V
Samples   400 /scan

Graphite
Captured Tue Aug 25 15:15:38 1887
Buffer 4, Rotated 0 degrees, Perspective off, Angstrom units

The surface of graphite as "viewed" with a scanning tunneling microscope, discussed in the essay in this chapter. This technique enables scientists to see small details on surfaces, with a resolution of about 2 Å. The contours seen here represent the arrangement of individual atoms on the crystal surface. *(Roger A. Freedman and Paul K. Hansma, University of California at Santa Barbara)*

In the previous chapter, we discussed why Newtonian mechanics must be replaced by Einstein's special theory of relativity when we are dealing with particles whose speeds are comparable to the speed of light. Although many problems were indeed resolved by the theory of relativity in the early part of the 20th century, many experimental and theoretical problems remained unsolved. Attempts to apply the laws of classical physics to explain the behavior of matter on the atomic level were totally unsuccessful. Various phenomena, such as blackbody radiation, the photoelectric effect, and the emission of sharp spectral lines by atoms in a gas discharge tube, could not be understood within the framework of classical physics. We shall describe these phenomena because of their importance in subsequent developments.

Another revolution took place in physics between 1900 and 1930. This was the era of a new and more general formulation called *quantum mechanics*. This new approach was highly successful in explaining the behavior of atoms, molecules, and nuclei. As with relativity, the quantum theory requires a modification of our ideas concerning the physical world.

The earliest and most basic ideas of quantum theory were introduced by Planck, and most of the subsequent mathematical developments, interpretations, and improvements were made by a number of distinguished physicists, including Einstein, Bohr, Schrödinger, de Broglie, Heisenberg, Born, and Dirac. An extensive study of quantum theory is certainly beyond the scope of this book. This chapter is simply an introduction to the underlying ideas of quantum theory and the wave-particle nature of matter. We shall also discuss some simple applications of quantum theory, including the photoelectric effect, the Compton effect, and x-rays.

913

Figure 29.1
The opening in the cavity of a body is a good approximation to a black body. As light enters the cavity through the small opening, part is reflected and part is absorbed on each reflection from the interior walls. After many reflections, essentially all of the incident energy is absorbed.

Wavelength, microns
(1 micron $= 10^{-6}$ m)

Figure 29.2
Intensity of blackbody radiation versus wavelength at three different temperatures. Note that the total radiation emitted (the area under a curve) increases with increasing temperature.

29.1   Blackbody Radiation and Planck's Hypothesis

An object at any temperature is known to emit radiation sometimes referred to as **thermal radiation.** The characteristics of this radiation depend on the temperature and properties of the object. At low temperatures, the wavelengths of the thermal radiation are mainly in the infrared region and hence are not observed by the eye. As the temperature of the object is increased, it eventually begins to glow red. At sufficiently high temperatures, it appears to be white, as in the glow of the hot tungsten filament of a lightbulb. A careful study of thermal radiation shows that it consists of a continuous distribution of wavelengths from the infrared, visible, and ultraviolet portions of the spectrum.

From a classical viewpoint, thermal radiation (electromagnetic waves) originates from accelerated charged particles near the surface of the object, which emit radiation much like small antennas. The thermally agitated charges can have a distribution of accelerations, which accounts for the continuous spectrum of radiation emitted by the object. By the end of the 19th century, it had become apparent that the classical theory of thermal radiation was inadequate. The basic problem was in understanding the observed distribution of wavelengths in the radiation emitted by a black body. By definition, a black body is an ideal system that absorbs *all* radiation incident on it. A good approximation to a black body is the inside of a hollow object, as shown in Figure 29.1. The nature of the radiation emitted through a small hole leading to the cavity depends only on the temperature of the cavity walls.

Experimental data for the distribution of energy for blackbody radiation at three different temperatures are shown in Figure 29.2. Note that the radiated energy varies with wavelength and temperature. As the temperature of the black body increases, the total amount of energy it emits increases. Also, as temperatures increase, the peak of the distribution shifts to shorter wavelengths. This shift was found to obey the following relationship, called **Wien's displacement law:**

$$\lambda_{max} \, T = 0.2898 \times 10^{-2} \text{ m} \cdot \text{K} \qquad \textbf{(29.1)}$$

where λ_{max} is the wavelength at which the curve peaks and T is the absolute temperature of the object emitting the radiation.

Early attempts to use classical ideas to explain the shape of the curves shown in Figure 29.2 failed. Figure 29.3 shows an experimental plot of the blackbody radiation spectrum (red curve) together with the theoretical picture of what this curve should look like based on classical theories (blue curve). At long wavelengths, classical theory is in good agreement with the experimental data. However, at short wavelengths there is major disagreement between theory and experiment. This can be seen by noting that as λ approaches zero, the classical theory predicts that the amount of energy being radiated should increase, that is, short-wavelength radiation should be most intense and all objects should have a blue glow. This is contrary to the experimental data, which show that as λ approaches zero, the amount of energy carried by short-wavelength radiation also approaches zero. This contradiction is often called the **ultraviolet catastrophe.**

Max Planck was born in Kiel and received his college education in Munich and Berlin. He joined the faculty at Munich in 1880 and five years later received a professorship at Kiel University. He replaced Kirchhoff at the University of Berlin in 1889 and remained there until 1926. He introduced the concept of a "quantum of action" (Planck's constant h) in an attempt to explain the spectral distribution of blackbody radiation, which laid the foundations for quantum theory. He was awarded the Nobel prize in 1918 for this discovery of the quantized nature of energy. The work leading to the "lucky" blackbody radiation formula was described by Planck in his Nobel prize acceptance speech (1920): "But even if the radiation formula proved to be perfectly correct, it would after all have been only an interpolation formula found by lucky guesswork and, thus, would have left us rather unsatisfied. I therefore strived from the day of its discovery, to give it a real physical interpretation and this led me to consider the relations between entropy and probability according to Boltzmann's ideas."

Planck's life was filled with personal tragedies. One of his sons was killed in action in World War I, and two daughters were lost in childbirth in the same period. His house was destroyed by bombs in World War II, and his son Erwin was executed by the Nazis in 1944, after being accused of plotting to assassinate Hitler.

Planck became president of the Kaiser Wilhelm Institute of Berlin in 1930. The institute was renamed the Max Planck Institute in his honor after World War II. Although Planck remained in Germany during the Hitler regime, he openly protested the Nazi treatment of his Jewish colleagues, and consequently was forced to resign his presidency in 1937. Following World War II, he was renamed the president of the Max Planck Institute. He spent the last two years of his life in Göttingen as an honored and respected scientist and humanitarian.

Max Planck
(1858–1947)

(Courtesy of AIP Niels Bohr Library, W. F. Meggers Collection)

In 1900, Planck developed a formula for blackbody radiation that was in complete agreement with experiments at all wavelengths. Planck's analysis led to a curve that follows the blue curve in Figure 29.3. In his theory, Planck made two assumptions, which at the time were bold and controversial, concerning the nature of the oscillating charges of the cavity walls:

1. The vibrating molecules that emitted the radiation could have only certain *discrete* amounts of energy, E_n, given by

$$E_n = nhf \tag{29.2}$$

where n is a positive integer called a **quantum number** and f is the frequency of vibration of the molecules. The energies of the molecule are said to be *quantized*, and the allowed energy states are called *quantum states*. The factor h is a constant, known as **Planck's constant**, given by

$$h = 6.626 \times 10^{-34} \text{ J} \cdot \text{s} \tag{29.3}$$

2. The molecules emit energy in discrete units of light energy called **quanta** (or **photons**, as they are now called). They do so by "jumping" from one

Figure 29.3
Comparison of the Planck theory with the classical theory for the distribution of blackbody radiation.

quantum state to another. If the quantum number changes by one unit, Equation 29.2 shows that the amount of energy radiated by the molecule equals hf. Hence, the energy of a light quantum, corresponding to the energy difference between two adjacent levels, is given by

$$E = hf \qquad\qquad (29.4)$$

The molecule will radiate or absorb energy only when it changes quantum states. If it remains in one quantum state, no energy is absorbed or emitted.

The key point in Planck's theory is the radical assumption of quantized energy states. This development marked the birth of the quantum theory. At that time, most scientists, including Planck, did not consider the quantum concept to be realistic. Hence, Planck and others continued to search for a more rational explanation of blackbody radiation. However, subsequent developments showed that a theory based on the quantum concept (rather than on classical concepts) had to be used to explain a number of phenomena at the atomic level.

Example 29.1   Thermal Radiation from the Human Body

The temperature of the skin is approximately 35°C. What is the wavelength at which the peak occurs in the radiation emitted from the skin?

Solution
From Wien's displacement law, we have

$$\lambda_{max} T = 0.2898 \times 10^{-2} \text{ m} \cdot \text{K}$$

Solving for λ_{max}, noting that 35°C corresponds to an absolute temperature of 308 K, we have

$$\lambda_{max} = \frac{0.2898 \times 10^{-2} \text{ m} \cdot \text{K}}{308 \text{ K}} = \boxed{940 \ \mu\text{m}}$$

This radiation is in the infrared region of the spectrum.

Exercise   (a) Find the wavelength corresponding to the peak of the radiation curve for the heating element of an electric oven at a temperature of 1200 K. Note that although this radiation peak lies in the infrared, there is enough visible radiation at this temperature to give the element a red glow. (b) Calculate the wavelength corresponding to the peak of the radiation curve for an object whose temperature is 5000 K, an approximate temperature for the surface of the Sun.

Answer   (a) 2.415 μm (b) 580 nm, in the visible region.

Example 29.2   The Quantized Oscillator

A 2-kg mass is attached to a massless spring of force constant $K = 25$ N/m. The spring is stretched 0.4 m from its equilibrium position and released. (a) Find the total energy and frequency of oscillation according to classical calculations. (b) Assume that the energy is quantized and find the quantum number, n, for the system. (c) How much energy would be carried away in a one-quantum change?

Solution

(a) The total energy of a simple harmonic oscillator having an amplitude A is $\frac{1}{2}KA^2$. Therefore,

$$E = \frac{1}{2}KA^2 = \frac{1}{2}(25 \text{ N/m})(0.4 \text{ m})^2 = \boxed{2.0 \text{ J}}$$

The frequency of oscillation is

$$f = \frac{1}{2\pi}\sqrt{\frac{K}{m}} = \frac{1}{2\pi}\sqrt{\frac{25 \text{ N/m}}{2 \text{ kg}}} = \boxed{0.56 \text{ Hz}}$$

(b) If the energy is quantized, we have $E_n = nhf$, and from the result of (a), we have

$$E_n = nhf = n(6.63 \times 10^{-34} \text{ J} \cdot \text{s})(0.56 \text{ Hz}) = 2.0 \text{ J}$$

Therefore,

$$n = \boxed{5.4 \times 10^{33}}$$

(c) The energy carried away in a one-quantum change of energy is

$$E = hf = (6.63 \times 10^{-34} \text{ J} \cdot \text{s})(0.56 \text{ Hz}) = \boxed{3.7 \times 10^{-34} \text{ J}}$$

The energy carried away by a one-quantum change in energy is such a small fraction of the total energy of the oscillator that we could not expect to see such a small change in the system. Thus, even though the decrease in energy of a spring-mass system is quantized and does decrease by small quantum jumps, our senses perceive the decrease as continuous. Quantum effects become important and measurable only on the submicroscopic level of atoms and molecules.

Example 29.3   The Energy of a "Yellow" Photon

Yellow light having a frequency of approximately 6×10^{14} Hz is the predominant frequency in sunlight. What is the energy carried by a quantum of this light?

Solution

The energy carried by one quantum of light is given by Equation 29.4:

$$E = hf = (6.63 \times 10^{-34} \text{ J} \cdot \text{s})(6 \times 10^{14} \text{ Hz}) = 4 \times 10^{-19} \text{ J} = \boxed{2.5 \text{ eV}}$$

29.2   The Photoelectric Effect

In the latter part of the 19th century, experiments showed that, when light is incident on certain metallic surfaces, electrons are emitted from the surfaces. This phenomenon is known as the **photoelectric effect,** and the emitted electrons are called **photoelectrons.** The first discovery of this phenomenon was made by Hertz, who was also the first to produce the electromagnetic waves predicted by Maxwell.

Light

Figure 29.4
Circuit diagram for observing the photoelectric effect. When light strikes the plate C, photoelectrons are ejected from the plate. Electrons collected at A constitute a current in the circuit.

Figure 29.4 is a schematic diagram of an apparatus in which the photoelectric effect can occur. An evacuated glass or quartz tube contains a metal plate, C, connected to the negative terminal of a battery, and a second metal plate, A, maintained at a positive potential by the battery. When the tube is kept in the dark, the galvanometer, G, reads zero, indicating that there is no current in the circuit. However, when monochromatic light of the appropriate wavelength shines on plate C, a current is detected by the galvanometer, indicating a flow of charges across the gap between C and A. The current associated with this process arises from electrons emitted from the negative plate and collected at the positive plate.

A plot of the photoelectric current versus the potential difference, V, between A and C is shown in Figure 29.5 for two light intensities. Note that for large values of V, the current reaches a maximum value, corresponding to the case where all photoelectrons are collected at A. In addition, the current increases as the incident light intensity increases, as you might expect. Finally, when V is negative, that is, when the battery in the circuit is reversed to make C positive and A negative, the photoelectrons are repelled by the negative plate A. Only those electrons having a kinetic energy greater than eV will reach A, where e is the charge on the electron. Furthermore, if V is less than or equal to V_0, called the **stopping potential**, no electrons will reach A and the current will be zero. The stopping potential is *independent* of the radiation intensity. The maximum kinetic energy of the photoelectrons is related to the stopping potential through the relation

$$KE_{max} = eV_0 \qquad \text{(29.5)}$$

Several features of the photoelectric effect could not be explained with classical physics or with the wave theory of light. The major observations that were not understood are as follows:

1. No electrons are emitted if the incident light frequency falls below some **cutoff frequency,** f_c, which depends on the material being illuminated. For example, in the case of sodium, $f_c = 5.50 \times 10^{14}$ Hz. This is inconsistent with the wave theory, which predicts that the photoelectric effect should occur at any frequency, provided the light intensity is high enough.
2. If the light frequency exceeds the cutoff frequency, a photoelectric effect is observed and the number of photoelectrons emitted is proportional to the light intensity. However, the maximum kinetic energy of the photoelectrons is independent of light intensity, a fact that cannot be explained by the concepts of classical physics.
3. The maximum kinetic energy of the photoelectrons increases with increasing light frequency.
4. Electrons are emitted from the surface almost instantaneously (less than 10^{-9} s after the surface is illuminated), even at low light intensities. Classically, it would be expected that the electrons would require some time to absorb the incident radiation before they acquire enough kinetic energy to escape from the metal.

A successful explanation of the photoelectric effect was given by Einstein in 1905, the same year he published his special theory of relativity. In his

Figure 29.5
Photoelectric current versus voltage for two light intensities. The current increases with intensity but reaches a saturation level for large values of V. At voltages equal to or less than $-V_0$, the current is zero.

photoelectric paper, for which he received the Nobel prize in 1921, Einstein extended Planck's concept of quantization to electromagnetic waves. He assumed that light (or any electromagnetic wave) of frequency f can be considered to be a stream of photons. Each photon has an energy E, given by

$$E = hf \qquad \text{(29.6)}$$

Energy of a photon

where h is Planck's constant. Thus, Einstein considered light to be much like a stream of particles traveling through space (rather than a wave), where each "particle" could be absorbed as a unit by an electron. Furthermore, Einstein argued that when the photon's energy is transferred to an electron in a metal, the energy acquired by the electron must be hf. However, the electron must also pass through the metal surface in order to be emitted, and some energy is required to overcome this barrier. The amount of energy, ϕ, required to escape the metal is known as the **work function** of the substance and is of the order of a few electron volts for metals. For example, the work function for zinc is about 3.0 eV. Hence, in order to conserve energy, the maximum kinetic energy of the ejected photoelectrons is the difference between the photon energy and the work function of the metal, or

$$KE_{\text{max}} = hf - \phi \qquad \text{(29.7)}$$

Photoelectric effect equation

That is, the excess energy $hf - \phi$ equals the maximum kinetic energy the liberated electron can have outside the surface.

With the photon theory of light, the features of the photoelectric effect that cannot be understood using classical concepts can be explained. These are briefly described in the order they were introduced earlier:

1. The fact that the photoelectric effect is not observed below a certain cutoff frequency follows from the fact that the energy of the photon must be greater than or equal to ϕ. If the energy of the incoming photon is not equal to or greater than ϕ, the electrons will never be ejected from the surface, regardless of the intensity of the light.
2. The fact that KE_{max} is independent of the light intensity can be understood with the following argument. If the light intensity is doubled, the number of photons is doubled, which doubles the number of photoelectrons emitted. However, their kinetic energy, which equals $hf - \phi$, depends only on the light frequency and the work function, not on the light intensity.
3. The fact that KE_{max} increases with increasing frequency is easily understood with Equation 29.7.
4. Finally, the fact that the electrons are emitted almost instantaneously is consistent with the particle theory of light, in which the incident energy appears in small packets and there is a one-to-one interaction between photons and electrons. This is in contrast to having the energy of the radiation distributed uniformly over a large area.

A final confirmation of Einstein's theory is a test of the prediction of a linear relationship between f and KE_{max}. Indeed, such a linear relationship is observed, as sketched in Figure 29.6. The slope of such a curve gives a value for h. The intercept on the horizontal axis gives the cutoff frequency, which is

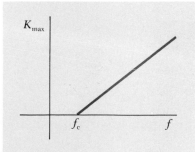

Figure 29.6
A sketch of KE_{max} versus frequency of incident light for photoelectrons in a typical photoelectric effect experiment. Photons with frequency less than f_c do not have sufficient energy to eject an electron from the metal.

related to the work function through the relation $f_c = \phi/h$. This corresponds to a **cutoff wavelength** of

$$\lambda_c = \frac{c}{f_c} = \frac{c}{\phi/h} = \frac{hc}{\phi} \qquad (29.8)$$

where c is the speed of light (3.00×10^8 m/s). Wavelengths *greater* than λ_c incident on a material with a work function ϕ do not result in the emission of photoelectrons.

Example 29.4   The Photoelectric Effect for Sodium

A sodium surface is illuminated with light of wavelength 300 nm. The work function for sodium is 2.46 eV. Find (a) the kinetic energy of the ejected photoelectrons and (b) the cutoff wavelength for sodium.

Solution

(a) The energy of the illuminating light beam is

$$E = hf = \frac{hc}{\lambda} = \frac{(6.63 \times 10^{-34} \text{ J·s})(3.00 \times 10^8 \text{ m/s})}{300 \times 10^{-9} \text{ m}}$$

$$= 6.63 \times 10^{-19} \text{ J} = \frac{6.63 \times 10^{-19} \text{ J}}{1.60 \times 10^{-19} \text{ J/eV}} = 4.14 \text{ eV}$$

where we have used the conversion 1 eV = 1.6×10^{-19} J. Using Equation 29.7 gives

$$KE_{max} = hf - \phi = 4.14 \text{ eV} - 2.46 \text{ eV} = \boxed{1.68 \text{ eV}}$$

(b) The cutoff wavelength can be calculated from Equation 29.8 after we convert ϕ from electron volts to joules:

$$\phi = 2.46 \text{ eV} = (2.46 \text{ eV})(1.6 \times 10^{-19} \text{ J/eV}) = 3.94 \times 10^{-19} \text{ J}$$

Hence

$$\lambda_c = \frac{hc}{\phi} = \frac{(6.63 \times 10^{-34} \text{ J·s})(3.00 \times 10^8 \text{ m/s})}{3.94 \times 10^{-19} \text{ J}}$$

$$= 5.05 \times 10^{-7} \text{ m} = \boxed{505 \text{ nm}}$$

This wavelength is in the green region of the visible spectrum.

29.3   Applications of the Photoelectric Effect

The photoelectric cell shown in Figure 29.4 acts much like a switch in an electric circuit in that it produces a current in an external circuit when light of sufficiently high frequency falls on the cell but does not allow a current in the dark. Many practical devices in our everyday lives depend on the photoelectric effect. One of the first practical uses was as the detector in the light meter of a camera. Light reflected from the object to be photographed strikes a photoelectric surface, causing it to emit electrons, which then pass through a

Figure 29.7
(a) When light strikes the phototube, the current in the circuit on the left energizes
the magnet, breaking the burglar alarm circuit on the right. (b) When the light
source is removed, the alarm circuit is closed, and the alarm sounds.

very sensitive ammeter. The magnitude of the current depends on the intensity of the light. Modern solid-state devices have now replaced light meters
that used the photoelectric effect.

A second example of the use of the photoelectric effect is the burglar
alarm. This device often uses ultraviolet rather than visible light in order to
make the presence of the beam less obvious. A beam of light passes from the
source to a photosensitive surface; the current produced is then amplified and
used to energize an electromagnet which attracts a metal rod, as in Figure
29.7a. If an intruder breaks the light beam, the electromagnet switches off and
the spring pulls the iron rod to the right (Fig. 29.7b). In this position, a
completed circuit allows current to pass and the alarm system is activated.

Figure 29.8 shows how the photoelectric effect is used to produce the
sound on a movie film. The sound track is located along the side of the film in
the form of an optical pattern of light and dark lines. A beam of light in the
projector is directed through the soundtrack toward a phototube. The variation in shading on the sound track varies the light intensity falling on the plate
of the phototube, thus changing the current in the circuit. This changing
current electrically simulates the original sound wave and reproduces it in the
speaker.

Figure 29.8
The shading on the sound
track varies the light intensity reaching the phototube
and hence the current to the
speaker.

Figure 29.9
X-ray diffraction pattern of NaCl.

29.4   X-Rays

In 1895 at the University of Wurzburg, Wilhelm Roentgen (1845–1923) was studying electrical discharges in low-pressure gases when he noted that a fluorescent screen glowed even when placed several meters from the gas discharge tube and even when black cardboard was placed between the tube and the screen. He concluded that the effect was caused by a mysterious type of radiation, which he called **x-rays** because of their unknown nature. Subsequent study showed that these rays traveled at or near the speed of light and that they could not be deflected by either electric or magnetic fields. This last fact indicated that x-rays did not consist of beams of charged particles, although the possibility that they were beams of uncharged particles remained.

In 1912 Max von Laue (1879–1960) suggested that one should be able to diffract x-rays by using the regular atomic spacings of a crystal lattice as a diffraction grating, just as visible light is diffracted by a ruled grating. Shortly thereafter, researchers demonstrated that such a diffraction pattern could be observed, similar to that shown in Figure 29.9 for NaCl. The wavelengths of the x-rays were then determined from the diffraction data and the known values of the spacing between atoms in the crystal. X-ray diffraction has proved to be an invaluable technique for understanding the structure of matter. We shall discuss this subject in more detail in the next section.

Typical x-ray wavelengths are about 0.1 nm, which is of the order of the atomic spacing in a solid. We now know that x-rays are a part of the electromagnetic spectrum and are characterized by frequencies higher than those of ultraviolet radiation and having the ability to penetrate most materials with relative ease.

X-rays are produced when high-speed electrons are suddenly decelerated, for example, when a metal target is struck by electrons that have been accelerated through a potential difference of several thousand volts. Figure 29.10a shows a schematic diagram of an x-ray tube. A current in the filament causes electrons to be boiled off, and these freed electrons are accelerated toward a dense metal target, such as tungsten, which is held at a higher potential than the filament.

Figure 29.11 represents a plot of x-ray intensity versus wavelength for the spectrum of radiation emitted by an x-ray tube. Note that there are two dis-

Figure 29.10
(a) Diagram of an x-ray tube.
(b) Photograph of an x-ray tube. *(Courtesy of GE Medical Systems)*

(a)

(b)

Figure 29.11
The x-ray spectrum of a metal target consists of a broad continuous spectrum plus a number of sharp lines, which are due to *characteristic x-rays*. The data shown were obtained when 35-keV electrons bombarded a molybdenum target. Note that $1 \text{ pm} = 10^{-12} \text{ m} = 10^{-3} \text{ nm}$.

Figure 29.12
An electron passing near a charged target atom experiences an acceleration and a photon is emitted in the process.

tinct patterns. One pattern is a continuous broad spectrum that depends on the voltage applied to the tube. Superimposed on this pattern is a series of sharp, intense lines that depend on the nature of the target material. The accelerating voltage must exceed a certain value, called the **threshold voltage,** in order to observe these sharp lines, which represent radiation emitted by the target atoms as their electrons undergo rearrangements. We shall discuss this further in Chapter 30. The continuous radiation is sometimes called **bremsstrahlung,** a German word meaning braking radiation. The term arises from the nature of the mechanism responsible for the radiation. That is, electrons emit radiation when they undergo a deceleration inside the target.

The deceleration of the electrons by the target produces an effect similar to an inverse photoelectric effect. In the photoelectric process, a quantum of radiant energy is absorbed by an electron in a metal and the electron gains enough energy to escape the metal. In the case of x-ray production, the inverse of this process occurs as shown in Figure 29.12. As an electron passes close to a positively charged nucleus contained in a target material, it is deflected from its path because of its electrical attraction to the nucleus, and hence experiences an acceleration. An analysis from classical physics shows that any charged particle will radiate energy in the form of electromagnetic radiation when it is accelerated. (An example of this is the production of electromagnetic waves by accelerated charges in a radio antenna as described in Chapter 23.) According to quantum theory, this radiation must appear in the form of photons. Since the radiated photon shown in Figure 29.12 carries energy, the electron must lose kinetic energy because of its encounter with the target nucleus. Let us consider an extreme example in which the electron loses all of its energy in a single collision. In this case, the initial energy of the electron

Figure 29.13
X-ray photograph of a human hand.

(eV) is transformed completely into the energy of the photon (hf_{max}). In equation form we have

$$eV = hf_{max} = \frac{hc}{\lambda_{min}} \qquad \text{(29.9)}$$

where eV is the energy of the electron after it has been accelerated through a potential difference of V volts and e is the charge on the electron. This says that the shortest-wavelength radiation that can be produced is

$$\lambda_{min} = \frac{hc}{eV} \qquad \text{(29.10)}$$

The reason that all the radiation produced does not have this particular wavelength is because many of the electrons are not stopped in a single collision. This results in the production of the continuous spectrum of wavelengths.

As noted earlier, x-rays are extremely penetrating and can produce burns or other complications if proper precautions are not taken by anyone exposed to them. Between 1930 and 1950, an x-ray device called a fluoroscope was widely used in shoe stores to examine the bone structure of the foot. Physicians used similar devices to examine the skeletal structure of their patients. Such devices are no longer in use since they are now known to be a health hazard. An x-ray photograph of a human hand is shown in Figure 29.13.

Example 29.5   The Minimum X-ray Wavelength

Calculate the minimum wavelength produced when electrons are accelerated through a potential difference of 100,000 V, a not-uncommon voltage for an x-ray tube.

Solution
From Equation 29.10, we have

$$\lambda_{min} = \frac{(6.63 \times 10^{-34} \text{ J} \cdot \text{s})(3.00 \times 10^8 \text{ m/s})}{(1.60 \times 10^{-19} \text{ C})(10^5 \text{ V})} = \boxed{1.24 \times 10^{-11} \text{ m}}$$

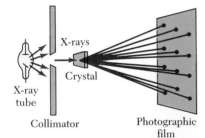

Figure 29.14
Schematic diagram of the technique used to observe the diffraction of x-rays by a single crystal. The array of spots formed on the film by the diffracted beams is called a Laue pattern.

*29.5   Diffraction of X-Rays by Crystals

In Chapter 27 we described how a diffraction grating can be used to measure the wavelength of light. In principle, the wavelength of *any* electromagnetic wave can be measured if a grating having the proper line spacing can be found. The spacing between lines must be approximately equal to the wavelength of the radiation to be measured. X-rays are electromagnetic waves with wavelengths of the order of 0.1 nm. It would be impossible to construct a grating with such a small spacing. As noted in the previous section, Max von Laue suggested that the regular array of atoms in a crystal could act as a three-dimensional grating for observing the diffraction of x-rays.

One experimental arrangement for observing x-ray diffraction is shown in Figure 29.14. A narrow beam of x-rays with a continuous range of wavelength is incident on a crystal such as sodium chloride. The diffracted radiation is very intense in certain directions, corresponding to constructive interference from waves reflected from layers of atoms in the crystal. The diffracted radiation is detected by a photographic film and forms an array of spots known as a "Laue pattern." The crystal structure is determined by analyzing the positions and intensities of the various spots in the pattern.

The arrangement of atoms in a crystal of NaCl is shown in Figure 29.15. The smaller red spheres represent Na$^+$ ions and the larger blue spheres represent Cl$^-$ ions. The spacing between successive Na$^+$ (or Cl$^-$) ions in this cubic structure, denoted by the symbol a in Figure 29.15, is approximately 0.563 nm.

A careful examination of the NaCl structure shows that the ions lie in various planes. The shaded areas in Figure 29.15 represent one example in which the atoms lie in equally spaced planes. Now suppose an x-ray beam is incident at grazing angle θ on one of the planes, as in Figure 29.16. The beam can be reflected from both the upper and lower plane of atoms. However, the geometric construction in Figure 29.16 shows that the beam reflected from the lower surface travels farther than the beam reflected from the upper surface by a distance of $2d \sin \theta$. The two portions of the reflected beam will combine to produce constructive interference when this path difference equals some integral multiple of the wavelength λ. The same is true for reflection from the entire family of parallel planes. (Note the similarity between this analysis and that used to describe thin film interference.) The condition for constructive interference is given by

$$2d \sin \theta = m\lambda \qquad (m = 1, 2, 3, \ldots) \qquad \textbf{(29.11)}$$

Bragg's law

This condition is known as **Bragg's law** after W. L. Bragg (1890–1971), who first derived the relationship. If the wavelength and diffraction angle are measured, Equation 29.11 can be used to calculate the spacing between atomic planes.

The method of x-ray diffraction to determine crystalline structures was thoroughly developed in England by W. H. Bragg, and his son W. L. Bragg, who shared a Nobel prize in 1915 for their work. Since then, thousands of

Figure 29.15
A model of the cubic crystalline structure of sodium chloride. The blue spheres represent the Cl$^-$ ions, and the red spheres represent the Na$^+$ ions. The length of the cube edge is $a = 0.563$ nm.

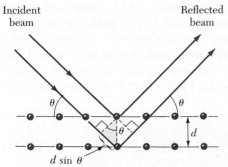

Incident beam   Reflected beam

Figure 29.16
A two-dimensional description of the reflection of an x-ray beam from two parallel crystalline planes separated by a distance d. The beam reflected from the lower plane travels farther than the one reflected from the upper plane by an amount equal to $2d \sin \theta$.

crystalline structures have been investigated. Recently, the technique of x-ray structural analysis has been used to unravel the mysteries of such complex organic systems as the important DNA molecule.

Example 29.6   Reflection from Calcite

If the spacing between certain planes in a crystal of calcite is 0.314 nm, find the angles of incidence at which first- and third-order interference will occur for x-rays of wavelength 0.07 nm.

Solution
For first-order interference, the value of m in Equation 29.11 is 1. Thus, the angle of incidence corresponding to this order of interference is found as follows:

$$\sin \theta = \frac{m\lambda}{2d} = \frac{(0.07 \text{ nm})}{2(0.314 \text{ nm})} = 0.111$$

$$\theta = \boxed{6.37°}$$

In third-order interference, $m = 3$, and we find

$$\sin \theta = \frac{m\lambda}{2d} = \frac{3(0.07 \text{ nm})}{2(0.314 \text{ nm})} = 0.334$$

$$\theta = \boxed{19.5°}$$

29.6   Compton Scattering

Further justification for the photon theory of light came from an experiment conducted by Arthur H. Compton in 1923. In his experiment, Compton directed a beam of x-rays of wavelength λ_0 toward a block of graphite. He found that the scattered x-rays had a slightly longer wavelength, λ, than the incident x-rays, and hence the energies of the scattered rays were lower. The amount of energy reduction depended on the angle at which the x-rays were scattered. The change in wavelength, $\Delta\lambda$, between a scattered x-ray and an incident x-ray is called the **Compton shift.**

In order to explain this effect, Compton assumed that if a photon behaves like a particle, its collision with other particles is similar to that between two billiard balls. Hence, both energy and momentum must be conserved. If the incident photon collides with an electron initially at rest, as in Figure 29.17, the photon transfers some of its energy and momentum to the electron. Consequently, the energy and frequency of the scattered photon are lowered and its wavelength increases. Applying relativistic energy and momentum conservation to the collision described in Figure 29.17, the shift in wavelength of the scattered photon is given by

The Compton shift formula

$$\Delta\lambda = \lambda - \lambda_0 = \frac{h}{m_0 c}(1 - \cos \theta) \qquad \textbf{(29.12)}$$

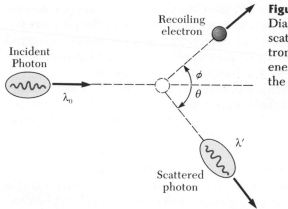

Figure 29.17
Diagram representing Compton scattering of a photon by an electron. The scattered photon has less energy (or longer wavelength) than the incident photon.

where m_0 is the rest mass of the electron and θ is the angle between the directions of the scattered and incident photons. The quantity h/m_0c is called the **Compton wavelength** and has a value $h/m_0c = 0.00243$ nm. Note that the Compton wavelength is very small relative to the wavelengths of visible light and hence would be very difficult to detect if visible light were used. Furthermore, note that the Compton shift depends on the scattering angle, θ, and not on the wavelength. Experimental results for x-rays scattered from various targets strongly support the photon concept.

Example 29.7 The Compton Shift for Carbon

X-rays of wavelength $\lambda_0 = 0.20$ nm are scattered from a block of carbon. The scattered x-rays are observed at an angle of 45° to the incident beam. Calculate the wavelength of the scattered x-rays at this angle.

Solution
The shift in wavelength of the scattered x-rays is given by Equation 29.12. Taking $\theta = 45°$, we find that

$$\Delta\lambda = \frac{h}{m_0 c} (1 - \cos \theta)$$

$$= \frac{6.63 \times 10^{-34} \text{ J·s}}{(9.11 \times 10^{-31} \text{ kg})(3.00 \times 10^8 \text{ m/s})} (1 - \cos 45°)$$

$$= 7.11 \times 10^{-13} \text{ m} = 0.000711 \text{ nm}$$

Hence, the wavelength of the scattered x-ray at this angle is

$$\lambda = \Delta\lambda + \lambda_0 = \boxed{0.200711 \text{ nm}}$$

29.7 Pair Production and Annihilation

In the photoelectric and Compton effects, the energy of a photon is transformed into kinetic and potential energy of an electron. When a photon interacts with matter through the photoelectric effect, an electron is removed from

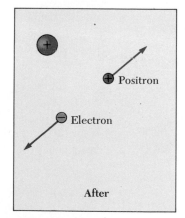

Figure 29.18
Representation of the process of pair production.

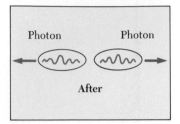

Figure 29.19
Representation of the process
of pair annihilation.

an atom and the photon disappears. In the Compton effect, a photon is scattered off an electron (or a nucleus) and loses some energy in the process. We shall now describe a process in which the energy of a photon is converted completely into rest mass. This is a striking verification of the equivalence of mass and energy as predicted by Einstein's theory of relativity.

A common process in which a photon creates matter is called **pair production,** illustrated in Figure 29.18. In this process, an electron and a positron are simultaneously produced, while the photon disappears. (Note that the positron is a positively charged particle having the same mass as an electron. The positron is often called the antiparticle of the electron.) In order for pair production to occur, energy, momentum, and charge must all be conserved during the process. Note that it is impossible for a photon to produce a single electron because the photon has zero charge, and charge would not be conserved in the process. The *minimum* energy that a photon must have to produce an electron-positron pair can be found using conservation of energy by equating the photon energy to the total rest energy of the pair. That is,

$$hf_{min} = 2m_0c^2 \qquad\qquad (29.13)$$

Since the rest energy of an electron is $m_0c^2 = 0.51$ MeV (see Chapter 28), the minimum energy required for pair production is 1.02 MeV. The wavelength of a photon carrying this much energy is 0.0012 nm. Photons with such short wavelengths are in the gamma-ray (or very short x-ray) region of the spectrum.

Pair production cannot occur in a vacuum but can only take place in the presence of a massive particle such as an atomic nucleus. The massive particle must participate in the interaction in order that energy and momentum be conserved simultaneously.

Pair annihilation is a process in which an electron-positron pair produces two photons, which is the inverse of pair production. Figure 29.19 is one example of pair annihilation in which an electron and positron initially at rest combine with each other, disappear, and create two photons. Since the initial momentum of the pair is zero, it is impossible to produce a single photon. Momentum can only be conserved if two photons are produced moving in opposite directions, both with the same momentum and energy. We shall discuss particles and their antiparticles further in Chapter 34.

29.8   Photons and Electromagnetic Waves

An explanation of a phenomenon such as the photoelectric effect presents very convincing evidence in support of the photon (or particle) concept of light. An obvious question that arises at this point is, "How can light be considered a photon when it exhibits wave-like properties?" On the one hand, we describe light in terms of photons having energy and momentum. On the other hand, we must also recognize that light and other electromagnetic waves exhibit interference and diffraction effects, which are consistent only with a wave interpretation. Which model is correct? Is light a wave or a particle? The answer depends on the specific phenomenon being observed. Some experiments can be better, or solely, explained on the basis of the photon concept, whereas others are best described, or can be described only, with a wave model. The end result is that *we must accept both models and admit that the true nature of light is not describable in terms of a single classical picture — light has a dual nature.* However, you should recognize that the same light beam that can eject photoelectrons from a metal can also be diffracted by a grating. In other words, *the photon theory and the wave theory of light complement each other.*

Light has a dual nature

We can perhaps understand why photons are compatible with electromagnetic waves in the following manner. We may suspect that long-wavelength radio waves do not exhibit particle characteristics. Consider, for instance, radio waves at a frequency of 2.5 MHz. The energy of a photon having this frequency is only about 10^{-8} eV. From a practical viewpoint, this energy is too small to be detected as a single photon. A sensitive radio receiver might require as many as 10^{10} of these photons to produce a detectable signal. Such a large number of photons would appear, on the average, as a continuous wave. With such a large number of photons reaching the detector every second, it would be unlikely that any graininess would appear in the detected signal. That is, we would not be able to detect the individual photons striking the antenna.

Now consider what happens as we go to higher frequencies, or shorter wavelengths. In the visible region, it is possible to observe both the photon and the wave characteristics of light. As we mentioned earlier, a light beam shows interference phenomena and at the same time can produce photoelectrons, which can be understood best by using Einstein's photon concept. At even higher frequencies and correspondingly shorter wavelengths, the momentum and energy of the photon increase. Consequently, the photon nature of light becomes more evident than its wave nature. For example, an x-ray photon is easily detected as a single event. However, as the wavelength decreases, wave effects, such as interference and diffraction, become more difficult to observe. Very indirect methods are required to detect the wave nature of very-high-frequency radiation, such as gamma rays.

All forms of electromagnetic radiation can be described from two points of view. At one extreme, electromagnetic waves describe the overall interference pattern formed by a large number of photons. At the other extreme, the photon description is natural when we are dealing with a highly energetic photon of very short wavelength. Hence,

light has a dual nature; it exhibits both wave and photon characteristics.

29.9   The Wave Properties of Particles

Students first introduced to the dual nature of light often find the concept very difficult to accept. In the world around us, we are accustomed to regarding such things as a thrown baseball solely as particles and such things as sound waves solely as forms of wave motion. Every large-scale observation can be interpreted by considering either a wave explanation or a particle explanation, but in the world of photons and electrons, such distinctions are not as sharply drawn. Even more disconcerting is the fact that, under certain conditions, *particles such as electrons also exhibit wave characteristics.*

In 1924, the French physicist Louis de Broglie wrote a doctoral dissertation in which his thesis was that *because photons have wave and particle char-*

acteristics, perhaps all forms of matter have wave as well as particle properties. This was a highly revolutionary idea with no experimental confirmation. However, the suggestion received immediate attention by the scientific community and played an important role in the subsequent development of quantum mechanics.

In Chapter 28, we found that the relationship between energy and momentum for a photon, which has a rest mass of zero, is $p = E/c$. We also know that the energy of a photon is

Energy of a photon

$$E = hf = \frac{hc}{\lambda} \tag{29.14}$$

Thus, the momentum of a photon can be expressed as

Momentum of a photon

$$p = \frac{E}{c} = \frac{hc}{c\lambda} = \frac{h}{\lambda} \tag{29.15}$$

From this equation we see that the photon wavelength can be specified by its momentum, or $\lambda = h/p$. de Broglie suggested that

de Broglie's hypothesis

material particles of momentum p should also have wave properties and a corresponding wavelength.

Because the momentum of a particle of mass m and velocity v is $p = mv$, the **de Broglie wavelength** of the particle is

de Broglie wavelength

$$\lambda = \frac{h}{p} = \frac{h}{mv} \tag{29.16}$$

Furthermore, in analogy with photons, de Broglie postulated that the frequencies of matter waves (that is, waves associated with real particles) obey the Einstein relation $E = hf$, so that

Frequency of a photon

$$f = \frac{E}{h} \tag{29.17}$$

The dual nature of matter is quite apparent in these two equations. That is, each equation contains both particle concepts (mv and E) and wave concepts (λ and f). The fact that these relationships are established experimentally for photons makes the de Broglie hypothesis that much easier to accept.

De Broglie's proposal that all particles exhibit both wave and particle properties was first regarded as pure speculation. If particles such as electrons had wave properties, then under the correct conditions they should exhibit interference phenomena. Three years later, in 1927, C. J. Davisson and L. Germer of the United States succeeded in measuring the wavelength of electrons. Their important discovery provided the first experimental confirmation of the matter waves proposed by de Broglie. In their original experiment, low-energy electrons (about 54 eV) were scattered from a single crystal of nickel. The electrons were scattered in preferred directions, as evidenced by intensity peaks at certain scattering angles. The results were explained by recognizing that the regularly spaced planes of atoms in the crystal acted as a diffraction grating for the electron waves, as discussed in Section 29.5.

The problem of understanding the dual nature of both matter and radiation is conceptually difficult because the two models seem to contradict each other. This problem as it applies to light was discussed earlier. Niels Bohr helped to resolve this problem in his principle of complementarity, which states that *the wave and particle models of either matter or radiation complement each other.* Neither model can be used exclusively to adequately describe matter or radiation. A complete understanding is obtained only if the two models are combined in a complementary manner.

Example 29.8 The Wavelength of an Electron

Calculate the de Broglie wavelength for an electron ($m = 9.11 \times 10^{-31}$ kg) moving with a speed of 10^7 m/s.

Solution
Equation 29.16 gives

$$\lambda = \frac{h}{mv} = \frac{6.63 \times 10^{-34} \text{ J} \cdot \text{s}}{(9.11 \times 10^{-31} \text{ kg})(10^7 \text{ m/s})} = \boxed{7.28 \times 10^{-11} \text{ m}}$$

This wavelength corresponds to that of x-rays in the electromagnetic spectrum.

Exercise Find the de Broglie wavelength of a proton ($m = 1.67 \times 10^{-27}$ kg) moving with a speed of 10^7 m/s.

Answer 3.97×10^{-14} m.

Example 29.9 The Wavelength of a Rock

A rock of mass 50 g is thrown with a speed of 40 m/s. What is the de Broglie wavelength of the rock?

Solution
From Equation 29.16, we have

$$\lambda = \frac{h}{mv} = \frac{6.63 \times 10^{-34} \text{ J} \cdot \text{s}}{(50 \times 10^{-3} \text{ kg})(40 \text{ m/s})} = \boxed{3.32 \times 10^{-34} \text{ m}}$$

Notice that this wavelength is much smaller than the size of any possible aperture through which the rock could pass. This means that we could not observe diffraction effects, and as a result the wave properties of large-scale objects cannot be observed.

Louis de Broglie (1892–1987), a French physicist, was awarded the Nobel prize in 1929 for his discovery of the wave nature of electrons. He was the scion of an aristocratic French family that produced marshals, ambassadors, foreign ministers, and at least one Duke, his older brother Maurice de Broglie. Louis de Broglie came rather late to theoretical physics, as he first studied history. Only after serving as a radio operator in World War I did he follow the lead of his older brother and begin his studies of physics. Maurice de Broglie was an outstanding experimental physicist in his own right and conducted experiments in the palatial family mansion in Paris. *(AIP Niels Bohr Library)*

▶ SPECIAL TOPIC
The Electron Microscope

A practical device that relies on the wave characteristics of electrons is the **electron microscope** (Fig. 29.20a), which is in many respects similar to an ordinary compound microscope. One important difference between the two is that the electron microscope has a much greater resolving power because electrons can be accelerated to very

Figure 29.20
(a) Diagram of an electron microscope. The "lenses" that control the electron beam are magnetic deflection coils.
(b) A scanning electron micrograph (SEM) of a Mediterranean fruitfly. (© *David Scharf/Peter Arnold, Inc.*)

high kinetic energies, giving them a very short wavelength. Any microscope is capable of detecting details that are comparable in size to the wavelength of the radiation used to illuminate the object. Typically, the wavelengths of electrons are about 100 times shorter than those of the visible light used in optical microscopes. As a result, electron microscopes are able to distinguish details about 100 times smaller.

In operation, a beam of electrons falls on a thin slice of the material to be examined. The section to be examined must be very thin, typically a few hundred angstroms, in order to minimize undesirable effects, such as absorption or scattering of the electrons. The electron beam is controlled by electrostatic or magnetic deflection, which acts on the charges to focus the beam to an image. Rather than examining the image through an eyepiece as in an ordinary microscope, a magnetic lens forms an image on a fluorescent screen. The fluorescent screen is necessary because the image produced would not otherwise be visible. An example of a photograph taken by an electron microscope is shown in Figure 29.20b. ◄

29.10   The Wave Function

De Broglie's revolutionary idea that particles should have a wave nature soon moved out of the realm of skepticism to the point where it was viewed as a necessary concept in understanding the subatomic world. In 1926, the Austrian-German physicist Erwin Schrödinger proposed a wave equation that described the manner in which matter waves change in space and time. The Schrödinger wave equation represents a key element in the theory of quantum mechanics. Its role is as important in quantum mechanics as that played by Newton's laws of motion in classical mechanics. Schrödinger's equation has been successfully applied to the hydrogen atom and to many other microscopic systems. Its importance in most aspects of modern physics cannot be overemphasized.

We shall not go through a mathematical derivation of Schrödinger's wave equation, nor shall we even state the equation here since it involves mathematical operations beyond the scope of this text. When we attempt to solve the Schrödinger equation, the basic entity we seek to determine is a quantity Ψ, called the **wave function.** Each particle is represented by a wave function Ψ that depends on both the position of the object and time. Once Ψ is found, what information does it give concerning the particle? To answer this question, let us consider an analogy with light.

In Chapter 26 we discussed Young's double-slit experiment and explained experimental observations of the interference pattern solely in terms of the wave nature of light. Let us now discuss this same experiment in terms of both the wave and particle nature of light.

First, recall from Chapter 23 that the intensity of a light beam is proportional to the square of the electric field strength E associated with the beam. That is, $I \propto E^2$. According to the wave model of light, there are certain points on the viewing screen where the net electric field is zero due to destructive interference of waves from the two slits. Because E is zero at these points, the intensity is also zero, and the screen is dark at these locations. Likewise, at points on the screen where constructive interference occurs, E is large, as is the intensity, hence these locations are bright.

Now consider the same experiment when light is viewed as having a particle nature. It seems reasonable to assume that the number of photons reaching a point on the screen per second increases as the intensity (brightness) increases. Thus, the number of photons which strikes a unit area on the screen each second is proportional to the square of the electric field, or $N \propto E^2$. Now let us consider the behavior of a single photon from a probabilistic viewpoint. What will be the fate of the photon as it moves through the slits in Young's experiment? From a probabilistic point of view, a photon has a high probability of striking the screen at a point where the intensity (and E^2) is high, and a low probability of striking the screen where the intensity is low.

When describing particles rather than photons, Ψ plays the role of the amplitude, rather than E. Using an analogy with the description of light, we make the following interpretations of Ψ for particles:

1. If a wave function Ψ is used to represent a large group of particles, the value of Ψ^2 at some location at a given time is proportional to the probability of finding a large number of particles at that location at that time.
2. If Ψ is a wave function used to describe a single particle, the value of Ψ^2 at some location at a given time is proportional to the probability of finding the particle at that location at that time.

29.11   The Uncertainty Principle

If you were to measure the position and velocity of a particle at any instant, you would always be faced with reducing the experimental uncertainties in your measurements as much as possible. According to classical mechanics, there is no fundamental barrier to an ultimate refinement of the apparatus and/or experimental procedures. That is, it would be possible, in principle, to make such measurements with arbitrarily small uncertainty or with infinite accuracy. Quantum theory predicts, however, that

it is fundamentally impossible to make simultaneous measurements of a particle's position and velocity with infinite accuracy.

This statement, known as the **uncertainty principle,** was first derived by Werner Heisenberg in 1927.

This photograph of Werner Heisenberg (1901–1976) was taken around 1924 (University of Hamburg). Heisenberg obtained his PhD in 1923 at the University of Munich where he studied under Arnold Sommerfeld and became an enthusiastic mountain climber and skier. Later, he worked as an assistant to Max Born at Göttingen and Niels Bohr in Copenhagen. While physicists such as de Broglie and Schrödinger tried to develop pictorialized models of the atom, Heisenberg, with the help of Born and Pascual Jordan, developed an abstract mathematical model called matrix mechanics to explain the wavelengths of spectral lines. The more successful wave mechanics by Schrödinger announced a few months later was shown to be equivalent to Heisenberg's approach. Heisenberg made many other significant contributions to physics including his famous uncertainty principle, for which he received the Nobel Prize in 1932, the prediction of two forms of molecular hydrogen, and theoretical models of the nucleus. During World War II he was director of the Max Planck Institute at Berlin where he was in charge of German research on atomic weapons. Following the war, he moved to West Germany and became director of the Max Planck Institute for Physics at Göttingen.

Werner Heisenberg (1901–1976)

Erwin Schrödinger (1887–1961) was an Austrian theoretical physicist best known as the creator of wave mechanics. He also produced important papers in the fields of statistical mechanics, color vision, and general relativity. Schrödinger did much to hasten the universal acceptance of quantum theory by demonstrating the mathematical equivalence between his wave mechanics and the more abstract matrix mechanics developed by Heisenberg.

Although Schrödinger's wave theory was generally based on clear physical ideas, one of its major problems in 1926 was the physical interpretation of the wave function Ψ. Schrödinger felt that the electron was ultimately a wave, Ψ was the vibration amplitude of this wave, and Ψ² was the electric charge density. Born, Bohr, Heisenberg, and others pointed out the problems with this interpretation and presented the currently accepted view that Ψ² is a probability, and that the electron is ultimately no more a wave than a particle. Schrödinger never accepted this view, but registered his "concern and disappointment" that this "transcendental, almost psychical interpretation" had become "universally accepted dogma."

In 1927 Schrödinger accepted the chair of theoretical physics at the University of Berlin where he formed a close friendship with Max Planck. In 1933, he left Germany and eventually settled at the Dublin Institute of Advanced Study where he spent 17 happy, creative years working on problems in general relativity, cosmology, and the application of quantum physics to biology. In 1956 he returned home to Austria and to his beloved Tirolean mountains, where he died in 1961.

Erwin Schrödinger
(1887–1961)

(AIP Niels Bohr Library, Photograph by Francis Simon)

Consider a particle moving along the x axis and suppose that Δx and Δp represent the uncertainty in the measured values of the particle's position and momentum, respectively, at some instant. The uncertainty principle says that the product $\Delta x \Delta p$ is never less than a number of the order of Planck's constant. More specifically,

Uncertainty principle

$$\Delta x \Delta p \geq \frac{h}{4\pi}$$

(29.18)

That is, *it is physically impossible to measure simultaneously the exact position and exact momentum of a particle.* If Δx is made very small, Δp will be large, and vice versa.

In order to understand the uncertainty principle, consider the following thought experiment. Suppose you wish to measure the position and momentum of an electron as accurately as possible. You might be able to do this by viewing the electron with a powerful microscope. In order for you to see the electron, and thus determine its location, at least one photon of light must bounce off the electron and pass through the microscope into your eye. This incident photon is shown moving toward the electron in Figure 29.21a. When the photon strikes the electron, as in Figure 29.21b, the photon transfers some of its energy and momentum to the electron. Thus, in the process of attempting to locate the electron very accurately (that is, by making Δx very small), we have caused a rather large uncertainty in its momentum. In other words, *the measurement procedure itself limits the accuracy to which we can determine position and momentum simultaneously.*

Let us analyze the collision between the photon and the electron by first noting that the incoming photon has momentum h/λ. As a result of the collision, the photon transfers part or all of its momentum to the electron. Thus, the uncertainty in the electron's momentum after the collision is at least as great as the momentum of the incoming photon. That is, $\Delta p = h/\lambda$. Furthermore, since light also has wave properties, we would expect the uncertainty in the position of the electron to be of the order of one wavelength of the light being used to view it, because of diffraction effects. Thus, $\Delta x = \lambda$. Multiplying these two uncertainties gives

$$\Delta x \, \Delta p = \lambda \left(\frac{h}{\lambda}\right) = h$$

This represents the minimum in the products of the uncertainties. Since the uncertainty can always be greater than this minimum, we have

$$\Delta x \, \Delta p \geqslant h$$

This agrees with Equation 29.18 (apart from a numerical factor introduced by Heisenberg's more precise analysis).

Another form of the uncertainty principle that applies to the simultaneous measurement of energy and time is

$$\Delta E \, \Delta t \geqslant \frac{h}{4\pi} \qquad \textbf{(29.19)}$$

where ΔE is the uncertainty in a measurement of the energy and Δt is the time it takes to make the measurement. It can be inferred from this relationship that the energy of a particle cannot be measured with complete precision in an infinitely short interval of time. Thus, when an electron is viewed as a particle, the uncertainty principle tells us that (a) its position and velocity cannot both be known precisely at the same time and (b) its energy can be uncertain (or may not be conserved) for a period of time given by $\Delta t = h/(4\pi \, \Delta E)$.

We can arrive at the uncertainty principle in the form of Equation 29.19 by returning to Figure 29.21. Recall that the uncertainty in the position of the electron being viewed by the microscope is of the order of the wavelength of the light being used to detect the electron. The photon used to detect the electron travels with a speed c; therefore the time it takes the photon to travel a distance Δx is given by

$$\Delta t \approx \frac{\Delta x}{c} \approx \frac{\lambda}{c}$$

As the photon collides with the electron, it can transfer all or part of its energy to the electron. Thus, the uncertainty in the energy transferred is

$$\Delta E \approx hf \approx \frac{hc}{\lambda}$$

The product of these two uncertainties is

$$\Delta E \, \Delta t \approx h$$

Again, the result of our approximate analysis agrees with Heisenberg's more accurate and detailed analysis within a small numerical factor.

Heisenberg's uncertainty principle enables us to better understand the dual wave-particle nature of light and matter. We have seen that the wave

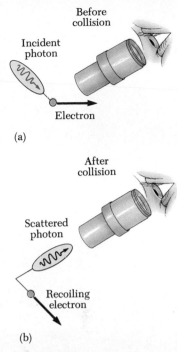

(a)

(b)

Figure 29.21
A thought experiment for viewing an electron with a powerful microscope. (a) The electron is viewed before colliding with the photon. (b) The electron recoils (is disturbed) as the result of the collision with the photon.

description is quite different from the particle description. Therefore, if an experiment is designed to reveal the particle character of an electron (such as the photoelectric effect), its wave character will become fuzzy. Likewise, if the experiment is designed to accurately measure the electron's wave properties (such as diffraction from a crystal), its particle character will become fuzzy.

Example 29.10   Locating an Electron

The speed of an electron is measured to be 5.00×10^3 m/s to an accuracy of 0.003%. Find the uncertainty in determining the position of this electron.

Solution
The momentum of the electron is

$$p = mv = (9.11 \times 10^{-31} \text{ kg})(5.00 \times 10^3 \text{ m/s}) = 4.56 \times 10^{-27} \text{ kg} \cdot \text{m/s}$$

Because the uncertainty in p is 0.003% of this value, we get

$$\Delta p = 0.00003p = (0.00003)(4.56 \times 10^{-27} \text{ kg} \cdot \text{m/s}) = 1.37 \times 10^{-31} \text{ kg} \cdot \text{m/s}$$

The uncertainty in position can now be calculated by using this value of Δp and Equation 29.18:

$$\Delta x \, \Delta p \geqslant \frac{h}{4\pi}$$

$$\Delta x \geqslant \frac{h}{4\pi \, \Delta p} = \frac{6.63 \times 10^{-34} \text{ J} \cdot \text{s}}{4\pi \, (1.37 \times 10^{-31} \text{ kg} \cdot \text{m/s})}$$

$$= 0.385 \times 10^{-3} \text{ m} = \boxed{0.385 \text{ mm}}$$

Example 29.11   Excited States of Atoms

As we shall see in the next chapter, electrons in atoms can be found in higher states of energy called excited states for short periods of time. If the average time that an electron exists in one of these states is 10^{-8} s, what is the minimum uncertainty in energy of the excited state?

Solution
From the uncertainty principle in the form of Equation 29.19, we find that the minimum uncertainty in energy is

$$\Delta E = \frac{h}{4\pi \Delta t} = \frac{(6.63 \times 10^{-34} \text{ J} \cdot \text{s})}{4\pi \, (10^{-8} \text{ s})} = 5.28 \times 10^{-27} \text{ J} = \boxed{3.30 \times 10^{-8} \text{ eV}}$$

Summary

The characteristics of **blackbody radiation** cannot be explained using classical concepts. The peak of a blackbody radiation curve is given by **Wien's displacement law:**

$$\lambda_{\max} T = 0.2898 \times 10^{-2} \text{ m} \cdot \text{K} \tag{29.1}$$

where λ_{max} is the wavelength at which the curve peaks and T is the absolute temperature of the object emitting the radiation.

Planck first introduced the quantum concept when he assumed that the vibrating molecules responsible for blackbody radiation could have only discrete amounts of energy given by

$$E_n = nhf \qquad \text{(29.2)}$$

where n is a positive integer called a **quantum number** and f is the frequency of vibration of the molecule.

The **photoelectric effect** is a process whereby electrons are ejected from a metal surface when light is incident on that surface. Einstein provided a successful explanation of this effect by extending Planck's quantum hypothesis to electromagnetic waves. In this model, light is viewed as a stream of particles called photons, each with energy $E = hf$, where f is the frequency and h is **Planck's constant.** The maximum kinetic energy of the ejected photoelectrons is

$$KE_{max} = hf - \phi \qquad \text{(29.7)}$$

where ϕ is the **work function** of the metal.

X-rays are produced when high-speed electrons are suddenly decelerated. When electrons are accelerated through a voltage V, the shortest-wavelength radiation that can be produced is

$$\lambda_{min} = \frac{hc}{eV} \qquad \text{(29.10)}$$

The regular array of atoms in a crystal can act as a diffraction grating for x-rays. The condition for constructive interference of the diffracted rays is given by **Bragg's law**

$$2d \sin \theta = m\lambda \qquad (m = 1, 2, 3, \ldots) \qquad \text{(29.11)}$$

X-rays from an incident beam are scattered at various angles by electrons in a target such as carbon. In such a scattering event, a shift in wavelength is observed for the scattered x-rays. This phenomenon is known as the **Compton shift.** Conservation of momentum applied to a photon-electron collision yields the following expression for the shift in wavelength of the scattered x-rays:

$$\Delta\lambda = \lambda - \lambda_0 = \frac{h}{m_0 c}(1 - \cos \theta) \qquad \text{(29.12)}$$

where m_0 is the rest mass of the electron, c is the speed of light, and θ is the scattering angle.

Pair production is a process in which the energy of a photon is converted into rest mass. In this process, the photon disappears as an electron-positron pair is created. Likewise, the energy of an electron-positron pair can be converted into electromagnetic radiation by the process of **pair annihilation.**

De Broglie proposed that all matter has both a particle and a wave nature. The **de Broglie wavelength** of any particle of mass m and speed v is

$$\lambda = \frac{h}{p} = \frac{h}{mv} \qquad\qquad \textbf{(29.16)}$$

De Broglie also proposed that the frequencies of the waves associated with particles obey the Einstein relationship, $E = hf$.

In the theory of **quantum mechanics,** each particle is described by a quantity Ψ called the **wave function.** The probability of finding the particle at a particular point at some instant is proportional to $|\Psi^2|$. Quantum mechanics is very successful in describing the behavior of atomic and molecular processes.

According to Heisenberg's **uncertainty principle,** it is impossible to measure simultaneously the exact position and exact momentum of a particle. If Δx is the uncertainty in the measured position and Δp the uncertainty in the momentum, the product $\Delta x \, \Delta p$ is given by

$$\Delta x \, \Delta p \geqslant \frac{h}{4\pi} \qquad\qquad \textbf{(29.18)}$$

or by

$$\Delta E \, \Delta t \geqslant \frac{h}{4\pi} \qquad\qquad \textbf{(29.19)}$$

where ΔE is the uncertainty in the energy of the particle and Δt is the uncertainty in the time it takes to measure the energy.

Additional Reading

A. Baker, *Modern Physics and Antiphysics*, Reading, MA, Addison-Wesley, 1970.

G. Binnig and H. Rohrer, "The Scanning Tunneling Microscope," *Sci. American*, August 1985, p. 50.

L. de Broglie, "The Revolution in Physics: A Non-Mathematical Survey of Quanta," Noonday Press, 1953.

W. C. Dampier, *A History of Science*, London, Cambridge University Press, 1977, Chap. 10.

G. Gamow, "The Principle of Uncertainty," *Sci. American*, January 1958, p. 51.

G. Gamow, *Thirty Years That Shook Physics*, New York, 1966. Doubleday, Anchor Books, 1966.

W. Heisenberg, *Physics and Beyond*, New York, Harper and Row, 1971 (a "biography" of quantum mechanics).

B. Hoffman, *Strange Story of the Quantum*, New York, Dover, 1959.

B. Wheaton, "Louis de Broglie and the Origin of Wave Mechanics," *The Physics Teacher*, May 1984, p. 297.

Questions

1. When wood is stacked on a special elevated grate in a fireplace (which is commercially available), there is formed beneath the grate a pocket of burning wood whose temperature is higher than that of the burning wood at the top of the stack. Explain how this device provides more heat to the room than a conventional fire does and thus increases the efficiency of the fireplace.

2. The brightest star in the constellation Lyra is the bluish star, Vega. The brightest star in Boötes is the reddish star, Arcturus. How do you account for the difference in color of the two stars?

3. All objects radiate energy. Why, then, are we not able to see all objects in a dark room?

4. What assumptions were made by Planck in dealing with the problem of blackbody radiation? Discuss the consequences of these assumptions.

5. Which has more energy, a photon of ultraviolet radiation or a photon of yellow light?

6. If the photoelectric effect is observed for one metal, can you conclude that the effect will also be observed for another metal under the same conditions?

7. What effect, if any, would you expect the temperature of a material to have on the ease with which electrons can be ejected from it in the photoelectric effect?

8. In the photoelectric effect, explain why the photocurrent depends on the intensity of the light source but not on the frequency.

9. In the photoelectric effect, explain why the stopping potential depends on the frequency of light but not on the intensity.

10. Why does the existence of cutoff frequency in the photoelectric effect favor a particle theory for light rather than a wave theory?

11. An x-ray photon is scattered by an electron. What happens to the frequency of the scattered photon relative to that of the incident photon?

12. If a photon is deflected by the Compton effect, can its wavelength ever become shorter?

13. What assumptions were made by Compton in dealing with the scattering of a photon from an electron?

14. Is light a wave or a particle? Support your answer by citing specific experimental evidence.

15. Suppose a photograph were made of a person's face using only a few photons. Would the result be simply a very faint image of the face? Discuss.

16. Why is an electron microscope more suitable than an optical microscope for "seeing" objects of an atomic size?

17. An electron and a proton are accelerated from rest through the same potential difference. Which particle has the longer wavelength?

18. If matter has a wave nature, why is this wave-like character not observable in our daily experiences?

19. Is an electron a particle or a wave? Support your answer by citing some experimental results.

20. Why is it impossible to simultaneously measure the position and velocity of a particle with infinite accuracy?

21. Suppose there are two electrons inside a hollow box. Can we determine which electron is which? Discuss this in terms of the attributes that uniquely identify each electron and the uncertainty principle.

22. Discuss whether the behavior of an electron is mainly wave-like or particle-like in each of the following situations: (a) traversing a circular orbit in a magnetic field; (b) absorbing a photon and being photoelectrically ejected from the surface of a metal; (c) forming an interference pattern.

Problems

Section 29.1   Blackbody Radiation and Planck's Hypothesis

1. Calculate the energy in electron volts of a photon in (a) the radio frequency range, 90 MHz, (b) the infrared range, 10^{13} Hz, and (c) the ultraviolet, 10^{16} Hz.

2. Calculate the energy in electron volts of a photon having a wavelength in (a) the microwave range, 5 cm, (b) the visible light range, 500 nm, and (c) the x-ray range, 10 nm.

3. The threshold of dark-adapted (scotopic) vision is 4.0×10^{-11} W/m^2 at a central wavelength of 500 nm. If light having this intensity and wavelength enters the eye and the pupil is open to its maximum diameter of 8.5 mm, how many photons per second enter the eye?

4. Figure 29.22 shows the spectrum of light emitted by a firefly. Determine the temperature of a blackbody that would emit radiation peaked at the same frequency. Based on your result, would you say firefly radiation is blackbody radiation?

Figure 29.22   Problem 4

5. If the surface temperature of the Sun is 5800 K, find the wavelength that corresponds to the maximum rate of energy emission from the Sun.

6. A certain light source is found to emit radiation with the peak value having a frequency of 10^{15} Hz. Find the temperature of the source.

7. A quantum of electromagnetic radiation has an energy of 2 keV. What is its wavelength?

8. A 0.5-kg mass falls from a height of 3 m. If all of the energy of this mass could be converted to visible light of wavelength 500 nm, how many photons would be produced?

9. Through what height would the object of Problem 8 have to fall in order to have the energy equivalent of one photon of 500-nm light?

10. A 1.5-kg mass vibrates at an amplitude of 3 cm on the end of a spring of spring constant 20 N/m. If the energy of the spring is quantized, (a) find its quantum number. (b) If n changes by 1, find the fractional change in energy of the spring.

Section 29.2   The Photoelectric Effect

11. From the scattering of sunlight, Thomson calculated the classical radius of the electron to have a value of 2.82×10^{-15} m. If sunlight having an intensity of 500 W/m² falls on a disk with this radius, estimate the time required to accumulate 1.0 eV of energy. Assume light to be a classical wave and that the light striking the disk is completely absorbed. How does your estimate compare with the observation that photoelectrons are promptly (within 10^{-9} s) emitted?

12. Ultraviolet light is incident normally on the surface of a certain substance. The binding energy of the electrons in this substance is 3.44 eV. The incident light has an intensity of 0.055 W/m². The electrons are photoelectrically emitted with a maximum speed of 4.2×10^5 m/s. How many electrons are emitted from a square centimeter of the surface? Assume 100% of the photons are absorbed.

13. A metal has a work function of 2×10^{-19} J. If yellow light of wavelength 600 nm falls on the surface of the metal, find (a) the maximum kinetic energy of the ejected electrons and (b) the cutoff wavelength for the metal.

14. When light of wavelength 350 nm falls on a potassium surface, electrons are emitted that have a maximum kinetic energy of 1.31 eV. Find (a) the work function of potassium, (b) the cutoff wavelength, and (c) the frequency corresponding to the cutoff wavelength.

15. When cesium metal is illuminated with light of wavelength 300 nm, the photoelectrons emitted have a maximum kinetic energy of 2.23 eV. Find (a) the work function of cesium and (b) the stopping potential if the incident light has a wavelength of 400 nm.

16. When a certain metal is illuminated with light of frequency 3×10^{15} Hz, a stopping potential of 7 V is required to stop the ejected electrons. What is the work function of this metal?

17. Consider the metals lithium, iron, and mercury, which have work functions of 2.3 eV, 3.9 eV, and 4.5 eV, respectively. If light of wavelength 300 nm is incident on each of these metals, determine (a) which metals exhibit the photoelectric effect and (b) the maximum kinetic energy for the photoelectrons for those that exhibit the effect.

18. When light of wavelength 253.7 nm falls on cesium, it is found that the stopping potential required is 3 V. If light of wavelength 435.8 nm is used, the stopping potential is 0.9 V. Use this information to plot a graph like that shown in Figure 29.6, and from the graph determine the cutoff frequency for cesium and its work function.

19. What wavelength light would have to fall on sodium (work function 2.3 eV) if it is to emit electrons with a maximum speed of 10^6 m/s?

20. Red light of wavelength 670.0 nm produces photoelectrons from a certain photosensitive material. Green light of wavelength 520.0 nm produces photoelectrons from the same material with 1.50 times the maximum kinetic energy. What is the material's work function?

Section 29.4   X-Rays

21. What is the shortest x-ray wavelength that can be produced with an accelerating voltage of 10 kV?

22. What minimum accelerating voltage would be required to produce an x-ray with a wavelength of 0.03 nm?

23. Calculate the minimum wavelength x-ray that can be produced when a target is struck by an electron that has been accelerated through a potential difference of (a) 15 kV and (b) 100 kV.

24. The extremes of the x-ray portion of the electromagnetic spectrum range from approximately 10 nm to 10^{-4} nm. Find the minimum accelerating voltages required to produce wavelengths at these two extremes.

Section 29.5   Diffraction of X-Rays by Crystals

25. The highest-quality diffraction gratings have about 25,000 grooves/cm. Suppose that such a grating were used in the standard transmission mode (Fig. 27.15) to analyze x-rays with a wavelength of 0.50 Å. At what angle θ would the first-order maximum appear? Comment on your result.

26. The spacing between planes of nickel atoms in a nickel crystal is 0.352 nm. At what angle does a second-order Bragg reflection occur in nickel for 11.3-keV x-rays?

27. The spacing between certain planes in a crystal is known to be 0.3 nm. Find the smallest angle at which interference will occur for x-rays of wavelength 0.07 nm.

28. Potassium iodide has an interplane spacing of $d = 0.296$ nm. A monochromatic x-ray beam shows a first-order diffraction maximum when the angle of incidence is 7.6°. Calculate the x-ray wavelength.

29. A monochromatic x-ray beam is incident on a NaCl crystal surface where $d = 0.353$ nm. The second-order maximum in the reflected beam is found when the angle between the incident beam and the surface is 20.5°. Determine the wavelength of the x-rays.

30. X-rays of wavelength 0.14 nm are reflected from a certain crystal, and the first-order maximum occurs at an angle of 14.4°. What value does this give for the interplane spacing of this crystal?

Section 29.6   Compton Scattering

31. X-rays of wavelength 0.20 nm are scattered from a block of carbon. If the scattered radiation is detected at 90° to the incident beam, find the Compton shift.

32. Show that the Compton wavelength has the numerical value 0.00243 nm.

33. X-rays of wavelength 0.071 nm undergo Compton scattering from free electrons in carbon. What is the wavelength of the photons, which are scattered at 90° relative to the incident direction?

34. X-rays are scattered from electrons in a carbon target. The measured wavelength shift is 0.0012 nm. Calculate the scattering angle.

35. A beam of 0.68-nm photons undergoes Compton scattering from free electrons. What are the energy and momentum of the photons that emerge at a 45° angle with respect to the incident beam?

36. A 0.45-nm x-ray photon is deflected through a 23° angle after scattering from a free electron. (a) What is the kinetic energy of the recoiling electron? (b) What is its speed?

37. After a 0.80-nm x-ray photon scatters from a free electron, the electron recoils with a speed equal to 1.4×10^6 m/s. (a) What was the Compton shift in the photon's wavelength? (b) Through what angle was the photon scattered?

38. A 0.0016-nm photon scatters from a free electron. For what (photon) scattering angle will the recoiling electron and scattered photon have the same kinetic energy?

Section 29.7   Pair Production and Annihilation

39. Find (a) the minimum energy of the photon required to produce a proton-antiproton pair and (b) the wavelength of this radiation.

40. Repeat Problem 39 for a neutron-antineutron pair.

41. How much total kinetic energy will an electron-positron pair have if produced by a photon of energy 3 MeV?

42. If an electron-positron pair is produced having a total kinetic energy of 2.5 MeV, find (a) the energy of the photon which produced the pair and (b) its frequency.

43. Two photons are produced when a proton and antiproton annihilate each other. What is the minimum frequency and corresponding wavelength of each proton?

Section 29.9   The Wave Properties of Particles

44. (a) If the wavelength of an electron is equal to 500 nm, how fast is it moving? (b) If the electron has a speed of 1.0×10^7 m/s, what is its wavelength?

45. Calculate the de Broglie wavelength of a 2000-kg car moving at 65 mi/h.

46. Calculate the de Broglie wavelength of a proton moving with a speed of (a) 10^4 m/s and (b) 10^7 m/s.

47. Calculate the de Broglie wavelength for an electron with kinetic energy of (a) 50 eV and (b) 50 keV.

48. The resolving power of a microscope is proportional to the wavelength used. If one wished to use a microscope to "see" an atom, a resolution of approximately 10^{-11} m (0.1 Å) would have to be obtained. (a) If electrons are used (electron microscope), what minimum kinetic energy is required for the electrons? (b) If photons are used what minimum photon energy is needed to obtain 10^{-11} m resolution?

49. Find the de Broglie wavelength of a ball having a mass of 0.2 kg just before it strikes the Earth after being dropped from a building 50 m tall.

50. Calculate the de Broglie wavelength of a proton that is accelerated through a potential difference of 10 MV.

51. Through what potential difference would an electron have to be accelerated to give it a de Broglie wavelength of 10^{-10} m?

52. Show that the de Broglie wavelength of a nonrelativistic electron accelerated from rest through a potential difference V is given by $\lambda = 1.228/\sqrt{V}$ nm, where V is in volts.

53. A monoenergetic beam of electrons is incident on a single slit of width 0.5 nm. A diffraction pattern is formed on a screen 20 cm from the slit. If the distance

between successive minima of the diffraction pattern is 2.1 cm, what is the energy of the incident electrons?

54. De Broglie postulated that the relationship $\lambda = h/p$ is valid for relativistic particles. What is the de Broglie wavelength for a (relativistic) electron whose kinetic energy is 3 MeV?

Section 29.11   The Uncertainty Principle

55. A ball of mass 50 g moves with a speed of 30 m/s. If its speed is measured to an accuracy of 0.1%, what is the minimum uncertainty in its position?

56. A 0.5-kg block rests on the icy surface of a frozen pond, which we can assume to be frictionless. If the location of the block is measured to a precision of 0.5 cm, what speed must the block acquire because of the measurement process?

57. The position of an electron is known to a precision of 10^{-8} cm. What is the minimum uncertainty in the measurement of the electron's velocity?

58. In the ground state of hydrogen the uncertainty in the position of the electron is roughly 0.1 nm. If the speed of the electron is of the order of the uncertainty in the speed, how fast is the electron moving?

59. The uncertainty of the momentum of electrons in a beam 1 m long is one part in 10^3. If the speed of the electrons is 6.4×10^6 m/s, what is the uncertainty in the position of the electron? Express the result as a fraction of the beam's length.

60. The energy of an electron in a particular atom is approximately 2 eV. How long would it take to measure this energy to a precision of 1%?

61. (a) Show that the kinetic energy of a nonrelativistic particle can be written in terms of its momentum as $KE = p^2/2m$. (b) Use the results of (a) to find the smallest kinetic energy of a proton confined to a nucleus of size 10^{-15} m.

Additional Problems

62. The current in a photocell is cut off by a retarding potential of 0.92 V for radiation of wavelength 250 nm. Find the work function for the material.

63. The material in a photocell has a work function of 2 eV. When a retarding potential is applied, the cutoff wavelength is found to be 350 nm. What is the value of the retarding potential?

64. A jungle hero of mass 70 kg swings at the end of a vine at a frequency of 0.5 Hz and has a speed of 2 m/s as he moves through the lowest point on his arc. (a) Assume the energy is quantized and find the quantum number n for the system. (b) Find the energy carried away in a one-quantum change in his energy.

65. How many photons are emitted per second by a 100-W sodium lamp if the wavelength of sodium light is 589.3 nm?

66. An x-ray tube is operated at 50,000 V. (a) Find the minimum wavelength radiation emitted by this tube. (b) If this radiation is directed at a crystal, the first-order maximum in the reflected radiation occurs when the angle of incidence is 2.5°. What is the spacing between reflecting planes in the cyrstal?

67. A light source of wavelength λ illuminates a metal and ejects photoelectrons with a maximum kinetic energy of 1 eV. A second light source of wavelength $\lambda/2$ ejects photoelectrons with a maximum kinetic energy of 4 eV. What is the work function of the metal?

68. An electron and a proton each have a thermal kinetic energy of $3kT/2$. Calculate the de Broglie wavelength of each particle when they are at a temperature of 2000 K. (Recall that k is Boltzmann's constant, which has the value $k = 1.38 \times 10^{-23}$ J/K.)

69. Photons of wavelength 450 nm are incident on a metal. The most energetic electrons ejected from the metal are bent into a circular arc of radius 20 cm by a magnetic field whose strength is 2×10^{-5} T. What is the work function of the metal?

70. In the Compton scattering event illustrated in Figure 29.17, the scattered photon has an energy of 120 keV and the recoiling electron has a kinetic energy of 40 keV. Find (a) the wavelength of the incident photon, (b) the angle θ at which the photon is scattered, and (c) the recoil angle of the electron. (*Hint:* Conserve both mass-energy and relativistic momentum.)

"Actually I started out in quantum mechanics, but somewhere along the way I took a wrong turn."

The Scanning Tunneling Microscope

The basic idea of quantum mechanics, that particles have properties of waves and vice versa, is among the strangest found anywhere in science. Because of this strangeness, and because quantum mechanics mostly deals with the very small, it might seem to you to have little practical application. But in fact one of the basic phenomena of quantum mechanics—the tunneling of a particle—is at the heart of a very practical device that is one of the most powerful microscopes ever built. This device, the *scanning tunneling microscope,* or *STM,* enables us to get highly detailed images of surfaces with resolution comparable to the size of a *single atom.* Such images promise to revolutionize our understanding of structures and processes on the atomic scale.

Figure 1 shows an example of what the STM can do. This is an image of the surface of a piece of graphite. Note the high quality of the image and the recognizable rings of carbon atoms. What makes this image so remarkable is that the *resolution* of the image—the size of the smallest detail that can be discerned—is about 2 Å (2×10^{-10} m). For an ordinary microscope, the resolution is limited by the wavelength of the waves used to make the image. Thus, an optical microscope has a resolution of no better than 2000 Å, about half the wavelength of visible light, and so could never show the detail displayed in Figure 1. Electron microscopes can have a resolution of 2 Å by using electron waves of this wavelength, given by the de Broglie formula $\lambda = h/p$. The electron momentum p required to give this wavelength is 10,000 eV/c, corresponding to an electron speed of 2% of the speed of light. Electrons traveling at this speed would penetrate into the interior of the piece of graphite in Figure 1 and so couldn't give us information about individual surface atoms.

The STM achieves its very fine resolution by using the basic idea shown in Figure 2. A conducting probe with a very sharp tip is brought near the surface to be studied. Because it is attracted to the positive ions in the surface, an electron in the surface has a lower total energy than would an electron in the empty space between surface and tip. The same thing is true for an electron in the tip, which is attracted to the positive ions in the tip. In classical Newtonian mechanics this would mean that electrons could not move between the surface and the tip, because they would lack the energy to escape either material. But because the electrons obey quantum mechanics, they can "tunnel" across the barrier of empty space between the tip and the surface. By applying a voltage between the surface and tip, the electrons can be made to tunnel preferentially from the surface to the tip. In this way the tip samples the distribution of electrons just above the surface.

Roger A. Freedman and
Paul K. Hansma
*University of California,
Santa Barbara*

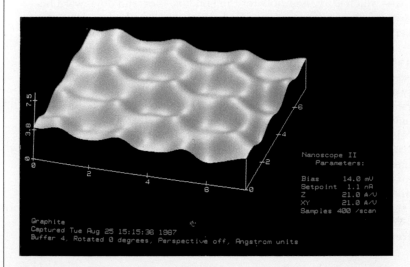

Figure 1
The surface of graphite as "viewed" with a scanning tunneling microscope. This technique enables scientists to see small details on surfaces with a lateral resolution of about 2 Å and a vertical resolution of 0.01 Å. The contours seen here represent the arrangement of individual atoms on the crystal surface. The scale is in Å. This image was obtained with a commercial STM, the Nanoscope II from Digital Instruments in Goleta, California.

Figure 2

A schematic view of an STM. The tip, shown as a rounded cone, is mounted on a piezoelectric x, y, z scanner. A scan of the tip over the sample can reveal contours of the surface down to the atomic level. An STM image is composed of a series of scans displaced laterally from each other. *(Based on a drawing from P. K. Hansma, V. B. Elings, O. Marti, and C. Bracker,* Science *242:209, 1988. Copyright 1988 by the AAAS.)*

z piezo

y piezo

x piezo

I

Because of the nature of tunneling, the STM is very sensitive to the distance z from tip to surface. The reason is that in the empty space between the tip and the surface, the electron wave function falls off exponentially with a decay length of order 1 Å, that is, the wave function decreases by $1/e$ over that distance. For distances z greater than 10 Å (that is, beyond a few atomic diameters), essentially no tunneling takes place. This exponential behavior causes the current of electrons tunneling from surface to tip to depend very strongly on z. This sensitivity is the basis of the operation of the STM: by monitoring the tunneling current as the tip is scanned over the surface, scientists obtain a sensitive measure of the topography of the electron distribution on the surface. The result of this scan is used to make images like that in Figure 1. In this way the STM can measure the height of surface features to within 0.01 Å, or approximately one one-hundredth of an atomic diameter!

You can see just how sensitive a probe the STM is by looking carefully at the image of the graphite surface in Figure 1. You may be able to see that of the six carbon atoms in each ring, three of the atoms *appear* lower than the other three. In fact all six atoms are at the same level, but have slightly different electron distributions. The three atoms that appear lower are bonded to other carbon atoms directly beneath them in the underlying atomic layer, and so their electron distributions—which are responsible for the bonding—extend downward beneath the surface. The atoms in the surface layer that appear higher do not lie directly over subsurface atoms, and hence are not bonded to carbon atoms beneath them. For these higher-appearing atoms, the electron distribution extends upward into the space above the surface. This extra electron density is what makes these electrons appear higher in Figure 1, since what the STM maps is the topography of the electron distribution.

Commercial STMs are now available and are speeding the use of scanning tunneling microscopy in a variety of applications (Figure 3). These include characterizing electrodes for electrochemistry (while the electrode is still in the electrolyte), characterizing the roughness of surfaces, measuring the quality of optical gratings, and even imaging replicas of biological structures.

Perhaps the most remarkable thing about the scanning tunneling microscope is that its operation is based on a quantum-mechanical phenomenon—tunneling—that was well understood in the 1920s, yet the first STM was not built until the 1980s. What other applications of quantum mechanics may yet be waiting to be discovered?

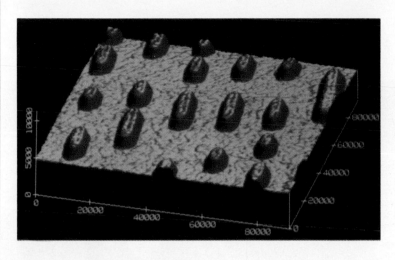

Figure 3

A technological application of the STM: bumps on a compact disk stamper. Five tracks are shown. The scale is in Å. This image was also obtained with the Nanoscope II. *(Figure from P. K. Hansma, V. B. Elings, O. Marti, and C. Bracker,* Science *242:209, 1988. Copyright 1988 by the AAAS.)*

Suggested Reading

G. Binnig, H. Rohrer, Ch. Gerber, and E. Weibel, *Physical Review Letters 49:57* (1982). The first description of the operation of a scanning tunneling microscope.

G. Binnig and H. Rohrer, *Scientific American,* August 1985, p. 50. A popular description of the STM and its applications.

G. Binnig and H. Rohrer, *Reviews of Modern Physics 59:615* (1987). The text of the lecture given on the occasion of the presentation of the 1986 Nobel Prize in Physics.

P. K. Hansma, V. B. Elings, O. Marti, and C. Bracker, *Science, 242:209* (1988). A review of how the STM and related devices are being applied to problems in biology and technology.

Essay Questions

1. The density of the Earth's atmosphere depends on altitude z according to the relation $\rho = \rho_0 e^{-z/z_0}$, where z_0, the "scale height" of the atmosphere, is 8430 meters. What is the scale height for the electron "atmosphere" above a conductor? Give both a formula and a numerical estimate.

2. Our discussion of STM operation was based on the assumption that the conducting probe has only a single tip. In fact a probe may have a number of protrusions on its end, each of which acts as a "tip." Hence it might be expected that such a probe would give multiple STM images of the surface, greatly complicating the analysis. Explain why in fact there is no problem with multiple images, provided that the multiple tips differ in proximity to the surface by 20 Å or more.

3. The STM, while using physical concepts which date from the 1920s, was not developed until the 1980s. Suggest some reasons why the STM was not invented half a century earlier.

4. Is was stated in the essay that for conventional microscopes, the resolution is limited by the wavelength of the waves used to make the image. To see whether this guideline applies to the STM, estimate the wavelength of an electron in the surface of a conductor and compare your estimate to the vertical resolution (about 0.01 Å) and lateral resolution (about 2 Å). Does the guideline apply to these two resolutions? Why or why not?

CHAPTER 30

Atomic Physics

This futuristic picture might represent an artist's version of atomic species being emitted by a solid. The atoms could be surface impurities emitted by thermal excitation, or they may even be ions emitted in the presence of a strong applied electric field. *(© Orion 1988/FPG International)*

A large portion of this chapter is concerned with the study of the hydrogen atom. Although the hydrogen atom is the simplest atomic system, it is an especially important system to understand for several reasons:

1. Much of what is learned about the hydrogen atom with its single electron can be extended to such single-electron ions as He^+ and Li^{2+}, which are hydrogen-like in their atomic structure.
2. The hydrogen atom is an ideal system for performing precise tests of theory against experiment and for improving our overall understanding of atomic structure.
3. The quantum numbers used to characterize the allowed states of hydrogen can be used to describe the allowed states of more complex atoms. This enables us to understand the periodic table of the elements, the explanation of which is one of the greatest triumphs of quantum mechanics.

In this chapter, we shall first discuss the Bohr model of hydrogen, which helps us understand many features of hydrogen but fails to explain many finer details of atomic structure. Following this discussion, we shall examine the hydrogen atom from the viewpoint of quantum mechanics and the quantum numbers used to characterize various atomic states. Additionally, we shall examine the physical significance of the quantum numbers and the effect of a magnetic field on certain quantum states. The Pauli exclusion principle is also presented in this chapter. This physical principle is extremely important in understanding the properties of complex atoms and the arrangement of elements in the periodic table. Finally, we shall apply our knowledge of atomic structure to describe the mechanisms involved in the operation of a laser.

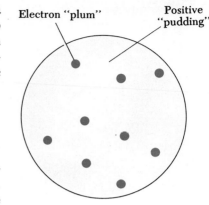

Figure 30.1
Thomson's "plum-pudding" model of the atom.

30.1 Early Models of the Atom

The model of the atom in the days of Newton was that of a tiny, hard, indestructible sphere. This model provided a good basis for the kinetic theory of gases. However, new models had to be devised when later experiments revealed the electrical nature of atoms. J. J. Thomson (1856–1940) suggested a model that described the atom as a volume of positive charge with electrons embedded throughout the volume, much like the seeds in a watermelon (Fig. 30.1). Thomson referred to this model as the "plum-pudding atom."

In 1911, Geiger and Marsden under the supervision of Ernest Rutherford (1871–1937) performed a critical experiment that showed that Thompson's plum-pudding model could not be correct. In this experiment, a beam of positively charged **alpha particles,** the nuclei of helium atoms, was projected into a thin metal foil, as in Figure 30.2a. The results of the experiment were astounding. It was found that most of the alpha particles passed through the foil as if it were empty space! Furthermore, many of the alpha particles that were deflected from their original direction of travel were scattered through very large angles. Some particles were even deflected backwards so as to reverse their direction of travel. When Rutherford was informed that some alpha particles were scattered backward, he wrote, "It was quite the most incredible event that has ever happened to me in my life. It was almost as incredible as if you fired a 15-inch shell at a piece of tissue paper and it came back and hit you."

Such large deflections were not expected on the basis of the plum-pudding model. According to this model, a positively charged alpha particle would never come close enough to a large enough volume of charge to cause any

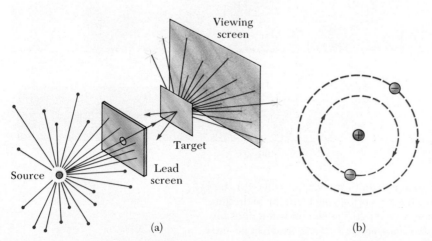

(a)   (b)

Figure 30.2
(a) Geiger and Marsden's technique for observing the scattering of alpha particles from a thin foil target. The source is a naturally occurring radioactive substance, such as radium. (b) Rutherford's planetary model of the atom.

Sir Joseph John Thomson (1856–1940), an English physicist and the recipient of the Nobel Prize in 1906. Thomson, usually considered the discoverer of the electron, opened up the field of subatomic particle physics with his extensive work dealing with the deflection of cathode rays (electrons) in an electric field. *(The Bettmann Archive)*

large-angle deflections. Rutherford explained these observations by assuming that the positive charge was concentrated in a region that was small relative to the size of the atom. He called this concentration of positive charge the **nucleus** of the atom. Any electrons belonging to the atom were assumed to be in the relatively large volume outside the nucleus. In order to explain why electrons in this outer region of the atom were not pulled into the nucleus, Rutherford developed a model similar to that of our Solar System. He viewed the electrons as particles moving in orbits about the positively charged nucleus in the same manner as the planets orbit the Sun, as in Figure 30.2b.

There are two basic difficulties with Rutherford's planetary model of the atom. As we shall see in the next section, an atom emits certain characteristic frequencies of electromagnetic radiation and no others. The Rutherford model is unable to explain this phenomenon. A second difficulty is that, according to classical theory, an accelerated charge must radiate energy in the form of electromagnetic waves. In the Rutherford model, the electrons undergo a centripetal acceleration because they move in circular paths about the nucleus. Therefore, an accelerating electron would rapidly lose its energy as it radiates, which would cause it to spiral quickly into the nucleus.

30.2   Atomic Spectra

Suppose an evacuated glass tube is filled with a gas such as neon, helium, or argon. If a potential difference between electrodes in the tube produces an electric current in the gas, the tube will emit light whose color is characteristic

Figure 30.3
Visible spectra (a) Line spectra produced by emission in the visible range for the elements hydrogen, mercury, and neon. (b) The absorption spectrum for hydrogen. The dark absorption lines occur at the same wavelengths as the emission lines for hydrogen shown in (a). *(Whitten, K. W., K. D. Gailey, and R. E. Davis, General Chemistry, 3rd ed., Saunders College Publishing, 1987).*

of the gas. If the emitted light is analyzed by passing it through a narrow slit and then through a spectroscope, a series of discrete lines is observed, each line corresponding to a different wavelength, or color. We refer to such a series of lines as a **line spectrum.** As you can see in Figure 30.3a, the wavelengths contained in a given line spectrum are characteristic of the element emitting the light. The simplest line spectrum is observed for atomic hydrogen. Other atoms such as mercury and neon emit completely different line spectra. Because no two elements emit the same line spectrum, this phenomenon represents a practical technique for identifying elements in a chemical substance.

The line spectrum of hydrogen includes a series of lines in the visible region of the spectrum, shown in Figure 30.4. Four of the most prominent lines in this region occur at the wavelengths 656.3 nm, 486.1 nm, 434.1 nm, and 410.2 nm. Later it was discovered that this series of lines continued into the ultraviolet region, ending at 364.6 nm, as shown in Figure 30.4.

In 1885, Johann Balmer (1825–1898) found that the wavelengths of these lines can be described by this simple empirical equation:

Figure 30.4
A series of spectral lines for atomic hydrogen. The prominent lines labeled are part of the Balmer series.

$$\frac{1}{\lambda} = R\left(\frac{1}{2^2} - \frac{1}{n^2}\right) \qquad \textbf{(30.1)}$$

Balmer series

where n may have integral values of 3, 4, 5, . . . and R is a constant, now called the **Rydberg constant.** If the wavelength is in meters, R has the value

$$R = 1.0973732 \times 10^7 \text{ m}^{-1} \qquad \textbf{(30.2)}$$

Rydberg constant

The first line in the Balmer series, at 656.3 nm, corresponds to $n = 3$ in Equation 30.1; the line of 486.1 nm corresponds to $n = 4$, and so on.

Balmer predicted that other series of lines might exist outside the visible region that would obey the equations given below. Several years later, these lines were indeed discovered. These spectra are called the Lyman, Paschen, and Brackett series after their discoverers. The wavelengths of the lines in these series can be calculated by the following empirical formulas:

$$\frac{1}{\lambda} = R\left(1 - \frac{1}{n^2}\right) \qquad n = 2, 3, 4, \ . \ . \ . \qquad \textbf{(30.3)}$$

Lyman series

$$\frac{1}{\lambda} = R\left(\frac{1}{3^2} - \frac{1}{n^2}\right) \qquad n = 4, 5, 6, \ . \ . \ . \qquad \textbf{(30.4)}$$

Paschen series

$$\frac{1}{\lambda} = R\left(\frac{1}{4^2} - \frac{1}{n^2}\right) \qquad n = 5, 6, 7, \ . \ . \ . \qquad \textbf{(30.5)}$$

Brackett series

In addition to emitting light at specific wavelengths, an element can also absorb light at specific wavelengths. The spectral lines corresponding to this process form what is known as an **absorption spectrum.** An absorption spectrum can be obtained by passing a continuous spectrum (one containing all wavelengths) through a vapor of the element being analyzed. The absorption spectrum consists of a series of dark lines superimposed on the otherwise continuous spectrum. It is found through experiments that each line in the absorption spectrum of a given element coincides with a line in the emission spectrum of the element. That is, if hydrogen is the absorbing vapor, dark lines will appear at the visible wavelengths 656.3 nm, 486.1 nm, 434.1 nm, and 410.2 nm, as shown in Figure 30.3b. (In practice, the hydrogen vapor must be very hot before it will absorb in the visible portion of the spectrum.)

The absorption spectrum of an element has many practical applications. For example, the continuous spectrum of radiation emitted by the Sun must pass through the cooler gases of the solar atmosphere and through the Earth's atmosphere. The various absorption lines observed in the solar spectrum have been used to identify elements in the solar atmosphere. It is interesting to note that, when the solar spectrum was first being studied, some lines were found that did not correspond to any known element. A new element had been discovered! Since the Greek word for Sun is *helios*, the new element was named helium. Scientists are able to examine the light from stars other than our Sun in this fashion, but elements other than those present on Earth have never been detected.

30.3 The Bohr Theory of Hydrogen

At the beginning of the 20th century, scientists were perplexed by the failure of classical physics in explaining the characteristics of atomic spectra. Why did hydrogen emit only certain lines in the visible part of the spectrum? Furthermore, why did hydrogen absorb only those wavelengths that it emitted? In 1913, the Danish scientist Niels Bohr (1885–1963) provided an explanation of atomic spectra that included some features contained in the currently accepted theory. Bohr's theory contained a combination of ideas from classical physics, Planck's original quantum theory, Einstein's photon theory of light, and Rutherford's model of the atom. Bohr's model of the hydrogen atom contains some classical features as well as some revolutionary postulates that could not be justified within the framework of classical physics. The Bohr model can be applied quite successfully to such hydrogen-like ions as singly ionized helium and doubly ionized lithium. However, the theory does not properly describe the spectra of more complex atoms and ions.

The basic postulates of the Bohr model of the hydrogen atom are as follows:

1. The electron moves in circular orbits about the nucleus (the planetary model of the atom) under the influence of the Coulomb force of attraction between the electron and the positively charged nucleus (Fig. 30.4).

2. The electron can exist only in very specific orbits; hence the states are quantized (Planck's quantum hypothesis). The allowed orbits are those for which the angular momentum of the electron about the nucleus is an integral multiple of $\hbar = h/2\pi$, where h is Planck's constant. The angular momentum of the electron is $I\omega$, where $I = mr^2$ and $\omega = v/r$. Thus, the angular momentum is $I\omega = (mr^2)(v/r) = mvr$. Applying the condition that the **angular momentum is quantized**, we have

Postulates of the Bohr model

$$mvr = \frac{nh}{2\pi} = n\hbar \qquad n = 1, 2, 3, \ldots \qquad \textbf{(30.6)}$$

Niels Bohr, a Danish physicist, proposed the first quantum model of the atom. He was an active participant in the early development of quantum mechanics and provided much of its philosophical framework. He made many other important contributions to theoretical nuclear physics, including the development of the liquid-drop model of the nucleus and work in nuclear fission. He was awarded the 1922 Nobel prize in physics for his investigation of the structure of atoms and of the radiation emanating from them.*

Bohr spent most of his life in Copenhagen, Denmark, and received his doctorate at the University of Copenhagen in 1911. The following year he traveled to England where he worked under J. J. Thomson at Cambridge, and then went to Manchester where he worked under Ernest Rutherford. He was married in 1912 and returned to the University of Copenhagen in 1916 as professor of physics. During the 1920s and 1930s, Bohr headed the Institute for Advanced Studies in Copenhagen, under the support of the Carlsberg brewery. (This was certainly the greatest benefit provided by beer to the field of theoretical physics.) The institute was a magnet for many of the world's best physicists and provided a forum for the exchange of ideas. Bohr, a firm believer in doing physics on a "person-to-person" basis, was always the initiator of probing questions, reflections, and discussions with his guests.

When Bohr visited the United States in 1939 to attend a scientific conference, he brought news that the fission of uranium had been discovered by Hahn and Strassman in Berlin. The results, confirmed by other scientists shortly thereafter, were the foundations of the atomic bomb developed in the United States during World War II. He returned to Denmark and was there during the German occupation in 1940. He escaped to Sweden in 1943 to avoid imprisonment, and helped to arrange the escape of many imperiled Danish citizens.

Although Bohr himself worked on the Manhattan project at Los Alamos until 1945, he strongly felt that openness between nations concerning nuclear weapons should be the first step in establishing their control. After the war, Bohr committed himself to many human issues including the development of peaceful uses of atomic energy. He was awarded the Atoms for Peace award in 1957. As John Archibald Wheeler summarizes, "Bohr was a great scientist. He was a great citizen of Denmark, of the World. He was a great human being."

* For several interesting articles concerning Bohr, read the special Niels Bohr centennial issue of *Physics Today*, October 1985.

Niels Bohr
(1885–1962)

3. When the electron is in one of its allowed orbits, it does not radiate energy; hence the atom is stable. Such stable orbits are called **stationary states.** (As discussed earlier, according to classical electricity and magnetism models, the accelerating electron must radiate electromagnetic waves and in doing so would quickly spiral into the nucleus and annihilate the atom.)

4. The atom radiates energy only when the electron "jumps" from one allowed stationary orbit to another. The frequency of the radiation obeys the condition

$$hf = E_i - E_f \qquad \textbf{(30.7)}$$

where E_i and E_f are the energies of the initial and final stationary states. This postulate states that the energy given off by an atom, $E_i - E_f$, is

Figure 30.5
Diagram representing Bohr's model of the hydrogen atom. In this model, the orbiting electron is allowed to be only in specific orbits of discrete radii.

carried away by a photon of energy hf. Note that this expression for the photon energy agrees with that proposed by Einstein in arriving at his photoelectric effect equation (Section 29.2).

With these assumptions, we shall now calculate the allowed energies of the hydrogen atom, which we can then use to calculate the wavelengths of the spectral lines emitted by the atom. We shall use the model pictured in Figure 30.5, in which the electron travels in a circular orbit of radius r with an orbit speed v.

The electrical potential energy of the atom is

$$PE = k\frac{q_1 q_2}{r} = k\frac{(-e)(e)}{r} = -k\frac{e^2}{r}$$

where k is the Coulomb constant. The total energy, E, of the atom is the sum of the kinetic energy of the electron (when the nucleus is at rest) and the potential energy:

$$E = KE + PE = \frac{1}{2}mv^2 - k\frac{e^2}{r} \tag{30.8}$$

Let us apply Newton's second law to the electron. We see that the electric force of attraction on the electron, ke^2/r^2, must equal ma_r, where $a_r = v^2/r$ is the centripetal acceleration of the electron:

$$k\frac{e^2}{r^2} = m\frac{v^2}{r} \tag{30.9}$$

From this equation, we see that the kinetic energy of the electron is

$$\frac{1}{2}mv^2 = \frac{ke^2}{2r} \tag{30.10}$$

We can combine this result with Equation 30.8 and express the **total energy** of the atom as

Total energy of the hydrogen atom

$$E = -\frac{ke^2}{2r} \tag{30.11}$$

An expression for r is obtained by solving Equations 30.6 and 30.9 for v and equating the results:

$$v^2 = \frac{n^2\hbar^2}{m^2 r^2} = \frac{ke^2}{mr}$$

Radii of the allowed orbits

$$r_n = \frac{n^2\hbar^2}{mke^2} \qquad n = 1, 2, 3, \ldots \tag{30.12}$$

This equation is based on the assumption that *the electron can exist only in certain allowed orbits*.

The orbit with the smallest radius, called the **Bohr radius**, a_0, corresponds to $n = 1$ and has the value

$$a_0 = \frac{\hbar^2}{mke^2} = 0.0529 \text{ nm}$$ **(30.13)**   Bohr radius

A general expression for the radius of any orbit in the hydrogen atom is obtained by substituting Equation 30.13 into Equation 30.12:

$$r_n = n^2 a_0 = n^2 (0.0529 \text{ nm})$$ **(30.14)**

The first three Bohr orbits for hydrogen are shown in Figure 30.6.

Equation 30.12 may be substituted into Equation 30.11 to give the following expression for the energies of the quantum states:

$$E_n = -\frac{mk^2e^4}{2\hbar^2}\left(\frac{1}{n^2}\right) \qquad n = 1, 2, 3, \ldots$$ **(30.15)**   Allowed energies of the hydrogen atom

If we insert numerical values into Equation 30.15, we have

$$E_n = -\frac{13.6}{n^2} \text{ eV}$$ **(30.16)**

The lowest stationary energy state, or **ground state,** corresponds to $n = 1$ and has an energy $E_1 = -mk^2e^4/2\hbar^2 = -13.6$ eV. The next state, corresponding to $n = 2$, has an energy $E_2 = E_1/4 = -3.4$ eV, and so on. An energy level diagram showing the energies of these stationary states and the corresponding quantum numbers is shown in Figure 30.7. The uppermost level shown, corresponding to $n \to \infty$, represents the state for which the electron is completely removed from the atom. In this case, $E = 0$ for $r = \infty$. The minimum energy required to ionize the atom, that is, to completely remove the electron, is called the **ionization energy.** The ionization energy for hydrogen is 13.6 eV.

Equations 30.7 and 30.15 and the fourth Bohr postulate show that if the electron jumps from one orbit, whose quantum number is n_i, to a second orbit, whose quantum number is n_f, it emits a photon of frequency f, given by

$$f = \frac{E_i - E_f}{h} = \frac{mk^2e^4}{4\pi\hbar^3}\left(\frac{1}{n_f^2} - \frac{1}{n_i^2}\right)$$ **(30.17)**

Finally, to compare this result with the empirical formulas for the various spectral series, we use the fact that $\lambda f = c$ and Equation 30.17 to get

$$\frac{1}{\lambda} = \frac{f}{c} = \frac{mk^2e^4}{4\pi c\hbar^3}\left(\frac{1}{n_f^2} - \frac{1}{n_i^2}\right)$$ **(30.18)**

A comparison of this result with Equation 30.1 gives the following expression for the Rydberg constant:

$$r_n = n^2 a_0 \qquad n = 1, 2, 3, \ldots$$

Figure 30.6
The first three circular orbits predicted by the Bohr model of the hydrogen atom.

Bohr's Correspondence Principle

In our study of relativity, we found that Newtonian mechanics cannot be used to describe phenomena that occur at speeds approaching the speed of light. Newtonian mechanics is a special case of relativistic mechanics and is usable only when v is much less than c. Similarly, *quantum mechanics is in agreement with classical physics when the quantum numbers are very large*. This principle, first set forth by Bohr, is called the **correspondence principle.**

For example, consider the hydrogen atom in an orbit for large values of n, say $n > 10,000$. For such large values of n, the energy differences between adjacent levels approach zero and the levels are nearly continuous. Consequently, the classical model is reasonably accurate in describing the system for large values of n. According to the classical picture, the frequency of the light emitted by the atom is equal to the frequency of revolution of the electron in its orbit about the nucleus. Calculations show that for $n > 10,000$, this frequency is different from that predicted by quantum mechanics by less than 0.015%.

30.4 Modification of the Bohr Theory

The Bohr theory of the hydrogen atom was a tremendous success in certain areas in that it explained several features of the spectra of hydrogen that had previously defied explanation. It accounted for the Balmer series and other series, it predicted a value for the Rydberg constant that is in excellent agreement with the experimental value, it derived an expression for the radius of the atom, and it predicted the energy levels of hydrogen. Although these successes were important to scientists, it is perhaps even more important that the Bohr theory gave us a model of what the atom looks like and how it behaves. Once a basic model is constructed, refinements and modifications can be made to enlarge upon the concept and to explain finer details.

The analysis used in the Bohr theory is also successful when applied to *hydrogen-like* atoms. An atom is said to be hydrogen-like when it contains only one electron. Examples are singly ionized helium, doubly ionized lithium, triply ionized beryllium, etc. The results of the Bohr theory for hydrogen can be extended to hydrogen-like atoms by substituting Ze^2 for e^2 in the hydrogen equations, where Z is the atomic number of the element. For example, Equations 30.15 and 30.18 become

$$E_n = -\frac{mk^2Z^2e^4}{2\hbar^2}\left(\frac{1}{n^2}\right) \qquad n = 1, 2, 3, \ldots \qquad \textbf{(30.21)}$$

and

$$\frac{1}{\lambda} = \frac{mk^2Z^2e^4}{4\pi c\hbar^3}\left(\frac{1}{n_f{}^2} - \frac{1}{n_i{}^2}\right) \qquad \textbf{(30.22)}$$

Although many attempts were made to extend the Bohr theory to more complex (multi-electron) atoms, the results were unsuccessful. Even today, there are only approximate methods available for treating multi-electron atoms.

Within a few months following the publication of Bohr's theory, Arnold Sommerfeld (1868–1951) extended the results to include elliptical orbits.

Table 30.1
Shell and Subshell Notations

n	Shell Symbol	ℓ	Subshell Symbol
1	K	0	s
2	L	1	p
3	M	2	d
4	N	3	f
5	O	4	g
6	P	5	h
. . .		. . .	

We shall examine his model briefly because much of the nomenclature used in this treatment is still in use today. When Bohr introduced the concept of quantization of angular momentum, the idea led to the introduction of the **principal quantum number** n, which determines the energy of the allowed states of hydrogen. Sommerfeld's theory retained n, but also introduced a new quantum number ℓ, called the **orbital quantum number,** where the values that ℓ can have range from 0 to $n - 1$, in integer steps. According to this model, an electron in any one of the allowed energy states of a hydrogen atom may move in any one of a number of orbits. For each value of n there are n possible orbits. Since $n = 1$ and $\ell = 0$ for the first energy level (ground state), there is only one possible orbit for this state. The second energy level, with $n = 2$, has two possible orbits corresponding to $\ell = 0$ and $\ell = 1$. The third energy level, with $n = 3$, has three possible orbits corresponding to $\ell = 0$, $\ell = 1$, and $\ell = 2$.

It became customary to use a specific nomenclature when referring to these states and orbits. **All states with the same principal quantum number are said to form a shell.** These shells are identified by the letters K, L, M, . . . , which designate the states for which $n = 1, 2, 3, \ldots$. Likewise, **the states having the same values of n and ℓ are said to form a subshell. The** letters $s, p, d, f, g, \ldots$ are used to designate the states for which $\ell = 0, 1, 2, 3,$ 4, ... These notations are summarized in Table 30.1.

You should note that the maximum number of electrons that are allowed in any given subshell is given by $2(2\ell + 1)$. For example, the p subshell ($\ell = 1$) is filled when it contains 6 electrons. This fact will be important to us later when we discuss the *Pauli exclusion principle.*

Another modification of the Bohr theory arose when it was discovered that the spectral lines of a gas are split into several closely spaced lines when the gas is placed in a strong magnetic field. (This is called the Zeeman effect after its discoverer.) Figure 30.9 shows a single spectral line being split into three closely spaced lines. This observation indicates that the energy of an electron is slightly modified when the atom is immersed in a magnetic field. In order to explain this observation, a new quantum number, m_ℓ, called the **orbital magnetic quantum number,** was introduced. The theory is in accord with experimental results when m_ℓ is restricted to values ranging from $-\ell$ to $+\ell$, in integer steps.

Finally, very high resolution spectrometers revealed that spectral lines of gases are in fact two very closely spaced lines even in the absence of an external magnetic field. This splitting was referred to as **fine structure.** In 1925, Samuel Goudsmit and George Uhlenbeck introduced the idea of an electron spinning about its own axis to explain the origin of fine structure. The

Figure 30.9
A single line (A) can split into three separate lines (B) in a magnetic field.

results of their work introduced yet another quantum number, m_s, called the **spin quantum number.** We shall withhold further discussion of this quantum number for a later section. It is interesting to note that each of the new concepts introduced into the original Bohr theory added new quantum numbers and improved the original model. However, the most profound step forward in our current understanding of atomic structure came with the development of quantum mechanics.

Example 30.3   Singly Ionized Helium

Singly ionized helium, He$^+$, a hydrogen-like system, has one electron in the 1s orbit when the electron is in its ground state. Find (a) the energy of the electron in the ground state and (b) the radius of the ground-state orbit.

Solution

(a) From Equation 30.21, the energy of a level whose principal quantum number is n is given by

$$E_n = -\frac{mk^2 Z^2 e^4}{2\hbar^2}\left(\frac{1}{n^2}\right)$$

This can be expressed in eV units as

$$E_n = -\frac{Z^2(13.6)}{n^2}\text{ eV}$$

Since $Z = 2$ for helium, and $n = 1$ in the ground state, we have

$$E_1 = -4(13.6)\text{ eV} = \boxed{-54.4\text{ eV}}$$

(b) The radius of the ground state orbit can be found with the help of Equation 30.12. This equation must be modified in the case of a hydrogen-like atom by substituting Ze^2 for e^2, to obtain

$$r_n = \frac{n^2\hbar^2}{mkZe^2} = \frac{n^2}{Z}(0.0529\text{ nm})$$

For our case, $n = 1$ and $Z = 2$, and the result is

$$r_1 = \boxed{0.0265\text{ nm}}$$

30.5   De Broglie Waves and the Hydrogen Atom

One of the postulates made by Bohr in this theory of the hydrogen atom was that angular momentum of the electron is quantized in units of $\hbar = h/2\pi$, or

$$mvr = n\hbar$$

For more than a decade following Bohr's publication, no one was able to explain why the angular momentum of the electron was restricted to these discrete values. Finally, de Broglie recognized a connection between his theory of the wave character of material properties and the quantization condition given above. De Broglie assumed that an electron orbit would be stable (allowed) only if it contained an integral number of electron wave-

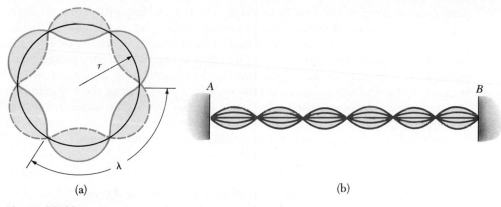

Figure 30.10
(a) Standing wave pattern for an electron wave in a stable orbit of hydrogen. There are three full wavelengths in this orbit. (b) Standing wave pattern for a vibrating stretched string fixed at its ends. This pattern has three full wavelengths.

lengths. Figure 30.10a demonstrates this point when three complete wavelengths are contained in one circumference of the orbit. Similar patterns can be drawn for orbits containing two wavelengths, four wavelengths, five wavelengths, and so forth. This situation is analogous to that of standing waves on a string discussed in Chapter 15. There we found that strings have preferred (resonant) frequencies of vibration. Figure 30.10b shows a standing wave pattern containing three wavelengths for a string fixed at each end. Now imagine that the vibrating string is removed from its supports at A and B and bent into a circular shape such that points A and B are brought together. The end result is a pattern like that shown in Figure 30.10a.

In general, the condition for a de Broglie standing wave in an electron orbit is that the circumference must contain an integral multiple of electron wavelengths. We can express this condition as

$$2\pi r = n\lambda \qquad (n = 1, 2, 3, \ldots) \qquad \text{(30.23)}$$

De Broglie's equation for the wavelength of an electron in terms of its momentum is

$$\lambda = \frac{h}{mv}$$

When this expression for λ is substituted into Equation 30.23, we find

$$2\pi r = \frac{nh}{mv}$$

or

$$mvr = n\hbar$$

This is precisely the quantization of angular momentum condition imposed by Bohr in his original theory of hydrogen.

The electron orbit shown in Figure 30.10a contains three complete wavelengths and corresponds to the case where the principal quantum number n equals three. The orbit with one complete wavelength in its circumference would correspond to the first Bohr orbit, $n = 1$; the orbit with two complete wavelengths would correspond to the second Bohr orbit, $n = 2$, and so forth.

By applying the wave theory of matter to electrons in atoms, de Broglie was able to explain the appearance of integers in the Bohr theory as a natural consequence of interference. This was the first convincing argument which suggested that the wave nature of matter was at the heart of the behavior of atomic systems. Although the analysis provided by de Broglie was a promising first step, gigantic strides were subsequently made with the development of Schrödinger's wave equation and its application to atomic systems.

30.6 Quantum Mechanics and the Hydrogen Atom

One of the first great achievements of quantum mechanics was the solution of the wave equation for the hydrogen atom. We shall not attempt to carry out this solution. Rather, we shall simply describe its properties and some of the implications with regard to atomic structure.

According to quantum mechanics, the energies of the allowed states are in exact agreement with the values obtained by the Bohr theory (Eq. 30.15), where the allowed energies depend only on the principal quantum number, n.

In addition to the principal quantum number, two other quantum numbers emerged from the solution, ℓ and m_ℓ. The quantum number ℓ is called the **orbital quantum number,** and m_ℓ is called the **orbital magnetic quantum number.** As pointed out in Section 30.4, these quantum numbers had already appeared in modifications made to the Bohr theory. The significance of quantum mechanics is that these quantum numbers and the restrictions placed on their values arose directly from the mathematics and not by some ad hoc assumptions to make the theory consistent with experimental observation. Because we shall need to make use of the various quantum numbers in the next several sections, the ranges of their values are repeated below.

The values of n can range from 1 to ∞.
The values of ℓ can range from 0 to $n - 1$.
The values of m_ℓ can range from $-\ell$ to ℓ in integer steps.

For example, if $n = 1$, only $\ell = 0$ and $m_\ell = 0$ are permitted. If $n = 2$, the value of ℓ may be 0 or 1; if $\ell = 0$, then $m_\ell = 0$, but if $\ell = 1$, then m_ℓ may be 1, 0, or -1. Table 30.2 summarizes the rules for determining the allowed values of ℓ and m_ℓ for a given value of n.

Table 30.2
Three Quantum Numbers for the Hydrogen Atom

Quantum Number	Name	Allowed Values	Number of Allowed States
n	Principal quantum number	1, 2, 3, . . .	Any number
ℓ	Orbital quantum number	0, 1, 2, . . . , $n - 1$	n
m_ℓ	Orbital magnetic quantum number	$-\ell, -\ell + 1, . . . ,$ 0, . . . , $\ell - 1, \ell$	$2\ell + 1$

States that violate the rules given in Table 30.2 cannot exist. For instance, one state that cannot exist is the $2d$ state, which would have $n = 2$ and $\ell = 2$. This state is not allowed because the highest allowed value of ℓ is $n - 1$, or 1 in this case. Thus, for $n = 2$, $2s$ and $2p$ are allowed states but $2d$, $2f$, . . . are not. For $n = 3$, the allowed states are $3s$, $3p$, and $3d$.

Example 30.4   The $n = 2$ Level of Hydrogen

Determine the number of states in the hydrogen atom corresponding to the principal quantum number $n = 2$ and calculate the energies of these states.

Solution
For $n = 2$, ℓ can have the values 0 and 1. For $\ell = 0$, m_ℓ can only be 0; for $\ell = 1$, m_ℓ can be -1, 0, or 1. Hence, we have a state designated as the $2s$ state associated with the quantum numbers $n = 2$, $\ell = 0$, and $m_\ell = 0$, and three states designated as $2p$ states for which the quantum numbers are $n = 2$, $\ell = 1$, $m_\ell = -1$; $n = 2$, $\ell = 1$, $m_\ell = 0$; and $n = 2$, $\ell = 1$, $m_\ell = 1$.

Because all of these states have the same principal quantum number, $n = 2$, they also have the same energy, which can be calculated using Equation 30.16, that is, $E_n = -13.6/n^2$. For $n = 2$, this gives

$$E_2 = -\frac{13.6}{2^2} \text{ eV} = \boxed{-3.4 \text{ eV}}$$

Exercise   How many possible states are there for the $n = 3$ level of hydrogen? For the $n = 4$ level?

Answers   9 states for $n = 3$, and 16 states for $n = 4$.

30.7   The Spin Magnetic Quantum Number

Example 30.4 was presented to give you some practice in manipulating quantum numbers. For example, we found that there are four states corresponding to the principal quantum number $n = 2$. As we shall see in this section, there actually are *eight* such states rather than four in this case. This can be explained by requiring a fourth quantum number for each state, called the **spin magnetic quantum number, m_s**.

As pointed out in Section 30.4, the need for this new quantum number first came about because of an unusual feature in the spectra of certain gases, such as sodium vapor. Close examination of one of the prominent lines of sodium shows that this line is, in fact, two very closely spaced lines. The wavelengths of these lines occur in the yellow region at 589.0 nm and 589.6 nm. In 1925, when this was first noticed, the theory of atomic structure had not been adequately developed to explain why there were two lines rather than one. To resolve this dilemma, Samual Goudsmidt and George Uhlenbeck, following a suggestion by the Austrian physicist Wolfgang Pauli, proposed that a new quantum number, called the spin quantum number, must be added to the set of required quantum numbers in describing a quantum state.

In order to describe the spin quantum number, it is convenient (but incorrect) to think of the electron as spinning on its axis as it orbits the nucleus, just as the Earth spins about its axis as it orbits the Sun. According to the theory, there are only two ways that the electron can spin as it orbits the nucleus, as

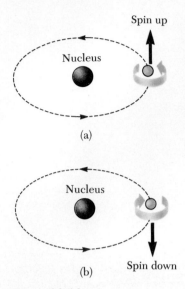

Spin up

Nucleus

(a)

Nucleus

(b)   Spin down

Figure 30.11
As an electron moves in its orbit about the nucleus, its spin can be either (a) up or (b) down.

shown in Figure 30.11. If the direction of spin is relative to its orbital motion, as shown in Figure 30.11a, the electron is said to have "spin up." If the direction of spin is relative to its orbital motion, as in Figure 30.11b, the electron is said to have "spin down." The energy of the electron is slightly different for the two spin directions. As it turns out, this energy difference properly accounts for the observed splitting of the yellow sodium lines. The quantum numbers associated with the spin of the electron are $m_s = \frac{1}{2}$ for the spin-up state and $m_s = -\frac{1}{2}$ for the spin-down state. As we shall see in the following example, this added quantum number doubles the number of allowed states specified by the quantum numbers n, ℓ, and m_ℓ.

The classical description of electron spin described above is incorrect because quantum mechanics tells us that, since the electron cannot be precisely located in space, it cannot be considered to be spinning as pictured in Figure 30.11.

Although the classical description of electron spin described above is helpful as a mental construct, it should not be taken too seriously. It is incorrect to view the electron as a charged, spinning object with a definite size. In spite of this conceptual difficulty, all experimental evidence supports the fact that an electron has an intrinsic property that can be described by the spin magnetic quantum number.

Example 30.5   Adding Electron Spin to Hydrogen

Determine the quantum numbers associated with the possible states in the hydrogen atom corresponding to the principal quantum number $n = 2$.

Solution
With the addition of the spin quantum number, we have the following possibilities:

n	ℓ	m_ℓ	m_s	Subshell	Shell	Number of Electrons in Subshell
2	0	0	$\frac{1}{2}$			
2	0	0	$-\frac{1}{2}$	$2s$	L	2
2	1	1	$\frac{1}{2}$			
2	1	1	$-\frac{1}{2}$			
2	1	0	$\frac{1}{2}$			
2	1	0	$-\frac{1}{2}$	$2p$	L	6
2	1	-1	$\frac{1}{2}$			
2	1	-1	$-\frac{1}{2}$			

Exercise   Show that for $n = 3$, there are 18 possible states. (This follows from the restrictions that the maximum number of electrons in the $3s$ state is 2, the maximum number in the $3p$ state is 6, and the maximum number in the $3d$ state is 10.)

30.8   Electron Clouds

The solution of the wave equation for the wave function Ψ, discussed in Section 29.10, yields a quantity dependent on the quantum numbers n, ℓ, and m_ℓ. Let us assume that we have found a value for Ψ and see what it may tell us about the hydrogen atom. Let us choose a value of $n = 1$ for the principal quantum number, which corresponds to the lowest energy state for hydrogen. For $n = 1$, the restrictions placed on the remaining quantum numbers are that $\ell = 0$ and $m_\ell = 0$. We can now substitute these values into our expression for Ψ.

The quantity $|\Psi|^2$ has great physical significance because it is proportional to *the probability of finding the electron at a given position*. Figure 30.12 gives the probability of finding the electron at various distances from the nucleus in the $1s$ state of hydrogen. Some useful and surprising information can be extracted from this curve. First, the curve peaks at a value of r equal to 0.0529 nm, the Bohr value of the radius of the first electron orbit in hydrogen. This means that the probability of finding the electron at this distance from the nucleus is a maximum. However, as the curve indicates, there is also a probability of finding the electron at other distances from the nucleus. In other words, the electron is not confined to a particular orbital distance from the nucleus, as assumed in the Bohr model. The electron may be found at various distances from the nucleus, but *the probability of finding it at a distance corresponding to the first Bohr orbit is a maximum*. Quantum mechanics also predicts that the wave function for the hydrogen atom in the ground state is spherically symmetric, hence the electron can be found in a spherical region surrounding the nucleus. This is in contrast to the Bohr theory that confines the position of the electron to points in a plane. This result is often interpreted by viewing the electron as a cloud surrounding the nucleus. An attempt at picturing this cloud-like behavior is shown in Figure 30.13. The densest regions of the cloud represent those locations where the electron is most likely to be found.

If a similar analysis is carried out for the $n = 2$, $\ell = 0$ state of hydrogen, a peak of the probability curve is found at $4a_0$. Likewise, for the $n = 3$, $\ell = 0$ state, the curve peaks at $9a_0$. Thus, quantum mechanics predicts a most probable electron location that is in agreement with the locations predicted by the Bohr theory.

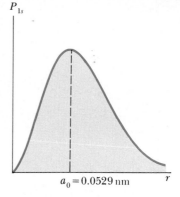

Figure 30.12
The probability of finding the electron versus distance from the nucleus for the hydrogen atom in the $1s$ (ground) state. Note that the probability has its maximum value when r equals the first Bohr radius, a_0.

30.9 The Exclusion Principle and the Periodic Table

Earlier we found that the state of an electron in an atom is specified by four quantum numbers: n, ℓ, m_ℓ, and m_s. For example, an electron in the ground state of hydrogen could have quantum numbers of $n = 1$, $\ell = 0$, $m_\ell = 0$, $m_s = \frac{1}{2}$. As it turns out, the state of an electron in any other atom may also be specified by this same set of quantum numbers. In fact, these four quantum numbers can be used to describe all the electronic states of an atom regardless of the number of electrons in its structure.

An obvious question that arises here is "How many electrons can have a particular set of quantum numbers?" This important question was answered by Pauli in 1925 in a powerful statement known as the **exclusion principle**:

> No two electrons in an atom can ever be in the same quantum state; that is, no two electrons in the same atom can have the same set of quantum numbers n, ℓ, m_ℓ, and m_s.

It is interesting to note that if this principle were not valid, every electron would end up in the lowest energy state of the atom and the chemical behavior of the elements would be grossly modified. Nature as we know it would not

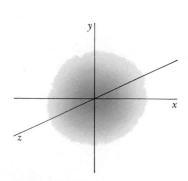

Figure 30.13
The spherical electron cloud for the hydrogen atom in its $1s$ state.

exist! In reality, we can view the electronic structure of complex atoms as a succession of filled levels increasing in energy, where the outermost electrons are primarily responsible for the chemical properties of the element.

As a general rule, the order of filling of an atom's subshell with electrons is as follows. Once one subshell is filled, the next electron goes into the vacant subshell that is lowest in energy. One can understand this principle by recognizing that if the atom was not in the lowest energy state available to it, it would radiate energy until it reached this state. Recall that a subshell is filled when it contains $2(2\ell + 1)$ electrons. This rule is based on the analysis of quantum numbers to be described below. Following this rule, shells and subshells are filled according to the pattern given in Table 30.3.

The exclusion principle can be illustrated by an examination of the electronic arrangement in a few of the lighter atoms.

Hydrogen has only one electron, which, in its ground state, can be described by either of two sets of quantum numbers: $1, 0, 0, \frac{1}{2}$, or $1, 0, 0, -\frac{1}{2}$. The electronic configuration of this atom is often designated as $1s^1$. The notation $1s$ means that we are referring to a state for which $n = 1$ and $\ell = 0$, and the superscript indicates that one electron is present in this level.

Neutral *helium* has two electrons. In the ground state, the quantum numbers for these two electrons are $1, 0, 0, \frac{1}{2}$ and $1, 0, 0, -\frac{1}{2}$. There are no other possible combinations of quantum numbers for this level, and we say that the **K** shell is filled. Helium is designated by the notation $1s^2$.

Neutral *lithium* has three electrons. In the ground state, two of these are in the $1s$ subshell and the third is in the $2s$ subshell because this subshell is slightly lower in energy than the $2p$ subshell. Hence, the electronic configuration for lithium is $1s^2 2s^1$.

A list of electronic ground-state configurations for a number of atoms is provided in Table 30.4. This listing will help us to gain an understanding of the periodic table of the elements given in the endpapers. In 1871 Dmitri Mendeleev (1834–1907), a Russian chemist, arranged the elements known at that time in a table according to their atomic weights and chemical similarities. The first table Mendeleev proposed contained many blank spaces, and he boldly

Table 30.3
The Number of Electrons in Filled Subshells and Shells

Shell	Subshell	Number of Electrons in Filled Subshell	Number of Electrons in Filled Shell
K ($n = 1$)	$s(\ell = 0)$	2	2
L ($n = 2$)	$s(\ell = 0)$ $p(\ell = 1)$	2 6	8
M ($n = 3$)	$s(\ell = 0)$ $p(\ell = 1)$ $d(\ell = 2)$	2 6 10	18
N ($n = 4$)	$s(\ell = 0)$ $p(\ell = 1)$ $d(\ell = 2)$ $f(\ell = 3)$	2 6 10 14	32

Table 30.4
Electronic Configuration of Some Elements

Z	Symbol	Ground-State Configuration	Ionization Energy (eV)	Z	Symbol	Ground-State Configuration	Ionization Energy (eV)
1	H	$1s^1$	13.595	19	K	[Ar] $4s^1$	4.339
2	He	$1s^2$	24.581	20	Ca	$4s^2$	6.111
				21	Sc	$3d4s^2$	6.54
3	Li	[He] $2s^1$	5.390	22	Ti	$3d^24s^2$	6.83
4	Be	$2s^2$	9.320	23	V	$3d^34s^2$	6.74
5	B	$2s^22p^1$	8.296	24	Cr	$3d^54s^1$	6.76
6	C	$2s^22p^2$	11.256	25	Mn	$3d^54s^2$	7.432
7	N	$2s^22p^3$	14.545	26	Fe	$3d^64s^2$	7.87
8	O	$2s^22p^4$	13.614	27	Co	$3d^74s^2$	7.86
9	F	$2s^22p^5$	17.418	28	Ni	$3d^84s^2$	7.633
10	Ne	$2s^22p^6$	21.559	29	Cu	$3d^{10}4s^1$	7.724
				30	Zn	$3d^{10}4s^2$	9.391
11	Na	[Ne] $3s^1$	5.138	31	Ga	$3d^{10}4s^24p^1$	6.00
12	Mg	$3s^2$	7.644	32	Ge	$3d^{10}4s^24p^2$	7.88
13	Al	$3s^23p^1$	5.984	33	As	$3d^{10}4s^24p^3$	9.81
14	Si	$3s^23p^2$	8.149	34	Se	$3d^{10}4s^24p^4$	9.75
15	P	$3s^23p^3$	10.484	35	Br	$3d^{10}4s^24p^5$	11.84
16	S	$3s^23p^4$	10.357	36	Kr	$3d^{10}4s^24p^6$	13.996
17	Cl	$3s^23p^5$	13.01				
18	Ar	$3s^23p^6$	15.755				

Note: The bracket notation is used as a shorthand method to avoid repetition in indicating inner-shell electrons. Thus, [He] represents $1s^2$, [Ne] represents $1s^22s^22p^6$, [Ar] represents $1s^22s^22p^63s^23p^6$, and so on.

stated that the gaps were there only because those elements had not yet been discovered. By noting the column in which these missing elements should be located, he was able to make rough predictions about their chemical properties. Within 20 years of this announcement, these elements were indeed discovered.

The elements in our current version of the periodic table are still arranged such that all those in a vertical column have similar chemical properties. For example, consider the elements in the last column: He (helium), Ne (neon), Ar (argon), Kr (krypton), Xe (xenon), and Rn (radon). The outstanding characteristic of these elements is that they do not normally take part in chemical reactions, that is, they do not join with other atoms to form molecules, and are therefore classified as being inert. Because of this aloofness, they are referred to as the noble gases. We can partially understand this behavior by looking at the electronic configurations shown in Table 30.4. The element helium is one in which the electronic configuration is $1s^2$. In other words, one shell is filled. Additionally, it is found that the electrons in this filled shell are considerably separated in energy from the next available level, the $2s$ level.

The electronic configuration for neon is $1s^22s^22p^6$. Again, the outer shell is filled and there is a large difference in energy between the $2p$ level and the $3s$ level. Argon has the configuration $1s^22s^22p^63s^23p^6$. Here, the $3p$ subshell is filled and there is a wide gap in energy between the $3p$ subshell and the $3d$ subshell. We could continue this procedure through all the noble gases, but the pattern remains the same. A noble gas is formed when either a shell or a subshell is filled and there is a large gap in energy before the next possible level is encountered.

Wolfgang Pauli and Niels Bohr watch a spinning top. *(Courtesy of AIP Niels Bohr Library, Margarethe Bohr Collection.)*

Wolfgang Pauli
(1900–1958)

(Photo taken by S. A. Goudsmit, AIP Niels Bohr Library)

Wolfgang Pauli was an extremely talented Austrian theoretical physicist who made important contributions in many areas of modern physics. At the age of 21, Pauli gained public recognition with a masterful review article on relativity, which is still considered to be one of the finest and most comprehensive introductions to the subject. Other major contributions were the discovery of the exclusion principle, the explanation of the connection between particle spin and statistics, theories of relativistic quantum electrodynamics, the neutrino hypothesis, and the hypothesis of nuclear spin. Pauli was a forceful and colorful character, well known for his witty and often caustic remarks directed at those who presented new theories in a less than perfectly clear manner. Pauli exerted great influence on his students and colleagues by forcing them with his sharp criticism to a deeper and clearer understanding. Victor Weisskopf, one of Pauli's famous students, has aptly described him as "the conscience of theoretical physics." Pauli's sharp sense of humor was also nicely captured by Weisskopf in the following anecdote:

"In a few weeks, Pauli asked me to come to Zurich. I came to the big door of his office, I knocked, and no answer. I knocked again and no answer. After about five minutes he said, rather roughly, "Who is it? Come in!" I opened the door, and here was Pauli—it was a very big office—at the other side of the room, at his desk, writing and writing. He said, "Who is this? First I must finish calculating." Again he let me wait for about five minutes and then: "Who is that?" "I am Weisskopf." "Uhh, Weisskopf, ja, you are my new assistant." Then he looked at me and said, "Now, you see I wanted to take Bethe, but Bethe works now on the solid state. Solid state I don't like, although I started it. This is why I took you." Then I said, "What can I do for you sir?" and he said "I shall give you right away a problem." He gave me a problem, some calculation, and then he said, "Go and work." So I went, and after 10 days or so, he came and said, "Well, show me what you have done." And I showed him. He looked at it and exclaimed: "I should have taken Bethe!"[1]

[1] From *Physics in the Twentieth Century: Selected Essays, My Life as a Physicist*, Victor F. Weisskopf, The MIT Press, 1972, p. 10.

The elements in the first column of the periodic table are called the alkali metals and are characterized by the fact that they are very chemically active. Referring to Table 30.4, we can understand why these elements interact so strongly with other elements. All of these alkali metals have a single outer electron in an *s* subshell. This electron is shielded from the nucleus by all the electrons in the inner shells. Thus, it is only loosely bound to the atom and can readily be accepted by other atoms to form molecules.

All the elements in the seventh column of the periodic table (called the halogens) are also very active chemically. Note that all these elements are lacking one electron in a subshell. Consequently, these elements readily accept electrons from other atoms and form molecules.

Example 30.6   The Quantum Number for the 2*p* Subshell

List the quantum numbers for electrons in the 2*p* subshell.

Solution

For this subshell, $n = 2$ and $\ell = 1$. The magnetic quantum number can have the values -1, 0, 1, and the spin quantum number is always $+\frac{1}{2}$ or $-\frac{1}{2}$. Thus, the six possibilities are

n	ℓ	m_ℓ	m_s
2	1	-1	$-\frac{1}{2}$
2	1	-1	$\frac{1}{2}$
2	1	0	$-\frac{1}{2}$
2	1	0	$\frac{1}{2}$
2	1	1	$-\frac{1}{2}$
2	1	1	$\frac{1}{2}$

30.10 Characteristic X-Rays

In Section 29.4, we discussed the manner in which x-rays are emitted when a metal target is bombarded by high-energy electrons. The x-ray spectrum typically consists of a broad continuous band and a series of sharp lines that are dependent on the type of metal used for the target, as shown in Figure 30.14. The presence of these lines, called **characteristic x-rays,** was discovered in 1908, but their origin remained unexplained until the details of atomic structure, particularly the shell structure of the atom, were developed.

The first step in the production of characteristic x-rays occurs when a bombarding electron collides with an electron in an inner shell of a target atom with sufficient energy to remove the electron from the atom. The vacancy created in the shell can now be filled when an electron in a higher level drops down into the lower energy level containing the vacancy. The time it takes for this to happen is very short, less than 10^{-9} s. This transition is accompanied by the emission of a photon whose energy will equal the difference in energy between the two levels. Typically, the energy of such transitions is greater than 1000 eV, and the emitted x-ray photons have wavelengths in the range of 0.01 nm to 1 nm.

Let us assume that the incoming electron has dislodged an atomic electron from the innermost shell, the K shell. If the vacancy is filled by an electron dropping from the next higher shell, the L shell, the photon emitted in the process is referred to as the K_α line on the curve of Figure 30.14. If the vacancy is filled by an electron dropping from the M shell, the line produced is called the K_β line.

Other characteristic x-ray lines are formed when electrons drop from upper levels to vacancies other than those in the K shell. For example, L lines are produced when vacancies in the L shell are filled by electrons dropping from higher shells. An L_α line is produced as an electron drops from the M shell to the L shell, and an L_β line is produced by a transition from the N shell to the L shell.

We can estimate the energy of the x-rays emitted by an atom as follows. Consider two electrons in the K shell of an atom whose atomic number is Z, where Z is the number of protons in the nucleus. Each electron partially shields the other from the charge of the nucleus, Ze, and so each electron is subject to an effective nuclear charge $Z_{\text{eff}} = (Z - 1)e$. We can now use a modified form of Equation 30.15 to estimate the energy of either electron in the K shell (with $n = 1$):

$$E_{\text{K}} = -mZ_{\text{eff}}^2 \frac{k^2 e^4}{2\hbar^2} = -Z_{\text{eff}}^2 E_0$$

Figure 30.14
The x-ray spectrum of a metal target consists of broad continuous spectrum plus a number of sharp lines which are due to *characteristic x-rays.* The data shown were obtained when 35-keV electrons bombarded a molybdenum target. Note that 1 pm = 10^{-12} m = 0.001 nm.

Figure 30.15
A Moseley plot. A straight line is obtained when $\sqrt{1/\lambda}$ is plotted versus Z for the K_α x-ray lines of a number of elements.

where E_0 is the ground state energy. Substituting $Z_{eff} = Z - 1$ gives

$$E_K = -(Z - 1)^2(13.6 \text{ eV}) \tag{30.24}$$

As we shall show in the following example, one can estimate the energy of an electron in an L or M shell in a similar fashion. Taking the energy difference between these two levels, one can then calculate the energy and wavelength of the emitted photon.

Example 30.7   Estimating the Energy of an X-Ray

Estimate the energy of the characteristic x-ray emitted from a tungsten target when an electron drops from an M shell ($n = 3$ state) to a vacancy in the K shell ($n = 1$ state).

Solution
The atomic number for tungsten is $Z = 74$. Using Equation 30.24, we see that the energy of the electron in the K shell state is approximately

$$E_K = -(74 - 1)^2(13.6 \text{ eV}) = -72,500 \text{ eV}$$

The electron in the M shell ($n = 3$) is subject to an effective nuclear charge that depends on the number of electrons in the $n = 1$ and $n = 2$ states, which shield the nucleus. Because there are eight electrons in the $n = 2$ state and one electron in the $n = 1$ state, roughly nine electrons shield the nucleus, and so $Z_{eff} = Z - 9$. Hence, the energy of an electron in the M shell ($n = 3$), following Equation 30.15, is equal to

$$E_M = -Z_{eff}^2 E_3 = -(Z - 9)^2 \frac{E_0}{3^2} = -(74 - 9)^2 \frac{(13.6 \text{ eV})}{9} = -6380 \text{ eV}$$

where E_3 is the energy of an electron in the $n = 3$ level of the hydrogen atom. Therefore, the emitted x-ray has an energy equal to $E_M - E_K = -6380 \text{ eV} - (-72,500 \text{ eV}) = 66,100 \text{ eV}$. Note that this energy difference is also equal to hf, where $hf = hc/\lambda$, and where λ, is the wavelength of the emitted x-ray.

Exercise   Calculate the wavelength of the emitted x-ray for this transition.

Answer   0.0188 nm.

In 1914, Henry G. J. Moseley plotted the Z values for a number of elements versus $\sqrt{1/\lambda}$, where λ is the wavelength of the K_α line for each element. He found that such a plot produced a straight line, as in Figure 30.15. This is consistent with our rough calculations of the energy levels based on Equation 30.24. From this plot, Moseley was able to determine the Z values of other elements, which provided a periodic chart in excellent agreement with the known chemical properties of the elements.

30.11   Atomic Transitions

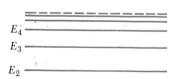

Figure 30.16
Energy level diagram of an atom with various allowed states. The lowest energy state E_1 is the ground state. All others are excited states.

In this section we shall look at some of the basic processes involved in an atomic system. It is necessary to understand these mechanisms before we can understand the operation of a laser, which we shall examine in the next section.

We have seen that an atom will emit radiation only at certain frequencies, which correspond to the energy separation between the various allowed

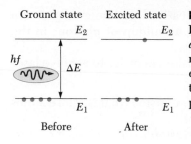

Ground state   Excited state

E_2 ——————— E_2 ———•———

hf ~~~~➤ ΔE

E_1 ——••••— E_1 ——•••——

Before   After

Figure 30.17
Diagram representing the process of *stimulated absorption* of a photon by an atom. The blue dots represent electrons in the various states. One electron is transferred from the ground state to the excited state when the atom absorbs one photon whose energy $hf = E_2 - E_1$.

states. Consider an atom with many allowed energy states, labeled E_1, E_2, E_3, . . . , as in Figure 30.16. When light is incident on the atom, only those photons whose energy, hf, matches the energy separation ΔE between two levels can be absorbed by the atom. A schematic diagram representing this **stimulated absorption process** is shown in Figure 30.17. At ordinary temperatures, most of the atoms are in the ground state. If a vessel containing many atoms of a gaseous element is illuminated with a light beam containing all possible photon frequencies (that is, a continuous spectrum), only those photons of energies $E_2 - E_1$, $E_3 - E_1$, $E_4 - E_1$, and so on, can be absorbed. As a result of this absorption, some atoms are raised to various allowed higher energy levels called **excited states**.

Once an atom is in an excited state, there is a certain probability that it will jump back to a lower level by emitting a photon, as shown in Figure 30.18. This process is known as **spontaneous emission**. In typical cases, an atom will remain in an excited state for only about 10^{-8} s.

Finally, there is a third process, which is of importance in lasers, known as **stimulated emission**. Suppose an atom is in an excited state E_2, as in Figure 30.19, and a photon of energy $hf = E_2 - E_1$ is incident on it. The incoming photon will increase the probability that the electron will return to the ground state and thereby emit a second photon having the same energy, hf. This process of speeding up atomic transitions to lower levels is called stimulated emission. Note that there are two identical photons that result from this process, the incident photon and the emitted photon. The emitted photon will be exactly in phase with the incident photon. These photons can, in turn, stimulate other atoms to emit photons in a chain of similar processes. The many photons produced in this fashion are the source of the intense, coherent light in a laser.

*30.12   Lasers and Holography

We have described how an incident photon can cause atomic transitions either upward (stimulated absorption) or downward (stimulated emission). Both processes are equally probable. When light is incident on a system of atoms, there is usually a net absorption of energy because there are many more atoms in the ground state than in excited states. That is, in a normal situation, there are more atoms in state E_1 ready to absorb photons than there are atoms in states E_2, E_3, . . . ready to emit photons. However, if the situation can be inverted so that there are more atoms in an excited state than in the ground state, a net emission of photons can result. Such a condition is called **population inversion**. This, in fact, is the fundamental principle involved in the

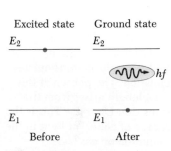

Excited state   Ground state

E_2 ——•——— E_2 ———————

~~~~➤ hf

E_1 ——————— E_1 ———•———

Before   After

Figure 30.18
Diagram representing the process of *spontaneous emission* of a photon by an atom that is initially in the excited state E_2. When the electron falls to the ground state, the atom emits a photon whose energy is $hf = E_2 - E_1$.

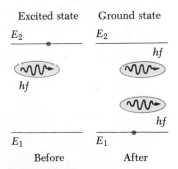

Excited state   Ground state

E_2 ——•——— E_2 ———————

                              hf

~~~~➤   ~~~~➤

hf ~~~~➤

                              hf

E_1 ——————— E_1 ———•———

Before   After

Figure 30.19
Diagram representing the process of *stimulated emission* of a photon by an incoming photon of energy hf. Initially, the atom is in the excited state. The incoming photon stimulates the atom to emit a second photon of energy $hf = E_2 - E_1$.

Figure 30.20
Energy level diagram for the neon atom, which emits photons at a wavelength of 632.8 nm through stimulated emission. The photon at this wavelength arises from the transition $E_3^\circ \rightarrow E_2$. This is the source of coherent light in the helium-neon gas laser.

operation of a **laser,** an acronym for $\underline{l}$ight $\underline{a}$mplification by $\underline{s}$timulated $\underline{e}$mission of $\underline{r}$adiation. The amplification corresponds to a buildup of photons in the system as the result of a chain reaction of events.

The following three conditions must be satisfied in order to achieve laser action:

1. The system must be in a state of population inversion (that is, more atoms in an excited state than in the ground state).
2. The excited state of the system must be a *metastable state,* which means its lifetime must be long compared with the usually short lifetimes of excited states. When such is the case, stimulated emission will occur before spontaneous emission.
3. The emitted photons must be confined in the system long enough to allow them to stimulate further emission from other excited atoms. This is achieved by the use of reflecting mirrors at the ends of the system. One end is made totally reflecting, and the other is slightly transparent to allow the laser beam to escape.

One device that exhibits stimulated emission of radiation is the helium-neon gas laser. The energy level diagram for the neon atom in this system is shown in Figure 30.20. The mixture of helium and neon is confined to a glass tube sealed at the ends by mirrors. An oscillator connected to the tube causes electrons to sweep through the tube, colliding with the atoms of the gas and raising them into excited states. Some neon atoms are excited to state E_3 through this process and also as a result of collisions with excited helium atoms. Stimulated emission occurs as the neon atoms make a transition to state E_2 and neighboring excited atoms are stimulated to emit at the same frequency and phase. This results in the production of coherent light at a wavelength of 632.8 nm. Figure 30.21 summarizes the steps in the production of a laser beam.

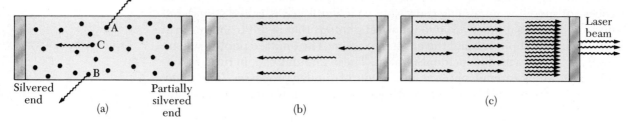

Figure 30.21
(a) Neon atoms in the gas are raised to a metastable excited state by collisions with electrons sweeping through the tube and by collisions with excited helium atoms. The excited neon atoms, shown as colored dots, eventually return to their ground state and emit photons. Some of the emitted photons emerge from the sides of the glass tube and are lost (atoms A and B). However, other photons are emitted along the axis of the tube (atom C). (b) A photon moving along the axis of the tube encounters other excited neon atoms and causes them to emit a photon. The two photons then stimulate other excited atoms, causing further emission. Thus, the number of photons multiplies as the process continues. (c) The photon beam is reflected from the left silvered end and moves back toward the partially silvered end at the right. The process of stimulated emission continues and the number of photons in the beam increases further. After many reflections between the two ends of the tube, a small fraction of the photons emerges from the partially silvered end. These are the photons that make up the coherent laser beam.

Since the development of the first laser in 1960, there has been a tremendous growth in laser technology. Lasers are now available that cover wavelengths in the infrared, visible, and ultraviolet regions. Applications include surgical "welding" of detached retinas, precision surveying and length measurement, a potential source for inducing nuclear fusion reactions, precision cutting of metals and other materials, and telephone communication along optical fibers. These and other applications are possible because of the unique characteristics of laser light. In addition to its being highly monochromatic and coherent, laser light is also highly directional and can be sharply focused to produce regions of extremely intense light energy. Three particular uses of lasers are discussed in the following section and in the Special Topics.

Holography

One of the most unusual and interesting applications of the laser is in the production of three-dimensional images of an object in a process called **holography**. Figure 30.22a shows how a hologram is made. Light from the laser is split into two parts by a half-silvered mirror at B. One part of the beam reflects off the object to be photographed and strikes an ordinary photographic film. The other half of the beam is diverged by lens L_2, reflects from mirrors M_1 and M_2, and finally strikes the film. The two beams overlap to form an extremely complicated interference pattern on the film. Such an interference pattern can be produced only if the phase relationship of the two waves is maintained constant throughout the exposure of the film. This condition is met if light from a laser is used because such light is coherent. The hologram records not only the intensity of the light but also the phase difference between the two halves of the laser beam. It is this last piece of information, the phase relationship, that causes the interference pattern that results in the three-dimensional image.

A hologram is best viewed by allowing coherent light to pass through the developed film as the viewer looks back along the direction from which the beam comes. Figure 30.22b is a photograph of a hologram made using a cylindrical film. A three-dimensional image of the object is seen such that as the viewer's head is moved, the perspective changes, as for an actual object. The applications of holography promise to be many and varied. For example, someday your television set may be replaced by one using holography.

Scientist checking the performance of an experimental laser-cutting device mounted on a robot arm. The laser is being used to cut through a metal plate. *(Courtesy of Philippe Plailly, Science Photo Library/Photo Researchers, Inc.)*

Figure 30.22
(a) Experimental arrangement for producing a hologram. (b) Photograph of a hologram which uses a cylindrical film. Note the detail of the Volkswagen image. *(Courtesy of CENCO.)*

(a)

(b)

Figure 30.23
A laser disc player.

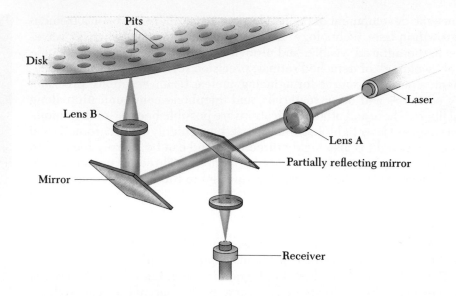

▶ **SPECIAL TOPIC**
Laser Disc Players

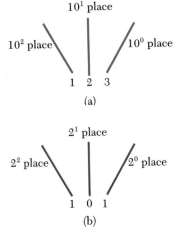

Figure 30.24
Weighting the digits in (a) the decimal system and (b) the binary system.

Figure 30.25
A byte consists of eight bits weighted as shown.

One of the most recent applications of the laser that has found its way into the life of college students is the laser disc player, a device that is rapidly replacing phonographs and tape players in ultra-high-fidelity sound reproduction. Figure 30.23 shows the important components of one of these devices. Light from a laser is sent through lens A toward a partially reflecting mirror. The portion of the light that passes through this mirror is redirected by another mirror through lens B toward a laser disc. The light reflected off this disc retraces the original path until it reaches the partially reflecting mirror, where a portion of it is reflected into a receiver. The receiver then uses this reflected signal to drive a loudspeaker and to recreate the information stored on the laser disc.

In order to understand how the light retrieves information from the laser disc, let us first explain the binary number system. Consider an ordinary decimal number such as 123, where each digit in the number has a place value associated with it. In this case, the number 123 contains 1 one hundred, 2 tens, and 3 ones. Figure 30.24a shows an alternative way of expressing this same information. Note that in the figure we say that the number 3 occupies the 10^0 location, which is equivalent to saying that it occupies the one's place because $10^0 = 1$. Also, the number 2 occupies the 10^1 location, and the number 1 occupies the 10^2 location. This procedure can be extended upward as high as you care to count, where the weight of each position would be determined by some power of the number system base, which is 10 for decimal numbers.

Each bit position of a binary number also carries a particular weight that determines its magnitude, except the place value of each bit is determined by some power of the base number 2. For example, Figure 30.24b shows a binary number 101 with its corresponding weight positions. The 1 on the right is in the $2^0 = 1$ location, the 0 is in the location having a place value of 2^1, and the 1 on the left is in the 2^2 position. Each digit in a binary number is referred to as a bit, and a collection of eight bits is called a byte. Thus, we can represent a byte with a box constructed as shown in Figure 30.25, where the weight of each bit in the box is indicated. The box says that the weight of the lowest order bit (the rightmost one) is 1 $(=2^0)$, and the weight of the highest order bit (the leftmost one) is 128 $(=2^7)$. Thus, referring to the box, we see that the binary number 101 is decimal 5. As an exercise, convert the following binary numbers to decimal numbers: 00101101, 11011100, and 11111111. (The answers are 45, 220, and 255, respectively.)

Modern electronic devices such as computers use binary numbers in the form of a series of on and off signals. To understand this, consider the binary number 00000001, which is the binary representation of the decimal number 1. This binary number could be manipulated by an electronic circuit by turning a switch "off" for seven equal intervals of time, then "on" for one of these time intervals.

The operation of a laser disc player is based upon the procedure just described. The amplitude of a sound signal is sampled at regular intervals of time as in Figure 30.26, and the loudness of the sound at each sampling is then converted to a binary number by an electronic circuit. For example, suppose the amplitude of the sound signal in Figure 30.26 at t_1, t_2, and t_3 is decimal 4, 5, 4, respectively. The circuitry will translate this to binary 00000100, 00000101, and 00000100. This pattern of binary numbers is then recorded by placing a series of pits and smooth places on a laser disc such that each pit is smaller than the dot over an "i" in this textbook. When laser light is reflected from one of these smooth places on the disc, it is reflected back to the detector in Figure 30.23, which interprets the presence of the reflection as an "on" signal, or as a 1 in binary. On the other hand, when the light strikes a pit on the rotating disc, it reflects in some random direction and does not return to the detector, which interprets the absence of a reflection as an "off" or a 0 in binary. Electrical equipment connected to the detector then drives a loudspeaker according to the size of the binary number it receives, thus reconstructing the original sound that was used to etch the disc. ◄

Figure 30.26
Sampling a sound signal whose amplitude varies with time.

►SPECIAL TOPIC
Lasers at the Supermarket

Lasers are becoming more and more a part of everyday life, as a visit to your local supermarket will verify. In most large supermarkets, the cashier will move your grocery items across the beam from a helium-neon laser that senses information on the Universal Product Code, or bar code, label on a package. These bar codes, one of which is shown in Figure 30.27, are printed such that the spacing and darkness of successive lines can be interpreted by a computer as representing specific numbers. The two thin lines at the far left are used to indicate to the computer that a new bar code is on the way to be interpreted. The next two lines, a heavy one and a thin one, are the basic code for the number 0. The number 0 is interpreted by the computer to indicate that the bar code for a grocery item is to follow. The next several lines are interpreted by the computer as the number 21140, a number that identifies the manufacturer of the item. Two more thin lines then appear that indicate that the bar code for the item is on its way. This pattern of lines is representative of the number 20786, which is the number assigned by the company to a can of green beans. As the beam of the laser moves across this bar code, the light is reflected back to a detector when a white space appears but is absorbed when a dark space appears. The detector senses these light and dark spaces as the numbers described above and sends this information to a computer that looks up the price for the product and rings it up on the cash register. ◄

Figure 30.27
The bar code on a can of green beans. (*Courtesy of Bill Schulz*)

*30.13   Fluorescence and Phosphorescence

When an atom absorbs a photon of energy and ends up in an excited state, it can return to the ground state via some intermediate states, as shown in Figure 30.28. The photons emitted by the atom will have lower energy, and therefore

Figure 30.28
The process of fluorescence.
(a) An atom absorbs a photon
of energy hf_1 and ends up in
an excited state, E_3. (b) The
atom emits a photon of energy
hf_2 when the electron moves
from an intermediate state,
E_2, back to the ground state.

lower frequency, than the absorbed photon. The process of converting ultraviolet light to visible light by this means is referred to as **fluorescence.**

The common fluorescent light, which makes use of this principle, works as follows. Electrons are produced in the tube as a filament at the end is heated to sufficiently high temperatures. The electrons are accelerated by an applied voltage, and this causes them to collide with atoms of mercury vapor present in the tube. As a result of the collisions, many mercury atoms are raised to excited states. As the excited atoms drop to their normal levels, some ultraviolet photons are emitted, and these strike a phosphor-coated screen. The screen absorbs these photons and emits visible light by means of fluorescence. Different phosphors on the screen emit light of different colors. "Cool white" fluorescent lights emit nearly all the visible colors and hence the light is very white. "Warm white" fluorescent lights have a phosphor that emits more red light and thereby produces a "warm" glow. It is interesting to note that the fluorescent lights above the meat counter in a grocery store are usually "warm white" to give the meat a redder color.

Two common fluorescent materials that you may have in your medicine cabinet are Murine eye drops and Pearl Drops toothpaste. If you were to use these products and then stand under a "black light," your eyes and teeth would glow with a beautiful yellow color. (A black light is simply a lamp that emits ultraviolet light along with some visible violet-blue light.)

Fluorescence analysis is often used to identify compounds. This is made possible by the fact that every compound has a "fingerprint" associated with the specific wavelength at which it fluoresces.

As we have seen, a fluorescent material emits visible light only when ultraviolet radiation bombards it. Another class of materials, called **phosphorescent** materials, continue to glow long after the illumination has been removed. An excited atom in a fluorescent material drops to its normal level in about 10^{-8} s, but an excited atom of a phosphorescent material may remain in an excited metastable state for periods ranging from a few seconds to several hours. Eventually, the atom will drop to its normal state and emit a visible photon. For this reason, phosphorescent materials emit light long after being placed in the dark. Paints made from these substances are often used to decorate the hands of watches and clocks and to outline doors and stairways in large buildings so that these exits will be visible if there is a power failure.

Summary

The **Bohr model** of the atom is successful in describing the spectra of atomic hydrogen and hydrogen-like ions. One of the basic assumptions of the model is that the electron can exist only in certain orbits such that its angular momentum, mvr, is an integral multiple of $h/2\pi = \hbar$, where h is Planck's constant. Assuming circular orbits and a Coulomb force of attraction between electron and proton, the energies of the quantum states for hydrogen are

$$E_n = -\frac{mk^2e^4}{2\hbar^2}\left(\frac{1}{n^2}\right) \qquad n = 1, 2, 3, \ldots \qquad \textbf{(30.15)}$$

where k is the Coulomb constant, e is the charge on the electron, and n is an integer called a **quantum number.**

If the electron in the hydrogen atom jumps from an orbit whose quantum number is n_i to an orbit whose quantum number is n_f, it emits a photon of frequency f, given by

$$f = \frac{mk^2e^4}{4\pi h^3}\left(\frac{1}{n_f{}^2} - \frac{1}{n_i{}^2}\right)$$

(30.17)

Bohr's **correspondence principle** states that quantum mechanics is in agreement with classical physics when the quantum numbers for a system are very large.

One of the many great successes of quantum mechanics is that the quantum numbers n, ℓ, and m_ℓ associated with atomic structure arise directly from the mathematics of the theory. The quantum number n is called the **principal quantum number,** ℓ is the **orbital quantum number,** and m_ℓ is the **orbital magnetic quantum number.** In addition, a fourth quantum number, called the **spin magnetic quantum number,** m_s, is needed to explain certain features of atomic structure.

An understanding of the periodic table of the elements became possible when Pauli formulated the **exclusion principle,** which states that no two electrons in an atom can ever be in the same quantum state, that is, no two electrons in the same atom can have the same set of quantum numbers, n, ℓ, m_ℓ, and m_s.

Characteristic x-rays are produced when a bombarding electron collides with an electron in an inner shell of an atom with sufficient energy to remove the electron from the atom. The vacancy thus created is filled when an electron in a higher level drops down into the level containing the vacancy.

Lasers are monochromatic, coherent light sources that work on the principle of **stimulated emission** of radiation from a system of atoms.

Additional Reading

E. N. Andrade, *Rutherford and the Nature of the Atom,* NY, Doubleday Anchor, 1964.

G. Gamow, *Mr. Tomkins Explores the Atom,* New York, Cambridge University Press, 1945.

W. E. Kock, *Lasers and Holography,* Garden City, NY, Doubleday, 1969.

A. V. La Rocca, *"Laser Applications in Manufacturing,"* *Sci. American,* March 1982, p. 94.

Lasers and Light, San Francisco, Freeman, 1969 (readings from *Scientific American*).

R. E. Latham, *"Holography in the Science Classroom,"* *The Physics Teacher,* October 1986, p. 395.

E. N. Leith, "White-Light Holograms," *Sci. American,* October 1976, p. 80.

R. Moore, *Niels Bohr, The Man, His Science, and the World They Changed,* NY, Knopf, 1966.

P. Schewe, "Lasers," *The Physics Teacher,* November 1981, p. 534.

H. Spetzler, "Interesting Example of Holographic Interferometry," *The Physics Teacher,* February 1986, p. 80.

C. L. Strong, "How to Make Holograms," *Sci. American,* February 1967, p. 122.

J. A. Wheeler, "Niels Bohr, the Man," *Physics Today,* October 1985, p. 66.

Questions

1. (a) Explain why the electrons in Rutherford's model of the atom cannot be at rest. (b) Discuss the interpretation of Rutherford's experiments if no large-angle deflection of the α particles had been seen.

2. Suppose that the electron in the hydrogen atom obeyed classical mechanics rather than quantum mechanics. Why would such a hypothetical atom emit a continuous spectrum rather than the observed line spectrum?

3. The Bohr theory of the hydrogen atom is based upon several assumptions. Discuss these assumptions and their significance. Do any of these assumptions contradict classical physics?

4. Does the light emitted by a neon sign constitute a continuous spectrum or only a few colors? Defend your answer.

5. Define the following terms: (a) energy level; (b) quantum jump; (c) ionization energy; (d) ground state.

6. "The binding energy (in electron volts) of an electron in an atom is numerically equal to the minimum voltage (ionization potential) necessary to ionize the atom by electrical discharge." Explain this statement.

7. Must an atom first be ionized before it can emit light? Discuss.

8. Can the electron in the ground state of hydrogen absorb a photon of energy (a) less than 13.6 eV and (b) greater than 13.6 eV? Explain.

9. Suppose an electron in the ground state of an atom has its first excited level at an energy of 3 eV above its ground-state value. What could happen to this electron if it were struck by an incoming electron having an energy of (a) 3 eV and (b) 4 eV?

10. Why aren't the relative intensities of the various spectral lines in a gas all the same?

11. What is excluded by the exclusion principle?

12. It was stated in the text that, if the exclusion principle were not valid, every electron would end up in the lowest energy state of the atom and the chemical behavior of the elements would be grossly modified. Explain this statement.

13. Why do lithium, potassium, and sodium exhibit similar chemical properties?

14. Which of the following elements would you predict to have properties most like those of silver: copper, cadmium, or palladium?

15. In what ways is the introduction of the quantum numbers via quantum mechanics more satisfying than their introduction through improvements on the Bohr theory?

16. List some ways in which quantum mechanics altered our view of the atom as pictured by the Bohr theory.

17. Explain how the fact that the angular momentum of an electron can be zero does not introduce any philosophical difficulties in the quantum mechanical view of the atom.

18. If matter has a wave nature, why is this not observable in our daily experiences?

19. Discuss why the term *electron clouds* is used to describe the electronic arrangement in the quantum mechanical view of the atom.

20. Why are the frequencies of characteristic x-rays dependent on the type of material used for the target?

21. In what sense do characteristic x-rays indicate that the electrons in the inner shells are tightly bound? Discuss.

22. Explain the difference between laser light and light from an incandescent lamp.

23. Why is *stimulated emission* so important in the operation of a laser? (Interestingly, the concept of stimulated emission was first discussed by Albert Einstein 35 years before the first successful laser.)

24. Why does a laser usually emit only one particular color of light rather than several colors or perhaps a continuous spectrum?

25. When a hologram is produced, the system (including light source, object, beam splitter, and so on) must be held motionless within a quarter of a wavelength. Why?

Problems

Section 30.2   Atomic Spectra

1. Use Equation 30.3 to calculate the wavelength of the first three lines in the Lyman series. On the basis of your calculation identify the region of the electromagnetic spectrum in which these lines appear.

2. Use Equation 30.1 to calculate the wavelength of the first three lines in the Balmer series for hydrogen.

3. Use Equation 30.4 to calculate the frequency of the emissions from hydrogen for the first two lines of the Paschen series. On the basis of your calculations identify the region of the electromagnetic spectrum in which these lines appear.

4. On the basis of your calculations for Problem 2, find three frequencies that will be absorbed if a continuous spectrum is passed through heated hydrogen gas.

5. Determine the wavelength of the following radiation of hydrogen: (a) the third-longest wavelength of the

Paschen series: (b) the longest wavelength of the Balmer series; (c) the shortest wavelength (series limit) of the Paschen series; (d) the next-to-longest wavelength of the Lyman series.

6. (a) Compute the shortest wavelength in each of these hydrogen spectral series: Lyman, Balmer, Paschen, and Brackett. (b) Compute the energy (in eV) of the highest-energy photon produced in each of these series.

7. (a) What value of n is associated with the Lyman series line in hydrogen whose wavelength is 94.96 nm? (b) Could this wavelength be associated with the Paschen or Brackett series?

Section 30.3   The Bohr Theory of Hydrogen

8. Show that
$$\frac{mk^2e^4}{2\hbar^2} = 13.6 \text{ eV}$$

9. Show that the velocity, **v**, of the electron in the first (ground state) Bohr orbit of the hydrogen atom is given by $\beta = v/c = 1/137$.

10. Consider a large number of hydrogen atoms, with electrons all initially in the $n = 4$ state. (a) How many different wavelengths would be observed in the emission spectrum of these atoms? (b) What is the longest wavelength that could be observed? To which series does it belong?

11. A photon is emitted from a hydrogen atom, which undergoes a transition from the $n = 3$ state to the $n = 2$ state. Calculate (a) the energy, (b) the wavelength, and (c) the frequency of the emitted photon.

12. How much energy is required to ionize hydrogen when it is in (a) the ground state and (b) the state for which $n = 3$?

13. A hydrogen atom initially in its ground state ($n = 1$) absorbs a photon and ends up in the state for which $n = 3$. (a) What is the energy of the absorbed photon? (b) If the atom returns to the ground state, what photon energies could the atom emit?

14. Use Equation 30.13 to show that the radius of the ground state orbit of hydrogen is 0.529×10^{-10} m.

15. Substitute numerical values into Equation 30.19 to determine a value for R.

16. Calculate the radius of the first, second, and third Bohr orbits of hydrogen.

17. Show that the speed of the electron in the nth Bohr orbit in hydrogen is given by
$$v_n = \frac{ke^2}{n\hbar}$$

18. A hydrogen atom is in its first excited state ($n = 2$). Using the Bohr theory of the atom, calculate (a) the

radius of the orbit, (b) the linear momentum of the electron, (c) the angular momentum of the electron, (d) the kinetic energy, (e) the potential energy, and (f) the total energy.

19. What wavelength radiation would ionize hydrogen when the atom is in its ground state?

20. Calculate the Coulomb force of attraction on the electron when it is in the ground state.

21. Calculate the angular momentum of the electron in the $n = 3$ state of hydrogen.

22. What is the energy of the photon that, when absorbed by a hydrogen atom, could cause (a) an electronic transition from the $n = 3$ state to the $n = 5$ state and (b) an electronic transition from the $n = 5$ state to the $n = 7$ state?

23. Four possible transitions for a hydrogen atom are listed below:

   (A) $n_i = 2$; $n_f = 5$ (B) $n_i = 5$; $n_f = 3$
   (C) $n_i = 7$; $n_f = 4$ (D) $n_i = 4$; $n_f = 7$

   (a) Which transition will emit the shortest wavelength photon? (b) For which transition will the atom gain the most energy? (c) For which transition(s) does the atom lose energy?

24. Find the shortest wavelength photon emitted in the Lyman series of hydrogen.

25. An electron is in the first Bohr orbit of hydrogen. Find (a) the speed of the electron, (b) the time required for the electron to circle the nucleus, and (c) the current in amps corresponding to the motion of the electron.

26. (a) Calculate the angular momentum of the Moon due to its orbital motion about the Earth. In your calculation, use 3.84×10^8 m as the average Earth-Moon distance and 2.36×10^6 s as the period of the Moon in its orbit. (b) Assume that the Moon is in its lowest energy state while orbiting the Earth and determine the corresponding quantum number. (c) By what fraction would the Earth-Moon radius have to be increased to increase the quantum number by 1?

27. Consider a hydrogen atom. (a) Calculate the frequency f of the $n = 2 \rightarrow n = 1$ transition and compare with the frequency f_{orb} of the electron orbital motion in the $n = 2$ state. (b) Make the same calculation for the $n = 10,000 \rightarrow n = 9999$ transition. Comment on the results.

Section 30.4   Modification of the Bohr Theory

28. Plot an energy level diagram like that of Figure 30.7 for singly ionized helium.

29. (a) Find the energy of the electron in the ground state of doubly ionized lithium, which has an atomic number $Z = 3$. (b) Find the radius of this ground state orbit.

30. Construct an energy level diagram for doubly ionized lithium (Li^{2+}), for which $Z = 3$.

31. What is the radius of the first Bohr orbit in (a) He^+ and (b) Be^{3+}?

32. What are the longest and shortest wavelengths in the "Lyman" series of a singly ionized helium atom? In what part of the spectrum will this series be found?

33. (a) Substitute numerical values into Equation 30.22 to find a value for the Rydberg constant for singly ionized helium, He^+. (b) Use the result of Part (a) to find the wavelength associated with a transition from the $n = 2$ state to the $n = 1$ state of He^+. (c) Identify the region of the electromagnetic spectrum associated with this transition.

34. Using the Sommerfeld model, (a) identify all the possible subshells that can exist for the P shell. (b) Find the maximum number of electrons that can occupy each subshell and (c) find the maximum number of electrons that can occupy the P shell.

Section 30.6 Quantum Mechanics and the Hydrogen Atom

Section 30.7 The Spin Magnetic Quantum Number

35. List the possible sets of quantum numbers for electrons in the $3d$ subshell.

36. List the possible sets of quantum numbers for electrons in the $3p$ subshell.

37. When the principal quantum number is $n = 4$, how many different values of (a) ℓ and (b) m_ℓ are possible?

Section 30.9 The Exclusion Principle and the Periodic Table

38. Verify the electronic configurations for the ground state given in Table 30.4 for (a) helium, (b) neon, and (c) argon.

39. (a) Write out the electronic configuration for the ground state for oxygen ($Z = 8$). (b) Write out the values for the set of quantum numbers n, ℓ, m_ℓ, and m_s for each of the electrons in oxygen.

40. How many different sets of quantum numbers are possible for an electron for which (a) $n = 1$, (b) $n = 2$, (c) $n = 3$, (d) $n = 4$, and (e) $n = 5$? Check your results to show that they agree with the general rule that the number of different sets of quantum numbers is equal to $2n^2$.

41. Suppose two electrons in the same system each have $\ell = 0$ and $n = 3$. (a) How many states would be possible if the exclusion principle were inoperative?

(b) List the possible states, taking the exclusion principle into account.

Section 30.10 Characteristic X-Rays

42. The K-shell ionization energy of copper is 8979 eV. The L-shell ionization energy is 951 eV. Determine the wavelength of the K_α emission line of copper. What must be the minimum voltage on an x-ray tube with a copper target to see the K_α line?

43. The K_α x-ray is the one emitted when an electron undergoes a transition from the L shell ($n = 2$) to the K shell ($n = 1$). Calculate the frequency of the K_α x-ray from a nickel target ($Z = 28$).

44. Use the method illustrated in Example 30.7 to calculate the wavelength of the x-ray emitted from a molybdenum target ($Z = 42$) when an electron undergoes a transition from the L shell ($n = 2$) to the K shell ($n = 1$).

45. The K series of the discrete spectrum of tungsten contains wavelengths of 0.0185 nm, 0.0209 nm, and 0.0215 nm. The K-shell ionization energy is 69.5 keV. Determine the ionization energies of the L, M, and N shells. Sketch the transitions.

Additional Problems

46. A hydrogen atom in the ground state absorbs ultraviolet light of wavelength 97.3 nm. Calculate the wavelengths of the possible emission lines. Sketch the energy levels and transitions.

47. A laser emits a 0.5-W beam of light at a wavelength of 640 nm. How many photons does the laser emit per second?

48. (a) How much energy is required to cause an electron in hydrogen to move from the $n = 1$ state to the $n = 2$ state? (b) If the electrons gain this energy by collision with hydrogen atoms at a high temperature, find the minimum temperature of the heated hydrogen gas. The thermal energy of the heated atoms is given by $3kT/2$.

49. (a) Calculate the radius of the third Bohr orbit in hydrogen. (b) Calculate the speed of an electron in this orbit. (c) From the de Broglie theory what is the wavelength of an electron that has this speed? (d) How does the wavelength of the electron compare with the circumference of the orbit for $n = 3$?

50. A laser used in a holography experiment has an average output power of 5 mW. The laser beam is actually a series of pulses of electromagnetic radiation at a wavelength of 632.8 nm, each having a duration of 25 ms. Calculate (a) the energy (in joules) radiated

with each pulse and (b) the number of photons in each pulse.

51. Find the potential energy and kinetic energy of an electron in the ground state of the hydrogen atom.

52. A hydrogen atom is in its ground state. Using the Bohr theory of the atom, calculate (a) the radius of the orbit, (b) the linear momentum of the electron, (c) the angular momentum of the electron, (d) the kinetic energy, (e) the potential energy, and (f) the total energy.

53. Mercury's ionization energy is 10.39 eV. The three longest wavelengths of the absorption spectrum of mercury are 253.7 nm, 185.0 nm, and 158.5 nm. (a) Construct an energy-level diagram for mercury. (b) Indicate all possible emission lines, starting from the highest energy level on the diagram. (c) Disregarding recoil, determine the minimum speed an electron must have to make an inelastic collision with a mercury atom.

54. Suppose the ionization energy of an atom is 4.1 eV. In this same atom, we observe emission lines with wavelengths 310 nm, 400 nm, and 1377.8 nm. Use this information to construct the energy-level diagram with the least number of levels. Assume the higher energy levels are closer together.

55. An electron has a de Broglie wavelength equal to the diameter of the hydrogen atom. (a) What is the kinetic energy of the electron? (b) How does this energy compare with the ground-state energy of the hydrogen atom?

56. In order for an electron to be confined to a nucleus, its de Broglie wavelength would have to be less than 10^{-14} m. (a) What would be the kinetic energy of an electron confined to this region? (b) On the basis of this result, would you expect to find an electron in a nucleus? Explain.

57. Use Bohr's model of the hydrogen atom to show that when the atom makes a transition from the state n to the state $n - 1$, the frequency of the emitted light is given by

$$f = \frac{2\pi^2 m k^2 e^4}{h^3}\left(\frac{2n-1}{(n-1)^2 n^2}\right)$$

58. Calculate the classical frequency for the light emitted by an atom. To do so, note that the frequency of revolution is $v/2\pi r$, where r is the Bohr radius. Show that as n approaches infinity in the equation of Problem 57, the expression given there varies as $1/n^3$ and reduces to the classical frequency. (This is an example of the correspondence principle, which requires that the classical and quantum models agree for large values of n.)

59. An electron of momentum p is at a distance r from a stationary proton. The electron has a kinetic energy

$KE = p^2/2m$ and potential energy $PE = -ke^2/r$. Its total energy is $E = KE + PE$. If the electron is bound to the proton to form a hydrogen atom, its average position is at the proton, but the uncertainty in its position is approximately equal to the radius r of its orbit. The electron's average momentum will be zero, but the uncertainty in its momentum will be given by the uncertainty principle. Treat the atom as a one-dimensional system in the following: (a) Estimate the uncertainty in the electron's kinetic, potential, and total energies in terms of r.

60. A dimensionless number that often appears in atomic physics is the *fine-structure constant* α, given by

$$\alpha = \frac{ke^2}{\hbar c}$$

where k is the Coulomb constant. (a) Obtain a numerical value for $1/\alpha$. (b) In terms of α, what is the ratio of the Bohr radius, a_0, to the Compton wavelength, $\lambda = h/m_0 c$? (c) In terms of α, what is the ratio of the reciprocal of the Rydberg constant, $1/R$, to the Bohr radius?

Lasers and Applications

Isaac D. Abella
The University of Chicago

Laser Principles

We encounter lasers in many applications in the home and workplace. The CD (compact disc) player in home audio systems incorporates a small laser disc-reader. The laser-printer in the office uses a laser beam to paint an image directly from the computer onto the xerographic surface, dispensing with the photographic lens of conventional copiers. The phone company uses laser communications devices on the fiber-optic network. The supermarket checkout counter is where most people see the laser beam in action. This enormous growth of laser applications was stimulated by many scientific and engineering advances that exploit some of the unique properties of laser light.

These properties derive from the distinctive way laser light is produced in contrast to the generation of ordinary light. Laser light originates from *energized* ("excited") atoms, ions, or molecules through a process of *stimulated emission* of radiation. The suitably prepared *active* laser medium is contained in an enclosure or *cavity* which organizes the normally random emission process into an intense, directional, monochromatic and coherent wave. The end *mirrors* provide the essential optical feedback which selectively builds up the stimulating wave along the tube axis. However, in an ordinary sodium vapor street lamp, for example, the energized atoms *spontaneously* emit in random directions and at irregular times, over a broad spectrum, resulting in isotropic illumination of incoherent light. Laser light has a well-defined phase, permitting a wide variety of applications based on interference or wave modulation.

Currently operating laser systems utilize a variety of atomic gases, solids, or liquids as the working laser substance. These devices are designed to emit either continuous or pulsed monochromatic beams, and operate over a broad range of the optical spectrum (ultraviolet, visible, infrared) with output powers from milliwatts (10^{-3}) to megawatts (10^6). The particular application determines the choice of laser system, wavelength, power level, or other relevant variables, since no one laser has all the desirable properties.

For a laser to operate successfully, several *atomic physics* conditions must be satisfied, as noted in Section 30.12. The requirement for population inversion, that is, more atoms in a particular *excited state* than in a lower state, essentially means that energy must be supplied from outside the system. Otherwise, atoms would eventually radiate, and develop an increasing probability for absorbing light. Finally, they fall to the lowest energy state and stop emitting altogether.

Therefore, all laser systems must be connected to external energy sources, usually electrical, as required by conservation of energy, to maintain this non-thermal equilibrium situation. For example, we can energize atoms by electron impact in gaseous discharges (so called "electrical pumping"). We can also supply energy to lamps whose light populates excited states by photon absorption ("optical pumping") for those solids or liquids which do not conduct electric charge. These pumping mechanisms tend to have low efficiency (ratio of laser energy output to electric energy supplied), typically a few percent, with the balance discharged as heat into cooling water or circulating air.

Controlling the electrical input into the laser system allows a variable laser energy output, which may be important in many applications. Thus, the argon ion laser system can emit up to about 10 W in the green optical beam by adjustment of the electric current in the argon gas, which in turn controls the degree of population inversion. Chemical lasers, on the other hand, operate without direct electrical input. Several highly reactive gases are mixed in the laser chamber, with the energy released in the ensuing reaction populating the excited levels in the molecule. In this case, the reactants need to be resupplied for the laser to operate for any length of time.

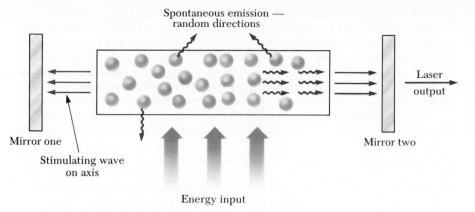

Figure 1
A schematic of a laser design. The tube contains atoms, which represent the active medium. An external source of energy (optical, electrical, etc.) is needed to "pump" the atoms to excited energy states. The parallel end mirrors provide the feedback of the stimulating wave.

Some laser systems have fluid media, containing dissolved dye molecules. The dye lasers are usually pumped to excited levels by an external laser. The advantage of this arrangement is that dye lasers can be continuously "tuned" over a wide range of wavelengths, using prisms or gratings, whereas the pump source has a fixed wavelength. Color variability is important for those cases where the laser is directed at materials whose absorption depends on wavelength. Thus, the laser can be tuned into exact coincidence with selected energy states. For example, blood does not absorb red light to any extent, which excludes red light use for most surgical applications on blood-rich tissue.

Recent developments in tunable solid state materials have permitted design of tunable lasers without the need for unstable dye molecules: most notable are sapphire crystals containing titanium ions. These materials are optically pumped by flashlamps or fixed wavelength lasers acting to populate the upper state directly.

A variety of laser systems are in general use today. They include the 1-mW helium-neon laser, usually appearing as a red beam at 632.8 nm (although yellow and green beams are available); the argon ion laser, which operates in the green or blue up to 10 W; the carbon dioxide gas laser, which emits in the infrared at 10 mm and can produce several hundred watts; and the neodymium YAG laser, a powerful solid-state optically pumped system which emits at 1.06 mm either continuous or pulsed. The recently perfected diode junction laser emits in the near infrared and operates by passage of current through the semiconductor material. The recombination radiation is essentially direct conversion of electrical energy to laser light and is a very efficient process. The diodes can emit up to 5 W and can be used to energize other laser materials.

Applications

We shall describe a few applications that should serve to illustrate the wide variety of laser uses. Other applications are discussed as Special Topics in this chapter. First, there is the use of lasers in precision long-range distance measurement (range-finding). It has become important, for astronomical and geophysical purposes, to measure as precisely as possible the distance from various points on the surface of the Earth to a point on the Moon's surface. To facilitate this, the Apollo astronauts set up a compact array, a 0.5-m square of reflector prisms on the Moon, which allows laser pulses directed from an Earth station to be retro-reflected to the same station. (See Section 24.9 on prism reflectors.) Using the known speed of light and the measured round-trip travel time of a 1-ns pulse, one can determine the Earth-Moon distance, 380,000 km, to a precision of better than 10 cm. Such information would be useful, for example, in making more reliable earthquake predictions and for learning more about the motions of the Earth-Moon system. This technique requires a high-power pulsed laser for its success, since a sufficient burst of photons must return to a collecting telescope on Earth and be detected. Variants of this method are also used to measure the distance to inaccessible points on the Earth as well as in military rangefinders.

Novel medical applications utilize the fact that the different laser wavelengths can be absorbed in specific biological tissues. A widespread eye condition, glaucoma, is manifested by a high fluid pressure in the eye, which can lead to destruction of the optic nerve. A simple laser operation (iridectomy) can "burn" open a tiny hole in a clogged membrane, relieving the destructive pressure. Along the same lines, a serious side effect of diabetes is neovascularization, the formation of weak blood vessels, which often leak blood into extremities. When this occurs in the eye, vision deteriorates (diabetic retinopathy) leading to blindness in diabetic patients. It is now possible to direct the green light from the argon ion laser through the clear eye lens and eye fluid, focus on the retina edges, and photo-coagulate the leaky vessels. These procedures have greatly reduced blindness in glaucoma and diabetes patients.

Laser surgery is now a practical reality. Infrared light at 10 mm from a carbon dioxide laser can cut through muscle tissue, primarily by heating and evaporating the water contained in cellular material. Laser power of about 100 W is required in this technique. The advantage of the "laser knife" over conventional methods is that laser radiation cuts and coagulates at the same time, leading to substantial reduction of blood loss. In addition, the technique virtually eliminates cell migration, which is very important in tumor removal. Furthermore, a laser beam can be trapped in fine glass-fiber light-guides (endoscopes) by means of total internal reflection (Section 24.9). The light fibers can be introduced through natural orifices, conducted around internal organs, and directed to specific interior body locations, eliminating the need for massive surgery. For example, bleeding in the gastrointestinal tract can be optically cauterized by fiber-optic endoscopes inserted through the mouth.

Finally, we describe an application to biological and medical research. It is often important to isolate and collect unusual cells for study and growth. A laser cell separator exploits the fact that specific cells can be tagged with fluorescent dyes. All cells are then dropped from a tiny charged nozzle and laser-scanned for

An argon laser passing through a cornea and lens during eye surgery.
(© Alexander Tsiaras, Science Source/Photo Researchers, Inc.)

the dye tag. If triggered by the correct light-emitting tag, a small voltage applied to parallel plates deflects the falling electrically charged cell into a collection beaker. This is an efficient method for extracting the proverbial needles from the haystack.

Suggested Reading

O'Shea, D., R. Callen, and W. T. Rhodes, *Introduction to Lasers and Their Application*, Reading, Mass., Addison-Wesley, 1977.

Seigman, A. E. *An Introduction to Lasers and Masers*, New York, McGraw-Hill, 1979.

Essay Problems

1. (a) In the case of the Earth-Moon laser range-finder, what is the round-trip time for a laser pulse? (b) What precision in timing is required, that is, how small a time change needs to be detectable, to be able to measure the distance to an error of 10 cm? What effect does the Earth's atmosphere have?

2. A laser beam $\lambda = 600$ nm is directed at the Moon from a laser tube 1 cm in diameter. Does the beam spread at all and what is the diameter of the "spot" on the Moon's surface? Does your answer demand a good strategy for successful lunar-array illumination and how would you do it?

3. Estimate how much chemical reagent is required to produce a chemical laser of 1-kW output, in moles/second? Pick a reasonable exothermic reaction rate in kcal/mole. Estimate how many kilograms each of hydrogen and chlorine gas would be needed to make an HCl laser operate for an hour at 1-kW output.

CHAPTER 31

Molecules and Solids

Natural calcite crystals, calcium carbonate.
(Brian Parker/Tom Stack and Associates)

The preceding chapter was concerned with the atomic and electronic structure of single atoms. However, except for the inert gases, single isolated atoms are generally not found in nature. Instead, elements combine to form chemical compounds that consist of individual atoms joined by chemical bonds.

In this chapter, we shall first describe the bonding mechanisms in molecules, and the various types of energy that a molecule can have. We shall then take the next logical step and show how molecules combine to form solids. Our study will enable us to explain the differences between insulating, metallic, and semiconducting solids. The chapter concludes with a discussion of semiconducting junctions and the operation of several semiconductor devices.

31.1 Molecular Bonds

Fundamentally, the bonding mechanisms in a molecule are primarily due to electrostatic forces between atoms (or ions). When two atoms are separated by a very large distance, the dominant force between them is attractive. As the atoms are brought closer together, both attractive and repulsive forces act. For small separations, repulsive forces between like charges begin to dominate, and the atoms are forced apart. At some distance of separation, called the **equilibrium separation**, the attractive and repulsive forces balance each

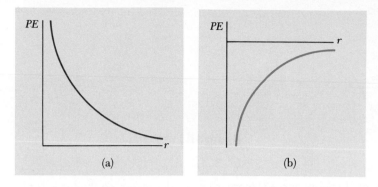

Figure 31.1
Plots of the potential energy versus particle separation for two charges when (a) the charges have the same sign and (b) the charges have opposite sign.

other. Although it is possible to describe the interaction between the atoms in a molecule in terms of the forces between them, it is more helpful to analyze the interaction in terms of potential energy diagrams. (By potential energy diagram, we mean a plot of the potential energy of the atoms versus their distance of separation.) Let us examine some simple potential energy diagrams to see what information they can reveal. First, consider two point charges q_1 and q_2 separated by a distance r. In Chapter 17 we found that the potential energy of a pair of charges is given by

$$PE = k\frac{q_1 q_2}{r} \tag{31.1}$$

Figure 31.1a is a plot of PE versus r for the cases in which both charges have the same sign and thus repel one another. Note that the characteristic shape of this curve is that the potential energy increases as r decreases. Figure 31.1b is a plot of PE versus r for the case in which the charges have opposite signs and thus attract one another. The shape of this curve is such that the potential energy decreases as r decreases.

With this background, let us now examine the general shape of a potential energy curve for a molecule containing two atoms whose nuclei are separated by a distance r, as in Figure 31.2. For large separations, the potential energy curve decreases as r decreases, corresponding to a net attractive force between the atoms. For small separations, the potential energy curve increases as r decreases, corresponding to a net repulsive force between the atoms. At the equilibrium separation, the potential energy has its minimum value, which corresponds to the binding energy of the molecule. The binding energy is a measure of how much energy would have to be added to the molecule to separate its atoms by an infinite distance.

The following example will involve a more realistic potential energy curve that is typically encountered in practice. This example also discusses the concept of activation energy.

Example 31.1   Activation Energy

In certain situations, the potential energy diagram for a molecule looks like that pictured in Figure 31.3. (a) Determine whether the force is attractive or repulsive at points A, B, and C on the curve.

Figure 31.2
The potential energy versus internuclear separation for two atoms.

Figure 31.3   (Example 31.1)

Solution

At point A the potential energy curve is increasing as r decreases, so the force is repulsive. At B the curve decreases as r decreases, so the force is attractive, and at C the curve again increases as r decreases, indicating a repulsive force. The fact that the force between the atoms is repulsive at point A means that the atoms will not come together without external assistance. Some additional energy must be added to the system to force the atoms over this potential energy barrier. The amount of energy that must be added to perform this task is called the *activation energy*.

The activation energy peak in the curve often arises because other bonds must be broken before the bond between the atoms can form. A common example of this is the formation of water from H_2 and O_2 molecules. Before the water bonds can be formed to make H_2O, the bonds between the H and O atoms of H_2 and O_2 molecules must first be broken. Passing a current (or spark) through gaseous H_2 and O_2 will break some of the bonds in the gas, thus providing the activation energy. When these freed H and O atoms combine to form water, there is a release of energy that can provide the activation energy to cause additional bonds to be broken and the subsequent formation of more water.

(b) Find the binding energy, the activation energy, and the equilibrium separation of the molecule.

Solution

From the graph, we find that the depth of the potential well, and thus the binding energy, is equal to 5.3 eV. Also, the graph shows that the activation energy is 4.2 eV, and the equilibrium separation is 0.33 nm.

31.2 Types of Molecular Bonds

A complete description of the binding mechanisms in molecules is a highly complex problem because it involves the mutual interactions of many particles. In this section, we shall discuss some simplified models in the following order of decreasing bond strength: the ionic bond, the covalent bond, the hydrogen bond, and the van der Waals bond. We shall also briefly describe the bonding mechanism in metals.

Ionic Bonds

Ionic bonds are due to the Coulomb attraction between oppositely charged ions. A familiar example of an ionically bonded molecule is sodium chloride, NaCl, which is common table salt. Sodium has an electronic configuration

Figure 31.4
Total energy versus the internuclear separation for the NaCl molecule.

$1s^2 2s^2 2p^6 3s^1$. Thus, the arrangement is such that the atom has 11 electrons, with 10 of them in closed shells or subshells. The electron in the outermost shell "sees" a charge of $+11e$ on the nucleus, but it is partially shielded from the nucleus by the inner core electrons, which have a charge of $-10e$. As a result, the electron in the outermost shell is attracted by a net charge of only $+1e$, so it is relatively easy to remove this electron from the atom. In fact, the energy required to ionize sodium to form Na^+ is 5.1 eV. Chlorine, which has an electronic configuration $1s^2 2s^2 2p^5$, is one electron short of a closed-shell structure. Because closed-shell configurations are energetically more favorable, the Cl^- ion is more stable than the neutral Cl atom. In fact, if neutral chlorine takes on an electron to become Cl^-, there will be an energy release of 3.7 eV. The energy released when an atom takes on an electron is called the **electron affinity**. Therefore, the energy required to combine Na^+ and Cl^- to form NaCl is $5.1 - 3.7 = 1.4$ eV. When the ions are brought close together, the electrostatic potential energy is greater than the 1.4 eV needed, and the ions coalesce to form NaCl.

The total energy versus the internuclear separation for NaCl is shown in Figure 31.4. Note that the total energy of the molecule has a minimum value of -4.2 eV at the equilibrium separation of about 0.24 nm. Thus, the total energy required to separate the NaCl molecule into neutral sodium and chlorine atoms is equal to 4.2 eV. Also, note that the energy at an infinite separation is 1.3 eV, which corresponds to the energy required to form the ions from neutral atoms. The equilibrium separation for an isolated NaCl molecule is 0.24 nm, compared to a value of 0.28 nm for solid NaCl. The equilibrium separation is greater in the solid state because the cubic arrangement of the ions in a NaCl crystal is such that each ion is surrounded by six nearest-neighbor ions of opposite charge.

When the ions in an isolated NaCl molecule are brought closer than their equilibrium separation, 0.24 nm, the electrons in the closed shells begin to overlap, which results in a repulsive force between the ions.

A ball and stick model of the molecule methane, CH_4, which has a tetrahedral geometry. The H—C—H bond angles in this structure are $109°$. *(Charles D. Winters)*

(a)

(b)   (c)

Crystalline solids. (a) A fabricated cylinder of nearly pure crystalline silicon (Si), approximately 10 inches long. Such crystals are cut into wafer form and processed to make various semiconductor devices such as integrated circuits and solar cells. (© *Charles D. Winters*) (b) Natural quartz (SiO_2) crystals, one of the most common minerals on Earth. Quartz crystals are used to make special lenses and prisms and in certain electronic applications. (*Courtesy of Ward's Natural Science*) (c) Amethyst crystals are an impure form of quartz crystals. Their color, which is due to iron impurities (Fe^{3+}), can range from pale lilac to deep purple, depending on the iron concentration. (*Courtesy of Ward's Natural Science*)

Covalent Bonds

A covalent bond between two atoms can be visualized as the sharing of electrons supplied by one or both atoms that form the molecule. Many diatomic molecules such as H_2, F_2, and CO owe their stability to covalent bonds. In the case of the H_2 molecule, the two electrons are equally shared between the nuclei, and form a so-called *molecular orbital.* To understand how this bond is formed, consider two H atoms approaching one another. Each has an electron in the ground state orbiting its nucleus, and each electron can have spin quantum numbers of either $+\frac{1}{2}$ or $-\frac{1}{2}$. If the two electrons have the same spin quantum number, a bond cannot form because this would violate the exclusion principle. However, if the spins are opposite, the two atoms are able to share their electrons. The two electrons are more likely to be found between the two nuclei, hence the electron density is large in this region. The formation of the molecular orbital from the *s* orbitals of the two hydrogen atoms is shown in Figure 31.5.

More complex stable molecules such as H_2O, CO_2, and CH_4 are also formed by covalent bonds. Consider methane, CH_4, a typical organic molecule shown schematically in Figure 31.6a. Note that four covalent bonds are formed between the carbon atom and each of the hydrogen atoms. The spatial electron distribution of the four covalent bonds is shown in Figure 31.6b. The four hydrogen nuclei are at the corners of a regular tetrahedron, with the carbon nucleus at the center. The energy required to break a bond between carbon and hydrogen is about 4.28 eV.

The Hydrogen Bond

Because hydrogen has only one electron, it is expected to form a covalent bond with only one other atom. However, in some molecules hydrogen forms a different type of bond between two atoms or ions, called a hydrogen bond.

Figure 31.5
The covalent bond formed by the two $1s$ electrons of the H_2 molecule. The depth of blue color is proportional to the probability of finding an electron in that location.

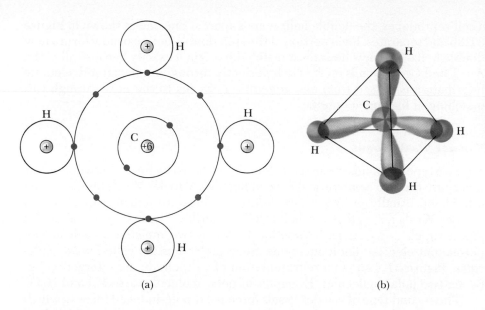

(a)   (b)

Figure 31.6
(a) Diagram of the four covalent
bonds in the CH_4 molecule.
(b) The electron distribution of
the four covalent bonds in the
CH_4 molecule. Note that the
carbon atom is at the center of a
tetrahedron with hydrogen
atoms at its corners.

One example of a hydrogen bond, shown in Figure 31.7, is the hydrogen
difluoride ion, $(HF_2)^-$. The two negative fluorine ions are bound by the positively charged proton between them. The bond is similar to the covalent bond
except that it is the proton associated with hydrogen that is being shared rather
than electrons. The hydrogen bond is relatively weak, with a binding energy of
about 0.1 eV. Although the bond is weak, it is the mechanism responsible for
linking giant biological molecules and polymers. For example, in the case of
the famous DNA molecule that has a double helix structure as shown in Figure
31.8a, hydrogen bonds link the turns of the helix. DNA, or deoxyribonucleic
acid, is the primary molecule in chromosomes, which are the carriers of genetic information. In Figure 31.8a, the hydrogen bonds connecting the double
helix are labeled A—T and C—G. These subunits of the helix, called nucleo-
tide base pairs, are composed of adenine (A), thymine (T), cytosine (C), and
guanine (G). Because of the structure of these units, hydrogen bonds can only
form between adenine and thymine and between cytosine and guanine. When

Figure 31.7
Hydrogen bonding in the
$(HF_2)^-$ molecule ion. The two
negative fluorine ions are
bound by the positively
charged proton between them.

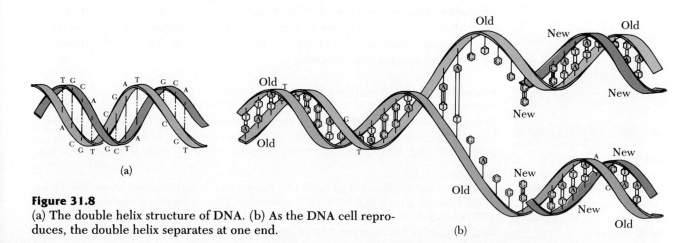

(a)   (b)

Figure 31.8
(a) The double helix structure of DNA. (b) As the DNA cell reproduces, the double helix separates at one end.

Figure 31.9
A representation of the dipole-dipole force between two polar molecules.

a cell reproduces, the double helix comes apart at one end as shown in Figure 31.8b and separates. Each portion of the split double helix can now form a new double helix. The new helix must match the original exactly because all of the A—T and C—G bonds must match perfectly throughout the strand. Genetic information is passed from one generation of cells to the next through this matching of hydrogen bonds.

Van der Waals Bonds

The two types of bonds discussed so far are produced by forces between atoms that cause them to combine and form molecules. **Van der Waals bonds,** on the other hand, usually occur *between molecules.* There are actually three types of van der Waals forces that we shall briefly describe. The first type, called the dipole-dipole force, is an interaction between two molecules each having permanent electric dipole moments. Such molecules are called *polar molecules.* Figure 31.9 shows a representation of a dipole-dipole interaction between two polar molecules. Examples of polar molecules are HCl and H_2O.

The second type of van der Waals force is a dipole-induced force in which a polar molecule induces a dipole moment in a nonpolar molecule. The third type of van der Waals force is a force of attraction between two nonpolar molecules. Even though a nonpolar molecule has no permanent dipole moment (when averaged over time), the separation of charges within a dynamic molecule at any instant may induce a dipole moment in a nearby molecule.

Van der Waals bonds are very weak and hence are easily broken by thermal motion of the molecules. However, at sufficiently low temperatures where thermal excitations are negligible, these bonds cause substances to condense into a liquid.

Metallic Bonds

Metallic bonds are generally weaker than ionic or covalent bonds. The valence electrons in a metal are relatively free to move throughout the material. There are a large number of such mobile electrons in a metal, typically one or two electrons per atom. The metal structure can be viewed as a "sea" or "gas" of nearly free electrons surrounded by a lattice of positive ions (Fig. 31.10). The binding mechanism in a metal is the attractive force between the positive ions and the electron gas. The free electrons are responsible for the high electrical and thermal conductivities of metals. For this reason, the free electrons are often called conduction electrons.

Metal ion

Electron gas

Figure 31.10
Schematic diagram of a metal. The colored area represents the electron gas, while the spheres represent the positive metal ion cores.

Example 31.2   Breaking Bonds in Water

In both the solid state and the liquid state, H_2O molecules are held in place by six bonds to their nearest-neighbor molecules. If the latent heat of fusion of water is 3.34×10^5 J/kg, find the average energy required to break each of the bonds in ice.

Solution
The molecular weight of H_2O is 18 (see Appendix C), and thus, we know that 18 kg of ice contains a total of 6.02×10^{26} molecules. Thus, 1 kg of ice contains $6.02 \times 10^{26}/18 = 3.34 \times 10^{25}$ molecules. If each of these is bound by six bonds, we have a total number of bonds in the 1 kg of ice of

$$\left(6 \frac{\text{bonds}}{\text{molecule}}\right)(3.34 \times 10^{25} \text{ molecules}) = 2.00 \times 10^{26} \text{ bonds}$$

The latent heat of fusion of ice is the energy required to break all of these bonds in one kg of ice. Thus, the energy per bond is

$$\frac{3.34 \times 10^5 \text{ J}}{2.00 \times 10^{26} \text{ bonds}} = 1.67 \times 10^{-21} \text{ J} = \boxed{0.01 \text{ eV}}$$

Example 31.3   Energy to Break Bonds (Methane)

If the average energy associated with a bond between carbon and hydrogen is 4.28 eV in methane (CH_4), find the total amount of energy required to break all the bonds in 1 kg of methane.

Solution
The molecular weight of methane (CH_4) is 16. Thus, the number of molecules in 1 kg of methane is

$$\frac{6.02 \times 10^{26}}{16} = 3.76 \times 10^{25} \text{ molecules}$$

and with four bonds per molecule (see Figure 31.6) the total number of bonds broken is

$$\left(4 \frac{\text{bonds}}{\text{molecule}}\right)(3.76 \times 10^{25} \text{ molecules}) = 1.50 \times 10^{26} \text{ bonds}$$

Thus, the total energy required is

$$\left(4.28 \frac{\text{eV}}{\text{bond}}\right)(1.50 \times 10^{26} \text{ bonds}) = 6.42 \times 10^{26} \text{ eV} = \boxed{1.03 \times 10^8 \text{ J}}$$

31.3   The Energy and Spectra of Molecules

As in the case of atoms, the structure and properties of molecules can be studied by examining the radiation they emit or absorb. A molecule emits radiation when it moves from an excited state to a state lower in energy, and it absorbs radiation when it moves from a low energy state to a state of higher energy. Much of the information about molecules comes through a study of their rotational and vibrational motions, and we shall study each of these separately.

Rotational Motion of a Molecule

Let us consider the rotation of a molecule about its center of mass. We shall confine our discussion to a diatomic molecule, although the basic ideas can be extended to more complex molecules. The diatomic molecule pictured in Figure 31.11 can have rotational energy associated with its rotation about either the z or the y axis. The x axis is not considered because the atoms are assumed to be points, and thus, they have no moment of inertia about this axis.

Figure 31.11
A diatomic molecule can rotate about the y or z axes.

If ω is the angular velocity of rotation about one of these axes, the rotational kinetic energy of the molecule can be expressed in the form

$$E_{rot} = \frac{1}{2} I\omega^2 \tag{31.2}$$

where I is the moment of inertia of the molecule about the axis under consideration. Equation 31.2 can be written in terms of the angular momentum, $I\omega$, of the molecule as

$$E_{rot} = \frac{1}{2} I\omega^2 = \frac{(I\omega)^2}{2I} \tag{31.3}$$

Wave mechanics restricts the angular momentum values to multiples of $\hbar$ such that the allowed values of the rotational angular momentum are given by

$$I\omega = \sqrt{J'(J'+1)}\,\hbar \qquad J' = 0, 1, 2, \ldots \tag{31.4}$$

where J' is an integer called the **rotational quantum number**. Substituting Equation 31.4 into Equation 31.3, we get an expression for the allowed values of the rotational kinetic energy:

$$E_{rot} = \frac{\hbar^2}{2I} J'(J'+1) \qquad J' = 0, 1, 2, \ldots \tag{31.5}$$

Rotational energy of a molecule

Thus, we see that the rotational energy of the molecule is quantized and depends on the moment of inertia of the molecule. The allowed rotational energies of a diatomic molecule are plotted in Figure 31.12. The radiation emitted as a molecule makes a transition from one rotational level to another lies in the microwave range of frequencies, which are of the order of 10^{11} Hz. The radiation emitted as a molecule undergoes such transitions can be used to find such important properties of a molecule as its moment of inertia and the spacing between atoms, as the following example shows.

Example 31.4   Rotation of Oxygen

The O_2 molecule is found to make a transition from the lowest rotational energy state ($J' = 0$) to its first excited state ($J' = 1$) when the molecule absorbs a photon of frequency 4.30×10^{10} Hz. The molecule is shown in Figure 31.13, where the mass of each atom is 2.67×10^{-26} kg and the distance of separation is r. (a) Find the moment of inertia of the oxygen molecule.

Solution
The energy of the oxygen molecule in the $J' = 1$ state is found from Equation 31.5 to be

$$E_1 = \frac{\hbar^2}{2I} J'(J'+1) = \frac{\hbar^2}{2I}(2) = \frac{\hbar^2}{I}$$

From the same equation, we see that the rotational energy is 0 in the $J' = 0$ state. Thus, the difference in energy between these two states is

$$\Delta E = E_1 - E_0 = \frac{\hbar^2}{I}$$

The molecule acquires the energy to make this transition from the incoming photon of energy hf. Thus, we have

$$
\begin{array}{rcl}
J & & \begin{array}{c}\text{Rotational} \\ \text{energy}\end{array} \\
6 & \underline{\quad\quad} & 42E_1 \\
\\
5 & \underline{\quad\quad} & 30E_1 \\
\\
4 & \underline{\quad\quad} & 20E_1 \\
\\
3 & \underline{\quad\quad} & 12E_1 \\
2 & \underline{\quad\quad} & 6E_1 \\
1 & \underline{\quad\quad} & 2E_1 \\
0 & \underline{\quad\quad} & 0
\end{array}
$$

Energy

Figure 31.12
Allowed rotational energies of a diatomic molecule calculated using Equation 31.5, where $E_1 = \hbar^2/2I$.

$$hf = \frac{\hbar^2}{I}$$

and from this we see that the moment of inertia can be found as follows.

$$I = \frac{h}{4\pi^2 f} = \frac{6.63 \times 10^{-34}\ \text{J}\cdot\text{s}}{4\pi^2 (4.30 \times 10^{10}\ \text{Hz})} = \boxed{3.91 \times 10^{-46}\ \text{kg}\cdot\text{m}^2}$$

(b) Find the distance of separation of the oxygen atoms in the molecule.

Solution

Since both atoms have the same mass, the center of mass of the molecule is located midway between the two atoms, thus $r_1 = r_2 = r/2$ in Figure 31.13. We shall consider the atoms to be point masses, so the moment of inertia about the center of mass is given by

$$I = \sum m_i r^2 = m r_1^2 + m r_2^2 = m\left(\frac{r}{2}\right)^2 + m\left(\frac{r}{2}\right)^2 = \frac{1}{2}mr^2$$

Solving for r, the distance of separation, we have

$$r = \left(\frac{2I}{m}\right)^{1/2} = \left[\frac{2(3.91 \times 10^{-46})\ \text{kg}\cdot\text{m}^2}{2.67 \times 10^{-26}\ \text{kg}}\right]^{1/2} = \boxed{1.71 \times 10^{-10}\ \text{m}}$$

Figure 31.13 (Example 31.4) A diatomic molecule with an axis shown through its center of mass. For O_2, $r_1 = r_2 = r/2$.

Example 31.5   The Rotation of CO

The CO molecule has a distance of separation between the C and O atoms given by $r = 1.13 \times 10^{-10}$ m. The mass of the carbon atom is 2.00×10^{-26} kg and the mass of the oxygen atom is 2.67×10^{-26} kg. **(a)** Find the center of mass of this molecule.

Solution

To find the center of mass of the CO molecule, we shall use a coordinate system like that shown in Figure 31.14 where the origin is selected at the position of the carbon atom. In Chapter 6, we found that the center of mass for a system such as this can be found as

$$x = \frac{\sum m_i x_i}{M} = \frac{(2.00 \times 10^{-26}\ \text{kg})(0) + (2.67 \times 10^{-26}\ \text{kg})(1.13 \times 10^{-10}\ \text{m})}{4.67 \times 10^{-26}\ \text{kg}}$$

$$x = 6.46 \times 10^{-11}\ \text{m} = \boxed{0.646\ \text{nm}}$$

Thus, we see that the center of mass is located 6.46×10^{-11} m from the carbon atom, which corresponds to 4.84×10^{-11} m from the oxygen atom.

(b) Find the moment of inertia of the CO molecule about its center of mass.

Solution

The moment of inertia is given by

$$I = \sum m_i r_i^2 = m_1 r_1^2 + m_2 r_2^2$$

where m_1 is taken to be the mass of the carbon atom, m_2 is the mass of the oxygen atom, r_1 is the distance from the center of mass to the carbon atom, and r_2 is the distance from the center of mass to the oxygen atom. Thus, we have

$$I = (2 \times 10^{-26}\ \text{kg})(6.46 \times 10^{-11}\ \text{m})^2 + (2.67 \times 10^{-26}\ \text{kg})(4.84 \times 10^{-11}\ \text{m})^2$$

$$= \boxed{1.46 \times 10^{-46}\ \text{kg}\cdot\text{m}^2}$$

Figure 31.14 (Example 31.5) Representation of the CO molecule.

(c) Find the energy of the CO molecule in the $J' = 1$ state and in the $J' = 0$ state.

Solution
For the $J' = 1$ state, Equation 31.5 gives

$$E_1 = \frac{\hbar^2}{2I} J'(J' + 1) = \frac{\hbar^2}{I} = \frac{(1.054 \times 10^{-34} \text{ J} \cdot \text{s})^2}{1.46 \times 10^{-46} \text{ kg} \cdot \text{m}^2} = \boxed{7.61 \times 10^{-23} \text{ J}}$$

The energy in the $J' = 0$ state is zero, as can be seen from Equation 31.5.

Exercise   Find the frequency of the photon that would be emitted as the molecule makes a transition from the $J' = 1$ state to the $J' = 0$ state.

Answer 1.15×10^{11} Hz. (The photon energy hf must be equal to the difference in energies of these two states.)

Vibrational Energy of Molecules

As we mentioned earlier, molecules may also have energy associated with their vibrational motion. A molecule is a flexible structure whose atoms are bound together as though they were connected by a spring, as shown in Figure 31.15a. Hence, the molecule can be set into vibration (through collisions with other molecules or the walls of a container) and acquire vibrational energy. The problem of finding an expression for the vibrational energy for an oscillator was one of the first to be solved in quantum mechanics. It was found that the energy of the oscillator is quantized, and the allowed energies are given by

Vibrational energies of a molecule

$$E_{\text{vib}} = \left(v + \frac{1}{2} \right) hf \qquad \textbf{(31.6)}$$

where v is an integer called the **vibrational quantum number**. At ordinary temperatures, most molecules have vibrational energies corresponding to the $v = 0$ state having the lowest energy. The vibrational energy of a molecule in

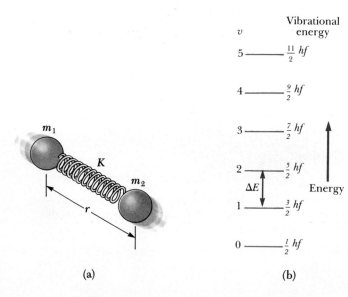

Figure 31.15
(a) Model of a diatomic molecule whose atoms are bonded by an imaginary spring of spring constant K. (b) Allowed vibrational energies of a diatomic molecule. Note that the spacings between adjacent levels are equal.

the $v = 0$ state is $E_{vib} = hf/2$, and is called the **zero-point energy.** In the first excited state, $v = 1$, and its energy is $3hf/2$, and so on. The allowed vibrational energies of a diatomic molecule are shown in Figure 31.15b. Note that the spacings between adjacent levels are equal, unlike the situation for the rotational levels. The frequency of the radiation emitted or absorbed during a vibrational transition is usually in the infrared region of the spectrum, with frequencies of the order of 10^{13} Hz.

31.4   Band Theory of Solids

An understanding of the behavior of semiconductors would not be possible without the techniques and ideas of quantum physics. Let us examine what happens to the energy levels of atoms as they are assembled to form a solid. We have already seen that the electrons of an isolated atom (one far removed from other atoms) can occupy only certain well-defined energy levels as shown in Figure 31.16a. Now suppose that two similar atoms are brought close together as in Figure 31.16b. An original single energy level common to both is split such that two energy levels will form. If six atoms are combined, the original energy level splits into six closely spaced energy levels, as in Figure 31.16c. Finally, imagine a system of N atoms being brought together as would be the case in a solid. The atomic energy levels are split into N closely spaced levels as in Figure 31.16d. In fact, if N is very large, as it is in a solid (of the order of 10^{23} atoms/cm^3), the spacing between these levels is so small that they can be considered to form a continuous band of energies.

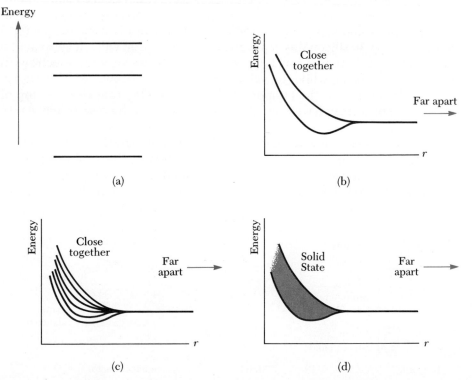

Figure 31.16
(a) Well-defined energy levels of an isolated atom. (b) The splitting of one of the levels when two atoms are brought together. (c) The splitting of one of the levels when six atoms are brought together. (d) The formation of a band of energies when many atoms are assembled to form a solid.

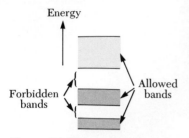

Figure 31.17
Energy level of a solid consists of a set of allowed bands (shaded) separated by forbidden bands.

By extending the discussion above, it is easy to understand that each atomic energy level of the individual atoms is replaced by a continuous band of energies. As a result, the energy level diagram for a solid is a set of bands as shown in Figure 31.17. The figure shows a series of allowed bands of energies separated by *forbidden bands*. These latter bands are called forbidden because electrons in the solid are not allowed to have these particular energies.

We can use this simple band theory to explain some of the electrical properties of solids. First, consider the energy bands for an insulator shown in Figure 31.18a. At ordinary temperatures, the insulator has a band of energy levels, called the *valence band,* which is completely filled with electrons. Above this in energy lies a completely empty band, called the *conduction band,* and between the valence and conduction bands is a wide energy gap. This gap is so wide (typically greater than 10 eV) that it is virtually impossible at ordinary temperatures to move electrons from the valence band into the conduction band. Since all the levels in the valence band are filled, an individual electron in this band is essentially immobile, even when an electric field is present. Hence, there is no current and the material behaves like an insulator. From another point of view, thermal agitation is insufficient to free an electron from its covalent bonds. Electrons in the valence band remain in a bound state as opposed to becoming "free" (moving into the conduction band).

The energy band structure for a metal is shown in Figure 31.18b. In this case, the uppermost band, which is the conduction band, is only partially filled. In this band the electrons have many levels to which they can easily move; therefore an electric current is readily established when an electric field is applied. Such conduction band electrons are often referred to as *free* electrons since they are not bound to any one atom in the structure.

In a semiconductor, the last filled energy band (the valence band) is completely occupied as in the case of an insulator. However, the energy gap that separates the valence band from the conduction band is very narrow for a semiconductor. In silicon this gap is about 1 eV in width, while in germanium it is about 0.7 eV. At very low temperatures, there are very few electrons in the conduction band, while the valence band is filled. However because the energy gap is small, as the temperature increases, electrons near the top of the valence band may have enough thermal energy to be able to move into the conduction band (Fig. 31.18c). These electrons can then participate in the conduction process if a voltage is applied across the material.

When an electron moves from the valence band into the conduction band, it leaves behind a vacant crystal site, or so-called hole, in the valence band. This hole (electron-deficient site) appears as a positive ion of charge $+e$. The

Figure 31.18
(a) An insulator has a filled valence band and an empty conduction band separated by a wide forbidden band. (b) The uppermost band of a metal, which is the conduction band, is partially filled with electrons. (c) A semiconductor at ordinary temperatures has a nearly filled valence band and a nearly empty conduction band. The forbidden band is very narrow, usually less than 1 eV.

hole also acts as a charge carrier in the sense that a valence electron from a nearby bond can transfer into the hole without becoming a free electron. Thus we see that there are both negative and positive charge carriers in a semiconductor.

When impurities are added to semiconductors, their band structures are modified as shown in Figure 31.19. For example, when an atom with five valence electrons, such as arsenic, is added to a semiconductor, four valence electrons participate in the covalent bonds and one electron is left over. This extra electron is nearly free and has an energy level in the band diagram that lies within the energy gap, just below the conduction band (Fig. 31.19a). Such an impurity in effect donates an electron to the structure and hence is referred to as a *donor atom*, and the levels they occupy are called *donor levels*. Since the energy spacing between the donor levels and the bottom of the conduction band is very small (typically, about 0.05 eV), a small amount of thermal energy will cause an electron in these levels to move into the conduction band. If the semiconductor is doped with atoms with three valence electrons, such as indium and aluminum, the three electrons form covalent bonds with neighboring atoms, leaving an electron deficiency, or hole. Such impurities are called *acceptors* and the levels they occupy are called *acceptor levels*. The energy levels of acceptors also lie within the energy gap, just above the valence band, as indicated in Figure 31.19b. Electrons from the valence band have enough thermal energy at room temperature to fill acceptor levels, leaving behind a hole in the valence band.

*31.5 Semiconductors

Electronic devices made from semiconducting materials are used in many facets of our everyday life. In this section, we shall describe the electrical properties of semiconductors and their role in the operation of electronic devices. In Chapter 18, we found that a good conductor contains a large number of free electrons that are able to move readily when a voltage is applied across the conductor. On the other hand, insulators have very few free electrons and hence are very poor conductors of electricity. Semiconductors are a class of materials in which the number of charge carriers is intermediate between those of insulators and those of conductors.

The most common semiconducting materials used in electronic devices are silicon and germanium. Each of these materials is characterized by the fact that the constituent atoms have four valence electrons. In the solid state, each atom shares one of its valence electrons with each of its four nearest neighbors. A two-dimensional model for this covalent bonding is illustrated in Figure 31.20.

As the temperature of the material is increased above room temperature, the atoms vibrate more vigorously about their equilibrium positions. In this process, some electrons acquire sufficient energy to break free of their bonds

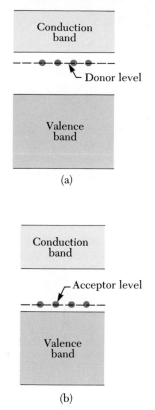

Figure 31.19
(a) Energy band diagram for a semiconductor containing donor impurities whose levels lie just below the bottom of the conduction band. (b) A semiconductor containing acceptor impurities has acceptor levels that are just above the top of the valence band.

Figure 31.20
A two-dimensional model of the bonding between atoms in a semiconductor. Each horizontal or vertical line between atoms represents a shared electron.

◯ = Semiconductor atom

○ = Semiconductor atoms

● = Impurity atom with
      five valence electrons

⊖ = Extra electron from
      impurity atom

Figure 31.21
A semiconductor material
containing an impurity atom
with five valence electrons.

○ = Semiconductor atoms

● = Impurity atom with
      three valence electrons

+ = Hole, or electron
      deficiency in a bond

Figure 31.22
Two-dimensional representation of a semiconductor containing an impurity with three
valence electrons.

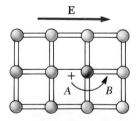

Figure 31.23
An electron originally at point
B jumps to fill the hole originally at point A. Thus, the
hole moves from A to B, in
the direction of the electric
field.

and are able to move freely through the material. These electrons can then participate in the conduction process if a voltage is applied across the material. This explains why the resistance of a pure semiconductor decreases with increasing temperature.

Doped Semiconductors

Pure semiconducting materials, such as silicon, are virtually never used in electronic devices. Instead, controlled amounts of impurities, such as arsenic or indium, are added. The electrical properties of the semiconductor are substantially altered by these impurities even if the number of impurity atoms added is relatively small. The process of adding impurities is called **doping.**

In order to understand the effect of doping on semiconductors, consider the situation in which an impurity atom, such as arsenic, with five valence electrons, is added to a semiconductor, such as silicon. Four of the arsenic valence electrons participate in the bonding between nearest-neighbor silicon atoms, and one electron is left over (Fig. 31.21). This extra electron is loosely bound and hence easily freed from the atom. When the electron is free, it contributes to conduction in the material. Thus, the impurity atom in effect donates an electron to the material and hence is referred to as a **donor** atom. Semiconductors doped with donor atoms are called **n-type semiconductors** because most of the charge carriers are electrons, whose charge is negative.

If a semiconductor is doped with atoms having three valence electrons, such as indium or aluminum, the three electrons form bonds with nearest-neighbor atoms. However, there is an electron deficiency in one of the bonds with a neighbor atom (Fig. 31.22). The vacancy in the lattice structure of the solid caused by this deficiency is referred to as a **hole.** Figure 31.23 shows how holes can contribute to conduction in a semiconductor. If an electric field is applied to the semiconductor, by connecting the material to a battery, say, electrons bound to atoms near the hole can jump over to the site of the hole. As a result, the hole migrates in the direction of the electric field as if it, the hole, were a positive charge carrier (shown by the curved arrow in Fig. 31.23). Because there are more holes than electrons, materials doped in this way are known as **p-type semiconductors.** The impurity atoms added are referred to as **acceptors.**

The *p-n* Junction

Most semiconductor devices, such as diodes and transistors, are fabricated using a semiconductor that is partly *p*-type and partly *n*-type. Figure 31.24 shows a semiconductor in which the left side has been made *p*-type and the right side has been made *n*-type. The boundary between the *p* and *n* regions is called a **p-n junction.** One of the outstanding features of this junction is that it is virtually swept free of charge carriers that move through it. Electrons in the *n* region diffuse through the junction into the *p* region. At the same time, holes from the *p* region diffuse through the junction into the *n* region. The electrons and holes in the junction region recombine as an electron moves into the vacant site where a hole exists. This recombination process of electrons and holes explains why this narrow region around the junction has no free charge carriers.

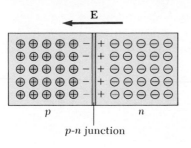

Figure 31.24
A semiconductor fabricated such that the left side is p-type and the right side is n-type. At the junction, an electric field is set up directed from right to left.

It might be guessed that this diffusion of electrons into the p region and of holes into the n region could continue until all the electrons and holes in these regions are neutralized. In order to understand why this does not occur, let us examine what happens when a hole and an electron recombine in the n region. The n region is electrically neutral because for every free electron in this region, there exists a positive charge on the atom from which it came. If an electron in the n region combines with a hole, effectively causing both to disappear, the positive charge present on the atom is no longer electrically balanced. Hence, the right side of the junction becomes positively charged (see Fig. 31.24). In a similar manner, the left side of the junction becomes negatively charged. An electric field is thus set up directed from right to left, as in Figure 31.24. The presence of this field prevents further diffusion of electrons and holes through the junction.

The p-n Junction Diode

The p-n device just described is called a **p-n junction diode.** It is a good conductor when a voltage applied to it attempts to drive current in one direction but a poor conductor if the voltage is reversed and attempts to drive current in the opposite direction.

Suppose a battery is connected across the junction, with the p side positive and the n side negative, as in Figure 31.25a. This situation is referred to as a **forward-biased junction.** This polarity provides an extra push for the electrons and holes in the form of an accelerating field, enabling them to overcome the opposing electric field at the junction. Thus, free electrons move readily from the n side to the p side of the junction, while holes move readily from the p side to the n side. This flow of charge across the junction constitutes a current.

Now consider what happens when the polarity of the battery is reversed, as in Figure 31.25b, where the p side is negative and the n side is positive. In this case, the junction is said to be **reverse-biased.** The electric field at the

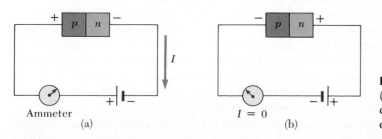

Figure 31.25
(a) A forward-biased diode carries a large current. (b) In a reverse-biased junction, the current is approximately zero.

Figure 31.26
(a) Current versus applied voltage for a *p-n* junction diode. (b) Circuit symbol for a diode.

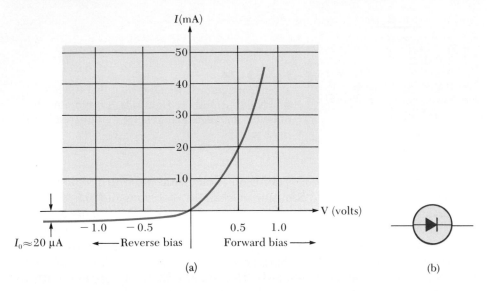

(a)

(b)

junction is now enhanced, and very few charges are able to cross the junction. This results in a very small current under reverse-bias conditions.

The current-voltage curve of a typical diode is shown in Figure 31.26a. Note that a large current passes through the diode under forward-bias conditions but there is virtually no current under reverse-bias conditions.

The circuit symbol for a diode is shown in Figure 31.26b. The arrow indicates the direction of the current through the diode.

The Diode as a Rectifier

The current from electric outlets in this country is an alternating current. However, the tubes and transistors in such electronic devices as your television set and stereo system require a direct current or voltage in order to operate properly. Therefore, it is necessary to change the alternating current from the outlet to a direct current. This process is called **rectification,** and the diode provides a simple and convenient means for achieving this goal.

We can understand how a diode rectifies a current by considering Figure 31.27a, which shows a diode placed in series with a resistor and an ac source. Because current can pass through the diode only in one direction, the alternating current in the resistor is reduced to the form shown in Figure 31.27b. The diode is said to act as a **half-wave rectifier** since there is current in the circuit during only half of each ac cycle.

When a diode is used as a half-wave rectifier, the current in the resistor is a pulsating direct current, as shown in Figure 31.27b. A closer approximation to a true direct current can be achieved by using two diodes and a transformer connected as shown in Figure 31.28a. Consider an instant at which the voltage at point A in the secondary of the transformer is higher than the voltage at B, which is higher than the voltage at C. At this instant, the current path in the circuit is from A through diode D_1, downward through the resistor and back to point B. Half a cycle later, the alternating voltage at the output of the secondary of the transformer is reversed so that point C is at a higher voltage than point B, which is at a higher voltage than point A. In this case, the current path

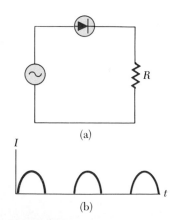

(a)

(b)

Figure 31.27
(a) A diode in series with a resistor allows current to flow during one half of the cycle. (b) The current versus time for the circuit in (a).

in the circuit is from C through diode D_2, again downward through the resistor, and back to point B. Thus there is current in the circuit during the complete cycle of the ac voltage, and the direction of the current is always downward through the resistor. The current in the resistor varies with time as shown in Figure 31.28b. In this situation, the pair of diodes is said to be acting as a **full-wave rectifier.** Note that the current in the resistor is still pulsating but is closer to being a true direct current than with the half-wave rectifier.

Filters

As stated earlier, the transistors and tubes in an electronic device such as a television or stereo need to be supplied with a smooth direct current or voltage. The output of a full-wave rectifier is not smooth enough. Various kinds of filters using inductors and capacitors have been developed to smooth out the pulsating output of a full-wave rectifier. In this section we shall briefly describe how a filter that uses only a single capacitor works.

Figure 31.29a shows a capacitor connected in parallel with the resistor in a full-wave rectifier circuit. In the absence of the capacitor, the current in the resistor would be as shown by the red curve in Figure 31.29b. When the circuit includes the capacitor, however, we have the following. As the current in the resistor increases with time, from A to B in Figure 31.29b, the capacitor charges such that its upper plate is positive. As the current through the resistor attempts to drop off between points B and C, the capacitor begins to discharge through the resistor. As a result, the current in the resistor never reaches zero. Instead, it follows the dashed path from B to D in Figure 31.29b. This results in a current in the resistor as shown in Figure 31.29c. Although a ripple still remains in the current, it is smooth enough for most applications.

The Junction Transistor

The discovery of the transistor by John Bardeen, Walter Brattain, and William Shockley in 1948 totally revolutionized the world of electronics. For this work, these three men shared a Nobel prize in 1956. By 1960, the transistor

Figure 31.28
(a) A full-wave rectifier circuit. (b) The current versus time through the resistor in (a).

Figure 31.29
(a) A full-wave rectifier circuit with a filter capacitor.
(b) The blue curve represents the current versus time with no filter capacitor, and the red curve represents the current when the circuit includes the capacitor.
(c) A graph of current versus time for the circuit shown in (a), that is, another rendering of the red curve in (b).

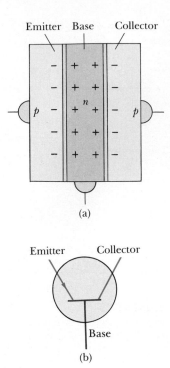

Figure 31.30
(a) The *pnp* transistor consists of an *n* region (base) sandwiched between two *p* regions (the emitter and collector). (b) The circuit symbol for the *pnp* transistor.

Figure 31.31
A bias voltage applied to the base as shown produces a small base current I_b, which is used to control the collector current I_c.

had replaced the vacuum tube in many electronic applications. The advent of the transistor created a multibillion dollar industry that produces such popular devices as pocket radios, handheld calculators, computers, television receivers, and electronic games.

The junction transistor consists of a semiconducting material with a very narrow *n* region sandwiched between two *p* regions. This configuration is the *pnp* **transistor.** Another configuration is the *npn* **transistor,** which consists of a *p* region sandwiched between two *n* regions. Because the operation of the two transistors is essentially the same, we shall describe only the *pnp* transistor.

The structure of the *pnp* transistor, together with its circuit symbol, is shown in Figure 31.30. The outer regions of the transistor are called the **emitter** and **collector,** and the narrow central region is called the **base.** Note that the configuration contains two junctions. One junction is the interface between the emitter and the base, and the other is between the base and the collector.

Suppose a voltage is applied to the transistor such that the emitter is at a higher potential than the collector. (This is accomplished with battery V_{ec} in Fig. 31.31.) If we think of the transistor as two diodes back to back, we see that the emitter-base junction is forward-biased and the base-collector junction is reverse-biased.

Because the *p*-type emitter is heavily doped relative to the base, nearly all of the current consists of holes moving across the emitter-base junction. Most of these holes do not recombine in the base because it is very narrow. The holes are finally accelerated across the reverse-biased base-collector junction, producing the current I_c in Figure 31.31.

Although only a small percentage of holes recombine in the base, those that do limit the emitter current to a small value because positive charge carriers accumulate in the base and prevent holes from flowing into this region. In order to prevent this limitation of current, some of the positive charge on the base must be drawn off; this is accomplished by connecting the base to a second battery, V_{eb} in Figure 31.31. Those positive charges that are not swept across the collector-base junction leave the base through this added pathway. This base current, I_b, is very small, but a small change in it can significantly change the collector current, I_c. For example, if the base current is increased, the holes find it easier to cross the emitter-base junction and there is a large increase in the collector current. Since the base current is very small, the collector current is approximately equal to the emitter current.

One common application of the transistor is to amplify a small, time-varying signal. The small voltage to be amplified is placed in series with the battery V_{eb} in Figure 31.31. The time-varying input signal produces a small variation in the base voltage. This results in a large change in collector current and hence a large change in voltage across the resistor.

The Integrated Circuit

The integrated circuit, invented independently by Jack Kilby at Texas Instruments in late 1958 and by Robert Noyce at Fairchild Camera and Instruments in early 1959, has been justly called "the most remarkable technology ever to hit mankind." Jack Kilby's first device is shown in Figure 31.32a. Integrated circuits have indeed started a second industrial revolution and are found at the heart of computers, watches, cameras, automobiles, aircraft, robots, space

(a)

(b)

Figure 31.32
(a) Jack Kilby's first integrated circuit, tested on September 12, 1958. *(Courtesy of Texas Instruments, Inc.)* (b) A 32-bit microprocessor integrated circuit containing over 200,000 components is about the size of a dime. *(Courtesy of A.T.&T. Bell Labs)*

vehicles, and all sorts of communication and switching networks. In simplest terms, an integrated circuit is a collection of interconnected transistors, diodes, resistors, and capacitors fabricated on a single piece of silicon. State of the art chips easily contain several hundred thousand components in a 1-cm^2 area (Fig. 31.32b).

It is interesting that integrated circuits were invented partly in an effort to achieve circuit miniaturization and partly in order to solve the interconnection problem spawned by the transistor. In the era of vacuum tubes, power and size considerations of individual components set modest limits on the number of components that could be interconnected in a given circuit. With the advent of the tiny, low-power, highly reliable transistor, design limits on the number of components disappeared and were replaced with the problem of wiring together hundreds of thousands of components. The magnitude of this problem can be appreciated when you consider that second generation computers (consisting of discrete transistors) contained several hundred thousand components requiring more than a million hand-soldered joints to be made and tested.

In addition to solving the interconnection problem, integrated circuits possess the advantages of miniaturization and fast response, two attributes critical for high-speed computers. The fast response is actually also a product of the miniaturization and close packing of components. This is so because the response time of a circuit depends on the time it takes for electrical signals traveling at about 3×10^8 m/s to pass from one component to another. This time is clearly reduced by the closer packing of components.

Summary

Two or more atoms combine to form molecules because of a net attractive force between the constituent atoms. The mechanisms responsible for the bonding in the molecule can be classified as follows:

1. **Ionic bond:** Certain molecules have ionic bonds because of the Coulomb attraction between oppositely charged ions. Sodium chloride (NaCl) is one example of an ionically bonded molecule.

2. **Covalent bond:** The covalent bond in a molecule is formed by the sharing of valence electrons of its constituent atoms. For example, the two electrons of the H_2 molecule are equally shared between the two nuclei.

3. **Hydrogen bond:** This type of bonding corresponds to the attraction of two negative ions by an intermediate hydrogen atom (a proton).

4. **Van der Waals Bond:** This is a weak electrostatic bond between atoms which do not form ionic or covalent bonds. It is responsible for the condensation of inert gas atoms and nonpolar molecules into the liquid phase.

Bonding mechanisms in solids can be classified in a manner similar to the schemes already described for molecules. For example, the Na^+ and Cl^- ions in NaCl form ionic bonds, while the carbon atoms in diamond form covalent bonds. Another type of bonding mechanism is the metallic bond, which is generally weaker than ionic or covalent bonds. The mechanism for the metallic bond in a metal is a net attractive force between the positive ion cores and the mobile valence electrons.

In a crystalline solid, the energy levels of the system form a set of bands. Electrons are allowed to occupy only certain bands. There are energy regions between the allowed bands corresponding to states which the electrons cannot occupy.

The properties of metals, insulators, and semiconductors can best be understood in terms of the band theory of solids. The conduction band of a metal is partially filled. Therefore, there are many electrons free to move throughout the metal and contribute to the conduction current. In an insulator at low temperatures, the valence band is completely filled with electrons, while the conduction band is empty. The region between the valence band and the conduction band is called the energy gap of the material. The energy gap for an insulator is of the order of 10 eV. Because this gap is large compared to the thermal energy at ordinary temperatures, very few electrons are thermally excited into the conduction band, which explains the small electric conductivity of an insulator.

A semiconductor is a material with a small energy gap, of the order of 1 eV, and a valence band which is filled at low temperatures. Because of their small energy gap, a significant number of electrons can be thermally excited from the valence band into the conduction band as the temperature increases. The band structures and electrical properties of a semiconductor can be modified by adding donor atoms with five valence electrons (such as arsenic), or by adding acceptor atoms with three valence electrons (such as indium). A semiconductor doped with donor impurity atoms is called an *n*-type semiconductor, while one doped with acceptor impurity atoms is called *p*-type. The energy levels of these impurity atoms falls within the energy gap of the material.

Additional Reading

P. W. Atkins, *Molecules,* Scientific American Library, New York, NY, W. H. Freeman, 1987.

M. L. Cohen, V. Heine, and J. C. Phillips, "Quantum Mechanics of Materials," *Sci. Amer.,* June 1982, p. 82.

P. C. Hinkle and R. E. McCarty, "How Cells Make ATP," *Sci Amer.,* March 1978, p. 104.

"Microelectronics," *Sci. Amer.,* September 1977.

Questions

1. The H_2 molecule is stable, yet the H_3 molecule is not. Explain why, using the Pauli exclusion principle.

2. When the emission spectrum is examined from a collection of diatomic molecules, many more lines are seen than is typical when the spectrum is from a collection of atoms. Why should this occur?

3. In general, must a photon have more energy to excite a rotational energy state than it does a vibrational energy state or vice versa? Defend your answer.

4. How can the analysis of the rotational spectrum of a molecule lead to an estimate of the size of that molecule?

5. Use the band theory of solids to discuss the difference between a material that is a good insulator and one that is a good conductor of electricity.

6. Would you expect the energy gap in a solid to increase or decrease with increasing temperature? Defend your answer.

7. Would you expect the energy gap in a solid to increase or decrease with increasing pressure? Defend your answer.

8. The resistivity of most metals increases with increasing temperature, whereas the resistivity of an intrinsic semiconductor decreases with increasing temperature. What explains this difference in behavior?

9. When a photon is absorbed by a semiconductor, an electron-hole pair is said to be created. Give a physical interpretation of this statement based on the energy band model of a solid.

10. An often used, and often incorrect, rule of thumb says that a metal is opaque, a semiconductor is translucent, and an insulator is transparent. Use the energy band model of a solid to explain how this rule of thumb might have arisen.

11. The energy of visible light ranges between 1.8 eV and 3.2 eV. Does this explain why silicon with an energy gap of 1.1 eV between the valence band and the conduction band appears black whereas diamond with an energy gap of 5.5 eV appears transparent?

12. Pentavalent atoms such as arsenic are donor atoms in a semiconductor such as silicon, while trivalent atoms such as indium are acceptors. Inspect the periodic table of the elements on the endpapers and determine what other elements would be considered as either donors or acceptors.

13. Suppose we could obtain an intrinsic semiconductor from a hexavalent substance. Discuss the types of materials that could serve as impurities to produce n- and p-type semiconductors from this material.

14. "A single piece of n-type semiconductor could be used as a resistor." Is this statement true? Discuss.

15. What is the resistance of a diode when connected in reverse bias?

16. Explain how a p-n junction diode operates as a rectifier.

17. Why is a pnp or npn "sandwich" (with the central region very thin) essential to the operation of a transistor?

18. It is often said that a transistor consists of two back-to-back diodes, with the base-emitter junction acting as one diode and the base-collector junction as the second. Which of these junctions is reverse biased, and which is forward biased?

Problems

Section 31.1 Molecular Bonds

1. Plot a potential energy diagram for an object moving at the end of a spring of spring constant $K = 150$ N/m. Recall that the potential energy for an object moving with simple harmonic motion is given by $PE = \frac{1}{2}Kx^2$.

2. Plot a graph of the gravitational potential energy between any pair of particles of masses m_1 and m_2. Recall that the expression for the gravitational potential energy is given by $PE = -Gm_1m_2/r$.

3. The separation between the K^+ and Cl^- ions in a KCl molecule is 2.8×10^{-10} m. Assuming the two ions act like point charges, (a) determine the force of attraction between them, and (b) determine the potential energy of attraction in eV.

4. Consider the potential energy diagram shown in Figure 31.33. Determine whether the force is positive, negative, or zero at the various points indicated.

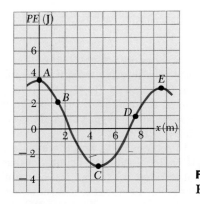

Figure 31.33
Problem 4

5. Consider the HCl molecule, which consists of a hydrogen atom of mass 1 u bound to a chlorine atom of mass 35 u. The equilibrium separation between the atoms is 0.128 nm, and it requires 0.15 eV of work to

increase or decrease this separation by 0.01 nm. (a) Recall that the potential energy of a simple harmonic oscillator is $PE = \frac{1}{2}Kx^2$ and determine the "spring constant" K for this molecule. (b) The classical frequency of vibration of a diatomic molecule as shown in Figure 31.15a is given by $f = \frac{1}{2\pi}\sqrt{\frac{K}{\mu}}$, where the "reduced mass" μ is defined by the expression $\mu = \frac{m_1 m_2}{m_1 + m_2}$. What is the classical frequency of vibration for the HCl molecule?

Section 31.2   Molecular Bonds: Ionic, Covalent, Hydrogen, Van der Waals, and Metallic

6. The ionization energy of K is 4.34 eV and the electron affinity of Cl is 3.7 eV. Find the energy required to form KCl.

7. The ionization energy of K is 4.34 eV and the electron affinity of F is 4.1 eV. Find the energy required to form KF.

8. Look up the heat of vaporization of water and calculate the energy per bond in liquid water. (We consider the water molecule to be loosely bound by six bonds to its nearest neighbor.)

9. How much methane could be broken into its components with an input energy of 3.63×10^8 J?

10. Ethane is an organic compound that is formed by two carbon atoms bound together, and with three hydrogen atoms bound to each carbon atom. The C—H bond energy is about 4.28 eV and the C—C energy bond is 3.60 eV. Find the energy input required to break existing bonds when 1 kg of ethane is burned.

Section 31.3   The Energy and Spectra of Molecules

11. Find the location of the center of mass of the HF molecule with respect to the fluorine atom if the equilibrium separation is 0.11 nm.

12. Find the location of the center of mass of the NO molecule with respect to the oxygen atom if the equilibrium separation is 0.114 nm.

13. Use the result of Problem 11 to find the moment of inertia of the HF molecule about its center of mass.

14. Use the result of Problem 12 to find the moment of inertia of the NO molecule about its center of mass.

15. The absorption of a photon changes the HF molecule's state of rotation from $J' = 2$ to $J' = 3$. Use the result of Problem 13 and determine the wavelength of the photon.

16. The NO molecule makes a transition from the $J' = 4$ rotational state to the $J' = 3$ state. Use the result of Problem 14 to determine the wavelength and the frequency of the photon emitted.

17. Find the frequency of the photon that would cause the N_2 molecule to undergo a transition from the $J' = 0$ state to the $J' = 1$ state. The equilibrium separation is about 0.11 nm.

18. Find the energy that would have to be carried away by a photon as the HBr molecule undergoes a transition from the $J' = 1$ state to the $J' = 0$ state. The equilibrium separation is about 0.11 nm.

19. Consider the HCl molecule described in Problem 5 and determine (a) the location of the center of mass of the molecule, (b) the moment of inertia of the molecule about its center of mass, (c) the energies of the two lowest rotational states of this molecule, and (d) the longest wavelength radiation the molecule can emit in a pure rotational transition.

20. The $J' = 5$ to $J' = 6$ rotational absorption line of a diatomic molecule occurs at a wavelength of 1.35 cm (in the vapor phase). (a) Calculate the wavelength and frequency of the $J' = 0$ to $J' = 1$ rotational absorption line. (b) Calculate the moment of inertia of the molecule.

21. The frequency absorbed by the HF molecule as it makes a transition from the $v = 0$ to the $v = 1$ state is 8.72×10^{13} Hz. Use this information to sketch a vibrational energy level diagram similar to that shown in Figure 31.15b.

22. Find the zero-point energy of HBr if it absorbs radiation of frequency 8.66×10^{13} Hz as it moves from the $v = 0$ state to the $v = 1$ state.

23. If the force constant of a vibrating HCl molecule is $K = 480$ N/m, what is the energy difference between successive vibrational energy levels of this molecule? What is the wavelength of the photon emitted when the molecule makes a transition from one vibrational state to the next lower state? (*Hint:* See Problem 5.)

24. The $v = 0$ to $v = 1$ vibrational transition of the HI molecule occurs at a frequency of 6.69×10^{13} Hz. The same transition for the NO molecule occurs at a frequency of 5.63×10^{13} Hz. (a) Calculate the effective force constant for each molecule. (b) Explain why the force constant of the NO molecule is so much larger than that of the HI molecule.

Section 31.5   Semiconductors

25. The energy gap for Si at 300 K is 1.14 eV. (a) Find the lowest frequency photon that will promote an electron from the valence band to the conduction band of silicon. (b) What is the wavelength of this photon?

26. The longest wavelength radiation that is absorbed by

a certain semiconductor is $1.85 \, \mu m$. What is the energy gap for this semiconductor?

27. Aluminum is added as an impurity to silicon. Is the resulting semiconductor n-type or p-type? Explain.

28. Arsenic is added as an impurity to silicon. Is the resulting semiconductor n-type or p-type? Explain.

29. Electrons from donor atoms of an n-type semiconductor can reach the conduction band by absorbing an energy of approximately 0.01 eV. (a) At what temperature does a particle have a thermal energy, kT, equivalent to 0.01 eV? (b) At what temperature would a particle have thermal energy equivalent to the energy gap in pure diamond, which is 7 eV?

30. A light-emitting diode (LED) made of the semiconductor GaAsP gives off red light ($\lambda = 650$ nm). Determine the band energy gap E_g in the semiconductor.

31. Most solar radiation has a wavelength of 10^{-6} m or less. What is the energy gap that a solar cell material should have to be able to absorb this radiation?

Additional Problems

32. The equilibrium separation of the atoms in a certain molecule is 0.27 nm, the depth of the potential well is 4.3 eV, and the activation energy is 3.8 eV. Sketch a graph which shows these features at the appropriate locations.

33. Propane is an organic compound that is formed as follows. Three carbon atoms are bonded together; two of the carbon atoms have three hydrogen atoms bonded to each of them and the third carbon atom has two hydrogen atoms bonded to it. The C—H bond energy is about 4.23 eV and the C—C bond energy is 3.60 eV. Find the energy required to break all the bonds in 1 kg of propane as it is burned.

34. Find the frequency of the photon that would make the HCl molecule undergo a transition from the $J' = 0$ to

the $J' = 1$ state. The equilibrium separation is about 0.12 nm.

35. In the oxygen molecule O_2, the separation between the atoms is 1.2×10^{-10} m. Treating the oxygen atoms as particles, determine the allowed values of the rotational energy for the $J' = 1$ and $J' = 2$ states.

36. The transition from the $v = 0$ to the $v = 1$ vibrational state in a CO molecule occurs when a photon of 4.67-μm wavelength is absorbed. (a) What is the force constant in a CO molecule? (b) Determine the amplitude of vibration of a CO molecule in the lowest ($v = 0$) vibrational state. (c) How does the amplitude found above compare to the equilibrium separation distance of 0.113 nm in the CO molecule?

37. The separation between the protons of the hydrogen molecule is 1.06×10^{-10} m. The mass of a hydrogen atom is 1.67×10^{-27} kg. (a) Find the hydrogen molecule's energies of rotation about an axis perpendicular to and bisecting the line joining the protons. (b) If the force constant for the hydrogen molecule is known to be 40.0 N/m, determine the vibrational energies of the molecule.

38. Assuming energy levels are excited thermally up to energies of the order of kT, determine the highest rotational and vibrational energy levels excited in a chlorine molecule at a temperature of 4.5 K, 450 K, and 4500 K. Take the mass of a chlorine atom to be 5.81×10^{-26} kg, the distance between the nuclei to be 2.5×10^{-10} m, and the spring constant to be 50 N/m.

39. The energy gap for germanium is 0.7 eV. (a) Find the lowest frequency photon that will promote an electron from the valence band to the conduction band of germanium. (b) What is the wavelength of this photon?

40. When the following impurities are added to pure germanium, does the material become n-type or p-type: (a) phosphorus, (b) gallium, (c) indium?

Ernest Rutherford (1871–1937), a physicist from New Zealand, was awarded the Nobel Prize in 1908 for discovering that atoms can be broken apart by alpha rays and for studying radioactivity. "On consideration, I realized that this scattering backward must be the result of a single collision, and when I made calculations I saw that it was impossible to get anything of that order of magnitude unless you took a system in which the greater part of the mass of the atom was concentrated in a minute nucleus. It was then that I had the idea of an atom with a minute massive center carrying a charge." *(Photo courtesy of AIP Niels Bohr Library)*

$$\text{Mass of proton} = 1.007276 \text{ u}$$

$$\text{Mass of neutron} = 1.008665 \text{ u}$$

$$\text{Mass of electron} = 0.000549 \text{ u}$$

For reasons that will become apparent later, note that the mass of the neutron is greater than the combined masses of the proton and the electron.

Because the rest energy of a particle is given by $E = m_0c^2$, it is often convenient to express the particle's mass in terms of its energy equivalence. For a proton, we have

$$E = m_0c^2 = (1.67 \times 10^{-27} \text{ kg})(3 \times 10^8 \text{ m/s})^2$$

$$= 1.50 \times 10^{-10} \text{ J} = 9.39 \times 10^8 \text{ eV} = 939 \text{ MeV}$$

Following this procedure, it is found that the rest energy of an electron is 0.511 MeV.

The rest masses of the proton, neutron, and electron are given in Table 32.1. The masses and some other properties of selected isotopes are provided in Appendix B.

Example 32.1   The Unified Mass Unit

Use Avogadro's number to show that 1 u = 1.66×10^{-27} kg.

Solution

We know that 12 kg of ^{12}C contains Avogadro's number of atoms. Avogadro's number, N_A, has the value of 6.02×10^{23} atoms/g·mole = 6.02×10^{26} atoms/kg·mole.

Thus, the mass of one carbon atom is

$$\text{Mass of one } ^{12}\text{C atom} = \frac{12 \text{ kg}}{6.02 \times 10^{26} \text{ atoms}} = 1.99 \times 10^{-26} \text{ kg}$$

Since one atom of ^{12}C is defined to have a mass of 12 u, we find that

$$1 \text{ u} = \frac{1.99 \times 10^{-26} \text{ kg}}{12} = 1.66 \times 10^{-27} \text{ kg}$$

The Size of Nuclei

The size and structure of nuclei were first investigated in the scattering experiments of Rutherford, discussed in Section 30.1. In these experiments, positively charged nuclei of helium atoms (alpha particles) were directed at a thin

Table 32.1
Rest Mass of the Proton, Neutron, and Electron in Various Units

Particle	Mass		
	kg	u	MeV/c^2
Proton	1.6726×10^{-27}	1.007276	938.28
Neutron	1.6750×10^{-27}	1.008665	939.57
Electron	9.109×10^{-31}	5.486×10^{-4}	0.511

piece of metal foil. As the alpha particles moved through the foil, they often passed near a nucleus of the metal. Because of the positive charge on both the incident particles and metal nuclei, particles were deflected from their straight-line path by the Coulomb repulsive force. In fact, some particles were even deflected back along the direction from which they had come. These particles were apparently moving directly toward a nucleus, on a head-on collision course.

Rutherford applied the principle of conservation of energy to the alpha particles and nuclei and found an expression for the nearest distance, d, that the particle can approach the nucleus when moving directly toward it before it is turned around by Coulomb repulsion. In such a head-on collision, the kinetic energy of the incoming alpha particle must be converted completely to electrical potential energy when the particle stops at the point of closest approach and turns around (Fig. 32.1). If we equate the initial kinetic energy of the alpha particle to the electrical potential energy of the system (alpha particle plus nucleus), we have

$$\frac{1}{2}mv^2 = k\frac{q_1 q_2}{r} = k\frac{(2e)(Ze)}{d}$$

Solving for d, the distance of closest approach, we get

$$d = \frac{4kZe^2}{mv^2}$$

From this expression, Rutherford found that the alpha particles approached nuclei to within 3.2×10^{-14} m when the foil was made of gold. Thus, the radius of the gold nucleus must be less than this value. For silver atoms, the distance of closest approach was found to be 2×10^{-14} m. From these results, Rutherford concluded that the positive charge in an atom is concentrated in a small sphere, which he called the nucleus, whose radius is no greater than 10^{-14} m.

Since the time of Rutherford's scattering experiments, a multitude of other experiments have shown that most nuclei are approximately spherical and have an average radius given by

$$r = r_0 A^{1/3} \tag{32.1}$$

where A is the mass number and r_0 is a constant equal to 1.2×10^{-15} m. Because the volume of a sphere is proportional to the cube of its radius, it follows from Equation 32.1 that the volume of a nucleus (assumed to be spherical) is directly proportional to A, the total number of nucleons. This suggests that *all nuclei have nearly the same density*, as shown in the following example.

Example 32.2 The Volume and Density of a Nucleus

Find (a) an approximate expression for the mass of a nucleus of mass number A, (b) an expression for the volume of this nucleus, and (c) an expression for its density.

Solution

(a) The mass of the proton is approximately equal to that of the neutron. Thus, if the mass of one of these particles is m_0, the mass of the nucleus is approximately Am_0.

Figure 32.1
An alpha particle on a head-on collision course with a nucleus of charge Ze. Because of the Coulomb repulsion between the like charges, the alpha particle will stop instantaneously at a distance d from the nucleus.

(b) Assuming the nucleus is spherical and using Equation 32.1, we find that the volume is given by

$$V = \frac{4}{3}\pi r^3 = \frac{4}{3}\pi r_0{}^3 A$$

(c) The nuclear density can be found as follows:

$$\rho_n = \frac{\text{mass}}{\text{volume}} = \frac{Am_0}{\frac{4}{3}\pi r_0{}^3 A} = \frac{3m_0}{4\pi r_0{}^3}$$

Taking $r_0 = 1.2 \times 10^{-15}$ m and $m_0 = 1.67 \times 10^{-27}$ kg, we find that

$$\rho_n = \frac{3(1.67 \times 10^{-27}\text{ kg})}{4\pi(1.2 \times 10^{-15}\text{ m})^3} = 2.3 \times 10^{17}\text{ kg/m}^3$$

Because the density of water is only 10^3 kg/m^3, the nuclear density is about 2.3×10^{14} times greater than the density of water!

___

Nuclear Stability

Since the nucleus consists of a closely packed collection of protons and neutrons, you might be surprised that it can exist, given that like charges (the protons) in close proximity exert very large repulsive electric forces on each other and these forces should cause the nucleus to fly apart. However, nuclei are stable because of the presence of a force that we have not mentioned previously, called the **strong nuclear force**. This force, which is a very short-range one, is an attractive force that acts between all nuclear particles. The protons attract each other via the nuclear force, and at the same time they repel each other through the Coulomb force. However, the attractive nuclear force is much stronger than the force of Coulomb repulsion. The nuclear force also acts between neutrons and neutrons and between neutrons and protons.

There are about 400 stable nuclei; hundreds of others have been observed, but they are unstable. (A stable nucleus is one that remains as a unit essentially forever. An unstable nucleus eventually breaks apart by radioactive decay as we shall discuss in a later section.) A plot of N versus Z for a number of stable nuclei is given in Figure 32.2. Note that light nuclei are most stable if they contain an equal number of protons and neutrons, that is, if $N = Z$. For example, the helium nucleus (two protons and two neutrons) is very stable. Also note that heavy nuclei are more stable if the number of neutrons exceeds the number of protons. This can be understood by noting that, as the number of protons increases, the strength of the Coulomb force increases, which tends to break the nucleus apart. As a result, more neutrons are needed to keep the nucleus stable since neutrons experience only the attractive nuclear forces. Eventually, the repulsive forces between protons cannot be compensated by the addition of more neutrons. This occurs when $Z = 83$. Elements that contain more than 83 protons do not have stable nuclei.

Figure 32.2
A plot of the neutron number N versus the proton number Z for the stable nuclei. The dashed straight line corresponding the condition $N = Z$ is called the *line of stability*.

32.2   Binding Energy

The total mass of a nucleus is always less than the sum of the masses of its individual nucleons. According to the Einstein mass-energy relationship, if the mass difference, Δm, is multiplied by c^2, we obtain the binding energy of the nucleus. In other words, *the total energy of the bound system (the nucleus) is less than the total energy of the separated nucleons.* Therefore, in order to separate a nucleus into protons and neutrons, energy must be delivered to the system.

Example 32.3 The Binding Energy of the Deuteron

Calculate the binding energy of the deuteron, which consists of a proton and a neutron, given that the mass of the deuteron is 2.014102 u.

Solution
We know that the proton and neutron masses are

$$m_p = 1.007825 \text{ u}$$

$$m_n = 1.008665 \text{ u}$$

Note that the masses used for the proton and deuteron in this example are actually those of the neutral atoms, as found in Appendix B. We are able to use atomic masses for these calculations since the electron masses cancel. Therefore,

$$m_p + m_n = 2.016490 \text{ u}$$

To calculate the mass difference, we subtract the deuteron mass from this value:

$$\Delta m = (m_p + m_n) - m_d$$
$$= 2.016490 \text{ u} - 2.014102 \text{ u}$$
$$= 0.002388 \text{ u}$$

Since 1 u corresponds to an equivalent energy of 931.50 MeV (that is, $1 \text{ u} \cdot c^2 = 931.50 \text{ MeV}$), the mass difference corresponds to the following binding energy:

$$E_b = (0.002388 \text{ u})(931.5 \text{ MeV/u}) = \boxed{2.224 \text{ MeV}}$$

This result tells us that, in order to separate a deuteron into its constituent parts (a proton and a neutron), it is necessary to add 2.224 MeV of energy to the deuteron. One way of supplying the deuteron with this energy is by bombarding it with energetic particles.

If the binding energy of a nucleus were zero, the nucleus would separate into its constituent protons and neutrons without the addition of any energy, that is, it would spontaneously break apart.

It is interesting to examine a plot of the binding energy per nucleon, E_b/A, as a function of mass number for various stable nuclei (Fig. 32.3). Note that except for the lighter nuclei, the average binding energy per nucleon is about 8 MeV. For the deuteron, the average binding energy per nucleon is $E_b/A = 2.224/2 \text{ MeV} = 1.112 \text{ MeV}$. Note that the curve in Figure 32.3 peaks in the vicinity of $A = 60$. That is, nuclei with mass numbers greater or less than 60 are not as strongly bound as those near the middle of the periodic table. As we shall see later, this fact allows energy to be released in fission and fusion reactions.

Note from Figure 32.3 that the average binding energy per nucleon is approximately constant for values of A greater than about 40. As we move to the right on the curve, we find that the number of nucleons in the nuclei increases. Thus, we might expect that an individual nucleon contained in a heavy nucleus would be tightly bound because there are many surrounding

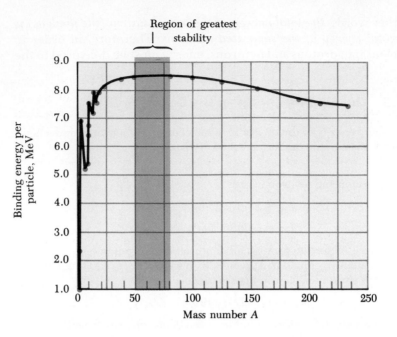

Figure 32.3
A plot of the binding energy per nucleon versus the mass number A for nuclei that are along the line of stability shown in Figure 32.2.

nucleons to hold it in place. Since the binding energy per nucleon can be considered as the energy needed to break these bonds, we might expect the curve in Figure 32.3 to continue to rise rather than level off. The fact that the curve does not continue to rise leads to the idea that the nuclear forces are *saturated*. This, in turn, suggests that an individual nucleon does not interact equally with all other nucleons in the nucleus but is influenced only by nearby nucleons.

32.3   Nuclear Models

Although the detailed nature of nuclear forces is still not well understood, several nuclear models have been proposed, and these are useful in understanding some experimental features concerning the nucleus. The models we shall discuss are (1) the liquid-drop model, (2) the independent-particle model, and (3) the collective model.

Liquid-Drop Model

The **liquid-drop model** of the nucleus, proposed by Bohr in 1936, treats the nucleons as if they were molecules in a drop of liquid. The nucleons interact strongly with each other and undergo frequent collisions as they "jiggle around" within the nucleus. This is analogous to the thermally agitated motion of molecules in a liquid.

The primary reason for the success of the liquid-drop model is that it accounts very well for the nuclear binding energy. As a first approximation we may write an equation for the binding energy as

$$E_b = C_1 A - C_2 A^{2/3} - C_3 \frac{Z(Z-1)}{A^{1/3}} \qquad (32.2)$$

where the C_1, C_2, and C_3 are proportionality constants. Let us discuss the significance of each term in this equation.

1. **The volume effect:** The first term, C_1A, is that part of the binding energy arising from the attraction between nucleons. As we showed earlier, the binding energy per nucleon is approximately constant. In order for the binding energy divided by A to be a constant, the binding energy must be proportional to A. This term is called the volume effect because the volume of the nucleus is also proportional to A.

2. **The surface effect:** Since many of the nucleons will be located on the surface of the drop, they will have fewer neighbors than those in the interior of the drop. Hence, these surface nucleons will reduce the binding energy by an amount proportional to the surface area of the drop, which varies as the square of the nuclear radius or as $A^{2/3}$.

3. **The Coulomb repulsion effect:** Each proton repels every other proton in the nucleus. The corresponding potential energy per pair of interacting particles is given by ke^2/r, where k is the Coulomb constant. The total Coulomb energy represents the work required to assemble Z protons from infinity to a sphere of volume V. The energy is proportional to the number of proton pairs $Z(Z-1)$ and is inversely proportional to the nuclear radius, which varies as $A^{1/3}$.

In a more complete equation, another term has to be added to the binding energy formula, a term which is significant only for heavy nuclei with a large excess of neutrons. We shall not consider this term here.

The three constants in Equation 32.2 can be adjusted to fit this expression to experimental data, with reasonable results. For nuclei with $A > 15$, the constants have the values

$$C_1 = 15.7 \text{ MeV} \qquad C_2 = 17.8 \text{ MeV} \qquad C_3 = 0.71 \text{ MeV}$$

Equation 32.2 describes the binding energy of the nucleus very well. Furthermore, it provides a qualitative description of the important process of nuclear fission to be discussed in Chapter 33. However, the liquid-drop model cannot account for such features as the existence of quantized energy levels in the nucleus. Such difficulties have led to the development of other nuclear models.

The Independent-Particle Model

We have mentioned some of the limitations of the liquid-drop model and its failure to explain some features of nuclear structure. In the **independent-particle model,** often called the *shell model,* the nuclear structure is viewed in terms of orbits and shells, in analogy with the electronic structure of the atom. In this model, the nucleons exist in well-defined orbits with quantized energy states. The quantized states occupied by the nucleons can be described by a set of quantum numbers. Like the electron, both the proton and the neutron have spin $\frac{1}{2}$ and obey the Pauli exclusion principle. As a result, each nuclear state can contain only two protons (or two neutrons) with opposite spin as shown in Figure 32.4. Because the protons are charged, the added Coulomb repulsion causes their energy levels to be somewhat higher than those of the neutrons.

On the basis of this model, it is easy to understand nuclear excited states in terms of nucleons moving from lower energy levels to higher levels. Also, it is

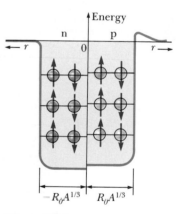

Figure 32.4
A nucleus containing 12 nucleons. The gray spheres represent protons, and the colored spheres represent neutrons. The energy levels for the protons are slightly higher than those for the neutrons because of the Coulomb potential of the protons. The difference in the levels increases as Z increases. Note that only two nucleons with opposite spin can occupy a given level, as required by the Pauli exclusion principle.

**Maria Goeppert-
Mayer**
(1906–1972)

*(Courtesy of Louise Barker/AIP
Niels Bohr Library)*

Maria Goeppert-Mayer was born in Germany and received a PhD in physics from Göttingen University. Her thesis work, which dealt with quantum mechanical effects in atoms, was encouraged by Paul Ehrenfest. She moved to the United States in 1930 after her husband received a professorship in chemistry at Johns Hopkins University. While at home raising two children, she and her husband wrote a book on statistical mechanics. Following the publication of the book, she was offered a lectureship in chemistry at Johns Hopkins University, but her presence as a woman at faculty functions was awkward.

In the late 1940s, she and her husband received appointments at the University of Chicago, but her position was without pay because of a strict nepotism rule. While in Chicago, she worked with Enrico Fermi and Edward Teller. During her collaboration with Teller, she became interested in understanding why certain elements in the periodic table were so abundant and stable. She eventually realized that the most stable elements were characterized by particular values of atomic and neutron numbers, which she called "magic numbers." She labored over a theoretical explanation of these numbers for about one year, and finally arrived at a solution during a conversation with Fermi. This resulted in a 1950 publication in which she described the "shell" model of the nucleus. (It is interesting to note that the paper confused many Russian scientists who translated "shell" as "grenade.") As often happens in scientific research, a similar model was simultaneously developed by Hans Jensen, a German scientist. Maria Goeppert-Mayer and Hans Jensen were awarded the Nobel prize in physics in 1963 for their extraordinary work in understanding the structure of the nucleus. When Goeppert-Mayer heard that she had won the Nobel prize, she said "Oh, how wonderful, I've always wanted to meet a king."

possible to understand why nuclei containing an even number of protons and neutrons are more stable than others. In fact, there are 160 such stable isotopes (called even-even nuclei) and only 4 stable isotopes with an odd number of both protons and neutrons. Any particular state is filled when it contains two protons (or two neutrons) with opposite spins. *An extra proton or neutron can be added to the nucleus only at the expense of increasing the energy of the nucleus.* This increase in energy leads to a nucleus that is less stable than the original nucleus.

Certain values of Z and N correspond to nuclei with unusually high stability. These values of Z and N, called **magic numbers,** are given by

$$Z \text{ or } N = 2, 8, 20, 28, 50, 82, 126$$

For example, the alpha particle (two protons and two neutrons), with $Z = 2$ and $N = 2$ is very stable.

The Collective Model

A third model of nuclear structure, known as the **collective model,** combines some features of both the liquid-drop model and the independent-particle model. The nucleus is considered to have some "extra" nucleons moving in quantized orbits, in addition to the filled core of nucleons. Deformations can be set up in the core as a result of a strong interaction between the core and the extra nucleons, thereby setting up vibrational and rotational motions as in the liquid-drop model. This collective model has been very successful in explaining many nuclear phenomena.

32.4   Radioactivity

The study of nuclear physics began when Becquerel discovered radioactivity in 1896, one year after the discovery of x-radiation. Becquerel found, quite by accident, that an ore containing uranium emits an invisible radiation that can penetrate a black paper wrapper and expose a photographic plate. Following several observations of this type under controlled conditions, he concluded that the radiation emitted by the ore was of a new type and occurred spontaneously. This process of spontaneous emission of radiation by uranium was soon called **radioactivity.** Subsequent experiments by other scientists showed that other substances were also radioactive. The most significant investigations of this type were conducted by Marie Curie and her husband, Pierre. After several years of careful and laborious chemical separation processes on tons of pitchblende, a radioactive ore, the Curies reported the discovery of two previously unknown elements, both of which were radioactive. These were named polonium and radium. Subsequent experiments, including Rutherford's famous work on alpha particle scattering, suggested that radioactivity was the result of the decay, or disintegration, of unstable nuclei.

There are three types of radiation that can be emitted by a radioactive substance: alpha (α) decay, in which the emitted particles are helium nuclei; beta (β) decay, in which the emitted "rays" are electrons; and gamma (γ) decay, in which the emitted "rays" are high-energy photons.

It is possible to distinguish these three forms of radiation by using the scheme described in Figure 32.5. The radiation from a radioactive sample is directed into a region in which there is a magnetic field, and the beam splits into three components, two bending in opposite directions and the third not changing direction. From this simple observation, it can be concluded that the radiation of the undeflected beam carries no charge (the gamma ray), the component deflected upward contains positively charged particles (alpha particles), and the component deflected downward contains negatively charged particles (electrons).

The three types of radiation have quite different penetrating powers. Alpha particles barely penetrate a sheet of paper, beta particles can penetrate a few millimeters of aluminum, and gamma rays can penetrate several centimeters of lead.

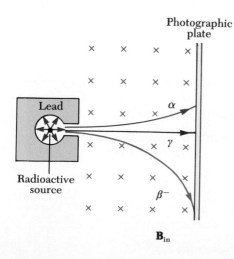

Figure 32.5
The radiation from a radioactive source, such as radium, can be separated into three components using a magnetic field to deflect the charged particles. The photographic plate at the right records the events.

Marie Curie
(1867–1934)

Marie Sklodowska Curie was born in Poland shortly after the unsuccessful Polish revolt against the Russians in 1863. Following her high school education, she worked diligently to help meet the educational expenses of her older brother and sister who had left for Paris. At the same time, she managed to save enough money for her own trip to Paris and entered the Sorbonne in 1891. Although she lived under very frugal conditions during this period (fainting once from hunger in the classroom), she graduated at the top of her class.

In 1895 she married the French chemist Pierre Curie, who was already known for the discovery of piezoelectricity. (A piezoelectric crystal exhibits a potential difference when it is under pressure.) Using piezoelectric materials to measure the activity of radioactive substances, she demonstrated the radioactive nature of the elements uranium and thorium. In 1898, she and her husband discovered a new radioactive element contained in uranium ore, which they called polonium, named after Madame Curie's native land. By the end of 1898, the Curies succeeded in isolating trace amounts of an even more radioactive element which they named radium. In an effort to produce weighable quantities of radium, they embarked on a painstaking effort of isolating radium from pitchblende, an ore rich in uranium. After four years of purifying and repurifying tons of ore, and using their own life savings to finance their work, the Curies succeeded in preparing about 0.1 g of radium. In 1903, Marie and Pierre Curie were awarded the Nobel prize in physics, which they shared with A. H. Becquerel, for their studies of radioactive substances.

After her husband's death in a tragic accident in 1906, Madame Curie took over his professorship at the Sorbonne. Unfortunately, she experienced prejudice in the scientific community because she was a woman. For example, after being nominated to the French Academy of Sciences, she was refused membership, losing by one vote.

In 1911, she was awarded a second Nobel prize in chemistry for the discovery of radium and polonium. She spent the last few decades of her life supervising the Paris Institute of Radium. Madame Curie died of leukemia caused by the years of exposure to radiation from radioactive substances.

The Decay Constant and Half-Life

If a radioactive sample contains N radioactive nuclei at some instant, it is found that the number of nuclei, ΔN, that decay in a small time interval Δt is proportional to N:

$$\Delta N = -\lambda N \Delta t \qquad \textbf{(32.3)}$$

where λ is a constant called the **decay constant** and the negative sign signifies that N decreases with time, that is, ΔN is negative. The value of λ for any isotope determines the rate at which that isotope will decay.

The **decay rate,** or **activity,** R, of a sample is defined as the number of decays per second. From Equation 32.3, we see that the decay rate is

$$R = -\frac{\Delta N}{\Delta t} = \lambda N \qquad \textbf{(32.4)}$$

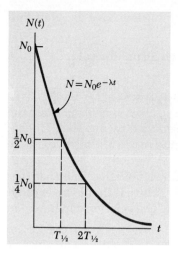

Figure 32.6
Plot of the exponential decay law for radioactive nuclei. The vertical axis represents the number of radioactive nuclei present at any time t, and the horizontal axis is time. The parameter $T_{1/2}$ is the half-life of the sample.

Thus we see that isotopes with a large value of λ decay at a rapid rate and those with a small λ value decay slowly.

A general decay curve for a radioactive sample is shown in Figure 32.6. It can be shown from Equation 32.4 (using calculus) that the number of nuclei present varies with time according to the expression

$$N = N_0 e^{-\lambda t} \tag{32.5}$$

where N is the number of radioactive nuclei present at time t, N_0 is the number present at time $t = 0$, and $e = 2.718 \ldots$ is the base of the natural logarithm. Processes that obey Equation 32.5 are sometimes said to undergo exponential decay.[1]

The unit of activity is the **curie** (Ci), defined as

$$1 \text{ Ci} \equiv 3.7 \times 10^{10} \text{ decays/s} \tag{32.6}$$

This number of decay events per second was selected as the original activity unit because it is the approximate activity of 1 g of radium. The SI unit of activity is the **becquerel** (Bq):

$$1 \text{ Bq} = 1 \text{ decay/s} \tag{32.7}$$

Therefore, $1 \text{ Ci} = 3.7 \times 10^{10}$ Bq. The most commonly used units of activity are the millicurie (10^{-3} Ci) and the microcurie (10^{-6} Ci).

One of the most common terms encountered in any discussion of radioactive materials is **half-life**, $T_{1/2}$. This is because all radioactive substances follow the same general decay pattern. After a certain interval of time, half of the original number of nuclei in a sample will have decayed; then in a second time

[1] Other examples of exponential decays were discussed in Chapter 19 in connection with *RC* circuits, and in Chapter 21 in connection with *RL* circuits.

interval equal to the first, half of those nuclei remaining will have decayed, and so on.

> The **half-life** is the time required for half of a given number of radioactive nuclei to decay.

Let us assume that the number of radioactive nuclei present at $t = 0$ is N_0, as in Figure 32.6. The number left after one half-life, $T_{1/2}$, passes is $N_0/2$. After the second half-life, the number remaining is again reduced by one half. Hence, after a time $2T_{1/2}$, the number remaining is $N_0/4$, and so forth. Half-lives range from about 10^{-22} s to 10^{21} years.

As noted earlier, the decay constant is large for radioactive materials that decay rapidly and small for those that decay slowly. Thus, it should be evident that the half-life and the decay constant are related. The relationship between them is[2]

Half-life

$$T_{1/2} = \frac{0.693}{\lambda} \qquad \text{(32.8)}$$

Example 32.4   How Many Nuclei Are Left?

Carbon-14, $^{14}_{6}C$, is a radioactive isotope of carbon that has a half-life of 5730 years. If you start with a sample of 1000 carbon-14 nuclei, how many will still be around in 22,920 years?

Solution

In 5730 years, half the sample will have decayed, leaving 500 carbon-14 nuclei remaining. In another 5730 years (for a total elapsed time of 11,460 years), the number will be reduced to 250 nuclei. After another 5730 years (total time 17,190 years), 125 remain. Finally, after four half-lives (22,920 years), only about 62 remain.

It should be noted here that these numbers represent ideal circumstances. Radioactive decay is an averaging process over a very large number of atoms, and the actual outcome depends on statistics. Our original sample in this example contained

---

[2] Equation 32.8 can be derived by starting with Equation 32.5, $N = N_0 e^{-\lambda t}$. After a time interval equal to one half-life, $t = T_{1/2}$, the number of radioactive nuclei remaining is $N = N_0/2$. Therefore,

$$\frac{N_0}{2} = N_0 e^{-\lambda T_{1/2}}$$

Dividing both sides by N_0 gives

$$\frac{1}{2} = e^{-\lambda T_{1/2}}$$

Taking the natural logarithm of both sides of this equation eliminates the exponential factor on the right, since $\ln e = 1$:

$$\ln \frac{1}{2} = -\lambda T_{1/2}$$

Since $\ln \frac{1}{2} = -0.693$ (use your calculator to check this), we have

$$-0.693 = -\lambda T_{1/2}$$

$$T_{1/2} = \frac{0.693}{\lambda}$$

only 1000 nuclei, certainly not a very large number. Thus, if we were actually to count the number remaining after one half-life for this small sample, it probably would not be 500. If the initial number is very large, the probability of getting extremely close to the predicted number remaining after one half-life increases greatly.

Example 32.5 The Activity of Radium

The half-life of the radioactive nucleus $^{226}_{88}\text{Ra}$ is 1.6×10^3 years. If a sample contains 3×10^{16} such nuclei, determine the activity.

Solution

First, let us calculate the decay constant, λ, using Equation 32.8 and the fact that

$$T_{1/2} = 1.6 \times 10^3 \text{ years} = (1.6 \times 10^3 \text{ years})(3.15 \times 10^7 \text{ s/year})$$
$$= 5.0 \times 10^{10} \text{ s}$$

Therefore,

$$\lambda = \frac{0.693}{T_{1/2}} = \frac{0.693}{5.0 \times 10^{10} \text{ s}} = 1.4 \times 10^{-11} \text{ s}^{-1}$$

We can calculate the activity of the sample at $t = 0$ using Equation 32.4 in the form $R_0 = \lambda N_0$, where R_0 is the decay rate at $t = 0$ and N_0 is the number of radioactive nuclei present at $t = 0$. Since $N_0 = 3 \times 10^{16}$, we have

$$R_0 = \lambda N_0 = (1.4 \times 10^{-11} \text{ s}^{-1})(3 \times 10^{16}) = 4.2 \times 10^5 \text{ decays/s}$$

Since 1 Ci $= 3.7 \times 10^{10}$ decays/s, the activity, or decay rate, is

$$R_0 = \boxed{11.3 \ \mu\text{Ci}}$$

32.5 The Decay Processes

As we stated in the previous section, a radioactive nucleus spontaneously decays via three processes: alpha decay, beta decay, and gamma decay. Let us discuss these in more detail.

Alpha Decay

If a nucleus emits an alpha particle (^4_2He), it loses two protons and two neutrons. Let us adopt the historical terminology of calling the nucleus before decay the **parent nucleus** and the nucleus remaining after decay the **daughter nucleus**. The nucleus $^{238}_{92}\text{U}$ decays by alpha emission, and it is easy to visualize the process in symbolic form as follows:

$$^{238}_{92}\text{U} \longrightarrow \ ^{234}_{90}\text{Th} + \ ^4_2\text{He} \qquad\qquad \textbf{(32.9)}$$

This says that a parent nucleus, $^{238}_{92}\text{U}$, emits an alpha particle, ^4_2He and thereby changes to a daughter nucleus, $^{234}_{90}\text{Th}$. Note the following facts about this reaction: (1) the atomic number (number of protons) on the left is the same as on the right ($92 = 90 + 2$) since charge must be conserved, and (2) the mass number (protons plus neutrons) on the left is the same as on the right ($238 = 234 + 4$). The half-life for ^{238}U decay is 4.47×10^9 years.

When one element changes into another, as in alpha decay, the process is called **transmutation.** In order for alpha emission to occur, the mass of the parent must be greater than the combined mass of the daughter and the alpha particle. In the decay process, this excess mass is converted into energy and appears in the form of kinetic energy in the daughter nucleus and the alpha particle. Most of the kinetic energy is carried away by the alpha particle because it is much less massive than the daughter nucleus. That is, because momentum must be conserved in the decay process, *the lighter alpha particle recoils with a much higher velocity than the daughter nucleus.* Generally, light particles carry off most of the energy in nuclear decays.

Example 32.6   The Alpha Decay of Radium

Radium, $^{226}_{88}$Ra, decays by alpha emission. What is the daughter element formed?

Solution
The decay can be written symbolically as

$$^{226}_{88}\text{Ra} \longrightarrow \text{X} + {^4_2}\text{He}$$

where X is the unknown daughter element. Requiring that the mass numbers and atomic numbers balance on the two sides of the arrow, we find that the daughter nucleus must have a mass number of 222 and an atomic number of 86:

$$^{226}_{88}\text{Ra} \longrightarrow {^{222}_{86}}\text{X} + {^4_2}\text{He}$$

The periodic table shows that the nucleus with an atomic number of 86 is radon, Rn. Thus, the process is

$$^{226}_{88}\text{Ra} \longrightarrow {^{222}_{86}}\text{Rn} + {^4_2}\text{He}$$

This decay is shown schematically in Figure 32.7.

Parent   Daughter α particle

$^{226}_{88}$Ra $^{222}_{86}$Rn $^4_2$He

Figure 32.7 (Example 32.6) Alpha decay of the $^{226}_{88}$Ra nucleus.

Example 32.7   The Energy Liberated When Radium Decays

In Example 32.6, we showed that the $^{226}_{88}$Ra nucleus undergoes alpha decay to $^{222}_{86}$Rn. Calculate the amount of energy liberated in this decay. Take the mass of $^{226}_{88}$Ra to be 226.025406 u, that of $^{222}_{86}$Rn to be 222.017574 u, and that of $^4_2$He to be 4.002603 u, as found in Appendix B.

Solution
After decay, the mass of the daughter, m_d, plus the mass of the alpha particle, m_α, is

$$m_d + m_\alpha = 222.017574 \text{ u} + 4.002603 \text{ u} = 226.020177 \text{ u}$$

Thus, calling the mass of the parent nucleus M_p, we find that the mass lost during decay is

$$\Delta m = M_p - (m_d + m_\alpha) = 226.025406 \text{ u} - 226.020177 \text{ u} = 0.005229 \text{ u}$$

Using the relationship 1 u = 931.5 MeV, we find that the energy liberated is

$$E = (0.005229 \text{ u})(931.50 \text{ MeV/u}) = \boxed{4.87 \text{ MeV}}$$

Beta Decay

When a radioactive nucleus undergoes beta decay, *the daughter nucleus contains the same number of nucleons as the parent nucleus but the atomic number is increased by 1.* A typical beta decay event is

$$^{14}_{6}\text{C} \longrightarrow \, ^{14}_{7}\text{N} + \, ^{0}_{-1}\text{e} \tag{32.10}$$

The superscripts and subscripts on the carbon and nitrogen nuclei follow our usual conventions, but those on the electron may need some explanation. The -1 indicates that the electron has a charge whose magnitude is equal to that of the proton but negative. The 0 used for the electron's mass number indicates that the mass of the electron is almost zero relative to that of carbon and nitrogen nuclei.

The emission of electrons from a *nucleus* is surprising because, in all our previous discussions, we stated that the nucleus is composed of protons and neutrons only. This apparent discrepancy can be explained by noting that the electron that is emitted is created in the nucleus by a process in which a neutron is transformed into a proton. This can be represented by the equation

$$^{1}_{0}\text{n} \longrightarrow \, ^{1}_{1}\text{p} + \, ^{0}_{-1}\text{e} \tag{32.11}$$

Let us consider the energy of the system of Equation 32.10 before and after decay. As with alpha decay, energy must be conserved in beta decay. The following example illustrates how to calculate how much energy is released in the beta decay of $^{14}_{6}\text{C}$.

Example 32.8 The Beta Decay of $^{14}_{6}\text{C}$

Find the energy liberated in the beta decay of $^{14}_{6}\text{C}$ to $^{14}_{7}\text{N}$ as represented by Equation 32.10. Refer to Appendix B for the atomic masses.

Solution
We find from Appendix B that $^{14}_{6}\text{C}$ has a mass of 14.003242 u and $^{14}_{7}\text{N}$ has a mass of 14.003074 u. Here, the mass difference between the initial and final states is[3]

$$\Delta m = 14.003242 \text{ u} - 14.003074 \text{ u} = 0.000168 \text{ u}$$

This corresponds to an energy release of

$$E = (0.000168 \text{ u})(931.50 \text{ MeV/u}) = \quad 0.156 \text{ MeV}$$

From Example 32.8, we see that the energy released in the beta decay of ^{14}C is approximately 0.16 MeV. As with alpha decay, we expect the electron to carry away virtually all of this as kinetic energy because apparently it is the lightest particle produced in the decay. However, as Figure 32.8 shows, only a small number of electrons have this maximum kinetic energy, represented as KE_{max} on the graph; most of the electrons emitted have kinetic energies less

Figure 32.8
A typical beta-decay spectrum.

---

[3] In beta decay, we must keep track of the electrons involved in the process. In the initial state, the neutral ^{14}C atom has six electrons. In the final state, the ^{14}N atom also has six electrons. Thus, we must conclude that ^{14}N is not neutral. However, if we include the electron emitted in the decay process, we effectively have a mass equivalent to a neutral ^{14}N atom.

Enrico Fermi
(1901 – 1954)

Enrico Fermi, an Italian-American physicist, received his doctorate from the University of Pisa in 1922 and did postdoctorate work in Germany under Max Born. He returned to Italy in 1924 and became a professor of physics at the University of Rome in 1926. He received the Nobel Prize for physics in 1938 for his work dealing with the production of transuranic radioactive elements by neutron bombardment.

Fermi first became interested in physics at the age of 14 after reading an old physics book in Latin. He had an excellent scholastic record and was able to recite Dante's *Divine Comedy* and much of Aristotle from memory. His great ability to solve problems in theoretical physics and his skill for simplifying very complex situations made him somewhat of an oracle. He was also a gifted experimentalist and teacher of physics. During one of his early lecture trips to the United States in the depths of the Depression, a car which he had purchased became disabled and he pulled into a nearby gas station. After Fermi repaired the car with ease, the station owner immediately offered him a job.

Fermi and his family immigrated to the United States and he became a naturalized citizen in 1944. Once in America, Fermi accepted a position at Columbia University and later became a professor at the University of Chicago. After the Manhattan Project was established, Fermi was commissioned to design and build a structure (called an atomic pile) in which a self-sustained chain reaction might occur. The structure, built in the squash court of the University of Chicago, contained uranium in combination with graphite blocks to slow the neutrons to thermal velocities. Cadmium rods inserted in the "pile" were used to absorb neutrons and control the reaction rate. History was made at 3:45 P.M. on December 2, 1942 as the cadmium rods were slowly withdrawn, and a self-sustained chain reaction was observed. Fermi's earthshaking achievement of the world's first nuclear reactor marked the beginning of the atomic age.

Fermi died of cancer in 1954 at the age of 53. One year later, the one-hundredth element was discovered and named *fermium* in his honor.

than this predicted value. If the daughter nucleus and the electron are not carrying away this liberated energy, then the requirement that energy is conserved leads to the question, "What accounts for the missing energy?" As an additional complication, further analysis of beta decay shows that the principles of conservation of both angular momentum and linear momentum appear to be violated!

In 1930, Pauli proposed that a third particle must be present to carry away the "missing" energy and to conserve momentum. Enrico Fermi later named this particle the **neutrino** (little neutral one) because it had to be electrically neutral and have little or no rest mass. Although it eluded detection for many years, the neutrino (symbol v) was finally detected experimentally in 1950.

1. It has zero electric charge.
2. It has a rest mass smaller than that of the electron, and in fact its rest mass may be zero (although recent experiments suggest that this may not be true).
3. It interacts very weakly with matter and is therefore very difficult to detect.

Properties of the neutrino

Thus, with the introduction of the neutrino, we are now able to represent the beta decay process of Equation 31.10 in its correct form:

$$^{14}_{6}C \longrightarrow ^{14}_{7}N + ^{0}_{-1}e + \bar{\nu} \tag{32.12}$$

where the bar in the symbol $\bar{\nu}$ indicates that this is an **antineutrino.** To explain what an antineutrino is, let us first consider the following decay.

$$^{12}_{7}N \longrightarrow ^{12}_{6}C + ^{0}_{1}e + \nu \tag{32.13}$$

Here we see that when ^{12}N decays into ^{12}C a particle is produced that is identical to the electron except that it has a positive charge of $+e$. This particle is called a positron. Because it is like the electron in all respects except charge, the positron is said to be **antiparticle** to the electron. We shall discuss antiparticles further in Chapter 34. For now, it suffices to say that *a neutrino is emitted in positron decay and an antineutrino is emitted in electron decay.*

Practical Uses of Radioactivity

Carbon Dating   The beta decay of ^{14}C given by Equation 32.12 is commonly used to date organic samples. Cosmic rays (high-energy particles from outer space) in the upper atmosphere cause nuclear reactions that create ^{14}C from ^{14}N. In fact, the ratio of ^{14}C to ^{12}C isotopic abundance in the carbon dioxide molecules of our atmosphere has a constant value of about 1.3×10^{-12} as determined by measuring carbon ratios in tree rings. All living organisms have the same ratio of ^{14}C to ^{12}C because they continuously exchange carbon dioxide with their surroundings. When an organism dies, however, it no longer absorbs ^{14}C from the atmosphere, and so the ratio of ^{14}C to ^{12}C decreases as the result of the beta decay of ^{14}C. It is therefore possible to measure the age of a material by measuring its activity per unit mass due to the decay of ^{14}C. Using carbon dating, samples of wood, charcoal, bone, and shell have been identified as having lived from 1000 to 25,000 years ago. This knowledge has helped us to reconstruct the history of living organisms—including humans—during this time span.

A particularly interesting example is the dating of the Dead Sea Scrolls. This group of manuscripts was first discovered by a shepherd in 1947. Translation showed them to be religious documents, including most of the books of the Old Testament. Because of their historical and religious significance, scholars wanted to know their age. Carbon dating applied to fragments of the scrolls and to the material in which they were wrapped established their age at about 1950 years.

Smoke Detectors   Smoke detectors are frequently used in homes and industry for fire protection. Most common devices are ionization-type smoke detectors which use radioactive materials (see Fig. 32.9). The detector consists of an ionization chamber, a sensitive current detector, and an alarm. A weak radioactive source ionizes the air in the chamber of the detector, which creates charged particles. A voltage is maintained between the plates inside the chamber, setting up a small but detectable current in the external circuit. As long as the current is maintained, the alarm is deactivated. However, if smoke drifts into the chamber of the detector, the ions become attached to the

Figure 32.9
An ionization-type smoke detector. Smoke entering the chamber reduces the detected current, causing the alarm to sound.

smoke particles. These heavier particles do not drift as readily as do the lighter ions, which causes a decrease in the detector current. The external circuit senses this decrease in current and sets off the alarm.

Radon Detecting Radioactivity can also affect our daily lives in harmful ways. Soon after the discovery of radium by the Curies, it was found that the air in contact with radium compounds becomes radioactive. It was shown that this radioactivity came from the radium itself, and the product was therefore called "radium emanation." Rutherford and Soddy succeeded in condensing this "emanation," confirming that it is a real substance — the inert, gaseous element now called radon, Rn. We now know that the air in uranium mines is radioactive because of the presence of radon gas. The mines must therefore be well ventilated to help protect the miners. The fear of radon pollution has now moved from uranium mines into our own homes. Since certain types of rock, soil, brick, and concrete contain small quantities of radium, some of the resulting radon gas finds its way into our homes and other buildings. The most serious problems arise from leakage of radon from the ground into the structure. One practical remedy is to exhaust the air through a pipe just above the underlying soil or gravel directly to the outdoors by means of a small fan or blower.

Example 32.9   Should We Report This to Homicide?

A 50-g sample of carbon is taken from the pelvis bone of a skeleton and is found to have a carbon-14 decay rate of 200 decays/min. It is known that carbon from a living organism has a decay rate of 15 decays/min·g and that ^{14}C has a half-life of 5730 y = 3.01×10^9 min. Find the age of the skeleton.

Solution
Let us start with Equation 32.5

$$N = N_0 e^{-\lambda t}$$

and multiply both sides by λ as

$$\lambda N = \lambda N_0 e^{-\lambda t}$$

But, from Equation 32.4, we see that this is the same as

$$R = R_0 e^{-\lambda t}$$

where R is the present activity, and R_0 was the activity when the skeleton was a part of a living organism. We are given that $R = 200$ decays/min, and we can find R_0 as

$$R_0 = \left(15 \frac{\text{decays}}{\text{min} \cdot \text{g}} \right) (50 \text{ g}) = 750 \frac{\text{decays}}{\text{min}}$$

The decay constant is found from Equation 32.8 as

$$\lambda = \frac{0.693}{T_{1/2}} = \frac{0.693}{3.01 \times 10^9 \text{ min}} = 2.30 \times 10^{-10} \text{ min}^{-1}$$

Thus, we make the following substitutions.

$$R = R_0 e^{-\lambda t}$$

$$200 \frac{\text{decays}}{\text{min}} = \left(750 \frac{\text{decays}}{\text{min}} \right) e^{-(2.30 \times 10^{-10} \text{ min}^{-1})t}$$

or

$$0.266 = e^{-(2.30 \times 10^{-10}\,\text{min}^{-1})t}$$

Now, we take the natural log of both sides of the equation, to give

$$\ln(0.266) = -(2.30 \times 10^{-10}\,\text{min}^{-1})t$$

$$-1.32 = -(2.30 \times 10^{-10}\,\text{min}^{-1})t$$

$$t = 5.74 \times 10^9\,\text{min} = \boxed{10,930\text{ y}}$$

Gamma Decay

Very often a nucleus that undergoes radioactive decay is left in an excited energy state. The nucleus can then undergo a second decay to a lower energy state, perhaps to the ground state, by emitting one or more photons. The process is very similar to the emission of light by an atom. An atom emits radiation to release some extra energy when an electron "jumps" from a state of high energy to a state of lower energy. Likewise, the nucleus uses essentially the same method to release any extra energy it may have following a decay or some other nuclear event. In nuclear de-excitation, the "jumps" that release energy are made by protons or neutrons in the nucleus as they move from a higher energy level to a lower level (see Section 32.3). The photons emitted in such a de-excitation process are called **gamma rays**, which have very high energy relative to the energy of visible light.

A nucleus may reach an excited state as the result of a violent collision with another particle. However, it is more common for a nucleus to be in an excited state as a result of alpha or beta decay. The following sequence of events represents a typical situation in which gamma decay occurs:

$$^{12}_{5}\text{B} \longrightarrow {}^{12}_{6}\text{C}^* + {}^{0}_{-1}\text{e} \qquad \textbf{(32.14a)}$$

$$^{12}_{6}\text{C}^* \longrightarrow {}^{12}_{6}\text{C} + \gamma \qquad \textbf{(32.14b)}$$

Equation 32.14a represents a beta decay in which ^{12}B decays to $^{12}\text{C}^*$, where the asterisk is used to indicate that the carbon nucleus is left in an excited state following the decay. The excited carbon nucleus then decays to the ground state by emitting a gamma ray, as indicated by Equation 32.14b. Note that gamma emission does not result in any change in either Z or A.

▶ SPECIAL TOPIC
Carbon-14 and the Shroud of Turin

Since the Middle Ages, many people have marveled at a 14-foot-long, yellowing piece of linen found in Turin, Italy, which is purported to be the burial shroud of Jesus Christ. The cloth bears a remarkable, full-size likeness of a crucified body, with wounds that could have been caused by a crown of thorns and a wound in the side that could have been the cause of death. The skepticism over the authenticity of the shroud has existed since its first public showing in 1354; in fact, a French bishop declared it to be a fraud at the time. Because of its controversial nature, religious bodies have taken a neutral stance on its authenticity.

The Shroud of Turin as it appears in a photographic negative image. *(The IMAGE Bank)*

In 1978, the bishop of Turin allowed the cloth to be subjected to scientific analysis, but notably missing from these tests was a carbon-14 dating. The reason for this omission was that, at the time, carbon dating techniques required a piece of cloth about the size of a handkerchief. In 1988, the process had been refined to the point that pieces as small as one inch on a side would be sufficient, and permission was granted to allow the dating to proceed. Three labs were selected for the testing, and each was given four pieces of material. One of these was a piece of the shroud while the other three were control pieces similar in appearance to the shroud.

The testing procedure consisted of burning the cloth to produce carbon dioxide, which was then converted chemically to graphite, and the graphite sample was subjected to carbon-14 analysis. The end result was that all three labs agreed amazingly well on the age of the shroud. The average of their results gave a date for the cloth of A.D. 1320 ± 60 years, with an absolute assurance that the cloth is no older than A.D. 1200. Thus, carbon-14 dating has unraveled the most important mystery concerning the shroud, but others remain. For example, the investigators have not as yet been able to explain how the image was imprinted. ◀

32.6 Natural Radioactivity

Radioactive nuclei are generally classified into two groups: (1) unstable nuclei found in nature, which give rise to what is called **natural radioactivity,** and (2) nuclei produced in the laboratory through nuclear reactions, which exhibit **artificial radioactivity.**

There are three series of naturally occurring radioactive nuclei (Table 32.2). Each series starts with a specific long-lived radioactive isotope whose half-life exceeds that of any of its descendants. The three natural series begin with the isotopes ^{238}U, ^{235}U, and ^{232}Th, and the corresponding stable end products are three isotopes of lead: ^{206}Pb, ^{207}Pb, and ^{208}Pb. The fourth series in Table 32.2 begins with ^{237}Np and has as its stable end product ^{209}Bi. The element ^{237}Np is a transuranic element (one having an atomic number greater than that of uranium) not found in nature and has a half-life of "only" 2.14×10^6 years.

Figure 32.10 shows the successive decays for the ^{232}Th series. Note that ^{232}Th first undergoes alpha decay to ^{228}Ra. Next, ^{228}Ra undergoes two successive β decays to ^{228}Th. The series continues and finally branches when it reaches ^{212}Bi. At this point, there are two decay possibilities. The end of the decay series is the stable isotope ^{208}Pb.

The two uranium series are somewhat more complex than the ^{232}Th series. Also, there are several naturally occurring radioactive isotopes, such as ^{14}C and ^{40}K, that are not part of either decay series.

Figure 32.10
Decay series beginning with ^{232}Th.

Table 32.2 The Four Radioactive Series			
Series	**Starting Isotope**	**Half-life (years)**	**Stable End Product**
Uranium	$^{238}_{92}$U	4.47×10^9	$^{206}_{82}$Pb
Actinium	$^{235}_{92}$U	7.04×10^8	$^{207}_{82}$Pb
Thorium	$^{232}_{90}$Th	1.41×10^{10}	$^{208}_{82}$Pb
Neptunium	$^{237}_{93}$Np	2.14×10^6	$^{209}_{83}$Bi

It is interesting to note that the existence of radioactive series in nature enables our environment to be constantly replenished with radioactive elements that would otherwise have disappeared long ago. For example, because the Solar System is about 5×10^9 years old, the supply of ^{226}Ra (whose half-life is only 1600 years) would have been depleted by radioactive decay long ago if it were not for the decay series that starts with ^{238}U, whose half-life is 4.47×10^9 years.

32.7 Nuclear Reactions

It is possible to change the structure of nuclei by bombarding them with energetic particles. Such collisions, which change the identity or properties of the target nuclei, are called **nuclear reactions.** Rutherford was the first to observe nuclear reactions in 1919, using naturally occurring radioactive sources for the bombarding particles. He found that protons were released when alpha particles were allowed to collide with nitrogen atoms. The process can be represented symbolically as

$$\ce{^4_2He + ^14_7N \longrightarrow X + ^1_1H} \tag{32.15}$$

This equation says that an alpha particle ($^4_2$He) strikes a nitrogen nucleus and produces an unknown product nucleus (X) and a proton ($^1_1$H). Balancing atomic numbers and mass numbers, as we did for radioactive decay, enables us to conclude that the unknown is characterized as $^{17}_8$X. Since the element whose atomic number is 8 is oxygen, we see that the reaction is

$$\ce{^4_2He + ^14_7N \longrightarrow ^17_8O + ^1_1H} \tag{32.16}$$

This nuclear reaction starts with two stable isotopes, helium and nitrogen, and produces two different stable isotopes, hydrogen and oxygen.

Since the time of Rutherford, thousands of nuclear reactions have been observed, particularly following the development of charged-particle accelerators in the 1930s.

Example 32.10 The Discovery of the Neutron

A nuclear reaction of significant historical note occurred in 1932 when Chadwick, in England, bombarded a beryllium target with alpha particles. Analysis of the experiment indicated that the following reaction occurred:

$$\ce{^4_2He + ^9_4Be \longrightarrow ^12_6C + X}$$

What is X in this reaction?

Solution
Balancing mass numbers and atomic numbers, we see that the unknown particle must be represented as $^1_0$X, that is, with a mass of 1 and zero charge. Hence, the particle X is the neutron, $^1_0$n. This experiment was the first to provide positive proof of the existence of neutrons.

Example 32.11   Synthetic Elements

(a) A beam of neutrons is directed at a target of $^{238}_{92}\text{U}$. The reaction products are a gamma ray and another isotope. What is the isotope? (b) This isotope is radioactive and emits a beta particle. Write the equation symbolizing this decay and identify the resulting isotope. (c) This isotope is also radioactive and decays by beta emission. What is the end product? (d) What is the significance of these reactions?

Solution
(a) Balancing input with output gives

$$^{1}_{0}\text{n} + {}^{238}_{92}\text{U} \longrightarrow {}^{239}_{92}\text{U} + \gamma$$

(b) The decay of ^{239}U by beta emission is

$$^{239}_{92}\text{U} \longrightarrow {}^{239}_{93}\text{Np} + {}^{0}_{-1}\text{e} + \bar{\nu}$$

(c) The decay of $^{239}_{93}\text{Np}$ by beta emission gives

$$^{239}_{93}\text{Np} \longrightarrow {}^{239}_{94}\text{Pu} + {}^{0}_{-1}\text{e} + \bar{\nu}$$

(d) The feature of these reactions that makes them of interest is the fact that uranium is the element with the greatest number of protons, 92, that exists in nature in any appreciable amount. The reactions in Parts (a), (b), and (c) do occur occasionally in nature so that minute traces of neptunium and plutonium are present. In 1940, however, researchers bombarded uranium with neutrons to produce plutonium and neptunium by the steps given above. These two elements were thus the first elements made in the laboratory, and by bombarding them with neutrons and other particles, the list of synthetic elements has been extended to include those up to atomic number 108 (and possibly 109).

32.8   Q Values

In the last section, we examined some nuclear reactions for which mass numbers and atomic numbers were balanced in the equations. We shall now consider the energy involved in these reactions, since energy is another important quantity that must be conserved.

Let us illustrate this procedure by analyzing the following nuclear reaction:

$$^{2}_{1}\text{H} + {}^{14}_{7}\text{N} \longrightarrow {}^{12}_{6}\text{C} + {}^{4}_{2}\text{He} \tag{32.17}$$

The total amount of mass on the left side of the equation is the sum of the mass of $^{2}_{1}\text{H}$ (2.014102 u) and the mass of $^{14}_{7}\text{N}$ (14.003074 u), which equals 16.017176 u. Similarly, the mass on the right side of the equation is the sum of the mass of $^{12}_{6}\text{C}$ (12.000000 u) plus the mass of $^{4}_{2}\text{He}$ (4.002603 u), for a total of 16.002603 u. Thus, the total mass before the reaction is greater than the total mass after the reaction. The mass difference in this reaction is equal to 16.017176 u − 16.002603 u = 0.014573 u. This "lost" mass is converted to the kinetic energy of the nuclei present after the reaction. In energy units, 0.014573 u is equivalent to 13.567 MeV of kinetic energy carried away by the carbon and helium nuclei.

The amount of energy required to balance the equation is called the *Q* value of the reaction. In Equation 32.17 the *Q* value is 13.567 MeV. Nuclear reactions in which there is a release of energy, that is, positive *Q* values, are said to be **exothermic reactions.**

The energy balance sheet is not complete, however. We must also consider the kinetic energy of the incident particle before the collision. As an example, let us assume that the deuteron in Equation 32.17 has a kinetic energy of 5 MeV. Adding this to our *Q* value, we find that the carbon and helium nuclei will have a total kinetic energy of 18.567 MeV following the reaction.

Now consider the reaction

$$^4_2\text{He} + ^{14}_7\text{N} \longrightarrow ^{17}_8\text{O} + ^1_1\text{H} \qquad \textbf{(32.18)}$$

Before the reaction, the total mass is the sum of the masses of the alpha particle and the nitrogen nucleus: 4.002603 u + 14.003074 u = 18.005677 u. After the reaction, the total mass is the sum of the masses of the oxygen nucleus and the proton: 16.999133 u + 1.007825 u = 18.006958 u. In this case, the total mass after the reaction is *greater* than the total mass before the reaction. The mass deficit is 0.001281 u, which is equivalent to an energy deficit of 1.193 MeV. This deficit is expressed by saying that the *Q* value of the reaction is a negative number, −1.193 MeV. Reactions with negative *Q* values are called **endothermic reactions.** Such reactions will not take place unless the incoming particle has at least enough kinetic energy to overcome the energy deficit.

At first, it might appear that the reaction in Equation 32.18 could take place if the incoming alpha particle had a kinetic energy of 1.193 MeV. In practice, however, the alpha particle must have more energy than this. If it had an energy of only 1.193 MeV, energy would be conserved but careful analysis would show that momentum was not. This can easily be understood by recognizing that the incoming alpha particle has some momentum before the reaction. However, if its kinetic energy was only 1.193 MeV, the products (oxygen and a proton) would be created with zero kinetic energy and, thus, zero momentum. It can be shown that, in order to conserve both energy and momentum, the incoming particle must have a minimum kinetic energy given by

$$KE_{\min} = \left(1 + \frac{m}{M}\right)|Q| \qquad \textbf{(32.19)}$$

where *m* is the mass of the incident particle, *M* is the mass of the target, and the absolute value of the *Q* value is to be used. For the reaction given by Equation 32.18, we find

$$KE_{\min} = \left(1 + \frac{4.002603}{14.003074}\right)|-1.193 \text{ MeV}| = 1.534 \text{ MeV}$$

This minimum value of the kinetic energy of the incoming particle is called the **threshold energy.** The nuclear reaction shown in Equation 32.18 will not occur if the incoming alpha particle has an energy less than 1.534 MeV but will occur if the kinetic energy is equal to or greater than 1.534 MeV.

Summary

Nuclei are represented symbolically as $^A_Z X$, where X represents the chemical symbol for the element. The quantity A is the **mass number**, which equals the total number of nucleons (neutrons plus protons) in the nucleus. The quantity Z is the **atomic number**, which equals the number of protons in the nucleus. Nuclei that contain the same number of protons but different numbers of neutrons are called **isotopes**. In other words, isotopes have the same Z value but different A values.

Most nuclei are approximately spherical, with an average radius given by

$$r = r_0 A^{1/3} \qquad \text{(32.1)}$$

where A is the mass number and r_0 is a constant equal to 1.2×10^{-15} m.

The total mass of a nucleus is always less than the sum of the masses of its individual nucleons. This mass difference, Δm, multiplied by c^2 gives the **binding energy** of the nucleus.

There are three different models of the nucleus. These are the **liquid-drop model**, the **independent-particle model**, and the **collective model**. Each has its own sphere of usefulness. For example, the liquid-drop model is quite effective in describing the binding energy of the nucleus.

The spontaneous emission of radiation by certain nuclei is called **radioactivity**. There are three processes by which a radioactive substance can decay: alpha (α) decay, in which the emitted particles are $^4_2$He nuclei; beta (β) decay, in which the emitted particles are electrons; and gamma (γ) decay, in which the emitted particles are high-energy photons.

The **decay rate**, or **activity**, R, of a sample is given by

$$R = \lambda N \qquad \text{(32.4)}$$

where N is the number of radioactive nuclei at some instant and λ is a constant for a given substance called the **decay constant**.

Nuclei in a radioactive substance decay such that the number of nuclei present varies with time according to the expression

$$N = N_0 e^{-\lambda t} \qquad \text{(32.5)}$$

where N is the number of radioactive nuclei present at time t, N_0 is the number at time $t = 0$, and $e = 2.718. \ldots$

The **half-life**, $T_{1/2}$, of a radioactive substance is the time required for half of a given number of radioactive nuclei to decay. The half-life is related to the decay constant as

$$T_{1/2} = \frac{0.693}{\lambda} \qquad \text{(32.8)}$$

If a nucleus decays by alpha emission, it loses two protons and two neutrons. A typical alpha decay is

$$^{238}_{92}\text{U} \longrightarrow ^{234}_{90}\text{Th} + ^{4}_{2}\text{He} \qquad \textbf{(32.9)}$$

Note that in this decay, as in all radioactive decay processes, the sum of the Z values on the left equals the sum of the Z values on the right; the same is true for the A values.

A typical beta decay is

$$^{14}_{6}\text{C} \longrightarrow ^{14}_{7}\text{N} + ^{0}_{-1}\text{e} + \bar{\nu}$$

When a nucleus decays by beta emission, a particle called an **antineutrino** ($\bar{\nu}$) is also emitted. In positron decay a **neutrino** (ν) is emitted. A neutrino has zero electric charge and a small rest mass (which may be zero) and interacts weakly with matter.

Nuclei are often in an excited state following radioactive decay and release their extra energy by emitting a high-energy photon called a **gamma ray** (γ). A typical gamma ray emission is

$$^{12}_{6}\text{C}^{*} \longrightarrow ^{12}_{6}\text{C} + \gamma \qquad \textbf{(32.14b)}$$

where the asterisk indicates that the carbon nucleus was in an excited state before gamma emission.

Nuclear reactions can occur when a bombarding particle strikes another nucleus. A typical nuclear reaction is

$$^{4}_{2}\text{He} + ^{14}_{7}\text{N} \longrightarrow ^{17}_{8}\text{O} + ^{1}_{1}\text{H} \qquad \textbf{(32.16)}$$

In this reaction, an alpha particle strikes a nitrogen nucleus, producing an oxygen nucleus and a proton. As in radioactive decay, atomic numbers and mass numbers balance on the two sides of the arrow.

Nuclear reactions in which there is a release of energy are said to be **exothermic reactions** and are characterized by positive Q values. Reactions with negative Q values, called **endothermic reactions,** cannot occur unless the incoming particle has at least enough kinetic energy to overcome the energy deficit. In order to conserve both energy and momentum, the incoming particle must have a minimum kinetic energy, called the **threshold energy,** given by

$$KE_{\min} = \left(1 + \frac{m}{M}\right)|Q| \qquad \textbf{(32.19)}$$

where m is the mass of the incident particle and M is the mass of the target atom.

Additional Reading

G. F. Bertsch, "Vibrations of the Atomic Nucleus," *Sci. American*, May 1983, p. 62.

H. Bethe, "What Holds the Nucleus Together?" *Sci. American*, September 1953, p. 201.

D. A. Bromley, "Nuclear Models," *Sci. American*, December 1978, p. 58.

B. N. Da Costa Andrade, *Rutherford and the Nature of the Atom.* Garden City, NY, Science Study Series, Doubleday, 1964.

S. A. Fetter and K. Tsipis, "Catastrophic Releases of Radioactivity," *Sci. American*, April 1981, p. 41.

O. Hahn, "The Discovery of Fission," *Sci. American*, February 1958, p. 76.

R. E. Hedges and J. Gowlett, "Radiocarbon Dating by Accelerator Mass Spectrometry," *Sci. American*, January 1986, p. 100.

R. E. M. Hedges, "Radioisotope Clocks in Archaeology," *Nature*, Vol. 281, No. 5725, 1979, p. 19.

M. Jacob and P. Landshoff, "The Inner Structure of the Proton," *Sci. American*, March 1980, p. 66.

J. G. Learned and D. Eichler, "A Deep-Sea Neutrino Telescope," *Sci. American*, February 1981, p. 138.

J. M. LoSecco, F. Reines, and D. Sinclair, "The Search for Proton Decay," *Sci. American*, June 1985, p. 54.

R. B. Marshak, "The Nuclear Force," *Sci. American*, March 1960, p. 98.

S. Weinberg, "The Decay of the Proton," *Sci. American*, June 1981, p. 64.

V. F. Weisskopf and E. P. Rosenbaum, "A Model of the Nucleus," *Sci. American*, December 1955, p. 261.

E. Witten, "New Ideas About Neutrino Masses," *The Physics Teacher*, February 1983, p. 78.

Questions

1. In Rutherford's experiment, assume that an alpha particle is headed directly toward the nucleus of an atom. Why does the alpha particle not make physical contact with the nucleus?

2. If the atom were indeed like the plum-pudding model, what results would Geiger and Marsden have obtained in their alpha-scattering experiment?

3. Estimate the mass of a pin head if it were composed entirely of densely packed nuclear matter.

4. How many protons are there in the nucleus $^{222}_{86}Rn$? How many neutrons? How many electrons are there in the neutral atom?

5. Element X has several isotopes. What do these isotopes have in common? How do they differ?

6. Why do heavier elements require more neutrons in order to maintain stability?

7. Would the liquid-drop or independent particle nuclear model be more appropriate to predict the behavior of a nucleus in a fission reaction? Which could better explain the γ-ray spectrum of an excited nucleus?

8. Explain the main differences between alpha, beta, and gamma rays.

9. If film is kept in a box, alpha particles from a radioactive source outside the box cannot expose the film but beta particles can. Explain.

10. An alpha particle has twice the charge of a beta particle. Why does the former deflect less than the latter when passing between electrically charged plates, assuming they both have the same speed?

11. If no more people were to be born, the law of population growth would strongly resemble the radioactive decay law. Discuss this statement.

12. If a nucleus has a half-life of one year, does this mean that it will be completely decayed after two years? Explain.

13. What fraction of a radioactive sample remains after two half-lives have elapsed?

14. "The more probable the decay, the shorter the half-life." Explain this statement.

15. Two samples of the same radioactive nuclide are prepared. Sample A has twice the initial activity of sample B. How does the half-life of A compare to the half-life of B? After each has passed through five half-lives, what is the ratio of their activities?

16. If a nucleus such as ^{226}Ra initially at rest undergoes alpha decay, which has more kinetic energy after the decay, the alpha particle or the daughter nucleus?

17. Use the analogy of a bullet and a rifle to explain why the recoiling nucleus carries off only a very small fraction of the disintegration energy in the α decay of a nucleus.

18. Explain why many heavy nuclei undergo alpha decay but do not spontaneously emit neutrons or protons.

19. Explain why neutron-rich isotopes are more likely to be β^- emitters, whereas neutron-deficient isotopes are more likely to be β^+ emitters.

20. Pick any beta decay process and show that the neutrino must have zero charge.

21. What is the difference between a neutrino and a photon?

22. Suppose it could be shown that the cosmic ray intensity was much greater 10,000 years ago. How would this affect present values of the age of ancient samples of once-living matter?

23. Why is carbon dating unable to provide accurate estimates of very old material?

24. If a nucleus captures a slow-moving neutron, the product is left in a highly excited state with an energy approximately 8.0 MeV above the ground state. Explain the source of the excitation energy.

25. Suppose a radioactive parent substance with a very

long half-life has a daughter with a very short half-life. Describe what happens, starting with a freshly purified sample of the parent substance.

26. The radioactive nucleus $^{226}_{88}$Ra has a half-life of about 1.6×10^3 years. Since the Solar System is about 5 billion years old, how can you explain why we still can find this nucleus in nature?

27. A free neutron undergoes beta decays with a half-life of about 15 min. Can a free proton undergo a similar decay?

Problems

Table 32.3 will be useful for many of these problems. A more complete list of atomic masses is given in Appendix B.

Table 32.3 Some Atomic Masses	
Element	**Atomic Mass (u)**
$^{4}_{2}$He	4.002603
$^{7}_{3}$Li	7.016004
$^{9}_{4}$Be	9.012182
$^{10}_{5}$B	10.012938
$^{12}_{6}$C	12.000000
$^{13}_{6}$C	13.003355
$^{14}_{7}$N	14.003074
$^{15}_{7}$N	15.000109
$^{15}_{8}$O	15.003065
$^{18}_{8}$O	17.999159
$^{18}_{9}$F	18.000937
$^{20}_{10}$Ne	19.992439
$^{23}_{11}$Na	22.989770
$^{23}_{12}$Mg	22.994127
$^{27}_{13}$Al	26.981541
$^{30}_{15}$P	29.978310
$^{40}_{20}$Ca	39.962591
$^{43}_{20}$Ca	42.958770
$^{56}_{26}$Fe	55.934939
$^{64}_{30}$Zn	63.929145
$^{64}_{29}$Cu	63.929599
$^{93}_{41}$Nb	92.906378
$^{197}_{79}$Au	196.966560
$^{202}_{80}$Hg	201.970632
$^{216}_{84}$Po	216.001790
$^{220}_{86}$Rn	220.011401
$^{238}_{92}$U	238.050786

Section 32.1   Some Properties of Nuclei

1. Find the radius of a nucleus of (a) $^{4}_{2}$He and (b) $^{238}_{92}$U.

2. Find the atomic number of a nucleus that has a radius 4.36×10^{-15} m.

3. Find the nucleus that has a radius approximately equal to one-half the radius of uranium, $^{238}_{92}$U.

4. Compare the nuclear radii of the following nuclides: $^{2}_{1}$H, $^{60}_{27}$Co, $^{197}_{79}$Au, $^{239}_{94}$Pu.

5. (a) Find the weight density of nuclear matter in lb/ft³. (See the results to Example 32.2.) (b) Find the weight in pounds that a sphere 5 cm in radius would have if it were composed completely of nuclear matter and located on the surface of the Earth.

6. Find the diameter of a sphere of nuclear matter that would have a mass equal to that of the Earth. Base your calculation on an Earth radius of 6.37×10^6 m and an average Earth density of 5.52×10^3 kg/m³.

7. Consider the hydrogen atom to be a sphere of radius equal to the Bohr radius, 0.53×10^{-10} m, and calculate the approximate value of the ratio of the nuclear density to the atomic density.

8. What would be the gravitational force between two golf balls (each of 4.3-cm diameter), 1 meter apart, if they were made of nuclear matter?

9. Use energy methods to calculate the distance of closest approach for a head-on collision between an alpha particle with an initial energy of 0.5 MeV and a gold nucleus (^{197}Au) at rest. Assume the gold nucleus remains at rest during the collision.

10. (a) Find the velocity an alpha particle must have in order to come within 3.2×10^{-14} m of a gold nucleus. (b) Find the energy of the alpha particle in MeV.

11. Use energy methods to find the distance of closest approach of a 0.5-MeV proton to a gold nucleus at rest.

12. The nucleus of an iron atom has a radius equal to 4.60×10^{-15} m. What must be the minimum speed of an α particle if it is to reach the nucleus? Disregard the effect of the outer electrons.

13. An α particle (Z = 2, mass 6.64×10^{-27} kg) approaches to within 1.0×10^{-14} m of a carbon nucleus (Z = 6). What is the (a) maximum Coulomb force on

the α particle; (b) acceleration of the α particle at this point; (c) potential energy of the α particle at this point?

Section 32.2   Binding Energy

14. Calculate the total binding energy for $^{20}_{10}\text{Ne}$.
15. Calculate the total binding energy for $^{40}_{20}\text{Ca}$.
16. Calculate the binding energy per nucleon for $^{93}_{41}\text{Nb}$.
17. Calculate the binding energy per nucleon for $^{197}_{79}\text{Au}$.
18. Compare the average binding energy per nucleon of $^{24}_{12}\text{Mg}$ and $^{85}_{37}\text{Rb}$.
19. A pair of nuclei for which $Z_1 = N_2$ and $Z_2 = N_1$ are called *mirror isobars* (the atomic and neutron numbers are interchangeable). Binding energy measurements on such pairs can be used to obtain evidence of the charge independence of nuclear forces. Charge independence means that the proton-proton, proton-neutron, and neutron-neutron forces are approximately equal. Calculate the difference in binding energy for the two mirror nuclei $^{15}_{8}\text{O}$ and $^{15}_{7}\text{N}$.
20. Determine the difference of the binding energy of $^{3}_{1}\text{H}$ and $^{3}_{2}\text{He}$.
21. Two isotopes having the same mass number are known as isobars. Calculate the difference in binding energy per nucleon for the isobars $^{23}_{11}\text{Na}$ and $^{23}_{12}\text{Mg}$. How do you account for this difference?
22. Calculate the binding energy of the last neutron in the $^{43}_{20}\text{Ca}$ nucleus. (*Hint:* You should compare the mass of $^{43}_{20}\text{Ca}$ with the mass of $^{42}_{20}\text{Ca}$ plus the mass of a neutron. The mass of $^{42}_{20}\text{Ca} = 41.958622$ u.)

Section 32.3   Nuclear Models

23. (a) In the liquid-drop model of the nucleus, why does the surface-effect term have a minus sign? (b) The binding energy of the nucleus increases as the volume-to-surface ratio increases. Calculate this ratio for both spherical and cubical shapes and explain which shape is more plausible for nuclei.
24. Use Equation 32.2 and the numerical values given for the constants C_1, C_2, and C_3 to calculate the binding energy for $^{64}_{29}\text{Cu}$.

Section 32.4   Radioactivity

25. The half-life of a radioactive sample is 30 min. If you start with a sample containing 3×10^{16} nuclei, how many of these nuclei remain after 10 min?
26. How long will it take for a sample of polonium of half-life 140 days to decay to 1/10 its original strength?

27. Suppose that you start with 10^{-3} g of a pure radioactive substance and 2 hours later determine that only 0.25×10^{-3} g of the substance remains. What is the half-life of this substance?
28. The half-life of an isotope of phosphorus is 14 days. If a sample contains 3×10^{16} such nuclei, determine its activity.
29. A drug tagged with $^{99}_{43}\text{Tc}$ (half-life = 6.05 h) is prepared for a patient. If the original activity of the sample was 1.1×10^4 Bq, what is its activity after it has sat on the shelf for 2.0 h?
30. The half-life of ^{131}I is 8.04 days. (a) Calculate the decay constant for this isotope. (b) Find the number of ^{131}I nuclei necessary to produce a sample with an activity of 0.5 μCi.
31. How many radioactive atoms are present in a sample that has an activity of 0.2 μCi and a half-life of 8.1 days?
32. A freshly prepared sample of a certain radioactive isotope has an activity of 10 mCi. After 4 h, the activity is 8 mCi. (a) Find the decay constant and half-life of the isotope. (b) How many atoms of the isotope were contained in the freshly prepared sample? (c) What is the sample's activity 30 h after it is prepared?
33. Tritium has a half-life of 12.33 years. What percentage of the nuclei in a tritium sample will decay in 5 years?
34. A radioactive sample contains 30.0×10^{-15} kg of $^{108}_{47}\text{Ag}$ (half-life = 2.42 min) at some instant. What is the activity of this sample in millicuries?
35. A building has become accidentally contaminated with radioactivity. The longest-lived material in the building is strontium-90 ($^{90}_{38}\text{Sr}$, atomic mass 89.9077). If the building initially contained 5.0 kg of this substance and the safe level is less than 10.0 counts/min, how long will the building be unsafe?

Section 32.5   The Decay Processes

36. Complete the following radioactive decay formulas:

$$^{212}_{83}\text{Bi} \longrightarrow ? + ^{4}_{2}\text{He}$$

$$^{95}_{36}\text{Kr} \longrightarrow ? + ^{0}_{-1}\text{e}$$

$$? \longrightarrow ^{4}_{2}\text{He} + ^{140}_{58}\text{Ce}$$

37. Complete the following radioactive decay formulas:

$$^{12}_{5}\text{B} \longrightarrow ? + ^{0}_{-1}\text{e}$$

$$^{234}_{90}\text{Th} \longrightarrow ^{230}_{88}\text{Ra} + ?$$

$$? \longrightarrow ^{14}_{7}\text{N} + ^{0}_{-1}\text{e}$$

38. Find the energy released in the alpha decay of $^{232}_{92}\text{U}$.

39. Find the energy released in the alpha decay of $^{238}_{92}$U. You will find the following mass values useful:

$$M(^{238}_{92}U) = 238.050786 \text{ u}$$

$$M(^{234}_{90}Th) = 234.043583 \text{ u}$$

$$M(^{4}_{2}He) = 4.002603 \text{ u}$$

40. Is it energetically possible for a $^{8}_{4}$Be nucleus to spontaneously decay into two alpha particles? Explain.

41. Determine which of the following suggested decays can occur spontaneously:

    (a) $^{40}_{20}Ca \longrightarrow ^{0}_{1}e + ^{40}_{19}K$ (b) $^{144}_{60}Nd \longrightarrow ^{4}_{2}He + ^{140}_{58}Ce$

42. Figure 32.11 shows the steps by which $^{235}_{92}$U decays to $^{207}_{82}$Pb. Enter the correct isotope symbol in each square.

Figure 32.11 Problem 42

43. $^{66}_{28}$Ni (mass = 65.9291 u) decays by β^- emission to $^{66}_{29}$Cu (mass = 65.9289 u). (a) Write the complete decay formula for this process. (b) Find the maximum kinetic energy of the emerging β^- particles.

44. A wooden artifact is found in an ancient tomb. Measuring the carbon-14 ($^{14}_{6}$C) activity indicates 60% of the activity in a fresh sample of wood taken from the same region. Assuming the same amount of carbon-14 was initially present in the wood from which the artifact was made, determine the age of the artifact.

45. A piece of charcoal used for cooking is found at the remains of an ancient campsite. A 1000-g sample of carbon from the wood is found to have an activity of 2000 decays per minute. Find the age of the charcoal.

Section 32.7   Nuclear Reactions

Section 32.8   *Q* Values

46. Find the *Q* value of the reaction discussed in Example 32.10.

47. The first nuclear reaction utilizing particle accelera-

tors was performed by Cockcroft and Walton. Accelerated protons were used to bombard lithium nuclei, producing the following reaction:

$$^{1}_{1}H + ^{7}_{3}Li \longrightarrow ^{4}_{2}He + ^{4}_{2}He$$

Since the masses of the particles involved in the reaction were well known, these results were used to obtain an early proof of the Einstein mass-energy relation. Calculate the *Q* value of the reaction.

48. Determine the *Q* values of the following reactions: (a) $^{27}_{13}Al + n \rightarrow ^{24}_{11}Na + ^{4}_{2}He$; (b) $^{11}_{5}B + ^{1}_{1}H \rightarrow ^{11}_{6}C + n$.

49. (a) Determine the product of the reaction $^{7}_{3}Li + ^{4}_{2}He \rightarrow ? + n$. (b) What is the *Q* value of the reaction?

50. Complete the following nuclear reactions:

$$? + ^{14}_{7}N \longrightarrow ^{1}_{1}H + ^{17}_{8}O$$

$$^{7}_{3}Li + ^{1}_{1}H \longrightarrow ^{4}_{2}He + ?$$

51. Complete the following nuclear reactions:

$$^{27}_{13}Al + ^{4}_{2}He \longrightarrow ? + ^{30}_{15}P$$

$$^{1}_{0}n + ? \longrightarrow ^{4}_{2}He + ^{7}_{3}Li$$

52. The first known reaction in which the product nucleus was radioactive (achieved in 1934) was one in which $^{27}_{13}$Al was bombarded with alpha particles. Produced in the reaction were a neutron and a product nucleus. (a) What was the product nucleus? (b) Find the *Q* value of the reaction.

53. When a $^{6}_{3}$Li nucleus is struck by a proton, an alpha particle and a product nucleus are released. (a) What is the product nucleus? (b) Find the *Q* value of the reaction.

54. Find the energy released in the fission reaction

$$n + ^{235}_{92}U \longrightarrow ^{98}_{40}Zr + ^{135}_{52}Te + 3n$$

The atomic masses of the fission products are: $^{98}_{40}$Zr, 97.9120 u; $^{135}_{52}$Te, 134.9087 u.

55. A beam of 6.61 MeV protons is incident on a target of $^{27}_{13}$Al. Those that collide produce the reaction

$$p + ^{27}_{13}Al \longrightarrow ^{27}_{14}Si + n$$

($^{27}_{14}$Si has a mass of 26.986721 u.) Neglect any recoil of the product nucleus and determine the kinetic energy of the emerging neutrons.

56. (a) Suppose $^{10}_{5}$B is struck by an alpha particle, releasing a proton and a product nucleus in the reaction. (a) What is the product nucleus? (b) An alpha particle and a product nucleus are produced when $^{13}_{6}$C is struck by a proton. What is the product nucleus?

57. (a) The reactions discussed in Problem 56 are often called inverse reactions. Why? (b) Show that the two reactions have the same absolute value of *Q*.

58. When ^{18}O is struck by a proton, ^{18}F and another particle are produced. (a) What is the other particle? (b) This reaction has a *Q* value of -2.453 MeV, and

ter, Lisa Meitner and Otto Frisch explained what had happened. The uranium nucleus had split into two nearly equal fragments after absorbing a neutron. Such an occurrence was of considerable interest to physicists attempting to understand the nucleus, but it was to have even more far-reaching consequences. Measurements showed that about 200 MeV of energy is released in each fission event, and this fact was to affect the course of human history.

The fission of ^{235}U by slow (low energy) neutrons can be represented by the equation

$$^1_0n + {}^{235}_{92}U \longrightarrow {}^{236}_{92}U^* \longrightarrow X + Y + \text{neutrons} \qquad (33.1)$$

where ^{236}U* is an intermediate state that lasts only for about 10^{-12} s before splitting into X and Y. The resulting nuclei, X and Y, are called **fission fragments**. There are many combinations of X and Y that satisfy the requirements of conservation of mass-energy and charge. In the fission of uranium, there are about 90 different daughter nuclei that can be formed. The process also results in the production of several neutrons per fission event, typically two or three. On the average 2.47 neutrons are released per event.

A typical reaction of this type is

$$^1_0n + {}^{235}_{92}U \longrightarrow {}^{141}_{56}Ba + {}^{92}_{36}Kr + 3\,{}^1_0n \qquad (33.2)$$

The fission fragments, barium, and krypton, and the released neutrons have a great deal of kinetic energy following the fission event.

The breakup of the uranium nucleus can be compared to what happens to a drop of water when excess energy is added to it. All the atoms in the drop have energy, but this energy is not great enough to break up the drop. However, if enough energy is added to set the drop vibrating, it will undergo elongation and compression until the amplitude of vibration becomes large enough to cause the drop to break apart. In the uranium nucleus, a similar process occurs (Fig. 33.1). The sequence of events is

1. The ^{235}U nucleus captures a thermal (slow-moving) neutron.
2. This capture results in the formation of ^{236}U*, and the excess energy of this nucleus causes it to undergo violent oscillations.
3. The ^{236}U* nucleus becomes highly distorted, and the force of repulsion between protons in the two halves of the dumbbell shape tends to increase the distortion.
4. The nucleus splits into two fragments, emitting several neutrons in the process.

Figure 33.1
The stages involved in a nuclear fission event as described by the liquid-drop model of the nucleus.

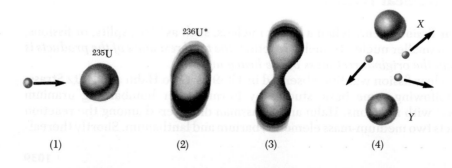

235U   236U*

X

Y

(1)   (2)   (3)   (4)

Let us estimate the disintegration energy, Q, released in a typical fission process. From Figure 32.3 we see that the binding per nucleon is about 7.6 MeV for heavy nuclei (those having a mass number approximately equal to 240) and about 8.5 MeV for nuclei of intermediate mass. This means that the nucleons in the fission fragments are more tightly bound and therefore have less mass than the nucleons in the original heavy nucleus. This decrease in mass per nucleon appears as released energy when fission occurs. The amount of energy released is $(8.5 - 7.6)$ MeV per nucleon. Assuming a total of 240 nucleons, we find that the energy released per fission event is

$$Q = (240 \text{ nucleons}) \left(8.5 \, \frac{\text{MeV}}{\text{nucleon}} - 7.6 \, \frac{\text{MeV}}{\text{nucleon}} \right) = 220 \text{ MeV}$$

This is indeed a very large amount of energy relative to the amount of energy released in chemical processes. For example, the energy released in the combustion of one molecule of the octane used in gasoline engines is about one millionth the energy released in a single fission event!

Example 33.1 The Fission of Uranium

Two other possible ways by which ^{235}U can undergo fission when bombarded with a neutron are (1) by the release of ^{140}Xe and ^{94}Sr as fission fragments and (2) by the release of ^{132}Sn and ^{101}Mo as fission fragments. In each case, neutrons are also released. Find the number of neutrons released in each of these events.

Solution
By balancing mass numbers and atomic numbers, we find that these reactions can be written

$$^{1}_{0}\text{n} + \, ^{235}_{92}\text{U} \longrightarrow \, ^{140}_{54}\text{Xe} + \, ^{94}_{38}\text{Sr} + 2 \, ^{1}_{0}\text{n}$$

$$^{1}_{0}\text{n} + \, ^{235}_{92}\text{U} \longrightarrow \, ^{132}_{50}\text{Sn} + \, ^{101}_{42}\text{Mo} + 3 \, ^{1}_{0}\text{n}$$

Thus, two neutrons are released in the first event and three in the second.

Example 33.2 The Energy Released in the Fission of ^{235}U

Calculate the total energy released if 1 kg of ^{235}U undergoes fission, taking the disintegration energy per event to be $Q = 208$ MeV (a more accurate value than the estimate given before).

Solution
We need to know the number of nuclei in 1 kg of uranium. Since $A = 235$, the number of nuclei is

$$N = \left(\frac{6.02 \times 10^{23} \text{ nuclei/mole}}{235 \text{ g/mole}} \right) (10^3 \text{ g}) = 2.56 \times 10^{24} \text{ nuclei}$$

Hence the disintegration energy

$$E = NQ = (2.56 \times 10^{24} \text{ nuclei}) \left(208 \, \frac{\text{MeV}}{\text{nucleus}} \right) = \boxed{5.32 \times 10^{26} \text{ MeV}}$$

Since 1 MeV is equivalent to 4.45×10^{-20} kWh, $E = 2.37 \times 10^7$ kWh. This is enough energy to keep a 100-W lightbulb burning for about 30,000 years. Thus, 1 kg of ^{235}U is a relatively large amount of fissionable material.

33.2   Nuclear Reactors

We have seen that, when ^{235}U undergoes fission, an average of about 2.5 neutrons are emitted per event. These neutrons can in turn trigger other nuclei to undergo fission, with the possibility of a chain reaction (Fig. 33.2). Calculations show that if the chain reaction is not controlled (that is, if it does not proceed slowly), it could result in a violent explosion, with the release of an enormous amount of energy, even from only 1 g of ^{235}U. If the energy in 1 kg of ^{235}U were released, it would be equivalent to detonating about 20,000 tons of TNT! This, of course, is the principle behind the first nuclear bomb, an uncontrolled fission reaction.

A nuclear reactor is a system designed to maintain what is called a **self-sustained chain reaction.** This important process was first achieved in 1942 by Fermi at the University of Chicago, with natural uranium as the fuel. Most reactors in operation today also use uranium as fuel. Natural uranium contains only about 0.7% of the ^{235}U isotope, with the remaining 99.3% being the ^{238}U isotope. This is important to the operation of a reactor because ^{238}U almost never undergoes fission. Instead, it tends to absorb neutrons, producing neptunium and plutonium. For this reason, reactor fuels must be artificially enriched to contain a few percent of the ^{235}U isotope.

Earlier, we mentioned that an average of about 2.5 neutrons are emitted in each fission event of ^{235}U. In order to achieve a self-sustained chain reaction, one of these neutrons, on the average, must be captured by another ^{235}U nucleus and cause it to undergo fission. A useful parameter for describing the level of reactor operation is the **reproduction constant,** K, defined as *the average number of neutrons from each fission event that will cause another event.* As we have seen, K can have a maximum value of 2.5 in the fission of uranium. However, in practice K is less than this because of several factors, which we shall soon discuss.

A self-sustained chain reaction is achieved when $K = 1$. Under this condition, the reactor is said to be **critical.** When K is less than unity, the reactor is subcritical and the reaction dies out. When K is greater than unity, the reactor is said to be supercritical and a runaway reaction occurs. In a nuclear reactor

Figure 33.2
A nuclear chain reaction.

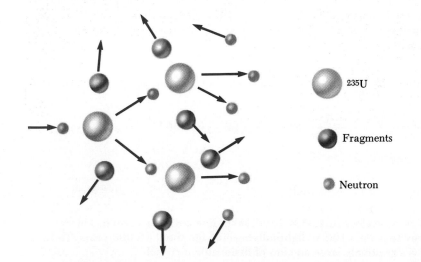

235U

Fragments

Neutron

used to furnish power to a utility company, it is necessary to maintain a value of *K* close to unity.

The basic design of a nuclear reactor is shown in Figure 33.3. The fuel elements consist of enriched uranium. The function of the remaining parts of the reactor and some aspects of its design will now be described.

Neutron Leakage

In any reactor, a fraction of the neutrons produced in fission will leak out of the core before inducing other fission events. If the fraction leaking out is too large, the reactor will not operate. The percentage lost is large if the reactor is very small because leakage is a function of the ratio of surface area to volume. Therefore, a critical feature of the design of a reactor is to choose the correct surface-area-to-volume ratio so that a sustained reaction can be achieved.

Regulating Neutron Energies

The neutrons released in fission events are very energetic, with kinetic energies of about 2 MeV. It is found that slow neutrons are far more likely than fast neutrons to produce fission events in ^{235}U. Therefore, in order for the chain reaction to continue, the neutrons must be slowed down. This is accomplished by surrounding the fuel with a **moderator** substance.

In order to understand how neutrons are slowed, consider a collision between a light object and a very massive one. In such an event, the light

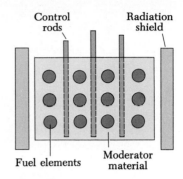

Figure 33.3
Cross-section of a reactor core surrounded by a radiation shield.

Painting of the world's first reactor. Because of wartime secrecy, there are no photographs of the completed reactor. The reactor was composed of layers of graphite interspersed with uranium. A self-sustained chain reaction was first achieved on December 2, 1942. Word of the success was telephoned immediately to Washington with this message: "The Italian navigator has landed in the New World and found the natives very friendly." The historic event took place in an improvised laboratory in the racquets court under the west stands of the University of Chicago's Stagg Field and the Italian navigator was Fermi. *(Courtesy of Chicago Historical Society)*

object rebounds from the collision with most of its original kinetic energy. However, if the collision is between objects whose masses are nearly the same, the incoming projectile will transfer a large percentage of its kinetic energy to the target. In the first nuclear reactor ever constructed, Fermi placed bricks of graphite (carbon) between the fuel elements. Carbon nuclei are about 12 times more massive than neutrons, but after several collisions with carbon nuclei, a neutron is slowed sufficiently to increase its likelihood of fission with ^{235}U. In this design the carbon is the moderator; most modern reactors use water as the moderator.

Neutron Capture

In the process of being slowed down, the neutrons may be captured by nuclei that do not undergo fission. The most common event of this type is neutron capture by ^{238}U. The probability of neutron capture by ^{238}U is very high when the neutrons have high kinetic energies and very low when they have low kinetic energies. Thus the slowing down of the neutrons by the moderator serves the dual purpose of making them available for reaction with ^{235}U and decreasing their chances of being captured by ^{238}U.

Control of Power Level

It is possible for a reactor to reach the critical stage ($K = 1$) after all the neutron losses described above are minimized. However, a method of control is needed to maintain a K value near unity. If K rises above this value, the heat produced in the runaway reaction would melt the reactor. To control the power level, control rods are inserted into the reactor core (see Fig. 33.3). These rods are made of materials, such as cadmium, that are very efficient in absorbing neutrons. By adjusting the number and position of these control rods in the reactor core, the K value can be varied and any power level within the design range of the reactor can be achieved.

A diagram of a pressurized-water reactor is shown in Figure 33.4. This type of reactor is commonly used in electric power plants in the United States. Fission events in the reactor core supply heat to the water contained in the

Figure 33.4
Main components of a pressurized-water reactor.

primary (closed) system, which is maintained at high pressure to keep it from boiling. This water also serves as the moderator. The hot water is pumped through a heat exchanger, and the heat is transferred to the water contained in the secondary system. The hot water in the secondary system is converted to steam, which drives a turbine-generator to create electric power. Note that the water in the secondary system is isolated from the water in the primary system in order to avoid contamination of the secondary water and steam by radioactive nuclei from the reactor core.

Reactor Safety[1]

The safety aspects of nuclear power reactors are often sensationalized by the media and misunderstood by the public. The 1979 near-disaster of Three Mile Island in Pennsylvania and the accident at the Chernobyl reactor in Russia rightfully focused attention on reactor safety. Yet the safety record in the United States is enviable. The records show no fatalities attributed to commercial nuclear power generation in the history of the United States nuclear industry.

Commercial reactors achieve safety through careful design and rigid operating procedures. Radiation exposure and the potential health risks associated with such exposure are controlled by three layers of containment. The fuel and radioactive fission products are contained inside the reactor vessel. Should this vessel rupture, the reactor building acts as a second containment structure to prevent radioactive material from contaminating the environment. Finally, the reactor facilities must be in a remote location so as not to expose the general public should radiation escape the reactor building.

According to the Oak Ridge National Laboratory Review, "the health risk of living within 8 km (5 miles) of a nuclear reactor for 50 years is no greater than the risk of smoking 1.4 cigarettes, drinking 0.5 liters of wine, traveling 240 km by car, flying 9600 km by jet, or having one chest x-ray in a hospital. Each of these activities is estimated to increase a person's chances of dying in any given year by one in a million."

Another potential danger in nuclear reactor operations is the possibility that the water flow could be interrupted. Even though the nuclear fission chain reaction would be stopped immediately, residual heat could build up in the reactor to the point of melting the fuel elements. The molten reactor core would melt to the bottom of the reactor vessel and conceivably melt its way into the ground below — the so-called "China syndrome." Although it might appear that this deep underground burial site would be an ideal safe-haven for this radioactive blob, there would be danger of a steam explosion should the molten mass encounter water. This nonnuclear explosion could spread radioactive material to the areas surrounding the power plant. To prevent such an unlikely chain of events, nuclear reactors are designed with emergency core cooling systems which require no power and automatically flood the reactor with water in the event of loss of coolant. The emergency cooling water moderates heat build-up in the core, which in turn prevents the occurrence of melting.

[1] The authors are grateful to Professor Gene Skluzacek of the University of Nebraska at Omaha for rewriting this section on reactor safety.

A continuing concern in nuclear fission reactors is the safe disposal of radioactive material when the reactor core is replaced. This waste material contains long-lived, highly radioactive isotopes and must be stored over long periods of time in such a way that there is no chance of environmental contamination. At present, sealing radioactive wastes in waterproof containers and burying them in deep salt mines seems to be the most promising solution.

Transportation of reactor fuel and reactor wastes pose additional safety risks. However, the danger of theft during transport (say by a terrorist group) is greatly exaggerated. The difficulties in handling highly radioactive material reduce the possibility of such events. Furthermore, neither the waste nor the fuel of nuclear power reactors can be used to construct a nuclear bomb.

Accidents during transportation of nuclear fuel could expose the public to harmful levels of radiation. The Department of Energy requires stringent crash tests on all containers used to transport nuclear materials. Container manufacturers must demonstrate that their containers will not rupture even in a high speed-collision.

The safety issues associated with nuclear power reactors are complex and often emotional. All sources of energy have risks associated with them. In each case, one must weigh the risks of each energy source against the benefits and availability of the source of energy.

▶SPECIAL TOPIC
Three Mile Island and Chernobyl

The two most publicized nuclear accidents have been those at the Three Mile Island nuclear plant in Middletown, Pennsylvania, and at the Soviet reactor at Chernobyl, Russia. Let us describe the events and consequences of these accidents.

The series of events that caused the Three Mile Island incident began at 4 A.M. on March 28, 1979, with a routine maintenance operation. Workers changing the water purifier in a part of the piping system at the plant inadvertently allowed some air into the pipe, which caused an interruption in the flow of water to the reactor. If such an event occurs, backup systems are installed to respond automatically, but in this case several things went wrong. Two pumps in the backup system were out of service for routine maintenance and, therefore, could not deliver water for cooling. Reactor operators did not notice the water-pump problem because warning lights were obscured with tags that had been placed on the pumps when they were taken out of service. The loss of cooling water to the reactor caused a pressure build-up in the reactor core. This caused a relief valve in the primary coolant loop to open automatically and release superheated steam. This automatic sequence led to further difficulties because the valve failed to close after venting, and a dangerous drop in pressure occurred inside the loop. At this point, the emergency cooling system was activated and pressure gauges in the control room indicated that the water level was being maintained such that the fuel rods were covered. This, however, was a false reading because a 1000 ft³ bubble of steam and hydrogen gas was pushing the water level up, leaving part of the core exposed. This caused the core to overheat to the point where temperatures inside the reactor vessel were so high that they could not be measured on instruments at the plant. When the situation was finally brought under control about three days later, it was found that the fuel rods had cracked, leading to contamination of the pressure vessel. The cleanup was a major problem and continues to be difficult to handle. Fortunately, there was no major release of radiation to the public. The radiation release was such that the public received no more than about 10% of what would be normally received from natural sources in one year. The subsequent threat to public health via cancers, if any, is too low to estimate or detect in any reliable way.

A far more serious accident occurred at the Soviet reactor at Chernobyl, Russia, in 1986. This disaster started when the cooling system failed, which caused an overheating and ultimate melting of the uranium oxide fuel rods. The next stage in the disaster occurred when the operators were late in flooding the core with water. The already hot core changed the water to steam, which in turn reacted with the graphite moderator in the core, the melted uranium, and other molten substances to produce a mixture of hydrogen, oxygen, methane, and carbon monoxide. These gases then interacted violently to produce a chemical explosion, which ruptured the walls of the plant. The plant was built without the added protection of a containment vessel, and enormous levels of radioactive materials were spewed into the atmosphere for several days.

Heroic measures were required to bring the fire and release of radioisotopes into the environment under control. Several tons of sand, clay, lead, and boron were dropped on the reactor from above by helicopters, and other workers tunneled under the plant to seal it from below with concrete, lead, and boron in an unsuccessful attempt to isolate the radioactive materials from the groundwater. Virtually all of these workers and pilots have now died from leukemia induced from the radiation dose received. At present an area of about 700 km (250 mi²) is so contaminated that it is a wasteland and will be so for decades. The final story of the death and devastation caused by the Chernobyl incident is yet to be written. It should be noted here that the Soviet reactor was old and out of date and lacked most of the safety features, such as a containment vessel, inherent to U.S. designs. A disaster of these proportions is very unlikely to happen in a properly designed power plant. ◀

33.3 Nuclear Fusion

Figure 32.3 shows that the binding energy for light nuclei (those having a mass number of less than 20) is much smaller than the binding energy for heavier nuclei. This suggests a possible process that is the reverse of fission. *When two light nuclei combine to form a heavier nucleus, the process is called nuclear fusion.* Because the mass of the final nucleus is less than the rest masses of the original nuclei, there is a loss of mass accompanied by a release of energy. It is important to recognize that although fusion power plants have not yet been developed, a great world-wide effort is under way to harness the energy from fusion reactions in the laboratory. Later we shall discuss the possibilities and advantages of this process for generating electric power.

Fusion in the Sun

All stars generate their energy through fusion processes. About 90% of the stars, including our own Sun, fuse hydrogen, while some older stars fuse helium or other heavier elements. Stars are born in regions of space containing vast clouds of dust and gas. Recent mathematical models of these clouds indicate that star formation is triggered by shock waves passing through a cloud. These shock waves are similar to sonic booms and are produced by events such as the explosion of a nearby star, called a supernova explosion. The shock wave compresses certain regions of the cloud, causing this region to collapse under

This photograph shows one of the most spectacular solar flares ever recorded; it spanned more than 588,000 km (367,000 mi) of the Sun's surface. The flare gives the distinct impression of a twisted sheet of gas in the process of unwinding itself. During solar flares, which may occur several times a week, an enormous amount of energy is released, partly in the form of electromagnetic energy. *(Courtesy of NASA)*

its own gravity. As the gas falls inward toward the center, the atoms gain speed, which causes the temperature of the gas to rise. Two conditions must be met before fusion reactions in the star can sustain its energy needs: (1) The temperature must be high enough (about 10^7 K for hydrogen) to allow the kinetic energy of the positively charged hydrogen nuclei to overcome their mutual Coulomb repulsion and collide, and (2) the density of nuclei must be high enough to ensure a high probability of collision.

When fusion reactions occur in a star, the energy liberated eventually becomes sufficient to prevent further collapse of the star under its own gravity. The star then continues to live out the remainder of its life under a balance between the inward force of gravity tending to cause it to collapse and the outward force due to thermal effects and radiation pressure. The proton-proton cycle is a series of three nuclear reactions that are believed to be the stages in the liberation of energy in our Sun and other stars rich in hydrogen. An overall view of the proton-proton cycle is that four protons combine to form an alpha particle and two positrons, with the release of 25 MeV of energy in the process.

The three steps in the proton-proton cycle are

$$\begin{aligned} {}^1_1\text{H} + {}^1_1\text{H} &\longrightarrow {}^2_1\text{H} + {}^0_1\text{e} + \nu \\ {}^1_1\text{H} + {}^2_1\text{H} &\longrightarrow {}^3_2\text{He} + \gamma \end{aligned}$$

(33.3)

This second reaction is followed by either

$${}^1_1\text{H} + {}^3_2\text{He} \longrightarrow {}^4_2\text{He} + {}^0_1\text{e} + \nu$$

or

$${}^3_2\text{He} + {}^3_2\text{He} \longrightarrow {}^4_2\text{He} + {}^1_1\text{H} + {}^1_1\text{H}$$

The energy liberated is carried primarily by gamma rays, positrons, and neutrinos, as can be seen from the reactions. The gamma rays are soon absorbed by the dense gas, thus raising its temperature. The positrons combine with electrons to produce gamma rays, which in turn are also absorbed by the gas within a few centimeters. The neutrinos, however, almost never interact with matter, and hence they escape from the star carrying about 2% of the energy generated with them. These energy-liberating fusion reactions are called **thermonuclear fusion reactions**. The hydrogen (fusion) bomb, first exploded in 1952, is an example of an uncontrolled thermonuclear fusion reaction.

Fusion Reactors

The enormous amount of energy released in fusion reactions suggests the possibility of harnessing this energy for useful purposes here on Earth. A great deal of effort is currently under way to develop a sustained and controllable thermonuclear reactor—a fusion power reactor. Controlled fusion is often called the ultimate energy source because of the availability of its source of fuel: water. For example, if deuterium were used as the fuel, 0.6 g of it could be extracted from 1 gal of water at a cost of about four cents. Such rates would make the fuel costs of even an inefficient reactor almost insignificant. An additional advantage of fusion reactors is that comparatively few radioactive by-products are formed. As noted in Equation 33.3, the end product of the fusion of hydrogen nuclei is safe, nonradioactive helium. Unfortunately, a thermonuclear reactor that can deliver a net power output over a reasonable time interval is not yet a reality, and many difficulties must be solved before a successful device is constructed.

We have seen that the Sun's energy is based, in part, upon a set of reactions in which ordinary hydrogen is converted to helium. Unfortunately, the proton-proton interaction is not suitable for use in a fusion reactor because the event requires very high pressures and densities. The process works in the Sun only because of the extremely high density of protons in the Sun's interior.

The fusion reactions that appear most promising in the construction of a fusion power reactor involve deuterium and tritium, which are isotopes of hydrogen. These reactions are

$$\begin{aligned} {}^{2}_{1}\text{H} + {}^{2}_{1}\text{H} &\longrightarrow {}^{3}_{2}\text{He} + {}^{1}_{0}\text{n} \qquad &Q = 3.27\ \text{MeV} \\ {}^{2}_{1}\text{H} + {}^{2}_{1}\text{H} &\longrightarrow {}^{3}_{1}\text{H} + {}^{1}_{1}\text{H} \qquad &Q = 4.03\ \text{MeV} \qquad \textbf{(33.4)} \\ {}^{2}_{1}\text{H} + {}^{3}_{1}\text{H} &\longrightarrow {}^{4}_{2}\text{He} + {}^{1}_{0}\text{n} \qquad &Q = 17.59\ \text{MeV} \end{aligned}$$

where the Q values refer to the amount of energy released per reaction. As noted earlier, deuterium is available in almost unlimited quantities from our lakes and oceans and is very inexpensive to extract. Tritium, however, is radioactive ($T_{1/2} = 12.3$ years) and undergoes beta decay to ^{3}He. For this reason, tritium does not occur naturally to any great extent and must be artificially produced.

One of the major problems in obtaining energy from nuclear fusion is the fact that the Coulomb repulsion force between two charged nuclei must be overcome before they can fuse. The fundamental problem then is to give the two nuclei enough kinetic energy to overcome this repulsive force. This can be accomplished by heating the fuel to extremely high temperatures (about 10^{8} K, far greater than the interior temperature of the Sun). As you might expect, such high temperatures are not easy to obtain in the laboratory or a power plant. At these high temperatures, the atoms are ionized, and the system consists of a collection of electrons and nuclei, commonly referred to as a plasma.

In addition to the high temperature requirements, there are two other critical factors that determine whether or not a thermonuclear reactor will be successful: **plasma ion density,** n, and **plasma confinement time,** τ, the time the interacting ions are maintained at a temperature equal to or greater than the temperature required for the reaction to proceed successfully. The density and confinement time must both be large enough to ensure that more fusion energy will be released than is required to heat the plasma.

Lawson's criterion states that a net power output in a fusion reactor is possible under the following conditions:

$$\begin{aligned} n\tau &\geqslant 10^{14}\ \text{s/cm}^3 \qquad &\text{Deuterium-tritium interaction} \\ && &\textbf{(33.5)} \quad \text{Lawson's criterion} \\ n\tau &\geqslant 10^{16}\ \text{s/cm}^3 \qquad &\text{Deuterium-deuterium interaction} \end{aligned}$$

The problem of plasma confinement time has yet to be solved. How can a plasma be confined at a temperature of 10^{8} K for times of the order of 1 s? The two basic techniques under investigation to confine plasmas are discussed following Example 33.3.

ter 31). As an energetic particle passes through the junction, electrons and holes are simultaneously created. The internal electric field sweeps the electrons toward the side of the junction connected to the positive side of the battery and the holes are swept toward the negative side. This creates a pulse of current that can be measured with any electronic counter. In a typical device, the duration of the pulse is about 10^{-7} s.

A **scintillation counter** (Fig. 33.8) usually uses a solid or liquid material whose atoms are easily excited by the incoming radiation. These excited atoms emit visible light when they return to the ground state. Common materials used as scintillators are crystals of sodium iodide and certain plastics. If such a material is attached to one end of a device called a photomultiplier tube, the photons emitted by the scintillator can be converted to an electric signal. The photomultiplier tube consists of several electrodes, called dynodes, whose potentials are increased in succession along the length of the tube, as shown in Figure 33.8. The top of the tube contains a photocathode, which emits elec-

Figure 33.8

Diagram of a scintillation counter connected to a photomultiplier tube.

trons by the photoelectric effect. An emitted electron striking the first dynode has sufficient kinetic energy to eject several other electrons. When these electrons are accelerated to the second dynode, many more electrons are ejected and an avalanche occurs. The end result is 1 million or more electrons striking the last dynode. Hence, one particle striking the scintillator produces a sizable electric pulse at the output of the photomultiplier tube, and this pulse is in turn sent to an electronic counter. A scintillator device is much more sensitive than a Geiger counter, mainly because of the higher density of the detecting medium. It is especially sensitive to gamma rays, which interact more weakly with matter than do charged particles.

The devices described so far make use of ionization processes induced by energetic particles. Various other devices can be used to view the tracks of charged particles directly. **A photographic emulsion** is the simplest example. A charged particle ionizes the atoms in a emulsion layer. The path of the particle corresponds to a family of points at which chemical changes have occurred in the emulsion. When the emulsion is developed, the particle's track becomes visible. Such devices, called **dosimeters,** are common in the film badges used in any environment where radiation levels must be monitored.

A cloud chamber contains a gas that has been supercooled to just below its usual condensation point. An energetic particle passing through ionizes the gas molecules along its path. These ions serve as centers for condensation of the supercooled gas. The track of the particle can be seen with the naked eye and can be photographed. A magnetic field can be applied to determine the sign and energy of the charges as they are deflected by the field.

A device called a **bubble chamber,** invented in 1952 by D. Glaser, makes use of a liquid (usually liquid hydrogen) maintained near its boiling point. Ions produced by incoming charged particles leave bubble tracks, which can be photographed. Because the density of the detecting medium of a bubble chamber is much higher than the density of a gas in a cloud chamber, the bubble chamber has a much higher sensitivity.

*33.6 Uses of Radiation

Practical applications of nuclear physics are extremely widespread in manufacturing processes, medicine, and biology. Even a brief discussion of all the possibilities would fill an entire book, and to keep such a book up to date would require a number of revisions each year. In this section, we shall present a few of these applications and some of the underlying theories supporting them.

Tracing

Radioactive particles can be used to trace chemicals participating in various reactions. One of the most valuable uses of radioactive tracers is in medicine. For example, ^{131}I is an artificially produced isotope of iodine (the natural, nonradioactive isotope is ^{127}I). Iodine, which is a necessary nutrient for our bodies, is obtained largely through the intake of iodized salt and seafood. The thyroid gland plays a major role in the distribution of iodine throughout the body. In order to evaluate the performance of the thyroid, the patient drinks a

Figure 33.9
A tracer technique for determining the condition of the human circulatory system.

① ²⁴NaCl solution injected

② Blood carries ²⁴NaCl to both legs

Site of constriction

③ High reading — good circulation

④ Low reading — poor circulation

very small amount of radioactive sodium iodide. Two hours later, the amount of iodine in the thyroid gland is determined by measuring the radiation intensity at the neck area.

A second medical application is indicated in Figure 33.9. Here a salt containing radioactive sodium is injected into a vein in the leg. The time at which the radioisotope arrives at another part of the body is detected with a radiation counter. The elapsed time is a good indication of the presence or absence of constrictions in the circulatory system.

The tracer technique is also useful in agricultural research. Suppose the best method of fertilizing a plant is to be determined. A certain material in the fertilizer, such as nitrogen, can be tagged with one of its radioactive isotopes. The fertilizer is then sprayed on one group of plants, sprinkled on the ground for a second group, and raked into the soil for a third. A Geiger counter is then used to track the nitrogen through the three types of plants.

Tracing techniques are as wide-ranging as human ingenuity can devise. Present applications range from checking the absorption of fluorine by teeth to checking contamination of food-processing equipment by cleansers to monitoring deterioration inside an automobile engine. In the latter case, a radioactive material is used in the manufacture of the pistons, and the oil is checked for radioactivity to determine the amount of wear on the pistons.

Activation Analysis

For centuries, a standard method of identifying the elements in a sample of material has been chemical analysis, which involves testing a portion of the material for reactions with various chemicals. A second method is spectral analysis, which utilizes the fact that, when excited, each element emits its own characteristic set of electromagnetic wavelengths. These methods are now supplemented by a third technique, **neutron activation analysis.** Both chemical and spectral methods have the disadvantage that a fairly large sample of the material must be destroyed for the analysis. In addition, extremely small quantities of an element may go undetected by either method. Activation analysis has an advantage over the other two methods in both of these respects.

When the material under investigation is irradiated with neutrons, nuclei in the material will absorb the neutrons and be changed to different isotopes. Most of these isotopes will be radioactive. For example, ^{65}Cu absorbs a neutron to become ^{66}Cu, which undergoes beta decay:

$$^{1}_{0}n + \, ^{65}_{29}Cu \longrightarrow \, ^{66}_{29}Cu \longrightarrow \, ^{66}_{30}Zn + \, ^{0}_{-1}e \qquad (33.7)$$

The presence of the copper can be deduced because it is known that ^{66}Cu has a half-life of 5.1 min and decays with the emission of beta particles having maximum energies of 2.63 and 1.59 MeV. Also emitted in the decay of ^{66}Zn is a gamma ray having an energy of 1.04 MeV. Thus, by examining the radiation emitted by a substance after it has been exposed to neutron irradiation, one can detect extremely small traces of an element.

Neutron activation analysis is used routinely by a number of industries, but the following nonroutine example of its use is of interest. Napoleon died on the island of St. Helena in 1821, supposedly of natural causes. Over the years, suspicion has existed that his death was not all that natural. After his death, his head was shaved and locks of his hair were sold as souvenirs. In 1961, the amount of arsenic in a sample of this hair was measured by neutron activation analysis. Unusually large quantities of arsenic were found in the hair. (Activation analysis is so sensitive that very small pieces of a single hair could be analyzed.) Results showed that the arsenic was fed to him irregularly. In fact, the arsenic concentration pattern corresponded to the fluctuations in the severity of Napoleon's illness as determined from historical records.

▶SPECIAL TOPIC
Medical Imaging Techniques

Computed Axial Tomography (CAT Scans)

The normal x-ray of a human body has two primary disadvantages when used as a source of clinical diagnosis. First, it is difficult to distinguish between various types of tissue in the body because they all have similar x-ray absorption properties. Second, a conventional x-ray absorption picture is indicative of the average amount of absorption along a particular direction in the body, leading to somewhat obscured pictures. To overcome these problems, a device called a CAT scanner was developed in England in 1973; it is capable of producing pictures of much greater clarity and detail than were previously obtainable.

The operation of a CAT scanner can be understood by considering the following hypothetical experiment. Suppose a box consists of four compartments, labeled A, B, C, and D as in Figure 33.10a. Each compartment has a different amount of absorbing material than any other compartment. What set of experimental procedures will enable us to determine the relative amounts of material in each compartment? The following steps outline one method that will provide this information. First, a beam of x-rays is passed through compartments A and C, as in Figure 33.10b. The intensity of the exiting radiation is reduced by absorption by some number that we assign as 8. (The number 8 could mean, for example, that the intensity of the exiting beam is reduced by eight-tenths of one percent from its initial value.) Since we do not know which of the compartments, A or C, was responsible for this reduction in intensity, half the loss is assigned to each compartment as in Figure 33.10c. Next, a beam of x-rays is passed

(a)

(b)

Figure 33.12
State-of-the-art magnetic resonance images taken of the human head. (a) is a sagittal view and (b) is a coronal view. The slice images are of 1 cm thickness and the result of a 128 × 128 point reconstruction. *(Photo Researchers, Inc.)*

is applied to align the magnetic moments, while a second, weak oscillating magnetic field is applied perpendicular to the dc field. When the frequency of the oscillating field is adjusted to match the precessional frequency of the magnetic moments, the nuclei will "flip" between the two spin states. These transitions result in a net absorption of energy by the spin system, which can be detected electronically.

In MRI, image reconstruction is obtained using spatially varying magnetic fields and a procedure for encoding each point in the sample being imaged. Some **MRI** images taken on a human head are shown in Figure 33.12. In practice, a computer-controlled pulse sequencing technique is used to produce signals that are captured by a suitable processing device. This signal is then subjected to appropriate mathematical manipulations to provide data for the final image. The main advantage of **MRI** over other imaging techniques in medical diagnostics is that it causes minimal damage to cellular structures. Photons associated with the rf signals used in **MRI** have energies of only about 10^{-7} eV. Since molecular bond strengths are much larger (of the order of 1 eV), the rf photons cause little cellular damage. In comparison, x-rays or γ-rays have energies ranging from 10^4 to 10^6 eV and can cause considerable cellular damage. ◀

Summary

In **nuclear fission** and **nuclear fusion,** the total rest mass of the products is always less than the original rest mass of the reactants. Nuclear fission occurs when a heavy nucleus splits, or fissions, into two smaller nuclei. In nuclear fusion, two light nuclei combine to form a heavier nucleus.

A **nuclear reactor** is a system designed to maintain what is called a self-sustaining chain reaction. Nuclear reactors using controlled fission events are currently being used to generate electric power.

Controlled fusion events offer the hope of plentiful supplies of energy in the future. The nuclear fusion reactor is considered by many scientists to be the ultimate energy source because its fuel is water. **Lawson's criterion** states that a fusion reactor will provide a net output power if the product of the plasma ion density, n, and the plasma confinement time, τ, satisfies the following relationships:

| $n\tau \geqslant 10^{14}$ s/cm^3 | Deuterium-tritium interaction |
| $n\tau \geqslant 10^{16}$ s/cm^3 | Deuterium-deuterium interaction |

Devices used to detect radiation are the Geiger counter, semiconductor diode detector, scintillation counter, photographic emulsion, cloud chamber, and bubble chamber.

Additional Reading

H. Bethe, "The Necessity of Fission Power," *Sci. American*, January 1976, p. 21.

B. L. Cohen, "The Disposal of Radioactive Wastes from Fission Reactors," *Sci. American*, June 1977, p. 21.

R. W. Conn, "The Magnetic Fusion Reactors," *Sci. American*, October 1983, p. 176.

L. Fermi, *Atoms in the Family: My Life with Enrico Fermi*, Chicago, University of Chicago Press, 1954.

H. P. Furth, "Progress Toward a Tokamak Fusion Reactor," *Sci. American*, August 1979, p. 50.

R. K. Lester, "Rethinking Nuclear Power," *Sci. American*, March 1986, p. 31.

H. W. Lewis, "The Safety of Fission Reactors," *Sci. American*, March 1980, p. 3.

P. R. Moran, R. J. Nickles, and J. A. Zagzebski, "The Physics of Medical Imaging," *Physics Today*, July 1983, p. 36.

I. I. Pykett, "NMR Imaging in Medicine," *Sci. American*, May 1982, p. 78.

M. M. Ter-Pogossian, M. E. Raichle, and B. E. Sobol, "Positron-Emission Tomography," *Sci. American*, October 1980, p. 170.

A. C. Upton, "The Biological Effects of Low-Level Ionizing Radiation," *Sci. American*, February 1982, p. 41.

Questions

1. Why is water a better shield against neutrons than lead or steel?

2. Explain the function of a moderator in a fission reactor.

3. In a fission reactor, nuclear reactions produce heat to drive a turbine generator. How is this heat provided?

4. A breeder reactor produces more fissionable fuel than it consumes. Does this conflict with the law of conservation of energy? Discuss.

5. Discuss the advantages and disadvantages of fission reactors from the point of view of safety, pollution, and resources. Make a comparison with power generated from the burning of fossil fuels.

6. Discuss the similarities and differences between fission and fusion.

7. Why would a fusion reactor produce less radioactive waste than a fission reactor?

8. What factors make a fusion reaction difficult to achieve?

9. Discuss the major problems associated with the development of a controlled fusion reactor.

10. Lawson's criterion states that the product of ion density and confinement time must exceed a certain number before a break-even fusion reaction can occur. Why should these two parameters determine the outcome?

11. Why is the temperature required for deuterium-tritium fusion less than that needed for deuterium-deuterium fusion? Estimate the relative importance of coulombic repulsion and nuclear attraction in each case.

12. Describe two techniques that are being pursued in an effort to obtain power from nuclear fusion.

13. Discuss the advantages and disadvantages of fusion power from the point of view of safety, pollution, and resources.

14. If two radioactive samples have the same activity measured in curies, will they necessarily create the same amount of damage in a medium? Explain.

15. Radiation can be used to sterilize such things as surgical equipment and packaged foods. Why do you suppose this works?

16. One method of treating cancer of the thyroid is to insert a small radioactive source directly into the

tumor. The radiation emitted by the source can destroy cancerous cells. Very often, the radioactive isotope $^{131}_{53}I$ is injected into the bloodstream in this treatment. Why do you suppose iodine is used?

17. Many radiopharmaceuticals must be made immediately prior to use. Why?

18. Is a Geiger counter sensitive to further radiation during the formation of a pulse? Discuss this question in relation to the maximum count rate.

19. The design of a photomultiplier tube might suggest

that any number of dynodes may be used to amplify a weak signal. What factors do you suppose would limit the amplification in this device?

20. Physicists have attempted to measure the solar neutrinos associated with nuclear processes in the Sun's interior (presumably the source of the Sun's energy). These studies involve placing a large volume of scintillator material into a deep mine shaft to reduce the background from cosmic rays. Explain why the neutrinos are not similarly affected.

Problems

Section 33.1   Nuclear Fission
Section 33.2   Nuclear Reactors

1. Find the energy released in the following fission reaction:

$$^{1}_{0}n + ^{235}_{92}U \longrightarrow ^{144}_{56}Ba + ^{89}_{36}Kr + 3\,^{1}_{0}n$$

2. Another fission reaction similar to the one in Problem 1 leads to the formation of ^{141}Ba and ^{92}Kr when ^{235}U absorbs a neutron. Write down this reaction. How many neutrons are released in this reaction?

3. Strontium-90 is a particularly dangerous fission product of ^{235}U because it is radioactive, and it substitutes for calcium in bones. What other direct fission products would accompany it in the neutron-induced fission of ^{235}U? (*Note:* This reaction may release 2, 3, or 4 free neutrons.)

4. Find the energy released in the following fission reaction:

$$^{1}_{0}n + ^{235}_{92}U \longrightarrow ^{88}_{38}Sr + ^{136}_{54}Xe + 12\,^{1}_{0}n$$

5. If the average energy released in a fission event is 208 MeV, find the total number of fission events required to keep a 100-W lightbulb burning for 1 hour.

6. In order to minimize neutron leakage from a reactor, the ratio of the surface area to the volume must be as small as possible. Assume that a sphere and a cube both have the same volume. Find the surface-to-volume ratio for (a) the sphere and (b) the cube. (c) Which of these shapes would have the minimum leakage?

7. How many grams of ^{235}U must undergo fission to operate a 1000-MW power plant for one day if the conversion efficiency is 30%? (Assume 208 MeV released per fission event.)

8. An all-electric home uses approximately 2000 kWh of electricity per month. How much ^{235}U would be required to provide this house with its energy needs for 1 year? (Assume 100% conversion efficiency and 208 MeV released per fission.)

9. It has been estimated that the Earth contains 10^9 tons of natural uranium that can be mined economically. Of this total, 0.7% is ^{235}U. If all the world's energy needs (7×10^{12} J/s) were to be supplied by ^{235}U fission, how long would this supply last?

10. The first atomic bomb released an energy equivalent to 20 kilotons of TNT. If 1 ton of TNT releases about 4×10^9 J, how much uranium was lost through fission in this bomb? (Assume 208 MeV released per fission.)

11. Suppose that the water exerts an average frictional drag of 10^5 N on a nuclear powered ship. How far can the ship travel per kilogram of fuel if the fuel consists of enriched uranium containing 1.7% of the fissionable isotope ^{235}U and the ship's engine has a efficiency of 20%? (Assume 208 MeV released per fission event.)

Section 33.3   Nuclear Fusion

12. Find the energy released in the fusion reaction

$$^{2}_{1}H + ^{2}_{1}H \longrightarrow ^{3}_{1}H + ^{1}_{1}H$$

13. Find the energy released in the fusion reaction

$$^{2}_{1}H + ^{3}_{1}H \longrightarrow ^{4}_{2}He + ^{1}_{0}n$$

14. If an all-electric home uses approximately 2000 kWh of electricity per month, how many fusion events of the type described in Problem 13 would be required to keep this house running for one year?

15. Another series of nuclear reactions that can produce energy in the interior of stars is the cycle described below. This process is most efficient when the central temperature in a star is above 16,000,000 K. Because the temperature at the center of the Sun is only 15,000,000 K, the cycle below produces less than 10% of the Sun's energy. (a) A high-energy proton is absorbed by ^{12}C. Another nucleus, A, is produced in the reaction, along with a gamma ray. Identify nucleus A. (b) Nucleus A decays through positron emission to form nucleus B. Identify nucleus B. (c) Nu-

cleus B absorbs a proton to produce nucleus C and a gamma ray. Identify nucleus C. (d) Nucleus C absorbs a proton to produce nucleus D and a gamma ray. Identify nucleus D. (e) Nucleus D decays through positron emission to produce nucleus E. Identify nucleus E. (f) Nucleus E absorbs a proton to produce nucleus F plus an alpha particle. What is nucleus F? (*Note:* If nucleus F is not ^{12}C, the nucleus you started with, you have done something wrong and you should review the sequence of events to find your error.)

16. When a star has exhausted its hydrogen fuel, it may fuse other nuclear fuels. At temperatures above 100,000,000 K, helium fusion can occur. Write the equation for the processes described below. (a) Two alpha particles fuse to produce a nucleus A and a gamma ray. What is nucleus A? (b) Nucleus A absorbs an alpha particle to produce a nucleus B and a gamma ray. What is nucleus B? (c) Find the total energy released in the reactions given in (a) and (b).

17. Of all the hydrogen nuclei in the ocean, 0.0156% are deuterium. (a) How much deuterium could be obtained from 1 gal of ordinary tap water? (b) If all of this deuterium could be converted to energy through reactions such as the first reaction in Equation 33.4, how much energy could be obtained from the 1 gal of water? (c) The energy released through the burning of 1 gal of gasoline is approximately 2×10^8 J. How many gallons of gasoline would have to be burned in order to produce the same energy as the 1 gal of water?

18. To understand why containment of a plasma is necessary, consider the rate at which a plasma would be lost if it were not contained. (a) Estimate the rms speed of deuterons in a plasma at 10^8 K. (b) Estimate the time such a plasma would remain in a cube 10 cm on an edge if no steps were taken to confine it.

19. The oceans have a volume of 317 million cubic miles and contain 1.32×10^{21} kg of water. Of all the hydrogen nuclei in this water, 0.0156% are deuterium. (a) If all of these deuterium nuclei were fused to helium via the first reaction in Equation 33.4, determine the total amount of energy which could be released. (b) Present world energy consumption is about 7×10^{12} W. If consumption were 100 times greater, how many years would the energy supply calculated in (a) last?

Section 33.4   Radiation Damage in Matter

20. Two workers using an industrial x-ray machine accidentally insert their hands in the x-ray beam for the same length of time. The first worker inserts one hand in the beam, and the second worker inserts both hands. Which worker receives the larger dose in rad?

21. A particular radioactive source produces 100 mrad of 2-MeV gamma rays per hour at a distance of 1.0 m. (a) How long could a person stand at this distance before accumulating an intolerable dose of 1 rem? (b) Assuming the gamma radiation is emitted isotropically, at what distance would a person receive a dose of 10 mrad/h from this source?

22. Assume that an x-ray technician takes an average of eight x-rays per day and receives a dose of 5 rem/year as a result. (a) Estimate the dose in rem per x-ray taken. (b) How does this result compare with the amount of low-level background radiation the technician is exposed to?

23. A person whose mass is 75 kg is exposed to a dose of 25 rad. How many joules of energy are deposited in the person's body?

24. In terms of biological damage, how many rad of heavy ions is equivalent to 100 rad of x-rays?

25. A 200-rad dose of radiation is administered to a patient in an effort to combat a cancerous growth. Assuming all of the energy deposited is absorbed by the growth, (a) calculate the amount of energy delivered. (b) Assuming the growth has a mass of 0.25 kg and a specific heat equal to that of water, calculate its temperature rise.

26. Estimate your mass and then estimate the temperature rise of your body if you were subjected to a 1000-rad whole-body radiation dose. (A dose this large is unquestionably fatal.)

27. A "clever" technician decides to heat some water for his coffee with an x-ray machine. If the machine produces 10 rad/s, how long will it take to raise the temperature of a cup of water by 50°C?

28. Calculate the radiation dose in rad supplied to 2 kg of water such that the energy deposited equals (a) the rest energy of the water and (b) the heat of vaporization of the water.

29. A patient swallows a radiopharmaceutical tagged with phosphorus-32($^{32}_{15}$P), a β^- emitter with a half-life of 14.3 days. The average kinetic energy of the emitted electrons is 700 keV. If the initial activity of the sample is 1.31 MBq, determine: (a) the number of electrons emitted in a 10-day period, (b) the total energy deposited in the body during the 10 days, and (c) the absorbed dose if the electrons are completely absorbed in 100 g of tissue.

Section 33.5   Radiation Detectors

30. In a Geiger tube, the voltage between the electrodes is typically 1 kV and the current pulse discharges a 5-pF capacitor. (a) What is the energy amplification of this device for a 0.5-MeV beta ray? (b) How many electrons are avalanched by the initial electron?

Figure 34.1
(a) Bubble-chamber tracks of electron-positron pairs produced by 300-MeV gamma rays striking a lead sheet. *(Courtesy of Lawrence Berkeley Laboratory, University of California.)* (b) Sketch of the pertinent pair-production events. Note that the positrons deflect upward while the electrons deflect downward in an applied magnetic field directed into the diagram.

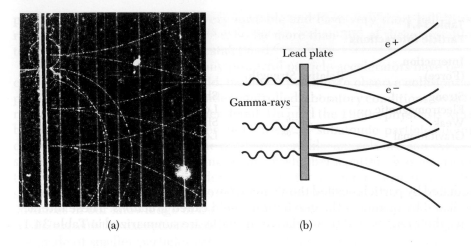

(a)   (b)

GeV.) In order to discriminate between positive and negative charges, the cloud chamber was placed in a magnetic field, causing moving charges to follow curved paths. Anderson noted that some of the electron-like tracks deflected in a direction corresponding to a positively charged particle.

Since Anderson's initial discovery, the positron has been observed in a number of experiments. Perhaps the most common process for producing positrons is **pair production.** In this process, a gamma ray with sufficiently high energy collides with a nucleus, and an electron-positron pair is created. Since the total rest energy of the electron-positron pair is $2m_0c^2 = 1.02$ MeV (where m_0 is the rest mass of the electron), the gamma ray must have at least this much energy to create an electron-positron pair. Thus, electromagnetic energy in the form of a gamma ray is transformed into mass in accordance with Einstein's famous relation $E = m_0c^2$. Figure 34.1 shows tracks of electron-positron pairs created by 300-MeV gamma rays striking a lead sheet.

A process that is the reverse of pair production can also occur. Under the proper conditions, an electron and positron can annihilate and produce two photons that have a combined energy of at least 1.02 MeV. The reaction can be expressed as

$$e + \bar{e} \longrightarrow 2\gamma$$

Likewise, a proton-antiproton pair can annihilate to produce two gamma rays; however, this event is very rare.

Practically every known elementary particle has an antiparticle. Among the exceptions are the photon and the neutral pion (π^0). (Note that the π^0 and η^0 are their own antiparticles!) Following the construction of high-energy accelerators in the 1950s, many other antiparticles were discovered. These included the antiproton ($\bar{p}$) discovered by Emilio Segre and Owen Chamberlain in 1955, and the antineutron ($\bar{n}$) discovered shortly thereafter.

34.4   Mesons and the Beginning of Particle Physics

The structure of matter as viewed by physicists in the mid-1930s was fairly simple. The building blocks of matter were considered to be the proton, the electron, and the neutron. Three other particles were known at the time: the gamma particle (photon), the neutrino, and the positron. These six particles were considered to be the fundamental constituents of matter. Although the accepted picture of the world was marvelously simple, no one was able to provide an answer to the following important question. If the nucleus of an atom contains many charged protons in close proximity that should strongly repel each other due to their like charges, what is the nature of the force that holds the nucleus together? Scientists recognized that this mysterious force must be much stronger than anything encountered in nature up to that time.

The first theory to explain the nature of the strong force was proposed in 1935 by the Japanese physicist Hideki Yukawa (1907–1981), an effort that later earned him the Nobel Prize. In order to understand Yukawa's theory, it is useful to first recall that *two atoms can form a covalent chemical bond by the exchange of electrons.* Similarly, in the modern views of electromagnetic interactions, *charged particles interact through the exchange of photons.* Yukawa used this same idea to explain the strong force by proposing a new particle whose exchange between nucleons in the nucleus produces the strong force with a range of about 10^{-15} m (the order of the nuclear diameter). Furthermore, he established that the range of the force is inversely proportional to the mass of this carrier particle, and predicted that the mass would be about 200 times the mass of the electron. Since the new particle would have a mass between that of the electron and proton, it was called a **meson** (from the Greek *meso*, meaning "middle").

In an effort to substantiate Yukawa's predictions, physicists began an experimental search for the meson by studying cosmic rays that enter the Earth's atmosphere from interstellar space. In 1937, Carl Anderson and his collaborators discovered a particle whose mass was 106 MeV/c^2, which is about 207 times the mass of the electron. However, subsequent experiments showed that the particle interacted very weakly with matter, and hence could not be the carrier of the strong force. The puzzling situation inspired several theoreticians to propose that there are actually two mesons with slightly different masses. This idea was confirmed in 1947 with the discovery in cosmic rays of the pi meson (π), or simply *pion*, by Cecil Frank Powell (1903–1969) and Guiseppe P. S. Occhialini (1907–   ). The lighter meson discovered earlier by Anderson, now called a *muon* (μ), has only weak and electromagnetic interaction and plays no role in the strong interaction.

The pion comes in three varieties, corresponding to three charge states: π^+, π^-, and π^0. The π^+ and π^- particles have masses of 139.6 MeV/c^2, while the π^0 has a mass of 135.0 MeV/c^2. Pions and muons are very unstable particles. For example, the π^- first decays into a muon and an antineutrino with a lifetime of about 2.6×10^{-8} s. (The lifetime is defined as the inverse of the decay rate, λ). The muon then decays into an electron, a neutrino, and an antineutrino with a lifetime equal to 2.2×10^{-6} s. The sequence of decays is

$$\pi^- \longrightarrow \mu^- + \bar{\nu}$$
$$\mu^- \longrightarrow e + \nu + \bar{\nu} \tag{34.1}$$

The interaction between two particles can be represented in a simple diagram called a *Feynman diagram*, developed by Richard P. Feynman (1918–1988). Figure 34.2 is a Feynman diagram for the case of the electromagnetic interaction between two electrons. In this simple case, a single photon acts as the carrier of the electromagnetic field between the electrons. The photon transfers energy and momentum from one electron to the other in

Figure 34.2
Feynman diagram representing how a photon mediates the electromagnetic force between two electrons.

34.7   Strange Particles and Strangeness

Many particles discovered in the 1950s were produced by the strong interaction of pions with protons and neutrons in the atmosphere. A group of these particles, namely the K, Λ, and Σ, were found to exhibit unusual properties in their production and decay, and hence were called *strange particles*. One unusual property is that they are always produced in pairs. For example, when a pion collides with a proton, two neutral strange particles are produced with high probability (Fig. 34.4) following the reaction

$$\pi^- + p \longrightarrow K^0 + \Lambda^0$$

On the other hand, the reaction $\pi^- + p \rightarrow K^0 + n$ never occurred, even though no known conservation laws were violated and the energy of the pion was sufficient to initiate the reaction. The second peculiar feature of strange particles is that although they are produced by the strong interaction at a high rate, they do not decay into strongly interacting particles at a very high rate as might be expected. Instead, they decay very slowly, which is characteristic of the weak interaction. Their half-lives are in the range 10^{-10} s to 10^{-8} s; strongly interacting particles have lifetimes of the order of 10^{-23} s.

Figure 34.4
This bubble-chamber photograph shows many events, and the inset represents a drawing of identified tracks. The strange particles Λ^0 and K^0 are formed (at the bottom) as the π^- interacts with a proton according to $\pi^- + p \rightarrow \Lambda^0 + K^0$. (Note that the neutral particles leave no tracks, as indicated by the dashed lines.) The Λ^0 and K^0 then decay according to $\Lambda^0 \rightarrow \pi^- + p$ and $K^0 \rightarrow \pi + \mu^- + \nu_\mu$. *(Courtesy of Lawrence Berkeley Laboratory, University of California, Photographic Services.)*

In order to explain these unusual properties of strange particles, a new conservation law called conservation of strangeness was introduced, together with a new quantum number S called the **strangeness**. The strangeness numbers for various particles are given in Table 34.2. The production of strange particles in pairs is explained by assigning $S = +1$ to one of the particles and $S = -1$ to the other. The nonstrange particles such as the π mesons, protons, and leptons are assigned strangeness $S = 0$. The **law of conservation of strangeness** can be stated as follows:

> Whenever a nuclear reaction or decay occurs, the sum of the strangeness numbers before the process must equal the sum of the strangeness numbers after the process.

The slow decay of strange particles can be explained by assuming that the strong and electromagnetic interactions obey the law of conservation of strangeness, while the weak interaction does not. Since the decay reaction involves the loss of one strange particle, it violates strangeness conservation, and hence proceeds slowly via the weak interaction.

Example 34.3 Is Strangeness Conserved?

(a) Determine whether the following reaction occurs on the basis of conservation of strangeness.

$$\pi^0 + n \longrightarrow K^+ + \Sigma^-$$

Solution
The initial state has a total strangeness $S = 0 + 0 = 0$. Since the strangeness of the K^+ is $S = +1$, and the strangeness of the Σ^- is $S = -1$, the total strangeness of the final state is $+1 - 1 = 0$. Thus, we see that strangeness is conserved and the reaction is allowed.

(b) Show that the following reaction does not conserve strangeness, and hence cannot occur.

$$\pi^- + p \longrightarrow \pi^- + \Sigma^+$$

Solution
In this case, the initial state has a total strangeness $S = 0 + 0 = 0$, while the final state has a total strangeness $S = 0 + (-1) = -1$. Thus strangeness is not conserved, so the reaction does not occur.

Exercise Show that the observed reaction $p + \pi^- \rightarrow K^0 + \Lambda^0$ obeys the law of conservation of strangeness.

34.8 The Eightfold Way

As we have seen, conserved quantities such as spin, baryon number, lepton number, and strangeness are labels we associate with particles. Many classification schemes have been proposed that group particles into families based on such labels. First, consider the first eight baryons listed in Table 34.2, all having a spin of one half. The family consists of the proton, the neutron, and six other particles. If we plot their strangeness versus their charge using a sloping

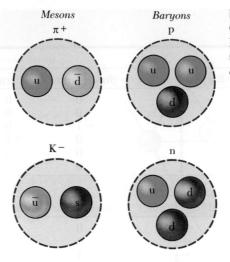

Mesons   *Baryons*

$\pi+$ p

$K-$ n

Figure 34.6
Quark compositions of several particles. Note that the mesons on the left contain two quarks, while the baryons on the right contain three quarks.

tains two u quarks and one d quark with $Q = +1$, $B = 1$, and $S = 0$. Note that just two of the quarks, u and d, are contained in all hadrons encountered in ordinary matter (protons and neutrons). The third quark, s, is needed only to construct strange particles with a strangeness number of either $+1$ or -1. For example, the neutral particle Λ^0 has uds as its quark composition, with $Q = 0$, $B = 1$, and $S = -1$. Figure 34.6 is a pictorial representation of the quark composition of several particles.

Charm and Other Recent Developments

Although the original quark model was highly successful in classifying particles into families, there were some discrepancies between predictions of the model and certain experimental decay rates. Consequently, a fourth quark was proposed by several physicists in 1967. They argued that if there are four leptons (as was thought at the time), then there should also be four quarks because of an underlying symmetry in nature. The fourth quark, designated by c, was given the new property, or quantum number, called **charm**. The *charmed* quark would have a charge $+2e/3$, but its new property of charm would distinguish it from the other three quarks. The new quark would have a charm of $C = +1$, while its antiquark would have a charm of $C = -1$ as indicated in Table 34.3. Charm, like strangeness, would be conserved in strong and electromagnetic interactions, but would not be conserved in weak interactions.

In 1974, a new heavy meson called the J/Ψ particle (or simply Ψ) was discovered independently by a group led by Burton Richter at the Stanford Linear Accelerator (SLAC) and another group led by Samuel Ting at the Brookhaven National Laboratory. Richter and Ting were awarded the Nobel Prize in 1976 for this work. The J/Ψ particle did not fit into the original three-quark model, but had the properties of a combination of a charmed quark and its antiquark ($c\bar{c}$). Its mass was much heavier than the other known mesons (~ 3100 MeV/c^2) and its lifetime was much longer than other strongly decaying particles. Soon thereafter, other related charmed mesons including D^+ and D^- were discovered corresponding to such quark combinations as $\bar{c}d$

Table 34.5
The Fundamental Particles and Some of Their Properties

Particle	Rest Energy	Charge
Quarks		
u	360 MeV	$+\frac{2}{3}e$
d	360 MeV	$-\frac{1}{3}e$
c	1500 MeV	$+\frac{2}{3}e$
s	540 MeV	$-\frac{1}{3}e$
t (?)	~100 Gev	$+\frac{2}{3}e$
b	5 GeV	$-\frac{1}{3}e$
Leptons		
e^-	511 keV	$-e$
μ^-	107 MeV	$-e$
τ^-	1784 MeV	$-e$
ν_e	<30 eV	0
ν_μ	<0.5 MeV	0
ν_τ	<250 MeV	0

and $c\overline{d}$, all of which have large masses and long lifetimes. In 1975, researchers at Stanford University reported strong evidence for the tau (τ) lepton, with a mass of 1784 MeV/c^2. Such discoveries led to more elaborate quark models, and the proposal of two new quarks, named *top (t)* and *bottom (b)*. (Some physicists prefer the whimsical names *truth* and *beauty*.) To dintinguish these quarks from the old ones, quantum numbers called *topness* and *bottomness* were assigned to these new particles, and are included in Table 34.3. In 1977, researchers at the Fermi National Laboratory under the direction of Leon Lederman reported the discovery of a very massive new meson, Y, whose composition is considered to be $b\overline{b}$.

At this point, you are probably wondering whether or not such discoveries will ever end. How many "building blocks" of matter really exist? At the present, physicists believe that the fundamental particles in nature include six quarks and six leptons (together with their antiparticles). Some of the properties of these particles are given in Table 34.5.

It should be noted that in spite of many extensive experimental efforts, no isolated quark has ever been observed. Physicists now believe that quarks are permanently confined inside ordinary particles because of an exceptionally strong force that prevents them from escaping. This force, called the "color" force, increases with separation distance (similar to the force of a spring); the properties of this force are discussed in the next section. The great strength of the force between quarks has been described by one author as follows:[3]

> Quarks are slaves of their own color charge, . . . bound like prisoners of a chain gang. . . . Any locksmith can break the chain between two prisoners, but no locksmith is expert enough to break the gluon chains between quarks. Quarks remain slaves forever.

[3] Harald Fritzsch, *Quarks, The Stuff of Matter*, London, Allen Lane, 1983.

34.10   The Standard Model

Shortly after the concept of quarks was proposed, scientists recognized that certain particles had quark compositions that were in violation of the Pauli exclusion principle. Recall that quarks are fermions with spins of 1/2, and hence are expected to follow the exclusion principle. One example is the Ω^- (*sss*) baryon, which contains three *s* quarks having parallel spins, giving it a total spin of 3/2. Other examples of baryons that have identical quarks with parallel spins are the Δ^{++} (*uuu*) and the Δ^- (*ddd*). To resolve this problem, it was suggested that quarks possess a new property called **color**. This property is similar in many respects to electric charge except that it occurs in three varieties (of color) called red, green, and blue. Another term that is used to distinguish among the six quarks is **flavor**. Thus, each flavor of quark can have three colors. (The whimsical names "color" and "flavor" should not be taken literally.) Of course, the antiquarks have the colors antired, antigreen, and antiblue. In order to satisfy the exclusion principle, all three quarks in a baryon must have different colors. A meson consists of a quark of one color and an antiquark of the corresponding anticolor. The result is that baryons and mesons are always colorless (or white). Furthermore, the new property of color increases the number of quarks by a factor of three.

Although the concept of color in the quark model was originally conceived to satisfy the exclusion principle, it also provided a better theory for explaining certain experimental results. For example, the modified theory correctly predicts the lifetime of the π^0 meson. The theory of how quarks interact with each other is called **quantum chromodynamics**, or **QCD**, because it is similar in its structure to quantum electrodynamics (the theory of interaction between electric charges). In QCD, the quark is said to carry a *color charge*, in analogy to electric charge. The strong force between quarks is often called the *color force*. As mentioned earlier, the strong interaction between hadrons is mediated by massless particles called **gluons** (analogous to photons for the electromagnetic force). According to the theory, there are eight gluons, six of which have color charge. Because of their color charge, quarks can attract each other and form composite particles. When a quark emits or absorbs a gluon, its color changes. For example, a blue quark that emits a gluon may become a red quark, while the red quark that absorbs this gluon becomes a blue quark. The color force between quarks is analogous to the electric force between charges—like colors repel and opposite colors attract. Therefore, two red quarks repel each other, but a red quark will be attracted to an antired quark. The attraction between quarks of opposite color to form a meson ($q\bar{q}$) is indicated in Figure 34.7a. Differently colored quarks also attract each other, but with less intensity than opposite colors of quark and antiquark. For example, a cluster of red, blue, and green quarks all attract each other to form baryons as indicated in Figure 34.7b. Thus, all baryons contain three quarks, each of which has a different color.

Recall that the weak force is believed to be mediated by the W^+, W^-, and Z^0 bosons (spin 1 particles). These particles are said to have *weak charge* just as a quark has color charge. Thus, each elementary particle can have mass, electric charge, color charge, and weak charge. Of course, one or more of these could be zero. Scientists now believe that the truly elementary particles are leptons and quarks, and the force mediators are the gluon, the photon, $W^\pm$,

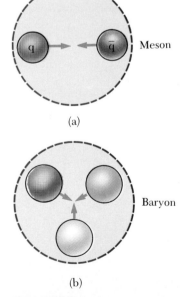

(a)

(b)

Figure 34.7
(a) A red quark is attracted to an antired quark. This forms a meson whose quark structure is ($q\bar{q}$). (b) Three different colored quarks attract each other to form a baryon.

Z^0, and the graviton. (Note that quarks and leptons have spin $\frac{1}{2}$ and hence are fermions, while the force mediators have spin 1 or higher, and are bosons.)

In 1979, Sheldon Glashow, Abdus Salam, and Steven Weinberg won a Nobel Prize for developing a theory that unified the electromagnetic and weak interactions. This so-called **electroweak theory** postulates that the weak and electromagnetic interactions have the same strength at very high particle energies. Thus, the two interactions are viewed as two different manifestations of a single unifying electroweak interaction. The photon and the three massive bosons ($W^\pm$ and Z^0) play a key role in the electroweak theory. The theory makes many concrete predictions, but perhaps the most spectacular is the prediction of the masses of the W and Z particles at about 82 GeV/c^2 and 93 GeV/c^2, respectively. The 1984 Nobel Prize was awarded to Carlo Rubbia and Simon van der Meer for their work leading to the discovery of these particles at just these energies at the CERN Laboratory in Geneva, Switzerland.

The combination of the electroweak theory and QCD for the strong interaction forms what is referred to in high-energy physics as the "Standard Model." Although the details of the Standard Model are complex, its essential ingredients can be summarized with the help of Figure 34.8. The strong force, mediated by gluons, holds quarks together to form composite particles such as protons, neutrons, and mesons. Leptons participate only in the electromagnetic and weak interactions. The electromagnetic force is mediated by photons, while the weak force is mediated by W and Z bosons. Note that all fundamental forces are mediated by spin 1 particles whose properties are given, to a large extent, by the symmetries involved in the theories. However, the Standard Model does not answer all questions. A major problem is why the photon has no mass while the W and Z bosons are massive particles. Because of their mass difference, the electromagnetic and weak forces are quite distinct at low energies, but become similar in nature at very high energies. This behavior as one goes from low to high energies, called *symmetry breaking*, leaves open the question of the origin of particle masses. In order to resolve

A computer reconstruction of the magnetic field distribution in the cross section of a prototype SSC magnet. In the green central area, the field is uniform to better than 0.02%. Other colors denote deviations of 0.04%, 0.06%, and so on. High quality field almost fills the whole beam chamber, ensuring that the proton beam will have excellent stability. *(Courtesy Brookhaven National Laboratory)*

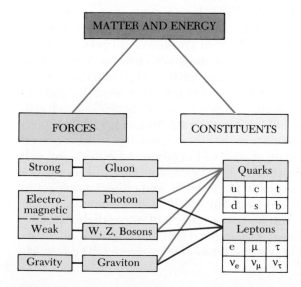

Figure 34.8
The Standard Model of particle physics.

Interaction halls

Future
interaction
halls

Tunnel cross section

Proton beam
(inside magnets)

Injector

Proton beam
(inside magnets)

Interaction
halls

Utility straight sections for beam-abort
and radio-frequency accelerating systems

Figure 34.9
The proposed layout of the
SSC, which is 52 mi in circumference. The experimental halls and other components on the ring are not to
scale. Access and service
points are placed at regular
intervals around the ring.

this problem, a hypothetical particle called the *Higgs boson* has been proposed, which provides a mechanism for breaking the electroweak symmetry.
The Standard Model, including the Higgs mechanism, provides a logically
consistent explanation of the massive nature of the W and Z bosons. Unfortunately, the Higgs boson has not yet been found, but physicists know that its
mass should be less than 1 TeV (10^{12} eV).

In order to determine whether the Higgs boson exists, two quarks of at
least 1 TeV of energy must collide, but calculations show that this requires
injecting 40 TeV of energy within the volume of a proton. The excess energy is
needed because of the quarks and gluons contained in the proton. Although no
existing accelerator can provide this energy, scientists in the United States are
currently planning to construct the world's largest and most powerful particle
accelerator called the Superconducting Super Collider, or SSC, which will
meet this need (Fig. 34.9). The SSC will accelerate protons to 20 TeV (two
protons traveling in opposite directions with this energy will give the required
40 TeV), will have a circumference of 82.944 km (about 52 miles), and will
cost about 4 billion dollars! The proton energies in the SSC will be twenty
times greater than is currently available at Fermilab, and physicists expect to
be able to explore distances down to about 10^{-18} m (one-thousandth the
diameter of a proton). Thus, it will be a microscope of unparalleled power.
Figure 34.10 shows the evolution of the various stages of matter that scientists
have been able to investigate with various types of microscopes. The Depart-

ment of Energy endorsed detailed research on the SSC in 1983, and federal funds were recently allocated to support research and development on the superconducting magnets to be used in the accelerator. The SSC is expected to be completed in the mid-1990s if federal funding and Congressional authorization are provided.

Following the success of the electroweak theory, scientists attempted to combine it with QCD in a *grand unification theory* known as GUT. In this model, the electroweak force was merged with the strong color force to form a grand unified force. One version of the theory considers leptons and quarks as members of the same family that are able to change into each other by exchanging an appropriate messenger particle. Many GUT theories predict that protons are unstable, and will decay with a half-life of about 10^{31} years. This is far greater than the age of the Universe, but as yet proton decays have not been observed.

34.11   The Cosmic Connection

As we have seen, the world around us is dominated by protons, electrons, neutrons, and neutrinos. Some of the other more exotic particles can be seen in cosmic rays. However, most of the new particles are produced using large, expensive machines that accelerate protons and electrons to energies in the GeV and TeV range. These energies are enormous when compared to the thermal energy in today's Universe. For example, kT is only about 1 keV at the center of the Sun. On the other hand, the temperature of the early Universe was high enough to reach energies of TeV and higher.

In this section we shall describe one of the most fascinating theories in all of science — the Big Bang theory of the creation of the Universe — and the experimental evidence that supports it. This theory of cosmology states that the Universe had a beginning, and further that the beginning was so cataclysmic that it is impossible to look back beyond the Creation. According to this theory, the Universe erupted from a point-like singularity about 15 to 20 billion years ago. The first few minutes after the Big Bang saw such extremes of energy that it is believed that all four interactions of physics were unified and that all matter melted down into an undifferentiated "quark soup."

The evolution of the four fundamental forces from the Big Bang to the present is shown in Figure 34.11. During the first 10^{-43} s (the ultra-hot epoch where $T \approx 10^{32}$ K), it is presumed that the strong, electroweak, and gravitational forces were joined to form a completely unified force. In the first 10^{-32} s following the Big Bang (the hot epoch where $T \approx 10^{29}$ K), gravity broke free of this unification while the strong and electroweak forces remained as one, described by a grand unification theory. This was a period when particle energies were so great ($> 10^{16}$ GeV) that very massive particles as well as quarks, leptons, and their antiparticles existed. Then the Universe rapidly expanded and cooled during the warm epoch when the temperatures ranged from 10^{29} to 10^{15} K (corresponding to energies ranging from 10^{16} to 10^2 GeV). During this epoch, the strong and electroweak forces parted company, and the grand unification scheme was broken. As the Universe continued to cool, about 10^{-10} s after the Big Bang, the electroweak force split in

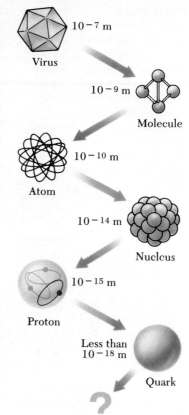

Figure 34.10
Looking at matter with various microscopes reveals structures ranging in size from the smallest living thing, a virus (at the top), down to a quark (at the bottom) which has not yet been observed as an isolated particle.

Figure 34.11
A brief history of the Universe from the Big Bang to the present. The four forces became distinguishable during the first microsecond. Following this, all the quarks combined to form the strongly interacting particles. However, the leptons remained separate, and they exist as individually observable particles to this day.

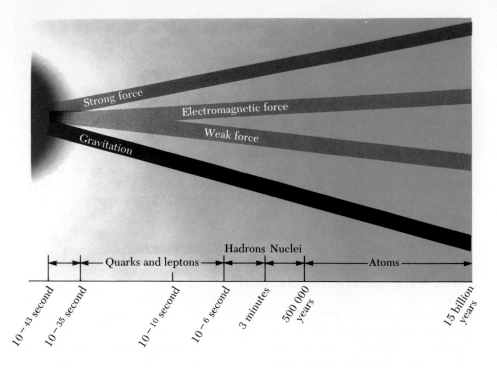

two. This resulted in the four forces in the Universe as we distinguish them today.

Until about 700,000 years after the Big Bang, the Universe was "radiation dominated"; ions absorbed and reemitted photons, thereby ensuring thermal equilibrium of radiation and matter. Energetic radiation also prevented matter from forming clumps or even neutral hydrogen atoms. When the Universe was about 700,000 years old, it had expanded and cooled to about 3000 K, and protons could bind to electrons to form neutral hydrogen atoms. Since neutral atoms do not appreciably scatter photons, the Universe suddenly became transparent to photons. The photons were frozen in time with an energy distribution characteristic of a 3000-K blackbody. Radiation no longer dominated the Universe and clumps of neutral matter steadily grew — first atoms, followed by molecules, gas clouds, stars, and finally galaxies.

Observation of Radiation from the Primordial Fireball

In 1965, Arno A. Penzias and Robert W. Wilson of Bell Labs were testing a sensitive microwave receiver and made an amazing discovery. A pesky signal producing a faint background "hiss" was interfering with their satellite communications experiments, and in spite of their valiant efforts, the signal remained. Ultimately, it would become clear that they were observing a microwave background radiation (at a wavelength of 7.35 cm) representing the left-over glow from the Big Bang.

The microwave horn that served as their receiving antenna is shown in Figure 34.12. The intensity of the detected signal remain unchanged as the antenna was pointed in different directions. The fact that the radiation had equal strengths in all directions suggested that the entire Universe was the source of this radiation. Booting a flock of pigeons from the 20-foot horn (bird droppings in the horn were a problem) and cooling the microwave detector both failed to remove the "spurious" signal. Through a casual conversation, Penzias and Wilson discovered that a group at Princeton had predicted the residual radiation from the Big Bang and were planning an experiment seeking to confirm the theory. The excitement in the scientific community was high when Penzias and Wilson announced that they had already observed an excess microwave background compatible with a 3-K blackbody source.

Because the measurements of Penzias and Wilson were taken at a single wavelength, they did not completely confirm the radiation as 3-K blackbody radiation. Subsequent experiments by other groups added intensity data at different wavelengths. The results confirm that the radiation is that of a blackbody at 2.9 K and represent, perhaps, the most clearcut evidence for the Big Bang theory. The 1978 Nobel prize in physics was awarded to Penzias and Wilson for their most important discovery.

Figure 34.12
Robert W. Wilson (left) and Arno A. Penzias with Bell Telephone Laboratories horn-reflector antenna. *(AT&T Bell Laboratories)*

34.12 Problems and Perspectives

While particle physicists have been exploring the realm of the very small, cosmologists have been exploring cosmic history back to the first microsecond of the Big Bang. The observation of events that occur when two particles collide in an accelerator is essential in reconstructing the early moments in cosmic history. Perhaps the key to understanding the early Universe is first to understand the world of elementary particles. Cosmologists and particle physicists now find they have many common goals and are joining hands to attempt to understand the physical world at its most fundamental level.

Our understanding of physics at short distances is far from complete. Particle physics is faced with many questions. Why is there so little antimatter in the Universe? Do neutrinos have a small rest mass, and if so, how do they contribute to the "dark matter" of the Universe? Is it possible to unify the strong and electroweak theories in a logical and consistent manner? Why do quarks and leptons form three similar but distinct families? Are muons the same as electrons (apart from their difference in mass), or do they have other subtle differences that have not been detected? Why are some particles charged and others neutral? Why do quarks carry a fractional charge? What determines the masses of the fundamental constituents? Can isolated quarks exist? The questions go on and on. Because of the rapid advances and new discoveries in the field of particle physics, by the time you read this book some of these questions will likely be resolved while others may emerge.

An important and obvious question that remains is whether leptons and quarks have a substructure. If they have a substructure, an infinite number of deeper structure levels could be envisioned. However, if leptons and quarks are indeed the ultimate constituents of matter, as physicists today tend to believe, we should be able to construct a final theory of the structure of matter as Einstein himself dreamed. In the view of many physicists, the end of the road is in sight, but it is anyone's guess as to how long it will take to reach that goal.

Summary

There are four fundamental forces in nature: strong (hadronic), electromagnetic, weak, and gravitational. The strong force is the force between nucleons that keeps the nucleus together. The weak force is responsible for beta decay. The electromagnetic and weak forces are now considered to be manifestations of a single force called the electroweak force.

An antiparticle and a particle have the same mass, but opposite charge. Furthermore other properties may have opposite values such as lepton number and baryon number. It is possible to produce particle-antiparticle pairs in nuclear reactions if the available energy is greater than $2m_0c^2$, where m_0 is the rest mass of the particle (or antiparticle).

Particles other than photons are classified as being *hadrons* or *leptons*. Hadrons interact primarily through the strong force. Hadrons have size and structure, and hence are not fundamental constituents of matter. There are two types of hadrons called *baryons* and *mesons*. Mesons have a baryon number zero and have either zero or integral spin. Baryons, which generally are the most massive particles, have a nonzero baryon number and a spin of 1/2 or 3/2. The neutron and proton are examples of baryons.

Leptons have no structure or size, and are considered to be truly elementary particles. Leptons interact only through the weak and electromagnetic forces. There are six leptons, the electron e^-, the muon μ^-, the tau τ^-, and their neutrinos ν_e, ν_μ, and ν_τ.

In all reactions and decays, quantities such as energy, linear momentum, angular momentum, electric charge, baryon number, and lepton number are strictly conserved. Certain particles have properties called *strangeness* and *charm*. These unusual properties are conserved only in those reactions and decays that occur via the strong force.

Recent theories in elementary particle physics have postulated that all hadrons are composed of smaller units known as *quarks*. Quarks have a fractional electric charge and a baryon number of 1/3. There are six flavors of quarks, up (*u*), down (*d*), strange (*s*), charmed (*c*), top (*t*), and bottom (*b*). All baryons contains three quarks, while all mesons contain one quark and one antiquark.

According to the theory of *quantum chromodynamics*, quarks have a property called *color*, and the strong force between quarks is referred to as the *color force*.

Every fundamental interaction is said to be mediated by the exchange of field particles. The electromagnetic interaction is mediated by the photon; the weak interaction is mediated by the $W^\pm$ and Z^0 bosons; the gravitational interaction is mediated by gravitons; the strong interaction is mediated by gluons.

Additional Reading

E. D. Bloom and G. J. Feldmann, "Quarkonium," *Sci. American*, May, 1982.

Frank Close, *The Cosmic Onion: Quarks and the Nature of the Universe*, The American Institute of Physics, 1986. A timely monograph on particle physics, including lively discussions of the Big Bang theory.

Harald Fritzsch, *Quarks, The Stuff of Matter*, London, Allen and Lane, 1983. An excellent introductory overview of elementary particle physics.

H. Harari, "The Structure of Quarks and Leptons," *Sci. American*, April, 1983.

J. David Jackson, Maury Tigner, and Stanley Wojcicki, "The Superconducting Supercollider," *Sci. American*, March, 1986.

Leon M. Lederman, "The Value of Fundamental Science," *Sci. American*, November, 1984.

N. B. Mistry, R. A. Poling, and E. H. Thorndike, "Particles with Naked Beauty," *Sci. American*, July, 1983.

Chris Quigg, "Elementary Particles and Forces," *Sci. American*, April, 1985.

James S. Trefil, *From Atoms to Quarks*, New York, Scribner, 1980. This is an excellent introduction to the world of particle physics.

Steven Weinberg, *The Discovery of Elementary Particles*, New York, Scientific American Library, W. H. Freeman and Company, 1983. This book emphasizes the important discoveries, experiments, and intellectual exercises which reshaped physics in the 20th century.

Steven Weinberg, "The Decay of the Proton," *Sci. American*, June, 1981.

Questions

1. Name the four fundamental interactions and the particles that mediate each interaction.
2. "All mesons are hadrons, but not all hadrons are mesons." Is this true? Discuss.
3. Describe the properties of baryons and mesons and the important differences between them.
4. Discuss the differences between hadrons and leptons.
5. Is the term "heavy lepton" a contradiction? Discuss.
6. Particles known as resonances have very short lifetimes, of the order of 10^{-23} s. From this information, would you guess they are hadrons or leptons? Explain.
7. The W and Z bosons were first produced at CERN in 1983 (by having a beam of protons and a beam of antiprotons meet at high energy). Why was this an important discovery?
8. Discuss the following conservation laws: energy, linear momentum, angular momentum, electric charge, baryon number, lepton number, and strangeness. Are all of these laws based on fundamental properties of nature? Explain.
9. The family of K mesons all decay into final states that contain no protons or neutrons. What is the baryon number of the K mesons?
10. An antibaryon interacts with a meson. Can a baryon be produced in such an interaction? Explain.
11. Identify the particle decays listed in Table 34.2 that occur by the weak interaction. Justify your answers.

12. Identify the particle decays listed in Table 34.2 that occur by the electromagnetic interaction. Justify your answers.
13. The Ξ^0 particle decays by the weak interaction according to the decay mode $\Xi^0 \rightarrow \Lambda^0 + \pi^0$. Would you expect this decay to be fast or slow? Explain.
14. Two protons in a nucleus interact via the strong interaction. Are they also subject to the weak interaction?
15. Discuss the quark model of hadrons, and describe the properties of quarks.
16. Discuss the essential features of the Standard Model of particle physics.
17. How many quarks are there in (a) a baryon, (b) an antibaryon, (c) a meson, and (d) an antimeson? How do you account for the fact that baryons have half-integral spins while mesons have spins of 0 or 1? (*Hint:* Quarks have spins of $\frac{1}{2}$.)
18. In the theory of quantum chromodynamics, quarks come in three colors. How would you justify the statement that "all baryons and mesons are colorless"?
19. What is the quark composition of the Ξ^- particle? (See Table 34.4.)
20. The sky is illuminated with a uniform background of radiation corresponding to a temperature of 2.9 K. Discuss the spectrum of this radiation, thought to be a remnant of the "Big Bang" as the Universe began.

Problems

Section 34.3 Positrons and Other Antiparticles

1. Two photons are produced when a proton and antiproton annihilate each other. What is the minimum frequency and corresponding wavelength of each photon?
2. A photon produces a proton-antiproton pair according to the reaction $\gamma \rightarrow p + \bar{p}$. What is the frequency of the photon? What is its wavelength?

Section 34.4 Mesons and the Beginning of Particle Physics

3. One of the mediators of the weak interaction is the Z^0 boson whose mass is 96 GeV/c^2. Use this information to find an approximate value for the range of the weak interaction.
4. Occasionally, high-energy muons will collide with electrons and produce two neutrinos according to the

reaction $\mu^+ + e \rightarrow 2\nu$. What kind of neutrinos are these?

5. When a high-energy proton or pion traveling near the speed of light collides with a nucleus, it travels an average distance of 3×10^{-15} m before interacting. From this information, estimate the time for the strong interaction to occur.

6. If a π^0 at rest decays into two γ s, what is the energy of each of the γ s?

Section 34.6   Conservation Laws
Section 34.7   Strange Particles and Strangeness

7. Identify the unknown particle on the left side of the reaction

$$? + p \longrightarrow n + \mu^+$$

8. The neutral ρ meson decays by the strong interaction into two pions according to $\rho^0 \rightarrow \pi^+ + \pi^-$ with a half-life of about 10^{-23} s. The neutral K meson also decays into two pions according to $K^0 \rightarrow \pi^+ + \pi^-$, but with a much longer half-life of about 10^{-10} s. How do you explain these observations?

9. Which of the following processes are allowed by the strong interaction, the electromagnetic interaction, the weak interaction, or no interaction at all?
   (a) $\pi^- + p \rightarrow 2\eta^0$
   (b) $K^- + n \rightarrow \Lambda^0 + \pi^-$
   (c) $K^- \rightarrow \pi^- + \pi^0$
   (d) $\Omega^- \rightarrow \Xi^- + \pi^0$
   (e) $\eta^0 \rightarrow 2\gamma$

10. For the following two reactions, the first may occur but the second cannot. Explain.

$$K^0 \rightarrow \pi^+ + \pi^- \quad \text{(can occur)}$$

$$\Lambda^0 \rightarrow \pi^+ + \pi^- \quad \text{(cannot occur)}$$

11. Determine whether or not strangeness is conserved in the following decays and reactions.
   (a) $\Lambda^0 \rightarrow p + \pi^-$
   (b) $\pi^- + p \rightarrow \Lambda^0 + K^0$
   (c) $\bar{p} + p \rightarrow \bar{\Lambda}^0 + \Lambda^0$
   (d) $\pi^- + p \rightarrow \pi^- + \Sigma^+$
   (e) $\Xi^- \rightarrow \Lambda^0 + \pi^-$
   (f) $\Xi^0 \rightarrow p + \pi^-$

12. Each of the following decays is forbidden. For each process, determine a conservation law that is violated.
   (a) $\mu^- \rightarrow e^- + \gamma$
   (b) $n \rightarrow p + e^- + \nu_e$
   (c) $\Lambda^0 \rightarrow p + \pi^0$
   (d) $p \rightarrow e^+ + \pi^0$
   (e) $\Xi^0 \rightarrow n + \pi^0$

13. Identify the conserved quantities in the following processes:
   (a) $\Xi^- \rightarrow \Lambda^0 + \mu^- + \bar{\nu}_\mu$
   (b) $K^0 \rightarrow 2\pi^0$
   (c) $K^- + p \rightarrow \Sigma^0 + n$
   (d) $\Sigma^0 \rightarrow \Lambda^0 + \gamma$
   (e) $e^+ + e^- \rightarrow \mu^+ + \mu^-$
   (f) $\bar{p} + n \rightarrow \bar{\Lambda}^0 + \Sigma^-$

14. Fill in the missing particle. Assume that (a) occurs via the strong interaction while (b) and (c) involve the weak interaction.
   (a) $K^+ + p \rightarrow \underline{\quad} + p$
   (b) $\Omega^- \rightarrow \underline{\quad} + \pi^-$
   (c) $K^+ \rightarrow \underline{\quad} + \mu^+ + \nu_\mu$

15. Each of the following reactions is forbidden. Determine a conservation law that is violated for each reaction.
   (a) $p + \bar{p} \rightarrow \mu^+ + e^-$
   (b) $\pi^- + p \rightarrow p + \pi^+$
   (c) $p + p \rightarrow p + \pi^+$
   (d) $p + p \rightarrow p + p + n$
   (e) $\gamma + p \rightarrow n + \pi^0$

16. (a) Show that baryon number and charge are conserved in the following reactions of a pion with a proton.

   (1) $\quad \pi^+ + p \rightarrow K^+ + \Sigma^+$

   (2) $\quad \pi^+ + p \rightarrow \pi^+ + \Sigma^+$

   (b) The first reaction is observed, but the second never occurs. Explain these observations.

17. The following reactions or decays involve one or more neutrinos. Supply the missing neutrinos.
   (a) $\pi^- \rightarrow \mu^- + ?$
   (b) $K^+ \rightarrow \mu^+ + ?$
   (c) $? + p \rightarrow n + e^+$
   (d) $? + n \rightarrow p + e^-$
   (e) $? + n \rightarrow p + \mu^-$
   (f) $\mu^- \rightarrow e^- + ? + ?$

18. Determine the type of neutrino or antineutrino involved in each of the following processes:
   (a) $\pi^+ \rightarrow \pi^0 + e^+ + ?$
   (b) $? + p \rightarrow \mu^- + p + \pi^+$
   (c) $\Lambda^0 \rightarrow p + \mu^- + ?$
   (d) $\tau^+ \rightarrow \mu^+ + ? + ?$

19. Determine which of the reactions below can occur. For those that cannot occur, determine the conservation law (or laws) that each violates.
   (a) $p \rightarrow \pi^+ + \pi^0$
   (b) $p + p \rightarrow p + p + \pi^0$
   (c) $p + p \rightarrow p + \pi^+$
   (d) $\pi^+ \rightarrow \mu^+ + \nu_\mu$
   (e) $n \rightarrow p + e^- + \bar{\nu}_e$
   (f) $\pi^+ \rightarrow \mu^+ + n$

20. Two protons approach each other with equal and opposite velocities. What is the minimum kinetic energy of each of the protons if they are to produce a π^+ meson at rest in the reaction,

$$p + p \longrightarrow p + n + \pi^+$$

21. A Σ^0 particle at rest decays according to

$$\Sigma^0 \longrightarrow \Lambda^0 + \gamma$$

Find the gamma-ray energy.

Section 34.8   Quarks — Finally

22. The quark compositions of the K^0 and Λ^0 particles are $d\bar{s}$ and uds, respectively. Show that the charge, baryon number, and strangeness of these particles equal the sums of these numbers for the quark constituents.

23. The quark composition of the proton is uud, while that of the neutron in udd. Show that the charge, baryon number, and strangeness of these particles equal the sums of these numbers for their quark constituents.

24. What is the electrical charge of the baryons with the quark compositions (a) $\overline{uud}$ and (b) $\overline{udd}$. What are these baryons called?

25. Analyze each of the reactions in terms of their constituent quarks:

(a) $\pi^- + p \rightarrow K^0 + \Lambda^0$

(b) $\pi^+ + p \rightarrow K^+ + \Sigma^+$

(c) $K^- + p \rightarrow K^+ + K^0 + \Omega^-$

(d) $p + p \rightarrow K^0 + p + \pi^+ + ?$

In the last reaction, identify the mystery particle.

26. Identify the particles corresponding to the following quark states: (a) uds; (b) $\bar{u}d$; (c) $\bar{s}d$; (d) ssd.

27. Neglect binding energies and estimate the mass of the u and d quarks from the mass of the proton and neutron.

APPENDIX A

Mathematical Review

A.1 Scientific Notation

Many quantities that scientists deal with often have very large or very small values. For example, the speed of light is about 300,000,000 m/s and the ink required to make the dot over an i in this textbook has a mass of about 0.000000001 kg. Obviously, it is very cumbersome to read, write, and keep track of numbers such as these. We avoid this problem by using a method dealing with powers of the number 10:

$$10^0 = 1$$

$$10^1 = 10$$

$$10^2 = 10 \times 10 = 100$$

$$10^3 = 10 \times 10 \times 10 = 1000$$

$$10^4 = 10 \times 10 \times 10 \times 10 = 10,000$$

$$10^5 = 10 \times 10 \times 10 \times 10 \times 10 = 100,000$$

and so on. The number of zeros corresponds to the power to which 10 is raised, called the **exponent** of 10. For example, the speed of light, 300,000,000 m/s, can be expressed as 3×10^8 m/s.

For numbers less than one, we note the following:

$$10^{-1} = \frac{1}{10} = 0.1$$

$$10^{-2} = \frac{1}{10 \times 10} = 0.01$$

$$10^{-3} = \frac{1}{10 \times 10 \times 10} = 0.001$$

$$10^{-4} = \frac{1}{10 \times 10 \times 10 \times 10} = 0.0001$$

$$10^{-5} = \frac{1}{10 \times 10 \times 10 \times 10 \times 10} = 0.00001$$

In these cases, the number of places the decimal point is to the left of the digit 1 equals the value of the (negative) exponent. Numbers that are expressed as some power of 10 multiplied by another number between 1 and 10 are said to be in **scientific notation.** For example, the scientific notation for 5,943,000,000 is 5.943×10^9 and that for 0.0000832 is 8.32×10^{-5}.

When numbers expressed in scientific notation are being multiplied, the following general rule is very useful:

$$10^n \times 10^m = 10^{n+m} \qquad\qquad \text{(A.1)}$$

where n and m can be *any* numbers (not necessarily integers). For example, $10^2 \times 10^5 = 10^7$. The rule also applies if one of the exponents is negative. For example, $10^3 \times 10^{-8} = 10^{-5}$.

When dividing numbers expressed in scientific notation, note that

$$\frac{10^n}{10^m} = 10^n \times 10^{-m} = 10^{n-m} \qquad \text{(A.2)}$$

Exercises

With help from the above rules, verify the answers to the following:

1. $86{,}400 = 8.64 \times 10^4$
2. $9{,}816{,}762.5 = 9.8167625 \times 10^6$
3. $0.0000000398 = 3.98 \times 10^{-8}$
4. $(4 \times 10^8)(9 \times 10^9) = 3.6 \times 10^{18}$
5. $(3 \times 10^7)(6 \times 10^{-12}) = 1.8 \times 10^{-4}$
6. $\dfrac{75 \times 10^{-11}}{5 \times 10^{-3}} = 1.5 \times 10^{-7}$
7. $\dfrac{(3 \times 10^6)(8 \times 10^{-2})}{(2 \times 10^{17})(6 \times 10^5)} = 2 \times 10^{-18}$

A.2   Algebra

A. Some Basic Rules

When algebraic operations are performed, the laws of arithmetic apply. Symbols such as x, y, and z are usually used to represent quantities that are not specified, what are called the **unknowns**.

First, consider the equation

$$8x = 32$$

If we wish to solve for x, we can divide (or multiply) each side of the equation by the same factor without destroying the equality. In this case, if we divide both sides by 8, we have

$$\frac{8x}{8} = \frac{32}{8}$$

$$x = 4$$

Next consider the equation

$$x + 2 = 8$$

In this type of expression, we can add or subtract the same quantity from each side. If we subtract 2 from each side, we get

$$x + 2 - 2 = 8 - 2$$

$$x = 6$$

In general, if $x + a = b$, then $x = b - a$.

Now consider the equation

$$\frac{x}{5} = 9$$

If we multiply each side by 5, we are left with x on the left by itself and 45 on the right:

$$\left(\frac{x}{5}\right)(5) = 9 \times 5$$

$$x = 45$$

In all cases, *whatever operation is performed on the left side of the equality must also be performed on the right side.*

The following rules for multiplying, dividing, adding, and subtracting fractions should be recalled, where a, b, and c are three numbers:

	Rule	Example
Multiplying	$\left(\dfrac{a}{b}\right)\left(\dfrac{c}{d}\right) = \dfrac{ac}{bd}$	$\left(\dfrac{2}{3}\right)\left(\dfrac{4}{5}\right) = \dfrac{8}{15}$
Dividing	$\dfrac{(a/b)}{(c/d)} = \dfrac{ad}{bc}$	$\dfrac{2/3}{4/5} = \dfrac{(2)(5)}{(4)(3)} = \dfrac{10}{12}$
Adding	$\dfrac{a}{b} \pm \dfrac{c}{d} = \dfrac{ad \pm bc}{bd}$	$\dfrac{2}{3} - \dfrac{4}{5} = \dfrac{(2)(5) - (4)(3)}{(3)(5)} = -\dfrac{2}{15}$

Exercises

In the following exercises, solve for x:

Answers

1. $a = \dfrac{1}{1+x}$ $\qquad x = \dfrac{1-a}{a}$

2. $3x - 5 = 13$ $\qquad x = 6$

3. $ax - 5 = bx + 2$ $\qquad x = \dfrac{7}{a-b}$

4. $\dfrac{5}{2x+6} = \dfrac{3}{4x+8}$ $\qquad x = -\dfrac{11}{7}$

B. Powers

When powers of a given quantity x are multiplied, the following rule applies:

$$x^n x^m = x^{n+m} \tag{A.3}$$

For example, $x^2 x^4 = x^{2+4} = x^6$.

Table A.1
Rules of Exponents

$$x^0 = 1$$
$$x^1 = x$$
$$x^n x^m = x^{n+m}$$
$$x^n/x^m = x^{n-m}$$
$$x^{1/n} = \sqrt[n]{x}$$
$$(x^n)^m = x^{nm}$$

When dividing the powers of a given quantity, note that

$$\frac{x^n}{x^m} = x^{n-m} \tag{A.4}$$

For example, $x^8/x^2 = x^{8-2} = x^6$.

A power that is a fraction, such as $\frac{1}{3}$, corresponds to a root as follows:

$$x^{1/n} = \sqrt[n]{x} \tag{A.5}$$

For example, $4^{1/3} = \sqrt[3]{4} = 1.5874$. (A scientific calculator is useful for such calculations.)

Finally, any quantity x^n that is raised to the mth power is

$$(x^n)^m = x^{nm} \tag{A.6}$$

Table A.1 summarizes the rules of exponents.

Exercises

Verify the following:

1. $3^2 \times 3^3 = 243$
2. $x^5 x^{-8} = x^{-3}$
3. $x^{10}/x^{-5} = x^{15}$
4. $5^{1/3} = 1.709975$ (Use your calculator.)
5. $60^{1/4} = 2.783158$ (Use your calculator.)
6. $(x^4)^3 = x^{12}$

C. Factoring

Some useful formulas for factoring an equation are

$$ax + ay + az = a(x + y + z) \qquad \text{common factor}$$
$$a^2 + 2ab + b^2 = (a + b)^2 \qquad \text{perfect square}$$
$$a^2 - b^2 = (a + b)(a - b) \qquad \text{differences of squares}$$

D. Quadratic Equations

The general form of a quadratic equation is

$$ax^2 + bx + c = 0 \tag{A.7}$$

where x is the unknown quantity and a, b, and c are numerical factors referred to as **coefficients** of the equation. This equation has two roots, given by

$$x = \frac{-b \pm \sqrt{b^2 - 4ac}}{2a} \tag{A.8}$$

If $b^2 \geq 4ac$, the roots will be real.

Example

The equation $x^2 + 5x + 4 = 0$ has the following roots corresponding to the two signs of the square-root term:

$$x = \frac{-5 \pm \sqrt{5^2 - (4)(1)(4)}}{2(1)} = \frac{-5 \pm \sqrt{9}}{2} = \frac{-5 \pm 3}{2}$$

that is,

$$x_+ = \frac{-5 + 3}{2} = \boxed{-1} \qquad x_- = \frac{-5 - 3}{2} = \boxed{-4}$$

where x_+ refers to the root corresponding to the positive sign and x_- refers to the root corresponding to the negative sign.

Figure A.1

Exercises

Solve the following quadratic equations:

	Answers	
1. $x^2 + 2x - 3 = 0$	$x_+ = 1$	$x_- = -3$
2. $2x^2 - 5x + 2 = 0$	$x_+ = 2$	$x_- = 1/2$
3. $2x^2 - 4x - 9 = 0$	$x_+ = 1 + \sqrt{22}/2$	$x_- = 1 - \sqrt{22}/2$

E. Linear Equations

A linear equation has the general form

$$y = ax + b \tag{A.9}$$

where a and b are constants. This equation is referred to as being linear because the graph of y versus x is a straight line, as shown in Figure A.1. The constant b, called the **intercept,** represents the value of y at which the straight line intersects the y-axis. The constant a is equal to the **slope** of the straight line and is also equal to the tangent of the angle that the line makes with the x-axis. If any two points on the straight line are specified by the coordinates (x_1, y_1) and (x_2, y_2), as in Figure A.1, then the *slope* of the straight line can be expressed

$$\text{Slope} = \frac{y_2 - y_1}{x_2 - x_1} = \frac{\Delta y}{\Delta x} \tag{A.10}$$

Note that a and b can have either positive or negative values. If $a > 0$, the straight line has a *positive* slope, as in Figure A.1. If $a < 0$, the straight line has a *negative* slope. In Figure A.1, both a and b are positive. Three other possible situations are shown in Figure A.2: $a > 0$, $b < 0$; $a < 0$, $b > 0$; and $a < 0$, $b < 0$.

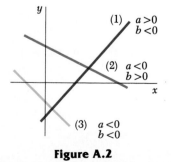

Figure A.2

Exercises

1. Draw graphs of the following straight lines:
   (a) $y = 5x + 3$ (b) $y = -2x + 4$ (c) $y = -3x - 6$
2. Find the slopes of the straight lines described in Exercise 1.
   Answers: (a) 5   (b) -2 (c) -3
3. Find the slopes of the straight lines that pass through the following sets of points:
   (a) $(0, -4)$ and $(4, 2)$, (b) $(0, 0)$ and $(2, -5)$, and (c) $(-5, 2)$ and $(4, -2)$
   Answers: (a) $3/2$ (b) $-5/2$ (c) $-4/9$

F. Solving Simultaneous Linear Equations

Consider an equation such as $3x + 5y = 15$, which has two unknowns, x and y. Such an equation does not have a unique solution. That is, $(x = 0, y = 3)$, $(x = 5, y = 0)$, and $(x = 2, y = 9/5)$ are all solutions to this equation.

If a problem has two unknowns, a unique solution is possible only if we have *two* independent equations. In general, if a problem has n unknowns, its solution requires n independent equations. In order to solve two simultaneous equations involving two unknowns, x and y, we solve one of the equations for x in terms of y and substitute this expression into the other equation.

Example

Solve the following two simultaneous equations:

$$(1) \quad 5x + y = -8 \qquad (2) \quad 2x - 2y = 4$$

Solution
From (2), $x = y + 2$. Substitution of this into (1) gives

$$5(y + 2) + y = -8$$

$$6y = -18$$

$$y = -3$$

$$x = y + 2 = \boxed{-1}$$

Alternate solution: Multiply each term in (1) by the factor 2 and add the result to (2):

$$10x + 2y = -16$$

$$\underline{2x - 2y = 4}$$

$$12x = -12$$

$$x = -1$$

$$y = x - 2 = \boxed{-3}$$

Two linear equations with two unknowns can also be solved by a graphical method. If the straight lines corresponding to the two equations are plotted in a conventional coordinate system, the intersection of the two lines represents the solution. For example, consider the two equations

$$x - y = 2$$
$$x - 2y = -1$$

These are plotted in Figure A.3. The intersection of the two lines has the coordinates $x = 5$, $y = 3$. This represents the solution to the equations. You should check this solution by the analytical technique discussed above.

Exercises

Solve the following pairs of simultaneous equations involving two unknowns:

Answers

1. $x + y = 8$ $x = 5, y = 3$
   $x - y = 2$
2. $98 - T = 10a$ $T = 65, a = 3.27$
   $T - 49 = 5a$
3. $6x + 2y = 6$ $x = 2, y = -3$
   $8x - 4y = 28$

G. Logarithms

Suppose that a quantity x is expressed as a power of some quantity a:

$$x = a^y \qquad \text{(A.11)}$$

The number a is called the **base** number. The **logarithm** of x with respect to the base a is equal to the exponent to which the base must be raised in order to satisfy the expression $x = a^y$:

$$y = \log_a x \qquad \text{(A.12)}$$

Conversely, the **antilogarithm** of y is the number x:

$$x = \text{antilog}_a y \qquad \text{(A.13)}$$

In practice, the two bases most often used are base 10, called the *common* logarithm base, and base $e = 2.718 \ldots$, called the *natural* logarithm base. When common logarithms are used,

$$y = \log_{10} x \qquad (\text{or } x = 10^y) \qquad \text{(A.14)}$$

When natural logarithms are used,

$$y = \ln_e x \qquad (\text{or } x = e^y) \qquad \text{(A.15)}$$

For example, $\log_{10} 52 = 1.716$, so that antilog$_{10}$ 1.716 $= 10^{1.716} = 52$. Likewise, $\ln_e 52 = 3.951$, so antiln$_e$ 3.951 $= e^{3.951} = 52$.

In general, note that you can convert between base 10 and base e with the equality

$$\ln_e x = (2.302585) \log_{10} x \qquad \text{(A.16)}$$

Finally, some useful properties of logarithms are

$$\log (ab) = \log a + \log b \qquad \textbf{(A.17)}$$

$$\log (a/b) = \log a - \log b \qquad \textbf{(A.18)}$$

$$\log (a^n) = n \log a \qquad \textbf{(A.19)}$$

A.3   Geometry

Table A.2 gives the areas and volumes for several geometric shapes used throughout this text:

Table A.2
Useful Information for Geometry

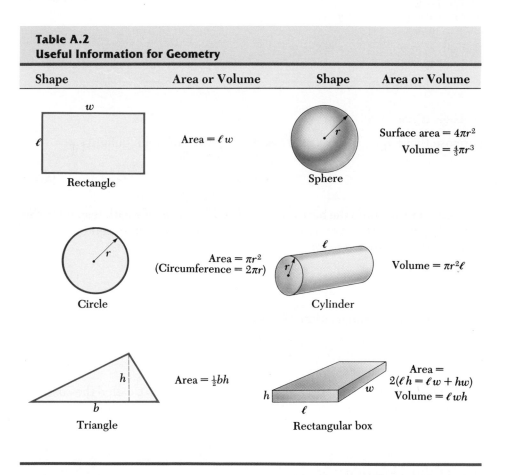

Shape	Area or Volume	Shape	Area or Volume
Rectangle	Area $= \ell w$	Sphere	Surface area $= 4\pi r^2$ Volume $= \frac{4}{3}\pi r^3$
Circle	Area $= \pi r^2$ (Circumference $= 2\pi r$)	Cylinder	Volume $= \pi r^2 \ell$
Triangle	Area $= \frac{1}{2}bh$	Rectangular box	Area $= 2(\ell h = \ell w + hw)$ Volume $= \ell wh$

A.4   Trigonometry

Some of the most basic facts concerning trigonometry are presented in Chapter 1, and we encourage you to study the material presented there if you are having trouble with this branch of mathematics. In addition to the discussion of Chapter 1, certain useful trig identities that can be of value to you follow.

$$\sin^2 \theta + \cos^2 \theta = 1 \tag{A.20}$$

$$\sin \theta = \cos(90° - \theta) \tag{A.21}$$

$$\cos \theta = \sin(90° - \theta) \tag{A.22}$$

$$\sin 2\theta = 2 \sin \theta \cos \theta \tag{A.23}$$

$$\cos 2\theta = \cos^2\theta - \sin^2\theta \tag{A.24}$$

$$\sin(\theta \pm \phi) = \sin \theta \cos \phi \pm \cos \theta \sin \phi \tag{A.25}$$

$$\cos(\theta \pm \phi) = \cos \theta \cos \phi \mp \sin \theta \sin \phi \tag{A.26}$$

APPENDIX B

An Abbreviated Table of Isotopes

Atomic Number, Z	Element	Symbol	Mass Number, A	Atomic Mass[a]	Percent Abundance, or Decay Mode if Radioactive[b]	Half-life if Radioactive
0	(Neutron)	n	1	1.008665	β^-	10.6 min
1	Hydrogen	H	1	1.007825	99.985	
	Deuterium	D	2	2.014102	0.015	
	Tritium	T	3	3.016049	β^-	12.33 y
2	Helium	He	3	3.016029	0.00014	
			4	4.002603	≈ 100	
3	Lithium	Li	6	6.015123	7.5	
			7	7.016005	92.5	
4	Beryllium	Be	7	7.016930	EC, γ	53.3 days
			8	8.005305	2α	6.7×10^{-17} s
			9	9.012183	100	
5	Boron	B	10	10.012938	19.8	
			11	11.009305	80.2	
6	Carbon	C	11	11.011433	β^+, EC	20.4 min
			12	12.000000	98.89	
			13	13.003355	1.11	
			14	14.003242	β^-	5730 y
7	Nitrogen	N	13	13.005739	β^+	9.96 min
			14	14.003074	99.63	
			15	15.000109	0.37	
8	Oxygen	O	15	15.003065	β^+, EC	122 s
			16	15.994915	99.76	
			18	17.999159	0.204	
9	Fluorine	F	19	18.998403	100	
10	Neon	Ne	20	19.992439	90.51	
			22	21.991384	9.22	
11	Sodium	Na	22	21.994435	β^+, EC, γ	2.602 y
			23	22.989770	100	
			24	23.990964	β^-, γ	15.0 h
12	Magnesium	Mg	24	23.985045	78.99	
13	Aluminum	Al	27	26.981541	100	
14	Silicon	Si	28	27.976928	92.23	
			31	30.975364	β^-, γ	2.62 h
15	Phosphorus	P	31	30.973763	100	
			32	31.973908	β^-	14.28 days
16	Sulfur	S	32	31.972072	95.0	
			35	34.969033	β^-	87.4 days

(Table continues)

Note: Data taken from *Chart of the Nuclides*, 12th ed., New York, General Electric, 1977, and from C. M. Lederer and V. S. Shirley (eds.), *Table of Isotopes*, 7th ed., New York, Wiley, 1978.

[a] Masses are those for the neutral atom, including the Z electrons.

[b] The process EC stands for "electron capture."

Atomic Number, Z	Element	Symbol	Mass Number, A	Atomic Mass[a]	Percent Abundance, or Decay Mode if Radioactive[b]	Half-life if Radioactive
17	Chlorine	Cl	35	34.968853	75.77	
			37	36.965903	24.23	
18	Argon	Ar	40	39.962383	99.60	
19	Potassium	K	39	38.963708	93.26	
			40	39.964000	β^-, EC, γ, β^+	1.28×10^9 y
20	Calcium	Ca	40	39.962591	96.94	
21	Scandium	Sc	45	44.955914	100	
22	Titanium	Ti	48	47.947947	73.7	
23	Vanadium	V	51	50.943963	99.75	
24	Chromium	Cr	52	51.940510	83.79	
25	Manganese	Mn	55	54.938046	100	
26	Iron	Fe	56	55.934939	91.8	
27	Cobalt	Co	59	58.933198	100	
			60	59.933820	β^-, γ	5.271 y
28	Nickel	Ni	58	57.935347	68.3	
			60	59.930789	26.1	
			64	63.927968	0.91	
29	Copper	Cu	63	62.929599	69.2	
			64	63.929766	β^-, β^+	12.7 h
			65	64.927792	30.8	
30	Zinc	Zn	64	63.929145	48.6	
			66	65.926035	27.9	
31	Gallium	Ga	69	68.925581	60.1	
32	Germanium	Ge	72	71.922080	27.4	
			74	73.921179	36.5	
33	Arsenic	As	75	74.921596	100	
34	Selenium	Se	80	79.916521	49.8	
35	Bromine	Br	79	78.918336	50.69	
36	Krypton	Kr	84	83.911506	57.0	
			89	88.917563	β^-	3.2 min
37	Rubidium	Rb	85	84.911800	72.17	
38	Strontium	Sr	86	85.909273	9.8	
			88	87.905625	82.6	
			90	89.907746	β^-	28.8 y
39	Yttrium	Y	89	88.905856	100	
40	Zirconium	Zr	90	89.904708	51.5	
41	Niobium	Nb	93	92.906378	100	
42	Molybdenum	Mo	98	97.905405	24.1	
43	Technetium	Tc	98	97.907210	β^-, γ	4.2×10^6 y
44	Ruthenium	Ru	102	101.904348	31.6	
45	Rhodium	Rh	103	102.90550	100	
46	Palladium	Pd	106	105.90348	27.3	
47	Silver	Ag	107	106.905095	51.83	
			109	108.904754	48.17	
48	Cadmium	Cd	114	113.903361	28.7	
49	Indium	In	115	114.90388	95.7; β^-	5.1×10^{14} y
50	Tin	Sn	120	119.902199	32.4	
51	Antimony	Sb	121	120.903824	57.3	
52	Tellurium	Te	130	129.906423	34.5; β^-	2×10^{21} y
53	Iodine	I	127	126.904477	100	
			131	130.906118	β^-, γ	8.04 days
54	Xenon	Xe	132	131.90415	26.9	
			136	135.90722	8.9	
55	Cesium	Cs	133	132.90543	100	

(Table continues)

Atomic Number, Z	Element	Symbol	Mass Number, A	Atomic Mass[a]	Percent Abundance, or Decay Mode if Radioactive[b]	Half-life if Radioactive
56	Barium	Ba	137	136.90582	11.2	
			138	137.90524	71.7	
			144	143.922673	β^-	11.9 s
57	Lanthanum	La	139	138.90636	99.911	
58	Cerium	Ce	140	139.90544	88.5	
59	Praesodymium	Pr	141	140.90766	100	
60	Neodymium	Nd	142	141.90773	27.2	
			144	143.910096	α, 23.8	2.1×10^{15} y
61	Promethium	Pm	145	144.91275	EC, α, γ	17.7 years
62	Samarium	Sm	152	151.91974	26.6	
63	Europium	Eu	153	152.92124	52.1	
64	Gadolinium	Gd	158	157.92411	24.8	
65	Terbium	Tb	159	158.92535	100	
66	Dysprosium	Dy	164	163.92918	28.1	
67	Holmium	Ho	165	164.93033	100	
68	Erbium	Er	166	165.93031	33.4	
69	Thulium	Tm	169	168.93423	100	
70	Ytterbium	Yb	174	173.93887	31.6	
71	Lutecium	Lu	175	174.94079	97.39	
72	Hafnium	Hf	180	179.94656	35.2	
73	Tantalum	Ta	181	180.94801	99.988	
74	Tungsten	W	184	183.95095	30.7	
75	Rhenium	Re	187	186.95577	62.60, β^-	4×10^{10} y
76	Osmium	Os	191	190.96094	β^-, γ	15.4 days
			192	191.96149	41.0	
77	Iridium	Ir	191	190.96060	37.3	
			193	192.96294	62.7	
78	Platinum	Pt	195	194.96479	33.8	
79	Gold	Au	197	196.96656	100	
80	Mercury	Hg	202	201.97063	29.8	
81	Thallium	Tl	205	204.97441	70.5	
			210	209.990069	β^-	1.3 min
82	Lead	Pb	204	203.973044	β^-, 1.48	1.4×10^{17} y
			206	205.97446	24.1	
			207	206.97589	22.1	
			208	207.97664	52.3	
			210	209.98418	α, β^-, γ	22.3 y
			211	210.98874	β^-, γ	36.1 min
			212	211.99188	β^-, γ	10.64 h
			214	213.99980	β^-, γ	26.8 min
83	Bismuth	Bi	209	208.98039	100	
			211	210.98726	α, β^-, γ	2.15 min
			214	213.998702	β^-, α	19.7 min
84	Polonium	Po	210	209.98286	α, γ	138.38 days
			214	213.99519	α, γ	164 μs
85	Astatine	At	218	218.00870	α, β^-	≈ 2 s
86	Radon	Rn	222	222.017574	α, γ	3.8235 days
87	Francium	Fr	223	223.019734	α, β^-, γ	21.8 min
88	Radium	Ra	226	226.025406	α, γ	1.60×10^3 y
			228	228.031069	β^-	5.76 years
89	Actinium	Ac	227	227.027751	α, β^-, γ	21.773 years
90	Thorium	Th	228	228.02873	α, γ	1.9131 years
			232	232.038054	100, α, γ	1.41×10^{10} y
91	Protactinium	Pa	231	231.035881	α, γ	3.28×10^4 y

(Table continues)

Atomic Number, Z	Element	Symbol	Mass Number, A	Atomic Mass[a]	Percent Abundance, or Decay Mode if Radioactive[b]	Half-life if Radioactive
92	Uranium	U	232	232.03714	α, γ	72 years
			233	233.039629	α, γ	1.592×10^5 y
			235	235.043925	0.72; α, γ	7.038×10^8 y
			236	236.045563	α, γ	2.342×10^7 y
			238	238.050786	99.275; α, γ	4.468×10^9 y
			239	239.054291	β^-, γ	23.5 min
93	Neptunium	Np	239	239.052932	β^-, γ	2.35 days
94	Plutonium	Pu	239	239.052158	α, γ	2.41×10^4 y
95	Americium	Am	243	243.061374	α, γ	7.37×10^3 y
96	Curium	Cm	245	245.065487	α, γ	8.5×10^3 y
97	Berkelium	Bk	247	247.07003	α, γ	1.4×10^3 y
98	Californium	Cf	249	249.074849	α, γ	351 years
99	Einsteinium	Es	254	254.08802	α, γ, β^-	276 days
100	Fermium	Fm	253	253.08518	EC, α, γ	3.0 days
101	Mendelevium	Md	255	255.0911	EC, α	27 min
102	Nobelium	No	255	255.0933	EC, α	3.1 min
103	Lawrencium	Lr	257	257.0998	α	≈ 35 s
104	Unnilquadium	Rf	261	261.1087	α	1.1 min
105	Unnilpentium	Ha	262	262.1138	α	0.7 min
106	Unnilhexium		263	263.1184	α	0.9 s
107	Unnilseptium		261		α	1 – 2 ms

APPENDIX C

Some Useful Tables

Table C.1
Mathematical Symbols Used in the Text and Their Meaning

Symbol	Meaning
$=$	is equal to
$\neq$	is not equal to
$\equiv$	is defined as
$\propto$	is proportional to
$>$	is greater than
$<$	is less than
$\gg$	is much greater than
$\ll$	is much less than
$\approx$	is approximately equal to
Δx	change in x
Σx_i	sum of all quantities x_i
$\|x\|$	magnitude of x (always a positive quantity)

Table C.2
Standard Abbreviations of Units

Abbreviation	Unit	Abbreviation	Unit
A	ampere	J	joule
Å	angstrom	K	kelvin
atm	atmosphere	kcal	kilocalorie
Btu	British thermal unit	kg	kilogram
C	coulomb	km	kilometer
°C	degree Celsius	kmol	kilomole
cal	calorie	lb	pound
cm	centimeter	m	meter
deg	degree (angle)	min	minute
eV	electron volt	N	newton
°F	degree Fahrenheit	rev	revolution
ft	foot	s	second
G	gauss	T	tesla
g	gram	u	atomic mass unit
H	henry	V	volt
h	hour	W	watt
hp	horsepower	Wb	weber
Hz	hertz	μm	micrometer
in.	inch	Ω	ohm

Table C.3
The Greek Alphabet

Alpha	A	α	Iota	I	ι	Rho	P	ρ
Beta	B	β	Kappa	K	κ	Sigma	Σ	σ
Gamma	Γ	γ	Lambda	Λ	λ	Tau	T	τ
Delta	Δ	δ	Mu	M	μ	Upsilon	Y	υ
Epsilon	E	ϵ	Nu	N	ν	Phi	Φ	ϕ
Zeta	Z	ζ	Xi	Ξ	ξ	Chi	X	χ
Eta	H	η	Omicron	O	o	Psi	Ψ	ψ
Theta	Θ	θ	Pi	Π	π	Omega	Ω	ω

Table C.4
Conversion Factors

Length
1 m = 39.37 in. = 3.281 ft
1 in. = 2.54 cm
1 km = 0.621 mi
1 mi = 5280 ft = 1.609 km
1 light-year = 9.461×10^{15} m
1 angstrom (Å) = 10^{-10} m

Mass
1 kg = 10^3 g = 6.85×10^{-2} slug
1 slug = 14.59 kg
1 u = 1.66×10^{-27} kg

Time
1 min = 60 s
1 h = 3600 s
1 day = 8.64×10^4 s
1 year = 365.242 days = 3.156×10^7 s

Volume
1 liter = 1000 cm^3 = 3.531×10^{-2} ft^3
1 ft^3 = 2.832×10^{-2} m^3
1 gallon = 3.786 liter = 231 in.^{3}

Angle
$180° = \pi$ rad
1 rad = 57.30°
1° = 60 min = 1.745×10^{-2} rad

Speed
1 km/h = 0.278 m/s = 0.621 mi/h
1 m/s = 2.237 mi/h = 3.281 ft/s
1 mi/h = 1.61 km/h = 0.447 m/s = 1.467 ft/s

Force
1 N = 0.2248 lb = 10^5 dynes
1 lb = 4.448 N
1 dyne = 10^{-5} N = 2.248×10^{-6} lb

Work and energy
1 J = 10^7 erg = 0.738 ft·lb = 0.239 cal
1 cal = 4.186 J
1 ft·lb = 1.356 J
1 Btu = 1.054×10^3 J = 252 cal
1 J = 6.24×10^{18} eV
1 eV = 1.602×10^{-19} J
1 kWh = 3.60×10^6 J

Pressure
1 atm = 1.013×10^5 N/m^2 (or Pa) = 14.70 lb/in.^{2}
1 Pa = 1 N/m^2 = 1.45×10^{-4} lb/in.^{2}
1 lb/in.^{2} = 6.895×10^3 N/m^2

Power
1 hp = 550 ft·lb/s = 0.746 kW
1 W = 1 J/s = 0.738 ft·lb/s
1 Btu/h = 0.293 W

APPENDIX D

SI Units

Table D.1
SI Base Units

Base Quantity	SI Base Unit	
	Name	Symbol
Length	Meter	m
Mass	Kilogram	kg
Time	Second	s
Electric current	Ampere	A
Temperature	Kelvin	K
Amount of substance	Mole	mol
Luminous intensity	Candela	cd

Table D.2
Derived SI Units

Quantity	Name	Symbol	Expression in Terms of Base Units	Expression in Terms of Other SI Units
Plane angle	Radian	rad	m/m	
Frequency	Hertz	Hz	s^{-1}	
Force	Newton	N	$kg \cdot m/s^2$	J/m
Pressure	Pascal	Pa	$kg/m \cdot s^2$	N/m^2
Energy: work	Joule	J	$kg \cdot m^2/s^2$	$N \cdot m$
Power	Watt	W	$kg \cdot m^2/s^3$	J/s
Electric charge	Coulomb	C	$A \cdot s$	
Electric potential (emf)	Volt	V	$kg \cdot m^2/A \cdot s^3$	W/A
Capacitance	Farad	F	$A^2 \cdot s^4/kg \cdot m^2$	C/A
Electric resistance	Ohm	Ω	$kg \cdot m^2/A^2 \cdot s^3$	V/A
Magnetic flux	Weber	Wb	$kg \cdot m^2/A \cdot s^2$	$V \cdot s$
Magnetic field intensity	Tesla	T	$kg/A \cdot s^2$	Wb/m^2
Inductance	Henry	H	$kg \cdot m^2/A^2 \cdot s^2$	Wb/A

Answers to Odd-Numbered Problems

Chapter 1

3. (a) MLT^{-2},   (b) kg m/s²

5. $h = \dfrac{4R^3}{3r^2}$, $L = \dfrac{[L^3]}{[L^2]} = [L]$

7. (a) 3.00×10^8 m/s,   (b) 2.9979×10^8 m/s,
   (c) 2.997925×10^8 m/s

9. (a) 797,   (b) 11,   (c) 17.8

11. (a) 22 cm,   (b) 67.9 cm²

13. 91.4 m

15. 3.16×10^9 y

17. 2.95×10^2 m, 2.95×10^8 cm³

19. 35.7 m²

21. 2.9 cm

23. (a) 1 mi/h = 1.609 km/h,   (b) 88.5 km/h,
   (c) 16.1 km/h

25. 6.71×10^8 mi

27. 10^{10} lb, 4.45×10^{10} N

29. 1.25×10^{10} lb, 4.17×10^7 head, (assuming 0.25 lb
per burger and 300 lb net of meat per head of cattle)

31. 4.4×10^4 pitches per season (assumes 30 pitches
per inning and 9-inning games)

33. 2.24 m

35. 8.06 cm

37. 8.60 m

39. (a) 1.5 m,   (b) 2.60 m

41. (a) 1 megaphone,   (b) 1 gigolo,
   (c) 1 decaration,   (d) 2 megacycles,
   (e) 1 terapin,   (f) 1 decadent,
   (g) 2 kilomockingbirds,   (h) 1 picoboo,
   (i) 1 nanogoat

43. $V = \left(0.579 \dfrac{ft^3}{s}\right) t + \left(1.19 \times 10^{-9} \dfrac{ft^3}{s^2}\right) t^2$

45. 417 ft (assumes 200 bills per inch)

47. 8.60 m

Chapter 2

1. 6.10 m/s

3. 2.4×10^4 s = 6.67 h

5. (a) 2.50 m/s,   (b) -2.27 m/s,   (c) 0

7. (a) 780 m,   (b) 1.86 m/s

9. (a) negative,   (b) positive,   (c) 0,   (d) 0

11. (b) 41.03 m/s, 41.008 m/s, 41.002 m/s,
   (c) 17 m/s

13. (a) 4.00 m/s,   (b) -0.50 m/s,
   (c) -1.00 m/s,   (d) 0

15. 0.75 m/s, -3.00 m/s, 3.00 m/s, 0

17. -1.33 m/s²

19. 8.60 m/s

21. (a) 5510 m,   (b) 20.8 m/s, 41.6 m/s, 20.8 m/s,
38.7 m/s

23. (a) 8.00 m/s²,   (b) 11.0 m/s²

25. 3.87 m/s

27. 2.07 s

29. 2.5 m/s²

31. (a) 20 s,   (b) No, the minimum distance to stop =
1000 m.

33. 1.50 m/s, 32.3 m

35. (a) 8.20 s,   (b) 134 m

37. 44.1 m

39. -14.0 m/s

41. 3.94 s

43. 941 m

45. (a) -6.26 m/s,   (b) 6.02 m/s,   (c) 1.25 s

47. -29.6 m/s, -9.80 m/s²

49. 4.16 m/s

51. 0.6 s

53. (a) 3.0 s,   (b) -15.2 m/s,   (c) -31.37 m/s,
-34.8 m/s

55. (a) 6.46 s after sports car starts,   (b) 334 ft,
   (c) v(sports car) = 103 ft/s, v(stock car) = 89.6 ft/s

57. (a) 4.74 s, 85.3 s,   (b) 200 ft/s,   (c) 18,500 ft/s
from starting point,   (d) 100 s

Chapter 3

1. 85 mi

3. (a) 5 units, at an angle of 53° below the x axis.
   (b) 5 units, 53° above x axis.

5. 421 ft, 3° below horizontal

7. 83 m, 33° north of west

9. 103 ft horizontally, 86.8 ft vertically

11. 42.7 yd

13. 61.8 m at 14.0° from original line of travel, or 25 m
at 36.9° from original line of travel

15. (a) 185.4 N at 77.8° from x axis,   (b) 185.4 N at
257.8°

17. 35.3 m downfield

19. 34.6 m/s

21. 29.4 m/s

23. 191 m/s

25. -0.944 km/h
27. (a) 22.3 s,   (b) 200 s
29. (a) (i) 1.2 m/s, (ii) 0,   (b) 0.96 m,   (c) 0.5 m/s
31. (a) $v_{bs} = 10.1$ m/s at 8.53° east of north,   (b) 45 m
33. (a) 14.5° north of west,   (b) 194 m/s
35. 15.3 m
37. 2017 s versus 1667 s in still water, for a 21% increase
39. 8.72 m at 53.4° south of east
41. 18 m on Moon, 7.89 m on Mars
43. 68.6 km/h
45. 14.0 m/s
47. 4.12 m
51. less than 265 m or more than 3476 m
53. 358 m at 2° south of east
55. 1.34 cm

Chapter 4

1. 19.6 N, 4.41 lb
3. 2380 N
5. (a) 534 N,   (b) 54.5 kg
7. (a) 12 N,   (b) 3 m/s²
9. 25 N
11. 312 N
13. (a) 789 N at 8.8° to right of forward direction,
    (b) 0.266 m/s² in direction of resultant force
15. 2160 N
17. 77.8 N in each
19. 166 N downward at 28.1° to left of vertical
21. 1.50 m
23. $\mathbf{F}_{ball} = -375$ N, $\mathbf{F}_{glove} = 375$ N (+ direction =
    direction of ball's velocity)
25. 23.1 m/s
27. 7.90 m/s
29. 551 N
31. 100.2 N, 203.8 N
33. 218 N
35. 236 N (upper rope), 118 N (lower rope)
37. (a) 36.8 N,   (b) 2.45 m/s²,   (c) 1.23 m
39. $\mu_k = 0.306$
41. 3.16 s
43. 56.7 N
45. 72.0 N
47. (a) 0.366 m/s². (b) No, the retarding force (friction
    plus tangential component of weight) exceeds the
    component of applied force along the incline.
49. (a) 98.6 m,   (b) 16.4 m
51. 19.0 m
53. 355 N
55. 5.72 m/s²
57. (a) -1.20 m/s²,   (b) $\mu_k = 0.122$,   (c) 45 m
59. 104 N
61. 3317 N at 2.68° from the 3000-N force
63. $F > 0$ for $t < 1.25$ s, $F = 0$ for $t = 1.25$ s, $F < 0$ for
    $t > 1.25$ s

65. 21.5 N
67. 7500 N, 50 m
69. 2260 N
71. $\mu_k = 0.4$
73. (a) 1730 N,   (b) 8.63°
75. 997 N, 796 N
77. (a) the friction force between box and truck bed,
    (b) 2.94 m/s²
79. (a) 8.93×10^3 N,   (b) 1.05 m/s²,   (c) 369 kg
81. (a) 706 N,   (b) 814 N,   (c) 706 N,   (d) 648 N
83. (a) 91.4 N,   (b) 48.6 N
85. (b) 9.8 N, 0.580 m/s²
87. (b) 5.75 m/s²,   (c) 17.4 N, 40.5 N
89. (a) 2 m/s²,   (b) 4 N, 6 N, and 8 N,   (c) m_2 on
    $m_3 = 8$ N, m_1 on $m_2 = 14$ N
91. (a) 304 N,   (b) 290 N,   (c) 152 N,   (d) 138 N
93. $a = g \sin \theta \left(\dfrac{m_1 - m_2}{m_1 + m_2} \right) - \mu g \cos \theta$,

    $T = \dfrac{2m_1 m_2 \, g \sin \theta}{(m_1 + m_2)}$

Chapter 5

1. 1.50×10^7 J
3. 30.6 m
5. 1.59×10^3 J
7. 4700 J
9. (a) 900 J,   (b) -900 J,   (c) $\mu_k = 0.383$
11. 9.07×10^5 J
13. (a) 1.20 J,   (b) 5.0 m/s,   (c) 6.30 J
15. 6.25×10^{-21} J
17. -2.40×10^5 J
19. (a) -168 J,   (b) -184 J,   (c) 500 J,
    (d) 148 J,   (e) 5.64 m/s
21. 2.04 m
23. (a) -19.6 J,   (b) 39.2 J,   (c) 0
25. $F = 6.30 \times 10^3$ N, $W = 2.59 \times 10^5$ J
27. (a) -147 J,   (b) -147 J,   (c) -147 J. The
    results should be the same because gravitational
    forces are conservative.
29. 44.3 m/s
31. 1.84 m
33. (a) 10.9 m/s,   (b) 11.6 m/s
35. (a) 8.85 m/s,   (b) 54.1%
37. 26.5 m/s
39. (a) 2.29 m/s,   (b) -15.6 J
41. 1.54 m
43. 7.10 m
45. (a) 7.50×10^4 J,   (b) 33.5 hp,   (c) 44.7 hp
47. 600 W
49. 1.84×10^3 W
51. (a) 0.408 m/s,   (b) 2.45×10^3 J
53. (a) 7.92 hp,   (b) 14.1 hp
55. (a) 22.5 J,   (b) 6.71 m/s

57. (a) 2.29 m/s, (b) 3.91 m/s, (c) 4.50 m/s
59. (a) 4π J, (b) 11.3 J, (c) 8.0 J
61. (a) 9.90 m/s, (b) 7.67 m/s
63. (a) 3.92×10^3 W, (b) 7.06×10^5 J
65. (a) 9.90 m/s, (b) -11.8 J
67. 9.80 m/s
69. (a) 4.43 m/s, (b) 1.47×10^5 N
71. (a) 400 J, (b) -294 J, (c) 5.94 m/s
73. (a) 1.31×10^5 J, (b) 871 W
75. (a) mgy, (b) mgy, (c) $mgy + \frac{1}{2}mv_0^2$
77. (a) 2.69 m/s^2, (b) 73.1 hp
79. (a) -5.60 J, (b) $\mu_k = 0.152$
81. (a) 0.225 J, (b) 0.363 J, (c) No, because the normal force and the force of friction change with the position of the object in the bowl.

Chapter 6

1. (a) 8.35×10^{-21} kg·m/s, (b) 4.50 kg·m/s, (c) 750 kg·m/s, (d) 1.78×10^{29} kg·m/s
3. 50.0 m/s
5. (a) 1.73 kg·m/s (horizontally), (b) 2.00 kg·m/s at an angle of 30° below horizontal
7. -5.4×10^5 N
9. (a) 1.35×10^4 N·s, (b) 9000 N
11. 2.40 N
13. (a) 9.00 m/s, (b) -15.0 m/s
15. 6.25×10^{-2} m/s
17. $\dfrac{v_{\text{large}}}{v_{\text{small}}} = \dfrac{1}{2}$
19. Boy moves west at 2.46 m/s.
21. 2.66 m/s
23. 2.50 m/s in direction car was originally moving
25. 3.0 m/s
27. 6 kg
29. -40 cm/s, 10.0 cm/s
31. (a) 6.0×10^5 m/s, 1.6×10^6 m/s, (b) $KE_{\text{alpha after}} = 1.20 \times 10^{-15}$ J, $KE_{\text{proton after}} = 2.14 \times 10^{-15}$ J, $KE_{\text{alpha before}} = 3.34 \times 10^{-15}$ J
33. 5.20×10^{-2} m/s
35. 0.816 m/s at 45° with respect to original direction of travel
37. 48,000 N
39. 0.398%
41. (0, 1.0 m)
43. ($\frac{1}{3}$ m, $\frac{5}{3}$ m)
45. (3.85 cm, 6.85 cm)
47. 2.78×10^{-14} m
49. (a) 20 m from ship, (b) He should unhook the engine from the rope and push the engine straight away from the ship.
51. 0.60 m/s
53. (a) $\frac{32}{3}$ m/s, $\frac{8}{3}$ m/s, (b) $\frac{8}{3}$ m/s, $-\frac{16}{3}$ m/s, (c) loses 7.11×10^{-3} J in (a) and 2.84×10^{-2} J in (b)

55. 1.09×10^3 N
57. (a) 0, 1.50 m/s, (b) -1.00 m/s, 1.50 m/s, (c) 1.00 m/s, 1.50 m/s
59. 240 s
61. 5.59 m/s
63. 0.98 m
65. (a) 8.00 N·s, (b) 5.33 m/s, 3.33 m/s
67. (a) 1.07 m/s at $\phi = 29.7°$, (b) 0.32 or 32%
69. 1.93 m
71. 529 m/s
73. (b) 2.14×10^{-1} m/s^2
75. (a) As the child walks to the right end of the boat, the boat moves left (toward pier). (b) 5.55 m, (c) No, his maximum reach is 6.55 m.
77. (a) $-2mv \sin \theta$, (b) 0, (c) $\dfrac{2mv \sin \theta}{t}$

Chapter 7

1. 0.524 rad, 0.785 rad, 1.05 rad, 3.14 rad, 4.71 rad, 6.28 rad
3. 2.15 m, 123 m, 773 m
5. 1.99×10^{-7} rad/s, 0.986 deg/day
7. 4.19×10^{-2} rad/s^2
9. (a) 8.22×10^2 rad/s^2, (b) 4.21×10^3 rad
11. (a) 1.75×10^{-1} rad/s^2, (b) 5.57 rev
13. 50.0 rev
15. -2.26×10^2 rad/s^2
17. (a) 1200 rad/s, (b) 25.0 s
19. (a) 0.40 rad/s, (b) 0
21. 36.5 rev
23. $v_{\text{small}} = 30.2$ m/s, $v_{\text{large}} = 33.9$ m/s, (b) 1.52×10^3 m/s^2, (c) 1.28×10^3 m/s^2
25. (a) 3.37×10^{-2} m/s^2, (b) 0
27. (a) 3.45×10^{-1} m/s^2, (b) 1.04 m/s, (c) 3.45×10^{-1} m/s^2, 0.937 m/s^2, 0.998 m/s^2, at 20.2° with respect to the direction of a_r.
29. 18.2 m/s
31. 11.7 m/s
33. (a) 2.49×10^4 N, (b) 12.1 m/s
35. (a) 1.46×10^2 m, (b) 5488 N
37. 68.6 N, 0.857 m/s^2
39. 3.13 m/s
41. 3.55×10^{22} N
43. 1.17×10^{-6} N
45. 8.87 m/s^2
47. (a) 1.63 m/s^2, (b) 1.68×10^3 m/s, (c) 108.3 min
49. The ratio of T^2 to r^3 is 2.98×10^{-19} s^2/m^3 for all cases
51. 1.49×10^{11} m (from Sun)
53. (a) 5310 s (88.5 mi), (b) 7.79 km/s, (c) 6.43×10^9 J
55. (a) -4.76×10^9 J, 5.68×10^2 N
57. (a) 1.99×10^{12} m/s^2, (b) 2.38×10^{11} N, (c) 1.39×10^{12} J

35. (a) 223 J/s,   (b) 1.32×10^{-2} J/s,
    (c) 4.50×10^{-2} J/s
37. 14.1 ft²·F°·h/BTU
39. 38.6 m³/day
41. 52.3 J/s (Thermopane), 1840 J/s (single pane)
43. 16.6 J/s
45. $\dfrac{P_h}{P_c} = 1.42$
47. 7.23×10^{-2} J/s·m·C°
49. 9.0 cm
51. 2.44 min
53. 38°C
55. 3.80×10^5 J (9.08×10^4 cal)
57. (a) system reaches 0°C and 3.78×10^{-4} kg of ice is melted,   (b) system reaches 100°C and 1.64 g of water is vaporized.
59. 44.5°C
61. (a) 2000 W,   (b) 4.47°C
63. 26.9 liters
65. 20.02 cm
67. 0.654 cal/g·C°
69. 109°C

Chapter 13

1. 804 J
3. (c) More work is done in process (a) because of the higher pressure during the expansion part of the process.
5. (a) 608 J,   (b) 0,   (c) −405 J,   (d) 0,
   (e) 203 J
7. 465 J
9. 2.86×10^5 Pa
11. (a) −152 J,   (b) −248 J
13. −720 J (from system)
15. (a) −180 J,   (b) −188 J
17. (a) 5.48×10^3 J,   (b) 5.48×10^3 J
19. (a) 12.0 kJ,   (b) −12.0 kJ
21. (a) 25.9 J,   (b) 8.96×10^5 J,   (c) 8.96×10^5 J
23. (a) 3.08×10^{-2} m³,   (b) -3.46×10^3 J,
    (c) -3.46×10^3 J
25. (a) 2300 J,   (b) 2300 J,   (c) $\dfrac{P_b}{P_a} = \dfrac{1}{2}$
27. (a) $Q < 0$, $W = 0$, $\Delta U < 0$,   (b) $Q > 0$, $W = 0$, $\Delta U > 0$
29. (a) $Eff = 0.067$,   (b) 350 J
31. (a) 667 J,   (b) 467 J
33. (a) 10,677 J,   (b) 0.533 s
35. $Eff = 48.8\%$
37. 546°C
39. (a) 9.10 kW,   (b) 1.19×10^4 J
41. (a) $Eff = 5.12\%$,   (b) 5.27×10^{12} J/h

43. (a) $Eff = 55.4\%$,   (b) $Eff_{Carnot} = 74.9\%$
45. 57.2 J/K
47. −612 J/K
49. (a)

Result	Possible Combinations	Total
All red	RRR	1
2R,1G	RRG, RGR, GRR	3
1R, 2G	RGG, GRG, GGR	3
All green	GGG	1

(b)

Result	Possible Combinations	Total
All red	RRRRR	1
4R, 1G	RRRRG, RRRGR, RRGRR, RGRRR, GRRRR	5
3R, 2G	RRRGG, RRGRG, RGRRG, GRRRG, RRGGR, RGRGR, GRRGR, RGGRR, GRGRR	10
2R, 3G	GGRRR, GGRGR, GRGGR, RGGGR, GGRRG, GRGRG, RGGRG, GRRGG, RGRGG, RRGGG	10
1R, 4G	RGGGG, GRGGG, GGRGG, GGGRG, GGGGR	5
All green	GGGGG	1

51. 0.477°C
53. The engine either violates the first law or it does not operate in a cyclic process (which would make it incapable of continuous operation). You should not invest.
55. (a) 3.45×10^3 J,   (b) 1.50×10^3 J,
    (c) 1.95×10^3 J
59. 29.3 J
61. 77.8 W
63. 17.9 C°
67. (a) $4P_0V_0$,   (b) $4P_0V_0$, 9.07×10^3 J
69. (a) 2.65×10^3 metric tons/day,
    (b) 7.74×10^6 dollars/y,   (c) 4×10^7 g/s

Chapter 14

1. (a) 49.0 N/m,   (b) 11.5 cm
3. (a) 110 N,   (b) The graph will be a straight line passing through the origin and with a slope of 1000 N/m.
5. 9.72 cm
7. 2.68 m/s

9. $\dfrac{PE_{0.4}}{PE_{0.2}} = 4$

11. (a) $PE = \dfrac{E}{4}$, $KE = \dfrac{3}{4}E$, (b) $x = \dfrac{A}{\sqrt{2}}$

13. 47.8 cm

15. 3.06 m/s

17. (a) 2.45×10^{-2} J, (b) 40.4 cm/s

19. (a) 0.153 J, (b) 0.783 m/s, (c) 17.5 m/s^2

21. (a) 0, (b) -4.33 m/s, (c) -5.00 m/s, (d) 0, (e) 5.00 m/s

23. 3.95 N/m

25. (a) quadrupled, (b) doubled, (c) doubled, (d) no change

27. 3.95 N/m

29. (a) at $t = 0$, $x = 20$ cm, (b) at $t = \dfrac{T}{8}$, $x = 14.1$ cm, (c) at $t = T/4$, $x = 0$, (d) at $t = 3T/8$, $x = -14.1$ cm, (e) at $t = T/2$, $x = -20$ cm, (f) at $t = 5T/8$, $x = -14.1$ cm, (g) at $t = 3T/4$, $x = 0$, (h) at $t = 7T/8$, $x = 14.1$ cm, (i) at $t = T$, $x = 20$ cm. When plotted, the graph is a cosine curve with amplitude 20 cm.

31. $x = (0.25 \text{ m}) \cos 2.47 \pi t$

33. (a) 250 N/m, (b) 0.281 s, 3.56 Hz, 22.4 rad/s, (c) 0.313 J, (d) 5.0 cm, (e) $v_{max} = 1.12$ m/s, $a_{max} = 25.0$ m/s^2, (f) 0.919 cm

35. 10.1 m/s^2

37. $\dfrac{T_E}{T_M} = \dfrac{1}{\sqrt{6}}$

39. (a) cut in half, (b) doubled

41. (a) 1.14×10^{-8} s, (b) 3.41 m

43. 0.082 m to 12.3 m

45. (a) 9 cm, (b) 20 cm, (c) 0.04 s, (d) 500 cm/s

47. (a) 8.31 min, (b) 1.28 s

49. 446 m

51. 80.0 N

53. (a) 30.0 N, (b) 25.8 m/s

55. 13.5 N

57. (a) 0, (b) 0.3 m

59. 127 m/s

61. (a) -500 N, -333 m/s^2, (b) -100 N, -66.7 m/s^2, (c) 0, 0, (d) 200 N, 133 m/s^2, (e) 500 N, 333 m/s^2

63. (a) 5.1×10^{-2} kg/m, (b) 19.6 m/s

65. 0.99 m

67. (a) 19.8 m/s, (b) 8.94 m

69. 3.14×10^6 Pa

71. (a) 100 m/s, (b) 374 J

73. (a) -28.0 J, (b) 0.446 m

75. (a) 588 N/m, (b) 0.70 m/s

77. $\mu = 0.115$

Chapter 15

1. (a) 347 m/s, (b) 387 m/s, (c) 436 m/s

3. 516 m

5. $\dfrac{\Delta \lambda}{\lambda} = 2.88\%$

7. 3.51×10^3 m/s

9. (a) 60 dB, (b) 70 dB

11. (a) 5×10^{-17} W, (b) 5×10^{-5} W

15. 5×10^4 m

17. (a) 7.96×10^{-2} W/m^2, (b) 109 dB, (c) 2.82 m

19. 30 m

21. 913 Hz

23. 32.1 m/s (72 mph)

25. (a) 537 Hz, (b) 465 Hz

27. 599 Hz

29. (a) 537 m/s, (b) 690 m/s

31. (a) 34.5 cm, (b) constructive

33. (a) 0.431 m, (b) 0.863 m

35. 0.089 m, 0.304 m, 0.518 m, 0.732 m, 0.947 m, 1.161 m

37. (a) 8×10^{-2} N, (b) 4.05×10^3 N

39. 813 N

41. (a) 3.36×10^5 N, (b) No, a wire this small would break under this tension.

43. 120 Hz

45. 0.352 Hz, 0.176 Hz, 0.117 Hz, etc.

47. 98.9 cm

49. (a) 0.66 m, (b) 0.33 m

51. (a) 0.55 m, (b) 317 Hz

53. 328 m/s

55. 438 Hz or 442 Hz

57. 0.297 m

59. 1.76 cm

61. (a) 362 Hz, (b) 287 Hz, (c) 0.953 m and 1.20 m

63. (a) 0.555 m, (b) 620 Hz

65. (a) decreases by factor of 2, (b) decreases by factor of $\sqrt{2}$, (c) increases by $\sqrt{2}$

67. 15.7 Hz

69. 2.82×10^8 m/s

71. 200 m/s

73. 32.9 m/s

75. 4 Hz

77. 1204 Hz

79. 10 cm/s

Chapter 16

1. (a) -80 μC, (b) 4.8 μC

3. 1.11×10^{-8} N

5. 5.13×10^5 N

7. 5.71×10^{13} C

9. -1.17×10^{-8} N (negative x direction)
11. 2.25×10^{-9} N/m
13. 3.90×10^{-7} N at $11.3°$
15. 7.45×10^{-7} C
17. 0.621 m
19. 1.45 m beyond the -3×10^{-9} C charge
21. 7.6×10^{5} N/C ($+x$ direction)
23. 1.44×10^{-7} N/C (toward electron)
25. 5.54×10^{11} N/C (away from proton)
27. (a) 2.0×10^{7} N/C to right,   (b) 40.0 N to left
29. 1.63×10^{5} N/C
31. (a) 5.27×10^{13} m/s²,   (b) 5.27×10^{5} m/s
33. (a) 6.13×10^{10} m/s²,   (b) 19.6 μs,
    (c) 11.7 m,   (d) 1.20×10^{-15} J
35. 10^{3} N/C in direction of electron's motion
37. 756 N/C at $70.1°$ above $-x$ axis
39. 0
41. 1880 N/C at $4.4°$ below x axis
43. 1.82 m to left of -2.5×10^{-6} C charge

47. $\dfrac{q_1}{q_2} = \dfrac{1}{3}$,   (b) $q_2 > 0$, $q_1 < 0$

51. (a) 0,   (b) -5 μC inside, $+5$ μC,   (c) 5 μC outside, 0 inside,   (d) 5 μC outside, 0 inside
53. (a) 0,   (b) 1.8×10^{6} N/C,   (c) 1.13×10^{5} N/C
55. (a) 5.27×10^{17} m/s²,   (b) 0.854 mm
57. 5.33×10^{-6} N at $22.2°$
59. (a) 24.2 N/C in $+x$ direction,   (b) 9.43 N/C at $63.5°$ above $-x$ axis
61. 5.25 μC
63. at $y = 9.79$ m
65. 4.43×10^{5} N/C
67. (a) -4.40×10^{15} m/s²,   (b) 9.09×10^{-10} s,
    (c) 1.82×10^{-3} m
69. (a) 1.09×10^{-8} C,   (b) 5.43×10^{-3} N

Chapter 17

1. (a) 6.40×10^{-19} J,   (b) -6.40×10^{-19} J
3. (a) -6.00×10^{-4} J,   (b) -50 V
5. 6.67×10^{4} V/m
7. 1.54×10^{6} J
9. 3.2×10^{-19} C
11. (a) 1.52×10^{5} m/s,   (b) 6.49×10^{6} m/s
13. (a) 9.22×10^{4} V,   (b) 1.69×10^{8} V
15. 2.00 m
17. 28.2 V
19. (a) 33.4 V,   (b) 195 V
21. 8.10×10^{-7} J
23. -9.09 J
25. (a) 5.93×10^{5} m/s,   (b) 1.38×10^{4} m/s
27. 6.75×10^{-3} J (4.22×10^{16} eV)
29. (a) 48 μC,   (b) 6 μC
31. 2.26×10^{-5} m²
33. 5.31×10^{-12} C

35. (a) 90.4 V,   (b) 9.04×10^{4} V/m
37. 17 μF
39. (a) 3.53 μF,   (b) 31.8 μC on each, 6.35 V, 2.65 V
41. (a) 18 μF,   (b) 1.78 μF
43. $Q_1 = 16$ μC, $Q_4 = 32$ μC, $Q_5 = 80$ μC, $Q_8 = 64$ μC
45. (a) 4 μF,   (b) $V_2 = 12$ V, $V_3 = 8$ V, $V_6 = 4$ V,
    $Q_2 = Q_3 = Q_6 = 24$ μC

47. Seven distinct values: $\dfrac{C}{3}, \dfrac{C}{2}, \dfrac{2C}{3}, C, \dfrac{3C}{2}, 2C, 3C$

49. (a) 9 μF,   (b) $V_5 = 6$ V, $V_6 = 4$ V, $V_{12} = 2$ V,
    $Q_5 = 30$ μC, $Q_6 = Q_{12} = 24$ μC
51. $Q_1 = \frac{10}{3}$ μC, $Q_2 = \frac{20}{3}$ μC
53. (a) 0.3 μF,   (b) 2.16×10^{-3} J
55. 2.55×10^{-11} J
57. 10.8 pF
59. Energy increases by a factor of 3.4.
61. (a) 8.13 nF,   (b) 2.40 kV
63. 5.63×10^{3} N/C
65. 4.22×10^{5} V
67. 6.25 μF
69. $\frac{4}{3}$ μF
71. 91.5 mV
73. $\kappa = 2.33$
75. 8000 V
77. $Q_4 = 177$ μC, $Q_2 = 88.6$ μC
79. $Q_6 = 750$ μC, $Q_2 = 250$ μC

Chapter 18

1. 45 C
3. 26.8 A
5. 3.75×10^{14} electrons/s
7. 0.159 A
9. 5.90×10^{28} electrons/m³
11. 1.57 A
13. 1.59×10^{-3} Ω
15. 1.69×10^{-4} m
17. 32.0 V, 0.16 V
19. 48.0 Ω
21. No, his claim violates Ohm's law. I is really 6 A.
23. $9R$
25. (a) 4.71×10^{-4} $\Omega \cdot$m,   (b) 30 Ω
27. (a) 2.82×10^{8} A,   (b) 1.76×10^{7} A
29. 10.8 Ω
31. $30.6°$C
33. 1.084×10^{-3} (C°)$^{-1}$
35. $51.3°$C
37. (a) 0.649 mV,   (b) 1.11 mV
39. 5 A, 24 Ω
41. 500 A
43. 1.75×10^{-8} m²
45. 36.1% increase
47. 1.55 A
49. 448 A

51. 13.2 cents

53. \$3.65/year (assumes 5 min of use per day)

55. I delivered = 15.9 A. Thus, use 20-A fuse.

57. (a) 18.8 A, (b) 7.46×10^6 J (2.07 kWh), (c) 16.6 cents

59. 3.24×10^5 C

61. 1.92 cents

63. 1.11×10^3 m

65. 3.77×10^{28} electrons/m^3

67. 0.4 μA

69. (a) 9.4×10^{-7} Ω, (b) 8.02×10^{-7} Ω

71. (a) 18 C, (b) 3.6 A

73. (a) 9.31 m, (b) 9.33×10^{-4} m

75. $\dfrac{\alpha_1}{\alpha_2} = -0.136$

Chapter 19

1. (a) 24 Ω, (b) 1.0 A

3. (a) $\frac{24}{11}$ Ω, (b) $I_4 = 6$ A, $I_8 = 3$ A, $I_{12} = 2$ A

5. There are seven distinct values possible: (1) use one alone, value $= R$; (2) use two in series, value $= 2R$; (3) three in series, value $= 3R$; (4) two in parallel, value $= \dfrac{R}{2}$; (5) three in parallel, value $= \dfrac{R}{3}$; (6) two in series with one in parallel, value $= \dfrac{2}{3}R$; (7) two in parallel with one in series, value $= \dfrac{3}{2}R$.

7. (a) 27.3 V, (b) $I_{10} = 2.73$ A, $I_{20} = 1.36$ A, $I_{30} = 0.910$ A

9. 15 Ω

11. 9.83 Ω

13. 5.13 Ω

15. 0.429 A

17. (a) $I_3 = 1.82$ A, $I_6 = 1.09$ A, $I_9 = 0.727$ A. (b) Same as (a).

19. 5 h

21. 2.4 V, 320 Ω

23. (a) 3 mA, (b) -19.0 V, (c) 4.5 V

25. In left branch, $I = 1$ A downward, and in center branch, $I = 1$ A down. $\mathcal{E} = 5$ V with polarity opposite to that shown in diagram.

27. $\mathcal{E}_1 = 6$ V, $\mathcal{E}_2 = 9$ V

29. 10.7 V

31. (a) 12 s, (b) 120 μC, (c) 1.0×10^{-5} A

33. $Q = 0.863Q_f$ or 86.3%

35. (a) 5×10^5 Ω

37. (a) 2.0 V, (b) 0.909 V, (c) 2.0 V (actually 1.9999976 V)

39. (a) 4.00 A, (b) 3.33 A, % error = 16.7%

41. 66.7 Ω

43. 400 Ω

45. 24 bulbs

47. (a) 0.101 W, (b) 10.1 W

49. (a) 8.0 A, (b) 120 V, (c) 0.8 A, (d) 576 W

51. (a) 1900 Ω, (b) 18,000 Ω, (c) 180,000 Ω

53. 15 Ω

55. $I_{10} = 1.5$ A, $I_2 = 4.5$ A, $\mathcal{E} = 33$ V

57. P(delivered by 12-V battery) = 6 W, power dissipated in 6-V battery = 3 W, power dissipated in resistors: $P_7 = 1.75$ W, $P_3 = 0.75$ W, $P_2 = 0.5$ W

59. 4.56 Ω

61. 7.5 Ω

63. 13.8 s

65. $\mathcal{E} = 1.8$ V, $r = 1$ Ω

67. 2.48 A

Chapter 20

1. 1.88×10^8 m/s

3. (a) 8.67×10^{-14} N, (b) 5.19×10^{13} m/s^2

5. (a) (a) to left, (b) into page, (c) out of page, (d) toward top of page, (e) into page, (f) out of page. (b) If the charge is negative, the direction of all forces above is reversed 180°.

7. (a) into page, (b) toward right, (c) toward bottom of page.

9. 10.8°

11. 2.09×10^{-2} T in $-y$ direction

13. 24.0 N

15. 1.0×10^{-2} N westward

17. (a) to the left, (b) into page, (c) out of page, (d) toward top of page, (e) into page, (f) out of page

19. 0.131 T downward

21. 0.245 A

23. (a) 9.0×10^{-3} N at 15° above horizontal in northward direction, (b) 2.33×10^{-3} N horizontal and westward

25. 0.277 A

27. 0.167 Ω

29. 1940 Ω

31. $R_1 = 3.33 \times 10^{-4}$ Ω, $R_2 = 2.998 \times 10^{-3}$ Ω, $R_3 = 2.997 \times 10^{-2}$ Ω

33. 1.39×10^{-2} T out of page

35. 8.65×10^{-8} s

37. 1.69×10^{-5} m

39. 5.0 T

41. 0.15 mm

43. 3.2 cm

45. (a) 10^{-5} T, (b) 2×10^{-6} T, (c) 5×10^{-7} T

47. 500 A

49. (a) 4×10^{-5} T, into page, (b) 5×10^{-6} T out of page, (c) 1.67×10^{-6} T out of page

51. (a) 10 μm, (b) It is highly unlikely that a wire that small could carry a 10-A current without melting.

53. (a) 2×10^{-4} N/m (attracted),   (b) 2×10^{-4} N/m (repelled)
55. 59.7 mA
57. 2.26×10^{-4} N, torque = 0
59. 38.2 mA
61. 8.44×10^{-5} T
63. 2.12 cm
65. (a) 5×10^{-7} T (out of page),   (b) 3.89×10^{-6} T in a plane parallel to the xy plane and at $31°$ clockwise from the y axis.
67. 0.331 T
69. (a) 30 A,   (b) 1.6×10^{-4} T (out of page)
71. (a) 2.85×10^{-6} T,   (b) 9.06×10^{-4} A
73. -2.70×10^{-5} N
75. 60 μA

Chapter 21

1. 7.07×10^{-2} T·m²
3. 7.71×10^{-1} T·m²
5. 2.96×10^{-5} T·m²
7. (a) 3.13×10^{-3} T·m²,   (b) 0
9. 6 V
11. 1.15×10^{5} V
13. (a) 25.2 mV,   (b) 3.16×10^{-2} Ω
15. 5.2×10^{-5} T
17. 33.9 mV
19. 78.5 μs
21. (a) 1.88×10^{-7} T·m²,   (b) 6.28×10^{-8} V
23. 12.0 mV
25. 2.0 mV (the western end is positive)
27. (a) 100 μV,   (b) 20 μA
29. 2.87 mV
31. (a) left to right,   (b) right to left
33. (a) right to left,   (b) right to left,   (c) left to right,   (d) left to right
35. left to right
37. (a) 0.157 mV, end B will be at the higher potential;   (b) 5.89 mV, end A will be at the higher potential.
39. 79.6 Hz, 240 V
41. 0.126 V
43. (a) 181 V,   (b) 6.03 A,
    (c) $\mathcal{E} = (181 \text{ V}) \sin (377 \text{ s}^{-1} t)$,
    (d) $i = (6.03 \text{ A}) \sin (377 \text{ s}^{-1} t)$
45. 76.4 V, 10.9 Ω
47. (a) 60.3 V,   (b) 56.9 V,   (c) 0.125 s
49. 256 turns
51. 1.88×10^{-4} T·m²/s
53. 2.4×10^{-7} T·m²
55. 12.0 mH
57. (a) 4 A,   (b) 2.52 A
59. 0.140 J
61. 1.85×10^{-3} T
63. 4.71 V

65. 9.67 V
67. This curve is a sine function with amplitude of 300 V and period of $\frac{\pi}{10}$ s.
69. 4.71×10^{-3} T·m²
73. 444 μV
75. (a) 52.5 V,   (b) 44.2 V,   (c) 11.5 N·m
77. 29 m/s

Chapter 22

1. (a) 141 V,   (b) 28.3 A,   (c) 20 A
3. (a) 14.7 Ω,   (b) 825 W
5. $R_1 = R_2 = 80.7$ Ω, $R_3 = 121$ Ω, $I = 3.64$ A
7. (a) 106 V,   (b) 60 Hz,   (c) 0,   (d) 3 A
11. 0.100 A
13. 0.224 A
15. (a) 0.188 Ω,   (b) 1.88×10^{4} Ω
17. (a) 184 Hz,   (b) 48.6 mA
19. (a) 31.8 mH,   (b) 120 Hz
21. 53.1 mH
23. (a) $17.4°$,   (b) voltage leads current
25. (a) 0.205 A,   (b) 154 V,   (c) 54.3 V,   (d) $90°$
27. 49.4 Hz
29. (a) 134 Ω,   (b) 0.748 A,   (c) $V_R = 22.5$ V, $V_L = 564$ V, $V_C = 662$ V,   (d) $-77°$
31. (a) 89.7 V,   (b) 108 V
33. (a) 146 V,   (b) 213 V,   (c) 179 V,   (d) 33.3 V
35. (a) 0,   (b) 11.2 mA
37. (a) 1440 Ω,   (b) 97.6 mA,   (c) $51.1°$,   (d) voltage leads current
39. 31.6 W
41. 35.6 W
43. (a) 1.23 A,   (b) 89.5 Ω
45. 1.82 pF
47. (a) 2.52×10^{3} Hz,   (b) 1.5 A,   (c) 0.967 A
49. 4.95×10^{-9} F to 5.07×10^{-8} F
51. 4.79×10^{-2} H
53. (b) $Q = 33.3$
55. 440 V
57. 780 V
59. (a) 1.67×10^{4} A,   (b) 10 A,   (c) 1.39×10^{5} W/m, 5×10^{-2} W/m,   (d) 167 lines, 0.1 line (or one line)
61. (a) 12.7 Hz,   (b) 3.18 A
63. 90.1 Hz
65. (a) 106 V,   (b) $-45°$
67. (a) 6 Ω,   (b) 11.9 mH
69. (a) 12 turns,   (b) 1.95 A,   (c) 0
71. (a) 0.670 A,   (b) $33.2°$,   (c) power factor = 0.837,   (d) 90.5 V,   (e) 22.5 W
73. 2.44 mH, 25.5 μF
75. (a) resistor and inductor,   (b) $R = 10$ Ω, $L = 29.7$ mH

77. $\dfrac{N_p}{N_s} = 31.6$

Chapter 23

1. 1.09 pF to 1.64 pF
3. 2.998×10^8 m/s
5. 5×10^{-7} T
7. (a) 10^{-7} T, 30 N/C,   (b) 2×10^{-8} T, 6 N/C
9. 1.01×10^3 N/C, 3.35×10^{-6} T
11. (a) 45.0 N/C,   (b) 2.69 W/m²
13. 438 N/C, 1.46×10^{-6} T
15. 3.33×10^3 m²
17. (a) 150 MHz,   (b) 15 MHz,   (c) 1.5 MHz
19. (a) 186 m to 556 m,   (b) 2.78 m to 3.41 m
21. 1.19 m
23. 0.276 ms
25. (a) 8.33 min,   (b) 1.34 s
27. 60 km
29. 720 N/C
31. 5.17 μH
33. 10^7 m^{-1}
35. 70 min
37. 29.4 W
39. (a) 6.67×10^{-16} T,   (b) 5.31×10^{-17} W/m²,
    (c) 1.67×10^{-14} W

Chapter 24

1. 2.27×10^8 m/s
3. (a) 2.08×10^4 rev/s, 4.17×10^4 rev/s,
    (b) 2.08×10^2 rev/s, 4.17×10^2 rev/s
5. 60 μs
7. (a) 1.94 m,   (b) 50° above horizontal (parallel to incident ray)
9. 2.00×10^8 m/s
11. 433 nm
13. 22.0°
15. 25.5°, 442 nm
17. 6.3 cm
19. first surface: $\theta_i = 30°$, $\theta_r = 19.5°$; second surface: $\theta_i = 19.5°$, $\theta_r = 30°$
21. 1.58×10^8 m/s
23. (a) first surface: $\theta_i = 30°$, $\theta_r = 19.5°$; second surface: $\theta_i = 40.5°$, $\theta_r = 77.1°$,   (b) first surface: θ (reflection) = 30° = angle of incidence; second surface: θ (reflection) = 40.5° = angle of incidence
25. 23.1°
27. (a) 1.98×10^8 m/s,   (b) 3.89×10^{-5} m
29. 11.3 m
31. 3.39 m
33. 245 nm
35. (a) $\delta_B = 57.5°$, $\delta_R = 54.64°$,   (b) 2.86°
37. (a) 31.3°,   (b) 44.2°,   (c) 49.8°
39. 48.5°

41. 24.4°
43. $n \geqslant \sqrt{2}$
45. 1.08 cm $< h <$ 1.17 cm
47. $n = 1.379$
49. 2.09×10^{-11} s
51. 2.11×10^8 m/s
53. 42.4°
55. 40°
57. $\phi = 55°$

59. $\dfrac{n_w}{n_b} = 0.890$

Chapter 25

3. 10 ft, 30 ft, 40 ft
5. (a) 40 cm in front of mirror, real, inverted, $M = -1$.   (b) No image is formed.   (c) 20 cm behind mirror, virtual, erect, $M = +2$.
7. (a) 13.3 cm behind mirror, virtual, erect, $M = +\frac{1}{3}$;   (b) 10 cm behind mirror, virtual, erect, $M = +\frac{1}{2}$;   (c) 6.67 cm behind mirror, virtual, erect, $M = +\frac{2}{3}$
9. (a) 2.22 cm,   (b) $M = 10$
11. 10 cm in front of mirror
13. $M = +0.130$
15. 48 cm
17. concave, $R = 40$ cm
19. 10 cm in front of mirror
21. 30 cm in front of mirror
23. (a) concave, with $f = 9.52$ cm,   (b) $M = -20$
25. (a) 1.5 m,   (b) 1.75 m
27. 52.5 cm
29. 9 cm inside the bowl
31. (a) 120 cm,   (b) -24 cm,   (c) -8 cm,   (d) -3.43 cm
33. (a) -13.3 cm,   (b) -8 cm,   (c) -4.8 cm,   (d) -2.67 cm
35. 16 mm
37. (a) 40 cm past lens, real, inverted, $M = -1$.
    (b) No image is formed. Rays are parallel.
    (c) 20 cm in front of lens, erect, virtual, $M = 2$.
39. (a) 6.5 m beyond lens, real, inverted.   (b) 6.0 m in front of lens, virtual, erect.
41. (a) 13.3 cm in front of lens, virtual, erect, $M = \frac{1}{3}$.
    (b) 10 cm in front of lens, virtual, erect, $M = \frac{1}{2}$.
    (c) 6.67 cm in front of lens, virtual, erect, $M = 0.667$
43. (a) -12 cm,   (b) -12 cm,   (c) -9 cm,   (d) -6 cm,   (e) -4 cm
45. 5.68 cm
47. -80 cm

49. $\dfrac{f}{9}$ beyond the focal point

51. either 36.2 cm or 13.8 cm from object; $M = -2.62$ for $s = 13.8$ cm; $M = -0.382$ for $s = 36.2$ cm

53. 30 cm in front of second lens, $M = -3$
55. (a) 9.26 cm in front of second lens, $M = 0.370$
57. from 0.232 m to 187 m
59. (a) 13.3 cm,   (b) $M = -5.90$,   (c) inverted
61. 25.3 cm to right of mirror, virtual, erect, $M = 8.05$
63. 10 cm
65. (a) 33.3 cm beyond lens, $M = -0.67$, real,
    inverted.   (b) 60 cm beyond lens, $M = -2$, real,
    inverted.   (c) 20 cm in front of lens, $M = 2$, erect
    and virtual
67. (a) 10 cm beyond second lens;   (b) real, erect and
    same size as object, $M = 1$;   (c) 20 cm beyond
    lens combination.
69. -30 cm
71. 1.5 cm
73. 32 cm beyond second surface (real image)

75. (a) $s = \dfrac{4f}{3}$,   (b) $s = \dfrac{3f}{4}$,   (c) $M_1 = -3$, $M_2 = 4$

Chapter 26

1. 433 nm
3. (a) 2.62 mm,   (b) 2.62 mm
5. 11.3 m
7. 2.61 m
9. 34.0 cm
11. (a) 55.7 m,   (b) 124 m
13. 99.8 m
15. 148 m
17. 91.9 nm
19. 85.4 nm (Other possible thicknesses are odd
    integral multiples of this.)
21. 290 nm, 580 nm, 870 nm
23. 290 nm
25. 647 nm
27. 324 nm, 417 nm, 584 nm
29. No, the wavelengths intensified are 276 nm,
    138 nm, 92 nm . . . .
31. (a) 123 nm,   (b) 81.6 nm
33. 147 bands
35. $4.23\ \mu m \leq t \leq 4.45\ \mu m$
39. (a) 5.12 cm,   (b) 0.8 mm
41. 0.227 mm
43. (a) 1.09 m,   (b) 1.70 mm
45. 56.3°
47. 31.2°
49. 48.4°
51. (a) $n = 1.11$,   (b) 42°
53. 60.5°
55. 58.1°
57. 4.22 mm
59. 681 nm
61. 432 nm

63. 6.89 units
65. 0.35 mm
67. 313 nm
69. 9.96×10^{-8} m

Chapter 27

1. 5.71 m
3. 177 m
5. $f/4.29$
7. $\frac{1}{128}$ s
9. (a) 33.3 cm,   (b) 3.0 diopters
11. -1.0 diopters
13. 17.0 diopters
15. 33.3 cm
17. Power of upper portion $= -1.0$ diopters. Power of
    lower portion $= 2.5$ diopters.
19. (a) $M = -1.22$,   (b) $\dfrac{\theta'}{\theta} = 6.08$
21. 4.07 cm
23. (a) $M = 2$,   (b) $M = 1$
25. 2.34 cm
27. 0.938 cm
29. 0.81 μm
31. $M = 15$
33. (a) 9.5 cm past the eyepiece.   (b) Probably not.
    When looking through a telescope, the eye is
    relaxed (focused at infinity) and would not properly
    focus on the bee. Also, the overall magnification of
    the bee is 0.1. Thus, the final image is 1/10 the size
    of the bee.
35. 157 miles
37. 0.818 m (32.2 in.)
39. (a) 1.03 cm,   (b) 26.8 cm
41. 38.3 cm
43. 1.06×10^{-5} rad
45. (a) 2.29×10^{-4} rad,   (b) 43.7 m
47. (a) 7.75×10^{-3} cm,   (b) 1.18 mm,   (c) 174 m
49. 499 nm
51. 654 fringes
53. 20 shifts
55. 6560 lines/cm
57. (a) 478.7 nm, 647.6 nm, 696.6 nm,   (b) 20.5°,
    28.30°, 30.66°
59. (a) 5.93°,   (b) 6.42°
61. 469 nm, 156 nm
63. 0.02 nm
65. (a) 101.5 cm,   (b) $M = 66.7$
67. 603 lines/mm
69. $M = 2.88$
71. 660 nm
73. (a) 1.96 cm,   (b) $f/3.27$,   (c) $f/9.8$

75. $n = 1.70$

77. 5.07 mm

Chapter 28

1. (a) $t_I = 1667$ s, $t_{II} = 2041$ s,   (b) $t_I' = 2500$ s, $t_{II}' = 2041$ s,   (c) $\Delta t = 85$ s

3. (a) 70 beats/min,   (b) 30.5 beats/min

5. $0.866c$

7. (a) 1.38 y,   (b) 1.31 lightyears

9. $0.866c$

11. 229.4 m

13. $0.927c$

15. (a) 20 m,   (b) 19 m,   (c) $0.312c$

17. ratio = 0.088

19. 3600 s $+ 3.2 \times 10^{-9}$ s

21. (a) 2.73×10^{-24} kg m/s,   (b) 1.58×10^{-22} kg m/s,   (c) 5.64×10^{-22} kg m/s

23. (a) $0.85c$,   (b) $0.85c$

25. $0.285c$

27. $0.927c$

29. 1.03×10^3 m

31. (a) 337 kg/m^3,   (b) 64 kg/m^3

33. $0.985c$

35. $+0.994c$

37. $0.982c$

39. 0.160 pulses/s

41. (a) $0.983c$,   (b) $0.695c$ to the right or $0.9992c$ to the left

43. (a) 939 MeV,   (b) 3006 MeV,   (c) 2067 MeV

45. $0.745c$

47. (a) 4.44×10^9 kg,   (b) 1.43×10^{13} y

49. 3.18×10^{-12} kg. No, a mass this small (out of 9 g) could not be detected.

51. (a) 5.09×10^4 MeV/c,   (b) $0.9998c$

53. (a) 4.92×10^{-13} J $= 3.07$ MeV,   (b) $0.986c$

55. $0.37c$

57. (a) 2.50 MeV/c,   (b) 4600 MeV/c

59. 1.22×10^{13} m

61. (a) 7.05 μs,   (b) 11,150 muons

Chapter 29

1. (a) 3.73×10^{-7} eV,   (b) 4.14×10^{-2} eV, (c) 41.4 eV

3. 5.71×10^3 photons

5. 500 nm

7. 0.622 nm

9. 8.19×10^{-20} m

11. 148 days

13. (a) 1.32×10^{-19} J (0.822 eV),   (b) 995 nm

15. 1.91 eV,   (b) 1.20 V

17. (a) lithium and iron,   (b) 1.84 eV and 0.244 eV, respectively

19. 242 nm

21. 0.124 nm

23. (a) 0.083 nm,   (b) 0.012 nm

25. 25.8 sec of arc

27. 6.7°

29. 0.124 nm

31. 2.43×10^{-3} nm

33. $\Delta\lambda = 7.34 \times 10^{-2}$ nm

35. 1.82 keV, 1.82 keV/c

37. (a) 2.87×10^{-3} nm,   (b) 100.5°

39. (a) 1878 MeV,   (b) 6.62×10^{-16} m

41. 1.98 MeV

43. 2.27×10^{23} Hz, 1.32×10^{-15} m

45. 1.14×10^{-38} m

47. (a) 0.174 nm,   (b) 5.37×10^{-12} m

49. 1.06×10^{-34} m

51. 151 V

53. 546 eV

55. 3.50×10^{-32} m

57. 5.79×10^5 m/s

59. 1 part in 10^8

61. (b) 5.21 MeV

63. 1.55 V

65. 2.96×10^{20} photons

67. 2.0 V

69. 1.35 eV

Chapter 30

1. 121.5 nm, 102.5 nm, 97.2 nm, far ultraviolet

3. 1.60×10^{14} Hz, 2.46×10^{14} Hz, infrared

5. (a) 1094 nm,   (b) 656.1 nm,   (c) 820.1 nm, (d) 102.5 nm

7. (a) $n = 5$.   (b) No, it is shorter than the series limit of these two series. (See Problem 30.6.)

11. (a) 1.89 eV,   (b) 658 nm,   (c) 4.56×10^{14} Hz

13. (a) 12.09 eV,   (b) 12.09 eV, 1.89 eV, 10.2 eV

15. 1.09243×10^7 m^{-1}

19. 91.4 nm

21. 3.15×10^{-34} J·s

23. (a) Transition B,   (b) Transition A, (c) Transitions B and C

25. (a) 2.19×10^6 m/s,   (b) 1.52×10^{-16} s, (c) 1.05×10^{-3} A

27. (a) 2.46×10^{15} Hz, $f_{orb} = 8.31 \times 10^{14}$ Hz. (b) 6565 Hz, $f_{orb} = 6649$ Hz. For large n, classical theory and quantum theory approach one another in their results.

29. (a) -122.4 eV,   (b) 1.76×10^{-11} m

31. (a) 0.265 Å,   (b) 0.132 Å

33. (a) 4.39×10^7 m^{-1},   (b) 30.3 nm,   (c) deep ultraviolet region

35.

n	ℓ	m_ℓ	m_s
3	2	+2	$+\frac{1}{2}$
3	2	+2	$-\frac{1}{2}$
3	2	+1	$+\frac{1}{2}$
3	2	+1	$-\frac{1}{2}$
3	2	0	$+\frac{1}{2}$
3	2	0	$-\frac{1}{2}$
3	2	−1	$+\frac{1}{2}$
3	2	−1	$-\frac{1}{2}$
3	2	−2	$+\frac{1}{2}$
3	2	−2	$-\frac{1}{2}$

(A total of 10 states.)

37. (a) 4 ($\ell = 3, 2, 1,$ and 0),
     (b) 7 ($-3, -2, -1, 0, 1, 2, 3$)
39. (a) $1s^2 2s^2 2p^4$ (b) The quantum numbers are

$1s^2$ state: $n = 1$	$\ell = 0$	$m_\ell = 0$	$m_s = \pm\frac{1}{2}$
$2s^2$ state: $n = 2$	$\ell = 0$	$m_\ell = 0$	$m_s = \pm\frac{1}{2}$
$2p^4$ state: $n = 2$	$\ell = 1$	$m_\ell = -1$	$m_s = \pm\frac{1}{2}$
		$m_\ell = 0$	$m_s = \pm\frac{1}{2}$

41. (a) 4,   (b) 2
43. 1.88×10^{18} Hz
45. L shell: 11.8 keV; M shell: 10.1 keV; N shell: 2.39 keV
47. 1.61×10^{18} photons
49. (a) 0.476 nm,   (b) 7.30×10^5 m/s,
     (c) 0.997 nm,   (d) $\lambda = \dfrac{\text{circumference}}{3}$
51. $PE = -27.2$ eV, $KE = 13.6$ eV
53. (a) The wavelengths of the emission lines are: 158.5 nm, 185 nm, 253.7 nm, 422 nm, 682 nm, 1109 nm.   (b) 1.31×10^6 m/s
55. (a) 135 eV,   (b) This is about 10 times the ground state energy of hydrogen.
59. (a) $\Delta p \approx \dfrac{\hbar}{r}$,   (b) $KE = \dfrac{\hbar^2}{2mr^2}$, $PE = -\dfrac{ke^2}{r}$,

$$E = \frac{\hbar^2}{2mr^2} - \frac{ke^2}{r}$$

Chapter 31

3. (a) 2.94×10^{-9} N, -5.14 eV
5. (a) 480 N/m,   (b) 8.68×10^{13} Hz
7. 0.24 eV
9. 3.52 kg
11. 5.54×10^{-3} nm from fluorine atom
13. 1.92×10^{-47} kg·m^2

15. 0.115 mm
17. 1.18×10^{11} Hz
19. (a) 0.1244 nm from hydrogen atom,
     (b) 2.64×10^{-47} kg·m^2,   (c) 4.17×10^{-22} J and 0 J,   (d) 0.477 mm
23. 0.359 eV, 3.46 μm
25. (a) 2.75×10^{14} Hz,   (b) 1.09 μm
27. p-type
29. (a) $-157°$C (116 K),   (b) 8.09×10^4 °C (8.12×10^4 K)
31. 1.24 eV
33. 9.07×10^7 J
35. 5.77×10^{-23} J, 1.73×10^{-22} J
37. (a) $(3.67 \times 10^{-3}$ eV$)J'(J' + 1)$ for $J' = 0, 1, 2, 3, \ldots,$ (b) $(v + \frac{1}{2})(0.144$ eV$)$ for $v = 0, 1, 2, 3, \ldots$
39. (a) 1.69×10^{14} Hz,   (b) 1.78 μm

Chapter 32

1. (a) 1.90×10^{-15} m,   (b) 7.4×10^{-15} m
3. any nucleus for which $A = 30$, for example $^{30}_{15}$P
5. (a) 1.4×10^{16} lb/ft^3,   (b) 2.6×10^{14} lb
7. $\dfrac{\rho_n}{\rho_a} = 8.6 \times 10^{13}$
9. 4.55×10^{-13} m
11. 2.28×10^{-13} m
13. (a) 27.6 N,   (b) 4.16×10^{27} m/s^2,
     (c) 1.73 MeV
15. 342 MeV
17. 7.92 MeV/nucleon
19. 3.54 MeV
21. 0.211 MeV/nucleon. The difference is largely due to increased coulomb repulsion caused by the extra proton in $^{23}_{12}$Mg.
23. (a) The surface-effect term must have a negative sign because the volume term over-approximates the binding energy by treating all nucleons as if they were completely surrounded by other nucleons. The nucleons on the surface of the "drop" are not totally surrounded and the surface term attempts to correct for this effect by subtracting off the excess which was included in the volume term.
     (b) $\left(\dfrac{V}{A}\right)_s = 0.207V^{1/3}$, $\left(\dfrac{V}{A}\right)_c = 0.167V^{1/3}$
25. 2.38×10^{16} nuclei
27. 1.0 h
29. 8.75×10^3 Bq
31. 7.47×10^9 nuclei
33. 24.5%
35. 1650 y
37. $^{12}_{6}$C, $^{4}_{2}$He, $^{14}_{6}$C

39. 4.28 MeV
41. (a) No, $\Delta m < 0$,   (b) Yes, $\Delta m > 0$
43. $^{66}_{28}\text{Ni} \rightarrow ^{66}_{29}\text{Cu} + _{-1}^{0}e + \bar{\nu}$,   (b) 187 keV
45. 16,600 y
47. 17.35 MeV
49. (a) $^{10}_{5}\text{B}$,   (b) -2.80 MeV
51. $^{1}_{0}\text{n}$, $^{10}_{5}\text{B}$
53. (a) $^{3}_{2}\text{He}$,   (b) 4.02 MeV
55. 0.989 MeV
57. (a) The second reaction has as its target and incident projectile the products formed in the first reaction.
59. 22.0 MeV
61. $Q < 0$. No.
63. 1.88 MeV
65. 0.055 mCi
67. 228 y
69. 1.60×10^8 nuclei
71. 12.2 mg

Chapter 33

1. 174 MeV
3. $^{144}_{54}\text{Xe}$ (2 neutrons released), $^{143}_{54}\text{Xe}$ (3 neutrons released), $^{142}_{54}\text{Xe}$ (4 neutrons released)
5. 1.08×10^{16} fissions
7. 3.38 kg
9. 2.46×10^3 y
11. 2900 km (1800 miles)
13. 17.6 MeV
15. (a) $^{13}_{7}\text{N}$,   (b) $^{13}_{6}\text{C}$,   (c) $^{14}_{7}\text{N}$,   (d) $^{15}_{7}\text{O}$,   (e) $^{15}_{7}\text{N}$,   (f) $^{12}_{6}\text{C}$
17. (a) 3.95×10^{22} nuclei,   (b) 1.03×10^{10} J   (c) 51.7 gallons
19. (a) 3.60×10^{30} J,   (b) 163 million years
21. (a) 10 h,   (b) 3.16 m
23. 18.8 J
25. (a) 2 J/kg,   (b) 4.78×10^{-4} °C
27. 24.2 days
29. (a) 9.00×10^{11} electrons,   (b) 1.01×10^{-1} J,   (c) 101 rad
31. (a) 10 electrons,   (b) 10^7 electrons
33. approximately 4100 electrons
35. (a) 3.72×10^{-24} kg·m/s,   (b) 1.78×10^{-10} m,   (c) The de Broglie wavelength is about the size of an atom (10^{-10} m) and about 10^5 times the size of a nucleus (10^{-15} m).
37. 26 collisions
39. 1.38×10^7 rad

Chapter 34

1. 2.26×10^{23} Hz, 1.32×10^{-15} m
3. 2.00×10^{-18} m
5. 10^{-23} s
7. $\bar{\nu}_\mu$

9. (a) Not allowed; violates conservation of baryon number,   (b) strong interaction,   (c) weak interaction,   (d) weak interaction,   (e) electromagnetic interaction.
11. (a) not conserved,   (b) conserved,   (c) conserved,   (d) not conserved,   (e) not conserved,   (f) not conserved
13. (a) Baryon number, charge, L_e, L_τ.   (b) Baryon number, L_e, L_τ, L_μ.   (c) Strangeness, charge, L_e, L_τ, L_μ.   (d) Baryon number, strangeness, charge, L_e, L_τ, L_μ.   (e) Baryon number, strangeness, charge, L_e, L_τ, L_μ.   (f) Baryon number, strangeness, charge, L_e, L_τ, L_μ.
15. (a) Violates conservation of L_e, L_μ,   (b) violates conservation of charge,   (c) violates conservation of baryon number,   (d) violates conservation of baryon number,   (e) violates conservation of charge
17. (a) $\bar{\nu}_\mu$,   (b) ν_μ,   (c) $\bar{\nu}_e$,   (d) ν_e,   (e) ν_μ,   (f) $\bar{\nu}_e$ and ν_μ
19. (a) Violates conservation of baryon number, (b) can occur,   (c) baryon number conservation violated,   (d) can occur,   (e) can occur,   (f) violates conservation of baryon number and muon-lepton number.
21. 74.4 MeV
23. (a)

	Proton	u	u	d	Total
Strangeness	0	0	0	0	0
Baryon number	1	$\frac{1}{3}$	$\frac{1}{3}$	$\frac{1}{3}$	1
Charge	e	$\frac{2}{3}e$	$\frac{2}{3}e$	$-\frac{1}{3}e$	e

(b)

	Neutron	u	d	d	Total
Strangeness	0	0	0	0	0
Baryon number	1	$\frac{1}{3}$	$\frac{1}{3}$	$\frac{1}{3}$	1
Charge	0	$\frac{2}{3}e$	$-\frac{1}{3}e$	$-\frac{1}{3}e$	0

25. (a) $\bar{u}d + uud \rightarrow d\bar{s} + uds$,
   (b) $u\bar{d} + uud \rightarrow u\bar{s} + uus$,
   (c) $\bar{u}s + uud \rightarrow u\bar{s} + d\bar{s} + sss$,
   (d) $uud + uud \rightarrow d\bar{s} + uud + u\bar{d} + ?$
   The quark combination of ? must be such as to balance the last equation for u, d, and s quarks.

$$u: 2 + 2 = 0 + 2 + 1 + ? \text{ (has 1 } u \text{ quark)}$$

$$d: 1 + 1 = 1 + 1 - 1 + ? \text{ (has 1 } d \text{ quark)}$$

$$s: 0 + 0 = -1 + 0 + 0 + ? \text{ (has 1 } s \text{ quark)}$$

Quark composite $= uds = \Lambda^0$.

27. $m_d = 313.6$ MeV/c^2, $m_u = 312.3$ MeV/c^2

Index

Periodic Table of the Elements*

Group I	Group II	Transition Elements						
H 1 1.0080 $1s^1$								
Li 3 6.94 $2s^1$	**Be** 4 9.012 $2s^2$							
Na 11 22.99 $3s^1$	**Mg** 12 24.31 $3s^2$							
K 19 39.102 $4s^1$	**Ca** 20 40.08 $4s^2$	**Sc** 21 44.96 $3d^14s^2$	**Ti** 22 47.90 $3d^24s^2$	**V** 23 50.94 $3d^34s^2$	**Cr** 24 51.996 $3d^54s^1$	**Mn** 25 54.94 $3d^54s^2$	**Fe** 26 55.85 $3d^64s^2$	**Co** 27 58.93 $3d^74s^2$
Rb 37 85.47 $5s^1$	**Sr** 38 87.62 $5s^2$	**Y** 39 88.906 $4d^15s^2$	**Zr** 40 91.22 $4d^25s^2$	**Nb** 41 92.91 $4d^45s^1$	**Mo** 42 95.94 $4d^55s^1$	**Tc** 43 (99) $4d^55s^2$	**Ru** 44 101.1 $4d^75s^1$	**Rh** 45 102.91 $4d^85s^1$
Cs 55 132.91 $6s^1$	**Ba** 56 137.34 $6s^2$	57–71‡	**Hf** 72 178.49 $5d^26s^2$	**Ta** 73 180.95 $5d^36s^2$	**W** 74 183.85 $5d^46s^2$	**Re** 75 186.2 $5d^56s^2$	**Os** 76 190.2 $5d^66s^2$	**Ir** 77 192.2 $5d^76s^2$
Fr 87 (223) $7s^1$	**Ra** 88 (226) $7s^2$	89–103‡	**Rf** 104 (261) $6d^27s^2$	**Ha** 105 (262) $6d^37s^2$	**Unh** 106 (263)	**Uns** 107 (261)		

Symbol — **Ca** 20 — Atomic number
Atomic mass† — 40.08
$4s^2$ — Electron configuration

‡ Lanthanide series

La 57	Ce 58	Pr 59	Nd 60	Pm 61	Sm 62
138.91 $5d^16s^2$	140.12 $5d^14f^16s^2$	140.91 $4f^36s^2$	144.24 $4f^46s^2$	(147) $4f^56s^2$	150.4 $4f^66s^2$

§ Actinide series

Ac 89	Th 90	Pa 91	U 92	Np 93	Pu 94
(227) $6d^17s^2$	(232) $6d^27s^2$	(231) $5f^2d^17s^2$	(238) $5f^36d^17s^2$	(239) $5f^46d^17s^2$	(239) $5f^66d^07s^2$

* Atomic mass values given are averaged over isotopes in the percentages in which they exist in nature.

† For an unstable element, mass number of the most stable known isotope is given in parentheses.